WORLD REGIONAL GEOGRAPHY

Global Patterns, Local Lives

Third Edition

LYDIA MIHELIČ PULSIPHER
University of Tennessee

ALEX A. PULSIPHER
Ph.D. Candidate
Graduate School of Geography
Clark University

with contributing author
Holly M. Hapke
East Carolina University

and the assistance of
Conrad Mac Goodwin
University of Tennessee

W. H. Freeman and Company
New York

Publisher: Sara Tenney
Acquisitions Editor: Jason Noe
Developmental Editor: Jennifer Van Hove
Marketing Manager: Scott Guile
Project Editor: Mary Louise Byrd
Cover and Text Designer: Blake Logan
Illustration Coordinator: Shawn Churchman
Maps: University of Tennessee, Cartographic Services Laboratory, Will Fontanez, Director;
  Anna Compton, Cartographer
Illustrations: Fine Line Illustrations
Photo Editor: Bianca Moscatelli
Photo Researcher: Elyse Rieder
Production Coordinator: Susan Wein
Editorial Assistant: Sharon Merritt
New Media and Supplements Editors: Elaine Palucki, Lisa Samols, Nick Tymoczko
Composition: Seven Worldwide Publishing Solutions and
  Sheridan Sellers, W. H. Freeman and Company, Electronic Publishing Center
Manufacturing: RR Donnelly Sons & Company

FRONTLINE/World is a trademark of the WGBH Educational Foundation.
Frontline is a registered trademark of the WGBH Educational Foundation.

Library of Congress Number: 2002105823

To the many Earthwatch volunteers who so competently contributed to our years of geography fieldwork on Montserrat, helping to excavate and decipher clues about lifeways in the Caribbean colonial past. Talking with Earthwatch volunteers and Montserratians is among the experiences that started us down the path of attempting to interpret the world at large.

ABOUT THE AUTHORS

LYDIA MIHELIČ PULSIPHER

is a cultural-historical geographer who studies the landscapes of ordinary people through the lens of archaeology and historical geography. She has contributed to several geography-related exhibits at the Smithsonian Museum of Natural History in Washington, D.C., including "Seeds of Change," which featured her research in the eastern Caribbean. Professor Pulsipher and her graduate students are now studying national identity issues in several European countries. She has taught cultural, gender, and Mesoamerican geography at the University of Tennessee at Knoxville since 1980, and, through her research, she has given many students their first experience in fieldwork abroad. Previously she taught at Hunter College and Dartmouth College. She received her B.A. from Macalester College, her M.A. from Tulane University, and her Ph.D. from Southern Illinois University.

ALEX A. PULSIPHER

is a Ph.D. candidate in geography at Clark University, where he is studying urban development and sustainability in the United States. In the early 1990s, Alex spent some time in South Asia working for a development research center and then went on to do an undergraduate thesis on the history of Hindu nationalism at Wesleyan University. Beginning in 1995, Alex worked full time on the research and writing of the first edition of this textbook. In 1999 and 2000, he traveled to South America, Southeast Asia, and South Asia, where he collected information for the second edition of the text and for the Web site. In 2000 and 2001, he returned to writing material and designing maps for the second edition.

While writing *World Regional Geography*, Lydia Pulsipher was assisted by her husband, **Conrad "Mac" Goodwin**, a historical archaeologist who specializes in sites created during the European colonial era in North America, the Caribbean, and the Pacific. He has particular expertise in the archaeology of agricultural systems, formal gardens, domestic landscapes, and urban spaces. He holds a research appointment in the Department of Anthropology at the University of Tennessee

(L. to R.) Alex A. Pulsipher, Lydia Mihelič Pulsipher, Conrad "Mac" Goodwin.

HOLLY M. HAPKE, Contributing Author,

is an economic-cultural geographer whose interests lie in the areas of political economy and development. She is particularly interested in the interaction of economic and cultural transformations and in local impacts of economic globalization. Professor Hapke has conducted field research in Kerala, India, where she has examined the impact of mechanization and commercialization on local fishing communities and women fish traders; she is currently studying the influence of gender, community identity, and ecological crisis on fisherfolk livelihoods in southern India. Another research project she is involved in investigates rural restructuring and the dynamics of Latino transnational migration in eastern North Carolina. Professor Hapke teaches geography at East Carolina University. She earned her B.A. from Hamline University and her M.A. and Ph.D. from Syracuse University.

BRIEF CONTENTS

CONTENTS

Doctors Rakesh Sachdeva and Seema Sachdeva, both from India, treat patients in Pikeville, Kentucky. (page 53) [Dena Potter/Appalachian News Express.]

In the Amazon Basin, colonists clear the rain forest for settlement and large-scale agriculture. (page 151) [Randall Hyman.]

French Muslim schoolgirls who became the center of a national debate over the wearing of religious garb in public schools. (page 204) [AFP/Getty Images]

CHAPTER 6
NORTH AFRICA AND SOUTHWEST ASIA   282

A once–affluent family huddles around a shelter built of their possessions. (page 344) [Oliver Jobard/Sipa.]

CHAPTER 8
SOUTH ASIA   388

A bicycle rickshaw and driver in Delhi, India. (page 400) [Lindsay Hebberd/ Woodfin Camp & Associates.]

CHAPTER 10
SOUTHEAST ASIA   500

I THE GEOGRAPHIC SETTING   503

Mass consumption in Indonesia. Notice the bag the young woman is carrying and the drink the young man is holding, as they pause amid manicured rice fields. (page 524) [Ian Loyd/Black Star.]

II CURRENT GEOGRAPHIC ISSUES   513

PREFACE

The scale of *World Regional Geography*, Third Edition, encompasses vast continents and global forces, but often its descriptive focus on individual lives has the most impact. Stories of people and families make the study of geography compelling. Students begin to grasp the complex patterns at work in the world today as they see how people are affected by, and respond to, economic, social, and political processes. Through these stories of individual lives, we hope to convey the impact of globalization, a major theme of the text.

To highlight global-to-local and interregional connections, the text includes a number of topics that have no borders: the war on terrorism, realignments in the global political order, interregional trade, the global economy, popular culture, the environment, and the Internet. Here, again, the focus on the individual person provides insight, offering local perspectives on these global trends.

World Regional Geography continues to supply a solid pedagogical program that aids and furthers student learning. For this edition we have worked diligently to equip students with the skills they need to think geographically by providing several new tools, described later in the preface. The map program has proved to be another outstanding tool, often combining information from multiple sources in insightful ways. The authors have carefully planned each map, working closely with Will Fontanez at the University of Tennessee Cartographic Services Laboratory.

For the third edition we also have made a number of enhancements, each explained below. We are especially excited to have the opportunity to provide students with a window into other cultures by making available ten short documentaries produced by the PBS series FRONTLINE/World. What's more, we now offer the additional flexibility of two versions of the text, one with subregions and one without.

Two people with different ethnic backgrounds celebrate a wedding in the United States. (page 64) [Stephen Zeigler.]

ONE VISION. TWO VERSIONS: WITH OR WITHOUT SUBREGIONAL COVERAGE

The new edition is now available in two versions to better serve the different needs of diverse faculty and curricula.

■ **World Regional Geography with Subregions, 581 pages.** The third edition continues to employ the consistent chapter structure for which this book has become known. The divisions in each chapter are now emphasized by roman numerals that distinguish chapter parts: I: The Geographic Setting; II: Current Geographic Issues; III: Subregions. The three major sections of each chapter conclude with review questions, so that each section can be studied independently.

To present topics in the most complete, yet concise, way possible, important current issues, such as conflict in Iraq and the War on Terrorism, and the conflict between the Israelis and the Palestinians, have been moved out of the subregional sections and are now covered in the main body of the text (Units I and II). The primary focus of the subregional units is now the descriptive characterization of particular countries and places.

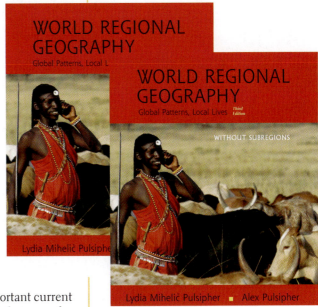

Two different versions of *World Regional Geography*, Third Edition, are available, one with subregional coverage and one without.

■ **World Regional Geography without Subregions, 450 pages.** The briefer version provides the same coverage as the version described above, except that the subregional sections have been omitted, creating a text that is 131 pages shorter. This is a useful alternative for teachers constrained by time.

VIDEO PARTNERSHIP WITH FRONTLINE/WORLD

Guatamalan women sifting coffee beans, taken during the filming of the FRONTLINE/World segment "Coffee Country." [Photo courtesy of Bill Kinzie Green, Mountain Coffee Roasters Foundation.]

W. H. Freeman and Company and the authors of *World Regional Geography* are proud to announce a new partnership with the groundbreaking PBS television series FRONTLINE/World. **Free to qualified adopters** and available in both VHS and DVD formats, these 10 video segments are seamlessly integrated into the text.

The new edition takes the pioneering "Personal Vignettes" in *World Regional Geography* to a new level by weaving the stories of individuals featured in the FRONTLINE/World video segments directly into the textbook chapters. The "Video Vignette" treatment and themes from the chapter will be useful to students as they watch, study, and discuss the FRONTLINE/World videos. This innovative multimedia package builds upon the book's purpose of putting a face on geography by giving students and instructors unprecedented and immediate access to the fascinating personal stories of people from all over the world.

The full set of videos and the supplements to support them are described below in the media section of the preface.

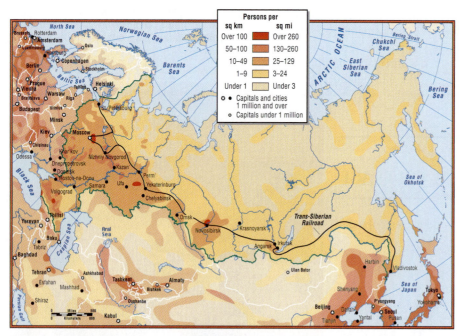

FIGURE 5.6 Population density in Russia and the newly independent states. (page 245)

PRACTICING GEOGRAPHY **READING MAPS** Compare the population patterns in this map with Figures 5.1, 5.2, and 5.5 to reach an understanding of the physical, social, historical, and political contexts for the patterns on this map. What are some physical explanations for the population patterns? What are the most important historical and social factors influencing these patterns? [Adapted from *Hammond Citation World Atlas*, 1996.]

NEW EXPANDED FOCUS ON "THINKING GEOGRAPHICALLY"

In this edition, the text builds on its innovative pedagogical program by increasing the focus on geographic methodology. To this end, we have added the following integrated text features:

■ **Stronger emphasis on map reading and spatial thinking skills in Chapter 1.** Prepares students from the outset to apply geographic analysis to maps, a skill they will use repeatedly throughout the text. The map-reading appendix from previous editions is now fully integrated into the chapter and has been expanded to include a more comprehensive introduction to the world of maps.

■ New "Practicing Geography: Reading Maps" critical thinking questions, embedded directly into the map captions. These questions encourage students to think more comparatively and analytically about spatially represented geographic concepts.

■ New "Practicing Geography: Comparing Local Landscapes" photo comparisons. The

PRACTICING GEOGRAPHY **COMPARING LOCAL LANDSCAPES**

Study the landscapes in these two photos. Photo A is from Beijing, China. Photo B was taken in Malacca, Malaysia. Can you determine what type of building appears in each landscape? What do the architectural features of the buildings tell you about what they are? What clue does the man's attire in Photo A provide? Finally, what cultural influences link the two buildings? Why? [A, Peter Sanders Photography; B, Kevin Bubriski.]

(page 469)

skills of photo interpretation are taught in Chapter 1 and emphasized in each subsequent chapter through the presentation of related images from two different regions. Students are asked to compare and contrast these different cultural landscapes, and are invited to use photo interpretation skills for all pictures in the book.

■ **Revised "Practicing Geography: Geographer in the Field" boxes.** In each chapter, a leading geographer discusses his or her methods of field research—how questions are posed and how they are answered—giving students a sense of how professional geographers contribute to our understanding of important global and local issues.

NEW MAPS

■ **Every chapter in the text includes six standardized thematic maps** that can be used for easy comparisons across regions. The six standardized maps in every chapter are

- A **global context map** at the beginning of every regional chapter that focuses on one issue, showing the region in relation to the rest of the world, and often emphasizing interregional connections
- A **large reference map** near the beginning of every regional chapter, showing landforms and political divisions
- A **climate map**
- A **population density map**

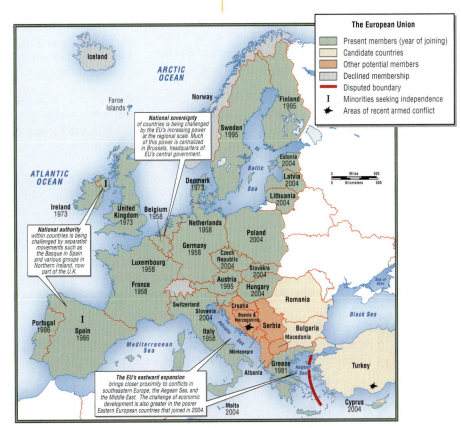

Figure 4.1 The European Union (EU). (page 179)

- An **economic issues map** that focuses on one or more key aspects of a region's economy, such as trade, debt, or gross domestic product, and usually illuminates economic linkages with other regions

- An **environmental issues map** that charts land use, pollution, acid rain, and threatened fisheries

■ **New maps for this edition include**

- **Updated thematic maps,** such as the most recent developments in the Israeli–Palestinian conflict, new members of the European Union, and tourism in Southeast Asia

- **Ten new political maps,** one for each region

- **Three new urban maps,** two for South America and one for Europe

NEW EXPERTISE:
Most Current Coverage Available

■ **New Contributing Author** We are delighted to welcome Holly Hapke, East Carolina University, as a new contributing author for this edition. Professor Hapke brings fresh insights and a new level of authority to the coverage of South Asia and economic and political geography in general. Holly's stamp is most apparent in Chapters 8 through 11, which treat South Asia, East Asia, Southeast Asia, and Oceania.

FIGURE 10.11 World Heritage Sites in the Greater Mekong Subregion. (page 519)
[Adapted from UN World Heritage Cities, UNESCO and Mekong River Commission.]

■ **Board of Regional Consultants** To give students the opportunity to explore the world's regions with those who know them best, we once again turn to a special group of the world's leading geographers to help identify the central issues and themes in their respective regions of expertise. The following are members of the board:

Chapter 2, North America: Stanley Brunn, *University of Kentucky.* Professor Brunn's teaching and research interests are in political, social, and urban geography, the human geographies of the twenty-first century, and the geographies of knowledge.

Chapter 3, Middle and South America: Dennis Conway, *Indiana University.* Professor Conway's research has centered on the development processes and prospects of Caribbean small islands, in general, and on the complex relationships between patterns of development, urbanization, and migration, in particular.

Chapter 4, Europe: John Agnew, *University of California, Los Angeles.* Professor Agnew's research interests include political geography, international political economy, European urbanization, and the changing economic geography of Italy.

Chapter 5, Russia and the Newly Independent States: Beth Mitchneck, *University of Arizona.* Professor Mitchneck's teaching and research interests are in economic and urban geography, local economic development policy, and migration processes and policies.

Chapter 6, North Africa and Southwest Asia: Donna Stuart, *Georgia State University.* Professor Stuart's primary research interests are comparative urbanization, urban spatial structure, urban landscape analysis, and historical preservation in the developing world. Much of her research focuses on the Middle East and Africa. She has studied and worked in Egypt periodically since 1988 and has traveled extensively throughout the region.

 Chapter 7, Sub-Saharan Africa: Abdi Samatar, *University of Minnesota.* Professor Samatar is interested in political economy, social change, and globalization in the developing world. Currently, he is researching democracy, leadership and the making of sustainable institutions in Africa, with a particular focus on Ethiopia and Somalia.

 Chapter 8, South Asia: Stuart Corbridge, *London School of Economics.* Professor Corbridge is a human geographer with teaching and research interests in development studies, India, and international political economy. He has conducted fieldwork in rural eastern India over a period of about twenty years, including work on forestry issues for the United Kingdom's Department for International Development.

 Chapter 9, East Asia (Japan): Roman Cybriwsky, *Temple University.* Professor. Cybriwsky's current research interests include urban-social geography, urban planning and development, and neighborhood social change. He also has a special interest in planning and urban design in big cities, and has worked extensively on related issues in New York City, Tokyo, Philadelphia, and elsewhere.

 Chapter 9, East Asia (China): Carolyn Cartier, *University of Southern California.* Professor Cartier teaches globalization and cultural and economic geographies in Asian cities, and currently conducts research on the space economy of Chinese urbanism and the gender and development aspects of transnational migration and tourism.

 Chapter 10, Southeast Asia: Richard Ulack, *University of Kentucky.* Professor Ulack's research and reaching interests are in the broad area of development, with emphasis on Southeast Asia and the South Pacific regions. His fieldwork has focused on population geography (internal migration and fertility), urbanization (especially of middle-sized cities), and, most recently, tourism.

 Chapter 11, Oceania: Robert Kuhlken, *Central Washington University.* A former regional planner and now a cultural geographer, Professor Kuhlken studies the traditional agricultural practices of Pacific Islanders.

In addition, two expert reviewers scrutinized the physical geography content to ensure consistent, high-quality coverage of the earth and the environment:

Gary Cummisk, *Dickinson State University*, reviewed every climate map for accuracy and clarity of presentation

Ken Orvis, *University of Tennessee*, reviewed text coverage of climate and the sculpting of the earth.

■ **Most Current Coverage Available** The new edition provides insightful coverage of important recent economic, political, and cultural events and patterns, including

- Domestic and global implications of the U.S. political, economic, and military stances
- The role of terrorism in the realignment of power—globally and locally
- The global phenomenon of labor outsourcing and its effect on countries importing and exporting jobs
- Growing turmoil in Southwest Asia and continuing conflicts in the Middle East
- The expansion of the European Union and its increased global influence

- Global disease management (HIV/AIDS and SARs)
- Cross-border aspects of environmental issues such as Mediterranean river basin management and carbon dioxide emissions

VIDEOS

Vladislav Dudakov came to Moscow as a soldier assigned to guard Lenin's tomb. Now, in a new Russia where entrepreneurship is encouraged, he manages a shop for an investor who has opened up a chain of Starbucks-style coffee shops. One of the unique challenges of his job is managing employees who are used to working for the government. He has to teach Western business values, just as he was taught them during his first Western-style job: working for McDonald's. When McDonald's came to Russia, Vladislav traded the life of a soldier for that of a businessman, working his way from floor sweeper to store manager: "I was different from my peers. They would say, 'Why don't we extend our break a bit? Why do we have to scrub so hard?' My answer was: you must fulfill the task. At McDonald's we were always earning achievement awards. . . . The American style of management became my education." Vladislav's eagerness to learn earned him a trip to McDonald's "Hamburger University" in the United States. Now, he encourages the waitresses at his coffee shop to forget old Soviet habits and to "smile, smile, smile." His success has made Vladislav an example of a new class of Muscovite: young, entrepreneurial, and newly affluent.

Hear more firsthand accounts of Russian capitalism in the Frontline World video "Moscow: Rich in Russia: The Brave New World of Young Capitalists and Tycoons."

(page 250)

New FRONTLINE/World Video Anthology (available in VHS or DVD format) Drawn from the acclaimed PBS series FRONTLINE/World, these ten video segments, each between 10 and 20 minutes long, concern matters both current and relevant. The videos naturally complement not only the subject matter of the text but also its unique approach to world regional geography. Taken together, the book and the videos are an especially engaging educational resource.

FRONTLINE/World *Instructor Video Guide.* The guide provides background information and offers ideas and resources for connecting the videos with classroom discussions and homework assignments.

FRONTLINE/World videos available on VHS or DVD include

Chapter 1   Geography: An Exploration of Connections
            India "Hole in the Wall: Opening the door to cyberspace"
Chapter 2   North America
            Mexico "A Death in the Desert: The fatal journey of a migrant worker"
Chapter 3   Middle and South America
            Guatemala/Mexico "Coffee Country: Can fair trade save the farm?"
Chapter 4   Europe
            Spain "The Lawless Sea: Investigating a notorious shipwreck"
Chapter 5   Russia and the Newly Independent States
            Moscow "Rich in Russia: The brave new world of young capitalists and tycoons"
Chapter 6   North Africa and Southwest Asia
            Iraq "The Road to Kirkuk: After Saddam's terror, can Kurds and Arabs live together?"
Chapter 7   Sub-Saharan Africa
            Nigeria "The Road North: What the Miss World riots reveal about a divided country"
Chapter 8   South Asia
            Bhutan "The Last Place: Television arrives in a Buddhist kingdom"
Chapter 9   East Asia
            Hong Kong "Chasing the Virus: Trying to stop the deadly SARS epidemic"
Chapter 10 Southeast Asia
            Cambodia "Pol Pot's Shadow: Searching for a mysterious executioner"

INSTRUCTOR SUPPLEMENTS

Instructor's CD-ROM, 0-7167-6264-1
To help instructors create their own Web sites and orchestrate dynamic lectures, the discs contain
- **All text images** in PowerPoint and JPEG formats with enlarged labels for better projection quality
- **PowerPoint lecture outlines** for each chapter
- **Test Bank** designed to match the pedagogical intent of the text and offering more than 2000 test questions (multiple-choice, short answer, matching, true/false, and essay) in a Word format that makes it easy to edit, add, and resequence questions

- **Set of 10 lecture questions per chapter** taken from the Test Bank for use with the personal response system (see below)

Instructor's Web Site (password-protected)
In addition to all the resources available on the Instructor's CD-ROM, this password-protected Web site also includes

- **Web Site Guide,** by Tim Oakes and Chris McMorran of the University of Colorado, provides ideas on how to incorporate the *World Regional Geography* media into your course

- **Instructor's Resource Manual,** by Jennifer Rogalsky, State University of New York, Geneseo, and Helen Ruth Aspaas, Virginia Commonwealth University, contains suggested lecture outlines, points to ponder for class discussion, and ideas for exercises and class projects available as chapter-by-chapter Word files to facilitate editing and printing

- Syllabus-posting service

Course Management
All Instructor and Student resources are also available via WebCT/Blackboard to enhance your course. W. H. Freeman and Company offers a course cartridge that populates your site with content tied directly to the book.

Personal Response System
Offered by W. H. Freeman and Company, in partnership with GTCO CalComp (formerly EduCue, "the leader in personal response systems")

The InterWrite™ personal response system is the user-friendly way to make your class time more efficient, interactive, and effective. This wireless, remote system (about the size of a television remote control) lets you pause to ask questions and instantly record student responses, as well as take attendance, direct students through lectures, gauge students' understanding of the material, and much more. For more information, contact your W. H. Freeman representative or contact prs@bfwpub.com

Overhead Transparency Set, 0-7167-6260-9
100 overhead transparencies showing maps from the text are available to adopters. All labels have been resized for easy readability.

Slide Set with Lecture Notes, 0-7167-0000-0
A set of more than 100 slides of *National Geographic* images, available to adopters, provides extra photographs not included in the text along with contextual lecture notes. Additional extension sets are available to continuing and new adopters (see next item).

Regional Slide Sets, 0-7167-0000-0
In addition to the master set, there are 5 region sets (2 regions per set). Available to continuing adopters.

Set 1: North America, Middle America, and South America

Set 2: Europe and Russian Federation, Belarus, Caucasia, and Central Asia

Set 3: North Africa, Southwest Asia, and Sub-Saharan Africa

Set 4: South Asia and East Asia

Set 5: Southeast Asia and Oceania

Test Bank, by Jason Dittmer, Georgia Southern University, and Andy Walter, West Georgia University
Available in Word files on the Instructor's Resource CD and on the password-protected Instructor's Web Site.

Online Quizzing, powered by Questionmark
Instructors can easily and securely quiz students using the on-line multiple-choice Sample Tests (30 questions per chapter). Students receive instant feedback and can take the quizzes multiple times. Instructors can go into a protected Web site to view results by quiz, student, or question, or can get weekly results via email in a simple spreadsheet with all quizzes compiled and graded.

STUDENT SUPPLEMENTS

Mapping Workbook and Study Guide, 0-7167-6261-7
Jennifer Rogalsky, State University of New York, Geneseo, and Helen Ruth Aspaas, Virginia Commonwealth University
This book of mapping exercises and study guide questions develops and hones student skills in geographic analysis within the context of the main themes of each chapter. Mapping exercises help the students understand and explain geographic patterns by employing the skills geographers themselves routinely use.

World Regional Geography Online at www.whfreeman.com/pulsipher3e
Web Site Authors: Tim Oakes and Chris McMorran, University of Colorado, Boulder

- Map Learning Exercises—Students use this interactive feature to identify and locate countries, cities, and the major geographic features of each region. These exercises make learning place locations fun and are instructive for future work.

- Thinking Geographically—Encourages critical reflections for each chapter on the following issues: linkages of trade, finance, tourism, and political movements that connect that region to daily life in the United States. These activities also allow students to explore a set of current issues, such as deforestation, human rights, or free trade, and experience how geography helps clarify our understanding of them. Linked Web sites are matched with a series of questions and/or brief activities that give students an opportunity to think about the ways they are connected to the places and people they read about in the text. Helps students focus on key geography concepts, such as scale, region, place, and interaction, by using these concepts to drive analysis of compelling issues.

- Working with Maps—Offers two sets of map-related exercises that develop student analytical abilities:

  Thematic Maps: Students can place various maps from the text side-by-side in order to compare and contrast data. Associated questions accompany each option.

  Animated Population Maps: Animated maps show how regional populations have changed or fluctuated with time. Related questions ask students how and why the changes may have occurred.

- Thinking Critically—Asks students to think critically about issues and examples highlighted in the textbook.

- Blank Outline Maps—Printable maps of the world, and of each region, for use in note-taking and/or exam review, as well as for preparing assigned exercises.

- Sample Tests—A self-quizzing feature (30 questions per chapter) enabling the student to review key text concepts and sharpen his or her ability to analyze geographic material for exam preparation. Answers (correct or incorrect) prompt feedback referring students to the specific section in the text where the question is covered.

- Flashcards—Matching exercises to teach vocabulary and definitions.

- Audio Pronunciation Guide—Spoken guide of place names, regional terms, and historical figures.

- World Recipes and Cuisines—From *International Home Cooking*, the United Nations International School Cookbook.

Student CD-ROM, 0-7167-6258-7
For students without convenient access to the Internet, the CD provides all the material from the Web site accessible through the same user-friendly navigation. Can be packaged with the text for an additional $3.95 net.

Rand McNally's Atlas of World Geography, 2005 Edition paperback, 176 pages (0-7167-7193-4)
This atlas contains:

- 52 physical, political, and thematic maps of the world and continents; 49 regional, physical, political, and thematic maps and dozens of metro area inset maps.

- Geographic facts and comparisons covering topics such as population, climate, and weather.

- A section on common geographic questions, a glossary of terms, and a comprehensive 14-page index.

ACKNOWLEDGMENTS

Many of our colleagues in geography have reviewed and commented on various drafts of individual chapters or groups of chapters, offering advice on content and perspective. Not only have they greatly improved the quality of the book, but many have offered words of encouragement that were especially needed during this long process.

Third Edition

Kathryn Alftine
California State University at Monterey Bay

Donna Arkowski
Pikes Peak Community College

Tim Bailey
Pittsburg State University

Brad Baltensperger
Michigan Technological University

Michele Barnaby
Pittsburg State University

Richard Benfield
Central Connecticut State University

Sarah Brooks
University of Illinois at Chicago

Jeffrey Bury
University of Colorado at Boulder

Michael Busby
Murray State University

Norman Carter
California State University at Long Beach

Cyrus Dawsey
Auburn University

Elizabeth Dunn
University of Colorado at Boulder

Margaret Foraker
Salisbury University

Robert Goodrich
University of Idaho

Steve Graves
California State University at Northridge

Ellen Hansen
Emporia State University

Sophia Harmes
Towson University

Mary Hayden
Pikes Peak Community College

R.D.K. Herman
Towson University

Samantha Kadar
California State University at Northridge

James Keese
California Polytechnic State University at San Luis Obispo

Phil Klein
University of Northern Colorado

Debra D. Kreitzer
Western Kentucky University

Soren Larsen
Georgia Southern University

Unna Lassiter
California State University at Long Beach

David Lee
Florida Atlantic University

Anthony Paul Mannion
Kansas State University

Leah Manos
Northwest Missouri State University

Susan Martin
Michigan Technological University

Luke Marzen
Auburn University

Chris Mayda
Eastern Michigan University

Michael Modica
San Jacinto College

Heather Nicol
State University of West Georgia

Thomas Paradis
Northern Arizona University

Amanda Rees
University of Wyoming

Arlene Rengert
West Chester University of Pennsylvania

B. F. Richason
St. Cloud State University

Deborah Salazar
Texas Tech University

Steven Schnell
Kutztown University

Kathleen Schroeder
Appalachian State University

Roger Selya
University of Cincinnati

Dean Sinclair
Northwester State University

Garrett Smith
Kennesaw State University

Jeffrey Smith
Kansas State University

Dean Stone
Scott Community College

Selima Sultana
Auburn University

Ray Sumner
Long Beach City College

Christopher Sutton
Western Illinois University

Harry Trendell
Kennesaw State University

Karen Trifonoff
Bloomsburg University

David Truly
Central Connecticut State University

Kelly Victor
Eastern Michigan University

Mark Welford
Georgia Southern University

Wendy Wolford
University of North Carolina at Chapel Hill

Laura Zeeman
Red Rocks Community College

First Edition

Helen Ruth Aspaas
Virginia Commonwealth University

Brad Bays
Oklahoma State University

Stanley Brunn
University of Kentucky

Altha Cravey
University of North Carolina at Chapel Hill

David Daniels
Central Missouri State University

Dydia DeLyser
Louisiana State University

James Doerner
University of Northern Colorado

Bryan Dorsey
Weber State University

Lorraine Dowler
Penn State University

Hari Garbharran
Middle Tennessee State University

Baher Ghosheh
Edinboro University of Pennsylvania

Janet Halpin
Chicago State University

Peter Halvorson
University of Connecticut

Michael Handley
Emporia State University

Robert Hoffpauir
California State University, Northridge

Glenn G. Hyman
International Center for Tropical Agriculture

David Keeling
Western Kentucky University

Thomas Klak
Miami University of Ohio

Darrell Kruger
Northeast Louisiana University

David Lanegran
Macalester College

David Lee
Florida Atlantic University

Calvin Masilela
West Virginia University

Janice Monk
University of Arizona

Heidi Nast
De Paul University

Katherine Nashleanas
University of Nebraska

Tim Oakes
University of Colorado, Boulder

Darren Purcell
Florida State University

Susan Roberts
University of Kentucky

Dennis Satterlee
Northeast Louisiana University

Kathleen Schroeder
Appalachian State University

Dona Stewart
Georgia State University

Ingolf Vogeler
University of Wisconsin, Eau Claire

Susan Walcott
Georgia State University

Second Edition

Helen Ruth Aspaas
Virginia Commonwealth University

Cynthia F. Atkins
Hopkinsville Community College

Timothy Bailey
Pittsburgh State University

Robert Maxwell Beavers
University of Northern Colorado

James E. Bell
University of Colorado, Boulder

Richard W. Benfield
Central Connecticut State University

John T. Bowen, Jr.
University of Wisconsin, Oshkosh

Stanley Brunn
University of Kentucky

Donald W. Buckwalter
Indiana University of Pennsylvania

Gary Cummisk
Dickinson State University

Roman Cybriwsky
Temple University

Cary W. de Wit
University of Alaska, Fairbanks

Ramesh Dhussa
Drake University

David M. Diggs
University of Northern Colorado

Jane H. Ehemann
Shippensburg University

Kim Elmore
University of North Carolina at Chapel Hill

Thomas Fogarty
University of Northern Iowa

James F. Fryman
University of Northern Iowa

Heidi Glaesel
Elon College

Ellen R. Hansen
Emporia State University

John E. Harmon
Central Connecticut State University

Michael Harrison
University of Southern Mississippi

Douglas Heffington
Middle Tennessee State University

Robert Hoffpauir
California State University, Northridge

Catherine Hooey
Pittsburgh State University

Doc Horsley
Southern Illinois University, Carbondale

David J. Keeling
Western Kentucky University

James Keese
California Polytechnic State University

Debra D. Kreitzer
Western Kentucky University

Jim LeBeau
Southern Illinois University, Carbondale

Howell C. Lloyd
Miami University of Ohio

Judith L. Meyer
Southwest Missouri State University

Judith C. Mimbs
University of Tennessee, Chattanooga

Monica Nyamwange
William Paterson University

Thomas Paradis
Northern Arizona University

Firooza Pauri
Emporia State University

Timothy C. Pitts
Edinboro University of Pennsylvania

William Preston
California Polytechnic State University

Gordon M. Riedesel
Syracuse University

Joella Robinson
Houston Community College

Steven M. Schnell
Northwest Missouri State University

Kathleen Schroeder
Appalachian State University

Dean Sinclair
Northwestern State University

Robert A. Sirk
Austin Peay State University

William D. Solecki
Montclair State University

Wei Song
University of Wisconsin, Parkside

William Reese Strong
University of North Alabama

Selima Sultana
Auburn University

Suzanne Traub-Metlay
Front Range Community College

David J. Truly
Central Connecticut State University

Alice L. Tym
University of Tennessee, Chattanooga

This book has been a family project many years in the making. I (Lydia) came to the discipline of geography at the age of five, when my immigrant father, Joe Mihelič, hung a world map over the breakfast table in our home in Coal City, Illinois, where he was pastor of the New Hope Presbyterian church (see my childhood map, page 3). We soon moved to the Mississippi Valley of eastern Iowa, where my father, then a professor in the Presbyterian theological seminary in Dubuque, continued his geography lessons on the passing landscapes whenever I accompanied him on Sunday trips to small country churches. Once we had an especially long lesson when the flooding Mississippi River trapped us for hours. My sons, Anthony and Alex, got their first doses of geography in the bedtime stories I used to tell them. For plots and settings, I drew on Caribbean colonial documents I was then reading for my dissertation. They first traveled abroad and learned about the hard labor of field geography when as mid-sized children they were expected to help with the archaeological and ethnographic research my colleagues and I were conducting on the eastern Caribbean island of Montserrat. Alex, who is a co-author on the second and third editions of this book, is now completing a Ph.D. in geography at Clark University in Massachusetts. Mac Goodwin, my husband, has given up several years of his career as a colonial sites archaeologist to help with the research, writing, and production of this book. It was my brother John Mihelic who first suggested that we write a book like this one, after he too came to appreciate geography. He has been a loyal cheerleader during the process, as have our extended family and friends in Knoxville, Montserrat, Slovenia, and beyond.

My graduate students and faculty colleagues in the Geography Department at the University of Tennessee have been generous in their support, serving as helpful impromptu sounding boards for ideas. Ken Orvis, especially, has advised us on the physical geography sections of both editions. Will Fontanez and the cartography shop staff unfailingly pursued the goal of beautiful informative maps, often cheerfully making adjustments when it was definitely not convenient. Special thanks are owed to Will and to Anna Compton, Andrew Wunderlich, Tom Wallin, Susan Carney, Jennifer Barnes, Hilary Martin, Brad Kreps, Will Albaugh, and Jacqueline McDermott.

Sara Tenney and Liz Widdicombe at W. H. Freeman and Company were the first to persuade us that together we could develop a new direction for *World Regional Geography*, one that included the latest thinking in geography written in an accessible style. In accomplishing this goal, we are especially indebted to our first developmental editor, Susan Moran, and to the W. H. Freeman staff for all they have done to ensure that this book is well written, beautifully designed, and well presented to the public. Susan has been unfailingly wise, temperate, insightful, and probing, yet also kind and patient. Jennifer Van Hove, who ably succeeded her on this third edition, is also a remarkably patient and sensitive editor who smoothes the way for her authors. She was assisted by Sharon Merritt.

We would also like to gratefully acknowledge the efforts of the following people at W. H. Freeman: Susan Brennan, acquisitions editor for the first edition, Jason Noe, for the second and third editions; Mary Louise Byrd, project editor for all three editions; Norma Roche, copy editor; Blake Logan, designer; Bill Page and Shawn Churchman, illustration coordinators; Susan Wein, production coordinator; Sheridan Sellers, W. H. Freeman and Company Electronic Publishing Center; Inge King, photo researcher for the first edition, Meg Kuhta for the second edition; Bianca Moscatelli and Elyse Rieder for the third edition; Elaine Palucki, managing editor for supplements and media; Nick Tymoczko, assistant editor; Lisa Samols, editorial assistant; and Eleanor Wedge and Martha Solonche, proofreaders for the text and maps. We are also grateful to the supplements authors who have created what we think are unusually useful, up-to-date, and labor-saving materials for instructors who use our book: Jennifer Rogalsky and Helen Ruth Aspaas, authors of the *Mapping Workbook and Study Guide* and *Instructor's Resource Manual*; Tim Oakes and Chris McMorran, authors of the Web site; and Andy Walter, and Jason Dittmer, authors of the Test Bank.

CHAPTER 1

GEOGRAPHY
An Exploration of Connections

WHERE IS IT? WHY IS IT THERE?

Where are you? You may be in a house or a library or sitting under a tree on a fine fall afternoon. You are probably in a community (perhaps a college or university), and you are in a country (perhaps the United States) and a region of the world (perhaps North America, Southeast Asia, or the Pacific). Why are you where you are? There are immediate answers, such as "I have an assignment to read." But there are also larger explanations, such as your belief in the value of an education, your career plans, or someone's willingness to sacrifice to pay your tuition. Even past social movements in Europe and America that opened up higher education to more than a fortunate few may help to explain why you are where you are.

The questions *where* and *why* and *how* are central to geography. Like anyone who has had to find the site of a party on a Saturday night, the location of the best grocery store, or the fastest and safest route home, geographers are interested in location, spatial relationships, and connections between the environment and people.

Different places have different sights, sounds, smells, and arrangements of features. Understanding what has contributed to the look and feel of a place, to the standard of living and customs of its people, and to the way people in one place relate to people in other places, near and far, helps to answer the question of why places are as they are. Seeking these answers is the special endeavor of geographers. Furthermore, geographers often think at several scales: for example, when choosing the best local situation for a new grocery store, a geographer might consider the socioeconomic circumstances of the local neighborhood as well as the best location from the perspective of the city at large. A potential site would also be considered in relation to national or even international transportation routes, possibly to determine cost-efficient connections to suppliers.

To make it easier to understand a geographer's many interests, please try this exercise. On a piece of blank paper, draw a map of your favorite childhood landscape. Relax, and let your mind recall the objects and experiences that were most important to you in that place. If the place was your neighborhood, you might start by drawing and labeling your home, then fill in other places you encountered regularly as you went about your life—such as your backyard and the objects in it, your best friend's home, your school. Don't worry about creating a work of art—just make a map that reflects your experiences as you remember them.

When you are finished, think about how the map reveals the ways in which your life was structured by space. Were there places you were not supposed to go, but did anyway? Why were some places off-limits? (And why were they so intriguing to you?) Does your map reveal, perhaps in a very subtle way, such emotions as fear, pleasure, or longing? Does it indicate your sex, your ethnicity, or the makeup of your family? Did you draw some places in greater detail than others? What is the **scale** of your map? That is, how much space did you decide to illustrate on the paper? Did you draw a house and yard, a neighborhood, or a square mile or more? The amount of space you covered with your map may represent the degree of freedom you had as a child to go about on your own, or how aware you were of the wider world around you. Did you give the map reader some clue about the scale of your map—how much space is represented?

In making your map and analyzing it, you have engaged in several aspects of geography:

- Landscape observation

- Description of the earth's surface and consideration of the natural environment

- Historical reconstruction of bygone places

- **Spatial analysis** (the study of how people, objects, or ideas are, or are not, related to one another across space)

- The use of different scales of analysis (your map probably shows the spatial features of your childhood at a detailed local scale)

- Cartography (the making of maps)

As you progress through this book, you will acquire geographic information and skills. Whether you are planning where to jog, looking at a photograph of an intriguing faraway landscape, thinking about investing in East Asian timber stocks, searching for a good place to market an idea, or trying to understand current local events in the context of world events, knowing how to practice geography will make the task easier and more engaging.

WHAT IS GEOGRAPHY?

Geography might be defined as the study of our planet's surface and the processes that shape it. Yet such a succinct definition does

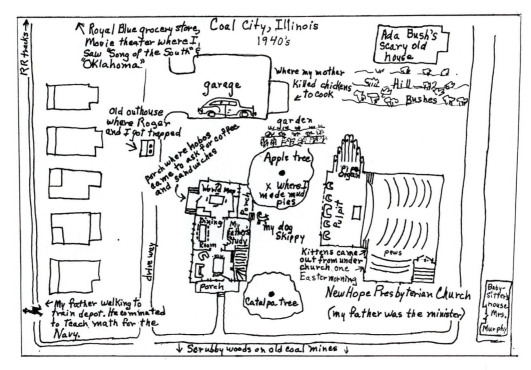

Lydia Pulsipher's childhood landscape map.

not begin to convey the fascinating interactions of physical and human forces that have given the earth its diversity of landscapes and ways of life. Geography, as an academic discipline, links the physical sciences—such as geology, physics, chemistry, biology, and botany—with the social sciences—such as anthropology, sociology, history, economics, and political science. **Physical geography** is the study of earth's physical processes to learn how they work, how they affect humans, and how they are affected by humans in return. **Human geography** is the study of various aspects of human life that create the distinctive landscapes and regions of the world. Physical and human geography are often tightly linked. For example, geographers might aim to understand

- how and why people came to occupy a particular place

- how they assess the physical aspects of that place (climate, landforms, resources) and then set about using and modifying them to suit their particular needs

- how they interact with other places, far and near

People are not always bound by their physical surroundings. Transport technology and communications media link people in one physical place to distant locations and resources, so people no longer need to design their lives around what is available locally. Thus the investigations of some geographers may seem to have little direct connection to the physical environment. For example, geographers may study such issues as political activism among homeowners, the spatial allocation of tax dollars, and socio-economic variables in housing choice.

Geographers are unified by a common interest in understanding the world in spatial terms. Just as historians study change over time, geographers study variation over spaces, large and small. Most of all, geographers are interested in the explanations for those variations. Among geographers' most important tools are maps, which they use to record and analyze spatial relationships, just as you recorded and analyzed the features that made up your favorite childhood landscape. The following section on map reading will help you to understand the maps in this book and elsewhere. Where appropriate, map captions will identify and explain the type of map being depicted.

READING MAPS

When geographers want to show spatial aspects of particular topics on a map—say, the locations and relative sizes of islands in the Caribbean Sea—they can begin by deciding at what scale the information should be mapped. The scale of a map represents the relationship between the distances shown on the map and the actual distances on the earth's surface. Sometimes a numerical ratio shows what one unit of measure on the map equals in the same units on the face of the earth: for example, 1:1,000,000, or the fraction 1/1,000,000, means that 1 inch or 1 centimeter on the map equals 1 million inches or 1 million centimeters on the face of the earth. Other maps may express scale using a phrase such as "1 centimeter equals 10 kilometers." Alternatively, a simple bar may express the information visually:

Thinking Geographically: ON THE WEB

A. The scale of this map is 1:3,000,000.

B. The scale of this map is 1:15,000,000.

C. The scale of this map is 1:45,000,000.

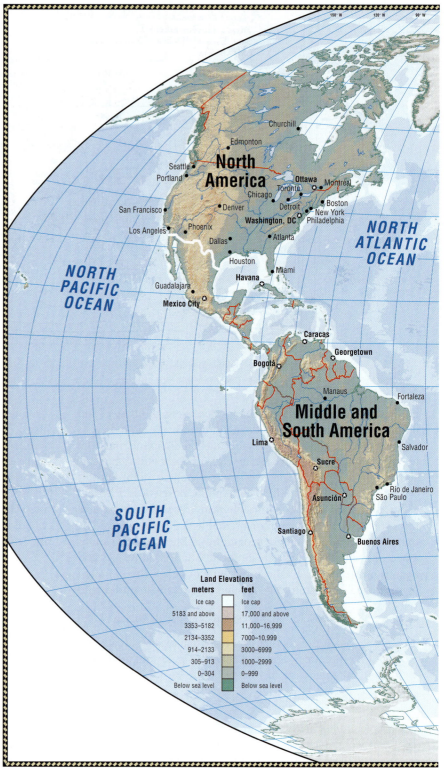

D. World map.

FIGURE 1.1 Examples of scale. A is a map of
Guadeloupe and Dominica in the eastern Caribbean.
The scale of 1:3,000,000 makes it possible to show towns,
a few roads, and a few landforms, but not much else. Map B is at a scale of 1:15,000,000. You can see much more
of the eastern Caribbean, but the only detail that can be shown is the shape of the islands and the location of the
capital cities. Map C, at a scale of 1:45,000,000, shows most of the Caribbean Sea and its general location between
Central and South America, but now the eastern Caribbean islands are too small to identify clearly. Map D, at a
scale of 1:100,000,000 is of the world. It has a colored elevation bar and a numerical scale in miles and kilometers.
[Adapted from *The Longman Atlas for Caribbean Examinations*, 2nd ed. (Essex, UK: Addison Wesley Longman, 1998), p. 4.]

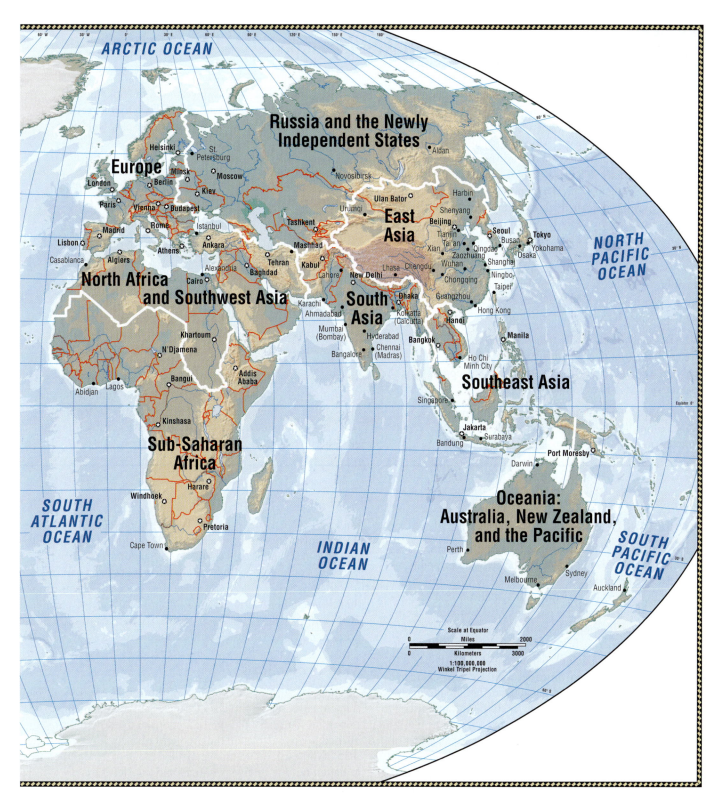

Notice that each of the maps (A, B, C) of the eastern Caribbean shown in Figure 1.1 is drawn using a different scale. As the amount of area shown on a map becomes larger, the amount of detail that can be displayed decreases.

The *title*, *caption*, and *legend* give basic information about the map. The title tells you the subject of the map, and the caption usually points out some features of the map that the author wants you to notice. The legend is the box that explains what the symbols and colors on the map represent (see, for example, Figure 1.6 on page 12, showing Cisco Systems offices worldwide). The map's scale is usually given in or near the map's legend, unless, as in Figure 1.1D, it is a world map. This world map, for example, has both a legend and a scale bar.

The detective work of photo interpretation

While maps are useful for translating vast distances or extensive quantities of information into a small, two-dimensional space, photos can be used to catalogue the vast amounts of physical and cultural data in a small space. Most geographers make use of photographs to help them understand or explain a geographic issue or depict the character of a place. It takes special skills to extract information from photographs. Here are some points to keep in mind as you look at pictures in this book or anywhere for that matter.

Make a mental inventory of everything in the picture that relates to geography. Most observable features will relate to geography in one way or another.

■ **Landforms:** Notice the lay of the land, and briefly describe the landform features. Speculate on any influences the landforms may have on climate or human activities.

■ **Vegetation:** Notice whether the visible vegetation indicates a wet or dry environment. Can you recognize species, such as a palm tree or an evergreen? Does the vegetation appear to be natural or disturbed by human use?

■ **Ambience:** What is the mood of the picture, time of day, temperature, weather, visible environmental issues? Is emotion apparent in the picture?

■ **Material culture:** What is indicated by the buildings, the architecture, or any tools you see? Do they reveal a distinctive design or reveal a particular level of technology or wealth?

■ **Activities and their possible meaning:** If there are people in the photo, what are they doing? How do their activities convey their level of technology or their living standards?

■ **Evidence about the economy:** Can you see evidence of the global economy or does the landscape appear to be influenced only by local issues? Is this picture of a wealthy or poverty-stricken place?

■ **Location:** Based on your observations, can you tell where the picture was taken? Can you narrow down the possible locations that might be depicted? Give your rationale.

Now, think of every possible statement you can make about the picture, taking note of any doubts you have. Your doubts may be useful in your detective work. You can use this system to analyze any of the photos in the book. Also, every chapter has special interpretation exercises.

Longitude and Latitude

Most maps contain *lines of latitude and longitude*, which enable us to establish a position on the map relative to other points on the globe. European cartographers developed latitude and longitude lines so that navigators far out at sea, with no visible landmarks and only the stars and the passage of time to orient them, could more easily locate themselves on a map. Lines of longitude (also called meridians) run from pole to pole; lines of latitude (also called parallels) run parallel to the equator (Figure 1.2).

Both latitude and longitude lines describe circles, so there are 360° in each circle of latitude and in each circle of longitude and 180° for each hemisphere. Each degree spans 60 minutes (designated with the symbol '), and each minute has 60 seconds (designated with the symbol "). Keep in mind that these are measures of relative space on a circle, not time, nor even real distance (because the circles of latitude get successively smaller to the north and south of the equator, until they become a virtual dot at the poles).

The *prime meridian*, 0° longitude, runs from the North Pole through Greenwich, England, to the South Pole. The half of the globe's surface west of the prime meridian is called the Western Hemisphere; the half to the east, the Eastern Hemisphere. The longitude lines both east and west of the prime meridian are labeled by their direction and distance in degrees from the prime meridian, from 1° to 180°. For example, 20 degrees east longitude would be written as 20°E. The longitude line at 180° runs through the Pacific Ocean and is used as the *international date line*, where the calendar day officially begins. The equator is 0° latitude; the poles are at 90° north and south of the equator.

Lines of longitude and latitude form a grid that can be used to designate the location of a place. In Figure 1.1A, you can see that the island of Marie Galante lies just south of the parallel at 16°N and just about 0.5° west of the 61st meridian. Hence, the position of Marie Galante's northernmost coast is 16°N, 61.5°W.

Map Projections

All maps must solve the problem of showing the spherical (ball-shaped) earth on a flat piece of paper. When reading a map, it is important to understand the limitations of the different strategies devised to do so. Imagine the problems of distortion if you drew a map of the earth on an orange, peeled the orange, and then tried to flatten out the orange-peel map and transfer it exactly to a flat piece of paper.

The various ways of showing the spherical surface of the earth on flat paper are called *projections*. All projections entail some distortion. For maps of small parts of the earth's surface, the distortion is minimal. Developing a projection for the whole surface of the earth that minimizes distortion is much more challenging. The *Mercator projection* (Figure 1.3A) is popular, but geographers rarely use it because of its gross distortion near the poles. To make his flat map, the Flemish cartographer Gerhardus

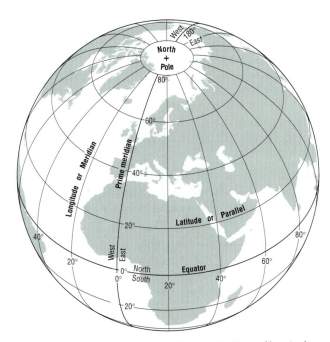

FIGURE 1.2 Summary of latitude and longitude. Lines of longitude (meridians) extend from pole to pole. The distance between them decreases steadily toward the poles, where they all meet. Lines of latitude (parallels) are equally spaced north and south of the equator and intersect the longitude lines at right angles. The only line of latitude that spans the complete circumference of the earth is the equator; all other lines of latitude describe ever smaller circles heading away from the equator. For example, the 60th lines of latitude (parallels) north and south are one-half the circumference of the equator. [Adapted from *The New Comparative World Atlas* (Maplewood, N.J.: Hammond, 1997), p. 6.]

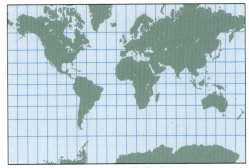

A. Mercator Projection

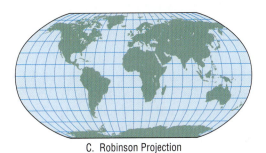

B. Goode's Interrupted Homolosine Projection

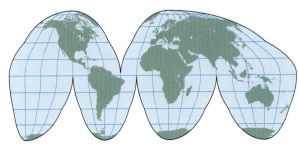

C. Robinson Projection

FIGURE 1.3 Three common map projections. [Mercator and Robinson projections adapted from *The New Comparative World Atlas* (Maplewood, N.J.: Hammond, 1997), pp. 6–7. Goode's interrupted homolosine projection adapted from *Goode's World Atlas*, 19th ed. (Chicago: Rand McNally, 1995), p. x.]

Mercator (1512–1594) stretched out the poles, depicting them as lines equal in length to the equator! Greenland, for example, appears about as large as Africa, even though it is only about one-sixth Africa's size. Nevertheless, Mercator's projection is still useful for navigation because it portrays the shapes of landmasses more or less accurately, and because a straight line between two points on this map gives the compass direction between them. Furthermore, this type of projection can be safely used for parts of the globe that are within 15 degrees south or north of the equator because the distortion in this range is minimal.

Goode's interrupted homolosine projection (Figure 1.3B) flattens the earth rather like an orange peel, thus preserving some of the size and shape of the landmasses. In this projection, the oceans get snipped up. The *Robinson projection* (Figure 1.3C) sacrifices accuracy at the poles for an uninterrupted view of land and ocean. In this book we often use the Robinson projection for world maps.

Maps are not unbiased. Most currently popular world projections reflect the European origins of modern cartography. Europe or North America is usually placed near the center of the map, where distortion is minimal; other population centers, such as East Asia, are placed at the highly distorted periphery. For less biased study of the modern world, we need world maps that center on different parts of the globe. For example, much of the world's economic activity is now taking place in and around Japan, Korea, China, Taiwan, and Southeast Asia. Discussions of this activity require maps that focus on these regions but still include the rest of the world in the periphery. Similarly, political developments in the Middle East or in Southwest Asia can best be depicted graphically on world maps that center on such places as Palestine or Iraq, so that the viewer can understand these places as they relate spatially to the wider world.

WHAT DO GEOGRAPHERS DO?

Geographers usually specialize in one or more fields of study, or *subdisciplines*. Table 1.1 gives examples of the different kinds of geographers who might be found in any college or university geography department or working for the government or a private firm. As you read through the descriptions, it is easy to see that cooperation among the various kinds of geographers would be useful in studying the interactions between people and places. For example, climate geographers, cultural geographers, and economic

TABLE 1.1   The subdisciplines of geography

Physical Geographers

Geomorphologists further our understanding of how landforms such as mountains, plains, river valleys, and beaches developed in the past and continue to change. Geographers specializing in geomorphology are often particularly interested in how people alter landforms through deforestation, cultivation, and the manipulation of watersheds and other environments. Geomorphologist Carol Harden has studied the eroding effects of rainfall on unpaved roads and paths in mountainous Ecuador by artificially creating a small rainstorm with a portable rainfall simulator and measuring the sediment displaced by the water.

Climatologists look at long-term weather patterns (climates), studying the interaction of climate, vegetation, and landforms. Recently, climatologists have begun to study the possibility that humans are altering climate on a worldwide scale by releasing carbon dioxide and other gases into the atmosphere, which results in global warming. The cause is thought to be the burning of wood and fossil fuels (oil, gas, and coal).

Biogeographers study the geography of life on earth: where plants, animals, and other living organisms are found today, where they were found in the past, and the circumstances that affected these distributions. They also examine how the distributions of ecosystems are shaped by human activity, climate change, plate tectonics, and other environmental and evolutionary processes. Biogeographer Sally Horn takes cores from the sediments of lakes in Central America and studies ancient pollen grains, charcoal fragments, and other material in the cores in order to chart the sequence of climatic conditions, fire history, and human impacts in lake basins over past thousands of years.

Human Geographers

Cultural geographers describe and seek to explain the spatial patterns and ecological relationships of culturally distinct groups of people. A cultural geographer might study how the religious or philosophical beliefs of a particular culture group influence the ways the members of the group practice agriculture, see themselves in relation to nature, organize their settlements, cope with long-term physical hardship and discrimination, or set down specific behavior rules for males and females from childhood through old age.

Human (also called **political** or **cultural**) **ecologists** combine the interests of biogeographers with those of cultural and political geographers. They examine the specific ways in which humans and physical environments (including all types of life-forms) interact. In Tanzania, political ecologist Richard Schroeder recently interviewed tourism operators, government and nongovernmental agency officials, foreign investors, and Masai activists to understand the nature of conflicts over wildlife management in northern Tanzania.

Historical geographers study long-term spatial processes, such as migrations; changes over time in human habitats; the effects of colonization; and the spread of technology, ideas, and other aspects of culture. They often use the techniques of biogeographers and cultural, political, and economic geographers in their analyses of past circumstances. Historical geographer Victoria Berry studies the history of medicinal plant use in the eastern Caribbean in order to establish how the many culture groups now living in this region contributed to plant introductions and customs of plant processing and use.

geographers are now cooperating to find out the spatial distribution of carbon dioxide emissions, which cultural practices might be changed to limit such emissions, and what the economic effects of such changes might be. Geographers interested in understanding the roots of terrorism are finding that historical geographers can shed light on how European colonialism set up situations that frustrated the development of democratizing customs, and that economic geographers can explain how the inequitable distribution of wealth and well-being can lead to disaffection and even violence, especially among the young. Cultural geographers specializing in the study of religion can examine the extreme circumstances under which fundamentalist religious beliefs can lead to the rationalization of terrorist violence. In order to communicate their findings to colleagues, individual geographers often read and write in several subdisciplines, and geography journals such as *Focus*, the *Geographical Review*, or the *Professional Geographer* typically contain articles on a wide array of topics.

Many geographers are led by their interest in complex spatial relationships and by the difficulties of analyzing any phenomenon solely on a global scale to specialize in a particular region of the world, or even in a small place within that region. They may study their chosen location from several different geographic perspectives: how people there make a living, how the religions they practice affect their use of resources, the ways in which they allocate power and resources among themselves. **Regional geography** is the analysis of the geographic characteristics of a particular place (the size and scale of that place can vary radically). The study of a particular place from a geographic perspective can reveal previously unappreciated connections among physical features and ways of life, as well as connections to other places. These links are a key to understanding the present (and the past) and are essential in planning for the future.

This book follows a "world regional" approach because experience shows that people new to geography find general knowledge about specific regions of the world to be the most interesting and useful introduction to the subject. Such knowledge is produced by all the types of geographers described in Table 1.1, and in this book we rely on all of them to provide a holistic perspective on each of the world's regions. Just what geographers mean by *region* is discussed next.

(Human Geographers continued)

Economic geographers look at the spatial aspects of economic activity, such as how resources are allocated and exchanged from place to place and what happens when these allocations change for some reason. They also examine how people interact with their environment as they go about earning a living. For example, an economic geographer might focus on where women tend to work within a city, why they work there, and how they get there. Economic geographers might also study the amount and type of space occupied by particular economic enterprises, such as popcorn vending, shopping malls, or the international cocaine trade.

Political geographers are interested in the spatial expressions of power and in the institutions people have devised to channel that power, such as representative democracies or authoritarian religious states. Often political power struggles are signified by landscape features such as the Great Wall of China, the Berlin Wall, or the burned villages of Bosnia. Political geographer Christine Drennon studies how Christian and Muslim people in the Balkans viewed themselves in ways that until recently allowed very different groups to coexist as neighbors and friends.

Urban geographers study spatial patterns and processes within cities and the ways in which urban areas interact with surrounding suburban and rural areas. In order to make sense of ever-changing urban landscapes, urban geographer Larry Ford likes to combine "lurking" in city streets, taking photos, and making sketches with compiling facts and figures from census data, tax rolls, and voting information.

Gender geographers offer a new perspective on the world by examining gender relations and the roles and status of men and women in society. They also explore how activities that are linked in some way to a person's sex are expressed spatially, and analyze the meanings of these spatial patterns. Geographer James Tyner has studied how government and private institutions in the Philippines have created gendered patterns of international labor migration. His subsequent work addresses the policy implications of gender and migration.

Technical Geographers

Cartographers specialize in depicting geographic knowledge graphically in maps. They are interested in both the science and the aesthetics of portraying visual information for the map reader. Although some maps are still drawn by hand, computers are increasingly the cartographer's chief tool. Virtually all of the maps in this book were created with a computer.

Remote sensing analysts discern patterns on the earth's surface by examining high-altitude photographs and other images collected from space by military and space programs. Remotely sensed information can be used to determine changes in land use, military installations, loss or increase in vegetative cover, and even below-soil-surface archaeological patterns.

Geographic Information Systems (GIS) analysts use computers and software to capture, store, integrate, analyze, and display large bodies of statistical information about spatial relationships. A GIS analyst might chart the best way to deliver food aid in famine situations to those who need it most. Recently, geographers have begun looking at complex cultural and social problems, such as welfare reform in the United States. GIS specialist Jennifer Rogalsky helps government officials to understand the spatial problems welfare recipients must resolve if they are to become self-sustaining. She analyzes the distribution of welfare recipient residences and the location of training, transport, and child-care facilities, as well as the location of potential jobs, to demonstrate whether it is indeed feasible for welfare mothers to work outside the home.

THE REGION AS A CONCEPT

Thinking Geographically: ON THE WEB

A **region** is a unit of the earth's surface that contains distinct patterns of physical features or of human activities. We could speak of a desert region, a region that produces table wine, or a region that at a particular time is characterized by ethnic violence. Geographers find that a precise definition of the concept of "region" is elusive. For one thing, it is rare for any two regions to be described by the same set of indicators. One assemblage of places might be considered a region because of its distinctive vegetative landscape and such cultural features as foods, dialects, and music; see if you can list traits that characterize and distinguish the southern and southwestern regions of the United States. Looking beyond the United States, a region might be defined primarily by its cold climate and sparse population, such as Siberia. Still another group of places might be considered a region primarily because of common historical experiences. Middle and South America are regions in large part

because they experienced colonization by Spain and Portugal after 1492.

Another problem in defining regions is that their boundaries are rarely crisp. The more closely we look at the border zones, the fuzzier the divisions appear. Take the case of the boundary between the United States and Mexico (Figure 1.4). Just where is the true regional boundary? What criteria should we use to establish this boundary? Although there is a clearly delineated political border between the two countries, it is not really a marker of separation between cultures or even between economies. In a wide band extending over both sides of the border, one can find a blend of Native American, Spanish colonial, Mexican, and Anglo-American cultural features: languages, place names, food customs, music, and family organization, to name but a few. And the economy of the border zone—which is based on agriculture, manufacturing, trade, and tourism—depends on interactions across a broad swath of territory. In fact, the Mexican national economy is becoming more closely connected to the economies of the United States and Canada than to those of its close neighbors

FIGURE 1.4 Language patterns along the border of the United States and Mexico. The map reveals that the legal boundary is not a cultural boundary. Along the border there is a wide band of cultural blending. In this blended zone, language, food, religion, and architecture show influences of both U.S. and Mexican culture. [Adapted from *The World Book Atlas* (Chicago: World Book, 1996), p. 89.]

in Middle America (the countries of Central America and the Caribbean).

Why, then, does this book place Mexico in Middle America? At this point in time, Mexico still has more in common overall with Middle America than with North America. Its colonial history and the resulting use of Spanish as its official language tie Mexico to Middle and South America, where the dominant language is Spanish (in Brazil, it is Portuguese). The dominant language of the United States and Canada is English (except in Québec, where French is primary and English secondary). These language patterns are symbolic of the larger cultural and histori-cal differences between the two regions, which will be discussed in Chapters 2 and 3. Even though Mexico has much in common with parts of the southwestern United States, on the whole the similarities do not yet override the differences. Note, however, that this situation is changing as the economic connections among the United States, Canada, and Mexico grow stronger and as more citizens from all three countries emigrate across the bor-ders. Indeed, in the not too distant future it might make sense to include Mexico in the same world region as the United States and Canada.

If regions are so difficult to define and describe and are so changeable, why do geographers use them? The obvious answer is that it is impossible to discuss the whole world at once, and so we seek a reasonable way to divide it into manageable parts. There is nothing sacred about the criteria or the boundaries we use; they simply need to be practical. In defining the world regions for this book, we have considered such factors as physical features, political boundaries, cultural characteristics, and what the future may hold. Geographers find that using multiple factors enables them to arrive at regional definitions with which most people can agree.

We have organized the material in this book into four levels, or scales: the global scale, the world regional scale, the subre-gional scale (Figure 1.5), and the local scale. At the **global scale,** explored in this chapter, the entire world is treated as a single area—a unity that is still new to all of us but is more and more relevant as we become used to thinking of our planet as a global system. We use the term **world region** for the largest divisions of the globe, such as East Asia, Southeast Asia, and North America. There are ten world regions, each of which is covered in a separate chapter. The world regions are then divided into **subregions,** which may be independent countries, groups of countries, or even parts of a single country. For example, the world region of North America contains several subregions. In Canada, the province of Québec is considered a subregion because of its distinct French heritage. In the United States, no one state stands out so distinctly, and there subregions tend to be groups of states, such as the Southwest or the Northeast. In several cases, subregions include territory in both the United States and Canada.

Most of us are familiar with geography at the **local scale,** meaning the places where we live and work, whether in a city, town, or rural area. Our local geography influences whether we commute to work or work from home, whether we have diverse opportunities or few. The options afforded by our local situation shape our lifestyles and the culture we create together. Through-out this book, life at the local scale is illustrated with vignettes about people in a wide range of places and cultural, economic, and political contexts. Often the lives of individuals reflect trends that can be tracked across several world regions or may even occur at the global scale.

The regions, subregions, and local places discussed in this book vary dramatically in size and complexity. A region can be relatively small, such as Europe, or very large, as in the case of East Asia. A subregion can be as large as a continent: Australia, for example, is a *subregion* of the huge, but sparsely populated, *world region* of Oceania. While Australia has an area of nearly 3 million square miles (7.8 million square kilometers), another subregion of Oceania is a small group of tiny islands known as Micronesia, which together cover only 270 square miles (700 square kilome-ters). Continuing with Oceania as an example, at the local scale, a place can be a backyard in Polynesia, a neighborhood in Honolulu, or a town in New Zealand.

In summary, regions have the following traits:

- A region is a unit of the earth's surface that contains distinct environmental or cultural patterns.

- Regions are constructed by people to help them define spaces for varying purposes; hence regional definitions are fluid.

- No two regions are described by the same set of indicators.

- Regions can vary greatly in size (scale), from part of a small island to the size of a continent, or even a hemisphere.

- The boundaries of regions are usually fuzzy and hard to agree upon.

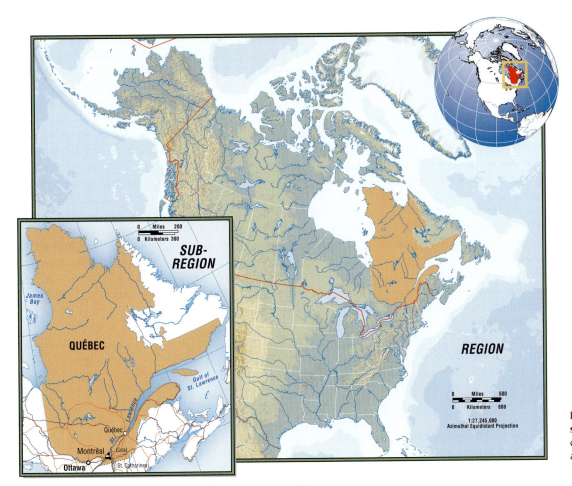

SUB-REGION

QUÉBEC

James Bay

Gulf of St. Lawrence

St. Lawrence

Québec

Montréal   Laval

Ottawa   St. Catharines

0 Miles 200
0 Kilometers 300

REGION

0 Miles 500
0 Kilometers 800

1:27,245,000
Azimuthal Equidistant Projection

FIGURE 1.5 Three regional scales: the world (upper right corner), a region (North America), and a subregion (Québec).

INTERREGIONAL LINKAGES AND GLOBALIZATION

Different regions (or subregions) often share some similarities, and they may have overlapping borders. Consequently, **contiguous regions**—those that lie next to or near each other—often interact in various ways. They may trade with each other, share labor pools of available workers, experience each other's economic recessions and booms, borrow new ideas from each other, and help each other in times of crisis.

Nonetheless, our present era is remarkable in that regions widely separated in space can have interdependent economic relationships that used to be possible only between close neighbors. For example, much of the clothing and tropical hardwoods used by North Americans come from Southeast Asia. Likewise, South Americans and Africans produce cash crops that end up on European tables. These connections between distant regions (**interregional linkages**) began to attain their present reach during the early stages of **European colonialism:** the practice of taking over the human and natural resources of often distant places in order to produce wealth for Europe. The European voyages to America in 1492 led Spain and Portugal to establish colonies in Middle and South America. The British, Portuguese,

Spanish, Dutch, and French founded colonies in North America and Asia in the sixteenth and seventeenth centuries and in the Pacific islands in the eighteenth century. By the end of the nineteenth century, most of Africa was divided into European colonies. Although most colonies are now independent countries, European colonization led to transglobal ties that still remain. For example, there are Muslims of Indonesian (Javanese) descent who have lived for generations in Suriname on the northern coast of South America (see the photo in Chapter 3 on page 167). Suriname and Indonesia were both colonies of the Dutch.

Wherever it took hold, European colonization changed local economies and landscapes. In the Caribbean and the islands of Southeast Asia, for example, forest cover was lost through plantation agriculture, logging, and mining. Sometimes valuable cash crops and the laborers to grow them were transferred to entirely new places; for example, nutmeg trees were sent from the Molucca islands of Southeast Asia to the British colony of Grenada in the Caribbean. There, with the help of enslaved labor from Africa, nutmeg plantations became an important source of income for Britain. Similarly, sugarcane, originally from Southeast Asia, became a widely grown cash crop in tropical America, and cacao (chocolate), from Middle America, became a cash crop in colonial West Africa. Many of the broad regional linkages we see today evolved from linkages that began hundreds of years ago, in large part under the influence of European colonialism.

Cisco Systems, Inc., is a global company that provides Internet networking at the local, regional, and global scales. Cisco sells the hardware and services that make it possible to link a company's computers together and then link the company to the World Wide Web. Cisco is young; it shipped its first product in 1984 and just 16 years later (in fiscal 2000) had revenues of $18.9 billion. Cisco has sales and service offices in more than 60 countries and sells its products in 115 countries (Figure 1.6). It trains thousands of technicians in its "Network Academies," and those technicians are hired either by Cisco or by its customers to provide service to its Internetworking systems worldwide. The corporate headquarters are in San Jose, California, and the company has operational centers in North Carolina, Texas, Massachusetts, the United Kingdom, and Australia. Cisco employs nearly 35,000 people worldwide.

Those who study cultural change have observed that the present revolution in information technology may rank with the two previous major technological revolutions in human history: the agricultural revolution and the industrial revolution. To be a revolution of the same magnitude, information technology would have to transform the ways in which people relate to their environments and to one another. Is that happening, and how inclusive is the transformation? Critics have noted a "digital divide" that separates those enjoying new opportunities via computers and the Internet from the billions of earth's inhabitants who have no access to new technologies and will likely be left further and further behind.

John Chambers, president and CEO of Cisco, says the transformation is real and spreading widely. He describes what he sees as a fundamental change in human society enabled by the open communication possible over the Internet. There is an emerging "Internet ecosystem," he says, that encourages cooperation and collaboration along lines that blur social and power differences. To businesspeople, this trend could mean less emphasis on cutthroat competition and more on prospering through interdependency. Cisco, like some other corporations and the United Nations, is attempting to close the digital divide by placing technology training facilities in a wide range of global locations (see the opening vignette in Chapter 6 and Figure 6.1).

John Chambers is not the first to claim that information technology will fundamentally change the ways people relate to one another. But to what extent are these claims believable, especially in the world of commerce? Couldn't this just be business hype, calculated to position a company to make more money? Answering these questions will take time and careful observation to see just where the advantages of, and profits from, information technology are allocated. The following Web sites are a good starting point for such an assessment:

- http://www.cisco.com/edu/emea
- http://cisco.netacad.net/public/academy/
- http://www.cisco.com/public/countries_languages.shtml
- http://www.netaid.org/projects/index_html
- http://www.undp.org/hdr2003
- http://cisco.netacad.net/public/news/Stats.html

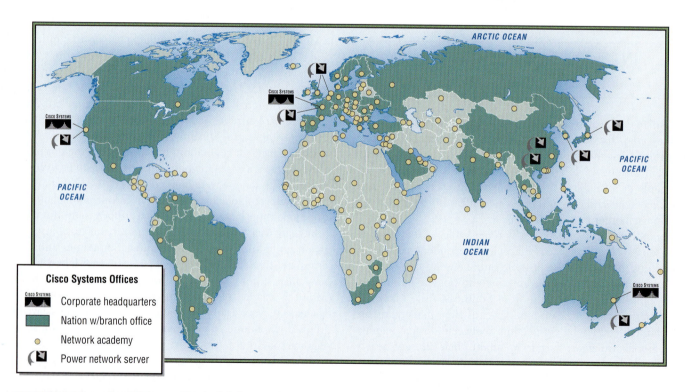

Cisco Systems Offices

CISCO SYSTEMS	Corporate headquarters
■	Nation w/branch office
○	Network academy
◪	Power network server

FIGURE 1.6 Interregional linkages: Cisco's global network. Cisco provides hardware and services for linking computers. This map shows several levels of Cisco's activities: multiple corporate headquarters, countries with branch offices, those with Network Academies for training, and those with power network servers. Can you offer an explanation for the clusters (or lack of clusters) in various parts of the world?

Today, because of rapid transportation and the speedy flow of electronic information, widely separated places can be so intimately linked that the details of their relationships can change from day to day. For example, fluctuations in the market price of cacao in London can alter overnight the food that a small cacao grower in West Africa can afford for his family. Construction workers in Saudi Arabia can be sending substantial amounts of their wages home to their families in Pakistan and Malaysia. Then, in a matter of days, hostilities in and around Saudi Arabia can send hundreds of thousands of workers scurrying home, not only disrupting construction in Saudi Arabia but also reducing tens of thousands of family budgets in Pakistan and Malaysia and limiting the fortunes of shopkeepers in their home villages. Paradoxically, as local places react to the **homogenization** (widespread sameness) that often comes with globalization, they often search for ways to show how very different they are from all other places. Furthermore, modern electronic communication can actually help groups of migrants to maintain an identity with home, even though they may be living far away.

The term **globalization** encompasses the many types of interregional linkages and flows and the changes they are bringing about. Here are some trends that have contributed to the globalization of societies around the world.

- As the volume of international trade grows, localities are becoming increasingly connected to this trade, but local producers and workers have little power over the circumstances of the new trade links.

- Businesses and corporations in rich countries have established manufacturing, sales, and service branches in poorer countries to be closer to their markets and to take advantage of low-cost resources, land, labor, and professional expertise. See, for example, the map (Figure 1.6) in the box "Cisco Systems" on the opposite page. Some transnational or multinational organizations also find lower taxes and less rigorous environmental and trade regulations than in their home countries.

- As manufacturing and some information processing jobs move to places where labor is cheaper, workers in the old locations are left unemployed and must prepare for other types of work by retraining or moving, while workers in the new location may be forced by need to work at very low wages and in inhumane conditions.

- International migration is surging as people seek opportunities in better locations.

- As people, goods, and capital flow across regional and national borders in increasing volumes, the world's nation-states must vie with international and supranational organizations to control the new flows and relationships.

- Increased cross-cultural contact is leading to greater homogenization of culture, especially through the spread of modern popular culture. But, at the same time, rapid communication is reinforcing cultural differences and enabling the development of distinct cultural identities.

- Technology now facilitates the rapid transfer of money, making it easier for legitimate firms to do business and for migrant workers to send money to their families. Meanwhile, illegal activities, such as drug trading and theft, also benefit from rapid international money transactions, in part because identities can be disguised more easily. However, access to technology, such as the Internet, is unevenly distributed, resulting in what is called "the **digital divide**" (see the box "Cisco Systems" and the Video Vignette "Hole in the Wall" below).

- Local environments increasingly show the effects of worldwide trade and resource consumption. Resources are moved from one region to serve the needs of another, environmental degradation in one place is often the result of demands for resources or goods in other places, and air and water pollution can flow across national borders.

Because globalization has multiple facets—economic, social, political, environmental—that influence one another, we will discuss globalization at many points in this book. Individuals can be harmed or empowered by its effects—and sometimes both at once. As you read the many references to globalization in the following sections, reflect on how you as an individual are part of the picture through the inexpensive products made by low-wage labor you buy at Wal-Mart, or through the factory or high-tech job lost by a relative when the company moved abroad.

 **Sugata Mitra:** "The hole in the wall gives us a method to create a door, if you like, through which large numbers of children can rush into this new arena. When that happens it will have changed our society forever."

Rajinder is an eight-year-old slum kid in New Delhi, India, who has a history of not doing well in school. Yet, in just a few minutes, he has leapt across the digital divide after encountering one of Sugata Mitra's kiosk computers. The **digital divide** is the gap between the small percentage of the world's population that has access to computers and the huge majority who do not.

Mitra is head of research and development for NIIT, a computer software and training company with offices around the world and annual sales of $300 million. NIIT was interested in developing kiosk computers for the global market—those that give passersby quick access to the Internet.

Supposing that kids, given access to information, would teach themselves to use unfamiliar technology, Mitra cut a hole in the wall of his office compound in New Delhi and installed a high-speed computer with Internet access facing an adjacent slum. As he had guessed, within hours the local kids were browsing the Internet. There "was a spiral of self-instruction," says Mitra, with one kid making a discovery, three witnessing it, saying, "Cool!" and then sharing three or four more discoveries while they explored together. In a matter of 5 hours, Rajinder had visited the Disney site, learned to use a drawing tool, and read news stories about the Taliban in Afghanistan. A young girl found a graphics program to help her father, a tailor, design the clothes he sews.

There are now more than 100 computers in slum neighborhoods across India. This is important, because although India is a leader in technological development, it is plagued by a digital divide. In the southern state of Karnataka, where Bangalore is located, better known as India's "Silicon State," and ranked by the United Nations as the world's fourth-best hub of technological innovation, 85 percent of people still don't have access to a computer, and 100,000 school-aged children don't go to school.

Find out more about Rajinder and the hundreds of Indian children who love the Internet. Watch the FRONTLINE/World video "Hole in the Wall."

THE ORGANIZATION OF THIS CHAPTER AND OF THE BOOK

The rest of this chapter introduces several subfields of human geography emphasized in this book: cultural/social, economic, population, environmental, and political. Basic concepts in physical geography are also included, and the organization shows how physical and human geography are interwoven. When you made the map of your favorite childhood landscape, you undoubtedly included material from several of these subfields of geography, without necessarily realizing you were doing so. In addition to this mapping experience at a local scale, reading more about the subfields should help you to understand how people in the various regions of the world have worked out their relationships with the local physical environment, with one another, and with the wider world.

The geographic subfields presented in this chapter will guide the description and discussion of each world region. Each chapter is organized along similar lines, and each covers the same general topics. The order of these topics changes slightly from chapter to chapter, simply because circumstances in a particular region may lend themselves to a particular arrangement of topics. However, if you wish to compare general conditions or specific topics in two or more world regions, you will be able to do so. Simply go to the Table of Contents and choose a topic, such as "Making Global Connections" or "The Geographic Setting." Under the latter, for example, in each regional chapter you will find discussions of "Physical Patterns," "Human Patterns over Time," and "Population Patterns." If you are interested in a cross-regional comparison of some specific issue, you can go to the Index and look up that issue. If, for example, you search for gender, or a more specific topic such as gender roles, you will find the page numbers for discussions of gender roles in every region of the world.

CULTURAL AND SOCIAL GEOGRAPHIC ISSUES

Culture is an important distinguishing characteristic of human societies. It comprises everything we use to live on earth that is not directly part of our biological inheritance. Culture is represented by the ideas, materials, and institutions that people have invented and passed on to subsequent generations. Among other things, culture includes language, music, gender roles, belief systems, and moral codes (for example, those prescribed in Confucianism, Islam, or Christianity). **Material culture** comprises all the things that people use: clothing, houses and office buildings, axes, guns, computers, earthmoving equipment (from hoes to work animals to bulldozers), books, musical instruments, domesticated plants and animals, and agricultural and food-processing equipment. **Institutions** are formal or informal associations among people. The family, in its many different forms, is an **informal institution.** So is a community, which can be made up of the people who share a physical space (a village or a neighborhood) or those who share only a belief system. The Roman Catholic community, for example, stretches across continents (see Figure 1.7 on page 16). **Formal institutions** include official religious organizations such as the Presbyterian Church (U.S.A.) or the Vatican; local, state, and national governments; nongovernmental organizations that provide philanthropic services, such as the Red Cross and Red Crescent; and specific business corporations.

In this section we explore the concept of culture groups and some of the cultural attributes, such as value systems and languages, that help to define them. We also examine gender roles and perceptions about race as cultural phenomena that vary over space and play an important role in human relationships.

CULTURE GROUPS

A group of people who share a set of beliefs, a way of life, a technology, and usually a place form a **culture group.** The term "culture group" is often used interchangeably with **ethnic group,** but these terms can become problematic. Like the concept of region, the concepts of culture and ethnicity are imprecise, especially as they are popularly used. In the modern era, for instance, as part of the globalization process, migrating people often move well beyond their customary cultural or ethnic boundaries to cities or to distant countries. In these new places, where they are often a minority, they take on new ways of life or even new beliefs, yet still identify with their culture of origin. The formerly nomadic Kurds in Southwest Asia, for example, are asserting their right to create a nation-state (see Figure 6.22) in the territory where they once lived, but which is now claimed by Syria, Iraq, Iran, and Turkey. We might picture a typical Kurd as a woman living in a certain type of tent and weaving the wool from her family's herd of sheep, and then learn that many Kurds who actively support the cause of the herders are now urban dwellers in Turkey, Iraq, or even London. They live in apartments and may work as clerks, housekeepers, taxi drivers, rug dealers, or engineers. Although these people think of themselves as ethnic Kurds and are so regarded in the larger society, they do not follow the traditional Kurdish way of life. Hence, we could argue that these urban Kurds both are and are not part of the Kurdish culture or ethnic group.

Another problem with the concept of culture is that it is often applied to a very large group that shares only the most general of characteristics. For example, one often hears the terms *American culture*, *African-American culture*, or *Asian culture*. In each case, the group referred to is far too large to share more than a few broad characteristics.

It might fairly be said, for example, that American culture (in both Canada and the United States) is characterized by beliefs that promote individual rights, autonomy, and responsibility; by high levels of consumption; by dependence on modern technology; and by a market economy based on credit and widely accessible middle and higher education. If we try to be more specific, we quickly run into disagreements. You can try this with your friends. Can you agree on a religious heritage, typical food preferences, or precise tastes in television programming that are common to all? And although we might all agree that we value individualism, people in the United States constantly debate just how independent the individual should be, at what stages in life, and at whose expense. Think for a moment about current debates over public school dress codes, whether the terminally ill have the right to choose "managed death," who should have the right to marry, or how much control women should have over their reproductive systems. In fact, U.S. culture encompasses many subcultures that share some of the core set of beliefs with the majority, but disagree over parts of the core and over a host of other matters. The same is true, in varying degrees, for all other regions of the world.

GLOBALIZATION AND CULTURAL CHANGE

There are indications that the diversity of culture is fading as trends and fads circle the globe via the many types of instant communication now available. American fast food, popular music, and clothing styles can now be found from Mongolia to Mozambique. At the same time, a wide variety of ethnic music, textiles, cuisines, and even modes of dress from distant places now grace the lives of consumers across the world. Thus, as globalization proceeds, especially as people migrate and ideas are diffused, some measure of **cultural homogeneity**—more overall similarity between culture groups—will occur.

Are we all drifting toward a common repertoire of material culture, and perhaps even similar ways of thinking? Possibly, but there are also countervailing trends. It is now possible for people to reinforce their feeling of **cultural identity** with a particular group through the same channels that are encouraging homogenization. We cite just two examples: the people of Aceh in Southeast Asia and the Nunavut in Canada.

Aceh is a province with a distinctive cultural identity located in northern Sumatra, an island within the country of Indonesia in

Southeast Asia. The people of Aceh desire greater **autonomy**—the right to control their own affairs and especially to retain control of recently discovered oil deposits that the central Indonesian government views as a national resource, not a local one. The Indonesian government has sent military troops to enforce its interests, with lethal consequences to the Acehese. Meanwhile, the Acehese have established several Web sites and Internet chat groups to build awareness among Acehese migrants worldwide of the political struggle and armed conflict in their homeland.

The Nunavut are a Native American people in northern Canada. They have set up several Web sites through which they teach those who have moved away about Nunavut tribal history and technology. These sites also inform tribal people about legal efforts to reestablish control over ancestral lands lost hundreds of years ago to British and French colonists.

These two examples show us that the ability to communicate easily over the Internet and to travel quickly may actually reaffirm cultural identity. These capabilities also enhance the conditions necessary for **multiculturalism:** the state of relating to, reflecting, or being adapted to several cultures. For example, the young man from Aceh who helped us to understand the perspective of the Acehese in Indonesia is now an artist in the United States, where he earns a living painting not only scenes from his tropical homeland, but also the mansions of U.S. corporate executives, as well as portraits of these executives.

CULTURAL MARKERS

Members of a particular culture group share features, such as language and common values, that help to define the group. These shared features are called **cultural markers.** The following sections examine the roles of values, religion, language, and material culture as cultural markers. Notice how colonization and modern communications are causing some cultural markers to disappear and others to become more dominant.

Values

Occasionally you will hear someone say, "After all is said and done, people are all alike" or "People ultimately all want the same thing." It is a heartwarming sentiment, but it is often wrong. We would be wise not to expect or even to want other people to be like us. People are not all alike, and that is in large part what makes the study of geography interesting. It is often more fruitful to look for the reasons behind differences among people than to search hungrily for similarities. **Cultural diversity** is one reason that humans are such successful and adaptable animals. The various cultures serve as a bank of possible strategies for responding to the social and physical challenges faced by the human species. The reasons for differences in behavior from one culture to the next are usually complex, but they are often related to differences in values.

Let us look at an example that contrasts the values held by modern urban individualistic culture with those held by rural community-oriented culture. In urban areas in many parts of the world, it is often acceptable, even desirable, for a woman to

Thinking Geographically: **ON THE WEB**

choose clothing that allows her to make a strong statement about who she is as an individual and how well off she is. One rainy afternoon, one of the authors of this book (Lydia Pulsipher) saw a beautiful middle-aged Asian woman walking alone down a fashionable street in Honolulu, Hawaii. She wore a white suit with matching white shoes, hat, and bag, and over it all an elegant, clear raincoat. Everyone noticed and admired her because she exemplified an ideal Honolulu woman: beautiful, self-assured, and rich enough to keep herself outstandingly well dressed. I guessed that she was a businesswoman.

But in her grandmother's village—whether it be in Japan, Korea, Taiwan, or rural Hawaii—people would regard such dress as outrageously immodest and dangerously antisocial. In a rural Asian community, her clothes would breach a widespread traditional ethic that no individual should stand out from the group. Furthermore, her costume exposed her body to open assessment and admiration by strangers of both sexes. In a village context, such behavior would signal that she lacked modesty. The fact that she walked alone down a public street—unaccompanied by her father, husband, or female relatives—would indicate that she was not a respectable woman. Thus a particular behavior may be admired when evaluated by one set of values, yet considered questionable or even despicable when evaluated by another.

If culture groups have different sets of values and standards, does that mean that there are no overarching human values or standards? This question increasingly worries geographers, who try to be sensitive both to the particularities of place and to larger issues of human rights. Those who lean too far in the direction of appreciating difference could be led to the tacit acceptance of inhumane behavior, such as the oppression of minorities and women or even torture and genocide. Acceptance of difference does not mean that we cannot make judgments about the value of certain extreme customs or points of view. Nonetheless, many geographers and others interested in human rights have observed that, although it is important to take a stand against cruelty, deciding when and where to take that stand is rarely easy.

Religion and Belief Systems

The religions of the world are formal and informal institutions that embody value systems. Most have roots deep in history, and many include a spiritual belief in a higher power (God, Yahweh, Allah) as the underpinning for their value systems. These days, religions often focus on reinterpreting age-old values for the modern world. Some formal religious institutions—such as Islam, Buddhism, and Christianity—*proselytize*; that is, they try to extend

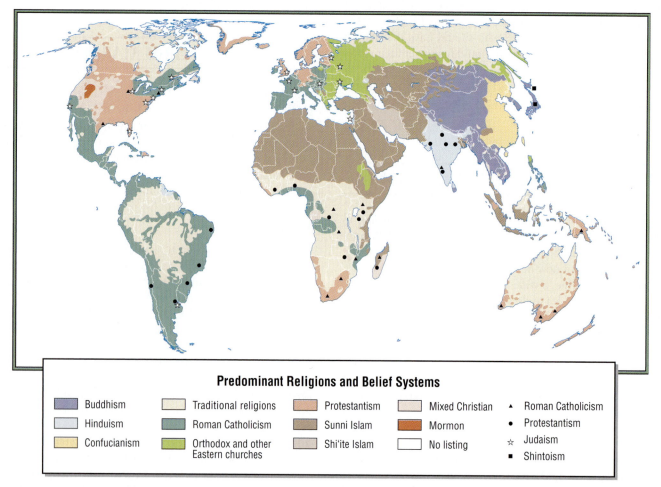

Predominant Religions and Belief Systems

FIGURE 1.7 Major religions around the world. Look at this map of the formal and informal religious traditions on earth and see if you can already offer explanations for some of the patterns you observe. [Adapted from *Oxford Atlas of the World* (New York: Oxford University Press, 1996), p. 27.]

This mosque, in Banja Luka, Serbia, is one of twenty-one mosques that were destroyed in the town during the 1992–1995 war. Rebuilding the mosque became a central goal of the Muslim community, as a symbol of their survival. [AFP/Getty Images.]

their influence by seeking converts. Others, such as Judaism and Hinduism, accept converts only reluctantly. Informal religions, often called *belief systems*, have no formal central doctrine and no policy on who may or may not be a practitioner.

Religious beliefs are often reflected in the landscape. For example, settlement patterns often demonstrate the central role of religion in community life: village buildings may be grouped around a mosque (see the photo on this page) or synagogue or an urban neighborhood organized around a Catholic church. In some places, religious rivalry is a major feature of the landscape. Certain spaces may be clearly delineated for the use of one group or another, as in Northern Ireland's Protestant and Catholic neighborhoods.

Religion has also been used to wield power. For example, religion served as a political instrument during the era of European colonization—as a way to impose a quick change of attitude on conquered people. Figure 1.7, which shows the distribution of the major religious traditions on earth today, demonstrates some of the religious consequences of colonization. Note the distribution of Catholicism in the parts of the Americas, Africa, and Southeast Asia colonized by European Catholic countries. Religion can also spread through trade contacts. In the seventh and eighth centuries, Islamic people used a combination of trade and political power (and less often, actual conquest) to extend their influence across North Africa, throughout Central Asia, and eventually into South and Southeast Asia. The distribution of major religions has changed many times over the course of history. Moreover, a world map is too small in scale to convey the complexity of actual spatial patterns, such as where two or more religious traditions intersect at the local level. Also, as the world's cultural traditions become increasingly mixed and urban life spreads, **secularism,** a way of life informed by values that do not derive from any one religious tradition, is spreading.

Language

Language is one of the most important criteria in delineating cultural regions. The modern global pattern of languages (shown in Figure 1.8) reflects the complexities of human interaction and isolation over several hundred thousand years. In reality, the map does not begin to depict the actual details of language distribution: between 2500 and 3500 languages are spoken on earth today—some by only a few dozen people in isolated places.

The geographic pattern of languages has continually shifted over time as people have interacted through trade and migration. The pattern changed most dramatically, however, after the age of European exploration and colonization began around 1500. From then on, the languages of European colonists often replaced the languages of the colonized people. This is why we find large patches of English, Spanish, Portuguese, and French in the Americas, Africa, Asia, and Oceania. In North America, European languages largely replaced Native American languages, but in Central and South America, Africa, and Asia, European and native tongues coexisted. Many people became bilingual or trilingual.

Today, with increasing trade and instantaneous global communication, just a few languages have become dominant. At the same time, other languages are becoming extinct because children no longer learn them. Arabic is an important **lingua franca,** or language of international trade, as are English, Spanish, and Chinese. Among these, English dominates, largely because the British colonial empire introduced English as a second language to many places around the globe. Now the need for a common world language in the computerized information age is rapidly pushing the world community to accept English as the primary lingua franca.

For many people, a particular language represents a cultural and geographic identity they wish to retain. For example, the newly independent country of Slovenia (once part of Yugoslavia in southeastern Europe) jealously guards its language, Slovene. Surrounded by more populous countries (Austria, Hungary, Italy, and Croatia), each using a different language, this small country of just 2 million people sees Slovene as one of only a few remaining cultural markers that give them identity. Slovenes worry that their language will vanish as their country becomes integrated into Europe, where French, German, and English are dominant. Indeed, Slovene children study all of these languages in school. As a result, Slovenes devote much energy to preserving their language. They have one of the highest rates of book publishing on earth and translate many foreign texts into Slovene. They even attempt to extend the use of Slovene beyond the borders of Slovenia by inviting the children and grandchildren of emigrants to return from the Americas, Western Europe, and Australia for heavily government-subsidized summer language courses. The Slovenes are an example of a culture group that is intensifying its distinct identity even as its connections to the wider world increase.

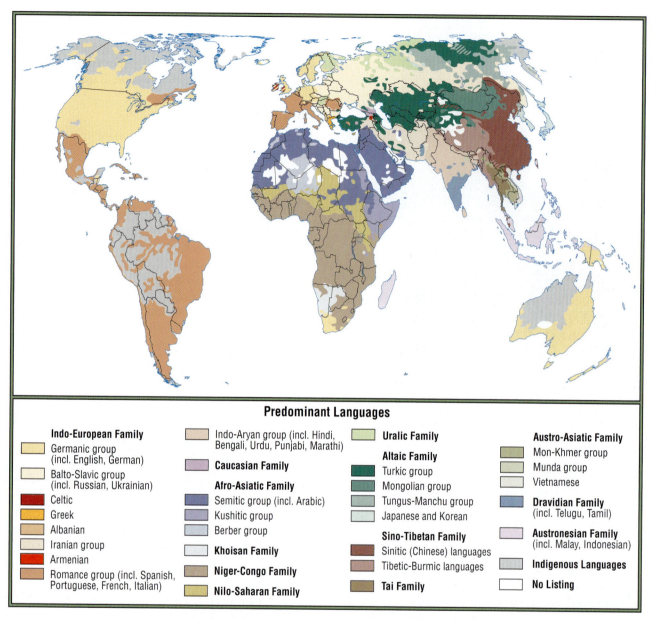

Predominant Languages

Indo-European Family
- Germanic group (incl. English, German)
- Balto-Slavic group (incl. Russian, Ukrainian)
- Celtic
- Greek
- Albanian
- Iranian group
- Armenian
- Romance group (incl. Spanish, Portuguese, French, Italian)

- Indo-Aryan group (incl. Hindi, Bengali, Urdu, Punjabi, Marathi)
- **Caucasian Family**

Afro-Asiatic Family
- Semitic group (incl. Arabic)
- Kushitic group
- Berber group
- **Khoisan Family**
- **Niger-Congo Family**
- **Nilo-Saharan Family**

- **Uralic Family**

Altaic Family
- Turkic group
- Mongolian group
- Tungus-Manchu group
- Japanese and Korean

Sino-Tibetan Family
- Sinitic (Chinese) languages
- Tibetic-Burmic languages
- **Tai Family**

Austro-Asiatic Family
- Mon-Khmer group
- Munda group
- Vietnamese
- **Dravidian Family** (incl. Telugu, Tamil)
- **Austronesian Family** (incl. Malay, Indonesian)
- **Indigenous Languages**
- **No Listing**

FIGURE 1.8 The world's major language families. Distinct languages (Spanish and Portuguese, for example) are part of a larger language group (Romance languages), which in turn is part of a language family (Indo-European). Does the category "indigenous languages" or its distribution raise any questions in your mind about how the mapmaker defined the term *indigenous*? [Adapted from *Oxford Atlas of the World* (New York: Oxford University Press, 1996), p. 27.]

Material Culture and Technology

As we have seen, *material culture* refers to the tangible items that a specific culture group produces or uses. A group's material culture reflects its **technology,** which is the integrated system of knowledge, skills, tools, and methods upon which a culture group bases its way of life. Material culture and technology help to define a culture—its resource base, standard of living, trading patterns, and belief systems. Archaeologists use the interconnectedness of material artifacts, technology, and culture to deduce—from an excavated assemblage of potsherds (bits of broken pottery), tools, animal bones, and other material remains—what was going on long ago in a particular place.

Housing provides an example of how material culture reveals a particular culture group's way of life, especially its values, technology, and resource use. The typical American suburban ranch house has a living room, a dining room, a kitchen, two baths, three bedrooms, a laundry room, a two-car garage, a front lawn, and a backyard. With its distinctive architecture, electrical and plumbing systems, and landscaping, it silently reveals a great deal about American family values. It reflects a nuclear family structure (mother, father, children)—a type that remains an ideal in American society but now constitutes less than 30 percent of American families. This sort of house is less suited to extended families that include aunts, uncles, and grandparents or to single-

This home in Canyon, California, belongs to the Cavin family. They are pictured (in the foreground) with all of their possessions. After looking over the assemblage, to what extent do you think the house and possessions are typical for a North American family? [Peter Menzel/Material World.]

parent families; in fact, it often has to be modified to accommodate these types of families. The ranch house reflects certain ideas about privacy (separate bedrooms) and gender roles (Mom's special spaces may be the kitchen and laundry room; Dad's, the TV room, the garage, and the tool shed). This house bespeaks a certain level of affluence and leisure, equipped as it is with labor-saving devices and conveniences and set apart on a green lawn requiring constant maintenance. It also symbolizes American ideas about private property, polite neighborliness, and mobility.

Because we in North America move so often in our quest for a better job and standard of living, yet insist on privately owned homes, a vast system of interchangeable dwellings has evolved across America, supported by an institution: the American home real estate market. Families moving across the North American continent can find similar houses with nearly the same floor plan and yard in just about any community.

An astute foreign observer could learn a great deal about Americans simply by "reading" the domestic material culture of their homes and surrounding landscapes. Were you that astute observer, you could look at housing in Japan, Mongolia, France, or the West Indies and learn much about each particular culture's notions of proper family structure, gender roles, intimacy rules, aesthetic values, property rights, building technology, use of resources, and trading patterns. Such an observer might even learn how modifications and remodeling of existing housing reflect cultural change.

Gender Issues

Geographers have begun to pay more attention to gender roles in different culture groups. In virtually all parts of the world, and for

at least tens of thousands of years, the biological fact of maleness and femaleness has been translated into specific roles for each sex. Although the activities assigned to men and to women can vary greatly from culture to culture and from era to era, there are some rather startling consistencies around the globe and over time. Men are usually expected to fulfill public roles, working outside the home—whether as traveling executives, animal herders, hunters, or government workers. Women are usually expected to fulfill private roles. They keep house, bear and rear the children, tend the elderly, plant the gardens, prepare the meals, fetch the water and firewood, and in some cultures, run the errands. In nearly all cultures women are defined as dependent on men—their fathers, husbands, or adult sons—even when the women may produce most of the family sustenance. Because their activities are focused on the home, women typically have less access to education and paid employment, and hence less access to wealth and political power. When they do work outside the home (as is the case increasingly in every region), women tend to fill lower-paid positions—whether as laborers, service workers, or professionals—and to retain their household duties.

Gender is both a biological and a cultural phenomenon. Men and women have different reproductive roles, and there are certain other physical differences as well. Men have larger muscles, can lift heavier weights, and can run faster than women (but not necessarily for longer periods). In some physical exercises, average women have more endurance and are capable of more precise movements than average men, and in populations that enjoy overall good health, women tend to outlive men by an average of 3 to 5 years. Women's physical capabilities are somewhat limited during pregnancy and nursing—and, for some, during menstruation—but their susceptibility to pregnancy is limited to about 30 years. Beyond about age 45, women are no longer subject to the physical limits imposed by reproduction, and most contribute in some significant way to the well-being of their adult children and grandchildren. A number of evolutionary biologists have postulated that the evolutionary function of menopause in midlife may be to give women the time and energy to help succeeding generations thrive—an idea sometimes labeled the *grandmother hypothesis*.

Although average physical gender differences exist, in most cultures they are amplified to carry greater social significance than the biological facts would warrant. The culturally defined differences between the sexes have enormous effects on the everyday lives of men, women, and children. Customary ideas about masculinity and femininity, proper gender roles, and sexual orientation are handed down from generation to generation within a particular culture group or society. Perhaps more than for any other culturally defined human characteristic, there is significant agreement from place to place and over time that gender is important.

The historical and modern global picture is a puzzlingly negative one for women. In nearly every culture, in every region of the world, and for a great deal of recorded history, women have been (and are) defined as second-class. It is hard to find exceptions, although the intensity of this second-class designation varies considerably. In Europe, Asia, the Americas, and Africa; on tiny islands in the Pacific; in cities and in villages, people of both sexes routinely accept the idea that males are more productive and intelligent than females. In nearly all cultures, families prefer boys over girls because, as adults, boys have greater earning capacity (in large part as a result of discrimination against females; Table 1.2), have more power in society, and will perpetuate the family name (because of patrilineal naming customs). Around the globe, females have less access to medical care, are more likely to die in infancy (see Figure 1.19 on page 41), start work earlier in their young lives than males, work longer hours, and eat less well. In the United States, when domestic violence occurs, statistics show that females are the victims 96 percent of the time. The actual conditions of women vary, of course; yet, in all regions, the evidence is overwhelming that women have an inferior status. The question of how and why women became subordinate to men has not yet been well explored because, oddly enough, few thought the question significant until recently.

There are growing challenges to traditional notions of femininity and ideas of female capabilities. In many countries around the world, females are now acquiring education at higher rates than males, a fact that should eventually make women as competitive as males for jobs. Unless discrimination persists, women should also earn higher pay. In addition, female Olympic athletes have demonstrated superlative strength, speed, and endurance. They make up more than a third of Olympic competitors, and in some countries (such as China), more than half (Figure 1.9).

Yet looking only at women's woes misses half the story. Men are also disadvantaged by strict gender expectations that may ride roughshod over personal preferences. Probably for most of human history, young men have borne the lion's share of onerous physical tasks and dangerous undertakings. Until recently it was usually young men who left home to migrate to distant, low-paying jobs. Overwhelmingly, it is young men who have died alone in faraway wars. It is boys who are taught to repress their feelings, although some cultures emphasize this more than others. Men who grow up with negative attitudes toward women have been taught those attitudes by their families (often female kin) and the larger society.

This book will return repeatedly to the question of gender disparities in an effort to investigate this most perplexing cross-cultural phenomenon. Our examination will also reveal that many societies are addressing gender inequalities and that in every country on earth, what it means to be a man or a woman is being renegotiated, at least in small ways.

Race

Just as ideas about gender roles affect life in all world regions, ideas about race affect human relationships everywhere on earth. However, according to the science of biology, all people now alive on earth are members of one species, *Homo sapiens sapiens*, and the popular markers of race (skin color, hair texture, face and body shapes) have no biological significance. For any supposed racial trait, such as skin color, there are wide variations within human groups. In addition, many invisible biological characteristics, such as blood types and DNA patterns, actually cut across skin color distributions and are shared by what are commonly viewed as different races. Over the last several thousand years, in fact, there has been such massive gene flow among human

TABLE 1.2   Comparisons of male and female income in countries where average education levels are higher for females than for males

Country	Female income (PPP[a] $U.S., 2001)	Male income (PPP $U.S., 2001)	Female income as percent of male income
Sweden	19,636	28,817	68
Australia	20,830	29,945	70
Belgium	15,835	35,601	44
United States	26,389	42,540	62
Austria	17,940	35,923	50
Poland	7,253	11,777	62
Israel	13,726	26,011	53
Barbados	11,852	19,496	61
Russia	5,609	8,795	64
United Kingdom	18,180	30,476	60
United Arab Emirates	6,041	28,223	21

[a]PPP = purchasing power parity.

Source: United Nations Human Development Report 2003 (New York: United Nations Development Programme), Table 23, "Gender-related development."

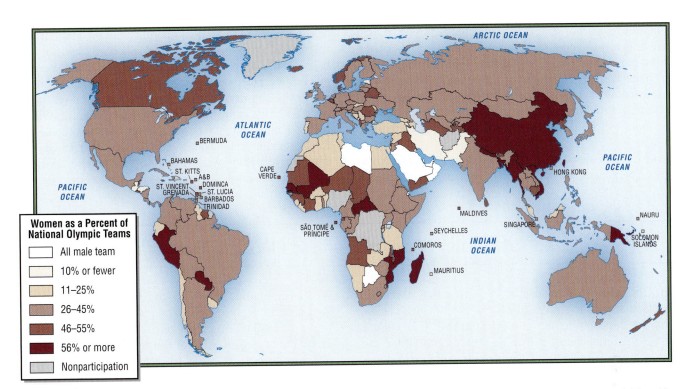

FIGURE 1.9 Women's participation in the 2000 Olympics. Traditional notions of femininity, female roles, and female strength, speed, and endurance are being strongly challenged by Olympic athletes. In the first modern Olympic Games in 1896, women were barred from participating. In 2000, women made up 38 percent of the Olympic competitors, and in some countries (China, for instance), more than 50 percent of the athletes were women. What might account for the varying representation of women in different parts of the world? [Adapted from Joni Seager, *The State of Women in the World Atlas* (New York: Penguin, 2003), pp. 50–51.]

populations that no modern group presents a discrete set of biological characteristics. Although we may look quite different, biologically speaking, we are all closely related.

It is likely that some of the easily visible features of particular human groups evolved to adapt them to environmental conditions, but precise explanations of just how these physical changes occurred are hard to come by. For example, biologists think that darker skin evolved in regions close to the equator, where sunlight is most intense. All humans need the nutrient vitamin D, and sunlight striking the skin helps the body to absorb vitamin D; however, too much of it can result in improper kidney functioning. Dark skin protects against too much vitamin D absorption in equatorial zones. In high latitudes, where the sun's rays are more dispersed, light skin allows for the absorption of sufficient vitamin D. Many physical characteristics, such as big ears, deep-set eyes, or high cheekbones, do not serve any apparent adaptive purpose, however; they are probably the result of random chance and ancient inbreeding within isolated groups. Similarly, there is no evidence that any so-called "race" has particularly high math ability or athletic ability; rather, these characteristics are present in individuals in all human populations and may become enhanced by cultural practices.

It may be comforting to learn that, biologically speaking, race is meaningless. But we cannot overlook the significant political and social import that race (often paired with culture or ethnicity) carries in some parts of the world. Race seems to have acquired more significance over the last several thousand years, especially when humans from different parts of the world began to encounter each other in situations of unequal power. For example, some researchers have suggested that European colonizers adopted **racism**—the negative assessment of unfamiliar, often darker-skinned people—to justify taking land and resources away from

In all major cities of the world, one will encounter a great diversity of people. Here, in Detroit, Michigan, people from several places around the world are being sworn in as U.S. citizens. [Jim West/Impact Visuals.]

these supposedly inferior beings. Still, it would be wrong to suggest that disparaging appraisals and exploitation of others are recent cultural innovations, or particularly European traits. The human animal has long committed atrocities against its own kind, often in the name of race or ethnicity, or even gender. Race and its implications in North America will be a focus in Chapter 2, and the topic will be discussed in several other world regions as well.

While recognizing all the ills that have emerged from racism and similar prejudices, we should not infer that human history has been marked primarily by conflict and exploitation. Actually, the opposite is probably true: humans have been so successful because of a strong inclination toward *altruism*, the willingness to sacrifice one's own well-being for the sake of others. Writ small, this altruism can be found in the sacrifices individuals make to help family, neighbors, and community. Writ large, it includes charitable giving to help anonymous people in need. It is probably our capacity for altruism that causes us such deep consternation over the relatively infrequent occurrences of inhumane behavior.

PHYSICAL GEOGRAPHY: PERSPECTIVES ON THE EARTH

Humans have always had to adapt to the physical environment, although the nature of the interactions between humans and their environments has changed over time and has varied from culture to culture. In this section we look at two components of the physical environment that are of particular importance to physical geographers: landforms and climate. We finish by examining the origins of agriculture as an example of how the interactions between humans and the environment can alter both the physical environment and human society.

LANDFORMS: The Sculpting of the Earth

Have you ever marveled at the landforms around you and wondered what awesome forces produced them? The processes that create mountain ranges, continents, and the deep ocean floor are some of the most powerful and slow-moving forces on earth. Originating deep beneath the earth's surface, these forces can move entire continents and often take hundreds of millions of years to do their work. Many of the earth's features, however, such as a beautiful waterfall or a dramatic rock formation, are formed by more rapid and delicate processes that take place on the surface of the earth. All of these forces are studied by *geomorphologists*, who focus on the processes that constantly shape and reshape the earth's surface.

Plate Tectonics

Two key ideas in physical geography are the Pangaea hypothesis and plate tectonics (Figure 1.10). The **Pangaea hypothesis**, first suggested by geophysicist Alfred Wegener in 1912, proposes that all the continents were once joined in a single vast continent called *Pangaea* (meaning "all lands"), which very slowly fragmented over time into the continents we know today. As one piece of evidence for his theory, Wegener pointed to the neat fit between the west coast of Africa and the east coast of South America. For decades, most scientists rejected Wegner's hypothesis, but we now know that the earth's continents have been assembled into supercontinents at least three different times, only to break apart again in what is called the *Wilson cycle*. Pangaea,

which existed from 200 million to 180 million years ago, was only the latest; it was preceded by Rodinia and Pannotia. In the distant future, our still-moving continents are predicted to coalesce once again into the supercontinent Pangaea Ultima. All of this activity is made possible by plate tectonics, a process of continental motion discovered in the 1960s, long after Wegener's time.

The premise of **plate tectonics** is that the earth's surface is composed of large plates that float on top of an underlying layer of molten rock. The plates are of two types. *Oceanic plates* are dense and relatively thin, and they form the floor beneath the oceans. *Continental plates* are thicker and less dense. Much of their surface rises above the oceans, forming continents. These massive plates move slowly, driven by the circulation of the underlying molten rock flowing from hot regions deep inside the earth to cooler surface regions and back. The creeping movement of tectonic plates created the continents we know today by fragmenting and separating Pangaea (see Figure 1.10).

Plate movements influence the shapes of major landforms, such as continental shorelines and mountain ranges. Mountain ranges arise mostly from the folding and warping of plates as they collide. The continents have piled up huge mountains on their leading edges as the plates carrying them collided with other plates. Hence, the theory of plate tectonics accounts for the long, linear mountain ranges extending from Alaska to Chile in the Western Hemisphere and from Southeast Asia to the European Alps in the Eastern Hemisphere. The highest mountain range in the world, the Himalayas of South Asia, was created when what is now India, at the northern end of the Indian-Australian Plate, ground into Eurasia, pushing up huge sections of thick continental crust. The only continent that lacks these long, linear mountain ranges is Africa. Often called the "plateau continent," Africa is believed to have been at the center of Pangaea and to have moved relatively little since the breakup.

Humans encounter tectonic forces most directly as earthquakes and volcanoes. Plates slipping past each other create the catastrophic shaking of the landscape we know as an **earthquake.** **Volcanoes** arise at plate boundaries or sometimes in the middle of a plate, where gases and molten rock (called magma) can rise to the earth's surface through fissures and holes in the plate. Volcanoes and earthquakes are particularly common around the edges of the

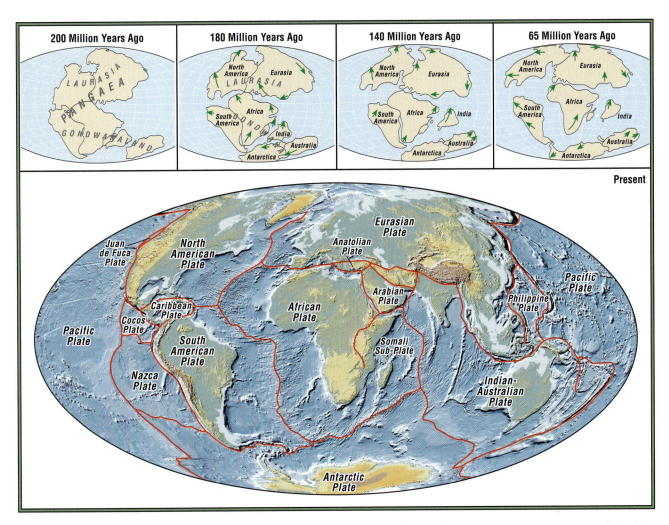

FIGURE 1.10 The breakup of Pangaea and resulting continental drift *(top)* **and the modern boundaries of the major tectonic plates** *(bottom).* Pangaea is only the latest of several global continents that have coalesced and then fragmented. [*Top:* Adapted from *Goode's World Atlas,* 19th ed. (Chicago: Rand McNally, 1995), p. 8. *Bottom:* Adapted from Frank Press and Raymond Siever, *Understanding Earth,* 2nd ed. (New York: W. H. Freeman, 1998), p. 509.]

Pacific Ocean, an area known as the **Ring of Fire** (Figure 1.11). In the Philippines, for example, the eruption of the Mount Pinatubo volcano in 1991 killed 550 people and ruined the livelihoods of 650,000 more; in addition, sulfur haze from the eruption may have influenced global climate patterns for several years.

Landscape Processes

The processes of plate tectonics are **internal processes,** driven by forces that originate deep beneath the surface of the earth. The landforms thus created have been further shaped by **external processes,** which are more familiar to us because we can observe them daily. One such process is **weathering,** the breakdown of rock exposed to the onslaught of sun, wind, rain, snow, ice, and the effects of life-forms. Exposure to these elements fractures rock and decomposes it into tiny pieces. These particles then become subject to another external process, **erosion.** In erosion, wind and water carry rock particles away and deposit them in new locations. The **deposition** of eroded material can raise and flatten the land around a river, where huge quantities of silt are spread about by

periodic flooding. As small valleys between hills are filled in by silt, a **floodplain** is created. Where rivers meet the sea, floodplains often fan out roughly in the shape of a triangle, creating a **delta.** External processes tend to smooth out the dramatic mountains and valleys created by internal processes.

Humans often contribute to external landscape processes through building, agriculture, and forestry. By altering the vegetative cover, agriculture and forestry expose the earth's surface to sunlight, wind, and rain—agents that increase weathering and erosion. Flooding becomes more common because the removal of vegetation limits the ability of the earth's surface to absorb rainwater. As erosion increases, rivers may fill with silt and deltas extend into the oceans. Building with concrete, asphalt, and steel often covers formerly wooded land with impervious surfaces. Again, flooding is the result, because rainwater runs over the surface to the lowest point instead of being absorbed into the ground. The physical effects of human activities vary in intensity from one culture to another depending on the tools used: mechanized earthmovers used to build roads change the earth's surface more rapidly and profoundly than machetes used to clear a path.

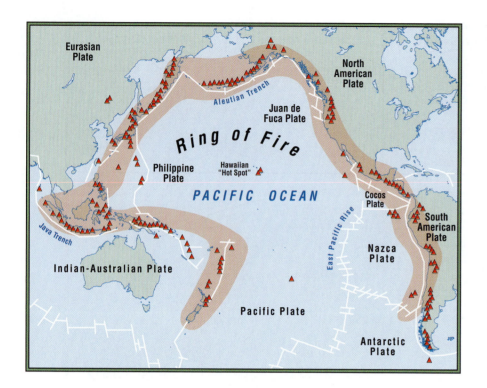

FIGURE 1.11 The Ring of Fire. Volcanic formations encircling the Pacific Basin form the so-called Ring of Fire, a zone of frequent earthquakes and volcanic eruptions. The red triangles represent active volcanoes. [Adapted from http://vulcan.wr.usgs.gov/Glossary/PlateTectonics/Maps/map_plate_tectonics_world.html.]

CLIMATE

The processes associated with climate are much more rapid than those that shape landforms. **Weather,** the short-term expression of climate, can change in a matter of minutes. **Climate** is the long-term balance of temperature and precipitation that keeps weather patterns fairly consistent from year to year. By this definition, the last major global climate change took place 15,000 years ago, when the glaciers of the last ice age began to melt.

Energy from the sun gives the earth a temperature range hospitable to life. The earth's atmosphere, oceans, and land surfaces absorb huge amounts of solar energy, and the atmosphere traps much of that energy at the earth's surface, insulating the earth from the deep cold of space. Solar energy is also the engine of climate. The most intense, direct sunlight falls in a broad band stretching about 30° north and south of the equator. The highest average temperatures on earth occur within this band. Moving away from the equator, sunlight becomes less intense, and average temperatures drop.

Temperature and Air Pressure

The wind and weather patterns we experience daily are largely a product of complex patterns of air temperature and air pressure. **Air pressure** can best be understood by thinking of air as existing in a particular unit of space—for example, a column of air above a square foot of the earth's surface. Air pressure is the amount of force exerted by that column on that square foot of surface. Air pressure and temperature are related:

1. The gas molecules in warm air are relatively far apart and are associated with low air pressure.

2. The gas molecules in cool air are relatively close together (dense) and are associated with high air pressure.

As a unit of cool air is warmed by the sun, the molecules move farther apart. The air becomes less dense and exerts less pressure. Air tends to move from areas of higher pressure to areas of lower pressure, creating wind. If you have ever been to the beach on a hot day, you may have noticed a cool breeze blowing in off the water. Land heats up (and cools down) faster than water, so on a hot day, the air over the land warms, rises, and becomes less dense than the air over the water, causing that cooler, denser air to flow inland. Often at night the breeze reverses direction, blowing from the now cooling land onto the now relatively warmer water.

These air movements have a continuous and important influence on global weather patterns. Over the course of a year, continents heat up and cool off much more rapidly than the oceans that surround them. Hence, the wind tends to blow from the ocean to the land during summer and from the land to the ocean during winter. It is almost as if the continents were breathing once a year, inhaling in summer and exhaling in winter.

Precipitation

Perhaps the most tangible way we experience changes in air temperature and density is through the falling of rain or snow. Precipitation occurs primarily because warm air holds more moisture than cool air. Warm air holds water vapor in the form of individual molecules. When this moist air rises to a higher altitude, its temperature drops, which reduces its ability to hold moisture. The moisture condenses into drops to form clouds and may eventually fall as rain or snow.

Several conditions that encourage moisture-laden air to rise influence the pattern of precipitation observed around the globe

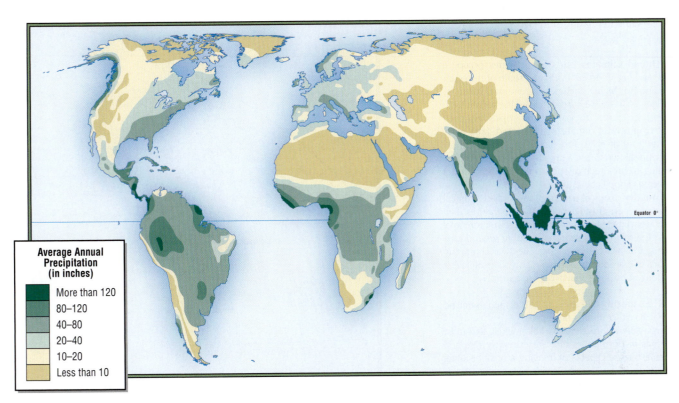

FIGURE 1.12 **World average annual precipitation.** Notice the concentration of precipitation along a wide irregular band on either side of the equator. [Adapted from *Goode's World Atlas,* 19th ed. (Chicago: Rand McNally, 1995), pp. 16–17.]

(Figure 1.12). When moisture-bearing air is forced to rise over mountain ranges, the air cools and the moisture condenses to produce rainfall. This process, known as **orographic rainfall,** is most common in coastal areas where wind blows moist air from above the ocean onto the land and up the side of a coastal mountain range (Figure 1.13). Most of the moisture falls as rain as the air is rising along the coastal side of the range. On the inland side, the descending air warms and ceases to drop its moisture. The dry side of a mountain range is called the **rain shadow.** Rain shadows may extend for hundreds of miles across the interiors of continents, as they do on the Mexican Plateau, east of California's Pacific coast, or north of the Himalayas of Eurasia.

A rain belt associated with the equator is primarily the result of moisture-laden tropical air being heated by the strong equatorial sunlight and rising to the point where it releases its moisture as rain. Equatorial areas in Africa, Southeast Asia, and South America are watered in this way. Neighboring nonequatorial areas also receive some of this moisture when seasonally shifting winds blow the rain belt north and south of the equator. The huge downpours of the Asian summer **monsoon** are an example. The Eurasian continental landmass heats up during the summer, causing the overlying air to expand, become less dense, and rise. The somewhat cooler, yet moist, air of the Indian Ocean is drawn (*Text continues on page* 28.)

FIGURE 1.13 **Orographic rainfall.** This type of rainfall is most common in coastal areas where wind blows moist air from above the ocean onto the land and up the side of a coastal mountain range. The air cools and drops most of its moisture on the windward side of the mountain. As the air descends on the leeward side, it warms and retains its remaining moisture; hence the leeward side often has a "rain shadow" where little rain falls. [Adapted from Frank Press and Raymond Siever, *Understanding Earth,* 2nd ed. (New York: W. H. Freeman, 1998), p. 291.]

FIGURE 1.14 Climate regions of the world.
Types of Climate Regions

Tropical Humid Climates (A). These climates occupy a wide band reaching 15° to 20° north and south of the equator, extending to higher latitudes when moderated by marine influences. Here we have simplified the variations to just two distinct climates: tropical wet and tropical wet/dry.

In the **tropical wet climate**, rain falls predictably every afternoon and usually just before dawn. The natural vegetation is the tropical rain forest, a broad-leafed evergreen forest consisting of hundreds of species of trees that form a several-layered canopy above the soil.

The **wet/dry tropical climate**, also called **tropical savanna**, experiences a wider range of temperatures than the tropical wet climate and may actually receive more total rainfall, but the rain comes seasonally and in great downpours, during the heat of the summer. The tropical forest in these regions is less vigorous and may show signs of having to survive long dry periods that occur unpredictably.

Arid and Semiarid Climates (B). Arid and semiarid climates may be either deserts or steppes.

Deserts generally receive very little rainfall, and most of that comes in downpours that are extremely rare and unpredictable, but are capable of bringing a brief, beautiful flourishing of desert life. Usually, deserts have sparse vegetation and almost no cloud cover, which leads to wide swings in temperature between day and night. Life is a struggle for both plants and animals because they must be able to survive heat stress during the day and freezing at night.

Steppes, such as the pampas of Argentina or the Great Basin of the American West, have climates similar to those of deserts, but more moderate. They usually receive about 10 inches more rain per year than deserts and are covered with grass and/or scrub.

Arid and semiarid climates are found primarily in two locations: the subtropics (slightly poleward of the tropics) and the midlatitudes. Subtropical deserts and steppes are found between 20° and 30° north and south latitudes, where high-pressure air descends in a belt around the planet. They are generally much warmer than midlatitudinal deserts and steppes, which are found farther toward the poles in the interiors of continents, often in the rain shadows of high mountains. Although soils are generally thin and unproductive in most deserts and steppes, some midlatitude steppes, such as the Great Plains of North America, have some of the thickest and richest soils in the world. The slightly colder temperatures and pronounced seasonality of these steppes keep down rates of decay and hence encourage the accumulation of organic matter in the soil over time.

Temperate Climates (C). In this book, we distinguish among just three temperate climates.

Midlatitude climates, such as those in southeastern North America and China, are moist all year and have short, mild winters and long, hot summers. A variant of the midlatitude climate is the **marine west coast climate**, such as that of western Europe, which is noted for fine drizzling rains.

Subtropical climates differ from midlatitude climates in that winters are dry.

Mediterranean climates have moderate temperatures but are dry in summer and wet in winter. Plants do not get moisture when temperature and evaporation rates are highest, so the plant species that live in this climate tend to be xerophytic (adapted to dry conditions), with scrubby,

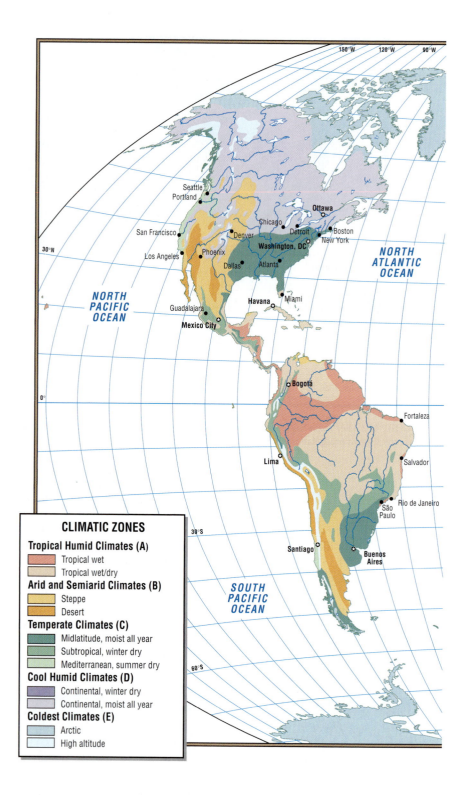

CLIMATIC ZONES

Tropical Humid Climates (A)
- Tropical wet
- Tropical wet/dry

Arid and Semiarid Climates (B)
- Steppe
- Desert

Temperate Climates (C)
- Midlatitude, moist all year
- Subtropical, winter dry
- Mediterranean, summer dry

Cool Humid Climates (D)
- Continental, winter dry
- Continental, moist all year

Coldest Climates (E)
- Arctic
- High altitude

shiny leaves capable of storing moisture. California, Portugal, northwestern Africa, southern Italy, Greece, and Turkey are examples of places with this climate type.

Cool Humid Climates (D). Stretching across the broad interiors of Eurasia and North America are continental climates, either **moist all year** (North America and north-central Eurasia) or with **dry winters** (northeastern Eurasia). Summers in cool humid climates are short, but can have very warm days. The natural vegetation of southern cool humid climates is broad-leafed deciduous and evergreen forest. Here the soil is deep and rich as a result of low temperatures

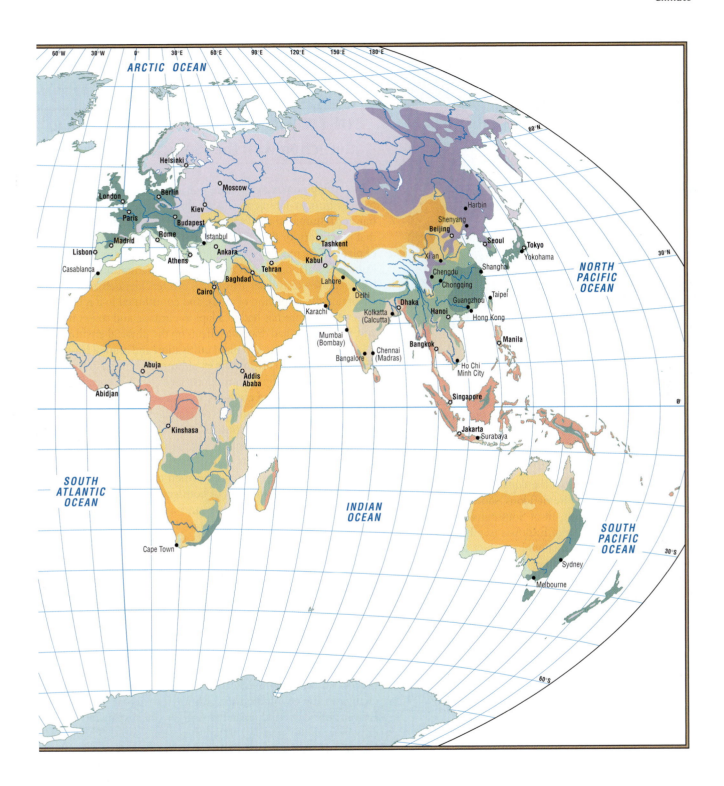

that inhibit decay, as in the midlatitude steppes. In the more northerly areas, winters are long and cold. Vast needle-leafed evergreen forests called *taiga* stretch across the cold interior. Soils here can be deep, but are not as rich as they are farther south; growing seasons are so short that cultivation is minimal.

Arctic and High-Altitude Climates (E). These climates are by far the coldest and are also among the driest. Although moisture is present, there is little evaporation because of the low temperatures.

The **Arctic climate** is often called *tundra*, after the low-lying vegetation that covers the ground; this dwarfed vegetation is a response to the 7 to

11 months of below-freezing temperatures. What little precipitation there is usually comes during the warmer months, and even this may fall as snow.

The **high-altitude** version of this climate is more widespread and subject to greater daily fluctuations in temperature. High-altitude microclimates can vary tremendously depending on factors such as available moisture, aspect, and vegetation cover. Ascending in altitude loosely mimics the climate changes found as one ascends to higher latitudes. These changes are known as temperature-altitude zones, a concept that will be visited when we study mountainous tropical regions such as the Andes and the Himalayas.

In the United States, people living along the Mississippi River floodplain are at risk nearly every year. When winter snows and early summer rains are particularly heavy, floodwaters rise over the banks and drown fields of soybeans and corn, destroy houses, and damage farm buildings—as happened at Wapello, Iowa, in 1993. People who live along the Yangtze and Yellow rivers in China and on the Brahmaputra Delta in Bangladesh are at even greater risk: in some years, thousands die as a result of summer floods. [Chris Stewart/Black Star.]

inland. The effect is so powerful that the equatorial rain belt is sucked onto the land, resulting in tremendous, sometimes catastrophic, rains throughout virtually all of South and Southeast Asia and much of coastal and interior East Asia (see Figures 8.3 and 8.4). Similar forces pull the equatorial rain belt south during the Southern Hemisphere's summer.

Much of the moisture that falls on North America and Eurasia is **frontal precipitation** caused by the interaction of large air masses of different temperatures and densities. These masses develop when air stays over a particular area long enough to take on the temperature of the land or sea beneath it. Often when we listen to a weather forecast we hear about warm fronts or cold fronts. A front is the zone where warm and cold air masses come into contact, and it is always named after the air mass whose leading edge is moving into an area. At a front, the warm air tends to rise over the cold air, carrying its clouds to a higher altitude, and rain or snow may follow.

Climate Regions

Geographers have several systems for classifying the world's climates that are based on the patterns of temperature and precipitation just examined. This book uses a modification of the widely known Köppen classification system, which divides the world into several types of climate regions, labeled A, B, C, D, and E on the climate map in Figure 1.14. As we look at each of the regions on this map and read the climate descriptions that accompany them, the importance of climate to vegetation becomes evident. Each regional chapter will contain a climate map; when reading these maps, you can refer back to the verbal

descriptions in Figure 1.14 if necessary. Keep in mind that the sharp boundaries on climate maps are in reality much more gradual transitions.

THE ORIGINS OF AGRICULTURE: Human Interaction with the Physical Environment

The development of agriculture provides a compelling illustration of how human interactions with the physical environment can transform human society and ways of life. Agriculture includes animal husbandry, or the raising of animals, as well as the cultivation of plants. The practice of agriculture has had long-term effects on human population growth and on rates of natural resource use. It has increased disparities in wealth, created hierarchies of power, facilitated the formation of cities, and contributed to a reliance on trade. Agriculture has had to adapt to climatic and soil conditions and, in turn, has had dramatic effects on the physical environment.

Where and when did plant cultivation and animal husbandry first develop? Very early ancestors of humans hunted animals and gathered plants and plant products (seeds, fruits, roots, fibers) for their food, shelter, and clothing. Early assumptions by prehistorians that males were the hunters and females the gatherers are now disputed. Successful hunting is more dependent on skill and intelligence than on strength, so women could have participated, and gathering, which was equally dependent on skill, was very likely something both men and women did. During the hunting and gathering stage, humans of both genders were manipulating the environment of wild plants and animals by reducing their numbers and changing their habitats. It is very likely that the transition from hunting and gathering to tending pastures and gardens was a gradual process arising from a long familiarity with the growth cycles and reproductive mechanisms of the plants and animals that humans liked to use.

Genetic studies support the view that at varying times between 8000 and 20,000 years ago, people in many different places around the globe (the Americas, Southwest Asia, Central Africa, Southeast Asia, and East Asia) independently learned to develop especially useful plants and animals through selective breeding, a process known as **domestication**. Another term for this time of change in the economic base of human society is the **Neolithic Revolution**: a period characterized by the expansion of agriculture and the making of polished stone tools. The map in Figure 1.15 shows several well-known centers of domestication. A continuing process of agricultural innovation also occurred in many places outside these centers. Sometimes **diffusion** was the mechanism that spread agriculture as farmers or herders moved into new areas, taking their tools, methods, or plants and animals with them. The geographer Carl Sauer noted in the 1950s that women were probably as active as men in plant domestication and the development of agriculture, and many others since have pursued this line of reasoning, noting the dominance of women in Native American agriculture in prehistoric times. In Greek and Roman agricultural mythology, deities in charge of agriculture and harvests were both female—Demeter, Ceres, Egesta, and Horta—and male—Saturn, Cronus.

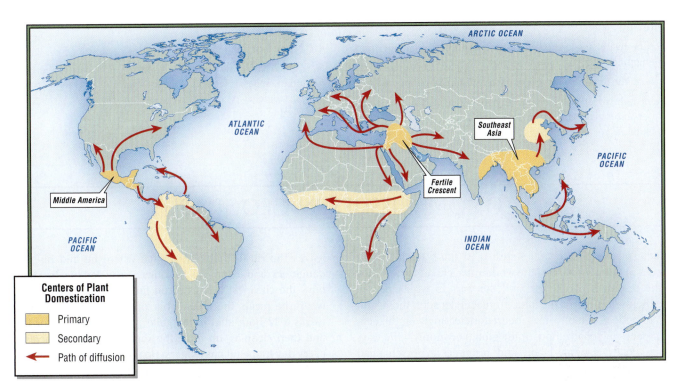

FIGURE 1.15 The origins of agriculture. Scientists have identified six main areas of the world (three primary and three secondary) where agriculture emerged independently. For lengthy periods, people in these different places tended plants and animals and selected for the genetic characteristics they valued. This knowledge then spread around the world. Domesticated plants and animals were then further adapted to new locations. This selection process continues in the present. [Adapted from Terry G. Jordan-Bychkov and Mona Domosh, *The Human Mosaic* (New York: Longman, 1999), pp. 108–109.]

Why did agriculture and animal husbandry develop in the first place? Certainly the desire for more secure food resources played a role. But Richard MacNeish, an archaeologist who has studied plant domestication in Mexico and Central America, suggests that the opportunity to trade was also a major motivation and may have been just as important as the simple desire to increase or improve food supplies. Many of the known locations of agricultural innovation lie near early trade centers. Is this just coincidental? People in such places would have had at least two reasons to pursue plant cultivation and animal husbandry: first, they would have had access to new information and new plants and animals brought by traders, and second, they would have needed products to trade with the people passing through. Perhaps, then, agriculture was at first a profitable hobby for hunters and gatherers that eventually, because of food security *and* market demand, grew into a "day job"—their primary source of sustenance.

Recent Insights on the Role of Agriculture in Human History. Agriculture is usually viewed as a major human achievement that made it possible to amass surplus stores of food for lean times and allowed some people to specialize in activities other than food procurement. While agriculture did eventually produce these effects, the picture is actually more complex than that. Anthropologist E. N. Anderson, writing about the beginnings of agriculture in China, suggests that agricultural production may have been the impetus for several developments now regarded as problems: rapid population growth, extreme social inequalities, environmental degradation, and even famine. He suggests that as groups turned to raising animals and plants for their own use or for trade, more labor was needed. As the population expanded to meet this need and more resources were used to produce food, natural habitats were destroyed, and hunting and gathering were gradually abandoned. The quality of human diets may have declined at first, as people abandoned foraging among diverse wild plants and began to eat primarily corn, wheat, or rice. Moreover, the wealth provided by trade was not shared equally by everyone, and some of those who specialized in nonagricultural activities began to amass wealth and power. Gradually, a small elite emerged atop a mass of much poorer people. The clearing associated with agriculture opened up the land to erosion and increased its vulnerability to drought and other natural disasters that could wipe out an entire harvest. Thus, as ever larger populations depended solely on agriculture, famine became more common.

Anderson's theory illustrates some points that are important for understanding modern geographic issues, especially general human well-being. The potential for human impact on the world's environments has increased markedly as fields of cultivated plants and pastures for cattle, sheep, and goats have replaced forests and grasslands. This trend of increasingly intense human impact has become even more pronounced over the past few centuries as the human population has doubled and redoubled. We apply ever-larger amounts of chemicals (hormones, fertilizers, pesticides) and irrigation water to keep our fields and pastures productive.

The invention of agriculture was an early step in the development of economic systems, all of which are based in some way on interactions between humans and their environments. Over the course of history, local economic systems became linked to increasingly distant systems. Let us now consider some current economic issues that have grown out of the resulting complex linkages.

Quick Review

1. Describe two characteristics that distinguish lines of latitude from lines of longitude.

2. What are the four different levels of scale explored in this book?

3. Which term describes the separation between people who enjoy access to opportunities via the computer and people who can't access new technology?

4. Name four cultural markers.

5. Which landform processes are linked primarily to the creation of mountain ranges, volcanoes, and earthquakes and which to rivers, floodplains, and deltas?

ECONOMIC ISSUES IN GEOGRAPHY

Geographers have long been interested in the ways in which economies in different parts of the world interact and in how they affect the resources people use and the ways in which people arrange themselves across the land. In recent decades, those who study economic geography have focused increasingly on the economic aspects of globalization, or the **global economy**—the ways in which goods, capital, labor, and resources are exchanged among distant and very different places. Here are a few examples of people who are part of that exchange and whose day-to-day lives are strongly affected by the global economy.

WORKERS IN THE GLOBAL ECONOMY

PERSONAL VIGNETTES Olivia lives in Soufrière on St. Lucia, an island in the Caribbean. Soufrière was once a quiet fishing village, but it now hosts cruise ship passengers several times a week. Olivia is 60. She, her daughter Anna, and three grandchildren live in a wooden house surrounded by a leafy green garden dotted with fruit and banana trees. Anna has a tiny shop at the side of the house, from which she sells matches, cigarettes, toilet paper, sugar, flour, canned food, soft drinks, locally grown spices, and preserves she and her mother make from the garden fruits. On days when the cruise ships dock, Olivia strolls down to the market shed on the beach with a basket of papaya, rolls of cocoa paste made from cacao beans picked in a neighbor's yard, local nutmeg, cloves, and cinnamon, and roasted peanuts packed up in tiny paper bags. She calls out to the passengers as they are rowed to shore, offering her spices and snacks for sale. In a good week she makes U.S. $50, and her daughter about U.S. $100 in the shop. Most days, Olivia tends the garden and some chickens, and her daughter constantly looks for other ways to earn a few dollars, making necklaces for tourists or taking in washing, so usually the family of five makes do with about U.S. $170 a week (U.S. $8840 per year). From this income they pay rent on the house, the electric bill, and school fees for the granddaughter who will go to high school in the capital next year. They also buy clothes for the children and whatever food (chiefly flour and sugar) they can't grow themselves.

Henry works in the galley on a luxurious four-masted German cruise ship that plies the Caribbean (including stops in Soufrière harbor) from December to June. Every November he leaves the rural Philippine village where he and his wife, two daughters, and a son live. He rides a crowded bus to Manila and buys the cheapest ticket he can find on a freighter to Panama, where he meets the cruise ship on December 1. All winter, he works a 12-hour shift in the galley, preparing food for the chefs and then cleaning up after the three sumptuous meals served every day to the 60 passengers. He receives U.S. $40 a day, or U.S. $3.33 an hour, plus meals, medical care if he needs it, and a hammock in the crew cabin to sleep in (crew in the world's merchant marine are paid according to the pay scale of their home country). He gets one day off every two weeks. He has little chance to spend his money, so by June he will have sent home U.S. $6800 to his wife. While Henry is gone, his wife tends their garden and takes care of the children and her elderly parents, who live downstairs in the new concrete home they finished building themselves 2 years ago. Occasionally, she clerks in the local general store where they buy food, kerosene, and most of their other goods, except clothes. Their total family cash income is about U.S. $9000 a year, which makes them exceptionally well off for their village. There is money in the bank, and all the children plan to attend technical college for at least 2 years.

Thirty-year-old Reyhan Purwanto sits on the dock in Riau, Indonesia, anxiously waiting with several other male and female Indonesians for the boat that will take them across the Strait of Malacca to Malaysia, where they plan to work illegally. They will join more than a million fellow Indonesians in Malaysia, some working legally, some not. All are attracted by Malaysia's booming economy, where average wages are four times higher than at home. This is Reyhan's second trip to Malaysia. His first was in 1996, when—after many years of working as a farm laborer on the island of Java, Indonesia, earning $600 per year—he signed up for an overseas employment program of the government's Ministry of Manpower. Although he was promised a 2-year work visa and a contract to work legally on a Malaysian oil palm plantation, upon arrival Reyhan found that his first 3 months' wages would go toward paying off his boat fare and that his visa was valid for only 2 months. Not one to give in easily, he soon escaped to the city of Malacca, where he secured a construction job earning U.S. $10 a day (about U.S. $2600 a year). On this wage, he was able to send enough money to his wife and two children in Indonesia to pay for their food, school fees, and a new roof for their house. In 1998, however, Reyhan was one of thousands of Indonesians deported by Malaysia, which was suffering growing unemployment due to the Asian financial crisis and wanted to create more jobs for locals by expelling foreign workers. The crisis affected Indonesia even more severely, and after several years of only part-time farm work, Reyhan is willing to give Malaysia another try.

Tanya is a 50-year-old grandmother with a son in high school and a married daughter. Tanya works at a fast-food store in North Carolina, making less than U.S. $6 an hour. She had been earning U.S. $8 an hour sewing shirts at a textile plant until it closed and moved to Indonesia. Her husband is a delivery truck driver for a snack-food company and has made enough over the years to build and improve a small home for them along a ridge outside Greenville. Between them, they make $27,000 a year; but from this income they must cover their mortgage and car loan, meet regular monthly expenses for food and utilities, and help their daughter, Rayna. She quit school after 11th grade and married a man who is now out of work and can't afford a car to look for a job. They and a baby live in the old mobile home at the back of the lot that Tanya and her family once lived in. With Tanya's new lower wage ($5000 less a year), there will not be enough money to pay the college tuition for her son, who had hoped to be an engineer and would be the first in the family to go past high school. For now, he is working at the local gas station. [*Adapted from Lydia Pulsipher's field notes, 1992–2000 (Olivia, Henry, and Tanya), and Alex Pulsipher's field notes, 2000 (Reyhan).*]

These people, living worlds apart, are all part of the global economy. Workers are paid at startlingly different rates for jobs that require about the same skill level. But varying costs of living and varying local standards of wealth make the difference between the relative affluence enjoyed by Henry's family and the near poverty and threatened hopes felt by Tanya's family, which actually has the highest income by far.

Most people want a better life than their parents had, but some have a better chance of achieving that goal than others.

Henry's children may have the best chance because their parents have some savings. Also, their father's worldwide travel has given him many valuable connections and the experience to know where and how to push his children into the best opportunities. Reyhan, by far the poorest, seems trapped by his status as an illegal worker, which robs him of many of his rights. Still, the higher pay he can earn in Malaysia offers him a possible way out of poverty.

Olivia and her family, though not well off, do not think of themselves as poor because they have what they need and because others around them live in similar circumstances. The tourist trade promises increased income, but it also means dependence on circumstances beyond their control—in an instant, the cruise-line companies can choose another port of call. Olivia's granddaughter dreams of a well-paid managerial job in tourism, but that requires a college degree, available only at the University of the West Indies in faraway Barbados. How will her family pay for that? With no degree, she will be competing with many other undertrained young people for low-paying waitress, kitchen, and maid jobs in resort hotels.

WHAT IS THE ECONOMY?

Earlier in this chapter, we observed that economic geography is the study of how people interact with their environment as they go about earning a living. The **economy** is the forum in which people make their living, and **resources** are what they use to do so. Some resources are tangible materials, such as mineral ores, timber, plants, and soil. Because they must be mined from the earth's surface or grown from its soil, they are called **extractive resources**. There are also **nonmaterial resources**, such as skills and brainpower. Often resources must be transformed to produce

From a distance, cruise ship passengers still see the village of Soufrière, St. Lucia, as a quaint place. When carried ashore by lighter, they encounter a small but vibrant entrepreneurial community. [Lydia Pulsipher.]

These construction workers have fled the economic crisis in Indonesia to work illegally in Malaysia. They are now under arrest and will be questioned by Malaysian immigration officials before being deported. Notice their faces and try to imagine their emotions and worries at this point in their lives. [S. Thinakaran/AP Photo.]

new commodities (such as refrigerators or sugar) or bodies of knowledge (such as books or computer software). **Extraction** (mining and agriculture) and **industrial production** are two types of economic activities, or *sectors* of the economy. A third is the **exchange** or **service sector:** the bartering and trading of resources, products, and people's services.

The **formal economy** includes all the activities that are recorded as part of a country's official production. Examples from the vignettes at the beginning of this section include Olivia's daughter in her shopkeeper's role; Henry as a ship employee and his wife when she clerks in the store; and Tanya, her husband, and their son. All are registered workers who earn recorded wages and presumably pay taxes to their governments. The activity of formal economies is measured by the **gross domestic product (GDP)**, a number that gives the total value, in monetary terms, of all goods and services officially recognized as produced in a country during a given year.

Many goods and services are produced outside formal markets, in the **informal economy.** Here, work is often traded for payment other than cash or for cash payment that is not reported to the government as taxable income. Only recently have economists begun paying attention to informal economies, yet it is estimated that one-third or more of the world's work falls into this category. Examples of workers in the informal economy include Olivia when she sells her goods to the tourists; Reyhan when he works illegally in Malaysia; and any members of the four families when they

contribute to their own or someone else's well-being through such unpaid services as housework, gardening, and elder and child care.

WHAT IS THE GLOBAL ECONOMY?

The global economy includes the parts of any country's economy that are involved in global flows of resources: extracted materials, manufactured products, money, and people. Most of us participate in the global economy every day. For example, this book was manufactured using paper made from trees cut down in Southeast Asia, North America, or Siberia and shipped to a paper mill in Oregon. It was printed in the United States, but many books are now printed in Asia because labor costs are lower there. This long-distance movement of resources and products has grown tremendously in the past 200 years, but it existed at least 2000 years ago, when silk fabrics, spices, food, and other goods were traded along Central Asian land routes that connected Rome with China or along sea routes stretching from East Africa to Arabia to India to East Asia.

Starting in about 1500, long-distance trade entered a period of expansion and change as Europeans began to take control of distant lands and people for economic gain. These acquired lands became known as **colonies.** Europeans began extracting resources from their colonies and organizing systems that processed those resources into higher-value goods to be traded back in Europe or wherever else there was a market. Sugarcane, for example, was grown on Caribbean, Brazilian, and Asian plantations, made into crude sugar and rum locally, and then sold in Europe and North America, where further refining often took place. The global economy grew as each region produced goods for export, rather than just for local consumption, and became increasingly dependent on imported food, clothing, machinery, and knowledge.

New wealth and ready access to global resources allowed Europeans to fund the **industrial revolution,** a series of innovations and ideas that changed the way goods were produced. One such innovation was the orderly timing and spacing of production steps and the use of specialized labor. Formerly, for example, one woman would produce the fiber (cotton or wool) for cloth, clean it, spin thread, weave the thread into cloth, and sew a shirt. The industrial production of shirts, by contrast, was organized as a set of tasks performed by different workers in separate spaces. Some workers specialized in producing the fiber, others in spinning, weaving, or sewing. Then, perhaps because workers were specialized in this way, they and their supervisors began to see ways to improve the efficiency and speed of their efforts. These innovations led to labor-saving improvements in tools and eventually to mechanized reaping, spinning, weaving, and sewing.

Mechanized production methods that made possible mass-produced goods of all sorts created a demand for many different kinds of raw materials. These materials were provided at low cost by the various European colonies in the Americas, Africa, and Asia. The colonies also became important markets for European manufactured goods. For example, sugar plantation managers in the British Caribbean bought British-made cast iron industrial equipment for crushing sugarcane and boiling cane juice. Hundreds of thousands of Caribbean slaves wore cotton cloth woven in England from cotton grown in colonies in America, Africa, and Asia. The sugar they produced was transported to European markets in

ships made in the British Isles of trees and minerals from various parts of the world. From just the example of sugar, we can see that during the European colonial and industrial eras (which overlapped in time) the extension of production, trade, and manufacturing networks to global dimensions increased markedly.

Until the early twentieth century, much of the activity of the global economy took place within the huge colonial empires amassed by a few European nations. However, as we will discuss later, economic and political changes brought an end to these empires by the 1960s, and almost all colonial territories are now independent countries. Nevertheless, the global economy continues to grow as the flows of resources and manufactured goods, linking now-independent countries, are sustained by private companies, many of which first developed during colonial times. **Multinational corporations** such as De Beers, Nabisco, Nike, Microsoft, Toyota, Cisco, and Bechtel operate across international borders, extracting resources from many places, producing products in factories carefully located to take advantage of cheap labor and transport facilities, and marketing their products wherever they can make an acceptable profit. One of the key characteristics of multinationals is that their global influence, wealth, and importance to local economies enable them to influence and manipulate the economic and political affairs of the countries in which they operate.

Although the vast majority of multinationals are still headquartered in the industrialized world, some are emerging in less wealthy former colonies. Moreover, while many multinationals still focus on a particular product or set of products, they are also increasingly important as conduits for the flow of **capital,** or investment money. For example, consider what was once a family-owned Mexico City cement company we will call MEXCRETE. It recently borrowed heavily from international banks to acquire controlling interests in cement industries in Texas, Mexico, the Philippines, and Thailand. The focus of the company is increasingly global, and MEXCRETE now hopes to outmaneuver its rivals by using

Among the first global economic institutions were Caribbean plantations like Old North Sound on Antigua. In the eighteenth century, thousands of sugar plantations in the British West Indies, subsidized by slaves who were forced to work without wages, provided huge sums of money to England and helped fund the industrial revolution. [Museum of Antigua and Barbuda.]

borrowed money to locate new cement plants that use cutting-edge technology in strategic places, such as Europe.

Such international investment has advantages and disadvantages. In this case, high-quality cement may be delivered more efficiently and cheaply to all the markets served. Competition may spur technological advancement. Jobs will be created in Southeast Asia, Mexico, and Texas. And the diverse new markets may vastly improve MEXCRETE's profits. But to maximize its profits, and if local laws allow, MEXCRETE may not pay its workers a living wage or provide a healthy workplace, and its cement plants may cause pollution. Competing smaller cement providers may fail, creating local unemployment. By taking the profits home to Mexico, MEXCRETE deprives local economies of capital. Moreover, MEXCRETE itself may fail, because its expansion resulted in a huge debt that requires regular payments. If it misses its payments, MEXCRETE's creditors could foreclose, putting thousands of jobs from Texas to Southeast Asia at risk.

THE DEBATE OVER FREE TRADE AND GLOBALIZATION

Free trade is the unrestricted international exchange of goods, services, and capital. There are differing views of the value of free trade and the globalization of markets. All governments presently impose some restrictions on trade to protect their own national economies from foreign competition. Restrictions take two main forms: tariffs and import quotas. **Tariffs** are taxes imposed on imported goods that have the effect of increasing the cost of those goods to the consumer. **Import quotas** set limits on the amount of a given good that may be imported over a period of time. These limits increase the price of the imported good by restricting the amount available, and they are intended to place imported goods at a price disadvantage when competing with domestic goods.

Other strategies to protect a country's trade include requiring the use of its own ships or other carriers to transport products being imported from abroad. Some countries also try to guard against unpredictable and potentially destructive movements of investment capital and the jobs, resources, and products that come or go with it. To do this, countries may impose **capital controls** that require investment capital and the profits it earns to stay in a country for a certain amount of time. Such controls may also require that companies funded by foreign capital have a certain percentage of domestic investors.

Such protections are a subject of contention. Proponents of free trade have successfully argued that the unrestricted movement of goods and services across national borders encourages efficiency and the production of higher-quality goods and services and gives consumers more product choices at lower prices. With free trade, companies can sell to larger markets and take advantage of mass production systems that lower costs further. If local economies obtain access to large markets, they can grow faster, thereby providing people with jobs and opportunities to raise their standard of living.

Restrictions on trade imposed by individual countries are now being reduced through the formation of regional trade blocs and the support of global institutions. **Regional trade blocs** are associations of neighboring countries that agree to lower trade barriers for one another. Examples are the North American Free Trade Agreement (NAFTA), the European Union (EU), the Southern Common Market (Mercosur) in South America, and the Association of Southeast Asian Nations (ASEAN). Examples of global institutions that support free trade are the **World Trade Organization (WTO)** and the **World Bank** (officially named the International Bank for Reconstruction and Development). The stated mission of the World Trade Organization is to lower trade barriers and establish ground rules for international trade. The World Bank is a lending institution that makes loans to countries that need money to pay for development projects. It may require a borrowing country to reorganize its national economy to achieve freer trade. For example, before approving a loan, the World Bank often requires a country to improve conditions for private enterprise by reducing and eventually removing government support for domestic industries and agriculture and by reducing government services (schools, health care, social services) in order to lower taxes. Such economic reorganization requirements are called **structural adjustment policies,** or **SAPs.**

Groups opposed to free trade argue that its gains are offset by the instability—even chaos—that comes with a less regulated global economy. They emphasize that even in more regulated times, the global economy has been prone to rapid cycles of growth and decline that can wreak havoc on smaller national economies. When controls and protections are removed for freer trade, cycles of growth and decline may increase in frequency, intensity, and duration. Labor unions and other workers' organizations point out that as corporations relocate factories to poorer countries where wages are lower, jobs are lost in richer countries. In the poorer countries, multinational corporations and governments often prevent workers from organizing themselves into unions that could bargain for **living wages** (wages that support a minimum healthy life). Environmentalists argue that in newly industrializing countries, which often don't have effective environmental protection laws, multinational corporations tend to use highly polluting, unsustainable production methods because such methods lower production costs and raise profits. Many fear that a "race to the bottom" in wages, working conditions, and environmental quality is under way as countries compete for profits and for the attention of potential investors.

In developing an opinion about free trade and globalization, you might consider how many of the things you own—computer, clothes, furniture, appliances, and car—were made outside the United States. These products are cheaper for you to buy, and your standard of living is higher, as a result of lower production costs and competition among many global producers. Also, as a result of free trade within the United States, you can travel unencumbered by border crossings at state lines, and you can hunt for a job and live in any part of the country. On the other hand, you or a relative may have lost a job because the company moved to another location where labor is cheaper. You may be concerned that the products you buy so cheaply were made under harsh conditions by underpaid workers, or that resources were used unsustainably in the manufacturing and transport processes. These are the kinds of issues to be considered in reaching a point of view on the question of free trade. What might be the advantages and drawbacks of less-free trade?

MEASURES OF DEVELOPMENT

Until recently, the term **development** was used to describe only economic changes that lead to better standards of living. These changes often accompany the greater productivity in agriculture and industry that comes from such technological advances as mechanization and computerization. Increasingly, however, the question **"Development for whom?"** is being asked, as it becomes clear that merely raising average national productivity will not necessarily improve average standards of living for most people. Often the benefits of increased productivity have gone primarily to those who are already economically well off, leaving the majority in circumstances that are little improved, or even worsened. Furthermore, development, as measured by average economic gains, often results in environmental side effects, such as air and water pollution, that reduce the quality of life for everyone. Some development experts (notably, for example, Nobel Prize–winning economist Amartya Sen) are advocating a broader definition of development that includes measures of **human well-being** (a healthy and socially rewarding standard of living) as well as measures of environmental quality.

GROSS DOMESTIC PRODUCT PER CAPITA

The most popular economic measure of development is **gross domestic product (GDP) per capita,** which is the total value of all goods and services produced in a country in a given year, divided by the number of people in the country. This figure (see column 2 of Table 1.3) is often used as a crude indicator of how well people are living in a given country. There are, however, several problems with GDP per capita as a measure of overall well-being. First, because it is an average figure, it can hide the fact that a country has a few fabulously rich people and a mass of abjectly poor people. For example, a GDP per capita of U.S. $20,000 would be meaningless if a few lived on millions per year and most lived on $5000 per year. Second, the purchasing power of currency varies widely around the globe, so a GDP of U.S. $5000 per capita in Jamaica might represent a middle-class standard of living, whereas that same amount in New York City could not buy even basic shelter. Because of these purchasing power variations, in this book we use GDP per capita figures that have been adjusted for **purchasing power parity (PPP).** PPP is the amount that the local currency equivalent of U.S. $1 will purchase in a given country. For example, according to *The Economist*, on April 24, 2003, a Big Mac at McDonald's in the United States cost U.S. $2.71; in Australia it cost the equivalent of just U.S. $1.86; and in China it cost U.S. $1.20. Of course, for the consumer in China, where per capita incomes average about $4000, this would be a very expensive hamburger.

A third problem with GDP per capita is that it measures only what goes on in the formal economy, whereas in some places the informal economy actually accounts for more activity. Recently, researchers who examined all types of societies and cultures have shown that, on average, women perform about 60 percent of all the work done on a daily basis, much of it unpaid. Nonetheless, only their paid work performed in the formal economy appears in the statistics. Statistics also neglect the contributions of millions of men and children who work in the informal economy as subsistence farmers or as seasonal laborers. For example, a recent worldwide study conducted by the United Nations uncovered the

The air that envelops Cairo, Egypt, is by some measures the most polluted of any urban area in the world. Environmental regulations either don't exist or are not enforced. Here, smoke and dust from cement factories cover newly cleaned surfaces within seconds. Much of the funding for Cairo's factories has come from international loans or from internationally controlled companies. The factories provide needed jobs for desperately poor local residents. [Mohamed El Dakhakhny/AP/Wide World Photos.]

fact that 250 million children under the age of 14 are employed in the informal economy, half of them full-time! Many of these working children live in Asia, but some work, virtually unnoticed, in the United States (see Chapter 2, page 110) and Europe.

The most important failing of GDP per capita as a measure for comparing countries is that it ignores all aspects of development other than economic ones. There is no way to tell from GDP per capita figures, for example, how fast a country is consuming its natural resources, or how well it is educating its young, or whether it is treating its citizens equitably, caring for the sick, or maintaining its environment. There are a number of movements to refine the definition of development to include these factors and others. Therefore, along with the traditional GDP per capita figure (see Table 1.3, column 2), we have chosen several other measures of development to use in this book: the United Nations Human Development Index (HDI), the United Nations Gender Development Index (GDI), and the United Nations Gender Empowerment Measure (GEM). Together, these measures reveal some of the subtleties and nuances of well-being and make comparisons between countries somewhat more valid. Because these more sensitive indexes are also more complex than the purely economic GDP per capita, they are all still being refined by the United Nations.

HUMAN WELL-BEING

The United Nations Human Development Index (HDI). The Human Development Index considers adjusted real income, which takes into account what people can buy with what they earn, as well as data on life expectancy at birth and educational attainment. Countries are measured on the HDI factors and ranked from highest (1) to lowest (175). The HDI provides no way to score a country directly on the equality of its distribution of income or purchasing power. These factors are indicated only indirectly by the information about health and education. The assumption is that a country providing widely available health and education services has a more equitable distribution of wealth than a country that provides low levels of access to health care and education. Also, notice that the ranks of countries based on HDI (see Table 1.3, column 3) are quite different from those based solely on GDP per capita (figures in parentheses in column 2). Japan, for example ranks 14th globally in GDP per capita, but because it provides so well for its citizens, it has a higher rank on HDI (9).

The United Nations Gender Development Index (GDI). The Gender Development Index looks at whether countries make basic literacy, health care, and access to income available to both women and men (see Table 1.3, column 4). Because of a lack of data, only 144 countries of the 175 ranked by HDI are in this index. GDI does not measure general social acceptance of the idea of gender equality, which is better measured by GEM.

The United Nations Gender Empowerment Measure (GEM). The United Nations devised the Gender Empowerment Measure to score and rank countries according to how well they enable participation by women in the political and economic life of the country (see Table 1.3, column 5). The indicators used include percentage of women holding parliamentary seats; percentage of administrators, managers, and professional and technical workers who are women; and women's GDP per capita. Although the GEM tells us something about the relative power of men and women in a society, the indicators now available are unsatisfactory. In most countries, very little data is collected separately on men and women, and in many cases data on women is missing altogether, making comparisons impossible. Because sufficient data was available to rank only 70 of the possible 175 countries in the year 2003, a GEM rank higher than an HDI rank may be merely the result of the small number of countries ranked on GEM relative to HDI. Also, a high rank does not indicate that a country is treating women and men equally, but only that the country is doing better than those ranked lower. Although women are half the world's population, nowhere on earth do women hold 50 percent of the parliamentary seats or have an average earning power even close to that of men.

The Sample Human Well-Being Table. Table 1.3 shows GDP per capita, HDI, GDI, and GEM rankings for four selected

TABLE 1.3   Sample human well-being table

Selected countries (1)	GDP per capita adjusted for PPP[a] in 2001 $U.S. (GDP ranking among 175 countries ranked from highest) (2)	Human Development Index (HDI) rankings among 175 countries[b] (3)	Gender Development Index (GDI) rankings of 144 countries (4)	Gender Empowerment Measure (GEM) rankings among 70 countries (5)
Japan	25,130 (14)	9 (high)	13	44
United States	34,320 (2)	7 (high)	5	10
United Arab Emirates	20,530 (23)	48 (high)	49	65
Barbados	15,560 (36)	27 (high)	27	20

[a]PPP = purchasing power parity.
[b]The high, medium, and low designations indicate where the country ranks among the 175 countries classified into just three categories.

Source: United Nations Human Development Report 2003 (New York: United Nations Development Programme).

countries: Japan, the United States, the United Arab Emirates (UAE), and Barbados. Notice how the rankings change in the four columns. The UAE rank drops significantly from 23rd in the GDP per capita column (2) to 48th in the HDI column (3); Japan and the UAE (and the United States to a lesser extent) show drastic drops from their GDI to GEM ranks; meanwhile, Barbados, with a relatively low GDP per capita, rises in rank significantly from the GDP, to the HDI and GDI, to the GEM columns. Perhaps you can already suggest some explanations for why these rankings differ so radically.

Each chapter in this book is designed to provide historical, demographic, cultural, social, economic, and political information that will help you to interpret the rankings for the countries in each world region. A human well-being table similar to this one, with a short discussion of the rankings, will be included in each chapter.

POPULATION PATTERNS

The changing levels of well-being associated with economic development have had dramatic effects on population growth. **Demography**, the study of population patterns and changes, is an important part of geographic analysis. Because geographers are concerned with the interaction between people and their environments, it is essential to know how many people there are on earth and how they are distributed, how fast their numbers are growing, how they make a living, and what their patterns of migration and consumption are.

GLOBAL PATTERNS OF POPULATION GROWTH

It took between 1 million and 2 million years, or at least 10,000 generations, for humans to evolve and to reach a population of 2 billion, which happened in about 1945. Then, in just 55 years, by the year 2000, the world's population more than tripled to 6.1 billion. What happened to make the population grow so very quickly in such a short time?

The explanation lies in changing relationships between humans and the environment. For most of human history, fluctuating food availability, natural hazards, and disease kept human death rates high, especially for infants. Periodically, there would even be crashes in human population—as happened in the 1300s throughout Europe and Asia during the pandemic known as the Black Death (Figure 1.16). Then an astonishing upsurge in human population began about 1500, at a time when the technological, industrial, and scientific revolutions were beginning in some parts of the world. These revolutions helped people to exploit land and resources, providing better diets, and improved health and disease control. Human life expectancy increased dramatically, and more and more people lived long enough to reproduce successfully, often many times over. The result was an exponential pattern of growth that is often called a J curve (see Figure 1.16), because the ever shorter periods between doubling and redoubling cause an abrupt upward swing in growth as depicted on a graph.

Today, the human population is growing in virtually all parts of the world—but more rapidly in some places than in others. Even if all couples agreed to have only one child, world population

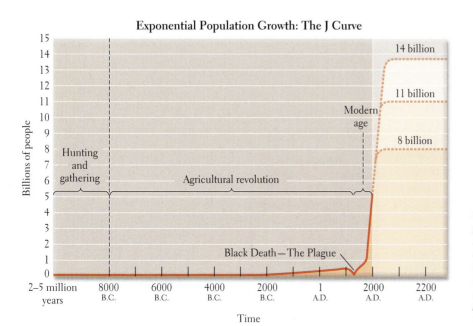

Exponential Population Growth: The J Curve

FIGURE 1.16 **The J curve depicts the exponential growth of the human population.** The curve's shape is a result of successive doublings of the population: it starts out nearly flat, but as doubling time shortens, the curve bends ever more sharply upward. Why do you think the Black Death is visible in the curve, but the millions killed in recent wars do not show up? [Adapted from G. Tyler Miller, Jr., *Living in the Environment*, 8th ed. (Belmont, Calif.: Wadsworth, 1994), p. 4.]

would probably grow to at least 8 billion before zero growth would set in. Growth would continue until then because there are presently so many people who have not yet reached the age of reproduction. Nevertheless, there are indications that the rate of growth is slowing globally. In 1993, the world growth rate was 1.7 percent. By 2003, it had decreased to 1.3 percent, but even this decreased rate still results in a doubling time of only 54 years. If present slower growth trends continue, world population may level off at somewhat over 9 billion before 2050, but this projection is contingent on couples in less developed countries having access to birth control technology and to the information on why fewer children would be advisable. Nine billion people will tax the earth's resources beyond imagining, especially if more and more people have lifestyles based on mass consumption, as is increasingly the case.

LOCAL VARIATIONS IN POPULATION DENSITY AND GROWTH

If the more than 6 billion people on earth today were evenly distributed across the land surface, they would produce an **average population density** of about 122 people per square mile (47 per square kilometer). As you can see from the population density map in Figure 1.17, people are not evenly distributed across the face of the earth. We can make certain general statements about their distribution. First, nearly 90 percent of all people live north of the equator, and most of them live between 20° N and 60° N. Even within that limited territory, people are concentrated on about 20 percent of the available land. For the most part, people choose to live in lowland regions—nearly 80 percent live below 1650 feet (500 meters)—in zones that have climates warm and wet enough to support agriculture. Many people live along rivers, and most people live fairly close to the sea, within about 300 miles (500 kilometers). In general, people are located where resources are available.

Nevertheless, many places with meager resources contain a great many people because the resources to support them can be garnered from elsewhere. If people have the means to pay for them, food, water, and clothing, raw materials for building and manufacturing and for producing electricity, and other material support can all be imported. An extreme example is Macao, the former Portuguese trading enclave in South China. The population density in Macao is the highest on earth: 57,600 per square mile (22,000 per square kilometer). Life there is sustained at a high standard not by local physical resources, but by trading connections enhanced by Macao's location on the edge of the huge and populous Chinese mainland. For many centuries, residents of cities around the world have relied on distant resources, and this is increasingly true for more and more of the world's people, no matter where they live.

Just as there is no easy correlation between population density and richness of resources, there is no easy correlation between density and poverty or wealth. Some rather densely occupied places, such as parts of Europe and Japan, are very wealthy. Other

Thinking Geographically: **ON THE WEB**

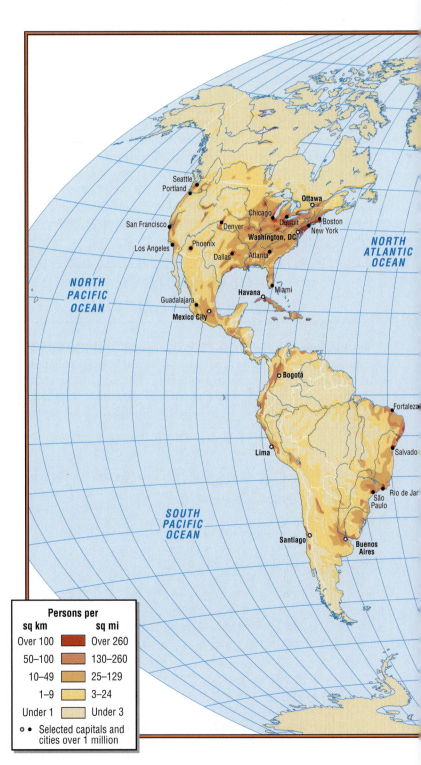

FIGURE 1.17 World population density. The text makes some general statements about patterns of world population density; for example, many people live north of the equator, in lowlands, and close to rivers and the sea. What are some social and physical explanations for these distribution patterns? [Adapted from *Hammond Citation World Atlas* (Maplewood, N.J.: Hammond, 1996); and *World Population Data Sheet, 2000* (Washington, D.C.: Population Reference Bureau).]

Persons per		
sq km		sq mi
Over 100		Over 260
50–100		130–260
10–49		25–129
1–9		3–24
Under 1		Under 3
○ ●	Selected capitals and cities over 1 million	

densely populated places, such as parts of India and Bangladesh, are desperately poor. To explain population density patterns today, we must look to cultural, social, and economic factors and to such historical events as experiences with colonialism. We also must identify present circumstances that attract migrants to some places and cause them to flee others.

Usually, the variable that is most important for understanding population growth in a region is the **rate of natural increase** (often shortened to **growth rate**). The rate of natural increase is the relationship between the number of people being born (**birth rate**) and the number dying (**death rate**) in a given population, without regard to the effects of migration. The rate of natural increase is usually expressed as a percentage. Take the example of Austria in Europe, which has 8,200,000 people. The annual birth rate is 9 per 1000 people, and the death rate is 9 per 1000 people. Therefore, the annual rate of natural increase is 0 per 1000

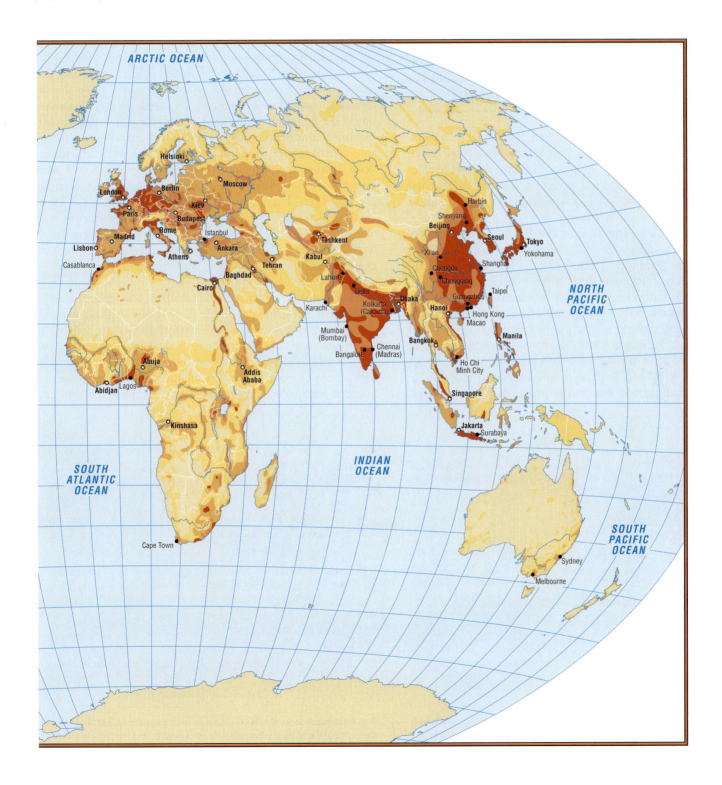

(9 − 9 = 0), or zero percent. (Actually, Austria is probably growing a tiny bit per year.) Just 16 percent of Austrians are under 15 years of age, and 16 percent are over 65 years of age. It is estimated that, under current conditions, Austria will take 2310 years to double its population. For the sake of comparison, consider Jordan in Southwest Asia, which has 5,500,000 people. The birth rate is 29 per 1000 and the death rate is 5 per 1000. The growth rate is thus 24 per 1000 (29 − 5 = 24), or 2.4 percent. At this rate, Jordan will double its population in just 29 years. As you might expect, the population of Jordan is very young: 40 percent are under 15 years of age, and just 5 percent are over 65.

AGE AND GENDER STRUCTURES

The **age distribution** (also known as the **age structure**) of a population is the proportion of the total population in each age group, and the **gender structure** is the proportion of males and females in each age group. The age and gender structures reflect past and present social conditions, and knowing these structures helps us to predict future population trends.

The **population pyramid** is a graph that depicts age and gender structures. Careful study of the population pyramids of places such as Austria and Jordan (Figure 1.18) reveals the age and gender distribution of the populations of those countries. As we have noted, most people in Jordan are very young, so they are clustered toward the bottom of the pyramid, with the largest groups in the age categories 0 through 9. The pyramid tapers off quickly as it rises above age 34, showing that in Jordan most people die before they reach old age. Only one category is needed for those over 80.

On the other hand, Austria's pyramid has an irregular shape, indicating that Austria has had experiences that alternately increased or decreased population growth. The narrow base indicates that there are now fewer people in the younger age categories than in young adulthood or middle age, and those over 70 greatly outnumber the youngest (ages 0 to 4). In the Austrian pyramid four age categories are used for those over 70. This age distribution tells us that many Austrians now live to an old age and that in the last several decades Austrian couples have been choosing to have only one child, or none. Austrians worry that their population will begin to decline and be weighted with elderly people who will need care and financial support from an ever-declining group of working-age people.

Population pyramids also reveal subtle gender differences within populations. Look closely at the right (female) and left (male) halves of the pyramids in Figure 1.18. Notice that in several age categories the sexes are not evenly balanced on either side of the line. In the Austrian pyramid, there are more women than men near the top (age 70 and older), reflecting the deaths among male soldiers in World War II and the as yet poorly understood trend in wealthy countries for women to live about 5 years longer than men. In Jordan, the gender imbalances occur at younger ages. There are about 220,000 women and about 269,000 men aged 25 to 29. These gender imbalances exist in other age groups as well; for example, among those under age 4, there are about 16,000 fewer females than males. Because gender-based research is so new, explanations for the imbalances are only now being proposed.

One possible explanation is that males and females may be treated differently throughout life (see the box "Missing Females in Population Statistics" on the next page). In societies afflicted by poverty or where the preference for sons is high, girls are sometimes fed less than boys and may die in early childhood more often. Because poor women frequently do not receive needed nutrition and medical care during pregnancy, more mothers tend

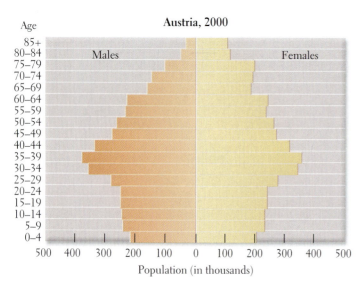

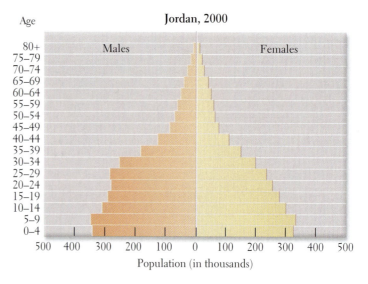

FIGURE 1.18 Population pyramids for Austria and Jordan. Careful study of the pyramids of places such as Austria and Jordan reveals contrasting age and gender distributions. Jordan's young population is clustered toward the bottom, while Austria's is not. Notice also that, on average, Austrians live longer than Jordanians. How would you account for these differences? Find the cases where there are gender imbalances in Austria and Jordan. How do they differ, and how would you explain these patterns? [Adapted from U.S. Bureau of the Census, International Data Base, at http://www.census.gov/ipc/www/idbpyr.html.]

to die during childbirth. In some places, rules requiring female seclusion from males make professional health care less accessible to women. All of these factors probably affect the survival rates of women and girls in Jordan. Another factor may be the availability of new technology that identifies the gender of an early-stage fetus.

Even surviving females may be invisible in statistics. Much data is collected without distinguishing between the sexes, thus obscuring important statistical differences. For example, research has shown that around the world, in every country from Sweden to Swaziland, females benefit less from development than do males. But development statistics rarely show gender differences, so the lesser well-being of females is not apparent.

Another aspect of the "missing female" problem is found in global statistics related to work. Almost universally, women's work has been so undervalued that it is virtually missing from national labor statistics, which deal only with the formal economy. Women's contributions as subsistence farmers, as homemakers, as domestic servants, as child-care providers, as volunteers, and as workers elsewhere in the informal economy have been considered mere pastimes, not real productive labor, so much of women's work has not been calculated into national income statistics. Even though the oversight has been documented for more than 20 years, the customs of statistics gathering are only beginning to change, and the missing female phenomenon continues.

AT THE GLOBAL SCALE Missing Females in Population Statistics

The worldwide cultural preference for sons over daughters is observable statistically in that between 60 million and 100 million females are missing from the world population. The normal ratio is for about 95 females to be born for every 100 males. Because boy babies are somewhat weaker than girls, the ratio evens out naturally within the first 5 years of life. Since 1900, however, the ratio of females to males at birth has been declining; in 1990, it was about 92.6 to 100 globally, and for parts of South and East Asia it was as low as 80 females for every 100 males. Just why the ratio

started to decline around 1900 has not yet been explained satisfactorily, but the likely reason is the nearly global preference for boys over girls, which seems to increase as couples chose to have fewer children. Some of the missing females were conceived but were never born because their parents chose to abort them. Others died at a young age because girls received less adequate health care and poorer nutrition than boys, and in some cases, parents committed female infanticide. Figure 1.19 shows the global pattern of abnormally high death rates for young females.

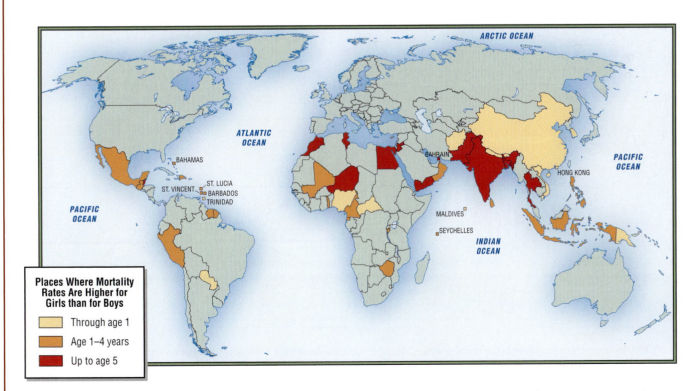

FIGURE 1.19 Young female mortality rates. In the countries shaded yellow, orange, and rust, the mortality rate for girls is abnormally high. The darker the color, the longer the risk to girls lasts. How might the explanations for these patterns vary from country to country? [Adapted from Joni Seager, *The State of Women in the World Atlas* (New York: Penguin, 1997), p. 35.]

POPULATION GROWTH RATES AND WEALTH

Although there is a wide range of variation, regions with slow population growth rates usually tend to be affluent, and regions with fast growth rates tend to have widespread poverty. The reasons that poor people tend to have more children than the rich are complicated. Again, the cases of Austria and Jordan serve as examples.

Austria has a GDP per capita (PPP) of $26,730. Its highly educated population is 100 percent literate and is employed in high-tech industry, business, research, teaching and writing, manufacturing, and upscale tourism. The large amounts of time, effort, and money required to prepare a child to compete in this economy cause many couples to choose not to have any children. Jordan, on the other hand, has a GDP per capita (PPP) of $3870. In Jordan, although many people live in urban areas (and 97 percent of the recorded GDP comes from industry and services), much everyday work is still done in the informal economy by hand, and each new child is seen as a potential contributor to the family well-being and income and as an eventual caregiver when the parents grow old. Not producing enough children who will live to adulthood is still a significant worry, because 33 children per 1000 die before they are 5 years old. (In Austria, only 5 children per 1000 die before they are 5.) It is not surprising, then, that only 56 percent of the women in Jordan use birth control (more than 67 percent do in Austria), in part because in Jordan birth control is less accessible, but also because the chance of losing a child to illness is much greater than in Austria. But the situation is changing in Jordan. Just 25 years ago, the GDP per capita was $993; infant mortality was 77 per 1000; and the average woman had eight children, whereas now she has only four.

The kinds of changes evident in Jordan are taking place in many countries. As agricultural production is mechanized, there is less need for agricultural labor and, instead, a demand for fewer, well-trained specialists. Furthermore, **subsistence lifestyles** —circumstances in which a family produces most of its own food, clothing, and shelter—are losing their appeal because cash is needed to buy such goods as television sets, bicycles, blue jeans, sneakers, T-shirts, meat, sugar, and canned goods. In **cash economies,** children become a drain on the family income. Each child must be educated in order to qualify for a good cash-paying job. There are often school fees, and children who are in school all day and doing their lessons in the evening are not able to earn money until they are 18 or 20 years old.

Societies that are going through the transition from subsistence to cash economies must learn all these consequences first-hand. Until recently, it took people one full generation or more to

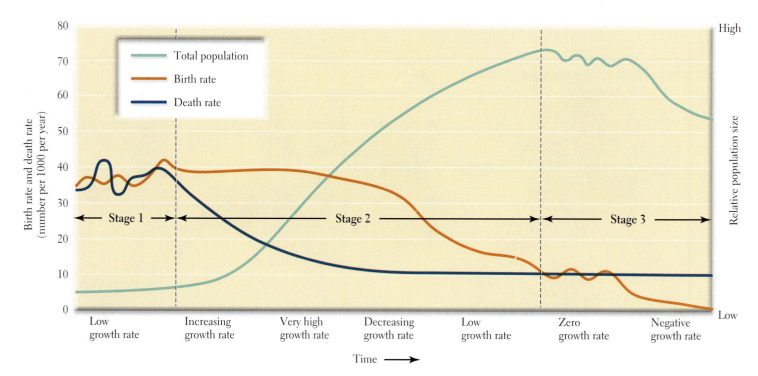

FIGURE 1.20 The demographic transition. In traditional societies, little touched by technology, food production, and health care developments, birth rates and death rates are usually high, at 35 or more births and deaths per 1000 people per year (left vertical axis), and population numbers (right vertical axis) remain low and stable (Stage 1). With advances in food production, education, and health, death rates usually drop rapidly (to 10 or fewer per 1000 per year), but strong cultural values regarding reproduction remain, often for generations, so birth rates drop much more slowly (hovering between 40 and 25 per 1000 per year), with the result that for decades or longer, the population grows significantly (Stage 2). When the understanding grows that it is no longer necessary to have large families because, due to the changed social and economic circumstances, most children will survive to adulthood and it is no longer necessary to produce a cadre of family labor (in fact, many children are a detriment to family well-being) population growth rates slow (Stage 3). At this point, demographers say that the society has gone through the **demographic transition.** [Adapted from G. Tyler Miller, Jr., *Living in the Environment*, 8th ed. (Belmont, Calif.: Wadsworth, 1994), p. 218.]

see the benefit of having fewer children and then to choose to have fewer. Now growth rates are dropping more quickly, perhaps as a result of better public education. When growth rates in a region slow, demographers say that the region has gone through the **demographic transition:** that is, a period of high growth rates has given way to a period of much lower growth rates, at the same time that major social and economic changes are taking place within the society (Figure 1.20).

HUMANS AND THE ENVIRONMENT

The long, continuous give-and-take between humans and their physical environment has resulted in many improvements in the circumstances of human life. At the same time, humans have had an enormous impact on physical environments in their efforts to obtain food and shelter, to travel, and to protect themselves from storms, droughts, flooding, and extreme cold and heat. Some of these effects on the environment are now regarded as harmful (and perhaps irreversible). They include ozone depletion, erosion, global warming, acid rain, deforestation, and chemical pollution of groundwater. High standards of living and mass consumption tend to alter the environment most profoundly, but all human ways of life—whether on small cultivation plots in the tropics or in huge industrial cities—have some environmental effects.

Mounting awareness of environmental impacts has prompted numerous proposals to limit the damage. Many of these plans are based on the use of advanced technology, such as catalytic converters to decrease harmful auto emissions and scrubbers on factory chimneys. Other proposals focus on reducing resource consumption. Finding ways to halt or even reverse environmental damage may be the greatest challenge our species has yet faced. Strategies are being devised in all parts of the world to create more sustainable ways of life, but all will require us to reduce our habits of mass resource consumption—habits that are most elaborately developed in rich countries.

SUSTAINABLE DEVELOPMENT

The United Nations defines **sustainable development** as the effort to improve present living standards in ways that will not jeopardize those of future generations. The concern is that by destroying resources (as in deforestation) or poisoning them (as in pollution of water and air supplies), we may be depriving future generations of the resources they will need for their well-being. Those who support the idea of sustainability hope to promote more environmentally and socially friendly development for all, but especially for the vast majority of earth's people who do not yet enjoy an acceptable level of well-being. Although sustainable development is considered desirable by many people, it is still a concept that is being refined, and no country or organization has yet devised a workable policy.

Geographers who study the interactions among development, human well-being, and the environment are called **political ecologists.** They examine how the power relationships in a society affect the ways in which development proceeds, whose needs it addresses, and how success is measured. Political ecologists have noted, for example, that a country may appear to be developing because its gross domestic product is rapidly increasing—but to achieve that increase, it may be using its resources in ways that cannot be sustained. For instance, the clearing of forests to grow oil palm trees might raise a Southeast Asian country's GDP through the sale of palm oil, but GDP does not take into account the resulting loss of forest resources, or the unsustainable use of soils that do not maintain their fertility when a single plant species replaces a multispecies forest. Also unaccounted for is the destruction of a way of life when forest dwellers are forced to migrate to cities, where their woodland skills are useless. The monetary value of national production may increase, and landowners may prosper, at the same time that quality of life for the former forest dwellers sinks abysmally and losses in other categories go unrecognized.

On the other hand, this same country could choose to measure its development by improvements in average human well-being and in present environmental quality. It could consider the

A young man in Yunnan Province, China, cultivates onions and cabbages early in the morning before heading off to a nearby school. Most families in China use intensive agricultural techniques to maximize garden production on small plots that have been sustainable for generations. [Leong Ka Tai/ Material World.]

potential for maintaining that human well-being and environmental quality into the future. A country using such measures would soon see that oil palm plantation development would fail to meet both its development and sustainability goals. The answer to the question "Development for whom?" would be that with oil palms, only a few would benefit, whereas the majority of citizens, the environment, and future generations would lose.

Developing sustainably is not easy. Robert Prescott-Allen, a Canadian specialist in sustainability, has produced an index that combines measures of human well-being with those of environmental sustainability. He found that even the countries that provide the highest overall standard of living for their citizens (Sweden, Finland, Norway, and Iceland) achieve this standard only at the expense of the environment. The following sections examine some of the issues of sustainability in agriculture, industry, and the growth of cities. We will then look at the issue of global warming, which is thought to be linked to unsustainable use of fossil fuel energy.

Sustainable Agriculture

Farming that meets human needs without poisoning the environment or using up water and soil resources is called **sustainable agriculture.** Food production on earth has increased remarkably, especially over the last several decades as many types of technology have been applied to agricultural production. In the 25 years between 1965 and 1990, total food production rose between 70 and 135 percent, depending on the region (Figure 1.21). But population also rose quickly during this period, so the gains were much less per capita, and in Africa, per capita food production actually decreased. By 2004, it appeared that growth in production was slowing. For wheat, growth rates slid from nearly 3 percent per year as of the late 1970s to just 1.78 percent by the late 1990s. Increasingly, scientists from many disciplines are saying that we will reach the limit of our productive capacity within the next 50 years, because as populations grow, environmental problems proliferate. There is some hope that technological advances, such as genetically modified (GM) crops and advanced irrigation techniques, will make present agricultural land more productive and unused land useful.

Just how sustainable are the world's present agricultural systems? The answer is unclear, but previously unrecognized side effects of development are just coming to light. The global map of soil degradation in Figure 1.22 shows that many of the most agriculturally productive parts of North America, Europe, and Asia have already suffered moderate to serious losses of soil through erosion. Moreover, many of the areas labeled as stable on this map may be unusable for agriculture because of extremes in temperature and moisture. Globally, soil degradation and other problems related to food production affect about 7 million square miles (2000 million hectares), putting the livelihoods of a billion people at risk.

The main causes of soil degradation are overgrazing, deforestation, and mismanagement of farmland. Irrigation often makes soil salty and infertile over time (see Figure 5.16, page 267). It also can deplete water resources, because irrigation depends on diverting water from rivers and streams or pumping it up from natural underground reservoirs, often at a rate that causes the sources to shrink or even dry up completely. Modern agricultural techniques pioneered in the United States and now spreading throughout the world—such as the use of fertilizers, pesticides, and herbicides—have caused massive die-offs of birds, insect pollinators, fish, and other life-forms.

The definition of sustainable agriculture requires that food production meet the needs of everyone on the planet. Yet about one-fifth of humanity subsists on a diet too low in total calories and vital nutrients to sustain adequate health and normal physical and mental development. At the same time, wealthy countries have experienced surplus food production that has depressed prices and contributed to farm failures. The problem of hunger, so far, is not that the world's agricultural systems cannot produce enough, but that food often does not reach hungry people. A further problem is that when rich countries send their surplus food to areas suffering famine, local farmers in the famine zone who can still produce are often forced out of business because the markets where they previously sold their produce are flooded for a time with free food. Many experts now agree that the most promising solution to hunger seems to be for individual countries and regions to develop their own plans for sustainable agricultural

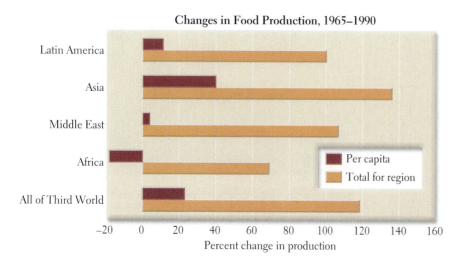

Changes in Food Production, 1965–1990

Latin America

Asia

Middle East

Africa

All of Third World

■ Per capita
☐ Total for region

−20   0   20   40   60   80   100   120   140   160
Percent change in production

FIGURE 1.21 Changes in food production between 1965 and 1990, by region. In the 25 years between 1965 and 1990, total food production rose between 70 (Africa) and 135 (Asia) percent, depending on the region. But population also rose quickly during this period, so the gains were much less per capita, and in Africa, per capita production actually decreased. [Adapted from John Bongaarts, "Can the growing human population feed itself?" in *Global Issues, 1995-96* (Guilford, Conn.: Dushkin/Brown & Benchmark, 1995), p. 119.]

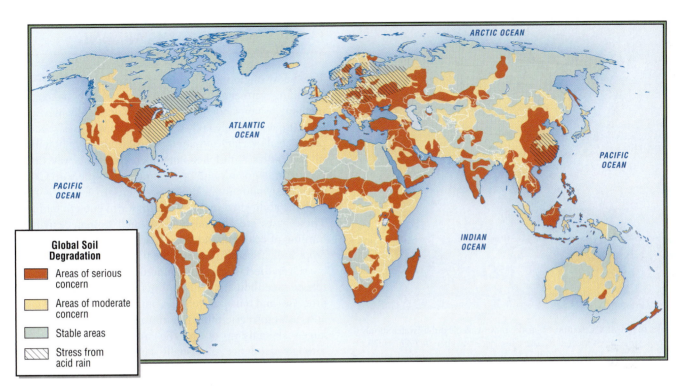

FIGURE 1.22 Global soil degradation. Many of the world's most productive agricultural lands have suffered soil degradation. Can you speculate as to why agriculture could not simply be shifted to some of the areas on this map labeled "stable"? [Adapted from John L. Allen, *Student Atlas of Environmental Issues* (Guilford, Conn.: Dushkin/McGraw-Hill, 1997), p. 109; Tom McKnight, *Physical Geography* (Upper Saddle River, N.J.: Prentice Hall, 1996), p. 159.]

development. Wealthier countries should help when asked, but in ways that will encourage sustainability, not dependence. In places like the United States that may mean cutting production.

In general, it is safe to say that up to now, there have been few truly sustainable agricultural solutions. On the one hand, mass production almost always results in environmental problems and in misallocations, and on the other hand, small-scale production that may be more environmentally sound cannot meet the huge food demands of growing urban populations. The agricultural successes and failures of various countries and agencies will be examined further in the chapters on world regions.

Sustainability and Urbanization

The world is fast becoming urbanized. In 1700, less than 10 percent of the world's total population—about 7 million people—lived in cities. Only 5 of those cities had populations as high as several hundred thousand people. By 2004, there were more than 400 cities of over 1 million and more than 24 cities with 10 million. By 2003, 47 percent of the world's population lived in cities. Although many people anticipate enjoying city life, it is the elusive ideal of life in a modern, wealthy city that they cherish. Most urban dwellers today cope with far more difficult realities.

Water and sanitation are cases in point. In rapidly growing cities in Asia, Africa, and Latin America, housing is often self-built of scavenged materials, and sanitation systems are absent. In many cities, working toilets are in short supply, and desperate citizens relieve themselves in poorly drained pit toilets or on city streets. Sewage and other wastewater is often pumped into a nearby river, swamp, or ocean, even from modern high-rise apartments and luxury hotels. This method of waste disposal causes serious health hazards and widespread ecological damage.

Increasingly, people must boil all the water they use, even for bathing, or must buy drinking water that has been purified. But the poorest urban inhabitants, who are often a majority of the population, lack the money to buy clean water or the fuel to boil it. Adults are often chronically ill, and waterborne diseases cause many children to die before the age of 5.

Building adequate wastewater collection and water purifying systems in cities already housing several million inhabitants is so costly that this option is rarely considered, yet new affordable sources of clean water are not available. As we will see in later chapters, clean water is only one of several problems faced by the world's urban dwellers. Technological and other advances will no doubt help alleviate some problems, but we are now far short of the sustainable ideal for the world's cities.

Changing Patterns of Resource Consumption

One of the most interesting revelations to come from the geographic analysis of rates of resource use is that as people move from rural agricultural work to industry or service sector jobs in cities, they begin to use more resources per capita. Moreover, they draw their resources from a wider and wider area. Water once fetched from nearby village wells may now be piped into urban apartments or shantytowns from hundreds of miles away. Clothing once made laboriously by hand at home is now purchased from manufacturers half a world away. Consumers in

In sections of Kolkatta (Calcutta), India, the only available public water supply comes from small pipes located at curbsides. Here a woman *(left center)* fills buckets for washing dishes and clothes from a public spigot (under her left hand). [Dilip Mehta, Contact/The Stock Market.]

Thinking Geographically: ON THE WEB

wealthy countries may have access to a variety of imported products at lower prices than they would pay for locally produced items. But there is a downside to accessing a wide net of resources: the tendency to overconsume. When water is piped into the house or yard, its very convenience encourages waste. When clothes can be purchased at low prices, closets become crowded. In fact, resource use has become so skewed toward the affluent with cash to pay that in any given year, the relatively rich minority (about 20 percent of the world's population) consumes more than 80 percent of the available world resources. Meanwhile, the poorest 80 percent of the population must be content with only 20 percent of the resources.

Another statistic related to resource consumption is that the wealthiest 20 percent of the world's population produces close to 90 percent of the world's hazardous wastes, and this same group also consumes well over 50 percent of the world's fossil fuel, metal, and paper resources. The economist E. F. Schumacher, in his book *Small Is Beautiful*, commented that "the problem passengers on spaceship Earth are the first class passengers." Schumacher's point is an important one: consumers in industrialized societies consume far more than their share of resources and produce most of the hazardous wastes and environmental degradation. But this observation does not recognize the full complexity of the situation. Although the people of the developing world presently consume much less as individuals than do those of the developed world, the developing world is already suffering from environmental deterioration. As populations in these countries increase, and as their consumption per capita grows closer to consumption rates in the developed world, negative environmental effects will also increase proportionately. A case in point is global warming.

GLOBAL WARMING

The theory of **global warming** proposes that the earth's climate is becoming warmer as atmospheric levels of carbon dioxide (CO_2), methane, water vapor, and other gases drastically increase. These gases are collectively known as "greenhouse gases" because their presence allows large amounts of heat from sunlight to be trapped in the earth's atmosphere in much the same way that heat is trapped in a greenhouse or a car parked in the sun. Greenhouse gases exist naturally in the atmosphere. In fact, it is their heat-trapping ability that makes the earth warm enough for life to exist. These gases are also released by such human activities as the burning of coal, oil, and other fossil fuels, the growing of certain crops such as paddy rice, and the use of nitrogen fertilizers.

Over the last several hundred years, human activities that release greenhouse gases have intensified dramatically. Industrial processes and transport vehicles burn large amounts of fossil fuels, and cash-crop agriculture uses nitrogen-based fertilizers. Even the large-scale raising of grazing animals contributes methane through flatulence. Unusually large quantities of greenhouse gases from these sources are accumulating in earth's atmosphere, and many scientists think that their presence has already led to a warming of the planet's climate.

At the same time, deforestation is removing growing plants (particularly trees) from landscapes around the world, especially in developing countries (forest cover in some developed countries, such as the United States, is actually increasing, although it is less diverse in species than the original forest). Living forests are a reservoir of carbon, which is continuously taken up, processed, and released back into the atmosphere by the plants. When trees are cut down and their biomass used for fuel or other purposes, the carbon reservoir is converted, either immediately or eventually, into carbon dioxide. Scientists at the World Resources Institute in Washington, D.C., say that as much as 30 percent of the buildup of CO_2 in the atmosphere results from the loss of trees and other forest organisms and their replacement with bare soil, ash, and eventually regrowth that is extremely carbon poor compared to the original forest.

Climatologists and other scientists are documenting global warming by examining evidence in tree rings, fossilized pollen and marine creatures, and glacial ice. This data indicates that the twentieth century was the warmest in 600 years and that the decade of the 1990s was the hottest since the late nineteenth century. It is estimated that average global temperatures will rise between 3°F and 9°F (about 5°C) by 2030, but it is not clear just what the consequences of such a rise in temperature will be. Certainly the effects will not be uniform across the globe. One prediction is that glaciers and the polar ice caps will melt—as they are already doing—causing a corresponding rise in sea level. Hundreds of millions of people in coastal areas and on low-lying islands could be displaced. Scientists also forecast a shift of warmer climate zones northward in the Northern Hemisphere and southward in the Southern Hemisphere. Such shifts might also displace huge numbers of people, as the zones where specific crops can grow would change dramatically. Animal and plant

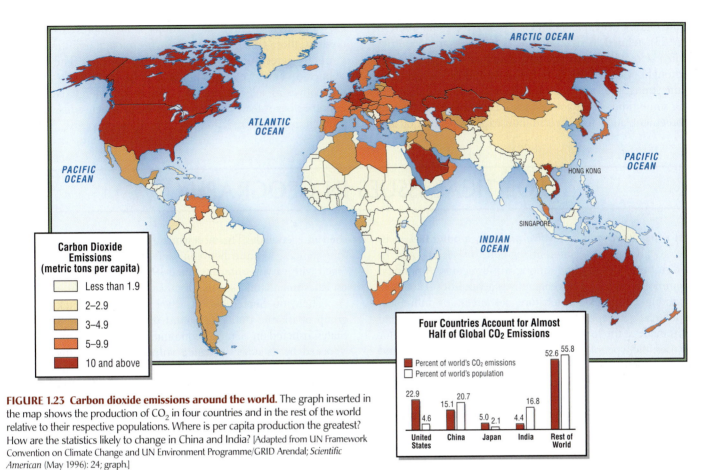

FIGURE 1.23 **Carbon dioxide emissions around the world.** The graph inserted in the map shows the production of CO_2 in four countries and in the rest of the world relative to their respective populations. Where is per capita production the greatest? How are the statistics likely to change in China and India? [Adapted from UN Framework Convention on Climate Change and UN Environment Programme/GRID Arendal; *Scientific American* (May 1996): 24; graph.]

species that could not adapt to the change rapidly would disappear. Another effect of global warming could be a shift in ocean currents, resulting in more chaotic and severe weather.

Most of the responsibility for CO_2 emissions rests on the shoulders of the industrialized countries of North America, Europe, northern Eurasia, and Australia, as well as the oil-producing countries of Southwest Asia (Figure 1.23). For the period 1859–1995, developed countries produced roughly 80 percent of the greenhouse gases from industrial sources, and developing countries produced 20 percent. But by 2000, the developing countries were catching up, accounting for nearly 40 percent of total CO_2 emissions. As developing nations industrialize over the next century and continue to cut down their forests, they will release more and more greenhouse gases every year (Figure 1.24). If present patterns hold, greenhouse gas contributions by the developing countries will exceed those of the developed world by 2040.

In 1992, members of the **Organization for Economic Cooperation and Development (OECD),** which includes the highly industrialized countries of North America, Europe, East Asia, and Oceania, drafted an agreement known as the Kyoto Protocol calling for scheduled reductions in CO_2 emissions by developed countries. The agreement also encourages OECD cooperation with developing countries to help them curtail their emissions as well. By 1997, 160 nations had signed the agreement, but the United

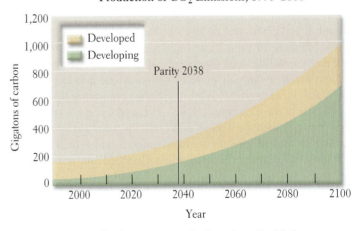

FIGURE 1.24 **Contributions to atmospheric carbon dioxide by developed and developing countries.** The carbon dioxide releases shown in this graph include both industrial emissions and amounts released as a result of deforestation. When both these sources are taken into account, the developing countries will exceed the developed countries in CO_2 production after 2038. How could developed countries best encourage lower emission rates in developing countries? [Adapted from Duncan Austin, José Goldemberg, and Gwen Parker, "Contributions to climate change: Are conventional metrics misleading the debate?" in *Climate Notes* (World Resource Institute Climate Protection Initiative, Oct. 1998), at http://www.wri.org/cpi/notes/metrics.html.]

States, the highest per capita producer of CO_2 (see graph insert in Figure 1.23), refused to sign, arguing that developing countries should first be required to reduce their emissions more stringently.

Although evidence for human-induced global warming is now quite strong, the issue will probably be a source of disagreement for some time. The Kyoto Protocol contains few provisions for enforcement, and a few affected industries—especially oil and gas—still argue that global warming is an unproved theory. In contrast, those concerned about environmental pollution see the currently agreed-upon reductions as far too low. They call for stepped-up energy conservation and for more research into such alternatives as solar, wind, and geothermal energy. Developing countries have steadfastly refused to reduce their emissions until the rich, industrialized, and largest emitters do so first.

POLITICAL ISSUES IN GEOGRAPHY

Political issues in geography revolve primarily around power: its exercise, its allocation to different segments of society, and its spatial distribution within and among world regions. A political issue commonly discussed in geography is **geopolitics**—the jockeying among countries for territory, resources, or influence. Other political geography topics include how public and private decisions made at the local level affect the allocation of space for various uses; the types of government that exist at local, state, and national levels and how they affect social organization, resource allocation, and land use; the ramifications of changing borders between countries; and the roles of international organizations focused on development, human well-being, and peacekeeping.

GEOPOLITICS

Geopolitics encompasses the strategies that countries use to ensure that their own interests are served in their relations with other countries. Geopolitics typified the **cold war era,** the period from 1946 to the early 1990s when the United States and its allies in Western Europe faced off against the Union of Soviet Socialist Republics (USSR) and its allies in Eastern Europe and Central Asia. Ideologically, the United States promoted a version of free-market capitalism and democracy, whereas the USSR and its allies favored a centrally planned economy and a socialist system in which citizens participated in government indirectly, through members of the Communist Party. The cold war grew into a contest that resulted in a race to attract the loyalties of unallied countries and to arm them. Eventually, because of the rivalry for their allegiance, the cold war influenced the internal and external policies of virtually every country on earth, often oversimplifying complex local issues to a contest of democracy verses communism.

In the post–cold war period of the 1990s, geopolitics shifted: the Soviet Union dissolved, creating many independent states, nearly all of which began to implement some democratic and free-market reforms. Globally, countries jockeyed for a better position in what appeared to some to be a new era of trade and amicable prosperity, rather than war. But throughout the 1990s, while the developed countries enjoyed unprecedented prosperity, there were many unresolved political conflicts in Asia, Africa, the Middle East, and Europe, which too often erupted into bloodshed and genocide. The most worrisome of these conflicts was the long-standing dispute between Israel and Palestine, which had raged with varying intensity since 1948, when Israel was created on Palestinian lands (see the discussion in Chapter 6).

Then, on September 11, 2001, a terrorist attack led by a group of Islamic militants against several targets in the United States ushered in a new geopolitical era that is still evolving. Shocked by its newly revealed vulnerability, the United States reacted by greatly increasing its emphasis on internal security while simultaneously embarking on a foreign policy that candidly placed U.S. security interests above all other concerns. Its first retaliatory step was a preemptive attack on Afghanistan, which was thought to harbor the terrorist organization Al Qaeda. Although Al Qaeda was curtailed in Afghanistan, the organization actually spread and appeared to flourish as more recruits were attracted in response to the aggressive U.S. policy.

Still searching for a more effective retaliatory move against terrorism, the United States waged a 2-month war against Iraq in the spring of 2003, even though a connection between Iraq and the September 11 attacks was not established. As a pretext, the George W. Bush administration claimed that Iraq had maintained an arsenal of deadly (possibly nuclear) weapons. As of late 2004, no such weapons had been found. Although many Iraqis at first seemed pleased to have been liberated from the rule of Saddam Hussein (see the discussion in Chapter 6), many became disenchanted with the U.S. occupation and administration, especially after graphic revelations that prisoners had been tortured by U.S. personnel in Iraqi prisons. Insurrections began in earnest in early 2004 and escalated into nearly daily suicide bombings. The war proved much more costly than anticipated by the United States. Thousands of Iraqi civilians died during and after the 2-month war. There were an estimated 9000–11,000 Iraqi civilian deaths by May 21, 2004; 802 U.S. soldiers were dead and some 10,000 soldiers injured by the same date.

Because of the size and the geopolitical power of the United States, the September 11 attacks (and especially the U.S. reactions to them) affected virtually all international relationships, public and private. Diplomatic affairs, trading partnerships, military alliances, development aid and private philanthropy initiatives, and the travel habits of private citizens all continue to be reevaluated and renegotiated. These numerous adjustments are directly or indirectly affecting the daily lives of billions of people, including all U.S. citizens.

NATIONS AND BORDERS

In this book, you will often (but not always) see regions defined by the countries within them. Countries are a common unit of geographic analysis, and their boundaries are a major feature of the maps in this book (note the small political boundary maps in each chapter). Why a country has its particular boundaries is a complex subject, but the following discussion addresses some of the determining factors.

The eighteenth-century French writer Jean-Jacques Rousseau believed that people realize their true potential by joining together to create a cohesive nation-state anchored to a particular territory. According to this view, a **nation,** per se, is not a political unit or an official country, but a group of people who share language, culture, and political identity: the Cherokee Nation, for example, or the Palestinians. Once those people formally establish themselves as occupying a particular territory and become a country, they are a **nation-state.** Japan and Slovenia are present examples of nation-states because each is a country and within each, the people (with, for the time being, only minor exceptions) share a common language, culture, and political identity. According to this definition, the United States and Australia are not nation-states because their diverse populations originated in virtually every corner of the globe and, to the extent that political identity is shared, in both countries it is based on a loosely defined recognition and affirmation of diversity. They are perhaps best called simply "countries."

Ideas about the nature of national identity spread from France to the rest of Europe, the Americas, and beyond. Cultural diversity usually prevented the establishment of nation-states in the former colonies of Europe, but eventually nearly all became independent countries. In some cases, these new countries were **pluralistic states** in which power was shared among several groups, each defined by common language, ethnicity, culture, or other characteristics. More typically, only one group, which had managed to gain the advantage over several others during the colonial period, achieved power when the state was created. In some cases, groups found that the territory they had customarily occupied was claimed by several new states, leaving them without a homeland. For example, the once nomadic Kurds, who ranged through parts of what are now Turkey, Syria, Iraq, and Iran, still do not have a recognized territory because all of that land has been claimed already. In this case and in many others around the world, groups are in conflict with national governments over rights to territory. The nation-state is theoretically a powerful way to bind together a large group of people in a particular place, but rarely is a substantial territory occupied by only a single group. With no clear way to divide once shared territory equitably among groups, the idea of the nation-state has often led to conflict.

The emergence of the idea of the nation-state was linked with the concept of **sovereignty,** which means that a country is self-governing (though not necessarily according to democratic principles) and can conduct its internal affairs as it sees fit without interference from outside. Sovereignty increased the

Some 200,000 Serbs, expelled from the Krajina region of Croatia, cross through Bosnia to reach Serbia during one of the many "ethnic cleansing" episodes of the conflict in former Yugoslavia. [Peter Turnley/CORBIS.]

importance of precise legal boundaries as demarcations of power and control. A country's borders mark the extent of its territory and hence its sovereignty. When we look at a map, we are viewing a spatial representation of the outcomes of struggles over control of territory. Some such struggles have been resolved only recently or are still under way. In the 1990s, for example, the map of Yugoslavia changed markedly as parts of that country became independent states. Then various ethnic groups within these new states began to struggle over territory they had previously occupied in common, and a devastating armed conflict ensued.

INTERNATIONAL COOPERATION

The idea of the nation-state has been recognized as deeply flawed, and there is reason to think that the idea of national sovereignty is also becoming less viable. Increasingly, people, goods, and capital are flowing freely across national borders, and this trend favors interdependence over national sovereignty. It also creates a need for some way to enforce laws governing business, trade, and human rights at the international level. When extreme human rights abuses became painfully obvious in the former Yugoslavia, first in Bosnia and then in Serbia and Kosovo, eventually the United States and Europe reluctantly decided that it was appropriate to intervene and thus breach the concept of sovereignty.

There is some evidence that the role of the individual state (country) may decline as international organizations play an increasing role in world affairs. The prime example of government-to-government international cooperation is the **United Nations (UN)**, an assembly of 185 member states that sponsors programs and agencies focusing on scientific research, humanitarian aid, economic development, general health and well-being, and peacekeeping assistance in hot spots around the world—such as Afghanistan, Angola, Bosnia-Herzegovina, Cambodia, East Timor, Iraq, Israel and Palestine, Kosovo, Serbia, Lebanon, Rwanda, Sierra Leone, and Haiti. Because most countries are unwilling to relinquish sovereignty, the United Nations has little legal power and can often enforce its rulings only through the power of persuasion. Even in its peacekeeping mission, there are no true UN forces, but rather troops from member states who wear UN designations on their uniforms and take orders from temporary UN commanders.

The World Bank (see page 34) and the **International Monetary Fund (IMF)** are financial institutions funded by the developed nations to help developing countries reorganize and formalize their economies in accordance with a free-market model.

These two institutions wield enormous power over global economic development through their lending activities. The World Trade Organization (WTO), mentioned earlier on page 34, oversees world trade practices, enforces trade regulations, and settles trade disputes among member countries.

Nongovernmental organizations (NGOs) are an increasingly important embodiment of globalization. They range from groups concerned primarily with local issues, such as East Tennessee's "Save Our Cumberland Mountains," to globally active transnational groups, such as the World Conservation Union and Médecins Sans Frontières (Doctors Without Borders). In such associations, individuals from widely differing backgrounds and locations agree on a political, economic, or environmental goal, such as saving endangered species or dispensing medical care to those who need it most. Sometimes the educational efforts of these organizations succeed in raising awareness of an issue among the general public. NGOs such as Amnesty International, which monitors and exposes human rights abuses, have prepared the way for formal agreements in the United Nations. Although these organizations often operate on a global scale (especially since the advent of the Internet), some, despite their global reach, resemble community-level action groups in that they solicit input from a wide range of individuals—a feature that political scientists consider essential to participatory democracy. On the other hand, there is some concern that huge international NGOs might wield such power that they could undermine democratic processes, especially in small countries.

Quick Review

1. What are three ways national governments protect their economies from international competition?

2. Describe the types of information that a population pyramid conveys.

3. Name two causes of soil degradation.

4. How is deforestation related to global warming?

5. According to Rousseau, is the United States a nation-state? Why?

Chapter Key Terms

age distribution or age structure 40	cultural diversity 15	"Development for whom?" 35
air pressure 24	cultural homogeneity 15	diffusion 28
autonomy 15	cultural identity 15	digital divide 13
average population density 38	cultural marker 15	domestication 28
birth rate 39	culture 14	earthquake 22
capital 33	culture group 14	economy 31
capital controls 34	death rate 39	erosion 23
cash economy 42	delta 23	ethnic group 14
climate 24	demographic transition 42	European colonialism 11
cold war era 48	demography 37	exchange or service sector 32
colony 32	deposition 23	external processes (geophysical) 23
contiguous regions 11	development 35	extraction 32

CHAPTER 2

NORTH AMERICA

PERSONAL VIGNETTE Mist curls off the Tennessee River just outside of Knoxville as Jesse and Iqbal settle in on a dock for some Saturday morning fishing. After a few minutes, Iqbal quietly puts down his pole and bows to the southeast, toward Mecca, observing his duty as a Muslim.

Jesse keeps fishing. He finds Iqbal's practice a bit strange, but is grateful for his company. The two men met along the river early one morning as Jesse was getting ready to fish and Iqbal was finishing his prayers. Iqbal joined in the fishing, and they have been meeting at least once a week ever since.

When the Levi Strauss plant moved to Mexico 2 years ago, Jesse lost his job. Now 56 years old, he works part time at a McDonald's, and even though the pay is low, he feels he has to stay there until something better comes along. His Social Security pension won't begin for years, and even then it won't meet all of his and his wife's expenses. His son has started a job in New Mexico as a schoolteacher and won't be available to help him financially in emergencies. All of these complications make these relaxing mornings with Iqbal even more special.

Iqbal finds his companion and everything else in the United States to be strange. Why, for example, does Jesse insist that they throw the fish they catch back into the river? His assertion that pollution in the river makes the fish unsafe to eat is hard to believe, especially because they look so big and healthy. Iqbal is amazed by the surrounding forested hills and the almost total absence of people on a river so close to a large city. In his home state of Kerala in India, waterways are crowded with boats and fishers and with people doing laundry. He is amazed by the smooth, clean streets and by the fact that Jesse, a mere restaurant "wallah," as he would be called in India, can afford a house and a lawn that even a well-paid doctor could scarcely afford in Kerala.

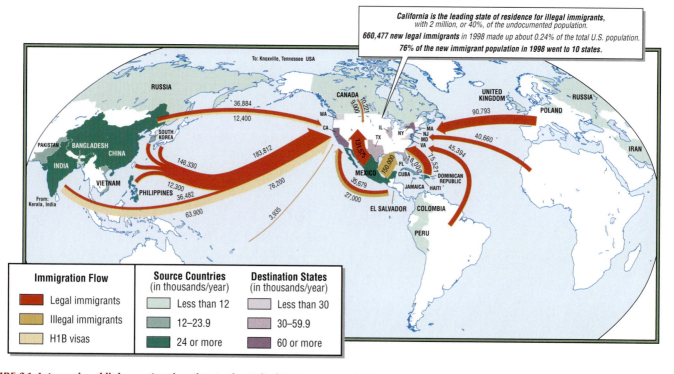

California is the leading state of residence for illegal immigrants, with 2 million, or 40%, of the undocumented population.
660,477 new legal immigrants in 1998 made up about 0.24% of the total U.S. population.
76% of the new immigrant population in 1998 went to 10 states.

Immigration Flow
- Legal immigrants
- Illegal immigrants
- H1B visas

Source Countries (in thousands/year)
- Less than 12
- 12–23.9
- 24 or more

Destination States (in thousands/year)
- Less than 30
- 30–59.9
- 60 or more

FIGURE 2.1 Interregional linkages: Immigration to the United States. The United States is being transformed by the immigrants it attracts from around the globe. Companies and universities recruit highly skilled workers to fill high-tech jobs. At the same time, a considerable flow of less skilled illegal immigrants arrives from Middle and South America to take jobs at the lower end of the pay scale. Immigration is also leading to an ethnic transformation. According to data from the 2000 U.S. census, ethnic "minorities" together constitute majorities in half the 100 largest U.S. cities. Census Bureau projections show that all minorities combined will constitute a majority of the U.S. population by 2060.

All this makes Iqbal determined to make the most of his special H1B temporary visa for high-tech workers. Although his new job as a network administrator at the university pays him less than fellow workers with less technical competence, he works diligently. His visa is good for 6 years, and if he loses his current job in the meantime, he may be forced to return to India, where the pay is one-tenth what he makes now.

Just as he is finishing his prayer, Iqbal's line goes taut. A long struggle produces an enormous carp. He turns to Jesse and asks, "But . . . must I throw this one back too?" "'Fraid so," Jesse answers. "If you're hungry, you could come on into work with me and get a Filet–O–Fish sandwich." [*Adapted from Alex Pulsipher's field notes and experiences, Knoxville, Tennessee, 1998–2001.*]

MAKING GLOBAL CONNECTIONS

Friendships like that of Iqbal and Jesse are forming more and more often across North America as ordinary Americans meet and spend time with some of the many migrants from distant countries coming there to learn or work, often only temporarily (Figure 2.1). Immigrants from Asia, Africa, and Middle and South America are bringing new customs and new international sensibilities to even the most remote communities. Rural villages in the mountains of North Carolina have Mexican grocery stores catering to farm laborers from Michoacan in central Mexico. Physicians from India live in eastern Kentucky hill-country towns and treat predominantly Appalachian patients. In Minneapolis, Somali women in Muslim veils lead labor protests for better working conditions in Minnesota's workplaces.

Americans are ambivalent about these new immigrants and guest workers and the major social and economic changes they represent. The family farm is no longer the basic unit of American society, and rural farm communities are shrinking. Manufacturing jobs that once provided a high standard of living are moving out of North America, leaving those who are ill-equipped to take the new high-tech jobs with little but low-wage service employment. A new but still unstable economy is emerging, based on sophisticated technology and the high-speed exchange of information, goods, and money at a global scale. There are rapid changes in employment levels and in the locations of jobs across the continent. Immigrants have a central role in these changes, some entering the job market at the low-skill end to take jobs that Americans no longer wish to fill, others entering at the high-skill end to take jobs for which there are too few qualified Americans. These immigrants are changing the face of America, increasing its multiethnic hue, and contributing significantly to its tax coffers.

It is likely that you are a resident of North America. You already have some impressions of this land and the changes it is undergoing. Please bring your own experience to the task of defining the evolving character of North America. Think of this chapter as a guide to enhance your understanding of what you already know and to construct, over a lifetime, an ever more complex understanding of the geography of North America. If you grew up in North America, perhaps your experiences with new American residents from faraway places will be part of what you bring to this task. You might want to glance at the map of your childhood landscape that you drew at the beginning of Chapter 1. As you read this chapter, you can place that map in its broader context. You may even see how features of your map—things you noticed as a young child—are related to larger patterns across North America, both physical and cultural.

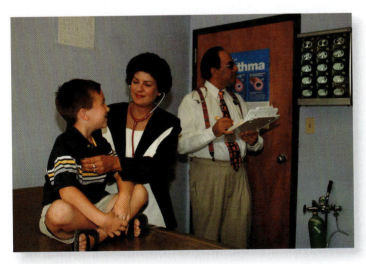

Doctors Rakesh Sachdeva and Seema Sachdeva examine Connor Mayhorn at their medical office, in Pikeville, Kentucky. The Sachdevas, both from India, moved to the United States in 1988 and to Pikeville in 1994. They both became U.S. citizens in September 2002. Rakesh Sachdeva is board certified in pediatrics and pediatric/gastroenterology, and Seema Sachdeva is board certified in pediatrics and pediatric emergency. Pikeville is a small city in eastern Kentucky that for years was underserved by doctors. [Dena Potter/*Appalachian News-Express.*]

Somali women have become prominent in the labor movement in Minnesota. Here, as members of the Hotel and Restaurant Employees Union, they are picketing the Hilton Hotel in downtown Minneapolis. As they urge a taxi driver not to cross their picket line, a security guard, hired by the hotel, is trying to convince him to pull in. Labor disputes often pit low-wage workers against each other. [Craig Borck/Pioneer Press.]

THEMES TO EXPLORE IN NORTH AMERICA

Some central patterns in North American life emerge in this chapter:

1. Mobility. North Americans value the ability to move and to observe different parts of their continent. This behavior is related to the individualism that characterizes North American lifestyles. Links to family and community are important, but migrating to a new geographic location is accepted as a way to achieve individual fulfillment and financial success. About one out of every three citizens of the United States between the ages 5 and 64 moved in the period from 1995 to 2000.

2. Changing population composition. North America is becoming less European and African and more Hispanic and Asian as new residents arrive from Middle and South America and from various parts of Asia. According to the 2000 U.S. census, Euro-Americans are now minorities in 50 of the 100 largest North American cities, and African-Americans, long the largest minority, are now outnumbered by Hispanic Americans.

3. Changing population distribution. Overall, there has been an internal shift in the distribution of the population from the northeast and middle of North America to the south and west.

4. Similar landscapes; regional differences. All major cities in Canada and the United States have similar suburbs, office towers, malls, ethnic restaurants, and home improvement centers. Yet, despite this sameness, there are physical differences among subregions and often stark cultural disparities within subregions. Over the past few decades, North Americans have become more, rather than less, culturally and economically diverse.

5. Wealth and poverty. Although Americans as a whole are prospering more than ever, the gap between rich and poor is widening. According to the 2000 U.S. census, there are 4 million millionaires (1000 times more than in 1900), but 37 million remain poor (13 percent of the population). People in Appalachia and across the southern tier of states are the most likely to be poor, and children are more likely to be poor than adults.

6. Increasing global awareness and interaction. On a personal scale, Americans can now maintain daily relationships with individuals in virtually any country on earth by using e-mail. And Americans must be agile in their response to shifts in the global economy as local jobs are alternately lost and created. North American reactions to events, such as terrorism or economic recession, can cause reverberations around the globe as individuals and governments struggle to find appropriate responses.

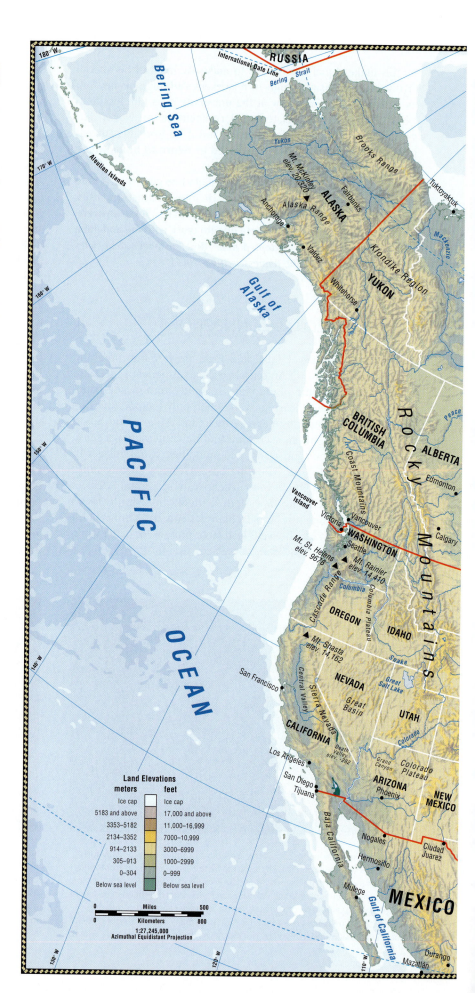

FIGURE 2.2 Regional map of North America.

ARCTIC OCEAN

Lincoln Sea

Greenland Sea

Beaufort Sea

Queen Elizabeth Islands

Ellesmere Island

Thule

Greenland

Denmark Strait

ICELAND

Faroe Islands
Tórshavn

Banks Island

Viscount Melville Sound

Resolute

Baffin Bay

Arctic Circle

Reykjavík

Victoria Island

Great Bear Lake

NUNAVUT

Baffin Island

Davis Strait

Narssarssuaq

Godthab

Dogrib Territory

NORTHWEST TERRITORIES

Yellowknife

Foxe Basin

Great Slave Lake

Cape Dorset

CANADA

Athabasca Lake

Reindeer Lake

Athabasca

SASKATCHEWAN

MANITOBA

Churchill

Hudson Bay

Labrador Coast

NEWFOUNDLAND

Labrador

Goose Bay

Saskatoon

Saskatchewan

James Bay

QUÉBEC

St. John's

Regina

Lake Winnipeg

ONTARIO

Canadian Shield

Island of Newfoundland

Gulf of St. Lawrence

Saint-Pierre

Winnipeg

Lake of the Woods

Lake Nipigon

St. Lawrence

PRINCE EDWARD ISLAND

Cape Breton Island

MONTANA

Great

Missouri

NORTH DAKOTA

MINNESOTA

Lake Superior

Québec

NEW BRUNSWICK

Fredericton

NOVA SCOTIA

Halifax

Yellowstone

Black Hills

SOUTH DAKOTA

Minneapolis

WISCONSIN

Lake Michigan

Lake Huron

Peterborough

Ottawa

Montréal

MAINE

VT NH

WYOMING

Front Range

Platte

Sand Hills

NEBRASKA

IOWA

MICHIGAN

Toronto

Lake Ontario

Hamilton

Lake Erie

NEW YORK

Mts.

MA

Boston

Chicago

Detroit

Cleveland

CT RI

Denver

Pikes Peak elev. 14,110

Plains

COLORADO

KANSAS

ILLINOIS

St. Louis

MISSOURI

INDIANA

OHIO

Ohio

Pittsburgh

PENNSYLVANIA

Baltimore

NJ

New York

Philadelphia

DE

Washington, DC

UNITED STATES

KENTUCKY

WEST VIRGINIA

VIRGINIA

OKLAHOMA

Arkansas

ARKANSAS

TENNESSEE

Tennessee

Appalachian

NORTH CAROLINA

Ouachita Mtns.

Red

Mississippi

SOUTH CAROLINA

Dallas

MISSISSIPPI

ALABAMA

Atlanta

GEORGIA

TEXAS

Pecos

Brazos

LOUISIANA

ATLANTIC OCEAN

Bermuda

Rio Grande

Houston

Tampa

FLORIDA

BAHAMAS

Tropic of Cancer

Monterrey

Rio Grande

Matamoros

Gulf of Mexico

Miami

Nassau

I   THE GEOGRAPHIC SETTING

Terms to Be Aware Of

The world region discussed in this chapter consists of Canada and the United States. The term *North America* is used to refer to both countries. Even though it is common on both sides of the border to call the people of Canada "Canadians" and people in the United States "Americans," this text will use the term *United States*, rather than "America," for the United States. The term *American*, then, describes citizens of, or patterns in, both countries, not just the United States.

Other terms relate to the growing cultural diversity in this region. For example, the text uses both **Hispanic** and **Latino**. *Hispanic* is a loose ethnic (not racial) term that refers to all Spanish-speaking people from Latin America and Spain. Hispanics may be black, white, Asian, or Native American. *Latino* is an alternative term for the same group. In Canada, the **Québecois**, or French Canadians, are an ethnic group that is distinct from the rest of Canada. They are the largest of an increasingly complex mix of minorities in that country, most of whom are still content to be called simply Canadians.

PHYSICAL PATTERNS

The continent of North America is a magnificent display of mountain peaks, ridges and valleys, expansive plains, long winding rivers, sparkling lakes, and extraordinarily long coastlines. You will find it easier to learn the physical geography of this large and complex territory if you break it into a few smaller segments, as described in the following section on landforms.

Landforms

The most dramatic and complex North American landform is the wide mass of mountains and basins in the West, known as the *North American Cordillera* (see Figure 2.2). It sweeps down from the Bering Strait in the far north, through Alaska, and all the way to the Isthmus of Tehuantepec in Mexico (see Figure 3.2 on page 115). The *Rocky Mountain zone* of Canada and the United States is a major part of the North American Cordillera. The less rugged Appalachian Mountains stretch along the eastern edge of North America from New Brunswick and Maine to Georgia. Both mountain ranges resulted from the collision of tectonic plates. About 200 million years ago, when the supercontinent Pangaea began to pull apart (see Figure 1.10 on page 23), the broad band of mountains along the western edge of the continent was created as the Pacific Plate pushed against the North American Plate. The Pacific Plate continues to slip to the northwest, and the North American Plate, to the southeast. These plate movements cause the earthquakes that are common along the Pacific coast of North America. The much older and hence more eroded and lower Appalachians resulted from earlier collisions of the North American Plate with North Africa.

Between these two mountain ranges lies a huge, irregularly shaped *central lowland* of undulating plains that stretch from the Arctic to the Gulf of Mexico. The central lowland was created by the deposition of material eroded from the mountains. This material was carried to the region by wind and rain and by the rivers flowing east and west into what is now the Mississippi drainage basin. These deposits can be several kilometers deep.

Over the last 2 million years, glaciers that formed during periodic ice ages covered the northern portion of the lowlands (as well as adjoining mountain ranges to the west and east) and extended south well into what is now the central United States. During the height of the most recent ice age (between 25,000 and 10,000 years ago) glaciers, sometimes as much as 2 miles (about 3 kilometers) thick, scoured the very old, exposed rock surface of the Canadian Shield in the far north and picked up sediment from lands to the south; then, as the glaciers melted, the sediment was dumped. The Great Lakes are depressions left by glacial scouring, as are the smaller lakes, ponds, and wetlands that dot Minnesota, Wisconsin, Michigan, and much of central Canada. In the upper Mississippi Valley, some of the sediment dumped by the melting glaciers has been picked up and redeposited by the wind. This wind-deposited layer of soil, called **loess**, is often many meters deep; it has proved particularly suitable for large-scale mechanized agriculture, but it remains susceptible to wind and water erosion.

East of the Appalachians, *a coastal lowland*, which is intermittent and narrow in the north but widens into a broad band toward the south, stretches from New Brunswick to Florida. From Florida, the wide coastal lowland sweeps to the west, joining the southern reaches of the central lowland along the Gulf of Mexico. In the southern reaches of the coastal zone, there is a low, flat transition zone between land and sea characterized by swamps, lagoons, and sandbars. This coastal lowland was formed from huge loads of silt deposited by North America's rivers during frequent floods. The Mississippi Delta, for example, has been extending to the south for more than 150 million years. It began at what is now the junction of the Mississippi and Ohio rivers and advanced 1000 miles (1600 kilometers) into the Gulf of Mexico, filling in much of the southern central lowland and creating the lands that are now Louisiana and Mississippi. Human activities such as deforestation, deep plowing, and heavy grazing have added to the silt load of the rivers, increased the likelihood of flooding, and extended the land into the Gulf of Mexico.

Climate

Every major type of climate on earth, except for the truly tropical, exists in North America. This huge expanse of land experiences widely different temperature patterns from north to south and different amounts of humidity from the dry continental interior to the moist marine fringes. North America's landforms contribute to its climatic variety by influencing the movement and interaction of air masses.

Look at the climate map of North America (Figure 2.3) and observe the thin band of light green along the Pacific coast. The mild Mediterranean climate there is cool and moist in the winter and warm and dry in the summer. Conditions in the north (around Seattle) are generally cooler and wetter than in the south (near Los Angeles). The west coast of North America, especially north of San Francisco, receives moderate to heavy rainfall (see Figure 1.12 on page 25) because moist air passes onto land from the Pacific Ocean. When this moist air encounters the coastal mountain ranges, it rises, cools, and dumps frequent rain along the windward (western) slopes (see the discussion of orographic rainfall and Figure 1.13 on page 25). Climates are much drier to the east of the coastal ranges because air masses tend to hold what moisture they have left as they move eastward across the Great Basin and Rocky Mountains, resulting in a rain shadow. In this arid region, many expensive dams and reservoirs for irrigation projects have been built to make agriculture and large-scale human habitation possible.

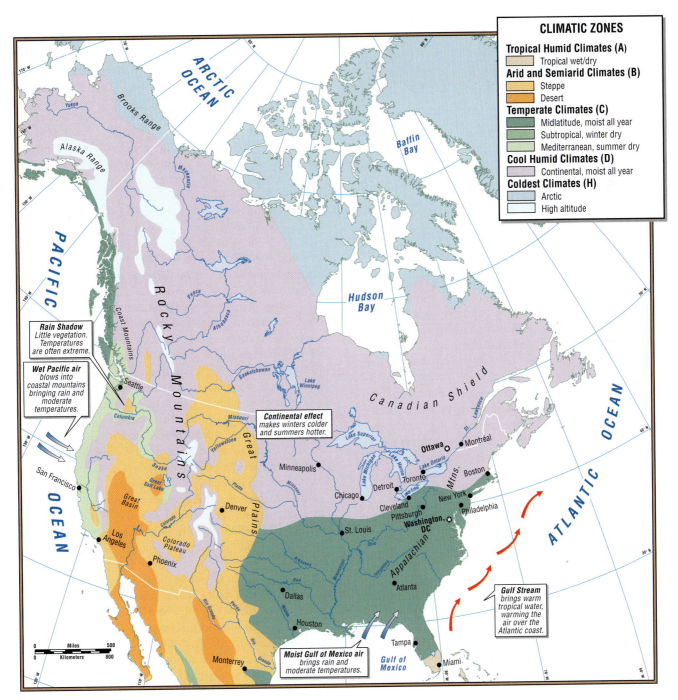

FIGURE 2.3 Climates of North America. Showing climate regions on a map can be misleading because the stark boundaries imply abrupt changes that rarely occur on the ground; instead, there are broad transition zones.

For example, the boundary between the dark green midlatitude climate and the lavender continental climate on this map should be a wide band of several hundred miles that shifts north or south from year to year.

Flood control upstream has decreased silt deposition rates in the Mississippi Delta wetlands. As a result, salt water intrudes into swamps along the edge of the delta and kills the cypress trees. [Gay Gomez.]

On the eastern side of the Rocky Mountains, the main source of moisture is the Gulf of Mexico. When the continent and the air masses above it are warming in the spring and summer and the air rises, moist air is pulled in from the Gulf. This air interacts with cooler, drier air masses moving into the central lowland from the north and west, often creating violent thunderstorms. Generally, the central part of North America is wettest in the east and south and driest in the north and west. Along the Atlantic coast, moisture is supplied by the Gulf Stream, a warm ocean current that flows up the eastern seaboard from the tropics. It brings a supply of warm, wet air that often creates high humidity and rain, and sometimes hurricanes.

The huge size of the North American continent creates large temperature variations in the continental interior. Because land heats up and cools off more rapidly than water, temperatures in the continental interior are hot in summer and cold in winter. Proximity to water generally modifies air temperature extremes, so temperatures are relatively moderate along the wide fringes of the continent adjacent to oceans.

The geographic distribution of the landforms and climates of North America has influenced the settlement patterns of the continent and the behavior, occupations, and problems of the people living there. We now discuss the historical geography of North America.

HUMAN PATTERNS OVER TIME

We can view the human history of North America as a series of successive arrivals and dispersals of people across the vast continent. These movements began with the original peopling of the continent from Eurasia, and they continued with waves of European immigrants, enslaved Africans, and their descendants, who settled the country from east to west from the 1600s on.

Today, immigrants from Asia and Latin America are arriving primarily in the West. The patterns of these movements have contributed to the formation of distinct subregions. Today, internal migration is still a defining characteristic of life for North Americans, who are among the most mobile people in the world.

The Peopling of North America

Thinking Geographically: ON THE WEB

Humans first came to North America from northwestern Asia between 25,000 and 14,000 years ago, during the last ice age. At that time, the polar ice caps were considerably thicker, sea levels were lower, and the Bering land bridge, a huge, low landmass more than 1000 miles (1600 kilometers) wide, connected Siberia to Alaska. Small bands of hunters crossed over into Alaska and traveled across the North American continent, south through Central America, and deep into South America, occupying virtually every climate region. Temperatures began to rise about 10,000 years ago, and as the ice caps melted, the land bridge sank below the rising sea.

Over thousands of years, the people settling in the Americas domesticated plants, created paths and roads, cleared forests, built permanent shelters, and sometimes created elaborate social systems. They developed a wide range of cultural traditions attuned to the circumstances in which they lived. The domestication of corn is thought to be closely linked to settled life and population growth in North America. Corn was introduced from Mexico into the southwestern desert about 3000 years ago along with other Mexican domesticates, particularly squash and beans. Even before corn arrived in the eastern woodlands, people had domesticated many native plants that produced nutritious seeds, among them *Cucurbita pepo* (a gourd) and *Helianthus annus* (sunflowers).

These foods provided surpluses that allowed some community members to specialize in activities other than agriculture, hunting, and gathering. The impressive harvests made possible large, citylike regional settlements. For example, by A.D. 200, such distinctive culture groups as the Anasazi, Hohokam, and Mongollon had emerged in the Southwest. These people built villages of attached adobe houses and irrigated their crops. By A.D. 1000, the urban settlement of Cahokia (in what is now central Illinois) covered 5 square miles (12 square kilometers) and was home to an estimated 30,000 people.

The Arrival of the Europeans. North America today bears little resemblance to the land that the Native Americans had created by A.D. 1500, after hundreds of generations of occupation. The climate and landforms are roughly the same, but virtually every other aspect of the continent's geography, from the vegetation to the cultural orientation of the population, has been transformed. The major reason for this transformation was the sweeping occupation by non-Native peoples of North America beginning in the seventeenth century.

From the early 1600s to the present, settlements have rolled over North America like an unstoppable tide, fed by masses of immigrants from Europe and smaller numbers from Africa, Asia, and elsewhere in the Americas. In the early seventeenth century, the British established colonies along the Atlantic coast in what is now Virginia (1607) and Massachusetts (1620). Over the next two centuries, colonists from northern Europe built villages, towns, port cities, and plantations up and down the east coast of North America. By the mid-1800s they had occupied most Native American lands throughout the Appalachian Mountains and into the central part of the continent.

The Disappearance of Native Americans. The rapid expansion of European settlement was assisted by the long biological and technological isolation of Native American populations from the rest of the world. Native Americans had little resistance to diseases carried by Europeans, such as measles and smallpox. By 1500, these diseases were no longer common killers in Europe, Africa, and Asia because the people there had built up immunity to them. Diseases transmitted by Europeans and Africans were particularly deadly to Native Americans, however, killing an average of 90 percent of any given group within 100 years of contact. Conflict between encroaching Europeans and resisting Native Americans also took a large toll. The Europeans were aided by technologically advanced weapons, including primitive firearms, steel swords, and lances, and by trained dogs and horses. Often the Native Americans had only stones, bows, and poisoned arrows, although by the late 1500s Native Americans in the Southwest had acquired horses from the Spanish and begun to breed them and use them in warfare against the Europeans.

Simple numbers reveal the general effect of European settlement on Native American populations in North America. There were roughly 18 million Native Americans on the continent in 1492. By 1542, after only a few incursions into North America by Spanish expeditions such as that led by Hernando de Soto, there were half that many. By 1907, slightly more than 400,000, or just 2 percent, remained.

The European Transformation

European settlement erased many of the landscapes familiar to Native Americans and imposed new ones that fit the needs of the new occupants. Those needs varied according to the opportunities offered by a particular physical environment and the cultural backgrounds of the settlers in that area. Hence distinct subregions, which still exist today, developed as settlement proceeded.

The Mid-Atlantic and Southern Settlements. European settlement of eastern North America began with the Spanish in Florida in 1565 and with the establishment of the British colony of Jamestown in Virginia in 1607. By the late 1600s, the colonies of Maryland, Virginia, the Carolinas, and Georgia were dependent on the cultivation of cash crops such as tobacco and rice, much of it grown on large holdings known as *plantations*. Because land was readily available even to the poorer settlers, large plantation owners could not easily secure the large, stable labor force necessary for their huge operations. The unfortunate solution was the use of enslaved African labor. African workers were first brought into North America at Jamestown in 1619. Although slavery was not widely accepted at first, within 50 years enslaved Africans were becoming the dominant labor force on some of the larger Southern plantations. By the start of the Civil War in 1861—a conflict that grew largely out of competition between the northern, mid-Atlantic, and Southern economies—slaves made up about one-third of the population throughout the Southern states and were concentrated where plantations were located.

In many ways, the plantation system was detrimental to the economic and social development of the South, and it encouraged stark race and class divisions. Much of the area's wealth was concentrated in the hands of a small class of plantation owners, who made up just 12 percent of Southerners in 1860. Taxes were kept low, and the planter elites invested their money in Europe and the more prosperous northern colonies, instead of in industry and social services at home. More than half of Southerners were poor white subsistence and small cash-crop farmers who consumed most of what they produced and had little money to invest. Because both slaves and poor whites were forced to live simply, their meager consumption did not provide a market for goods, so there were few trading towns. Plantations tended to be self-sufficient. Some food was produced, a few staple commodities were imported, and off-season labor was used to produce or repair equipment such as wagons and tools, breed draft animals, and sew garments for the plantation's residents. As a result, plantations generated few of the independent spin-off enterprises—such as transport and repair services, shops, garment making, and small manufacturing—that could have invigorated the regional economy. The plantation economy declined after the Civil War (1861–1865), and the South sank deeper into poverty, remaining economically and socially underdeveloped well into the twentieth century.

The Southern plantation era had many lingering effects; place names and architecture are examples. Among its most enduring and significant effects is that in the twenty-first century, the largest concentrations of African-Americans are still found in the southeastern states where the slave-based plantation economy was centered (see Figure 2.16 on page 86).

The Northern Settlements. Throughout the seventeenth and eighteenth centuries, relatively poor subsistence farmers dominated agriculture in the northern colonies of New England and southeastern Canada. In contrast to the Southern colonies, there were no plantations and few slaves, and not many cash crops were exported. Instead, farmers often augmented their incomes with exports of timber and animal pelts. Some communities were completely dependent on the rich fishing grounds of the Grand Banks, located off the shores of Newfoundland and Maine. By the late seventeenth century (1600s), a few industries were already producing goods both for local consumption and for export. By the eighteenth century, metalworks and pottery, glass, and textile factories were turning out products in Massachusetts, Connecticut, and Rhode Island. They supplied markets in North America, but also exported to British plantations in the Caribbean.

Industry began to flourish in the early nineteenth century, when Francis Lowell designed a water-powered loom. By 1822, "factory girls" were working in the textile mills of Lowell, Massachusetts, and living in company-owned housing. Southern New England, especially the region around Boston, became the center of manufacturing in North America, drawing on male and female immigrant labor from French Canada and Europe. It retained that status as a manufacturing center into the second half of the nineteenth century.

The Economic Core. New England and southeastern Canada were eventually surpassed in numbers of people and in wealth by the mid-Atlantic colonies of New York, New Jersey, and Pennsylvania. This region benefited from more fertile soils, a slightly warmer climate, multiple deepwater harbors, and better access to the resources of the interior. By the end of the Revolutionary War (in 1783), the mid-Atlantic region was on its way to becoming the **economic core,** or the dominant economic region, of North America.

Both agriculture and manufacturing boomed in the early nineteenth century, drawing immigrants from much of northwestern Europe. The land they tilled produced grain in such vast quantities that the region had been called the "bread colonies" in pre-Revolutionary times. The tremendous success of agriculture laid the groundwork for industrial development. Food-processing industries packaged meat and turned grain into cereal, flour, and bread. As farmers prospered, they bought mechanized equipment, appliances, and consumer goods, much of it made in nearby cities. The richest farmers put up money to build the factories. Port cities such as New York, Philadelphia, and Baltimore prospered because they were well positioned to trade industrial products to the vast continental interior. This relatively small, early core region yielded almost a third of U.S. industrial products on the eve of the Civil War (1860).

Throughout the nineteenth century, the borders of the economic core expanded as connections with interior areas to the west deepened. Rivers, canals, railways, and roads connected the big cities of the mid-Atlantic coast with the entire Great Lakes area, including the area that became the states of Ohio, Michigan, Indiana, Illinois, and Wisconsin (Figure 2.4). In this new expansion of the economic core, the early settlers were farmers who supplied food to growing urban populations. By the mid-nineteenth century, its economy was increasingly based on the steel industry, which diffused westward from the eastern end of the economic core to the Great Lakes industrial cities of Cleveland, Lorraine, Gary, and Chicago. Local deposits of coal, iron ore, and (later) oil provided power and raw materials for the steel industry. That industry further stimulated the mining of coal and iron ore, as well as the invention, patenting, and manufacture of a wide range of goods needed on farms and in urban households: tools, plows, carriages, pottery, iron cooking and heating stoves, and many others. As mechanization proceeded, heavy farm and railroad equipment, including steam engines, became important products of this region.

By the late nineteenth century, the economic core stretched from the Atlantic to Chicago and from Toronto to Cincinnati. This rapidly industrializing region dominated the other parts of North America economically and politically well into the

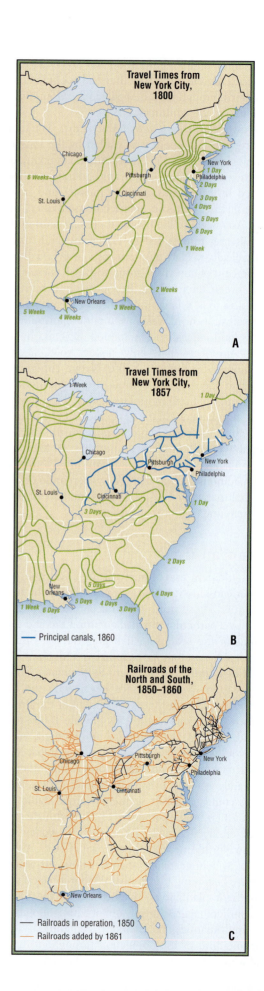

FIGURE 2.4 Nineteenth-century transportation. With the building of the railroads in the decade before the Civil War, the mobility of people and goods increased dramatically across the eastern half of the United States. In 1800 (Map A), it took a day to travel from New York City to Philadelphia and a week to go to Pittsburgh. By 1857 (Map B), the travel time from New York to Philadelphia was only 2 or 3 hours, and to Pittsburgh less than a day. People could travel farther, easily reaching principal cities along the Mississippi River, and the travel was less expensive and onerous. Because of their ability to move troops and materials quickly, the railroads (Map C) played a significant role in the North's victory in the Civil War. [Adapted from James A. Henretta, W. Elliot Brownlee, David Brody, and Susan Ware, *America's History,* 2nd ed. (New York: Worth, 1993), pp. 400–401.]

twentieth century. Most other areas produced food and raw materials for the core's markets and depended on the core's factories for manufactured goods. (See pages 98–100 for more discussion of the rise and decline of this region.)

Expansion West of the Mississippi

The east-to-west trend of settlement continued. By the mid-nineteenth century, the land in the densely settled eastern parts of the continent had become too expensive for new immigrants. By the 1840s, immigrant farmers from central and northern Europe, as well as some older settlers, were pushing their way west across the Mississippi River.

The Great Plains. Much of the land west of the Mississippi was dry grassland, or prairie, which seemed desolate and strange to people whose cultural roots were in the well-watered fields and forests of northwestern Europe and the eastern United States. Some settlers interested in crop farming founded homesteads on the Great Plains and adapted to the lack of trees and water (average annual rainfall is 20 inches or less; see Figure 1.12 on page 25). The soil usually proved amazingly fertile in wet years. But the naturally arid character of this land eventually created an ecological disaster for Great Plains farmers. After 10 dry years, a series of devastating dust storms in the 1930s blew away topsoil by the ton; animals died, and entire crops were lost. This hardship was made worse by the economic depression of the 1930s, and many farm families packed up what they could and headed west to California and other states on the Pacific coast.

Settlement of the Western Great Plains and Beyond. While the Great Plains filled up, other settlers were alerted to the possibilities farther west. The Louisiana Purchase in 1803 had doubled the size of the United States, and the Lewis and Clark expedition of 1804–1806 explored this new acquisition, reaching the Pacific coast by traveling along the Columbia River in the Oregon country. Settlers were drawn to the valleys of the Rocky Mountains (especially the Great Basin) and to the well-watered and fertile coastal zones of the Oregon country and California. The great rush to Oregon began in the 1840s and lasted until the 1860s. Over that 20-year period, more than 350,000 people walked the Oregon Trail, carrying their possessions in wagons that wore 3-foot-deep ruts still visible today. Although farming was the major occupation in the Oregon country, settlers also turned their attention to the abundance of trees, which filled the rising demand for building materials. Soon logging became the dominant industry in the Pacific Northwest as vast stands of enormous redwoods, Douglas firs, spruces, and many other species were clear-cut throughout the region. (**Clear-cutting** is the cutting down of all trees on a given plot of land, regardless of age, health, or species.)

Among the most successful new settlers in the West were the Mormons, a group of utopian Christians also known as the Church of Jesus Christ of Latter-day Saints. The Mormons began a trek across the continent in the 1830s from western New York. They demonstrated that the semiarid Great Basin region could be farmed with the use of irrigation. Mormons remain an important cultural influence in the Great Basin states.

Settlement in the Southwest. The Southwest was first colonized by Spaniards from Mexico at the end of the sixteenth century. They established sheep ranches and missions in the area that today forms central and southern California, southern Arizona, New Mexico, and Texas. Their settlements were few and weakly connected; therefore, as settlers from the United States expanded into the region, Mexico found it increasingly difficult to retain its claimed territory. By 1850, nearly all of the Southwest was under U.S. control, and the new state of California in particular was attracting farmers and laborers. By the twentieth century, a vibrant agricultural economy was developing in California, where the mild Mediterranean climate made it possible to grow vegetables almost year-round. With the advent of refrigerated railroad cars in the early twentieth century, these vegetables could be sent to the major population centers of the East. Massive government-sponsored irrigation schemes greatly increased the amount of arable land. These projects diverted whole rivers to supply fields and urban developments hundreds of miles away, totally transforming the natural course of water drainage in southern California and much of the West.

European Settlement and Native Americans

Thinking Geographically: ON THE WEB

Native Americans living in the eastern part of the continent who survived early European incursions often occupied land that Europeans wished to use. These Native Americans were continually pushed westward. Eventually those who remained were settled on relatively small reservations with few resources on the western side of the Mississippi River. The best known of these organized relocation events was the Trail of Tears. In the late 1830s, European settlers pushing into Georgia and across the Appalachian Mountains hoped to acquire land at low cost. The Creek and Cherokee, whose lands the settlers wanted, were by then assimilated to European ways. The Cherokee had built roads, schools, and churches, and most earned a living by farming or raising cattle. Some owned ferries and grain mills and lived in substantial brick homes of European style. The European settlers prevailed on President Andrew Jackson to sign the Indian Removal Act, passed by Congress in 1830. The Cherokee resisted by filing suits that reached the U.S. Supreme Court twice, but in May of 1838, General Winfield Scott and 7000 soldiers of the U.S. Army began the removal of the Cherokee to Oklahoma. The route they took became known as the Trail of Tears because of the hardships they endured.

Many different reservations were created both east and west of the Mississippi for Native Americans, usually with insufficient attention paid to their needs and customs. Then, as European settlers occupied the Great Plains and prairies, many of the reservations set aside for Native Americans, especially in the West, were further shrunk or relocated on ever less desirable land to make room for these newcomers.

Native American populations are now expanding, after plunging from 18 million to less than 400,000 between 1500 and 1900. In 2000, there were almost 2.5 million Native Americans in North America, most living in the United States, where they constitute less than 1 percent of the total population. Ironically, North Americans today often identify with the Native peoples: many now claim some Native American blood, making it difficult to determine who should be called a Native American and just how many there are.

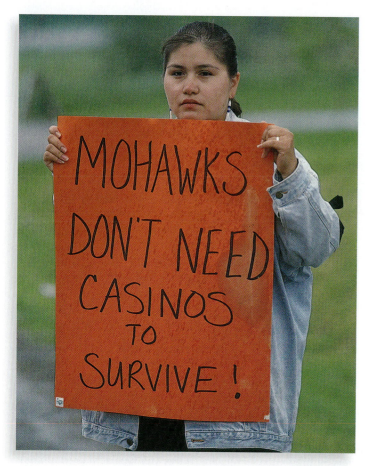

In the early 1990s, Canadian Mohawks who viewed gambling as cultural pollution strongly opposed the operation of a casino by the Mohawks across the border in New York State. The protesting Canadian Mohawks destroyed some of the U.S. machines, causing a bitter dispute within the general Mohawk Nation. [Sarah Leen/Matrix.]

Reservations cover just over 2 percent of the United States, but more than 20 percent of Canada, since the creation of the Nunavut territory in 1999 and the Dogrib territory in 2003 in Canada's far north (see Figure 2.30 on page 105). The Nunavut and Dogrib stand out as having won the right to control their own resources and the right to home rule, including an elected legislative assembly, a public service system, and a court. (It is unusual for Native groups in Canada to have legal control over their territory.) Most Native American reservations in North America have insufficient resources to support their populations at the standard of living enjoyed by other citizens. After centuries of mistreatment, many Native Americans are more familiar with poverty than with true Native ways. Severely demoralized by the lack of opportunity, some have turned to alcohol, and many live as impoverished wards of the state, totally dependent on meager welfare money. Alcohol consumption often leads to family violence, and in the United States, Native American suicide and homicide rates are almost twice the national average.

For many years, Native young people faced a difficult choice: they could remain in poverty with family and community on the reservation, or they could seek a more affluent way of life far from

home and kin. Outside their communities, many young people lose touch with traditional ways and encounter considerable prejudice. In the 1990s, several tribes found avenues to greater affluence on the reservations. Some tribes have sought and received substantial monetary compensation from the government for past losses. Others have opened gambling casinos designed to lure non-Native gamblers. Loopholes in the old treaties allow casinos on reservations in states where local law would otherwise forbid them. Although these few roads to prosperity (especially casinos) can produce enormous income, the stress on weakened tribal structure can be intense. In some cases, graft has diverted funds to a few elites. In others, economic development has been accompanied by environmental pollution, as in the case of the many oil and gas wells and coal and uranium mines located on reservations.

Some tribes have been successful in taking creative approaches to development. The Choctaw in Mississippi have developed factories that produce plastic utensils for McDonald's restaurants, electrical wiring for automobiles, and greeting cards. They employ not only their own people but also 1000 non-Native Americans who come onto the reservation to work. The Salish and Kootenai tribes of Montana have reduced their unemployment to levels below those in nearby communities by creating jobs for themselves in farming, tourism, and recreation. In the Great Plains states, Native American populations are growing as migrants return to their homelands to take up new occupations (see the discussion later in this chapter on page 103). In Canada, Native Americans have a higher standard of living than their U.S. counterparts, and the government spends much more per capita on them. But with the exception of the new Nunavut and Dogrib territories in northern Canada, Canadian tribes have less control over the resources on their lands, which are often developed for use by non-Native Americans (see Figure 2.30 on page 105).

The Changing Social Composition of North America

The distinctive subregions of North America created by European-led settlement still remain. However, the boundaries and distinctive characteristics of these subregions, both rural and urban, are becoming increasingly blurred as the process of change continues. In the coming section on geographic issues, we will see that the economic core no longer dominates North America, as industry has spread to other parts of the continent. Some regions that were once dependent on agriculture, logging, or mineral extraction now have high-tech industries as well. In cultural terms, once-distinct ethnic groups have become assimilated as succeeding generations married outside their groups and moved to new locations. In Los Angeles, for example, one-third of all U.S.-born middle-class Latinos and more than one-fourth of all U.S.-born middle-class Asians marry someone of another ethnic group (see the photo on page 64). Non-European immigrants from Middle and South America, India, Pakistan, Southeast Asia, East Asia, Oceania, and all parts of Africa are finding a wide variety of social and economic niches as they add to the cultural diversity of their new homeland. Before we look more closely at these emerging patterns, however, some information about the changing population of North America will be useful.

POPULATION PATTERNS

The population map of North America (Figure 2.5) shows the uneven distribution of the more than 323 million people who live on the continent. Canadians, who account for just over one-tenth (31.6 million) of the population of North America (the United States has 291.5 million), live primarily in southeastern Canada, close to the border with the United States. Sixty percent of Canadians live in southern Ontario in the Great Lakes region and in Québec along the St. Lawrence River. The greatest concentration of people in the United States is not far south of the border with Canada, in the northeastern part of the United States. A quadrant marked by Boston and Washington, D.C., along the

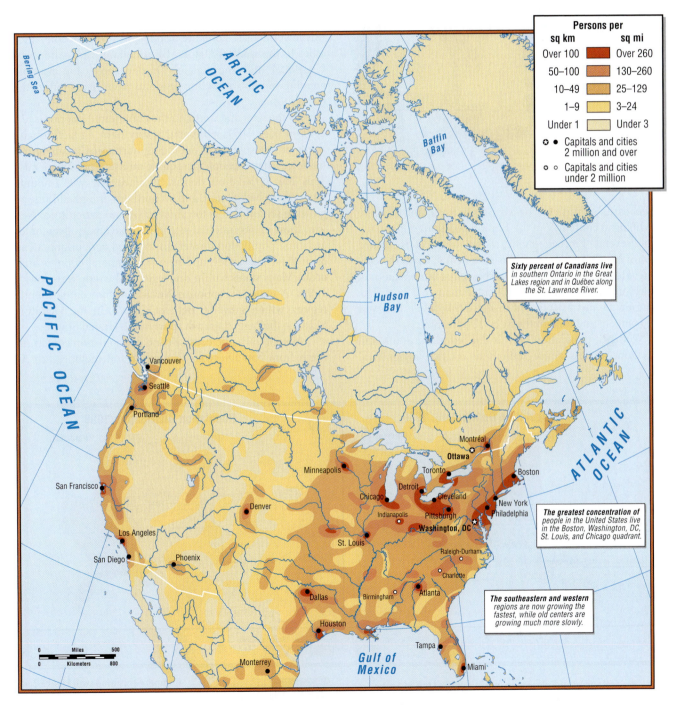

Sixty percent of Canadians live in southern Ontario in the Great Lakes region and in Québec along the St. Lawrence River.

The greatest concentration of people in the United States live in the Boston, Washington, DC, St. Louis, and Chicago quadrant.

The southeastern and western regions are now growing the fastest, while old centers are growing much more slowly.

Persons per

sq km	sq mi
Over 100	Over 260
50–100	130–260
10–49	25–129
1–9	3–24
Under 1	Under 3

⊛ ● Capitals and cities 2 million and over

⊛ ○ Capitals and cities under 2 million

FIGURE 2.5 Population distribution in North America. [Adapted from *Hammond Citation World Atlas* (Maplewood, N.J.: Hammond, 1996).]

In Southern California, near Los Angeles, in June 2003, an American woman, Dr. Molina Patel, whose family originated in the state of Gujarat, India, married another American, Thomas Osmand, whose family originated in Australia and the United Kingdom. They decided on a Hindu wedding complete with a *mandap* (a ceremonial platform with four pillars supporting a canopy) and joyous processions in which the wedding guests participated; but their wedding was an abbreviated version of a typical Gujarati ceremony, which can last for days. [Stephen Zeigler.]

land that is dotted with many medium-sized cities, most with populations under 2 million (see Figure 2.5). On average, in this landscape of large fields, woodlands, and grasslands, rural densities are 3 to 129 people per square mile (1 to 50 per square kilometer). Rural densities in the Middle West are actually declining as farms are bought by corporations and young people leave to seek urban careers (notice the change in the area labeled Midwest between 1900 and 1999 in Figure 2.6). Midwestern cities are growing modestly and becoming more ethnically diverse, with rising populations of Hispanics and Asians in such places as Indianapolis, St. Louis, and Chicago. Meanwhile, the rural landscapes in the southern portion of this lower-density zone are also losing young people, and many cities to the south—Atlanta, Birmingham, Knoxville, Nashville, Charlotte, and the Raleigh-Durham area, for example—are growing rapidly, attracting residents from many different ethnic backgrounds.

To the west and north, in the continental interior (see Figure 2.30 on page 105), settlement is very light. In fact, roughly two-thirds of this subregion, including northern Canada and Alaska, has fewer than 3 people per square mile or 1 per square kilometer. This low population density is correlated with mountainous topography, lack of rain, and, in northern or high-altitude zones, a growing season that is too short to sustain agriculture. Nevertheless, these environmental restrictions have been overcome; there are some population clusters in irrigated agricultural areas, such as in the Utah Valley, and near rich mineral deposits and resort areas. Many states within this region grew only very slowly in the 1990s. A major exception is Nevada, at the southern end of the region, which grew 66 percent (mostly in the vicinity of Las Vegas)—the biggest growth spurt in the United States.

Along the Pacific coast, a band of population centers stretches north from San Diego to Vancouver and includes Los Angeles,

Atlantic seaboard, St. Louis on the Mississippi River, and Chicago on Lake Michigan contains 7 of the 12 most populous states in the country and several of its largest cities. Population is concentrated in this part of North America because, historically, it was the economic core. However, the 2000 census shows that other regions of the country are now growing much faster than this one. The Northeast grew by just 5.5 percent in the 1990s, while the West grew by 20 percent and the Southeast by 17 percent.

The Northeast population core is flanked to the south and west by a zone of less dense settlement on rich agricultural

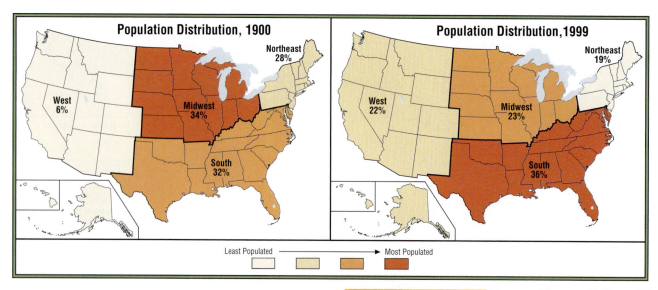

Population Distribution, 1900 — Northeast 28%, West 6%, Midwest 34%, South 32%

Population Distribution, 1999 — Northeast 19%, West 22%, Midwest 23%, South 36%

Least Populated ———→ Most Populated

FIGURE 2.6 Percentage of population by region in the United States, 1900 and 1999. The population of the United States has shifted to the South and West over the past 100 years. [Adapted from "America's diversity and growth," *Population Bulletin* (June 2000): 12.]

PRACTICING GEOGRAPHY READING MAPS Compare this map with Figure 2.3. What climatic conditions led to the relative underpopulation of the western United States? Consider also the effects of shifting employment opportunities (see Figure 2.10 on page 74).

San Francisco, Portland, and Seattle. These are all port cities engaged in air and maritime trade around the **Pacific Rim** or **Basin** (all the countries that border the Pacific Ocean on the west and east) and with countries of the Atlantic community via the Panama Canal. These Pacific cities are flanked by irrigated agricultural zones that supply a major portion of the fresh produce for North America. Over the past several decades, these cities also have become centers of innovation in computer technology.

During the twentieth century, there have been clear shifts in the U.S. population from the Northeast to the South and West (Figure 2.6). Because Americans are among the most mobile people on earth, this kind of redistribution of people is common. Every year, in the United States, almost one-fifth of the population relocates. Some are changing jobs, others are attending school or retiring to a warmer climate, others are merely moving across town or to the suburbs and countryside. Still others are arriving from outside the region: immigrants enter North America at the rate of about 3000 per day. New residents usually contribute to the local economy by finding work, by buying or renting homes, by becoming taxpayers and consumers, and by increasing the number of representatives in state, provincial, and federal legislatures. They also demand more services and increase the costs of city administration, so gaining and losing residents has important political consequences for cities.

People's perceptions of which regions are desirable places to live change over time, so the movement of people ebbs and flows. Why have the South and West of the United States been attracting new residents over the past two decades? There are many factors at work. Internal migrants are often lured by opportunities for employment. For example, many farm towns in the Middle West (the large central farming region of North America), once magnets for European immigrants, are now losing population as farms consolidate, labor needs decrease, and young people choose better-paying careers in cities. Some towns in the Middle West are attempting to recruit new immigrants from Middle and South America and central Europe to work in farm-related industries, just as such towns sought immigrants from Europe in the nineteenth century. Southern cities such as Atlanta, Georgia, and Birmingham, Alabama, are growing more rapidly than cities in the old economic core. These cities and others farther west have sprouted satellite or "edge" cities around their peripheries based on businesses dealing with technology and international trade. In addition, the warm climates of the South and Southwest are attracting the recently retired. You will find additional explanations for these population movements as you read the following sections on geographic issues and individual subregions.

The rate of natural increase in North America (0.5 percent per year) is low, less than one-third the rate of the rest of the Americas (1.7 percent). Still, according to the Population Reference Bureau, North Americans are adding to their numbers through births and immigration fast enough to reach 460 million by 2050. Many of the important social issues now being debated in North America, and discussed below, are reflected in changing population patterns—issues such as legal and illegal immigration, voting patterns among new immigrants, the language of instruction in public schools, the aging of the population, availability and use of birth control, the social costs and benefits of low birth rates, changing family structures, and gender and ethnic diversity.

Recent History: The National and Global Geopolitical Repercussions of the Terrorist Attacks of September 11, 2001

A recent event in the history of North America bears mentioning at this point because it appears to be having a deep and lasting influence on how North Americans relate to one another and to the rest of the world. While it is still too early to assess the true depth or duration of the impacts of this event, some of its repercussions are discussed here, as well as elsewhere in this chapter and the book.

On September 11, 2001, around 9:00 A.M. on a bright and sunny morning, two passenger jets, carrying 147 people, were purposely flown into the World Trade Center towers in New York City. A third passenger jet struck the Pentagon, and a fourth plane crashed in rural Pennsylvania. All four planes had been commandeered by suicide hijackers, who steered three of the aircraft into their intended targets. The passengers on the Pennsylvania plane were able to force it down before it reached its unknown target in Washington, D.C. Within the hour, the World Trade Center towers collapsed, killing 2792 people. At the Pentagon, 184 were killed, including those on board the plane, and 40 died in Pennsylvania. Altogether, 3016 people died. Soon it was learned that the attacks were masterminded by a radical Muslim organization known as Al Qaeda, led by a dissident Saudi Arabian, Osama bin Laden, who had allied himself with, and lived among, a faction of Muslim extremists known as the Taliban, then in control of Afghanistan.

Security Issues Become Paramount. A wave of fear swept the United States during and after the attacks. That fear grew first out of immediate worries about the safety of loved ones, but also out of the sudden realization that the United States was vulnerable. In the nearly 60 years since the Japanese had bombed Pearl Harbor in December 1941, U.S. citizens had come to believe that on the North American continent they were safe from attack. On September 11, it became clear that the very openness of U.S. society and its physically unguarded borders allowed multiple opportunities for intrusion by those who might do harm. Several years after the attack, it remains difficult for Americans to agree on which security strategies are appropriate, given the strong value placed on personal freedom in North American society. The borders of North America are now somewhat less penetrable (but see the description of the long, unfortified U.S.–Canadian border on page 68). A cabinet-level Department of Homeland Security has been established and a color-coded (yellow, orange, red) national threat condition advisory system implemented, but the Patriot Act that was quickly passed after the 9/11 attacks has been deemed by a diverse coalition of interest groups, such as the American Civil Liberties Union (ACLU) and the National Rife Association (NRA), to be too restrictive and abusive of individual liberty.

Even more divisive has been the issue of what strategies the United States should implement on the international level. Should the United States focus primarily on its own interests and take aggressive stands against supposed enemies and possible terrorists, or might the better strategy be to ask the rest of the world to join with the United States in designing ways to alleviate the festering resentment and outrage (justified or not) that had led to the 9/11 events? National debate was cut short when, in the fall of 2001, the administration of George W. Bush, with the advice and consent of Congress, launched what was called a War on Terrorism. The first target was Afghanistan, thought to be host to the elusive Al Qaeda network; the aim was to capture bin Laden (as of Fall 2004, his whereabouts were still unknown). Then, in the spring of 2003, based on the unsubstantiated claims that Saddam Hussein of Iraq may have been involved in the 9/11 events and was seeking to acquire materials for nuclear weapons, President Bush brought the War on Terrorism to Iraq. The United Nations and all major U.S. allies, with the exception of the United Kingdom, advised against this attack and did not cooperate in it. A few small countries reliant on U.S. aid agreed to help and sent several thousand soldiers to serve under U.S. commanders, but overwhelmingly, world opinion was against U.S. policy in Iraq. This opinion only solidified when, after its initial success in removing Hussein from power, the United States found achieving its goals of establishing order and democracy in Iraq increasingly difficult due to increasing Iraqi resistance against the U.S. occupation. In order to defuse the escalating violence, the United States then asked for help from the United Nations and other allies.

The National Identity of the United States Is Redefined. Among the longer-term results of the 9/11 attacks was that citizens of the United States were forced to reevaluate what the United States stood for in their own eyes and in the eyes of the world at large. As they were long accustomed to being warmly regarded abroad, the mere fact of the attacks was a terrible shock; understanding of what had motivated the attackers emerged only slowly over several years and remains incomplete.

Shortly after the terrorist attacks, it was learned that the hijackers were Muslim extremists from the Middle East, especially Saudi Arabia and Egypt, and that they were bent on undermining the power of the West—both Europe and the United States—and on stopping the global diffusion of Western culture and economic influence. This, alone, was difficult for most people in the United States to understand because they had long assumed that most people around the world sought to emulate their way of living and thinking. The realization that, while preparing their attacks, many of the hijackers had lived, worked, and studied among Americans, all the while developing harsh antagonisms, led at first to an anti-Muslim (and anti-foreign) backlash and to some acts of violence against innocent Muslims and other people with apparent foreign origins. The U.S. government itself used the Patriot Act to incarcerate without due process hundreds of Muslim citizens and legal residents as well as illegal aliens. But across the country, leaders and ordinary citizens quickly stepped forward to protest such reprisals as being counter to American values. Even before September 11, a redefinition of diversity in the United States had begun when, during the summer of 2001, analyses of the 2000 census revealed that the United States was no longer primarily a nation of blacks and whites and Christians and Jews. Census-related research showed that

Hispanic and Asian immigrants were changing the cultural palette of communities across the country, with Hispanics just about to exceed African-Americans as the largest minority group.

After September 11, Americans again learned to expand their understanding of diversity when surveys revealed that Muslims and Jews were present in about equal numbers in the United States (6 million to 7 million of each), many of them native-born. Jews had long been accepted as having an important role in U.S. society, but Muslims had not. In the months and years after the 9/11 events, newspaper articles from across the country revealed that U.S. residents of all backgrounds were reflecting on how they might best adjust to an emerging multicultural national identity. At the same time, immigrants realized that they had an obligation to show they were assimilating, at least in part, to American ways of life. New and old Americans began to express solidarity across cultural and religious lines by flying the flag or wearing red, white, and blue ribbons and by singing patriotic songs.

During this reevaluation process, a countertrend was developing at the international level. Immediately after the attacks of September 11, the international community extended warm sympathy to the United States. Many leaders said that their people saw the attack on the United States as an attack on themselves. But soon the United States began to lose support as the prevailing voices of U.S. national leaders became those who advocated violent retribution. The international reputation of the United States as a beacon of freedom and democracy deteriorated, first as a result of the decision to go to war in Afghanistan with the intent of eliminating the Taliban and Al Qaeda. Many saw this as a rash step that would endanger innocent lives and that might lead to subsequent U.S. attacks against Muslim countries at large. The decision in 2003 to go to war against Iraq without United Nations support and without the wholehearted agreement of allies, as well as the decision not to treat captured prisoners according to the **Geneva Conventions** (treaties that protect the rights of prisoners of war), further undermined the reputation of the United States. By 2004, opinion polls around the world revealed little support for the U.S. War on Terrorism and increasing criticism of the United States in general, though most of those polled distinguished between the U.S. government and the people of the United States.

The Economy Is Shaken. The 9/11 attacks had immediate negative effects on the American economy, which had already shown signs of a recession in the spring of that year. The recession deepened as investors lost confidence, and hundreds of thousands of people lost jobs: in New York City alone 79,000 people lost their jobs as a direct result of the destruction of the World Trade Center. Nearly all North America's cities experienced a precipitous decline in economic prospects. Some of the decline was simply due to uncertainty about the future, but some was due to direct losses of business. Perhaps the most important of these losses was a reduction in air travel. Because the hijackers had been able to breach airport security easily and had used fully loaded commercial jets as missiles, Americans suddenly cancelled planned trips and reevaluated the role air travel would play for them in the future. By mid-2004, air traffic was still down from what it had been in the summer of 2001.

Within weeks of September 11, several airlines, in already shaky financial condition, required federal aid to avoid bankruptcy. Those cities and businesses dependent on a freely traveling public began to suffer. Las Vegas, San Francisco, Miami, and Honolulu, all dependent on vacation tourism for income, experienced drastic reductions in tourist arrivals. And there were other ripple effects. Companies that had placed a premium on face-to-face relationships with customers stopped flying staff to visit clients and began to use conference calls instead. As air transport business shifted to ground systems, interstate highways and trains became temporarily overloaded. Industrial managers dependent on on-time delivery of parts, resources, and services had to stop production as they waited for deliveries. Drastically increased security measures raised business costs in time and money.

Ultimately, the costs of the wars in Afghanistan and Iraq were made harder to bear because an economic slump, only partially related to 9/11, continued. By August of 2004, despite tax cuts meant to increase job creation in the United States, 1.7 million U.S. jobs had been lost, many going to less expensive workers abroad, and the creation of replacement jobs fell far below forecasts. And it was revealed that even high-paying technology jobs were now vulnerable to "outsourcing," as employers began to take advantage of technologically skilled workers in India, China, and central Europe. (**Outsourcing** is a strategy whereby employers seek only short-term contracts with workers provided by third-party companies. Usually these jobs pay less than permanent jobs and do not carry health and retirement benefits.)

The general slowing of the U.S. economy affected the global economy. Unemployment quickly spread to distant countries that supplied products to U.S. consumers. For example, a decrease in garment purchases in the United States in November 2001 led to a decrease in orders to suppliers in Pakistan and elsewhere. Anti-U.S. sentiment built when Pakistani workers were laid off just as their government agreed to help the United States with the war it was waging in neighboring Afghanistan.

Dependence on Oil Is Questioned. The 9/11 attacks, connected as they were to Arab extremists, reminded Americans of how dependent they are on oil from Arab countries. As of 2003, 20 percent of the oil used in the United States came from Arab producers, 40 percent from the **Organization of Petroleum Exporting Countries (OPEC)**. Were that supply to be interrupted, oil production by the United States and its allies would not be sufficient to quickly take up the slack. Nor were U.S. reserves substantial enough to last for long. Some speculated that the time was ripe to develop alternative energy resources, such as solar, geothermal, and hydrogen, and to steeply increase energy conservation. But consistently high oil prices are necessary to give sufficient impetus to the developers of these alternative energy sources. With shrinking economies, oil prices fell drastically in the months following September 11. Only after a year of war in Iraq did oil prices rise sufficiently to interest consumers in alternatives to petroleum-based energy. Hybrid cars that run on a combination of gasoline (oil-derived) and computer-controlled rechargeable batteries increased in popularity to such an extent that in 2004, consumers had to wait months to buy one.

Those who say that geographic relationships have been changed forever in the post-9/11 world are probably right. Four years later, the effects on internal and external U.S. affairs remain widespread indeed. Because these effects touch diplomatic relations, long-standing alliances, the price and availability of petroleum products, investment prospects, jobs, political fortunes, transport costs, food supplies, and much more, they reaffirm the need for continuing geographic analysis.

Quick Review

1. During the last ice age, what force created the depressions that would become the Great Lakes?

2. Which body of water contributes the most to the moist climate of the south-central United States?

3. Describe in broad terms the landform features of North America.

4. Describe shifts in the distribution of population in the United States.

5. List three ways in which the 9/11 attacks affected the U.S. economy and three examples of how U.S. national identity was redefined.

II CURRENT GEOGRAPHIC ISSUES

The residents of North America live in what is often considered a privileged part of the world. Nonetheless, the region faces complex problems posed by a rapidly changing regional and global economy, an increasingly diverse population, and rising environmental concerns. The events following the 9/11 attacks further complicate the situation. Responding appropriately to the sense that the United States, if not Canada as well, now faces serious internal and external security risks is a difficult proposition for North Americans, with their cultural diversity and their constitutional rights to freedoms such as travel. Some people fear that the region may not be able to find solutions to these challenges while also maintaining the high standard of living currently enjoyed by the vast majority of its residents. North America's integration with the global economy is visible in nearly every landscape of the region, yet just how the global economy and post-9/11 geopolitics will affect jobs, incomes, and daily life is far from clear. In this section we examine a series of social, economic, political, and environmental issues as they affect both Canada and the United States.

FIGURE 2.7 Political map of North America.

RELATIONSHIPS BETWEEN CANADA
AND THE UNITED STATES

Citizens of Canada and the United States share many characteristics and concerns. Indeed, in the minds of many people (especially those in the United States), the two countries are one. Yet that is hardly the case. In fact, Canadians often make the point that their national identity is focused on noting how different Canada is from the United States. Here, we develop a perspective on relations between these two countries. Three key words characterize the interaction between Canada and the United States: *asymmetries*, *similarities*, and *interdependencies*.

Asymmetries

We start with the concept of *asymmetry*, which means "lack of balance." The two countries occupy about the same amount of space, but much of Canada's territory is sparsely inhabited cold country in the north and west of the continent. In nearly every other way, the United States is much larger than Canada. For example, the U.S. population, as we have seen, is nearly ten times the Canadian population. Although Canada's economy is one of the largest and most productive in the world (producing U.S. $923 billion in goods and services in 2002), the U.S. economy is more than ten times larger ($10.4 trillion in 2002).

There is also asymmetry in international affairs. The United States is an economic, military, and political superpower with a world leadership role that preoccupies it. Since 1990, U.S. for-

eign policy has focused primarily on relations with Russia, Europe, China, Japan, Southeast Asia, Middle and South America, and more recently, Southwest Asia (the Middle East). Canada is only an afterthought in U.S. foreign policy, in part because the country is so close and so secure an ally. But for Canada, managing its relationship with the United States is the foreign policy priority. As a result, Canadians focus much more on events and circumstances in the United States and how they should react to them than vice versa. These asymmetries are cause for discontent in Canada, and as the importance of transborder issues such as migration, free trade, environmental pollution, and security against terrorism increases, people in the United States may have to pay closer attention.

Similarities

Despite these asymmetries, the United States and Canada have much in common. Both are former British colonies, and from this experience they have developed comparable political traditions. Both are federations (of states or provinces), and both are representative democracies; their legal systems are also alike. Not the least of the features they share is a 4200-mile (6720-kilometer) border, which, despite heightened security against terrorism since 9/11, is still the longest unfortified border in the world. Official checkpoints on highways have been increased; the border on the U.S. side is now under the jurisdiction of the Department of Homeland Security, rather than the Immigration and Naturalization Service; and there are now 1000 border guards, rather than the 300 deployed back in 2000. Still, for thousands of miles, there is not even a fence. In some locations the international boundary runs through bedrooms and backyards. Thousands of rural residents pass out of one country and into the other many times in the course of a day simply by going about their usual business of farming, hunting, or housekeeping. While protecting against terrorist threats is important to North Americans, preserving the casually efficient way in which Canadians and U.S. citizens have traditionally traveled back and forth across their common border is also important to them.

Well beyond the border country, Canada and the United States share many other landscape similarities. Their cities and suburbs look much the same. The billboards that line their highways and freeways advertise the same brand names. Shopping malls have followed suburbia into the countryside, drawing millions of affluent buyers away from old urban centers. The same fast-food franchises dispense the same meals of questionable nutrition. And the two countries share similar patterns of ethnic diversity that developed, as we shall see shortly, in nearly identical stages of immigration from abroad.

Interdependencies

Canada and the United States are perhaps most intimately connected by their long-standing economic relationship. By 2002, that relationship had evolved into a two-way trade flow of U.S. $387.2 billion annually (Figure 2.8). Each country is the other's largest trading partner. Canada sells 84 percent of its exports to the United States and buys 73 percent of its imports from the United States. The United States, in turn, sells 22 percent of its exports to Canada and buys 19 percent of its imports from Canada. Notice that asymmetry exists even in the realm of interdependencies: Canada's smaller economy is much more dependent on the United States than the reverse. Nonetheless, if Canada were to disappear tomorrow, as many as a million

American jobs would be threatened. The asymmetry of this relationship will undoubtedly persist; because of its size, large population, and giant economy, the United States is likely to remain the dominant partner. As former Canadian Prime Minister Pierre Trudeau once told the U.S. Congress, "Living next to you is in some ways like sleeping with an elephant: No matter how friendly and even-tempered the beast, one is affected by every twitch and grunt."

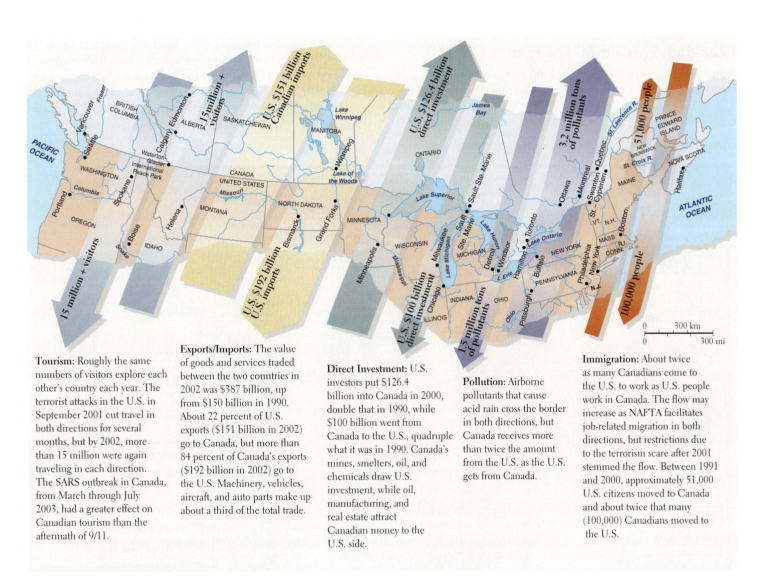

Tourism: Roughly the same numbers of visitors explore each other's country each year. The terrorist attacks in the U.S. in September 2001 cut travel in both directions for several months, but by 2002, more than 15 million were again traveling in each direction. The SARS outbreak in Canada, from March through July 2003, had a greater effect on Canadian tourism than the aftermath of 9/11.

Exports/Imports: The value of goods and services traded between the two countries in 2002 was $387 billion, up from $150 billion in 1990. About 22 percent of U.S. exports ($151 billion in 2002) go to Canada, but more than 84 percent of Canada's exports ($192 billion in 2002) go to the U.S. Machinery, vehicles, aircraft, and auto parts make up about a third of the total trade.

Direct Investment: U.S. investors put $126.4 billion into Canada in 2000, double that in 1990, while $100 billion went from Canada to the U.S., quadruple what it was in 1990. Canada's mines, smelters, oil, and chemicals draw U.S. investment, while oil, manufacturing, and real estate attract Canadian money to the U.S. side.

Pollution: Airborne pollutants that cause acid rain cross the border in both directions, but Canada receives more than twice the amount from the U.S. as the U.S. gets from Canada.

Immigration: About twice as many Canadians come to the U.S. to work as U.S. people work in Canada. The flow may increase as NAFTA facilitates job-related migration in both directions, but restrictions due to the terrorism scare after 2001 stemmed the flow. Between 1991 and 2000, approximately 51,000 U.S. citizens moved to Canada and about twice that many (100,000) Canadians moved to the U.S.

FIGURE 2.8 Transfers of tourists, goods, investment, pollution, and immigrants between the United States and Canada. Canada and the United States have the world's largest trading relationship. The flows of goods, money, and people across the long Canada–U.S. border are essential to both countries, but because of its relatively small population and economy, Canada is more reliant on the United States than the reverse. Restrictions on travel and residency permits since September 11, 2001, have affected Canadians more than U.S. citizens. [Adapted from *National Geographic* (February 1990): 106–107; augmented with data from Office of Travel and Tourism Industries [http://tinet.ita.doc.gov/view/f-2000-04-001/index.html?ti_cart_cookie= 20030901.194057.17337], and International Travel Forecasts: Fourth Quarter Update, 2003: http://ftp.canadatourism.com/ctxUploads/en_publications/InternationalTravel ForcastsQ4.pdf; British Columbia Chamber of Commerce: http://www.bcchamber.org/ publications/dilemmabudget3.html; Canadian Embassy, Washington, D.C.: http://www. canadianembassy.org/trade/wltr-en.asp; *CIA Factbook*, 2002, 2003; The Green Lane, Acid Rain: http://www.ec.gc.ca/acidrain/acid health.html; *Population Today*, vol. 30, no. 2, February–March, 2002.]

ECONOMIC AND POLITICAL ISSUES

The economic and political systems of Canada and the United States have much in common. Both countries evolved from societies based primarily on family farms. After an era of industrialization, both now have primarily service-based economies with important technology sectors. Politically, the two countries have similar governments and face similar issues. Yet they often take very different approaches to social issues, such as unemployment, and social services, such as health care.

North America's Changing Agricultural Economy

North America benefits from an abundant supply of food, and exports of agricultural products were once the backbone of the American economy (Figure 2.9). Today, North America's importance as a producer of food for the continent and for foreign consumers remains high; most farm income derives from sales within the country, but exports account for 20 to 30 percent of American farm income. Nonetheless, the overall economic role of agriculture within North America has been shrinking for many years. It now accounts for just 1.7 percent of the region's gross domestic product—this despite the fact that American agriculture is highly productive per worker and per acre.

Agriculture employed 90 percent of the American workforce in 1790 and 50 percent in 1880. European immigrants set up thousands of highly productive family-owned farms, spreading over much of the United States and southern Canada. These farms provided the majority of food and fiber for domestic consumption and the majority of all exports until 1910. Currently, agriculture employs less than 2 percent of the North American workforce because today's highly mechanized farms need few workers. Instead, these farms require heavy capital investment in land and machinery—often a half-million dollars or more per farm. As farms became mechanized, they became substantially more productive, as measured in yield per acre and per worker. But one result of increased yields is declining prices for farm

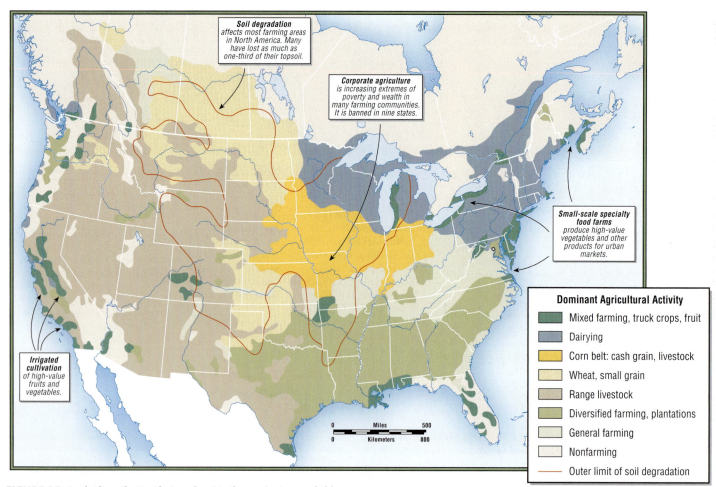

Soil degradation affects most farming areas in North America. Many have lost as much as one-third of their topsoil.

Corporate agriculture is increasing extremes of poverty and wealth in many farming communities. It is banned in nine states.

Small-scale specialty food farms produce high-value vegetables and other products for urban markets.

Irrigated cultivation of high-value fruits and vegetables.

Dominant Agricultural Activity

- Mixed farming, truck crops, fruit
- Dairying
- Corn belt: cash grain, livestock
- Wheat, small grain
- Range livestock
- Diversified farming, plantations
- General farming
- Nonfarming
- Outer limit of soil degradation

FIGURE 2.9 Agriculture in North America. North America is remarkable in that some type of agriculture is possible throughout much of the continent. The major exceptions are the northern reaches of Canada and Alaska and the dry mountain and basin region (the continental interior) lying between the Great Plains and the Pacific coastal zone. However, even some marginal areas, such as southern California, southern Arizona, and the Utah Valley, are cultivated with the help of irrigation. [Adapted from Arthur Getis and Judith Getis, eds., *The United States and Canada: The Land and the People* (Dubuque, Iowa: Wm. C. Brown, 1995), p. 165.]

The rapid industrialization of the South is evident in this recent air photo of the Saturn plant in Springhill, Tennessee. It was laid down over the rural landscape south of Nashville. Old field patterns are still clearly visible within the confines of the new plant. [Saturn Corporation.]

products, making it increasingly difficult for small farms to break even. A successful farm now requires so much investment that large **agribusiness** corporations, with their ready access to cash, have an advantage over individual farmers, especially because they can engage in all aspects of food production: farming, marketing, research, processing, transport, and delivery to customers. Corporate agriculture provides a wide variety of food at low prices for North Americans; nonetheless, in many rural areas, it has depressed local economies and created social problems. Communities in such places as rural Iowa, Nebraska, and Kansas were once made up of farming families with similar incomes, similar social standing, and a commitment to the region. By 2000, farm communities were increasingly composed of a few wealthy farmer-managers amid a majority of poor, often migrant, Hispanic or Asian laborers working for low wages on corporate farms and in food-processing plants (see the box "Meat Packing in the Great Plains" on page 104). The result is increasing class disparity, with low levels of education and well-being for many people, and less functional local governments. Under the assumption that the family farm has inherent value, the U.S. federal government and some states are increasing efforts to protect family farms and the rural communities of which they are a part. The states of Iowa, Kansas, Minnesota, Missouri, Nebraska, North and South Dakota, Oklahoma, and Wisconsin all have laws that restrict corporate involvement in agriculture. Family-owned corporations are permitted only if family members live on the farm.

Corporate agriculture relies heavily on the use of highly modified seeds and chemical fertilizers, pesticides, and herbicides to increase crop yields. These farming methods have the potential to contaminate food and pollute environments. Compared with owner-operators, the small, salaried labor forces of today's farms

have a lesser stake in the long-term sustainability of farming. It is easier for them to ignore such environmental effects as erosion and stream pollution. Many North American farming areas had lost as much as one-third of their topsoil (see Figure 1.22 on page 45) even before corporate farms took over. The fear is that with the profit motive now paramount and with no incentive to preserve the home farm for future generations, corporate agriculture will hasten the demise of America's productive topsoil. Using the **low-till** cultivation method, which entails plowing only once every few years, is one strategy for reducing the loss of topsoil. However, low-till methods require increased use of chemicals to control weeds and pests. An equally controversial effort to reduce dependence on chemicals involves **genetic engineering,** the scientific modification of crops to create varieties resistant to pests and diseases. In Europe and other parts of the world to which North American farms have traditionally exported their crops, genetic manipulation of food crops is seen as potentially so unsafe that in a number of cases, Europe has refused to import U.S. agricultural products.

Changing Transport Networks and the North American Economy

North America is going through dramatic transformations, many of them related to changes in technology and trade. Traditional manufacturing is fading in economic importance, just as agriculture did earlier. A vibrant new service economy, based increasingly on technology, is emerging. The old economic core (see Figure 2.27 on page 99) no longer dominates, as new centers of production spring up across the continent. The diffusion of production has been aided in part by improved networks for transportation and information exchange.

The Manske cousins, who own adjoining farms in Wisconsin, employ contour furrows to curb erosion. They also grow alternating rows of corn (their cash crop) and alfalfa (which provides hay for their dairy cows) to help control runoff. [Jim Richardson/Richardson Photography.]

The development of the inexpensive mass-produced automobile in the 1920s changed North American transport dramatically. Soon trucks were delivering cargo more quickly and conveniently than the railroads could, and miles of railroad track decreased between 1930 and 1980. The **Interstate Highway System,** a 45,000-mile (72,000-kilometer) network of high-speed, multilane roads, was begun in the 1950s and completed in 1990. Because this network was connected to the vast system of local roads, it introduced previously unimagined flexibility, speed, and low cost to the delivery of manufactured products. Thus the highway system made possible the dispersal of industry into nonurban locales across the country, away from the old industrial core in the Northeast. The availability of quick delivery virtually anywhere in the country stimulated service and manufacturing industries in the hinterland, where labor, land, and living costs were cheaper.

The 9/11 attacks, however, raised questions about the vulnerability of this dispersed industrial production system. If availability of oil from the Arab countries decreased and fuel prices rose, the vast and efficient air and ground transportation system might suffer. Would dispersed factories lose their competitive edge if the highway-based delivery system became overloaded or if air service became more costly and less frequent? Conversely, might a dispersed pattern of industrial distribution be less susceptible to attack and therefore worth keeping?

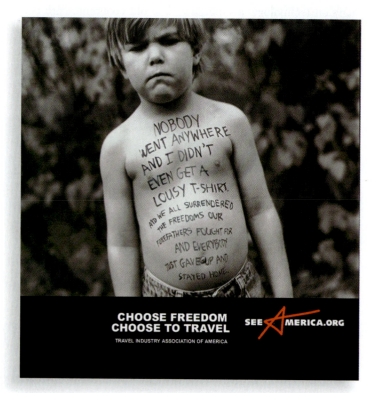

In the fall of 2001, in an effort to stem the decline of passenger travel, the Travel Industry Association of America launched a campaign to attract reluctant fliers back to the sky. This trade association allied its campaign, "Freedom to Travel," with those warning that overly strict security measures would abridge the basic liberties Americans hold dear. [Courtesy of the Travel Industry Association of America.]

Air transport, which came into its own after World War II, also served economic growth in North America, but its primary niche is business travel. Despite the increasing use of instant communication via the Internet, face-to-face contact is still seen as essential to American business culture. To address North America's pattern of widely dispersed industries in numerous medium-sized cities and to facilitate one-day travel between any two points, air service is organized as a **hub-and-spoke network.** Hubs are strategically located airports used as collection and transfer points for passengers and cargo. Airlines schedule many flights to arrive at a hub within a few minutes of one another. Passengers and cargo then are shifted to flights departing to other hubs or intended final destinations. Although delays can affect the synchronization of the system, causing passengers to miss a connection, other connecting flights are often only a few hours away. Flying is a way of life in North America. In 1999, North American airlines carried nearly 640 million passengers, more than double the continental population. A Gallup survey in 2000 revealed that 80 percent of the U.S. adult population had flown at least once, more than one-third of them in the previous 12 months.

The air transport system of North America was among the industries most deeply affected by the 9/11 attacks and by fears of further terrorism. Post-9/11, the airline industry suffered the sharpest decline in revenues in its history (approximately 38 percent), and 90,000 employees were furloughed. A year later, domestic passenger traffic was still off 22 percent, and international traffic by 37 percent. But passenger traffic rebounded steadily, and by the summer of 2004 it was forecast to equal travel rates in summer 2001.

The New Service and Technology Economy

Since the 1970s, the economic base of North America has shifted from manufacturing to a broad and varied **service sector** (economic activity that involves the sale of services). By 2002, the employment picture in North America had changed in three important ways. Occupational structure had changed: routine, low-skill, mass-production industrial jobs were less available due to mechanization and because firms moved such operations abroad. Education requirements had increased: knowledge-intensive jobs requiring specialized professional training in technology and management were growing so fast that positions went unfilled. The workforce had become diversified: the old model of a male head of household, earning one salary and supporting a wife and family, had virtually disappeared.

Decline in Manufacturing Employment. As labor unions won higher pay and better working conditions for most workers in the industrial Northeast (the economic core), those benefits added to the costs of production. For this reason and others, such as the cost of keeping equipment technologically current, the Northeast began to lose its advantageous position. By the 1960s, the geography of manufacturing was changing: many companies were moving their factories elsewhere to take advantage of cheaper labor that was less unionized. At first, the southeastern United States was a major destination for these industries because the lack of labor unions there resulted in lower wages, and the warm climate meant lower energy costs than elsewhere in the country. By the 1980s, however, another geographic shift was under way. Especially after the passage of the North American Free Trade Agreement (NAFTA) in 1994, certain industries, such as clothing, electronic assembly, and auto parts manufacturers, were moving farther south to Mexico or overseas. In these places labor was vastly cheaper, and laws mandating environmental protection and safe and healthy workplaces were absent or less strictly enforced—a further source of savings.

Another factor in the decline of manufacturing employment is automation. The steel industry provides an illustration. In 1980, huge steel plants, most of them in the economic core, employed more than 500,000 workers. At that time, it took about 10 man-hours and cost about $1000 to produce 1 ton of steel. Spurred by more efficient foreign competitors, the North American steel industry applied new technology in the 1980s and 1990s to lower production costs, improve efficiency, and increase production.

The steel industry as a whole now employs less than half the workers it did in 1980. Revamped plants in the late 1990s produced 1 ton of steel with just 3 man-hours of labor (a 60 percent reduction) at a cost of just $200 per ton. In addition, new technology has allowed the use of cheap scrap metal from many sources (such as junked cars and refrigerators) in smaller operations (minimills) that are located in less urbanized areas (principally the Southeast). In 2000, these minimills produced steel at the rate of just 0.44 man-hour per ton (a more than 90 percent reduction since 1980) and at a cost of about $165 per ton.

This example helps us to understand the changing employment picture in North America: far fewer people are now producing more of a given product at far lower cost than was the case 20 years ago. So, although the share of the gross domestic product (GDP) produced by manufacturing has declined, the actual level and value of industrial production has not, at least in certain very important sectors, such as steel. It should be noted, however, that the balance is precarious, as the U.S. economy, especially the low-wage service sector, continues to expand.

Growth of the Service Sector. As factories moved outside the United States and manufacturing employment declined, the economic base of North America shifted increasingly to a broad and varied service sector. This sector is **bimodal** in that some jobs require advanced training or higher education and are high-paying, while others require few skills and pay very low wages. As of the year 2000, in both Canada and the United States, about three-fourths of the GDP and a majority of the jobs were in the service sector. In Canada, 70 percent of workers—and in the United States, 80 percent—now work in transportation and utilities, wholesale and retail trade, health, leisure, maintenance, government, and education. In all of these categories there are high-paying jobs, but low-paying jobs are common. Wal-Mart and Manpower Inc. are the largest employers in the United States. Service jobs are often connected in some way to international trade, entailing the processing, transport, and trading of agricultural and manufactured products and information that originate outside North America; hence international events can shrink or expand the numbers of these jobs.

An important subcategory of the service sector involves the creation, processing, and communication of information—what is often labeled the **knowledge economy** (or k-economy). The knowledge economy includes workers who manage information, such as those employed in finance, publishing, print and visual media, higher education, research and development, and professional aspects of health care. Industries that rely on the use of computers and the Internet to process and transport information—banks, software companies, medical technology companies, publishing houses—are increasingly called **information technology (IT)** industries (Figure 2.10). They are freer to locate where they wish than were the manufacturing industries of the old economic core, which depended on massive amounts of steel and energy, especially from coal. By contrast, IT industries depend mostly on highly skilled thinkers and technicians (as well as on electrical energy), and hence they are often located near major universities and research institutions. Examples of such IT centers include Toronto's Spadina Bus district, near Ryerson Polytechnic; California's Silicon Valley, near Stanford University; the Route 128 corridor near Boston's Massachusetts Institute of Technology (MIT) and Harvard University; and the Research Triangle near the University of North Carolina. There are IT centers in hundreds of other towns and cities near institutions of higher learning. Patches of IT activity also exist around government research labs such as Oak Ridge in Tennessee and Brookhaven on Long Island, New York.

The evolving economy based on information technology relies on a continuous stream of innovation and is dependent on the establishment of networks of individuals, communities, corporate organizations, and government. By the late 1990s, yet another technological change, the Internet, was expanding these networks. The **Internet** is a computer network that allows for the electronic transfer of all kinds of information virtually instantaneously to any place in the world—information that used to be carried via mail over roads, rails, water, and air. The Internet is emerging as an economic force more rapidly in North America than in any other world region. It was here that the Internet was first widely available, and, for the time being, it is where the largest block of its users live (36 percent of the world's Internet users are in the United States). The *United Nations Human Development Report 2003* indicates that there are 50 Internet users for every 100 people in the United States and 47 users per 100 in Canada. In comparison, Europe as a whole has about 33 Internet users per 100 people.

The geography of the "new economy" of information technology bears some striking resemblances to the hierarchies of the "old economy" of industrial production. As we have just observed, over the past several hundred years, North America laid down a network of roads, canals, railroads, highways, telegraph and telephone lines, and airports to move resources, goods, people, and information. There was a geographic hierarchy of places connected by transport and communications systems. Usually the large cities and highly productive industries were connected early and well. Less influential outlying towns and enterprises were connected later, and rural areas later still. Today, those places first "wired" for information technology tend to be large and wealthy

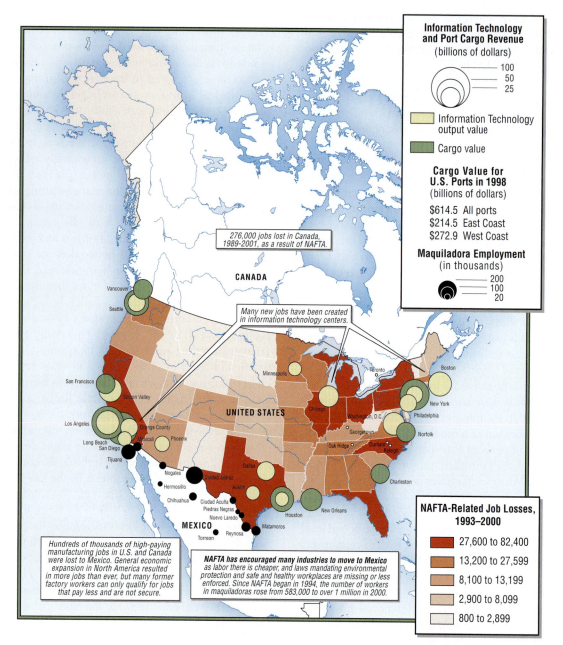

**Information Technology
and Port Cargo Revenue**
(billions of dollars)

100
50
25

Information Technology
output value

Cargo value

**Cargo Value for
U.S. Ports in 1998**
(billions of dollars)

$614.5 All ports
$214.5 East Coast
$272.9 West Coast

Maquiladora Employment
(in thousands)

200
100
20

*276,000 jobs lost in Canada,
1989-2001, as a result of NAFTA.*

CANADA

*Many new jobs have been created
in information technology centers.*

Vancouver
Seattle

San Francisco
Silicon Valley
Los Angeles
Orange County
Long Beach   Mexicali
San Diego
Tijuana
Nogales
Hermosillo
Chihuahua   Ciudad Juárez
          Ciudad Acuña
          Piedras Negras
          Nuevo Laredo
MEXICO   Reynosa   Matamoros
Torreon

UNITED STATES

Minneapolis
Chicago
Phoenix
Dallas
Austin
Houston   New Orleans
Toronto   Boston
New York
Washington, D.C.   Philadelphia
Georgetown   Norfolk
Oak Ridge   Durham
        Raleigh
        Charleston

*Hundreds of thousands of high-paying
manufacturing jobs in U.S. and Canada
were lost to Mexico. General economic
expansion in North America resulted
in more jobs than ever, but many former
factory workers can only qualify for jobs
that pay less and are not secure.*

NAFTA has encouraged many industries to move to Mexico
as labor there is cheaper, and laws mandating environmental
protection and safe and healthy workplaces are missing or less
enforced. Since NAFTA began in 1994, the number of workers
in maquiladoras rose from 583,000 to over 1 million in 2000.

**NAFTA-Related Job Losses,
1993–2000**

27,600 to 82,400
13,200 to 27,599
8,100 to 13,199
2,900 to 8,099
800 to 2,899

FIGURE 2.10 Economic issues: North America. Trading and industrial
patterns in the United States changed during the 1990s. For the first time,
U.S. trade with Asia grew faster than trade with Europe (see Figure 2.11), and
New York dropped to the third largest port in value of cargo handled, behind
Los Angeles and Long Beach, California. Information technology increased
dramatically in value of output, while manufacturing and related jobs
decreased—in part due to the effects of NAFTA. [http://www.ratical.org/co-
globalize/NAFTA@7/index.html.]

Thinking Globally: ON THE WEB

cities, major government research laboratories, and universities
on the cutting edge of technology innovations. A spatial **digital
divide** has developed that leaves small, rural, and poor places out
of the loop. Some have waited years for Internet connections, but
increasingly rural middle-class homes are online. There are still
important portions of the population, however, that are left out:
the poor, the elderly, women, and minorities are less likely to have
a computer and to be able to afford Internet connections. For
them, the public library or the county courthouse may be the
only access. Once the information technology infrastructure is in

place, however, distance from centers is far less of a hindrance to
participation in the new IT economy than it used to be in the
industrial economy.

Globalization and the Economy

Throughout the previous discussion, globalization—the process
of forging worldwide economic linkages, sometimes by country-
to-country agreements, sometimes by general reductions in barriers to trade—has been recognized as a force driving economic

change in North America. North America has a wide trading network and, because of its size and wealth, wields great power in the international organizations that affect global economic relations, such as the United Nations, the World Bank, the International Monetary Fund (IMF), and the World Trade Organization (WTO).

In the past, trade barriers have been important aids to North American development. For example, upon achieving independence from Britain in 1776, the new U.S. government imposed tariffs and quotas on imports and gave subsidies to domestic producers to protect fledgling domestic industries and commercial agriculture. As a result, the economic core region flourished. But if tariffs and quotas are kept in place too long, they can result in stagnation. Without competition, quality may suffer and prices will remain high. The United States is now an active partner in the World Trade Organization, which seeks to lower trade barriers and increase competition. Nonetheless, some tariffs and quotas against low-priced manufactured goods from developing countries (such as textiles, shoes, and agricultural products) remain in effect in the United States. The stated role of these tariffs is to prevent less expensive foreign-produced versions of these goods from driving North American producers out of business, but they have the added effect of shutting poor countries out of lucrative markets just when these fledgling economies need market opportunities, as the United States once did. U.S. tariffs aimed at developing economies are what spur the public protests against the United States and the European Union that often accompany WTO meetings. The September 2003 meeting in Cancún, Mexico, was shut down by such protests. **Government subsidies** —payments that cover part of the production costs of some farm products—also help North American producers to sell their goods for less than their foreign competitors, even those in very poor countries. Nonetheless, global competition is stiff, and less developed countries and world regions have some advantages, such as lower wages, fewer financial and environmental controls, and lower taxes. Increasingly, workers in North America, like those in developing countries, are noting that globalization is a mixed bag, offering both opportunities and risks for the individual worker.

The Role of Regional Trade Agreements in Global Trade: NAFTA. The United States and Canada have long had free trade unrestricted by tariffs or quotas between the states and provinces. In recent years, Canada and the United States have promoted the idea of free trade in the whole of North America because tariffs and quotas are now seen as constraints on economic growth that raise prices for consumers, discourage competition among producers, and restrict the movement of workers. The process formally began with the Canada–U.S. Free Trade Agreement of 1989. The creation in 1994 of the North American Free Trade Agreement (NAFTA) brought in Mexico as well. Today, NAFTA is the world's second largest trading bloc, both in population size and in dollar value of trade (the European Union is the largest).

AT THE GLOBAL SCALE Wal-Mart

One of the most common places where North Americans encounter the global economy is at the local Wal-Mart store, where they can buy a wide range of products manufactured abroad. The selection is so good and the prices so low that in nearly every market, Wal-Mart is driving local merchants and manufacturers out of business.

Based in Bentonville, Arkansas, Wal-Mart is the world's largest retailer and the world's largest company in terms of sales ($256 billion as of January 31, 2004). It is the largest private-sector employer in the United States (1.3 million employees) and in Mexico; the company also has stores in Canada, Brazil, Germany, Japan, Argentina, Puerto Rico, Korea, Britain, and China. More than 100 million people visit Wal-Mart stores annually; in the United States, eight out of ten households shop at a Wal-Mart store weekly.

Criticism of large "big box" retailers such as Wal-Mart is growing. First of all, the jobs associated with them provide low wages and few or no health benefits, and unionization is discouraged. Stores are usually located on relatively inexpensive land at the edges of suburban communities, where they take up open space and farmland, cause environmental disruption, and contribute to sprawl and automobile dependence. Some cities, such as Phoenix, Arizona, welcome the stores: in the Phoenix–Scottsdale vicinity (population 1.3 million), there are 6 Super-Wal-Marts and 14 regular stores. But in the small New England state of Vermont (population 600,000), which already has 4 stores, expansion plans for 7 new stores are under fire. In May 2004, the National Trust for Historic Preservation designated the State of Vermont one of America's 11 most endangered historic places because huge retailers like Wal-Mart are putting out of business the hundreds of small stores that have for years been crucial to the special lifestyle of Vermont's many villages.

According to *The Economist* (April 15, 2004), "The company is skilled at obtaining its products cheaply, and the emergence of China as a centre of low-cost production is playing to its strengths." Each year Wal-Mart buys nearly $8 billion of goods directly from China, and the company's suppliers purchase an equal amount from China. This translates into big savings for Wal-Mart shoppers and profits for the company through large-volume sales.

[*Sources*: http://www.walmartstores.com/wmstore/wmstores/HomePage.jsp; "Wal-Mart and the World," PBS broadcast on *Now with Bill Moyers* 12-19-2003; Fran Spielman, "Wal-Mart makes pledges on hiring minorities, locals," *Chicago-Sun-Times*, May 26, 2004; Herbert G. McCann, Associated Press, "City OKs Wal-Mart on W. Side, rejects S. Side store," *Chicago-Sun-Times*, May 26, 2004; "National Trust Names the State of Vermont One of America's 11 Most Endangered Historic Places"; Sana Siwolop, "Wal-Mart's Mixed Success Where Land Is Costly," NYTimes.com, May 26, 2004; The Economist online, April 15, 2004.]

A major long-term goal of NAFTA is to increase the level of trade among Canada, the United States, and Mexico, though not necessarily to balance that trade (the Canadian and Mexican economies, at $900 billion plus per year each, are both about one-tenth the size of the U.S. economy). By the time of its full implementation in 2009, NAFTA will have largely integrated the economies of Canada, the United States, and Mexico by removing or reducing most trade barriers. Establishing uniform standards is part of this process. For example, Mexican, Canadian, and U.S. trucks eventually will be equipped with brakes and pollution control devices that meet agreed-upon standards so that they can move freely through all three countries.

It is not possible at this time to assess the overall effects of NAFTA with any accuracy, partly because data have not been collected for a long enough time, and partly because it is difficult to know whether perceived changes are due to NAFTA or to other changes in world trade. Furthermore, the effects of NAFTA have not been uniform across North America. For example, although the value of U.S. trade with Canada and Mexico increased 179 percent between 1990 and 2000, the benefits varied from state to state (and province to province) and over time. The long-term effect of NAFTA on employment is also obscure. Some evidence indicates that about 440,000 U.S. jobs have been lost because of the movement of industries to Mexico and the closing of firms that could not compete with companies in Mexico (see Figure 2.10 on page 74); more extreme estimates say that close to 800,000 jobs have been lost. At the same time, perhaps between 660,000 and 1,000,000 U.S. jobs have been created by economic opportunities forged by NAFTA and other aspects of globalization. Canada and Mexico show similar mixed and hard-to-fathom pictures of job loss and gain due to NAFTA. While Canada's trade with the United States and Mexico has boomed, internal trade has fallen, and trade with the rest of the world has been less than it was. Both phenomena have eliminated jobs in Canada. Moreover, although some new NAFTA-related jobs pay up to 18 percent higher than the average American wage, those jobs are usually in different locations from those that were lost, and the people who take them tend to be younger and more highly skilled than those who lost jobs. Often the new jobs taken by former factory workers are short-term contract jobs or low-skill jobs that pay minimum wage and carry no benefits, such as the fast-food job taken by Jesse in the vignette that opens this chapter.

Even as the benefits and drawbacks of NAFTA are being assessed, there is talk of extending NAFTA to the entire Western Hemisphere under what would be called the Free Trade Area of the Americas (FTAA). Such an agreement would have drawbacks and benefits as well, and some countries, such as Brazil, are wary of being overwhelmed by the U.S. economy. This emerging trade agreement is discussed in Chapter 3 (see page 139).

The Asian Link. NAFTA is only one way in which the North American economy is becoming globalized. The lowering of trade barriers has encouraged the growth of trade between North America and Asia (Figure 2.11). During the early 1990s, growth in trade with Asia surpassed growth in trade with Europe. Although U.S.–Asian trade slowed during the Asian recession of the late 1990s, the growing importance of this region to the U.S. economy is obvious.

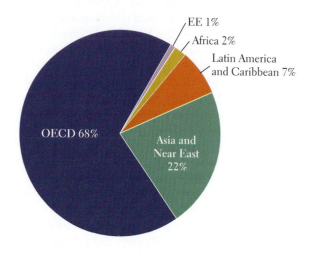

(a) Regional Distribution of U.S. Total Trade, 1998

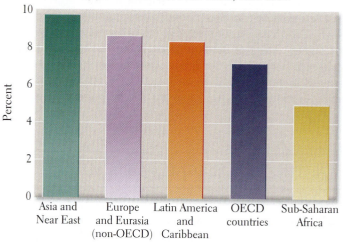

(b) Growth in U.S. Total Trade, 1990–2000

FIGURE 2.11 U.S. trade with selected regions, 1998. Although the bulk of U.S. trade is with OECD countries in the 1990s, U.S. trade with Asia as a whole grew faster than with any other world region. [Data from USAID: http://www.usaid.gov/economic_growth/trdweb/overview.html; http://www.usaid.gov/economic_growth/trdweb/Images.]

Many Japanese automotive plants dot the landscape of rural Middle America east of the Mississippi River. Japanese companies sought plant locations in North America to be near their most important pool of car buyers: commuting North Americans. They chose rural locations to save money on land, taxes, and utility costs, but sought property that was close to the interstate highway system and not far from major cities. In such places, Japanese companies found a ready, inexpensive labor force among those who value life in rural America, yet do not wish to farm. Many are willing to commute 20 miles (32 kilometers) more for secure jobs that pay reasonably well and include health and retirement benefit packages.

The Toyota Camry plant constructed during the 1980s in Georgetown, Kentucky, is an example of such an enterprise. Georgetown lies north of Lexington in the heart of the Kentucky Bluegrass country. The Toyota site is adjacent to I-75 and I-64, which form a major Interstate Highway junction in the nation's midsection. The plant is Toyota's largest outside Japan and employs 7500 people, who turn out nearly 500,000 vehicles and engines per year (not all Camrys). Several dozen smaller Japanese plants in central and western Kentucky produce parts for the main plant.

The Toyota plant has brought social and physical change to Georgetown. The arrival of Japanese executives and their families brought new ethnic diversity and required adjustments in the public schools, such as the teaching of Japanese and of English as a second language. New religious institutions were built for those Japanese who practice Buddhism. The utility supply (water and electricity) had to be increased greatly to accommodate the new plant and labor force, and new businesses were established to cater to those workers.

New Competition from Developing Economies for IT Jobs. By the early 2000s, the North American economy was affected by yet another aspect of globalization. Information technology jobs were being outsourced, meaning that a range of jobs, from software programming to telephone-based customer and technology support services, were shifted to lower-cost zones outside North America. By mid-2003, 500,000 jobs had been outsourced, and forecasts are that by 2020, 3.3 million more IT jobs will follow. Newly developed IT centers in India, China, Southeast Asia, Central Europe, and Russia, where there are large pools of highly trained, English-speaking young people who will work for less than 15 percent of their American counterparts' wages, have won contracts from North American firms such as Microsoft and Oracle. Some advocates of outsourcing argue that rather than depleting jobs, the trend will actually help job creation in North America by saving corporations money, which will then be reinvested in new ventures. The truth of this argument remains to be seen. One North American beneficiary of U.S. outsourcing is Canada, where the U.S. firm Allmerica has realized 20 percent savings on software development expenses. The firm reports that outsourcing to Canada has four advantages: the work is done more cheaply, travel and communications costs are lower than in more distant locations such as India, China, or Europe; there are few cultural differences among workers; and there are no disruptive time zone changes.

Canadian and U.S. Responses to Economic Change

The U.S. and Canadian governments have responded differently to the displacement of workers by economic change. For many decades, Canada has protected its working population from job losses or declines in income by enacting strict unemployment insurance laws. It also has spent more than the United States on social programs that lessen the financial burdens of working people in times of economic crisis. Although these policies have made the financial lives of working people more secure, they have also made Canada slightly less attractive than the United States to new businesses. For example, Canadian employers must pay somewhat higher taxes and sometimes higher wages, and they are required to be more restrained in firing workers. Consequently, they are more hesitant to create new jobs, and this contributes to Canada's unemployment rate (usually several percentage points higher than that in the United States) and its slightly lower rate of economic growth. On the other hand, there are significant social problems associated with the U.S. approach, which provides little job protection or government unemployment assistance. Many poor rural and urban areas in the United States that have lost agricultural and manufacturing jobs have experienced increases in ill health, violent crime, drug abuse, and family disintegration—all of which are social problems connected with declining incomes and the lack of a social safety net.

One important part of a strong social safety net is health care. The Canadian health-care system covers 100 percent of the population. It is heavily subsidized by the government and is less expensive per capita and more effective in preventing illness than the largely private health-care system in the United States. The United States spends 14 percent of its gross domestic product on health care; Canada spends only 9.3 percent. Health expenditures per capita in Canada are just a little over half those in the United States. Nevertheless, Canada does better than the United States on most indicators of overall health, such as infant and maternal mortality and overall life expectancy (Table 2.1). Medicines are also notably less expensive in Canada. Having guaranteed health care removes considerable pressure from those who work for low wages, or who lose their jobs, even temporarily.

TABLE 2.1 Health-related indexes for Canada and the United States

Country	Health-care costs as percentage of GDP, 2002	Percentage of population fully insured	Deaths per 1000	Infant mortality per 1000 live births	Maternal mortality per 100,000 live births	Life expectancy at birth (years)	Health expenditures per capita (PPP U.S.$), 2000
Canada	9.3	100	7	5.3	6	79	2534
United States	14	66	9	6.9	12	77	4499

Sources: United Nations Human Development Report 2003, 2002 World Population Data Sheet, Population Reference Bureau, from http://www.prb.org/pdf/WorldPopulationDS03_Eng.pdf; 2002 *Women of Our World,* Population Reference Bureau, from http://www.data.worldpop.org;

OECD Health Data 99 (Paris: Organization for Economic Cooperation and Development, 1999) and subsequent updates, in *Health Care Spending in 23 Countries,* http://www.thirdworldtraveler.com and http://www.thirdworldtraveler.com/Health/O_Canada.html.

Systems of Government: Shared Ideals, Different Trajectories

Canada and the United States have similar democratic systems of government, but there are differences between the two countries in the way power is divided between the federal government and provincial or state governments. There are also differences in the way the division of power has changed since each country achieved independence. Both countries have a federal government, in which a union of states recognizes the sovereignty of a central authority while retaining certain residual powers of government. In both Canada and the United States, the federal government has an executive branch, a legislature, and a judiciary. In Canada, however, the executive branch is more closely bound to the legislature, and the Canadian federal government has more and stronger powers (at least constitutionally) than does the U.S. federal government. Over the years, both the Canadian and U.S. federal governments have moved away from the original intentions of their constitutions. Canada's originally strong federal government has become somewhat weaker, whereas the initially more limited federal government in the United States has extended its powers. The main force behind the weakening of Canada's central government has been demands by provinces for greater autonomy over local affairs. The most assertive effort for provincial control has come from the French-speaking province of Québec, where many people feel overwhelmed by the rest of Canada. In the past decades, several referenda on political independence for Québec have narrowly failed (see page 98).

The U.S. federal government's original source of power was its mandate to regulate trade between states. Over time, this mandate has been interpreted ever more broadly. By the late twentieth century, the federal government was affecting life at the state and local levels primarily through its ability to dispense federal tax monies. The federal government has been able to influence local policy through grants for school systems, federally assisted housing, mortgage programs, health care, and environmental protection, as well as through the establishment of military bases, anti-crime legislation, and the building of Interstate Highways. Only 10 percent of the funding for the Interstate Highway System came from the states linked by that system; the remaining 90 percent was supplied by the federal government in accordance with its mandate to facilitate interstate commerce. Money for these various federal programs is withheld if state and local governments do not conform to federal standards. This practice has made some poorer states dependent on the federal government, but it also has encouraged some state and local governments to enact more enlightened laws than they might have done otherwise. In the 1960s, federally funded programs promoted civil rights for southern African-American citizens by providing enriched educational programs for minority children. Eventually, federal money was made available for programs promoting adult literacy, employment training, job opportunities, school vouchers, voter registration, and community development for disadvantaged African-American and Hispanic communities. Several U.S. Supreme Court decisions also promoted racial integration in the schools and the workplace.

Gender in National Politics and Economics

There are some powerful political contradictions in North America with regard to gender. Women apparently decided the U.S. presidential election of 1996, voting overwhelmingly for Bill Clinton, and such women's issues as family leave, equal pay, day care, and reproductive rights have been forced into the national debate. However, men still hold 70 percent of the top executive positions in business and government, and they earn about 30 percent more, on average, than women in those positions. Of the 535 people elected to the U.S. Congress in 2002, just 74, or 14 percent, are women (up from 12.5 percent in 1998). At the state level, 22 percent of legislators are women; the six states with the highest percentage of women legislators are all located in the West and Southwest. In the Canadian Houses of Parliament, 23.6 percent of the members are women, and about 10 percent of provincial elected officials are women. A joint study in 1998 by U.S. and Canadian political scientists showed that in the United States, women are most likely to be elected in rural districts where politics are conservative and competition for positions is weak. In Canada, on the other hand, women tend to be elected in urban districts where the electorate is multicultural and liberal on gender issues. Consequently, urban women are underrepresented at the state level in the United States, and rural women are underrepresented at the provincial level in Canada. Both the United States and Canada are well behind several other countries in their percentages of women in national legislatures: Cuba (36), South Africa (30), Mozambique (30), Norway (36.4), and Sweden (45.3).

Within the broader workforce, according to the *United Nations Human Development Report* of 2003, North American women earned only 62 percent of what men earned, but this was up from 59 percent in 1970. Women tend to work more hours than men, however, and their longer working hours contribute to their gain in earnings. In the United States, it is estimated that if women and men earned equal wages, the poverty rate would be cut in half. Women are now about half the labor force, with most working for male managers. However, by 2003, nearly 46 percent of all businesses were at least 50 percent owned by a woman or women, and such businesses employed 18 million people and contributed $2.3 trillion to the $10.4 trillion U.S. economy. Despite these advances by women-owned businesses, they tend to be much smaller and less secure than those in which the dominant control is held by men, in part because it is harder for women business owners to obtain loans and to land contracts.

In education, Canadian women have attained virtually the same level as men in most categories. In the United States, although women students now outnumber men in higher education, they are coming from behind: 25 percent of U.S. women aged 25 and over hold an undergraduate degree, while 29 percent of men do. This imbalance is likely to disappear, because by 2000, women were earning slightly more undergraduate degrees than were men. The disparity in earnings between men and women persists, however, as do disparities in workers' compensation and retirement benefits.

SOCIOCULTURAL ISSUES

North America is undergoing rapid sociocultural change, and much of this change has geographic aspects. In this section we discuss some especially widespread changes that are likely to persist into the future. First we examine a set of issues—urbanization, immigration, race and ethnicity, and religion—that stem from the fact that North Americans are an increasingly diverse people living in concentrated settlements where their differences can make it difficult to find common ground. Then we discuss two issues—the American family and aging—that illustrate changing social organization in this large, complex, and prosperous region.

Urbanization

More than 80 percent of North Americans now live in **metropolitan areas,** meaning cities of 50,000 or more plus their surrounding suburbs. In the nineteenth and early twentieth centuries, North American cities consisted of dense inner cities and less dense urban peripheries. Since the mid-twentieth century, central cities have been losing population and investment while urban peripheries have been growing rapidly. This development began in the early 1900s, when workers were drawn to the peripheries of crowded cities (the suburbs) by the opportunity to raise their families in single-family homes with large lots in secure and pleasant surroundings. Some continued to travel to the inner city to work, usually on streetcars, then increasingly in private cars. After World War II, factories and firms seeking less expensive land on which to develop, as well as better access to transport facilities (highways and airports), followed their workers to the suburbs. They have been joined there in the past two or three decades by the high-tech businesses of the new IT economy. Eventually, huge tracts of inner-city land that once held manufacturing industries were abandoned, and cities lost the tax revenue once generated there. These old industrial sites, often contaminated with chemicals and covered with obsolete structures, are called **brownfields** because their degraded condition poses obstacles to redevelopment.

Also left behind in the inner cities were the least skilled and least educated citizens, many of whom are members of racial and ethnic minorities. In the early 1990s, 70 of the 100 largest U.S. cities had white majorities, but by 2000 almost half of the largest U.S. cities had nonwhite majorities. The majority population in these cities is a mixture of African-Americans, Asians, Hispanics who are nonwhite, and other groups who identify themselves as nonwhite. As whites and some African-Americans moved to the suburbs and urban fringe, immigrants (especially Hispanics) took their places in the inner cities. Often those people who remain in the inner cities are in great need of the very services—medical, social support (including churches, synagogues, and mosques), and educational—that have moved with more affluent people to the suburbs.

Geographers James Fonseca and David Wong have observed that urban and suburban densities continually fluctuate. **Dense nodes** (small regions with extremely dense populations) may appear in urban peripheries where high-tech industries attract thousands of skilled workers. These workers occupy suburban high-rise office and apartment buildings as well as condominiums and single-family homes. Nodes can also reappear in old urban centers and then regenerate into larger dense settlements when businesses and new industries locate on old industrial brownfields that have been rehabilitated. In this way, portions of the old industrial Northeast (Baltimore, Maryland, and Hackensack, New Jersey, for example) are becoming densely populated again, not so much by population growth as by population redistribution. People are moving to the city to take newly available jobs and because of new preferences for urban living. This movement is called **New Urbanism,** and it is the result of increased appreciation for the many advantages of urban life: less dependence on automobiles, the balanced development of housing and jobs, reduced or eliminated commuting time, the cultural diversity of residents, chance meetings with friends and neighbors as one goes about daily business, café life, and less urban sprawl. **Gentrification** of old urban residential districts is increasingly common as primarily well-educated people invest substantial sums of money in renovating old houses and apartments, often displacing poor residents in the process. Gentrification contributes to the nodes of density found in inner cities, but these nodes may also follow streets and avenues extending out of the inner-city hub, in a pattern resembling spokes on a wheel. The rim of the wheel may be formed by the interstate beltways that girdle many American cities.

During the 1990s, a significant number of people moved from densely occupied cities to lightly settled areas beyond the urban fringe (farther out than the suburbs). Most of these new

In Megalopolis, there is a continuing battle between urban developers and those who want to preserve landscapes such as wetlands. In 1969, the Hackensack (New Jersey) Meadowlands Development Commission was charged with overseeing development while protecting nearly 20,000 acres of wetlands. Today railroad yards, toxic waste sites, highways, power plants, and other industrial developments have eaten up more than 11,000 of those wetland acres. [Melissa Farlow/National Geographic Image Collection.]

Here and there in every world city there are free spirits who occupy marginal land where they create unique spaces for themselves and others to enjoy. Here you see two such efforts; and in both cases, high-rise skylines are just out of sight. See if you can identify which is a place in Asia and which is in a Western country. Are there similarities between the two pictures? Can you judge the standard of living enjoyed by the inhabitants? [A, Mac Goodwin; B, Lydia Pulsipher.]

rural residents remain employed in urban areas; others work at home via computers or as skilled craftspeople. Many have chosen to lower their standard of living somewhat for the simpler pleasures of rural life. Very few have taken up farming, although some raise a small cash crop, such as strawberries or fresh produce, or raise a few animals as a sideline.

The growth of cities and the spread of their suburbs have resulted in two common urban phenomena: the megalopolis and urban sprawl. A **megalopolis** is created when several cities grow to the extent that their edges meet and coalesce. "Megalopolis" is the name chosen by French geographer Jean Gottman in the 1960s to describe the 500-mile-long (800-kilometer-long) band of urbanization that stretches across portions of ten states from Boston through New York City, Philadelphia, and Baltimore, terminating south of Washington, D.C. Other megalopolis formations in North America include the San Francisco Bay area, the region around Chicago, and the Los Angeles to San Diego area. **Urban sprawl** results when lightly populated, car-dependent suburbs encroach on agricultural land.

Urban Sprawl.   Among the many consequences of industrialization and urbanization in the economic core and elsewhere across North America is the invasion of farmland, forest, grassland, and desert by bulldozers preparing the way for suburban development: residences, malls, and discount outlets. Typically, real estate developers find rural land cheaper to develop than city properties, even when inner-city property is occupied only by defunct factories and abandoned houses. Rural land is taxed at a lower rate (at least at the beginning of the development process), is less polluted, and requires less preparation for development than inner-city land. However, as so vividly described by James Kunstler in his book *The Geography of Nowhere* (1993), huge tracts of suburban housing and mall development are gobbling up farm and wilderness areas around the peripheries of North American cities. Residents of these areas require automobiles to get through all aspects of daily life; anticipated time with family may instead be taken up with long commutes to jobs and schools, while boredom and isolation lead to domestic disintegration. (The video *Home Economics: A Documentary of Suburbia*, by Jenny Cool, illustrates these problems in the case of urban sprawl into Antelope Valley, east of Los Angeles.) Increasing air pollution and rising traffic deaths are two additional results of dependence on vehicles. And urban sprawl can be especially problematic for farmers. Even though fertile land is abundant in North America, some of the best land is located close to urban areas because cities and towns were often intentionally founded near rich farmland.

PERSONAL VIGNETTE   Mark Greene's family has been farming in Pittsford, New York (Figure 2.12), near the Canadian border, since 1812, but until 1997 the prospects of his 400-acre farm remaining in business for another generation looked dim. As land prices rose, so did property taxes, and the Greene family found that it could not meet its tax payments. New suburban homes were sprouting up on what had been neighboring farms, and the homeowners sometimes pushed local officials to halt normal farm practices, such as noisy nighttime harvesting or planting, spreading smelly manure, or importing bees to pollinate fruit trees. In 1997, however, Pittsford issued $410 million in bonds so that it could pay Greene and six other farmers for promises that they would not sell their 1200 acres to developers, but would continue to farm them.

Advocates of farmland preservation argue that farms provide more than food and fiber: they also provide environmental benefits, soul-soothing scenery, economic diversity, and especially, tax

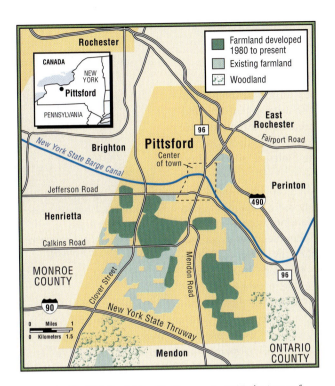

FIGURE 2.12 The Pittsford, New York, area. In 1997, the town of Pittsford, New York, decided that farms were such a positive influence on the community that it was willing to pay farmers to continue farming rather than sell their land to developers. [Adapted from *The New York Times* (March 20, 1997): D19.]

savings, as they require fewer government services than suburban residential development. Saving prime farmland from development also helps the nation as a whole by reducing the need for food imports and by protecting against volatility in food prices. In the United States, urban sprawl eats up 2 acres a minute—a million acres a year—including 400,000 acres of land that is especially well suited for high-quality specialty crops. Urban sprawl is now an issue in virtually all parts of North America; however, some geographic features, such as mountains and bodies of water, can force cities to concentrate rather than sprawl. Open flat or rolling spaces are most vulnerable. An urban sprawl index that ranks cities on the severity of the problem can be found at www.smartgrowthamerica.com.

Immigration

As noted earlier in this chapter, immigration has played a central role in populating both the United States and Canada. Most Americans have roots in some other part of the globe. Figure 2.13 shows the proportion of immigrants to the United States from different areas of the world and how that distribution has changed over time.

Once in North America, immigrants from various parts of the world have tended to settle in particular parts of the continent. Early arrivals were particularly attracted to rural agricultural locations that offered affordable land. Because immigrants from the British Isles and France occupied eastern Canada and the northeastern United States, later waves of immigrants from

Thinking Geographically: ON THE WEB

Europe tended to settle farther west. Scandinavians went to the upper Midwest—to Minnesota and the surrounding states. Many Germans settled in a wedge across the central Midwest, from Ohio to Missouri and in Nebraska and the Dakotas. Other Germans went to southern Appalachia and to south Texas. When early immigrants from Europe sent for friends and relatives, word of an attractive American location would spread to adjacent villages in their homeland. Immigrants from other parts of the world still do the same thing. As a result of this phenomenon, which geographers have dubbed **chain migration,** people from particular places tend to be concentrated in certain urban neighborhoods or rural communities. For example, shrimp fishermen from coastal Vietnam have, since the 1960s, congregated along the south Texas coast, where they can continue to ply their trade.

Canada's general immigration pattern is similar to that of the United States, but specific settlement patterns differ. The English, Irish, and Scots tended to settle in the Maritime Provinces bordering the Atlantic coast, around the Great Lakes, and on farms

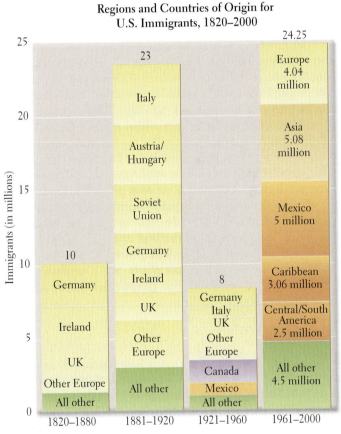

Regions and Countries of Origin for U.S. Immigrants, 1820–2000

FIGURE 2.13 National origins of U.S. immigrants, in 40-year periods, 1820–2000. [Adapted from Philip Martin and Elizabeth Midgley, "Immigration to the United States: Journey to an uncertain destination," *Population Bulletin* 49 (September 1994): 25; Martha Farnsworth Riche, "America's diversity and growth," *Population Bulletin* 55 (June 2000): 11; Table 2, "Immigration by region and selected country of last residence: Fiscal years 1820–1998," *1998 Statistical Yearbook of the Immigration and Naturalization Service,* p. 10; Table 2, "Immigration by region and selected country of last residence: Fiscal years 1961–2000," *2002 Statistical Yearbook of the Immigration and Naturalization Service:* http://www.bcis.gov/graphics/shared/aboutus/statistics/IMM02yrbk/IMMExcel/table2.xls.]

PRACTICING GEOGRAPHY   Where, Why, and So What?

Stanley D. Brunn
Department of Geography,
University of Kentucky

I conduct geographic research in ways similar to geographers in the past, by seeking to answer questions we have about places, landscapes, environments, and regions. Just as the early geographers approached their quest for learning by exploring unfamiliar territory, I discover and examine fascinating questions about my hometown and region, the United States, and the world. Geographers are formally instructed to become familiar with various methodologies; these include methods for collecting data during fieldwork, conducting social surveys, studying archival materials, and utilizing geographic and nongeographic databases for qualitative and quantitative analyses, including GIS. While for some geographers, research is "technique driven," I embark on research with a series of "where, why, and so what?" kinds of questions. The salient questions, databases, and methods may come from reading articles in professional journals, listening to presentations at meetings, or engaging in conversations with others. To answer the pertinent questions, I will often read a wide variety of sources, geographic and nongeographic, about the issue at hand or about a particular methodology or about some useful construct, theory, or model. I find it immensely valuable to collaborate with junior and senior colleagues, and those in other disciplines, on topics of mutual interest. It is at the interfaces of human and environmental geography and of the social sciences and humanities that I find much intellectual comfort and challenge.

Currently I am engaged in several research projects, some dealing with places and regions near my residence, others about the United States, and still others global. One project investigates the immobility of the elderly in rural Appalachian Kentucky. We live in a highly mobile society, yet most people don't move every 5 years. While the mobility literature heavily focuses on those that move and why, I think we need to find out why people choose not to move. Many residents of marginal and declining rural areas have a strong sense of place, which includes deep family ties, limited interaction with distant places, and a satisfaction with permanence. Survey information gathered from family reunions will help answer pertinent questions. To what extent the findings from eastern Kentucky are applicable elsewhere in the United States, I simply don't know.

The geopolitical impacts of 9/11 interest a number of geographers. One project I am pursuing examines how magazines published by major U.S. religions depicted Muslims and Arabs in the months before 9/11 and in the months after. Were there any articles, editorials, letters, photos, and maps about Muslims? What information did they convey and how was it presented? Were there any differences in the content of magazines produced for traditional mainline Protestant denominations and those focusing on evangelism? Or those serving Catholics and Jews? A content analysis of these magazines will provide some answers.

In January 2004 I attended an interdisciplinary conference on globalization at the University of California, Riverside. The major corporation mentioned at this event was Wal-Mart. I learned that there has not been much research conducted on this giant corporation by scholars in the social sciences and humanities, so I am currently editing a volume with contributors from North America, Asia, and Europe. The geographic, social, and cultural impacts of this Arkansas-based corporation on small towns and cities in the United States and cities in the above continents are significant, fascinating, and worthy of scholarly inquiry.

There are countless geographic problems that one can study in one's backyard, one's homeland, and at global levels and a variety of methods and techniques available to examine them. Geographic research is full of adventure for the inquiring mind and the geographer in all of us.

---

in the Prairie Provinces. The French settled primarily along the lower St. Lawrence River, and their descendants remain concentrated in and near Québec. Canada never had a slave-based economy, so Canadians with some African ancestry have roots among those who escaped to Canada from the United States during slavery or who later came voluntarily from the United States, the Caribbean, or Africa. They tend to live in the cities of Ontario and Québec. During most of the twentieth century, Canada encouraged immigration to fill up its empty lands. After farmers from the British Isles arrived, others from Eastern Europe, including Russia and Ukraine, settled mostly on the prairies. By the twenty-first century, as in the United States, the immigrants' places of origin had shifted away from Europe to Latin America,

the Caribbean, and Asia. By the 1990s, 55 percent of Canada's new immigrants came from Asia. Many of these Asian immigrants are educated and affluent, and they are particularly attracted to Canada's Pacific coast. Still, while North America used to have a pattern of regionally homogenous ethnic cities (for example, Boston was Italian and Irish; Pittsburgh had lots of Slavic people), North American cities are increasingly characterized by wide ethnic diversity.

Diversity and Immigration. In both Canada and the United States, the relatively recent influx of immigrants from areas other than Europe has challenged the long-held assumption that the dominant culture of North America would forever be derived from Europe, with but a few small minorities of Native Americans,

African-Americans, Hispanics, and Asians. Figure 2.14 shows the home countries of the 794,000 legal immigrants who came to New York City from 1990 to 1996. Like most new immigrants, they chose neighborhoods where earlier arrivals from their homelands had settled. The result is an ethnic patchwork across the boroughs of New York City.

Of the two countries, Canada seems to be the more receptive to newcomers. In recent decades, Canadians have accepted with equanimity many refugees from Asia and Africa. More than in the United States, emigrants to Canada are allowed, even encouraged, to retain their culture, use their native languages in schools, and maintain fairly strict (and usually voluntary) residential segregation. Toronto, Ontario, provides a model of how Canadians accommodate ethnicity (see the box "Ethnicity in Toronto" on page 87). In the United States, immigration and cultural diversity are topics of considerable public debate. The census of 2000

FIGURE 2.14 Ethnic neighborhoods of New York City. New York City's newest residents come from around the world, and they have created discrete neighborhoods within the city. [Adapted from *The New York Times Magazine* (September 17, 2000): 44.]

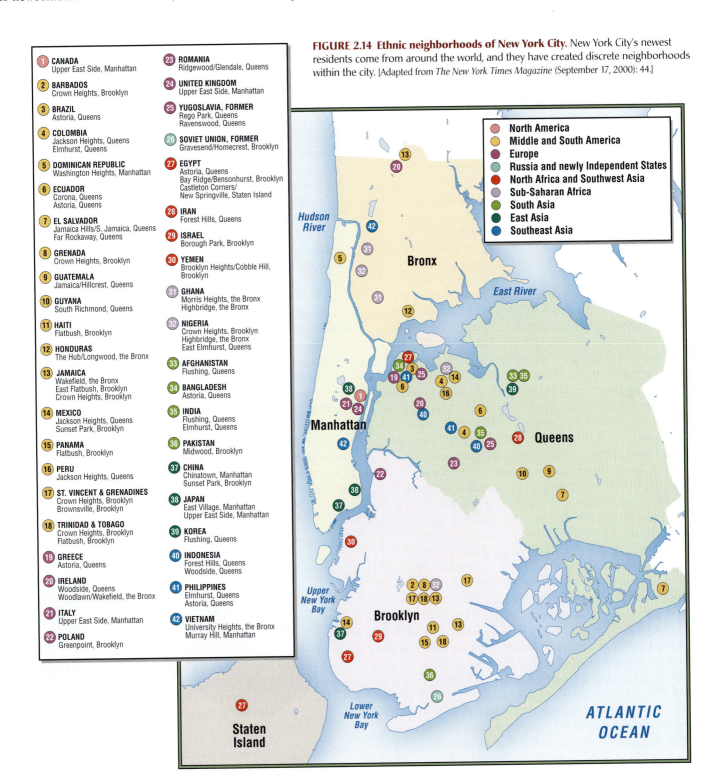

1 **CANADA**
Upper East Side, Manhattan

2 **BARBADOS**
Crown Heights, Brooklyn

3 **BRAZIL**
Astoria, Queens

4 **COLOMBIA**
Jackson Heights, Queens
Elmhurst, Queens

5 **DOMINICAN REPUBLIC**
Washington Heights, Manhattan

6 **ECUADOR**
Corona, Queens
Astoria, Queens

7 **EL SALVADOR**
Jamaica Hills/S. Jamaica, Queens
Far Rockaway, Queens

8 **GRENADA**
Crown Heights, Brooklyn

9 **GUATEMALA**
Jamaica/Hillcrest, Queens

10 **GUYANA**
South Richmond, Queens

11 **HAITI**
Flatbush, Brooklyn

12 **HONDURAS**
The Hub/Longwood, the Bronx

13 **JAMAICA**
Wakefield, the Bronx
East Flatbush, Brooklyn
Crown Heights, Brooklyn

14 **MEXICO**
Jackson Heights, Queens
Sunset Park, Brooklyn

15 **PANAMA**
Flatbush, Brooklyn

16 **PERU**
Jackson Heights, Queens

17 **ST. VINCENT & GRENADINES**
Crown Heights, Brooklyn
Brownsville, Brooklyn

18 **TRINIDAD & TOBAGO**
Crown Heights, Brooklyn
Flatbush, Brooklyn

19 **GREECE**
Astoria, Queens

20 **IRELAND**
Woodside, Queens
Woodlawn/Wakefield, the Bronx

21 **ITALY**
Upper East Side, Manhattan

22 **POLAND**
Greenpoint, Brooklyn

23 **ROMANIA**
Ridgewood/Glendale, Queens

24 **UNITED KINGDOM**
Upper East Side, Manhattan

25 **YUGOSLAVIA, FORMER**
Rego Park, Queens
Ravenswood, Queens

26 **SOVIET UNION, FORMER**
Gravesend/Homecrest, Brooklyn

27 **EGYPT**
Astoria, Queens
Bay Ridge/Bensonhurst, Brooklyn
Castleton Corners/
New Springville, Staten Island

28 **IRAN**
Forest Hills, Queens

29 **ISRAEL**
Borough Park, Brooklyn

30 **YEMEN**
Brooklyn Heights/Cobble Hill,
Brooklyn

31 **GHANA**
Morris Heights, the Bronx
Highbridge, the Bronx

32 **NIGERIA**
Crown Heights, Brooklyn
Highbridge, the Bronx
East Elmhurst, Queens

33 **AFGHANISTAN**
Flushing, Queens

34 **BANGLADESH**
Astoria, Queens

35 **INDIA**
Flushing, Queens
Elmhurst, Queens

36 **PAKISTAN**
Midwood, Brooklyn

37 **CHINA**
Chinatown, Manhattan
Sunset Park, Brooklyn

38 **JAPAN**
East Village, Manhattan
Upper East Side, Manhattan

39 **KOREA**
Flushing, Queens

40 **INDONESIA**
Forest Hills, Queens
Woodside, Queens

41 **PHILIPPINES**
Elmhurst, Queens
Astoria, Queens

42 **VIETNAM**
University Heights, the Bronx
Murray Hill, Manhattan

● North America
● Middle and South America
● Europe
● Russia and newly Independent States
● North Africa and Southwest Asia
● Sub-Saharan Africa
● South Asia
● East Asia
● Southeast Asia

Hudson River

Bronx

East River

Manhattan

Queens

Brooklyn

Upper New York Bay

Lower New York Bay

Staten Island

ATLANTIC OCEAN

revealed that whites are now a minority in half the nation's 100 largest cities and that immigrants from Middle and South America and Asia have moved into the small cities and towns of the Middle West, the South, and the West Coast. Some U.S. residents believe that future immigration should be controlled, but others feel that immigration is the strength of North America. Many church organizations take special pains to sponsor new immigrants, especially refugees.

Do New Immigrants Cost U.S. Taxpayers Too Much Money? Many people believe that immigrants cost U.S. taxpayers money because they use public services such as food stamps and welfare. Several studies have shown that in the long run, immigrants contribute more to the U.S. economy than they cost. Most immigrants start to work and pay taxes soon after their arrival. Even immigrants who draw on public services tend to do so only in the very first few years after they arrive. For example, a study in Los Angeles County in the mid-1990s showed that new immigrants made up 25 percent of the population (most were from Mexico and various parts of Asia). These 2.3 million people paid $4.3 billion in taxes in 1991–1992. Yet they received just $947 million in county services, less than one-fourth of the taxes they paid.

In fact, immigrants play important roles as taxpayers. Perhaps most noteworthy is their role in supporting the aged. As the U.S. population ages, more people will be drawing Social Security, and the money to pay their benefits will come out of the pockets of young and middle-aged workers. Yet the base of native-born young workers is shrinking because those people now reaching retirement age had relatively few children. The Social Security contributions of new immigrant workers, who tend to be young adults, will be essential to older U.S. residents, and will substantially alleviate the tax burden of native-born working adults. Interestingly, a 2004 study by the National Institutes of Health (NIH) reports that immigrants are healthier and live longer than native U.S. residents and hence are themselves less of a drain on the health-care and social service systems. NIH attributes this difference to a stronger work ethic, a healthier lifestyle, and more nutritious eating patterns than in U.S. society at large.

Do Immigrants Take Jobs Away from U.S. Citizens? Some people in the United States, particularly professional people, are occasionally in competition with highly trained immigrants for jobs. Usually, however, this competition is not a big problem, because there are not enough trained native-born people to fill these positions. The computer technology industry, for example, regularly recruits abroad in such places as India, where there is a surplus of high-tech workers (see the vignette at the opening of this chapter). Another developing trend is the outsourcing of high-tech jobs to overseas locations, as described earlier (page 77). Certainly, the least educated, least skilled American workers may find themselves competing with immigrants for jobs, and it is possible that the large pool of immigrant labor drives down the wages of some Americans. Usually, however, immigrants with little education fill the very lowest-paid service and agricultural jobs, which U.S. workers have already rejected. Although such immigrants are often willing workers, the wages they earn may be too low to support a decent way of life, making assimilation hard.

Are Too Many Immigrants Being Admitted to the United States? Several circumstances have caused some people to call for curtailing immigration into the United States. First, after a lull at mid-century, the immigration rate has picked up again (see Figure 2.13 on page 81). In the 1990s, the rate of immigration appeared to increase faster than ever before (although this increase may be partly a result of changes in census coverage): 10.6 million legal immigrants entered the United States, and 6.4 million children were born to them. Immigrants and their children accounted for 78 percent of population growth in the 1990s. At present rates, by the year 2050, the U.S. population will be over 400 million, nearly one-fourth larger than it would be if all immigration stopped now.

Second, illegal immigration reached unprecedented levels during the 1980s and 1990s. The Census Bureau estimates that the illegal immigrant population in the United States is about 8.7 million, with 200,000–400,000 more entering each year. While this trend is worrisome because illegal immigrants tend to lack skills, their numbers are uncontrolled, and they are not screened for criminal background or terrorist intentions, the vast majority do not engage in criminal activity. Since September of 2001, border controls have tightened significantly; hence many fewer illegal immigrants are entering. Furthermore, because they must hide from the authorities, illegal immigrants do not partake of public programs, so their drain on government services is negligible. In reality, illegal immigrants contribute to overall productivity by subsidizing the cost of products and services with their very low wages.

For some people, an added concern is that so many of the new immigrants are from Latin America and Asia. Figure 2.15 shows the projected change in the ethnic composition of the U.S. population from 1996 to 2050. Some Euro- and African-Americans worry that they will lose their sense of home in a United States of America where the largest minority is Hispanic, where Spanish is a prominent second language, and where another large minority is from Asia. New religious landscapes may emerge as well, with Hindu temples and Muslim mosques prominent in some places. Will the U.S. national identity be diluted by so many newcomers from different backgrounds? Or can a new national identity be constructed that will be more useful in the emerging global society?

U.S. Population by Race and Ethnicity, 1996 and 2050

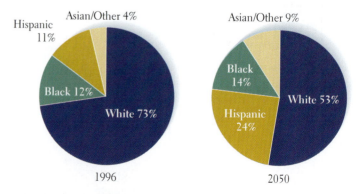

FIGURE 2.15 The changing U.S. ethnic composition. [Adapted from Jorge del Pinal and Audrey Singer, "Generations of diversity: Latinos in the United States," *Population Bulletin* 52 (October 1997): 14.]

Sheriff's deputy Michael Walsh works along the Arizona–Mexico border. Talking with a reporter about his recent encounter with two bereaved Mexican young men holding the body of their relative, Walsh said, *"Mostly you just find skeletons in the desert. This time there was a lot more emotion, family emotion. . . . Obviously they were close, they were crying. You have a name, you know that he had a brother, a cousin, a family in Mexico. . . ."* Deputy Walsh's voice trails off.

Matias Garcia, age 29, died after walking 32 miles through the desert. He was a Zapotec Indian who lived near Oaxaca (southern Mexico) with his wife and three children, as well as his parents, younger brothers, and several cousins. Matias's cash crop of chili peppers was ruined by a spring frost just as they were ripening, leaving him in debt. So he reluctantly decided to risk a trip to the United States to work in some vineyards where he had worked on and off since he was a teenager. From there he could send money home to his family to keep the house in repair and to send his children to school. But it takes money just to cross the border and it took until May for Matias, his younger brother, and a cousin to save the necessary amount.

May is one of the hottest and driest months in this part of the Americas, and since the NAFTA agreement of 1994, the border has been more carefully patrolled. This meant that Matias and the two younger men would attempt to cross the less patrolled but more perilous Arizona desert on a route to the United States that has come to be known as the "Devil's Highway." The men tried to avoid the worst of the heat by walking at night, but they didn't reach the highway on the Arizona side by dawn. Their water ran out, the sun turned especially hot, and Matias began having seizures. His brother and cousin carried him, desperately looking for the highway, but he died shortly before they found it and could flag down someone to call for help. That's where Deputy Walsh found the two men grieving over Matias's body.

Find out more about Matias Garcia, why he migrated, how he died, and what his death has meant back in his village. Watch the FRONTLINE/World video "Mexico, A Death in the Desert."

One answer to these questions is that new immigrants from non-European places are adding talent, entrepreneurial vigor, and creativity to the U.S. cultural mix. The new immigrants appear to have a work ethic every bit as strong as that of their European predecessors. For example, Harvard's Joint Center for Housing Studies reported in 2003 that 50 percent of all immigrants owned their own homes, and that immigrants were acquiring homes at a faster rate than native-born Americans. Home ownership is commonly thought to point toward responsible participation in civil society. Ethnic diversity is already contributing significantly to U.S. culture by enriching friendships, entertainment, business opportunities at home and abroad, and political perspectives. Furthermore, U.S. leadership in international affairs is enhanced by a less monolithic Euro-American presence at conferences and negotiating sessions.

Race and Ethnicity in North America

In any discussion of race and ethnicity in North America, it is important to refer to the material on race in Chapter 1, where it was noted that "race" has no biological meaning and is a socially constructed concept (see pages 20–22). That is, to greater and lesser extents around the world, people have decided to invest meaning in skin color and other superficial features. The same may be said for **ethnicity,** which is the cultural counterpart to race in that people may also ascribe overwhelming significance to cultural characteristics, such as religion or family structure or gender customs. Happily, because race and ethnicity are social constructs, people can also decide that those concepts, having no basis in scientific fact, are obsolete and irrelevant. Unfortunately, dispensing with racial and ethnic prejudice and its social consequences can take time. Race and ethnicity arise as issues in all regions of the world.

Although North America is ethnically and racially diverse, in this region the term "race" usually has been used in reference to deeply embedded discrimination against African-Americans. Prejudice and lack of opportunity have clearly hampered the ability of African-Americans to reach social and economic parity with Americans of other ethnic backgrounds. Unfortunately, despite the removal of legal barriers to equality, African-Americans as a group still experience lower life expectancies, higher infant mortality rates, lower levels of academic achievement, higher poverty rates, and greater unemployment than do other Americans, although Hispanics of all racial backgrounds have similar difficulties.

In recent years, there has been a tendency to include Hispanics, Asians, Native Americans, and Pacific Islanders in discussions of race and ethnicity because these other minority groups are growing as a proportion of the population. Figure 2.16 shows the distribution of ethnic minorities in the United States in 2000. In June 2001, African-Americans, whose numbers are growing only slowly, were overtaken by Hispanics as the largest minority group in the United States. Hispanics (of all racial identities), because of a higher birth rate and a high immigration rate, increased by 58 percent in the 1990s. Asians, who in 2000 made up only 3.6 percent of the U.S. population, nonetheless increased by 48 percent.

Figure 2.17 shows the discrepancies in income among Asians, whites, Native Americans, Hispanics, and African-Americans in the United States. There is considerable public debate over the causes and significance of these racial and ethnic differences in income. Over the past few decades, many African-Americans and Hispanics have joined the middle class and achieved success in the highest ranks of government and business. In particular, African-Caribbean emigrants to North America, especially Afro-Cubans and English-speaking West Indians, have been outstandingly successful, yet their roots in slavery and past discrimination are very similar to those of other African-Americans. Numerous surveys show that Americans of all backgrounds favor equal opportunities for groups that have experienced past discrimination. Nonetheless, many middle-class African-Americans, Hispanics, and Asians report experiencing both overt and covert discrimination that affects them economically.

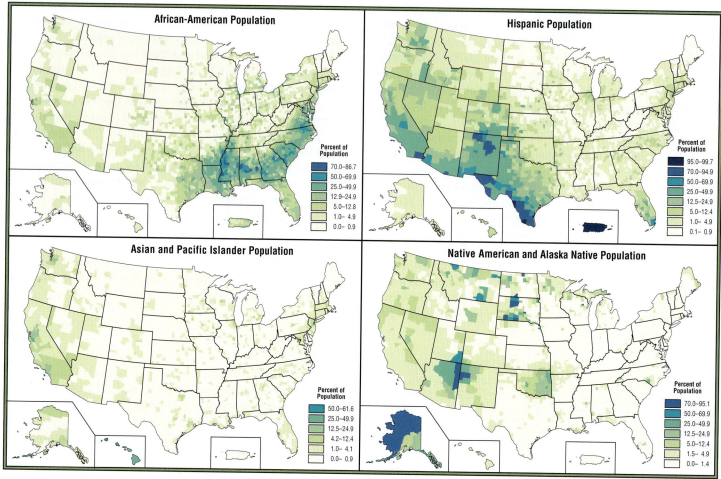

FIGURE 2.16 Minority population distribution, 2000. While representatives of the four major minority groups in the United States are to be found in nearly every part of the country, the distribution is uneven: African-Americans are concentrated in the Southeast, and Hispanic people in the Southwest. Asians are in greatest concentration in California, and Native American people are concentrated near the reservations onto which they were forced in the nineteenth century. Native Alaskans are most heavily concentrated in northern Alaska. Can you offer explanations for these patterns? [U.S. Bureau of the Census, Mapping Census 2000: The Geography of U.S. Diversity.]

Is there anything besides the prejudice of others that may be holding back African-Americans, Hispanics, and Native Americans? The experts disagree, but there is a consensus that poverty and low social status breed the perception among their victims that there is no hope for them and hence no point in trying to succeed. Furthermore, the poorest Americans, no matter what their backgrounds, have little opportunity to learn the basic skills for success at home or in school. The increasing number of poor single-parent families is also thought to play a role. For example, 50 years ago, African-American families were primarily two-parent units or extended families, and fathers and grandfathers influenced the development of their children. In the 1990s, nearly 75 percent of African-American children were born to single mothers, and often the fathers of those children were not active in their support and upbringing. The causes of this change are complex and not yet well understood. They are undoubtedly related to employment discrimination, especially against African-American men, and to a welfare system that issues funds overwhelmingly to single mothers. Thus fathers are encouraged to be

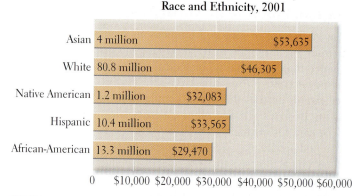

Median Household Income by Race and Ethnicity, 2001

Race/Ethnicity	Population	Median Income
Asian	4 million	$53,635
White	80.8 million	$46,305
Native American	1.2 million	$32,083
Hispanic	10.4 million	$33,565
African-American	13.3 million	$29,470

0    $10,000 $20,000 $30,000 $40,000 $50,000 $60,000

FIGURE 2.17 U.S. median household income by race and ethnicity, 2001. The "Asian" category includes Pacific Islanders and excludes Hispanics; "White" and "African-American" exclude Hispanics; and "Native American" includes Eskimos and Aleuts, but excludes Hispanics. [Adapted from Kelvin M. Pollard and William P. O'Hare, "America's racial and ethnic minorities," *Population Bulletin* 54 (September 1999): 22, 24–27, 36, updated according to U.S. Census Bureau Historical Income Tables-Households, Table H-16, see http://www.census.gov/hhes/income/histinc/h16.html.]

absent from their families, leaving the enormous responsibilities of both child rearing and breadwinning in the hands of often undereducated young mothers.

There is growing evidence that the pattern of poverty, underachievement, and poor health experienced by African-Americans in particular is part of a larger class problem, rather than simply a race problem. For example, there are many poor whites and Hispanic Americans of all races in the same predicament. They, too, are experiencing a growing number of single-parent families, absent fathers, low achievement, and poor health. The crux of the problem may well be the growing economic disparity and spatial separation between poor and rich Americans. In both the United States and Canada, the increasingly prosperous middle class, of whatever race or ethnicity, has fled to the suburbs. As geographer John B. Strait found in his study of the concentration of poverty in particular neighborhoods of Atlanta, Georgia, the poor are increasingly isolated in very poor neighborhoods; hence rich and poor rarely encounter each other in daily life. This isolation makes it difficult for the affluent of all racial and ethnic backgrounds to see how they might help individuals move out of poverty, and the very poor rarely have the chance to witness models of success.

Evidence for this class-based explanation is the increasing range of opportunity that exists for middle- and upper-class African-Americans, Hispanics, and Asian-Americans with education and skills. Privileged Americans of all racial backgrounds are beginning to share middle- and upper-class neighborhoods, workplaces, places of worship, and marriages, with markedly decreasing attention paid to matters of skin color. Meanwhile, the poor of all races see the material evidence of the success of others all around them, but have little inkling of the life choices that have made that success possible.

Nonetheless, it would be wrong to conclude that blocks to success for middle-class minority Americans have been completely removed. Robert Fallows, who studies poverty amid wealth in North America, writes that in the new highly paid IT professions, Hispanics and African-Americans are conspicuously absent, despite a broadly multicultural (but overwhelmingly male) workplace. People in the information technology world often have office mates with black, brown, or white skin tones, but they tend to be highly trained immigrants from such places as China, Malaysia, Poland, Colombia, and India, rather than Hispanics and African-Americans.

AT THE GLOBAL SCALE   Ethnicity in Toronto

The lines dividing Toronto's ethnic neighborhoods are neither strict nor exclusionary, but when a city bus moves through town, Italians get off at particular stops, Chinese at others, and Greeks at yet others. Many of the neighborhoods have been rehabilitated and upgraded, or gentrified, to attract the middle-class children and grandchildren of the much poorer laborers who arrived earlier in the twentieth century. There are about 400,000 Torontonians of Italian descent, many of whom live in Corso Italia (the Little Italy of Toronto), an area of shops, sidewalk cafés, and elegant homes and apartments. The Chinese community of 350,000 is the fastest growing. Many Chinese who arrived in the 1990s are prosperous businesspeople from Hong Kong who feared the consequences of China's resumption of control of Hong Kong in June 1997. Koreans own many of the fresh produce and grocery stores throughout the city, and Koreatown is home to herb shops and import emporiums. The Greek community revolves around the National Bank of Greece and the Greek Orthodox Church. Jews and people from the Caribbean are sprinkled in small enclaves throughout the city, while Polish people often settle together and specialize in owning bakeries and butcher shops. All in all, there are said to be at least 80 different ethnic neighborhoods in the city.

[*Source:* Adapted from Richard Conniff, "Toronto," *National Geographic* 189(6) (June 1996): 120–139.]

Throughout Toronto, signs on city streets are written in English, the lingua franca, as well as in the predominant language of the neighborhood—in this case, Vietnamese. [David R. Frazier/Tony Stone Images.]

Religion

Because so many early immigrants to North America were Christian in their home countries, Christianity is presently the predominant religious affiliation claimed by North Americans. Nonetheless, virtually every medium-sized city has at least one synagogue, mosque, and Buddhist temple; legally, one is free to accept whatever creed one likes, or none at all. In some localities, adherents of Islam, Judaism, or Buddhism are numerous enough to constitute a prominent cultural influence. Religion in North America has some interesting and unusual geographic patterns (Figure 2.18). Two of these are the distributions of particular faiths and the regional differences in the role of religion in daily life.

There are many versions of Christianity in North America, and their distributions are closely linked to the settlement patterns of the immigrants who brought them. The map in Figure 2.18 shows the dominance of Roman Catholicism in regions where Hispanic, French, Irish, and Italian people have settled—in southern Louisiana, the Southwest, and the far Northeast in the United States and in Québec and other parts of Canada. Lutheranism is dominant where Scandinavian people have settled, primarily in Minnesota and the eastern Dakotas. Baptists dominate the religious landscapes of the South. Generally speaking, Southern Baptists tend to take the teachings of the Bible more literally than many other Christian denominations, so this part of the United States has come to be called the Bible Belt. Christianity—especially the Baptist version—is such an important part of community life in the South that newcomers to the region are asked almost daily what church they attend. (This question is rarely asked in most other parts of the continent.)

Those who reply "none" or something other than "Baptist" may be left with the feeling that they have disappointed the questioner. In all parts of the United States, Mormons and Pentecostals are among the fastest-growing Christian groups, perhaps because they offer firm rules for living to those who have recently immigrated or have moved to cities and cut ties with older Christian traditions.

The proper relationship of religion and politics has long been a controversial issue in the United States, in large part because the framers of its Constitution supported the idea that church and state should remain separate. Even now, most people try to avoid mixing the two in daily life and conversation. In the past decade or two, however, some of the more conservative Protestants have pushed for prayer in the public schools, posting Christian religious symbols in public places, teaching the biblical version of creation instead of evolution, banning abortion as a method of birth control, and keeping women primarily in the role of home-makers. Although the policies of conservative Christians have met with the most success in the South, their goals are shared by people scattered across the country.

New immigrants have brought their own faiths and belief systems, and they are contributing to the debate about religion and public life. Muslims, for example, are challenging the policies of public schools and local governments regarding public holidays, food served in school cafeterias, and the content of textbooks on such topics as sex education and the religious heritage of America. The long-term outcome of these conflicts in American religious and political life is not yet apparent, but recent national surveys indicate that a substantial majority of Americans favor the continued separation of church and state and favor personal choice in reproductive and mating matters.

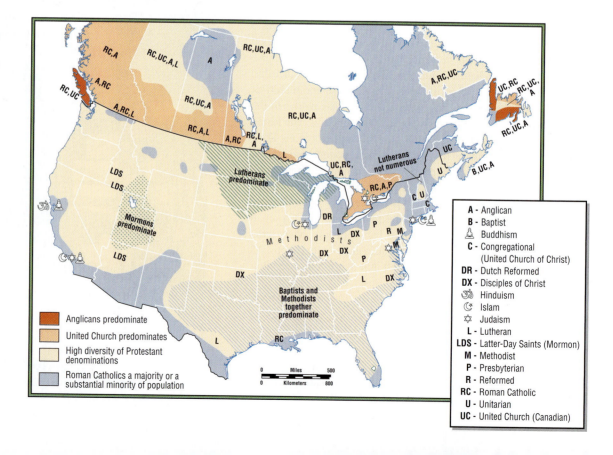

FIGURE 2.18 Religious affiliations across North America. [Adapted from Jerome Fellmann, Arthur Getis, and Judith Getis, *Human Geography* (Dubuque, Iowa: Brown & Benchmark, 1997), p. 164.]

Gender and the American Family

The family has repeatedly been identified as the institution most in need of shoring up in today's fast-changing and ever more impersonal world. A century ago, most Americans lived with members of several generations in extended families. Family members pooled their incomes and shared chores. Aunts, uncles, cousins, siblings, and grandparents were almost as likely to provide daily care for a child as its mother and father. Humans have long preferred to reside in communities of kinfolk, but in industrialized countries the extended family was replaced by the **nuclear family.** The nuclear family, consisting of a married father and mother and their children, is actually a rather recent invention of the industrial age.

Beginning after World War I, and especially after World War II, many young people left their large kin groups on the farm and migrated to distant cities, where they established new nuclear families. Soon suburbia, with its many similar single-family homes, seemed to provide the perfect domestic space for the emerging nuclear family. This small, compact family suited industry and business, too, because it had no firm ties to other relatives and hence was portable. An industry could draw on a large labor pool that was willing and able to move. Many North Americans born since 1950 moved as often as five to ten times before reaching adulthood. The grandparents, aunts, and uncles who were left behind missed helping to raise the younger generation, and they had no one to look after them in old age. Institutional care for the elderly proliferated.

In the 1970s, the whole system began to come apart. It was a hardship to move so often. Suburban sprawl meant onerous commutes to jobs for men and long, lonely days at home for women. Women began to want their own careers, and rising consumption patterns made their incomes increasingly useful to family economies. By the 1980s, 70 percent of the females born between 1947 and 1964 were in the workforce, compared with 30 percent of

Many North American suburbs, like this one in Maryland, contain dozens of houses built according to just two or three plans. Paint color, trim, the position of the house on the lot, and landscaping detail are used to give some variety to the neighborhood. How is (or was) your community similar to or different from this one? [Frank Fisher/Gamma-Liaison.]

their mothers' generation. But once employed, women could not easily move to a new location with an upwardly mobile husband. Nor could working women manage all of the family's housework and child care as well as a job. Some married men began to handle part of the household management and child care, and the demand for commercial child care grew sharply. With husband and wife both spending long hours at work, many people of both sexes found that their social life increasingly revolved around work, while family life receded in importance. With kinfolk no longer around to shore up the marital bond, and with the new possibility of self-support available to women in unhappy marriages, divorce rates rose drastically. By 2001, the U.S. divorce rate had reached nearly 50 percent; in Canada, the divorce rate was about 40 percent.

By the late 1980s, the nuclear family, which had been dominant for only a few decades, was less and less representative of the American family. As Figure 2.19 shows, the percentage of nuclear families continually shrank. According to the 2000 census, only 24 percent of U.S. households were nuclear families; the most common household type—almost 29 percent—was a married couple *without* children, while the fastest-growing household type, a single person living alone, had reached 25 percent. More Americans than ever before are living alone; the majority are over the age of 45 and were once part of a nuclear family that dissolved due to divorce or death. In 2000, U.S. Census Bureau analysts reported that there is no longer a typical American household, only an increasing diversity of forms.

Some of these new forms do not necessarily represent an improvement over the nuclear family. By the late 1990s, one-fourth of U.S. children were born to unmarried women, and more than one-fourth lived in single-parent households. Although most single parents do a good job of rearing their children, the responsibilities can be overwhelming, especially because single-

Household Composition, 1970–2000 (percent)

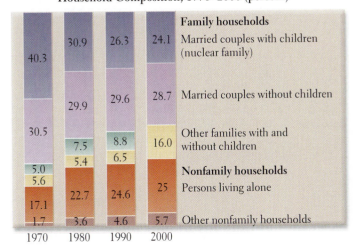

				Family households
40.3	30.9	26.3	24.1	Married couples with children (nuclear family)
30.5	29.9	29.6	28.7	Married couples without children
5.0	7.5	8.8	16.0	Other families with and without children
5.6	5.4	6.5		**Nonfamily households**
17.1	22.7	24.6	25	Persons living alone
1.7	3.6	4.6	5.7	Other nonfamily households
1970	1980	1990	2000	

FIGURE 2.19 Changing U.S. family compositions, 1970–2000.
[Adapted from *U.S. Census Bureau Current Population Survey, America's Families and Living Arrangements* (Washington, D.C.: U.S. Bureau of the Census, June 2001), see http://www.census.gov/prod/2001pubs/p20–537.pdf.]

parent families tend to be hampered by economic hardship and lack of education. The vast majority of single-parent households are headed by young women, and the incomes of single female-headed households are, on average, 33 percent lower than those of families headed by single males. In 2000, the U.S. Census Bureau reported that children are more likely than adults to be poor: in that year; 20 percent (one in five) of U.S. children lived in poverty, whereas only 11.4 percent of people aged 18 to 64 did. In Canada, just 14.7 percent of children were poor; in Sweden, by comparison, 2.4 percent of children lived in poverty; in Ireland, 12.4 percent; in Spain, 12.4 percent; and in Poland, 12.7 percent.

Aging

The populations of Canada and the United States are aging, meaning that the numbers of people over 65 are growing more rapidly than the numbers of those under 15. In most societies, people between youth and old age (15 and 65 years) must support those who are younger or older. The age structure of a population tells us how great their burden is likely to be. Wealthier, developed societies tend to have lower birth rates, and their members tend to have longer life spans. Thus, over time, the proportion of those younger than 15 becomes smaller, while the proportion of those over 65 becomes larger. During the twentieth century, the number of older Americans grew rapidly. In 1900, 1 in 25 individuals was over the age of 65; by 1994, 1 in 8 was. By 2050, when most of the current readers of this book will be over 65, 1 in 5 Americans will be elderly.

In North America, the number of elderly people will shoot up especially fast between the years 2010 and 2030. This predicted spurt is the result of a marked jump in birth rate that took place in the years after World War II, from 1947 to 1964. The so-called **baby boomers** born in those years constitute the largest age group in North America, as indicated by the wide band through the middle of the population pyramids shown in Figure 2.20. In the years from 2010 to 2030, this group will reach age 65 and retire. During this period, payments from Social Security and private pensions will be high, and medical costs

will leap upward. The boomers had fewer children than their parents had, so there will be fewer people of working age to pay the taxes and pension fund contributions necessary to meet these costs (although, as mentioned earlier, the contributions of new young adult immigrants will help). Moreover, people in this smaller population will have fewer brothers and sisters with whom to share responsibility for the daily personal care and companionship needed by their elderly kin. Most families will not be able to afford assisted living and residential care, which already (in 2004) costs from $30,000 to $60,000 per year for one person. We might expect that once the boomers begin to retire in large numbers, the makeup of many households and the spaces in which they live will reflect people's efforts to find humane and economical means to care for elderly family members at home.

The maps in Figure 2.21 show where the elderly were concentrated in the 1990s and how these patterns are expected to look by 2020. In rural areas (through the Middle West and Northeast), the young have departed in large numbers for other parts of the country, leaving aging parents behind. In other areas—Florida, especially, but also all across the southern United States—the elderly are the new residents, attracted by the warm, pleasant climate of the Sunbelt and gated communities that appeal to the security-conscious wealthy elderly. California has the largest number of elderly people, but they make up only a small percentage of the total population because there are so many young people in that state.

The problems presented by aging populations reveal a paradox. On the one hand, it is widely agreed that population growth should be reduced to lessen the environmental impact of human life on earth, especially that of the societies that consume the most. On the other hand, slower population growth means that as more of earth's citizens grow old, there will be fewer young, working-age people to keep the economy going and to provide the financial and physical help the elderly require, unless immigrants can be persuaded to assist.

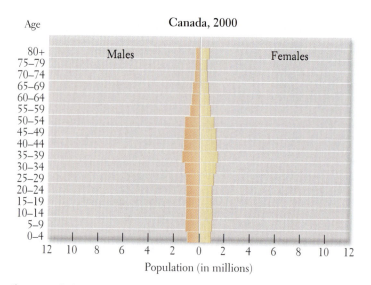

FIGURE 2.20 Population pyramids for the United States and Canada. The "baby boomers," born between 1947 and 1964, constitute the largest age group in North America, as indicated by the wide band through the middle of these population pyramids. [Adapted from "Population pyramids of the United States, 2000" and "Population pyramids of Canada, 2000" (Washington, D.C.: U.S. Bureau of the Census), International Data Base, at http://www.census.gov/ipc/www/idbpyr.html.]

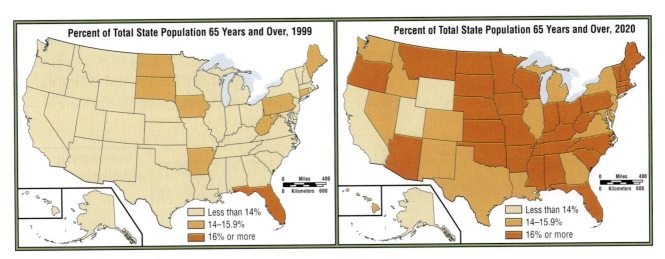

FIGURE 2.21 Changing distribution of the elderly in the United States. The proportion of elderly people in the population is expected to increase across the country in the coming decades. In 1999, Florida had the highest proportion of elderly people, but by 2020 more than half the states will have similar proportions of 16 percent or more. The actual numbers of elderly in California and Florida are masked on these maps by the fact that these states also have many young people. How might daily life for aging Americans differ between a retirement community in Arizona and a small, depopulated town in South Dakota? Who cares for, or pays for the care of, very old retirement community residents once they can no longer live on their own? [Data from U.S. Bureau of the Census, 2000 census.]

ENVIRONMENTAL ISSUES

All human activity has environmental consequences, but some activities of North Americans, especially those in the United States, are particularly damaging. The production of hazardous waste is an example. **Hazardous waste** is produced by nuclear power generation and weapons manufacture, by mineral mining and drilling, by waste incinerators, and by industry in general, which produces 80 percent of all liquid hazardous waste. Many other harmful wastes are not yet universally defined or regulated by law, such as chemical wastes from small businesses and private homes, including oven cleaners, drain, toilet, and window cleaners, weed killers, and gasoline. With a population of roughly 290 million, each year the United States generates five times the amount of hazardous waste generated by the entire European Union (with a population of more than 300 million). Canada generates much less hazardous waste per capita each year than the United States, although its citizens generate several times the global average.

The disposal of hazardous waste within the United States has a geographic pattern. Sociologist Robert D. Bullard has shown that a disproportionate amount of hazardous waste is disposed of in the South and in locations inhabited by poor Native American, African-American, and Hispanic people. Bullard writes that "nationally, 60 percent of African-Americans and 50 percent of Hispanics live in communities with at least one uncontrolled toxic-waste site." Military bases are also major sources of hazardous waste, generating more wastes than the top five U.S. chemical companies combined.

Dumping of hazardous waste is only one way in which human actions can damage the environment. Consider the many consequences of building and living in a typical North American suburban home. Habitat for wild creatures and plants is lost on the site of the home and its lawn and on the landfill where its solid waste is disposed of. Lawn fertilizer runoff pollutes nearby streams. Resources for building and maintaining the home and lawn, such as wood and petroleum products, may be drawn from very distant places, where their extraction causes environmental damage. The rest of this section focuses on a few environmental consequences of the North American lifestyle: air pollution, loss of habitat for plants and animals, and depletion of water resources.

Air Pollution

With only 5 percent of the world's population, North America produces 26 percent of the greenhouse gases released globally by human activity. This large share can be traced to North America's high consumption of fossil fuels, which is in turn related to its oil-dependent industrial and agricultural processes, the heating of its homes and offices, and its dependency on automobiles (see the discussion of U.S. contributions to global warming in Chapter 1, pages 46–48).

Smog is a combination of industrial air pollution and car exhaust that frequently hovers over many North American cities, causing a variety of health problems for their inhabitants. In Los Angeles the intensity of the smog is due in large part to the city's warm land temperatures and West Coast seaside location, which often results in a **thermal inversion**—a warm mass of stagnant air that is trapped beneath cooler air blowing in off the ocean. The inversion is held in place, often for days, by the mountains that surround the city.

The burning of fossil fuels also contributes to **acid rain** (and acid snow) because it releases sulfur dioxide and nitrogen oxides into the air. Acid rain is created when these gases dissolve in falling rain and make it acidic. Acid rain can kill certain forest trees and, when concentrated in lakes and streams and snow cover, can destroy fish and wildlife. The eastern half of the continent, from the Gulf Coast to Newfoundland and including the entire eastern seaboard, is greatly affected by acid rain. The United States, with its larger population and more extensive industry, is responsible for the vast majority of acid rain. Due to continental weather patterns, however, the area most affected by acid rain is along the U.S.–Canadian border (Figure 2.22).

Loss of Habitat for Plants and Animals

Throughout North America, many plants and animals are in danger of becoming extinct because humans are overusing them or

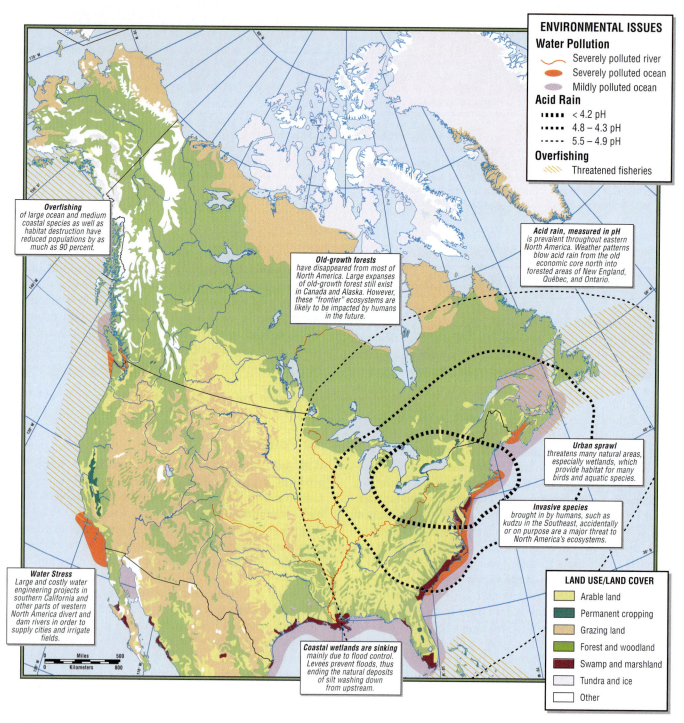

FIGURE 2.22 Environmental issues: North America. The environmental consequences of modern life in North America are varied and overlapping. Coastal zones are especially affected by pollution and other negative environmental impacts generated by urban and agricultural activities. Note that one result of air pollution is acid rain. Acidity is measured in terms of pH, with 7 representing neutral. A pH less than 7 is acidic: a pH of 5 is approximately the acidity of black coffee; a pH of 4, that of tomatoes. As you can see, eastern North America experiences acid rain with a pH of 5.5 to 4.2, the approximate acidity of tomato juice—sufficient to kill fish, snails, and crustaceans.

destroying the environments in which they live. Of all countries in the world, as of 2003, the United States had both the *largest number* (20,892) of well-understood plant and animal species and the *highest percentage* (33 percent) threatened by extinction. Earlier in North American history, millions of acres of forests and grasslands were cleared to make way for farms. Now the last bits of natural land near cities are disappearing to make way for residential developments, highways, golf courses, office complexes, and shopping centers. The destruction of wetlands is especially significant because they are important reproductive zones for many

bird and aquatic species. Overuse of ocean resources has reached such a critical state that scientists now think it is likely that humans will force even the great whales into extinction.

Water Resource Depletion

Many people who live in the humid eastern part of North America find it difficult to believe that water is becoming scarce. But as populations grow and per capita water usage increases, it has become necessary to look farther and farther afield for sufficient

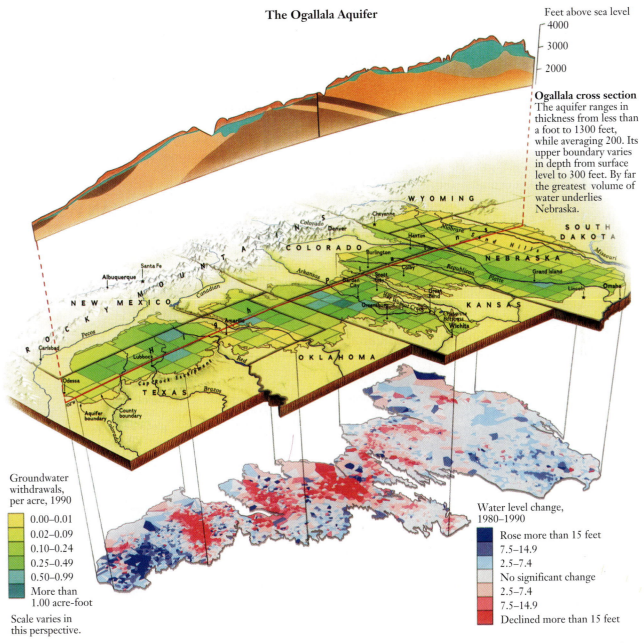

The Ogallala Aquifer

Ogallala cross section The aquifer ranges in thickness from less than a foot to 1300 feet, while averaging 200. Its upper boundary varies in depth from surface level to 300 feet. By far the greatest volume of water underlies Nebraska.

Groundwater withdrawals, per acre, 1990

- 0.00–0.01
- 0.02–0.09
- 0.10–0.24
- 0.25–0.49
- 0.50–0.99
- More than 1.00 acre-foot

Scale varies in this perspective.

Water level change, 1980–1990

- Rose more than 15 feet
- 7.5–14.9
- 2.5–7.4
- No significant change
- 2.5–7.4
- 7.5–14.9
- Declined more than 15 feet

FIGURE 2.23 The Ogallala aquifer. From the 1940s through the 1980s, the aquifer lost an average of 10 feet (3 meters) of water overall, and more than 100 feet (30 meters) of water in some parts of Texas. During the 1980s the decline was less due to abundant rain and snow, but this is an area where the climate fluctuates from moderately moist to very dry. A drought began in mid-1992 and continued until late 1996. Large agribusiness firms pumped water from the aquifer for irrigation to supplement precipitation.

As a result, water levels declined an average of 1.35 feet per year during the mid-1990s. [Graphic adapted from *National Geographic* (March 1993): 84–85, with supplemental information from High Plains Underground Water Conservation District 1, Lubbock, Texas: http://www.hpwd.com/ogallala/ogallala.asp; Erin O'Brian, Biological and Agricultural Engineering, National Science Foundation Research Experience for Undergraduates, Kansas State University, 2001.]

water resources. New York City, for example, obtains most of its water from the distant Adirondack Mountains in upstate New York. Atlanta, Georgia, with over 4 million residents, is located in the usually moist Southeast, but absorbs water so insatiably that downstream users in Alabama and Florida have sued Atlanta to get their rights to water that no longer makes it into their rivers. In an interview with the Associated Press, Jim Campbell, an attorney who represents Alabama in talks with Georgia and Florida over allocation of water to all three states, noted, "This whole subject of finite water supply is something that's new to the Southeast." Disputes over water rights are difficult to resolve; for a look at how this topic is being addressed in other regions of the world, see discussions of water and water rights in Chapters 4, 5, 6, 8, 9, and 10.

In North America, water becomes increasingly precious the farther west one goes. On the Great Plains, rainfall in any given year may be too sparse to support healthy crops and animals. To make farming more secure and predictable, taxpayers across the continent have subsidized the building of stock tanks and reservoirs. Irrigation has also increased in recent decades, drawing on "fossil water" that has been stored over the millennia in natural underground reservoirs called **aquifers.** The **Ogallala aquifer** (Figure 2.23, page 93) underlying the Great Plains is the largest such body of water in North America. In parts of the aquifer, water is being pumped out at rates that exceed natural replenishment by 10 to 40 times. Although local precipitation plays a small role in recharging the aquifer, it received most of its water from highland streams to the west. Because of geologic change, access to much of this stream flow has been cut off. Now it may take as long as 35 years for water from the highlands to reach the Ogallala.

Irrigation is also important in Southern California, which supplies much of the fresh fruit and vegetables consumed in the United States, but at a high cost to taxpayers. Billions of dollars of federal and state funds have paid for massive water engineering projects that bring in enormous quantities of water from hundreds of miles away in Washington and Oregon, northern California, Colorado, and Arizona, at times pumping it up and over mountain ranges. Irrigation in Southern California also deprives Mexico of this much-needed resource. The mouth of the Colorado River, which used to be navigable, is now dry and sandy, and a mere trickle gets to Mexico, which also would like to use more water for irrigation.

Increasingly, citizens in western North America are recognizing that the use of scarce water for irrigated agriculture is unsustainable and uneconomical. Conflicts over moving water from wet regions to dry ones, or from sparsely inhabited to urban areas, have halted some new water projects, raised awareness of the need to conserve water, and resulted in some reduction in the use of water for irrigation. Nevertheless, government subsidies keep water artificially low in price, and past successes in harnessing new water supplies provide a disincentive to change.

MEASURES OF HUMAN WELL-BEING

Both Canada and the United States consistently rank high on global scales of well-being, yet a comparison of these two wealthy countries makes it clear why old ways of making such compar-

TABLE 2.2   Human well-being rankings of Canada, the United States, and other developed countries, 2003

Country	GDP per capita, adjusted for PPP[a] in 2001 $U.S. (GDP ranking among 175 countries, ranked from highest) (2)	Human Development Index (HDI) global rankings, 2003 (3)	Gender Empowerment Measure (GEM) global rankings, 2003 (4)	Gender Development Index (GDI) global rankings, 2003 (5)	Women's average GDP per capita as a percentage of men's average GDP per capita (PPP 2001 $U.S.) (6)
United States	34,342 (2)	7	10	5	62
Canada	27,130 (9)	8	9	6	63
Japan	25,130 (14)	9	44	13	45
Germany	25,350 (13)	18	8	15	57
Sweden	24,180 (18)	3	3	3	68
France	23,990 (20)	17	(31 in 1996)[b]	17	(63 in 1998)[b]
United Arab Emirates	20,530 (23)	48	65	49	21
Australia	25,370 (12)	4	11	4	70
Mexico	8,430 (58)	55	42	52	38
World	7,376	—	—	—	(51 in 1998)[b]

[a]PPP = purchasing power parity.   [b]Data not available in 2003.

Source: United Nations Human Development Report 2003.

isons are misleading. If we looked only at gross domestic product (GDP) per capita for the two countries, we would see that Canada's GDP per capita is U.S. $27,130 (ninth highest in the world) and that that of the United States is U.S. $34,342 (second highest). From these numbers, we might conclude that Canadians are doing significantly less well than people in the United States. But remember that GDP per capita ignores all measures of well-being other than income and is only an average for the entire country. In fact, despite its significantly lower GDP per capita, Canada ranks nearly the same as the United States on many indicators of well-being and, as we have seen, exceeds the United States in gender empowerment, life expectancy, and infant and maternal survival rates (see pages 77–78).

Table 2.2 compares Canada, the United States, and other selected developed countries in terms of GDP per capita and several other measures developed by the United Nations (see Chapter 1, pages 36–37). On all the UN measures included here,

Canada is neck and neck with the United States. The United Nations Human Development Index (HDI; column 3 in the table), combines three components—life expectancy at birth, educational attainment, and adjusted real income—to rank 175 countries on how well their citizens achieve basic human capabilities. On this scale, Canada ranks eighth and the United States seventh. On the Gender Empowerment Measure (GEM, column 4), which measures opportunities for women to participate as equals with men, Canada ranks ninth (Iceland is first in this category), and the United States tenth. On the Gender Development Index (GDI, column 5), which measures women's access to health care, income, and education, Canada ranks sixth in the world and the United States fifth. Finally, in both the United States and Canada, women, on average, are poorer than men (column 6), earning just 62 to 63 percent of what men earn, despite the fact that younger women tend to have more education than their male age-mates (see page 78).

Quick Review

1. List the physical and human geography similarities and differences between Canada and the United States.

2. Is the goal of NAFTA to increase trade among the United States, Canada, and Mexico, or to balance trade?

3. Which country, Canada or the United States, spends more per capita on health care?

4. Describe how patterns of immigration to the United States have ebbed and flowed since 1820.

5. Give several reasons why developers are often attracted to rural land for their projects.

6. Which characteristic of the nuclear family was especially suited to business after the industrial revolution?

III   SUBREGIONS OF NORTH AMERICA

When geographers try to understand patterns of human geography in a region as large and varied as North America, they usually impose some sort of subregional order on the whole (Figure 2.24). In this section we divide North America into subregions, and then sketch in the features that give each subregion its "character of place." Keep in mind, however, that, as we observed in Chapter 1, geographers often have trouble reaching consensus on just where regional boundaries should be drawn.

NEW ENGLAND AND CANADA'S ATLANTIC PROVINCES

Among the earliest parts of North America to be settled by Europeans were New England and Canada's Atlantic Provinces (Figure 2.25, page 97). Of all the regions of North America, this one may maintain the strongest connection with the past, and it holds a certain cultural prestige and reputation as North America's cultural hearth. Philosophically, New England laid the foundation for religious freedom in North America with a strong conviction that eventually became part of the U.S. Constitution: that

there should be no established church and no requirement that public officials hold any particular religious beliefs. New England is especially noted as the source of the classic American village style, arranged around a town square, and of such early American house styles as Cape Cod and other wood-framed houses. Many domestic material culture styles also originated in New England; in all classes of American homes, popular interior furnishings that are copies of New England-designed furniture and accessories, such as lamps, candlesticks, dishes, silverware, and textile patterns, can be found.

Economically, northern New England and the Atlantic Provinces of Canada are relatively poor. They still depend on fishing, which has become severely depressed, and on timber, vegetable crops, dairying, and other extractive industries. In recent decades, southern New England has used its celebrated "Yankee ingenuity" to reinvent itself as an information technology center reliant on the intellectual skills of those attracted by its many universities. Although New England has long been the site of several large cities (Boston, Providence, Hartford), it is now becoming both more urban and more ethnically diverse as it shifts its focus, at least partially, to the new IT economy. Alongside the new economy, tourism remains of enduring importance to all of

Thinking Geographically: ON THE WEB

FIGURE 2.24 Subregions of North America. There are many schemes for dividing North America into subregions. The scheme we use here is partly indebted to Joel Garreau and his successful book, *The Nine Nations of North America* (1981), in which he noted that state and provincial boundaries are particularly useless for sketching the regional characteristics of the continent. He proposed a set of regions that cut not only across state boundaries but also across national boundaries, to include parts of Mexico and the Caribbean. [Adapted from Joel Garreau, *The Nine Nations of North America* (Boston: Houghton Mifflin, 1981).]

New England as Americans from all regions make pilgrimages to sites of early settlement and important historic events.

Many of the continuities between the present and past economies of New England and the Atlantic Provinces derive from the region's geography. During the last ice age, glaciers scraped away much of the topsoil, leaving behind picturesque mountains with rounded, bare summits—the source of some of the finest building stone in North America. But without much topsoil, the land is only marginally productive. Although many farmers settled here, they struggled to survive. Today, although much of the landscape is rural and looks agricultural, farming is not particularly productive, so many areas try to capitalize on their rural and village ambiance and historic heritage by enticing tourists, especially winter skiers, as well as retirees who will build homes and stay for a decade or more.

After being cleared in the days of early settlement, New England's evergreen and hardwood deciduous forests have slowly returned to fill in the fields abandoned by struggling farmers. These second-growth forests are now being clear-cut by logging companies—a practice driven in part by the fear that most of

the trees will be lost eventually to a severe and spreading infestation of the eastern spruce budworm. In rural areas, paper milling from wood is still supplying jobs as other blue-collar occupations die out.

Abundant fish were the major attraction that drew the first wave of Europeans to New England. Throughout the 1500s, hundreds of fishermen from Europe's Atlantic coast came to the Grand Banks, off the shores of Newfoundland and Maine, to take huge catches of cod and other fish. Over the last century, the fish stocks of the Grand Banks have been badly depleted by modern fishing vessels (some from outside North America), which use enormous mechanized nets able to bring in valuable species. These nets damage the fragile sea bottom and kill many unmarketable sea creatures that are nonetheless essential to the marine ecosystem. The fishing industry in this region has also been hurt by competition from fish farming, a rapidly growing industry in many parts of North America. There is a global connection here: farmed fish are fed with millions of tons per year of wild oceanic fish (10 million tons in 1997). The harvesting of these wild fish throughout the world ocean to feed farmed fish for markets in

North America poses a threat to oceanic resources not only off the shores of New England and the Atlantic Provinces, but also in countless places where the poor are dependent on fishing for their basic nutrition.

Many parts of southern New England now thrive on service- and knowledge-based industries that require skilled and educated workers: insurance, banking, high-tech engineering, and genetic and medical research, to name but a few. New England's considerable human resources derive in large part from the strong emphasis placed on education, hard work, and philanthropy by the earliest Puritan settlers, who established many high-quality schools, colleges, and universities. The city of Boston has some of the nation's foremost institutions of higher learning (Harvard, the Massachusetts Institute of Technology, Boston University, Boston College, Northeastern University). Boston has capitalized on its supply of university graduates to become North America's second most important high-technology center, after California's Silicon Valley.

At present, New England finds itself confronted by some startling discontinuities with its past. As its remaining agricultural enterprises and rural industries become increasingly mechanized and larger in scale, many rural New Englanders are flocking to the cities for work. Once in the cities, they become aware that New England is not the Anglo-American stronghold they had thought it to be. Native Americans, African-Americans, and immigrants from Portugal and southern Europe have long shared New England with those from the British Isles and elsewhere. But today, in such places as Middletown, Connecticut, and the suburbs of Boston, corner groceries are owned by Koreans and Mexicans; Jamaicans sell meat patties and hot wings; restaurants serve food from Thailand; and schoolteachers are Filipino, Brazilian, and West Indian. The blossoming cultural diversity of New England and the entrepreneurial skills of new migrants are helping New England to keep pace with change across the continent. These trends are less obvious in the Atlantic Provinces of Canada.

QUÉBEC

Québec is the most culturally distinct formal region in Canada or the United States (Figure 2.26). For more than 300 years, a substantial portion of the population has been French-speaking. French Canadians are now in the majority and are struggling to resolve Québec's relationship with the rest of Canada.

In the seventeenth century, France encouraged its citizens to settle in Canada. By 1760, there were 65,000 French settlers in Canada. Most lived along the St. Lawrence River, on long, narrow strips of land stretching back from the river's edge. This **long-lot system** gave the settlers access to resources from both the river and the land: fishing and river-borne trade, the fertile soil of the river's floodplain, and, on higher ground, the interior forest, where they could hunt. Because of the orientation of the long lots to the riverside, early French colonists commented that one could travel along the St. Lawrence and see every house in Canada. Later, the long-lot system was repeated inland, so that today narrow farms also stretch back from roads that parallel the river, forming a second tier of long lots.

Through the first half of the twentieth century, Québec remained a relatively poor land of farmers who grew only enough food to feed their families. After World War II, Québec's economy grew steadily, propelled by increasing demand for the natural resources of northern Québec, such as timber, iron ore, and hydroelectric power. Many of the cities of the St. Lawrence River valley prospered from the processing and transport of these resources, although most of these enterprises were in the hands of Anglo-Canadians (those whose ancestors came from the British Isles). Québec's new prosperity was most visible in the rapid growth of Montréal. This city's interior location, near the confluence of several rivers, had always made it an attractive site for British entrepreneurs interested in exporting the natural resources of Québec.

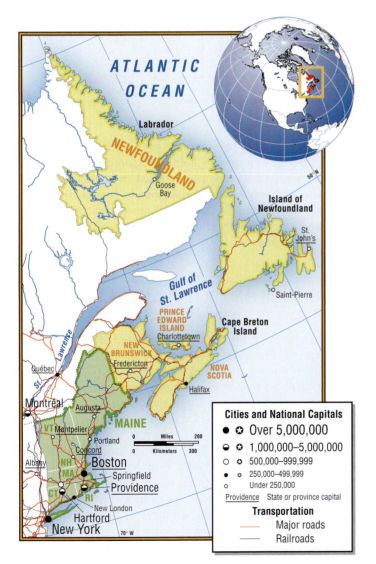

FIGURE 2.25 The New England and Atlantic Provinces subregion.
[Map adapted from Joel Garreau, *The Nine Nations of North America* (Boston: Houghton Mifflin, 1981).]

FIGURE 2.26 The Québec subregion.
The population of Québec is concentrated along the St. Lawrence River, and a majority of Québecois live in cities. Native Americans from various ethnic groups live throughout Québec, but one-fourth of them live in far northern and eastern rural Québec (see insert at right). [Adapted from Joel Garreau, *The Nine Nations of North America* (Boston: Houghton Mifflin, 1981); "Map of Indian Nations of Québec," Indian and Northern Affairs Canada http://www.ainc-inac.gc.ca/qc/map/index_e.html.]

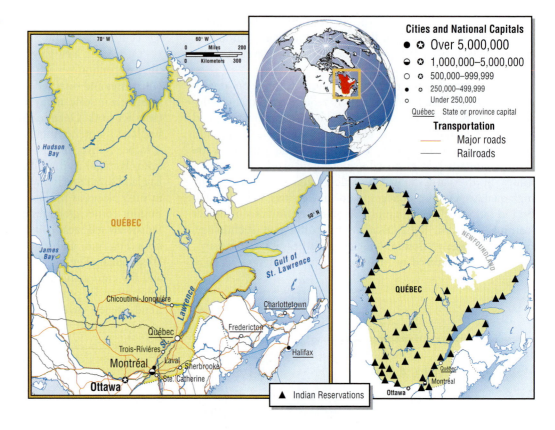

As many French Canadians moved into the cities, the so-called Quiet Revolution began. With increasingly better access to education and training, the Québecois were for the first time able to challenge English speakers for higher-paying jobs and power. Gradually, their conservative Catholic agricultural society gave way to a more cosmopolitan one that was increasingly resentful of discrimination at the hands of English speakers, which took the form of social denigration of French culture as well as blocked access to education and economic participation. In the 1970s, support for increased autonomy and outright national independence grew. At that time, the province of Québec passed laws that heavily favored the French language in education, government, and business. In response, many English-speaking natives of Québec relocated to Ontario and elsewhere. Québecois themselves began to fear for Québec's economic survival if the province left Canada. Nevertheless, a referendum on independence failed only narrowly in 1996.

Québec's northern portions extend into areas around Hudson Bay that are rich in timber and mineral deposits (iron ore, copper, and oil). These resources are hard to reach in the remote, difficult terrain of the Canadian Shield, a vast expanse of undulating coniferous forests and tablelands dotted with small lakes and wetlands. Although these are the native lands of the Algonquin-speaking Cree people, the Québec provincial government has legal control of the resources. In the 1960s, that government became interested in developing hydroelectric power in the vicinity of James Bay (part of Hudson Bay) in order to run mineral-processing plants, sawmills, and paper mills. The Cree protested the clear-cutting of their forests and the changes to community life brought by outsiders. In the 1990s, further hydro-electric projects were put on hold in response to Cree protests that the enormous shallow lakes created by the dams flooded sacred ancestral hunting and burial grounds. Nonetheless, development of resources in Cree lands by Euro-Canadians continues, and some electric power from Cree country is sold to the northeastern United States.

THE OLD ECONOMIC CORE

Southern Ontario and the north-central part of the United States, from Illinois to New York (Figure 2.27), were once the heart of North American iron- and steel-based heavy industry. This region is still industrial, but it no longer dominates the continent economically as it once did. Since 1970, many of its communities have struggled to redefine themselves in the aftermath of plant closings.

The old economic core represents less than 5 percent of the total land area of the United States and Canada, yet as recently as 1975, its industries produced more than 70 percent of the continent's steel and a similar percentage of its motor vehicles and parts. This concentrated output was made possible by the availability of energy and mineral resources in the region, or just beyond its boundaries, to supply the great steel mills and automotive plants. Coal came from Appalachia and southern Illinois, oil and gas came from Pennsylvania, and iron ore came from the great Mesabi Iron Range in Minnesota and similar deposits at Red Lake in Ontario (see Figure 2.29 on page 102).

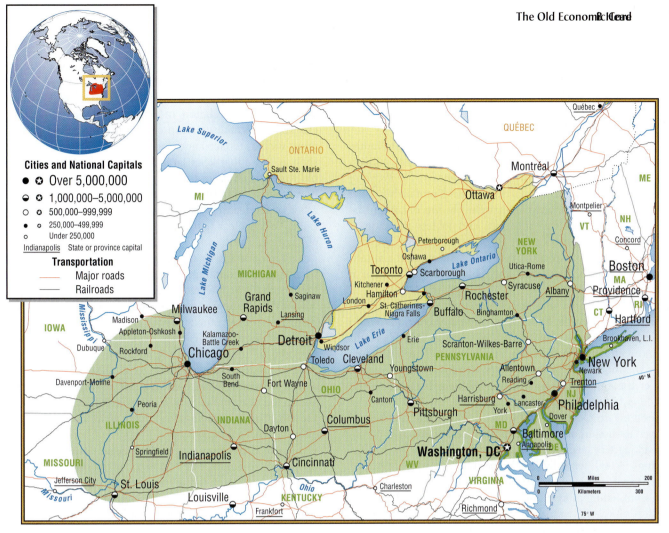

FIGURE 2.27 The old economic core subregion. [Adapted from Joel Garreau, *The Nine Nations of North America* (Boston: Houghton Mifflin, 1981).]

By the 1990s, much of the region was in decline or undergoing economic reorganization. As the map in Figure 2.28 shows, industrial and manufacturing jobs can now be found in the South, Middle West, and Pacific Northwest, well beyond the old industrial heartland. Brand-new state-of-the-art automobile factories built by Japanese and American firms are situated in rural landscapes in Kentucky and Tennessee (see the discussion on pages 76–77). In addition, industrial resources are coming from different places. Much coal now comes from Wyoming, and Canada's coal comes from British Columbia and Alberta, although Appalachian production has remained steady. Steel, the mainstay of the automotive and construction industries, can be more cheaply derived from scrap metal or purchased abroad from Brazil, Japan, or Europe. Hence huge, outdated factories sit empty or underused throughout much of the economic core, and thousands of aging factory workers have retrained or have entered an early, impoverished retirement. In recent times, some people have taken to calling this obsolete industrialized region the **Rust Belt.**

It would be wrong to think of the economic core, even during its heyday, as solely a region of factories and mines. In between the great industrial cities of this region are thousands of acres of some of the best farmland in North America. Some of the

A working-class neighborhood of Youngstown, Ohio, as viewed from the Republic Steel plant, now closed, which once employed 10,000 people. Early-twentieth-century frame houses surround a large Roman Catholic church. North Star Steel now operates minimill facilities on the Republic Steel site and employs 250 workers. [Richard Kalvar/Magnum Photos.]

appliance and food-related industries that grew as offshoots of the farming industry remain in the old economic core. Many others, such as the meat-packing industry, have consolidated or moved to the South, where labor unions scarcely exist and safety regulations are more lax.

The loss of millions of manufacturing jobs in the old industrial core has had serious consequences. Sociologist William Julius Wilson, in his book *When Work Disappears: The World of the New Urban Poor*, takes special note of the demoralizing effect of plant closings on working-class families in the old industrial cities. "Neighborhoods that are poor and jobless are entirely different from neighborhoods that are poor and working," he says. "Work is not simply a way to make a living and support a family. It also constitutes a framework for daily behavior because it imposes discipline."

Industrial jobs drew millions of rural men and women, both black and white, from the South to the industrial core after World War II. It is their sons and daughters who are now without work and without funds to retrain and relocate. They constitute some of the more than 35 million U.S. citizens living in poverty. Consider the case of one inner-city black neighborhood on the west side of Chicago, called North Lawndale. In 1960, two large factories there employed 57,000 workers, and a large retail chain employed thousands of secretaries and office workers in its corporate headquarters nearby. One factory closed in the late 1960s, removing 14,000 jobs; by 1974, the retail headquarters had moved downtown; and by 1984, the other large factory had closed, eliminating a breathtaking 43,000 jobs. North Lawndale began to disintegrate with the first loss of jobs. Because no one had money to spend anymore, thousands of service jobs disappeared, and with them, many middle-class families. By the early 1990s, the housing stock had deteriorated. As families and businesses left to look for opportunities elsewhere, landlords abandoned buildings, and financial institutions saw little reason to support reinvestment. By 2003, North Lawndale had shrunk to just 50,000 people, but with the help of a family foundation and local civic groups, the North Lawndale Industrial Development Team was actively recruiting industry, providing a custom-trained labor force, and maintaining an informative Web site (http://www.nlidt.com/index.html).

THE AMERICAN SOUTH (THE SOUTHEAST)

The regional boundaries of the American South are perhaps fuzzier and based more on a perceived state of mind and way of life than is the case for most other U.S. regions. This region, in fact, covers only the southeastern part of the country, not the whole of the southern United States (Figure 2.28). Somewhere east of Austin, Texas, the American South grades into the Southwest, a region with noticeably different environmental and cultural features. But what characteristics best define the South? And where should we draw its borders?

Within this region there is a complex of features that many people would identify as southern: comfort food; bluegrass, jazz

and blues music; the open friendliness of the people; a variety of dialects; country Baptist churches; conservative political stances on gun ownership, women's rights, environmental protection, and the role of religion in public life; the rolling hills and crooked roads; the early onset of spring; the field patterns and crops such as tobacco and cotton; and rural settlement patterns, such as those that keep large kin groups together in clusters of old wooden cabins, or lines of mobile homes, or fine newly built houses. But southern cultural features are hard to measure: only a few of the features listed here are present in any given place; there are few clear spatial distributions, and the patterns are not contiguous. Furthermore, arguments ensue if one tries to define just which accents, or what recipes, or which styles of music, or what kinds of friendliness are "southern." Some places located in the South have few recognizable southern qualities. Western Missouri, Oklahoma, and Texas have strong ties to the Great Plains; western Kentucky has the look of the Midwest. Miami, on the far southern tip of Florida, with its cosmopolitan Latino culture and its trade and immigration ties to the Caribbean and South America, seems to have lost all vestiges of the Old South and is redefining what it means to be southern. New Orleans, also in the heart of the South, has a unique set of cultural roots—French, Cajun, Creole—that set it apart from other southern cities.

North Americans and foreign visitors hold many outdated images of the South. The region today is a different place than it was during the civil rights efforts of the 1960s, let alone during the Civil War. It is true that significant racial segregation still persists in that, by custom, black people and white people tend to live and worship separately, but, in fact, many more whites and blacks share neighborhoods in the South than in the North; and in both rural and urban settings, schools are integrated as a result of residential patterns as well as busing plans and magnet schools that draw white children into black neighborhoods to access especially good science, music, and math programs. The workplace is now integrated, and African-Americans are in supervisory and administrative positions, especially in government and educational institutions. Black officials have been elected by white constituencies across the South. In the 1990s, Hispanic immigrants added another dimension to the South. Now in most southern cities much of the landscaping work, restaurant service, road maintenance, and construction work is done by Spanish-speaking workers from Mexico and Central America. These Hispanic people are not just seasonal or transient workers; many are sinking roots in the region by establishing businesses and buying homes.

Poverty is still a problem: the South has the nation's highest concentration of families living below the poverty line. On the other hand, most southerners, black, white, or Hispanic, are able to maintain a substantially better standard of living than their parents did. That older generation toiled as illiterate laborers on plantations or on poor, eroded, hilly farms in Appalachia, or fled the region for jobs in the industrial North. In the decade of the 2000s, the vast majority of southerners work in the service sector of the economy. Those who left for factory jobs in the North are being attracted back to the region by jobs, business opportunities, lower taxes, lower costs of living, a milder climate, and safer, more spacious, and friendlier neighborhoods than they could afford in such places as New York, Illinois, or California. The Population Reference Bureau reports that in the decade of the 1990s, black migration to the South grew dramatically: 3.5 million people who identified themselves as black moved to the South from the Northeast, the Midwest, and the West. Seven of the ten metropolitan areas that gained the most black migrants between 1990 and 1996 were in the South, with Atlanta leading in gains: 160,000 for the period.

The South is steadily improving its position as a growth region. The federally funded interstate highway system opened the region to auto and truck transport. Inexpensive industrial locations, close to arterial highways, have drawn many businesses to the South, including automobile and modular home manufacturing, food processing, forest-based building products, light-metals processing, high-tech electronic assembly, furniture manufacturing, and high-end crafts production (for example, the making of fine quilts or stringed instruments such as the dulcimer). Wal-Mart (see page 75) has its headquarters there. More recently,

tourists and retirees by the hundreds of thousands have been driving south on the Interstate Highway System, many attracted by the bucolic rural landscapes and warm temperatures, the scenery of the national parks, outlet mall shopping, and recreational theme parks.

Southern agriculture is now mechanized, and is at once more diversified and specialized than ever before. Many cash crops other than the traditional tobacco, cotton, and rice are now grown in the South. Strawberries, blueberries, peaches, tomatoes, mushrooms, herbs, vegetables, and wines from local vineyards are often produced for urban consumers on small holdings by part-time farmers who may also have jobs in nearby factories or service agencies. Most of the country's broiler chickens are now produced by huge operations throughout the South. Interestingly, the laborers on these factory-like chicken farms are often not local. Rather, they are refugee immigrants from Russia, Vietnam, Haiti, Honduras, the Philippines, and Ukraine who are willing to take these low-wage jobs, at least for a while. They often live in communities of prefabricated houses, another major product of the South.

FIGURE 2.28 The American South subregion. [Adapted from Joel Garreau, *The Nine Nations of North America* (Boston: Houghton Mifflin, 1981).]

THE GREAT PLAINS BREADBASKET

The Great Plains (Figure 2.29) receives its nickname, the Breadbasket, from the immense quantities of grain—wheat, corn, sorghum, barley, and oats—it produces, much of which is exported to Europe and other parts of the world. The gently undulating prairies give the region a certain visual regularity; its weather and climate, in contrast, can be extremely unpredictable, making life precarious at times. More than a hundred tornadoes may strike across the region in a single night, taking lives and demolishing whole towns. In some places the environmental challenges of farming are so great that the land is being abandoned.

European settlers began to stake claims on the Great Plains in the 1860s, and for the most part, they had to adjust to harsher climates than they were used to. Summers in the middle of the continent, even as far north as the Dakotas, can be oppressively hot and humid. Winters can be terribly cold and dry, or so snowy that it may be difficult for farmers to get from the house to the barns. Year-to-year unpredictability is also demonstrated by variable rainfall patterns. Summer rains in some years are plentiful; in other years, hardly a drop falls, and a year's labor may come to naught as the crops wither.

The people of the plains have learned to adapt to the harsh weather and unpredictable rainfall. As in yet drier regions to the west, Great Plains farmers have taken to irrigating their crops regularly with water pumped from deep aquifers (see the photo on page 103). They have chosen to grow crops—wheat, corn, oats, barley, sorghum, sugar beets, sunflowers—that suit the Great Plains environment and world market demand. To select the most

FIGURE 2.29 The Great Plains Breadbasket subregion. Gently undulating prairies give this region a certain visual regularity, but its weather and climate can be extremely unpredictable, making farming a challenge. [Adapted from Joel Garreau, *The Nine Nations of North America* (Boston: Houghton Mifflin, 1981).]

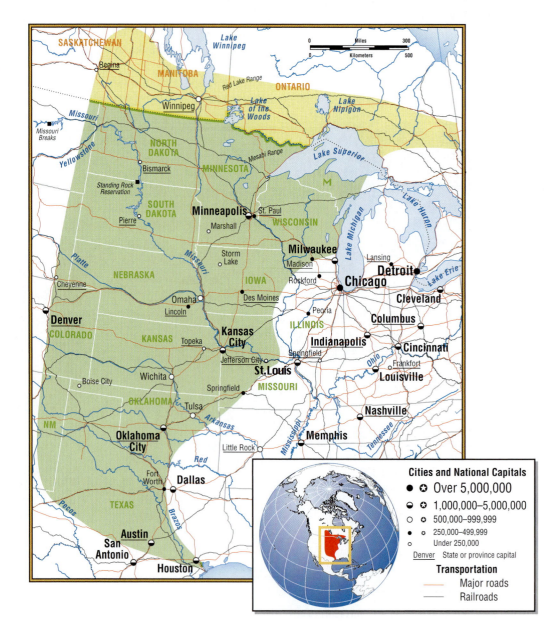

Ever notice dark green circles below you as you fly over the Great Plains on the way to the Pacific coast? Central pivot irrigators create these great circles in the landscapes of the Great Plains and farther west, as in this semiarid environment near Tuscarora, Nevada. The water supports crops that would not grow otherwise. Often a well at the center of each circle taps an underground aquifer, which is then depleted at an unsustainable rate. When abandoned, the circles fade to a brownish green. [Alex S. MacLean/ Landslides.]

profitable crops and find the best time to sell them, plains farmers keep close tabs on the global commodities markets, usually with computers installed in barns, homes, or even the cabs of the huge farm machines they use to work their land. Their primary crop, however, remains wheat, much of it genetically selected to resist frost damage. Winter wheat, sown in the fall and reaped in the early summer, is grown on the southern and central plains. Fast-growing summer wheat is grown in the north, where the winters are too severe for young wheat plants to survive. Most wheat is harvested by traveling teams of combines that start harvesting in the south in June and move north over the course of the summer.

Cattle raising is another important activity on the Great Plains. Rather than being turned out onto the open range, as in the past, cattle now are raised in fenced pastures and then shipped to feedlots, where they are fattened for market on a diet rich in sorghum and corn. Cattle (and other livestock such as hogs and turkeys) are slaughtered and processed for market in small plants across the plains (see the box "Meat Packing in the Great Plains" on page 104). Erosion is a serious problem on the plains: soil is disappearing more than 16 times faster than it can form. Grain fields and pasture grasses do not hold the soil as the original dense prairie grasses once did. The sharp hooves of the cattle loosen the soil so that it is more easily carried away by wind and water erosion. Experts estimate that each pound of steak produced in a feedlot results in 35 pounds of eroded soil, so cattle raising is by no means a sustainable economic activity.

Women and men have worked equally hard on the Great Plains since the first settlers arrived. For generations, women have borne and raised large families, grown, preserved, and cooked most of the food, and often managed the bookkeeping. They have helped their male family members care for farm stock, milk cows, drive farm equipment, and fix motors. Many women taught in the country schools and organized the church-related social functions that drew widely dispersed farm families together once or twice a week. It is common now to find farm families in which several members of both sexes have college degrees. Increasingly, farms are jointly managed by several family members, while others work in a nearby town, not only to earn salaries but also to gain health and retirement benefits for their families.

Population patterns on the Great Plains are changing. The few cities around the periphery of the region (Denver, Minneapolis, St. Louis, Dallas, Kansas City) are growing as young people leave the small towns on the prairies. Mechanization has reduced the number of jobs in agriculture and encouraged the consolidation of ownership. Increasingly, corporations own farms in the Great Plains. Individuals who still farm may own several large farms in different locales. They often choose to live in cities, traveling to their farms seasonally. As a result of this depopulation, thousands of small prairie towns are dying out. The 2000 census revealed that 60 percent of the U.S. counties in the Great Plains had lost population. An area equal to the Louisiana Purchase (900,000 square miles, or 2.3 million square kilometers) now has fewer than 6 people per square mile. Those left behind are struggling to continue to provide schools and opportunities for their own children and, in some towns, for the children of poor Hispanic or Asian immigrants attracted by low-paying meat-packing or crop-processing jobs.

A countertrend in some parts of the Great Plains is that Native Americans are coming home. While hundreds of Great Plains counties are losing population, those with significant numbers of Native Americans are the only counties growing. Although still only a fraction of the total plains population, the overall Native American population grew 12 to 23 percent between 1990 and 2000 in the Great Plains states of North and South Dakota, Montana, Nebraska, and Kansas. Many Native Americans are returning to work in newly established gambling casinos like the one at Standing Rock, in North Dakota. The casino is the county's largest employer, with 376 jobs, and has expanded six times since 1993. Myron Gutmann, a University of Texas professor, says these and other statistics suggest that European agricultural settlement on the Great Plains may have been an experiment that is ending. In hundreds of counties, native grasses and wildflowers, prairie dogs, and bison have made comebacks. Cattle ranches and grain farms still prevail in the middle and southern plains, but in the northern plains, there are now 300,000 bison — which, unlike cattle, require little management. Mike Faith, a Sioux trained at an accredited Native American college, manages a bison herd at Standing Rock Reservation, not far from where Sitting Bull was killed in 1890. Bison meat, which is low in cholesterol, may help to alleviate the high rate of diabetes among Native Americans. Faith adds, "Just having these animals around, knowing what they meant to our ancestors, and bringing kids out to connect with them has been a big plus."

THE CONTINENTAL INTERIOR

Among the most striking features of the continental interior (Figure 2.30) are its huge size, its physical diversity, and its very low population density. This is a land of extreme physical environments, characterized by rugged terrain, frigid temperatures, and lack of water (compare the maps in Figures 2.2, 2.3, and 2.5 with that in 2.30 to gain an appreciation of these characteristics). These physical features restrict many economic enterprises and account for the low population density: fewer than 2 people per square mile (1 person per square kilometer) in most parts of the region.

Increasing numbers of people are migrating to this region, some to take advantage of its open spaces and often dramatic scenery, others to exploit its considerable natural resources. These two groups often find themselves in conflict with each other and with the indigenous people of the continental interior. In fact, this region is one of the most intense battlegrounds in North America between environmentalists and resource developers.

Physically, there are four distinct zones within the continental interior: the **Canadian Shield,** the frigid and rugged lands of Alaska, the Rocky Mountains, and the Great Basin. The Canadian Shield is a vast glaciated territory lying north of the Great Plains that is characterized by thin or nonexistent soils, innumerable lakes, and large meandering rivers. The rugged lands of Alaska lie to the northwest of the shield. The shield and Alaska have northern coniferous (**boreal**) and subarctic (**taiga**) forests along their southern portions. Farther north, the forests give way to the **tundra,** a region of winters so long and cold that the ground is permanently frozen several feet below the surface. Shallow-rooted, ground-hugging plants such as mosses, lichens, dwarf trees, and some grasses are the only vegetation. The Rocky Mountains stretch in a wide belt from southeastern Alaska to New Mexico. The highest areas are generally treeless, with glaciers or

AT THE LOCAL SCALE   Meat Packing in the Great Plains

In the 1970s, a meat-packing job in the Great Plains provided a stable annual income of $30,000 or more. The workforce was unionized, and virtually all workers were descendants of German, Slavic, or Scandinavian immigrants. But in the 1980s, a number of highly unionized meat-packing companies closed their doors. Other nonunionized plants have opened, often in isolated small towns in Iowa, Nebraska, and Minnesota. The labor is supplied not by local residents, but by immigrants from Mexico, Central America, Laos, and Vietnam. In 1982, for example, there were only 28 non-English-speaking students in all of the schools in Storm Lake, Iowa. By the fall of 1996, the first language of 47 percent of the kindergarten class was Spanish, Tai Dam, Lao, Cambodian [Khmer], German, Korean, or Chinese. The meat-packing companies now pay $6.00 an hour, which yields an income, after taxes, of less than $12,000 per year. And union-won work rules are gone. Working hours are long or short at the convenience of the packing house manager, and those who protest may be summarily fired.

Many of the Latino and Asian workers are refugees from war in their home countries, and most of the Asians have spent a decade or more in refugee camps in Southeast Asia before coming to the United States. These immigrant workers have difficulty affording housing on their wages, and midwesterners are reluctant to rent to them. Hence, many live in makeshift housing, like the Laotian families in Storm Lake, Iowa, who occupy a series of old railroad cars and shanties. Down the road, Latino workers live in two dilapidated trailer parks.

A few hours away from Storm Lake by car, in Marshall, Minnesota, Roberto Trevino is the personnel director at the Heartland Company, a turkey-processing plant. He supervises 500 workers, 70 percent of whom are from Latin America and Asia, along with a few from Somalia (also war refugees). All wear white smocks and caps and labor with dangerous machinery in icy temperatures.

They slaughter, carve, trim, and package 32,000 turkeys per day and ship them under more than 60 different brand names.

Trevino is the college-educated son of Hispanic migrant farm workers and feels that his company is providing the immigrants with a stepping-stone. "If you are new in this country . . . you take the jobs Americans don't want and you may not get ahead. But you do it for your kids," he says. Yet he admits that these jobs are far from stable and do not provide an income sufficient to raise a family.

The Reverend Tom Lo Van, a Laotian Lutheran pastor, sees little chance that Laotian youth will prosper from their parents' toil. "This new generation is worse off," he says. "Our kids have no self-identity, no sense of belonging . . . no role models. Eighty percent of [them] drop out of high school."

On June 22, 2003, an article in the *Des Moines Register* documented further how globalization is complicating the raising and processing of meat in the North American Great Plains. A Taiwanese family firm recently bought the Hospers, Iowa, meat packing plant (near Storm Lake) and joined with a northwestern Iowa meat producers' confederation to supply meat to customers in Taiwan, Korea, and Japan. The Iowa meat producers were happy for the deal because they can no longer afford to independently grow, process, and market their meat in the global marketplace. Now, with a guaranteed market in Asia, there is suddenly a demand for much more Iowa meat. But the laborers in the packing plants are still poor migrants primarily from Latin America and Asia, working now for about $7.00 an hour.

[*Source:* Adapted from Marc Cooper, "The heartland's raw deal: How meatpacking is creating a new immigrant underclass," *The Nation* (February 3, 1997): 1–18; updated in 2003, http://rwor.org/a/v19/920–29/920/storm.htm; http://lombardi.wctc.net/~hchao/;   http://desmoinesregister.com/business/stories/c4789013/21538112.html.]

FIGURE 2.30 The continental interior subregion. [Adapted from Joel Garreau, *The Nine Nations of North America* (Boston: Houghton Mifflin, 1981).]

tundralike vegetation, while forests line the rock-strewn slopes on the lower elevations. Between the Rockies and the Pacific coastal zone is the **Great Basin** (see Figure 2.2, pages 54–55), a dry region of widely spaced mountains covered mainly by desert scrub and a few woodlands.

The continental interior has the greatest concentration of Native Americans in North America. Most of them live on reservations. In the United States and southern Canada, the reservations are usually not part of their original native lands.

Many of those who live in the vast tundra and northern forests of the Canadian Shield, however, still occupy their original territory. The Nunavut, for example, recently won rights to their territory after 30 years of negotiation. They are now able to hunt and fish and generally maintain the ways of their ancestors, although most of them now use snowmobiles and modern rifles (see page 62, for more on the Nunavut). In September 2003, the Dogrib people, a Canadian Native American group of 3500, reclaimed 15,000 square miles of their land just south of the Arctic Circle, and were given $100 million in compensation. The Dogrib, who have a strong sense of cultural heritage and a strong entrepreneurial streak as well, will manage the resources of their ancestral lands, which may include oil, gas, gold, and diamonds.

Although much of the continental interior remains sparsely inhabited, nonindigenous people have settled in considerable density in a few places. Where irrigation is possible, agriculture has been expanding—in the Utah Valley, in the lowlands along the Snake and Columbia rivers of Idaho, Oregon, and Washington, and as far north as the Peace River district in the Canadian province of Alberta. There are many cattle and sheep ranches throughout much of the Great Basin. Overgrazing and erosion are problems there; more serious problems include groundwater pollution from chemical fertilizers and the malodorous effluent

The Grand Coulee Dam is one of 264 dams on the Columbia River and its tributaries. Dams like this one have nearly destroyed the Pacific Northwest's salmon industry, while simultaneously providing irrigation water to what was once a high desert and generating substantial amounts of affordable electric power. [Jim Richardson/National Geographic Image Collection.]

from huge feedlots where large numbers of beef cattle are fattened for market. Efforts by the U.S. government to curb abuses on federal land, which is often leased to ranchers in extensive holdings, have not been very successful.

More than half the land in the continental interior is federally owned. In Utah, Arizona, and Nevada, 75 percent of the land is held by the U.S. federal government or state governments, and much of this land is leased to mining or logging operators. Mining and oil drilling are by far the largest industries. Since the mid-nineteenth century, the wide range of mineral resources in the region has supported most of its major permanent settlements. But even in these remote towns, life has been governed by the global economy—which has brought alternating booms and busts, depending on the world market prices for minerals. In recent times, the most stable mineral enterprises have been oil-drilling operations along the northern coast of Alaska. The Trans-Alaska Pipeline runs southward for 800 miles through mountains and tundra, terminating at the port of Valdez in southern Alaska. The pipeline, which often runs above ground to avoid shifting as the earth above the permafrost freezes and thaws, constitutes a major ecological disruption, interfering with caribou migrations and always posing the threat of oil spills. A giant spill from the oil tanker *Exxon Valdez* in 1989 devastated 1100 miles of Alaskan shoreline, killed much wildlife, and ruined native livelihoods and commercial fishing.

Increasing numbers of immigrants and vacationers flock to the continental interior for its open spaces and natural beauty. The many national parks in the region attract millions of North American tourists; seasonal and permanent residents come to work in such parks as Rocky Mountain, Yellowstone, Glacier, and Grand Teton. Others come to work in mineral mining. Towns such as Laramie, Wyoming, and Calgary, Alberta, have swelled with both seasonal and permanent residents who require services of all kinds. In the United States, environmental groups have pressured the federal government to set aside more land for parks and wilderness preserves. These groups also want to limit or eliminate activities such as mining and logging, which they see as damaging to the environment and the scenery. A switch to more recreational and preservation-oriented uses would change, and probably lessen, employment opportunities throughout the region. The increasing numbers of visitors would continue to place stress on natural areas, particularly on water resources.

THE PACIFIC NORTHWEST

Once a fairly isolated region, the Pacific Northwest (Figure 2.31) is now at the center of debates about how North America should deal with environmental and development issues. The economy in this region is shifting from logging, fishing, and farming to information technology industries. As this happens, its residents' attitudes about their environment are also changing. Forests that were once valued primarily for their timber are now also prized for their recreational value, especially their natural beauty and their wildlife, both of which are threatened by heavy logging.

Energy, needed to run the technology and other industries, is now more than ever a crucial resource for the region.

The physical geography of this long coastal strip consists of mountains and valleys. Most of the agriculture, as well as the largest cities, are located in the southern part, in a series of valleys and lowlands lying between two long, rugged mountain zones extending north and south. Throughout the region, the climate is wet. Winds blowing in from the Pacific Ocean bring moist and relatively warm air inland, where it is pushed up over the mountains, resulting in copious orographic rainfall and snowfall. The close proximity of the ocean gives this region a milder climate than is found at similar latitudes farther inland. In this rainy, temperate zone, enormous trees and vast forests have flourished for eons. The balmy climate, along with the spectacular scenery, attracts many vacationing or relocating Canadians and U.S. citizens.

Pacific Northwest logging provides most of the construction lumber and an increasing amount of the paper used in North America. Lumber and wood products are also important exports to Asia, and the lumber industry is responsible directly and indirectly for hundreds of thousands of jobs in the region. As the forests have shrunk, environmentalists have harshly criticized the logging industry for such practices as clear-cutting, the cheapest and most widely used method of harvesting timber. Clear-cutting destroys wild animal and plant habitats, and it leaves the land susceptible to erosion and the adjacent forest susceptible to pests. The trees that grow back after clear-cutting tend to be of a single species, allowing harmful insects and infections to spread from tree to tree more easily than in a multispecies forest. There are many reasons to criticize clear-cutting, but an alternative logging method that will preserve forest diversity and not be unduly costly in time and money is not easy to come by. The battle lines have been drawn between those who make a living from logging or related activities and those who make a living from occupations that tout the beauty of Pacific Northwest forests—people who are often advocates of strict environmental protection.

PERSONAL VIGNETTE In the early 1990s, in the northern California town of Quincy (slightly east of the border of the Pacific Northwest), several advocates of logging and supporters of environmental preservation decided to meet in neutral territory—the public library—for regular discussions of their differences. Every week, burly loggers sat down with a collection of local environmental advocates. After some years of discussion and conflict with the U.S. Forest Service, which favored clear-cutting, the Quincy Library Group was able to draw up a plan aimed at protecting forest habitats. The plan endorsed managing rather than eliminating fires (fires are needed to clear out highly combustible underbrush) and controlling logging to some extent. The Library Group has also sponsored successful federal legislation to protect old-growth timber. More important, they have established the precedent of citizen input into planning by the U.S. Forest Service, a democratic policy that has changed the dynamics of forestry management in several adjacent subregions.

FIGURE 2.31 The Pacific Northwest subregion. [Adapted from Joel Garreau, *The Nine Nations of North America* (Boston: Houghton Mifflin, 1981).]

Several large hydroelectric dams in the Pacific Northwest have attracted industries in need of cheap electricity, particularly aluminum smelting and associated manufacturing industries in aerospace and defense. These industries are major employers, but their demand for labor is erratic and periodic layoffs are common. In addition, the hydroelectric dams have been criticized by another major employer in the region, the fishing industry. The dams block the seasonal migrations of salmon to and from their spawning grounds, so some of the region's most valuable fish cannot produce young.

Given the environmental impacts of the logging and hydroelectric industries, it is not surprising that many people are enthusiastic about the increasing role of information technology within the major urban areas of the Pacific Northwest: San Francisco, Portland, Seattle, and Vancouver. IT companies, for the most part, are clean and efficient, and they produce high-priced finished goods that are easy to transport. With Silicon Valley just outside San Francisco and with a growing number of IT companies around Seattle, the Pacific Northwest is the world leader in information technology (the failure of many new technology firms in the late 1990s did not seriously affect the dominance of IT in this region). But IT firms are particularly dependent on secure and steady sources of electricity. Hence, although IT employees are often vociferous supporters of environmental preservation and restoration, their jobs depend on electric power sources—whether water, gas, oil, coal, or nuclear fuels—that have significant negative environmental impacts.

The Pacific Northwest often leads the rest of the United States and Canada in adjusting to emerging realities. This role is illustrated by the region's growing connections to Asia. The adjustment of a predominantly Euro-American society to large numbers of Asian immigrants has been well under way along the Pacific coast for many years. This transition has been accompanied by the growing importance of the countries around the Pacific Rim as trading partners. Asia's economic downturn in the last half of the 1990s hindered the ability of the Pacific Northwest to market forest products there—particularly to Japan, its largest market. West Coast ports became clogged with timber shipments, and jobs were lost. With some measure of economic recovery in Asia, the market rebounded somewhat by 2000. But by then competitors from Europe and South America were also selling timber in Asia, taking over part of the market and bringing overall prices down. In the short term, the North American economy absorbed most of the excess timber for the flourishing housing market. The timber trade with Asia is now recovering, even though many parts of Asia also produce wood.

SOUTHERN CALIFORNIA AND THE SOUTHWEST

Southern California and the Southwest (Figure 2.32) are united primarily by their warm, dry climate, which has attracted many migrants from across America; by their long and deep ties to Mexico; and by the promise that the North American Free Trade Agreement will make the subregion a nexus of development.

The varied landscapes of the Southwest include the Pacific coastal zone, the hills and interior valleys of Southern California, the widely spaced mountains and dramatic mesas and canyons of southern Arizona and New Mexico, and the gentle coastal plain of south Texas. The common physical characteristics of these landscapes are their warm average year-round temperatures and arid climate. Most areas receive less than 20 inches of annual rainfall, and the dominant vegetation is scrub, bunchgrass, and widely spaced trees. Because of the region's aridity, ranching is the most widespread form of land use, but other forms are more important economically and in numbers of employees.

Where irrigation is possible, farmers take advantage of the nearly year-round growing season to cultivate fruits and vegetables. California's Central Valley is the leading producer of fruits, vegetables, and wine in the United States and is one of the most valuable agricultural districts in the world. Most of this land is devoted to such crops as grapes, tomatoes, peppers, lettuce, and other vegetables, which are shipped in refrigerated containers throughout the United States and Canada. Many of these crops are produced on vast, plantation-like farms. Migrant Mexican workers, often illegal immigrants, supply most of the labor. These workers, sometimes mere children (see the photo on page 110), lower the cost of North American food by working for very low wages.

Although agriculture is important, the bulk of the region's economy is nonagricultural. The coastal zones of Southern California and Texas are home to oil drilling and refining and associated chemical industries, as well as other industries that

Los Angeles keeps expanding farther and farther into the dry, desertlike foothills of the mountains surrounding it. Such development destroys the natural vegetation that holds the soil in place, so large landslides and mud slides may occur when the snowcap melts in the spring or during the occasional heavy rains coming in off the Pacific. In this new development, the lots have been graded to impede runoff. [Alex S. MacLean/Landslides.]

Thinking Geographically: ON THE WEB

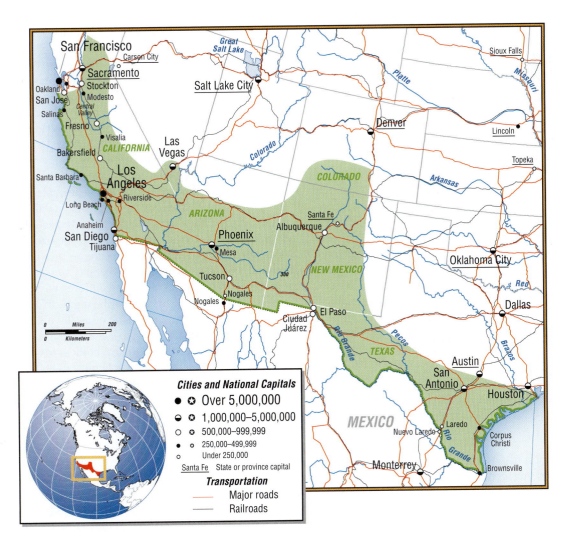

FIGURE 2.32 The Southern California and Southwest subregion. [Adapted from Joel Garreau, *The Nine Nations of North America* (Boston: Houghton Mifflin, 1981).]

need the cheap transport provided by the ocean. The region's access to the Pacific Ocean along the California coast and to the Gulf of Mexico along the Texas coast, its mild climate, its sunny weather, and its spectacular scenery have made the region a popular location for a wide range of service industries, and it is these activities that employ the vast majority of workers. For example, the U.S. film and television industry has a strong presence in Southern California, drawing writers, actors, producers, investors, and support staff from all over the world. Los Angeles—with its major trade, transport, media, and finance industries, as well as important research facilities—recently surpassed New York as the most populous city in the United States. Like San Francisco, Seattle, and other Pacific coastal cities, Los Angeles is strategically located for trade with Pacific Rim countries, especially in Asia. Los Angeles, rather than New York, is now the largest port in the United States (see Figure 2.10 on page 74).

On the eastern flank of this region, cities such as Austin, Texas, have attracted information technology industries, lured by research activities connected to the University of Texas, by the warm climate, and by the laid-back, folksy lifestyle of central Texas. Austin has close to 3 million people and grew by more than

30 percent from 1990 to 2000. Across the region, the climate draws large numbers of vacationers and retirees; entire semi-planned communities of 10,000 or 20,000 people have grown up in the deserts of New Mexico and Arizona in the space of just a few years.

In the decade of the 2000s, residents of Southern California and the Southwest must worry about energy costs along with congestion and smog. Questions about the region's rate of energy consumption were raised when prices suddenly escalated and supplies became precarious in California in the early part of the decade. In that state, the costs of electricity to consumers had been held artificially low by state regulation of, and subsidies to, utility companies. The low cost of electricity encouraged its ever-increasing use for air conditioning, nighttime commercial lighting, produce refrigeration, and computer-related equipment. State attempts at deregulation failed to take into account the fact that much of the region's power was purchased from wholesale producers outside the region that had been selling energy to the state-regulated utility companies at inflated prices. By early 2001, costs of electricity in some cases rose to ten times what they had been a year earlier. At the same time, supplies of power decreased, forcing rolling blackouts (sequenced reductions in power to

Thinking Geographically: ON THE WEB

José Madrid, 11, picks green chilies in New Mexico. "I'm not good at math, but I'm good at money," he says. Like many child migrant workers in the United States, he goes to school only intermittently. [Eric Draper/AP/Wide World Photos.]

are becoming increasingly interdependent as NAFTA fosters increased trade (see the discussion on pages 75–76).

In the 1970s, U.S., Canadian, European, and Asian companies began to set up factories, known as **maquiladoras,** in Mexican towns just across the border from U.S. towns, in such places as Ciudad Juárez (El Paso), Nuevo Laredo (Laredo), Nogales (Nogales), and Tijuana (San Diego). Maquiladoras produce manufactured goods for sale primarily in the United States and Canada, taking advantage of cheaper Mexican wages, cheaper land and resources, lower taxes, and weaker environmental regulations. (The maquiladora phenomenon is illustrated in Figure 2.10 on page 74 and is discussed in greater detail in Chapter 3.) These factories are a key part of a larger transborder economic network that stretches across North America. In fact, the U.S.–Mexican border has been one of the world's most **permeable national borders,** meaning that people and goods flowed across it easily (though not without controversy). This permeability was greatly increased by the North American Free Trade Agreement of 1994. There are now as many as 200 million legal border crossings each year. According to a U.S. Immigration and Naturalization Service study in 1998, illegal border crossings then were less than 10 percent of this figure, at about 150,000 per year. Since the terrorist attacks of 2001, the border has become much less permeable and the trip more dangerous. Smugglers who bring workers across the more heavily guarded border charge high fees, often cheat their passengers, and in a number of cases have left whole truckloads of people to die in the desert heat, without food or water.

Several contentious issues surround the estimated 8 million to 9 million illegal immigrants currently residing in the United States, 4.5 million to 5 million of whom are Mexicans. Many of these people remain in California and the Southwest, where, in addition to contributing vitality to the local economy, they keep down the wages of low-skilled workers. Inadvertently, they have also increased tension over the status of the English language in the United States. Although English speakers in the Southwest far outnumber Spanish speakers, many people fear that English could be challenged in much the same way as it has been challenged by French Canadians in Québec. Accordingly, Arizona and California recently became the first states to make English the official language of government. Some bilingual programs, installed in the 1970s to help migrant children make the transition from Spanish to English, are being abolished by those who feel it is better for those children to learn English as quickly as possible. Other educators think that migrant children do better in math and science and enjoy higher self-esteem if they can study for a period in their native language.

ensure that there is not a general blackout). A wide range of businesses and industries simply could not continue under these conditions, and worries about the region's economic future escalated. In the short term, the electricity supply problem has been alleviated by a range of strategies, including controls on prices and reduction of nonessential use. It would be wrong to conclude that energy is in critically short supply; the immediate problem was more one of corruption among some private generators and distributors that deliberately controlled the supply to drive up prices. But like the rest of the United States, Southern California and the Southwest have been consuming energy at rates that are not ultimately sustainable, affordable, or safe for the environment.

Much of this region was originally a colony of Spain, and this area has maintained the Spanish language, a distinctive Hispanic culture, and other connections with Mexico (see Figure 1.4 on page 10). Today, Hispanic culture is gaining prominence in the Southwest and spreading beyond it. As we have seen, large numbers of immigrants are arriving from Mexico (see Figure 2.13 on page 81), and the economies of the United States and Mexico

Quick Review

1. Name a few of the natural resources that encouraged Europeans to settle in New England.

2. What is the evidence that European agricultural settlement in the Great Plains is an experiment that is ending?

3. How are Asian and Hispanic immigrants contributing to the economies of various U.S. regions?

4. Name two environmental issues currently being debated in the Pacific Northwest.

5. Which European countries colonized which portions of North America?

REFLECTIONS ON NORTH AMERICA

It is not hard to rhapsodize about North America: the sheer size of the continent, its incredible wealth in natural and human resources, its superior productive capacities, and its powerful political position in the world are attributes enjoyed by no other world region. North America enjoys this prosperity, privilege, and power as a result of fortunate circumstances, the hard work of its inhabitants, the diversity and creativity of its settlers, and astute planning on the part of early founders, especially the writers of the U.S. Constitution. Perhaps the most important factor in North America's success has been its democratic governments and supporting institutions, which allow for individual freedom and flexibility as times and circumstances change.

Yet life has not been good to everyone in North America, nor has the influence of North America on other parts of the world always been benign. Native Americans, enslaved Africans and their descendants, and other ethnic minorities suffered as Canada and the United States were being created out of what had been lands long occupied by indigenous peoples. Settlers of all origins and their descendants had, and continue to have, a significant negative impact on the continent's environments, and the standard of living expected by all North Americans promises to increase the strain on those environments. Furthermore, as Canada and the United States have developed into wealthy world powers, their impact on environments and people elsewhere around the world, through trade and cultural diffusion, has increased.

There is no guarantee that North America will continue its leadership role into the future; in fact, in the post-9/11 world, challenges to that leadership are common in every corner of the globe. Some critics name the recent U.S. tendency to use military force before adequate consultation with allies as the major cause for concern. Others cite insensitive trading policies or mean treatment of employees in the overseas workplaces of U.S.-based corporations such as Wal-Mart or Nike. North American models for development—based on democracy and on assumptions of rich and inexhaustible resources—are being challenged as inappropriate for much of the rest of the world. As we shall see in subsequent chapters, societies elsewhere are beginning to prosper without following North American examples, and sometimes without first installing democratic institutions. Environmental concerns, increasingly a part of the North American consciousness (if not yet practice) are not central in many developing countries. Rather, material prosperity is often the chief goal, just as it still is for many in North America. Many people around the world are eager to bring the North American material miracle to their lands, regardless of negative environmental impacts.

The closer formal economic association of Canada, the United States, and Mexico through NAFTA has strengthened the already strong economic relationships among these three countries. NAFTA may turn out to be just the beginning of new alignments between North America and countries in Middle and South America as well as in Europe and Asia. As we shall see in Chapter 3, over the past few centuries Middle and South America have had experiences very different from those of North America, yet recent social and economic changes in that region have been dramatic and, despite the disruptions of 9/11, have brought closer ties to North America. Perhaps all of the Americas will eventually be much more integrated economically and, possibly, socially and politically.

Chapter Key Terms

acid rain   91
agribusiness   71
aquifer   94
baby boomer   90
bimodal   73
boreal forest   104
brownfields   79
Canadian Shield   104
chain migration   81
clear-cutting   61
dense nodes   79
digital divide   74
economic core   60
ethnicity   86
genetic engineering   71
Geneva Conventions   66
gentrification   79

government subsidies   75
Great Basin   105
hazardous waste   91
Hispanic   56
hub-and-spoke network   72
information technology (IT)   73
Internet   73
Interstate Highway System   72
knowledge economy   73
Latino   56
loess   56
long-lot system   97
low-till   71
maquiladoras   110
megalopolis   80
metropolitan areas   79
New Urbanism   79

nuclear family   89
Ogallala aquifer   94
Organization of Petroleum Exporting Countries (OPEC)   67
outsourcing   67
Pacific Rim (Basin)   65
permeable national borders   110
Québecois   56
Rust Belt   99
service sector   72
smog   91
taiga   104
thermal inversion   91
tundra   104
urban sprawl   80

CHAPTER 3

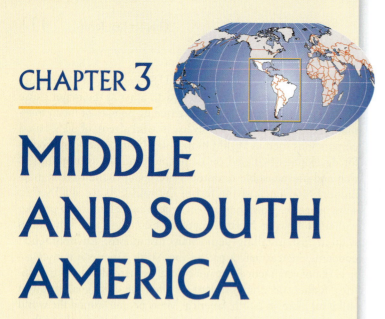

MIDDLE AND SOUTH AMERICA

On a trip through the Ecuadorian Amazon in 1999, Alex Pulsipher looked into the geography of the region's rapidly expanding oil industry and its effects on local populations.

PERSONAL VIGNETTE The boat trip down the Aguarico River in Ecuador takes one into a world of magnificent trees, river canoes, and houses built high up on stilts to avoid floods and pests and to catch the breeze. I was there to visit the Secoya (Figure 3.1, inset), a group of 350 indigenous people in the Ecuadorian Amazon who are currently negotiating with the U.S. oil company Occidental Petroleum over its plans to drill for oil on Secoya lands. Ecuador's government sees the revenues from oil production as essential to paying off its huge national debt. The Secoya wish to protect themselves from the pollution and cultural disruption that come

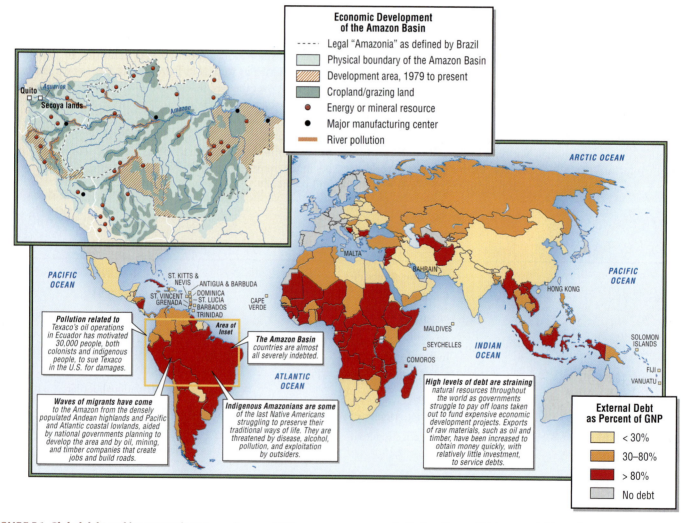

FIGURE 3.1 Global debt and its connections to environmental issues. Economic development always has environmental effects, but environmental damage often increases when countries have moderate to high levels of external debt (that is, debt owed by a country to nonresidents that is repayable in foreign currency). When external debt is high, natural resources may be extracted at unsustainable rates and sold in order to make debt payments. Such patterns of high external debt and unsustainable resource use can be found in the countries that share the Amazon Basin, as well as in Africa, Russia, and Asia. [Debt data from *United Nations HDI Report 2000*; Amazon inset adapted from John L. Allen, *Student Atlas of Environmental Issues* (Guilford, Conn.: Dushkin/McGraw-Hill, 1997), p. 82.]

A worker samples one of the several hundred open waste pits that Texaco left behind in the Ecuadorian Amazon. Wildlife and livestock trying to drink from the pits often are poisoned or fall into the pits. After heavy rains, the pits overflow, polluting nearby streams and wells. [Alex Pulsipher.]

with oil development. As Colon Piaguaje, chief of the Secoya, puts it, "A slow death will occur. Water will be poorer. Trees will be cut. We will lose our culture, our language, alcoholism will increase, as will marriages to outsiders, and eventually we will disperse to other areas. There is no other way. We will negotiate, but these things will happen." Given all the impending changes, Chief Piaguaje is asking Occidental to use the highest environmental standards in the industry and to establish a fund to pay for the educational and health needs of the Secoya people.

Chief Piaguaje bases his assertions on what has happened in parts of the Ecuadorian Amazon that have already experienced several decades of oil development. The U.S. company Texaco was the first major oil developer to establish operations in Ecuador. From 1964 to 1992, its pipelines and waste ponds leaked almost 17 million gallons of oil into the Amazon Basin—50 percent more than the 11 million gallons the oil tanker *Exxon Valdez* spilled into Prince William Sound in Alaska. Although Texaco sold its operations to the government and left Ecuador in 1992, its oil wastes continue to leak into the environment from hundreds of open pits. In addition, many people have come from the densely populated highlands to establish farms along the roads Texaco built to service its oil wells. In 1993, some 30,000 people, both colonists and indigenous people, sued Texaco in New York State, where Texaco (now called ChevronTexaco) is headquartered, for damages from the pollution. They complain that oil contamination has contributed to their skin problems, respiratory ailments, stomach diseases, and dying crops and animals.

As an alternative to oil development, many inhabitants of the Ecuadorian Amazon argue for locally based "sustainable devel-opment" that won't damage the land. Such development might include programs to help the Secoya market their agricultural produce in the cities or abroad and to start small, environmentally friendly businesses. Others believe that oil development would not be as problematic if effective environmental and cultural protections were in place and if a share of the revenue went to the local community.

Occidental offered the Secoya $20,000 for the right to extract oil, and then upped that offer to $90,000, but Chief Piaguaje rejected that, too, saying it was far too little to compensate for the billions of barrels of oil extracted and all the changes that would follow in the course of large-scale oil extraction. In 2002 the Ecuadorian suit against Texaco was dismissed by the U.S. Second Court of Appeals, which said Ecuador would be a more convenient venue for the suit. For the time being, nothing seems likely to stop the "slow death" of the Secoya that worries Chief Piaguaje.

MAKING GLOBAL CONNECTIONS

The rich resources of the Americas have attracted outsiders since the first voyage of Christopher Columbus in 1492. Europe's encounter with the Americas marked a major expansion of the global economy. Within a few years, the Americas were producing gold and silver that eventually financed much of the industrial revolution in Europe. Some crops domesticated by Native Americans became cash crops mass-produced for foreign markets with low-paid or slave labor. For several hundred years, Middle and South America occupied a peripheral position in global trade, supplying primarily raw materials. Recently, however, this region began to forge stronger global connections by forming trade blocs within the region, by sending migrants abroad who retained attachments to home, by attracting outside investment, and by developing trade partnerships with Japan, China, and European countries.

Middle and South America differ from North America in several ways. Physically, this world region is larger and exhibits a more varied mosaic of environments (Figure 3.2). Culturally, it is more complex than the continent to the north. Large Native American populations influence daily life just by their presence as well as through their legacies of crop plants, languages, religious beliefs, material culture, and customs. Immigrant groups from Europe, Asia, and Africa have also left distinctive marks, often because circumstances and isolation encouraged a certain cultural conservatism. And this ethnic variety is made yet more complex by highly stratified social systems based on class, race, and gender. Politically, too, the region is more complex: it includes more than three dozen independent countries that implement different models of self-government with different degrees of success. Economically, levels of well-being vary more markedly than they do in North America.

Despite the distinct features of the many parts of Middle and South America, there are regional commonalities: widespread use of Spanish, cohesive multigenerational families, and a strong social and political role for the Roman Catholic Church are just a few examples. Most of these commonalities originate from many of the countries' shared experience as colonies of Spain or Portugal. Colonialism introduced a version of capitalism that did not provide for development in the region, but instead made it a producer of raw materials for development elsewhere—chiefly Europe. This focus on production for export greatly altered local landscapes, just as oil extraction is altering the Ecuadorian Amazon today. Complex mosaics of Native American lands once devoted to multiple uses were transformed into vast stretches of a single crop, such as sugarcane or cotton. Sometimes the surface of the land was stripped away to mine silver, copper, or gold. In both colonial and postcolonial times, as internal elites allied first with European and then with U.S. private investors to continue to exploit local lands and extract resources, Native Americans were denied equal rights, held in subservient class positions, and discriminated against as *indios*—the lowest of the low.

Nonetheless, in every country of the region, people like Chief Colon Piaguaje in Ecuador are making efforts to change the patterns of the colonial and postcolonial eras, and successes are not hard to find. Self-help projects and efforts to democratize everyday life are evident in countless villages and city neighborhoods. The military, which in the past was often called upon to maintain order (for good or ill), is now much less evident. The economies of most countries are being reorganized to encourage economic growth in markets free of government controls. Such **economic restructuring,** enforced by international lending agencies, was deemed necessary because of the huge **external debts** (debts owed to nonresidents that are repayable in foreign currency) that many countries incurred to finance large development projects (see Figure 3.1). The poor have been hit hard by these mandates: they lost jobs as programs were cut and industries were streamlined, and they have been greatly affected by rising prices. Yet many have managed to survive using informal networks and communal self-help strategies. In confronting the hardships wrought by restructuring, many people of the region are realizing that development is not just a matter of economic growth, sleek skyscrapers, and massive dam projects. To be judged successful, development must change the lives of the majority for the better.

THEMES TO EXPLORE IN MIDDLE AND SOUTH AMERICA

You will encounter a few major themes repeatedly in this chapter:

1. **Cultural diversity.**   The region of Middle and South America is notable for its cultural variety and richness.

2. **Increasing regional integration of trade.**   The countries of the region are forming economic links with one another by lowering old colonial trade barriers, and are in the process of establishing the Free Trade Agreement of the Americas (FTAA), a trade bloc that would include all of the Americas.

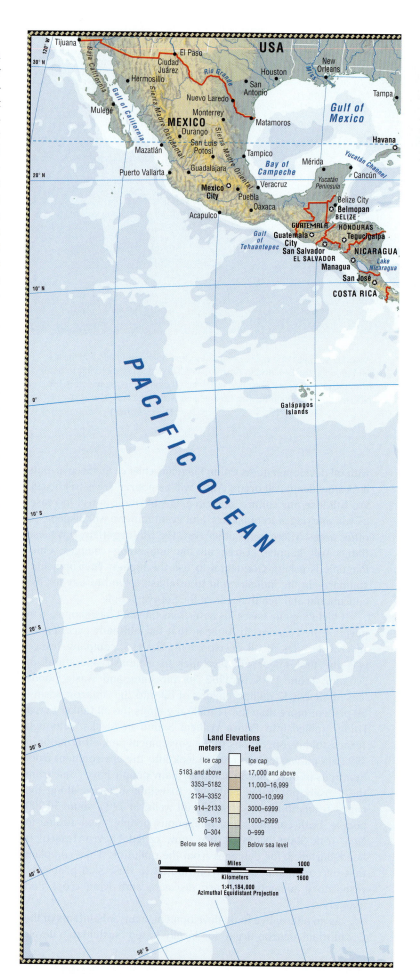

FIGURE 3.2 Regional map of Middle and South America.

Jacksonville
Orlando
Miami
BAHAMAS
Bermuda
Tropic of Cancer
GREATER ANTILLES
CUBA
DOMINICAN
REPUBLIC
JAMAICA   HAITI   *Hispaniola*   San Juan
Santo   Puerto
Domingo   Rico
Montserrat
LESSER ANTILLES
DOMINICA
ST. LUCIA
ST. VINCENT   BARBADOS
GRENADA
TOBAGO
TRINIDAD
Caribbean Sea
Gulf of
Venezuela
Barranquilla   Maracaibo   Caracas
Colón   Panama
Canal   Valencia
Panama   Lake
Maracaibo   VENEZUELA
PANAMA   Medellín   *Orinoco*   Georgetown   Paramaribo
Gulf of   Bogotá   Cayenne
Panama   *Llanos*   GUYANA
Cali   SURINAME   Fr.
Guiana
COLOMBIA   *Guiana Highlands*
Quito
ECUADOR
Guayaquil   *Negro*   Manaus   *Amazon*   Belém
Iquitos   *Solimões*   Equator
Marañón   *Amazon Basin*   Fortaleza
Piura
Madeira   B R A Z I L
PERU   *Xingu*   *São Francisco*   Recife
Araguaia   *Brazilian Highlands*
Callao   Lima   *Tocantins*
Cuzco   *Ucayali*   *Mato*
Andes   *Grosso*   Salvador da Bahia
BOLIVIA   Brasília
Lake   La Paz
Titicaca   Cochabamba   Belo
Altiplano   Sucre   Santa Cruz   Horizonte
Potosí   *Paraná*
Antofagasta   PARAGUAY
Atacama   Asunción   Rio de Janeiro
Desert   Curitiba   São Paulo
CHILE   Tropic of Capricorn
Mtns.   *Paraná*
Córdoba   Pôrto Alegre
Valparaíso   Rosario
Santiago   URUGUAY
Buenos Aires   Montevideo
ARGENTINA   La Plata   *Rio de la Plata*
Pampas
Puerto Montt   *Monte*
Verde
Patagonia
Comodoro Rivadavia
ATLANTIC OCEAN
Falkland
Islands
Straits of Magellan   Stanley
Punta Arenas
Tierra
del Fuego

3. Raw materials production. Many countries still rely on income from the export of agricultural products and extracted resources, though manufacturing for export is a rapidly growing economic sector.

4. Highly stratified social systems. Political and economic power remains concentrated in the hands of a small, rich minority, often of European descent. The majority, made up of people of Native American, mestizo, and African descent, is poor and has relatively little political influence. The disparity between rich and poor in this region is the widest on earth and may be increasing.

5. Rural-to-urban and international migration. In all parts of the region, large numbers of people are migrating from the countryside to cities, where traditional patterns of life are being transformed. Temporary emigration to North America and Europe is an alternative strategy for the young and adventuresome. Most international migrants originate in Mexico, the Caribbean, or Central America, but increasingly they come from rural and urban South America as well. The money and goods they send home form a significant portion of national incomes.

6. The extended family. Despite declining birth rates, large multigenerational families, with defined roles for men and women, continue to predominate in the region and serve as a source of mutual support during hard times.

7. Outside influences. Throughout most of the twentieth century, the dominant outside influence in the region has been the United States. Today, multinational corporations from several world regions are increasing their presence and competing with U.S. corporations for shares in regional trade and commerce.

I   THE GEOGRAPHIC SETTING

Terms to Be Aware Of

In this book, **Middle America** refers to Mexico, the narrow ribbon of land south of Mexico that makes up Central America, and the islands of the Caribbean. **South America** refers to the continent south of Central America. For several reasons, we usually don't use the term *Latin America* in this book, even though it is the term most often used by others for what we call Middle and South America. Latin America is so called because it was colonized, for the most part, by Spain and Portugal, whose cultures are thought of as having Latin (Roman) origins. (Actually, the cultures of Spain and Portugal are the products of many non-Latin influences as well.) Therefore, when *Latin* is used to refer to the Americas, it serves as a permanent reminder of the region's former colonial status. Furthermore, the designation ignores the other cultures present in the region, most notably those of the various Native American groups, but also those of non-Latin people who arrived during and after the colonial period: Africans, Dutch, Germans, British, Chinese, Japanese, and others from elsewhere in Asia. Nor does it acknowledge the new, distinctly American cultures—often called **Creole cultures**—that have been created from these many strands.

The terms *New World* and *Old World* are also problematic. Is the New World so designated because it is actually newer in some way than the Old World, or only because it was new to Europeans in 1492? In this book, we occasionally use the term *Old World* to mean an entity other than the Americas, which includes virtually the entire rest of the world. This designation is useful at times because Eurasia, Africa, and Oceania have interacted in many ways over the last 20,000 years and more, exchanging cultural attributes, and it is not always clear where an idea or item of material culture originated within that huge territory.

PHYSICAL PATTERNS

You can see from the map in Figure 3.2 that the region of Middle and South America extends south from the midlatitudes of the Northern Hemisphere all the way across the equator through the Southern Hemisphere, nearly to Antarctica. This expanse is part of the reason for the wide range of climates in the region; another contributing factor is its variation in altitude. Tectonic forces have shaped the primary landforms of this huge territory to form an overall pattern of highlands to the west and lowlands to the east.

Landforms

Highlands. A grandly curving and nearly continuous chain of mountains stretches along the western edge of the American continents for more than 10,000 miles (16,000 kilometers) from Alaska to Tierra del Fuego, at the southern tip of South America (see Figures 1.1D, 2.2, and 3.2). This long mountain chain, known as the Sierra Madre in Mexico, by various local names in Central America, and as the Andes in South America, was formed by a process called **subduction,** in which the edge of one tectonic plate descends under the edge of another. In this case, the eastward-moving oceanic plates—the Cocos Plate and the Nazca Plate—are plunging beneath the three continental plates—the North American Plate, the Caribbean Plate, and the South American Plate—at a lengthy subduction zone running thousands of miles along the Pacific (western) coast (see Figure 1.10 on page 23). The overriding plate crumples to create mountain chains, often developing fissures that allow molten rock from beneath the earth's crust to ascend to the surface and form volcanoes (see Figure 1.11 on page 24). These highlands constitute a major barrier to transportation and communication throughout their length, but in northern and central South America, where the population is dense, they pose a special challenge.

The chain of low and high mountainous islands in the eastern Caribbean is similarly volcanic in origin, created as the Atlantic Plate thrusts under the eastern edge of the Caribbean Plate. It is not unusual for volcanoes to erupt in this active tectonic zone. On the island of Montserrat, for example, people have been coping with an active volcano since July 1995. The eruptions have been violent **pyroclastic flows**—blasts of superheated rocks, ash, and gas that move with great speed and force down the side of a mountain. Twenty people were killed one morning in their gardens, and eventually the capital of Plymouth was destroyed, along with 20 settlements in the southern two-thirds of the island. About 7000 people have left the island, but 4000 others, relocated to the north of the island, are struggling to adapt to life in a still beautiful but greatly changed place.

Lowlands. A look at the regional map in Figure 3.2 will show that lowlands extend over most of the land to the east of the western mountains. In Mexico, however, the Sierra Madre is divided into a broad U. Within the U sits a plateau, and to the east of the Sierra Madre, a coastal plain borders the Gulf of Mexico. Farther south in Central America, wide aprons of sloping land descend to the Caribbean coast. In South America, a huge wedge of lowlands, widest in the north, stretches from the Andes east to the Atlantic Ocean. These South American lowlands are interrupted in the northeast and the southeast by two modest

On the day in July 1999 when this photo was taken, the Soufrière volcano on Montserrat was quiet, but you can see the ash deposits left by pyroclastic flows on the left and right sides of the mountain. Virtually all the houses in the foreground are in the current restricted zone and cannot be occupied. If you are wondering why people reside close to volcanoes, despite the danger, consider that these zones have many attractive features: rich soils, plenty of orographic rainfall, lush vegetation, moderate temperatures, and beautiful, dramatic landscapes. Volcanologists predict that the Soufrière volcano will remain moderately active for several years. [Mac Goodwin.]

highland zones: the Guiana Highlands and the Brazilian Highlands. Elsewhere in the lowlands, grasslands cover huge, flat expanses, including the llanos of Venezuela and the pampas of Argentina. The latter has become one of the region's most productive agricultural zones.

The largest feature of the South American lowlands is the Amazon Basin, drained by the Amazon River (see Figure 3.2 on page 115). This basin lies within Brazil and its western neighbors, but it has global biological significance: it contains the earth's largest expanse of tropical rain forest, 20 percent of the earth's fresh water, and more than 100,000 species of plants and animals. The basin's rivers are so deep that ocean liners can steam 2300 miles (3700 kilometers) upriver, all the way to Iquitos, jokingly referred to as Peru's "Atlantic seaport." The vast Amazon River system starts as streams high in the Andes. These streams eventually unite as rivers that flow eastward toward the Atlantic, which are joined along the way by rivers flowing from the Guiana Highlands to the northeast and the Brazilian Highlands to the southeast. Once these rivers reach the flat land of the Amazon Plain, their velocity slows abruptly, and fine soil particles, or **silt**, sink to the riverbed. When the rivers flood, silt and organic material transported by the floodwaters renew the soil of the surrounding areas, nourishing millions of acres of tropical forest. Not all of the Amazon Basin is rain forest, however. Variations in rainfall, cloud cover, wind patterns, and soil types, as well as human activity, can create the conditions for seasonally dry deciduous tropical forest or even grassland.

The interior reaches of the Amazon Basin are home to some of the last remaining relatively undisturbed Native American cultures. When Europeans first came to these tropical wetlands, there were few ways to exploit them for profit, so for several centuries they remained largely unexplored by Europeans. The Amazon Basin was home to perhaps 2 million hunters and gatherers and subsistence cultivators. In the nineteenth century, European settlement and commercial exploitation (mining, lumbering, and some ranching and plantation cultivation) began in earnest, and soon the native people were in decline. By 1900, there were 230 known ethnic groups in the Amazon Basin, often referred to as tribes. Since then, 87 entire tribes have become extinct, while in some isolated tribes, such as the Secoya, only a few hundred members survive. Many died as the result of introduced diseases or mistreatment while laboring on plantations or in mines. Recently, others have been forced to leave by encroachment on their lands and losses of habitat and environmental quality caused by logging, ranching, oil exploration, commercial agriculture, and new settlements.

Climate

From the steaming jungles of the Amazon and the Caribbean to the high, glacier-capped peaks of the Andes to the parched moonscape of the Atacama Desert, the climatic variety of Middle and South America is astounding (Figure 3.3). This variety results from several factors. The wide range of temperatures reflects the great distance the landmass spans on either side of the equator: northern Mexico lies at 33°N latitude, while Tierra del Fuego, at the southern tip of South America, lies at 55°S latitude. The region's long mountainous spine creates tremendous changes in

CLIMATIC ZONES

Tropical Humid Climates (A)
- Tropical wet
- Tropical wet/dry

Arid and Semiarid Climates (B)
- Steppe
- Desert

Temperate Climates (C)
- Midlatitude, moist all year
- Subtropical, winter dry
- Mediterranean, summer dry

Cool Humid Climates (D)
- Continental, moist all year

Coldest Climates (H)
- High altitude

Northeast trade winds bring heavy seasonal rains to Central America, parts of the Caribbean, and the Amazon Basin of South America.

Seasonal winds bring rains to west coast of Central America.

Peru Current brings cold surface waters. Air above is very dry, contributing to arid coastal climate. *El Niño* brings warm water instead of cold every few years, impacting climates worldwide.

Rain Shadow Andes block the northeast and southeast trade winds off the Atlantic.

Globe-encircling eastward-blowing winds (also called westerlies) bring steady cold rains. The rainfall supports forests much like those of the Pacific Northwest of North America.

Southeast trade winds bring rain to the eastern midlatitudes of South America.

Rain Shadow Andes block rains coming from the west.

FIGURE 3.3 Climates of Middle and South America. After reading about the circulation of air and moisture in this region and looking at this climate map, can you offer an explanation for the existence of the Atacama and Patagonian desert regions? After analyzing Figure 3.3, see if you can estimate where the various temperature-altitude zones would be found on this map.

altitude that also contribute to the wide range of temperatures. In addition, global patterns of wind and ocean currents result in a distinct pattern of precipitation.

Temperature-Altitude Zones. Four main **temperature-altitude zones** are commonly recognized by Middle and South Americans (Figure 3.4). As altitude increases, the temperature of the air decreases by about 1°F per 300 feet (1°C per 165 meters) of elevation. Thus temperatures are warmest in the lowlands, which are known in Spanish as the **tierra caliente,** or "hot lands," and extend up to about 3000 feet (1000 meters) in elevation. In some parts of the region, these lowlands cover wide expanses. Where moisture is adequate, tropical rain forests thrive, as well as

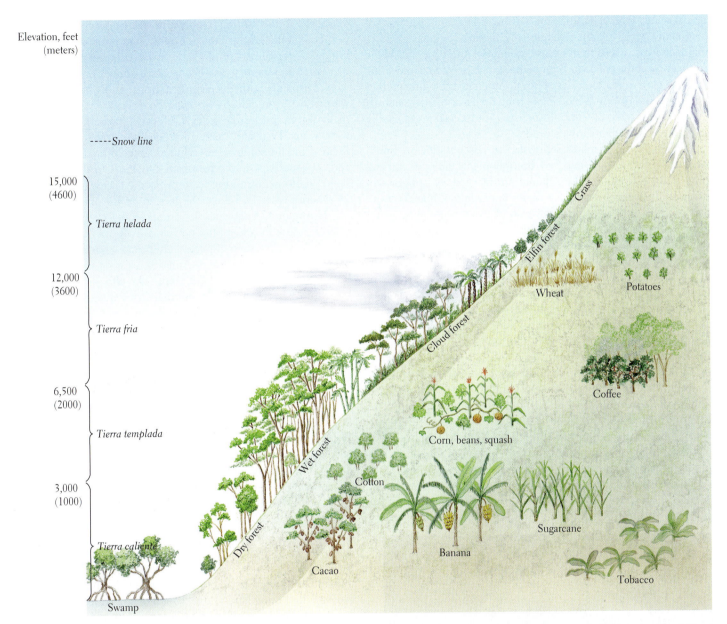

Elevation, feet (meters)

- - - - Snow line

15,000 (4600)

Tierra helada

12,000 (3600)

Tierra fria

6,500 (2000)

Tierra templada

3,000 (1000)

Tierra caliente

Swamp

Dry forest

Wet forest

Cloud forest

Elfin forest

Grass

Wheat

Potatoes

Coffee

Corn, beans, squash

Cotton

Cacao

Banana

Sugarcane

Tobacco

FIGURE 3.4 Temperature–altitude zones of Middle and South America. Each zone is suited for growing specific crops, and the natural vegetation changes with altitude as well. [Illustration by Tomo Narashima, based on fieldwork and a drawing by Lydia Pulsipher.]

a wide range of tropical crops, such as bananas, sugarcane, cacao, and pineapples. Many coastal areas of the *tierra caliente*, such as northeastern Brazil, have become zones of plantation agriculture that support populations of considerable size.

Between 3000 and 6500 feet (1000 to 2000 meters) are the cooler *tierra templada* ("temperate lands"). The year-round springlike climate of this zone drew large numbers of Native Americans in the distant past and, later, Europeans. Here such crops as corn, beans, squash, various green vegetables, wheat, and coffee are grown. Between 6500 and 12,000 feet (2000 to 3600 meters) are the *tierra fria* ("cool lands"). Many midlatitude crops—such as wheat, fruit trees, and root vegetables (potatoes, onions, and carrots)—and cool-weather vegetables—such as cabbage and broccoli—do very well at this altitude. Several modern

population centers are in this zone, including Mexico City and Quito, Ecuador. Above 12,000 feet (3600 meters) are the *tierra helada* ("frozen lands"). At the lowest reaches of this zone, some grains and root vegetables are cultivated, and some animals—such as llamas, sheep, and guinea pigs—are raised for food and fiber. Higher up, vegetation is almost absent, and mountaintops emerge from under snow and glaciers. The remarkable feature of such tropical mountain zones is that in a single day of strenuous hiking, one can encounter most of the climate types known on earth.

Precipitation. The pattern of precipitation throughout the region (see Figure 1.12 on page 25) is influenced by the interaction of global wind patterns, topographic barriers such as mountains, and nearby ocean currents. The northeast **trade winds**

(tropical winds that blow from the northeast and southeast toward the equator) sweep off the Atlantic (see Figure 3.3 on page 118), bringing heavy seasonal rains to eastern Central America, parts of the Caribbean, and the Amazon Basin. The southeast trade winds bring rain to the eastern midlatitudes of South America, including parts of the Amazon Basin. Winds blowing in from the Pacific bring seasonal rain to the west coast of Central America, but mountains block the rain from reaching the Caribbean side.

The Andes are a major influence on precipitation in South America (see Figures 1.12 and 3.3). Rains borne by the northeast and southeast trade winds off the Atlantic are blocked, creating a rain shadow on the western side of the Andes in northern Chile and western Peru. Southern Chile is in the path of eastward-trending winds that sweep north from Antarctica, bringing steady cold rains that support forests similar to those of the Pacific Northwest of North America. The Andes block this flow of wet, cool air and divert it to the north, thus creating an extensive rain shadow on the eastern side of the mountains along the southeastern coast of Argentina (Patagonia).

The pattern of precipitation is also influenced by the adjacent oceans and their currents. Along the west coasts of Peru and Chile, the cold surface waters of the Peru Current bring cold air that is unable to carry much moisture. The combined effects of the Peru Current and the central Andes rain shadow have created what is possibly the world's driest desert, the Atacama Desert of northern Chile.

El Niño. An interesting and only partly understood aspect of the Peru Current is its tendency to change its position every few years. When this happens, warm water flows eastward from the western Pacific, bringing warm water and rain, instead of cold water and drought, to parts of the west coast of South America. Among other consequences, this change in the current causes fish catches to fall drastically as the nutrient-poor warm water replaces the nutrient-rich cold water of the Peru Current. The event was named **El Niño,** or "the Christ Child," by Peruvian fishermen, who noticed that when it occurs, it reaches its peak sometime in December, when Hispanic cultures celebrate Christmas. El Niño has worldwide effects, periodically bringing cold air and drought to normally warm and humid western Oceania, torrential rains to the usually dry Peruvian coast, unpredictable weather patterns to Mexico and the southwestern United States, droughts to the Amazon, and perhaps fewer hurricanes to the Caribbean. The El Niño phenomenon in the western Pacific is covered in Chapter 11, where Figure 11.5 illustrates its transpacific effects.

Hurricanes. Large storm systems form every year in the Atlantic, close to Africa (and occasionally in the Pacific as well). They sometimes sweep into the Caribbean Sea or onto the shores of North, Middle, and South America, causing major wind and flood damage to buildings and crops. The effects of these storms have always been significant, but their consequences are increasing as coastal population density increases.

HUMAN PATTERNS OVER TIME

The conquest of the Americas by Europeans set in motion a series of changes that helped create the ways of life found in Middle and South America today. That conquest wiped out much of Native American civilization and set up colonial regimes in its place. It introduced many new cultural influences to the Americas and led to the lopsided distribution of power and wealth that continues to this day.

The Peopling of Middle and South America

Between 13,000 and perhaps 40,000 years ago, groups of hunters and gatherers from Asia spread throughout North America after crossing the Bering land bridge, which once connected northeastern Asia to Alaska. Many of these groups ventured south through Mexico and Central America. Anthropologists now think that in only a few thousand years, these people managed to adapt to a wide range of ecosystems: the dry mountains and valleys of northern Mexico, the tropical rain forests and upland zones of Central America, the vast Amazon Basin, the Andes Mountains, and the cold, windy grasslands near the tip of South America. Recent archaeological evidence from the Monte Verde site in the southern Andes indicates occupation perhaps as much as 30,000 years ago.

By 1492, there were 50 million to 100 million people in Middle and South America. In some places, even rural population densities were high enough to threaten sustainability. These people altered the landscape in many ways. They modified drainage to irrigate crops, constructed raised fields in lowlands, terraced hillsides, built paved walkways across swamps and even mountains, constructed cities with sewer systems and freshwater aqueducts, and raised huge earthen and stone ceremonial structures that rivaled the Pyramids of Egypt. They also perfected a productive system of **shifting cultivation** that is still common in wet, hot regions in Central America and the Amazon Basin.

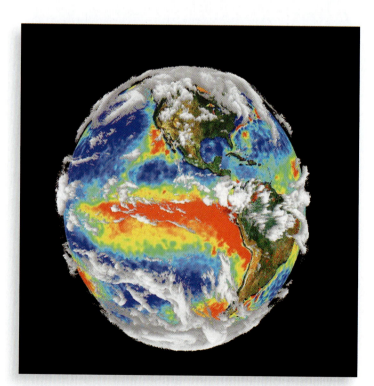

El Niño (in red) flares across the Pacific in this image created from satellite data. [Rudolf B. Husar, *National Geographic* (October 2000): 90–91.]

Thinking Geographically: ON THE WEB

Small plots were cleared in forestlands, the brush was dried and burned to release nutrients to the soil, and the clearings were planted with multiple crop species. Each plot was used for only 2 or 3 years and then abandoned for many years of regrowth. This system was (and is) highly productive per unit of land and labor if there is sufficient land for long periods of regrowth. Shifting cultivation produces abundant food for the cultivators, but little surplus for marketing. If population pressure increased to the point that the same plot had to be used before it had fully regrown, yields would have decreased drastically, and forest resources would have been depleted.

These various preconquest activities significantly changed the habitats of land and water species, some of which became extinct; in addition, some animals were hunted to extinction. Then the trauma of the Spanish conquest and of the ensuing 500 years of European dominance further affected physical environments and obliterated many remarkable accomplishments of Native American cultures, some of which are only now coming to light through archaeological research.

Although they lacked the wheel and gunpowder, the **Aztecs** of the high central valley of Mexico had some technologies (such as urban water and sewage systems) and levels of social organization (such as highly organized marketing systems) that rivaled or surpassed those of civilizations of the time in Asia and Europe. Recent historians have concluded that, on the whole, all social classes of Aztecs lived better and more comfortably than did their contemporaries in Europe.

The largest pre-Columbian state in the Americas was that of the **Incas,** whose domain stretched from southern Colombia to northern Chile and Argentina. The main population clusters were in the Andes highlands. The cooler temperatures at these high altitudes eliminated the diseases of the tropical lowlands, yet proximity to the equator guaranteed mild winters and long growing seasons. For several hundred years, until the Spanish arrived in the 1500s, the Inca Empire was one of the most efficiently managed in the history of the world. The ruling class divided the population into a hierarchy of family groups: communities of 10, 50, 100, and all the way up to 40,000 families. Each unit was strictly controlled by a leader who reported to his superior. Highly organized systems of cooperative and reciprocal labor were used to construct paved road systems linking high Andean communities and to build great stone centers in the highlands. Incan agriculture was advanced, particularly in the development of staple crops such as the numerous varieties of potatoes that were adapted to grow in different upland environments.

The Conquest

The European conquest of Middle and South America was one of the most significant events in human history, rapidly altering landscapes and cultures and ending the lives of millions of people. Europeans instigated the conquest after learning of the Americas from Christopher Columbus following his first voyage in 1492. Most of the early colonizers came from Spain and Portugal, on Europe's Iberian Peninsula. By the 1530s, a mere 40 years after Columbus's arrival, all major population centers in the Americas had been conquered and were rapidly being transformed by Iber-

ian colonial policies (Figure 3.5). The superior military technology of the Iberians speeded the process considerably, but the major factor that explains the swiftness and completeness of the conquest was the vulnerability of Native Americans to diseases carried by the Europeans. Epidemics of diseases such as smallpox and measles killed nine out of ten Native Americans, who lacked the immunity to these diseases that most Europeans had developed. In about 150 years, the total population of the Americas was reduced by more than 90 percent, to just 5.6 million.

Columbus established the first Spanish colony in 1492 on the Caribbean island of Hispaniola, now occupied by Haiti and the Dominican Republic. This initial seat of empire expanded to include the rest of the Greater Antilles—Cuba, Puerto Rico, and Jamaica. At first, Native Americans were forced to work on plantations and in mines. Their populations soon plummeted, however, as a result of disease, malnutrition, and brutality. To obtain a new supply of labor, the Spanish initiated the first shipments of enslaved Africans to the Americas in the early 1500s.

Roman Catholic diplomacy prevented conflict between Spain and Portugal over the lands of the Americas. The Treaty of Tordesillas of 1494 divided the Americas at approximately 46°W longitude. Portugal took all lands to the east and eventually acquired much of what is today Brazil; Spain took all lands to the west.

The first part of the mainland to be conquered was Mexico, home to several advanced Native American civilizations, most notably the Aztecs. The Spanish, unsuccessful in their first attempt to capture the Aztec capital of Tenochtitlán, succeeded a few months later when a smallpox epidemic they inadvertently brought with them was in full sway. The Spanish demolished the Aztec capital in 1521, including its grand temples, public spaces, causeways, residences, and aqueducts. Mexico City was built on its ruins, to become one of two main seats of the Spanish Empire in the Americas. Called the Viceroyalty of New Spain, this part of the empire extended from Panama in the south all the way to what is now San Francisco on the northern California coast. Wealth from the gold and silver mines of Mexico flowed through the port of Veracruz on the Caribbean and from there on to Spain.

The conquest of the Incas bore a remarkable resemblance to that of the Aztecs in that a tiny band of Spaniards was able to capture the capital city of an extensive empire. Their leader, Francisco Pizarro, himself admitted that his campaign received its greatest assistance from a smallpox epidemic (brought unintentionally by earlier Spanish scouts) that preceded his arrival.

Out of the ruins of the Inca Empire in South America, the Spanish created the Viceroyalty of Peru, which originally encompassed all of South America except Portuguese Brazil. The newly constructed capital of Lima flourished on the trade of the huge silver mines established in the highlands of Bolivia at Potosí.

The conquest of Brazil by the Portuguese was similar to the Spanish conquest of other areas, in that land was seized and people were killed or enslaved; but it differed in some key respects. Brazil was apparently only sparsely populated by people who lived in small, impermanent villages. There were no highly organized urban cultures, as in parts of Middle America and the Andes. Most Atlantic coastal cultures were annihilated early on, and the populations of the huge Amazon Basin declined sharply as contagious diseases spread through trading. It proved difficult to extract

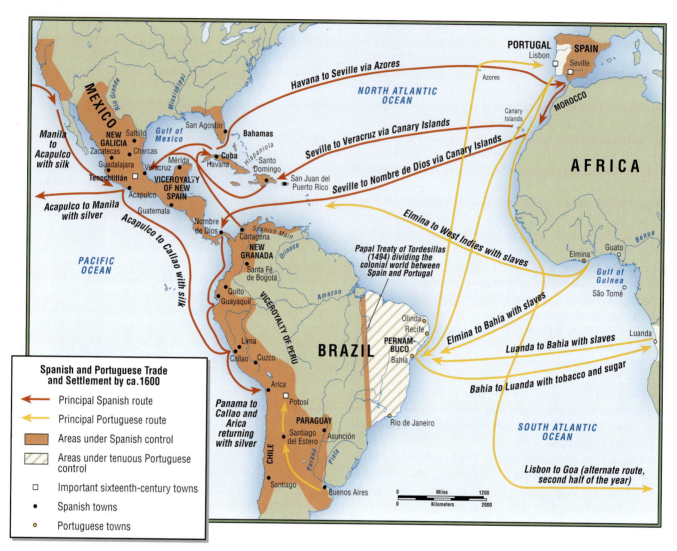

FIGURE 3.5 Spanish and Portuguese trade routes and territories in the Americas circa 1600. The major trade routes from Spain to its colonies led to the two major centers of its empire, Mexico and Peru. The Spanish colonies could trade only with Spain, but there were direct trade routes from Portuguese colonies in Brazil to Portuguese outposts in Africa. Many millions of Africans were enslaved and traded to Brazilian plantation and mine owners. [Adapted from *Hammond Times Concise Atlas of World History* (Maplewood, N.J.: Hammond, 1994), pp. 66–67.]

quick wealth from tropical forests, however, so most of the Amazon was left alone for several centuries. Instead, the Portuguese focused on extracting mineral wealth, especially gold and precious gems, from the Brazilian Highlands and on establishing plantations along the Atlantic coast. Only in the nineteenth and twentieth centuries did commercial interest in the Amazon Basin increase to the point that serious colonization took place. Today, indigenous Amazonians are some of the last Native Americans still struggling to preserve their own ways of life.

A Global Exchange of Crops and Animals

Beginning in the earliest days of conquest, a number of plants and animals were exchanged among the Americas, Europe, Africa, and Asia. Many plants from the Old World are essential to agriculture in Middle and South America today: rice, sugarcane, bananas, citrus, melons, onions, apples, wheat, barley, and oats are just a few examples. When disease decimated the native pop-

ulations of the Americas, the colonists turned the abandoned land into pasture for herd animals imported from Europe, including sheep, goats, oxen, cattle, donkeys, horses, and mules. European draft animals helped fill in Native American irrigation canals, drain lakes, and plow the relics of complex Indian gardens into huge one-crop fields of sugarcane or wheat. Surviving Native Americans living on the plains of Mexico and Argentina adopted European-introduced horses, using them to hunt large game. Others learned to herd sheep and adapted their fleece to ancient spinning and weaving technologies.

Plants first domesticated by Native Americans have changed diets everywhere and have become essential components of agricultural economies around the globe. In Europe, for example, the potato had so improved the diet of the poor by 1750 that it fueled a population explosion. Manioc, though less nutritious, played a similar role in West Africa. Corn is a widely grown garden crop in Africa, and peanuts and cacao (chocolate) are essential cash crops to many African farmers. Peppers are a cash crop in China, as

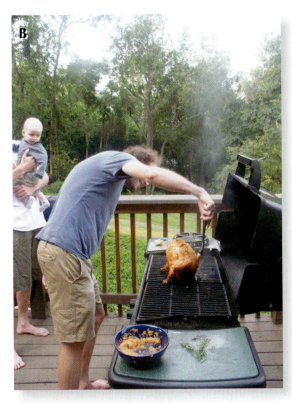

COMPARING LOCAL LANDSCAPES

Meat! Barbequed meat is a favorite throughout the Americas. Here, in two different contexts, you can see Americans cooking for friends using a method first described by the Europeans who reached the Caribbean with Columbus. The word *barbeque* is probably a Native American term from the island Caribbean.

(A) This woman is in South America, cooking an animal long associated with that region. Can you identify the animal? From the background, can you tell where she is cooking? (B) This fellow is cooking in another American context. Where do you think he is? What can you guess about the circumstances of the people? What is the origin of the animal being cooked? [A, © 2004 Julia Ng; B, Mac Goodwin.]

pineapples are in the Pacific, and tomatoes have transformed Mediterranean cuisine. Table 3.1 lists some of the more common globally used plants from Middle and South America and their sites of probable domestication.

The Legacy of Underdevelopment

Today, despite Middle and South America's potentially productive environment, perhaps 40 percent of the people are poor and lack access to such basic resources as land, adequate food and shelter, and basic education. A small elite class, however, enjoys levels of affluence equivalent to those of the wealthy in the United States. In part, these inequalities result from the lingering effects of colonization: economic policies that still favor the export of raw materials and foster the dominance of outside investors who spend their profits elsewhere rather than reinvesting them within the region. Other factors leading to inequality are official corruption at all levels of government and social attitudes and policies that stifle local entrepreneurial development and upward social mobility. All these factors have resulted in an undereducated and underskilled populace and an infrastructure that is inadequate by modern standards. For these reasons, it is dif-

ficult for the region to compete in the global marketplace, which increasingly demands an educated labor force, good transport networks, and technological capabilities.

The first colonial enterprises to be established were based on the extraction of raw materials, including gold, silver, and other minerals, timber, and various agricultural products, such as cotton, indigo, tobacco, and sugar. These activities were part of an emerging policy of **mercantilism** by which the rulers of Spain and Portugal and, later, England, France, and Holland sought to increase the power and wealth of their realms by managing all aspects of production, transport, and trade. Resources were mined or grown in the colonies at the lowest possible cost, brought home to Europe, and manufactured into trade goods (cotton into cloth, indigo into dye, sugar into rum). These finished products were then sold to the colonies and elsewhere in the world markets. Money flowed into Europe as payment for the manufactured goods, enriching the merchants, and taxes on the trade brought revenues to the national treasuries. A *merchant marine* (a fleet of privately owned trading ships) was maintained to transport resources, manufactured goods, and, once the slave trade began in earnest in the 1600s, people. Nearly all commercial activity was focused solely on enriching the **colonizing,** or **mother, country.** The colonies of Spain in the Americas, for

TABLE 3.1 Major domesticated plants originating in the Americas and now used commercially around the world

Type	Common name	Scientific name	Place of origin
Seeds	Amaranth	*Amaranthus cruentus*	Southern Mexico, Guatemala
	Beans	*Phaseolus* (4 species)	Southern Mexico
	Maize (corn)	*Zea mays*	Valleys of Mexico
	Peanut	*Arachis hypogaea*	Central lowlands of South America
	Quinoa	*Chenopodium quinoa*	Andes of Chile and Peru
	Sunflower	*Helianthus annuus*	Southwestern and southeastern North America
Tubers	Manioc (cassava)	*Manihot esculenta*	Lowlands of Middle and South America
	Potato (numerous varieties)	*Solanum tuberosum*	Lake Titicaca region of Andes
	Sweet potato	*Ipomoea batatas*	South America
	Tannia	*Xanthosoma sagittifolium*	Lowland tropical America
Vegetables	Chayote (christophene)	*Sechium edule*	Southern Mexico, Guatemala
	Peppers (sweet and hot)	*Capsicum* (various species)	Many parts of Middle and South America
	Squash (including pumpkin)	*Cucurbita* (4 species)	Tropical and subtropical America
	Tomatillo (husk tomato)	*Physalis ixocarpa*	Mexico, Guatemala
	Tomato (numerous varieties)	*Lycopersicon esculentum*	Highland South America
Fruit	Avocado	*Persea americana*	Southern Mexico, Guatemala
	Cacao (chocolate)	*Theobroma cacao*	Southern Mexico, Guatemala
	Papaya	*Carica papaya*	Southern Mexico, Guatemala
	Passion fruit	*Passiflora edulis*	Central South America
	Pineapple	*Ananas comosus*	Central South America
	Prickly pear cactus (tuna)	*Opuntia* (several species)	Tropical and subtropical America
	Strawberry (commercial berry)	*Fragaria* (various species)	Genetic cross of Chile berry + wild berry from North America
	Vanilla	*Vanilla planifolia*	Southern Mexico, Guatemala, perhaps Caribbean
Ceremonial and drug plants	Coca (cocaine)	*Erythroxylon coca*	Eastern Andes of Ecuador, Peru, and Bolivia
	Tobacco	*Nicotiana tabacum*	Tropical America

Source: B. Kermath and L. Pulsipher, "Guide to Food Plants Now Used in the Americas," forthcoming 2005.

example, were allowed to trade only with Spain—not with one another. If businesspeople in Peru wanted to trade legally with their counterparts in Mexico, for instance, the goods first had to cross the Atlantic in Spanish ships, be taxed in Spain, and then be reshipped back to Mexico in Spanish ships. These cumbersome and expensive restrictions nurtured a huge underground economy and institutionalized dishonesty as merchants and customers often tried to bribe their way around the trade laws. Though some trade took place, the building of formal trading institutions was blocked, and the numerous restrictions led to considerable resentment of the Spanish authorities.

In the early nineteenth century, wars of independence transformed the region (Figure 3.6). The modern countries of Middle and South America emerged, and Spain was left with only a few colonies in the Caribbean. Many supporters of these revolutions were not true reformists, but rather **Creoles** (people of usually

European descent born in the Americas) who wished to consolidate their control of economic assets in the colonies. Also active in the effort to overthrow the Spanish Empire were relatively wealthy **mestizos** (people of mixed European and Native American descent), whose access to the profits of the colonial system had been restricted due to their race. Once these pseudo-revolutionary leaders came to power, they emulated their colonial predecessors. They became a new elite that controlled the state and monopolized economic opportunity, doing little to expand economic development or the people's access to political power. The mestizos who participated strove to emulate European ways and deny their native roots.

During the twentieth century, some countries in the region began to experiment with radical ways of fostering development, as we shall see in the section on economic and political issues. Today, the economies of Middle and South America are much more complex and technologically sophisticated than they once were. Nevertheless, the colonial pattern of dominance by an elite and dependence on extraction of raw materials remains, contributing to a persistent pattern of underdevelopment, unsustainable use of resources, and unequal distribution of wealth.

POPULATION PATTERNS

The events of the last 500 years have greatly affected the present distribution and character of human settlement in Middle and South America. Conquest and colonization brought the swift demise of Native Americans; soon they were more than replaced by millions of people from Europe, Africa, and Asia. It would be difficult to find a case elsewhere in human history that rivaled this massive shift in human population. Today, the patterns of human settlement continue to change, but now these changes are being generated from within the region, rather than coming from outside. The population continues to climb, but primarily because of high birth rates rather than immigration. At present, the major migration trend is internal: rural-to-urban migration is taking place everywhere in the region and transforming traditional ways of life. A second, growing migration trend is international: many people from the Caribbean, Mexico, and Central America are leaving their countries, temporarily or permanently, to seek opportunities elsewhere. In the last 20 years, the United States has been the preferred destination for hundreds of thousands of these transnational migrants, but Europe

Persons per

sq km	sq mi
Over 100	Over 260
50–100	130–260
10–49	25–129
1–9	3–24
Under 1	Under 3

✪ ● Capitals and cities 1 million and over
✪ Capitals under 1 million

FIGURE 3.7 Population density in Middle and South America. In Middle and South America, as in most places on earth, there is no consistent relationship between population density patterns and landforms. Some of the highest densities are in highland areas (*tierra templada*). Elsewhere, high densities are found in lowland zones along the Pacific coast of Central America and the Atlantic coast of South America. Most of these coastal lowland concentrations, in *tierra caliente*, are near important seaports. There are also many empty areas in the region. What are some of the explanations for why these lands remain so lightly settled? Are they likely to stay that way? [Adapted from *Hammond Citation World Atlas* (Maplewood, N.J.: Hammond, 1996.]

is increasingly popular as a destination. Since 1996, several hundred thousand Peruvians and Ecuadorians have migrated to Spain and Portugal in search of jobs within the European Union.

Population Numbers and Distribution

As of 2003, 540 million people were living in Middle and South America—close to ten times the population of the region in 1492.

This number is about 200 million more than presently live in North America.

Population Distribution. The population density map in Figure 3.7 reveals a very unequal distribution of people in Middle and South America. It is often thought that people concentrate where certain physical landforms are common, but that is only partly true. If you compare Figure 3.7 with the regional map on pages 114–115, you will see that there is no obvious, consistent relationship between population density patterns and landforms. Some of the highest densities, such as those around Mexico City and in Colombia and Ecuador, are in highland areas. Elsewhere, high concentrations are found in lowland zones along the Pacific coast of Central America and especially along the Atlantic coast of South America. In tropical and subtropical zones, the cool uplands (*tierra templada*) are particularly pleasant and healthful and were comparatively densely occupied even before the European conquest. Most of the coastal lowland concentrations, in *tierra caliente*, are near important seaports. Despite the hot climate, people are attracted to these port cities by the possibility of jobs in the vibrant and varied economy and by the interesting social life. Living near the sea is also attractive because the water modifies the heat and humidity, making the coast relatively cooler and breezier than lowlands in the interior.

Figure 3.7 also reveals the lands in the region that are relatively empty of human occupants, especially in South America. Cold and windy Patagonia at the far south end of South America has very few people, as do the desert regions of Chile and Peru along the Pacific coast. The vast Amazon Basin is also only lightly settled. Despite recent efforts to develop and populate this wet tropical zone, it seems able to truly sustain only hunting and gathering, shifting cultivation, and light forestry. Although some people think that advancements in technology will make it possible to use and settle the Amazon environments more intensively in the future, so far no such strategies have proved sustainable for more than a few years.

Population Growth. Twentieth-century rates of natural population increase have been high in Middle and South America. (Recall from Chapter 1 that the rate of natural increase is population growth resulting from births alone, not counting growth resulting from immigration.) Although rates of natural increase are now declining (Figure 3.8), at present rates, the region's population could double in just 41 years (in North America, doubling will take 140 years). Rapid population doubling rates are a disturbing prospect for a region in which the majority of people already suffer from a low standard of living. Any gains in well-being might be checked by the costs of supplying more and more new people with houses, schools, and hospitals.

There are a number of reasons why rates of natural increase in this region have remained high when compared with those of North America, Europe, and East Asia, or even with world averages. One is that until recently, most people in Middle and South America lived in agricultural areas. Children were seen as sources of wealth because they could do useful work at a young age and eventually would care for their aging elders. High infant death rates encouraged parents to have many children to be sure of raising at least a few to adulthood. In addition, the Catholic church has discouraged systematic family planning. Further, as will be discussed later, the cultural mores of *machismo* and *marianismo* reinforced the idea that both men and women validate themselves as adults by reproducing prolifically. When medical care in the region improved modestly beginning in the 1930s, death rates began to decline rapidly. By 1975, death rates were about one-third what they had been in 1900. Typically birth rates do not decrease as quickly as death rates, primarily because of deeply ingrained values such as those we have just described. Hence population growth was especially rapid between 1940 and 1975.

By the 1980s, for a number of reasons, Middle and South Americans were beginning to limit family size, and population growth rates started to fall. Now the region is beginning to undergo demographic transition (see Figure 1.20 on page 42). Between 1975 and 2001, the rate of natural increase for the entire region fell from about 2.5 percent to 1.9 percent—still a high rate of growth when compared with those of North America, Europe, and the East Asian countries of China, Japan, and the Koreas (all at 0.7 percent or less in 2003). Figure 3.8 compares the natural increase rates of selected countries in the region for 1975–2001 and their projected rates for 2001–2015.

In 2003, the Caribbean as a whole had the lowest rate of natural increase in the region (1.2 percent), with Cuba (0.5 percent), Trinidad and Tobago (0.6 percent), and Barbados (0.6 percent) having the very lowest rates, about the same as North America (0.5 percent). Several factors explain these low rates. Cuba, Trinidad and Tobago, and Barbados provide both women and men with

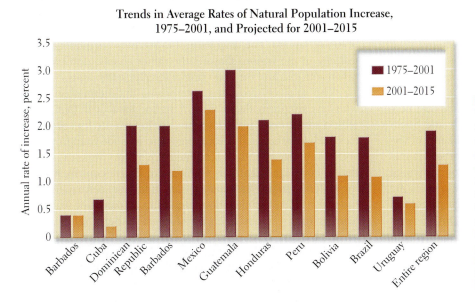

Trends in Average Rates of Natural Population Increase, 1975–2001, and Projected for 2001–2015

FIGURE 3.8 Trends in natural population increases, 1975–2015. Rates of natural increase have declined steadily throughout the region and are projected to continue to do so into the future, but in many countries natural population increase remains high enough to outstrip efforts at improving standards of living. In this region a majority of people already suffer from a low standard of living. Resources that might have gone toward improving their lives will be used instead to supply more and more new (young) people with the basics for survival. Note that the rates of natural increase given here are for extended periods and, therefore, do not match figures for a given year. [Adapted from *United Nations Human Development Report 2003,* Table 5, United Nations Development Programme.]

education, meaningful work, and basic health care. Infant mortality rates are low, so people can expect to see their one or two children grow to adulthood. Barbados prospers from a diversified economy of manufacturing, computerized information processing, tourism-related services, **remittances** (money sent home by migrants), and agriculture. Its skilled citizens have many economic, civic, and social options for constructing interesting lives that make large families less attractive. Trinidad and Tobago, while less prosperous, has oil-related industries and a thriving business community. Cuba has made a strong effort to improve social welfare and encourage smaller families over the past 35 years.

In Mexico and Central America, poverty is more widespread and women have less access to education and employment than in the Caribbean; both factors are associated with high rates of natural increase (2.4 percent for these subregions). South America, also afflicted with high rates of poverty, has a rate of just 1.5 percent. Within South America, however, there is considerable variation: in the year 2003, the rate of natural increase in Bolivia was 2.3 percent, in Argentina 1.2 percent, in Paraguay 2.7 percent, and in Uruguay 0.6 percent. Again, this variation can be partially explained by differing standards of living and access to education and jobs, especially for women. A wild card in population projections for this region is the rising incidence of HIV-AIDS, which is discussed in the section on human well-being (page 156).

Migration and Urbanization

Migration is perhaps the most important social force at work in the world today. Why? Because people are migrating at unprecedented rates, and because migration often abruptly introduces large numbers of people from rural areas into new, usually crowded, cities. Residence in urban places may or may not provide rural immigrants with more opportunities, but it will surely expose them to values and ways of life that contrast sharply with those they have known. Since the early 1970s, the region of Middle and South America has led the world in rates of migration.

Geographers recognize the importance of economic and social factors in initiating and sustaining migration. People are drawn to migrate by the possibility of better job opportunities. Often they hear of those opportunities from networks of friends and family who have migrated earlier and report increased prosperity. Communication between these new migrants and their home communities, including the sending of remittances, stimulates still more rural-to-urban migration. As a result, cities throughout the region have grown remarkably quickly. More than 75 percent of the people in the region now live in towns with populations of at least 2000; but increasingly, in Middle and South America and elsewhere, one city or one metropolitan area of several contiguous cities is vastly larger than all the others, sometimes accounting for one-fourth or more of the country's total population. Such cities are called **primate cities;** examples are Mexico City (with 21 million people and about 20 percent of Mexico's population; see the box on Mexico City on the next page); Managua, Nicaragua (20 percent); Lima, Peru (30 percent); Santiago, Chile (34 percent); and Buenos Aires, Argentina (38 percent).

The concentration of people into just one or two primate cities in a country leads to *uneven spatial development and urban bias* (government policies and social values that favor urban areas). Wealth and power are concentrated in one place, while distant rural areas, and even other towns and cities, have difficulty competing for talent, investment funds, industries, and government services. Many provincial cities languish as their most educated youth leave for the main city. Primate cities are usually found in underindustrialized countries. They attract people even though they do not provide enough jobs, housing, or services for the flood of hopeful migrants.

Because the rush to the cities has not been accompanied by a general rise in productivity and prosperity, no city in the region is prepared for the influx. Spending on urban infrastructure (such as roads and sanitation) and on social services (such as schools and hospitals) has failed to keep pace with the inflow of people, and the signs of urban decay are everywhere. (For an exception to this pattern, see the box "Curitiba, Brazil: A Model of Urban Planning" on page 131.) Although there are wealthy neighborhoods in the cities, they are often heavily guarded because of their proximity to unsightly, uncontrolled shantytowns built by the poor. These self-help settlements—called **colonias, barrios, favelas,** or **barriadas**—can arise spontaneously. Because most recent urban migrants can't afford to buy or lease land, they become squatters who invade a piece of property after dark and set up simple dwellings by morning. Once established, these communities are extremely difficult to dislodge, because the impoverished are such a huge portion of the urban population that even those in positions of power will not challenge them directly. The result is a city landscape that is remarkably different from the common U.S. pattern of a poor inner city surrounded by affluent suburbs, with relatively clear spatial separation of classes, races, and family styles. Rather, housing types are intermingled. Inner-city neighborhoods of vastly different socioeconomic classes nestle close to one another, and the suburban fringe is also a mixture of wealthy and poor neighborhoods. The wealthy neighborhoods are similar in architectural style and formality to those of North America, but are usually ringed with high fences or walls. The poor neighborhoods, which predominate, are often built of unconventional materials and laid out in an uncontrolled and seemingly chaotic pattern. Figure 3.10 (page 130) shows the distribution of favelas in Vitória, Brazil, a city about 300 miles northeast of Rio de Janeiro. The diagram next to this map is a model of a typical city in the region, showing the rather chaotic distribution of urban activities and the outer ring of squatter settlements.

The squatters are often enterprising and admirable people who have simply made the most of the bad hand dealt them. Such settlers sometimes organize themselves to pressure for social services such as schools, day care, part-time jobs for young mothers, and sewer systems. Some farsighted urban governments, such as that in Fortaleza (Ceará) in northeastern Brazil, contribute building materials so that the squatters can build more permanent structures with at least crude indoor plumbing. Expanses of shacks and lean-tos have thereby been transformed through self-help into modest but livable suburban places. Moreover, these communities can be centers of pride and support for their residents, where community work, music, folk belief systems, and crafts flourish. For

AT THE LOCAL SCALE   Mexico City

Mexico City is one of the world's largest primate cities (Figure 3.9). It has drawn so many migrants because it is a vibrant center of cultural heritage with large universities and museums. And it is a world center of international banking and financial institutions and home to the headquarters of several Mexican multinational corporations. Mexico City and adjacent cities such as Cuernavaca, Toluca, and Puebla are also home to thousands of smaller formal and informal enterprises that employ millions from the continuous stream of migrants arriving from the countryside. The city's physical situation, its high population density, and its rapid modernization have all contributed to major environmental crises. Automobiles and trucks crowd the streets, often creating gridlock traffic jams that take hours to clear. The snarl of motor vehicles contributes tons of CO_2 pollution to the air, which is prevented from dispersing by the surrounding mountains. Although the *tierra fria* climate of Mexico City once made it a place of sparkling clear vistas, heavy brown smog now obstructs views on most days and causes asthma and related respiratory maladies in the citizenry (see photo).

The millions of poor migrants coming to the city have had to build their own housing, which may be unsafely constructed and usually lacks adequate sanitary plumbing—another source of pollution. Furthermore, the location of the Mexico City **conurbation** (multiple cities tightly linked; see Figure 3.9 and inset map) on a drained lake bed in a tectonically active zone makes it vulnerable to **subsidence** (sinking land), disastrous earthquake damage, and even volcanic hazards. But still the people come, because for all its faults, the city promises opportunities and rich life experiences.

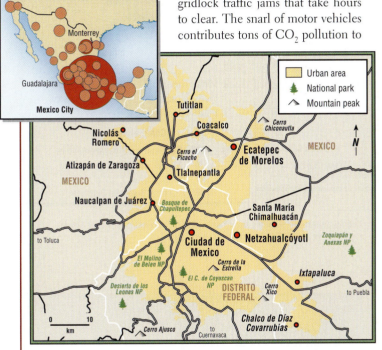

FIGURE 3.9 Map of Mexico City.

Once famed for its clear air and beautiful mountainous surroundings, Mexico City is now swathed in foggy pollution generated by motor vehicles and industry. The mountains trap and concentrate the pollution. [Stuart Franklin/Magnum Photos.]

example, many of the most prestigious steel bands of Port of Spain, Trinidad, have their homes in such shantytowns. In Brazil, the favelas are home to samba clubs (musical ensembles) and folk religious movements such as Umbanda, Batuque, and Condomble (see the discussion and photo on pages 174–175).

Interestingly, rural women are just as likely to migrate to the city as rural men. This is especially true when employment is available in foreign-owned factories that produce goods for export, where women are preferred because they are a low-cost, usually passive labor force. Rural development projects can actually encourage female migration because, due to mechanization, these projects often end up decreasing available jobs. Women may be disproportionately affected because, despite their long experience working in agriculture, they are rarely considered for training as farm equipment drivers and mechanics. In urban areas, unskilled women migrants can usually find work as domestic servants. Low wages, however, force them to live in the households where they work, where they are often subject to sexual overtures, yet are themselves blamed if they fail to adhere to rigid standards of behavior. Male urban migrants, on the other hand, tend to depend on short-term day work, most often in low-skill jobs in construction, maintenance, small-scale manufacturing, and petty commerce. They must compete with a throng of other eager workers, and most are periodically unemployed. Many make some cash in the informal economy, doing such work as street vending, running errands, cleaning, washing cars, and recycling trash; some may engage in crime. The loss of family ties and village life is sorely felt by both men and women, and the chances for recreating normal family life in the urban context are extremely low. Opportunities to meet and marry mates of their

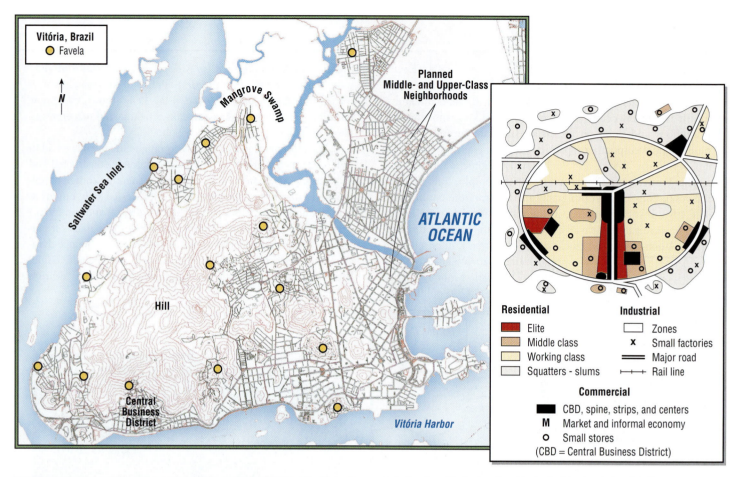

FIGURE 3.10 The pattern of squatter settlements in Vitória, Brazil, and Crowley's model of urban land use in mainland Middle and South America. [Vitória map adapted from fieldwork map by Eric Spears; Crowley's model of

Latin American city structure from *Yearbook of the Association of Pacific Coast Geographers* 57 (1995): 28; printed with permission.]

own background are limited, decent housing is beyond the reach of most couples, and the chances of having a child as a single parent are very high. Usually, the burden of single parenthood falls to the woman.

PERSONAL VIGNETTE   One squatter settlement in Fortaleza, Brazil, lies along the Atlantic coast just north of a line of high-rise tourist hotels. One day, I took a stroll past the settlement and was invited by a resident to join him on his porch, just visible in the photograph behind the red Coca-Cola chest. He maintained a small refreshment stand, and his wife ran a beauty parlor that catered to women from the area. The man explained that they had come to the city 5 years before, after being forced to leave the drought-plagued interior when they lost the right to cultivate their rented land. The landowner had sold it to the state government, which was building a dam to create an irrigation reservoir. Here in Fortaleza, they had slowly constructed the building they used for home and work from resources they collected along the beach, and eventually they were able to purchase the roofing tiles that gave it an air of permanency. [*Lydia M. Pulsipher, personal interview, August 1987.*]

Squatter communities, like this one in Fortaleza, Brazil, sometimes become permanently established with the help of government funding for cinder blocks, roofing tiles, and plumbing. [Lydia Pulsipher.]

AT THE LOCAL SCALE   Curitiba, Brazil: A Model of Urban Planning

The southern Brazilian city of Curitiba, capital of the grain-producing state of Paraná, has gained international renown for its environmentally friendly urban planning. Like most other Brazilian cities, Curitiba has mushroomed, doubling its population in just 20 years to more than 1.3 million people. But Curitiba is unusual in that it has carefully oriented its expansion around an integrated public transport system. Minibuses bring people from their neighborhoods to terminals where they meet express buses to all parts of the city. Being able to get to work quickly and cheaply has helped Curitiba's poor to find and keep jobs. The city's streets and its many parks and green spaces are kept spotlessly clean and decorated with flowers. The city also has a decentralized public health service. A trash recycling program encourages people in the informal economy to collect specific kinds of trash and sell them to recycling companies.

One goal is to keep migrants from swamping Curitiba's well-designed, environmentally sensitive urban environment with shantytowns. The city first tries to stem the flow by offering free bus tickets back home to new migrants. Twenty-five thousand people have used the tickets. To accommodate those migrants who come to take what are often short-term jobs in Curitiba's industrial sector, the city is building rural satellite towns, called *vilas rural*, where people can live and maintain their agricultural skills by farming small plots when they are between industrial jobs. In this way, they will be able to feed themselves and possibly some urbanites as well. Financed by the World Bank and the Inter-American Development Bank, 5 *vilas rural* have been built, 15 are under construction, and 60 more are planned. The strategy is to accommodate a significant proportion of Paraná's landless farmers in these urban fringe communities, where the advantages of both rural and urban life can be enjoyed by those of meager means.

[*Sources:* http://www.dismantle.org/curitiba.htm; http://www.pbs.org/frontlineworld/fellows/brazil1203.]

It always takes some resourcefulness to move from one place to another. Hence it is usually those who already have some advantages of education and unusual ambition who migrate to cities (where their talents may, however, go to waste). Thus the rural sending communities are deprived of young adults in whom they have invested years of nurture and education. This loss, often referred to as **brain drain,** happens at several scales in Middle and South America: there is rural-to-urban migration from villages to regional towns and from towns and small cities to primate cities, as well as international migration from the many countries in the region to North America and Europe. On the other hand, families often encourage their youth to migrate so that they can benefit from the remittances the migrants send home. Remittances and gifts of consumer goods are now thought to constitute a significant portion of the GDP of some countries. The money goes to pay for medicine, clothing, food, school fees and books, housing, community projects, and entrepreneurial ventures. The brain drain is therefore offset somewhat by the remittances, goods, and services that migrants provide to their home communities (for more on the effects of remittances in Mexico, see page 161). These contributions can bring considerable prestige to returning migrants, who, with their skills enhanced by the overseas sojourn, are in a good position to become local leaders. Thus migration has a mix of consequences for sending communities—some positive, some negative.

The receiving societies, whether cities such as Rio de Janeiro or Mexico City or countries such as the United States, garner considerable benefit from immigration: a huge, inexpensive labor pool and a concentration of skilled professionals. For example, in 2000, the American Medical Association reported that one-fourth of the doctors practicing in the United States were immigrants, as were about 50 percent of medical students. Of the practicing doctors who were immigrants, more than 5 percent were from Middle and South America. The United States saves money when other countries supply the early education for these physicians (and sometimes even provide government scholarships for higher education), who may then spend their most productive years treating patients in the United States. And so it is with migrants generally: most of those who migrate are people with above-average education and with the initiative and creativity to take a long, risky journey into unfamiliar territory—all characteristics badly needed in the places they leave.

Quick Review

1. What are some of the known effects of El Niño in this region?

2. Describe the characteristics of the four main temperature-altitude zones.

3. What are a few of the reasons why the rate of population growth is still relatively high in this region?

4. Describe the effects of migration on both the sending and host societies.

PRACTICING GEOGRAPHY | How does return migration affect the West Indies and Caribbean?

Migration "off the island" has been a survival strategy for generations of West Indians and Caribbean people and continues to be a flexible response to opportunities at home and abroad today. Many islanders circulate temporarily, returning to their homes. Others, who initially circulate as a temporary strategy, eventually emigrate permanently to join overseas diasporas across the Caribbean region and North America and globally across Europe, Australia, and beyond. These networks of extended families have become global social systems that help members live between two or three worlds, taking advantage of the best opportunities available in transnational spaces. They form ties that have potent value to their home societies—remittances, the return migration of skilled émigrés, transnational business ventures, information circulation, cross-cultural exchanges of music, art, and cultural performances, and improved transnational or interregional communication and political alliances.

I've been studying return migration effects, the effects of remittances, and transnational networks in West Indian and Caribbean communities, in part because previous research has been somewhat negative and disparaging. Recent work (including some of my contributions) has begun to change the conventional wisdom. Research conducted in Barbados and St. Lucia by Robert Potter, who is interviewing young returning nationals, has found that their adaptation experiences are not only complex but contradictory. A new cohort of return migrants is reversing the brain drain, but the scale of their effects is unknown, and there is evidence that some returned migrants migrate again. They may constitute a major return flow. Because they bring back educational and financial capital as well as material wealth, their social and economic effects are likely to be greater than their demographic proportion would indicate.

A better understanding of these complex relationships can be accomplished only by in-depth interviews of returnees, remittance recipients, and possibly migrant donors if they have returned to their island homes. Conducting these interviews in Trinidad, Tobago, or Grenada means that I need to involve local collaborators, who help me with the interviews and use their local connections to open doors and find potential respondents through their social networks of family and friends. *Snowball sampling* of young or retired returnees is the only option because none of the local governments, their statistical offices, or universities have records of these returning nationals. Once my collaborators and I have identified the first wave of returnees and interviewed them, then they identify others they know, and our interview schedule can continue to grow until it is large enough to be a representative sample, and we can generalize about trends, patterns, and processes.

Dennis Conway
Department of Geography, Indiana University

For such studies as these, I seek out local university personnel as co-investigators, and local university graduates to help me with the open-ended, in-depth interviews. I find that female interviewers get the best responses (and fewest rejections) from my West Indian respondents. This coming year, I will be conducting interviews of young returning nationals in Trinidad and Tobago, helped by two Trinidadians affiliated with the University of the West Indies at St. Augustine and a St. Lucian-born interviewer. Together, we intend to add to the comparative regional framework of primary data collection (via in-depth interviews) already under way in Barbados, St. Lucia, and Grenada, and add Trinidad and Tobago's ethnic plural society to the mix of contemporary West Indian societal milieus under scrutiny. Insights into the transnational flexibility of these respondents' lives will help us understand some of the contemporary changes afoot in these small but dynamic island societies—the "transnational geographies" of West Indian/Caribbean migrant families and their extended family networks, in particular.

II CURRENT GEOGRAPHIC ISSUES

Historically, power and wealth in Middle and South America have been concentrated in the hands of a few. The economic modernization of most countries and their transformation from rural to urban societies over the last century have not changed that reality. The rewards of urbanization and industrialization have spread only thinly to the masses of poor and not-so-poor, who instead have had to absorb the effects of large government debts incurred in the 1970s and 1980s.

ECONOMIC AND POLITICAL ISSUES

Although this region is not as poor on average as sub-Saharan Africa, South Asia, or Southeast Asia, it has the widest gap between rich and poor. Except for a few relatively well-off Caribbean countries, such as Barbados and Cuba (see pages 158–159), by 2000 the richest 20 percent of the population was between 12 and 29 times richer than the poorest 20 percent (Table 3.2), and nearly 40 percent of the population was living in poverty in 2004. This **income disparity** is troubling in many ways. From a moral standpoint, many people argue that it is wrong for wealth and resources to be concentrated among a tiny number of people while so many have so little. From an economic perspective, poverty prevents people from contributing to economies with their own purchasing power or with their skills, which tend to remain at a low level. Thus, wherever large numbers of people sink into poverty, the potential for a nation's growth diminishes. Political instability also increases because impoverished people may rebel violently and be violently repressed by governments. In the not so distant past—in Bolivia, Peru, Chile, Brazil, Argentina, Colombia, Guatemala, Nicaragua, Honduras, El Salvador, Mexico, and Haiti—hundreds of thousands died in repeated waves of government repression, revolution, counterrevolution, and coups d'état.

It has been argued that globalization is reducing the importance of income disparity in the region as increases in international trade provide more jobs that raise the income of the poorest. Certain aspects of globalization have had some positive effects. Lowered trade barriers expand opportunities for entrepreneurs, who then create jobs. In addition, multinational corporations are investing in the region (an activity called **foreign direct investment** or **FDI**), building factories and hiring semiskilled workers and some local managers. Markets for products from the region, such as meat and building materials, are being created in Asia and Europe, another process that increases economic growth. This trend toward globalization is especially important in expanding opportunities for the more educated and affluent minority. Highly educated, urban, technologically proficient businesspeople (sometimes called *technocrats*) are enjoying new wealth, prestige, and power. And significant numbers of the less educated poor have also benefited from jobs created by globalization. However, for them, job security has proved to be precarious: many have gained and then lost jobs as businesses, large and small, open and close, expand and contract. Furthermore, as private foreign direct investment has increased, government-funded programs and aid from the United States and Europe have contracted, under the assumption that market economies will eventually make everyone self-supporting. Multilateral development aid, such as that from the World Bank, usually requires governments to streamline their economies to attract investors (in accordance with SAPs, discussed below); hence governments have dramatically cut their programs of assistance to the poor, hoping the trickle-down effect will suffice.

TABLE 3.2 Income disparity in selected countries

Country	Ratio of wealth of richest 20% to wealth of poorest 20%[a]	
	1987	1998–2000[b]
Bolivia	9:1	12:1
Brazil	26:1	29:1
Chile	17:1	19:1
Colombia	20:1	20:1
Guatemala	30:1	16:1
Mexico	16:1	17:1
Peru	12:1	12:1
China	8:1	8:1
France	6:1	6:1
Jordan	6:1	6:1
Philippines	10:1	10:1
Thailand	8:1	8:1
Turkey	8:1	8:1
Canada	5:1	5:1
United States	9:1	9:1

Note: The UN used data from 1987 and from 1998 to 2000 on either income or consumption to calculate an approximate representation of how much richer the wealthiest 20 percent of the population is than the poorest 20 percent in selected countries. The lower the ratio, the more equitable the distribution of wealth in that country. Middle and South America have wider disparities than all other world regions. Notice that in Brazil the disparity has increased since 1987, while in Guatemala it has lessened, but is still high. Can you account for the general pattern, and for the difference between these two countries?

[a]Decimals rounded up or down.

[b]Survey years fall within this range.

Source: United Nations Human Development Report 2000, Table 13; 2003, Table 13.

Phases of Economic Development: Globalization and Income Disparity

The current economic and political situation in Middle and South America is deeply rooted in its history of dependent relationships with external forces: colonial mother countries, the economic interests of Europe and America, and foreign and multinational corporations. This history can be divided into three major economic phases: the early extractive phase, the import substitution industrialization phase, and the structural adjustment phase of the present. All three phases have been characterized by various levels of external influence, and all have helped entrench wide income disparities.

The Early Extractive Phase. The **early extractive phase,** beginning with the European conquest and lasting until the early twentieth century, was characterized by colonial policies such as mercantilism that resulted in unequal trade. A relatively small flow of foreign investment and manufactured goods came into the region, and a vast flow of resources left for Europe and beyond. Foreign money to fund enterprises aimed at extracting resources for export (farms, plantations, mines, sawmills, roads to ports, and, more recently, industries and railways) came first from Europeans—the Spanish and Portuguese, then other Western Europeans—and later from North Americans and other international sources. Investors kept the costs of these enterprises low by acquiring local resources at bargain prices, while the profits were usually banked abroad. The region as a whole was at a global economic disadvantage because it sold raw materials and did not manufacture even essential items, such as farm tools and household utensils, which had to be purchased from Europe and North America at relatively high prices. Moreover, the vast majority of the population received little benefit from the extraction of the region's resources because wages remained low (or, in the case of slave economies, nonexistent). Employment opportunities were scarce, and chances to move up the social ladder were few.

It is useful to compare this pattern with the economic development of North America at the time. In North America after the American Revolution, the profits from industries based on the extraction of raw materials went to owners who tended to live in the region and who wanted to bring the industrial revolution to their homeland. Hence they invested their money locally in industries that processed North America's raw materials into more valuable finished goods that were then bought by American workers or exported. This investment provided the foundation for North America's present greater economic stability and higher living standards. In contrast, the export of raw materials remained the basis of Middle and South America's economy.

A number of economic institutions arose to feed raw materials to Europe and North America. Large rural estates called **haciendas** were first granted to colonists as a reward for claiming territory and people for Spain, then passed down through the families of those colonists. Hacienda laborers were virtually feudal serfs, while the owner often lived in a distant city or in Europe. Generally, haciendas produced several types of crops, but used only a fraction of their potential agricultural land. The low cost of labor and the ability of the system to control large areas of land and large groups of people is what kept the haciendas in business.

Plantations, which were large factory farms growing a single crop such as sugar, coffee, cotton, or (more recently) bananas, were first developed by English, French, and Dutch colonizers on Caribbean islands in the 1600s. Northeastern Brazil had some early sugar plantations, and they became more common in South America by the late nineteenth century. Owners of plantations made larger investments in equipment and labor than did hacienda owners. Unlike haciendas, which were often established in the continental interior in a variety of climates, plantations were for the most part situated in tropical coastal areas with year-round growing seasons. Their coastal location also gave easier access to global markets via ocean transport.

As markets for meat, hides, and wool grew in Europe and North America, the **livestock ranch** emerged, specializing in raising cattle and sheep. Today, commercial ranches are found in the drier grasslands and savannas of South America, Central America, and northern Mexico, and even in the wet tropics on cleared rain-forest lands.

Mining was another early extractive industry. Mines had to be located wherever the desired minerals (at first primarily gold and silver) were found. Labor had to be moved to the mine, and the minerals had to be partially refined on site to reduce the costs of transport. Important mines were located on the island of Hispaniola, in north central Mexico, in the Andes, in the Brazilian Highlands, and in many other locations as well. Today oil and gas have been added to the mineral extraction industry, but rich mines throughout the region continue to produce gold, silver, copper, tin, precious gems, titanium, bauxite, and tungsten.

FIGURE 3.11 Political map of Middle and South America. Small island countries in the Eastern Caribbean are not included.

The mines, ranches, plantations, and haciendas of the early extractive phase did not contribute to the development of the region because of low wages and because the owners generally invested their earnings outside the region. By the late nineteenth and early twentieth centuries, when nearly all the countries of Middle and South America were independent, wealthy European and North American private investors had purchased many of the extractive industries in the region, thus ensuring that the pattern of profits leaving the region would persist. These investments came to be known as *neocolonial* because the power these foreign investors wielded resembled that of colonial officials. (**Neocolonialism** refers to modern efforts by dominant countries to further their own aims by controlling economic and political affairs in other countries.)

The Import Substitution Industrialization Phase. In the early twentieth century, waves of popular political protest arose against the domination of the economy and society by foreign businesses and a few wealthy citizens that characterized the extractive phase. Many governments—Mexico and Argentina most prominent among them—proclaimed themselves socialist democracies and enacted a set of policies that came to be known as **import substitution industrialization (ISI).** The goal of these policies was to keep money and resources within the region in order to foster economic self-sufficiency. First, national governments replaced foreigners as the main source of investment

ing industries would grow large enough to replace the extractive raw materials industries as the backbone of the economy.

Although ISI strategies still survive in a few countries, and are being reconsidered in a few, they failed to bring about a thorough transformation of the industrial sector. In the ISI phase, some progress was made toward reducing income disparities as governments increased spending on public health, education, and infrastructure and enacted reforms that broke up some large land holdings and distributed them among landless farmers. But the goal of lifting large numbers of people out of poverty through jobs and general economic growth was never realized. The state-owned manufacturing sectors upon which the success of ISI depended were never able to expand sufficiently to provide jobs for the migrants flooding into the cities. The number of people able to afford manufactured goods did not increase significantly, either within any country or in the region as a whole. The region might have had enough consumers to sustain manufacturing industries if trade and transport barriers between countries had not restricted trade and raised prices for the consumer. However, state-owned manufacturing industries could not compete in price or quality with U.S.- and European-made products, and tariff barriers, applied to goods produced by neighboring countries as well as those from outside the region, were required to protect locally made goods from higher-quality imports. Not all state-owned corporations were losing propositions, however: Brazil—with its aircraft, armament, and auto industries—and Mexico—with its oil and gas industries—both experienced successes in their ISI development. Most critics conclude, however, that had these industries been efficiently managed as private corporations, they would have been yet more successful.

For all these reasons, state-owned manufacturing industries—often unprofitable and inefficient—failed as incubators of innovation and expansion and as suppliers of employment on a wide scale. As a result, the region remained ever more dependent on the export of raw materials.

The Debt Crisis. Ultimately, a global financial crisis brought the ISI phase to an end. Starting in the early 1950s, the global economy enjoyed a period of economic expansion, during which the prices for raw materials held steady. But that period of prosperity ended in the 1970s due to increases in oil prices and decreases in the prices of raw materials on global markets. While earnings from exports were falling, governments and private interests continued to pursue expansive plans to modernize and industrialize their national economies. They paid for these projects by borrowing millions of dollars from major international banks, most of which were in North America or Europe. These banks had surplus cash to lend at the time and encouraged this borrowing. The beginning of a global recession in 1980 put a halt to such ambitious plans. Dragged down by the recession, the Middle and South American economies could not meet their targets for growth. Thus the governments were unable to repay their loans (see Figure 3.1 on page 112). The damage to Middle and South America was made worse by the fact that the biggest borrowers—such as Mexico, Brazil, and Argentina—also had the largest economies in the region. Hence huge external debts now burdened the very countries that had been the most likely to grow (Figure 3.12).

These precipitation tanks for the separation of aluminum are part of the C.V.G. Venalum Company's holdings in Venezuela. C.V.G. is the largest aluminum smelter in Middle and South America. The company began operations in 1973, primarily to make products for export. The Venezuelan government owns about 80 percent of the company (10 to 20 percent of this 80 percent is owned by C.V.G. workers), and the remaining 20 percent is foreign-owned, primarily by Japanese companies. Since 1999, C.V.G. has offered additional shares to foreign investors, and companies from the United States, France, and the United Kingdom have expressed interest. [Mario Corvetto/Evergreen Photo Alliance.]

money in the region. States either bought or seized (usually with some payment to owners) the most profitable extractive industries, then used their profits to create manufacturing industries that could supply the goods once purchased from Europe and North America. To encourage the local population to buy manufactured goods from local suppliers, governments placed high tariffs on imported manufactured goods. The money and resources kept within each country were expected to provide the basis for further industrial development that would create numerous well-paying jobs and raise living standards for the majority of people. National governments further hoped that these new manufactur-

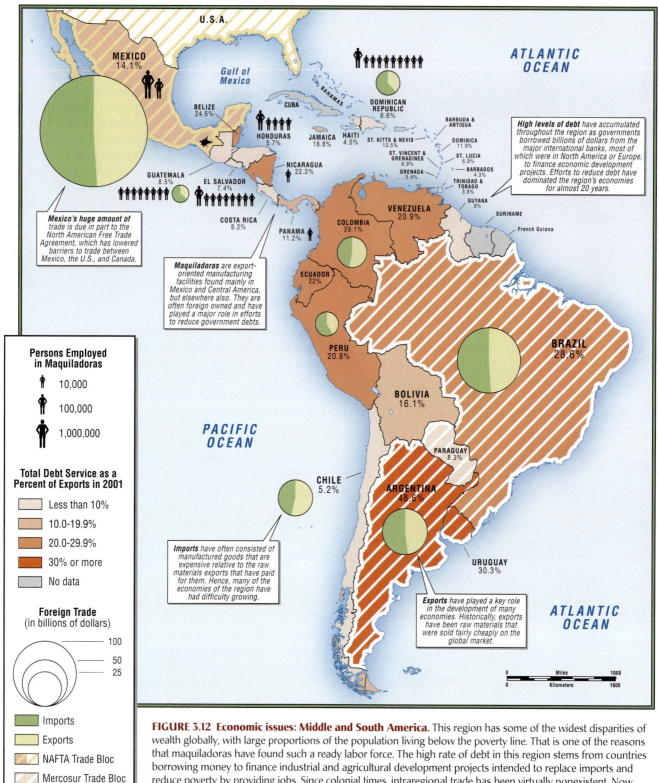

Persons Employed in Maquiladoras

👤	10,000
👤	100,000
👤	1,000,000

Total Debt Service as a Percent of Exports in 2001

- Less than 10%
- 10.0-19.9%
- 20.0-29.9%
- 30% or more
- No data

Foreign Trade (in billions of dollars)

- 100
- 50
- 25

- Imports
- Exports
- NAFTA Trade Bloc
- Mercosur Trade Bloc

Mexico's huge amount of trade is due in part to the North American Free Trade Agreement, which has lowered barriers to trade between Mexico, the U.S., and Canada.

Maquiladoras are export-oriented manufacturing facilities found mainly in Mexico and Central America, but elsewhere also. They are often foreign owned and have played a major role in efforts to reduce government debts.

High levels of debt have accumulated throughout the region as governments borrowed billions of dollars from the major international banks, most of which were in North America or Europe, to finance economic development projects. Efforts to reduce debt have dominated the region's economies for almost 20 years.

Imports have often consisted of manufactured goods that are expensive relative to the raw materials exports that have paid for them. Hence, many of the economies of the region have had difficulty growing.

Exports have played a key role in the development of many economies. Historically, exports have been raw materials that were sold fairly cheaply on the global market.

FIGURE 3.12 Economic issues: Middle and South America. This region has some of the widest disparities of wealth globally, with large proportions of the population living below the poverty line. That is one of the reasons that maquiladoras have found such a ready labor force. The high rate of debt in this region stems from countries borrowing money to finance industrial and agricultural development projects intended to replace imports and reduce poverty by providing jobs. Since colonial times, intraregional trade has been virtually nonexistent. Now trade blocs, such as NAFTA and Mercosur, are facilitating such trade. Debt is presented as total debt service (repayment of public and private loans) as a percentage of a country's exports of goods and services. In 2001, debt for all developing countries was 11 percent, whereas that for just Middle and South America was 19.7 percent, according to UN figures. [Adapted from *United Nations Human Development Report 2003*, Table 16.]

The Current Structural Adjustment Phase. When the advanced industrial economies of the United States, Europe, and Japan rebounded from the global recession of 1980–1983, influential experts recommended that developing countries around the world adopt the same economic policies that seemed to be so successful in those advanced economies. These policies were labeled "structural adjustment" because they mandated profound changes in the organization of economies. For example, they advocated export-oriented manufacturing as a more robust way to achieve development than import substitution industries, and the success of Japanese and Pacific Rim export economies in the 1980s reinforced this view. Free trade and privatization, rather than government-funded industry, were considered the soundest ways to achieve economic expansion.

This chapter gives the most detailed explanation of such structural adjustment policies or programs (SAPs), and should be referred to when SAPs are mentioned in succeeding chapters. In virtually every region of the world, such policies have become highly influential and controversial.

The structural adjustment phase was launched in Middle and South America in 1982, when a drop in the price of oil left Mexico unable to make its payments on a huge national debt. Following Mexico's default, international banks refused to lend to the debtor nations of Middle and South America unless they first agreed to **structural adjustment programs (SAPs)**, which were designed to free up money for loan repayment through a number of methods. SAPs were developed and enforced by a global financial institution known as the International Monetary Fund (IMF), acting on behalf of the major international banks—the lending agencies. The goal of the IMF was to promote free-market policies in order to generate cash for loan repayments. In order to obtain further loans, governments were required to begin removing tariffs on imported goods of all types and to sell state-owned raw materials and manufacturing industries to private investors—often multinational corporations located in Asia, North America, and Europe. SAPs also reversed the trend during the ISI era toward the expansion of government social programs. Again, to generate funds for debt repayment, governments were required to fire many civil servants and drastically reduce spending on public health, education, and infrastructure. Advocates of structural adjustment often associate these policies with **neoliberalism,** the idea that social justice—meeting basic nutritional, shelter, and educational needs for everyone—can best be achieved through free-market economic development, which presumably will create living-wage jobs for the poor. SAPs also encourage the (preferably unregulated) expansion of industries that produce goods for export. In most developing countries, these export industries are still based on the extraction of raw materials because they are seen as a reliable way to earn the money necessary first to repay national debts quickly and then to invest in job-creating manufacturing and service industries.

Outcomes of SAPs. Without a doubt, SAPs have increased economic activity in Middle and South America. Multinational corporations, for example, have driven expansion in the mining, agriculture, forestry, and fishing industries. And the U.S. corporations Cargill, Del Monte, and Dole are major players in the region, producing fresh fruits, flowers, and vegetables for U.S. supermarkets regardless of season. These corporations provide pesticides, fertilizers, and agricultural technologies for agribusinesses in the region. Asian corporations are moving into timber extraction in Brazil and Guyana. Partnerships between foreign companies and local businesspeople operate the largely unregulated gold mining, forestry, and other extractive industries of the Amazon Basin territories of Brazil, Ecuador, Guyana, and Bolivia. Not surprisingly, because limiting regulation of all types is an important part of structural adjustment, the very industries that have expanded as a result of SAPs are those that raise worries about environmental impacts, worker safety, and sustainability (see discussion on pages 149–153).

A number of countries, such as Mexico, have attempted to move beyond extractive industries by establishing **Export Processing Zones (EPZs)**, also known as **free trade zones.** EPZs are specially created legal spaces or industrial parks within a country where duties and taxes are not charged in order to attract foreign-owned factories. These factories are allowed to assemble products, often textiles or electronics, strictly for export back to the owner's country. EPZs exist in nearly all countries on the American mainland and on Caribbean islands, but by far the largest of the EPZs is the conglomeration of assembly factories, called **maquiladoras,** located along the Mexican side of the U.S.–Mexican border (see Table 3.4 and the discussion on pages 160–162). Because the taxes paid by these industries are low, the main benefit to the host country is the employment of local labor, which helps to ease the country's high rate of unemployment. EPZs now compete rather fiercely with one another.

Have SAPs resolved the debt crisis and reduced income disparity? The answer is yes and no. SAPs seem to have played a role in the slow economic recovery of the region since the start of the debt crisis in 1982. The burden of debt has fallen throughout the region, but it is still very high and will continue to be so for many years (see Figure 3.12). SAPs have encouraged greater business and industrial efficiency, and some countries (Mexico, Brazil, Chile) experienced high rates of economic growth during the 1990s. But for most, growth slowed in the 2000s. During the 1990s, exports increased dramatically, especially in countries, such as Mexico, that are hosts to maquiladora-like assembly factories. However, poor working conditions in these new export-oriented industries, and their general disregard for workers' well-being and the environment, have created problems for the communities in which they are located. Moreover, except for Mexico and Brazil, most of the region's exports are still raw materials; as a result, many economies remain vulnerable to rapid price changes on the global market. Thus, while the debt crisis seems to be easing, it could reemerge.

The fact is, SAPs were never intended to decrease income disparity directly, but there was the implication that rising employment would do so automatically, and that the new buying power generated by the new jobs would create demands for more goods and services, leading to still more jobs. Well-being was expected to trickle down to the poorest, and hence social justice would be achieved. The reality is that SAPs have contributed to increases in the gap between rich and poor. Chile's reforms, for example, have produced national economic growth rates as high as 7 percent in some recent years—but they have also resulted in some of the highest rates of income disparity in the world. Between 1994 and 1996, the average yearly income for the 40 percent of the Chilean population classified as poor was $1440 per capita, while the average yearly income for the top 10 percent of the population was $48,000 per capita. For the region as a whole, the number of people living in poverty has increased since SAPs were introduced: between 1980 and 1990, virtually every country in the region experienced a drop in purchasing power. In many countries, the true value of workers' wages is now half, or less than half, of what it was in 1980. Although levels of public funding for education and health care have recently increased in many countries, the years of SAP-mandated cuts in education have left the region poorly equipped to move into the technological age. In Mexico, Brazil, Ecuador, and Argentina, annual debt payments still exceed the amounts spent on health and education combined. In most of the region, fewer than 75 percent of those eligible are enrolled in high school; in Ecuador and Venezuela, that number is just 50 percent. Finally, incomes have declined significantly for those in the middle class, especially former government employees who lost their jobs as a result of SAPs. People who have lost income have increasingly turned to the informal economy to make a living.

By the year 2000, the IMF and experts at the United Nations and in North America and Europe were acknowledging that IMF loan programs and subsequent structural adjustment programs had worked undue hardships on the poorest. Moves were initiated to alleviate those hardships by forgiving the overwhelming debt of the poorest countries worldwide. In an effort to address the debt and environmental crises all at one time, some environmentalists proposed forgiving a certain amount of debt for every tract of land placed in environmentally protected status. And, in 2001, Harvard economist Jeffrey Sachs, an architect of SAPs in the 1990s, did an about-face and began advocating increased aid from rich to poor countries, especially those where structural adjustment had increased poverty.

The Informal Economy

PERSONAL VIGNETTE   Maria del Rosario Valdez lost her government job as an elevator operator in Mexico City in the 1980s. Unable to find formal employment, she created her own "informal" business: selling secondhand clothing to poor people. Every Thursday evening, she boarded a bus for Texas filled with people doing similar business—buying and selling toys, auto parts, cosmetics, kitchenware. All along the way, wherever the bus stopped, the passengers themselves interacted with others in the informal economy—people selling tacos, pastries, cassettes, ice cream. For centuries, such low-profile businesspeople have operated outside the law throughout the Americas, paying no taxes but supporting their families.

By early morning, the bus reached the border. Maria took a cab to a huge warehouse, where she chose items she knew would sell from the towers of clothing collected from all over the United States. Some of the clothes were new—unsalable overstock. She filled boxes and suitcases with about 500 pounds of clothing, for which she paid about $150. Later she paid another $150 in bribes to Mexican customs officials. Once back in Mexico City, early Saturday, Maria went immediately to her stall in the market. There

---

AT THE LOCAL SCALE   The Informal Economy in Peru

Street vending is the most visible part of the informal economy in most cities of Middle and South America. The vendors sell useful items such as vegetables, spices, cooked snack foods, sunglasses, and umbrellas. Some sell luxury goods at prices lower than those in shops—products such as perfume, cosmetics, and handwoven rugs. Street vendors use public streets and sell to the public, so the governments in most countries try to formalize and control this activity to protect public health, ensure orderly urban spaces, and collect taxes.

The informal economy is a lifesaver in Peru, which has been hit by unprecedented economic recession, losses in real wages, and underemployment for up to 70 percent of urban workers. It is thought to employ a majority of the working urban population. By the year 2000, perhaps as many as 68 percent of urban workers were in the informal sector, and they generated an estimated 42 percent of the country's total gross domestic product. Most of them work as street vendors; up to a fourth of the working population is so employed.

Geographer Maureen Hays-Mitchell, who studies street vending in Peruvian cities, found a vibrant, highly organized system in which the vendors assess the market and the risks entailed in this semi-illegal activity. She found street vending in Peru to be a highly rational sector of the informal economy, fraught with intense competition for particular urban spaces and requiring continuous strategizing for market advantages. She also found that

street vending not only was an important source of sustenance for the vendors, but also improved the overall quality of urban life by making goods available in convenient places and at affordable prices, something the inefficient formal retailing system seemed unable to do.

Street vendors specialize in particular products and carefully place their stands where they can attract the most customers for the products they sell. Vendors sell food and small gifts in front of hospitals during visiting hours; candy, games, and toys in front of schools; shoeshines and newspapers near hotels and restaurants; lotion, hats, and bikinis near beaches. Few opportunities are left unexploited: movie patrons can buy comics and magazines while waiting in line; outside jails, visitors, guards, and even prisoners can buy food, souvenirs, and handicrafts produced by inmates. The most popular locations, though, are along the edges of streets surrounding a city's central retailing district, the entrances to specific buildings and stores, and the intersections of major streets. A single city block may contain, on average, from 15 to 30 street vendors.

A desirable location has itself become a commodity for sale, though the vendors have no legal right to any particular space. Hays-Mitchell observed vendors painting the outlines of their stands on the street itself. The operators have sometimes sold the rights to use such spaces to new vendors for several hundred dollars.

she bundled her leftover unsold clothing (to be hawked door-to-door in the poor barrios by another woman) and laid out her new purchases. She sold for 10 or more hours each day until Thursday, when she left for the border once again.

Unfortunately, after the terrorist attacks of September 11, 2001, vendors like Maria del Rosario Valdez found that border delays made their carefully timed trips to and from the United States impossible, and many have ceased to participate in this part of the Mexican informal economy. [*Adapted from Randall Hansis, The Latin Americans: Understanding Their Legacy (New York: McGraw-Hill, 1997), pp. 178–179; NPR reports, 2003.*]

The **informal economy** (described in greater detail in Chapter 1) consists of any and all economic activity that is not legal in the sense that taxes are not paid to the government and the work does not appear in official statistics. People working in the informal economy are often small-scale entrepreneurs who see an opportunity to earn some money in the market economy (see the box "The Informal Economy in Peru" on page 138). As in the case of Maria's business described in the vignette, this entrepreneurship often entails recycling used items and, in some instances, can serve as an incubator for businesses that might eventually provide legitimate jobs for family and friends. Many people argue, however, that informal workers are just treading water, making too little to ever expand their businesses significantly. Moreover, the bribes they have to pay (some have labeled these bribes "informal taxes") are much less beneficial to the economy as a whole than the taxes paid by legitimate businesses. Work in the informal economy is also risky because there is no protection of workers' health and safety and no retirement or disability benefits. In many ways, the growth of the informal economy is a symptom of the increasing disparity between those with substantial access to wealth and power and those with very little.

Regional Trade and Trade Agreements

The growth in international trade and foreign investment encouraged in part by SAPs has been joined by a growth in regional free trade agreements, which reduce tariffs and other barriers to trade among a group of neighboring countries. The two most successful free trade agreements within the region are NAFTA and Mercosur (see Figure 3.12). The **North American Free Trade Agreement (NAFTA)**, created in 1994, is by far the larger of the two. It links the economies of Mexico, the United States, and Canada in a free trade bloc containing more than 420 million people, with a total annual economy worth at least $12 trillion. **Mercosur** is older, having been created in 1991, and smaller. It links the economies of Brazil, Argentina, Uruguay, Paraguay, and Chile (an associate member) to create a common market with 237 million potential consumers and an economy worth more than $1.9 trillion per year. There are a growing number of other free trade agreements within the region, some of them spurred in part by competition between the European Union (EU) and the United States to expand their influence in the region. Traditionally, the United States and Canada have traded and invested more heavily in the northern parts of Middle and South America—as reflected by Mexico's position within NAFTA and by

the many U.S. economic ties to Central America and the Caribbean. Two reasons for North America's greater interest in trade with the northernmost countries of Middle and South America are that these countries are the closest, so transport costs are lower, and that this part of the region is seen as strategically important, especially to the United States.

In the 1990s, the United States began pushing for the creation of a NAFTA-like trade bloc for all of the Americas, hoping to at least partially check European trading power in the region. In April 2001 (before the 9/11 attacks), government leaders from all countries in the Western Hemisphere, except Cuba, signed the Free Trade Agreement of the Americas (FTAA). This agreement calls for the removal of trade barriers among members by the year 2005. While the current administrations of the United States, Canada, and Mexico view FTAA as a logical expansion of their NAFTA agreement, the actual implementation of such an ambitious free trade plan is unlikely to take place by 2005, in part because of security issues raised by the 9/11 terrorist attacks, and perhaps more importantly because small countries in the region fear open competition with well-established corporations in the United States, Canada, and Mexico.

The record of regional free trade agreements so far is mixed. Over the period from 1990 to 1994, NAFTA and related agreements that preceded it increased exports from Mexico to the United States by 137 percent. Trade among the Mercosur countries grew by 400 percent in the period from 1990 to 1998 but then underwent a decline in the early 2000s. When trade grows, the benefits usually are not spread evenly among regions or among sectors of society. In Mexico, the benefits of NAFTA have been concentrated in the northern states that border the United States, where the growth of maquiladoras has been the most intense. Still, most of those who work in the maquiladoras are paid only marginally above the going national wage rate, and living conditions are so difficult that their actual standard of living rises only slightly, if at all (see the discussion on pages 160–163).

Producers within the Mercosur sphere differ in their views of the trade agreement. Chilean grain farmers fear competition from cheaper Argentine and Uruguayan wheat and corn, while Chilean fruit growers and winemakers see greater access to buyers in neighboring Brazil as beneficial. In Argentina, Mercosur has brought greater efficiency and productivity as a result of increased competition, but it has also brought high unemployment because, to be competitive, excess labor had to be cut. Brazil favors a customs union in which there would be no tariffs between Mercosur countries, but there would be high external tariffs on goods entering from outside Mercosur. These external tariffs would secure regional markets for manufactured goods produced in member countries, but external tariffs also would hamper the ability of Argentina and Uruguay to attract foreign industrial capital.

Benefits to all Mercosur members would grow if improved roads and railroads could be developed. There are now few links across the Andes, and individual countries within Mercosur have only modest transport networks, including air service. Mercosur meetings by themselves are a benefit because they provide occasions for discussing regional issues, such as improvements in transport, as well as in power grids, waterways, and pipelines. On the whole, Mercosur seems likely to help member economies protect themselves from unpredictable global markets and to build more competitive and profitable industries; but so far, trade within Mercosur accounts for less than one-fifth of the members' total trade.

Global Free Trade Issues as Seen from Middle and South America

 **Reporter Sam Quinones:** "In the current crisis peasant coffee growers have to learn the Starbucks lesson and focus on quality. Consumers meanwhile have to be willing to pay extra for the best coffee, searching out regional coffees, the way they do with wine."

The Guatemalan coffee farmer frowns down at a slick package of Green Mountain organic coffee. It has his picture on the front. It's the first time he has ever seen how the coffee he grows is marketed to upscale consumers in North America. He has never tasted coffee himself, preferring sweet soda drinks. Only a harsh, low-quality coffee is sold in his village and virtually no one drinks it.

Guatemala used to be covered with coffee plantations (fincas), but many are now abandoned; the workers have migrated to Mexico and the United States seeking some way to support their families. The owners have gone broke because there is a glut in the global coffee market. Too many poor small tropical farmers on 10 or 12 acres are trying to make a living in coffee. Most are paid only 7 to 20 cents a pound by the itinerant coffee trader (called a coyote) who is their only access to market.

When you buy a $1.00 cup of coffee, the grower gets just 10 to 12 cents, the trader 3 cents, the shipper 4 cents, but the roaster (most are large multinational companies) gets 65 to 70 cents. The retailer typically gets 10 to 15 cents.

Green Mountain and other fair trade coffee marketers are trying to change that lopsided allocation of profits by going directly to the growers to do their buying. This direct-to-market approach means growers can get as much as $1.26 a pound for their work, but they have to produce high-quality coffee without chemical pesticides and fertilizers—coffee that passes the high standards of professional coffee tasters. So farmers, themselves, must learn to judge coffee by its taste. The goal is to teach them to be coffee connoisseurs, no longer producers of low-quality beans.

For more on how fair trade affects the coffee market and the lives of growers, see the FRONTLINE/World video "Coffee Country."

In September 2003, what may be temporary brakes were applied to the expansion of free trade in the Americas. At the Cancún, Mexico, meetings of the World Trade Organization (WTO), which seeks to promote free trade globally, farmers from across Central America demonstrated, and a coalition of more than 22 developing countries took a firm stand against any further free trade talks until twin protectionist issues could be resolved: tariffs and subsidies.

Calling themselves the G22 (the 22 included such countries as Brazil, Argentina, Chile, Costa Rica, Jamaica, Pakistan, India, Thailand, Philippines, Bulgaria, and Botswana), these countries challenged the powerful G8 (Group of Eight: the United States, Canada, France, Germany, Italy, Japan, Russia, and the United Kingdom) to stop the hyprocrisy of promoting free trade for others while practicing protectionist policies themselves. Farmers in the G22 and in many other countries say they are shut out of both G8 markets and their own domestic markets by G8 policies. Tariffs imposed by the industrial economies (G8) on farm products from G22 countries increase the market prices of those products, making them uncompetitive. Moreover, subsidies to farmers in rich G8 countries help them sell their products on the global market at less than cost. These low-priced agricultural imports from wealthy countries put farmers from poor countries out of business in their own homelands. One consequence is increasing dependence on imported food and rising food prices for people who are already malnourished. Brazil, a leader among the G22 and a nation afflicted by serious debt (see Figure 3.1 on page 112), says the United States must stop protecting its steel and orange juice industries if it wants Brazil, a leading producer of both products, to join in any free trade agreements.

Yet another focus of free trade protests is the dominant role now played by multinational corporations in agriculture in the G22 countries and the little sympathy shown by these corporations for low-wage agricultural workers (see the vignette below).

Agriculture and Contested Space

The agricultural lands of Middle and South America (Figure 3.13) are an example of what geographers call **contested space**: various groups are in conflict over the rights to use specific territory as they see fit.

PERSONAL VIGNETTE   Aguilar Busto Rosalino used to work on a hacienda, where he had a plot for his own subsistence and in return worked 3 days a week for the hacienda. Now he rises well before dawn and goes to work on the banana plantation that took over the hacienda. From 5:00 a.m. to 6:00 p.m., stopping only for a half-hour lunch break, he places plastic bags containing pesticide around bunches of young bananas. He prefers this work to his last assignment of applying a more powerful pesticide, work that left him and 10,000 other plantation workers sterile. He works very hard because he is paid according to how many bananas he treats. Usually he earns between $5.00 and $14.50 a day. Right now he is working for a plantation in Costa Rica that supplies bananas to Del Monte, but he thinks that in a few months he will be working for another plantation nearby. It is common practice for these banana operations to fire their workers every 3 months so that they can avoid paying the employee benefits that Costa Rican law mandates. Although Aguilar makes barely enough to live on, he has no plans to protest for higher wages, because he knows that he would be put on a blacklist of people that the plantations agree they will not hire.

[*Adapted from Andrew Wheat, "Toxic bananas," Multinational Monitor 17, no. 9 (September 1996): 6–7.*]

Agricultural Zones

1	Mixed farming
2	Irrigated market-oriented areas
3	Cash grain
4	Shifting cultivation
5	Cattle or sheep ranching
6	Plantation agriculture
7	Low-tech subsistence farming
8	Nonfarming

FIGURE 3.13 Agricultural zones in Middle and South America.
Subsistence farming (7) remains widespread, but this traditional mode of agriculture is losing ground to modern, mechanized, export-oriented agriculture (2, 3, 5, 6). The displaced farmers often join the stream of migration to urban areas and North America. [Adapted from Edward F. Bergman, *Human Geography–Cultures, Connections, and Landscapes* (Englewood Cliffs, N.J.: Prentice Hall, 1995), p. 194.]

Thinking Geographically: ON THE WEB

At the Personal Scale. Throughout Middle and South America, increasing numbers of rural people find themselves in situations not unlike that of Aguilar Busto Rosalino. For centuries, while people labored for very low wages on haciendas, the owner would at least allow them a bit of land to grow a small garden or some cash crops. In recent years, however, governments have encouraged a shift to large-scale, absentee-owned, export-oriented farms and plantations like the one where Aguilar now works. This primarily export-oriented agriculture is favored because it has the potential to earn large amounts of the foreign currency needed to pay off debts, and because it mass-produces food for urban populations. SAPs have encouraged governments to relax their restrictions on foreign ownership, so it is easier for foreign multinationals such as Del Monte to dominate the most profitable agribusinesses. These companies are usually highly efficient, state-of-the-art operations. No longer are bits of land available for subsistence plots or sharecropping parcels for laborers. Meanwhile, smaller farmers

are squeezed out because they cannot afford the latest production techniques and hence can't sell their bananas or other produce at the lower prices of the more modern operations. In addition, governments subject to SAPs have cut subsidies and other policies helpful to small farmers. Increasingly, the only new land available to farmers of limited income is in forested areas, where soil fertility lasts only a few years, or in hilly areas, where soil erosion is intense. If they cannot find new land, farmers migrate to cities or work as wage laborers on plantations under conditions like those experienced by Aguilar Busto Rosalino in Costa Rica.

These trends in rural agriculture have sparked resistance in a number of places. In Brazil, 65 percent of the land is owned by wealthy farmers who make up just 2 percent of the population. Since 1985, more than 2 million small-scale farmers have been pressed to sell their land to large corporate farms specializing in the production of cattle and other agricultural products for export. Such large farms are popular with the government because they earn money that helps pay off the country's debt, but the now land-less farmers have been forced to migrate. To help these farmers, organizations such as the National Movement of Landless Rural Workers began taking over unused portions of some large farms, saying that it is wrong for agricultural land to lie unused while there are people living in extreme poverty. Since the mid-1980s, the landless movement has coordinated the occupation of more than 51 million acres (21 million hectares) of land (an area about the size of Kansas). Some 250,000 families have gained land titles, while the elite owners have been paid off by the Brazilian government and have moved elsewhere. Public opinion polls show that 77 percent of Brazilians support the landless movement.

At the Community Scale. The community of Filho de Sepe is using the space of a large old underproductive hacienda to create an entirely new way of life for formerly low-wage field hands. Recently founded, Filho de Sepe is one of a thousand such new towns now dotting the Brazilian interior. There, 376 families operate their own school and medical clinic and grow organic rice for sale as well as vegetables for their own needs. The town is a magnet for other Brazilians seeking agricultural innovations, and farmers from North America also visit to learn about the latest in drip irrigation, wheatgrass crops, and wetlands restoration.

At the National Scale. In the southern Mexican state of Chiapas, agricultural activists have mobilized armed opposition to entrenched systems that have left them poor and powerless. The Zapatista rebellion by Native American agricultural workers (named for the Mexican revolutionary hero Emiliano Zapata) began on the day the North American Free Trade Agreement took effect in 1994. Although the Mexican government redistributed some hacienda lands to poor farmers earlier in the century in land reform programs, most land in Chiapas is still held by a wealthy few who grow cash crops for export. The rest of the population farms tiny plots on unfertile hillsides. In the year 2000, about three-fourths of the rural population was malnourished, and one-third of the children did not attend school. The Zapatista farmers in Chiapas view NAFTA as a threat because it has diverted the Mexican federal government from land reform to the support of large-scale export agriculture. Unlike Brazil, the Mexican government used the army to repress the rebellion.

By the time of the WTO trade talks in Cancún in September 2003, thousands of landless former farmers from Chiapas had migrated to work in tourism in Cancún. They live in a squatter settlement west of town, and most do not support the Chiapas farmers' protest against the WTO. "To me, it's not worthwhile to protest," said Maria Arguez, who came to Cancún from a small town and now works selling cookies for the Mrs. Fields chain. "We all have to work." Antiglobalization leaders say, however, that tourism in Cancún should not be held up as an example of the merits of free trade. "It's not a question of how many people have work in Cancún," said Rafael Alegria, a Honduran leader of farmers who came to Cancún to protest. "It's a question of how many people don't have any work at all." As many as 3 million Mexicans have lost employment due to shifts in the economy brought on by NAFTA. Many of those people were farmers.

The three instances of contested agricultural land just discussed have all arisen out of circumstances of gross inequality: masses of underpaid workers laboring on the large holdings of elites. Only rarely is the contest resolved in favor of the workers, as appears to have happened in Filho de Sepe. The situation in Chiapas and Costa Rica is more typical: expanding large-scale mechanized commercial agriculture or ranching displaces low-wage agricultural laborers. Yet modernization of agriculture is needed. In fact, there are many types of successful agriculture in the region, as a look at Figure 3.13 will confirm. Some of these farms have been around for a long time. For example, a wide belt of commercial grain production, similar to that found in the midwestern United States, stretches through the Argentine pampas. Zones of modern mixed farming, including the production of meat, vegetables, and specialty foods for sale in urban areas, surround the major urban centers of southern Brazil, Argentina, and Uruguay, as well as south central Chile. Irrigated agriculture is practiced along the dry, narrow coastal desert plain of Peru and northern Chile, as well as on the other side of the Andes in the Argentine pampas. These various types of market-based agriculture on large holdings have long been essential to the trade of many countries and often help to feed local urbanites, thus decreasing the amount of food that must be imported. But one result of the recent expansion of market-based mechanized agriculture into lands formerly in small subsistence plots is that significant numbers of traditional farmers are displaced. Families are split up as members of all ages migrate to nearby cities, to other countries, or to North America to find work.

Is Democracy Rising?

For the first time in a long while, most countries in the region have multiparty political systems and democratically elected governments. Cuba is the main exception. In the last 25 years, there have been repeated peaceful democratic transfers of power in countries once dominated by monolithic alliances composed of the military, wealthy rural landowners, wealthy urban entrepreneurs, foreign corporations, and even foreign governments such as the United States. Some of the elected governments, however, are repeatedly threatened with coups d'état because of policies that are unpopular with the masses or with powerful elites—Venezuela and Bolivia were so threatened in the early 2000s. Change has been especially dramatic in Mexico, where the huge entrenched political machine of the Institutional Revolutionary Party (PRI), which had ruled the country for 71 years, was unseated in a 2000 election by a coalition led by the maverick businessman and reformer Vicente Fox. Although President Fox does not control the Mexican Congress, where the PRI is still the majority, he has managed to pass a number of reform-oriented laws, such as a freedom of information act, legislation to reform and update financial institutions, and legislation that will enable universal health care.

In Venezuela, where the profits from the country's rich oil deposits have not trickled down—one-third of the population lives on $2.15 (PPP) a day, and 80 percent of the population is reported to live in poverty—the landslide election of Hugo Chavez as president in 1998 appeared to advance the cause of democracy. Chavez called for fundamental restructuring of the governmental system along more egalitarian lines. Although President Chavez originally had the support of the middle class, that support waned when he failed to take the kinds of actions that would improve the business climate, increase the potential for job creation in the private sector, and increase the potential for wealth accumulation. Rather, his administration chose the opposite of SAP policies: it increased public spending on social programs that have improved the lives of the huge underclass by providing jobs, community health care, and subsidized food. Chavez was reelected in 2000, then was briefly deposed in a coup in 2001, but was quickly reinstated when there was a groundswell of popular support among workers. His position was sustained in a referendum in 2004. The Chavez-led government has tried unorthodox strategies to alleviate the effects of poverty. For example, it has contracted with Cuba, which has a surplus of medical personnel, to supply qualified doctors to community clinics (see page 159). By 2001, infant mortality rates and some other health indicators were improving. Nonetheless, despite Venezuela's considerable oil wealth, the overall economy has lost productive capacity. Perhaps most telling is that Venezuela's per capita GDP (PPP) was $2000 lower in 2001 than it was in 1977!

In Brazil, in 2002, the election of the socialist-leaning Luiz Inácio Lula da Silva was expected to bring radical changes in development policies. While the disruption of free market policies under President da Silva has been only slight, some policies, such as that on Internet expansion (discussed below), promise to further democratic processes.

In much of the region, democracy is still fragile and not particularly *transparent*—meaning that official decisions are not open to

public input and review. Policy is formulated without sufficient attention to its potential effects on the poor or those who lack strong political voices. Officials may remain in power not because they have broad public support, but because they are backed by powerful people, who may even be drug czars or otherwise corrupt.

Corruption. High-level corruption undermines democratic institutions by setting a damaging example for the rest of society. Virtually every country in the region has had a serious scandal in the past few years involving high-level officials performing million-dollar favors for friends and families or stealing money from taxpayers outright. The international banks in Miami, Florida, are well known as depositories for such stolen funds, so indirectly, North American financial institutions have benefited from corruption, while Middle and South American countries have been robbed of much-needed investment capital. For years, in Mexico, banks regularly and openly recorded massive transfers of money into foreign bank accounts at the end of every 6-year presidential term. Former Mexican president Carlos Salinas fled the country at the end of his term in 1996 with millions of dollars. Some people argue that SAPs are increasing pressure on leaders to be financially accountable, but noticeable effects have yet to appear.

The international drug trade is a major factor contributing to corruption, violence, and the subversion of the democratic process throughout Middle and South America. Middle America and

FIGURE 3.14 Interregional linkages: Cocaine sources and traffic in the Americas. South America is a major source of cocaine and other illegal drugs for the U.S. market. Drug production and trade is a major factor in corruption and crime in the region. Although drugs are illegal, poor farmers in remote locations now grow cash crops of coca for cocaine, poppies for heroin, and marijuana because these crops provide a better income for their families than more traditional crops. Drug production is now so great in the region that it outstrips the available U.S. market and much is traded within Middle and South America; hence, addiction there is increasing. [Adapted from George F. Rengert, *The Geography of Illegal Drugs* (Boulder, Colo.: Westview Press, 1996), p. 16.]

northwestern South America are the primary sources for the raw materials of the drug trade to North America—coca for cocaine, poppies for heroin, and marijuana (Figure 3.14). Many growers are small-scale farmers of Native American or mestizo origin in remote locations, who can make a better income for their families from these plants than from other cash crops. Although drugs are illegal in all of Middle and South America, in countries such as Colombia, Venezuela, Bolivia, and Mexico, public figures from the local police on up to the highest officials are paid to turn a blind eye to the industry. Because law enforcement is lax in drug-producing areas, drug lords can fund their own private armies to protect drug transport routes and force the cooperation of local citizens.

For more than a decade, the United States has sponsored a "war on drugs" that is controversial in the region. This effort is intended to stem the flow of cocaine, heroin, and other narcotics into the United States from Middle and South America by stopping production abroad, intercepting drug traffic, and prosecuting users within the United States (in order to reduce the demand for illegal drugs). An important part of the strategy to impede the supply is to destroy drug crops using chemical defoliants. Figure 3.14 shows the flow of cocaine to North America from Middle and South America.

The effort to stop the production of drugs in Middle and South America has led to the largest U.S. military presence in the region in history. The U.S. effort is focused on supplying intelligence, military equipment, and training to military forces in the region. Many leaders in Middle and South America question the wisdom of the U.S. war on drugs. For example, President Andres Pastrana of Colombia expressed the fear that just as countries such as Colombia, Peru, and Bolivia are trying to turn away from long-standing tendencies toward military rule, U.S. aid to national armies encourages military leaders to maintain a high political profile and promotes a gun-based popular culture. Colombia is now the biggest recipient of U.S. aid, globally, after Egypt and Israel, and U.S. military aid to Middle and South America is now about equal to U.S. aid for education and other social programs in the region. The U.S. government takes the position that in waging the war on drugs in Middle and South America, it is making the security forces of legitimately elected governments more professional and equipping them to combat the guerrillas and militias who protect drug-producing underworld figures and force local cultivators to grow illegal drug plants. Critics say that social programs would do more to get at the root causes of the drug-based economy.

Destroying enough crops to stop the flow of drugs into the United States will be a challenging task. By the U.S. government's own estimates, drug production worldwide is so abundant that the entire U.S. consumption could be supplied by 1 percent of the worldwide drug crop. Oscar Arias, former president of Costa Rica and winner of the 1987 Nobel Peace Prize, asked rhetorically in a letter to the *New York Times* (August 29, 2000), "Haven't the last 20 years shown us that as long as there is a market for drugs in the United States, somebody will find a way to get them there?" This view would seem to advocate some method for decreasing the demand for illegal drugs in the United States. One solution that is favored by only a tiny minority of people in the United States is to declare drug use legal and then regulate the industry, much as is done with alcoholic beverages. Canada is experimenting with legal-

izing the use of marijuana for medical purposes. The most promising research on drug use in the United States finds that strong social programs that provide broad-based emotional and physical support for children and youth are the most likely to reduce the market for drugs, but in the last 10 years, money for such programs has been cut, not expanded. Meanwhile, drug production in Middle and South America is now so great that it outstrips the available U.S. market, and much of that production is traded within Middle and South America; hence addiction there is increasing.

Foreign Involvement in the Region's Politics. Interventions by powers outside the region have compromised democracy, even though sometimes their stated aim was to enhance it. Although the former Soviet Union, Britain, France, and other European countries have wielded influence in the region during the twentieth century, the most active foreign power has been the United States. That country proclaimed the Monroe Doctrine in 1823 to warn Europeans that the United States would allow no further colonization in the Americas. Successive U.S. administrations have interpreted the Monroe Doctrine more broadly to mean that the United States itself had the right to intervene in the affairs of countries in the Americas, ostensibly to make them safe for democracy, but in reality to protect U.S. national security and economic interests. One result was the installation of several U.S.-backed dictators in the 1930s through the 1950s, including the Somozas in Nicaragua, Batista in Cuba, and Papa Doc and Baby Doc Duvalier in Haiti. Later, the United States supported armed insurgencies for the purpose of preventing or reversing communist takeovers in Cuba (1961), the Dominican Republic (1965), and Nicaragua (1980s). In Chile, the elected government of Salvador Allende was overthrown by the U.S. Central Intelligence Agency (CIA) on September 11, 1973. Official CIA documents made public in 2003, 30 years after the fact, confirmed what was long suspected: the CIA had laid plans for the eventual coup as early as 1970. General Augusto Pinochet was installed as a military dictator with U.S. approval. He ruled for 17 years, and his regime imprisoned and killed thousands of Chileans who protested the loss of democracy.

In Panama, corruption linked to the Panama Canal and to the international drug trade led the United States to invade the country in 1989 in order to put a halt to the activities of the renegade Panamanian general, and then president, Manuel Noriega. Noriega had shadowy connections with both drug dealing and CIA covert operations in the region. He is now in jail in Florida. Since the end of the cold war, the United States has for the most part curtailed direct intervention in the region's politics. The United States is still highly influential, but is now more likely to wield its influence by enacting economic policies.

Can Internet Technology Further Democracy? Democracy is strongest where people are educated, healthy, and economically secure. Thus the process of strengthening democracy goes hand in hand with efforts to create more socially and economically equal societies. But ideas about how to reach a more egalitarian status quo differ. Neoliberals see free-market economic development and globalization as ultimately furthering social justice and democratic participation by creating jobs, opening up entrepreneurial opportunities, and putting people in touch with one another, especially via the Internet. On the other hand, critics of globalization and free-market economics think that these changes are likely to increase income disparities and undermine democratic participation by decreasing environmental quality, lowering wages, causing local job losses, and pitting the poorest laborers against one another. They argue that egalitarian conditions are more likely to arise out of local control, government support of social services, and regulation of aspects of the economy such as basic utilities and services, banks, energy brokers, communications infrastructure, stock exchanges, and businesses.

If use of the Internet is to enhance democratic participation, as neoliberal theoreticians suggest, then access to this new technology is crucial. A serious **digital divide** exists in Middle and South America. A recent United Nations study of global Internet use showed that users in the region were just 3.2 percent of the population in 2000, but that rate was already four times the 1998 rate. By the mid-2000s, rates had doubled again, and were highest in the Caribbean, Costa Rica, and Argentina. Overall, the region is the most advanced in technological achievement outside the highly developed regions (North America, Europe, Japan and Korea in East Asia, and Oceania). A 2000 study by *Wired* magazine identified those countries worldwide that were leaders in technological innovation. Although no countries in Middle or South America were then leaders, Mexico, Argentina, Costa Rica, Chile, and Brazil were considered to have excellent potential.

Brazil has two world-class technology hubs in the environs of São Paulo and is attempting to cover the whole country with broadband fiber optic cable networks for telecommunications and Internet service (Figure 3.15). Despite its technological superiority, there are worries that because Brazil is such a stratified society, the vast poor majority will not have access to the Internet. In January 2003, international NGOs promoting democracy and civil society challenged the newly elected president of Brazil, Luiz Inácio Lula da Silva, to reform Brazil's Internet administration. Brazilian Internet activists complained that its management was bureaucratic, worked behind closed doors, inadequately addressed access for the poor, and did not account for the millions of dollars raised in the sale of Brazilian (br) Internet addresses. The Lula da Silva government decided to support the transition to a more transparent and egalitarian Internet governance structure for Brazil. Now Brazil is entering into partnerships with China and India to create its own version of a digital TV system. This effort is part of a growing movement among nonaligned countries to develop a range of cutting-edge technologies in the developing world. (Here the word *nonaligned*, a cold war term, no longer refers to those countries that are neither communist or capitalist, but to countries that seek to distance themselves a bit from control by the rich industrialized countries of North America, Europe, and Asia—the G8.) Mexico has recently launched a program (described on page 163) to give its citizens access to training and higher education via the Internet, a movement intended to stem the flow of migration to the United States.

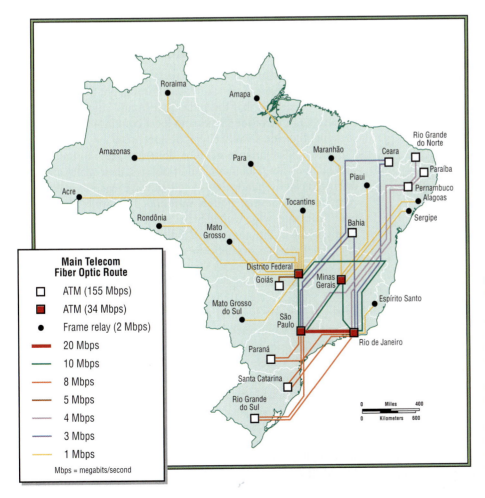

Main Telecom Fiber Optic Route

☐ ATM (155 Mbps)
■ ATM (34 Mbps)
● Frame relay (2 Mbps)
━━ 20 Mbps
━━ 10 Mbps
━━ 8 Mbps
━━ 5 Mbps
━━ 4 Mbps
━━ 3 Mbps
━━ 1 Mbps

Mbps = megabits/second

FIGURE 3.15 Telecom routes in Brazil. Brazil is leapfrogging old-fashioned copper wire telephone systems by installing high-speed, broadband fiber optic cable networks for telecommunications and Internet connectivity that are on a par with systems in the United States and Europe. The country is facing a major problem, however: disparity of distribution and access to the routes. The routes connect only the major cities and business districts; 95 percent of Brazil's population cannot gain access to the system. ATM in the key stands for asynchronous transfer mode, a means of very high speed digital (voice, video, image, and data) communications, and Brazil is hoping to lead a telecommunications revolution in the developing world. [Adapted from *Red Herring* (December 19, 2000): 81.]

SOCIOCULTURAL ISSUES

Under colonialism, a series of social structures evolved that guided daily life—standard ways of organizing the family, the community, and the economy. They included rules for gender roles, rules for race relations, and ways of religious observance. These social structures are still widely in place, but they are changing in response to urbanization, economic development, and the diffusion of ideas from outside the region. The results are varied. In the worst cases, change is leading to the breakdown of family life and the abandonment of children; in the best cases, to a new sense of initiative on the part of women, men, and the poor.

Cultural Diversity

The region of Middle and South America is one of the most culturally complex on earth. Contributing to this diversity are the many indigenous groups present when the Europeans arrived and the many cultures that were introduced during and after the colonial period. In the Caribbean, the Guianas (Guyana, Suriname, and French Guiana), and Brazil, the arrival of a relatively small number of European newcomers from Spain, England, Ireland, France, Holland, and Portugal resulted in the annihilation of indigenous cultures, and these regions then became populated almost entirely by various people from the Old World. From 1500 to the early 1800s, some 10 million Africans were brought to plantations in the islands and coastal zones of Middle and South America. After the emancipation of African slaves in the Caribbean in the 1830s, more than half a million Asians were brought there from India, Pakistan, and China as indentured agricultural workers. Their cultural impact remains most visible in Trinidad and Tobago, Jamaica, and the Guianas. In some parts of Mexico and in Guatemala, Ecuador, and other Andean zones, indigenous people have remained numerous, and to the unpracticed eye, may appear somewhat unaffected by colonization.

Mestizos are now the majority in Mexico, Central America, and much of South America. In some places, such as Argentina, Chile, and southern Brazil, Euro-Americans are now so numerous that they dominate the landscapes and ways of life. The Japanese, though a tiny minority everywhere in the region, are increasingly influential in innovative agriculture and industry in Brazil, the Caribbean, and Peru. In some ways, diversity is increasing as the media and trade introduce new influences into the region. Yet culture groups are becoming less distinct in urban areas, where various groups live in close proximity. The increasing contact between people of widely different backgrounds is accelerating the rate of **acculturation** (cultural borrowing) and **assimilation** (loss of old ways and adoption of new ones). In the big cities, such as Mexico City, Lima, and Rio de Janeiro, there is a blend of many different strains, so that no one cultural component remains unaffected by the others.

Race and the Social Significance of Skin Color

People from Middle and South America, and especially those from Brazil, often proudly make the public claim that race and color are of less consequence there than in North America, and they are right in certain ways. As discussed in Chapter 2, race remains socially significant in North America. In Middle and South America it is possible to effectively "erase" the color of one's skin through assimilation. For example, in Brazil, a Native American or a dark-skinned person of African or any other background may, by acquiring an education, a good job, a substantial income, the right accent, and a high-status mate, become recognized as having wiped away his skin color. Because of this possibility, race and class designations became more and more complex over the generations, and social class is no longer assigned on the basis of race or skin color alone. Today, the terms *clear-skinned* for "light" and *black, mulatto,* or *pardo* for "brown" tend to be merely descriptive and do not necessarily carry social meaning. The dark skin of an upper-class person would not be thought of as racially significant, but that person's family, wealth, education, and occupation would be very significant.

Nevertheless, the ability to erase the significance of skin color through accomplishments is not quite the same as color having no significance at all. This point is illustrated by the fact that overall, those who are poor, less well educated, and of lower social standing tend to have darker skins than those who are educated and well-off. And, while there are poor light-skinned people throughout the region (often the descendants of migrants from central Europe over the last century), most light-skinned people are middle and upper class. These observations seem to indicate that race and color have not yet disappeared as social factors in Middle and South America.

The Family and Gender Roles

The **extended family** is the basic social institution in all the societies of Middle and South America. Throughout this region, it is generally accepted that the individual should subvert his or her interests to those of the family and local community, and that by so doing, individual well-being is best secured.

The spatial arrangement of domestic life illustrates these strong family ties. Families of adult siblings, their mates and children, and elderly grandparents frequently live together in domestic compounds of several houses surrounded by walls. Such compounds are rare in the United States and Canada, where the single-family home is the norm. In Middle and South America, social groups out together in public are most likely to be family members of several generations, rather than unrelated groups of single young adults or of married couples, as would be the case in Europe or the United States. A woman's best friends are likely to be her female relatives. A man's business or social circle will almost certainly include male family members or long-standing family friends.

Gender roles have their roots in the extended family as well. Throughout Middle and South America, the Catholic Church was instrumental in the colonizing process and in the establishment of official and popular mores. The church remains influential in daily life (see page 148). The Virgin Mary, the mother of Jesus, is held up as the model for women to follow. Through their adoration of the Virgin, they absorb a set of values, known as *marianismo,* that puts a priority on chastity and service to the family. The ideal woman is the day-to-day manager of the house and of the family's well-being, training her sons to enter the wider world and her daughters to serve within the home. Over the course of her life, a woman's power increases as her skills and sacrifices for the good of all are recognized and enshrined in family lore.

Her husband, the titular family head, is expected to provide most of his income to his family, but he usually operates in the outside world, working and tending to his social network, which is deemed just as essential to the family's prosperity and status in the community as his work. The rules for how men are to contribute to family stability and well-being are less clearly spelled out than are the rules for women. Men have a good bit more autonomy and freedom to shape their lives than women do, simply because they are expected to move about the community and establish relationships, both economic and personal. In addition, as is the case in most societies, there is an overt double sexual standard for males and females. While expecting strict fidelity in mind and body from his wife, a man is freer to associate with the opposite sex. Males measure themselves by the model of **machismo,** which considers manliness to consist of honor, respectability, and the ability to father children and be attractive to women, to be an engaging raconteur in social situations, and to be the master of the household. Under traditional rules of machismo, the ability to acquire money was secondary to the other symbols of maleness. Now a new market-oriented culture also prizes visible affluence as a desirable male attribute.

Many factors are transforming the family and gender roles. For one thing, couples are now having only two or three children, instead of five or more. As discussed earlier, with infant mortality declining steeply, it is no longer necessary to have many children simply to see a few survive to adulthood. Because people still marry early, most parents are free of child-raising responsibilities by the time they are 40. For men, this change may require an adjustment of the machismo idea that manliness is defined in part by high fertility. For women, there is the problem of the empty-nest syndrome: after the children are grown, 30 or more years of active life loom, to be filled in some way. As it happens, economic change throughout the region has provided a solution. Increasingly, women, despite little formal education, have been able to find factory jobs or other employment that puts to use the skills they perfected while supervising a family: management of time and people, maintenance of equipment, organizational skills, the ability to anticipate problems, and long-range planning. For more and more women of all ages, employment outside the home (often in a distant city) is a way to gain a measure of independence and also contribute to the needs of the extended family. Many families simply bend to accommodate these new situations; others lose many members to migration, and some disintegrate.

Children in Poverty

Despite often difficult conditions, the vast majority of parents in Middle and South America, even those in the worst urban slums, heap loving care on their children, providing all that they possibly can and sacrificing so that their children might have a better life. Still, many families need the income that even young children can generate. It is estimated that one-third of the children in Middle and South America are economically active. In the countryside, children work in agriculture, logging, and mining. But city children also work, often on the street selling items such as chewing gum, flowers, and flip-flops (rubber sandals), cleaning cars, and performing small services. It is understandable that children are asked to help alleviate a family's poverty by doing some work, paid or unpaid. Nonetheless, in societies that traditionally place such high value on the family and children, it is difficult to understand recent news stories about homeless and abandoned children in such countries as Mexico, Guatemala, Colombia, and Brazil. The United Nations reports that in Mexico alone, there are 15,000 homeless children; in Guatemala City, 5000. Some of these reports may be overblown: a study in the year 2000 in São Paulo, Brazil, South America's largest city, revealed that just 609 children were homeless there, one-tenth of what had been thought to be the case. Nonetheless, throughout the region, many children live on the streets of cities, and thousands have been pushed out of the home at a tender age by overburdened parents. These children must fend for themselves, sleeping in doorways, hustling for money or food. Travelers to the region may experience the extreme distress of eating in an outdoor restaurant ringed with sturdy iron fences through which stretch the skinny arms of children begging to share part of their dinner. Reports of abandoned children are increasing around the world, and the explanations for this phenomenon are numerous.

In Middle and South America, structural factors are largely responsible for the difficult conditions under which families must operate. As discussed previously, rural-to-urban migration by young adults is common. Many inexperienced young women migrants are equipped for only the most menial jobs and rarely earn a living wage. They may soon ally themselves with a man to gain some measure of security, or even just a place to live. Soon,

A Nicaraguan boy sniffs glue from a baby-food jar in the Oriental Market in Managua, while his friend pulls his own jar out of his shirt. [Richard Sennott/ *Minneapolis–St. Paul Star Tribune.*]

children result. The man may already have several informal mating relationships, so his unpredictable income is stretched too thin to support several households adequately. Women are commonly expected to undertake the daily care of children, and in the case of young migrant mothers, they must also keep working. With no extended families to help them with child care, and with no older, solid role models—aunts, grandmothers—to enforce such traditional values as chastity, sobriety, and good parenting, their children may grow up neglected, malnourished, and unruly, with lonely, dysfunctional mothers who may turn to drugs or alcohol for solace. Recent reports tell of children who turn to brain-damaging glue sniffing to ease their anxiety (see the photo on page 147).

Religion in Contemporary Life

The Roman Catholic Church remains one of the most influential institutions throughout Middle and South America. From the beginning of the colonial era, the church was the major partner of the Spanish and Portuguese colonial governments. It received extensive lands and resources from those governments, and sent thousands of missionary priests to convert Native Americans. For many centuries, the Catholic Church encouraged working people to accept their low status, be obedient to authority, and postpone their rewards until heaven. Thus, while serving to unify the region and spread European culture, the church also reinforced class differences. Furthermore, the church ignored those teachings of Christ that admonish the privileged to share their wealth and attend to the needs of the poor. Nonetheless, poor people throughout Spanish and Portuguese America embraced the faith and still make up the majority of church members. They have put their own ethnic spin on Catholicism, creating multiple folk versions of the religion, with music, liturgy, and interpretations of Scripture that vary greatly from European ones.

The Catholic Church began to see its power erode in the nineteenth century in places such as Mexico. **Populist movements**—popularly based efforts seeking relief for the poor—seized and redistributed church lands and canceled the high fees that the clergy had been charging for simple rites of passage such as baptisms, weddings, and funerals. Over the years, the church became less obviously connected to the elite and more attentive to the needs of the poor. In the mid-twentieth century, the church began ordaining Native American and African-American clergy as a matter of course. Common people gained the right to read the Bible in their own languages and to replace Gregorian chants with ethnic music. After the Second Vatican Council in the 1960s, women were allowed to perform certain rites during Mass.

Just as the Catholic Church began to change its stance vis-à-vis the common people, a more radical movement known as **liberation theology**, which sought to reform the church from the ground up, emerged in the 1970s. Begun by a small group of priests and activists, the movement portrays Jesus Christ as a social revolutionary who symbolically spoke out for the redistribution of wealth when he divided the loaves and fishes among the multitude. Poverty is viewed as the problem of an entire society rather than as a personal failing. The perpetuation of gross inequality

and political repression is viewed as sinful, and social and economic reform is promoted as liberation from evil.

At its height, the liberation theology movement had more than 3 million adherents in Brazil alone and was the most articulate movement for region-wide social change. Today, the movement is somewhat reduced in strength. On the one hand, some of its positions have been adopted by the Vatican; on the other, in such places as Guatemala and El Salvador, it became the target of state-sponsored attacks that vilified its participants as communist conspirators. The liberation theology movement has also had to compete with newly emerging evangelical Protestant movements that also have many attractions for the poor.

Evangelical Protestantism has diffused into Middle and South America from North America. It is now the fastest growing religious movement in the region; already about 10 percent of the population, or more than 50 million people, are adherents. Like liberation theology, it appeals to the rural and urban poor—that segment of society most in need of hope for a better future. Some evangelical Protestants teach a "gospel of success," stressing that those who are true believers and give themselves to Christ and to a new life of hard work and clean living will experience prosperity of the body (wealth) as well as of the soul. (For a discussion of this same movement in Africa, see Chapter 7.) The movement is *charismatic*, meaning that it focuses on personal salvation and empowerment of the individual through miraculous healing and transformation, rather than on general social change aimed at benefiting all citizens. Individual adherents are thought to become "reborn." The movement has no central authority, consisting of a host of small, independent congregations. Leadership is often passed around so that both men and women in the congregations develop organizational and public speaking skills. A number of studies have shown a real change in the lives of individuals formerly beset with dejection, lethargy, and vice. After a few years of active church membership, an individual may seize educational and entrepreneurial opportunities, and indeed achieve success, aided by his or her belief in the gospel of Christ. There is considerable evidence that the gospel of success is an important impetus to the emergence of a middle class, and some theologians have noted that this version of Christianity meshes nicely with the neoliberal idea that social justice can be achieved in a trickle-down fashion through the success of the market economy. Others point out that general societal improvements are unlikely to spring from a religious movement that encourages personal well-being but does not mobilize the faithful to reform society on behalf of the poor and marginalized.

The geographer David Clawson, who has studied Protestants in Middle and South America, has provided a map showing the percentages of people throughout the region who are Protestant (Figure 3.16). The extent to which these geographic patterns reflect the spread of the evangelical version of Protestantism is not yet well understood. Evangelical Protestantism is growing in such formerly Catholic countries as Brazil and Chile. Brazil now has 34 million Protestants (75 percent of the total for the region), and in Chile, nearly a third of the more wealthy and literate people are Protestants. The evangelical movement is also strong in the Caribbean, Mexico, and Middle America, among both the poor

FIGURE 3.16 Percentage of Protestants in the populations of Middle and South America. It is difficult to establish the actual number of Protestants in the region because only a few countries (Chile and Brazil) include religious affiliation questions on their census forms. The data on this map, by geographer David Clawson, differs only slightly from information collected by other researchers. Indications are that Protestants of all sorts are growing as a percentage of the population, but in all mainland countries and most islands, are still a minority. [Adapted from David Clawson, *Latin America and the Caribbean* (Dubuque, Iowa: Wm. C. Brown, 1997), p. 210.]

and middle classes. The revival tents of North American evangelical pastors are a not uncommon sight in the urban fringe of cities throughout the region.

ENVIRONMENTAL ISSUES

Environments in Middle and South America were among the first to inspire concern about the use and misuse of the earth's resources. In the 1970s, scholars such as anthropologist Emilio Moran warned of an impending crisis in the tropical rain forests of the Americas. Beginning in that decade, construction of the Trans-Amazon Highway provided migrant farmers with access to the rain forest. They followed the new road into the Brazilian Amazon and began clearing its forests to grow crops. Scholars now know that even in prehistory, each and every human settlement has had consequences for environments across the Americas. Today's impacts, however, are particularly severe because the population has doubled and redoubled at the same time that per capita consumption of resources within the region and exports out of the region have increased dramatically.

Four World Regions Are Involved in Amazon Forest Extraction

During the 1980s and early 1990s, forests were cleared in Middle and South America at the rate of 30.4 million acres (12.3 million hectares) per year—an area about the size of Ohio. The clearing continues today; close to 75 percent of the forest lost annually in

FIGURE 3.17 Environmental issues: Middle and South America. This region was among the first on earth to recognize environmental deterioration, but there has been little agreement on the causes or the cures, partly because any one environmental problem may involve numerous local, state, and national political entities. (An explanation of pH is given in the legend of Figure 2.22 on page 92.)

PRACTICING GEOGRAPHY READING MAPS Compare this map with Figures 3.7 and 3.13. Do you see any connection between the patterns of available resources and the patterns of highly polluted areas? Looking at the environmental issues map, can you tell where the most urban areas are? Could you make a guess about where the most industrialized cities are? How?

this region is in Brazil, primarily within the Amazon Basin (Figure 3.17).

Amazon environments, though vast, face multiple threats. Primary among them are the logging of valuable hardwoods (especially mahogany), the clearing of land to raise cattle and to grow tropical cash crops, and the extraction of oil and gas from beneath the forests. All of these activities are focused on the global export market and connect investors, producers, and consumers across several world regions. The logging of Amazon hardwoods is often carried out by local entrepreneurs and landless loggers working for large landowners, who are encouraged by government policy. Investment capital comes from Asian multinational companies that have turned to the Amazon forests after having logged up to 50 percent of the tropical forests in Southeast Asia. Logging firms from Burma, Indonesia, Malaysia, the Philippines, and Korea are purchasing forestlands in Brazil after already having secured rights to Amazon Basin forests in the neighboring countries of Peru, Colombia, Venezuela, and the Guianas. A single tree can be worth $6000 to the barefoot men who fell the tree, cut it up, and raft it to a transshipment point, but the same tree will be worth $300,000 once it is turned into furniture or paneling in the United States or Europe. The World Wildlife Fund estimates that 50,000 such trees ended up in the U.S. market in 2002. According to an industry spokesperson, U.S. consumers typically do not inquire about the origins of the wood in the furniture they buy.

The governments of Peru and Brazil encourage surplus urban populations to occupy cheap land along the newly built logging roads. As a result, landless settlers from the highlands of the Andes and from southern and eastern Brazil clear yet more land for use as subsistence cultivation plots. After a few years of farming the poor rain-forest soils, these inexperienced settlers often abandon the land, now eroded and depleted of nutrients, and move on to new plots. Ranchers may buy the worn-out land from these failed small farmers to use as cattle pastures. As discussed in the vignette at the opening of this chapter, oil and gas extraction by multinational petroleum corporations is polluting the water and soil on remote Native American lands. The access roads built for oil extraction also encourage migrants, who inadvertently bring in deadly diseases to forest dwellers who rarely encounter people from the outside world.

The days of unopposed exploitation may be waning. Until the mid-1980s, Brazilian governmental policies encouraged forest clearing, considering it development of unused land. In particular, those who developed forestland did not have to pay taxes on it. Eventually, the Brazilian government changed its tax-relief policies in response to environmental concerns. Although they come late, environmental regulations are now being developed in all the countries of the Amazon. While Brazil now strictly controls the mahogany trade, illegal logging is on the rise. The pace of clear-cutting for pasture and small farming is slowing, but the rapid growth of urban settlements and suburbs, the increased use of forest timber as fuelwood, and the clearing of forest tracts for the large-scale production of cash crops such as soybeans all continue to threaten Amazon forestlands.

The destruction of the Amazon rain forest has global implications. Scientists now understand that the absorption of carbon dioxide and the release of free oxygen is one of the most important functions of the world's forests. Many scientists agree that the huge tracts of forested land in the Amazon Basin are essential to this process. The loss of these forests may be contributing to global warming (see the discussion on pages 46–48 in Chapter 1). One positive note is that reforestation is more widespread than previously thought. The tropical climate encourages fairly rapid rates of regrowth: abandoned pastures and farmland can regain a mature, although less diverse, forest in 15 to 20 years. Planting fast-growing trees in forest plantations is one way to maintain the exchange of carbon dioxide and oxygen, and the use of wood from such plantations may save some natural forests from cutting in the future. However, forest plantations fall far short of replicating the complex biodiversity of the rain forest.

The forests in other parts of Middle and South America are also at risk. About 65 percent of the forest clearing in Central America is intended to create pastures for beef cattle grown primarily for the U.S. fast-food industry. The forests of Central America (Costa Rica, El Salvador, Nicaragua, and Panama) and the Caribbean (Cuba, the Dominican Republic, and Jamaica) have been among the main targets of clearing for ranching and for export crops such as sugar, cotton, and tobacco in the past and bananas today. If deforestation continues in these areas at the present rates, the natural forest cover will be entirely gone in 20 years.

The forests of Middle and South America are another example of what geographers call contested space (see the discussion on page 140). Various groups—some new arrivals, some longtime inhabitants—are in conflict over the right to use the space as they see fit. For example, one well-financed government research agency in Brazil (IMPA, in Manaus) promotes the use of forest products, while another less well-financed government agency (FUNAE, also in Manaus) works to undo the resulting damage

In the Brazilian state of Rondônia, in the Amazon Basin, colonists from other areas clear the rain forest for settlement and large-scale agriculture. New roads, like the one at the upper left, provide the settlers with access. Some of the wood is harvested for local use or sold, but much is wasted. [Randall Hyman.]

and protect indigenous tribes from the encroachment of development. Meanwhile, in Manaus and elsewhere in the Amazon, ecotourism entrepreneurs take their clients to resorts in the rain forest, where luxury and spectacle often take priority over environmental sustainability (see the box "Ecotourism in the Ecuadorian Amazon" on page 153).

The Environment and Economic Development

In the past, governments in the region argued that economic development was so desperately needed that environmental regulations were an unaffordable luxury. Now there are increasing attempts to take a middle ground: embracing economic development as necessary to raise standards of living, yet hoping to minimize its negative effects on the environment. Grassroots environmental movements are active throughout the region, from the neighborhood level on up to the national level, often assisted by international nongovernmental organizations. Most such movements monitor, evaluate, and often challenge economic development projects via public protest, political negotiation, legal action, or public education.

Two cases of conflict between economic development and environmenal concerns are discussed here. Elsewhere, the text describes two locally designed projects that are sensitive to environmental concerns: one fostering urban quality of life in Curitiba, Brazil (see the box on page 131), and one reviving ancient cultivation systems in the Peruvian Andes (see the box on page 170).

Case 1: The Paraná-Medio Hydroelectric Project. Hydroelectric dams provide energy that can be used for industrialization and to provide electricity to urban areas. Such a dam has been proposed for the swampy central Paraná River basin in Argentina. The Paraná-Medio hydroelectric dam project is privately funded by a Metairie, Louisiana, firm that will receive a 50-year concession for electricity sales to Buenos Aires. The dam reservoir would turn a large wetlands ecosystem into the world's second largest artificial lake. An assessment firm based in the United States has said that the environmental effects would be negligible, but some Brazilian and Argentine scientists say that the lake would inundate the habitat of more than 1100 animal species. This disruption could harm 300 species of fish, fishing tourism, and the livelihood of fishers. Upstream flooding would eliminate seasonal grazing and rice farming and kill many valuable trees.

The public debate has persuaded the government to reevaluate the project and to include more public participation in the planning effort. However, energy is crucial to most kinds of economic development and is presently in short supply in much of Middle and South America. Figure 3.18 compares electricity use per capita in 1980 and in 2000 for several world regions. Middle and South America use only about one-tenth the electricity used in North America. Hydroelectric power is favored because it does not produce air pollution as coal or gas does, but the energy so produced is not "clean" because there are so many negative environmental and social consequences. The Paraná dam is only one of a score of privately funded hydroelectric dam projects being proposed for South America's major river systems.

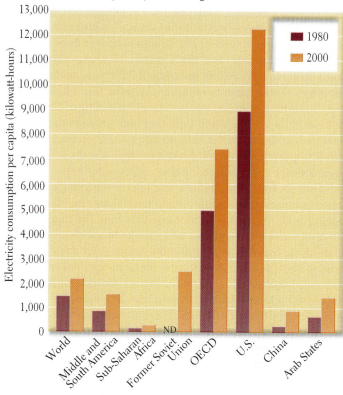

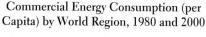

FIGURE 3.18 Electricity consumption (per capita) by world region, 1980 and 2000. Notice the differences in electricity consumption for all regions between 1980 and 2000. Also, notice that Middle and South America use only about one-tenth the energy per capita used in the United States, less than a quarter of that used in OECD (developed) countries combined, and about the same as Arab countries. Can you account for the change in use over the 20-year period and for the fact that this region uses so little energy? Can you anticipate future energy use patterns in various parts of this region? [*United Nations Energy Statistics Yearbook, 1977; United Nations Human Development Report 2000,* Table MDG5.]

Case 2: Shrimp Farming in Ecuador. The debt crisis that swept through the region beginning in the 1980s created tremendous pressure on countries to increase their exports so that they could pay off their debts. Environmental degradation is often one result. During the 1980s and 1990s, Ecuador became a major exporter of shrimp, but in the process it lost 80 to 90 percent of the mangrove swamp forests that once covered much of its coast on the Pacific Ocean. Mangroves are unique ecosystems that provide shelter to birds, fish, mollusks, crustaceans, and other species while also protecting coastal areas from erosion by sea waves. People along Ecuador's coast have traditionally used the mangroves as a source of firewood, clams, fish, and crabs without significantly altering the ecosystem. Because the mangroves flood periodically, those who use them move around constantly to take advantage of resources that shift depending on how much water is present. Shrimp farming requires that the mangroves be cut to make way for large, stable ponds in which shrimp are raised. The industry is highly profitable, and it has spread rapidly. The mangroves were acquired cheaply because traditional subsistence users never laid legal claim to them. They viewed the mangroves

AT THE LOCAL SCALE Ecotourism in the Ecuadorian Amazon

In a world where human efforts to prosper, or merely survive, continually threaten natural environments, one new strategy for both using and respecting nature is **ecotourism.** Ideally, ecotourism offers both natural and cultural travel experiences in unfamiliar environments. It encourages sustainable use and conservation of resources while providing a livelihood to local people and the broader host community. Ecotourism is now the most rapidly growing segment of the global tourism and travel industry, which itself is the world's largest industry, with $3.5 trillion spent annually. On a trip through the Ecuadorian Amazon in 1998, Alex Pulsipher saw several sides of the growing ecotourism industry.

PERSONAL VIGNETTE Puerto Misahualli is a small river town that is currently enjoying an economic boom due to the many European, North American, and other foreign travelers who come there, much as I did, looking for experiences that will bring them closer to the now legendary rain forests of the Amazon. The array of ecotourism offerings is impressive. One capable-looking indigenous man offered to be my personal guide for as long as I wanted as we traveled by boat and on foot, camping out in "untouched forest teeming with wildlife." His guarantee that we would eat monkeys and birds did not seem to promise the nonintrusive, sustainable experience I was after. A well-known "eco-lodge" nearby offered plush hotel-style rooms with a river view, a chlorinated swimming pool, a fancy restaurant serving "international cuisine," and a private 740-acre (300-hectare) nature reserve. A wall topped with broken glass separated the lodge from the surrounding community. It seemed more like a fortified beach resort than an eco-lodge.

The solar-powered Yachana Lodge, in contrast, had simple rooms, local cuisine, and a knowledgeable resident naturalist who was a veteran of many campaigns to preserve Ecuador's wilderness. Profits from the lodge funded a local clinic and various programs that teach agricultural methods to protect the fragile Amazon soils while increasing earnings from surplus produce. The "local" cultivators were actually poor migrants from Ecuador's densely populated cool highlands who had been given free land in the Amazon by the government—but no training for how to farm in the lowland rain forest. Many of their attempts have resulted in extensive soil erosion and deforestation. I soon learned from people working with the nonprofit group running the Yachana Lodge—the Foundation for Integrated Education and Development (FUNEDESIN)—that the lodge is earning just barely enough to sustain the clinic and agricultural programs.

This story touches on two themes related to environmental issues in Middle and South America: one is the tensions between economic development and environmental preservation and the other is the ways in which poverty is linked to environmental degradation. Ecotourism often turns out to be little different from other kinds of tourism that damage the environment and give little back to the surrounding community. Still, there is the potential to use the profits of ecotourism to benefit local communities and the environment.

as commonly held grounds. Local and international environmental advocates attempted to halt the encroachment of shrimp farms on mangroves, but bribes to officials by shrimp farmers made it possible for them to claim yet more mangrove land, and now it is nearly gone.

These two cases illustrate how local people in Middle and South America must learn to be their own environmental advocates in the tension-filled process of economic development. So far, success in halting unsustainable exploitation has been elusive. Collaboration between local indigenous groups and international conservation groups is just beginning to result in stronger environmental legislation and more sustainable development policies.

MEASURES OF HUMAN WELL-BEING

PERSONAL VIGNETTE Miss Eliza arises at six in the morning and prepares tea, bread, and fruit for her two daughters and her elderly Aunt Hettie, whose two-room wooden house she shares. Aunt Hettie helps ready the children for school, and Miss Eliza rushes off with a "head-load" of vegetables and spices to catch the jitney to Montego Bay, Jamaica. There, in the central market, she hustles all day, finding just the right breadfruit for a good customer, putting together a basket of spices and condiments (nutmeg, pimento, cinnamon, fiery hot sauce) for some tourists off a cruise ship, negotiating with a fellow higgler (a person who buys small surpluses to resell) for some special cuttings of dasheen to plant in her home garden. During all this time, she doesn't think once about the fact that her yearly income for herself, Aunt Hettie, and the two children is just $4000. Rather, every moment of her life is spent dealing with the realities of having very little. [*Lydia M. Pulsipher, fieldwork, 1992.*]

As geographers, we need ways to compare the well-being of people like Miss Eliza with that of people in other places. Table 3.3 provides you the opportunity to do just that, according to several different indices.

You will remember from the explanation in Chapter 1 that gross domestic product (GDP) per capita figures are often used as a crude indicator of well-being. In column 2 of Table 3.3, the countries in Middle and South America are compared with Japan, the United States, and the UAE in GDP per capita. You can see

TABLE 3.3   Human well-being rankings of countries in Middle and South America and other selected countries, 2003

Country (1)	GDP per capita, adjusted for PPP[a] in 2001 $U.S. (GDP ranking among 175 countries, ranked from highest) (2)	Human Development Index (HDI) global rankings, 2003[b] (3)	Gender Empowerment Measure (GEM) global rankings, 2003 (4)	Female literacy (percentage), 2001 (5)	Male literacy (percentage), 2001 (6)
Selected countries for comparison					
Japan	25,130 (14)	9 (high)	44	99	99
United States	34,342 (2)	3 (high)	10	99	99
United Arab Emirates	20,530 (23)	48 (high)	65	80	75
Caribbean					
Antigua and Barbuda	10,170 (49)	56 (medium)	—	88[e]	90[e]
Bahamas	16,270 (34)	49 (high)	18	96	95
Barbados	15,560 (36)	27 (high)	20	99.7	99.7
Cuba	5,259 (90)	52 (high)	25[c]	97	97
Dominica	5,520 (86)	68 (medium)	—	94[d]	94[d]
Dominican Republic	7,020 (68)	94 (medium)	37	84	84
Grenada	6,740 (70)	93 (medium	—	98[d]	98[d]
Guadeloupe	9,000 (1997 est.)[d]	—	—	90[d]	90[d]
Haiti	1,860 (140)	150 (low)	71[c]	50	53
Jamaica	3,720 (105)	78 (medium)	—	91	83
Martinique	10,700 (2001 est.)[d]	—	—	93[d]	92[d]
Netherlands Antilles	11,460 (2002 est.)[d]	—	—	93[d]	98[d]
Puerto Rico	11,500 (2002 est.)[d]	—	—	94[d]	94[d]
St. Kitts and Nevis	11,300 (46)	51 (high)	—	98	97
St. Lucia	5,260 (88)	71 (medium)	—	69	65
St. Vincent and the Grenadines	5,330 (87)	80 (medium)	—	96	96
Trinidad and Tobago	9,100 (55)	54 (high)	22	98	99

[a]PPP = purchasing power parity.
[b]The high, medium, and low designations indicate where the country ranks among the 175 countries classified by the United Nations; the only country in the Western Hemisphere to rank low is Haiti.
[c]Data from *United Nations Human Development Report 1998* (data not available in 2003).

immediately that they all lag far behind those three countries. These GDP figures have been adjusted for differences in prices and other factors to obtain internationally comparable indications of purchasing power parity (PPP) in U.S. dollars. Nonetheless, even these adjusted figures can be misleading, because they are averages that mask a very wide disparity of wealth in Middle and South America and because they ignore aspects of well-being other than income. From these figures, we can learn that purchasing power is very likely limited throughout the region. But purchasing power is not the best measure of well-being.

In column 3 of Table 3.3, the United Nations Human Development Index (HDI) adds two other components to GDP per capita (PPP)—life expectancy at birth and educational attainment—to arrive at a ranking of 175 countries that is sensitive to more factors than just income. Life expectancy figures show the overall level of health, while educational attainment is representative of the emphasis on human resource development in a country. You can see that some countries in Middle and South America rank close to the UAE (48) on the HDI index despite having much lower GDP per capita figures: the Bahamas (49),

TABLE 3.3 (*Continued*)

Country (1)	GDP per capita, adjusted for PPP[a] in 2001 $U.S. (GDP ranking among 175 countries, ranked from highest) (2)	Human Development Index (HDI) global rankings, 2003[b] (3)	Gender Empowerment Measure (GEM) global rankings, 2003 (4)	Female literacy (percentage), 2001 (5)	Male literacy (percentage), 2001 (6)
Mexico and Central America					
Mexico	8,430 (58)	55 (high)	42	89	93
Belize	5,690 (83)	67 (medium)	47	93	93
Costa Rica	9,460 (51)	42 (high)	19	96	95
El Salvador	5,260 (90)	105 (medium)	54	76	82
Guatemala	4,400 (97)	119 (medium)	35[c]	62	76
Honduras	2,830 (116)	115 (medium)	60	76	75
Nicaragua	2,450 (123)	121 (medium)	—	67	66
Panama	5,750 (82)	59 (medium)	50	91	93
South America					
Argentina	11,320 (45)	34 (high)	—	97	97
Bolivia	2,300 (126)	114 (medium)	38	80	92
Brazil	7,360 (64)	65 (medium)	68[c]	87	87
Chile	9,190 (53)	43 (high)	52	96	96
Colombia	7,040 (67)	64 (medium)	46	92	92
Ecuador	3,280 (109)	97 (medium)	49	90	93
French Guiana	6,000	—	—	82[d]	84[d]
Guyana	4,690 (94)	92 (medium)	—	98	99
Paraguay	5,210 (91)	84 (medium)	59	93	94
Peru	4,570 (96)	82 (medium)	39	86	95
Suriname	4,599 (95)	77 (medium)	—	91[d]	95[d]
Uruguay	8,400 (59)	40 (high)	43	98	97
Venezuela	5,670 (84)	69 (medium)	56	92	93

[d]Data from *CIA World Factbook*, 2003, which estimates or cites figures for variable years.
Source: United Nations Human Development Report 2003.

Barbados (27), Costa Rica (42), Argentina (34), Chile (43), and Uruguay (40). Their rankings are close to or higher than the UAE's partly because education is more available to both sexes and across classes in Middle and South America than in Southwest Asia, where women are secluded. In addition, life expectancy in the Caribbean can be as high as 80 years for women, whereas in the UAE it is 77. Nonetheless, investment in basic and secondary education in Middle and South America is not sufficient to prepare most people for skilled jobs, and poor health care is holding down life expectancy to age 70 for the region as a whole. The HDI rankings also indicate that poverty is especially deep in Haiti (150), Guatemala (119), Honduras (115), Nicaragua (121), and Bolivia (114). In these countries the burden of a low income is not eased by government social programs, as it is in most Caribbean countries (except Haiti).

The United Nations Gender Empowerment Measure (GEM) (Table 3.3, column 4) measures the extent to which females have opportunities to participate in economic and political life and takes into account female per capita GDP. Only 70 of 175 countries are ranked on the GEM, and figures are missing for a number of Caribbean countries, but those available show that four Caribbean countries plus Costa Rica rank relatively high on this scale: the Bahamas (18), Barbados (20), Trinidad and Tobago (22), Cuba (25), and Costa Rica (19). In all cases, their GEM ranks are higher than their HDI ranks. When a country's GEM rank is significantly higher than its HDI rank, it can mean that despite economic problems, the society is comparatively open to female participation in public life: education, jobs outside the home, entrepreneurship, leadership in business, and government. In the Caribbean, women have long been powerful family members while working outside the home and leading community organizations. Increasingly they serve in government and usually have higher educational attainments than men. Many other countries in Middle and South America, however, rank relatively low on the GEM, such as Brazil (68), Chile (52), Paraguay (59), and Venezuela (56). Still, nearly all do better than the relatively rich but gender-stratified UAE (65), where education and health services for women and opportunities for paid employment and political participation have been limited.

Returning to the case of Miss Eliza in Jamaica, we can now see that her income of $4000 for a family of four ($1000 per person) is considerably lower than the average per capita income for that country ($3720). However, it is also true that social support customs can compensate for low income, as exemplified by Miss Eliza's looking after and feeding her aged aunt, who in return provides a house and child care for Miss Eliza. It is through such informal reciprocal exchanges that people manage to survive and even thrive despite low incomes and declining tax-supported public services.

A Rising Threat to Human Well-Being

The global epidemic of HIV-AIDS is now taking a serious toll throughout Middle and South America and can be expected to affect well-being statistics negatively. The UN reported in 2003 that more than 600,000 women and at least 60,000 of their children in the region were living with the virus. It is safe to assume that at least 600,000 men are also infected. The Caribbean is thought to have one of the highest HIV rates on earth, but its high education rates and the relatively high status of women may result in the quick adoption of safe-sex measures. Haiti has the highest rate of infection in the region at 6 percent of the population, the Bahamas next highest at 3.5 percent. The Venezuelan journalist Silvana Paternostro, in her book *In the Land of God and Man: Confronting Our Sexual Culture* (1998), has written that cultural practices in the region probably contribute to the rate of infection. Cultural mores discouraging the discussion of sex, as well as the low status of women in much of the region, militate against safe-sex practices. Men are customarily not expected to be monogamous, and there are many sex workers, male and female. The participation of husbands in the sex trade means that married women are especially vulnerable to being infected and to infecting any children they bear. Condom use is only beginning to be accepted, and a wife would be loath to ask her husband to use a condom.

As is the case elsewhere around the globe, HIV-AIDS tends to claim a country's youngest, best educated, and most productive citizens. Places such as Barbados, which have state-of-the-art health-care facilities, are quickly overrun with patients who have the means to travel and pay for treatment. But for the vast majority, treatment is beyond their means.

Quick Review

1. What does the term *debt crisis* mean, and why were SAPs expected to relieve the crisis?

2. Name a few of the forces that act to perpetuate income disparity.

3. How might the spread of the Internet reinforce democratic institutions?

4. What is the difference between acculturation and assimilation?

5. How does the destruction of the Amazon rain forest have a global impact?

6. What are some factors that can make a country's actual standard of living higher than indicated by its GDP per capita?

III   SUBREGIONS OF MIDDLE AND SOUTH AMERICA

This tour of the subregions of Middle and South America focuses primarily on the themes of cultural diversity, economic disparity, and environmental deterioration. These themes are reflected somewhat differently from place to place. For every subregion, examples of connections to the global economy are given.

THE CARIBBEAN

Despite the Caribbean's strong record of fostering human well-being, visitors to the region often erroneously conclude that it is

desperately poor. Typically, short-term visitors are isolated in tourist enclaves or on cruise ships, where they glimpse only tiny swatches of the island landscapes. From that vantage point, it is hard to see beyond the tiny houses and garden plots or quaint, narrow streets to the statistical facts and social relationships that make life in the Caribbean so different from what it first appears to be.

Only in the past 50 years have most of the island societies emerged from colonial status to become independent, self-governing states (Figure 3.19). These islands, with the exception of Haiti and parts of the Dominican Republic, are no longer the poverty-stricken places they were 30 years ago. Rather, they are

FIGURE 3.19 The Caribbean subregion. These tiny island countries have the highest standards of living in the entire region of Middle and South America.

managing to provide a reasonably high quality of life for their people. Children go to school, and literacy rates for people under 70 years of age average close to 95 percent. There is basic health care: mothers receive prenatal care, nearly all babies are born in hospitals, and infant mortality rates are low (much lower than what they were even in the 1970s). Life expectancy is in the 70s for most islands. And people are choosing to have fewer children; the overall rate of population increase for the Caribbean is the lowest in Middle and South America. A number of Caribbean islands rank high on the Human Development Index, and some do particularly well on the Gender Empowerment Measure (see Table 3.3). Returned emigrants often say that the quality of life on many Caribbean islands actually exceeds that of far more materially endowed societies because life is enhanced by strong community and family support. And, of course, there is the healthful and beautiful physical environment.

Beyond the beaches and quaint villages, there are local civic organizations such as the Rotary and Lions clubs, libraries, chambers of commerce, gourmet cooking clubs, garden societies, community clean-up associations, and active churches. These organizations practice participatory democracy daily: citizens meet to educate one another about social and environmental issues, and they continually design and implement solutions to local problems. The progress indicated by the demographic statistics is the result of hard work and civic responsibility, as well as aid from the old colonial powers and from Canada (and in Puerto Rico, from the United States).

Island ministers of government continually search for ways to turn former plantation economies, once managed from Europe and North America, into more self-directed, self-sufficient, and flexible entities that can adapt quickly to the perpetually changing markets of the global economy. On the one hand, they are tempted to specialize; on the other, they are wary of the dependency and vulnerability that too much specialization can bring. When plantation cultivation of sugar, cotton, and copra (dried coconut meat) died out in the 1950s and 1960s, some islands turned to producing "dessert" crops, such as bananas and coffee, which they sold for high prices under special agreements with the countries that once held them as colonies. Now these protections are disappearing as the European Union makes such agreements illegal in Europe. Other island countries turned to the processing of their special resources (petroleum in Trinidad and Tobago, bauxite in Jamaica), the assembly and finishing of such high-tech products as computer chips and pharmaceuticals (in St. Kitts and Nevis), or the processing of computerized data (in Barbados). A number of islands combined one or more of these strategies with tourism development.

For some islands, tourism and related activities contribute as much as 80 percent of the gross national product (GNP), as was the case in Antigua and Barbuda, Barbados, the Bahamas, and St. Martin in the 1990s. Heavy borrowing to build hotels, airports, water systems, and shopping centers for tourists has left some islands with huge debts to pay off. Furthermore, there is a special stress that comes from dealing perpetually with hordes of strangers, especially when they dress too skimpily, act in ways that violate local mores, and make unflattering assumptions about one's state of well-being—concluding erroneously that poverty and illiteracy are widespread. Every year in the mid-1990s, Jamaica (population 2.6 million) received visitors numbering twice its population; Antigua and Barbuda (population 66,000) hosted more than 4 times its population, and St. Martin (population 20,000), 20 times its population.

Host countries consider tourism difficult to regulate, in part because it is a complex multinational industry. Tourism markets are controlled by external interests such as travel agencies, airlines, and the outside investors needed to fund the construction of hotels, resorts, restaurants, and golf courses. The rapid growth of the industry in an unregulated environment has led to overbuilding and to poorly developed hotel accommodations in such places as St. Croix, St. Thomas, Antigua, and St. Martin. The recent rapid growth of the cruise-line industry, operated mostly out of Puerto Rico and Miami, has caused overcrowding as well. Caribbean cruise ships now routinely bring several thousand passengers into small port cities, such as Soufrière, St. Lucia, or St. Johns, Antigua. Plans by corporations such as Disney Enterprises to build even larger cruise ships, equipped with elaborate entertainment facilities and carrying even more passengers, are troublesome for small islands such as Grenada and St. Lucia, which would be unable to adequately accommodate either the enormous ships in their harbors or the large numbers of day visitors

such liners would discharge. Entrepreneurs on some islands are targeting new types of tourists, who might stay longer and show a more informed interest in island societies. Nature tourism, sport tourism, and ecotourism are three variations that better suit the Caribbean environment and have been promoted as low-density alternatives to the "sand, sea, and sun" menu that presently attracts the majority of visitors to the region.

Cuba and Puerto Rico Compared

Cuba and Puerto Rico are interesting to compare because they shared a common history until the 1950s, then diverged starkly. North American interests dominated both islands after the end of Spanish rule around 1900. U.S. investors owned plantations, resort hotels, and other businesses on both islands. To protect their interests, the U.S. government influenced dictatorial local regimes to keep labor organization at a minimum and social unrest under control. By the 1950s, poverty was widespread in both Cuba and Puerto Rico. Infant mortality was at 50 per 1000 or higher, and most people worked as agricultural laborers. Then each island took a different course toward social transformation. In 1959, Cuba experienced a communist revolution under the leadership of Fidel Castro; meanwhile, Puerto Rico experienced a more gradual capitalist metamorphosis into a high-tech manufacturing center.

Since Castro seized control, Cuba has dramatically improved the well-being of its general population in all physical categories of measurement (life expectancy, literacy, and infant mortality). By 1990, it had a solid human well-being record despite its relatively low GDP per capita (see Table 3.3). Cuba thus proved that, once the requisite investment in human capital is made, poverty

Varadero, one of Cuba's new tourism areas, is on the island's north coast, about 100 miles (160 kilometers) east of Havana. [David Alan Harvey/Magnum Photos.]

and social problems are not as intractable as some people had thought. Unfortunately, Cuba's successes are not replicable elsewhere in the region, first, because it has been politically repressive in the extreme, forcing out the aristocracy and jailing and executing dissidents. Second, Cuba has been inordinately dependent on outside help to achieve its social revolution. Until 1991, the Soviet Union was Cuba's chief sponsor. It provided cheap fuel and generous foreign aid and bought Cuba's main export crop, sugar, at artificially high prices. With the demise of the Soviet Union, Cuba's economy declined sharply during what became known as the Special Period, when all Cubans were asked to make sacrifices and cut back their consumer spending. To survive this economic crisis, Cuba opened its economy to outside investment, especially European capital, to help redevelop its tourism industry. Canada has invested in joint biotech research on hepatitis and cancer drugs with Cuban medical scientists, who receive particularly good training—among the best in the region. Biotech exports accounted for $130 million in 2000 and are expected to grow. Cuban doctors are loaned (for a price) to other countries in the region (such as Venezuela; see page 143) to improve basic health care. Health and spa tourism is also a growing industry, with more than 5000 customers per year arriving for face-lifts, tummy tucks, and medical therapy.

Political relations between the Castro regime and successive U.S. administrations have remained on cold war terms since 1961. A recent example is the Helms–Burton Act, adopted in 1996 by the U.S. Congress under pressure from upper-class Cubans who fled to Florida after the Cuban revolution. This law attempted to stop all trade with Cuba in the hope that the hardships brought on by the ensuing economic crisis would lead to popular support for an end to Castro's leadership. Other governments in Europe and America, regarding the Helms–Burton Act as inconsistent with international law, continue to trade with and invest in Cuba, especially in tourism. Cuban tourism grossed $2 billion in 2002, hosting 1.68 million visitors, mostly from the European market. Nonetheless, Cuba's most natural trading partner would be the United States, and many U.S. businesspeople are eager for a repeal of the Helms–Burton Act.

In the 1950s, Puerto Rico began Operation Bootstrap, a government-aided program to transform the island's economy from its traditional sugar plantation base to modern industrialism. Many international industries took advantage of generous tax-free guarantees and subsidies to locate plants on the island. Since 1965, this industrial sector has shifted from light to heavy manufacturing, from assembly plants to petroleum processing and pharmaceutical manufacturing. It is believed that the Puerto Rican seaboard is heavily polluted as a consequence of the chemicals released by these more recent industries.

Because Puerto Rico is a commonwealth within the United States and its people are U.S. citizens, many Puerto Ricans migrate to work in the United States. The manufacturing jobs in Puerto Rico and remittances from people working on the mainland have greatly improved living conditions on the island, but social investment by the U.S. government has also upgraded the standard of living. Many Puerto Ricans receive some sort of support from the federal government, including retirement benefits and health care. The infant mortality rate is now 10.6 (Cuba's is just 6), and life expectancy is 77 (Cuba's is 76). The connection with the United States helps Puerto Rico's tourism and light-industry economy, but outside of San Juan, with its skyscraper tourist hotels, Puerto Rico's landscape reflects stagnation: few interesting or well-paid jobs, an inadequate transport system, little development at the community level, mediocre schools, and few opportunities for advanced training.

Statehood for Puerto Rico, which some people desire, would mean the end of tax holidays for U.S. companies and hence the end of Puerto Rico's ability to attract assembly plants, some of which are already moving to cheaper labor markets in Asia. In addition, many fear that statehood would eventually bring cultural assimilation and the loss of the island's Spanish linguistic and cultural heritage.

Haiti and Barbados Compared

Haiti and Barbados present another study in contrasts. During the colonial era, both were European possessions with plantation economies, Haiti a colony of France and Barbados a colony of Britain. Yet they have had very different experiences, and today they are far apart in economic and social well-being. Haiti, though not without useful resources, is the poorest nation in the Americas, with a ranking on the United Nations Human Development Index of 150 out of 175. Barbados has the highest HDI ranking in the entire region (27), even though the island has much less space, a higher population density, and few resources other than its limestone soil, its beaches, and its people.

Haiti was the richest plantation economy in the Caribbean by the end of the eighteenth century. When Haitian slaves revolted against the brutality of the French planters in 1804, Haiti became the first colony in Middle and South America to achieve independence. Haiti's early promise was lost, however, when the former-slave reformist leaders were overthrown by other former slaves who were violent and corrupt militarists. They neither reformed the exploitative plantation economy nor sought a new economic base. Under a long series of incompetent authoritarian governments, the people sank into abject poverty, while the land was badly damaged by particularly wasteful and unprofitable plantation cultivation. In the mid-twentieth century, when other only recently independent Caribbean islands, such as Barbados, were reorganizing their economies and achieving various levels of prosperity, in Haiti a class-based reign of terror under François "Papa Doc" Duvalier (1957–1986) pitted the mulatto elite against the black poor. Today, Haiti remains overwhelmingly rural, with widespread illiteracy and high infant mortality. Its lands are deforested and eroded and are subject to disastrous flooding. Efforts to establish democracy have repeatedly devolved into violence. Since the early 1990s, the United Nations has maintained peacekeeping troops in Haiti, and several humanitarian aid organizations run programs there. Multinational corporations have opened maquiladora-like assembly plants, employing primarily young women, but the plants are not yet flourishing. Although minerals such as bauxite, copper, and tin exist in Haiti, cost-effective development of these resources is not possible.

Far more prosperous Barbados has fewer natural resources and is more than twice as crowded as Haiti. Barbados has 166 square miles (430 square kilometers) and 1500 people per square mile (618 per square kilometer), whereas Haiti has 10,700 square miles (27,800 square kilometers) and 700 people per square mile (268 per square kilometer). Barbados's present prosperity is explained by the fact that its citizens successfully pressured the British government to invest in the people and infrastructure of its colony before giving it independence in 1966. Although both Haiti and Barbados began the twentieth century with large, illiterate, agricultural populations, their development paths have diverged sharply. In 2001, two-thirds of Haitians are still agricultural workers, and less than half are able to read and write. Barbados, on the other hand, now has a diversified economy that includes tourism, sugar production, remittances from migrants, information processing, offshore financial services, and modern industries that sell products throughout the Caribbean—including cement, food (hot sauce) and beverages (rum), furniture, beachware, customized fashions, footwear, and music. Barbadians hold jobs that demand skilled, literate employees; they are well educated and well fed, and most are homeowners. Meanwhile, the Barbadian government and private businesspeople constantly seek new employment options for the citizens and occupy a central role in Caribbean economic and social development.

MEXICO

Mexico today is working toward becoming a reasonably well-managed, middle-income democracy (see the earlier discussion on page 142). Most of the efforts to achieve this goal, both private and governmental, are focused in some way on Mexico's relationship with the United States, its main trade partner. The integration of the Mexican, U.S., and Canadian economies has proceeded rapidly since NAFTA became official in the mid-1990s. Two of the most obvious geographic markers of this integration are interconnecting highway systems between the three countries and, in Mexico, the phenomenon of the maquiladoras, discussed below (see Figure 3.20).

Mexico's formal economy has modernized rapidly in recent decades. Its components are mechanized agriculture, a vastly expanded manufacturing sector, petroleum extraction and refining, tourism, a service and information sector, and the remittances of millions of migrants working abroad. Mexico's large and varied informal economy employs an estimated 16 million people at least part-time—about half of the total working population of 33 million. These workers contribute at least 13 percent of the GDP through such activities as subsistence agriculture, fine craft making, and many types of services, such as hairdressing, running errands, tutoring, trash recycling, and street vending. The informal economy also includes a flourishing black market and the crime sector.

The Geographic Distribution of Economic Sectors

Mexico's economic sectors have a distinct geographic distribution pattern. Just 12 percent of Mexico is good agricultural land, and much of that is on hillsides. The biggest agricultural units are large corporate farms, aided by government marketing, research, and irrigation programs, which have largely replaced the old, inefficient haciendas and ranches. These farms produce high-quality meat and produce—beef, chicken, peaches, nuts, asparagus, chickpeas, citrus, melons, peppers, winter vegetables, oilseeds, strawberries, and raspberries—as well as sorghum for animal feed. Large corporate farms located in such places as the state of Sonora, in the arid north close to the U.S. border, send produce to the North American market. In the tropical coastal lowlands farther south, subsistence cultivation on small plots is much more common, but here too tropical fruits and products such as jute, sisal, and tequila are produced for export as well as for internal consumption. Mexico's convoluted upland landscapes provide many small and specialized niches where a wide variety of crops are grown, mostly on small farms: coffee, corn, tomatoes, sesame seeds, hot peppers, and flowers, among others. About 24 percent of Mexico's population is employed in agriculture, though it produces only 5 percent of Mexico's GDP.

The petroleum refining industry is located along the Gulf coastal plain, and tourism facilities are found primarily along the Caribbean and Pacific coastal plains. Manufacturing, which employs about 21 percent of the registered workforce and accounts for 27 percent of the GDP, is concentrated along the U.S. border in such cities as Tijuana, Mexicali, Ciudad Juárez, and Reynosa, where numerous U.S. firms operate maquiladoras; some of these factories are also found in the general vicinity of Mexico City and on the Yucatán Peninsula. The service sector employs 55 percent of the population and produces 68 percent of the GDP. Although it includes restaurants, hotels, financial institutions, consulting firms, and museums, this sector is dominated by small operations that cater to everyday needs: hair salons, tortilla vendors, tailoring shops, shoeshine and shoe repair services, Internet cafés, cleaning businesses, and so forth. Most such service jobs are concentrated in cities. The informal economy, which can include agricultural, manufacturing, and service activities, is everywhere, but it is most varied and noticeable in the big cities.

NAFTA and the Maquiladora Phenomenon

Maquiladoras are foreign-owned plants that hire people to assemble manufactured goods, usually from imported parts. The materials are imported duty-free, and the goods are returned to the country of origin for sale there or elsewhere. Nine out of ten such products are imported from and sold back to the United States. There are thousands of maquiladoras along the Mexican side of the U.S.–Mexican border, as well as elsewhere in the country (Figure 3.20; see also Figure 3.12 on page 136). Many maquiladoras are branches of American, European, or Asian corporations, though Mexicans hold the majority interest in about half of them. The Mexican side of the border is a desirable location for these factories because it provides an inexpensive labor pool within easy reach of the United States. In 2003, Mexican workers earned an average of about $6 to $8 a day. Furthermore, Mexico has far fewer regulations than the United States covering worker safety, fringe benefits, and environmental protection, and it charges much lower taxes on industries than the United States. These

FIGURE 3.20 The Mexico subregion. Mexico today is a middle-income democracy focused on industrialization. There are thousands of maquiladoras along the Mexican side of the U.S.–Mexican border, as well as elsewhere in the country. A complex road system is necessary to link the maquiladoras with sources of labor and with markets in the United States and Canada. In Mexico, the main artery of Mexican–North American trade is known as the "NAFTA Highway, " running from Mexico City to Ciudad Juárez/El Paso on the Mexican–U.S. border. It links to U.S. Interstate Highways 25 (north-south) and 10 (east-west). There is now talk of building "NAFTA Highways" in the United States to handle the increased traffic between Mexico, the United States, and Canada.

conditions attracted a considerable number of maquiladoras even before NAFTA took effect in 1994, and thereafter, the number of maquiladoras increased dramatically (Table 3.4, column 2). Total maquiladora employment grew until 2001, but by 2003 it had contracted by more than 200,000 jobs (see Table 3.4, column 3) as multinational firms found even less expensive labor forces in Asia (see the vignette on Orbalin Hernandez on pages 162–163). As contraction continued in 2004, the cities along the U.S. border were beginning to experience declining populations.

Studies in the late 1990s showed that more than half of the maquiladora laborers are women, many of them young and unmarried. Often living for the first time without the protection of their families, these young women have been subjected to high rates of criminal assault. Ciudad Juárez has been the site of serial killings that have taken the lives of dozens of young women, who have often been attacked or kidnapped while traveling to and from work. The film *La Señorita Extraviada* (2002) documents the failure of Mexican officials to investigate the killings and chronicles the efforts of women on both sides of the border to demand an accounting. Maquiladora laborers, male and female, migrate from rural locations across Mexico, and the money (remittances) they send home is crucial to their families' welfare. Because of their financial contributions, female workers enjoy status in the family that earlier generations of women did not have. Recent studies have shown that the majority do not use maquiladora work as a precursor to migration into the United States, as has been often claimed; if these migrants fail to find maquiladora work, they tend to return to their home villages.

Although many Mexicans are pleased to have the maquiladoras and the jobs they bring, the country is increasingly facing negative side effects. The border area has developed serious groundwater and air pollution problems because of the high

TABLE 3.4   Growth of the maquiladora sector

Year (1)	Number of maquiladoras (2)	Number of employees (3)
1968	79	17,000
1974	455	75,974
1981	605	130,973
1988	1400	369,489
1990	2042	486,000
1997	2676	897,354
1998	2983	1,008,021
1999	3500	1,096,000
2000	—	1,300,000
2001	(2882)[a]	1,081,526
2002	3800	1,100,000
2003	—	1,074,000[b]

[a]Counts only maquiladoras in border states.
[b]By mid-2003, there were 226,000 layoffs from the high in 2000. Maquiladoras are streamlining in order to compete in the global economy, and some maquiladora jobs are now going to Asia, especially China.

In Nuevo Laredo, Mexico, workers in this sewing factory, leased by R. G. Barry Corporation of Pickerington, Ohio, are making brand-name slippers for markets in North America and Europe. [Paul S. Howell/Gamma-Liaison.]

concentration of poorly regulated factories and the unplanned worker shantytowns that have sprung up around them. Social problems are also developing among those crowded into unsanitary living conditions. Family violence is a growing problem, brought on by the tensions of migration, changing gender roles, and wages that are too low to afford a decent standard of living. Women often can find work more easily than men because they are more compliant and will work for less pay. But men, displaced from agriculture and local factory work, are increasingly employed in maquiladoras, filling traditional female jobs such as sewing garments or assembling electronic devices. The schools in the border cities are inadequate for the increasing numbers of children. In Ciudad Juárez in 2001, for example, there were just ten day-care centers serving 180,000 women workers. A further problem has to do with fragmented governmental authority in the long border zone where maquiladoras dominate employment. There are more than 13 paired U.S.–Mexican cities along the border, and the multiplicity of governments makes it difficult to address common problems. Because NAFTA purports to be a trade agreement among equals, cooperation along the border to improve the lives of maquiladora workers is now gaining emphasis. The U.S.–Mexico Border XXI Project has joint task forces focusing on water and air pollution mitigation, hazardous waste cleanup and prevention, environmental studies, and emergency preparedness. In the next few years, Mexico hopes to provide 93 percent of the border population with clean drinking water, and 16 water-related projects are currently under way along the border, some in Mexico, some in the United States.

PERSONAL VIGNETTE   Orbalin Hernandez has just returned to his self-built shelter in Mexicali. He smiles as he enters the yard, and his wife breathes a sigh of relief. Recently fired for taking off his safety goggles while loading TV screens onto trucks at Thompson Electronics, he went to the personnel office to ask for his job back. Because there were no previous problems with him, he was rehired at his old salary of $300 a month ($1.88 an hour). Thompson, a French-owned electronics firm, took advantage of NAFTA and moved in 2001 to Mexicali from Scranton, Pennsylvania. In Scranton its 1100 workers had been paid an average of $20.00 an hour. With overtime, some Scranton workers made as much as $80,000 a year, but now 20 percent of them are unemployed, and many feel bitter toward the Mexicali workers.

Orbalin's salary at Thompson-Mexicali is barely enough for him to support his wife, Mariestelle, and four kids. They are from a farming community in Tabasco State, where, for people with no high school education, wages were $60 a month, so the Hernandez family is grateful for the job. Thompson is known for treating its Mexican workers well. It pays into a housing fund, supports local educational facilities, and provides bus transport for workers to and from home. But a recent study at the Autonomous University in Mexico City found that over the last 20 years of the maquiladora boom, Mexican wages have lost 81 percent of their buying power. A typical maquiladora worker has to work more than an hour to buy a kilo (2.2 pounds) of rice, whereas a dockworker in Los Angeles can buy that rice after just 3 minutes of work, and an undocumented minimum-wage worker in L.A. after 12 minutes of work.

Nonetheless, Mexicali's workers are now being told that their wages are too high. Because of the recession of the early 2000s in the United States, sales of maquiladora products are off, and the firms are cutting costs. Fourteen Mexicali plants have closed and moved to Asia; in China, workers with the same skills as Orbalin are earning just $2.80 a day. Those firms remaining in Mexicali have cut wages and reduced benefits. Some firms have started hiring temporary workers, who are then laid off just before they acquire permanent status under Mexican law, at the end of 90 days. [*Source: NPR reports by John Idste August 14, 2001, August 25, 2003, August 16, 2003, and by Gary Hadden, August 27, 2003; David Bacon, "Anti-China campaign hides maquiladora wage cuts," February 2, 2003, http://csf. colorado.edu/forums/labor-rap/current-discussion/msg01050.html.*]

Migration

Migration is a strategy often used to alleviate the problems of poverty. Rapid rural-to-urban migration in Mexico is legendary: 75 percent of Mexicans (79 million people) live in cities, the majority of them in large cities. The emigration of Mexicans to North America also has an enormous impact on Mexico as a nation. On the one hand, the skills of some of the the the best and brightest are lost, at least temporarily; on the other, some benefits are gained from their industry.

Unlike many Europeans, who usually cut their ties with family and native land when they came to North America, Mexicans often remain loyal to their home villages. Their migrations are usually undertaken with the express purpose of helping out their families and communities. A paper by geographer Dennis Conway and anthropologist Jeffrey H. Cohen reports that couples who migrate from Native American villages in Mexico often work just long enough at menial jobs in the United States to save a substantial nest egg. Then they return home to build a house (usually a family self-help project) and buy appliances. Afterward, the family may live off their nest egg for several years, while one or both members of the couple renders volunteer community service to the home village—perhaps building schools, serving on the town council, improving the town water system, or refurbishing the town's public spaces. When the money runs out, the couple, or just the husband, may migrate again to save another nest egg. Often these civic-minded migrants cycle other members of their family through a menial U.S. job, such as dishwashing, thus keeping the job open for themselves when they need it again. Their bosses and coworkers in the United States never guess that such unprepossessing minimum-wage workers are actually influential and public-spirited citizens in their home villages, who use migration and hard work to enhance community living standards and participatory democracy. All this is possible because Mexican communities tend to be self-reliant and because the Mexican economy is much less affluent than that of the United States. Hence, a few thousand dollars saved in the United States will accomplish a great deal in rural Mexico.

Does migration have to be the main route to advancement for Mexico's youth? Maybe not; most migrants seek opportunities they wish were available at home. One month after he was elected president in January 2001, Vicente Fox founded "e-Mexico," designed to bring the Internet to 10,000 communities by 2006. Fox thinks the Internet will be a way for Mexicans to leap out of poverty by gaining access to training and higher education. In the tiny mountain town of Santa Ana de Allende, where there is one telephone for 1400 people, Oracio Covarrubias uses the Internet to study at Tec de Monterrey Virtual University. He gets readings, assignments, access to study groups, and conferences with his advisor all via the Internet. Already students in Santa Ana are opting to stay in town, use their Internet training to establish businesses, and work for local change rather than migrating to the United States. This Internet initiative in Mexico resembles thousands of others elsewhere (see, for example, the opening vignette about Turkey in Chapter 6, page 282).

CENTRAL AMERICA

Central America's wealth is in its soil. Industry accounts for only about 20 percent of the gross domestic product, and the region is not rich in mineral resources. Thus the seven countries of Central America (Figure 3.21) remain dependent on the production of their plantations, ranches, and small farms. Here, the disparity of income seen in the region as a whole takes the form of disparity in the distribution of usable land.

Most of the Central American isthmus consists of three physical zones that are not well connected with one another: the narrow Pacific coast, the highland interior, and the long, sloping, rain-washed Caribbean coastal region (see Figure 3.2 on pages 114–115). Along the Pacific coast, mestizo (**ladino** is the local term) laborers work on large plantations that grow sugar, cotton, and bananas and other tropical fruits; coffee is grown in the hills in back of the coast. In the highland interior of Guatemala, Honduras, and Nicaragua, where Native American people have traditionally made their living from subsistence agriculture, cattle ranching and commercial agriculture have recently displaced subsistence farmers. Similarly, the humid Caribbean coastal region,

The pattern of tiny farm plots (minifundios) in El Salvador is visible in this photo. Some are the result of land redistribution efforts, but because of high population density, the minifundios are often too small to support a family. [Tomasz Tomaszewski/National Geographic Image Collection.]

FIGURE 3.21 The Central America subregion. Central America is the least industrialized part of Middle and South America. In Guatemala and Honduras, more than 50 percent of the population is still engaged in agriculture, and these countries, plus Nicaragua and El Salvador, are the poorest. Costa Rica is considered a developed country. Belize and Panama are relatively well off.

long sparsely populated with Native American and African-Caribbean subsistence farmers, is increasingly dominated by commercial agriculture, forestry, tourism development, and resettlement projects for displaced small farmers from the highlands.

In Central America, the majority of people are Native Americans or ladinos, and about half still live in small villages. In these villages, most people make a sparse living by cultivating cash crops as well as their own food on land they own or rent, or as sharecroppers, and by working seasonally as laborers on large farms and plantations (see the story of Aguilar Busto Rosalino in the vignette on page 140). The people of this region have experienced centuries of hardship, including long hours of labor at low wages and the loss of most of their farmlands to large landholders. In both rural and urban areas, infrastructure development has lagged.

Roads are primitive and few. Most people lack clean water, sanitation, health care, protection from poisoning by agricultural chemicals, and basic education. Often they do not have access to enough arable land to meet their basic needs. The majority of the land is held in huge tracts—ranches, plantations, and haciendas—owned by a few families. As a result, on the small bits of farmland available to the poor for growing their own food and cash crops, local densities may be 1000 people or more per square mile.

Although there is a general hunger for land among the poor majority, only tiny El Salvador is truly densely populated, with an average of 817 people per square mile (314 per square kilometer) (Table 3.5, column 4). In fact, arable land there is so scarce that people desperate for land have slipped over the border to cultivate land in Honduras.

TABLE 3.5 Population data on Mexico and Central America, 2003

Country (1)	Population (2)	Population density per square mile (3)	Rate of natural increase (percentage) (4)	Literacy rate (percentage) (5)
Mexico	104,900,000	139	2.4	91
Belize	300,000	31	2.3	93
Guatemala	12,400,000	294	2.6	69
Honduras	6,900,000	159	2.9	76
El Salvador	6,600,000	817	2.3	79
Nicaragua	5,500,000	109	2.7	67
Costa Rica	4,200,000	211	1.4	96
Panama	3,000,000	102	1.8	92

Source: Population Reference Bureau, Population Data Sheet, 2003; *United Nations Human Development Report 2003,* Table 1.

AT THE LOCAL SCALE   Panama and the Panama Canal

Since early Spanish colonial days, Panama has served as a conduit for moving goods between the Caribbean and the Pacific. In the early days, much of the trip across the isthmus was made by mules and other draft animals. Originally part of Colombia, Panama was created as an independent state under pressure from the United States, which, in the 1890s, wanted to finance the construction of a canal across the isthmus. The Panama Canal was built primarily with West Indian labor, and many West Indians stayed, making Panama's north coast a distinctly Caribbean place. The country remained a virtual colony of the United States, which managed the canal and maintained a large military presence there. For many years, those Panamanians employed in the Canal Zone or in canal-related occupations lived a way of life that was heavily influenced by U.S. culture and the U.S. economy;

Panama is relatively wealthy for Central America (see Table 3.3 on pages 154–155), and the canal is the reason. Away from the canal, however, poverty is much more common.

In 1977, under pressure from Panama, the United States agreed to turn the canal over to Panama in 1999 and remove itself as a dominant presence. Interestingly, the turnover of the canal came at a time when the canal was becoming obsolete. It is no longer large enough to accommodate the huge cargo container or cruise ships of the modern era. Thus traffic through the canal is no longer growing as quickly as it had been. Plans for updating the canal have been delayed by a number of considerations, one of which is that there is no alternative route to use during repairs, except around the tip of South America. Panama's hope of earning revenue from the canal may not be fulfilled over the long term.

Costa Rica is an exception to these extreme patterns in Central America of elite monopoly, mass poverty, and rapid population growth (as are, to some extent, Belize and Panama) (Table 3.5 and the box "Panama and the Panama Canal" on this page). The huge disparities in wealth between colonists and laborers did not develop in Costa Rica, chiefly because the fairly small native population died out soon after the conquest. Without a captive labor supply, the European immigrants to Costa Rica set up small but productive family farms that they worked themselves, not unlike early North American family farms. Costa Rica's democratic traditions stretch back to the nineteenth century, and it has unusually enlightened elected officials as well as one of the region's soundest economies. Population growth is low for the region at 1.4 percent per year, and investment in schools and education has been high. As a result, Costa Rica has Central America's highest literacy rates, and on many scales of comparison, including GDP per capita, the country stands out for its high living standards (see Table 3.3 on pages 154–155). Thus Costa Rica has often been hailed as a beacon for the more troubled nations of Central America.

Environmental Concerns

Although poor farmers may use unwise practices in wrenching enough to feed their families from the tiny plots they are allowed, and are often blamed by the news media for environmental degradation, most environmental problems in Central America are caused by large-scale agriculture and cattle ranching. In some places, commercial farmers have overused chemicals to control weeds and pests. Each year, hundreds of children die from allergies and other effects of chemical poisoning. Land management by the large landowners is also a problem. Clearing enough land for large-scale agriculture and cattle grazing leads to tremendous amounts of erosion. In Honduras, the reservoir for a large hydroelectric dam built only a few years ago has been nearly filled with silt eroded from surrounding cleared land. Its electricity output must now be supplemented with generators run on imported oil.

Costa Rica has been a leader in the environmental movement in Central America. In the 1980s, the country established wetland parks along the Caribbean coast and encouraged ecotourism, while at the same time paying some attention to the potentially negative environmental side effects of tourism. Costa Rica supports scientific research through several international study centers in its central highlands and lowland rain forests, where students from throughout the hemisphere study tropical environments. Elsewhere in Central America, support for national parks is growing. With the help of international NGOs such as the Audubon Society and the U.S. National Park Service, 175 small parks have been established.

Civil Conflict

Frustrated with governments unresponsive to their plight, Native Americans and other rural people in Guatemala, Honduras, Nicaragua, and El Salvador began organizing protest movements in the 1970s. In some cases, they had the help of liberation theology advocates and Marxist revolutionaries from outside the region. Despite reasonable requests, they met with stiff resistance from wealthy elites assisted by national military forces. The consequence was the destructive civil wars that plagued Guatemala, Honduras, El Salvador, and Nicaragua in recent decades. In the 1980s, protests were particularly strong in Guatemala, which has the largest Native American population in Central America. During these dying days of the cold war, the United States backed and armed military dictatorships because it was convinced that the revolutionaries posed a communist-inspired threat to the United States. U.S.-equipped national armies then killed many thousands of Native protesters and drove 150,000 into exile in Mexico. In time, the protesters responded with guerrilla tactics. Rigoberta Menchu is one Native American woman whose family was killed. She won the 1992 Nobel Peace Prize for her efforts to stop government violence against her people. Her autobiographical account attracted public attention to the carnage and was important in awakening worldwide concern. Eventually, after a number of regional peacemakers joined Menchu in bringing international pressure to bear on the Guatemalan government, the Guatemalan Peace Accord was signed in September 1996.

A Case Study of Civil Conflict: Nicaragua

Until recently, Nicaragua showed landownership patterns characteristic of the region: a tiny elite held the usual monopoly on land, while the mass of laborers lived in poverty. Then a socialist revolution attempted to change the distribution of land and reorient society along more egalitarian lines. The effort was thwarted, partly because of internal turmoil, but also because Nicaragua became a focus of U.S. worry over communist expansion in the Americas.

North American investors have had an interest in the coffee and fruit plantations of the Nicaraguan Pacific coastal plain since early in the twentieth century. Several times in the early 1900s, the United States sent in the Marines to quell labor protests that threatened Nicaragua's export economy. This U.S. interference helped the wealthy Somoza family to establish its members as dictators in Nicaragua in the 1930s. The particularly brutal Somoza regime was finally ousted by the Marxist-leaning Sandinista revolution of 1979. The Sandinistas, who eventually won several national elections, embarked on a program of land and agricultural reform, redistributing thousands of acres and downplaying ranching. They also improved basic education and health services. Soon, however, a debilitating war with right-wing counterinsurgents (known as contras) undid most of the social progress.

The contras were a mixed group of dissidents, outlaws, and adventurers from outside the region; they were backed by local elites and by the communist-wary United States during the Reagan administration. A trade embargo imposed by the United States further contributed to the ruin of the economy. By the end of the 1980s, Nicaragua was one of the poorest nations in the Western Hemisphere. In national elections in 1990, a coalition led by Violetta Chamorro defeated the Sandinistas, who had lost popularity as a result of their own incomplete reform programs and public weariness with Sandinista–contra violence. Chamorro intended to privatize both rural and urban economies, but was forced by public pressure to continue some redistribution of land. Since 1997, several free elections in which 75 percent of the eligible citizens voted have resulted in moderate governments that continue to find it difficult to bring Nicaragua any measure of prosperity.

THE NORTHERN ANDES AND CARIBBEAN COAST

The five countries in the northernmost part of South America share a Caribbean coastline and extend into a remote interior of wide river basins and humid uplands (Figure 3.22). The Guianas resemble Caribbean countries in that they were once traditional plantation colonies worked with slave and indentured labor; and

FIGURE 3.22 The northern Andes and Caribbean coast subregion.

In Suriname, Javanese Muslim women, descendants of indentured servants brought by the Dutch in the 1850s, mark the end of Ramadan, a major religious observance. [Robert Caputo/Aurora.]

today their multicultural societies are made up of descendants of African, East Indian, Pakistani, Southeast Asian, Dutch, French, and English settlers. Venezuela and Colombia share a Spanish colonial past and are predominantly mestizo, with a small upper class of primarily European heritage and a small population of African derivation in the western and Caribbean lowlands. In all the countries of this subregion, small indigenous populations survive, mainly in the lowland Orinoco and Amazon basins, where they hunt, gather, and grow subsistence crops.

The Guianas

To the north of Brazil lie the three small countries known collectively as the Guianas: Guyana, Suriname, and French Guiana. Guyana gained independence from Britain in 1966, and Suriname from the Netherlands in 1975. French Guiana, on the other hand, is not independent, but rather is considered part of France. Today, the common colonial heritage of these three countries is still visible in both the economy and the people. Sugar, rice, and banana plantations established by the Europeans in the coastal areas continue to be economically important, but logging and the mining of gold, diamonds, and bauxite in the resource-rich highlands are now the leading economic activities. The population descends mainly from laborers who once worked the plantations. These laborers formed two major cultural groups: Africans, brought in as slaves from 1620 to the early 1800s, and South and Southeast Asians, brought in as indentured servants after the abolition of slavery. The descendants of Asian indentured servants are Hindus and Muslims (see the photo on this page). Many of them became small-plot rice farmers or owners of small businesses. Those of African descent are Christian and are both agricultural and urban workers. Politics in the Guianas are complicated by the social and cultural differences between citizens of Asian and African derivation.

Venezuela

Venezuela has long been potentially one of the wealthier countries in South America, primarily because it holds large oil deposits and is an active member of the Organization of Petroleum Exporting Countries (OPEC). Oil has been the backbone of the economy since the mid-twentieth century. Venezuela not only is among the top suppliers of oil to the United States, but also is a major supplier to Japan and Europe. The U.S. invasion of Iraq in 2003 illustrates how geopolitics can link the fates of distant regions: Venezuelans feared that the invasion would result in the takeover of Iraqi oil management by the United States and Britain. This could have caused Venezuela to lose market share in the United States, and the result in already economically strapped Venezuela could have been civil disorder.

Venezuela's oil resource was nationalized in the 1970s, with the idea that the profits would fund public programs to help the poor majority. (Now, foreign firms control 25 percent of production.) Oil has not generated widespread prosperity, however, even during times of relatively high oil prices. Most of the wealth has been retained by the 20 percent of the population that forms the middle and upper classes—those mainly of European descent. Taxes on this wealth have been kept low, so oil has not generated enough government revenue to fund badly needed improvements in education, transport, and communications. As a result, Venezuela is not building a base for general economic and social advancement. Although the capital city of Caracas is bedecked with gleaming skyscrapers, modern freeways, and universities, it is also surrounded by poor shantytowns that lack access to clean drinking water and sanitation and are served by substandard schools. A third of the population lives in deep poverty, on $2.15 a day per capita.

Debt also dogs Venezuela. Despite periodically high oil prices in the 1980s, Venezuela accumulated a large debt through corruption and poor management of development programs such as land reform. In the 1980s, 42.9 percent of all farms covered only 1 percent of the arable land, while the largest 3 percent accounted for as much as 77 percent of the arable land. But large sums aimed at fixing this situation resulted in little change, and the funds evaporated. The country had planned to pay off the debt with future oil profits, but a slump in oil prices during the early 1990s left the country still deeper in debt. Structural adjustment programs (SAPs), intended to free money for debt repayment, reduced the role of government in industry and in social programs. Many government jobs disappeared, and the poor lost rent and food subsidies and access to education. Living standards fell even further, and violent riots were one result.

Since the mid-1990s, Venezuela's leftist populist president, Hugo Chavez, has encouraged the poor with suggestions that social welfare programs will return and the rich will lose some of their advantages through higher taxes (and perhaps even through land expropriation). His populist statements, however, have discouraged outsiders contemplating investment in Venezuela's industries and mechanized agriculture, thereby hindering economic expansion. Nevertheless, Chavez's populist policies have survived, in part because a rebound in oil prices in the late 1990s and early 2000s brought much-needed capital back into the Venezuelan government coffers.

Another problem looming for Venezuela is the drug-related civil war in neighboring Colombia. The discovery of large cocaine caches in Venezuela during several raids by U.S. and international police cast doubt on Venezuela's commitment to curtail corruption.

Colombia

An ongoing civil war in Colombia has displaced more than 1.5 million people over the past decade. The current wave of violence is part of a long string of conflicts that have arisen out of inequalities in the country's social order. Although Colombia is today the world's second largest exporter of coffee and a major exporter of oil and coal, a small proportion of the population, mostly of European descent, has received most of the income by keeping wages low and resisting the payment of taxes. In addition, the wealthy have not cooperated with efforts to redistribute some of their extensive landholdings. On one side of the civil war are revolutionary guerrilla bands seeking government-sponsored reforms for the poor. On the other side are the private armies of the wealthy, who have little faith in the will and ability of the ill-equipped Colombian military to defeat the guerrillas. The guerrilla group FARC controls a large part of the Colombian interior south of Bogotá, where it is said to encourage the cultivation of more than 100,000 acres (40,000 hectares) of coca.

The current hostilities are still basically driven by unequal access to wealth, power, and land, but they are complicated by the fact that, to raise money, all the warring parties participate in the drug trade to varying degrees. This trade depends on the ancient Andean coca plant, whose leaves have traditionally been chewed as a fatigue and hunger suppressant and as a mildly invigorating tonic and mood enhancer. Today, coca leaves are processed into the much stronger cocaine (a crystalline extract of coca leaves noted for its euphoric, stimulating, and addicting effects). The coca growers are usually poor farmers in remote areas, who make a somewhat better income from coca than they would from other crops. However, most of the profits are reaped by competing drug-smuggling rings. These drug rings pay the farmers relatively little, and they bribe, intimidate, and murder Colombian government officials who seek to reduce or eradicate the drug trade.

There are some signs that this era of drug-related violence in Colombia will abate, due in part, perhaps, to training and equipment supplied by the United States, but also to local initiatives. Significantly, ordinary Colombians have organized the "No Mas" movement, a civic foundation that works to stop the killing. Five million demonstrators marched in Bogotá in October 1999. In addition, there are many efforts to find sources of income for rural people other than coca. In the mountains of southwestern Colombia, 1200 farm families have formed a cooperative that produces a line of 20 products—preserved fruits, sauces, and candies—marketed especially to Hispanics in North America and Europe. The cooperative also focuses on increasing child and adult education and on enhancing such skills as running an effec-

tive business meeting and improving the sales of products through market research. Elsewhere in the region, rural people working with the food scientists at the International Center for Tropical Agriculture have developed new varieties of corn that will be more productive and nutritious. These farmers now have enough corn to feed their families, raise chickens, and sell corn in the marketplace. These activities diversify the diets of rural families, bring in cash, and make the new varieties of corn available to other farmers.

THE CENTRAL ANDES

The central Andes, which includes the countries of Ecuador, Peru, and Bolivia, is the poorest subregion in Middle and South America (Figure 3.23). On the eve of the Spanish conquest, it was the home of the Inca Empire. The legacy of the Incas is reflected in numerous ruins, such as those at Machu Picchu in Peru, in thousands of miles of trade routes, in ruined cities that remain incompletely explored, and especially in the roughly one-half of the population that is Native American—the largest proportion in South America.

After its fall to the Spanish, the central Andes, like other parts of the Americas, went into a long decline. During this time, a tiny group of landowners prospered while the vast majority lived in poverty. Most of the twentieth century has been marked by failed efforts at social change and the violence resulting from those failures, but the growing political involvement of the large indigenous population may help achieve greater social equity. In exploring the central Andes, we will look at three main areas within the subregion: the economically dynamic coastal lowlands, the poor and largely Native American highlands, and the resource-rich and environmentally threatened western Amazonian lowlands.

Environments, Settlement Patterns, and Production in the Central Andes

Only Ecuador and Peru have coastal lowlands; Bolivia is now landlocked, after having lost what is called the War of the Pacific in 1884, in which Bolivia's mineral-rich Atacama Desert coastline was annexed by Chile. The coastal populations of Ecuador and Peru are mainly mestizo and African. The Pacific coast in this subregion is a zone of productive agricultural land and large and modern cosmopolitan cities, including Peru's capital and commercial center, Lima, and Ecuador's leading industrial center, Guayaquil. Although the climate is often dry, especially in Peru, plantations and other agricultural enterprises that produce crops for export thrive along the coast with the help of irrigation. The production of export crops such as bananas, cotton, tobacco, grapes, citrus, apples, and sugarcane has increased dramatically. Shrimp farming and ocean fishing grounds nourished by the nutrient-rich Peru Current sustain a vital export-oriented fishing industry. Food production for local consumption can be precarious, however, and actually declined 14 percent in Peru between 1985 and 1990. Overall, dependence on food imports (chiefly

FIGURE 3.23 The central Andes subregion. The central Andes, which includes the countries of Ecuador, Peru, and Bolivia, is the poorest subregion in Middle and South America. Only Ecuador and Peru have coastal lowlands; Bolivia is landlocked. Bolivia has two capitals: La Paz is the seat of the government, and Sucre is the seat of the judiciary.

cereals) is increasing, even as exports of food commodities increase. This situation is largely a reflection of the increasing integration of global food markets: consolidated farms are producing for the export market rather than for local consumption, as small farms used to do; meanwhile, demand for what cannot be grown locally is increasing.

Like other parts of Middle and South America, this area has had to tighten its belt in response to government debts acquired through borrowing to fund development. Structural adjustment programs have mandated the privatization of state-run industries and the streamlining of government. These SAP reforms have resulted in job losses and social turmoil along with stronger economic growth. Governments have dramatically increased prices for gasoline, electricity, and transport in an effort to raise funds to pay off debts. Such policies hurt the urban poor, whose low wages cannot cover the increases. In rural coastal areas, SAPs have removed government assistance to small-scale farmers (such as low rents, price supports for staple crops, and subsidies for tools). Small farms are now being replaced by corporate farms and high-tech operations such as shrimp farms, most of which, ironically, cause environmental degradation (see the discussion of shrimp farming on page 152) and yet receive government assistance, at least in the start-up phase. In Ecuador, protests by working people brought down the government in 1997, and the same thing might have then happened in Peru had it not been for suppression by the military. Eventually, popular uprisings forced the resignation of the president of Peru in 2000 and the president of Bolivia in 2003 (see the following section). Ecuador declared bankruptcy in 1998 because it couldn't repay its debts to the international finan-

cial community, and the economy and society are still in a state of crisis and uncertainty.

The highlands, known as the **altiplano,** are the home of the bulk of the region's indigenous people, who for centuries worked on large haciendas and in rich mines (copper, lead, and zinc in Peru; tin, bauxite, lead, and zinc in Bolivia) owned by a small group of people of European descent. Most of the Native Americans today obtain their incomes from subsistence cultivation of a wide array of vegetables acclimated to the cool climate: more than 50 varieties of potatoes, for example, are cultivated in the altiplano. Surpluses are traded on market days in towns throughout the altiplano. Sheep herding, wool processing and weaving, and (for some men) mining also contribute to family incomes. Recently, foreign and local scientists have explored imaginative new (and ancient) cultivation techniques and crops that may improve standards of living (see the box "Learning from the Incas" on page 170). In Peru, agronomist Angel Valledolid oversees 7500 acres (3000 hectares) of *nuña,* "popping beans," an ancient crop plant that grows without pesticides. Roasted, seasoned, and salted or covered with chocolate, the popped beans are being marketed as a cocktail snack in Europe and North America. If successful in the global marketplace, the popping beans could improve the incomes of 200,000 Peruvian cultivators and their families. In the last few decades, highland Peruvians who cannot get access to land or other means of support have been moving east into Amazon lowland regions.

The Amazonian lowlands of Ecuador, Peru, and Bolivia have traditionally been the home of scattered groups of indigenous people. In recent years, this area has seen rapid and often destructive

AT THE LOCAL SCALE   Learning from the Incas: Agricultural Restoration in the Peruvian Highlands

Around the ancient Incan capital of Cuzco, in the Peruvian highlands, terraces and irrigation canals once supported crops that fed hundreds of thousands of people on what would otherwise have been barren mountain slopes. Much of this ancient infrastructure still remains, but it was not being used until recently. Some of it is now being restored, thanks mainly to more than two decades of careful research by the British rural development specialist and archaeologist Ann Kendall. Kendall recognized that Incan agricultural technologies were more advanced than archaeologists had thought. For example, on steep slopes, the construction of terraces improves the efficient recycling of nutri-

ents by inhibiting erosion and trapping moisture, thus making high-yield organic cultivation possible. In particular, Kendall found that the Incan use of clay as a flexible semiwet sealant on both terraces and canals was extremely well adapted to the earthquake-prone environment of the Peruvian highlands because it doesn't break the way concrete does and because cracks tend to reseal themselves. So far, reviving these traditional methods has proved highly productive, allowing some highland people who had migrated to lowland cities to return home and take up farming. One canal and its associated terrace system can irrigate enough land to support more than 2000 people.

development of natural resources. National governments have encouraged private extraction, first of trees and then of minerals and agricultural products, and these activities have severely damaged the home territories of Native Americans. At the forefront of this process today are oil and mining companies (see the vignette that opens this chapter). Roads—such as the one cut across the Andes in Peru from Lima to Pucallpa on the headwaters of the Amazon—have been built to gain access to lumber and minerals, but they have also opened up this region to waves of migrants from the highlands and coastal lowlands. Accompanying these migrants are diseases that decimate indigenous people in the Amazon Basin, who lack resistance to them.

Social Inequalities and Social Unrest

In the twentieth century, various strategies for lessening the subregion's gross social inequalities were tried, with varying degrees of success. In the 1950s, Bolivia was the first country to attempt reform when the government took over the tin mines in the altiplano after a revolution in which miners' unions played a key role. Wages were raised and conditions improved for several decades, but then world prices for tin declined because plastic packaging was replacing tin cans just as tin mining was intensifying around the globe. As tin prices fell, illegal drugs replaced tin as Bolivia's chief export. As in Colombia, the military and the government publicly condemn the drug trade, yet members of both are extensively involved, as are local cultivators who grow poppies for heroin or coca for cocaine.

In the 1970s, Peru's military dictatorship sought to improve the lives of indigenous people by transforming the haciendas of the elite into state-run communes. The experiment was not popular and soon inspired a resistance movement among indigenous small farmers, which was taken over in 1980 by a communist guerrilla movement called the Sendero Luminoso (Shining Path). Followers of Sendero Luminoso forced cooperation by killing farmers if they participated in the capitalist economy in any way or were associated with the central government, whether as traders, community officials, participants in development projects, or even just voters. Eventually, the farmers banded together to fight the Sendero Luminoso with the aid of the military. The

level of violence has decreased for several years now, but issues of affordable access to good agricultural land and of democratic participation for the majority remain. The Peruvian government, hoping to avoid reigniting past grievances, supplies highland communities with subsidies and protections from the harsher aspects of SAP reforms that are straining the lowlands.

The Native Americans of South America have visibly increased their participation in national politics. They played a part in forcing the resignation of Peru's president, Alberto Fujimori, in the autumn of 2000 after a series of corruption scandals. In Ecuador, CONAIE, a federation representing all of Ecuador's indigenous groups, has been a strong force fighting for the land rights of Native Americans. CONAIE's support was central to the success of the protests against SAPs that brought down Ecuador's government in 1997. Native American demands for an increased voice in national policy continued into 2000, as members of the federation blocked off roads to many remote highland areas in Ecuador in an effort to gain national and international attention for their protest.

In Bolivia, native people mounted massive protests against the weak conservative coalition government of President Gonzalo Sanchez de Lozada. The trigger for these protests was a plan to export natural gas from Bolivia to the United States via a pipeline to a port in Chile. In the 1980s, Bolivia embraced the free-market model and sold off state-owned industries to largely foreign interests. These policies, urged on Bolivia by the World Bank and IMF to reduce its debt, enriched a few, such as President Lozada, but also brought high unemployment and a drop in real per capita income for Bolivia's majority Indian population. Protestors interviewed by *New York Times* reporter Larry Rohter said they saw an "unbroken line" between the rapacious Spanish colonial policies of the past and modern movements linked to globalization and free trade, such as U.S. and Chilean designs on Bolivia's gas; the U.S. effort to curb the drug trade by squelching coca growing, which had resulted in high rural unemployment; and SAPs that eliminated social programs and encouraged the sale of Bolivia's assets to foreign investors. In one weekend of protests, the Bolivian military shot to death 50 unarmed protestors who were demanding the president's resignation. President Lozada eventually resigned in October 2003.

FIGURE 3.24 The southern cone subregion.

THE SOUTHERN CONE

The countries of Chile, Paraguay, Uruguay, and Argentina (Figure 3.24) have diverse physical environments but remarkably similar histories. The southern cone experienced little European settlement during the Spanish Empire, but in the late nineteenth and early twentieth centuries, European immigrants—mainly Germans, Italians, and Irish—were drawn there by temperate climates, economic opportunities, and the prospect of land-ownership. These immigrants swamped the surviving Native American populations throughout most of the region. Paraguay is the only country of the four that has a predominantly mestizo population.

The Economies of the Southern Cone

Agriculture has long been a leading economic sector in the southern cone. The primary agricultural zone is the pampas, an area of extensive grasslands and highly fertile soils in northern Argentina and Uruguay. The region is famous for its grain and cattle. Sheep raising dominates in Argentina's drier, less fertile southern zone, Patagonia. On the other side of the Andes, in Chile's central zone, a Mediterranean climate supports large-scale fruit production that caters to the winter markets of the Northern Hemisphere. Chile also benefits from considerable mineral wealth, especially copper.

Although the agricultural and mineral exports of the southern cone have created considerable wealth, fluctuating prices for raw materials on the global market have periodically sent the economies of this subregion into a sudden downturn. The desire for economic stability was a major impetus for industrialization and urbanization in the mid-twentieth century. At first, the new industries were based on processing agricultural and mineral raw materials. Later, state policies supported diversification into import substitution industries. However, as discussed on pages 134–135, inefficiencies, corruption, and small local markets prevented industry from becoming a leading sector for the region.

Economic policy in the southern cone has been a source of conflict within the subregion for decades (see the discussion of the Mercosur trade agreement on page 139). Despite their considerable resources, these countries have always had substantial impoverished populations that were, nonetheless, well enough educated to demand attention to their grievances. In response, each country developed mechanisms for redistributing wealth—government subsidies for jobs, food, rent, basic health care, and transport, for example—that did alleviate poverty to some degree, but did not contribute to viable development or the spread of democratic institutions. When global prices for raw materials fell in the 1970s and produced an economic downturn for the region, with accompanying deprivation, people clamored for more fundamental changes. In response, throughout the 1970s and into the early 1980s, military leaders fearful of civil unrest, with the support of the elite, took control of governments and waged bloody campaigns of repression against striking workers and the socialist and communist opposition. In the years of state-sanctioned violence that followed, known as the "dirty wars," 3000 people were killed in Chile and 14,000 to 18,000 in Argentina, and tens of thousands more were jailed and tortured for their political beliefs (see page 144 for a discussion of the overthrow of President Salvador Allende in Chile in 1973).

The International Monetary Fund has consistently applied pressure for repayment of the massive debts accumulated under both civilian and military governments from the 1960s through the 1990s. These financial pressures have effectively shifted countries away from poverty alleviation policies and toward free-market economic reform, which it was hoped would bring economic growth and expansion sufficient to make debt repayment possible. The record of success is mixed. Chile's national economy has witnessed growth of its export-oriented industry, mining, and agriculture, whereas Argentina has not been able to rid itself of its debt burden and defaulted in December 2001. An IMF package of structural reforms and loans to the Argentine government held off the crisis, but did not resolve it. Argentina's debt-induced social disarray and disquiet mirrors Ecuador's (see pages 169–170).

Buenos Aires: A Primate City

The primary urban center in the southern cone is Buenos Aires, the capital of Argentina and one of the world's largest cities. Forty percent of the country's people live in this city. Buenos Aires boasts premier shopping streets, elegant urban landscapes, and dozens of international banks, yet it has weathered six decades of decline that have left it with environmental degradation, empty factories, social conflict, severe poverty, and declining human well-being. Some people argue that the downward slide in quality of life in Buenos Aires is an unavoidable result of the restructuring required to create an economy that will be competitive in the global arena. Argentina's leaders contend that greater integration with the global economy will help to reverse decades of malaise. They want Buenos Aires to be seen as a world city—a center with pools of skilled labor that attract major investment capital, a sophisticated city with a beautiful skyline and a powerful sense of place. An impatient local population living in rundown apartments and on wages too low to afford basic nutrition, however, gives more importance to obtaining such basics as decent housing, better food, and modernized transportation.

BRAZIL

The observant visitor to Brazil (Figure 3.25) is quickly caught up in the country's physical complexity and in the richly exuberant, multicultural quality of its society. But its landscapes also plainly show, more obviously than those of North America, the environmental effects of both colonialism and recent underplanned economic development. Brazil's 176 million people live in a highly stratified society made up of a small, very wealthy elite, a modest but rising middle class, and a majority that lives below, or just barely above, the poverty line. In Brazil's megacities of São Paulo and Rio de Janeiro, vast shantytowns, high crime rates, and homeless street children are signs of the gross inequities in opportunity and well-being. The richest 20 percent of Brazil's population has 29 times the wealth of the poorest 20 percent—one of the widest disparities on earth (see Table 3.2 on page 133). Travelers find themselves caught up and delighted by the flamboyant creativity and elegance of the Brazilian people, yet sobered by the obvious hardships under which they labor.

Brazil's three main physical features are the Amazon Basin, the Mato Grosso, and the Brazilian Highlands (see Figures 3.2 and 3.3). The huge Amazon Basin covers the northern two-thirds of the country and is described elsewhere in this chapter. The Mato Grosso is a seasonally wet/dry interior lowland south of the Amazon, with a convoluted surface. It was once covered with grasses and scrubby trees adapted to long dry periods, but in the twentieth century it was extensively cleared for subsistence and commercial agriculture. The southern third of Brazil is mostly occupied by the Brazilian Highlands, a variegated plateau that rises abruptly just behind the Atlantic seaboard 500 miles (800 kilometers) south of the mouth of the Amazon. The northern por-

FIGURE 3.25 The Brazil subregion. The Brazilian population has been concentrated along the Atlantic seaboard since colonial days, but due to assertive national policy, settlement is now creeping inland.

tion of the plateau is arid; the southern part receives considerably more rainfall. Settlement in northeastern Brazil is concentrated in a narrow band along the Atlantic seaboard and south through Salvador (see Figure 3.7). Settlement extends deeper inland in Brazil's temperate zone, near Rio de Janeiro, São Paulo, Curitiba, and Pôrto Alegre.

Brazil is about the same size as the United States, and the Brazilian economy is the largest in Middle and South America—the eighth largest in the world. The resources available for development in Brazil are the envy of most nations, but management of those resources in Brazil's best interests has been a challenge. Local oil now provides for 60 percent of Brazil's needs, and more has recently been discovered west of Manaus in the Amazon. Gold, silver, and precious gems have been important resources since colonial days, but it is industrial minerals—titanium, manganese, chromite, tungsten, and especially iron ore—that are most valuable today. These minerals are found in many parts of the Brazilian Highlands. Hydroelectric power is widely available because of the many rivers and natural waterfalls that descend from the highlands.

There has been large-scale agriculture in Brazil for 400 years. Today the pattern is European-style farming of fruits and vegetables in the south; in the northeast, large commercial farms growing primarily sugarcane, tobacco, cotton, cassava, and oil palm; in the interior dry zones, ranching. Production of soybeans, pork, and beef has some potential to expand in the Mato Grosso, but most of the available land is in the tropical Amazon, where soils are fragile. Agriculture there requires special and sustained attention and more investment of time, research, labor, and skill than has been anticipated by entrepreneurs, both recent and historic. Overall, agricultural exports from Brazil are increasing and exceed imports, but agriculture as a proportion of Brazil's economy and as an employer of its labor force (23 percent) is rapidly losing out to industry (24 percent) and services (56 percent).

Brazil is the most highly industrialized country in South America. Most of its industry—steel, motor vehicles, aeronautics, appliances, chemicals, textiles, and shoes—is concentrated in a triangle formed by the huge southeastern cities of São Paulo, Rio de Janeiro, and Belo Horizonte, where hydroelectric power has been cheap. Until the 1990s, the vast majority of Brazil's mining and industrial operations were developed using government funding, and many are still owned and run by the government. In the 1980s and 1990s, elected governments adopted structural adjustment policies and privatized many formerly government-held industries and businesses (television stations, power plants, mines, and agricultural land) in an effort to make them more efficient and profitable. Many of these firms were sold to foreign investors at bargain-basement prices. In 1997 alone, direct foreign investors spent $23 billion on Brazilian properties. Privatizing industry often results in greater productivity, because the focus is on efficiency for profitability, but the sale of these industries may be hurting Brazil in the long run. Now any profits that might materialize will go not to the public, who paid for the development of these industries, but to the new private owners, who paid much less than the firms were worth and who may well invest any profits outside of Brazil.

Urbanization

Brazil has a number of large and well-known cities: Rio de Janeiro, São Paulo, Belo Horizonte, Salvador, Recife, Fortaleza, Belém, Manaus, and Brasília. All but Manaus, Brasília, and Belo Horizonte are on the Atlantic perimeter of the country. During the global economic depression of the 1930s, farm workers throughout the country began migrating into urban areas as world prices for Brazil's agricultural exports (jute, sugar, coffee, cacao, rubber) fell. By 1960, Brazil was 22 percent urban. Since then, efforts to modernize and mechanize agriculture have reduced

labor requirements further. Agricultural change pushed people off the land; during and after the 1960s, the chance of employment in huge factories being built with government money pulled them into the cities. Brazil's competitive edge globally was its cheap labor, and the military governments of the time thought that the country's then mostly government-owned industries could continue to pay very low wages for some time because there was such a surplus of workers to draw on from the low-wage countryside. They were right, but they had to quell many protests by workers who could not live decently on their wages.

By 1995, 77 percent of Brazil's population was urban, and at least one-third of the urban dwellers, many of them unemployed, were living in favelas, the Brazilian urban shantytowns (see the map of Vitória in Figure 3.10 on page 130). The poverty in the hardest-hit cities in the northeast rivals that of Haiti, the poorest country in the Americas. Brazilians, however, are intrepid at finding ways to make life worth living. Favela dwellers are famous for their efforts to create strong community life and support for those in distress. An example is a religious movement, known as Umbanda, that thrives in all the Atlantic coastal cities from Belém to São Paulo (see the photo of a similar movement, Batuque, below). Umbanda and related groups are centered around a male or female spiritual leader. The leader welcomes celebrants into a neighborhood center and invokes the spirits to help people cope with health problems and the ordinary stresses and strains of a life of urban poverty. Drumming, dancing, spirit possession, and psychological support are the central focus. Umbanda grew out of an older African-Brazilian belief system

The two women in lace shawls are in a trance, having been possessed by the spirits of Oxala and Yemanja during a Batuque ritual. They are greeted and supported by mediums called *filhas de santos*. Batuque ceremonies, like Umbanda and Condomble ceremonies, use mesmerizing drum rhythms to invoke various spirits, including Christian saints. These ceremonies regularly unite large groups of people in a common uplifting experience. [Jacques Jangoux, Brazil.]

called Condomble, and similar movements (Voodoo, Santeria, Obeah) are to be found elsewhere in the Americas, including the Caribbean, the United States, and Canada. In Brazil, Umbanda appeals to an increasingly wide spectrum of the populace. There are many centers with adherents of various African, European, and Native American backgrounds.

Rio de Janeiro and other cities in Brazil's industrial heartland in the southeast have districts that are elegant and futuristic show-places, resplendent with the very latest technology and graced with buildings that would put New York, Singapore, Kuala Lumpur, or even Tokyo to shame. The stores are filled with artis-tically designed consumer goods at attractive prices for those with desirable foreign currency. Brazilian planners now acknowledge, however, that in the rush to develop these modern urban land-scapes, developers neglected to underwrite the parallel develop-ment of a sufficient urban infrastructure (for an exception, see the box "Curitiba, Brazil: A Model of Urban Planning" on page 131). In short supply are sanitation and water systems, up-to-date elec-trical wiring, transportation facilities, schools, housing, and med-ical facilities—all necessary to sustain modern business, industry, and a socially healthy urban population. But the Brazilians should not be overly criticized for what in hindsight seems like an obvious error. During the 1970s and 1980s, development theory set forth by the World Bank and other financial institutions held that investment in what were called *urban growth poles*, and the accompanying industrialization and modernization, would deliver development and stimulate economic growth. Social transformation and infrastructure development would be natu-rally occurring side effects. The debt crisis of the 1980s, which ate up profits with *skyrocketing inflation* (a rapid rise in consumer prices due to the falling value on the international market of the Brazilian currency), put an end to such optimistic plans and policies. Only very recently are all parties beginning to recognize that planned development of human capital through educa-tion, health services, and community development, whether pri-vately or publicly funded, is a primary part of building strong economies.

Brasília

Brasília, the new capital of Brazil, is an intriguing example of the effort to lead development with urban growth poles. Built in the state of Goiás in just 3 years, beginning in 1957, Brasília lies about 600 miles (1000 kilometers) inland from the glamorous old administrative center of Rio de Janeiro. The official explanation of this move was that the city, built in this remote location, would serve as a **forward capital**—it would draw migrants and invest-ment for the development of the western highland territories, the Mato Grosso, and eventually, the Amazon Basin. The scholar William Schurz suggested an alternative explanation for Brasília's location: moving the capital so far away from the centers of Brazil-ian society was the most efficient way to trim the badly swollen and highly inefficient government bureaucracy. In any case, sym-bolism figured more prominently than practicality in the design. The city was laid out to look from the air like a swept-wing jet plane. There was to be no central business district, but rather shopping zones in each residential area and one large mall. Pedestrian traffic was limited to a few grand promenades; people were expected to move even short distances in cars and taxis. Pub-lic buildings were designed for maximum visual and ceremonial drama, but safety was an afterthought.

After people had spent several decades actually using this urban landscape, they made all sorts of interesting changes to the formal design. At the Parliament, legislative staff and messengers created footpaths where they needed them: through flower beds and—with little steps notched in the dirt—up and over land-scaped banks. Thus they efficiently connected the administration buildings, bypassing the sweeping promenades. Commercial dis-tricts were retrofitted in and around hotel complexes. Urban workers who couldn't afford taxi fares wore direct-route footpaths across great green swards of grass. Little hints of the informal economy that characterizes life in the old cities of Brazil began to show up—a fruit vendor here, a manicurist there. And the shanty-towns, which the planners had tried hard to eliminate, began to rise relentlessly around the perimeter. Today, life in the somewhat spiffed-up shantytowns is so much more interesting than life in the sterile environment of Brasília that tour buses take visitors to see them and shop there. Overall, in the 40 years since its con-struction, Brasília's success as a forward capital has been limited. Although Brasília has drawn poor laborers, the entire province of Goiás still has only 4.5 million people, just 6 percent of the coun-try's total, and their average income is only half that of the coun-try as a whole. Population and investment remain centered in the Atlantic coastal cities.

Quick Review

1. What are some primary differences between Haiti and Barbados and between Cuba and Puerto Rico?

2. Describe some of the effects maquiladoras have on the environment.

3. Why did the United States support the contras in Nicaragua?

4. What is the Sendero Luminoso?

5. Account for the cultural diversity of Middle and South America by citing the worldwide origins of the ancestors of modern inhabitants.

REFLECTIONS ON MIDDLE AND SOUTH AMERICA

Middle and South America were the first major non-European lands to be colonized by Europeans. In a number of ways, European colonialism in the Americas launched the modern global economic system. It was in the Americas that large-scale extractive practices were inaugurated. Raw materials were shipped at low prices to distant locales where they were turned into high-priced products, while the profits went to Europe. Local people were diverted from producing food and other necessities for themselves to working as low-paid or enslaved labor in the fields and mines. Rules governing private property were set down in societies that had long practiced communal land rights. In the process, the landscapes and settlement patterns of the Americas were reorganized as the focus shifted from producing for local and regional economies to producing for global markets.

After the colonial era, this region continued to serve as a testing ground for economic theories, both capitalist and socialist. Middle and South America tried out such capitalist ideas as mechanized agriculture for export, rural-to-urban migration as a solution to rural poverty, and government borrowing and financing for large-scale development projects. More socialist experiments included government-sponsored industrialization to create jobs and produce substitutes for imports, as well as broad government subsidies to address the basic needs of the poor. External factors, however, intervened to thwart any such optimistic plans for economic expansion and societal transformation. The global recession of the early 1980s left the region's governments in debt, and their indebtedness forced free-market, structural transformations on them so that wealth inequities widened and social disparities increased. The rapidly growing numbers of urban poor had to resort to self-help solutions for their shelter; they sought meager livings in the informal sector, and with the reduction of public health and welfare services, many resorted to community and political activism.

Despite the hardships of the last 20 years—the increased harshness of everyday life for the rural and urban poor, especially women, children, and the elderly—there are signs that the situation is not beyond retrieval. There is increasing recognition that to be judged successful, development must first change the lives of the majority for the better. Middle and South Americans are beginning to build on their strengths and to invent solutions to problems common in many places around the world. Curitiba, Brazil, seeks simple but humane solutions to rapid urbanization and to the pollution and social disruption it brings. Degraded agricultural lands are being rehabilitated in the Peruvian highlands by the revival of ancient indigenous practices. Environmental groups are addressing the root causes of deforestation: inequitable domestic distribution of lands and resources, unsustainable uses of forestlands such as ranching and cash-crop agriculture, and demand for forest products in distant markets. Increasingly, regional trade organizations are emerging with sufficient strength and motivation to negotiate for the good of the region rather than for parochial interests.

As you read other chapters of this book, it might be useful to reflect again on the geographic issues we have discussed in this chapter. For example, notice how Europe's situation today—its wealth, its position as a world leader, and its emerging commitment to help its former colonies—is related in part to its colonizing experiences, which began in the Americas. You will see that Africa and South Asia under colonial rule experienced some conditions similar to those in the Americas; more recently, they too have felt the sting of SAPs. Southeast Asia, Australia, New Zealand, and Oceania too have had comparable experiences as colonies; but in recent decades their "colonizers" have included wealthy Asian capitalists. In the former Soviet Union and Central Asia, outside investors intent on exploiting oil and forest resources are reminiscent of the conquistadors and their successors in the Americas.

Chapter Key Terms

acculturation 146	Creole cultures 116	forward capital 175
altiplano 169	digital divide 145	free trade zones 137
assimilation 146	early extractive phase 133	hacienda 134
Aztecs 121	economic restructuring 114	import substitution industrialization (ISI) 134
barriadas 128	ecotourism 153	
barrios 128	El Niño 120	Incas 121
brain drain 131	evangelical Protestantism 148	income disparity 133
colonias 128	Export Processing Zones (EPZs) 137	informal economy 139
colonizing country 123	extended family 146	ladino 163
contested space 140	external debts 114	liberation theology 148
conurbation 129	favelas 128	livestock ranch 134
Creole 124	foreign direct investment (FDI) 133	machismo 147

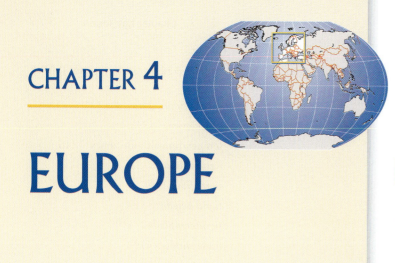
On a recent visit with his extended family in Ribnica, Slovenia, Alex Pulsipher had an experience that illustrates several ongoing transitions in Europe. Slovenia is a small Slavic country in southeastern Europe that was once the northernmost province of the country of Yugoslavia. Slovenia became independent in 1991.

their desire to seem cosmopolitan enough to gain acceptance in the European Union—the primary transnational institution taking shape in Europe.

MAKING GLOBAL CONNECTIONS

Slovenes typically have mixed feelings about the countries to the south (Croatia, Bosnia and Herzegovina, Serbia and Montenegro, Macedonia), even though Slovenia shared the former Yugoslavia with those countries until 1991. Slovenes, as citizens of the wealthiest and most educated province of the former Yugoslavia, the closest to western Europe, and now the first to join the European Union (in 2004), feel both affinity and distaste for the people of the south. Many, while desiring to distance themselves from their southern neighbors and viewing them as unworthy of living in Slovenia, also acknowledge that they are possessors of a rich and ancient kindred culture.

Slovene sentiments toward the south echo similar attitudes of countries in North and West Europe about the central European countries, such as Slovenia, that joined the **European Union (EU)** in 2004. The European Union, a supranational organiza-

PERSONAL VIGNETTE I was out one night with my cousin Maya when we passed some young men blasting unusual music on their car radio. It sounded almost Middle Eastern, like folk music, and very different from the U.S./Euro-Pop playing in the local bars. "What's that?" I asked. "South music," Maya said. "Those men are from the south" (by which she meant Croatia, Bosnia and Herzegovina, Serbia and Montenegro, and Macedonia—also once provinces of Yugoslavia). "They come here to take the jobs on farms and construction projects that pay too little for Slovenes."

We next stopped at Ribnica's only disco, where the DJ was playing typical U.S./Euro-Pop. When I asked him to play some South music, a look of haughty disapproval came over his face. "No! Only Slovene music here!" A few moments later, however, he must have had a change of heart, because the mournful yet rhythmic sounds of the south filled the room. The place erupted! The men who had been quietly drinking and chatting at the bar were suddenly throwing their arms up, spilling beer on one another, and happily singing along with the lyrics. A group raced to the dance floor, where they joined hands to form a semicircle, performing intricate footwork as their formation slowly rotated. But then, just as abruptly, the DJ put on Britney Spears, and the emotion-filled moment ended.

In the days that followed, I asked a number of Slovenes why, if South music was so popular, the DJ did not play more of it. The answers I received helped me see how Slovenes are motivated on the one hand by their love of Slavic culture and on the other by

For many years Albanians were welcome in Slovenia only if they stayed there temporarily as guest workers. This Slovene woman's companion is a man from Albania who migrated to a rural Slovene community where he and his parents own a popular ice cream store. The little boy is his son, and the woman now acts as his mother. Young people who have grown up since Slovene independence (1991) find it easier than their parents to adjust to living together with people of different ethnicities. [Lydia Pulsipher.]

tion including most of the countries of West, South, and North Europe, unites member countries in a single economy within which people, goods, and money can move freely (Figure 4.1). Many western Europeans fear that expansion of the EU into central Europe, and eventually into eastern Europe, will bring new political tensions and a throng of alien immigrants. At the same time, many people who support the idea of European unity harbor the fear that if the EU succeeds, cultural homogenization will result: regional culture of the kind that produced the South music so enjoyed by the Slovenes will disappear.

FIGURE 4.1 The European Union (EU). The EU, with the initial goal of economic integration, has led the global movement toward greater regional cooperation. It is older and more deeply integrated than its closest competitor, the North American Free Trade Agreement (NAFTA), which was largely a response to the EU. NAFTA is concerned mainly with eliminating trade restrictions between Canada, the United States, and Mexico. The EU goes beyond this goal to include the free movement of EU citizens across international borders within the EU, a common EU currency (the euro), and increasing levels of policy coordination in civil, judicial, economic, military, and foreign affairs. NAFTA is unlikely to achieve this level of integration any time soon, which could give the EU certain global advantages. Can you list some? [http://europa.eu.int/abc/index_en.htm.]

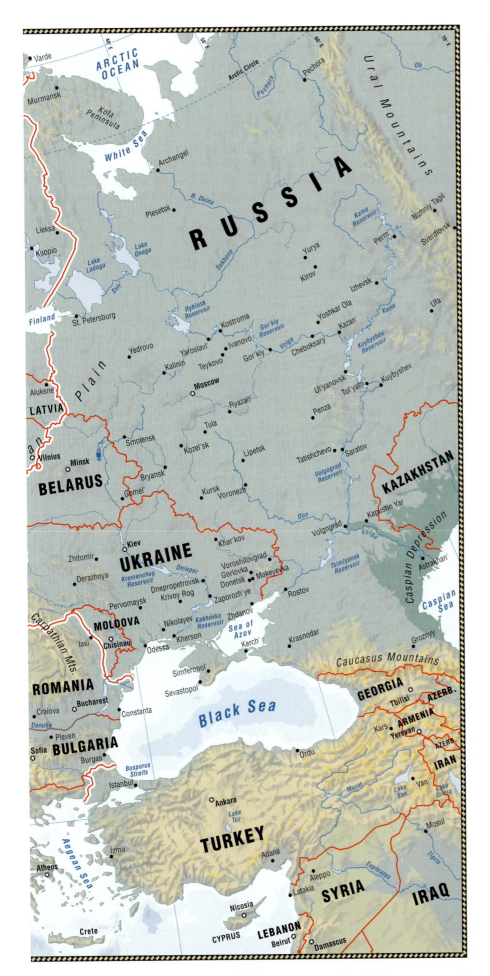

THEMES TO EXPLORE IN EUROPE

The following ideas are developed in this chapter.

1. **Economic union.** The lure of greater prosperity first drew the countries of western Europe into economic union, and now eight more countries in East and Central Europe have joined them. These 25 countries form the European Union, the world's strongest and most elaborate trading alliance (see Figure 4.1).

2. **Countertrends that resist change.** Despite, or even because of, increasing economic integration, some Europeans feel an increasing desire to establish a sense of national identity and to stem the tide of rapid cultural change that may lead to cultural homogenization. Worried that the European Union may change their way of life, some are wary of a wave of immigrants—either from neighboring countries or from outside Europe. Others are opposed to regulatory aspects of the European Union.

3. **Large role for government.** European governments play an active role in shaping urban and rural environments by providing housing, education, health care, transportation, and other services to increase overall human well-being. There are, however, regional variations in the degree to which national governments provide these services. These differences contribute to diversity in Europe and are reflected in human well-being indicators.

4. **Regional differences in economic development and well-being.** One long-term goal of the European Union has been to foster economic growth in those EU subregions that are poorer and less developed, primarily those in southern Europe. The new member states in East and Central Europe, many just emerging from a half-century of control by the Soviet Union, are poorer still. All are trying, with different levels of success, to improve well-being and enter the free-market economy.

5. **Effects of globalization.** Europe's overall prosperity arises in part from the many benefits it has obtained—and still obtains—from its access to peoples and resources around the globe, in many cases from its former colonies. At the same time, the globalizing economy has placed some of Europe's workers and products in stiff competition with those in developing parts of the world. Like the United States, Europe is losing jobs to other world regions.

6. **Environmental activism.** Europe's dense population and its high rates of consumption are contributing to air and water pollution. Environmental activism is growing in Europe.

FIGURE 4.2 Regional map of Europe.

I   THE GEOGRAPHIC SETTING

Terms to Be Aware Of

The end of the cold war (page 189) changed political alignments across Europe and raised questions about just where the borders of Europe lie. Hence special care is needed in designating the various parts of the region (Figure 4.2); the reader should remember that regions are just a convenient tool in geographic analysis and that their borders frequently change. In this book, Europe is divided into the following subregions: *North Europe* includes Iceland, Denmark (including Greenland and the Faroe Islands), Sweden, Norway, Finland, Estonia, Latvia, and Lithuania. The last three countries, formerly part of the Soviet Union, are new additions to North Europe because they share many economic and some cultural characteristics with that region, and North Europe seems willing to absorb them. *West Europe* includes the United Kingdom, the Republic of Ireland, France, Belgium, Luxembourg, the Netherlands, Germany, Switzerland, and Austria. *South Europe* includes Portugal, Spain, Italy, Greece, Malta, and Cyprus. *East and Central Europe*, the largest subregion, contains (1) those countries of central Europe that were formerly within the Soviet Russian sphere of influence and practiced a form of communism: Poland, the Czech Republic, Slovakia, Hungary, Albania, Romania, and Bulgaria; (2) all the countries once known as Yugoslavia, also communist, but independent of Soviet control: Slovenia, Croatia, Bosnia and Herzegovina, Serbia and Montenegro, and Macedonia, and (3) Albania. As we shall see, some of the East and Central European countries have recently joined the EU; others have not.

The actual physical location of a country—east, west, north, south—is less important in this subregional scheme than such other characteristics as historical, economic, political, and cultural alignments. For example, the countries lying between the Adriatic and the Black Sea (once called the *Balkans* but now usually referred to as *southeastern Europe*) are not considered part of South Europe, despite their southerly location. Their cultural history and recent political experience as communist countries have linked them more to East and Central Europe, which is the designation they prefer.

For convenience, we occasionally use the term *Western Europe* to refer to all the countries that were not part of the experiment with communism in the Soviet Union and Yugoslavia. That is, *Western Europe* is used for the combined subregions of North Europe (except Estonia, Latvia, and Lithuania), West Europe (except former East Germany), and South Europe. *Eastern Europe* will be used to designate all the countries once part of, or allied with, the former Soviet Union, as well as Albania, and those once a part of the former Yugoslavia. Occasionally, when it is necessary to identify only those countries once known as the *Balkans* (Albania, Bosnia and Herzegovina, Bulgaria, Croatia, Macedonia, Romania, and Serbia and Montenegro), the term *southeastern Europe* is used instead.

In the future, the borders of Europe may expand farther. Although just how far Europeans will be willing to stretch their borders is a topic of great debate, Romania and Bulgaria and other countries of southeastern Europe (Croatia, Bosnia and Herzegovina, Serbia) will probably join the EU within a few years. Turkey, now considered part of the Southwest Asia region, is officially an EU applicant. Ukraine, Russia, and Georgia are also likely to join in the distant future. Some observers suggest that the countries of North Africa and some countries in Central Asia that already trade heavily with Europe may seek some sort of formal association with the European Union. Figure 4.1 shows the present extent of the European Union and explores some ways in which Europe's economic sphere of influence may expand in the future.

PHYSICAL PATTERNS

Europe is a region of peninsulas upon peninsulas. The entire European region is one giant peninsula extending off the Eurasian continent, and the whole of its very long coastline is itself festooned with peninsular appendages, large and small. Norway and Sweden share one of the larger appendages. The Iberian Peninsula (shared by Portugal and Spain), Italy, and Greece are other examples of large peninsulas. One result of these many fingers jutting into oceans and seas is that much of Europe feels the climate-moderating effect of the large bodies of water that surround it.

Landforms

Although European landforms are fairly complex, the basic pattern is not hard to learn. The three basic landforms are mountains, uplands, and lowlands, all of which stretch roughly east-west in wide bands. Look at Figure 4.2 and notice first the beige color representing mountains. Europe's largest mountain chain stretches west to east through the middle of the continent, from southern France through Austria and Slovakia and curving southeast into Romania. The Alps are the highest and most central part of this formation. This network of mountains is mainly the result of pressure from the collision of the African Plate, which is moving northward, and the Eurasian Plate, which is moving to the southeast. Europe lies on the westernmost extension of the Eurasian Plate. South of the main formation, mountains extend into the peninsulas of Iberia and Italy and along the Adriatic Sea through Greece. The northernmost mountainous formation is shared by Scotland, Norway, and Sweden. These mountains are old (about the age of the Appalachians in North America) and worn down by glaciers.

Extending northward from the central mountain zone is a band of low-lying hills and plateaus (on the map in Figure 4.2, shown in a dusty mustard color curving from Dijon, through Frankfurt, to Krakow). These uplands form a transition zone between the high mountains and the North European Plain, the most extensive landform in Europe. The plain begins along the Atlantic coast in western France and stretches in a wide band around the northern flank of the main European peninsula, reaching across the English Channel and North Sea to take in southern England in the British Isles, southern Sweden, and most of Finland. The plain continues east through Poland, then broadens to the south and north to include all the land east to the Ural Mountains.

The coastal zones of the North European Plain are densely populated all the way east through Poland. Crossed by many rivers and holding considerable mineral deposits, this coastal lowland is an area of large industrial cities and densely occupied rural areas. Over the past thousand years, especially in the Netherlands, people have transformed the natural seaside marshes and vast river deltas into farmland, pastures, and urban areas by building dikes and draining the land with wind-powered pumps.

The rivers of Europe link its interior to the surrounding seas. Several of these rivers are navigable well into the upland zone, and Europeans have built large industrial cities on their banks. The Rhine carries more traffic than any other European river, and the course it has cut through the Alps and uplands to the North Sea also serves as the route for railways and motorways. The area where the Rhine flows into the North Sea is considered the economic core of Europe, and it is here that Rotterdam, Europe's largest port, is located. The larger and much longer Danube flows from Germany to the southeast, connecting the center of Europe with the Black Sea and passing the important and ancient cities of Vienna, Budapest, and Belgrade. As the EU expands to the east, the economic and environmental roles of the Danube River basin are getting increased attention.

Vegetation

Europe was once covered by forests and grasslands. Today, old forests exist only in scattered areas, primarily on the more rugged mountain slopes, in the most northern parts of Scandinavia. In some places in central and southeastern Europe, forests are regenerating where agricultural lands have been abandoned and where reforestation is under way. The dominant vegetation in Europe is now crops and pasture grass. Vast areas are covered with industrial sites, railways, roadways, parking lots, canals, cities and their suburbs, and parks.

Climate

Europe has three main climate types: temperate midlatitude, Mediterranean, and humid continental (Figure 4.3). The temperate midlatitude climate dominates in northwestern Europe, where the influence of the Atlantic Ocean is very strong. A broad warm-water current called the North Atlantic Drift, which is really just the easternmost end of the Gulf Stream (see Chapter 2, page 58), brings large amounts of warm water that has traveled from the Gulf of Mexico along the eastern coast of North America and across the North Atlantic toward Europe. The air above the North Atlantic Drift is warm and wet, and eastward-blowing winds push this air over northwestern Europe and the North European Plain, bringing moderate temperatures and rain deep into the Eurasian continent. These factors create a climate that, though still fairly cool, is much warmer than elsewhere in the world at similar latitudes. To minimize the effects of precipitation runoff, people in these areas have developed elaborate drainage systems for

houses and communities (steep roofs, rain gutters, storm sewers, drain fields, and canals), and they grow crops, such as potatoes, beets, turnips, and cabbages, that thrive in cool, wet conditions.

Farther to the south, the Mediterranean climate of warm, dry summers and mild, rainy winters prevails. In the summer, warm, dry air from North Africa shifts north over the Mediterranean, bringing high temperatures, clear skies, and dry periods as far north as the Alps. Crops grown in this climate, such as olives, citrus fruits, and wheat, must be drought resistant or irrigated. In the fall, this warm, dry air shifts to the south and is replaced by cooler temperatures, and in the west, by thunderstorms sweeping in off the Atlantic. For short periods in winter, the northern Mediterranean climate can resemble the temperate midlatitude climate of northwestern Europe, but overall, the climate here is mild. Along the Mediterranean coast, houses by the sea are often open and airy to afford comfort in the hot, sunny, rainless Mediterranean summers. Tourists, escaping the wet, cloudy climate

In the summer of 2003, hot, dry air from North Africa got stuck over the European continent and stayed there through July and August. Temperatures in England exceeded 100°F (38°C), breaking all records for some locations. On the Continent, temperatures exceeded century-old records. So unprepared were Europeans for these extremely hot, dry conditions that deaths soared; in France alone, as many as 11,000 died. [AP/Wide World Photos.]

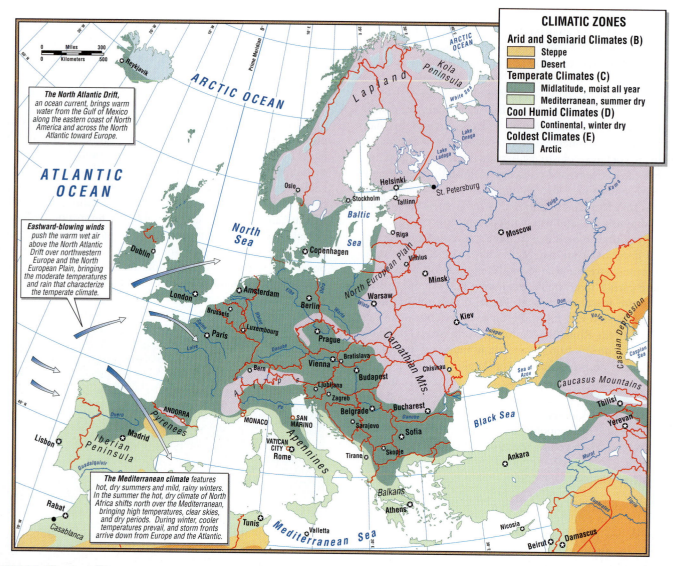

The North Atlantic Drift, an ocean current, brings warm water from the Gulf of Mexico along the eastern coast of North America and across the North Atlantic toward Europe.

Eastward-blowing winds push the warm wet air above the North Atlantic Drift over northwestern Europe and the North European Plain, bringing the moderate temperatures and rain that characterize the temperate climate.

The Mediterranean climate features hot, dry summers and mild, rainy winters. In the summer the hot, dry climate of North Africa shifts north over the Mediterranean, bringing high temperatures, clear skies, and dry periods. During winter, cooler temperatures prevail, and storm fronts arrive down from Europe and the Atlantic.

CLIMATIC ZONES

Arid and Semiarid Climates (B)
Steppe
Desert

Temperate Climates (C)
Midlatitude, moist all year
Mediterranean, summer dry

Cool Humid Climates (D)
Continental, winter dry

Coldest Climates (E)
Arctic

FIGURE 4.3 Climates of Europe.

of northwestern Europe, crowd the scenic areas, and especially the beaches, of Mediterranean shores.

In eastern Europe, the moderating influences of the Atlantic Ocean and the Mediterranean Sea are less or absent, and the climate is more extreme. In this region of humid continental climate (eastern Poland and Germany, Slovakia, Hungary, and most of Romania, as well as inland and eastern parts of Scandinavia), summers are fairly hot, and winters are longer and colder the farther north or the deeper into the interior of the continent one goes. Here, houses tend to be well insulated, with small windows and low ceilings. Crops must be adapted to much shorter growing seasons.

HUMAN PATTERNS OVER TIME

Some grand claims have been made in the name of Europe, and geographers have been among both the most vociferous supporters and the strongest critics of these ideas, as the following two quotations attest.

Europe has been a great teacher of the world. Almost every vital political principle active in the world today had its origin in Europe or its offspring, European North America. . . . [T]he same is true of the arts. Even though other parts of the world have produced rich folk arts, the culture of the West has become dominant.

[George F. Hepner and Jesse O. McKee (eds.),
World Regional Geography: A Global Approach
(Eagan, Minn.: West Publishing Co., 1992), p. 144.]

[There is] the [unfortunate] notion that European civilization— "The West"—has had some unique historical advantage, some special quality of race or culture or environment or mind or spirit, which gives this [particular] human community a permanent superiority over all other communities, at all times in history and down to the present.

[James Blaut, *The Colonizer's Model of the World*
(New York: Guilford Press, 1993), p. 1.]

We are so used to hearing the praises of Western (European) culture that for many people it is hard to spot the realities that they omit. You may have read the passages above idly and said to yourself, "So?" The first quote fails to recognize that Europe has been as much a learner from the world as a teacher of it, and that many "European" ideas and technologies were adopted from non-European sources. For example, the concept of the peace treaty, so vital to current global stability, was first documented not in Europe, but in ancient Egypt. Many people also criticize the assumption that the ability to dominate militarily, politically, economically, or culturally is evidence of overall superiority.

It is true that over the last 500 years (in fact, not a very long time), Europe has influenced how much of the rest of the world trades, fights, thinks, and governs itself. Attempts to explain this influence have ranged from arguments that Europeans are somehow a superior breed of humans to assertions that Europe's many bays, peninsulas, and navigable rivers have promoted commerce to a greater extent there than elsewhere. European capitalism is often cited as the crucial development that fueled Europe's geopolitical prominence. We may never have a single satisfying answer to the question of why Europe gained the leading role it continues to play. Still, it is worth taking a look at the broad history of this area and considering a few of the developments that have made Europe so influential over the past five centuries.

Sources of European Culture

Starting about 10,000 years ago, the practice of agriculture and animal husbandry gradually spread into Europe from the Tigris and Euphrates river valleys in Southwest Asia (also known as Mesopotamia) and from farther east in Central Asia and beyond (see Figure 1.15 on page 29). The cultivation of wheat, barley, and numerous vegetables and fruits and the keeping of cattle, pigs, sheep, and goats came to Europe from the east and south (from Central and Southwest Asia and North Africa), as did pottery making, weaving, mining, metalworking, and, most important, mathematics. All these innovations opened the way for a wider range of economic activity, especially trade.

The first European civilizations were ancient Greece (800–86 B.C.) and Rome (100 B.C.–A.D. 500). The innovations of these societies and their extensive borrowings from other culture groups formed some of the most important cultural legacies of Europe. Located in southern Europe, Greece and Rome interacted more with the Mediterranean rim, Southwest Asia, and North Africa than with the rest of modern Europe, which was then relatively poor and thinly populated. Greek artists, philosophers, and mathematicians were fascinated with the workings of both the natural world and human societies. Later European traditions of science, art, and literature were heavily based on Greek ideas derived from yet earlier systems of thought. The Romans, perhaps the greatest borrowers of Greek culture, also left important legacies in Europe. Many Europeans (and Middle and South Americans) today speak Romance languages, such as Spanish, Portuguese, Italian, and French, which are largely derived from Latin, the language of the Roman Empire. The Roman notion of individual ownership of private property also influenced the development of Europe, as did the Roman practices used in colonizing new lands. After a military conquest, Romans would secure control by establishing large plantation-like farms and communities of settlers transplanted from the heartland. By the second century A.D., hundreds of Roman towns dotted Europe. Roman systems of colonization were later used when European states laid claim to territory in the Americas and other places.

The influence of Islamic civilization on Europe is often overlooked. North African Muslims, called Moors in Europe, had a profound influence on language, music, food customs, architecture, and belief systems in Spain, which they ruled for 700 years, starting in A.D. 711. Similarly, the Muslims of the Ottoman Empire left a deep imprint on various parts of southeastern Europe (the Balkans) and Greece, which they ruled from the 1500s through the early 1900s. These Muslim rulers brought to Europe many textile and tempered-metal trade goods, food crops, architectural principles, and technologies from Arabia, China, India, and Africa. Muslim scholars preserved much learning from Greece, Egypt, and other ancient civilizations and brought Europe the Arabic numbering system as well as significant advances in medicine. Arab mathematicians, perhaps building on ideas picked up in India, brought algebra and algorithms—both essential elements of modern engineering and architecture—to Europe. Among the modern applications of these mathematical techniques are the design of computer hardware and software programs.

Growth and Expansion of Continental Europe and the Rise of Urban Life

As the Roman Empire declined in the fifth century A.D., a social system known as **feudalism,** which arose during the **medieval period** (450–1400), evolved from the need to provide some stability and order and to defend rural areas against raiders: rival elites, local bandits, Vikings from Scandinavia, and nomads from the Eurasian steppes. The objective of feudalism was to have a sufficient number of heavily armed, professional fighting men, or knights, ready to defend the farmers, or serfs, who cultivated plots of land for them. Over time, some of these knights became an elite class of warrior-aristocrats, called the nobility, who were bound together by a complex web of allegiances obligating them to assist one another in times of war. The often lavish lifestyles and elaborate fortifications (castles) of the wealthier knights were supported by the labors of the serfs, who were legally barred from leaving the lands they cultivated for their protectors. Most serfs lived in poverty outside the castle walls. During later colonial times, aspects of feudalism were brought to the Americas, where the Spanish crown expropriated land from Native Americans and granted it to colonists, along with the right to treat the native inhabitants as serfs.

While rural life followed established feudal patterns, in Europe's towns, protected by elaborate defensive walls and independent of feudal knights, new institutions began to develop that would influence settlement and commerce in modern Europe. Located along trade routes, these towns sheltered artisans and merchants who, through innovation, created the main institutions of modern European capitalism: markets, banks, insurance companies, and corporations. Documents called town charters set forth the rights of town citizens and provided the basis for the

European notion of civil rights, which were eventually extended to all citizens. These strong new social institutions became the basis for Europe's claim to being a distinct civilization. They allowed Europe's townsfolk to moderate the extreme divisions of status and wealth of the feudal system and to establish a pace of technological and social change, as well as contacts with the outside world, that left the feudal hinterland literally in the Dark Ages. This division in European thought between the "exciting, creative, cultured, safe urban" and the "behind-the-times, rude, dangerous rural" still shapes many of our attitudes toward economic development. In former European colonies, for example, urban development often receives more support than rural development, even though the majority of the population may be rural.

An outgrowth of European town life was the **Renaissance** (French for "rebirth"), a broad cultural movement that began in Italy in the fourteenth century (1300s) and was inspired by the older Greek, Roman, and Islamic civilizations. Renaissance thinkers turned their attention to science, politics, commerce, and the arts. By developing humanism, a philosophy that emphasizes the dignity and worth of the individual, they provided the foundation of modern European culture.

The liberating effects of the new urban institutions and the Renaissance led to a further transformation of life in Europe. One aspect of life so affected was religion. Since late Roman times, the Catholic Church had dominated religion, politics, and daily life throughout much of Europe. In the sixteenth century, however, a movement known as the **Protestant Reformation** arose in the towns of the North European Plain, Scandi-

navia, and the British Isles. The reformers challenged Catholic practices that stifled public participation in religious discussions, such as the Catholic tradition of printing the Bible and holding church services only in Latin, a language that none but a tiny educated minority understood. The Reformation coincided with the invention of the European version of the printing press, which facilitated widespread diffusion of reformist ideas and stimulated the development of written versions of local languages. In addition to translating the Bible and holding services in the languages of the people, the Protestants also promoted public literacy, individual responsibility, and more open public debate of social issues.

Europe's **Age of Exploration** was a direct outgrowth of the greater openness of the Renaissance, and it began a period of accelerated global commerce and cultural exchange. The globalization of modern international trade had its origins in the Age of Exploration. In the fifteenth and sixteenth centuries, Portugal took advantage of Renaissance advances in navigation, shipbuilding, and commerce to round the Horn of Africa, to set up a trading empire in Asia, and eventually, to establish a colony in Brazil. Spain, beginning with the first voyage of Christopher Columbus in 1492, founded a vast and profitable American-Pacific mercantile empire. By the seventeenth century, however, England and Holland (now called the Netherlands, and its people, the Dutch) had seized the initiative from Spain and Portugal. They perfected **mercantilism,** a strategy to increase a country's power and wealth not only by acquiring colonies with their human and natural resources, but also by managing all aspects of production, trans-

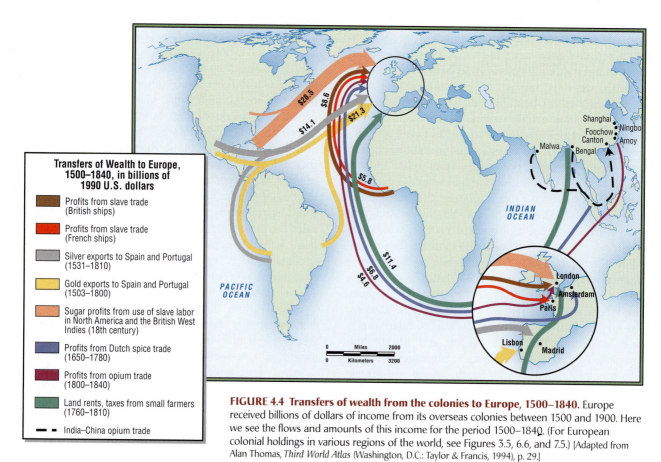

FIGURE 4.4 Transfers of wealth from the colonies to Europe, 1500–1840. Europe received billions of dollars of income from its overseas colonies between 1500 and 1900. Here we see the flows and amounts of this income for the period 1500–1840. (For European colonial holdings in various regions of the world, see Figures 3.5, 6.6, and 7.5.) [Adapted from Alan Thomas, *Third World Atlas* (Washington, D.C.: Taylor & Francis, 1994), p. 29.]

port, and trade for the colonizer's own benefit and to the colony's detriment (Figure 4.4). Mercantilism supported the industrial revolution in Europe by supplying cheap resources for new factories and markets for European manufactured goods. By the nineteenth century, the Spanish and Portuguese empires had weakened, partly because they still relied on archaic Greco-Roman colonial institutions, such as the plantation and the hacienda. In addition, Spain and Portugal maintained a strong alliance with the Catholic Church, which discouraged social and technological change. Their empires were soon overshadowed by those of England and Holland. The Dutch and the British, influenced by the Protestant emphasis on individualism and innovation, used somewhat more efficient methods of colonial expansion and control, and soon extended their influence into Asia and Africa. Regardless of which European powers dominated, by the twentieth cenury, European colonial systems had strongly influenced nearly every part of the world.

The Evolution of European Cities

Roman towns—some quite large, but most with only 2000 or so inhabitants—were sprinkled across Europe by A.D. 200. During the medieval period (A.D. 450–1400), many of these towns fell into ruin, but cities in northern Italy developed important trading links, by land and by sea (Figure 4.5, Map A).

After 1500, Europe's increasing wealth was particularly evident in the flourishing cities of northwestern Europe, to which the colonies of England and Holland in the Americas, Asia, and eventually Africa brought enormous infusions of cash to be invested in industrialization. Wealth and investment capital were transferred to cities across Europe (see Figure 4.5, Map B).

By the mid-1700s, industrial activities such as mining, milling, and manufacturing led to substantial migration of workers from rural areas to factory towns in England, Holland, Belgium, France, and Germany. The profits generated from the sale of new products on the expanding global market brought yet more wealth to this part of Europe. Some cities, such as Paris, London, and Vienna, were elaborately rebuilt to reflect their roles as centers of empires (see Figure 4.5, Map C).

London and Paris, each of which had a million people by 1800, are now Europe's largest cities and are considered **world cities** in that London is a main global center of finance, and Paris is a cultural center that is highly influential in global consumption patterns, from food to fashion to tourism. (For a discussion of modern patterns of urbanization, see page 192.)

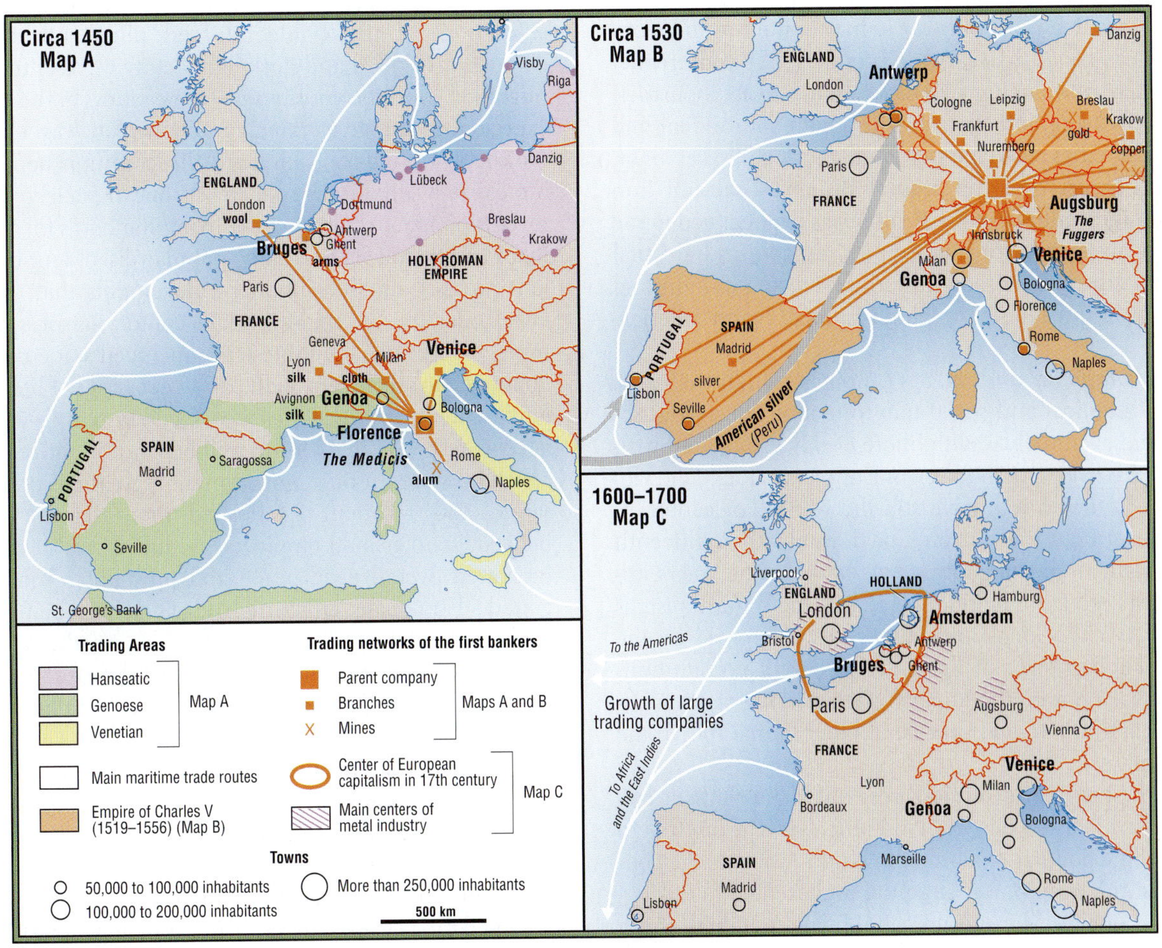

FIGURE 4.5 The development of cities in Europe and their changing links with the outside world.

An Age of Revolutions

Colonialism produced great wealth for Europe, at the expense of people and resources in its overseas colonies. This wealth helped fund two of the most dramatic transformations in a region characterized by rebirth and innovation: the industrial and the democratic revolutions.

The Industrial Revolution. Europe's industrial revolution was intimately connected with colonial expansion and the Age of Exploration. In particular, Britain's ascendancy as the leading industrial power of the nineteenth century had its origins partly in the expansion of its empire. In the sixteenth century (1500s), Britain was an island of only modest wealth and resources. In the seventeenth century, however, Britain developed a small but growing trading empire in the Caribbean, North America, and South Asia, which provided access to a wide range of raw materials and to markets for British goods. Sugar, produced by British colonies in the Caribbean, was an especially important trade crop (see Figure 4.4). Sugar production was a complex process requiring major investments in equipment and labor and long-term management of planting, harvesting, processing, storage, transport, and marketing—activities that contributed to the development of a set of skills needed in the industrial revolution. So, as the availability of sugar stimulated the demand for sweet foods, the mass production of sugar not only generated enormous wealth that helped fund industrialization in Britain, but also provided a model for ordering the manufacturing process sequentially.

By the late eighteenth century, Britain was straining to meet a demand for more goods than it could produce. The country met the challenge by introducing mechanization into its industries, first in textile weaving and then in the production of coal and steel. Eventually, cities with factories producing textiles, ceramics, tools, and machinery grew up around Britain's coal and iron fields. By the nineteenth century, Britain was a global economic power with a huge and growing empire, expanding industrial capabilities, and the world's most powerful navy, and its industrial technologies were spreading throughout Europe, North America, and elsewhere.

The Democratic Revolution. As Europe was industrializing, it experienced political and social transformations that redistributed power, and eventually wealth, more evenly throughout society. However, the road to democracy in Europe was rocky and violent, just as it is now in many parts of the world. For centuries, Europe's power structure had been feudal. But by the eighteenth century, especially in the kingdoms of western Europe, the political elite was expanding to include merchants and industrialists as well as nobles. While kings and churches were forced by the expanding financial power of the elites to accept constitutions that restricted their power, the impoverished working-class majority, both rural and urban, had no vote or other formal role in the political system.

In 1789, the French Revolution arose out of the conflicts created by extreme disparities of wealth in French society. Inspired in part by news of the popular revolution in North America, the poor rose up against the royal power structure and executed the king, Louis XVI, the queen, Marie Antoinette, many members of the nobility, and many people suspected of being royalist sympathiz-

ers. The French Revolution led to the first major inclusion of the common people in the political process in Europe. As the populace became more involved in governing, the idea of the nation was born, partly as a way to unify people despite divisive wealth and privilege disparities (see Chapter 1, page 49).

Throughout the nineteenth century, the idea of allegiance to the state, or **nationalism,** spread throughout much of Europe, transforming the political structure. Individual kingdoms, in the case of France, or collections of kingdoms, in the case of Germany and Italy, tried to establish the notion that all their people together formed a nation and that loyalty to that nation should supersede all other loyalties. The elite found this rallying call to nationalism an effective way to deflect attention away from the poor living conditions the majority of people experienced. Eventually the whole map of Europe was configured as a mosaic of small nation-states, but the reality was that virtually none of these political units was made up of a single cultural group; all contained numerous minorities that did not fit into the nation-state model. France had the Bretons and Basques, Spain the Basques, Galicians, and Catalans, Italy the Slovenes in the northeast and the Sicilians in the south, Austria the Tyroleans, and all of Europe had the Roma (Gypsy) and Jewish minorities. So internal loyalty within the nation-states was tenuous, and there were numerous efforts to expel, or even annihilate, those who were not considered sufficiently French or German or Italian.

While Europe was undergoing the industrial and democratic revolutions of the eighteenth and nineteenth centuries, the demand for skilled and unskilled factory workers triggered steady migration from the countryside to the cities. In the cities, workers experienced unsafe working conditions, low wages, and dirty and crowded housing. So, although the common people played an increasingly important role in European society, the industrial revolution failed at first to raise living standards for the vast majority of Europeans. Periodically, popular discontent erupted in violent protests and revolutionary movements that threatened the civic order. By the mid-twentieth century, as prosperity became more widespread, the protesters convinced some governments to guarantee to all citizens such basic necessities of life as education, employment, and health care. A social system in which the state accepts responsibility for the well-being of its citizens is known as a **welfare state.** In time, government regulations on wages, hours, safety, and vacations established more harmonious relations between workers and employers, reduced wealth disparities, and increased overall civic peace and prosperity. Today, though in general Europeans enjoy high levels of well-being and a state-supported safety net meets their basic needs, just how much support the welfare state provides has been resolved in different ways across Europe; we will discuss these variations starting on page 206.

Democratic institutions in Europe, such as constitutions, elected parliaments, and impartial courts, developed only slowly and unevenly. The right to elect leaders came first to some European men in the early nineteenth century, and, after considerable delay and political agitation, to women, who in Switzerland obtained the vote only in 1971. In East and Central Europe, democratic institutions were not adopted until the 1990s. When assessing progress toward democracy in other world regions, it is helpful to remember this long and difficult path to democracy in Europe.

Europe's Difficult Twentieth Century

World Wars, Cold War, and Decolonization. Despite the region's many advances in industry and politics, between 1914 and 1945 two horribly destructive world wars removed Europe from its position as the dominant region of the world. By the mid-twentieth century, Europe lay in ruins, millions had died, and Europe still lacked a system of collective security that could prevent war between its rival nations. The defeat of Germany, seen as the instigator of both world wars, resulted in a number of enduring changes in Europe. After the end of World War II, in 1945, Germany was divided into two parts. West Germany became an independent democracy allied with the rest of Western Europe, especially Britain and France, and with the United States. Through the Marshall Plan, the United States provided financial assistance to rebuild Western Europe's basic facilities, such as roads, housing, and schools. Western European countries continued their free-market economic system of privately owned businesses that adjusted prices and output to match the demands of the market.

East Germany and the rest of Eastern Europe (Poland, Czechoslovakia, Hungary, Romania, and Bulgaria) fell under the control of the Soviet Union (Union of Soviet Socialist Republics, or USSR), a revolutionary communist state that had emerged out of the old Russian Empire during World War I. The line between East and West Germany was part of the **iron curtain,** a fortified border zone that separated Western and Eastern Europe. The Soviet Union integrated Eastern Europe into its sphere of so-called

communist states, which employed a form of socialism in which the state owned all farms, industry, land, and buildings. In contrast to the market economies of Western Europe, these economies were "centrally planned": a central bureaucracy dictated prices and output with the stated aim of allocating goods equitably across society according to need.

The division of Europe laid the foundation for the **cold war** between the United States and the Soviet Union, an era lasting from 1945 to 1991, during which the entire world became a stage on which these two superpowers competed for dominance. Once-powerful Europe became subject to the geopolitical manipulations of the two superpowers. Yet another manifestation of Europe's decline was that it could no longer control its colonial empires. By the 1960s, most former European colonies were independent and had entered a difficult period, which still persists, of finding ways to thrive on their own.

Europe's Rebirth and Integration. In the decades after World War II, economic reconstruction proceeded rapidly in the free-market democracies of Western Europe, thanks in part to U.S. postwar aid policies. But progress was much slower in socialist Eastern Europe, where the Soviet Union wielded strong influence in much the same way that colonizers such as the British, French, and Dutch once controlled the resources and people of the Americas, Africa, and Asia. Eastern European state-run industries were inefficient and highly polluting, and public debate and citizen participation were discouraged. Local attempts at reform were squelched by the Soviets in 1953 and 1956 in Hungary and in 1967 in Czechoslovakia. Then, in the early 1980s, political and economic reforms in Eastern Europe began with labor protests in Poland. By the end of the 1980s, much of Eastern Europe had abandoned socialism. This change hastened the economic and

Strasbourg, at the confluence of the Ill River with the Rhine at the edge of Alsace wine country, is the seventh largest city in France. The medieval towers of the Ponts Couverts are all that remain of fortifications that surrounded the old "free" city, originally founded in 12 B.C. as a Roman camp. Today Strasbourg, a major attraction for tourists from around the world, is also home to the European Parliament, where representatives of the EU meet; the Council of Europe, which is concerned with social, cultural, educational, environmental, and human rights matters throughout the region; and the European Court of Human Rights, which is responsible for individuals' rights in all member countries. [Mac Goodwin.]

political collapse of the Soviet Union in 1991 because, through its quasi-colonial system, the Soviet Union had depended greatly on the resources and skilled labor of Eastern Europe.

Meanwhile, some of the free-market democracies of Western Europe began lowering barriers to trade among themselves. This process of increasing economic links had begun in the 1950s, and it led eventually to the establishment of the European Union (see Figure 4.1 on page 179). The EU encourages **economic integration**: the free movement of people, goods, money, and ideas among member countries. EU nations also exhibit some degree of political integration: there is a directly elected European Parliament with limited powers, and a constitution for the EU has been developed, but not yet adopted. Because Europe increasingly acts as a single large unit, encompasses a large and affluent population, and is pioneering innovative social programs, it is challenging the dominance of the United States in world affairs. Later in this chapter, however, we shall see that as the EU expands, primarily into the poorer countries of East and Central Europe, and takes on a larger global role, it is encountering many challenges of its own.

Ethnic Cleansing in Southeastern Europe. The peaceful absorption of former Eastern bloc states into the European Union was arrested by bloody conflict in southeastern Europe in the 1990s.

After World War II, the country of Yugoslavia was formed from six territories (see Figure 4.21 on page 231). In five of them—Serbia, Croatia, Slovenia, Macedonia, and Montenegro—a single ethnic group was dominant, although other ethnic minorities were present in most of the five. The sixth territory, Bosnia and Herzegovina, did not have a majority ethnic group. Rather, most of its people had been in Bosnia long enough to think of themselves as simply Bosnian, but they were aware of having historical roots as Croatians or Serbians. Many of these people were either Catholic or Orthodox Christians. There was also a third "ethnic" group, consisting of Muslims who had converted from Christianity generations ago and were largely secular in their religious practices.

In 1991, the first free elections ever held in Yugoslavia resulted in declarations of independence first by Slovenia, then by Croatia and Macedonia, and finally by Bosnia and Herzegovina. Although Slovenia and Macedonia were allowed to separate relatively peacefully, the now much smaller Yugoslavia, composed of Serbia and Montenegro, fought protracted wars to retain Croatia and Bosnia and Herzegovina. Both countries had Serb-populated provinces, which Serbia argued should be part of one Serbian-led country. To rid the coveted provinces of non-Serbs, some Serbs instigated genocidal **ethnic cleansing** campaigns against civilians in Bosnia and Herzegovina, in Croatia, and later in Kosovo (a province of Serbia). Their goal was to create ethnically "pure" nation-states, or independent countries consisting of just one nationality.

Only after agonizing delays did this systematic violence gain the attention of EU countries and the United States, both of which sent military peacekeepers. Peace accords were signed in Dayton, Ohio, in 1995. By this time, the results of the war were devastating, and Serbia itself was in shambles. It lost territory it had illegally claimed, and its economy was ruined. Unemploy-

ment reached 60 percent during the war, and inflation was at 20 percent. Personal income was cut in half between 1990 and 1996. In Bosnia and Herzegovina alone, 200,000 people, 5 percent of the population, had died in the war; many were the victims of Serbian war crimes. The cosmopolitan city of Sarajevo, the site of the Winter Olympics in 1984, was destroyed, as were countless villages.

The conflict in the former Yugoslavia caught most of its citizens off guard. As the personal vignette below illustrates, they once got along with people of different ethnicities, but then found that rapidly, and for reasons they still do not understand, the situation changed. Now they struggle to reconstruct life in their old multicultural communities. If sentiments like those of the Djukanovic family are common, as many observers say they are, it is difficult to understand why there has been so much violence between ethnic groups.

PERSONAL VIGNETTE   Mitra Djukanovic and her husband, Dusan, are Christian Serbian farmers who defied Serb leaders in 1996 by returning to their home village of Krtova, Bosnia and Herzegovina, in a territory now administered by Muslims. As they prepare for the coming hard winter, they tell their story amid bags of potatoes, walnuts, pickled peppers, and flour as drying meat smokes over a smoldering tub.

Dusan speaks: "I thought it was time to come back. I stopped worrying about living with Muslims. I lived with them most of my life. . . . My grandfather told me about the old times, and there was never trouble between us and the Muslims here. We all just worked a lot and had nothing. It was equal. Until this war, I never had trouble either. If there is a Muslim government, so what?" [*Adapted from Mike O'Connor, "Serbs go home to Bosnia village, defying leaders," New York Times (International) (December 8, 1996): 1.*]

POPULATION PATTERNS

Enumerating just how many people live in Europe is complex because, as a result of recent political changes, the region has no universally agreed-upon eastern border. According to the way we define Europe in this book (see the regional map on pages 180–181), there are about 540 million Europeans, and they are distributed unevenly. The population map (Figure 4.6) shows the densest settlement in Europe stretching in a disconnected band from the United Kingdom and north coastal France east through the Netherlands and central Germany all the way to Warsaw and Bucharest. (This band of density continues into Ukraine, as we shall see in Chapter 5.) Northern Italy is another zone of density, and pockets along the coasts of Portugal, Spain, southern France, Sicily, and southern Italy, as well as in southeastern Europe, are also densely populated. A glance at the global population map in Figure 1.17 (pages 38–39) will show that, overall, Europe is one of the most densely occupied regions on earth. You might assume that at

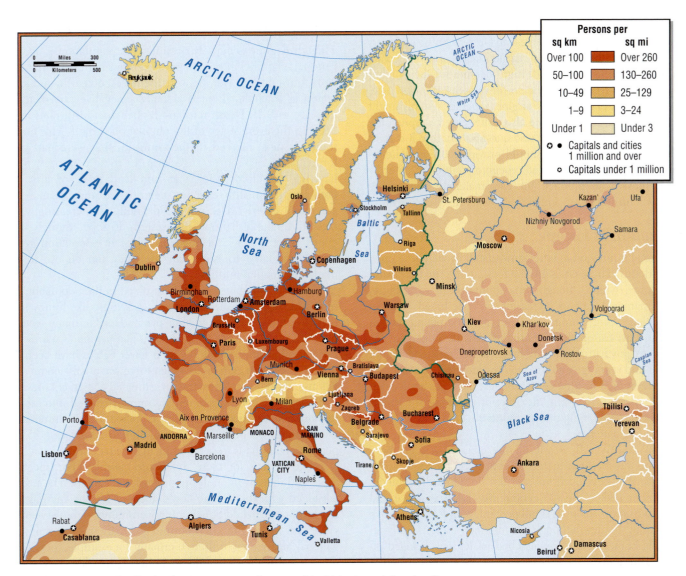

FIGURE 4.6 Population distribution in Europe. Europe has one of the highest population densities on earth. [Adapted from *Hammond Citation World Atlas* (Maplewood, N.J.: Hammond, 1996).]

such high population densities, there would be insufficient resources to go around. How do Europeans maintain their high standards of living?

Population Density and Access to Resources

Population density alone tells us little about how well people are doing in a particular place. If people have access to adequate resources, and if those resources are allocated fairly among the population, they may live well even at high densities. A region's **resource base** is the selection of raw materials available for domestic use and industrial development: coal, petroleum, iron ore, cotton and wool fiber, food, soil, and water, for example. Europe depleted many of its own resources, especially forests and minerals, early in the industrial revolution. Nonetheless, Europe is now both densely populated and wealthy. This is possible because for hundreds of years it has had access to a resource base

that reaches far beyond Europe. Figure 4.4 on page 186 shows transfers of wealth to Europe from the sixteenth to the nineteenth centuries. Since about 1500, colonies in the Americas, Africa, and Asia have contributed tremendous wealth to Europe in forest and mineral resources, in agricultural products extracted at low costs, with enslaved or very low-paid labor. This wealth has added considerably to the overall standard of living of Europeans and to the ability of Europe to develop an industrial economy. Even ordinary laborers in Europe have benefited. In contrast, those far away in former colonies who worked to create wealth for Europeans have often lived in poverty.

Consider the case of the European chocolate industry. The raw material for fine chocolates is cacao beans, which are grown in the tropics of the Americas, Africa, and Asia. Most cacao farmers are small producers who barely eke out a living and have little or no power to negotiate the prices they are paid. Meanwhile, European cacao traders are wealthy from the earnings they make in the world cacao market; European workers in chocolate factories also live well in comparison to cacao farmers. Some observers would argue that the European cacao traders and chocolate factory workers are rewarded with high incomes for their higher levels of skill and technology and for developing the chocolate market in the first place. Critics of global market systems might counter that the cacao farmers are also quite skilled, despite little access to schooling. They might also question whether it is right that education and technology should provide such overwhelming benefits to those who trade and process a product, in comparison with those who produce the raw materials. (See pages 215–216 for a further discussion of the global role of chocolate.)

Urbanization in Europe

Today, Europe is a region of cities surrounded by well-developed rural hinterlands. These cities are the focus of the modern European economy, which, though long grounded in agriculture, trade, and manufacturing, is increasingly service based. Even in sparsely settled North Europe, 73 percent of the people live in urban areas. East Europe is least urbanized, with 60 percent living in cities. Many European cities began as trading centers more than a thousand years ago and still bear the architectural marks of medieval life in their historic centers. These old cities are located either on navigable rivers in the interior or along the coasts because water transport figured prominently (and still does) in Europe's trading patterns.

Since World War II, nearly all the cities in Europe have expanded around their perimeters in concentric circles of apartment blocks. Well-developed rail and bus lines link these blocks to one another and to the old central city. Land is scarce and expensive in Europe, so only a small percentage of Europeans live in single-family homes, although the number is growing. Even single-family homes tend to be attached or densely arranged on small lots, surrounded by walls or fences to ensure privacy and to protect small gardens. Rarely does one see the sweeping lawns that North Americans spend so much effort grooming. Publicly funded transportation is widely available, so many people prefer to live in apartments near city centers. They can either walk or take public transport to their jobs and to shopping centers, and they can easily do without a car. Public transport also links the countryside to urban centers, and many Europeans commute daily from their ancestral villages to work in nearby cities. Much of the income of these urban workers is invested in manicuring these villages.

Life in Europe's many large cities—London has 7.6 million people in its metropolitan area, Paris 9.8 million, Madrid 5.1 million, Essen 6.5 million, and Berlin 3.3 million—can be quite intimate and personal as friends congregate in sidewalk restaurants, ancient pubs, and elegant public parks. Although deteriorating housing and slums do exist, substantial public spending on social welfare systems, sanitation, water, utilities, education, housing, and public transportation maintains a high standard of urban living in Europe—a standard that many feel, in the quality of daily life, exceeds that in the United States. Tourists from around the world are drawn to Europe's large cities by their art, music, museums, architectural heritage, outdoor spaces, and generally pleasant walkable ambience.

Population Growth Patterns

Though Europe is densely occupied, its population is aging, as European families choose to have ever fewer children. In fact, in 2004, birth rates in Europe were the lowest on earth, and for the region as a whole, there is actually a negative rate of natural increase (−0.2). Death rates are higher than birth rates in part because the average age of the population is high. This circumstance is most marked in the countries that were part of the former Soviet Union or Yugoslavia. In these countries, people have had few children for years, and death rates are fairly high, both because health care has deteriorated and because there are many elderly. The one-child family is increasingly common throughout Europe. In western Europe, though immigrants are a major, and needed, source of population growth, once assimilated to European life, they too choose to have markedly smaller families.

The declining birth rate is illustrated in the population pyramids of European countries, which look more like lumpy towers than pyramids. The population pyramid of Germany (reunited in 1990) is an example (Figure 4.7). The pyramid's narrow base indicates that since about 1975, far fewer babies have been born than in the 1950s and 1960s, when there was a baby boom across Europe. By 2000, in Germany, an estimated 25 percent of women and men were choosing to remain unmarried well into their 30s (although unmarried parenthood is increasingly common, with one European child in four born to unmarried parents); 35 to 40 percent of Germans were choosing to have no children at all.

The reasons for these trends are complex. For one thing, more and more women desire professional careers. The need for advanced education alone could account for late marriage and low birth rates; in addition, many governments make very few provisions for working mothers beyond giving them some months off with pay after giving birth. In Germany, for example, there is little day care available for children under 3. In addition, in places such as Austria, Italy, and Slovenia, school days are short, ending at 2 p.m., or no lunches are provided, so mothers must be home by noon or early afternoon. Hence even many married women are choosing not to become mothers because they would have to settle for part-time jobs. Within the European Union, to encourage higher birth rates, there is a move to give one parent (mother or father) a full year off with reduced pay after a child is born or adopted, but this provision is not yet fully operational.

Two results of the small number of births in Germany are that the population pyramid continues to narrow at the base, and that the vast majority of Germans are over the age of 30. If these trends continue, as the older generations die, the population will eventually settle into a stable age structure, as Sweden's population has already done (see Figure 4.7). Sweden's population pyramid resembles an elongated box, with only slight variations from birth through middle age, tapering at the top after age 60 for both males and females. The more rapid death rate for males after age 64 is a common pattern throughout the developed world, but it is not yet well understood.

A stable population with a low birth rate has several consequences. Although families with few children have extra money for luxury spending, new consumers are not being produced at the

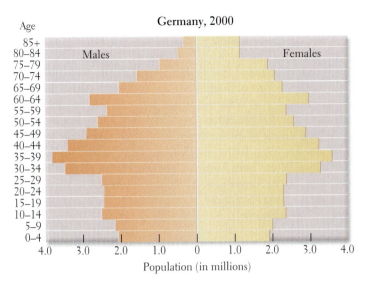

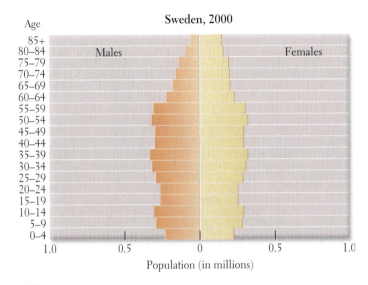

FIGURE 4.7 Population pyramids for Germany and Sweden. Notice that the population scales are different for the two countries. [Adapted from "Population Pyramids of Germany" and "Population Pyramids of Sweden" (Washington, D.C.: U.S. Bureau of the Census), International Data Base, 2003.]

rate that affluent elderly consumers are dying. Hence, in time, markets will contract unless immigrants can be attracted. Demand for new workers, especially highly skilled ones, may go unmet, again unless immigration supplies a solution, as is increasingly the case in Germany. The number of younger people available to provide expensive and time-consuming health care for the elderly, either personally or through tax payments, is small; currently, for example, there are just two German workers for every retiree. By comparison, in the United States, there were three workers for every retiree in 2001 (but by 2020, there will be just two).

Quick Review

1. What means have people in the Netherlands used to transform their landscape from marshes to farmland and cities?

2. What factors account for the fact that the North European Plain is warmer than other places at comparable latitudes?

3. What are the historical roots of the colonizing strategies that Europeans used around the world?

For centuries, economic growth has been the hallmark of progress, and the success of states has been measured by the ever-increasing wealth of their economies. But with declining and aging populations, economic growth in Europe may be difficult to maintain. Immigration provides one solution, but Europeans are reluctant to absorb large numbers of immigrants, especially from distant parts of the world whose cultures are very different (see the discussion on pages 201–205).

4. Where did funding for Britain's industrial revolution come from?

5. Part of the iron curtain separated two halves of which European country?

6. How do European cities differ from U.S. cities in form and in the ways people use them?

II CURRENT GEOGRAPHIC ISSUES

At the start of the twenty-first century, the social, economic, and political geography of Europe (Figure 4.8) is in a state of flux. This circumstance is the result of three major changes that occurred during the 1990s: the demise of the Soviet Union, the end of the cold war, and the effort toward economic and political integration that has produced the European Union. These developments, especially the emergence of the EU, could ultimately bring greater peace, prosperity, and world leadership to Europe, but many problems and tensions still have to be resolved.

ECONOMIC AND POLITICAL ISSUES

Almost all economic and political issues in Europe today are linked to the European Union in one way or another. At the end of World War II, European leaders felt that closer economic ties would prevent the hostilities that had led to two world wars. The first major step in achieving this economic unity took place in 1958, when Belgium, Luxembourg, the

FIGURE 4.8 Political map of Europe.

1 CROATIA
2 BOSNIA & HERZEGOVINA
3 MONTENEGRO
4 MACEDONIA

invited to join. But even as the EU expands, it must deal with changes fueled by advancing technology, the globalizing economy, and geopolitical shifts related primarily to the U.S. response to the terrorism threat.

It should be noted that Turkey, which in this book continues to be treated as part of Southwest Asia and North Africa, became an important ally of Western Europe and a member of the North Atlantic Treaty Organization (NATO) during the cold war. It has had a long history of interaction with Europe, and many millions of Turks are now living in Europe. But Turkey's culture is markedly different from Europe's. For economic and political reasons, the present Turkish government is interested in joining the European Union, but there is opposition to this idea, both in Turkey and in Europe.

Goals of the European Union

The original plan at the founding of the EEC, which remains the core idea of the European Union, was simply to work toward a level of economic and social integration that would make possible the free flow of goods and people across national borders. But two other roles for the EU have emerged. With the increasing importance of the global economy and the geopolitical maneuvering that followed the terrorist attacks on the United States on September 11, 2001 (first the war in Afghanistan and, more importantly, the U.S. attack on Iraq), some Europeans feel that the EU should become both a global economic power competing with the United States and Japan and a politically and militarily independent counterforce to the United States.

The Economic and Social Integrative Role of the European Union. Individual European countries have far smaller populations than their competitors in North America and Asia. With smaller markets for their products, companies in small countries earn lower profits. Previously, European businesses could sell their products outside their home countries, as is common in the global economy, but the extra costs imposed by tariffs, currency exchanges, and border regulations sapped their earnings. The EU solves this problem by joining European national economies into a common market, thus providing the firms in any one country with access to a much larger market and the potential for larger profits through **economies of scale**—reductions in the unit costs of production that occur when goods or services are produced in large amounts, resulting in a rise in profits per unit.

There are now close to 495 million people in the European Union (out of a total of 540 million in the whole of Europe)—roughly 200 million more than live in the United States. Overall, the countries of the EU are wealthy. The combined exports of the EU countries are 40 percent of the world's total, and their average gross domestic product (GDP, PPP) per capita is comparable to that of the United States (see Table 4.1 on page 212). Yet some countries are notably wealthier than others (see Figure 4.9 and Table 4.1). One aim of the EU is to promote the equitable distribution of economic activity, opportunity, human well-being, and environmental quality across Europe while at the same time building institutions that respect the many different regional identities within Europe. EU funds are raised through an annual 1.27 percent tax on the gross national product (GNP) of all members.

Netherlands, France, Italy, and West Germany formed the **European Economic Community (EEC).** The members of the EEC agreed to eliminate certain tariffs against one another and to promote mutual trade and cooperation. Five episodes of expansion followed: Denmark, Ireland, and the United Kingdom joined in 1973; Greece joined in 1981; Portugal and Spain joined in 1986; and Austria, Finland, and Sweden joined in 1995. In 1992, a treaty enlarging the concept of the EEC to that of a European Union was signed at Maastricht in the Netherlands. And in 2004, the EU accepted ten new members: Cyprus, the Czech Republic, Estonia, Hungary, Latvia, Lithuania, Malta, Poland, Slovakia, and Slovenia. Three more countries—Romania, Bulgaria, and Turkey—have formally applied for membership; future potential members are Croatia, Bosnia and Herzegovina, Serbia and Montenegro, Macedonia, and Albania.

As you look at the maps of Europe in Figures 4.1 and 4.9, you will notice that there is one lone holdout in the center of Europe and two on its northern periphery. Switzerland has long treasured its neutral role in world politics, and as yet, the Swiss have decided to remain outside the EU. Norway and Iceland have done the same for similar reasons. All three are wealthy countries and are concerned about losing control, primarily over their domestic economic affairs. Eventually, countries outside Europe (in addition to Turkey), such as Russia and Ukraine, may be

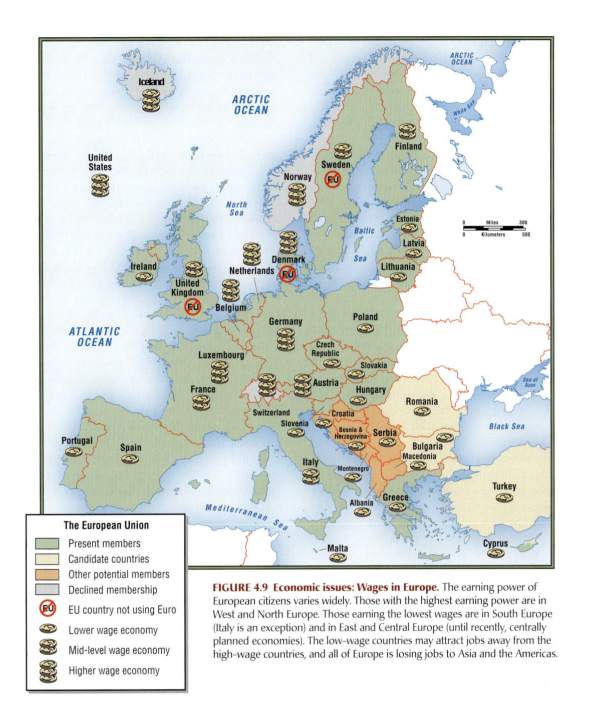

FIGURE 4.9 Economic issues: Wages in Europe. The earning power of European citizens varies widely. Those with the highest earning power are in West and North Europe. Those earning the lowest wages are in South Europe (Italy is an exception) and in East and Central Europe (until recently, centrally planned economies). The low-wage countries may attract jobs away from the high-wage countries, and all of Europe is losing jobs to Asia and the Americas.

Although financial allotments change from year to year, some member countries—such as Germany, the Netherlands, Austria, Sweden, the United Kingdom, France, and Italy—contribute more than they receive back in grants, while others—Greece, Ireland, Portugal, Spain, Malta, Cyprus, the eight countries from central Europe admitted in 2004, and to a lesser extent, Finland and Denmark—tend to receive more than they contribute.

Since 1993, the EU's agenda has expanded to include the creation of a common European currency, the defense of Europe's interests in international forums, and negotiation of EU-wide agreements on human rights and social justice. The EU is also experimenting with various forms of political unification (including the creation of a common European military). One

day, the people of Europe may become citizens of the European Union and live under the common EU constitution presently being considered.

A Common European Currency. On January 1, 2002, the **euro** became the official currency in twelve of the then fifteen EU countries. Three EU members—the United Kingdom, Denmark, and Sweden—had earlier elected not to use the euro because they feared losing control over their own national economies and over trade relations beyond Europe. In addition, they preferred to keep their national currencies alive as a patriotic symbol. Although the viability of the euro is not yet entirely certain, it is expected that a single regional currency will bring gains in efficiency; for example, the common currency will eliminate the currency exchange fees charged to international travelers. Countries that use the euro have a greater voice in the creation of EU economic policies; those that do not use it (including all ten countries that joined the EU in 2004) face the economic uncertainty of using a currency whose value fluctuates relative to the value of the euro. In the United Kingdom, for example, when the pound rises in value relative to the euro, British exporters are put at a disadvantage when selling in Europe because their goods become more expensive to consumers using the euro.

The European Union in the Global Economy. As mentioned above, a second goal of European economic unity is to provide member nations with a competitive edge in the global economy. To be more competitive with the United States and Japan, where technologies in some industries are more advanced and labor is cheaper, the EU is using fund pools such as the European Regional Redevelopment Fund (ERDF) to update its members' technologies and infrastructure. European industries are becoming more efficient, producing at a higher level with fewer workers. Now, however, Europe's firms must face the much more difficult problem of competing with developing economies in Asia, Africa, and South America, where production costs, including labor and raw materials, are much cheaper and environmental regulations are less stringent, hence less costly to meet. Will improving efficiency be enough to enable Europeans to compete in the wider world while maintaining their high standards of income and safety for workers?

Managing industrial location within Europe is another way in which the EU tries to remain competitive at the global scale. If factories can be relocated from the wealthiest countries to the EU's relatively poorer, lower-wage countries (see Figure 4.9), the costs of production will fall. The EU's economic leaders are Germany, France, the United Kingdom, the Netherlands, Belgium, Austria, and Luxembourg in West Europe, as well as Denmark, Sweden, and Finland in North Europe. Italy is the only such leader in South Europe (Figure 4.10). In all these countries, workers and managers are highly paid by world standards; these high wages, along with related expenses such as taxes and worker benefits, drive up production costs. The EU has encouraged its economic leaders to locate new facilities in the poorer, lower-wage countries of Portugal, Spain, Ireland, and Greece, and now in the new member countries of central Europe. Promoters of the EU hope that these efforts will help poorer European countries to prosper and that the removal of tariffs, the elimination of border regulations, and the use of the euro as the common currency will keep the costs of doing business in the EU low enough to restrain European companies from moving to Mexico, Southeast Asia, or China, where costs are lower still. This effort will undoubtedly mean that fewer jobs will be created in Europe's wealthiest countries.

Like the United States, the EU exerts a powerful influence in the global trading system, especially in the World Trade Organization (WTO), where it often negotiates privileged access to world markets for European firms and farmers and for former European colonies. Generally, the EU employs protectionist measures that favor European farmers at the expense of both European consumers, who must then pay higher prices for foreign goods, and producers outside Europe in both rich and poor countries. For exam-

ple, the EU has restricted the international trade of **genetically modified organisms (GMOs)**—plants or animals whose genetic code has been modified to make them more attractive as commodities. GMOs may be resistant to disease, produce greater yields, have a longer shelf life, or be more nutritious. The EU has generally opposed the free trade of GMO products for two reasons. First, many Europeans believe that these products may pose health, safety, and environmental dangers that are not yet understood. Second, curtailing the use of GMOs keeps American and other producers of GMOs from competing in the European market. The EU's opposition has scaled back the global trade of GMOs and sharply curtailed the profits of farmers who produce them.

The European Union as a Geopolitical Counterforce to the United States. A third goal of the EU has emerged since 1991. With the demise of the Soviet Union and the end of the cold war, it appeared that the United States would assume global leadership in political as well as economic matters. The United States retains its country-level dominance economically: with a GDP of $10.98 trillion at the end of 2003, the U.S. economy is three times larger (as well as more efficient) than that of its nearest competitor, Japan. But the combined EU economy is now about equal to that of the United States, and it promises to grow even more competitive. More important, the U.S. reaction to the terrorist attacks of 2001 eventually led to a sharp decline in the political prestige of the United States, especially among the leaders of Europe. The choice of U.S. leaders to resort to preemptive military action in Iraq led European leaders (with the exception of the British, and for a while, the Spanish and Italians) to adopt a global intermediary and peacemaking role in order to curtail the United States' future reliance on such military solutions.

The North Atlantic Treaty Organization (NATO) may also take on a new form of geopolitical leadership. NATO nations (originally the United States, Canada, Western Europe, and Turkey) cooperated militarily and learned to share authority during the cold war, and this experience was an important precursor to European economic union. Now, with the breakup of the Soviet Union, NATO has tried to reconfigure itself as a stabilizing institution that fosters the cooperation needed to expand the EU. Indeed, membership in NATO is considered a stepping-stone to membership in the EU, especially for countries once allied with the Soviet Union. By 2004, nine such countries— Poland, the Czech Republic, Hungary, Bulgaria, Estonia, Latvia, Lithuania, Romania, Slovakia—plus Slovenia had joined NATO. All but Romania and Bulgaria also had joined the EU by 2004.

As cold war tensions fade, and as strengthening global economic and political forces bring the world closer together, NATO's role will doubtless change further. Until the United States adopted a militarily assertive posture after September 11, 2001, most observers expected to see NATO expand its role as a peacekeeper to hot spots in adjacent regions, such as Russia (Chechnya) and Southwest Asia (Israel and Palestine), as it had done in southeastern Europe (the Balkans) in the 1990s and in Central Asia (Afghanistan) in 2003. But when the United States and Britain attacked Iraq (Poland, Spain, and Italy also sent troops), despite strong European dissent, and without the assistance of NATO, the future viability of this once powerful alliance

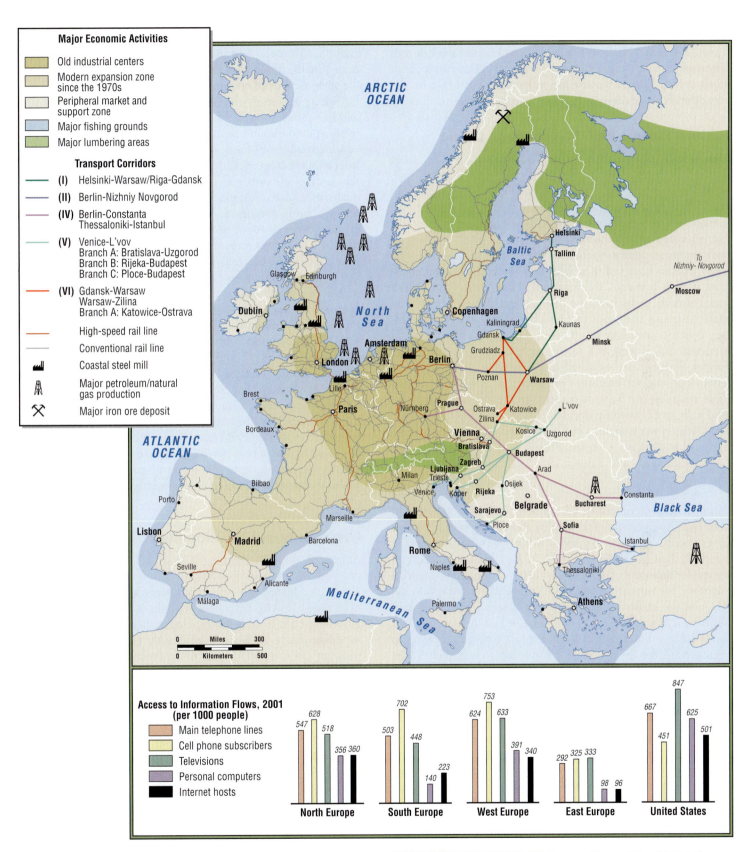

FIGURE 4.10 Europe's principal industrial centers. [Adapted from William H. Berentsen, *Contemporary Europe: A Geographic Analysis* (New York: Wiley, 1997), p. 164; Terry G. Jordan-Bychov and Bella Bychova Jordan, *The European Culture Area: A Systematic Geography* (Lanham, Md.: Rowman & Littlefield, 2000), p. 300; Linda McCarthy and Darrick Danta, Chapter 5, in Stanley D. Brunn, Jack F. Williams, and Donald J. Zeigler, eds., *Cities of the World*, 3rd ed. (Lanham, Md.: Rowman Littlefield, 2003).]

PRACTICING GEOGRAPHY   READING MAPS   **Compare Figures 4.9 and 4.10.** Is there any correlation between changes in industrial location and the wage scales of those countries involved in those changes?

was called into question. Because the war in Iraq did not go smoothly for the U.S.-British alliance, and because the threat of terrorism remains strong, it is possible that NATO, or a successor alliance with Europe playing the leading role, will take on the very difficult job of addressing global security issues.

Eastern Europe and the European Union

Membership in the EU became especially attractive to countries in East and Central Europe after the demise of the Soviet Union, when the economy and the socialist safety net began to deteriorate. The general move toward more open and competitive market systems has placed many of these countries in jeopardy. Former state-owned industries are now forced to survive on their own revenues; hence operations have been streamlined, and many workers have lost their jobs. In some of the poorest countries, such as Romania, Bulgaria, and Serbia, there is a threat of social turmoil, including a rise in organized crime. Although the social programs and funds that benefit EU members are appealing, it is the possibility of attracting foreign investment as EU members that is seen as a main source of economic salvation. Foreign investors are expected to provide funds for upgrading technology and for creating new industries (and jobs) so that these countries can begin to compete in the global marketplace. The path to EU membership for Poland, the Czech Republic, and Hungary was smoothed by German and French investors, for Slovenia, by Austrian, German, French, and Italian investors; for Latvia, Lithuania, and Estonia, Swedish and Finnish investors paved the way.

Standards for EU membership, however, are exacting. To be considered for membership, all ten of the countries admitted in 2004 had to achieve political stability and democratically elected governments. Each country had to adjust its constitution to EU standards that guarantee the rule of law, human rights, and respect for minorities. Each also had to have a functioning market economy, the capacity to cope with competition from within the EU, and the ability to take on the financial and administrative obligations of membership. The EU helped all the new member states achieve these standards, and in all countries aspiring to membership, the EU maintains commissions to help them adjust administrative structures, modify legal systems, and create banks and other financial institutions that will support a market economy. Finally, no country may become an EU member without the agreement of all current members.

Many people in older EU member countries fear that jobs will leave their countries and go to the new, poorer, lower-wage member countries in East and Central Europe, or that skilled and educated migrants from the new member states will move throughout Europe, taking jobs from locals. Some also fear that standards of all sorts will be lowered and that cheap products made in eastern Europe will flood EU markets, competing unfairly and putting a further strain on their countries' economies. By the same token, East and Central Europeans have their own reservations about joining the EU, especially because of what appear to be overzealous regulatory efforts. In the summer of 2003, Slovenia was abuzz with the news that their cucumbers would henceforth have to be a precise size and shape to be accepted for market. And, sure enough, in short order, the "EU Protocol for distinctness, uniformity and stability for *Cucumis sativus* L. (cucumber)" was adopted and published. Regulated cucumbers may be only the beginning, as exemplified by the worries of Hungarian sausage lover Dork Zygotian (see the box "The Hungarian Sausage Experience and the EU" below). The idea that European unification will erase distinguishing cultural features, encourage boring homogenization, and even diminish well-being comes up again and again throughout Europe. One result has been the rise of conservative political movements that foster a sort of "circling the wagons" mentality. European cultural conservatism in France is discussed on pages 204 and 217.

Future Paths of EU Organizational Development

At present, the EU governing body in Brussels, Belgium, makes decisions on the basis of consensus: issues are discussed and proposals are adjusted until all members agree on them. Supporters of a strong EU believe it can be strengthened by giving the central governing body greater power. Under their proposals, decisions would be made in Brussels by directly elected delegates from member countries on the basis of majority rule.

CULTURAL INSIGHT   The Hungarian Sausage Experience and the EU

The Hungarian writer, humorist, and self-professed anthropologist Dork Zygotian (probably a pseudonym) assesses the prospects of European union from the perspective of a sausage lover. Zygotian writes that when you bite into a Hungarian sausage (*kolbasz*), "great torrents of paprika colored grease and juice should explode into the atmosphere around you. If you eat more than two, you should expect to bite on some piece of bone or possibly find a tooth or hair sometime during your meal. There will be a large yellow gelatinous bit somewhere in your sausage that you should not be able to identify." This is all part of the tasty Hungarian *kolbasz* experience.

But now Hungary, long a part of Eastern Europe, has joined the EU, and Zygotian worries about the effect of membership on his country's sausages, which he rates as Europe's best. He fears that overzealous EU standards of cleanliness and purity will kill the special flavors of Hungarian sausages. Worse, "Eurofication" (his term) may well leave Hungarians unable to afford their own beloved *kolbasz* because prices in the small, poorer countries will rise to match those of wealthy Germany, France, the Netherlands, and Switzerland.

Many citizens of EU member countries, especially the smaller ones, are wary of this proposed change because they fear losing control of decisions important to their own countries. They prefer a more flexible arrangement in which groups of countries with common interests can band together to work out individual agreements. Under this plan, a single country would no longer have the power to block projects or policies, as it can now. A bloc of EU countries, for example, could purchase arms for a European army that would be staffed by soldiers only from the countries that agreed to this program. The other member countries would not have to participate, but they could not block the arrangement either, and might lose control over how that army would be used.

Economic Change in Europe

The EU powerfully influences all aspects of life in Europe as it works for progressive changes in European societies. Here we discuss changes in the manufacturing, service, and agricultural sectors of European economies.

Deindustrialization. Throughout the nineteenth century, heavy industry based on coal and steel production grew increasingly important in Europe. Industrial activity was concentrated in the British Midlands and along the Rhine and other major rivers of the North European Plain, close to Europe's main coal fields. In the twentieth century, industry began to diversify and spread into North and South Europe. Figure 4.10 shows these patterns. Over the past several decades, Europe's industries have modernized and become distinctly more efficient, with the result that production has increased while the demand for labor has declined. In the original 15 EU countries in 1970, 30 percent of the workers were in manufacturing; by 1994, only 20 percent were. Such changes are often labeled *deindustrialization*. It is important to note, however, that while there has been a loss of industrial jobs at the local level, this loss is mostly due to increased efficiency and the shifting of industrial location to less expensive areas in southern and eastern Europe, not to the overall loss of industry to other world regions (though a few factories have moved to Asia). Although the EU's easing of trade and migration restrictions has done much to make industrial location more flexible, the switch away from coal to oil, gas, and nuclear power as sources of energy and the expansion of transport systems, especially highways, have also been factors in the shifting location of factories to the south and east.

Europe's Growing Service Economies. As industrial jobs decrease, most Europeans (about 70 percent) find jobs in the service economy, which meets the many needs of governments and businesses for advice, information, testing, licensing, strategic planning (often called **producer services**), communication, and finance. As Europe's economies become better integrated regionwide, new jobs are created. Growth in services to the private

---

PRACTICING GEOGRAPHY   The Development of the EU

John Agnew
University of California, Los Angeles

I am primarily a political geographer. Typically I have used both quantitative methods such as looking at trends in election results and qualitative methods like interviews. I have found that living in places known to be strongholds of particular political groupings or bellwethers of political change is a great way to answer the questions I'm researching. For example, in 1988–1989 I spent long periods of time in the towns of Lucca and Pistoia in central Italy investigating the organization of and support for the Christian Democratic and Communist parties [*Place and Politics in Modern Italy*, University of Chicago Press, 2002].

Italy has long been seen by northern Europeans (and Americans) as backward or underdeveloped compared to themselves. I have learned that this is false, not because Italy is as virtuous as the others claim to be, but because they are all much the same as one another in so many ways. Cultural and economic differences between places are important, but not as great as often portrayed. The same goes for making comparisons between Europe and other parts of the world. I have learned to resist seeing places in terms of some "scale" of development with where I come from at the top end. Undoubtedly, Europe is still one of the world's major geopolitical centers, all the more so if the EU becomes a more integrated and effective organization. But such challenges as population decline, foreign immigration, deindustrialization, and increasing alienation from U.S. foreign policy will challenge most

of the region. Europe's relatively privileged place in the world is not guaranteed.

I am also interested in the geopolitics of European unification, particularly the expansion of the European Union into central Europe and the possible membership of Turkey. I have spent considerable time examining EU documents and interviewing EU officials in Brussels [e.g., "How Many Europes?" *European Urban and Regional Studies* (2001)]. Critical research questions concern the viability of a much-expanded European Union. Where does the project end geographically—with Russia and Turkey or across the Mediterranean? Can the EU encompass countries at different levels of economic development and varying political histories? What are the possibilities for a European citizenship that effectively competes with national identities? How can culturally distinctive immigrants fit into this process? Will Europe establish itself as an independent geopolitical center, competing with, rather than allied to, the United States?

sector has drawn hundreds of thousands of new employees to the main European cities, including those in South Europe and East and Central Europe. Financial services located in London and serving the entire world play a huge role in the British economy, and many transnational companies are headquartered in London. London's prominence in the global economy is a main reason that Britain has elected to not adopt the euro as its currency, because this step would place the British economy under the control of the European Central Bank.

Service jobs in the government sector are also numerous because European countries provide many tax-supported social services to their citizens. Another engine driving the creation of service jobs is the European Union itself, which has generated hundreds of thousands of clerical, administrative, and advisory jobs, especially in Brussels, Luxembourg, and Strasbourg, France. International organizations associated with the United Nations and headquartered in Europe (chiefly in Geneva, Switzerland) also employ thousands of permanent and contract workers. OPEC and the International Atomic Energy Agency (IAEA), both headquartered in Vienna, Austria, are also important sources of employment.

Telecommunications and information technology are well advanced in Europe generally, with North Europe (Finland, Sweden, Norway, and Iceland) leading the way; and European cell phones can be used worldwide. (Most U.S. cell phones cannot.) Personal computer use is also growing throughout Europe. UN statistics for 2001 show, however, that there were fewer homes and businesses with computers in Europe (about 43 per 100 people) than in the United States (where there were about 62 computers for every 100 people). In South Europe and East and Central Europe, where personal computer ownership is lower, public computer facilities in cafés and libraries are common, and surfing the Internet is popular, especially among schoolchildren. Overall, access to information is high across Europe, and improving rapidly in East and Central Europe (see Figure 4.10).

Another important component of Europe's service economy is tourism, both regional and global. Europe is itself the most popular tourist destination on earth, and Europeans are enthusiastic travelers, visiting one another's countries frequently. The long vacation is a European institution: most workers get at least 4 weeks off from the job every year. In the last several decades, taking several week-long trips has become popular, and this custom has increased the demand for urban hotels, rural guest houses, car rentals, and airline service. Not surprisingly, tourism is a substantial part of the European economy. One job in eight in the European Union is related to tourism, and the industry generates 13.5 percent of the EU's gross domestic product and 15 percent of its taxes.

A small percentage of Europeans have long been trekkers to exotic world locations, but now many more have the money to travel internationally, visiting Southeast Asia, India, Tibet, China, North America, Central America, and the Caribbean. Cuba alone by the early 2000s was attracting close to a million Europeans yearly.

Energy Resources. At the same time as Europe's economy has become oriented to services, its energy use has steadily moved away from coal to petroleum, natural gas, and nuclear power.

Europe imports much of its oil and gas from outside the region, buying natural gas from Algeria and Russia and much of its oil from the Middle East. Within Europe itself, there are large oil and gas deposits in the North Sea and under the Netherlands. The North Sea deposit, controlled by Britain and Norway, now supplies much of the energy used in Germany and Britain, and the Netherlands ships natural gas to most of West Europe. The use of nuclear power to generate electricity has been more common in Europe than in North America—especially in France, where it accounts for 78 percent of the electricity generated, compared to only 20 percent in the United States. Although many European countries relied on nuclear power to achieve energy self-sufficiency, most western European countries are now phasing out their use of nuclear energy. This movement is partly a response to public concerns about safety. The safety risks were dramatically illustrated by the Chernobyl accident in Ukraine in 1986, when an explosion sent a cloud of radiation drifting across western Europe. In addition, the desire for national self-sufficiency in energy is waning as the European Union fosters interdependence.

The Transportation Infrastructure. One goal of the EU has been to extend transportation networks to draw all of Europe together. Europeans have typically favored fast rail networks for both passengers and cargo, rather than multilane highways for cars and trucks (see Figure 4.10). In 1994, the Eurotunnel under the English Channel, linking England and France, continued this reliance on rail transport. What was once a ferry crossing involving long waits is now a 20-minute rail trip for passengers, autos, and cargo. But even though fast rail transport remains important and new lines are planned, there is a noticeable trend toward less energy-efficient but more flexible motorized vehicles. Since the mid-1990s, the EU has been building **Corridor Five,** a major transportation roadway across southern Europe, stretching east 2000 miles (3500 kilometers) from Barcelona, Spain, through Slovenia and eventually to Kiev, Ukraine.

Agriculture in Europe

Although, on average, only about 2 percent of Europeans are now engaged in full-time farming, modern farming methods employing mechanization and chemical fertilizers, pesticides, and herbicides have greatly increased production. As a result, most Europeans are eating as well as or better than ever before, and food is absorbing a smaller percentage of household budgets. Europeans like the idea of being self-sufficient in food, and the European Community Agricultural Policy (CAP) aids farmers with tariffs, **subsidies** (payments to farmers to lower their costs of production), and price supports on some agricultural commodities. **Price supports** are a method of maintaining high prices by setting legal minimum prices for certain crops that are being over-produced. Or farmers may be paid not to farm in order to cut supplies of crops so that prices will remain higher. These measures not only are expensive—payments to farmers are the largest expense category in the EU budget—but also raise food costs for the consumer, encourage farmers to overproduce (to collect more payments), and make European agricultural products uncompetitive on the world market. However, these measures do ensure a steady food supply in Europe and provide a decent living standard for farmers. Although many other countries, including the United States, Canada, and Japan, also use agricultural tariffs and subsidies, they are unpopular internationally because the tariffs lock out producers from poorer countries and the subsidies encourage overproduction, causing a glut of farm products. European countries sometimes sell this overproduction cheaply on the world market. This practice, called **dumping,** lowers global prices and hurts producers of the same commodities elsewhere in the world.

Farms in Europe have tended to be smaller than those in the United States. The average European farm is 45 acres (18 hectares), less than one-tenth the size of the average U.S. farm of 487 acres (197 hectares). Geographer Ingolf Vogeler points out that small family farms are disappearing in Europe, just as they did several decades ago in the United States, and that the trend is toward larger, more profitable farms with very few laborers. In part, this consolidation is driven by the use of expensive agricultural machinery, a practice that requires more acres in production to be profitable. However, mechanization also raises living standards for the farm families who remain after consolidation. Hard labor is reduced, the workday is shortened, and greater profits can be made. Moreover, farmers have time to learn scientific techniques that improve all aspects of farming, from soil fertility to marketing.

In East and Central Europe, agriculture developed somewhat differently in that the small family farm has not been the model for some time. When communist governments gained power in the mid-twentieth century, in most parts of the region, they consolidated small, privately owned farms into large collectives. These farms were worked and managed by a large labor force under the supervision of the state. After the breakup of the Soviet Union, the farms, many reclaimed by their original owners, were rented to large corporations, which in turn further mechanized the farms and laid off all but a few laborers. Rural poverty rose and small towns declined as farm workers left for the cities. In the more open markets of present-day Europe, these larger, more profitable corporate farms could give the smaller family farms of western Europe stiff competition, especially as eastern European countries join the EU. In some places, however, such as southeastern Poland and Slovenia, farms remained small and less modernized. In Poland, close to 40 percent of the population is still employed in agriculture, farmers continue to use draft animals and outdated techniques, and they produce crops that do not meet the standards of the EU. In wealthier Slovenia, 11 percent of workers are still in agriculture and even the tiniest farms have a tractor, but production is still inefficient and farm incomes comparatively low. The quandary is, how shall the farmers of East and Central Europe be treated now that they are in the EU? Were they to receive CAP payments like those typically paid to EU farmers, they would quickly bankrupt the EU budget, and their improved production might lead to price-lowering surpluses. On the other hand, failure to support farmers in the new EU member states could leave entire regions in a second-class, impoverished state, fueling resentment and a flood of ex-farmers into cities across Europe.

SOCIOCULTURAL ISSUES

Although the European Union was conceived primarily to promote economic cooperation and free trade, the economic integration that has resulted from its programs has social implications. In this section, we examine how the European Union is affecting attitudes toward immigrants, gender roles, and the evolution of social welfare programs. At the same time, religion and language are largely fading as divisive issues.

Open Borders Bring Immigrants, Cultural Complexity, Needed Taxpayers, and Worries

Few Europeans mourn the demise of strict regulations at the borders between countries. Not long ago, border crossings meant tension-filled moments for travelers as armed border guards closely inspected passports, vehicles, and luggage, often with a rude edginess. Then, in the 1990s, the EU and many of its neighbors approved the **Schengen Accord,** an agreement for free movement across common borders. Now, even on crossings into West Europe from East and Central European countries such as Slovenia, Poland, and Hungary, guards are usually scrupulously polite, uttering mild pleasantries while barely examining passports. One has to request a country stamp on one's passport as a souvenir.

Open borders have eased life throughout the region, facilitating trade, tourism, and social relationships. Open borders were also intended to bring new workers to European cities and towns. But, while workers are needed, there has been ambivalence in all European countries about freely admitting so many new people, even those from other parts of Europe. Europeans have long prided themselves on being culturally distinct from one another. Even the inhabitants of Germany and Austria, who share a language, a similar history, and numerous customs, see themselves as having distinct national identities. Little wonder, then, that each time the European Union has expanded, there have been fears that the cultural mix was getting too rich, too complex, and that the cohesiveness of particular countries would be lost. Now, in addition to people from East and Central Europe—a part of Europe that has always seemed alien to western Europeans—immigrants are arriving from around the globe. They are coming from Russia, Central Asia, China, North and sub-Saharan Africa, South and Southeast Asia, and South America and the Caribbean (Figure 4.11). Some, such as those from Turkey and North Africa, come legally as **guest workers,** who are expected to stay just a few years—fulfilling Europe's need for workers in certain sectors—and then go home. Others are refugees from the world's trouble spots—Afghanistan, Albania, Congo (Kinshasa), Haiti, Iraq, Kosovo, Rwanda, Serbia, Sudan—who are supported by country governments and community groups. Still others are illegal. But here is the dilemma: while Europeans fear that the presence of large numbers of immigrants, especially from outside Europe, will make them feel like strangers in their own lands, Europeans also desperately need these mostly young new residents to fill jobs at all levels of the economy and to lend vibrancy, creativity, and

Thinking Geographically: ON THE WEB

Migration into Western Europe, 1990–2000

- High-income countries (foreign population)
- Massive population movement

Iceland (3000)

ARCTIC OCEAN

Sweden (477,300)

Finland (91,100)

Norway (150,000)

Estonia (265,400)

Baltic Sea

Denmark (258,600)

Netherlands (667,800)

Latvia (29,300)

Ireland (126,500)

United Kingdom (2.6 mil)

Belgium (861,600)

ATLANTIC OCEAN

Germany (7.3 mil)

Czech Republic (201,000)

Lux. (164,700)

France (3.3 mil)

Switz. (1.2 mil)

Austria (761,400)

Hungary (110,000)

Slovenia (42,300)

Portugal (208,200)

Spain (895,700)

Italy (1.5 mil)

Black Sea

Sea of Azov

Greece (161,100)

Mediterranean Sea

Other former Soviet bloc countries: 2.4 million in the West, 64% to Germany

Turkey: 2 million guest workers, most in Germany

North Africa: 2 million immigrants from Morocco, Algeria, and Tunisia, most in France

0 Miles 300
0 Kilometers 500

FIGURE 4.11 Interregional linkages: Migration into western Europe. Migration to western Europe increased in the 1990s and continued to increase into the 2000s, becoming a crucial issue in EU debates. The numbers in parentheses show the numbers of immigrants living in western European countries in 2000. [Adapted from *National Geographic* (May 1993): 102.]

new young families to an aging Europe. In particular, they are needed to contribute to the tax coffers that will pay the pensions of the many Europeans about to retire.

The new arrivals are evident in schools, the workplace, sports arenas, and religious institutions. Schools in Switzerland may have students from Sri Lanka, Bosnia, Algeria, Russia, China, and Iran. Workplaces in Austria and Germany may have migrants from Turkey, Iran, Ukraine, and Kazakhstan. Soccer teams are made up of players originally from South Africa, Argentina, Zimbabwe, and Trinidad (see the box "Soccer: The Most Popular Sport in the World" on page 203). The presence of so many culture groups raises questions of national identity: Is Germany no longer a German place? Will France's self-image suffer if children of French-North African descent read textbooks that deal frankly with France's long repressive colonial regime in North Africa? Should the predominant religion of these children, Islam, have equal footing with Catholicism, the traditional religion of choice in France (see the photo on the left on page 204)? Questions of public welfare also arise as countries try to accommodate

unfamiliar value systems and family types in social services such as family counseling, housing, and health care.

European host countries have very different policies regarding immigrant workers. Some, like the Netherlands, have lenient laws that allow immigrants from former colonial territories to stay indefinitely and to receive subsidized housing and other social welfare services equivalent to those available for citizens. France provides its immigrants with less adequate housing blocks in informally segregated working-class neighborhoods, and the atmosphere is often hostile. In what are extreme cases in locations widely dispersed across Europe, nationalist feelings have produced violence against immigrants from Turkey, North Africa, the Caribbean, South Asia, and elsewhere. In Germany, Austria, Italy, and France, minority right-wing parties now openly speak of forcing immigrants to leave, but a wider informal movement favors accommodation and acceptance. Ultrarightist feelings bubble to the surface every few years in some part of the region, however, so officials at the community, country, and EU level must be continually cognizant of the need to educate the public about the advantages of bringing in new young workers who will stay and contribute their talents and money.

Citizenship. Until recently, achieving legal citizenship in a European country was very difficult for outsiders, especially those of non-European heritage. Germany, a country with a particularly large group of immigrants (in the late 1990s, 8 percent of its 82 million people were foreign-born), has had especially stringent rules. Even if they were born in Germany, the children and grandchildren of Turkish or North African immigrant workers were not considered citizens, and most were rejected if they applied for citizenship. In 1994, of the more than 2 million people of Turkish origin who lived in Germany and who ran more than 30,000 businesses, fewer than 1 percent had acquired citizenship. But attitudes and policies have changed. As of January

AT THE GLOBAL SCALE Soccer: The Most Popular Sport in the World

In Europe, soccer teams are a major component of national identity. They bring together citizens of vastly different cultural, racial, and class backgrounds to enjoy a common experience. Countries go to great lengths to recruit the best players for their teams, often importing them from South America, South Asia, Africa, and eastern Europe and granting them instant citizenship. Cities in Europe cultivate winning soccer teams in much the same way as cities in the United States support their professional basketball or football teams. Feelings can run so high at soccer matches that violence among the fans erupts easily.

Soccer, known as football in most countries except the United States, is a global game, played regularly by more than 240 million people, as reported in 2000 by the world soccer federation, Fédération Internationale de Football Association (FIFA).

During the 1999 women's World Cup matches played in the United States, it is possible that the entire population of the world could have seen some portion of the matches on worldwide television.

The oldest known precursors to soccer may have been kickball games played in China about 2500 years ago, but similar ball games were known in ancient times in other parts of the world (Figure 4.12). The Romans carried a version to England, where the game evolved. It emerged as the modern game of soccer in 1863, when the rules of the game were formalized. The first international match was played between teams from England and Scotland in 1872. British sailors then carried the game to mainland Europe, India, South America, and the South Pacific.

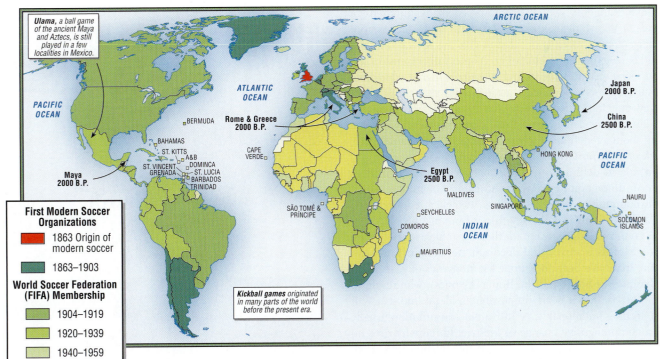

FIGURE 4.12 The origin and spread of soccer. Notice the color sequence of the spread of soccer, from red in England through dark green and then lighter greens to yellows. How might you account for the light yellow shades for Russia and Central Asian countries? Notice the locations of ancient versions of kickball.

These French schoolgirls, from a secular family (their mother is Muslim of North African origin; father, French Jewish), have decided to be observant Muslims and to wear the *hijab* to school. They were expelled for this and have become the center of a national discussion in France about civil liberties and freedom of expression and religion. France later passed a law against all types of religious symbolism in schools. [AFP/Getty Images.]

2000, all children born in Germany are citizens, and people born elsewhere who have lived in Germany for 8 years can apply to become citizens. The law is retroactive to 1975.

France's citizenship laws are somewhat less generous than Germany's, although it has long been a culturally diverse country. The United Kingdom, which has absorbed many hundreds of thousands of West Indians, Indians, Pakistanis, and Africans from its former colonies, recently decided to give the several hundred thousand citizens of its few remaining colonies full citizenship and rights to social welfare benefits and subsidized higher education. As a result of these new policies, an increasing number of Europeans can now trace their ancestry to China, Nigeria, Thailand, Brazil, Jamaica, or Turkey, and these demographic changes are bringing about a redefinition of what it means to be a European.

Rules for Assimilation. In Europe, race and skin color play less of a role in defining differences between people than does culture. An immigrant from Asia or Africa may be fully accepted into the community if he or she has wholeheartedly adopted European ways. This is especially true for immigrants skilled in the host country's language and for those who have mastered the finer points of its manners and decorum. Yet certain European minorities that have been in Europe for thousands of years, such as the Basques in Spain and the Roma (Gypsies), who reside in many countries, find it nearly impossible to blend into society if they retain their traditional ways. In Europe, **assimilation** usually means a comprehensive change of lifestyle. Immigrants give up the culture of home—language, dress, family relationships, food, customs, mores, and even religion—and take up instead the ways of their adopted country.

Geographer Eva Humbeck studied how Thai women who come to Germany explicitly to marry German men grapple with the demands of assimilation. Since 1990, about a thousand Thai brides have arrived in Germany every year. Although they

willingly take on the role of housewife as opposed to career woman, they usually live isolated in apartments and sorely miss their female relatives—who, in Thai culture, are a woman's best friends. Most feel obliged to drop all Thai ways of doing things because they sense that Germans view Thai culture as inferior. Few Thai wives know one another, and many confess to daydreaming about the Thai hospitality and family conviviality they once enjoyed.

Note that the degree of adjustment expected of immigrants in Europe differs from that expected of immigrants in North America, where the norm is only **acculturation;** that is, enough adaptation to the host culture for members of the minority culture to function effectively and be self-supporting. In the United States and Canada, immigrants are expected to learn the language and to abide by the laws of the land. Canada even encourages immigrants to have schools that teach in their own languages, provided they also teach English, or, in Québec, French. Overall, in North America, immigrants are free to keep their native cultures and cuisine. Those who retain native dress and particularly exotic customs may encounter a certain amount of informal prejudice or even discrimination, but they will also find that the laws protect their right to be different, and that some American customs—international street festivals featuring ethnic food and music, for example—actually encourage and reward a measure of cultural retention.

The changes in citizenship laws in Europe may herald a growing acceptance of outsiders. Cross-cultural marriage is increasingly common, and the multiethnic offspring of these mar-

Food habits are changing rapidly in central Europe as a result of market economies and entry into the EU. Central Europeans resisted trying alien ethnic cuisines for a time, saying, "We prefer to eat our own food." But that opinion is no longer shared by many. People are eating out more often, and their selection of restaurants is blossoming. Chinese food seems to be especially popular; restaurants serving dim sum and Peking duck are springing up even in remote small towns. [Lydia Pulsipher.]

Turkish migrants relax in Berlin's Tiergarten, the city's central park. A soccer game is under way. In Berlin, by the mid-1990s, one of every eight residents was foreign born. [Gerd Ludwig/National Geographic Image Collection.]

riages will undoubtedly influence attitudes in their generation. Furthermore, official EU agencies are encouraging public dialogue through frequent forums on cultural integration held at the local and international levels. The Council of Europe has drawn up guidelines for the legal acceptance of cultural diversity by new EU member states from East and Central Europe, and this project to get Europeans in places such as Serbia and Kosovo to reject ethnic prejudice has forced western Europeans to reflect on their own record as well.

European Ideas about Gender

Gender roles in Europe have changed significantly from the days when most women married young and worked in the home raising large families and tending to various agricultural duties. A large percentage of Europeans live in cities (73 percent), marry late, and have only one child; and increasing numbers of European women are working outside the home. In all but a few European countries, women are seeking advanced training more often than men—an indication that European women may eventually exceed men in job qualifications. But will this change translate into higher earnings and more policy-making roles for women?

Distinct differences remain between men and women, both in public perception and in people's day-to-day lives. As is the case nearly everywhere on earth, Europeans generally accept the notion that men and women have different innate abilities that determine the functions they should perform. European public opinion among both women and men holds that women are less able than men to perform the types of work typically done by men and that men are less skilled at domestic duties. These views continue to influence European social institutions. In most cases, men have greater social status, hold more managerial positions, earn higher pay, and have greater autonomy in daily life (more freedom of movement across space, for example) than do women. Despite these gender differences, women increasingly work outside the home as factory laborers, service workers, teachers, textile producers, architects, physicians, and professors.

Often women who work outside the home are also expected to do most of the domestic work, a situation called the **double day**. United Nations research shows that throughout Europe, women's workdays (including time spent in housework and child care) are 5 to 9 hours longer than men's. Women burdened by the double day generally operate with somewhat less efficiency in a paying job than do men, and they tend to choose employment closer to home that offers more flexibility in the hours and skills required. These more flexible jobs almost always offer lower pay and less chance for advancement, but not necessarily fewer working hours, than typical male jobs. In the former Eastern bloc countries, women continue to work outside the home and accept nearly full responsibility for the home and all domestic duties. Here, the double day is especially taxing for women because labor-saving devices are not yet widely available: laundry is often done by hand, kitchens are rudimentary, and food must be purchased daily because small refrigerators will hold only a day's food.

Despite strong EU emphasis on policies encouraging gender equality, the political influence and economic well-being of European women lag far behind that of European men. As Figure 4.13 illustrates, in France only 11.7 percent of elected representatives in the parliament are women, and in Hungary, under 10 percent

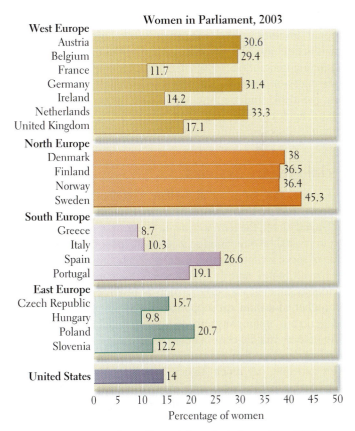

FIGURE 4.13 Women in parliament for selected countries, 2003. European Union policies emphasize gender equality, but women, who are about half the adult population, do not have anywhere near their fair share of representation in parliaments; hence their influence on legislation is seriously restricted. Notice that some regions are closer to equity than others, and also notice how the United States compares. [Data from *United Nations Human Development Report 2003*, United Nations Development Programme, Table 23.]

are women. Only in North Europe do women come anywhere close to filling 50 percent of the seats in the parliament or legislature; Sweden has the highest percentage at 45.3. Women are also poorly represented in government bureaucracies. They serve mostly in the lower ranks, which gives them little voice in the formation of national policies. Women have to rely on men to push for such legal measures as the right to work, equal job opportunities, and equal wages and fringe benefits. Progress has been slow. For example, female unemployment in the EU in 2001 was 152 percent of male unemployment, and 23 percent of women worked part-time, as opposed to only 4 percent of men. Throughout Europe, women are paid less on average than men. By 2000, women in Europe were acquiring university and technical education at much higher rates than men (for example, 15 percent more than men in Spain, 74 percent more in Iceland, 44 percent more in Poland), yet in the same year European women earned notably less than men (43 percent less in Spain, 17 percent less in Iceland, 20 percent less in Poland).

One might question why, given Europe's overall prosperity and progressive attitudes, women have not achieved more equality with men. The explanation is that although EU policies have officially ended legal discrimination against women, old patriarchal attitudes persist. Some analysts point out that European women face obstacles to collaboration and the development of solidarity: geographic separation, language differences, multiple ethnic and national loyalties, varied education levels, and long-standing social hierarchies. This situation is changing as the EU opens up entrepreneurial options for women and encourages international conferences on women's interests. It is noteworthy that, despite their political underrepresentation, on average, women in Europe hold a greater proportion of elected positions than do women in any other world region.

Social Welfare Systems

In Europe, the term **social welfare** describes elaborate, tax-supported systems that serve all citizens in one way or another. In the United States, the term *welfare* has a narrower definition, referring to limited tax-supported services to the poor or disabled. The value of social welfare is more widely accepted in Europe than in the United States, and in Europe such programs include many services to middle-class citizens, such as health care, free or low-cost higher education, tax-supported housing, disability payments, and generous unemployment and pension benefits. (Although some of these middle-class benefits also exist in the United States and Canada, there they are not considered welfare, per se.) Europeans generally pay much higher taxes than Americans, and in return they expect a wide range of services and safety nets. Still, Europeans do not agree on the goals of these welfare systems, or on just how generous they should be. Some argue that Europe can no longer afford high taxes if it is to remain competitive in the global market; others maintain that Europe's high standards of living are the direct result of the social contract to take care of basic human needs for all. The debate has been resolved differently in different parts of Europe, with the result that there is clear regional variation in approaches to welfare. These regional differences have become

a source of conflict because, with newly open borders, unequal benefits may encourage those in need to flock to a country with a generous welfare system and overburden the taxpayers there. Conversely, a country with a less generous welfare system could find its best and brightest workers emigrating to more generous countries.

Using the work of the British sociologist Crescy Cannan and others, we have classified European welfare systems into five basic categories (Figure 4.14). Each category makes certain assumptions about gender roles. All assume that the typical family has two parents and that the male is the sole or primary breadwinner, even though in most parts of Europe, as in the United States, such families are in the minority. Whether married with children or single, most women take employment outside the home to support themselves and their families.

1. **Social democratic welfare systems** are well developed in all of Scandinavia, but especially so in Sweden and Iceland. These systems attempt to achieve equality across gender and class lines by providing generous health, education, housing, and care benefits to all citizens from cradle to grave. Citizens pay high taxes to provide an environment, from prenatal care on, in which every child's physical and social potential is realized. Child care is widely available, not just to help women enter the labor market, but also to provide the state with a mechanism for instilling such values as polite behavior, good nutrition, the work ethic, and good study habits in its citizens. Such early childhood training is a key feature of this system because it is meant to ensure that, in adulthood, every citizen will be able to contribute to his or her highest capabilities, and that social problems will not develop. Although gender equality is a stated goal, traditional gender roles are officially emphasized throughout this lifelong support system, and women are not yet equal to men in the workplace and in public forums.

2. **Conservative welfare systems** are found in France, Germany, the Netherlands, Austria, and Switzerland (all in West Europe). They seek to provide a minimum standard of living for all citizens. Hence, the state assists those in need, but it does not see its mission as one of assisting upward mobility. For example, the state would not normally make a college education accessible to the children of poor families; although a university education is often free, strict entrance requirements are hard for the poor to meet. On the other hand, many state-supported services, such as health care and pension plans, are available to all economic classes. Like other public institutions in these countries, the welfare system reinforces the so-called "housewife contract" by assuming that women will stay home and take care of children. As discussed earlier, half-day school schedules and the lack of school lunches often compel mothers to stay at home. Those mothers who do work outside the home often choose part-time work or jobs with flexible hours. Even in France, where women are likely to work full-time and there is support for the equality of men and women in the workplace, the gender pay gap persists (French women earn,

Welfare Systems
- Social Democratic
- Conservative
- Former Communist
- Modest
- Rudimentary

FIGURE 4.14 European social welfare systems. Within Europe, there is clear regional variation in approaches to social welfare. This map recognizes five basic categories of social welfare systems, described in the text; it should be taken only as an informed approximation of the patterns.

on average, 23 percent less than French men), and gender equality does not extend to domestic duties.

3. **Modest welfare systems,** presently found only in the United Kingdom (West Europe), are designed to encourage individual responsibility and the work ethic. Welfare is thought to encourage dependency, whereas the ideal citizen is completely self-reliant. This type of welfare system has been evolving in the United Kingdom since conservative reforms in the 1980s and 1990s, and it is roughly similar to the welfare system in Canada (the U.S. system is less generous). Benefits are modest, just enough for those who qualify to maintain a minimally adequate standard of living. Recipients are often stigmatized in jokes and conversation as being lazy and "on the dole." The state claims no direct interest in the quality of its citizens' daily lives; how people live is thought to be a matter of individual choice in a free market. Women are free to work outside the home or not, but the state supports neither option. Critics point out that, in reality, women

cannot enter the labor market on the same terms as men. For example, women in the United Kingdom earn only about 74 percent of what men earn, and they occupy the lowest-paying factory jobs and the lower echelons of the business and professional sectors—yet they are also expected to be the unpaid caretakers of children and the elderly. Only minimal state-supported child care and elder care are provided, and their availability is unpredictable.

4. **Rudimentary welfare systems** are found primarily in South Europe—Portugal, Spain, Italy, and Greece—and in Ireland. These countries do not accept the idea that citizens have inherent rights to government-sponsored welfare support. Local governments provide some services or income for those in need, but the availability of such services varies widely, even within a country. The traditional extended family and community are still common in these countries. and the state assumes that when people are in need, their relatives and friends will intervene to provide

needed services, financial support, and care for the young, old, and disabled. The state also assumes that women are available to provide daily child care and other social services for free, despite the fact that in these countries, women actually form the bulk of the flexible, low-wage workforce, and few remain available to provide these services. Finally, there is the official assumption that the large informal (hidden, or gray-market) economy will provide some sort of employment for anyone who needs it. Some of these attitudes are now changing, especially in Italy, which experienced an economic boom in the 1990s. Young urban Italian women are very likely to work full-time outside the home, and they are much more likely than men to be highly trained; nevertheless, their average income is still less than half that of Italian men.

5. **Postcommunist welfare systems** prevail in the countries of the old Soviet bloc (Poland, the Czech Republic, Slovakia, Hungary, Romania, Bulgaria) and the countries that were formerly part of Yugoslavia (Slovenia, Croatia, Bosnia and Herzegovina, Macedonia, and Serbia and Montenegro). During the communist era, these systems were, in theory, comprehensive. Although the bureaucracy administering benefits was inefficient and the programs unevenly funded, the ideal resembled the cradle-to-grave social democratic system in Scandinavia, except that women were pressured to work outside the home. Benefits often extended to nearly free apartments and health care and to meals provided on the job at nominal cost to the worker. The big change in the postcommunist era is that state funding of these welfare systems has collapsed and people must cope with the loss of apartments, free meals, and other benefits of the welfare safety net at the same time as jobs are being lost. In the transition to a market economy, many people are going without such basic necessities as food, heating fuel, and health care; there is thus a temptation to migrate to western Europe, where the basics are still provided.

As the European Union evolves and expands, Europe's social welfare systems will probably become more similar to one another, but it is still unclear just which models will prevail. Some observers suggest that the dominance of the German economy in Europe could mean that the conservative German model will spread. Others think that the entry of Scandinavian countries into the EU will inspire a shift to more activist welfare policies with more equal treatment of women. As the economies of East and Central Europe adjust to the postcommunist era, state-provided benefits are continuing to shrink and may reach the level of the modest welfare system in the United Kingdom.

ENVIRONMENTAL ISSUES

Europe's environment has changed dramatically over the past 10,000 years as a result of human activities. Nearly all the original forests are gone; some have been gone for more than a thousand years. Many of Europe's seemingly natural landscapes are largely the creation of people who have changed the landforms, the drainage systems, and the vegetation cover repeatedly over time. An extreme case is the Netherlands, where almost no natural landscapes are left (see pages 183 and 216–217). Environmental issues in Europe often focus less on setting aside pristine natural areas (of which there are few) than on establishing livable environments for future generations.

Public awareness about the environment has increased over the past 30 years, and in all European countries there are **Green** (environmentally conscious) political parties that influence policies at the national level as well as within the EU. In many ways, European lifestyles result in lower resource consumption than those elsewhere in the developed world, especially North America. Europeans live in smaller spaces, their yards often contain vegetable gardens rather than great expanses of mowed lawns, their cars are smaller and more fuel efficient, public transportation is widely used, and people walk or bike to many of their appointments. Nevertheless, these practices are related more to high population density and social customs in the region than to widespread explicit support for Green principles. Despite the successes of Green politicians, their concerns consistently take a backseat to issues of economic growth and social welfare. Europe's air, seas, and rivers remain some of the most polluted in the world.

Pollution of the Seas

Europe is surrounded by seas: the Baltic Sea, the North Sea, the Atlantic Ocean, the Mediterranean Sea, and the Black Sea. Any pollutants that enter Europe's interior rivers, streams, and canals eventually reach those surrounding seas (Figure 4.15). The Atlantic Ocean and the North Sea are able to disperse most pollutants dumped into them along European shores because they are closely connected to the circulating flow of the world ocean. In contrast, the Baltic, Mediterranean, and Black seas are nearly landlocked bodies of water that do not have the power to flush themselves out quickly: all three are prone to accumulate pollution. The countries surrounding these seas are at different stages of development, which makes it difficult for them to cooperate on solutions.

The Mediterranean Sea is a receptacle for multiple pollutants. Municipal and rural sewage (75 percent of it untreated), eroded sediments, agricultural chemicals, industrial waste, nuclear contamination, and oil spills are pouring into the water from adjacent lands. Many of the chemical pollutants add nitrogen to the water, which causes vast tracts of algae to bloom. These algal blooms render the sea inhospitable to natural life-forms and make some beloved seaside resorts unsafe for swimmers.

Pollution in the Mediterranean is exacerbated by the fact that it has just one tiny opening to the world ocean. About 80 years are required for a complete exchange of water to flush out pollutants. Atlantic seawater flows in through the narrow Straits of Gibraltar and moves eastward, evaporating as it goes. Evaporation concentrates the natural salts in the water and, as a result, the water at the far eastern end of the Mediterranean is more saline and heavier than the water entering from the west. This heavier water sinks,

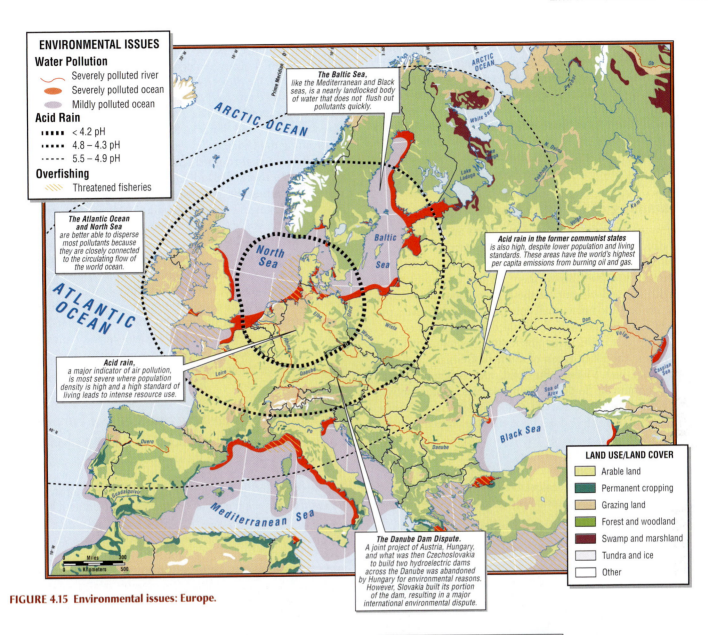

ENVIRONMENTAL ISSUES

Water Pollution
~~~ Severely polluted river
⬬ Severely polluted ocean
⬭ Mildly polluted ocean

Acid Rain
▪▪▪▪ < 4.2 pH
▪ ▪ ▪ 4.8 – 4.3 pH
- - - 5.5 – 4.9 pH

Overfishing
⫽⫽⫽ Threatened fisheries

The Baltic Sea, like the Mediterranean and Black seas, is a nearly landlocked body of water that does not flush out pollutants quickly.

The Atlantic Ocean and North Sea are better able to disperse most pollutants because they are closely connected to the circulating flow of the world ocean.

Acid rain in the former communist states is also high, despite lower population and living standards. These areas have the world's highest per capita emissions from burning oil and gas.

Acid rain, a major indicator of air pollution, is most severe where population density is high and a high standard of living leads to intense resource use.

The Danube Dam Dispute. A joint project of Austria, Hungary, and what was then Czechoslovakia to build two hydroelectric dams across the Danube was abandoned by Hungary for environmental reasons. However, Slovakia built its portion of the dam, resulting in a major international environmental dispute.

LAND USE/LAND COVER
▢ Arable land
▢ Permanent cropping
▢ Grazing land
▢ Forest and woodland
▢ Swamp and marshland
▢ Tundra and ice
▢ Other

FIGURE 4.15 Environmental issues: Europe.

flows back westward at a lower depth, and finally exits the Mediterranean at Gibraltar, many decades later. The ecology of the Mediterranean is attuned to this lengthy cycle, but the balance has been upset by the 320 million people now living in countries surrounding the sea.

Relatively rich and industrialized Europe shares the Mediterranean Basin with relatively poor North Africa and Southwest Asia. Although most of the water pollution is generated by Europe, populations to the south and east are increasing rapidly, and soon their sheer numbers will also pose an environmental threat. In 1995, in an effort to address water pollution problems, the EU inaugurated a cooperative effort with its southern neighbors to decide what kind of development is feasible for the welfare of the Mediterranean region. The developing countries of North Africa and the eastern Mediterranean, however, argue that they are being asked to make sacrifices for the environment that Europe was unwilling to make during its development era. They say that Europe has sounded the alarm only *after* Europeans have substantially polluted the water.

FRONTLINE WORLD™ VIDEO VIGNETTE
stories from a small planet

Belen Piniero: "All of us who have been cleaning these beaches, we know that even if we get everything clean, our land will be dead for many years. Maybe our children will be able to enjoy the beaches again. I don't know," she continued, but stopped short as tears welled in her eyes. "Sorry, but I don't want to talk any more."

It's December of 2002. Belen is one of hundreds of volunteers who wade through the thick black oil that covers everything for 350 miles of Spanish coastline. Millions of fish and birds are smothering in the gunk; dead and dying animals lie all about. Help is slow to come, so the people are inventing their own methods for cleaning the poisonous oil by hand. It is a hopeless and depressing task, and nerves are frayed.

In November 2002, a Greek-owned oil tanker, the *Prestige,* had broken apart in a heavy storm and sunk, spilling 20 million gallons of highly toxic crude oil (two times the amount dumped by the *Exxon Valdez* off Alaska in 1989). For the people of the

United States, this oil spill was a blip in the nightly news, coming during the buildup for war in Iraq. For the people of Galicia, the oil spill was an enduring nightmare. Was the disaster unavoidable, just one of those unfortunate times when a ship gets caught in a storm?

Watch the FRONTLINE/World video "The Lawless Sea" to learn the facts.

The Damming of Europe's Rivers

Europe's rivers are home to the region's oldest and largest cities, which today use those rivers both as a depository for municipal waste and as a source of energy. Many of Europe's rivers have hydroelectric dams built across them. It has been difficult to manage the use and cleanup of these rivers rationally because Europe is divided into many small countries, and many rivers run through, or form the borders of, several different countries. The Danube River is a particularly interesting environmental case study. It rises in the Black Forest of southern Germany and flows east and south past Vienna, Budapest, and Belgrade, passing through or forming the border of seven countries before emptying into the Black Sea. Before the fall of the Soviet Union, Austria, Hungary, and what was then Czechoslovakia had agreed jointly to build two hydroelectric dams across the river; these dams would have worked in tandem to flush huge volumes of water through electricity-generating turbines twice daily. The project was pushed ahead without an accurate calculation of the environmental impacts along what had been a relatively unspoiled part of the river.

Then, in the 1990s, the political climate in Eastern Europe shifted, and in Hungary, remarkably, a new reformist government withdrew from the project, dismantled the newly built Hungarian portion of the dams, and even partially restored the landscape. Meanwhile, Czechoslovakia had split into two countries, the Czech Republic and Slovakia. Slovakia persevered in the project by diverting the Danube and sending it through a canal and turbines. The original riverbed has dried up, and species that once occupied the river's wetlands are declining. Agricultural chemicals are polluting the reservoir created upstream of the Slovak dam and destroying its aquatic life. Although Hungary sued Slovakia in the World Court, the court eventually refused to make a judgment, ruling that the two countries would have to settle the case between themselves. Hungary argues that shared resources, such as rivers, can be exploited only by mutual consent, and that pollution of a common resource constitutes a breach of human rights. Slovakia cites customary European law that gives upstream users water rights over downstream users, a legal concept that has led to many a Wild West movie plot in the United States and to tensions in several parts of the world where this European concept of water rights applies.

Air Pollution in Eastern and Central Europe

Europe produces about a quarter of the world's carbon dioxide emissions by burning fossil fuels. Although there is significant air pollution across much of Europe, it is particularly heavy over the North European Plain. This is a region of heavy industry, dense transport routes, high population densities, and affluent lifestyles,

PRACTICING GEOGRAPHY   COMPARING LOCAL LANDSCAPES

Nuclear power in everyday lives. Nuclear power plays a role in everyday lives in many parts of the world. Here are two pictures related to issues of nuclear contamination. In Photo A, a bride-to-be is picking flowers for her wedding. Describe what you see and comment on the mood and the undercurrent of meaning conveyed. Where do you think this picture was taken? Which countries in Europe are likely locations? Photo B tells another story related to nuclear power. Here, a family man in Central Asia plays with his children and tries to distract himself from concerns about his work in a nuclear power plant. What sort of concerns is he likely to have, given the situation depicted here? [A, Joe Klamar/AFP/Getty Images; B, Paul Lowe/Panos Pictures.]

all of which lead to intense fossil fuel use and the production of CO_2 and other airborne pollutants that cause acid rain. Acid rain of different intensities is a threat over wide areas of Europe, even in East and Central Europe, where the population is not so dense and living standards are much lower (see Figure 4.15).

According to geographer Brent Yarnal, the high level of air pollution in the former communist states of East and Central Europe is due to the world's highest per capita emissions from burning oil and gas. Under Soviet influence, those countries used energy at per capita rates averaging 50 to 150 percent higher than those in the United States (one of the world's largest per capita users of energy). In Poland in the late 1980s, air quality in the industrial centers was so poor that young children were relocated to rural environments to preserve their health. After the breakup of the Soviet Union, reformers thought that market economies would improve energy efficiency and reduce emissions in the former Soviet bloc countries, and in Bulgaria, the least efficient and most polluting of the lot, pollution did decrease briefly after 1990. Yarnal found, however, that in Bulgaria and elsewhere, there is likely to be a long lag between economic reforms and the reduction of air pollution. The reasons for this lag are several: government subsidies for energy use encourage wasteful practices, reformist groups that might push for environmental improvements are not well organized, and antipollution activists find it hard to mobilize public support for regulations that might slow economic growth. Nonetheless, the public obviously has a stake in reducing pollution: it causes declines in public health, lowers crop yields, and affects water quality and availability. Stopping government subsidies that lower the costs of fossil fuels to consumers and encourage wasteful use is an important step. In the early 1990s, the World Bank had some success with this idea in six countries in East and Central Europe: Bulgaria, the Czech Republic, Slovakia, Hungary, Poland, and Romania. By 1999, all but Hungary had succeeded in further emissions reductions.

Why are environmental problems so serious in the countries of East and Central Europe? Part of the reason may be the Marxist theories and policies promoted by the Soviet Union, which portrayed nature as existing only to serve human needs. The Soviet leader Joseph Stalin once said, "We cannot expect charity from Nature. We must tear it from her." Another reason may be that, with only superficial democratic institutions in place during the Soviet era, there was little opportunity for public outrage at pollution to be channeled into constructive political activism and change.

MEASURES OF HUMAN WELL-BEING

We have already observed that there is considerable disparity in wealth across Europe. Table 4.1 compares the well-being of people in Europe with that of people outside the region, in the United States, Japan, and the United Arab Emirates. The three indices are those introduced in Chapter 1 and used throughout this book.

In the category of gross domestic product per capita (see Table 4.1, column 2), you can see that most countries in North and West Europe compare favorably with the United States, Japan, and the UAE. The only exceptions are Estonia, Latvia, and Lithuania, which separated from the Soviet Union in the early 1990s. These three countries have notably lower per capita GDPs than their neighbors in North Europe. South Europe is less affluent than either North or West Europe, and its per capita GDPs hover around $20,000. The countries of East and Central Europe are, for the most part, way down the scale in GDP per capita, with the notable exception of Slovenia. On this scale, several of these countries fall well below several Middle and South American countries (see Table 3.3 on pages 154–155). Yet despite these low GDP per capita figures, there has not been great disparity in wealth between classes in East and Central Europe until recently, because the communist system specifically sought to even out the distribution of wealth. As a result, GDP per capita figures for this region are a somewhat more meaningful measure of real well-being than they are in Middle and South America and elsewhere, where class disparities are huge. Even in the relatively wealthy country of Slovenia, the richest 20 percent of the population is just four times more wealthy than the poorest 20 percent. In Middle and South America, the richest 20 percent can be 20 to 30 times as wealthy as the poorest 20 percent.

The United Nations Human Development Index (HDI) (see Table 4.1, column 3) combines three components—life expectancy at birth, educational attainment, and adjusted real income—to arrive at a ranking of 175 countries that is sensitive to more factors than just income. In many parts of Europe, state subsidies keep the prices of necessities low and make salaries go further. All countries in North, West, and South Europe (with the exception of Estonia, Latvia, and Lithuania) rank in the top 24 on the global HDI. Iceland, Norway, and Sweden are at the top, undoubtedly because these countries have comprehensive social welfare systems. In contrast, the countries of East and Central Europe rank relatively low (from 29 to 95) because, as the communist system failed, most state-run firms closed or cut jobs, social welfare programs went bankrupt, and environmental quality declined. People lost not only their jobs, but also their health care, housing, and food subsidies. Those retaining their jobs saw their salaries rendered meaningless pittances as a result of high inflation rates. Air and water pollution increased, personal stress over uncertainties mounted, overall quality of life declined drastically, and life expectancies plummeted.

The Gender Empowerment Measure (GEM) (see Table 4.1, column 4) ranks countries by the extent to which females have opportunities to participate in economic and political life. It evaluates how women are employed in the economy and the extent to which women participate in society, especially as elected officials. The GEM figures show that North and West Europe, with the notable exceptions of Estonia, Latvia, and Lithuania, do better than most countries in empowering women. Estonia, Latvia, and Lithuania rank low in comparison because they were long part of the Soviet Union and retain many customs and institutions that restrict women. In South Europe, the GEM rankings are lower than those of North and West Europe. Spain, Portugal, and Italy are changing rapidly as they enjoy an economic boom. Women,

TABLE 4.1 Human well-being rankings of countries in Europe and other selected countries

Country (1)	GDP per capita, adjusted for PPP[a] in 2001 $U.S. (2)	Human Development Index (HDI) global rankings, 2003[b] (3)	Gender Empowerment Measure (GEM) global rankings, 2003[c] (4)
Selected countries for comparison			
Japan	25,130	9 (high)	44
Kuwait	18,700	46 (high)	75[d]
United Arab Emirates	—	48 (high)	65
United States	34,320	7 (high)	10
North Europe			
Denmark*	29,000	11 (high)	4
Estonia*	10,170	41 (high)	33
Finland*	24,430	14 (high)	5
Iceland	29,990	2 (high)	1
Latvia*	7,730	50 (high)	30
Lithuania*	8,470	45 (high)	48
Norway	29,620	1 (high)	2
Sweden*	24,180	3 (high)	3
West Europe			
Austria*	26,730	16 (high)	7
Belgium*	25,520	6 (high)	15
France*	23,990	17 (high)	31[e]
Germany*	25,350	18 (high)	8
Ireland*	32,410	12 (high)	16
Luxembourg*	53,780	15 (high)	14 (1998)
Netherlands*	27,190	5 (high)	6
Switzerland	28,100	10 (high)	13
United Kingdom*	24,160	13 (high)	17
South Europe			
Cyprus*	21,190	25 (high)	34
Greece*	17,440	24 (high)	40
Italy*	24,670	21 (high)	32
Malta*	13,160	33 (high)	—
Portugal*	18,150	23 (high)	21
Spain*	20,150	19 (high)	14

who had long been overeducated for the lower-echelon jobs they held, are being encouraged to take new, more responsible positions; hence, the GEM rankings for these countries have risen and are likely to rise further. Nevertheless, Greece has one of the lowest rankings in Europe. In much of East and Central Europe, useful statistics have not been collected, in part because of underfunded statistics departments and in part because the problem of gender inequity has gone unrecognized. Specific studies report that, compared to men, a disproportionate number of even well-educated young women in East and Central Europe are unemployed, while their older female relatives are forced into early, inadequately funded retirement (see page 229).

TABLE 4.1 (Continued)

Country (1)	GDP per capita, adjusted for PPP[a] in 2001 $U.S. (2)	Human Development Index (HDI) global rankings, 2003[b] (3)	Gender Empowerment Measure (GEM) global rankings, 2003[c] (4)
East and Central Europe			
Albania	3,680	95 (medium)	—
Bosnia and Herzegovina	5,970	66 (medium)	—
Bulgaria	6,890	57 (medium)	43[e]
Croatia	9,170	47 (high)	36
Czech Republic*	14,720	32 (high)	28
Hungary*	12,340	38 (high)	41
Macedonia	6,110	60 (medium)	—
Poland*	9,450	35 (high)	25
Romania	5,830	72 (medium)	53
Serbia and Montenegro	2,370[f]	—	
Slovakia*	11,960	39 (high)	24
Slovenia*	17,130	29 (high)	27
World	7,376		

[a]PPP = purchasing power parity.
[b]The high and medium designations indicate where the country ranks among the 175 countries classified into three categories (high, medium, low) by the United Nations.
[c]Rankings are for 2003 unless otherwise noted.

[d]Ranking for 1996.
[e]Ranking for 1998.
[f]2002 estimate (PPP), *CIA World Factbook*, 2003.
*Designates a member of the European Union.

Source: United Nations Human Development Report 2003.

Quick Review

1. What are some of the advantages for EU members that adopt the euro?

2. How does restricting trade of GMOs benefit the EU economically?

3. Distinguish between acculturation and assimilation. Which of these is expected of immigrants to Europe?

4. Which type of social welfare system provides a minimum standard of living for all citizens while not accepting the obligation to support the personal development of all citizens?

5. At present, does most of the pollution in the Mediterranean flow from Europe or from North Africa and Southwest Asia?

III   SUBREGIONS OF EUROPE

In this section we look at the subregions of Europe as separate units, examining in greater detail some of the issues that affect them. The subregion in which a particular country is included is based on location, on convention, and on our best judgment of the situation in Europe at this time. Good arguments for other choices could be made.

Every subregion of Europe is undergoing changes. These changes result from the end of cold war politics, the struggles of East and Central European countries to establish national identities, moves toward greater economic and political cooperation throughout the region, and global shifts in economic and political power.

WEST EUROPE

Despite their economic success, the countries of West Europe (Figure 4.16)—the United Kingdom, the Republic of Ireland, France, Germany, Belgium, Luxembourg, the Netherlands, Austria, and Switzerland—are confronting several issues that will affect their development and overall well-being:

1. The continuing trend in Europe toward economic and political union

FIGURE 4.16 The West Europe subregion.

2. Long-standing conflicts, such as that in Northern Ireland, and social tensions, such as those generated by the influx of immigrants from outside the subregion

3. The persistence of high unemployment despite general economic growth

4. The need to increase the global competitiveness of the subregion's agricultural, industrial, and service sectors while at the same time maintaining costly, but highly valued, social welfare programs

Benelux

Belgium, the Netherlands, and Luxembourg—often collectively called the Low Countries or Benelux—are all densely populated countries that have achieved very high standards of living. These three countries are well located for trade: they lie close to North Europe and the British Isles and are adjacent to the commercially active Rhine Delta and the industrial heart of Europe. The coastal location of Benelux and its great port cities of Antwerp, Rotterdam, and Amsterdam give these countries easy access to the global marketplace. The Benelux countries have long played a central role in the European Union. Brussels, Belgium, is considered the EU capital, with most EU headquarters located there, and this EU activity draws other business to Brussels.

International trade has long been at the heart of the economies of Belgium and the Netherlands. Both were active colonizers of tropical zones: Belgium in Africa and the Netherlands in the Caribbean and Southeast Asia. Their economies benefited from the wealth extracted from their colonies and from global trade in tropical products such as spices, cacao beans, fruit, wood, and minerals. Private companies based in Benelux still maintain advantageous relationships with the former colonies, which supply raw materials for European industries. The Benelux countries occasionally find themselves embroiled in conflicts related to both their colonial past and pressures to integrate with their neighbors in the new EU era; the European "chocolate standards war" is a case in point.

The Chocolate Standards War. In Europe, chocolate is an important luxury product that sells at very high prices. Godiva, for example, a brand loved by American chocoholics, is a Belgian company. Chocolate first enters Europe in the form of dried cacao beans, which are grown mostly by small producers in Africa. Belgium, France, and six other EU nations agree that real chocolate (so labeled) must be made only of cocoa butter derived from cacao beans, with no oil or fat additives. But the United Kingdom, Denmark, and several other countries routinely add cheaper oils to their mass-produced candy. The EU Commission, charged with defining common manufacturing standards for the entire EU, is trying to reconcile these different national chocolate traditions. Some contend that with open borders, the cheaper chocolate made with oil additives will have an unfair advantage over pure chocolate, both in the EU and abroad. Globally, six multinational firms, controlling well over half of all chocolate sales, side with the United Kingdom in arguing that chocolate recipes should not be standardized. The purists, on the other hand, argue that European chocolate will be more competitive in global markets if it is a gourmet product that maintains the highest standards of purity.

This European chocolate debate has global implications. The West African countries that produce cacao beans favor high standards of purity and the prohibition of cheaper oils in manufactured chocolate products. Ghana, Cameroon, and Côte d'Ivoire are former European colonies that depend on cacao for

These two photographs of the Netherlands, taken from the air, show intensive human use and manipulation of natural land and water interactions in coastal zones. The photograph at left shows part of the port of Rotterdam, looking northwest. Cranes transfer products between river vessels and oceangoing ships. The oil "tank farms" figure prominently in Dutch chemical industries. The photograph at right shows fields of flower bulbs on raised, drained lands close to the coast. Economically, the flowers, which are sold locally, are merely a by-product of the bulbs, which are marketed internationally. Dunes along the coast, planted with trees for stabilization, are the last defense against the sea. [Aeroview, Rotterdam.]

at least a quarter of their export earnings. They estimate that the use of even as little as 5 percent non-cocoa oils in the EU will mean a 10 percent drop in world cacao consumption and will devastate their economies. These three African countries have petitioned their former colonizers to maintain the pure chocolate standards in accordance with agreements they made to help African economies rebound from colonial exploitation. The chocolate story is taken up again in Chapter 7, where it is noted that African producers have little control over the market in which they must operate.

The Netherlands: A Human-Made Place. The largest Benelux nation, the Netherlands, is particularly noted for having reclaimed land that was previously under the sea (Figure 4.17). Today, its landscape is almost entirely a human construct. As populations grew during and after the Middle Ages, people created more living space by filling in a large natural coastal wetland. To

protect themselves from devastating North Sea surges, they built dikes, dug drainage canals, pumped water with windmills, and constructed artificial dunes along the ocean. Today, a train trip through the Netherlands between Amsterdam and Rotterdam takes one past raised rectangular fields crisply edged with narrow drainage ditches and wider transport canals. The fields are filled with commercial flower beds of crocuses, tulips, daffodils, and hyacinths (see the photo at the right on page 215); there are also grazing cattle and vegetable gardens, which feed the primarily urban population. Despite the almost complete absence of pockets of wilderness, the landscape has an open, bucolic ambience. In one vista, a train traveler can see huge cranes in the distance at the port of Amsterdam. To one side are high-tech office buildings and a new satellite town. In the foreground, one might see an elderly woman pushing her bicycle beside a sparkling blue, reed-lined canal, the rear basket stuffed with leeks and flowers.

FIGURE 4.17 Land reclamation areas in the Netherlands. The Netherlands is noted for having a landscape that is almost entirely human-made. As populations grew, people created more living space by filling in a large natural coastal wetland. Filling in wetlands is now prohibited in many parts of the world because of the known negative environmental impacts. [Adapted from William H. Berentsen, *Contemporary Europe: A Geographic Analysis* (New York: Wiley, 1997), p. 317.]0

The Netherlands has 15 million people, who enjoy a high standard of living. It is the most densely settled country in Europe (with the exception of Malta), and this density has consequences. There is no land left for the kind of high-quality suburban expansion preferred in the Netherlands without intruding on agricultural space. And there is not nearly enough space for recreation; bucolic as they appear, transport canals and carefully controlled raised fields are not a venue for picnics and soccer games. People now travel great distances to reach their jobs, and these long commutes add to air pollution and miserable traffic jams. For even weekend getaways, people usually leave the Netherlands for a neighboring country that has some more natural, less manicured areas. The choice to maintain agricultural space is largely a psychological and environmental one, as agriculture accounts for only 1 percent of both the GDP and employment.

France

France has the shape of an irregular hexagon bounded by the Atlantic on the north and west, mountains on the southwest and southeast, Mediterranean beaches on the south, and the lowlands of Belgium on the northeast (see Figure 4.16). The capital city of Paris lies in the north of this hexagon. It is the heart of the cultural and economic life of France, attracting worldwide admiration. Paris, as one of Europe's two world cities (London is the other), has an economy tied to the provision of highly specialized services (accounting, advertising, design, financial, scientific).

Paris is often called the City of Light because of the elegance of its architecture and urban design and its reputation as a world-class culinary, cultural, and, now, technology center. France's population is concentrated in Paris: together with its suburbs, the city has more than 11 million inhabitants, roughly a fifth of the country's total. It is also the hub of a well-integrated water, rail, and road transport system. Its central location and transport links attract a disproportionate share of trade and businesses. Firms that manufacture luxury and high-fashion items, high-tech and automotive industries, and distributors and processors of France's fine farm produce draw a steady stream of workers. In the 1970s, however, French planners decided that Paris was large enough, and they began diverting development to other parts of the country, especially to Toulouse and farther south to the Mediterranean.

To the west and southwest of Paris, toward the Atlantic, are less densely settled lowland basins that are used primarily for agriculture. France has the largest agricultural output in the EU and, globally, is second only to the United States in agricultural exports. Its climate is mild and humid, though drier to the south. Throughout the country, farmers have found profitable specialties: wheat (France is ranked fifth in world production), grapes (the Bordeaux region is a leading exporter of French wines), cheese (second in world production), fruit, olives, sugar beets and other vegetables, and sunflower seed oil (third in world production). Despite France's leadership in agricultural production, agriculture accounts for only 3 percent of the national GDP and employs only 3 percent of the population. This paradox of agriculture being highly productive while its contribution to the national economy and employment is decreasing needs explaining. Modernization has made ever higher agricultural production possible with ever lower labor inputs; meanwhile, the manufacturing sector quickly grew very large, and now the high-end service sector is growing at leaps and bounds, both in France and throughout western Europe. French industry now accounts for 26 percent of both GDP and employment, and services account for 71 percent.

The Mediterranean coast is the site of France's leading port, Marseille, and of the famous French Riviera. Here, development is booming. Tourism, in particular, has inspired the building of marinas, condominiums, and parks that threaten to occupy every inch of waterfront. The downside of this wealthy, densely populated region is that it produces large amounts of the pollutants that contribute to Mediterranean environmental problems (see the discussion on pages 208–209).

France derived considerable benefit and wealth from its large overseas empire, which it ruled until the mid-twentieth century. There were French colonies in the Caribbean, North America, North Africa, sub-Saharan Africa, Southeast Asia, and the Pacific. Many of the citizens of these former colonies now live in France and bring to it their skills and a multicultural flavor. Oddly enough, the French take pride in being cosmopolitan, but they are also very protective of their distinctive European culture and wish to guard what they consider to be its purity and uniqueness. The French expect immigrants to assimilate to French culture, even to the point of forcing people to give up religious dress (see the photo on the left on page 204).

Few would quarrel with the idea that French culture has a certain cachet and that France is an arbiter of taste. People of many different cultural backgrounds accept the idea that the French language is particularly refined. French style as exemplified in clothing, jewelry, perfume, and interior design is highly pursued, and French culinary arts remain among the most cultivated of cooking traditions. Its many cultural attributes have made France the leading tourist destination on earth, with 11 percent of global tourism arrivals. In 2002, France attracted 77 million international tourists, more than the population of the entire country (59.8 million). Other Europeans, Turks, people from the Americas, and Japanese form the majority of tourists in France.

France's strong cultural identity is both a blessing and a curse. Many people in France are deeply concerned about what they perceive as a decrease in France's world prestige as its industrial competitiveness declines, unemployment rises, social welfare benefits decrease, and French culture is diluted by an influx of migrants from former colonies. Such feelings have resulted in increasing popular support for the National Front, France's xenophobic right-wing party. But others in France are searching for a multicultural identity in the potentially more egalitarian global community that France helped to create by being a model for democracy. From a purely practical point of view, because it has a low birth rate, France needs immigrants to keep its economy humming, its technology on the cutting edge, and its tax coffers full.

Germany

The most famous image of Germany in recent times is that of a happy crowd dismantling the Berlin Wall in 1989. This important symbol of the end of Soviet influence in East and Central Europe had particular significance for Germany, which for 40 years had been divided into two unequal parts. Russian troops occupied the smaller, eastern third of Germany at the end of World War II; by 1949, the Soviets and their German counterparts had created East Germany in this area. Over the next decade or so, perhaps as many as 3 million East Germans fled to the other part of the country, now called West Germany. To retain the remaining East German population, the Soviets literally walled off the border in the summer of 1961. East German skilled labor, mineral resources, and industrial capacities were used to buttress the socialist economies of the Soviet bloc and to support the military aims of the Soviet Union and its allies.

Because Germany was regarded as the perpetrator of two world wars, West Germany had to tread a careful path after 1945: seeing to its own economic and social reconstruction and rebuilding a prosperous industrial base, yet not seeming to become too powerful economically or politically within the European community. For the most part, West Germany played this complex role successfully. Since the early 1980s, it has been a leader in building the European Union, and with the EU's largest economy, it has borne the greatest financial burden of the European unification process.

After the fall of the Berlin Wall, the two Germanys were reunited. But East Germany came home with a suitcase full of troubles that are expensive to fix. Its industries were outdated, inefficient, polluting, and, in large part, irredeemable. Its infrastructure of roads, bridges, dams, hydroelectric plants, nuclear energy plants, waterways, and buildings did not meet the standards of West Europe, and much of it had to be remodeled or dismantled and replaced. Its industrial products were not competitive in world markets, and East German workers, though considered highly competent in the Soviet sphere, were undereducated and underskilled by West European standards. West German taxpayers must foot the cost of addressing these problems.

Reunified Germany is Europe's most populous country (83 million people) and is a global leader in industry and trade, but the costs of absorbing the poor eastern zone have dragged Germany down in many rankings within Europe. Unemployment rose sharply (especially among women) in the 1990s, then declined to between 9 and 10 percent in the early 2000s. Not only have there been layoffs in the east (where unemployment is nearly double the national average), but German workers, accustomed to high wages and generous benefits, are losing jobs as firms mechanize or move overseas, some to the United States.

The German multinational corporation DaimlerChrysler, headquartered in Stuttgart, is an example of a German firm that has moved operations out of Germany. In the late 1990s, Daimler-Benz bought Chrysler, an American auto company based in Michigan, and built a Mercedes-Benz plant in Tuscaloosa, Alabama. In 2000, it announced plans to expand the Alabama plant and add 2000 jobs. By 2001, globally, Daimler-Chrysler had 467,000 employees (241,000 in Germany, 124,000 in the United States, 12,200 in Mexico City). Then, in 2003, the company reduced its total workforce by 100,000 while increasing its activities in central Europe (Poland, Hungary) and Asia (India,

Japan, Korea, China, Taiwan, Malaysia), not just because factory costs are lower in those places, but also because the market for luxury cars such as Mercedes-Benz is likely to soar in Asia.

The United Kingdom of Great Britain and Northern Ireland, and the Republic of Ireland

The British Isles, located off the northwestern coast of the main European peninsula, are occupied by two countries: (1) the United Kingdom of Great Britain (England, Scotland, and Wales) and Northern Ireland, often called simply Britain or the United Kingdom (U.K.), and (2) the Republic of Ireland. The Republic of Ireland (not to be confused with Northern Ireland) was once a colony of Britain, and the two countries have long held contrasting positions in the world. Britain is a powerful industrialized country that has experienced a decline in its global influence and wealth. Ireland is a once poor agricultural country that now has one of Europe's fastest-growing (but still small) economies.

Ireland. The only physical resources in the Republic of Ireland are its soil, abundant rain, and beautiful landscapes. This combination has contributed to a continued dependence on agriculture and tourism, while industrialization has lagged. As a result, the Irish people were for a long time the poorest in West Europe; and over the last two centuries some 7 million Irish migrated to find a better life. In the 1990s, however, a remarkable turnaround began. Ireland attracted foreign manufacturing companies by offering cheap, well-educated labor, accessible air transport, access to EU markets, low taxes, and other financial incentives. These foreign firms (such as Glaxo, Merck, Norsk Hydro, and Phillips Petroleum) brought in small industries with specialties such as training software, food and drink processing, and the manufacture of pharmaceuticals, chemicals, and high-end giftware and textiles. The economy grew so quickly that Irish labor was in short supply. To supply the 200,000 workers required, Ireland invited its diaspora back, and other foreign workers were recruited in many locales. In 2000, 42,000 people immigrated to Ireland, 18,000 of whom were returning emigrants. Now, Czech workers are packing meat in Ireland, and Filipino nurses are working in most Irish hospitals. The cost of living has risen sharply, however, and this rise could affect the country's ability to attract yet more foreign investment. Ireland did not favor the admission of East and Central European countries into the EU because it feared that those countries would attract investment away from Ireland by offering even cheaper labor pools.

Northern Ireland began to emerge in the seventeenth century, when Protestant England conquered the whole of Catholic Ireland. England removed or killed many of the indigenous people and settled Lowland Scots and English farmers on the vacated land. The remaining Irish resisted English rule with guerrilla warfare for nearly 300 years, until the Republic of Ireland gained independence from the United Kingdom in 1921. However, six counties in the northeastern corner of Ireland remained part of the United Kingdom and became known as Northern Ireland. Here Protestant majorities held political control, and the minority Catholics experienced economic and social discrimination. Catholic nationalists unsuccessfully lobbied for a united Ireland by constitutional means. Other Catholic groups, the most radical being the Irish Republican Army (IRA), resorted to violence, often against British peacekeeping forces, who were seen as supporting the Protestants. The Protestants reciprocated with more violence. More than 3000 people had been killed in the Northern Ireland conflict by 1995.

A peace accord was finally reached in 1998. Both of the opposing groups were tired of seeing the development of the entire island—especially of Northern Ireland—blighted by the persistent violence. In May 1998, voters in both Northern Ireland and the Republic of Ireland overwhelmingly approved the peace accord (71 percent in Northern Ireland, more than 90 percent in the Republic of Ireland). Although voter approval is a major step forward, problems of implementation remain. Since 1998, violence has periodically risen and then abated as elements within the IRA have refused to disarm.

The United Kingdom. The United Kingdom, in contrast to Ireland, has been operating from a position of power for many hundreds of years. Like Ireland, the United Kingdom has a mild, wet climate and a robust agricultural sector—based, in its case, on grazing animals. The usable land is extensive, and the mountains contain mineral resources, particularly coal and iron. Beginning in the seventeenth century, however, Britain no longer depended on just its own resources. After colonizing Ireland, Britain extended its empire to the Americas, Africa, and Asia. These colonies gave it access to enormous resources of labor, agricultural products, minerals, and timber. These foreign resources plus its own were sufficient to make Britain the leader of the industrial revolution in the early eighteenth century. By the nineteenth century, the British Empire covered nearly a quarter of the earth's surface. As a result, British culture was diffused far and wide, and English became the lingua franca of the world (displacing French).

Britain's widespread international affiliations, set up during the colonial era, positioned it to become a center of international finance as the global economy evolved. Today, London is Europe's leading financial center, and it handles 31 percent of global foreign currency exchange, more than twice as much as New York City, its closest rival. But although the United Kingdom remains Europe's financial center, it is no longer Europe's industrial leader. At least since World War II, and some say earlier, the United Kingdom has been sliding down from its high rank in the world economy toward the position of an average European nation. British manufactured goods are no longer competitive with goods from other world regions where labor and production costs are cheaper. The discovery of oil and gas reserves in the North Sea gave the United Kingdom a cheaper and cleaner source of energy for industry than the coal on which it had long depended, but it did not stop the economic decline. Cities such as Liverpool and Manchester in the old industrial heartland and Belfast in Northern Ireland experienced long depressions.

Beginning in the 1970s, a series of conservative governments tried to make the United Kingdom more competitive by instituting two decades of budget cutbacks, reducing social spending, selling off unprofitable government-run firms, and tightening education budgets. Although these measures were not so labeled, they were the same as the structural adjustment policies (SAPs) implemented in many other world regions. Unemployment rose. Eventually, foreign investment began to pour into the United Kingdom because the skills of its labor force were high compared to the wages workers would accept after years of cutbacks.

Like the Republic of Ireland, Britain has now successfully established technology industries in parts of the country that previously were not industrial—for example, near Cambridge and west of London in a region called Silicon Vale. But the service sector now dominates the economy (74 percent of both the GDP and the labor force). In the last decade, new, but not high-paying, jobs have been created in health care, food services, sales, financial management, insurance, communications, tourism, and entertainment. Even the movie industry has discovered that

"Mixed relationships are more accepted here [in London] than in the U.S.," says Joanne Evans as she spends a Sunday afternoon with her friend Neil Williams. Overall, Britain is a far more integrated society than the United States. Marriage across ethnic lines occurs often, and 74 percent of white Britons say they would not mind a close relative marrying someone of another color or ethnicity. [Jodi Cobb/National Geographic Image Collection.]

England is a cheap and pleasant place for film editing and production. Britain's past experience with a worldwide colonial empire has left it well positioned to provide superior financial and business advisory services (producer services) in a globalizing economy. These jobs, however, tend to go to young, educated, multilingual city dwellers, many of them from outside the United Kingdom, not to the middle-aged, unemployed ironworkers and miners left in the old industrial heartland. Because the EU allows the free movement of EU citizens, there are now more than 400,000 foreign nationals working in Greater London alone, many of them from other parts of Europe.

Politically, the United Kingdom consists of the countries of Scotland, Wales, England, and Northern Ireland. Until recently, it was governed by a single parliament with delegates from the four countries, but England, with the largest population by far, held 85 percent of the votes. In national debates, the will of the English majority usually prevailed. In 1997, Scotland and Wales voted to institute their own elected governing bodies—a parliament for Scotland and a national assembly for Wales. This decision for home rule is seen as an example of **devolution,** the weakening of a formerly tightly unified state. It springs both from dislike of the structural adjustment policies of recent U.K. conservative governments—which brought high unemployment and losses of social programs to both Scotland and Wales—and from rising feelings that the wishes of the voters in both countries were consistently overwhelmed by the English majority in the British parliament. It remains to be seen whether this change will result in the eventual independence of Scotland and Wales from the United Kingdom or merely the delegation of some powers to what amount to subassemblies within the United Kingdom. Much will depend on decisions within the EU regarding allocation of political power and development funds.

The Jamaican poet and humorist Louise Bennett used to joke that Britain was being "colonized in reverse." She was referring to the many immigrants from the former colonies who now live in the United Kingdom. Indeed, it would be hard to overemphasize the international spirit of the United Kingdom. London, for example, has become an intellectual center for debate in the Muslim world. Salman Rushdie, the controversial Indian Muslim writer, lives there; many bookstores carry Muslim literature and political treatises in a variety of languages; Israelis and Palestinians talk privately about peace; and the Saudi Arabian government has a strong presence. On a leisurely stroll through London's Kensington Gardens, you will discover thousands of people from all over the Islamic world who now live, work, and raise their families in London. They are joined by many other people from the British Commonwealth: Indian, Malaysian, African, and West Indian families pushing baby carriages and playing games with older children. Impromptu soccer games may have team members from a dozen or more countries (see the box "Soccer: The Most Popular Sport in the World" on page 203). A 2000 survey of ethnic diversity in Britain noted that Britons of African and West Indian heritage contribute £5 billion per year to the economy, and that over 200 languages are spoken in London daily.

SOUTH EUROPE

The Mediterranean region of South Europe was once unrivaled in wealth and power. First Athens in Greece and then Rome in Italy were imperial seats whose influence extended widely. In the medieval period, from the fifth to the fifteenth centuries, the Italian cities of Venice, Florence, and Genoa were major trading centers with contacts stretching across the Indian Ocean through Central Asia to eastern China. During the sixteenth century, Spain and Portugal developed large colonial empires, primarily in the Americas but also in Africa and Southeast Asia. But times changed. As Europe's own economy became more global with the development of colonies by England, France, and the Netherlands in distant world regions, the Mediterranean ceased to be a center of trade. South Europe declined, and the industrial revolution reached the subregion relatively late. Even today, much of the subregion remains poor in comparison to the rest of the continent: only the countries of East and Central Europe are poorer (see Table 4.1, pages 212–213). In this section, we focus on Spain and Italy, contrasting the parts of each country that are now prospering with the parts that have remained poor (Figure 4.18).

In all six countries of South Europe—Portugal, Spain, Italy, Malta, Cyprus, and Greece—agriculture was long the predominant occupation. Farmers often lived in poverty, using simple tools to produce crops for local consumption. Since the 1970s, however, there has been an effort to modernize and mechanize agriculture across the subregion. The focus has been on irrigating large holdings to grow commercial crops. Irrigation has increased output and opened up new semiarid locations for cultivation, but not everyone welcomes the abundance of produce flowing from South Europe. Today, both Spain and Italy mass-produce vegetables and fruits in quantities that exceed demand in the European Union. The surplus has driven down prices, hurting smaller family farmers across Europe and creating ill feelings toward the European Union (see the personal vignette about Vera Kuzmic on pages 228–229). As in the western United States, irrigation schemes result in water shortages, depletion of reservoirs and natural aquifers, and increases in soil salinity over time. Farmers in northern Spain and Portugal complain because water originating there is transported by canal to irrigate more arid regions in the south. And despite extensive improvement projects, agriculture as a whole (measured as a percentage of GDP and as a percentage of labor force) has declined in importance as manufacturing, tourism, and other services have increased.

For several decades, middle-class tourist money has brought new life to the economy of South Europe. Masses of western and northern Europeans flock to see the monuments of past civilizations and to enjoy the south's warm climate and beautiful beaches. But tourist money has done little to elevate the lives of poor rural people. Although some localities have grown prosperous as resort sites, many others have been bypassed. In Greece, for example, income from tourism has been confined to scenic coastal and island locations, and many interior rural areas remain

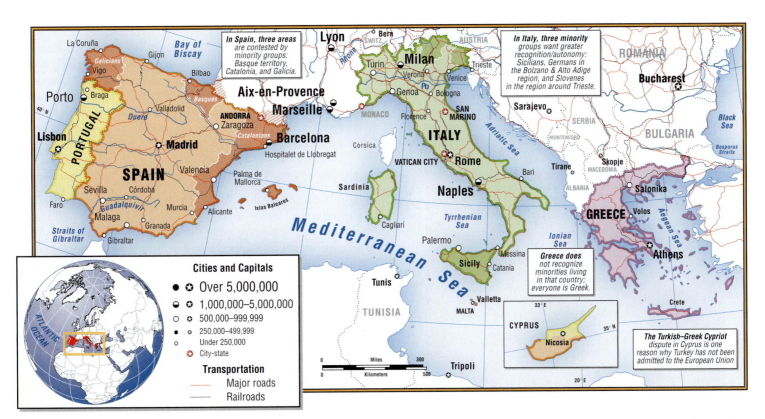

FIGURE 4.18 The South Europe subregion. The Mediterranean region of South Europe was once a major trading center unrivaled in wealth and power. Today, however, much of South Europe is poor in comparison to the rest of the continent.

poor and isolated. In much of South Europe, the very qualities that attract tourists are being threatened by the sheer numbers of visitors and the effects they have on coastal environments and on rural culture.

Spain

In the last quarter of the twentieth century, Spain emerged from centuries of underdevelopment. Today, it is poised to take advantage of opportunities in a more integrated Europe, and its future is bright. Just 40 years ago, however, Spain was known as the Old Man of Europe. By the beginning of the twentieth century, its once vast empire in the Americas and Asia was all but gone. This loss was followed by years of civil war and then a military dictatorship led by Francisco Franco. Even Spain's physical location was no longer an advantage, as commercial activity had shifted away from the Mediterranean to West Europe. Mountains (the Pyrenees) kept Spain isolated from the more prosperous parts of Europe, making travel and commerce difficult.

In 1975, Spain made a cautious transition to democracy, and in 1986 it joined the EU. Since then, growth has accelerated with the help of EU funds aimed at bringing Spain into economic and social harmony with the rest of the EU through investment in infrastructure and in human resources. As a result of these improvements, foreign businesses are investing in industry and factories have been modernized. The cities of Barcelona and Madrid are

now centers of population, wealth, and industry. Both are now linked to the rest of Europe by a network of roads, airports, and the high-speed Train à Grande Vitesse. Indeed, Spain's standing in Europe has so improved that one occasionally sees articles proclaiming it a world-class cultural center. Such was the case when on August 10, 2003, a *New York Times* piece began with the headline "Spain Is the New France." The author, Arthur Lubow, noted that chefs and wine connoisseurs across Europe, as well as art critics and movie analysts, were saying "Spain is rising, France is resting."

Nonetheless, some segments of Spanish society have not participated in this new creativity and prosperity. Many small farmers remain poor, particularly in the northwestern and southwestern provinces, which are arid and have few resources. Even in the relatively prosperous zones of Barcelona and Madrid, workers have been laid off as a result of factory modernizations. In the mid-2000s, unemployment hovered at 11.3 percent (Europe's highest) and affected women twice as much as men. Unemployed Spanish workers were willing to work for significantly lower wages than were paid in North and West Europe.

Spain's low wage scales are one reason that it has attracted foreign investment from around the world. Other attractions include democratic institutions able to withstand crises (such as the terrorist bombing in Madrid in March 2004), an educated and skilled workforce (partly the result of EU funds for training), and the absence of stringent environmental and workplace regulations. Additionally, Spain has good ports on both the Atlantic

Courtyard of the Parador de San Francisco in the Alhambra, Granada, Spain. Muslim traders and conquerors (known as Moors) came to southern Spain in the 700s and remained there until the 1400s. They deeply influenced all aspects of Spanish life: language, music, settlement patterns, architecture, principles of governance, and gender roles. [Mark D. Phillips/Photo Researchers.]

and the Mediterranean. Spain now has an industrial base that includes food and beverage processing and the production of automobiles and auto parts (Ford, DaimlerChrysler, Citroën, Opel, Renault, MDI, and Volkswagen all have factories in Spain), chemicals, metals, machine tools, and textiles. An interesting indicator of Spain's growth is its increasing investment in its former colonies in Latin America. Spanish banks have lent more than U.S. $7 billion to Cuba, Mexico, and various countries in South America. Between 1996 and 2000, Spanish corporations invested $14 billion in hotels, factories, utilities, and banks in Latin America. Spain's telecommunications giant, Telefonica, holds more than U.S. $5 billion in Latin American telecommunications companies.

Regional disparities of wealth in Spain contribute to high levels of ethnic rivalry. Two of Spain's most industrialized and now wealthiest regions—the Basque country in the central north (adjacent to France) and Catalonia in the east (also bordering France)—have especially strong ethnic identities. Their inhabitants often speak of secession as a response to long years of repression by the Spanish government; in addition, the Basques would like to avoid supporting Spain's poorer districts. A Basque separatist movement periodically resorts to violence. Meanwhile, Galicia, in the far northwestern corner of Spain, also has a strong cultural identity, and the preservation of the Galician language is a popular issue there. All three of these ethnic enclaves emphasize their distinctive languages as markers of identity and chafe against the use of Castilian Spanish as Spain's official language.

Italy

Italy is by far the wealthiest country in South Europe and is, in fact, the seventh largest economy in the world. Over the millennia, Italy has contributed greatly to European culture and prosperity. In Roman times, the Po Valley in the north was an uninhabited, mosquito-infested swamp, but the people of northern Italy eventually converted it into a productive agricultural region. By the thirteenth century, the northern trading center of Venice on the Adriatic had commercial links reaching around the Mediterranean and into Africa and Arabia, to the Malabar Coast of India, and all the way to China. Marco Polo, from a wealthy trading family in Venice, took a trip through Central Asia to China during the late 1200s and returned after some 20 years to give medieval Europeans their first impressions of life and commerce in China. The wider view of the world that Polo brought home contributed to the European Renaissance in northern Italy (see discussion on page 186), which, after the turbulence of the medieval period, revived optimism about the power of reason to improve the human condition. Wealthy families (such as the Medicis) patronized the arts and literature and invested in beautifying public and private spaces. Aided by the European version of the printing press (already in use in China), Italy spread its enlightened ideas about philosophy, art, and architecture throughout Europe. Italy's prominence faded, however, as Spain, England, Portugal, and the Netherlands grew rich on their colonies in the Americas and Asia. During the 1700s and 1800s, Italy was overrun by a long series of wars between France, Spain, and Austria.

Progress was rocky for Italy's ordinary people, who were buffeted by the political ambitions of politicians and religious leaders (the pope, the religious leader of the Catholic Church, remains headquartered in the Vatican in Rome). Wide disparities in wealth and power left whole sections of the country lawless and in decline. This was especially true of the southern half of the peninsula, Sardinia, and Sicily, which were collectively known as "the south." Absentee landlords and central government tax collectors from northern Italy hired henchmen (known as *mafiosi*) to collect fees from the cultivators who worked the land; meanwhile, the people of the south were given no chance for self-government. The mafiosi were the mechanism whereby the power structure in the industrializing north controlled and profited from the agricultural south.

Now, in the early twenty-first century, northern Italy has one of the most vibrant economies in the world, producing products renowned for their quality and design: Ferrari automobiles, Giorgio Armani clothing, Olivetti office equipment, and high-quality musical instruments. From Turin, Fiat automobiles pro-

Nestled within 10 miles of Mount Vesuvius, a volcano overdue for an eruption, is Naples, founded by Greeks in 600 B.C. as Neapolis. Many people consider it to be the gateway to southern Italy. This beautiful, vibrant city of more than a million inhabitants has many of the problems of the south: violence, drugs, and corruption. Yet it also houses what is perhaps Italy's finest opera house, luxurious palaces and gardens, and the Piazza del Plebiscito, a central square overlooking the Bay of Naples that serves as the background for thousands of wedding portraits for southern Italy's brides. [David A. Harvey/National Geographic Image Collection.]

duced in a fully robotic plant are shipped throughout Europe; in Milan, one of the largest cities in Italy, fashion designers have surpassed even their rivals in Paris. As elsewhere in Europe, the service sector, including such activities as sales, advertising, communications, banking, international design, and construction, is growing faster than industry. Tourism accounts for 67 percent of the GDP and 63 percent of the labor force. And yet, for all of Italy's stylish success as an industrial and services leader in Europe and the world, it could be yet more sleek and efficient. Estimates are that 20 to 25 percent of Italian economic activity is off the books and untaxed—part of the so-called informal economy.

Italy's regional disparities in wealth are apparent when one travels south of the city of Naples. Archaic agricultural practices, repressive politics, deforestation, eroding steep slopes, and an arid climate have made the south less and less productive over time. The government's episodic efforts to promote infrastructure development and industry are apparent, but across the landscape, unfinished roads and public buildings and empty factory hulks (called "cathedrals of the desert") attest to the difficulties these efforts have encountered. The multiple causes of poverty in Italy's south include the ways in which local people have been handicapped by the failure of government to provide education, stigmatized as backward and incapable of self-government, and plagued with misguided and corrupt development schemes imposed from outside.

Italy has been a democracy since World War II, and it is a charter member of the EU. Yet because of its domestic politics, Italy is known as the "Bad Boy" of Europe. This designation is to

be taken humorously and derives in large part from unfair stereotypes of Italians as clever operators on the fringes of legality. But it also derives from the fact that Italians vote governments in and out in rapid succession and are not above electing leaders who have been mixed up in shady dealings. In 2001, Prime Minister Silvio Berlusconi took office while under investigation for money laundering, perjury, and bribery. Others around him were found guilty, but Berlusconi managed to use his position as leader of the parliamentary majority to have legislation passed that helped his cause. Italians have devised ways of living with such official indiscretions without descending into national crises. During the dozens of governmental emergencies since 1950, a quasi-government based on informal relationships—the *sottogoverno*, or "undergovernment"—has taken over whenever a predicament has developed, and shortly daily life has proceeded apparently unimpaired.

Many of the peculiarities of Italy arise from its strong faith in institutions based on the family or on personal relationships. Most businesses are family-run, including several of the largest corporations of northern Italy. And, as discussed earlier in this chapter (see pages 207–208), the rudimentary welfare system found in Italy and other countries of South Europe assumes that strong families and personal networks will provide what the state cannot.

NORTH EUROPE

North Europe (Figure 4.19) traditionally has been defined as the countries of Scandinavia—Iceland, Denmark, Norway, Sweden, Finland—and their various dependencies. Greenland and the Faroe Islands are territories of Denmark; the island of Svalbard in the Arctic Ocean is part of Norway. North Europeans themselves are now including the three Baltic states of Estonia, Latvia, and Lithuania in their region, and so we do so here. All three were part of the Soviet Union until September 1991, and these small Baltic states have many remaining links to Russia, Belarus, and other parts of the former Soviet Union. Although their cultures are not Scandinavian, they are attempting to reorient their economies and societies in varying degrees to North Europe and the West.

The countries of North Europe are linked by their locations on the North Atlantic, the North Sea, and the Baltic Sea. Most citizens can drive to a seacoast within a few hours. The main cities of the region—Copenhagen, Oslo, Stockholm, and Helsinki—are vibrant ports that have long been centers of shipbuilding, fishing, and the warehousing of goods. They are also home to legal and financial institutions related to maritime trade.

Scandinavia

The southern parts of Scandinavia are where most people live and where economic activity is concentrated (see Figure 4.6 on page 191 and Figure 4.10 on page 197). These same warmer lowlands of southern Scandinavia are also dedicated to agriculture. Denmark, although a small country, produces most of North Europe's poultry, pork, dairy products, wheat, sugar beets, barley,

FIGURE 4.19 The North Europe subregion. The countries of North Europe are linked by their intimate connections with the sea.

and rye. Despite Denmark's agricultural successes, two-thirds of the Danish economy is nonagricultural, consisting of the service sector (financial services and education, for example), high-tech manufacturing, construction and building trades, and fisheries. Sweden is the most industrialized nation in North Europe. The Swedes produce most of North Europe's transport equipment and two highly esteemed automobiles, the Saab and the Volvo, as well as furniture and housewares famed for their spare, elegant design. Helsinki, Finland, is home to some of the world's most successful information technology companies, such as TietoEnator, Nokia, and Novo Group.

The northern part of Sweden and almost all of Norway are covered by mountainous terrain, while Finland is a low-lying land dotted with glacial lakes, much like northern Canada. Sweden and Finland contain most of Europe's remaining original forests. These well-managed forests produce timber and wood pulp. Norway, which has less usable forestland, had been considered poorly endowed with natural resources in comparison with

the rest of Scandinavia, but the discovery of gas and oil under the North Sea in the 1960s and 1970s has been a windfall to that country. The exploitation of these resources dwarfs all other aspects of Norway's economy. Norway is able to supply its own energy needs through hydropower, and it exports oil and gas to the rest of the EU. It is now one of Europe's wealthiest countries, with a high level of human well-being (see Table 4.1 on page 212).

The fishing grounds of the North Sea, the North Atlantic, and the southern Arctic Ocean are also an important resource for the countries of North Europe (as well as for several other countries on the Atlantic). In recent decades, however, overexploitation has severely reduced fish stocks. Rights of access to the fishing grounds remain a cause of dispute, but in 1994, the EU created a joint 200-mile (320-kilometer) coastal exclusive economic zone (EEZ) that ensures equal access and sets fishing quotas for member states. In January 2001, EU members plus Norway and Iceland agreed to an annual 3-month hiatus in cod fishing and in the taking of juvenile fish of all species to feed salmon in commercial farms. The agreement applies to a 40,000-square-mile (103,600-square-kilometer) swath of the North Sea lying between the British Isles, Denmark, and Norway. This conservation effort was aimed at giving North Sea fish populations a chance to rebound, but by 2003 the cod were thought to be in even worse shape, and a longer hiatus of as much as 12 years was called for.

Two important characteristics distinguish the Scandinavian countries from other European nations. The first is their strong social welfare systems. The second is the extent to which Scandinavian countries have achieved equality of participation and well-being for men and women. Iceland, Norway, Sweden, Denmark, and Finland hold the top five positions in the United Nations Gender Empowerment Measure (GEM) rankings.

Social Welfare in Sweden. Sweden's cradle-to-grave social welfare system (see page 206) is founded on three beliefs. First is the idea of security: that all people are entitled to a safe, secure, and predictable way of life, free from discomfort and unpleasantness. Second, the appropriate life is ordered, self-sufficient, and quiet, not marred by efforts to stand out above others. Third, when the first two concepts are practiced as they should be, the ideal society, *folkhem* ("people's home"), is achieved. These three concepts help to explain why Swedes are willing to pay for a social welfare system that consumes two-thirds of their government's annual budget (as of 1993) and provides stability and a safety net for every one of the country's 9 million people.

Equal Rights in Norway. Unlike most societies, Norway has directed much of its most innovative work on gender equality toward advancing male rights in traditional women's arenas. Men, for example, now have the right to at least a month of paternity leave to stay home with a newborn or adopted child. This policy recognizes a father's responsibility in child rearing and related family obligations; in addition, it provides a chance for father and child to bond during the early days of life. Even when addressing male violence against women, Norway recognizes in its public discussions that this global problem stems in part from cultural customs that suffocate sensitivity in individual men. For example, male children's games that include derision of boys who are not violent are now openly discouraged in Norway.

At the same time, equal rights for women are closer to being a reality in Norway than in any other country. In 1986, Gro Harlem Brundtland, a physician, became the country's first woman prime minister, and she appointed women to 44 percent of all cabinet posts. Norway became the first country in modern times to have such a high proportion of women in important government policy-making positions. She was reelected and served in the post for 10 years. By 2003, women made up 36.4 percent of Norway's parliament (third only to Sweden's 45 percent and Finland's 36.5 percent); they held nearly 33 percent of elected positions in municipal government and more than 40 percent of posts in county government. The growing policy-making power of women is perhaps best seen in the fact that in the 1990s, women were leaders of Norway's three major political parties.

It is not just in politics that gender equality is sought. Norway was the first country in the world to have an equal-status

Apartment complexes, such as the Tokarp development in the south-central Swedish town of Jönköping, reflect the Swedish sense of modernist elegant simplicity. [Tomasz Tomaszewski/National Geographic Image Collection.]

ombudsperson whose job it is to enforce the 1979 Equal Status Act. This act has a "60–40 rule" requiring a minimum of 40 percent representation by each sex on all public boards and committees. Even though the 40 percent rule has not yet been achieved in all cases, female representation averages over 35 percent for state and municipal agencies. In addition, employment ads are required to be gender-neutral, and all advertising must be nondiscriminatory. For example, an advertisement for a car cannot depict a bikini-clad male or female as an eye-catcher. Norwegians are no prudes: nude bodies can appear, but only in ads for products that are directly related to the body in question.

The Baltic States

The three small Baltic states of Latvia, Estonia, and Lithuania are situated together on the east coast of the Baltic Sea. Culturally, these countries are distinct from one another and from Scandinavia, with different languages, myths, histories, and music. After World War II, all three shared the experience of being forced to become Soviet republics. Then, in 1991, they gained their independence when the Soviet Union collapsed. Despite the recent connection to Russia, their cultural ties are traditionally with West and North Europe. Ethnic Estonians, who make up 62 percent of the people in Estonia, are nominally Protestant, with strong links to Finland; they view themselves as Scandinavians. Ethnic Latvians, who are just 50 percent of the population of Latvia, are mostly Lutheran; they see themselves as part of the old German maritime trade tradition. Lithuanians are primarily Roman Catholic and also see themselves as a part of West Europe.

To secure its dominance in the Baltic states, the Soviet Union settled many ethnic Russians in the main cities of all three countries. Today, Russians make up about a third of the population in both Estonia and Latvia, and they wish to keep strong ties with Russia. In both countries, the indigenous populations are aging because they reproduce at very low rates, whereas the Russian minorities are having many children and could become the dominant culture group within a few decades. Poles, not Russians, are the dominant minority in Lithuania (somewhat under 20 percent), and ethnic Lithuanians will remain the dominant majority for some time to come. Ironically, although the Russian and Polish minorities were established in the Baltic states to secure the influence of the Soviet Union, under EU rules their rights as minorities had to be guaranteed before these countries could join the European Union.

Of all the countries of North Europe, the Baltic states have the most precarious economic outlook. Under Soviet rule, the Russians expropriated their agricultural and industrial facilities and used them primarily for Russia's benefit. Until 1991, 90 percent of the Baltic states' trade was with other Soviet republics. Since then, in reorganizing their economies, the three countries have moved in different directions. Estonia has a light industrial base that is relatively easy to reorient. It has undertaken the most radical economic change, moving toward a market economy and increasing its trade with the West, especially Finland and Germany. Of the three, Estonia is the only one to experience

Riga, Latvia's capital, is spread along the Daugava River. Once known as the Paris of the Baltic, the city strives to regain its former economic prosperity. The return to private enterprise has helped many small entrepreneurs, but privatization of industrial production is proceeding more slowly. [Torleif Svensson/Corbis Stock Market.]

Gray snow floats down from the stacks of the Kunda cement factory in Estonia. As in most other former Soviet nations, environmental degradation in Estonia resulted from the Russian emphasis on industrial output as the key to economic growth and prosperity. [Larry C. Price.]

real economic growth, and its accomplishments have attracted foreign aid and private investors. Heavily industrialized Latvia and Lithuania now trade more with Germany and the West than with Russia. They have had to overcome low standards of quality, and many of their factories remain out of date and heavily polluting.

The ecological consequences of industrialization under the Soviets are widespread. One prominent example in Estonia is the oil shale production facility near Kohtla-Järve in the northeastern corner of the country. Smoldering hills of shale residue add to the noxious smoke from the plants. Oily acids seep into the groundwater and into the Baltic, causing skin ulcers on coastal fish. It is possible to ignite the chemicals floating on well water on nearby farms. The photo at the left shows the level of ground and air pollution near a cement factory in Estonia.

The Baltic states see national security as their major problem because their strategic position along the Baltic Sea has long been coveted by Russia, which has few easy outlets to the world ocean. In this regard, the status of the Russian **exclave** of Kaliningrad (old East Prussia) is crucial. Kaliningrad is called an exclave because, although it is an actual part of Russia, it lies far from that country along the Baltic, between Lithuania and Poland (see Figure 4.2 on pages 180–181. A relatively ice-free port, Kaliningrad is headquarters for the Russian Baltic fleet. Russia's strategic interest in Lithuania is unlikely to diminish, and Lithuania fears that changing power relationships in Russia could result in a reinvasion of its territory. The countries of Western Europe, unwilling to commit to military support of Lithuania, hope to resolve differences with the Russians without antagonizing them.

EAST AND CENTRAL EUROPE

No other part of Europe is in a more profound period of change than East and Central Europe (Figure 4.20). In the euphoria at the end of the cold war, many people believed that democracy and market forces would quickly turn around the region's centrally planned economies, which were sluggish, highly polluting, and inefficient. Instead, some parts of the region have experienced violent political turmoil, and virtually all experienced at least temporary drops in standards of living. It appears that two tiers of countries are emerging. In the 1990s, those countries physically closest to western Europe—Poland, the Czech Republic, Slovakia, Hungary, and Slovenia—elected new governments democratically, made rapid progress toward market economies, and began to attract foreign investment. Access to consumer goods improved markedly. In 2004, all of these countries joined the European Union, but conforming to the wide range of economic, environmental, and social requirements set forth by the EU has not been easy.

The countries of the former Eastern bloc that are experiencing the greatest economic and social difficulties are Albania, Bosnia and Herzegovina, Bulgaria, Croatia, Macedonia, Romania, and Serbia and Montenegro (including Kosovo). For these countries,

FIGURE 4.20 The East and Central Europe subregion. Two tiers are emerging in East and Central Europe. Poland, the Czech Republic, Slovakia, Hungary, and Slovenia have elected new governments democratically, made rapid progress toward market economies, and joined the European Union. Albania, Bosnia and Herzegovina, Bulgaria, Croatia, Macedonia, Romania, and Serbia and Montenegro (including Kosovo) have had a more difficult time adjusting to democracy and privatization.

adjusting to democracy and privatization has proved difficult. Plagued by corruption and economic chaos, many governments have relaxed the pace of reforms as they struggle merely to maintain civil peace. Meanwhile, most of the countries that were once provinces in the old Yugoslavia (Bosnia and Herzegovina, Croatia, Macedonia, and Serbia and Montenegro (including Kosovo) became mired in a series of armed ethnic conflicts that resulted in European and U.S. military intervention. Not surprisingly, many people in these countries are looking back nostalgically to the communist era, when jobs were stable, health care and education were free, and there was a strong, authoritative government. Still, many observers are optimistic about even the poorest and most politically unstable parts of East and Central Europe. The countries in this region have useful natural resources for both agriculture and industry, and all have large, skilled, yet inexpensive workforces that are beginning to attract foreign investment. Germany, France, the Netherlands, Italy, the United Kingdom, Russia, and the United States are their most important trade partners, but the combination changes from country to country; investors from Hungary, the Czech Republic, and Slovenia are also beginning to invest in their neighbors to the south.

The Difficulties of Economic Transition

The shift away from central planning toward a free-market economy has been rocky. One problem has been that, because democratic institutions are still immature, old political bosses and high-level bureaucrats have been able to claim benefits for themselves at the expense of other citizens, as demonstrated in the following discussion of agricultural reform. In the Czech Republic, Slovenia, Poland, and elsewhere, small farmers have found it difficult to adjust to the requirements of a market economy.

Agricultural Reform in the Czech Republic and Slovenia. In the Czech Republic in the 1990s, land that had been expropriated after World War II and turned into state farms and cooperatives was either reorganized into private owner-operated cooperatives or restored to its original owners and their descendants. The latter group, rather than farming the land themselves, usually leased their regained land, either in small parcels to those who wished to become family farmers or in large parcels to the new private cooperatives. The private cooperatives are owned by a group of workers who labor on the farm and who share its profits. Eventually, 1.25 million acres (500,000 hectares), representing 30 percent of the country's agricultural land, went to individual private farmers, large and small. These private owner-operated cooperatives now account for the majority of the agricultural land in the Czech Republic.

German economist Achim Schlüter, who studies Czech agriculture, discovered that land redistribution was only the beginning of the reform process. Successful privatized farms require assets in addition to land—such as animals, machinery, and buildings—and they need managers with marketing connections and agricultural expertise. Former agricultural bureaucrats, who had that expertise and understood the old system, were able to make quick, informed decisions that allowed them to farm profitably in the new era. The former bureaucrats became executives of large privatized farm cooperatives, while, Schlüter discovered, new small family farmers soon gave up because they had little understanding of market economies and were ill-equipped to negotiate the best terms for themselves. Consequently, there has been a trend toward large quasi-corporate farms, which are preferred by the EU economic advisors who readied the Czech Republic for entry into the European Union. Larger, more efficient farms are preferred in part because overproduction, which could contribute to falling commodity prices across Europe, is easier to control on such farms. EU advisors have suggested that the smaller family farms find more lucrative uses for their land, such as farm tourism. They say that "dude farms," similar to dude ranches in the western United States, or living history demonstrations of traditional food cultivation and preparation techniques could attract affluent urban families from all over Europe. In the Czech Republic itself, where parkland is scarce and city people often live in cramped quarters, farm tourism could become a popular form of weekend and summer recreation.

In Slovenia, unlike most of the rest of East and Central Europe, farms were not collectivized, so it has not been necessary to redistribute land. The problem instead is that farms are too small for efficient production. The average farm size is just 8.75 acres (3.5 hectares), and agriculture accounts for less than 3 percent of GDP. Although it has much rich farmland, Slovenia is a net importer of food, mostly from EU countries such as Italy, Spain, and Austria. Nonetheless, the new emphasis on private entrepreneurship has encouraged some Slovene farmers to seek a niche in the domestic market. The case of Vera Kuzmic is illustrative.

PERSONAL VIGNETTE   Vera Kuzmic (a pseudonym) lives a 2-hour drive south of Ljubljana, Slovenia's capital. Her family has farmed 12.5 acres (5 hectares) of fruit trees near the Croatian border for generations. During the socialist era, she and her husband worked in urban industrial jobs, commuting daily, and did their farm work in the late afternoon and on weekends. After Slovenia became independent in 1991, Mrs. Kuzmic was the first to lose her job when market reforms bankrupted many Slovene factories. The Kuzmic family decided to try earning its living in vegetable market gardening because vegetable farming could be more responsive to market changes than fruit tree cultivation. By 2000, the adult children and Mr. Kuzmic worked on the land, and Mrs. Kuzmic was in charge of marketing their produce and that of neighbors she had convinced to grow vegetables as well.

Mrs. Kuzmic secured market space in a suburban shopping center in Ljubljana, where she and one employee maintained a small but orderly vegetable and fruit stall. Every day but Sunday, she loaded her station wagon with fresh produce and drove to the shopping center. Her produce had to compete with much less expensive Italian-grown produce sold elsewhere in the same shopping center—all of it produced on large corporate farms in northern Italy and trucked in daily. But Mrs. Kuzmic gained market share by bringing her customers special orders daily and by guaranteeing that no pesticides or herbicides were used

Vera Kuzmic in her market stall in Ljubljana, Slovenia. [Lydia Pulsipher.]

on the fields, only animal manure. For a while, her special customer services and her organically grown produce kept her in business. But when Slovenia joined the EU in 2004, more had to be done to compete with produce growers and marketers across Europe.

Anticipating the challenges to come, the Kuzmics' daughter Lili completed a marketing degree at the University of Ljubljana. The family incorporated their business, and Lili is now its Ljubljana-based director, while Mrs. Kuzmic manages the farm. Lili's market research shows that it would be wisest to diversify. So, while continuing to focus on the expanding professional population in Ljubljana, whose food habits are changing and who are willing to pay extra for fine vegetables and fruits, the farm also produces gourmet preserves, marmalades, and spreads, and it sells as much as possible at the farm to save on transport costs. In a recently built banquet facility, Mrs. Kuzmic prepares special dinners for bus-excursion groups interested in traditional dishes made from home-grown crops. [Adapted from Lydia Pulsipher's conversations with Vera Kuzmic and Dusan Kramberger, 1993 through 2004.]

Challenges to Industrial Reform. Much of East and Central Europe's considerable industrial base is still burdened by legacies of the socialist era: a reliance on heavy industries that use energy and other resources inefficiently, and hence are highly polluting; a dependence on imported raw materials, and an emphasis on coal and steel production rather than on consumer-oriented manufacturing and service industries. Furthermore, when consumer products have been a focus, there has been too little attention to good design and efficient production. An example illustrating industrial pollution is Upper Silesia, Poland's leading coal-producing area. There forests have succumbed to acid rain, soils yield contaminated crops, and water

pollution is at deadly levels. Residents have experienced birth defects, high rates of cancer, and lowered life expectancies. Industrial pollution has been one of the greatest obstacles to entry into the European Union, but the EU also requires evidence that prospective members can compete successfully in the EU marketplace, something that few East and Central European countries are yet able to do.

Corruption has also inhibited industrial reform. During the last years of the Soviet Union and during the post–cold war era, corruption mushroomed. In some countries, involvement by organized crime plagued the sale of formerly state-owned industries, especially in military sectors. One mobster based in Budapest reportedly took over virtually the entire Hungarian armaments industry and sold weapons globally. For a time, post-communist politicians had connections to organized crime groups, and some foreign businesses eager to invest in East and Central Europe reluctantly paid bribes to those groups. But in recent years, the wish to join the EU has encouraged most countries to control organized crime. While industry in this subregion of Europe is still not particularly successful, better design has been encouraged, and regional trade shows are full of attractive, if not yet widely produced, goods. High-end artisan-made products, such as handmade designer shoes, luxury baby clothes, modern kitchens, prefabricated houses, and home furnishings appear to be the greatest successes.

Continuing Hardship. Throughout East and Central Europe, people suffered directly from the privatization or collapse of state-owned industries. Payrolls were unmet, and firms went bankrupt, resulting in widespread layoffs. Tax collection fell far behind schedule, and public funds for essential services were depleted. In reaction, scientists, technicians, artists, writers, and other highly skilled people moved to wealthier parts of the world.

The economic transition has been particularly hard on women, who were more vulnerable than men to job loss. As state-run firms were sold off and made more efficient, women—who occupied the lowest ranks—were the first to be laid off, despite the fact that in many cases they were a family's sole support. By 2004, most governments had official agencies that looked after the interests of women, who still suffered higher unemployment than men. Even young women, who by 2004 were completing higher-education degrees in larger numbers than young men, continued to find employment more slowly than men and to be paid only 80 percent as much as similarly qualified men. Many older women were choosing to found small private businesses, while younger women were taking their skills and migrating to western Europe, New Zealand, Australia, or North America. Since all countries in East and Central Europe have very low or negative population growth rates, losing these young future mothers is a serious blow.

Progress in Economic Transition

For all the signs of trouble, however, the governments of East and Central Europe are taking "free market-friendly" steps to help their economies grow more quickly. One strategy is to encourage foreign investment. Virtually all the countries of East and Central

The Fishermen's Bastion and its outdoor restaurant is one of Budapest's many tourist destinations. The structure was completed in 1905 on the site of the old fish market in memory of the fishermen who helped to defend the city during the Middle Ages. [Leo de Wys.]

moted as tourist attractions. The cities of East and Central Europe are some of Europe's most distinguished municipalities, and they are becoming magnets for tourists from America, Japan, and western Europe. One example is Budapest, the capital of Hungary. Situated on the Danube River and built on the remnants of the second- to third-century Roman city of Aquincum, Budapest was one of the seats of the Austro-Hungarian Empire, which ruled central Europe until World War I. By 2000, Budapest had reemerged as one of Europe's finest metropolitan centers, filled with architectural treasures, museums, concert halls, art galleries, discos, cabarets, theaters, and several universities of distinction. An added attraction is that Hungarians have revived their love affair with jazz, called "the music of the imperialists" during the communist era. Since the change to a market economy, dozens of jazz clubs have opened. An interesting example of links with tourism in western Europe is the two-day round-trip Danube cruise that now runs between Vienna and Budapest.

New Experiences with Democracy

The advent of democracy has reduced some of the tensions associated with the economic transition because people have been able to vote against governments whose reforms created high levels of unemployment and inflation. Most countries have now had several rounds of parliamentary and local elections, and voters are becoming accustomed to Western-style political campaigns. However, the tendency to remove reformers before they can accomplish anything significant has dampened the enthusiasm of potential foreign investors who are waiting for real change before they offer their capital. On the other hand, in many cases, even crudely practiced democracy is leading to long-term economic growth and stability by encouraging governments to take a greater interest in the financial security of the population as a whole. For example, in Hungary and the Czech Republic, largely because of pressure from voters and from the EU, there are plans for new national pension systems to replace defunct communist pension plans. Viable pension systems are important on two counts: they will give financial stability to the large cohort that labored under communism and lost their safety nets, and they will relieve younger workers who would have had to bear the burden of supporting aged parents.

Southeastern Europe: Understanding an Armed Conflict

A conflict that erupted in southeastern Europe (the Balkans) in 1991 took the lives of more than 250,000 people (see the earlier discussion of this conflict on page 190). This mountainous region north of Greece is home to many ethnic groups of differing religious faiths. While many in the media have asserted that violence and ethnic hatred is an age-old way of life in this region, a deeper look reveals greater complexity. Although ethnic tensions have existed for generations, they have greatly intensified in recent decades, suggesting that more immediate political and economic factors contributed to the conflict.

Many of the so-called indigenous ethnic groups of southeastern Europe were themselves invaders over the last 1200 years or

Europe actively recruit foreign investors and offer them a variety of incentives. Although there have been a number of false starts by investors unaware of the difficulties they would face, there have also been some successes. An example is Poland's joint venture with the French multinational glass company Saint-Gobain to construct one of the world's most modern glass factories. The factory is being constructed in the south, near the Czech border, in what has been Poland's most polluted region, Silesia. Significantly, it will be one of the most environmentally safe plants of its kind in Europe, with emissions well below the standards set by Germany, the industry leader. All waste glass will be completely recycled.

Tourism is increasingly recognized as an important part of a well-rounded market-based economy. The remnants of medieval forts, castles, and other buildings are being refurbished and pro-

so, coming from central Europe, Russia, Central Asia, and beyond. These disparate groups jointly occupied the Balkan Peninsula with groups who had been there even longer, and none of them were particularly concerned with precise land rights. For the most part, the various ethnic groups managed to live together peacefully, intermarrying, blending, and realigning into new groups. Historian Charles Ingrao argues that coexistence was even easier in places such as Bosnia, where three or more groups living side by side prevented any one group from dominating the others. More recently, external wars exacerbated local ethnic tensions, as when the Nazis made an ally of Croatia in World War II, pitting it against ethnic Serbs and others. Nevertheless, over the last several hundred years, ethnic rivalries were no worse than those in the rest of Europe.

Marshal Josip Broz Tito, Yugoslavia's founder and leader until his death in 1980, tried through the power of his personal leadership to gloss over ethnic differences by encouraging pan-Yugoslav nationalism. Although there was ethnic peace during Tito's rule, there was no national dialogue about the benefits of ethnic diversity. Other ethnic groups were made uneasy by the domination of the Serbs, the most populous ethnic group in the military and in much of the federal government bureaucracy. Throughout the 1980s, resentment of the Serb-dominated central government increased as the centrally planned economy of the country deteriorated. By the late 1980s, in order to divert public attention away from this economic deterioration, President Slobodan Milosevič, himself a Serb, claimed that Serbs were threatened by Albanians in Kosovo. When Slovenia and Croatia, two of Yugoslavia's wealthiest provinces, voted to declare independence in 1991 (Figure 4.21), they were responding to a deteriorating Yugoslav economy, exasperation with a government dominated by Serbs, and growing apprehension about burgeoning nationalism in Serbia. Partly in response to Serbian nationalism, Slovenes and Croatians themselves became more nationalistic, and a spiral into competing nationalisms threatened.

Given the scale of the interethnic violence that occurred in the following years, it is not surprising that many, especially outsiders, have seen ethnic hatred as the overriding cause of the

Dates of Secession from Yugoslavia

Slovenia	1991
Croatia	1991
Macedonia	1991
Bosnia & Herzegovina	1992
Serbia	2003
Montenegro	2003

FIGURE 4.21 The former Yugoslavia. The Union of the South Slavs (Yugoslavia) began to take shape between the two world wars and emerged as a viable country after World War II. It was held together by the charisma of its first leader, Josip Broz Tito, a war hero who led the resistance against the German invaders. After Tito's death in 1980, the country disintegrated. There are now five countries in place of the old Yugoslavia. [Adapted from "A survey of the Balkans," *The Economist* (January 24, 1998): 4.]

conflict, but Serbian geopolitics and an earlier lack of guidance by Marshal Tito on multicultural affairs were the main forces responsible. The Serb-dominated government and military of Yugoslavia allowed the province of Slovenia to separate after just a short skirmish, in part because of its location close to the heart of Europe and far from Serbia and in part because there were no significant enclaves of ethnic Serbs in Slovenia. Nonetheless, Slovenia was a major loss to Yugoslavia because it had been by far the most modernized and productive of the provinces. The Yugoslav government then feared that unless it made a strong show of force in dealing with the Croats and the Bosnians, the rest of the non-Serb populations of Yugoslavia would also move to secede and take valuable territory, resources, and skilled people with them. This fear became a reality when Croatia and Bosnia and Herzegovina attempted to secede from Yugoslavia. Despite the undoubted culpability of the Serbs in starting the conflict and in perpetrating much of the brutality, they cannot be blamed exclusively for all the interethnic violence that has taken place. While suffering ethnic cleansing, Croats and Bosnians have themselves used brutality and ethnic cleansing in retaliation, as did Albanians (also known as Kosovars) later.

Conflict sprang up again in the late 1990s in the province of Kosovo, the southern area of Serbia dominated by ethnic Albanians. Throughout the 1990s, Milosevič's government in Serbia had been removing Albanians (Kosovars) from government positions and barring them from using public services. Increasingly, the Serbian government used arrests and violence to quell Kosovar activists. In response, the guerrilla Kosovo Liberation Army was formed, supported by expatriates of Albanian ethnicity. They also attempted to secede from Yugoslavia — perhaps to join the country of Albania, which lies just to the west. In spring 1999, Serbia forced most of the majority Albanian population out of the province of Kosovo. Several weeks later, in their first-ever formal military action, NATO forces bombed Serbia in an effort to stop massacres and prolonged ethnic cleansing in Kosovo. The violence de-escalated, but hostilities between Kosovar Albanians and Serbs continued in 2002, fueled by covert aid to both sides from agents outside the region, some inspired by nationalism, some by Islamic fundamentalism. In November 2001, elections were held in Kosovo under UN supervision.

The conflicts in and around Serbia have devastated the economies of southeastern Europe. They destroyed the infrastructure of commerce, interrupting trade and tourism for all countries in the region. Dozens of bridges were bombed, and the Danube River, a major transport conduit through Europe to the Black Sea, was blocked with wreckage. In 2000, an election in Serbia resoundingly removed Slobodan Milosevič from power, and in 2002 he was brought to trial for war crimes in the International Court of Justice (World Court) in The Hague. To forestall any future troubles, the European Union and the countries of southeastern Europe had earlier agreed, once Milosevič was gone, to enact the Balkan Stability Pact. Funded by the European Union, this pact is promoting recovery of the region through public and private investment and through social training to enhance the public acceptance of multicultural societies

and strengthen informal democratic institutions. In the fall of 2000, the country now known as "Serbia and Montenegro" was admitted to the United Nations.

REFLECTIONS ON EUROPE

Europe's position as a center of world economic and political power seems less formidable when the region is viewed up close. Europe is not a monolith. Its geographic patterns are highly variable, and its borders are fluid: in a few years the Southwest Asian country of Turkey may join the European Union. The distribution of wealth, social welfare policies, approaches to environmental issues, treatment of outsiders, and participation of women in public and private affairs are just some of the aspects of life in Europe that reflect the region's geographic variation.

Many of the issues confronting people in Europe are similar to those encountered elsewhere around the world. For example, the effort at economic integration is exposing European countries to the same problems of structural adjustment now faced by so many, much poorer, developing countries. The vicissitudes of dealing with an integrated world economy are apparent on every level. Europe's problems, although generally less severe and played out in more affluent circumstances, are no less potentially disruptive than those faced by Latin America, Asia, and Africa. Europeans must answer the same questions: How can a country retain jobs for its people when there are qualified workers willing to work for less within relatively easy reach? How can more satisfying jobs be created? How can a society help the poor and unemployed to a better life while at the same time lessening the drain on public funds by welfare programs? How can poor parts of the region be stimulated to develop, and how can wealth be more equitably distributed? And how can development everywhere be made to conform to the requirements of an increasingly stressed environment? No less complex are the cultural questions of how countries with very different sets of traditions and mores can find sufficient common ground to cooperate economically and thus achieve greater prosperity for everyone. For all its current troubles, Europe's head start on development, its long-established world leadership, and the heightened awareness of global economic issues gained by its own efforts at integration all position Europe to retain its global influence.

Europe's questioning of U.S. foreign policy after September 11, 2001, was an important chapter in post–cold war geopolitics, and it will be interesting to see the extent to which this challenge to U.S. power actually affects global relationships over the long term. To people in the Americas at large, Europe's current preoccupation with economic integration should prove instructive. Many of the same issues must be confronted if the economic integration of the Americas is to proceed: the free movement of people, the loss of jobs to low-wage areas, the need to address gross disparities in wealth, and the need to strengthen and coordinate social and environmental policies, to name just a few.

The future of the relationship between Europe and the world region we will cover in Chapter 5—the Russian Federation, Belarus, Caucasia, and Central Asia—is not clear. Will these two regions eventually integrate their economies and societies and become one large Europe stretching to the Pacific with 800 million people? Will Russia's very serious economic and social problems make it too unattractive a partner for Europe, or will its oil resources make it attractive no matter what? The next chapter will discuss these and other geographic issues.

Quick Review

1. Comment on the role of tourism in the various European countries.

2. In the United Kingdom, which recent events are seen as examples of devolution?

3. What is meant by the statement that two tiers of countries are emerging in East and Central Europe?

4. Why has one particular European country sometimes been labeled the "Bad Boy of Europe"?

5. What are some of the unique features of gender politics in Norway?

6. What are the roots of the United Kingdom's present status as a place of multiethnic complexity?

Chapter Key Terms

acculturation   204

Age of Exploration   186

assimilation   204

cold war   189

Corridor Five   200

democratic institutions   188

devolution   220

double day   205

dumping   201

economic integration   190

economies of scale   194

ethnic cleansing   190

euro   195

European Economic Community (EEC)   194

European Union (EU)   178

exclave   226

feudalism   185

genetically modified organisms (GMOs)   196

Green   208

guest workers   201

iron curtain   189

medieval period   185

mercantilism   186

nationalism   188

price supports   200

producer services   199

Protestant Reformation   186

Renaissance   186

resource base   191

Schengen Accord   201

social welfare   206

subsidies   200

welfare state   188

world cities   187

CHAPTER 5

RUSSIA
and the Newly Independent States

PERSONAL VIGNETTE When he was a teenager, Boris K. used to race go-carts, and he even produced films for a while. Still a young man, Boris now owns several kiosks—which makes him, by most Russian standards, a gangster. Kiosks are mobile, modular stores, usually consisting of a large steel box about twice the size of a dumpster, with windows and a display case. They are usually crammed with an array of goods ranging from vodka and candy to phone cards, lottery tickets, newspapers, and magazines to toilet seats. Some kiosks are open 24 hours a day, but they can be locked up as securely as a safe. Kiosks can be loaded onto trucks and transported anywhere overnight. This portability allows their operators to take advantage of nearly any public space with commercial potential and to evade the authorities. The places where kiosks congregate are called *tolkuchkas*, or "places of pushing."

Kiosks are a highly visible component of Russia's move to a market economy and more open access to information—but why the gangster connection? Nearly all kiosks operate at least partially in the informal economy, dodging taxes and selling some sort of bootlegged or suspect product: counterfeit CDs, fake name brands, stolen property. Many are run by toughs who strong-arm other vendors or demand protection money. Some kiosks are in dirty, rubble-strewn squares, others next to pleasant parks and subway stations. Nearly all are temporary, disappearing and reappearing depending on the needs of those who own and supply them, on who is buying what, and on current government policy. Kiosks are also a major source of public information, selling newspapers and magazines subject to government censure as well as underground pamphlets critical of the government, so they incur government scrutiny for that reason as well.

In the spring of 2004, during a government crackdown on the press, Russia's Interagency Anti-terrorism Commission recommended that no kiosks be allowed closer than 25 meters to subway station entrances because of the danger that explosives could be hidden in them. Noting that 70 percent of Moscow's newspapers are sold in kiosks near subway stations, the Union of Publishers and Press Distributors said that if this policy were implemented, news distributors would go bankrupt, and 4000 jobs, as well as $110 million in trade, would be lost annually. Most significantly, public access to information would be severely restricted.

Many kiosks have names painted on the outside, usually in bright colors and in English. The names reflect their somewhat rebellious and capitalistic spirit: "Freedom," "My Valentine," "What You Want?" As Boris puts it, kiosks and their *tolkuchkas* are "spontaneous rackets," very responsive to the market trends of the moment. [*Adapted from Mark Kramer,* Travels with a Hungry Bear: A Journey into the Russian Heartland *(Boston: Houghton Mifflin, 1996), pp. 262–268;* Moscow Travelers Yellow Pages, *http://www.infoservices.com/ moscow/509.htm, 7/26, 2004;* The Moscow News, *http://english.mn.ru/ english/issue.php?2004-19-31.*]

MAKING GLOBAL CONNECTIONS

Boris lives in a region that has attempted to change its political and economic systems entirely in just a few short years. The change has offered opportunities to many people, like Boris, but has brought hardships and uncertainty as well. Until a few years ago, the **Union of Soviet Socialist Republics (USSR,** also known as the **Soviet Union)** was the largest political unit on earth, stretching from eastern Europe all the way across Central and East Asia to the Pacific Ocean, covering one-sixth of the earth's surface. In December 1991, the Soviet Union ceased to exist. In its place stood fifteen independent countries joined in a loose economic alliance. The largest and most powerful was Russia. But soon, three of the fifteen—Latvia, Lithuania, and Estonia—having chafed under Russian domination for many years, abandoned their close association with Russia. In this book, they are discussed in Chapter 4 along with the European countries with which they are negotiating new relationships.

Imported (probably smuggled) VCRs and TVs are sold in informal, non-tax-paying open markets like this one in Moscow. They are tended by men affecting, if not actually living, a "mafioso" lifestyle. Police are paid to look the other way. [Gerd Ludwig/National Geographic Image Collection.]

Thus the region we describe in this chapter presently consists of Russia and eleven loosely allied countries: Belarus, Moldova, and Ukraine to the west of Russia; to the southeast, Georgia, Armenia, and Azerbaijan, referred to jointly as the republics of Caucasia; and to the south, the five republics of Central Asia: Kazakhstan, Kyrgyzstan, Tajikistan, Uzbekistan, and Turkmenistan. In Chapter 1, we discussed some problems with the concept of region in geographic analysis. Russia and the newly independent states are in a part of the world where regional analysis is particularly difficult to implement at this time, precisely because the major changes taking place may ultimately result in new regional orientations for different parts of the region. Some republics in the east may eventually join Europe, while the Central Asian republics may align themselves with neighbors in Southwest or South Asia with which they share cultural traditions and economic interests. The far eastern parts of Russia are already finding common trading ground with East Asia and Oceania. The map in Figure 5.1 shows these potential regional realignments.

The breakup of the Soviet Union triggered a transformation in the region's political and economic systems. Authoritarian government gave way to experiments with democracy and an economy controlled by government bureaucrats was replaced by one that encourages its citizens to establish their own businesses, but gives mixed messages to those who would try entrepreneur-

ship. Boris K.'s roving kiosk is an example of the freewheeling experimentation going on in this region. The questionable legality of his operation illustrates the pitfalls and temptations of an economic system that is only beginning to develop adequate controls. His fondness for English phrases suggests the allure of the West's wealth and mass culture. The Russian government's tangential attack on the press with its threat to close those kiosks most accessible to subway stations illustrates how it still uses rumor and menacing gestures to keep the public unsure of their freedoms.

The demise of the Soviet Union in the early 1990s required a period of adjustment and reorganization around the globe because the cold war had affected so many economic and political relationships worldwide. During the cold war, the West (Western Europe and North America and their allies) and the East (the Soviet Union and its allies) were locked in a struggle for power and influence. Although the confrontation between East and West diminished drastically in the 1990s, it is not yet clear just what role the old components of the former Soviet Union will play in world affairs. Now, in the aftermath of the terrorist attacks of September 11, 2001, the U.S. response to these attacks by Muslim extremists has instigated further changes in global alignments. Russia, itself the target of repeated bombings in 2003 and 2004, for the most part is aligning itself with Europe and to some extent with the United States. The Central Asian republics appear

FIGURE 5.1 Russia and the newly independent states. The region of Russia and the newly independent states (green) is in flux as all of the component countries rethink their geopolitical positions vis-à-vis one another and adjacent regions. For example, Ukraine and Moldova are interested in a closer association with Europe but want to maintain connections to Russia as well. The text boxes point out that the newly independent states, and also Far Eastern Russia, are developing new relationships with their neighbors on the periphery while still retaining some sort of connection to European Russia.

to be finding common ground with oil-producing Muslim nations in Southwest Asia, but many points of contention remain, including debates over the role religion should play in government.

Russia's sphere of influence has diminished because it no longer has the economic means to court alliances by giving loans and assistance and through its role as the dominant trade partner among its allies; yet Russia remains prominent in the region because of its size, world standing, and historical precedence. The newly independent states maintain strong economic ties to Russia even as they forge new trading relationships with the outside world. Russia itself has had to make major adjustments. In the past, its economic system separated it from the West, and its satellite countries in Europe, the Caucasus, and Central Asia supplied food for its populace and raw materials for its industries. Now Russia must compete for resources and trading partners in the worldwide marketplace.

Today, internal economic, political, and social systems are in flux, and important regions once under the Soviet sphere of influence have become independent. Since the early 1990s, Russia, Belarus, Caucasia, and Central Asia have been overrun with Western businesspeople and lawyers, rock stars, franchise hawkers, international drug dealers and gun merchants, American academics and environmentalists, computer and cell phone vendors, multinational oil companies, U.S. Chamber of Commerce delegations—all missionaries in their own way for capitalism and the market economy. Even fundamentalist Christian and Muslim missionaries are seeking converts in many cities. On the one hand, the changes have caused great hardship for many, especially those over age 50, who have suffered dramatic drops in income and well-being. People feel anxious, particularly because it is unclear whether these changes will pay off. On the other hand, the people living in the former Soviet Union are experiencing the exhilaration of new opportunities. The changes offer them greater freedom of expression and of movement, the satisfaction and challenge of entrepreneurship, the right to practice religion openly, and the availability of consumer products in far greater variety and quality than ever before.

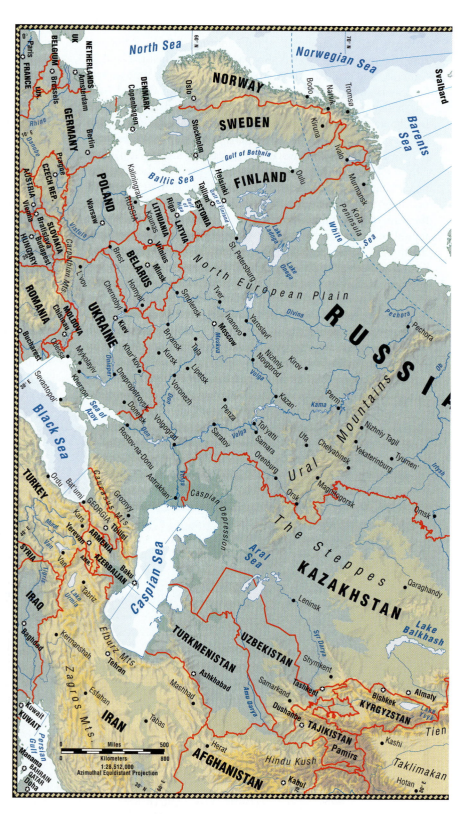

FIGURE 5.2 Regional map of Russia and the newly independent states.

THEMES TO EXPLORE IN RUSSIA AND THE NEWLY INDEPENDENT STATES

Several themes appear throughout this chapter, many of which are associated with the changing internal and external relationships just mentioned.

1. The move to a market economy.   The most remarkable change in this region is the effort to move rapidly from a centrally planned economy to a market economy based on private enterprise. This change,

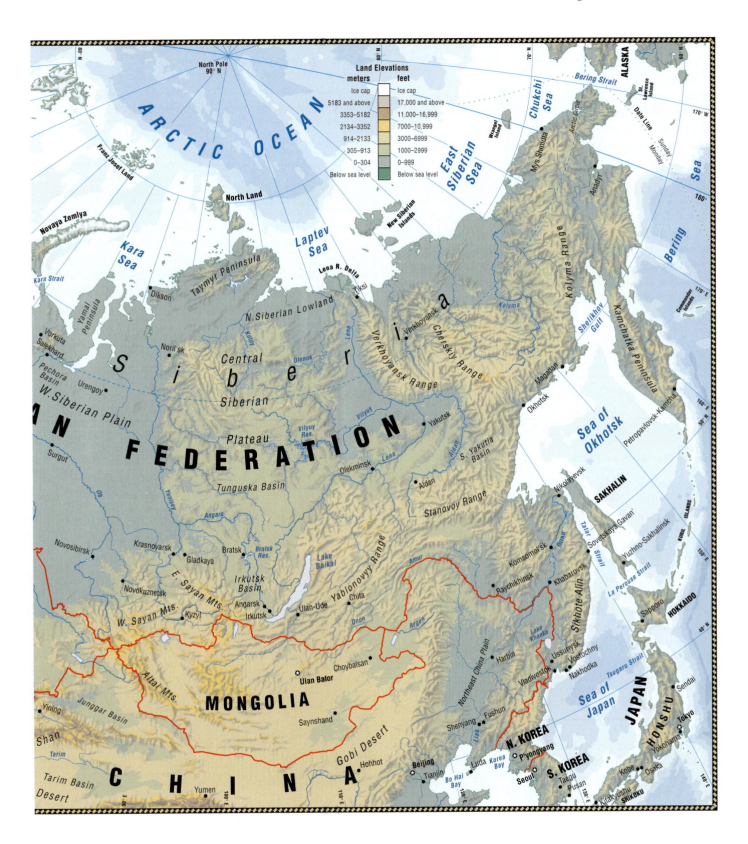

under way since the late 1980s, is uneven in its general success across the region. Wider choices in lifestyle, especially for the young and the educated, are balanced against the loss of jobs and social services and the decline in overall quality of life for large segments of the population.

2. New alignments of people and trade. The breakup of the Soviet Union into smaller, more autonomous political units has resulted in population movements, new allocations of resources, and new trading patterns. These changes have left some parts of the old union isolated and disconnected from former trading partners, from needed resources, and from the transportation infrastructure.

3. The slow emergence of democracy.   Market liberalization did not lead automatically to more popular participation in government, as Westerners had assumed. The newly independent states have all had to forge governing institutions in unstable circumstances, and while regular and orderly elections are increasingly the norm at all levels, from local to federal, the democratization process is uneven—successful in some places, much less so in others.

4. The reconfiguration of the institutions of a civil society.   The Soviet Union provided many social and civil safeguards for its citizens, but it did so in an authoritarian manner. During the disruptions of the transition to a market economy, crime and corruption have flourished. Progress in building trained bureaucracies and civic-minded volunteer and service organizations is uneven across the region.

5. Changing identities.   Regional unity on political, economic, and social issues, once imposed by the centralized Soviet system, no longer exists. In place of this unity, alliances are being forged among new, smaller countries and among ethnic, social, and religious groups. Russia and the newly independent states are also each looking outward to new relationships with Europe, with Southwest Asia, and with China and Japan.

6. Environmental challenges.   The region has long faced a difficult physical environment: a huge continental territory with difficult topography and a harsh climate, few useful rivers, large expanses of permafrost, little agricultural land, and few ice-free ports (see Figures 5.2 and 5.3). During Soviet rule, efforts to make this difficult landscape prosperous resulted in many unwise decisions that led to serious environmental degradation. Now the environment is under considerable stress, but it is getting somewhat more public attention, both because of the recently recognized public health effects of environmental pollution and because some pollution originating in this region has international implications.

‖   THE GEOGRAPHIC SETTING

Terms to Be Aware Of

There is no entirely satisfactory new name for the former Soviet Union. The formal name used in this chapter is *Russia and the newly independent states.* Russia is still closely associated economically with all of these states, but they are independent countries, and their governments are legally separate from Russia's. Russia itself includes more than 30 (mostly) ethnic *internal republics*—such places as Chechnya, Tatarstan, and Tuva (also spelled Tyva)—which constitute about a tenth of its territory and a sixth of its population. Occasionally, the word *transition* will be used to refer to the period from the collapse of the Soviet Union in 1991 to the present (the mid-2000s), during which many economic, political, and social changes have been taking place.

PHYSICAL PATTERNS

The physical features of Russia and the newly independent states vary greatly over the huge territory they encompass. The region bears some resemblance to North America in size, topography, climate, and vegetation.

Landforms

Look at the regional map (Figure 5.2) to get a feeling for the landforms of this region. Moving west to east, there is first the eastern extension of the North European Plain, then the Ural Mountains, then the West Siberian Plain, followed by an upland zone called the Central Siberian Plateau, and finally, in the far

Workers at an oil-extracting plant in the town of Raduzhny struggle with their equipment in the cold environment of western Siberia. Raduzhny is in the northeastern district of the Khanty-Mansiysk Autonomous Okrug (territory). Khanty-Mansiysk is the richest oil region in Russia, producing about 58 percent of all oil in the country and about 5.4 percent of the world's oil supply. Some 40 joint-stock companies (foreign and domestic combinations) operate there. The climate is severe: snow covers the ground from November to June, and the average temperature in January is between 0°F and −10°F. Temperatures may reach 65°F in July, the warmest month. [V. Kolpakov/TRIP.]

east, a mountainous zone bordering the Pacific. To the south of these territories, from west to east, there is an irregular border of semiarid grasslands (**steppes**), barren uplands, and high mountains that separates the region from the southern reaches of the Eurasian continent.

The eastern extension of the North European Plain rolls low and flat from the Carpathian Mountains in Ukraine and Romania 1200 miles (about 2000 kilometers) to the Ural Mountains. This part of the region is called European Russia because the Ural Mountains are traditionally considered part of the border between Europe and Asia. *European Russia* is the most densely settled part of the entire region and is its agricultural and industrial core. Its most important river is the Volga, which flows into the Caspian Sea. The Volga River is a major transport route; it is connected—via canals, lakes, and natural tributaries—to many parts of the North European Plain, to the Baltic and White seas in the north, and to the Black Sea in the southwest.

The Ural Mountains extend in a fairly straight line south from the Arctic Ocean into Kazakhstan. The Urals are not much of a barrier to humans or to nature: there are several easy passes across the mountains, and winds carry moisture all the way from the Atlantic and Baltic across the Urals and into Siberia. Much of the Urals' once-dense forests have been felled to build and fuel

the new industrial cities, such as Yekaterinburg and Chelyabinsk, established in the Urals during the twentieth century.

The West Siberian Plain, lying east of the Urals, is the largest plain in the world. A vast, mostly marshy lowland, it is drained by the Ob River and its tributaries, which flow north into the Arctic Ocean. Long, bitter winters mean that there is a layer of permanently frozen soil (**permafrost**) just a few feet beneath the surface (see photo and caption on this page). Because water doesn't sink through this layer, swamps and wetlands form in the summer months, providing habitats for many migratory birds. Many species breed in the **tundra**—a treeless permafrost zone—and fly south for the winter. This plain has some of the world's largest reserves of oil and natural gas, although their extraction is made difficult by the harsh terrain and climate.

The Central Siberian Plateau and the Pacific mountain zone are farther to the east; together they equal the size of the United States. Still farther east, on the Pacific coast, especially on the Kamchatka Peninsula, there are many active volcanoes. These are created as the Pacific Plate sinks under the Eurasian Plate, causing gas and molten rock to rise to the surface through fissures and holes. Kamchatka, as yet only lightly populated, is a haven for wildlife.

South of the steppes that run in an irregular band from near the Caspian Sea to the Pacific is a wide, curving band of

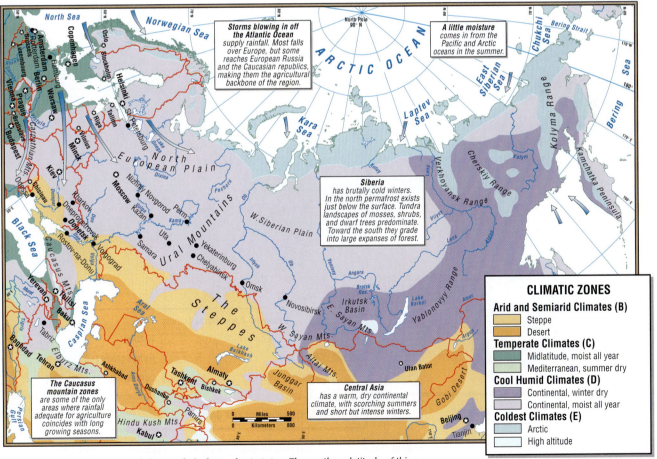

FIGURE 5.3 Climates of Russia and the newly independent states. The northern latitude of this region and its massive size create continental "breathing" that brings a long, cold exhalation of air from the Arctic in the winter and a short summer inhalation that brings little warm, moist oceanic air.

mountains. The mountains lying at least partly within this region include the Caucasus, Elburz, Hindu Kush, Tien Shan, and Altai. Their rugged terrain has not deterred cultural influences that have penetrated north and south for literally tens of thousands of years. People have persistently crisscrossed the mountains, exchanging plants (apples, onions, citrus fruits, wheat), animals (horses, sheep, cattle), technologies (cultivation, animal tending, portable shelter construction, rug and tapestry weaving), and religious belief systems (principally Islam and Buddhism, but also Christianity and Judaism).

Climate and Vegetation

Other than Antarctica, no place on earth has as harsh a climate as the northern part of the Eurasian landmass occupied by Russia (Figure 5.3). Here is a prime example of a continental climate, in which the winters are long and cold, with only brief hours of daylight; the summers are short and cool to hot, with long days. The short summers restrict the crops that can be grown and curtail overall agricultural production. The northerly location means that much of the land has underlying permafrost and hence is boggy and uncultivable. The seasonal continental "breathing" (described in Chapter 1, page 24) is dominated in winter by a long, dry, frigid exhalation of air that flows out of the arctic zones. A short inhalation in summer brings in only modest amounts of moist, warm oceanic air from the west and lesser amounts from the east. The mountain ranges along the region's southern edge block access to warm, wet air from the Indian Ocean far to the south.

Most rainfall in the region comes from storms that blow in from the Atlantic Ocean far to the west. By the time these initially rain-bearing air masses arrive, most of their moisture has been squeezed out over Europe. Nevertheless, sufficient rain reaches the fertile lands of Ukraine, European Russia, and the Caucasian republics to make this area the agricultural backbone of the region (Figure 5.4). However, this area is not large. In fact, the region has no large zone that is ideal for agriculture, with the

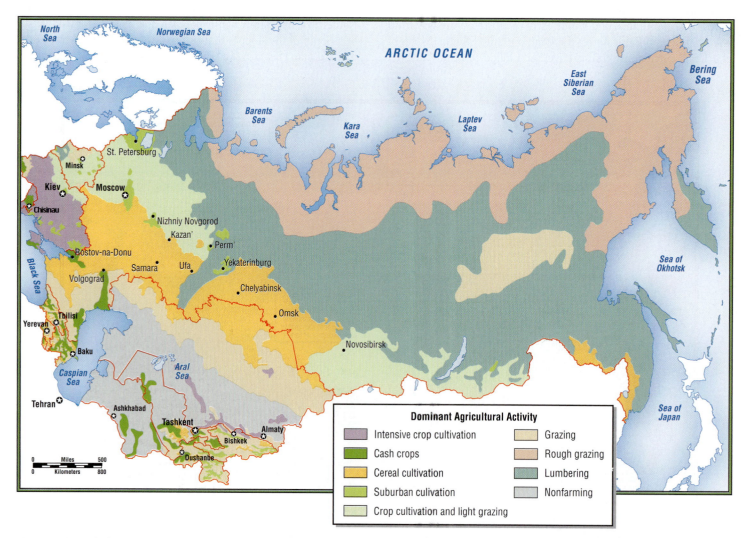

FIGURE 5.4 Agriculture in Russia and the newly independent states. Agriculture in this part of the world has always been a difficult proposition, partly because of the cold climate and short growing seasons, and partly because soil fertility or lack of rainfall are problems in all but a few places (Ukraine, Moldova, and the Caucasus). [Adapted from Robin Milner-Gulland with Nikolai Dejevsky, *Cultural Atlas of Russia and the Former Soviet Union*, rev. ed. (New York: Checkmark Books, 1998), pp. 186–187, 198–199, 204–205, 216–217.]

requisite fertile soils, long growing seasons, and sufficient moisture. Across the region, agriculture is usually precarious at best, requiring expensive inputs of labor, water, and fertilizer. The Caucasian mountain zones are some of the only areas in the region where rainfall adequate for agriculture coincides with a relatively warm climate and long growing seasons. Before the breakup of the Soviet Union, these tiny areas were important sources of fruits and vegetables for the whole region. The best soils are in a region stretching from Moscow south toward the Black and Caspian seas, including much of Ukraine. These black soils, of a type called **chernozem,** are as fertile as those in the American corn and wheat belt. Here the natural vegetation is open woodland and steppe. The soils in the northern part of European Russia are generally not particularly fertile, and growing seasons are very short. The natural vegetation is mixed and boreal coniferous forest.

East of the Urals, the lands of Siberia receive moderate precipitation (primarily from the east), but experience long, cold winters. Huge expanses of Siberia are covered with **taiga** (northern coniferous forest), which stretches to the Pacific. Agriculture is generally not possible because of the cold climate and poorly drained soils, though reindeer are tended in the far north beyond the taiga. Here, in the tundra, a landscape of mosses, lichens, shrubs, and dwarf trees grows above permafrost that is only a foot or so below the surface. Plants are small and do not withstand much grazing or trampling.

East of the Caucasus Mountains, the lands of Central Asia have semiarid to arid climates influenced by their location in the middle of a very large continent. The summers are scorching and short, the winters intense. These lands support grassland (steppe) vegetation. Large-scale cultivation is rarely possible without irrigation, and herding is practiced widely.

HUMAN PATTERNS OVER TIME

The core of the entire region has long been European Russia, the most densely populated area and the homeland of the ethnic Russians. From this center, the Russians conquered a large area inhabited by a variety of other ethnic groups. These conquered territories remained under Russian control as part of the Soviet Union, which attempted to create an integrated social and economic unit out of the disparate territories. The breakup of the Soviet Union has reversed this gradual process of Russian expansion for the first time in centuries.

The Rise of the Russian Empire

For thousands of years, the politically dominant people in the entire region were nomadic herders who lived on the meat and milk of their herds of sheep, horses, and other grazing animals. Their movements across the wide grasslands stretching from the Black Sea to Lake Baikal followed the changing seasons. Nomads would often take advantage of their superior horsemanship and hunting skills to plunder settled communities. To defend themselves, permanently settled peoples gathered in fortified towns.

The towns arose in two main areas: the dry lands of Central Asia and the moister forests of Ukraine and Russia. As early as 5000 years ago, Central Asia was supporting settled communities, and sometimes even large empires, which were enriched by irrigated croplands and by trade along the famed Silk Road, the ancient trading route between China and the Mediterranean (see Figure 9.19). About 1500 years ago, the **Slavs,** a group of farmers who originated between the Dnieper and Vistula rivers, in what is now Poland, Ukraine, and Belarus, moved in an easterly direction, founding numerous settlements, including the towns of Kyiv (Kiev) and eventually Moscow. By A.D. 600, Slavic trading towns were located along all of the rivers west of the Ural Mountains. Distinctive Slavic ethnicity emerged during a long peaceful period (A.D. 720–860) when an especially lucrative trade route was developed between Scandinavia, Constantinople, and Baghdad by way of the Volga River. This trade route, combined with the introduction of Christianity via Constantinople in about A.D. 1000, was instrumental in the development of powerful kingdoms in what are now Ukraine and European Russia. Slavic art and architectural traditions were influenced by Greek Christian missionaries, who also introduced the Cyrillic alphabet. The influences of these missionaries on landscapes in the region are still visible, especially in religious structures. Kiev, especially, but also Moscow, became well-organized urban commercial centers. In the eleventh century,

The archaeologist Jeannine Davis-Kimball, working in the steppes on the border of northern Kazakhstan, found the burial ground of a 14-year-old young woman with bowed legs, possibly caused by a life on horseback. Artifacts recovered included arrowheads (1), a sword (2), a stone amulet (3), and shells (4). These finds are part of the evidence that between two and three thousand years ago, nomadic women participated fully in the economic life of their society (herding, hunting, fighting, and playing games—often on horseback), as is the case today in nomadic cultures of Tibet and Mongolia (see the photo in Chapter 9, page 484). [Courtesy of Dr. Jeannine Davis-Kimball.]

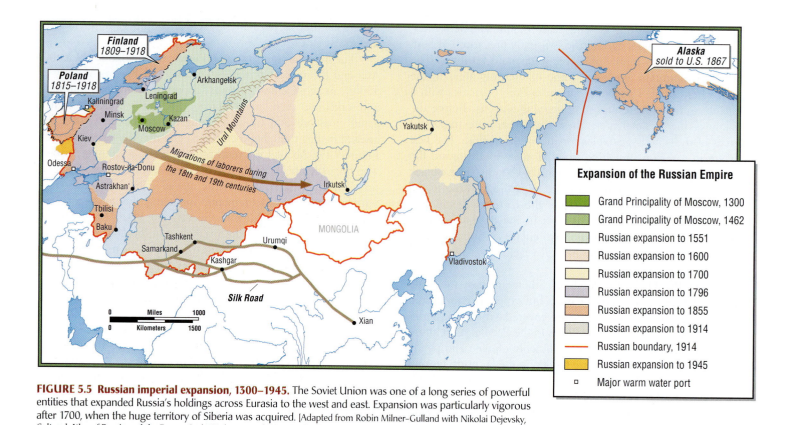

FIGURE 5.5 Russian imperial expansion, 1300–1945. The Soviet Union was one of a long series of powerful entities that expanded Russia's holdings across Eurasia to the west and east. Expansion was particularly vigorous after 1700, when the huge territory of Siberia was acquired. [Adapted from Robin Milner-Gulland with Nikolai Dejevsky, *Cultural Atlas of Russia and the Former Soviet Union*, rev. ed. (New York: Checkmark Books, 1998), pp. 56, 74, 128–129, 177.]

at the height of its glory, Kiev and its surrounding lands formed the largest and most populous state in Europe.

The threat of nomadic invasion remained, however, and the Mongol armies of Genghis Khan conquered these forested lands in the twelfth century. The **Mongols** were a loose confederation of nomadic pastoral people centered in eastern Central Asia and descended from much earlier nomadic groups (see the photo on page 241). By the thirteenth century, the Mongols had conquered an empire containing the rich Silk Road trade route stretching from Europe to the Pacific. Moscow's rulers became tax gatherers for the Mongols, dominating neighboring kingdoms and eventually growing powerful enough to challenge local Mongol rule. Ivan the Terrible's conquest of the Mongols in 1552 at Kazan' (Figure 5.5) is a clear marker of the beginning of the expansion of the Russian Empire.

By the beginning of the seventeenth century, Russians centered in Moscow had conquered many former Mongol territories, integrating them into their empire, which by then stretched eastward from the Baltic over the thinly populated northern Eurasian landmass toward the Pacific (see Figure 5.5). The outlying parts of the empire, south and east of the Ural Mountains, contained huge deposits of mineral resources and were inhabited by many non-Russian peoples. The first major non-Slavic area to be annexed was western Siberia (1598–1689). Although the Russians used many methods of conquest and administration developed by the Mongols, their expansion into Siberia in some ways resembled the spread of European colonial powers throughout Asia and the Americas. For example, much like Britain, France, and Spain, the Russian imperial colonists appropriated Siberian resources for

their own use and invested little in local development. They upheld European notions of private property at the expense of indigenous patterns of communal land use. Perhaps most significant was that Russian expansion into Siberia in the eighteenth and nineteenth centuries made way for massive migrations of laborers from Russia, who soon far outnumbered indigenous Siberians. By the mid-nineteenth century, Russia had also conquered Central Asia to gain control of cotton, its major export crop. Cotton was in high demand in global markets because the American Civil War was disrupting cotton exports from plantations in the United States.

The Russian Empire had great extremes of wealth and poverty. It was ruled by a **czar** (derived from "Caesar," the title of the Roman emperors), who, along with a tiny aristocracy, lived in splendor while the vast majority of the people lived in poverty. Many Russians were **serfs,** who were legally bound to live and farm on land owned by a lord. If the land was sold, they were transferred with it. Even though serfdom ended legally in the mid-nineteenth century, the brutal inequalities of Russian society persisted into the twentieth century, fueling opposition to the czar. By the early twentieth century, a number of violent uprisings had occurred.

A Communist Revolution

In 1917, at the height of Russian suffering during World War I, Czar Nicholas II was overthrown. Within a few months, a faction of revolutionaries called the **Bolsheviks** came to power. Though not very numerous, they were a highly disciplined organization inspired by a common ideology, **communism.** Based largely on the writings of the German revolutionary philosopher Karl Marx, communism criticized the societies of Europe as inherently flawed in that the **capitalists**—a wealthy minority who owned the factories, farms, businesses, and other means of production—dominated the impoverished, propertyless majority of people. With no other options, those people had to work for low wages that undervalued their labor. Communism called on workers to unite to overthrow the capitalists and establish a completely egalitarian society with no government and no money, in which people would work out of a commitment to the common good, sharing whatever they produced.

The Bolshevik leader, Vladimir Lenin, declared that the people of the former Russian Empire needed a transition period of centrally orchestrated drastic change in order to realize the ideals of communism. Accordingly, Lenin's Bolsheviks formed the **Communist party,** which set up a powerful government that ruled the former Russian Empire for the next 70 years. This government was centered in Moscow, which had also been the center of power during most of the Russian Empire. Ethnic Russians held most key government positions, even in non-Russian areas. Lenin was responsible for the New Economic Program, which was a hybrid of capitalism and socialism. Under Lenin, government management of the economy did not work well, and soon production was not meeting demand.

After Lenin's death, Joseph Stalin's 26-year rule of the Soviet Union as party chairman and premier (1922–1953) brought a mixture of revolutionary change and despotic brutality that largely set the course for the rest of the Soviet Union's history. Stalin sought to cure the mismatch between supply and demand by creating a centrally planned **command economy** in which the state owned all real estate and the means of production, while government bureaucrats in Moscow directed all economic activity. They decided where factories would be located, what and how much they would produce, who would manage them, where the products would be distributed, and what they should cost. Such a government-controlled economic system was termed **socialism.** The idea was that under a socialist system, the economy would grow quickly and hasten the transition to the idealized communist state in which everyone shared equally. The notion that the old Russian Empire was in a state of transition toward communism, but not yet there, was reflected in the new name chosen for the country: the Union of Soviet Socialist Republics.

Stalin used the powers of the command economy with fervor and cruelty. To increase agricultural production, he forced farmers to join large government-run collectives that were presumed to be more efficient than smaller plots farmed by families. Those who resisted collectivization were relocated or executed. Despite his reshaping of agriculture, Stalin saw increasing industrial production as the true key to achieving economic growth. Accordingly, he ordered massive government investments in gigantic economic development projects, such as factories, dams, and chemical plants, some of which are still the largest of their kind in the world. The labor was supplied largely by former farmers. Eventually, government-controlled companies monopolized every sector of the economy, from agriculture to mining to clothing design and production.

This strategy of government control resulted in both significant successes and massive failures. For those millions of farmers who peacefully went to work in urban industries, their wages brought higher standards of living. And the schools provided for their children made social mobility possible for future generations. During the Great Depression of the 1930s, the Soviet Union's industrial productivity grew steadily even while the economies of other countries stagnated. As was the case under the czar, however, much production was geared toward heavy industry (the manufacture of machines and transport equipment) and supplying the military with armaments, while little attention was paid to the demand for consumer goods and services that would have dramatically improved daily life for the Russian people. In Europe and North America during this same time, everyday consumers, supplied with even modest wages, were becoming a major source of economic growth for the whole economy (remember Henry Ford's idea that all American families should own a car). Meanwhile, Moscow gave scant consideration to the effects of heavy industrial development on either the natural environment or the health of Soviet citizens. During and after the Stalin regime, the Soviets perpetrated some of the world's worst industrial disasters (especially the unsafe disposal of hazardous materials) and created polluted and unlivable cities.

The most destructive aspect of Stalin's rule, however, was his ruthless accumulation of personal power through fear. Sometimes called the Marxist czar, Stalin used the secret police, starvation, and mass executions to silence almost all opposition to his rule. The lucky ones were those merely sentenced to labor camps in remote Siberia. Stalin's atrocities, which resulted in the deaths of at least 20 million people, were a major reason that the Soviet Union had no real political revolution. The climate of fear squelched civil society, and the Soviet Union continued nearly to the end to be governed by authoritarian power structures similar to those used by the czars.

The cold war, which pitted the Soviet Union against the United States in global geopolitical rivalry, further sapped the Soviet Union's resources. In an attempt to match the global military power of the United States, the Soviets diverted ever more resources to their military at the expense of much-needed economic and social development. Internationally, the Soviet Union spent scarce funds to promote the communist model (centrally planned and funded economic development) in such far-flung countries as China, Mongolia, North Korea, Cuba, Vietnam, Nicaragua, and various African nations. Closer to home, especially in eastern Europe but also in Caucasia and Central Asia, the Soviet Union maintained its pervasive influence through the use of political, economic, and military coercion. Nonetheless, by the late 1960s, the economies and political systems of countries under Soviet influence were steadily drifting toward the free-market democratic model advocated by the United States and its allies. Then, in 1979, a war in Afghanistan, launched to prop up a regime favorable to continued Soviet prominence in Central Asia, severely drained morale in the Soviet Union as a whole. The Soviets were badly beaten in Afghanistan by highly motivated Afghan freedom fighters, many of whom were influenced by a very conservative branch of Islam (see discussions of the mujahedeen in

Chapter 8). This defeat, as well as social and political changes already under way in the region, made the 1980s a time of rapid and monumental change in the Soviet Union.

The Post-Soviet Years

In 1985, a reform-minded leader of the Soviet Union, Mikhail Gorbachev, responded to pressures for change by opening up public discussions of social and economic problems. This innovation, labeled **glasnost,** ultimately brought about the end of the Soviet Union. Gorbachev's efforts to revitalize the Soviet economy through **perestroika,** or restructuring, resulted in little real change. But when he also began to democratize decision making throughout the Soviet Union in the late 1980s, long-silenced resentment of the government in Moscow boiled over and long-suppressed political and interethnic tensions emerged. Independence movements surfaced first in the Baltic states and quickly spread. Gorbachev's liberalizing policies were strongly opposed by those who favored centralized control. After a failed military coup in August 1991, the Soviet Union dissolved in an atmosphere of economic and political chaos.

In the post-Soviet years since 1991, one massive state has been replaced by 12 independent countries, each with its own agenda for reforming the failed systems inherited from the Soviet Union. Russia is the chief successor of the Soviet Union and, although much smaller now, is still the largest country in the world, nearly twice the size of Canada (the second largest). Russia's efforts to maintain its global influence have foundered because the Russian leadership has been consistently unable to solve major internal problems. The transition to a free-market economy has proceeded without much planning, leaving once centrally planned economies drifting dangerously. In some parts of the region, the already sad record of environmental abuse has worsened due to lax supervision. The record of public participation in unfamiliar democratic processes is spotty. For a time, there was more open debate on a wide range of issues, elections were held on schedule, and Russia's economic relationships with the newly independent states were renegotiated in a somewhat more open forum than they were in the old Soviet Union. But, in 2004, after a spate of deadly terrorist attacks, President Vladimir Putin called for constitutional amendments that severely limited democratic participation and freedom of the press.

POPULATION PATTERNS

European Russia's portion of the North European Plain is much less densely settled than is the western portion in Belgium and the Netherlands (Figure 5.6). Nonetheless, within this region, European Russia is the most heavily settled zone, with an average density of 22 people per square mile (57 per square kilometer). By way of comparison, the United States has an average density of 78 per square mile (102 per square kilometer). A broad area of moderately dense population forms a wedge shape, with the base of the wedge stretching between Odessa in southern Ukraine on the Black Sea north to St. Petersburg on the Baltic Sea (see Figure 5.6). The irregular point of the wedge stretches

beyond European Russia to the eastern industrial cities of Yekaterinburg and Chelyabinsk and on to Omsk and Novosibirsk, the largest city in Siberia. This area of relatively high population density corresponds roughly to the distribution of usable agricultural soils, but the correlation of usable soil with population density does not hold everywhere: some industrial cities east of the Ural Mountains are on particularly infertile land.

Beyond Novosibirsk, light settlement roughly follows the irregular pattern of industrial development (including the processing of such minerals as coal, uranium, natural gas, and petroleum) across Siberia. These activities, in turn, are linked to the course of the Trans-Siberian Railroad to the Pacific (see Figure 5.10 on page 253. The eastern industrial cities of Krasnoyarsk, Angarsk, and Irkutsk, all lying on the Trans-Siberian Railroad, each have several hundred thousand people. So Siberia, which is seen as a desolate, lonely land, is nonetheless a region where nearly 90 percent of the people are concentrated in a few large urban areas. But the large cities don't dispel the impression of isolation: beyond the cities, eastern Russia has an extremely light population density of just a few people per square mile.

In the west, a secondary spur of dense settlement extends south from the main wedge in European Russia into the Caucasus region between the Black Sea and the Caspian Sea. Another patch of relatively dense settlement is centered on Tashkent and Almaty and along major rivers in the Central Asian republics. Here the development of irrigated agriculture (especially the cultivation of cotton) and mineral extraction resulted in patches of high rural density, fueled partly by ethnic Russian immigration.

Recent Population Changes

Before its breakup, the Soviet Union was considered a developed region with fairly high standards of living and well-being. In the early 1990s, however, the well-being of the citizens of Russia and the newly independent states deteriorated significantly (Table 5.1, page 246). Not only did infants and children die in large numbers (column 7), but adult life spans also fell for a few years. In Russia, male life expectancy declined from 63.9 years in 1990 to 59 by 2003 (see Table 5.1, column 5). In the same period, female life expectancy dropped from 74.4 years to 72. No other country has experienced a gap of 13 years between average male and female life expectancies during peacetime. This gap indicates significant stress in the lives of the region's males. Male life expectancy in Russia (59 years) is now the shortest in any industrialized country. On the other hand, in Caucasia—despite armed conflict, severe economic reversals, and drastic reductions in living standards—both males and females have a significantly longer life span than they do in Russia or Central Asia (see Table 5.1, columns 5 and 6). This pattern was noticed well before the post-Soviet transition, and it may be partly a result of longer agricultural growing seasons, a more nutritious diet, little consumption of alcohol (many people are Muslim and hence do not drink), and a genetic predisposition to long life.

The apparent cause of declining life expectancies is the physical and mental distress caused by lost jobs and social disruption, most notably reflected in high rates of alcoholism and suicide. By 1996, divorce rates—already high before the Soviet

Thinking Geographically: ON THE WEB

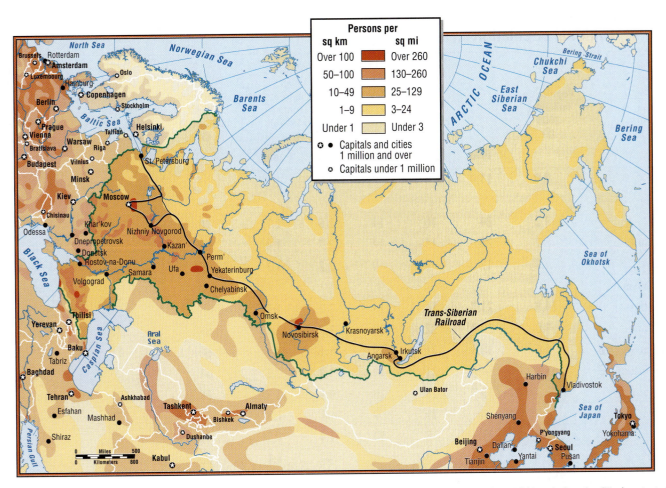

FIGURE 5.6 Population density in Russia and the newly independent states.

PRACTICING GEOGRAPHY   READING MAPS   Compare the population patterns in this map with Figures 5.1, 5.2, and 5.5 to reach an understanding of the physical, social, historical, and political contexts for the patterns on this map. What are some physical explanations for the population patterns? What are the most important historical and social factors influencing these patterns? [Adapted from *Hammond Citation World Atlas*, 1996.]

Union dissolved—reached more than 6 out of every 10 marriages in Belarus, Russia, and Ukraine. Alcohol abuse shortens and saddens many lives in the region, especially those of men. Male suicide rates in Russia, Belarus, and Ukraine are now the highest on earth. After 1991, many people began to suffer nutritional deficiencies caused by sharply falling incomes and by scarcities, some of which were brought on by conflict in Russia's internal republics and in Caucasia. Daily per capita intake of calories was markedly below that of the United States and Europe. The Canadian geographer James Bater has shown that per capita consumption of basic foodstuffs declined throughout Russia and the newly independent states in the mid-1990s. Production of food declined overall after 1990, increasing only in those areas that had previously grown only nonfood cash crops. In the Soviet era, for example, Uzbekistan grew cotton and purchased wheat and other food from Russia, but by the late 1990s, Uzbekistan was producing vegetables, fruit, grain, and livestock and exporting small amounts of food. Overall, during the upheaval of the 1990s, people were eating fewer vegetables; less meat, eggs, and dairy products; and even less bread, long a staple of the diet in this region. The cost of food took up as much as half of family budgets. Though entrepreneur-ial farmers (especially from Ukraine and the Caucasus region) now supply attractive fresh food to the best city markets, ordinary people can afford these luxuries only rarely. Urban residents who have access to a bit of land plant gardens, and since 1991, the option to lease land is more widely available. Urban gardens have long been an important source of nourishment in Russia and the other independent states, but there is a limit to what city people can produce with little time to cultivate, in the small spaces available, during short growing seasons.

Environmental pollution also plays a large, if poorly understood, role in untimely illness and death (see page 266, in the Environmental Issues section). A third or more of the population may be affected by noxious airborne emissions; heavy metal pollution of the ground, air, and water; and radiation contamination. In Central Asia, the buildup of chemicals used in cotton cultivation has threatened the health of agricultural workers and helped to pollute river waters and the Caspian and Aral seas (see pages 266–267). In the industrial cities of European Russia and Siberia, careless handling of industrial waste and inadequate water and sewage treatment endanger health through the pollution of air, water, soil, and food.

TABLE 5.1 National population statistics

Country (1)	Population (millions), 2003 (2)	Births per 1000 (3)			Fertility rate (number of children per woman aged 15–45) (4)		Life expectancy at birth, male, 2003 (5)	Life expectancy at birth female, 2003 (6)	Infant mortality rate (deaths per 1000 live births) (7)	
		1990	2001	2003	1990	2003			1990	2003
Belarus	9.9	13.9	9	9	1.9	1.3	63	75	11.9	9
Caucasia										
Armenia	3.2	24	10	14	2.6	1.7	70	74	18.6	36
Azerbaijan	8.2	26	15	14	2.7	1.9	68	75	23.0	13
Georgia	4.7	17	9	9	2.2	1.1	75	80	15.9	15
Central Asia										
Kazakhstan	14.8	21.7	14	15	2.7	1.8	58	71	26.4	19
Kyrgyzstan	5.0	29.3	22	20	3.7	2.4	65	72	30.0	23
Tajikistan	6.6	38.8	21	19	5.1	2.4	66	71	40.7	19
Turkmenistan	5.7	34.2	21	19	4.2	2.2	63	70	45.2	25
Uzbekistan	25.7	33.7	23	20	4.1	2.5	68	73	34.6	20
Moldova	4.3	—	11	9	—	1.3	65	72	—	16
Russia	145.5	13.4	8	10	1.9	1.3	59	72	17.4	15
Ukraine	47.8	—	8	8	—	1.1	62	74	—	11
Total	281.4									
For comparison										
United States	291.5	16	15	14	2.0	2.0	74	80	9.7	6.9
Europe	540	13	10	10	1.7	1.4	70	78	12	8

Sources: 2003 *World Population Data Sheet* (Washington, D.C.: Population Reference Bureau);
Population Bulletin 49 (December 1996): 48–49.

Declining birth rates are another significant change in population patterns. In the Soviet era, birth rates and fertility rates in Russia and Belarus were already low compared to those in the United States and Europe; since the early 1990s, these rates have dropped still further (see Table 5.1, columns 3 and 4). Now the traditionally higher birth rates in the Caucasian and Central Asian republics are also dropping, ranging from just 9 per 1000 in Georgia to 20 per 1000 in Kyrgyzstan and Uzbekistan. Surveys show that people are choosing to have fewer children mostly out of concern for the gloomy economic prospects of the near future. In 1992, 75 percent of Russian women who had decided against childbearing cited insufficient income as the reason. Because contraceptives are in short supply and of poor quality, abortion is the method of choice for birth control.

By 2003, Russia, Belarus, and Ukraine all had death rates at least 1.5 times their birth rates, and their populations were actually declining (although very slowly). Population pyramids for several of the republics (Figure 5.7) show the overall population trends and reflect differences in patterns of family structure and fertility. The pyramids for Belarus and Russia resemble those of European countries (for example, Germany, as shown in Figure 4.7 on page 193). They are significantly narrower at the bottom, indicating that birth rates in the last several decades have declined sharply. The narrower point at the top for males in all four pyramids depicts their much shorter average life span (see Table 5.1). The bases of the pyramids for Kazakhstan and Kyrgyzstan, in Central Asia, are narrowing as a drop in birth rates accompanies urbanization. But both pyramids still have substantial bases, reflecting the higher fertility rates common in a more rural, primarily Muslim society with a patriarchal family structure. All of the pyramids have a noticeable "waist" at the 55–59 age group, the result of high death rates and low birth rates during and just after World War II.

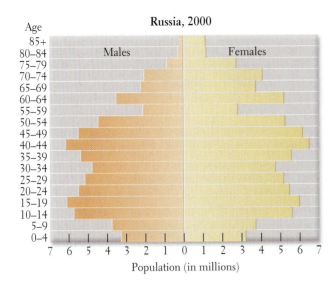

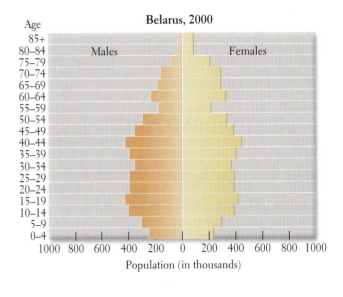

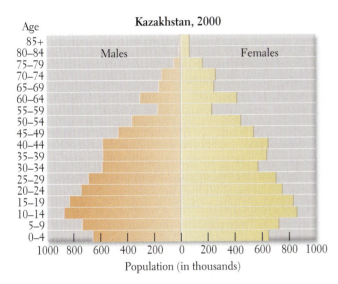

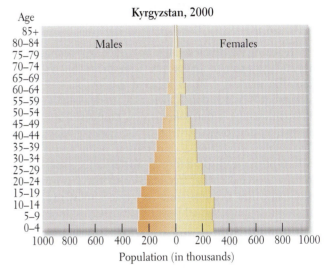

FIGURE 5.7 Population pyramids for Russia, Belarus, Kazakhstan, and Kyrgyzstan. Note that the pyramid for Russia is at a different scale (millions) from the other three (thousands), but this difference does not unduly affect the pyramid shape. [Adapted from U.S. Bureau of the Census, International Data Base, at http://www.census.gov/cgi-bin/ipc/idbpyry.pl.]

Quick Review

1. What is permafrost, and how does it affect the hydrogeography of Siberia?

2. Describe an important early trade route (circa A.D. 1000) through European Russia and explain what regions it connected.

3. How can the drop in male life expectancy in Russia since 1991 be explained?

4. Distinguish between communism and a command economy. How were these two systems related in the Soviet Union?

II   CURRENT GEOGRAPHIC ISSUES

The goals of the Soviet experiment begun in 1917 were unique in human history: to reform quickly and totally both a human society and its social, political, and economic systems. Now, in the aftermath of the Soviet Union's collapse, a second unique experiment is under way: the transformation of a communist society and a centrally planned economy into its antithesis—a society based on democracy and a free-market economy. But there is a lack of agreement on the wisdom of this experiment. At this point, the process has only begun, and for many, life is proceeding amid economic, political, and social uncertainty.

Many prefer less extreme reforms and the maintenance of a strong central government; such is the position of the present president of Russia, Vladimir Putin, who occasionally exercises autocratic control over the economy, the political process, and powerful new entrepreneurs.

ECONOMIC AND POLITICAL ISSUES

PERSONAL VIGNETTE   The kindly, gray-haired teacher tapped the blackboard under the words *profit* and *inventory* as the first-graders struggled to pronounce the Russian words. They were reviewing the story of Misha, a bear who opens a honey, berry, and nut store in the forest. He soon outsmarts the overpriced government-run Golden Beehive Cooperative, becoming first a prosperous bear and eventually the finance minister of the forest. [*Adapted from Sarah Koenig, "In Russia, teaching tiny capitalists to compete," New York Times (February 9, 1997).*]

Teaching economics in the first grade may seem a bit surprising, and the practice may not be widespread, but this story illustrates the marked shift in focus that took place in the 1990s throughout Russia and the newly independent states (Figure 5.8). One real change is that schools now emphasize the teaching of free-market economics instead of the ideas of Marx and Lenin. This shift in education is but one of many ambitious market reforms pursued in some former Soviet states, especially Russia. To fully appreciate what free-market reforms have meant to this region and the circumstances reformers must overcome, it is important to know something of the Soviet institutions that were previously in place.

The Former Command Economy

As we have seen, although the long-term goal of the command (or centrally controlled) economy was to achieve communism, its shorter-term goals were to end the severe deprivation suffered by so many under the czars and to distribute the benefits of economic development to all regions. To some extent, these goals were met: the Soviet economy grew rapidly until the 1960s, and abject poverty was largely conquered. Such basic necessities as housing, food, health care, and transportation were provided to all for free or at low cost—a remarkable accomplishment. Nonetheless, the command economy allocated goods and services inefficiently. Extreme scarcities of some goods and gluts of others made it clear that the Soviet economy was less efficient than market economies in allocating resources.

In a free-market economy, no single person or group decides what (and how much) will be produced or what prices will be. Production and pricing decisions are made independently by the managers of many privately owned competing companies, based on their judgments of what the market will bear. Hence, if one manager errs in predicting demand for a product, the mistake primarily affects his or her own company—which may fail or see the price of its stock fall—but the economy as a whole is not affected. In the Soviet command economy, however, bureaucrats set production goals for the whole country, which meant that even small miscalculations of production types and amounts created country-wide shortages or gluts. Shortages of food and raw materials occurred regularly throughout the Soviet era. Clothing was unimaginative, and most consumer goods—for example, washing machines, automobiles, and small appliances such as vacuum cleaners—were of poor quality and were available only at high cost to a privileged few. (Televisions and small refrigerators were an exception: they became widely available in the 1960s, and by the 1970s most urban households had these two appliances.)

Inefficient production methods were also typical in the command economy. In large part, this problem resulted from a lack of competition among producers. In free-market economies, competition among producers of goods usually encourages those producers to supply higher-quality goods at prices lower than their competitors because these are the goods that consumers tend to purchase. Because each Soviet factory tended to have a monopoly on the production of a particular product, Soviet goods became infamous for their high cost and low quality compared to similar goods produced in competitive free-market countries.

Another major problem of the Soviet Union was its failure to keep up with the rest of the industrialized world in technological and managerial innovation. Soviet engineers did develop advanced military and space exploration technologies, such as various nuclear weapons, world-renowned fighter jets, the first satellite (launched in 1957), and the first manned spacecraft (piloted by Yuri Gagarin in 1961). The Soviets also made technological breakthroughs in ferrous (iron) metallurgy and were pioneers in mainframe computer technology (later abandoned when personal computers became available). But in the civilian economy, Moscow's bureaucrats often settled for outdated technologies to produce consumer goods. The equipment typically used more resources, produced more pollution, and was less productive than equipment used in capitalist, free-market economies. At the same time, Soviet leaders paid little attention to the need for rewarding hard work and innovation in the work-

FIGURE 5.8 Political map of Russia and the newly independent states.

force. Rather than granting promotions and privileges to brilliant innovators, these rewards went instead to those who had loyalty to and personal connections with the Communist party.

Soviet Regional Development Schemes. A unique feature of Soviet central planning was that industrial projects were often located to serve political, rather than economic, purposes. For several reasons, the Soviet Union encouraged economic development in the farthest reaches of its territory. First, its leaders wanted to bring the higher standards of living enjoyed in industrialized European Russia to all parts of the region. The central planners and their military advisors also thought that dispersing industrial centers throughout the country would make them safer because an enemy would have to attack widely separated sites. In addition, there was some effort to locate new developments close to necessary resources. Thus the metal-processing firm Norilsk Nickel was situated near mineral beds in far northern Siberia (see the box "Norilsk: A City in Transition to a Global, Free-Market Economy" on page 262), and commercial cotton cultivation was linked to rivers in Central Asia that provided water for irrigation (see Figure 5.16 on page 267). Another factor in industrial location was the need to buttress Russia's claims to distant territories. Soviet leaders thought that economic development projects would placate remote ethnic minorities whose resources were being diverted to national uses and that the migration of ethnic Russian managers into minority areas would promote conformity with the dominant socialist ethic and with modernized ways of life.

In a market economy, many of the locations the Soviets selected for production centers would be seen as highly inefficient, mainly because it costs more to ship a product from a remote factory to a market center than it does to ship it from a nearby factory. Moreover, the far-flung industries were never sufficiently linked with ground transport to encourage voluntary extension of settlement and growth of local markets of the sort that fostered the spread of development in western North America. When the Soviet Union fell apart, the Soviet system of industrial location contributed to the drastic drop in national production because, once true costs were calculated, it was often too costly to transport goods from these factories to market at all.

Transport Issues. In locating industries in far-off corners of the region, the command economy ignored the tremendous transport problems posed by this strategy, given the region's huge size and challenging physical geography. Russia's rivers run north/south, while its primary transport needs are east/west. Ocean ports are few, remote, and often blocked with ice. Throughout the modern era, the development of land transport systems has been limited by long winters, permafrost, swampy forestlands, and complex upland landscapes, especially in Siberia. All of these conditions, plus a certain amount of bureaucratic inertia, conspire against road building; hence the former Soviet Union, more than two and a half times the size of the United States, has less than one-sixth the number of hard-surface roads and virtually no multilane highways. Most of the roads are in European Russia. In eastern Russia and Central Asia, existing roads are still of such poor quality that cities are linked primarily by rail and air. Nearly every Russian city has an airport, but the Trans-Siberian Railroad is the chief connection between Moscow

and Vladivostok, the main port city on the Russian Pacific coast (see Figures 5.6 and 5.10). The rail trip takes several days, and may be experienced in elegance or at lesser levels of comfort depending on the class of travel purchased. To drive the distance on roads would be practically impossible because, as writer Alexander Blakely reports, the Trans-Siberian Highway runs 3500 miles from Moscow to Lake Baikal, but then peters out. After a thousand miles of forest, the road picks up again and covers the last thousand miles to Vladivostok. He suggests thinking of the Trans-Siberian Highway as two very long roads with a thousand miles of forest between them, but notes that rumor has it the road will be completed in a few years. In European Russia, rail service is regular, and the network is dense, as is the case in western and central Europe. Feeder rails branching off main lines connect to industrial cities and resource supplies. The rail, air, and highway systems are supplemented by oil and gas pipelines built in the late twentieth century. During the 1990s, the Russian state transport agency, Rosavtodor, tried to assess the locations of the most likely markets for Russian products. As a result, Rosavtodor has recently increased road links to Europe by 10 percent, and has laid plans to build roads to the Central Asian republics, China, Mongolia, and the Pacific coast.

Reform in the Post-Soviet Era

Russia's economic reforms have been ambitious, but haphazard. Many key institutions of the command economy have been dismantled, but so far this change has enriched only a small group of wealthy and politically connected individuals, many of whom are corrupt. The lives of the majority have become more difficult in some respects, and certainly more unpredictable.

Two key economic reforms have been privatization and the lifting of price controls. **Privatization** (sometimes called **marketization**) is the selling of industries formerly owned and operated by the government to private companies or individuals, who, it is hoped, will operate them more efficiently in response to the free market. Estimates vary, but it appears that by 2000, approximately 70 percent of Russia's economy was in private hands, a significant change from the 100 percent state-owned economy of 1991. Even the remaining state-owned industries have changed the way they operate to follow free-market rules. In the old command economy, the government set the prices at which goods could be bought and sold. Prices on certain basic necessities such as bread, milk, and specific vegetables were often set lower than production costs, simply to make these goods affordable to all. Now prices are determined by supply and demand.

The lifting of price controls has had unexpected consequences. When price controls were removed and goods sold at their market prices, the prices skyrocketed, and the managers or new private owners gained wealth. But the drastically higher prices forced many people to use their savings to pay for simple necessities. One consequence was that individuals who would have liked to become small business owners lacked the capital to launch these enterprises.

Nor has privatization had the intended outcome. Among the first enterprises privatized were the most lucrative ones: those producing raw materials for export, such as oil, timber, or metals. Officials in the administrative hierarchy allowed buildings, factories, and access to resources to be sold for a fraction of their real value in return for kickbacks or bribes. Those who managed to acquire ownership of resources such as oil or timber, or the rights to their development, profited immensely. These Russian versions of "robber barons" are called **oligarchs** because their money affords them political power. Now that so much industry has been privatized, the profits collect in the pockets of the oligarchs instead of funding government social services or pension programs. These programs have become underfunded just as the need for them is increasing due to increasing unemployment. Many workers have lost their jobs as private companies have reduced their workforces to save money and operate more efficiently.

Vladislav Dudakov came to Moscow as a soldier assigned to guard Lenin's tomb. Now, in a new Russia where entrepreneurship is encouraged, he manages a shop for an investor who has opened up a chain of Starbucks-style coffee shops. One of the unique challenges of his job is managing employees who are used to working for the government. He has to teach Western business values, just as he was taught them during his first Western-style job: working for McDonald's. When McDonald's came to Russia, Vladislav traded the life of a soldier for that of a businessman, working his way from floor sweeper to store manager: "I was different from my peers. They would say, 'Why don't we extend our break a bit? Why do we have to scrub so hard?' My answer was: you must fulfill the task. At McDonald's we were always earning achievement awards. . . . The American style of management became my education." Vladislav's eagerness to learn earned him a trip to McDonald's "Hamburger University" in the United States. Now, he encourages the waitresses at his coffee shop to forget old Soviet habits and to "smile, smile, smile." His success has made Vladislav an example of a new class of Muscovite: young, entrepreneurial, and newly affluent.

Hear more firsthand accounts of Russian capitalism in the Frontline World video "Moscow: Rich in Russia: The Brave New World of Young Capitalists and Tycoons."

Foreign Direct Investment. The largest sources of capital for development in Russia are outside the country, primarily in Europe, Asia, and the United States. A report in 2003 by Elena Rogacheva and Julia Mikerova showed that during the 1990s, after a brief upsurge, foreign investors lost interest in Russia, in large part because of the lack of transparency and rampant corruption. Instead, they put large sums into the countries of East and Central Europe, which have much smaller but more predictable economies. The largest receptacle for foreign investment is the energy sector (oil and gas), which receives five times the investment dollars of the food and drink sector (the next largest). But energy is also the sector that attracts the most government interference, partly because in the early 1990s a few oli-

garchs managed to make off with huge profits, and several have attempted to turn their ill-gotten wealth into political power. Russia has some of the world's largest oil and gas reserves, which promise to be important sources of income for the country in the future. There was a slight upsurge in FDI in Russia in 2002 to U.S. $3.5 billion, but it dropped to U.S. $1.1 billion in 2003, probably because of the government's move against some of the oligarchs, especially Yukos Oil in the energy sector and broadcast media firms. Nonetheless, foreign investors can be expected to resume their interest in Russia's natural resources when the intentions of the central government are made clear.

The Growing Informal Economy. The informal economy—that part of the economy that operates outside official regulation and often involves illegal activities—is a major problem throughout the former Soviet Union because governments are unable to collect tax revenue on this economic activity. The informal economy in this region encompasses a large share of the total economic activity, and the taxes generated by the formal economy are not sufficient on their own to cover government obligations to the public. To some extent, the new informal economy is an extension of the old one that flourished under communism. The "black market" of that time was based on currency exchange and the sale of hard-to-find luxuries. For example, in the 1970s savvy Western tourists could enjoy a vacation on the Black Sea paid for by a pair or two of smuggled Levi's blue jeans and some Swiss chocolate bars. Today, however, the range of informal ventures involves more people and more occupations than did the old black market because many people who have lost stable jobs due to privatization now depend on the informal economy for their livelihoods. Often they operate out of their homes, selling cooked foods, making vodka in their bathtubs, assembling telephones, or pirating and selling computer software, just to name a few activities. Indeed, so much of the economy is in the informal sector that in many countries of this region, people may be better off than official gross domestic product (GDP) figures suggest. For example, the true agricultural output is underreported because so many farmers, still employed on declining collective farms, now earn a considerable portion of their income from the private gardens they cultivate on the fringes of those farms.

To demonstrate the scope of the lost tax revenue, consider the effect of the underground economy on vodka production. Homemade and covertly produced vodka now accounts for 70 percent of the Russian market, drastically reducing sales of the legitimate vodka that was once a large source of tax revenue for the government. In the late 1990s, the average Russian was estimated to drink between 25 and 45 liters of vodka a year, and even with 45 percent of sales then untaxed, vodka produced nearly as much tax revenue as oil. The loss of tax revenue means that the Russian government has not had sufficient money to pay the salaries of workers in the remaining state agencies. Pensions have gone underpaid, and many essential public services such as electricity and garbage collection have been insufficiently funded. In June 2000, the government set up a state watchdog agency to oversee and perhaps return state control to the alcohol industry so that taxes could again be levied and collected. Tax dodging

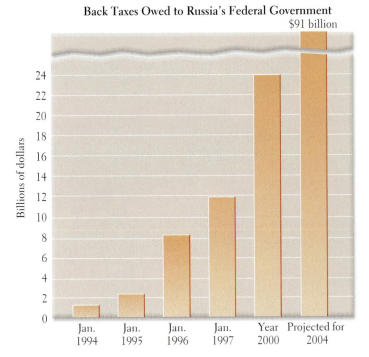

Back Taxes Owed to Russia's Federal Government

$91 billion

Billions of dollars

| Jan. 1994 | Jan. 1995 | Jan. 1996 | Jan. 1997 | Year 2000 | Projected for 2004 |

FIGURE 5.9 Tax debt in Russia. The unpredictable state of the Russian economy motivated some successful entrepreneurs to bank their profits outside the region, so the actual sum lost to taxation may exceed the figures indicated here. Tax evasion is now common grounds for legal action, even against the very influential oligarchs. [Adapted from *New York Times* (February 19, 1997): A8; *The Russia Journal* (June 17–23, 2000): 12.]

(Figure 5.9), which does not carry the stigma it does in other free-market countries, has reduced tax revenues even from businesses in the formal economy. Only in the last several years have tax collections increased markedly. Officially, the government says that as of 2002, 95 percent of known taxes due were collected, yet economists speculate that tax collectors are retrieving only about half the actual taxes due.

Debt Crisis. Russia's large foreign debt dominates the country's financial relationships with other countries. The money is owed to private commercial lenders, foreign countries (Germany is Russia's biggest creditor by far), and the IMF and the World Bank. The debt mounted after 1991 when the government had to borrow because tax collections waned at the same time that exports lagged, industries closed, and unemployment rose. In 1998, Russia's debt was a staggering 90 percent of its GDP (U.S. debt at the time was about 10 percent of its GDP). Two-thirds of this debt was incurred by the Soviet Union prior to 1991, and theoretically, the newly independent states should help Russia repay this money because some of it was borrowed for projects within their borders. But they have no surpluses to dedicate to this cause and little incentive to do so. In addition, the war in Chechnya (see pages 258–259) has added to Russia's debt at the rate of $200 million a year.

In early 2000, Russia defaulted on its debt payments and then requested debt rescheduling with all its creditors. Western lenders wanted to make sure that any rescheduling or forgiveness of debt was tied to Russia's commitment to achieve a free-market econ-

omy, reduce theft and corruption, and encourage Russian entrepreneurs to reinvest their profits in Russia, not simply keep them in foreign banks. For its part, Russia worried that its annual obligatory payments, if not rescheduled, would undermine its economic growth. In 2000, private lenders agreed to cut 36 percent off the $32 billion Russia owed them and to extend payments over a 30-year period. As of 2004, Russia was paying off its debt (down from 90 percent of GDP to just 28 percent), and its economy, after many years of decline, was growing at a rate of about 4 percent per year. Russia's growth was helped by the fact that elites were no longer sending all their savings out to foreign banks, but rather were investing in Russian projects. It is interesting to note that a number of foreign countries owe more than $100 billion to Russia, as the successor state of the Soviet Union. This money was loaned by the Soviets to their allies, such as Vietnam, Cuba, Mongolia, Mozambique, Angola, Algeria, and Yemen, but is unlikely to ever be repaid.

Market Reforms and International Trade. For decades, Russia had a colonial relationship with its associates in Eastern Europe, the Caucasus, and Central Asia, appropriating their resources in uneven exchanges that left some republics poorer and more dependent on Russia than they would have been otherwise. Today, despite Russia's struggle to maintain what is left of these arrangements, several former Soviet republics have already drifted far out of the sphere of Russian influence. The Baltic republics of Latvia, Lithuania, and Estonia and some of the central European countries that were formerly within the Soviet sphere of influence (Poland, the Czech Republic, Slovakia, and Hungary) are already so much more closely tied to Europe that they are not discussed in this chapter. Belarus retains close economic ties with Russia; the two countries trade freely without tariffs or customs duties, and there is talk of their eventually

In an effort to supplement their government pensions (many of which have not been paid for months), some older citizens here at Petrovskiy Zavod sell food to Trans-Siberian Railroad passengers, using old baby carriages as carts. [Gerd Ludwig/National Geographic Image Collection.]

reuniting politically. Ukraine and Moldova would like to be invited to join the European Union, but an invitation is not likely to come soon. Both countries retain systemic ties to Russia.

Other countries in the region remain economically connected with one another and with Russia by arrangements that may not last. For example, the cotton growers of the Central Asian republics currently sell their product to textile mills located within Russia, which exports cotton cloth back to Central Asia. In time, Central Asia will most likely build its own cotton mills and reap the profits of exporting finished textiles to the global marketplace. Similarly, Central Asian countries, whose oil and gas deposits and industries are unevenly distributed across the subregion, currently sell energy resources to Russia at prices below global market price, but this arrangement is unlikely to continue. In the future, they are likely to sell to the global market and to use their oil and gas resources to gain geopolitical advantages. Multinational energy companies are now pressuring Central Asian countries to bypass Russia and send their oil instead to the world market through new delivery systems that would run through Caucasia, Turkey, Afghanistan and Pakistan, or China. Russia has its own initiatives regarding future trade. Exports to the United States, the European Union, and China are increasing and have the potential to grow rapidly. Most Russian exports are raw materials, but fine finished goods and refined gasoline are beginning to make a mark (see the vignette on LUKOIL on this page).

Obstacles to International Trade. Russia and the newly independent states seek to gain wealth by increasing exports to the world market. To reach that common goal, they must establish new international trade relationships. But there are some roadblocks on their way to prosperity. First, they are accustomed to trading with one another under terms set by central planners. Each part of the region specialized in—and often held monopolies on—the production of specific products, such as oil, heavy equipment, cotton, electronics, and food crops. With new national borders now separating these regions, each country must establish new trade agreements to ensure sufficient food and consumer goods for its populace and adequate resources for its industries. Second, all must improve their products and manage their production costs to compete globally with producers in Europe, America, and Asia. Most former Soviet countries lack marketing expertise, and they must acquire these skills rapidly. Many of these countries now maintain Web sites on the Internet to familiarize the public with their products and to seek investors.

The underdeveloped transport system has created further obstacles for the newly independent states as they try to trade with partners in the global economy. Transportation routes are not always in ideal locations for emerging patterns of foreign trade. For example, Kazakhstan and Uzbekistan, independent since 1991, are increasing their direct trade with the outside world, but they are landlocked and still have only feeder rail connections with what are now Russian railroads (Figure 5.10). Their oil and gas pipelines flow to Russia (see Figure 5.11), not to other customers, and their air links, long focused on Moscow, have had to be refocused on Europe and Asia.

The Vostochny International Container Services (VICS) at Vostochny Port, in the Russian Far East, connects with the Trans-Siberian Railroad, a high-capacity, double-track electric line that reaches Central Asia and Europe. The combination offers the shortest route from East Asia to central China, Central Asia, Russia, and Europe at competitive rates and transit times. A shipment from Shanghai to Finland, for example, takes 23 days, roughly half as long as sea routes. [Courtesy of Vostochny International Container Services.]

PERSONAL VIGNETTE   It is September 26, 2003, and exciting changes are afoot at Paramgit Kumar's Manhattan gas station. He had managed the gas station for Getty Petroleum Marketing for several years, before Getty was purchased by Moscow-based LUKOIL. Now he is one of the first to join LUKOIL's plan to rebrand the old Getty stations—to make it obvious that Russian oil products are sold here. The Russian president, Vladimir Putin, in the United States for a meeting with President George W. Bush, stopped by to officially open Kumar's station, which sports a bright white and red logo.

One measure of the increasingly global influence wielded by Russia's oil and gas industries is their expansion into retail marketing in North America. In 2000, LUKOIL, Russia's largest oil company, purchased 72 percent of Getty Petroleum Marketing, a chain of 1300 gas stations in 13 eastern U.S. states. This purchase was the first public acquisition by a Russian corporation of a U.S. firm. Although the deal is small ($71 million was paid), it is important psychologically and could lead to other Russian investments in the United States. LUKOIL plans to expand the chain to 3000 stations, more than doubling the number of its U.S. stations.

Originally, LUKOIL was going to retain the Getty brand name, but in the aftermath of September 11, 2001, Presidents Bush and Putin began discussing ways to decrease U.S. dependence on oil from Southwest Asia (Saudi Arabia, the UAE, Kuwait) and to increase U.S. purchases of oil from Russia. Unexpectedly, due to post-9/11 geopolitics, a Russian identity for U.S. gas stations turned into a marketing advantage.

In the 21st century, transport of goods and people will become easier in the region as links open to the west.

The heaviest concentrations of industrial sites are inside Russia's border partly because they are near natural resources. Russia sold the products to other Soviet republics.

Transport problems plagued the USSR due to huge distances and challenging physical geography. The Trans-Siberian Railroad is the chief link between Moscow and the east.

Industries were located in remote places during Soviet times for a variety of political and economic reasons.

Industrial Regions

- Timber and wood products
- ▲ Ferrous ores and metals
- ▪ Nonferrous metals
- Industrial area
- ▬ Main trunk line, Trans-Siberian Railroad
- — Other railroads

Miles 500
Kilometers 800

FIGURE 5.10 Economic issues: Principal industrial areas and land transport routes of Russia and the newly independent states. The industrial, mining, and transport infrastructure is concentrated in European Russia and adjacent areas. The main trunk of the Trans-Siberian Railroad and its spurs link industrial and mining centers all the way to the Pacific, but the frequency of these centers decreases with distance from the borders of European Russia. [Adapted from Robin Milner-Gulland with Nikolai Dejevsky, *Cultural Atlas of Russia and the Former Soviet Union*, rev. ed. (New York: Checkmark Books, 1998), pp. 186–187, 198–199, 204–205, 216–217; and http://www.travelcenter.com.au/russia/images/trans-sib-map-v3.jpg.]

To facilitate this business venture, at the Russian (supply) end, LUKOIL converted a former Soviet military base at Murmansk on the Barents Sea into a processing facility capable of storing and shipping a million barrels per day to the United States. At the U.S. (demand) end, it made low-cost loans available to station managers like Kumar, who will need capital to make the costly upgrades necessary for the rebranding. [*Adapted from John Lofstock, "The Russians are coming," Convenience Store News, October 12, 2003, http://www.csnews.com/csnews/reports_analysis/oilwatch_display.jsp?vnu_content_id=1990275.*]

Supplying Oil and Gas to the World

Crude oil and natural gas are the most lucrative exports of the entire region. In the post-Soviet era, a tug-of-war has developed between Russia and multinational energy companies over the rights to develop Central Asian oil and gas (see the discussion on page 280). Despite the large oil reserves recently discovered in Central Asia, Russia on its own has the largest reserves of the region and remains the prime oil exporter in the region (Figure 5.11). In 2002, Russia was itself consuming the equivalent of about one-third of the oil it produced. In the same year, Russia was the largest producer of natural gas (the United States was second) and itself consumed about two-thirds of this production.

In Russia, oil companies and Gazprom (Russia's joint government and private gas extracting and exporting monopoly) already account for more than half of federal tax receipts. But much potential revenue has escaped because of the uncontrolled way in which Russia's oil and gas resources were privatized. Russia is now tightening its control over the oil and gas sector, strengthening rules to ensure that the industry pays all its taxes regularly. The possibility of reinstituting state ownership of oil and gas is mentioned from time to time, but that discussion alarms other potential private investors in the region. As world oil and gas production and prices rose and access to global markets increased into the mid-2000s, tax proceeds also increased, helping the budgets of Russia and of the newly independent states.

Greater Integration with Europe and the United States

Greater integration of the economies of Europe and the United States with those of Russia and the newly independent states would serve the interests of all concerned, yet there are worries all

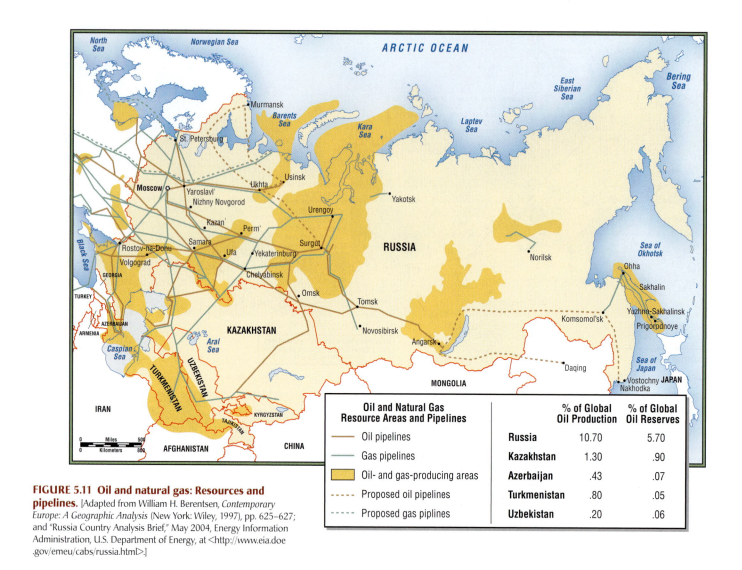

FIGURE 5.11 Oil and natural gas: Resources and pipelines. [Adapted from William H. Berentsen, *Contemporary Europe: A Geographic Analysis* (New York: Wiley, 1997), pp. 625–627; and "Russia Country Analysis Brief," May 2004, Energy Information Administration, U.S. Department of Energy, at <http://www.eia.doe.gov/emeu/cabs/russia.html>.]

Oil and Natural Gas Resource Areas and Pipelines	% of Global Oil Production	% of Global Oil Reserves
Russia	10.70	5.70
Kazakhstan	1.30	.90
Azerbaijan	.43	.07
Turkmenistan	.80	.05
Uzbekistan	.20	.06

Legend: Oil pipelines; Gas pipelines; Oil- and gas-producing areas; Proposed oil pipelines; Proposed gas piplines

around as well. The oil and gas trade with Europe, the United States, and Asia does much to improve the economies of Russia and the newly independent states and to connect them with the inner circle of industrialized nations. At the same time, the United States and Europe, motivated by the unspoken worry that some of these countries might return to foe status if their economic and political burdens are not eased, are seeking ways to bring them all into closer formal association with established trading institutions. Negotiations are under way to make Russia part of the World Trade Organization, and Russia has become a member of what was once the Group of Seven, an organization of the most highly industrialized nations, making that group the **Group of Eight (G8).** In part, such invitations are intended to reassure Russia. Its former allies Latvia, Lithuania, Estonia, Poland, the Czech Republic, Slovakia, and Hungary had all joined the European Union as of May 2004, and Poland, the Czech Republic, Slovakia, Hungary, Bulgaria, and Romania are now members of the North Atlantic Treaty Organization (NATO), the old cold war military alliance against the Soviet Union. Russia also craves closer association with the European Union because the EU has

been so successful in reorganizing the economic and political face of Europe and wields great influence globally.

The EU is also tempting the newly independent states with possible trade and guest worker agreements in exchange for compliance with European trade-related standards. These standards include regulations on personnel training, the personal safety of workers, and product liability; requirements for carrying various kinds of insurance; rules governing finance, bankruptcy, and contract law; environmental codes on pollution and endangered species; border control standards; and licensing standards of various kinds. Once these standards are met, WTO and NATO membership can be considered. Those hoping to join eventually include Ukraine, Moldova, various Caucasian republics, and Russia itself. Their desire to eventually be part of the European Union makes these countries willing to work toward conformity with European standards. Europe sees upgrading conditions in such places as Ukraine, Russia, and the countries of Caucasia as a way of developing a buffer zone between itself and unbridled illegal immigration from Central and East Asia. It is thought that each year, hundreds of thousands of illegal immigrants make their

PRACTICING GEOGRAPHY

How do the new local governments in Russia balance their budgets?

I started out studying the Soviet Union and am now a Russian specialist. At the time of the breakup of the former Soviet Union, I knew immediately I wanted to continue to study contemporary phenomena in the region. If anything, the former USSR is now a more attractive place to conduct research because so many new research topics and methods are available under the new government. During the Soviet period, it was impossible to interview and survey local government officials and to view materials about their budgets and policies. It was forbidden to copy government documents and, in some places, even to read budgets.

One of my recent projects was a study of local economic development policies. During the Soviet period, most of the policy making on development issues was done by the central government in Moscow. Imagine if policy makers in Washington, D.C., had the authority and the responsibility to make investment decisions in every city in the United States and to determine policy priorities in your hometown—with very little input from the local population and politicians! Soviet local governments were not responsible for funding their own budgets, but in the post-Soviet period, cities had to learn how to balance budgets and finance their own policies.

In the early 1990s, the first "democratic" elections were held to elect public officials, who were for the first time directly accountable to their constituents. Suddenly, these officials, with little or no funds at their disposal, had to steer and fund local economic development initiative. I developed a research project to survey and interview newly elected and other government officials to find out what they considered their major policy problems and priorities.

With nearly 700 responses from 70 different regions, we now better understand the policy-making environment in Russian cities in the mid-1990s. We found that policy makers agreed that their major policy problems relate to industrial and economic decline and that they are still beholden to federal officials in deciding policy priorities and resolving problems. However, public officials also felt that business leaders now played an important role in defining and resolving problems. We are puzzled by the apparent lack of geographic and spatial variation in the identification of policy problems and their resolution—everyone seemed to voice the same concerns—but we suspect that the similarities are in part due to past emphasis on central planning, which caused similar problems across the region.

Beth Mitchneck
Department of Geography,
University of Arizona

way into Europe via Russia and the newly independent states from such trouble zones as Afghanistan, Iraq, Iran, Ethiopia, Congo, and Somalia. Europe, which buys 20 percent of its natural gas from Russia, is also interested in negotiating lower gas prices.

Political Reforms in the Post-Soviet Era

Attempts to foster democracy in Russia and the newly independent states have been slower and more difficult than economic reform. Many observers in the West expected that the introduction of a market economy would go hand in hand with democratization, but this has not happened. Although several countries have held elections in which there were competing candidates for office, forms of authoritarian control remain. For example, Belarus, Russia, and the Central Asian republics have very strong presidencies. In these countries, elected representative assemblies often act as rubber stamps for presidents and exercise only limited influence on policy. Across the region, even though officials are elected, they sometimes are overly influenced by oligarchs that wield influence because of their enormous wealth.

To be fair, it is not clear whether a majority of the people in the region are prepared to participate more fully in the decision-making process: these people have never had their own voice in governance, and the last 80 years have not encouraged civic responsibility. Even today, ordinary citizens tend to think of political leaders as patriarchs who can dispense favors to individual citizens and produce simple solutions to complex problems. Furthermore, many citizens are now so concerned with daily survival and security that they have little energy for regular political participation. Parliamentary elections in December 2003 brought out just 50 percent of the eligible voters (in the United States in 2000, 60 percent of those eligible voted). Ralph Clem and Peter Craumer noted that the Russian electorate is increasingly differentiated geographically: older, rural, less educated Russians tend to vote in higher numbers and favor socialist parties, while urban, younger, better educated and white-collar Russians favor reform parties, but are less likely to vote at all. However, public participation in civil society in the form of demonstrations for or against Kremlin policy is now common. In November 2003, when President Putin ordered the arrest of the oil oligarch Mikhail Khodorkovsky, public demonstrators openly protested. Their banners pointed out that Khodorkovsky was most likely jailed not because of tax evasion, as the Kremlin claimed, but because he publicly challenged the political future of President Putin. Protesters in front of the Kremlin also pointed out that the arrest had a potential economic effect because it chilled the climate for outside investors.

The Media and Political Reform. In the Soviet era, all communications media were under government control. There was no free press, and public criticism of the government was punishable. Yet it was journalists, who risked punishment by engaging in increasingly vocal critiques of public officials and policies, who were instrumental in bringing an end to the Soviet Union. Many observers thought that an independent press would bring about democratic practices, which would in turn ensure the formal establishment of free speech and public debate. Between 1991 and the early 2000s, the communications industry was a center of privatization, and several media tycoons emerged. Privately owned newspapers and television stations regularly criticized various leaders of Russia and the other republics and their policies. It appeared that a free press was developing, and for a time, such leaders as President Vladimir Putin openly lauded a free press as vital to the growth of civil society. Nonetheless, Russian citizens are ambivalent about just how free the press is, or should be. When media tycoon Vladimir Gusinsky, whose newspapers and broadcast facilities constantly criticized the government, was arrested and detained in June 2000, a survey on Moscow's streets showed little outrage among ordinary citizens. Most did not see the arrest as a government effort to squelch free expression because Gusinsky himself was an oligarch. After his detention (Gusinsky now lives in England), his media facilities were acquired by the government. Local and international observers report that critical analysis of the government is now rare. During the parliamentary election of 2003, the media openly supported Putin supporters and lambasted his opponents. Putin's party won an overwhelming majority of the parliamentary seats.

Closely related to the development of a free communications industry is the availability of communications technology to the general public (Figure 5.12). Television sets were widely available in the Soviet Union, but programming was tightly controlled by the government and was free of advertisements. Now there are advertisements, programming is less tightly controlled, and many people receive European and other stations via satellite dishes. Access to telephone lines is still limited, but mobile phones have made instant personal communication possible in even the most remote cities of Siberia and Central Asia, where they are essential to many new enterprises. Mobile service, however, remains expensive and unreliable, and it is not available to most people.

Mobile phone companies are hoping to help the newly independent states leapfrog into the global economy by facilitating Internet commerce. Throughout the region, however, personal computers are still rare—Russia has one-tenth the computers per capita that the United States has. Nonetheless, information technology is available, and demand is growing. Commercial use of the Internet in the newly independent states is hampered by

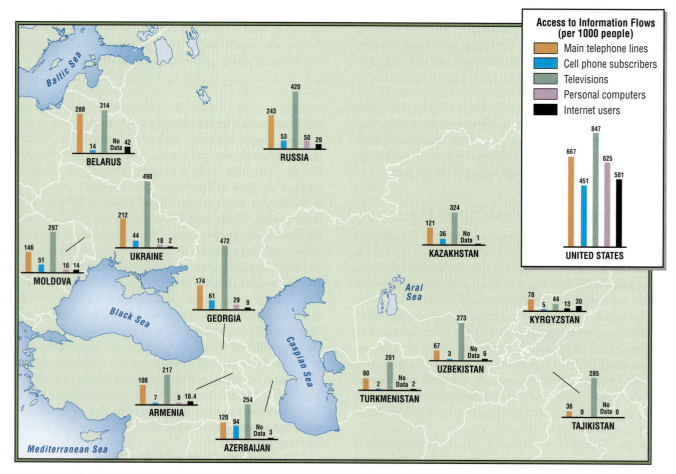

FIGURE 5.12 Access to information flows: Russia and the newly independent states. Access to information throughout the region is low compared with that in more developed countries. [*United Nations Human Development Report 2003*, United Nations Development Programme, Table 11, MGD 10.]

the lack of easy access for all and a lack of legislation to control the integrity of all Internet transactions. The Russian Academy of Science is participating in the framing of such legislation, and already Russia's leading newspapers have free access to Internet lines.

The Military in the Post-Soviet Era. The future role of the military is one of the most important political issues facing the region today. The military was once the most privileged sector of society, and its officers had access to goods and services that ordinary citizens could only dream about. Now, after huge funding cuts, the Russian military has been reduced by hundreds of thousands. In 2000, there were just 1.5 million troops, and further cuts were proposed by President Putin. Military personnel returning home from far-flung posts all across Central Asia and Siberia actually pose a threat to civil society. With few skills, few job prospects, little decent housing for their families, and sometimes not even adequate clothing, these former soldiers form a huge reservoir of discontent. Nonetheless, although it was a botched coup by top military commanders against President Gorbachev in August 1991 that led to the formal dissolution of the Soviet Union, further military rebellions have not occurred. If Russia is invited to join NATO, some of the unemployed military personnel may find work with that organization, or as United Nations peacekeepers. A more immediate concern is that the former Soviet nuclear arsenal, much of which remains deployed in the newly independent states, could fall into the wrong hands if military cuts are poorly managed. There is some evidence that nuclear expertise and even some warheads may have been smuggled out of the region to such countries as North Korea, Iran, and Iraq, whose governments are widely distrusted by the international community.

The Political Status of Women. Although granted equal rights in the Soviet Union's constitution, women never held much power in government and hence had little say in policy. In 1990, women accounted for 30 percent of Communist party membership, but they made up just 6 percent of the governing Central Committee, and generally women did not have a strong political voice. Since the fall of the Soviet Union, women have not fared well in the election process. While once they were appointed by party officials, now they are elected by the people, and as a result they have lost political representation in the parliaments of Russia and the Caucasian republics, dropping from 20 to 40 percent in Soviet times to just 3 to 10.5 percent in 2003 (Figure 5.13). Women in Belarus, Tajikistan, and Turkmenistan have fared slightly better.

Nonetheless, by the late 1990s, especially in Caucasia and the Central Asian republics, newly aware women were finding ways to influence policy by working through nongovernmental organizations (NGOs) and such agencies as the United Nations Development Program (UNDP). Hence formal representation in parliament is no longer the only, or even the best, measure of women's political activity and influence. Still, support for women's political movements is not widespread, despite strong evidence of sex discrimination. Women fear that working for women's rights and well-being would be seen as antimale or as a rejection of traditional feminine roles, and this is itself a sign of their low political status.

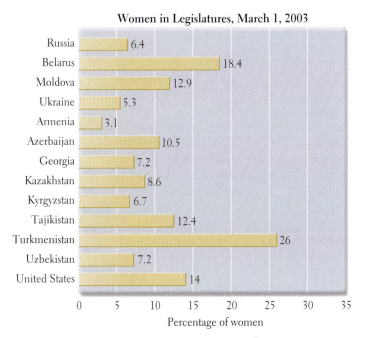

Women in Legislatures, March 1, 2003

Country	Percentage
Russia	6.4
Belarus	18.4
Moldova	12.9
Ukraine	5.3
Armenia	3.1
Azerbaijan	10.5
Georgia	7.2
Kazakhstan	8.6
Kyrgyzstan	6.7
Tajikistan	12.4
Turkmenistan	26
Uzbekistan	7.2
United States	14

Percentage of women

FIGURE 5.13 Women legislators. Percentage of women legislators in Russia and the newly independent states (with the United States for comparison). Belarus and Turkmenistan stand out as having the most women lawmakers, but both are authoritarian societies in which true democratic participation is rare. [Adapted from *United Nations Human Development Report 2003*, United Nations Development Programme, Gender Empowerment Measure, Table 23.]

Russia's Internal Republics

Russia's centuries-long history of expansion into neighboring lands has left it with an exceptionally complex political geography. More than 75 percent of the country is divided fairly systematically into provinces (called **oblasts**), but Russia also contains more than 30 **internal republics** and more than 10 so-called **autonomous regions** (Figure 5.14). The internal republics and autonomous regions have been set aside as the homelands of non-Slavic peoples. They are only somewhat analogous to counties, states, or reservations in the United States. Russia and these internal republics and autonomous regions together are often referred to as the **Russian Federation.**

The territory covered by the Russian Federation was gradually claimed over the last 500 years by the European Russian state (see Figure 5.5 on page 242). As the Russian czars and the Soviets pushed the borders of European Russia eastward toward the Pacific Ocean, they conquered a number of small non-Russian areas, which became republics in the Russian Federation. Many of the republics have significant ethnic minority populations that trace their origins to peoples who spoke such languages as German, Turkish, Finnish, Ugric, or Persian. The lands of the Yakutiya (now Sakha) in Siberia cover large, resource-rich areas that are of special interest to Russia. The peoples of some of the internal republics, particularly the more southerly ones, are followers of Islam, the dominant religion in Central Asia as well as Southwest Asia (see Chapter 6). The peoples of most other republics are Christian, but those of Kalmykia, between the Volga

FIGURE 5.14 Russia's political subdivisions and Russian population as a percent of total population in independent states. Russia with all its internal republics and autonomous regions is called formally the Russian Federation. Here, the provinces are outlined in white, and 20 of Russia's more than 30 internal republics are outlined in green. The ethnic character of many of the internal republics has been changed by the policy of central planning, known as Russification. The pie charts for the newly independent states show that, in these countries, Russians form significant minorities. [Adapted from James H. Bater, *Russia and the Post-Soviet Scene* (London: Arnold, 1996), pp. 280–281; and Graham Smith, *The Post Soviet States* (London: Arnold, 2000), p. 75.]

and Don rivers, are Buddhist, and some groups have animist beliefs that predate any of the organized religions.

Even under the czars, assimilation of all minorities to Russian (Slavic) ways (called Russification) was the central policy. During the Soviet era, ethnic Russians and other Slavs were settled in the internal republics in order to modernize and civilize what were viewed as culturally marginal people. Slavic immigrants often outnumbered native people, and they tended to receive the best jobs and housing. Resistance to Russification often enhanced local ethnic identities.

The internal republics vary greatly in size and have a bewildering array of political relationships to the Russian state. Their status is being continually renegotiated republic by republic. Russia has tried to avoid formally addressing the political status of the internal republics because if these ethnic enclaves were to demand greater autonomy, or even independence, the territorial integrity of the vast Russian Federation would be threatened. Shortly after the breakup of the Soviet Union, several internal republics did demand greater autonomy, and two of them, Tatarstan and Chechnya, declared outright independence. Whereas Tatarstan has since been placated by significant efforts

to increase its economic and political viability and autonomy, Chechnya was much more resistant. Russia responded to the Chechen rebellion with a major show of military force that has led to the worst bloodshed of the post-Soviet era. For the time being, most of the other ethnic enclaves are content to stay closely associated with Russia.

The Conflict in Chechnya. Chechnya, a small internal republic on the northern flanks of the Caucasus Mountains (number 5 in Figure 5.14; see also Figures 5.17 and 5.21), has a green, fertile, hilly landscape reminiscent of the Cumberland Mountains in eastern Tennessee. It is home to 800,000 people, about 130,000 of whom have been displaced by war since 1991. The Chechens converted to the Sunni branch of Islam in the 1700s; ever since then, Islam has served as an important symbol of Chechen identity and of resistance against the Orthodox Christian Russians, who annexed Chechnya in the nineteenth century. After the Russian Revolution in 1917, the Chechens and other groups in the Caucasus formed the Republic of Mountain Peoples, hoping to separate themselves from Russia, but the Soviets abolished the republic in 1924. Chechens were forced onto collective farms in the 1930s, and

during World War II they were accused of collaborating with the Germans.

In 1991, as the Soviet Union was dissolving, Chechnya declared itself an independent state. To the Russians, this declaration represented a dangerous precedent because they feared that Chechen independence could spark similar demands by other cultural enclaves throughout Russia. In addition, Russia wished to retain the region's agricultural, oil, and gas wealth. There are also suspicions that Russia may wish to use Chechen territory for the transport of oil or gas to Europe. Since 1991, Chechen guerrillas seeking independence have repeatedly challenged the Russian army, which has responded by bombing the capital of Groznyy and carrying out other reprisals that have left thousands of civilians homeless or dead. Chechens have claimed responsibility for taking hostages in a Moscow theater in 2002, a massive suicide bomb in Groznyy in January 2003, and the terrorist bombings of two passenger planes near Moscow and of a large grade school in North Ossetia in 2004. Although many Chechens claim to be fighting simply for independence and the right to their own resources, Russia claims that the factions that engage in terrorist bombings have links to Muslim extremist groups in Central Asia—perhaps even to global terrorists—and hence represent a threat to Russia's long-term security. Russia's ineffective attempts to address the troubles in Chechnya early on, before people resorted to terrorism, have raised doubts about Russia's commitment to human rights and its overall ability to address internal political dissent.

SOCIOCULTURAL ISSUES

When the winds of change began to blow through the Soviet Union in the 1980s, political and economic repercussions were expected. But few anticipated that the transition to a market economy and away from central control of most aspects of life would happen so speedily and be so unstructured and so disruptive of social stability. On the one hand, the new freedoms have encouraged self-expression and individual initiative for some people, as well as political participation and cultural and religious revival. On the other hand, the collapse of structure once provided by the socialist state has resulted in the loss not only of livelihoods, but also of housing, food, health care, and civil order. In the wake of uncontrolled change, widespread corruption took root.

Cultural Dominance of Russians and Other Slavs

The Russians always regarded themselves as the heart of the Soviet Union, a point of view that originated with the czars when they extended the Russian Empire over adjacent regions. Throughout the twentieth century, Russians felt closest culturally to the European Slavic republics of Ukraine, Moldova, and Belarus and to the Baltic states, whereas the Caucasian and Central Asian republics were of interest primarily for the resources they possessed and their strategic location. The distinctive languages, religion (Islam), and cultural origins of Caucasia and Central Asia set these republics apart. The Soviets considered this distinctiveness to be a liability to the cohesiveness of the Soviet Union, yet the resources these republics contained were indispensable. To ensure compliance with regional economic plans, to provide a skilled labor force, and to acculturate minorities to Slavic (especially Russian) customs and attitudes, the Soviets resettled Slavic people in all the peripheral republics (now the newly independent states), just as they did in the internal republics. Ethnic Slavic technicians, teachers, and professionals occupied the choice positions throughout the region. Russians and other Slavs entered all the republics of the Soviet Union in such large numbers that in some locations, they were significant and powerful minorities (see Figure 5.14, page 258).

Across the region, Russian culture dominated all aspects of public life. For example, although more than 40 legally recognized languages were spoken throughout the Soviet Union, Russian was the language of official business and was taught in the schools. Local languages were forced to use the Cyrillic (Russian) alphabet. By the 1980s, the use of minority languages was in decline everywhere in the Soviet Union. Ancient local customs of non-Russian peoples—including religion, family organization, domestic architecture, manner of dress, farming, and diet—were considered outmoded and were suppressed. In the school curriculum, Russian culture was presented as the norm. The Russians were especially intolerant of the Muslim religion in the internal republics and in Caucasia and Central Asia.

Until the mid-1970s, migration within the Soviet Union consisted primarily of ethnic Slavs engaged in Russification, while comparatively few people from Eastern Europe, Caucasia, Central Asia, and the internal republics moved to European Russia. Those from the periphery that did move did so temporarily for education and training. Then, in the 1980s, just preceding the demise of the Soviet Union, migration between Russia and its neighbors slowed and shifted into a new pattern: return migration. In the 1990s, as many as a third of the Russian and other Slavic emigrants returned home to Russia from the newly independent Caucasian and Central Asian states. Conversely, ethnic minorities in Russia went home to the Caucasian and Central Asian states or to internal ethnic enclaves. Most Russians in the newly independent European states (the Baltic states, Belarus, Ukraine, and Moldova) did not return to Russia. Instead, they remained in their host counties, where they have become a significant minority and where they saw a brighter future than in Russia.

Cultural Revival in the Post-Soviet Era

Since the achievement of independence and the departure of many Russians, the newly independent countries and the internal republics have to varying extents started to reassert their pre-Soviet cultural identities. Folk and religious practices that predate the Soviet era and Russian domination have been revived in response to the relaxation of controls by the central government. Belarus has shown the least tendency to reassert its culture, and remains the most Russian of the newly independent states.

One aspect of cultural revival is the resurgence of interest in religion. Under the Soviets, religious practice was discouraged because religious beliefs were thought to inhibit the commitment of the people to revolutionary change. In European Russia (as well as in Georgia and Armenia in Caucasia), most people have some ancestral connection to Orthodox Christianity, and those with Jewish heritage form a sizable minority. In the case of both Orthodox Christians and Jews, religious observance increased markedly in the 1990s, and sanctuaries were rebuilt and restored. A spectacular example of restoration is Christ the Savior Cathedral in Moscow (see the photos on the next page). In Russia's internal ethnic republics—such as Tatarstan, Chechnya, Dagestan, and Ingushetiya—and in Azerbaijan in Caucasia, many people are Muslims. Among these people, and among Muslims in the Central Asian republics, observance of Islam is now more open and Muslim identity more politically important.

A countertrend to the robust revival of Orthodox Christianity is the modest spread of Christian evangelical sects from the United States (Southern Baptists, Adventists, Pentecostals). The magnitude of this spread may be exaggerated by the missionaries themselves, who count their success by numbers of converts. The movement may also be ephemeral, lasting only as long as the money the missionaries bring keeps flowing. However, in European Russia, as in Latin America and Africa, these sects emphasize that with faith comes economic success and that salvation of the soul helps the individual face the rigors of everyday life. Some new capitalists see this emphasis on individualism as a particularly useful message. After years of relying on the state to meet their basic needs, people must now rely on their own resources, often for the first time. As the following vignette illustrates, these teachings can be particularly comforting, not only during times of hardship but also as individuals adjust to new prosperity.

PERSONAL VIGNETTE Valerii, age 35, once a government research scientist, is now a new capitalist. He makes a comfortable living importing and exporting goods in the informal economy. Though he has to bribe officials and pay protection money to the *reketiry* (racketeers or mobsters), his income places his family of three—himself; his wife, Nina, age 30; and their son, Mikhail, age 10—at a much higher economic level than their longtime friends. Nina is the only woman among them who does not work outside the home. Their new wealth, the precariousness of their position in the informal economy, and the fact that their friends do not share their prosperity cause Nina and Valerii to be uneasy. In search of values that will guide them in these new circumstances, both have recently been baptized in a Christian evangelical sect. They say they chose this particular religious group because it promotes modesty, honesty, and commitment to hard work. [*Adapted from Timo Piirainen,* Towards a New Social Order in Russia: Transforming Structures and Everyday Life *(Aldershot, U.K., and Brookfield, Vt.: Dartmouth Press, 1997), pp. 171–179.*]

In the Central Asian republics, the return to religious practices is often a subject of contention. Here most people have a Muslim

Geographer Dmitri Sidorov traced the history of the Cathedral of Christ the Savior in Moscow from the 1810s to the present. The photo at the left shows the cathedral as it is being destroyed by the Bolsheviks in 1931 to make way for a planned, but never built, Palace of the Soviets. The photo at the right shows the restored cathedral in 1997. The restoration of the cathedral is often publicized as signaling the dismantling of the antireligious Soviet state and the beginning of a new era in Russian history. [*Left:* Dmitri Sidorov/ E. I. Kirichenko/V. Mikosha. *Right:* Dmitri Sidorov.]

heritage, but some local leaders, schooled in Soviet theory, view traditional Muslim religious practices as obstacles to social and economic reform. Other devout Muslim Central Asian leaders worry that the extremists among modern Islamic movements in the region may be agents of religious states such as Iran, Afghanistan, and Saudi Arabia and hence may endanger the security of Central Asian countries. Between 1992 and 1997, Tajikistan fought a civil war against Islamic insurgents with links to the Taliban in Afghanistan. In 2000, Uzbekistan and Kyrgyzstan joined to eliminate an extremist Islamic movement. But many religious leaders and human rights groups say that the fervor to eliminate radical insurgents has resulted in persecution of ordinary devout Muslims, especially men. In Uzbekistan, men convicted of no more than possessing a religious pamphlet have died in prison, and young men who met to play soccer and eat together have been arrested and accused of planning a jihad (holy war). At their trials, defendants have shown bruises and torn-out fingernails. Human Rights Watch, which monitors rights abuses worldwide, reports that at least 4000 Muslim men have been arrested and detained in Uzbekistan alone.

Unemployment and Loss of Safety Net

Widespread unemployment and underemployment has been common throughout the region for more than a decade and is still increasing. Nearly all families are affected by job loss in some way. In fact, the number of those who no longer have paid employment is probably much higher than the official unemployment figures. Numerous state firms simply no longer operate and have no money to pay employees who are still listed as workers. It is estimated that in the mid-1990s, three-fifths of the Russian labor force was not being paid in full and on schedule. For years, many families lived on their savings, but by now, those savings have been depleted. By 2003, official unemployment for the region ran from 2.1 percent in Belarus to 20 percent and higher in Caucasia and Central Asia. The rate of *underemployment*, referring to people who aren't working enough hours to make a decent living, runs as high as 25 percent in some parts of the region. Hence many people have barely enough to survive, yet have time on their hands to reflect on their predicament. Of special concern is the high youth (ages 15–24) unemployment rate across the region, estimated as 25 percent or higher in 2001, up 9 percentage points or more since 1990. In the United States, by comparison, a youth unemployment rate of 10 percent is considered a major economic and social problem.

It is especially devastating to lose a job in the former Soviet Union because work once provided many benefits beyond a salary. For many decades the social welfare system was organized around the workplace. The job was an individual's link not only to income, but also to necessities of life and general social services. The state-owned companies provided housing, two main meals a day in the company cafeteria, health care, and day care for children. Work was the center of community and social life. Thus, for many workers, when the job ended, there was no safety net and no social support group. With so many industries now retrenching, in some cases whole cities face not just widespread unemployment, but also the loss of many social benefits for their people. Such is the case in the city of Norilsk in the far north of the Central Siberian Plateau (see the box "Norilsk: A City in Transition to a Global, Free-Market Economy," page 262).

AT THE LOCAL SCALE   Norilsk: A City in Transition to a Global, Free-Market Economy

Norilsk lies in Siberia, 200 miles (320 kilometers) north of the Arctic Circle (see Figure 5.2, page 237), where year-round average temperatures are below freezing. The city sits atop a rich deposit of minerals: 35 percent of the world's nickel supply, 10 percent of its copper, 40 percent of its platinum. The city's only industry is Norilsk Nickel, Russia's largest metal company. In 1998, Norilsk Nickel employed 110,000 workers out of a total population of about 270,000, and more than 18,000 of the 110,000 workers were employed in rendering social services (clinics, sports clubs, day-care centers, cafeterias) to the rest of the workers and growing food. Norilsk was important both to Russia and to the global economy; platinum, for example, is essential in the computer industry. But Norilsk is also a site of major industrial pollution and natural habitat destruction: male life expectancy sank to just 50 years, despite sports clubs and health care.

In 1994, a private firm acquired 38 percent of the company and instituted cost-cutting measures in all branches, especially the social services division. By 1997, social services still absorbed all the plant's profits, an estimated U.S. $260 million, or about U.S. $2000 per worker. By 2004, reports regarding Norilsk were mixed: reorganization made Norilsk Nickel solvent enough to pay a dividend to stockholders, and Norilsk won first place in a nationwide competition for the most socially oriented company, despite laying off 20,000 people. The much reduced labor force at first threatened to strike if an improved contract were not agreed to; but then they postponed the strike until 2007. Perhaps they were hoping to stick around and see what would become of Norilsk Nickel's efforts to be a successful multinational corporation. It is already Russia's largest multinational, owning 51 percent of Stillwater Mining in the United States, 20 percent of Goldfields Ltd. in South Africa and Noriment Ltd. in London, and other investments in Belgium and Switzerland.

[*Sources: The Economist* (January 10, 1998): 59; "Russia's cooked books," *The Economist* (September 9, 2000); Norilsk Nickel Web page, http://www.nornik.ru/index.html/english/default.htm; Marina Kamayeva, "Russia's Norilsk Nickel upgrades production facilities," *Business Information Service for the Newly Independent States* (Washington, D.C.: U.S. Department of Commerce, May 19, 2000), http://www.bisnis.doc.gov/bisnis/isa/000601nickel.htm; Norilsk Nickel site, 2003: http://www.nornik.ru/index.jsp?lang=E; United Nations Conference on Trade and Development, World Investment Report, 2004, "The Shift Toward Services" (New York, 2004), p. 74.]

The mountains that range over Norilsk were constructed from wastes produced by the mining and smelting operations around the city. [Randy Olson/National Geographic Image Collection.]

The loss of jobs and status has been especially difficult for many men, and, as discussed on pages 244–245, some have turned to drinking and drug use. Death by alcohol poisoning in Russia rose 25 percent in the 1990s and is thought to have killed more Russians than died during World War II. In May 2003, the Russian Interior Ministry reported that a poll of 5000 youths between ages 15 and 24 showed that Russian children tend to start smoking at age eleven-and-a half, drinking alcohol at age 13, and taking drugs at about age 14. Eighty percent of those polled reported a regular intake of alcohol.

Gender and Opportunity in Free-Market Russia

Soviet policy encouraged all women to work for wages outside the home, and most women in Russia and the newly independent states were, and are, laborers in factories and in the fields. By the

1970s, 90 percent of able-bodied women in Russia were working full-time, the highest rate of female paid employment in the world. Leninist theory had always argued that women should contribute fully to national development, although the traditional attitude that women are the keepers of the home persisted. Unlike men, most women put in long days, working in a factory or office or on a farm for 8 hours and then doing laundry and housework without the aid of household appliances. And because of shortages, they often had to stand in long lines to procure food and clothing for their families, as well.

By the 1990s, the female labor force in Russia was, on average, better educated than the male labor force (the same pattern was emerging in Muslim Central Asia, but there, overall educational attainment was lower). In Russia, the best-educated women commonly hold jobs as economists, accountants, scientists, and technicians. Yet, despite their generally higher qualifications, women are unlikely to hold senior supervisory positions, and as recently as 2003, the wages of women workers averaged 36 percent less than those of men. Although three in four physicians and one in three engineers were women, these occupations do not have high status in Russia, and pay can be lower than in factories. On the other hand, executive physicians and supervisory engineers, who earn high salaries, are routinely male. Unfair as the Russian gender gap in pay may seem, it is not an unusual one. For comparison, in the United States in 2003, women's earnings overall were 38 percent less than men's.

Market reforms reduced the number of jobs available to all citizens, and in the reshuffling, women were laid off in larger numbers than men; women with children were laid off first. Mikhail Gorbachev, Russia's president until 1992, publicly stated that the first role of Soviet women was domestic; hence they could best serve their country by returning to their homes and leaving the increasingly scarce jobs to men. By the late 1990s, 70 percent of the registered unemployed were women. But many, if not most, of the women left jobless were—due to illness, death, or divorce—the sole support of their families, which often included three generations: themselves, their parents, and their children. In the Russian Federation, two out of three marriages end in divorce, and the life expectancy for males is low. Finding a new job is complicated by blatant gender-biased hiring and promotion practices. Job advertisements routinely specify gender, asking for an "attractive female receptionist" or a "male account executive" (see the box "The Trade in Women" below).

Corruption and Social Instability

Rampant corruption poses a dangerous threat to social stability in Russia and the newly independent states. Since independence in 1991, gangsters have taken over portions of the economy in every republic, and literally every citizen has to deal daily with the effects of corruption. In the Caucasian republic of Georgia, until the election of reformist President Mikhail Saakashvili in 2004, the government bureaucracy was nearly nonfunctional until a bribe was paid. Someone who wanted to construct a building or open a shop could spend months waiting in lines to obtain the necessary permits—or shorten the process with a few well-placed bribes. Additionally, because laws were poorly written and contradictory, the most capable and ambitious people simply ignored restrictions and regulations. Many adopted an attitude of noncompliance with all regulations—even those that might have clearly served the common good, such as speed limits, building codes, sanitary regulations, and limits on the transport and disposal of nuclear waste.

It was a short step from such ingrained systemic corruption to protection rackets, outright robbery, and violence, especially after cheap weapons became readily available in 1991.

AT THE REGIONAL SCALE   The Trade in Women

Among the less savory entrepreneurial activities in the new Russian market economy are those connected with the "marketing" of women. As observed in other chapters of this book (see especially Chapter 10), there is a growing demand for females as a commodity in global markets. Women from European Russia and Ukraine are deemed especially desirable. A relatively benign part of this market is the phenomenon of Internet-based mail-order bride services. A woman in her late teens or early twenties pays about $20 to be included in an agency's catalog of pictures and discriptions (one Internet agency advertises 30,000 women listed). She is then interviewed by the prospective groom, who travels—usually to Russia or Ukraine—to choose from the women he has selected from the catalog.

A more unsavory practice is the hijacking of unsuspecting females for sex work outside Russia and the newly independent states. In 2000, The Economist estimated that 300,000 such women are smuggled into the European Union yearly, many being held first at an intermediary stop in central Europe. The sex business in Europe is thought to generate $9 billion a year, and the supply side seems to be dominated by Russian-speaking gangs. Usually, the women in this market, desperate for a job, sign up to work as domestic servants or waitresses in Europe, only to find upon arrival that they are being held by force and are expected to work as strippers, dancers, and prostitutes. Most often, the women kidnapped into sex work have no well-connected relatives or friends who will report them missing. Once ensnared, the women may be sold for a few thousand dollars to brothels in parts of the world where their European looks will be considered exotic. They are usually unprotected from sexually transmitted diseases.

[Sources: CBSNews.com, July 7, 2003: http://www.cbsnews.com/stories/2003/07/05/politics/main561828.shtml; "Trafficking in women: In the shadows," The Economist (August 26, 2000): 38–39; two studies by The Foundation of Women's Forum/Stiftelsen Kvinnoforum, Stockholm, August 1998; and The Angel Coalition Web site, 2004: http://www.angelcoalition.org/history.html.]

In the mid-1990s, an American Fulbright scholar, John G. Stewart, was posted to teach public administration ethics in Georgia. He found himself learning from his students about the roots of corruption in daily life. They reminded him that even before rampant inflation, all government salaries were ridiculously low; by the mid-1990s, civil servants routinely supplemented salaries too low to live on by accepting bribes. Stewart's students wondered how they, as future public administrators, could practice ethical values and survive in the real world. Since finishing the public administration course, his 28 Georgian students have set up a support group to help one another implement ethical procedures and to find legitimate supplementary employment, so that bribes need not become part of their income.

In November 2003, Georgia's president, Edvard Shevardnadze, was forced to resign because corruption infused his government. A few months later, Mikhail Saakashvili, a 35-year-old Georgian with a Columbia University law degree, was elected president on a platform stating that corruption was robbing Georgian children of a future.

Impatience with the Present, Nostalgia for the Past

It is understandable that Russians would be impatient with the slow pace of political and economic reforms and with the job losses, declining services, and widening disparities of wealth they brought. In 2001, geographers Grigory Ioffe and Tatyana Nefedova noted that Russia "is beginning to resemble an archipelago with islands of vibrant economic life [where the most modernized cities are] immersed in a sea of [rural] stagnation and decay." Declining rural population has meant the loss of services in rural areas (such as repair and postal services and landline telephone service as cell phones take over) and the abandonment of farmland and villages. The effect is stark rural-urban disparities of

"Everything rots, everything dies" are the sentiments expressed by this Azerbaijani woman, referring to her garden near the Baku oil fields.
[Reza Deghati/National Geographic Image Collection.]

wealth and well-being that leave those on the periphery in the worst shape. Not surprisingly, nostalgia for the past, when at least the basics of life could be counted on, is frequently expressed.

Nostalgia can play a political role. Public desire to alleviate the immediate problems of lost jobs and income, obsolete skills, inadequate health care, deteriorating shelter, and declining nutrition may push elected lawmakers and officials to abandon reforms before they have had time to be effective, but doing so may thwart the development of democratic institutions and a more market-oriented economy before they have had a chance to succeed.

Although many, if not most, people experience some nostalgia for the past, many of these same people also find the present era exciting. They enjoy the challenge of a more open society with better access to information, and they are personally doing better, at least in some ways (income, career options), than during Soviet times. It would be wrong to conclude that the majority are dissatisfied with the changes since 1991.

ENVIRONMENTAL ISSUES

There is nothing particularly Soviet about the idea that nature should be subservient to human economic desires. Capitalist economies operate largely under the same premise, although, as a result of public pressure, regulations safeguard against the worst abuses. Nonetheless, a comment by Soviet leader Joseph Stalin helps us understand what was happening in the Soviet Union: "We cannot expect charity from Nature. We must tear it from her." The Soviets believed that nature had no value unless it could be exploited by humans for its resources. Soviet leaders like Stalin took the extreme position that nature was the servant of industrial and agricultural progress, and the grander the evidence of human domination over nature, the better. Hence the Soviets built huge industrial facilities. Now Russia and the newly independent states have some of the worst environmental problems in the world.

The dismissal of environmental issues continued right through to the end of the Soviet regime. With all the financial problems that developed as the Soviet Union collapsed in 1991, environmental problems continued to be relegated to the back burner. As one Russian environmentalist recently put it: "When people become more involved with their stomachs, they forget about ecology." Attracting potential investors from Europe, America, and Asia has been the highest priority for Russia and all the newly independent states because this money will create jobs. But such development will also increase industrial production and resource exploitation. In fact, investors may be attracted to the region precisely because legal protections for the environment are lax, saving them money in the short term. Some have bribed officials to ignore environmental degradation. Although awareness that pollution ultimately inhibits development is growing, leaders typically have not responded to complaints about pollution or other environmental problems because they were being paid by the polluters. At the beginning of the twenty-first century in the former Soviet Union, more than 35 million people (15 percent of the population) live in areas where the air is dangerous to breathe, where birth defects such as missing limbs or hands are rampant, and where, by some estimates, only one-third of all schoolchildren enjoy good health.

Pollution controls are further complicated because instead of one government in Moscow coordinating economic development, there are now 12 independent countries with widely differing policies, and none of them has the money to correct even a few of the past environmental abuses. Notice in Figure 5.15 that environmental degradation (acid rain, water and air pollution by petrochemicals and agricultural chemicals) crosses international borders, making the resolution of pollution issues especially difficult.

Resource Extraction and Environmental Degradation

Russia and the newly independent states have considerable deposits of natural resources (see Figures 5.10 and 5.11). As mentioned, Russia alone has the world's largest natural gas reserves, major oil deposits, and forests that stretch across the northern reaches of the continent. Russia also has major deposits of coal

and industrial minerals such as iron ore, lead, mercury, copper, nickel, platinum, and gold. It is the third-largest producer of hydropower in the world. The Central Asian republics share substantial deposits of oil and gas, which are apparently centered on the Caspian Sea and extend east toward China (see Figure 5.11 on page 254). The five countries that border the Caspian (Russia, Azerbaijan, Iran, Turkmenistan, and Kazakhstan), as well as Uzbekistan, are competing for rights to tap these fossil fuels and transport them to the world market with the help of several multinational energy conglomerates.

Resource extraction has brought environmental effects as abundant as the resources themselves. Pollution and environmental degradation are associated with the mining and industrial processing of minerals (including oil and gas) and with the generation of the energy necessary for mineral processing. In Siberia, some of the world's worst inland oil spills have contaminated lakes, rivers, and wildlife that provide sustenance for indigenous populations of hunters and fishers. Hydroelectric dams are the main source of energy for the enormous industrial complexes in Russia and the newly independent states, but the flooding of large

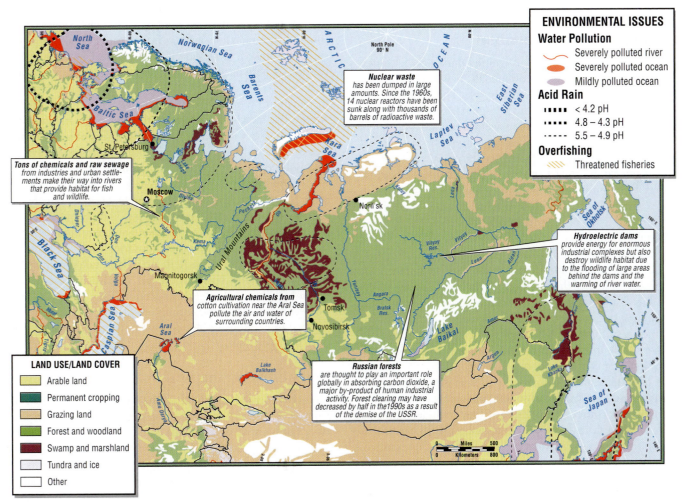

FIGURE 5.15 Environmental issues: Russia and the newly independent states. Like most industrialized countries, the countries in this region have a wide range of environmental problems caused by the burning of fossil fuels, unwise disposal of hazardous wastes, and uncontrolled use of natural resources. Notice that environmental degradation crosses international borders, making the resolution of pollution issues especially difficult. These countries have inherited many problems from the Soviet era, when the negative consequences to the environment of Soviet development policies were downplayed. (The pH of acid rain is explained in Figure 2.22 on page 92.)

areas behind the dams destroys wildlife habitat, and **thermal pollution** occurs when water warmed by the turbines to temperatures that will kill plant and animal life is returned to the rivers below the dams.

For a while, after the Soviet Union fell in 1991, general pollution levels fell and environmental degradation slowed, simply because the economy slowed markedly. Factories were closing, people's incomes had shrunk, and consumption fell. Gasoline was scarce and expensive. For example, forest clearing in Siberia decreased by as much as half because domestic demand for wood declined while at the same time (but for unrelated reasons) international prices for wood declined. This period of low demand gave some respite to Russian forest resources. But environmentalists worry that as the Russian economy rebounds, demand for all resources will grow, and indeed, air and water pollution levels are already on the rise. In the case of Russia's timber resources, as global wood prices rise again, the economically strapped Russian government is issuing contracts to private foreign concessionaires for rapid and unsustainable clear-cutting, especially in Siberia, where environmental monitoring is difficult due to the size of the area.

Urban and Industrial Pollution

Urban and industrial pollution was ignored during Soviet times. Cities expanded quickly to accommodate new industries and workers being relocated from the countryside. The dense concentration of workers alone was enough to generate lethal levels of many pollutants. Even today, urban sewer systems are rare, and of those that exist, few actually process and purify the sewage that enters them. Many large apartment blocks rely on the subsoil under and around the block to absorb sewage. Moreover, because cities were often built with residential areas located adjacent to industries producing harmful by-products, many people are exposed to high levels of industrial pollution. The city of Magnitogorsk in the Urals is one example among many of the effects of extreme industrial pollution on residents. The largest steel mill in the world is the basis of the city's economy and simultaneously causes misery for tens of thousands of Magnitogorsk citizens. One in three has respiratory problems, such as asthma or bronchitis, from breathing smoke and airborne chemicals. In all urban areas, air pollution resulting from the burning of fossil fuels is skyrocketing as more and more people purchase cars and as the industrial and transport sectors of the economy heat up.

It is often difficult to link urban pollution directly to health problems because the sources of contamination are diffuse. Often called **nonpoint sources of pollution,** they include untreated automobile exhaust, raw sewage, and agricultural chemicals that drain from fields into urban water supplies. Moscow, for example, is located at the center of a large industrial area, where infant mortality and birth defects are unexpectedly high. Researchers are convinced that these effects result from a complex mixture of pollution that is difficult to trace, but includes all the nonpoint sources just mentioned.

Rural pollution can often be just as severe as urban pollution. People in the countryside are subjected both to urban-generated airborne emissions that drift into rural areas and to fertilizer and pesticide pollution from agriculture. East of the Urals, industrial cities send tons of chemicals and raw sewage into the rivers that traverse Siberia and provide habitat for fish and wildlife.

Nuclear Pollution

Nuclear pollution in Russia and the newly independent states is the worst in the world, and the potential exists for its effects to spread well beyond the borders of the region. During the cold war, in the thinly populated deserts of Kazakhstan, the Soviet military set off almost 500 nuclear explosions. Local populations, mostly poor Kazakh herders, suffered radiation sickness and birth defects but were not told the cause until 1989. The most famous nuclear disaster in the world occurred in north-central Ukraine in 1986, when one of four nuclear reactors at Chernobyl exploded. The explosion severely contaminated a vast area in northern Ukraine, southern Belarus, and Russia and spread a cloud of radiation over much of eastern Europe and Scandinavia. As a direct result of this incident, 5000 people died, 30,000 were disabled, and 100,000 were evacuated from their homes. Yet several Chernobyl-style reactors still operate in Ukraine, Russia, and Lithuania, and increasingly all are safety hazards.

Even the pollution released by Chernobyl pales in comparison with the radiation leaking from former Soviet military sites such as Tomsk-7 (a closed city east of the Urals, now renamed Seversk), where the soil alone holds 20 times the amount of radiation released by Chernobyl. These facilities have been linked to radiation pollution recorded thousands of miles away in the Arctic. The radiation was brought there by rivers and by migrating ducks, who carry it within their bodies. The Arctic Ocean and the Sea of Okhotsk (see Figure 5.15) in the northwestern Pacific are also polluted with nuclear waste dumped at sea. Although the Soviet government signed an international antidumping treaty, it sank 14 nuclear reactors and dumped thousands of barrels of radioactive waste in the world's oceans.

Russia and Kazakhstan have sought to earn money by taking in the nuclear waste of other countries eager to be rid of it. According to experts in the field, no reliably safe system for storing nuclear waste has been found; yet the Russians and Kazakhs claim that they will safely and economically store the imported nuclear waste, using the earnings to clean up their own nuclear waste dumps. Reliable environmental impact studies, however, have not been done.

Irrigation and the Aral Sea

Once the fourth-largest lake in the world, the Aral Sea is disappearing as a result of large-scale irrigation projects in Central Asia. For millions of years, this landlocked inland sea was fed by the Syr Darya and Amu Darya rivers, which brought snowmelt from the lofty Hindu Kush and Tien Shan mountains to the northeast. In 1918, the Soviet leadership decreed that water diverted from the two rivers would irrigate millions of acres of land in Kazakhstan and Uzbekistan, which would be used to grow cotton. This Soviet plan to meet Russian textile needs was so successful that by 1937, Russia had become an exporter of cotton, or "white gold." In 1956, millions more acres of cotton were planted in Turkmenistan, irrigated with Amu Darya water

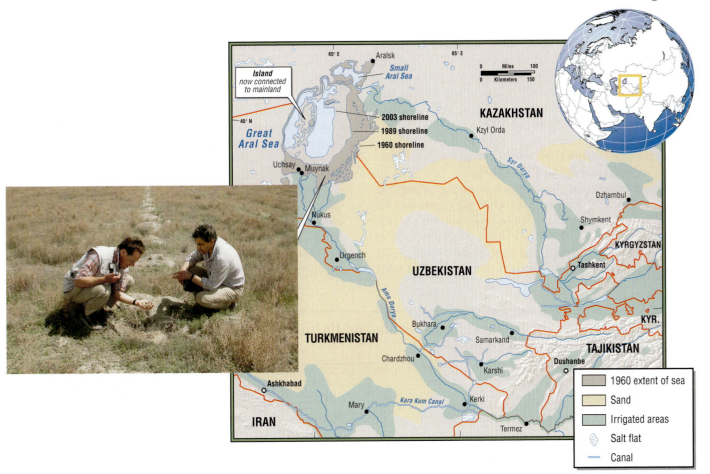

FIGURE 5.16 The decline and disappearance of the Aral Sea. Once the fourth-largest lake in the world, the Aral Sea is disappearing as a result of large-scale irrigation projects in Central Asia. Aral Sea shrinkage has caused climate change, air and water pollution, desertification, and chronic illnesses. To help restore the damaged ecology, university professors from Germany and Kazakhstan are attempting to get vegetation to grow on the now dry and dusty lake bed (see photo). [Adapted from *National Geographic* (February 1990): 72, 80–81; satellite images of the Aral Sea, 2003: http://www.redtailcanyon.com/items/16517.aspx?imageId=32940; and http://www.uni-bielefeld.de/biologie/Oekologie/figure1.html. Photo courtesy of Dr. Walter Wucherer and Professor Siegmar Breckle.]

diverted into an open, 850-mile-long (1368-kilometer-long) desert canal. So much water was lost through evaporation that within 4 years, the Aral Sea had shrunk measurably. Yet planners, enticed by the income the cotton generated, irrigated still more land. By the early 1980s, no water at all was reaching the Aral Sea. By early 1993, the sea had shrunk by more than 50 percent and had lost 75 percent of its volume (Figure 5.16). If irrigation continues, the sea will disappear entirely in a decade or two.

Shrinkage of the Aral Sea may have caused changes in climate and human health. Large bodies of water have a moderating effect on the surrounding land; since shrinkage of the sea became noticeable in the 1960s, the country around it has become drier, and there is some evidence that the summers are 2°F to 3°F (1.1°C to 1.6°C) hotter, while winters have become cooler and longer. The growing season is as much as 3 weeks shorter. Winds sweeping across the landscape now pick up the newly exposed salt and seabed sediment, creating poisonous dust storms that deposit a deadly cargo of salty sediment in people's lungs. Drinking water is heavily polluted with agricultural chemicals. In rural northern Turkmenistan near the Amu Darya, a majority of the population suffers from chronic illnesses. At the south end of the Aral Sea, in Uzbekistan, 69 of every 100 people report chronic illness, and in some villages, life expectancy is 38 years (the national average is 68 years).

What is being done to restore the Aral Sea? On the whole, very little; but university professors from Germany and Kazakhstan are attempting to grow vegetation on the dry and dusty lake bed (see inset photo in Figure 5.16). Efforts to stop irrigated agriculture, however, have not succeeded because it generates so much income. Drawing from both rivers, Uzbekistan, which irrigates 16 percent of its total cropland, is the world's fifth-largest cotton grower. Cotton is the country's leading hard currency earner (accounting for one-third of exports) and employs about 40 percent of the national labor force. In Turkmenistan and Kazakhstan, the situations differ in detail, but are comparable. Kazakhstan allows far more environmental activism and public input than its neighbors, but even here the people's desire for a better standard of living outweighs the desire to halt environmental degradation.

MEASURES OF HUMAN WELL-BEING

As discussed in the section on population patterns, average levels of well-being sank across the region after the breakup of the Soviet Union. Disparities in well-being increased as the transition to a market economy created opportunities for a few and troubles for many. But by 1998, living standards began to improve, and by 2001, in Russia, the per capita GDP was one and one-half times higher than in 1995. GDP was up by similar proportions in nearly all of the newly independent states as well (Table 5.2, column 2). Armenia, Uzbekistan, and Tajikistan recorded only modest gains between 1995 and 2003. Notice, however, that with the exception of Belarus, Russia and the newly independent countries all have GDP per capita figures well below the world average of $7376 (2003).

The GDP figures also show disparities among subregions, with Russia and Belarus each having a GDP per capita roughly three times (or more) that of some other countries in the region (Moldova, Armenia, Tajikistan, Uzbekistan, Kyrgyzstan). After 1991, Russia (and Belarus) retained most of the former Soviet Union's industrial capacity and much of its transport and trade infrastructure (banks, company headquarters, government agencies that supply statistics and customs services and that collect taxes), and these assets contributed to the disparities. The gaps between countries are narrowing, however. Kazakhstan, for example, had more than doubled its GDP by 2003 and had risen 17 points in rank on the United Nations Human Development Index (HDI) (see Table 5.2, column 3).

The GDP figures given here have been adjusted for differences in prices and other factors to obtain internationally comparable indications of purchasing power parity (PPP) in U.S. dollars. What the GDP figures do not reveal is that there is also disparity within each republic. The former communist system's goal of equalizing the distribution of wealth is now being discarded as a few become very rich in the new market economy while the majority experience a decline in living standards, at least tem-

TABLE 5.2   Human well-being rankings of Russia and the newly independent states and other selected countries

Country (1)	GDP per capita, adjusted for PPP[a] in 2001 $U.S. (2)		Human Development Index (HDI) global rankings, 2003[b] (3)		Gender Empowerment Measure (GEM) and Gender Development Index (GDI) global rankings, 2003 (4)	
	1995	2001	1998	2003	GEM	GDI
Selected countries for comparison						
Japan	21,930	25,130	8 (high)	9 (high)	44	13
United States	26,966	34,320	4 (high)	7 (high)	10	5
United Arab Emirates	17,400	20,530	—	48 (high)	65	49
World	5,990	7,376				
Russia and the Eurasian Republics						
Russian Federation	4,531	7,100	72 (medium)	63 (medium)	57	56
Belarus	4,398	7,620	68 (medium)	53 (high)	—	48
Moldova	1,547	2,150	113 (medium)	108 (medium)	51	87
Ukraine	2,361	4,350	102 (medium)	75 (medium)	61	63
Caucasia						
Georgia	1,389	2,560	108 (medium)	88 (medium)	62	—
Armenia	2,208	2,650	99 (medium)	100 (medium)	—	78
Azerbaijan	1,463	3,090	110 (medium)	89 (medium)	—	—
Central Asian Republics						
Kazakhstan	3,037	6,500	93 (medium)	76 (medium)	—	62
Turkmenistan	2,345	4,320	103 (medium)	87 (medium)	—	—
Uzbekistan	2,376	2,460	104 (medium)	101 (medium)	—	79
Tajikistan	943	1,170	118 (medium)	113 (medium)	—	92
Kyrgyzstan	1,927	2,750	109 (medium)	102 (medium)	—	—

[a]PPP = purchasing power parity.
[b]The high and medium designations indicate where the country ranks among the 175 countries classified into three categories (high, medium, low) by the United Nations.

Sources: United Nations Human Development Report 1998, United Nations Development Programme, Tables 1–3; *2003*, Tables 1, 22, 23.

porarily. Well-being remained as high as it did largely because citizens of this region helped one another to find shelter, food, and personal assistance. Certainly purchasing power was severely limited throughout the region, but purchasing power is not the best measure of well-being because it ignores aspects of well-being other than income.

The HDI tries to assess actual well-being by looking at GDP plus educational attainment and life expectancy (the lower the HDI number, the higher the rank). By 2003, the HDI ranking for nearly all of the republics was better than it had been in 1998; two rankings in Caucasia jumped 20 points or more (Table 5.2, column 3). Gains in Central Asia were more modest. These apparent improvements may be partly due to increasing production beginning in the late 1990s, but also to changes in statistical reporting.

The United Nations Gender Empowerment Measure (GEM) is available only for the Russian Federation, Moldova, Ukraine, and Georgia. Gender Development Index (GDI) figures are somewhat more complete (Table 5.2, column 4). They are not comparable to GEM, however, because they ignore the actual participation in government and society by females and measure only the extent to which males and females have access to the same basics, such as food, health care, and income. As is the case in all parts of the world, in no part of this region are women approaching equality with men in respect to income.

Quick Review

1. What financial problem does the informal economy pose to governments in this region?

2. From what source do oligarchs derive their power?

3. What is the political relationship between Russia and Chechnya?

4. How do Soviet ideas about resource exploitation still affect the level of environmental pollution today?

III SUBREGIONS OF RUSSIA AND THE NEWLY INDEPENDENT STATES

In this section, we discuss Russia first because it is the largest country in the region and because it has dominated the entire region for many hundreds of years in what most scholars now recognize as a quasi-colonial manner. The newly independent countries of Belarus, Moldova, and Ukraine are discussed next because of their location and their close social, cultural, and economic associations with European Russia. They are followed by the Caucasian countries of Georgia, Armenia, and Azerbaijan. Finally we discuss the countries of Central Asia: Kazakhstan, Kyrgyzstan, Tajikistan, Turkmenistan, and Uzbekistan. The themes covered for the various subregions echo those already discussed for the region as a whole. Here you will see what life is like in the various parts of this unusually large region.

RUSSIA

In the post-Soviet era, Russia remains influential, even as the newly independent states have begun to construct their own economies and political identities. Russia is still the largest country on earth—nearly twice the size of Canada, the United States, or China. It is also a leader in population; with 145.5 million people, it ranks sixth in the world. It also ranks high in natural wealth. For convenience, we have divided Russia into three parts: European Russia, Siberian Russia, and the Russian Far East.

European (or Western) Russia

European Russia is that area of Russia that shares the eastern part of the North European Plain with Latvia, Lithuania, Estonia, Belarus, Ukraine, and Moldova (Figure 5.17). It is usually considered the heart of Russia because it is here that early Slavic peoples established what became the Russian Empire, with its center in Moscow. Although it occupies only about one-fifth the total territory of present-day Russia, European Russia has most of the industry, the best agricultural land, and about 70 percent of the population of the Russian Federation (105 million out of 145.528 million).

The vast majority of western Russians live in cities, their parents and grandparents having left rural areas when Stalin collectivized agriculture and established heavy industry. Most of these people went to work in one of four major industrial regions chosen by central planners for accessibility and location of crucial minerals. One industrial region is centered on Moscow, another in the Ural Mountains and foothills (part of this zone lies in Siberian Russia), a third along the Volga River from Kazan' to Volgograd, and the fourth just north of the Black Sea extending into Ukraine around Donetsk.

The most heavily occupied part of European Russia—stretching south from St. Petersburg on the Baltic to the Russian Caucasus and east to the Urals—coincides with that part of the continental interior that has the most amenable climate (see Figure 5.6 on page 245). Even so, because the region lies so far to the north (Moscow is at a latitude 100 miles [160 kilometers] north of Edmonton, Canada) and in a continental interior, the winters are rather long and harsh and the summers short and mild.

Urban Life and Patterns. Moscow remains at the heart of Russian life. In the post-Soviet era, this city of about 9 million people (not particularly large by world standards) has the best selection of goods and food and the most exciting nightlife in Russia. But prices are higher here than in most Russian cities (in 2003, the newspaper *Pravda* reported that prices in Moscow were rising at the rate of about 15 percent per year). The criminals are more concentrated, more innovative, and more violent in Moscow. Entrepreneurs are more active, the market economy is more developed, and government jobs are particularly

Cities and Capitals

● ✪	Over 5,000,000
◒ ✪	1,000,000–5,000,000
○ ✪	500,000–999,999
•	250,000–499,999
○	Under 250,000

Transportation

Major roads
Trans-Siberian Railroad
Other railroads

1 Adygeya
2 Karachay-Cherkesskaya
3 Kabardino-Balkarskaya
4 North Ossetia
5 Chechnya & Ingushetiya
6 Dagestan
7 Kalmykiya

FIGURE 5.17 European Russia. Included here for reference are the Russian internal republics of Caucasia (see also Figure 5.21).

Kaliningrad's busy port is the only year-round ice-free port on Russia's Baltic coast. The channel you see here is 24 nautical miles long and connects the Baltic Sea with the port of Kaliningrad. With business booming, what used to be farmland along the channel is now crowded with cranes, warehouses, and ground transport facilities. Container ships are too big to pass in the narrow channel, so the ships come and go according to a detailed one-way schedule. [Courtesy of Nikolay Yagunov/Baltfinn-Kaliningrad Maritime Agency.]

PERSONAL VIGNETTE Natasha is an engineer in Moscow. She has managed to keep her job and the benefits it carries, but inflation has so diminished her buying power that in order to feed her family, she sells used household items and secondhand clothes in a street bazaar on the weekends. "Everyone is learning the ropes of this capitalism business," she laughs. "But it can get to be a heavy load. I've never worked so hard before!" Asked about her customers, Natasha says, "Many are former officials and high-level bureaucrats who just can't afford the basics for their families any longer. Old people love the warm sweaters. Some of them who shop with me have to eat in soup kitchens."

"But, you know," she continues, brightening and changing the subject, "the bales of used clothes I buy now come from the United States. A guy drives a carload from a container ship in Amsterdam harbor every 2 weeks. And there are lots of sturdy clothes, especially for children, some quite new with the price tags still on; but [her voice registering disappointment in American styles] only occasionally is there a really fashionable item for a woman." Moscow women are noted for making creative, even extreme, fashion statements. [*A composite story based on work by Alessandra Stanley, David Remnick, and David Lempert.*]

The privatization of real estate in Moscow gives an interesting insight into pre- and post-Soviet circumstances. During the communist era, all housing was state-owned, and there was a general housing shortage. Thousands of families were always waiting for housing, some for more than 10 years. Most apartments were built after the mid-1950s in large, shoddily constructed blocks; the units included one to three rooms and housed one to four (or more) people. But older buildings in the center city contained much more spacious, if bedraggled, dwellings. In 1991, the government allowed residents of Moscow to acquire, free, about 250 square feet (23 square meters) of usable living space per person—about the size of a typical American living room. Most families simply took ownership of the apartments in which they had been living; by 1995, a majority of Muscovite families owned their apartments. A lively real estate market quickly emerged. Because of the housing shortage, apartments in the most desirable neighborhoods commanded very high prices. The demand was fueled by a rising wealthy elite that was willing to pay as much as U.S. $100,000 to $250,000 for large remodeled apartments in the center city. Rents also skyrocketed. Those who could find alternative housing for themselves made tidy sums simply by renting out their center-city apartments. The demand for commercial space for private businesses reduced the number of dwelling units, thus exacerbating the housing shortage.

In Soviet times, most Russian cities were surrounded by many small parcels of land on which urbanites maintained small garden plots and sometimes second dwellings (called dachas). Long important in urban middle-class Russian life, these modest holdings provided city dwellers with a place to relax on weekends and cultivate vegetables and fruits to supplement purchased food. The dacha, often only a tiny toolshed (though some are quite large and well appointed), might accommodate occasional overnight stays. Moscow, however, had relatively few of these small gardens until recently. Rather, many productive socialist

precarious. Moscow's economy is privatizing rapidly. By 1992, just one year after the end of the Soviet Union, more than a quarter of Moscow's labor force of 4.4 million was in nonstate employment; by 1994, more than half the jobs were in the private sector. But after 1996, the growth of the private sector slowed, and by 2002, in an effort to quell social unrest, the government began to expand public employment. In Moscow, the retail sector and small-scale services to the public—such as street food vending and sales of inexpensive clothes, home furnishings, and small appliances—created the most private-sector jobs (often in the informal economy). Today, for individuals adjusting to the loss of "cradle-to-grave" security, life in Moscow can be painful, frightening, exhilarating, or all three, as the following account of the experiences of Natasha and her customers illustrates.

Muscovites, like many Europeans, often journey to the urban fringe on weekends, where they cultivate small plots. Here, members of a cooperative gardening society are sowing potatoes. The small house in the background is where they store tools and where they may spend the night in order to extend their gardening hours. [TASS/Sovfoto.]

farms surrounded the city and supplied food to the city as well as to distant locations within the Soviet Union. Since privatization, land use has changed markedly, though local farms still supply milk, meat, eggs, and potatoes to the city. Moscow's now wealthier and more numerous middle and upper classes have invested in substantial suburban dwellings on land formerly farmed collectively. Suburbs are beginning to ring the city, and urban garden plots are appearing. Given the instability of jobs in Moscow and other large cities, garden plots are likely to be important supplements to family nutrition. An interesting question is, who will do the family gardening?

St. Petersburg, renamed Leningrad during most of the communist period (1924–1991), is European Russia's second-largest city, with about 5 million people. It is a shipbuilding and industrial city located at the eastern end of the Gulf of Finland, on the Baltic Sea. Czar Peter the Great ordered the city built in 1703 as part of his effort to Europeanize Russia; from 1713 to 1918, it served as the Russian capital. The city has a rich architectural heritage, including many palaces and public parks. Since 1991, it has been undergoing a renaissance: the transport system is being refurbished, urban malls are proliferating, and historic sites are being renovated. The czars' Winter Palace, now called the Hermitage Museum, houses one of the world's most important art collections. Urban planners working in St. Petersburg maintain an elaborate Web site with numerous pictures that can be reached at http://www.spb.ru/eng/.

Russia East of the Urals

To the east of European Russia lies a vast territory stretching from the Ural Mountains to the Pacific. It is usually divided into three physical zones: the western plain, the middle plateau, and the far eastern mountains (see the section on landforms earlier in this chapter, on pages 238–239). Siberia has been the popular term for this area, but for our purposes here, Siberian Russia refers to

only about half the landmass east of the Urals: it encompasses the West Siberian Plain beyond the Urals and about half the Central Siberian Plateau (Figure 5.18) The rest of the landmass east to the Pacific is here called the Russian Far East.

Siberian Russia

Siberia is so cold that 60 percent of the land is permafrost. This vast central portion of Russia is home to just 33 million people, many of whom moved into the region from west of the Urals. Settlement is concentrated in cities in the somewhat warmer southern quarter. For millennia, Siberia was a land of wetlands and quiet, majestic forests. But during the Soviet era, it became dotted with bleak urban landscapes and industrial squalor—the legacy of the Soviet effort to establish industries to exploit Siberia's valuable mineral, fish, game, and timber resources (see the box "Norilsk: A City in Transition to a Global, Free-Market Economy" on page 262).

Novosibirsk, Siberia's twentieth-century capital and financial center, is the largest Russian city east of the Urals (1.4 million people). Its location illustrates the Soviet policy of claiming space (and resources in that space) by distributing industrial cities across its vast landmass. Novosibirsk lies more than 1600 miles (2500 kilometers) east, as the crow flies, from Moscow. Although there are few natural resources in its immediate environs, this commercial center has the highest concentration of industry between the Urals and the Pacific. During the cold war, it was a center for strategic industries such as optics and weaponry and contained more than 200 heavy-industry plants. In the post-Soviet era, these factories have become infamous for obsolescence and high rates of pollution, but they also represent a potential for future profit if privatized and reorganized. With its banks and financial services, Novosibirsk has become a center for outside investors interested in acquiring factories in the nearby surrounding region. In the 1990s, 400 joint ventures were established between Russians and Chinese, German, or South Korean partners. Many of them failed, but the successful ones were those that placed an emphasis on investing in Novosibirsk's human resources (rather than just its factory facilities) by offering retraining and education. An example of a still mostly state-owned industry that seems to be making a successful transition to a market economy is Novosibirsk Instruments. This enterprise once produced artillery for the Soviet army, then moved into telescopes for civilian consumers in the 1980s, and most recently has focused on providing advanced amateur astronomers around the world with good-quality telescopes; see, for example the Web site at http://www.telescopes.ru/links. A link at this site makes clear that Novosibirsk is still a center of weapons manufacture and raises the question of who around the world is buying the wide variety of devices displayed.

The Russian Far East

The Russian Far East is an extensive territory of mountain plateaus with long coastlines on the Pacific and Arctic oceans (Figure 5.19). Here, almost 90 percent of the land is permafrost; the coastal volcanic mountains stop the warmer Pacific air from moderating the Arctic cold that spreads deep into the continent.

FIGURE 5.18 Siberian Russia.

In area, this subregion makes up slightly over one-third of the entire country of Russia and is nearly two-thirds the size of the United States. If the population of 8.16 million were evenly distributed across the land, there would be just 2.3 persons per square mile (1 per square kilometer), but 75 percent of the people live in just a few cities. The residents of this subregion are primarily immigrants and exiles—some were sent to Siberia for perceived misdeeds in the Soviet era. But here, where serfdom never intruded, there is a surprising spirit of independence that is rare elsewhere in Russia. The earlier migrants from western Russia worked in timber and mineral extraction enterprises and in isolated industrial enclaves. Since the 1980s, many immigrants have headed for cities along the Pacific coast, especially Nakhodka and Vladivostok near North Korea.

The interior of the Russian Far East has many resources, but has not been developed because of its distance from European Russia and its difficult environments. Only 1 percent of the territory, mostly along the Pacific coast and in the Amur River drainage basin, is suitable for agriculture (see Figure 5.4 on page 240). Relatively unexploited stands of timber cover 45 percent of the area; the tree roots grow slowly in the sun-warmed, but not very fertile, sediment above the permafrost. Soviet central planners, and now private investors, have been attracted by the region's resources of timber, coal, natural gas, oil, tin, antimony, gold, diamonds, iron, and other minerals. Based on recent oil discoveries near Sakhalin Island, estimates are that 50 percent of Russia's oil reserves are in the Russian Far East. China, Japan, and Russia have plans to increase both oil and timber extraction and to develop fruit and grain production in the basin of the Amur River, which is the world's largest remaining unbridged and undammed river.

The Russian Far East has great potential for linkages with other world regions. In recent decades, port facilities on the Pacific coast have made the region accessible to such countries as Japan, Korea, and China (see photo on page 252). Japan, a rich country that is nonetheless poor in natural resources other than timber, accounts for 20 percent of the trade, and that portion is likely to

FIGURE 5.19 The Russian Far East. The Russian Far East makes up slightly over one-third of the entire country of Russia. Soviet planners and now private investors have been attracted to the region by its timber, coal, natural gas, oil, and other natural resources. It is also an area of wilderness that supports many endangered species, like the Siberian tiger, Far Eastern leopard, and the Kamchatka snow sheep.

increase. China, which badly needs more energy resources, and Japan are vying for the right to build an oil pipeline from Angarsk in Siberia to either Daqing in China or Nakhodka on the Russian Pacific coast, the closest port to Japan (see Figure 5.19; see also Figure 5.11 on page 254). China's proposed line would be ready in just a few years, but Japan's would give Russia access to much wider markets in the western Pacific. Russia is also considering a rail link across China to North Korea and Vladivostok. Some geographers and economists forecast that, because of its rich resource reserves and its location so far from the Russian core, the Russian Far East is likely to integrate economically with other Pacific Rim nations, and then perhaps eventually detach itself politically from European Russia.

BELARUS, MOLDOVA, AND UKRAINE

Sandwiched between Russia and the Eastern European countries that were once part of the Soviet Union's sphere of influence are the newly independent states of Belarus, Moldova, and Ukraine

(Figure 5.20). Each of these primarily agricultural countries is the home of a distinct Slavic people. Because of their position between Russia and the rest of Europe, these countries have found themselves faced with the choice of maintaining their closest ties with Russia or turning toward Europe.

Belarus is a globe-shaped country surrounded by Latvia, Lithuania, Poland, Ukraine, and Russia. In size and terrain, Belarus resembles Minnesota: its flat, glaciated landscape is strewn with forests and dotted with thousands of small lakes, streams, and marshes that are replenished by abundant rainfall. During the twentieth century, much of the land was cleared and drained for agriculture, mostly on collective farms. The stony soils are not particularly rich; nor, other than a little oil, are there many known useful resources or minerals beneath their surface. Belarus absorbed a large amount of radiation contamination after the Chernobyl nuclear accident in Ukraine; 20 percent of its agricultural land and 15 percent of its forestland were rendered unusable for the foreseeable future.

Belarus was forced into independence by the collapse of the Soviet Union in 1991 and now actively seeks to reintegrate with Russia, with which it shares cultural, economic, and political features. During the twentieth century, Belarus was rather thoroughly Russified—molded to accept Russian values and perspectives—by a relatively small but influential group of Russian workers and bureaucrats. Although 80 percent of the population of 10 million is Belarus (a Slavic group) and only 13 percent is Russian, the Russian language predominates, and Belarus culture survives primarily in museums and historical festivals. The urban concrete landscape looks and feels like Russia of a decade ago. The Belarus economy remains dominated by state firms that sell to Russia; only a few retail shops are now private. Its important petrochemical industry depends largely on Russian oil and gas. The Belarussian desire to reunite with Russia is welcomed by the remaining Russian communists as well as by the military—who see it as a useful buffer state against the growing European Union

and the extension of NATO membership to central European states such as Poland and Hungary—and by all those who wish to see the former Soviet Empire reconstituted. By 2001, Belarus and Russia had agreed to remove all customs and trading fees imposed on each other, and the countries appeared to be united for all practical purposes.

Moldova and Ukraine are two closely related countries that share the low-lying territory between Belarus and the Black Sea. They enjoy a warmer climate than Belarus and Russia, and, with good agricultural resources, they have the potential to increase agricultural production significantly. Ukraine, which has rich soils, produces sugar (from beets), grapes for wine, sunflower oil and seeds, flax, nuts and fruits, meat, and dairy products. It is comparable to France in size, population, and agricultural output. Agriculture accounts for one-fourth of both GDP and employment in Ukraine. Moldova, lying to the west and bordering Romania, is much smaller (4.3 million people) than Ukraine (48.5 million people), but produces many of the same products. It is now pursuing the possibility of emphasizing wine, nuts, and

FIGURE 5.20 The Belarus, Moldova, and Ukraine subregion.

fruits to be marketed not only to Russia but also to the Baltic states, which lie too far north to produce such items, and to Europe via Romania.

As of 2000, much of Ukraine was still collectivized because government officials were wary of making changes that could jeopardize food supplies. But private individuals are changing Ukrainian agriculture on their own. Many families grow up to half their own food supply in small family gardens that have proliferated in the years since independence, especially in urban areas. Many analysts believe that Ukraine—which supplied food to the Soviet Union, especially after World War II, and which continues to trade heavily with Russia—could enjoy a niche in the agricultural economy of Europe, giving France some serious competition in the production of vegetables, fruits, and wines. To do so, however, Ukraine would have to make many adjustments to meet EU standards. It would also need to refurbish its collective farms, which, for the most part, have outdated equipment (see the following vignette) and produce less efficiently than farms in western Europe. For the time being, therefore, Ukraine will probably continue to trade its agricultural products mostly with Russia.

PERSONAL VIGNETTE   Those Ukrainian farmers who spent their lives on collective farms have had difficulty making the transition to being private farmers. Vasyl Speiko, age 35, is a farmer in western Ukraine. After spending years on the Bukovyna collective farm, he recently obtained 12 acres (5 hectares) of his own. So far, the only equipment he can afford is a horse and wagon, but he is optimistic: "I want to know what it's like to be a free man." In working his land, he benefits from the counsel of his aged father, who lost his 61 acres (25 hectares) of farmland to the state collective in 1948. [*Mike Edwards, "After the Soviet Union's collapse—A broken empire," National Geographic (March 1993): 49–52.*]

The EU's interest in Ukraine and Moldova is related not so much to their future membership, but to illegal migration into Europe that passes through these two countries from Sri Lanka, Bangladesh, Afghanistan, Chechnya, and West Africa. The EU is now providing financial aid, especially to Ukraine, to tighten border security.

CAUCASIA: GEORGIA, ARMENIA, AND AZERBAIJAN

Caucasia is located around the rugged spine of the Caucasus Mountains, which stretch from the Black Sea to the Caspian Sea. It includes a piece of the Russian Federation on the northern flank of the mountains as well as the three independent countries of Georgia, Armenia, and Azerbaijan, which occupy the southern flank of the mountains in what is known as Transcaucasia. Transcaucasia is a band of subtropical intermountain valleys and high volcanic plateaus that drop to low coastal plains near the Black and Caspian seas. Whereas much of the land in the rest of Russia and the newly independent states is arid or cold, these treasured pieces of rugged mountain slopes and narrow plains are

Farmlands like these in southwestern Ukraine (the Ivano-Frankivs'k region) evoke images of French Impressionist paintings and display the richness of soil and climate that made Ukraine a prime agricultural producer for Soviet Russia. [Robert Semeniuk/Corbis Stock Market.]

blessed with warm temperatures and abundant moisture from the Black and Caspian seas. In these favorable conditions, farmers grow crops, such as citrus fruits and even bananas, that can be produced in no other part of the vast region. Before 1991, most of the Soviet Union's citrus and tea came from Georgia, as did much of its grapes and wine.

In a mountainous space the size of California between the Black Sea and the Caspian Sea live more than 50 ethnic groups: Armenians, Chechens, Ossetians, Karachays, Abkhazians,

Aziza Mustafa Zadeh is an Azerbaijani singer and jazz pianist, popular in Europe and the eastern Mediterranean. Her version of jazz is inspired by *mugam*, an ancient, mesmerizing modal system of traditional music from Caucasia. [Courtesy of Aziza Mustafa Zadeh.]

Georgians, and Tatars, to name but a few (Figure 5.21). The groups vary widely in size from a few hundred (for example, the Ginukh) to over 6 million (the Turkic Azerbaijanis). Some are Orthodox Christians, many are Muslims, some are Jews, Gypsies pass through, and some retain ancient elements of local animistic religions. All these groups, including Chechens and others who live in Russian Caucasia (see Figure 5.14 on page 258), are remnants of ancient migrations. For thousands of years, Caucasia was a stopping point for nomadic peoples moving between the Central Asian steppes, the Mediterranean, and Europe. Other ethnic enclaves were created more recently by Soviet-instigated relocation and then return of troublesome minorities. Today, many Caucasians maintain ties to Europe and North America, where hundreds of thousands of emigrants from the region live. Armenians may be the best known of these groups. Like native people in North America (the Nunavut) and Southeast Asia (the Acehese), some of the ethnic groups in Caucasia are using the Internet as a way to gain publicity for their unique geographic location and complex historical heritage. A good example is the Web site maintained by the Abkhazian Republic (http://www.abkhazia.org/). Abkhazia is part of Georgia.

Members of a single ethnic group may live in several different Caucasian republics, and because of their strong ethnic loyalties, this pattern can result in persistent conflict. After the Soviet collapse in 1991, the three culturally distinct republics of Georgia, Azerbaijan, and Armenia obtained independence as nation-states (see pages 263–264 on the role of corruption in Georgia's rocky path to democracy). Then, very quickly, several other ethnic groups within these already tiny states took up arms to obtain their own independent territories. The Abkhazians and the South Ossetians took up arms against Georgia, and the Christian Armenians in Nagorno Karabakh (an exclave of Armenia located inside Azerbaijan) fought against the Muslim majority. In the conflict over Nagorno Karabakh, 15,000 people died before a truce was signed in 1994. Although the ethnic strife across Caucasia appears to have local causes, it is often instigated by larger powers—Turkey, Russia, Iran—seeking access to strategic military installations and agricultural and mineral resources.

Access to oil and gas reserves also has been a source of conflict in Caucasia. Estimates of Caspian oil reserves vary widely, from 28 billion to 200 billion barrels. A major issue is how to get the oil and gas out of Caucasia safely and into the world market. Pipelines, trucks, and ocean tankers would all require passage through hostile territory or across difficult terrain. A consortium of Western oil companies has considered several pipeline routes to the Black Sea. Transit through Chechnya is problematic because of its continuing guerrilla war with Russia. Transit through Georgia and Russia exists now, and new lines are under construction. The United States and Europe favor a pipeline across Caucasia and Turkey. Turkey would use some of the energy supplied, but most would be sold to Europe.

THE CENTRAL ASIAN REPUBLICS

Since the fall of 1991, following nearly 12 decades of Russian colonialism, five independent nations have emerged in Central Asia: Kazakhstan, Kyrgyzstan, Tajikistan, Turkmenistan, and Uzbekistan (Figure 5.22). Consistent with the mosaic of cultures that is found here, each of these new nations has traditions that are recognizably different from those of the others. Yet all of them draw on the deep, common traditions of this ancient region. For example, most people speak different Turkic languages that are often mutually intelligible; the Tajiks, however, speak a Persian language. After decades of efforts to align their cultures and economies with those of the Soviet Union, the countries of Central Asia are poised for change brought on by political independence, by the rise of ethnic and religious (Islamic) sensitivities, and especially by the chance to profit from the exploitation of recently discovered oil and gas reserves. The revival of Muslim traditions, the emergence of oil as a major resource, and the need for new economic partners have led these new Central Asian countries to consider renewed association

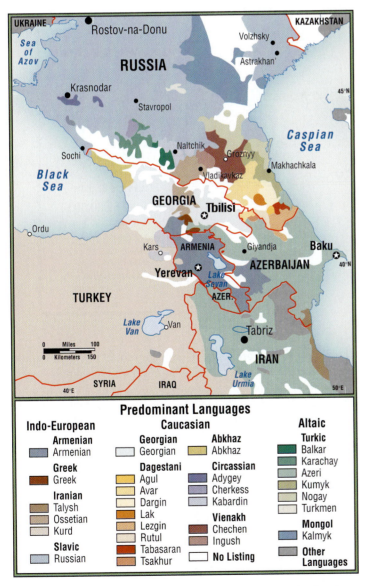

FIGURE 5.21 The Caucasian subregion. This map illustrates the diversity of ethnolinguistic groups in the Caucasian subregion and adjacent culturally related neighbors. [Adapted from http://www.geocities.com/southbeach/marina/6150/ethno.jpg.]

FIGURE 5.22 The Central Asian subregion.

Cities and Capitals

- ● ✪ Over 5,000,000
- ◒ ✪ 1,000,000–5,000,000
- ○ ✪ 500,000–999,999
- ● ○ 250,000–499,999
- ○   Under 250,000

Transportation
—— Major roads
—— Railroads

with the countries of Southwest Asia and with Afghanistan, Pakistan, and western China. It is not unlikely that in the future, regional analysis will be applied to a set of countries stretching from the Mediterranean across Central Asia to western China.

The Physical Setting

Central Asia is situated in the center of the Eurasian continent, in the rain shadow of the lofty mountains that lie to the south in Iran, Afghanistan, and Pakistan. Its dry continental climate is a reflection of this location. What rain there is falls mainly in the north, on the Kazakh plain, where wide grasslands support limited agriculture and large herds of sheep, goats, and horses. In the south are deserts crossed by rivers carrying glacial meltwater from the high peaks still farther to the south. These rivers are tapped to irrigate huge fields of cotton and smaller fields of wheat and other grains (see the discussion of irrigation on pages 266–267). In many areas of the south, gardens are attached to houses and are walled to protect against drying winds. The gardens nurture tree crops (plums, pistachios, apricots, and apples, as well as onions and garlic, were first domesticated in Central Asia) and many types of melons.

Uzbekistan and Turkmenistan are largely low-lying plains. Kazakhstan has plains in the north and uplands in the south that grade into high mountains. Kyrgyzstan and Tajikistan lie high in

the Hindu Kush, Pamir, and Tien Shan mountains to the southeast of Kazakhstan. These two countries have exceedingly rugged landscapes. Tajikistan's elevations go from near sea level to more than 22,000 feet (6705 meters), and those in Kyrgyzstan are similar. Both offer little in economic potential; Tajikistan is the poorest country in the entire region and has undergone five changes in government and a civil war since 1991.

Central Asia in World History

Civilization flourished in Central Asia long before it did in lands to the north. The ancient Silk Road, a continuously shifting ribbon of trade routes connecting China with the fringes of Europe, operated for thousands of years, diffusing ideas and technology from place to place (see Figure 5.5, page 242, and Figure 9.19, page 485). Modern versions of ancient trading cities still dot the land: two examples are Bukhara, renowned as a center of Islamic learning and culture, and Samarkand, which has been in existence for

One of the most remarkable cultural traditions of Central Asia is carpet weaving, an art practiced primarily by women and children. The carpets probably originated as the highly portable and useful furnishings of nomadic yurts and *gers* (tents); there are many different types and styles. Each region and culture group has distinctive patterns, with particular combinations of colors. The practice of weaving rugs stretches from interior China to Caucasia, Turkey, and the Balkans. The rug pictured here is an Ersari from Turkmenistan. [Courtesy of Charles W. Jacobsen, Inc., Syracuse, N.Y.]

An important episode in the story of Russian domination of Central Asia was the decision to invade Afghanistan in 1979 to prop up a pro-Soviet puppet government there that was encountering increasing popular resistance. The Russian invasion inspired many disparate Afghan interests to join in opposition, and it served as a cause around which anti-Russian sentiment in Central Asia could solidify. The Russians were eventually defeated in 1989 by a shaky coalition of Afghans and Afghan sympathizers (NATO reported 15,000 Soviet soldiers lost and over 1 million Afghans dead as a result of the 10-year war). With Russia weakened, the Central Asian republics gained independence.

In Central Asia, the transition since independence in 1991 has been rocky, with some conditions of daily life deteriorating significantly. Russia has sought to maintain a strong influence, but it is being challenged by growing international interest in the region's oil and gas resources. Transportation and services have deteriorated even from what they were in Soviet times. Poverty is now more widespread, while a few have prospered; and health standards are the lowest of any in the former Soviet Union. Although elections are held and free markets are opening up opportunities for entrepreneurship, government bureaucracies remain authoritarian and patriarchal, and elections have been hijacked by fraudulent vote counts. Most Central Asian governments fear Islamic fundamentalism; hence even moderate Muslims endure repression. The revival of Islam, although influential, is subdued, and believers live in fear (see pages 260–261).

A particularly troubling legacy of change in the modern era survives in the complex mountainous borderlands where Central Asia touches China, Pakistan, and Afghanistan. There, Soviet efforts to develop commercial agriculture and industry and to supply a labor force for this development resulted in the rearrangement of ethnic groups that previously had not been overly antagonistic. There are now, for example, new pockets of Tajiks in

5000 years. Commerce along the Silk Road diminished by the fifteenth century as traders shifted to shipping goods by sea. Central Asia then entered a long period of stagnation until, in the mid-nineteenth century, czarist Russia developed an interest in the region's major export crop, cotton.

The Russian imperial agents built modern mechanized textile mills, which quickly replaced small-scale textile firms that had employed people to do traditional cotton and silk weaving. The new mills were often located in entirely modern Russian towns built alongside railroad lines. Consequently, they benefited few Central Asians. In rural areas, cotton fields often replaced the wheat fields that had fed local people. Occasional famines struck whenever people could not afford imported food or when supplies were interrupted.

The Soviet and Post-Soviet Eras

Although Central Asia had been under Russian control since the mid-1800s, Russian domination intensified during the Soviet era. Only a few Central Asians actually joined the Communist party; those that did were thoroughly Russified. The practice of Islam, though officially tolerated, was undermined by the Communist government's promotion of atheism and by restrictions on Islamic worship, pilgrimages to Mecca, and access to mosques.

Once Kazakhstan gained independence, in 1991, ethnic Kazakhs returned home from self-imposed exile in Mongolia. Kazakhs and Mongolians share a cultural history of nomadic herding. Their portable yurt homes are visible in the background. [Gerd Ludwig/National Geographic Image Collection.]

Uzbekistan and Afghanistan, Afghans in Tajikistan, and Kazakhs in Turkmenistan. Much of the cultural patchwork of this region is ancient and can be accounted for partially by the complexity of the mountainous terrain, which kept groups cut off from one another. But modernization created new patterns of settlement. The forced movement of large portions of particular ethnic groups broke up families and clans and changed the groups' economic arrangements and trading patterns. The Soviet era displaced people, purposely destroyed traditional economies, and did not prepare Central Asians to govern themselves. When Soviet authoritarian control abruptly diminished in the 1990s, armed conflict erupted.

Geopolitics in the Central Asian Oil Fields

In this chapter, we have repeatedly noted that Central Asian oil and gas reserves have for some time attracted the attention of wealthy energy-consuming nations. Maneuverings to gain the rights to develop, transport, and sell (or buy) these resources have influenced international politics for many years. Countries with designs on Central Asian oil include Russia, Turkey, Iraq, China, India, Pakistan, Iran, the United Kingdom, and of course, the biggest energy consumer of all, the United States. Meanwhile, the holders of these considerable oil reserves, the new Central Asian republics, are not powerful or experienced enough to stave off the covetous advances of world powers, nor do they have the capital to develop the resources themselves, though their need for oil income is great. Hence they are between a rock and a hard place; suspicion abounds as leaders get bought off.

As of the mid-2000s, contention over Central Asian oil and gas hinges on pipeline routes. Russia favors a route it could control through Chechnya in northern Caucasia. Iran has already built a pipeline from its border with Turkmenistan south to the Persian Gulf. The United States favors two routes, one from Turkmenistan through Afghanistan and Pakistan to the Indian Ocean, and a second already under construction from Baku, Azerbaijan, on the Caspian Sea through Georgia to a Turkish seaport on the Mediterranean. China promotes an eastbound route via Kazakhstan and Kyrgyzstan. In 2004, presumably to protect its interests, the United States had 500 specially trained troops stationed in Georgia and another 3000 troops in Kyrgyzstan, near the capital, Bishkek.

The Free Market in Central Asia

Despite the overall slow pace of change and the difficulties of transition, the Central Asian republics appear to be finding their "capitalist legs" fairly quickly. As communism fades and the Russians leave, Central Asians are taking up old market-based skills that have been part of their heritage since the ancient Silk Road days, when trade and long-distance travel were the heart of the economy. For example, a largely contraband trade in consumer goods (clothes, electronics, cars) and illegal substances from drugs to guns is blossoming with Iran, Afghanistan, and China (see references to the revival of the Silk Road and the ancient market at Kashgar in Chapter 9). There are also opportunities to render such services as food, transport, and accommodations to tourists and traveling business-people who hope to set up legitimate enterprises in the region (one such effort is described in the vignette about Fahreeden below). Meanwhile, all five republics are negotiating with competing multinational oil companies to market Central Asian oil and gas to their best advantage. Iran, certain that shoppers will once again materialize from across the steppes, as they have for thousands of years, is establishing a free-trade zone along its northern border with Turkmenistan.

PERSONAL VIGNETTE   Fahreeden is a taxi driver who operates out of Samarkand, Uzbekistan, in a dusty and battered old Fiat that he starts by hot-wiring the ignition. The tires are worn down to the steel fibers, and the brakes no longer work. His client today is an American lawyer pursuing an interest in religious history. The destination is the train station in Bukhara, Fahreeden's hometown. Bukhara is 4 hours away, a distance that will bring three flat tires and five halts by highway police, who expect bribes (baksheesh) at each stop. Fahreeden and the lawyer have agreed upon a fee that includes the anticipated police bribes, tire repairs, and a generous luncheon worthy of a Silk Road merchant.

The dusty ribbon of road through the semiarid landscape is conveniently punctuated with shops dedicated to tire repair. There are no signs advertising this service, just a bald tire sitting upright along the roadside. As you wait for your tire to be repaired, you may be treated to tea and conversation with the family of the repair person. Eventually, due to missed trains and changed plans, the lawyer will spend the night in the taxi driver's home, where, to the guest's surprise, the walls and floors are covered with beautiful and expensive handmade carpets. Dinner, prepared by Fahreeden's wife, Miriam, is shish kebab, fresh fruit, tomato salad, round loaves of bread, rice pilaf, tea, and lassi (a yogurt-based drink, taken from a common bowl). The guest will sleep in splendor on the same soft rugs on which his meal was graciously served. [From the travels of Ron Leadbetter, personal communication, Autumn 1996.]

Fahreeden's efforts to adapt to the post-Soviet era have led him to reinvent ancient ways of life in Central Asia. As was the case for many families in the days of the ancient Silk Road trade, his income is based on accommodating travelers and merchants and steering them safely through a large, arid, and potentially dangerous territory. Given a modicum of peace, he is likely to survive, and even thrive.

Quick Review

1. Explain why European Russia is still considered the "heart of Russia."

2. Why has the development of the resources of the Russian Far East lagged in the past and now taken off?

3. Which physical attributes contribute to the agricultural capabilities of Ukraine and Moldova?

4. What are the ethnic, economic, and physical features of Caucasia, including that part that lies in Russia, that make this subregion of special interest and concern to the Russian people?

In Kyrgyzstan, Minavar Salijanova took advantage of a United Nations microcredit loan of 5000 SOM (about U.S. $100.00) to buy 60 chickens. Each day the chickens produced 60 eggs, which she sold in the market for about 6 cents each. Every 3 months she bought more chickens, and now has 147. She has also bought some goats and is expanding the business. "The program has improved our vision for life," she says, and neighbors have begun to follow her example. [Staton R. Winter/New York Times Pictures.]

REFLECTIONS ON RUSSIA AND THE NEWLY INDEPENDENT STATES

The region comprising the Russian Federation, Belarus, Ukraine, Moldova, Caucasia, and Central Asia has gone through dramatic changes over the last decade. No other set of countries has ever attempted such a rapid and peaceful transformation to a free-market economy. As the command economy disappears and localities direct their own economic affairs, regional and local differences and disparities are bound to increase. Although Russia retains its leadership position, its power and influence have weakened. Some of its closest allies in eastern Europe are no longer in its sphere of influence and are now closely connected with the increasingly powerful European Union. The Central Asian republics are likewise finding common cause with their neighbors to the west, south, and east. As time passes, it may no longer be appropriate to combine these particular political units into one region, but just what the future holds is unclear.

It is likely that central governments will remain especially strong in most republics and that participatory democracies will evolve only slowly. Although some individuals are eager to learn from the West, there is widespread and understandable resistance to rapid culture change and to excessive influence from abroad. Most republics in the region will probably have an ambivalent relationship with the West for some years to come, embracing many aspects of market economies and some aspects of democracy with enthusiasm, yet mourning the loss of certain civilities and welfare guarantees of the old system that made possible a feeling of egalitarian well-being. Rising crime, violence, and fraud, associated either with the breakdown of the old system or with the introduction of a market economy, have alienated many, especially in Russia, from Western models of development. In the Central Asian republics, the nuances of the market economy may be sorted out more rapidly because of the long heritage of entrepreneurial trade in the region. On the other hand, the religious values of a reviving Islamic tradition in Central Asia appear to be the nexus of anticapitalist and anti-Western feeling.

All parts of the region will be hampered for years to come by the aging and inefficient industrial infrastructure and by the severe environmental pollution that has accompanied industrialization. While pollution levels may recede as the industrial infrastructure is first shut down and then modernized, if central control is eroded, the rapid private development of the rich resource base will probably lead to increased pollution.

The region's people have a particularly challenging future as they deal with all that the move toward a market economy entails. On the one hand, jobs and social safety nets formerly associated with the workplace have been lost, skills have become obsolete, and changing gender roles are disorienting to both men and women. On the other hand, exhilarating opportunities exist for greater self-expression, for the possibility of entrepreneurial activities, and for the material well-being that market economies can provide.

Chapter Key Terms

CHAPTER 6

NORTH AFRICA AND SOUTHWEST ASIA

[*Adapted from Douglas Frantz, "Poverty forces new methods for educating Turkish youth," New York Times (July 1, 2001): 7; updated, 2004, with material from http://www.tegv.org/english/foundation.htm.*]

MAKING GLOBAL CONNECTIONS

As the world map in Figure 6.1 shows, this region has been slow to adopt technology; but Ibrahim Betil's efforts at educational reform in Turkey, and those of the 2000 volunteers in the foundation he originated, are among the many types of social change that are taking place in the region of North Africa and Southwest Asia. Part of what unifies this region is the religion of Islam, which guides the daily lives of the vast majority of the region's inhabitants and has traditionally guided its educational and legal systems and other aspects of public life as well. In Turkey and elsewhere in the region, a tension exists between those who campaign for the teaching of Islamic values in public schools and those, like Mr. Betil, who strictly exclude religious teaching from the classroom. The debates sparked by educational reform in Turkey are part of wider debates over how these societies should respond to many issues. What should be the public role of religion and of religiously inspired leadership? To what extent should women be brought into full participation in society and government? How important is broad-based democratic consensus? What should be the response to cultural influences from outside the region, especially from Europe and the United States (influences that some people fear are undermining traditional religious beliefs and ways of life)? The degree to which these debates can be held in public varies from country to country. In Turkey the debate is relatively open as it is in Jordan, Israel, Qatar, and even Egypt. Elsewhere in the region debating these issues is risky (Algeria, Morocco, Iraq, Iran, Syria, Yemen), or even seditious (Saudi Arabia).

The 21 countries in the region (see Figure 6.11, page 298) achieved political independence only in the twentieth century, many of them not until after World War II. Although none were outright colonies of Europe, most were forced to submit to some type of formal control by a European country (and more recently by the United States) that resembled colonialism (see definition in Chapter 1). These foreign countries tried to appropriate the region's resources (especially oil and gas) at very low prices and to influence their economies, government officials, and cultural institutions, such as schools, universities, government systems, military establishments, and even religious practices. As a result, many people in the region mistrust Europe and the United States, and they see a need to protect their cultures from outside influences.

This region (Figure 6.2) has long been subject to misrepresentation and unwarranted generalizations, both typical by-products of colonialism. Too often, outsiders see single issues as defining the entire region: the Arab-Israeli conflict, for example, the isolated revival of particularly extreme versions of Islam, or the U.S. invasions of Iraq in 1991 and 2003. Some outsiders have such a sketchy mental image of this region that they may erroneously equate life in Morocco with that in Saudi Arabia, Sudan, or Iran. In fact, despite a common arid climate and a general religious preference

PERSONAL VIGNETTE

In the spring of 2001, Ibrahim Betil, a former bank executive, was standing outside one of his mobile classrooms in rural southeastern Turkey when a teenage shepherd approached, trailed by his flock of several dozen sheep. The young shepherd asked Mr. Betil if he would take care of the sheep so that the shepherd could see the computers that his friends had described with awe.

It was not the first time Mr. Betil had filled in as a substitute shepherd. Betil is pioneering a nonprofit educational effort in Turkey, where 30 percent of the population is under the age of 15 and only one child in five attends school beyond the age of 14. International agencies report that lack of education is the chief cause of poverty in Turkey. Yet, because of economic troubles, the Turkish department of education has recently had to cut back its financial support of schools.

Betil's private Education Volunteers Foundation, funded by individuals and corporations and staffed mostly by volunteers, is focusing on what has been called leapfrogging: an effort to improve the prospects for poor urban and rural youth quickly and dramatically by sharply upgrading their training in math, science, and language skills in classrooms furnished with abundant computers. The free classes, which are meant to supplement public schools, emphasize stimulating students' curiosity. There are now 62 of the foundation's schools and 3 mobile classrooms in Turkey (see Figure 6.1, inset). The program hopes to reach a million students in the next five years.

As Ibrahim Betil hands the care of the sheep back to the young shepherd, he reflects that, with the computers, attracting a million students will be no problem. Once they see their names on the computer screen, they are hooked, and the natural inclination of children to explore takes over from there.

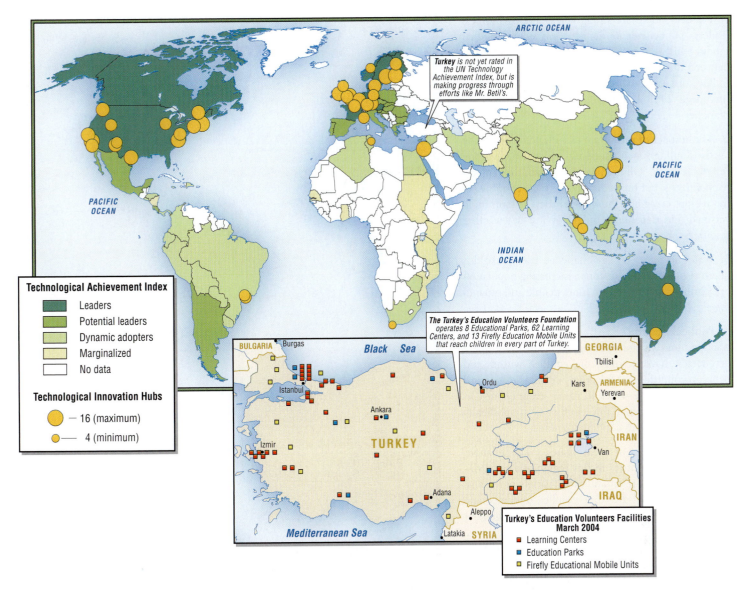

FIGURE 6.1 The geography of technological innovation and achievement. To become an innovative global technology hub, as defined by the United Nations (shown as a range of circle sizes on the map), a location must have each of the following elements: (1) a university or research center that can train skilled workers or develop new technologies; (2) an established company that provides expertise and stability; (3) local entrepreneurial drive to start new ventures; and (4) available venture capital to help take an idea to market. (Tunisia and Israel qualify.) Turkey does not yet have all these elements in place, but the Education Volunteers (EV) Foundation developed by Mr. Betil helps the country make the transformation into the information age. By 2005, EV plans to have learning centers in every part of the country. [Adapted from *United Nations Human Development Report 2001*, Map 2.1, p. 45, at http://www.undp.org/hdr2001/chaptertwo.pdf. Inset from a September 2004 facilities map provided by the Education Volunteers (EV) Foundation of Turkey.]

for Islam, there is much geographic and cultural variety throughout this region. Oil is present in abundance in some countries but totally absent in others. In some parts of the region, nearly everyone lives in a city; in other parts, rural life is dominant. In some countries, civil laws are heavily influenced by religious law; others have secular legal systems. Women may be active in public life and prominent in government, or they may lead secluded domestic lives and have few educational opportunities. And ethnicity varies greatly within countries and across the region.

THEMES TO EXPLORE IN NORTH AFRICA AND SOUTHWEST ASIA

Several themes appear throughout this chapter:

1. The role of Islamic culture. A major source of debate in the region is the degree of influence Islamic beliefs should have on government, law, gender roles, and social behavior. Although, in many countries there is popular acceptance of a religious influence in government, there are strong voices among business leaders, politicians, journalists, and educators that modify this stand. To some extent the debate is between the two main branches of Islam, Sunni and Shi'ite (discussed in the section on religion in daily life, page 298), but there are moderates in both branches.

2. The impact of oil.   The oil-rich countries of the Persian Gulf are major suppliers of fossil fuels to the rest of the world. Oil has provided wealth for a minority of citizens, but the economies of most countries in the region, especially the oil countries, remain undiversified and dependent on oil or just a few other resources.

3. Water scarcity.   A key challenge is to make this desert region inhabitable for an increasing population by obtaining more water for more people and expanding agriculture.

4. Political conflict and violence.   Rivalry for resources and territory and agitation for more direct democracy have resulted in conflict within and between countries that frequently devolves into terrorism. Terrorists have struck in nearly every country in the region. Too often this conflict is simplistically attributed to religious contentions alone.

5. Global impacts of local issues.   The relationships between countries in this region and countries around the world are difficult, for a variety of reasons: the escalating Israel–Palestine conflict, undemocratic governments within the region, the mismanagement of oil wealth, and related extreme reactions to oppression. Autocratic and unresponsive governments established locally and under the influence of Europe and the United States have bred violence as a political strategy. On several occasions, this violence has spread beyond the boundaries of the region to Europe, Africa, Southeast Asia, and the United States.

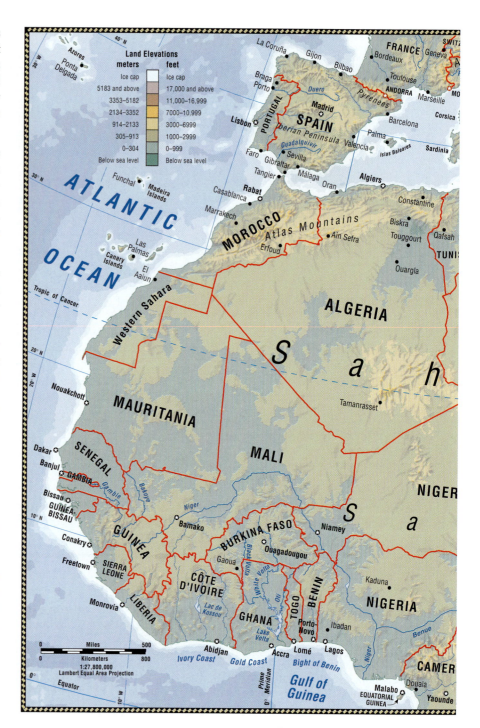

FIGURE 6.2 Regional map of North Africa and Southwest Asia.

I   THE GEOGRAPHIC SETTING

Terms to Be Aware Of

The common term *Middle East*, often used for the eastern Mediterranean countries of the region, is not used in this chapter because, as the well-known Palestinian scholar Edward Said pointed out, it reflects the Eurocentric tendency to lump all of Asia together, differentiating it only by its distance (near, middle, far) and direction (east) from western Europe. To a Japanese person the region lies to the far west, and to a Russian, to the south. Also, the term *Middle East* as popularly used does not usually include the western sections of North Africa (the Maghreb, see pages 319–321) or the eastern portions of Southwest Asia—Iran, for example—all of which we include in this region. On the other

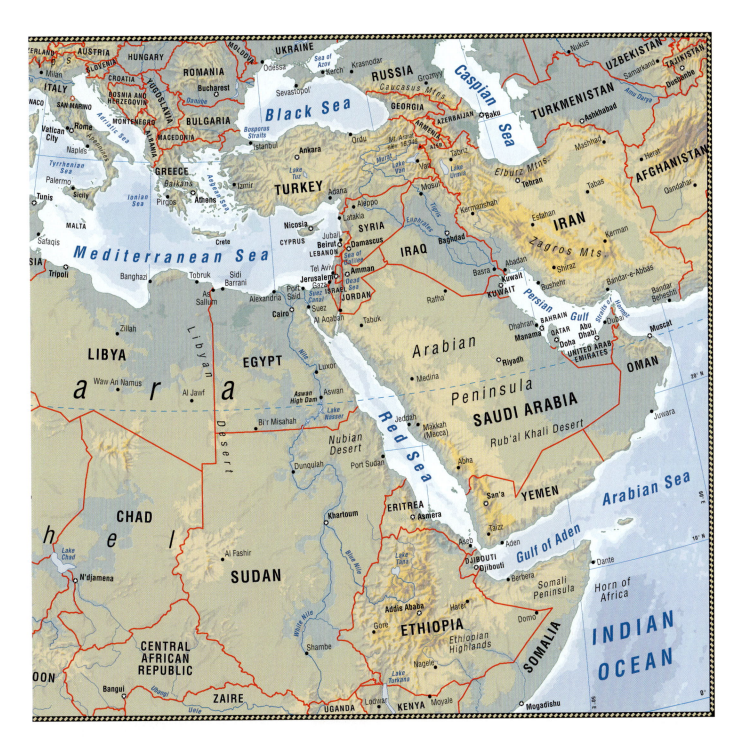

hand, the reader should know that some people who live in the region do use the term *Middle East* themselves, as do the press, broadcast media, some historians, and some archaeologists. Although it is now common in the media to refer to an *Islamic World*, this term is not used in this book because it carries a false implication that there is a cohesive Islamic community, the members of which could be expected to participate in common practices and have the same values. This is not the case, and the term is therefore misleading. Although all Muslims accept the Five Pillars of Islam, there are many versions of Islamic practice in day-to-day life. The variability of Islam is explored in a box with that title on page 300. Furthermore, non-Muslims constitute significant minority communities throughout the region. *Arab World* is not used because it excludes those, such as Iranians, Kurds, Turks, and Israelis, who are not Arabs.

PHYSICAL PATTERNS

Landforms and climate are particularly closely related in this region. The climate is dry and hot in the vast stretches of relatively low, flat land and somewhat moister where mountains are able to capture rainfall.

Climate

No other region in the world is as pervasively dry as North Africa and Southwest Asia (Figure 6.3). A global belt of dry air that circles the planet between roughly 20° and 30° north and south latitudes cuts through this region, creating desert climates in the Sahara of North Africa, the Arabian Peninsula, Iraq, and Iran. The Sahara's size and southerly location make it a particularly hot desert; in some places, temperatures can reach 130°F (54°C) in the shade at midday. In the nearly total absence of heat-retaining water or vegetation, nighttime temperatures can drop quickly to below freezing. Nevertheless, in even the driest zones, humans survive at scattered oases where they maintain groves of drought-resistant date palms and some irrigated field crops. Traditionally, desert inhabitants have worn light-colored, loose, flowing robes that reflect the sunlight during the day and provide insulation against the cold at night. Many people continue this practice.

On the uplands and margins of the deserts on both the African and Eurasian continents, enough rain falls to nurture grass, some trees, and limited agriculture. Such is the case in western Morocco and northern Algeria, Turkey, northern Iraq, and Iran. Generally too dry for cultivation, the rest of the area has long been the prime herding lands of the region's nomads, such as the Kurds of Southwest Asia, the Berbers in North Africa, and the Bedouin of the steppes and deserts on the Arabian Peninsula. Recently, the restrictions of a dry climate have been overcome with irrigation, but the general aridity of the region means sources of irrigation water are scarce.

Climate change has repeatedly affected human settlement over the millennia. Grazing animals depicted in caves like this one at Tassilli'n Agger, Algeria, in the heart of the Sahara, would not survive there under present conditions. [Pierre Boulat/Woodfin Camp and Associates.]

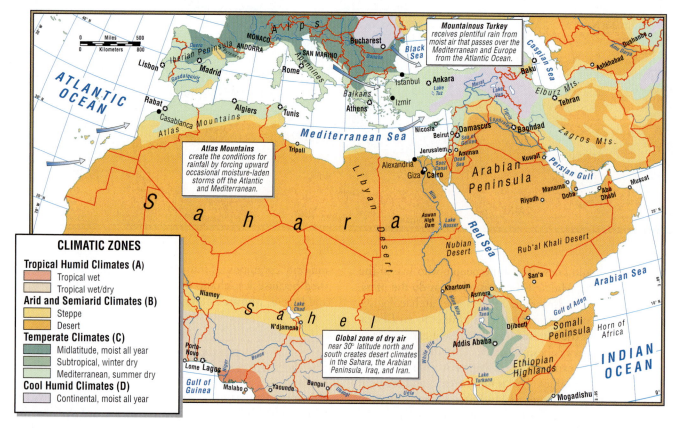

FIGURE 6.3 Climates of North Africa and Southwest Asia. To better understand how climate and landform patterns are related, compare this map with the map in Figure 6.2.

Landforms and Vegetation

The undulating surfaces of desert and steppe lands cover most of North Africa and Saudi Arabia (compare Figures 6.2 and 6.3). Fringes of mountains, located here and there across the region, capture moisture and make possible the survival and even flourishing of humans, plants, and animals. In northwestern Africa, the Atlas Mountains stretch from Morocco on the Atlantic to Tunis on the Mediterranean. They block and lift moisture-laden winds from the Atlantic, creating the conditions for rainfall of more than 50 inches (127 centimeters) a year in some places. Snowfall is sufficient in some Atlas Mountain locations to support a skiing industry. Behind these mountains, to the south and east, spreads the great Sahara Desert, where rain falls rarely and unpredictably.

Africa and Southwest Asia are separated by a rift formed between two tectonic plates—the African Plate and the Arabian Plate—that are moving away from each other (see Figure 1.10 on page 23). The rift, which began to form about 12 million years ago, is now occupied by the Red Sea. The Arabian Peninsula lies to the east of this rift. There, mountains bordering the rift in the southwest corner rise to 12,000 feet (3658 meters), and the precipitation they capture from moisture-laden monsoon winds crossing Central Africa is the primary source of rain for the entire peninsula.

Behind these mountains, to the east, lies the great desert Rub'al Khali. Like the Sahara, it is virtually devoid of vegetation. Strong winds drive sand into every crevice and move huge dunes across the landscape. The dunes of the Rub'al Khali are the world's largest, some reaching more than 2000 feet (610 meters) in height. To protect themselves from the persistent winds, blowing sands, and temperatures that vary radically from day to night, the few remaining desert dwellers still line their tents with animal skins, rugs, and tapestries.

The landforms of Southwest Asia are more complex than those of North Africa and Arabia. As the Arabian Plate, carrying the great desert peninsula of Arabia, rotates slowly to the northeast, away from the African Plate, it is colliding with the Eurasian Plate and pushing up the widely spaced mountains and plateaus of Turkey and the Zagros Mountains in Iran. Turkey, Iran, and adjacent lands are often the sites of earthquakes related to these tectonic plate movements (see Figure 1.10). Mountainous Turkey receives considerable rainfall from the moisture masses that pass over Europe from the Atlantic Ocean. But by the time this air reaches the mountains of Iran, to the southeast of Turkey, only a little moisture makes it over these mountains and hence, the interior of Iran is very dry.

There are only three major rivers in this entire region, and all have attracted human settlement for thousands of years. The Nile flows north from the central East African highlands across mostly arid Sudan and desert Egypt to a large delta on the Mediterranean. The Euphrates and Tigris rivers both begin in the mountains of Turkey and flow southeast to the swamps of the Shatt al Arab estuary terminating in the Persian Gulf. A fourth and much smaller river, the Jordan, starts as snow melt in the uplands of southern Lebanon and flows 223 miles (359 kilometers) through the Sea of Galilee to the Dead Sea. Most other streams (wadis) in the region flow only after the infrequent and usually light rains that fall between November and April. (See the photo above.)

A wadi is a dry riverbed in the desert. This is Wadi Rum in Jordan.
[Adam Woolfitt/Woodfin Camp & Associates.]

Although North Africa and Southwest Asia were home to some of the very earliest agricultural societies, agriculture today is possible only in a few places. In areas along the Mediterranean coast, where farmers can count on rain during the winter, they grow citrus fruits, grapes, olives, and many vegetables, often using supplemental irrigation. In the valleys of the major rivers, seasonal flooding and modern irrigation methods provide water for such crops as cotton, wheat, barley, and vegetables. In occasional oases, and in a few places where groundwater is pumped to the surface with the help of technologically advanced equipment, such crops as dates, melons, apricots, and vegetables are grown.

HUMAN PATTERNS OVER TIME

Important advances in agriculture and in the organization of human society took place long ago in this part of the world: it was the location of early centers of crop agriculture, urban settlement, and civilization. Later, three of the world's great religions were born here: Judaism, Christianity, and Islam. Islam became the region's dominant religion and the center of its private and public life. For many centuries, beginning about 1300 years ago (A.D. 700), the region influenced Europe and the rest of the world through its advanced learning, effective trading strategies, and refined culture. As we shall see, Muslim empires waxed and waned, and waxed again. In more recent times, however, especially after World War I, Western influences (Europe and the United States) and general modernization are challenging traditional mores regarding religion, gender roles, and the role of the state in society.

Agricultural Beginnings

Between 10,000 and 8000 years ago, nomadic peoples of this region began some of the earliest known agricultural communities in the world. These communities were located in an arc formed by the uplands of the Tigris and Euphrates river systems (in modern Turkey and Iraq) and the Zagros Mountains of modern Iran. This zone is often called the **Fertile Crescent** (Figure 6.4). There, people found bountiful environments: plentiful fresh water with fish; open forests and grasslands; abundant wild grains; and goats, sheep, wild cattle, and other large animals. As these early people adjusted to the environment, their emerging skills in domesticating plants and animals allowed them to build more elaborate settlements and societies that were eventually based on widespread irrigated agriculture in lowland locations along the base of the mountains and along the Tigris and Euphrates.

The Emergence of Gender Roles. Some researchers think that the dawning of agriculture may mark the transition to markedly different roles for men and women. Archaeologist Ian Hodder reports that at the 9000-year-old site Çatalhöyük, in south-central Turkey, there is little evidence of gender differences. Families were small and men and women performed similar chores in daily life, had similar status and power, and both sexes played key roles in social and religious life (*Archaeology*, January 2004). After the development of agriculture, as wealth and property became more important in human society, scholars think that a concern with family lines of descent and inheritance emerged, which led in turn to the idea that women's bodies needed to be controlled so as to prevent them becoming pregnant by a number of men and thus confusing lines of inheritance.

Other Sites of Early Cultivation. Agriculture was also an early practice in the Nile Valley 6500 years ago, in the Maghreb (northwestern Africa) at about the same time, and in the eastern mountains in Persia (modern Iran). In adjacent steppe zones,

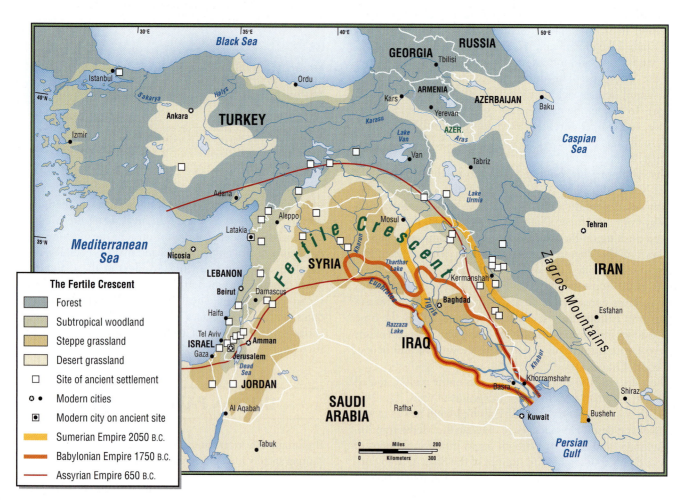

FIGURE 6.4 The Fertile Crescent. The Fertile Crescent, where people first domesticated plants in this region, extends from the eastern Mediterranean coast into the mountains of southern Turkey and curves to include the lands along the Tigris and Euphrates rivers and the Zagros Mountains. Three major empires developed successively in the eastern extent of this crescent: the Sumerian, Babylonian, and Assyrian. [Adapted from Bruce Smith, *The Emergence of Agriculture* (New York: Scientific American Library, 1995), p. 50.]

societies of nomadic animal herders traded meat, milk, hides, and other animal products for the grain and manufactured goods of the settled areas, which eventually took on urban qualities: dense settlement, specialized occupations, centralized government, and concentrations of wealth. The city of Sumer (in modern southern Iraq), for example, existed 5000 years ago and gradually extended its influence over the surrounding territory. The Sumerians developed wheeled vehicles, oar-driven ships, and irrigation technology. It is believed that this culture also pioneered the idea that people could specialize in livelihood activities, and that this specialization led ultimately to central authorities and government bureaucracies. At times, nomadic tribes would band together, sweep over settlements with devastating cavalry raids, and set themselves up as a ruling class. These ruling nomads tended to adopt the culture of the sedentary peoples they conquered, and after a few generations they became almost indistinguishable from them. Usually rulers tried to expand their domains through a combination of military aggression, administrative consolidation, and trading acumen. These strategies occasionally led to the creation of vast but unwieldy empires that did not last long. In time, the conquerors themselves became vulnerable to defeat by new waves of nomadic peoples. Thus, the civilization of Sumer was succeeded by the Babylonian and Assyrian empires, the latter reaching its zenith about 3000 years ago.

The Coming of Judaism, Christianity, and Islam

The early religions of this region were based on a belief in many gods linked to natural phenomena, but several thousand years ago **monotheistic** belief systems—those based on one god—emerged. Judaism, Christianity, and Islam have their origins in the eastern Mediterranean and are monotheisms.

Judaism was founded approximately 2000 years before Christianity. According to tradition, it was begun by the patriarch Abraham, who led his followers from Mesopotamia (modern Iraq) to the shores of the eastern Mediterranean (modern Israel and Palestine). Jewish religious history is recorded in the Torah (the Old Testament of the Bible) and is characterized by the belief in one God, Yaweh; a strong ethical code summarized in the Ten Commandments; and an enduring ethnic identity. After the Jews rebelled against the Roman Empire at the Masada fortress on the Dead Sea in A.D. 73, they were expelled from the eastern Mediterranean and migrated to other lands in a movement known as the **Diaspora** (the dispersion of an originally homogeneous people). Many Jews dispersed across North Africa, and some crossed to the Iberian Peninsula in Europe. Other groups of Jews made their way into eastern and central Europe; and still others, to various parts of Asia and eventually to the Americas.

The eastern Mediterranean is the ancient home of those who spoke Semitic languages, including the peoples of Aram (present-day Syria), Assyria in the upper valley of the Tigris River, Babylonia, and Phoenicia. Modern peoples speaking Semitic languages include the Arabs and Jews. The Jews of today base their claims to Palestine on the biblical kingdoms and on their expulsion from the region beginning in A.D. 73 until the fall of the Roman

Empire in A.D. 476. Many Jews scattered by the Diaspora encountered persistent discrimination through the ages, especially in eastern Europe, where occasionally whole villages would be murdered in **pogroms** (episodes of ethnic cleansing). This deadly oppression culminated in the Nazi era before and during World War II, when more than 6 million Jews across Europe were rounded up, imprisoned, made to work as slaves, and eventually killed in gas chambers in what has become known as the Holocaust. When the Jews left the eastern Mediterranean more than 1500 years ago, other non-Jewish Semitic people remained, many of whom converted to Islam shortly after the Prophet Muhammad founded the religion around A.D. 622. Palestinians descended from these peoples and from other Arab groups who joined them over the centuries.

Christianity is based on the teachings of Jesus of Nazareth, a Jew, also known as Christ, who gathered followers in the area of Palestine in the eastern Mediterranean. He taught that there is one God, whose relationship to humans is primarily one of love and support but who will judge those who do evil. Both Jewish (religious) and Roman (governmental) authorities of the time saw Jesus as a dangerous challenge to their power. After his execution in about A.D. 32, his teachings spread and became known as Christianity. By A.D. 400, Christianity was the official religion of the Roman Empire. In A.D. 1054, Christianity split into the Western tradition associated with Rome and the Eastern, or Orthodox, tradition headquartered in Constantinople (Istanbul). Christianity in its various forms gained its strongest foothold in Europe, and Orthodox Christianity was introduced by Greeks to Russia in about A.D. 1000. However, with the spread of Islam after A.D. 622, only remnants of Christianity remained in the eastern Mediterranean (Maronites), in North Africa (Copts), and in enduring pockets elsewhere in Asia (such as the Malabar Coast of India). During the eleventh, twelfth and thirteenth centuries, European Christians launched various military ventures, known as the Crusades, to retake what they regarded as the Holy Land (eastern Mediterranean) from the Muslims (see below). Christianity was later spread through European colonization and missionary activity to the Americas, and parts of Africa and Asia.

Islam is now the overwhelmingly dominant religion in the region of North Africa and Southwest Asia. Islam emerged in the seventh century A.D. when the archangel Gabriel revealed the principles of the religion to the Prophet Muhammad. Muhammad was a merchant and caravan manager in the small trading town of Makkah (Mecca) on the Arabian Peninsula near the Red Sea. Islam was, and is, considered an outgrowth of Judaism and Christianity. Followers of Islam, called Muslims, believe that Muhammad was the final and most important of a long series of revered prophets, which included Abraham, Moses, and Jesus. They also believe that the **Qur'an** (or **Koran**), the holy book of Islam, contains the words God (**Allah**) revealed to Muhammad. (*Allah* is the Arabic word for "God." Thus Arabic-speaking Christians as well as Muslims use the word *Allah* in their prayers.) In his early years as a messenger of God, Muhammad instructed his followers to pray toward Jerusalem, the central holy place for Jews and Christians (now Muslims pray toward Makkah). This, and the fact that Jerusalem is the place from which the Prophet

Muhammad is believed to have ascended into heaven, make the city particularly holy to Muslims around the world. All Muslims are encouraged to undertake the **hajj,** the pilgrimage to the cities of Makkah and Medina (sites of the Prophet Muhammad's mosque and burial place), at least once in a lifetime. Although there have been several powerful Muslim empires, the religion of Islam, per se, has virtually no religious hierarchy or central administration. The world's 1 billion Muslims communicate directly with God, not through a clergyman. As a result, and this is an important point, the interpretation of Islam varies widely within and among countries and from individual to individual.

The Spread of Islam

The nomads of the Arabian Peninsula, the Bedouin, were among the first converts to Islam. They were already spreading the faith and creating a vast Islamic sphere of influence by the time of Muhammad's death, in A.D. 632. Over the next century, Muslim armies extended an Arab-Islamic empire (here defined as control of politics and trade) over most of Southwest Asia through Iran and then over North Africa, as well as into Spain (Figure 6.5).

The new empire established bonds between government and Islam that continued after the empire's demise. Islamic rules of conduct guided such matters as taxation policy and particularly the system of law, which was administered by religious specialists. The leaders of the empire often claimed a blood relationship with Muhammad. Arabs remained the dominant elite as Islam spread, but they placed local non-Arabs and non-Muslims as administrators of newly taken territory. Thus, they allowed local rule but collected taxes that supported the empire. They were particularly tolerant of Jews, Christians, and other monotheists.

While Europe languished in the medieval period (A.D. 450–1400), Muslim scholars traveled extensively throughout Asia and Africa, and they made important contributions in the fields of history, mathematics, geography, medicine, and other academic disciplines that flourished in Alexandria, Baghdad, Damascus, Fez, and Toledo (in Spain). These centers of world learning arose in vibrant economies that benefited from the early Islamic development of financial institutions and practices, such as banks, trusts, checks, and receipts. From India, Muslims imported the useful mathematical concept of the zero, further advancing trade, record keeping, and scientific knowledge in the region. Islamic

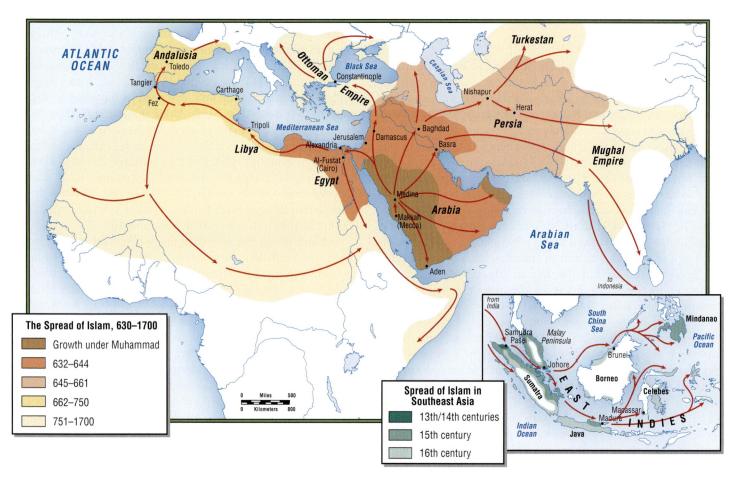

FIGURE 6.5 The spread of Islam, 630–1700. In the first 120 years following the death of the Prophet Muhammad in 632, Islam spread primarily by conquest. Over the next several centuries, Islam was carried to distant lands by both traders and armies. [Adapted from Richard Overy, ed., *The Times History of the World* (London: Times Books, 1999), pp. 98–99.]

PRACTICING GEOGRAPHY COMPARING LOCAL LANDSCAPES

Think of these kitchen counters in Mexico and Indonesia as tiny landscapes that reveal a great deal about cultural connections. Photo A shows the ingredients for a Mexican mole. Photo B shows the ingredients for an Indonesian curry dish, sayur lodeh. The similarities they display are in large part the result of the Arab connection. Arab traders linked many parts of

Asia (for example, China, the Philippines, Indonesia, Burma, and India) with the Mediterranean. The Spanish completed the connection by bringing new food crops from the Americas to the Mediterranean. Examine the two pictures and see if you can identify the origin of the ingredients in each. You can use Table 3.1 (page 124) to help you a bit. [*A* and *B*, Ignacio Urquiza.]

medicine drew on a wide variety of traditions that had their origins from China to West Africa; Islamic medical practices, such as medical record keeping, and regular peer review of physicians existed already in the eleventh century and remained superior to those in Europe until the nineteenth century.

By the tenth century, people throughout the Arab-Islamic empire's former territories had gradually converted to Islam and many adopted the Arabic language. But then the Arab-Islamic empire began to break apart into smaller units as outlying provinces, often ruled by a local Islamic dynasty, eluded the control of central authorities. From the eleventh to the fifteenth centuries, Mongols from eastern Central Asia and other peoples conquered parts of Muslim territory. Meanwhile, beginning in the 1200s, nomadic herders in western Anatolia (Turkey) had

begun to forge the Ottoman Empire, the greatest Islamic empire the world has ever known. By the fifteenth century, the Ottoman Muslims had defeated the Christian Byzantine Empire (the successor to the Roman Empire) and taken over its capital, Constantinople, which they renamed Istanbul. Very soon the Ottomans controlled most of the eastern Mediterranean, Egypt, and Mesopotamia, and by the late 1400s, they had extended their control to much of southeastern and central Europe and had nearly taken Vienna. Ironically, at about the same time that the Ottoman Muslims grew stronger (1400s), the Arab Muslims lost their control of the Iberian Peninsula.

Figure 6.5 shows that during these various empires, the Islamic faithful were found from Spain in the east, deep into North Africa, in Turkey and southeastern Europe (the Balkans), and across Central Asia to the Indus Valley and northern India. Muslim traders carried the Islamic faith still farther, by land across Central Asia to western China and by sea to Indonesia and Malaysia, the southern Philippines, and even parts of southeastern China. These areas, and those of the former Arab-Islamic empire, are still predominantly Islamic today, with the exception of Spain, some parts of the Balkans, and southeastern China, where, nonetheless, relics of Islamic culture remain.

The Ottoman Empire, like the Arab-Islamic empire before it, supported religious tolerance within its borders. Jews and Christians were allowed to practice their religion, although there were economic and social advantages to converting to Islam. The Ottomans prospered from the productivity of the empire's many different peoples. Drawing on trading networks that stretched from central Europe to Algeria to the Indian Ocean, Istanbul became a cosmopolitan capital, outshining most European cities until the nineteenth century.

The origins of cultural practices in Islam are often obscure and misunderstood. For instance, many people believe that the Muslim religion instituted restrictions on women, such as **seclusion** (the requirement that a woman stay out of public view) and veiling (the custom of covering the body with a loose dress and the head—and in some places, the face—with a scarf). There is ample evidence, however, that these practices predate Islam by thousands of years. Nor are the antiforeign attitudes associated with some movements in Islam today inherent in that religion. On the contrary, some past Islamic empires, such as the Ottoman, were known for their cosmopolitan culture, tolerance, and even appreciation and adoption of foreign ways. Antiforeign sentiment is a more recent phenomenon, coming after years of European and U.S. interference in the affairs of the region.

Western Domination and State Formation

The Ottoman Empire ultimately withered in the face of a Europe made powerful by the industrial revolution. Throughout the nineteenth century, North African lands changed from Ottoman to European control. At first, European influence was exercised primarily through control of trade and finance. North Africa became a source of raw materials for Europe in a trading relationship dominated by European merchants, who sought the agricultural products of the coast—the cotton of Egypt, and the phosphates, manganese, hides, and wool of the western Mediterranean countries. In 1830, France became the first European country to exercise direct control of a North African territory when its military gained control of Algeria in the western Mediterranean. France eventually administered that land almost as though it were a part of France—although Muslims were not allowed the benefits of full citizenship. In the last two decades of the nineteenth century, France, Britain, and Italy raced to occupy several other North African countries, motivated in part by European power politics—the fear that a rival would step in first. Britain gained control of Egypt (1882) and then Sudan; France took over Tunisia (1881) and Morocco; and Italy moved into Libya. In these countries, Europeans set up a form of dependence that they termed a protectorate: local rulers remained in place, but European officials made the important decisions.

In response to increased European influence, and in the hope of checking European encroachment while at the same time building more modern national identities, some countries, including some still under Ottoman control, adopted European institutions—systems of administration and taxation, trade, and legal systems—and gave up traditional local, ethnic, and more religiously based governing customs. This strategy of partial Europeanization backfired; these secular institutions were further developed and strengthened under formal European control, and popular (local) participation in governance was squelched by the Europeans who felt such participation would undermine their control.

Another result of European influence was that traditional, more egalitarian systems of landownership were undercut to the disadvantage of small cultivators. European officials privatized tribal collective lands and gave title to just the ethnic (tribal) leaders in return for their political support. Thus European domination disempowered the many and encouraged the concentration of wealth and power in the hands of a relative few.

World War I (1914–1918) brought the death of the Ottoman Empire. The Ottomans had allied with Germany and after the war ended in defeat for that country, the League of Nations (a precursor to the United Nations) allotted almost all former Ottoman territories at the eastern end of the Mediterranean to France and Britain for supervision (Figure 6.6). Only Turkey was recognized as an independent country at that time. Syria, and what is now Lebanon, became mandated territories of France (dependencies thought incapable of self-rule). Palestine and Iraq, out of which the kingdom of Jordan was carved, became mandated territories of Britain. In the Arabian Peninsula, Bedouin tribes were consolidated under the control of Sheik Ibn Saud in 1932, after which point Saudi Arabia began to emerge as a state according to the Western definition of that term. The smaller states bordering the peninsula (Yemen, Oman, the United Arab Emirates, Qatar, Bahrain, and Kuwait), which formed from ethnic groups independent of Ibn Saud, were British protectorates after World War I and eventually each became an independent country.

The aftermath of World War II led to further effects on the political development of the North Africa and Southwest Asia region. Foreign influence remained strong in many countries as the U.S.–USSR cold war developed. As the United States gained status as a world power, it supported autocratic local leaders who were most sympathetic to U.S. policies (as opposed to USSR policies) and most likely to maintain a friendly attitude toward U.S. business interests. In Iran and Saudi Arabia, where vast oil deposits became especially lucrative by mid-century, European and U.S. oil companies played a key role in deciding who ruled. Oil profits were undertaxed and most of the profits were gathered by a small elite and spent on opulent living or invested abroad. Although some oil revenue went toward building roads, hospitals and schools—facilities that had not previously existed—too little was invested in opening up opportunities for the masses. Wealthy sheiks often were generous to their subjects, but there were few institutionalized methods for ensuring that profits from the

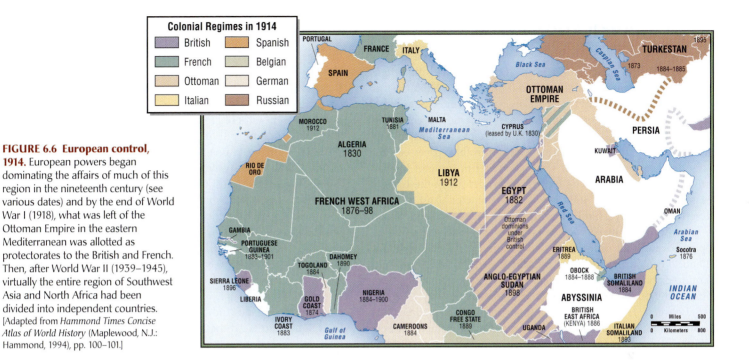

FIGURE 6.6 European control, 1914. European powers began dominating the affairs of much of this region in the nineteenth century (see various dates) and by the end of World War I (1918), what was left of the Ottoman Empire in the eastern Mediterranean was allotted as protectorates to the British and French. Then, after World War II (1939–1945), virtually the entire region of Southwest Asia and North Africa had been divided into independent countries. [Adapted from *Hammond Times Concise Atlas of World History* (Maplewood, N.J.: Hammond, 1994), pp. 100–101.]

region's primary natural resource asset were made available for the general welfare. Private oil wealth was invested abroad in shopping malls, resorts, banks, and department stores in places such as the United States, Australia, and the United Kingdom. The region remained dependent on Europe and North America for the technology needed to exploit oil and to begin the mechanization of manufacturing, transportation, and agriculture. During this period, disparity in wealth between the elite and the majority of the people increased dramatically. Also, the atrocities of World War II led to the formation of the state of Israel and to the resulting Israeli–Palestinian dispute that has monopolized the politics of the region for so long.

The creation of the state of Israel on Palestinian lands. In 1947, after World War II, the Western powers were searching for a place to settle the tens of thousands of European Jewish survivors of Nazi death camps who were still in European refugee centers. Although some Jews were taken in by England, France, the United States, and countries in South America, no country stepped forward to offer a home to all Jews. Meanwhile, the idea of European Jews migrating to the ancestral homeland once known as Israel had been gaining popularity among Jews since it was proposed in nineteenth-century Austria. A small group of Jews, known as **Zionists,** had begun to purchase land here and there from wealthy Palestinian landholders and establish communal settlements called *kibbutzim.* Jewish and Arab populations lived intermingled. Most of the people displaced by the Zionist land purchases were poor farmers and herders, who were accustomed to using bits of land held by village landlords for their fields, pastures, and houses. In 1917, the British government issued the Balfour Declaration, which committed it to support the establishment of "a national home for the Jewish people" in

Palestine, but only if the civil and religious rights of non-Jewish communities in Palestine could be assured (see Figure 6.15A on page 311). Land purchases by European Jews continued. It should be noted that the Balfour Declaration was itself a breach of previous agreements the British had with Arab leaders to assure them independent control over Arab lands. This breach is still cited as cause for distrusting the West.

By 1946, strong sentiment had built among the world's Jews that a Jewish homeland should be created in Palestine, and many Jews already in Palestine took up arms to convince the British to stand by the Balfour Declaration. The Palestinian Arabs and their primarily Muslim supporters in the region fiercely objected to the formation of the state of Israel, out of fear that Palestinians would lose claim to the land they had long owned and held title to under the Ottomans and British, and would be denied a voice in their own governance. Notice that the struggle between Zionists and Palestinian Arabs was not over religion but over land, and this remains the case, with water now also an issue.

The Palestinians were not unified, and they lacked connections in, and knowledge of, the West. By the late 1940s, British, European, and U.S. sentiment in the aftermath of World War II sided with the Jews, and in 1947 the United Nations voted to recommend a partition plan that would have divided the area of Palestine between Jews and Arabs with Jerusalem internationalized. Though neither Arabs nor Jews had the right to vote for or against the UN resolution, they both reacted. The Arabs rejected this plan because they would have lost land, and the World Zionist Organization (Jews) accepted it, but only with reservations. When the British left in early May 1948, Jewish leaders immediately created a Jewish state from part of Palestinian land (see further discussion on pages 309–311).

POPULATION PATTERNS

The great size of this region and its aridity have had important implications for human settlement and interaction. Although the region as a whole is nearly twice as large as the United States, the areas that are useful for agriculture and settlement are tiny by comparison. The land supports 100 million more people than the United States, and population densities in livable spaces can be quite high, over 100,000 per sqare kilometer in some of Egypt's urban neighborhoods. (The highest urban density in the United States does not exceed 4000 per square kilometer.) Vast tracts of desert are virtually uninhabited. In the population map (Figure 6.7), most people live along coastal zones, in river valleys, and in upland zones that capture orographic rainfall. Efforts are being made to extend livable space into the desert, but doing so may require extremely costly and environmentally questionable solutions to the problems of water scarcity and soil infertility.

Between 1980 and 2000, human fertility rates across the region fell dramatically. In the 1960s, when most women married shortly after reaching puberty, some countries had average fertility rates of close to nine children per woman. By 2001, as literacy rates rose (though they are still below the world average; see Table 6.2, page 318), and the average woman did not marry until age 21, the average fertility rate for the region was reduced to just under four children per woman. Nonetheless, this rate is still higher than the world average (2.8; only Africa's is higher); and in Iraq and Saudi

Arabia, fertility rates are still five children per woman and higher. One result is that well over 40 percent of the population is under the age of 15 in Iraq, Jordan, Palestine, Saudi Arabia, Syria, Yemen, and Sudan. Population will continue to grow rapidly as these children reach reproductive age, even if each couple has only two children.

At present rates of growth, the population of the region, now about 400 million, will double to 800 million by about the year 2030. It is difficult to imagine how the area will support the additional people. Fresh water is already in extremely short supply, and most countries must import food at great cost. Furthermore, 32 million more jobs would have to be created within the next decade to employ the added population—compared to just 4.5 million needed in Europe, where population is growing very slowly. If women start entering the workforce, even more new jobs will be needed. As Figure 6.8 shows, by the year 2001, considerably less than 50 percent of women across the region (except for Morocco, Israel, Qatar, and Turkey) worked outside the home (this figure excludes the large number of women who work on their family farms). If the status of women changes more quickly than anticipated, the demand for jobs will increase even beyond the anticipated 32 million.

Many inhabitants of the region believe that the industrialized world (Europe, Russia, North America, Japan, and parts of Oceania) is overly concerned with population growth in the developing world, and too little concerned with its own high rates of resource consumption. It is thought that the industrialized world

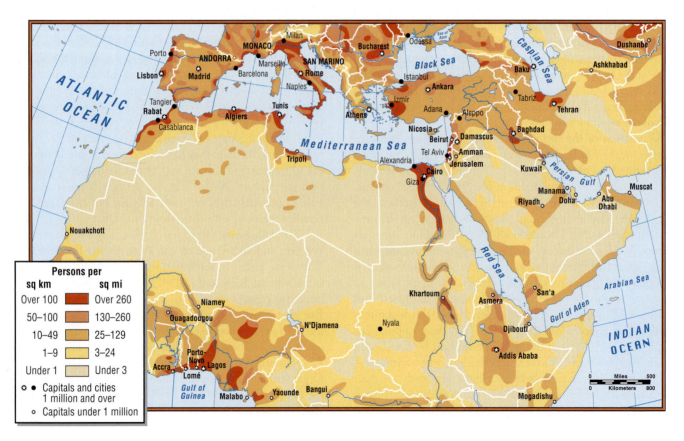

FIGURE 6.7 Population density in North Africa and Southwest Asia. [Adapted from *Hammond Citation World Atlas* (Maplewood, N.J.: Hammond, 1996).]

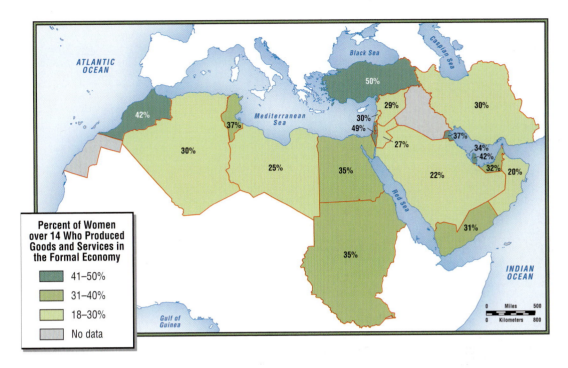

FIGURE 6.8 Percentage of women who are wage-earning workers. The percentage of women earning wages in this region is slowly but steadily increasing. [Adapted from Joni Seager, *Women in the World: An International Atlas* (New York: Viking Penguin, 1997), pp. 66–67, and *United Nations Human Development Report 2003* (New York: United Nations Development Programme), Table 25.]

promotes population control elsewhere so it can continue to consume per capita more than a fair share of the world's resources. On the other hand, some leaders in the region (especially in Algeria, Egypt, Iran, Jordan, Morocco, Oman, Qatar, Sudan, Tunisia, Turkey, and Yemen) see the wisdom of population control and encourage family planning. Yet, despite this official attitude and despite drastic declines in fertility, by 2003 in the region as a whole, only about half the women were using modern methods of contraception. Most specialists in Islamic law say that limiting family size is acceptable for a wide variety of reasons, so long as the motive is not to shirk parenthood altogether.

Gender Roles and Population Growth

In societies where fertility rates remain high despite a decline in labor-intensive agriculture of the sort that depends on children as a labor supply, gender inequality may partially explain why many couples are still choosing to have so many children. In urban contexts, especially where women have opportunities outside the home—to obtain knowledge, wages (see Figure 6.8), or prestige on the job, or to participate in civil society—they almost always choose to have fewer children. Throughout the region, wherever men have a much higher education level than women (see Table 6.2), greater access to employment, and greater power to make family and community decisions (see the discussion of the family on page 299), children are an important source of family involvement and power for women. Where women are secluded in the home and undereducated (see Table 6.2, page 318), they have little chance to accumulate wealth or to enhance the prospects and reputation of the family in ways other than having children.

Although women contribute their labor and goodwill to the reproduction, health, and well-being of families, one result of the lesser ability of women to contribute financially is a deeply entrenched cultural preference for sons. Because of their wider opportunities, sons are better positioned to increase the family wealth and social standing. Sons are expected to support their parents financially in their retirement (daughters are expected to render personal care). The strong preference for sons means that families sometimes continue having children until they have a sufficient number of sons, or young females simply may not survive, as is indicated by missing females in the 0- to 4-year age cohort of the population pyramids shown in Figure 6.9 and discussed in the legend. (The issue of missing females is discussed at greater length in Chapter 8.)

Migration and Urbanization

Emigration is increasingly common across North Africa and Southwest Asia. Because jobs are hard to find, especially for workers with more than a basic education, many millions emigrate to other parts of the world in search of work. Most of those leaving the region are young men, because here women typically do not travel widely, or on their own. Several million are guest workers in Europe, where, in 2004, Turkish guest workers alone numbered more than 3.3 million. Of the millions of emigrants, many intend to return home, in part because it is difficult to gain citizenship in European countries. In the meantime, remittances (earnings sent home) to their families in North Africa, Turkey, or Lebanon significantly increase local standards of living.

In some parts of the region, immigration rather than emigration is the trend. Over the past 30 years, the oil-rich states of the Arabian Peninsula have recruited large numbers of guest workers. When oil revenues poured in after price increases in the 1970s, the money was used to rapidly modernize some aspects of life. The labor force in these countries was too small and either too undereducated or otherwise ill-suited to fill the jobs created. As the infrastructure for a modern economy was being completed in the 1990s, there was a marked shift in migration trends. For example, in Saudi Arabia the demand for skilled guest workers

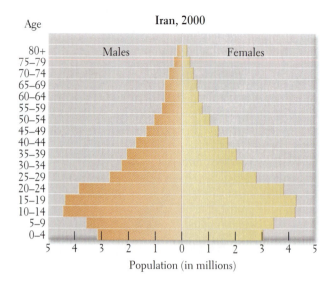

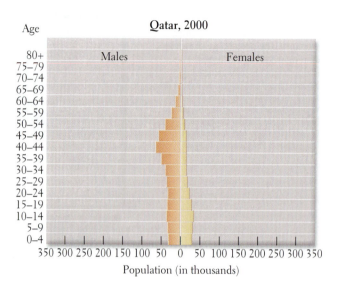

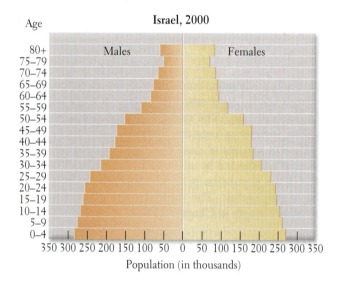

FIGURE 6.9 Population pyramids for Iran, Qatar, and Israel. The population pyramid of Iran is at a different scale (millions) from those of Israel and Qatar (thousands). Observe that Iran has had significantly fewer births over the last 10 years; that Qatar's pyramid is skewed by the presence of numerous male guest workers in the 25- to 54-year age groups; and note also that all three pyramids show missing females, especially in the age groups under 24. This is most easily observed by drawing lines from the ends of the male and female age bars to the scale at the bottom of the pyramid and comparing the numbers. [Adapted from "Population Pyramids for Iran," "Population Pyramids for Qatar," and "Population Pyramids for Israel" (Washington, D.C.: U.S. Bureau of the Census, International Data Base, May 2000.]

increased and although many young Saudis with education certificates are unemployed, they still lack the skills in demand. By 2004, according to Saudi Arabia's own agency for labor relations, immigrants made up 88 percent of the labor force. They filled jobs in all skill levels, teaching in schools and colleges, building roads, government buildings, housing, schools, universities and technical colleges, water desalinizing plants, oil and gas production facilities, and modern irrigation systems—and they run the systems once completed.

Guest workers come from all over the world, but employers prefer Muslims. Several hundred thousand workers—some technically skilled, many willing to do manual labor—arrived from Palestinian refugee camps in Lebanon and Syria, others from Egypt. In the 1980s, 1.5 million came from Pakistan alone. Many, especially female domestic workers, came from Muslim countries in South and Southeast Asia, where chances for employment are limited; they were attracted to Saudi Arabia by the relatively high wages and the chance to make the pilgrimage to the holy cities of Makkah and Medina.

Immigrant workers on the Arabian Peninsula are only temporary residents. During and after the Gulf War in 1991, millions of foreign workers fled or were summarily expelled from the Gulf countries. Although some have returned, some Arabian Peninsula governments are undertaking efforts to better prepare their own citizens to take over jobs, especially at the managerial levels. As those trained in Europe and America return from their studies abroad, some take jobs that immigrants once performed, but others wait many years for a high-status job. When migrant workers are forced back to refugee camps in Syria or to rural villages in Indonesia, Turkey, Mexico, and Egypt, the loss of income often means a return to grinding poverty for their families.

Israel also has many immigrants, but those arriving are Jews settling permanently in Israel, many of whom are fleeing persecution. They began coming to Palestine before and after World War I; and huge numbers liberated from Nazi concentration camps arrived when Israel was created in 1948, just after World War II. Most recently, immigration surged in the 1990s, when more than 700,000 Jews were allowed to enter from the former Soviet Union. Another 25,000 fled the Ethiopian civil war. Despite already crowded conditions, Israel encourages Jewish immigrants, partly because its constitution grants sanctuary to Jews and partly because it believes that a large Jewish population is important for political weight in the region.

Internal migration from rural villages to urban areas has also been an important population pattern across the region. Until recently, most people lived in small settlements; but by 2000, more than 60 percent of the region's people lived in urban areas,

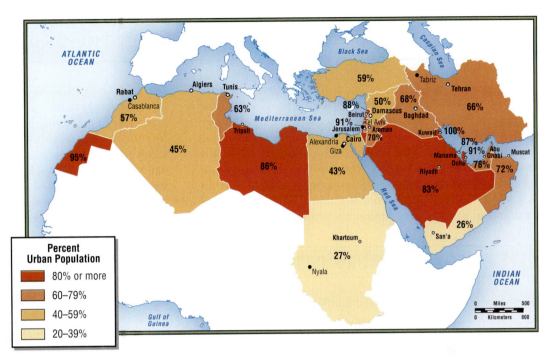

FIGURE 6.10 Percentage of urban population for each country in North Africa and Southwest Asia. By 2000, more than 60 percent of the region's people lived in urban areas, although the pattern varied greatly across the region. There are now more than 24 cities with more than 1 million people. [Data from *2003 World Population Data Sheet* (Washington, D.C.: Population Reference Bureau).]

although the pattern varied greatly across the region (Figure 6.10). There are now more than 120 cities with populations of at least 100,000; 24 of them have more than 1 million people. Contrary to patterns in other regions, where the poor build rings of shantytowns on the perimeters of relatively modern cities, poor urban migrants in this region often occupy the medieval interiors of old cities. Here streets are narrow pedestrian pathways and there is little plumbing, sewage disposal, or clean water. The ancient dwellings, dating back 500 years or more, are worthy of historic preservation, but the inhabitants are far too poor to provide even routine maintenance.

Refugees

There are hundreds of thousands of refugees in North Africa, the eastern Mediterranean, Turkey, Iraq, and Iran. Usually they seek to escape wars or environmental disasters such as earthquakes or long-term drought. When Israel was created, as many as 2 million Palestinians who were living there were displaced to refugee camps in Lebanon, Syria, Jordan, the West Bank, and the Gaza Strip. Conditions in these Palestinian refugee camps, which are no better than shantytowns, are extremely difficult. In the Gaza Strip, the population density is close to 8000 per square mile (3000 per square kilometer), per capita income is just $600 a year, the average age of the population is 15, and women have, on average, 6 children each. However, as of 2004, the country with the

world's largest international refugee population (2.21 million) was Iran, which sheltered 1.3 million Afghans and at least 600,000 Iraqis displaced by recent regional conflicts (see pages 329–331). Although some Iraqis and Afghans returned home, continuing violence and instability in both home countries convinced many to stay in Iran. Across the region, even more people are refugees within their home countries, unable to occupy their homes, usually because of civil unrest. Sudan had at least 2 million internally displaced persons living in camps in 2003 (see the discussion on pages 322–323). In 2004, the situation worsened when the government allowed (perhaps even encouraged) savage attacks by Arab-speaking Muslim vigilantes from the north against Arab-speaking Muslim black Africans in the southwest (Darfur).

In this region, refugee camps often become semipermanent communities of stateless people, in which whole generations are born, mature, and die. The residents may show enormous ingenuity in creating a community and an informal economy under adverse conditions. Nevertheless, the cost in social disorder is high. Children rarely receive enough schooling. Disillusionment is widespread. Years of hopelessness, extreme hardship, and lack of employment take their toll on youths and adults alike, leading some to do violence against those they see as responsible for their suffering. Moreover, even when international organizations contribute money to support a basic level of living for refugees, these displaced people constitute a huge drain on the resources of their host countries.

Quick Review

1. The term *Middle East*, though not used in this text, is commonly used to describe Southwest Asia relative to which other world region?

2. What global weather pattern creates the arid atmosphere of the Sahara?

3. Name the area defined by the sources of the Tigris and Euphrates rivers and the Zagros Mountains that is well known for early successes with plant- and animal-based agricutlure.

4. On what historical event do the Jews base their claim to Palestine?

5. Name the various physical and social causes for the large numbers of refugees in this region.

II   CURRENT GEOGRAPHIC ISSUES

The region of North Africa and Southwest Asia has sharp differences in physical landscape, level of economic development, political and economic systems, and social relations. Although many social institutions are shared among the countries in this region, notably those based on the practice of Islam, many countries have differing views on such issues as the role of the Islamic religion in daily life, proper roles for men and women, democracy as a political model, the pace and direction of economic development, the importance of oil, and rights to scarce water supplies.

SOCIOCULTURAL ISSUES

The countries of North Africa and Southwest Asia (Figure 6.11) are experiencing broad social changes. Here we examine a few examples that illustrate these changes: religion, family, gender roles and female seclusion, the lives of children, and cultural diversity as reflected in language.

Religion in Daily Life

More than 93 percent of the people in this region subscribe to Islam. The Five Pillars of Islamic Practice (see the box on the next page) embody the central teachings of Islam. All but the first pillar are things to do in daily life, rather than articles of faith. (Pregnant women and others who are poor, young, or sick are exempted from the physically demanding pillars.) Islamic practice in this region is thus a consistent part of daily life in ways rarely experienced in the West, where most people live by a personal ethical code but the formal practice of religion is often set aside for certain times (the Sabbath) and spaces (churches, synagogues, and mosques).

Saudi Arabia occupies a prestigious position in Islam. It is the site of two of Islam's three holy shrines, or sanctuaries: Makkah, the birthplace of the Prophet Muhammad and of Islam; and Medina, site of the Prophet's mosque and his burial place.

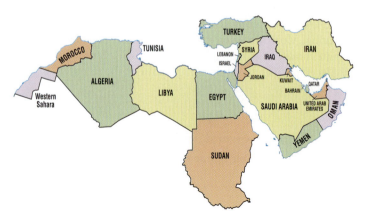

FIGURE 6.11 Political map of North Africa and Southwest Asia.

(Jerusalem is the third holy site.) The fifth pillar of Islam, that all Muslims should make a pilgrimage (hajj) to the two holy cities at least once in a lifetime, has placed Makkah and Medina at the heart of Muslim religious geography. Each year, a large private-sector service industry organizes and oversees the five- to seven-day hajj for more than 2.5 million foreign visitors. Most pilgrims visit during the month of Dhu al-Hijja on the Muslim lunar calendar. Although oil now overshadows the hajj as a source of national income, the event is economically, as well as spiritually, important to Saudi Arabia.

Beyond the Five Pillars, Islamic religious law, called **shari'a** ("the correct path"), guides daily life according to the principles of the Qur'an. Some Muslims believe that no other legal code is necessary in an Islamic society. There are, however, numerous interpretations of just what behavior meets the requirements of shari'a and what does not. Insofar as the interpretation of shari'a is concerned, the Islamic community is split into two major groups: **Sunni** Muslims, who today account for 85 percent of the world community of Islam; and **Shi'ite** (or **Shi'a**) Muslims, who are found primarily in Iran, but also southern Iraq and the south of Lebanon. The Sunni–Shi'ite split dates from just after the death of Muhammad when there were already divisions of opinion over interpretations of the Qur'an. In 661, the Shi'ite-favored Caliph Ali was assassinated, an act which gave the split significance that endures in the present. More important, today, within each of these groups, there are many different interpretations of shari'a, and devoutly observant Muslims, Shi'ite and Sunni, can be found living a host of different lifestyles (see the box "The Variability of Islam," page 300).

Role of Islam in Society

In Muslim communities across the world, a major debate is unfolding on the role of religion in society, and although the debate has led to violence when Muslim extremists have sought to overthrow governments they feel do not pay proper attention to religion (for example, Iran and Algeria, see pages 329 and 321–322), for the most part the debate is taking place in civil society. Typically, Islam has not recognized a separation of religion and the state as this concept is understood in the West. Rather there have been varying mixtures of government and religion. In several of the countries in this region—Saudi Arabia, Yemen, the United Arab Emirates (UAE), Oman, and Iran, for example—the state is the defender and even the enforcer of the religious principles of Islam. In these **theocratic states,** Islam is the officially accepted religion, the leaders must be Muslim and are divinely guided, and the legal system is based on conservative interpretations of shari'a that claim to hark back to the time of Muhammad (the accuracy of their position is disputed by more liberal Muslims). Other countries—Algeria, Egypt, Morocco, Iraq, Turkey, and Tunisia—have declared themselves to be **secular states.** In these countries, theoretically, there is no state religion and no direct influence of religion on affairs of state. In practice, how-

CULTURAL INSIGHT The Five Pillars of Islamic Practice

1. A testimony of belief in Allah as the only God and in Muhammad as his Messenger (Prophet).

2. Daily prayer at one or more of five times during the day (daybreak, noon, midafternoon, sunset, and evening). Prayer, although an individual activity, is encouraged to be done in a group and in a mosque.

3. Obligatory fasting (no food, drink, or smoking) during the daylight hours of the month-long Ramadan, followed by a light celebratory family meal.

4. Obligatory almsgiving (*zakat*) in the form of a progressive "tax" of at least 2.5 percent that increases as wealth increases. The

alms are given to Muslims in need. *Zakat* is based on the recognition of the injustice of economic inequity; although it is usually an individual act, the practice of government-enforced *zakat* is returning in certain Islamic republics.

5. Pilgrimage (hajj) to the Islamic holy places, especially Makkah (Mecca), during the twelfth month of the Islamic calendar. Rituals shared with the devout of all backgrounds, from around the world, reinforce the concept of *ummah*, the transcultural community of believers.

[*Source:* Carolyn Fluehr-Lobban, I*slamic Society in Practice* (Gainesville: University of Florida Press, 1994).]

ever, religion plays a public role even in the secular states. Most political leaders are at least nominally Muslim, and Islamic ideas play a role in governmental affairs. Israel is not a theocratic state, but Judaism strongly influences policy, and Lebanon is a multi-religious state. Still, here as elsewhere on earth, there is great variation in the stringency with which people practice their religion. Many people who consider themselves Muslim do not strictly observe the Five Pillars or shari‘a, and they prefer that Islam play a limited role in public life. The same can be said for many Christians and Jews in the region, many of whom trace their heritage back to pre-Islam days.

Family and Group Values

In North Africa and Southwest Asia the traditional multigeneration patriarchal family is still very much the norm. Nonetheless, patterns are changing: families are becoming smaller and no longer do several generations of one family always share the same household.

In traditional Islamic culture, the family was both a physical space and a functional grouping. Physically, the family space was usually a walled compound focused inward, where food, shelter, and companionship were provided (see the photo of Kalaa Sghrira, Tunisia, and the photo of a house in Jeddah on pages 301–302). The size and details of these compounds varied according to social class and geographic location. The family group consisted of kin that spanned several generations, including elderly parents, as well as adult siblings and cousins and their children. A system of interlocking duties, obligations, and benefits, often assigned by gender and age, provided a role for each individual and solidarity for the whole family. All accomplishments or misdeeds became part of a family's heritage. The responsibilities of family membership were enforced through informal social pressures that ensured no one became an obvious shirker. Whether one received a meal or something as grand as a university education from pooled resources, the recipient knew that some measure of repayment would come due, eventually. This informal contract between generations ensured the flow of financial remittances to elderly kinfolk; and although these remittances were formally expected only from sons, daughters who were able sent them, too. These concepts of the home and family and attendant obligations

A rare glimpse of the Grand Mosque in Makkah (Mecca), packed with devout worshipers, at the height of the pilgrimage season. [Mohamed Lounes, Gamma-Liaison.]

are still very much part of value systems in the region, but everyday practice is changing as people move into apartments in cities and away from strict village social pressures.

Thinking Geographically about Gender Roles and Gender Spaces

Carefully specified gender roles are common in many cultures and often there is a spatial component to these roles. In this region, the differences between male and female roles are reflected in the organization of space within the home and within the larger society; but practice varies widely from place to place.

In both rural and urban settings, men and boys go forth into **public spaces**—the town square, shops, the market. Here, men not only make a living, they also continuously transact alliances with other men to negotiate arrangements that will advance the interests of their particular families. It is through such networks,

AT THE GLOBAL SCALE The Variability of Islam

When they envision Islam, most Westerners refer only to what they have learned about Islam from Western journalists, who tend to suggest that the religion is uniform and extremist in its views. In fact there is no "Islamic World" and no central Muslim authority; rather, there are many versions of Islam beyond the two main divisions of Shi'ite and Sunni. There are many interpretations of the Qur'an, and several sets of religious law—not surprising, given that there are 1.3 billion Muslims in the world. In fact, Muslims in the region of Southwest Asia and North Africa make up less than a third of the world population of Muslims (see Figure 1.7 on page 16). Here are stories of how just three Muslims live out their faith.

Ebrahim Moosa was inspired to learn more about his Muslim faith when he was insulted. When a fundamentalist Christian friend told him that Islam was false and Muhammad a fraud, Moosa was hurt by this accusation. As a result, Moosa spent years studying in *madrasas* (Muslim schools) in Asia and learning Arabic so he could read and interpret the Qur'an himself. Later he traveled worldwide as a reporter for a Muslim magazine. His studies and experiences persuaded him that there are multiple versions of Islam. Now a professor of religious studies at Duke University in North Carolina, he is an activist for a progressive Islam—the idea that Muslims must find a modern path that eschews racism and sexism, one that works for human rights but does not succumb to Western secularism and materialist values.

In the United States, Saleh Abazeed plays football for the Fordson High Tractors, a high school team in Dearborn, Michigan, trying to reach the state championship with a 10–2 record. All but one of the 53 players on Abazeed's team are Muslims, most them the sons of immigrants from Iraq, Lebanon, Syria, and Egypt. In 2003, when the holy month of Ramadan fell during the football season, Abazeed and others of his teammates chose to

observe the required fast, even on game day: no food or water from sunup to sundown. Abstaining from food and drink is only part of the observance of Ramadan, which for Muslims everywhere entails a broad spiritual commitment to avoid bad deeds and thoughts. After the game, once the sun was down, the whole team broke fast together.

In Istanbul, Turkey, Nasiye Wadud (a pseudonym) hopes to become a physicist, but she is presently blocked from continuing her studies by a government ban on head scarves. The government of Turkey, which is secular, banned the scarves and other displays of religiosity in all public institutions, when an Islamic political party gained 20 percent of the vote in a national election. Nasiye chooses to wear a scarf and a long, light-weight coat whenever in public because these garments signal that she is a devout and modest Muslim. She and her female friends are active in a student-led movement to eliminate the ban on scarves and to allow other practices of Islam, such as daily prayers, within the walls of the university, but they do not regard themselves as militants. Nasiye thus finds herself in the quandary of seeking a professional and public role as an educated woman (even a feminist) while also serving, by her choice of dress, as a symbol of militant Islam—a movement that supports the idea that women should stay out of the public eye and play only familial roles.

[Adapted from material drawn from *The Detroit News* (November 5, 2003), at http://www.detnews.com/2003/highschools/0311/05/e01-316648.htm; "Morning Edition," National Public Radio (October 31, 2003); *The Washington Post National Weekly Edition* (January 19–25, 2004): 10–11; Yonat Shimron, *Raleigh News and Observer* (November 30, 2003); "Frontline," Portraits of ordinary Muslims, at http://www.pbs.org/wgbh/pages/frontline/shows/muslims/portraits/turkey.html and http://www.pbs.org/wgbh/pages/frontline/shows/muslims/themes/.]

which span a range of social strata, that people get the best price for an appliance or a car, find a mate for a son or daughter, obtain a job or admittance to a professional school, find some scarce item, or unsnarl a particularly nasty bureaucratic problem. For any favors they receive, they incur future obligations. If men are successful at making arrangements, they also garner considerable respect and prestige within their own families.

Women used to inhabit primarily secluded **private spaces.** In the traditional family dwellings described above, the courtyard was usually a private, female space within the home (see the photo of the old city of Kalaa Sghrira, Tunisia, on the next page). Only male relatives could enter. For the upper classes in urban places, female space was an upstairs set of rooms with latticework or shutters at the windows. Here it was possible to look out at street life and to enjoy breezes passing through the lattice, without being seen by people on the street (see the photo of a house in the historic district of Jeddah, Saudi Arabia, on page 302). Today the majority live in apartments, yet even here there is a demarcation

of public and private space. Most apartments have one or two formal reception areas for nonfamily visitors, with family rooms deeper into the dwelling reserved for sleeping, socializing, watching TV, cooking, and eating.

Now, both women and men go forth into public spaces, but just how women enter these spaces remains an issue and customs vary not only from country to country, but also from rural to urban setting, and by social class. In some parts of the region, particularly in Saudi Arabia, the rules remain strict. People believe that women should not be in public except when on important business, and it is the custom that women be accompanied by a male relative in public. In some parts of the region (Egypt, for example), some classes may enforce seclusion more strictly than others. Some affluent, urban women may observe seclusion more than do rural women, because although rural women are traditional in their outlook, they have many tasks that they perform outside the home: agricultural work, carrying water, gathering firewood, and marketing.

In Kalaa Sghrira, Tunisia, founded in 670, streets are narrow and traditional courtyard houses focus family life inward. As in many crowded cities throughout the region, houses seldom rise more than one or two stories, and the skyline is broken only by the minaret towers of nearby mosques. [Roger Wood/CORBIS.]

vignette about professional women in Yemen, page 326). Figure 6.12 compares the various levels of restrictions on women across the region. Some restrictions are officially sanctioned, and some are the result of informal social pressures. The primary issue is that when a woman enters public space her honor must be maintained; hence, in much of this region, it is important that her street clothing be modest. However, what she wears in the privacy of homes among family and friends may be quite different. When going to a party, for example, a woman may wear a long coat and shawl to get there and then shed them on arrival, revealing a stylish outfit for her friends to admire (see the box "The Veil" on page 303). Also, to maintain honor, it is common for women in public to travel in groups, or for a single woman to take a younger sibling along on errands.

There is considerable controversy over the origin and validity of female seclusion as a Muslim custom. Scholars of Islam, male and female, say that these ideas predate Islam by hundreds or even thousands of years. They do not derive from the teachings of the Prophet Muhammad, who seems to have advocated equal treatment of males and females and to have helped with domestic chores. His first wife, Khadija, was an independent businesswoman whose counsel he often sought. Although Muhammad did suggest that both sexes dress modestly, the Qur'an contains only the barest hint about veiling and nothing about seclusion. Modern scholars of Islam state that the most restrictive interpretations of the Prophet Muhammad's sayings were influenced by ancient and severely limiting customs of non-Islamic cultures that were conquered by Arab Muslims in the first few centuries after the Prophet's death. Islam, in fact, emerged as a reformist

In the more secular Islamic societies—such as Morocco, Tunis, Libya, Egypt, Turkey, Lebanon, and Iraq—women regularly engage in activities that place them in public spaces. And, increasingly, female doctors, lawyers, teachers, and businesswomen are found in even the most conservative societies (see the

FIGURE 6.12 Variability in the restrictions on women. There is considerable variation country to country in the restrictions placed on women in the region. Women's rights are perhaps most strongly protected in Turkey and Israel where equality for women is constitutionally guaranteed. But religious fundamentalists opposing these guarantees are working to repeal them. [Adapted from Joni Seager, *The Penguin Atlas of Women in the World*, Completely Revised and Updated (New York: Viking Penguin, 2003), pp. 14–15.]

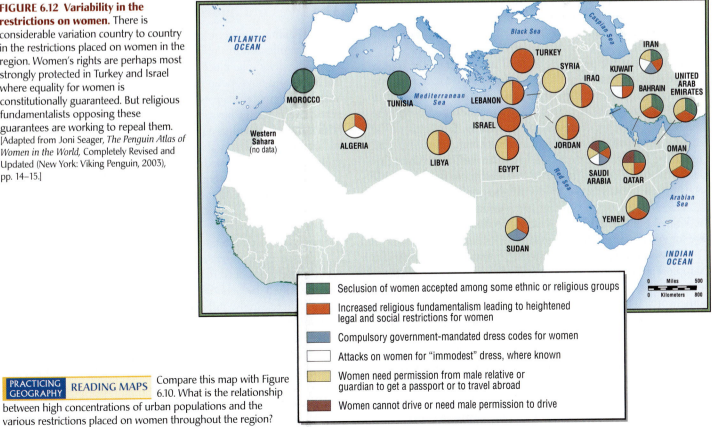

Seclusion of women accepted among some ethnic or religious groups

Increased religious fundamentalism leading to heightened legal and social restrictions for women

Compulsory government-mandated dress codes for women

Attacks on women for "immodest" dress, where known

Women need permission from male relative or guardian to get a passport or to travel abroad

Women cannot drive or need male permission to drive

PRACTICING GEOGRAPHY   READING MAPS   Compare this map with Figure 6.10. What is the relationship between high concentrations of urban populations and the various restrictions placed on women throughout the region?

This house in Jeddah, Saudi Arabia, exhibits numerous versions of louvered and latticed bays that allow women to observe life on the streets from seclusion. Such architectural details are found wherever Islam has been an influence. [Hubertus Kanus/Photo Researchers.]

movement in a troubled and violent time, and rather than enforcing restrictions on people, it was focused on bringing peace, stability and humane treatment to troubled societies. It is also important to note that Muslim scholars have shown that in some parts of the region, interpretations of Islamic law regarding female behavior have become markedly more conservative and restrictive just during the last half of the twentieth century!

People from Western cultures tend to exaggerate the importance of such customs as veiling, simply because to them veiling seems so exotic. Also, it is important to remember that European women had to wait more than a thousand years to acquire property-holding rights that were guaranteed by the Qur'an 1300 years ago (A.D. 700). The Qur'an allows a woman, married or not, to manage her own property and keep her wealth or a monthly salary for herself. By contrast, in Britain, it was not until 1870 that laws stopped the custom of transferring to her husband's control all the wealth a woman brought to a marriage. In some parts of the United States similar state laws were changed only after pressure for equal rights in the 1960s!

Many countries across the region (Tunisia and Morocco are examples) are improving the status of women by enacting legislation allowing women to initiate divorces and prohibiting men from divorcing their wives without cause or a court proceeding. In some places, contrary to the teachings of the Prophet Muhammad, women were forced to marry and to undergo virginity tests, and they were left unprotected in cases of domestic violence. Legislatures across the region are being petitioned by activist women to criminalize "honor killings" in which a woman or girl who is perceived to have dishonored her family in some way, or has been a victim of rape, is actually killed by her own relatives. Such acts, thought to restore family honor, are strictly forbidden by Jewish, Muslim, and Christian doctrine. In 1999, the new king of Morocco, Mohammed VI, began to implement a program that encourages participation by women in civil society and reforms the legal code to grant women legal rights. Tunisia recognizes the equal right of either spouse to seek divorce. It also requires mutual spousal obligation in household management and child care, and it no longer permits preteen girls to be married.

Marriage customs are a source of contention in this region. Although the Qur'an guarantees women equality with respect to divorce, in some countries (all the Arab republics, Morocco, Algeria, Syria, Jordan, Iran, and Iraq, are examples), Muslim men are legally permitted to take more than one wife, a custom called **polygyny.** But polygyny, although allowed under certain conditions by the Qur'an, is not encouraged, nor is it a common modern practice. It is estimated, for example, that less than 4 percent of males in North Africa have more than one wife (see Figure 7.21, page 368). The occurrence of polygyny is low partly because the Qur'an, which limits the number of wives to four, requires that all must be treated with scrupulous equality. Urbanization and modernization are also important factors. When agriculture was the main economic activity, multiple wives with several children each may have been more productive economically, but urban life with its small living spaces and cash requirements favors smaller families. According to the Ibn Khaldun Center for Development Studies in Cairo, democratization in North Africa and Southwest Asia led to the successful bans against polygyny in Tunisia, Lebanon, and Palestine.

The Lives of Children

Two sweeping statements can be made about the lives of children in the Islamic cultures of North Africa and Southwest Asia. First, their daily lives take place overwhelmingly within the family circle. Girls and boys spend their time within the family compound with adult female relatives and with siblings and cousins of both sexes. Even teenage boys in most parts of the region identify more with family than with age peers. Only the poorest of children play in urban streets. In rural areas, before puberty, girls often have considerable **spatial freedom**—the ability to move about in public space—as they go about their chores in the village. The U.S. geographer Cindi Katz found that until puberty, rural Sudanese Muslim girls have considerably more spatial freedom than do girls of similar ages in the urban United States, where they now are rarely allowed to range through their own neighborhoods.

The second observation is that the lives of children in this region, like those of so many around the world, are increasingly circumscribed by school and television. Most children go to school (increasingly, girls go for more than a few years, many boys for a decade or more), and in most urban areas even the poorest families have access to a television. Furthermore, the TV set is often on all day, in part because it provides a window on the world for secluded women. TV may serve either to reinforce traditional culture or as a vehicle for secular values, depending on which channels are watched. Since September 2001, the expo-

CULTURAL INSIGHT The Veil

The veil allows a woman to remain secluded when she enters a public space, and thus it actually increases the space she may occupy with honor preserved. Some garments totally cover the body, including the face; some allow the eyes to be seen; and some are a mere head covering. Among women who have worn the all-encompassing chador (full body veil), there are those who find that it offers an intriguing anonymity and satisfying privacy from the gaze of strangers. Women who wear some version of religious dress such as the *hijab*, a long, loose dress and a scarf that leaves the face exposed, or just the head scarf with blue jeans and a T-shirt, speak of liking the signal it sends that here is a devout Muslim woman. They add that when women don religious garb, they are reminding men to be religiously observant as well. Paradoxically, many women throughout the region wear high-fashion Western clothes under the chador or *hijab*, or at special gatherings that may be women only.

Veils may be bought in shopping malls across the region. But for women who think of the veil as providing a chance for a fashion statement, Wafeya Sadek runs an Islamic designer-veil studio out of her home in Cairo. Women come from across the city to buy a custom-made veil. A best-seller is a leopard-print veil, favored by young women for daywear. Some of Sadek's designs for evening wear seem to call for the type of bold individualism that is frowned upon by conservative Muslims: for example, a close-fitting black bonnet topped with a wildly colorful cockscomb of velvet. Operating in the informal economy, Madame Sadek counts on word-of-mouth advertising.

In the United States, Islamic clothing is a new fashion category. There are now about 8 million Muslims in the country, and Sadiya Shaikh, together with her mom and a friend, started Flippant, a design house in California that appeals to the young Muslim women who wish to be both stylish and religiously observant. In the photo, two young women enjoy a moment while modeling dresses from Flippant.

[*Sources:* Amy Docker Marcus, *Wall Street Journal* (May 1, 1997); I-chun Che, "Clothing line offers modesty and beauty for Muslim women," *The Sun*, Sunnyvale, CA (October 15, 2003), at http://www.svcn.com/archives/sunnyvalesun/20031015/sv-cover.shtml.]

[Jim Gensheimer, *San Jose Mercury News*.]

sure of children (and women) to broadcast news has taken a decided leap. (See the box "The Evolving Role of the Press" on page 305.)

Language and Diversity

Arabic is now the official language in all countries of this region except Iran, Turkey, and Israel (where Farsi, Turkish, and Hebrew, respectively, are spoken), but this uniformity of language masks considerable cultural diversity. There are numerous minorities within the region who have their own non-Arabic languages: Berbers, Tuareg, Sudanese, Nilotics, and Nubians, to name a few in North Africa; Kurds and Turkomans in Southwest Asia. There are also many dialects of Arabic that indicate deep cultural variations across the region. Yet, in this era of modern communications, dialects are disappearing as standardized Arabic becomes the language used by broadcasters. More importantly, Arabic is replacing the languages of many minorities as they become assimilated into mass media culture through radio, television, and movies. Nonetheless, the widespread use of Arabic facilitates region-wide communication and pan-Muslim solidarity. Meanwhile, the use of French and English as second languages, also very common especially in urban areas, and the dominance of English on the Internet, is contributing to the decrease of language diversity and to the intrusion of outside cultural influences. Figure 6.13 shows the distribution of languages in North Africa and Southwest Asia.

Islam in a Globalizing World

Many Muslims see modern culture, much of it originating in Europe and America, as undermining important values, such as the duty of the individual to the family and community. Hence, some Muslims object to the liberalization of women's roles, especially to their being active outside the home. Many Muslims also lament the global spread of Western culture with its open sexuality and consumerism and its tendency toward hedonism. They worry that such ways lead to family instability, alcoholism, drug addiction, and street crime by youth who have been poorly raised by inattentive parents. Consumerism is seen as leading to selfishness and to widening gaps between rich and poor. (Remember, a basic pillar of Islam is the duty to address the injustice of poverty.) Some social commentators argue that the basic concerns of Muslims about loss of a sense of responsibility to family and the obsession with materialism are similar to the concerns of people from other religions around the world (Baptists in the U.S. South, Hindu nationalists in India, for example) who worry about the erosion of traditional values and ethical systems.

More conservative Muslims, known as **Islamists** (also called Islamic fundamentalists), harken back to what they feel are more accurate interpretations of the Qur'an. They favor a simple, prayerful life focused on family and community, traditional gender roles, and respect for the elderly. The different Islamist factions vary greatly in the perspectives and fervor they bring to their causes. The majority simply seek to moderate Western influence; for example, most countries want to connect their people to the computer age, but they are reluctant to accept some of what they see as negative Western influences, such as pornography and con-

sumerism, that come with connecting to the Internet. The proper role of the press is also hotly debated, with some feeling that a free press is essential and others that the press should not go beyond simple statements about events (see the box "The Evolving Role of the Press" on the next page).

Moderate Abdullahi An-na'im, a Sudanese professor of religious studies at Emory University in Georgia, United States, calls for an open debate in Islam that includes a reevaluation of the ancient shari'a interpretations of the Qu'ran. He notes that the shari'a were conditioned by the times in which they emerged—within several hundred years of Muhammad's life. They were progressive for the time—giving rights to women and minorities, which was not yet the norm in Europe or elsewhere. But newly interpreted in light of today's circumstances, the Qur'an could become the basis of a modern society, where, for example, human rights and gender equity are recognized as central to a civil society.

ECONOMIC AND POLITICAL ISSUES

There are major economic and political barriers to peace and prosperity within the region today. The oil wealth so prominent in Western minds is limited to a few elites; most people are low-wage urban workers or relatively poor farmers or herders. The economic base of the region is unstable because oil and agricultural commodities, the main resources in the region, are subject to wide price fluctuations on world markets. New bases of economic development are badly needed, but a more diverse range of industries is just beginning to emerge and is in need of investment. Meanwhile, hard times loom for many poorer non-oil-producing countries as large national debts are forcing governments to restructure their economies in the ways commonly associated

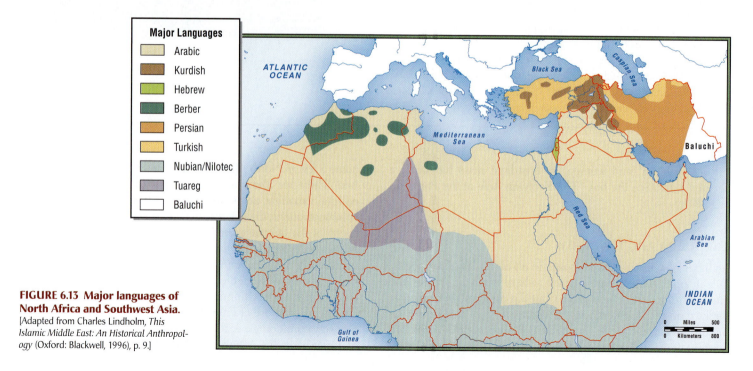

FIGURE 6.13 Major languages of North Africa and Southwest Asia.
[Adapted from Charles Lindholm, *This Islamic Middle East: An Historical Anthropology* (Oxford: Blackwell, 1996), p. 9.]

Major Languages

- Arabic
- Kurdish
- Hebrew
- Berber
- Persian
- Turkish
- Nubian/Nilotec
- Tuareg
- Baluchi

AT THE REGIONAL SCALE   The Evolving Role of the Press

The press, both print and broadcast, did not play a major role in this region until very recently. The few respected newspapers in major cities rarely took strong oppositional stands to government policies. In many cases the press was actually controlled by sitting governments and served primarily as government information organs, informing the public about events and policies but exercising little critical analysis. Such is still the case in Saudi Arabia, discussed first; some exceptions to this general pattern follow.

In Saudi Arabia, there are more than a dozen newspapers on the newsstands every morning, and all are owned or controlled by the Saud royal family. The Saudi printed press is said to be the most influential in the Arab world. Yet all Saudi newspapers are constrained by the proviso that nothing critical may be said about Islam, the royal family, or the Saudi government, and there is no investigative reporting. When accidents happen or some malfeasance of a public official is revealed, the story is blandly reported with little or no effort to explore the causes of events or their future effects, or to hold responsible officials accountable. Occasionally, critical public discourse will begin about an issue and then be preemptively shut down. In March 2002, for example, a fire in a girls' school killed 15 and injured 50. Soon it was revealed that rescuers and firefighters were held back by men from the Commission for the Promotion of Virtue—the country's government-paid religious police—because the girls had fled to the balconies without their *abiyas* (veils and robes). For a few days the Saudi press covered the public outcry and called for investigations of the religious police, but then the Minister of the Interior informed the editors that the stories were to stop, and they did immediately. Many Saudis watch the news broadcast on the Al Jazeera TV network in neighboring Qatar (see below) via their illegal satellite TV hookups. The Saudi paper *Arab News* may be read online in English at http://www.arabnews.com/.

[*Source*: Lawrence Wright, "The Kingdom of Silence," *The New Yorker* (January 5, 2004): 48–73.]

In Egypt, Hisham Kassem has published the *Cairo Times*, an independent English-language weekly, since 1997. It is non-partisan and meant to inform readers about current affairs in Egypt, but because the *Cairo Times* often carries articles critical of the Egyptian government, Kassem lost his license to publish in Egypt and has had to register the *Cairo Times* in the state of Delaware in the United States. Each week, he prepares the current issue in Delaware, then flies with the files to Cyprus, where the weekly is printed. He then carries the weekly into Egypt as luggage, where it must clear the censors, who can refuse to let him sell a particular issue. Many times he has been approached by the Egyptian government to become a pro-government "snitch," as he puts it, but he persists with his critical stance saying a free press is essential to reforming what he calls the failed regimes of the Middle East. The *Cairo Times* survives in part because it is a major advertising medium for Egyptian and Middle Eastern markets. It may be read online at http://www.cairotimes.com/.

[*Source*: "Freshair," interview by Terry Gross with Hisham Kassem, National Public Radio (February 4, 2004).]

An important change in the role of the Arab press in this region occurred in the aftermath of 9/11 when Al Jazeera, a satellite TV network, founded in 1996 and funded by the Emir of Qatar and other Arab moderates, suddenly became the main source of news in the Arab-speaking world. Al Jazeera's programming represents a major change in that it provides wide-ranging Arab-language documentaries and uncensored debates on primarily political issues to a loyal and well-educated audience. Because Al Jazeera questioned the U.S. stance on terrorism, it was quickly denounced by the U.S. government and by some commentators as a mouthpiece for Islamic fundamentalism. However, Al Jazeera regularly stands up to puritanical religious leaders, and it also reports on corruption by high officials in Egypt and Saudi Arabia, Turkey, and Palestine. Many international journalists now see the network as an increasingly important world-class broadcaster. Al Jazeera may be read in English online at http://english.aljazeera.net/HomePage.

[*Source*: Michael Moran, senior producer for special projects, MSNBC, at http://www.msnbc.com/news/643471.asp?cp1=1.]

---

with structural adjustment policies (SAPs): the streamlining of production, and the cutting of jobs and social services.

The Oil Economy

This region contains some of the world's largest known petroleum (oil and gas) reserves. (Russia and Central Asia may prove to have yet larger reserves.) These deposits are located mainly around the Persian Gulf (Figure 6.14); thus, the petroleum wealth so often associated with the region as a whole is concentrated mostly in countries that border the Gulf: Saudi Arabia, Kuwait, Iran, Iraq, Oman, Qatar, and the United Arab Emirates. Oil and gas are also found in the North African countries of Algeria, Tunisia, Libya, and Sudan. Because the region possesses few other natural resources in amounts that are useful for national development, oil gets a great deal of attention.

Early in the twentieth century, European and North American companies were the first to exploit the region's oil reserves. These companies paid governments a small royalty for the right to explore and drill for oil. The oil was processed at on-site oil refineries owned by the foreign companies and shipped primarily to Europe, the United States, and eventually other places, like Japan. The governments of the region did not assume control of their oil reserves until the 1970s, when they declared all oil resources and industries to be the property of the state. Even before this, in the 1960s, the oil-producing countries organized a

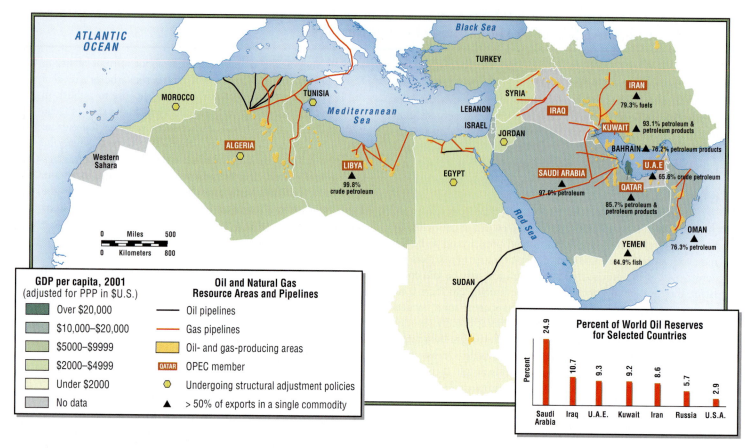

FIGURE 6.14 Economic issues: North Africa and Southwest Asia.
[Adapted from Rafic Boustani and Philippe Farques, *The Atlas of the Arab World–Geopolitics and Society* (New York: Facts on File, 1990), pp. 85, 88, 89; Richard Overy, ed., *The Times History of the World* (London: Times Books, 1999), p. 304; and *Hammond Atlas of the Middle East*, revised (Union, N.J.: Hammond, 2001), pp. 8-9.]

cartel, a group of producers that is strong enough to control production and set prices for its products. The oil cartel is called **OPEC**—the **Organization of Petroleum Exporting Countries.** OPEC now includes all the oil states marked with a red box on Figure 6.14, plus Venezuela, Indonesia, and Nigeria. (Gabon and Ecuador have left the organization). OPEC members cooperate to periodically restrict oil production, thereby significantly raising the price of oil on world markets. Note that world petroleum prices result from a complicated set of factors. OPEC countries often fail to reach agreement on the management of prices (see discussion of the Gulf War, 1990–1991 on page 330 and the Iraq War, begun in 2003, on page 309 and in Chapter 2 on page 67), and non-OPEC producers can influence petroleum supplies and, hence, prices. Consumers, by reducing demand, can also affect prices. Furthermore, geopolitical events can have a major effect on oil prices, as demonstrated by the aftermath of September 11, 2001, when oil prices rose and fell a number of times.

After the major Arab and non-Arab oil-producing countries raised oil prices dramatically in 1973, oil income in Saudi Arabia alone shot up from U.S. $2.7 billion in 1971 to U.S. $110 billion in 1981. Yet because the OPEC countries did not invest much of their oil wealth at home in basic human resources (such as providing broad, quality education and health care, for example), they have been unable to improve their economic bases. Like their poorer neighbors, they have remained dependent on the industrialized world for their technology, manufactured goods, and skilled labor. Nevertheless, oil has brought significant benefits to the region as a whole. Those countries that had significant amounts of oil (Saudi Arabia, the UAE, and Kuwait) sharply increased investment in some (but not all) parts of their long-neglected national infrastructures. They invested heavily in roads, airports, new cities, irrigated agriculture, and petrochemical industries, but much less in education, social services, housing, and health care. Even people from non-oil-producing countries were able to share in the wealth: many of the laborers on these projects came from Egypt, Jordan, Syria, and Turkey, none of which are major oil producers.

Over the next several decades demand for energy will surely rise as technology spreads; if oil supplies dwindle worldwide, oil prices will rise. At some point in the future, the depletion of oil resources and the development of new sources of energy will force OPEC countries to find other ways of earning an income or risk economic ruin.

The Traditional Economy: Agriculture and Herding

Despite the abundance of oil and gas along the Persian Gulf and elsewhere, much of the region remains poor and highly dependent on agriculture and herding. It often appears from production

An irrigated wheat field in the Saudi Arabian desert. [Ray Ellis/Photo Researchers.]

(GDP) and export figures that economies in this region are dominated by industries and services, but in several of the largest countries in this region, substantial portions of the population are still employed in agriculture and depend upon their own efforts for most of their daily nutrition. Nonetheless, the actual market value of their production is very low. For example, in Morocco, industry accounts for 33 percent of GDP and agriculture for only 15 percent, yet 50 percent of the labor force works in agriculture producing this 15 percent of the GDP. In Egypt, agriculture accounts for only 17 percent of GDP, yet 49 percent of the people work in agriculture. In Turkey, 18 percent of the GDP is in agriculture, while agriculture employs 45 percent of the people. Herding remains an important aspect of agricultural life and land use in Morocco (47 percent of the land is used for herding), Algeria, Egypt, Turkey, and Iran, but in none of these countries is herding a significant income earner when compared to industry and services. When measured in dollar value, agriculture is not productive, but when measured in value to diets and family budgets, farming and herding are essential to national economies.

Agriculture. Crop agriculture in the region has, until recently, been confined to areas with a Mediterranean climate (dry summers, moister winters)—places like moist river valleys, coastal zones, and upland areas where mountain ridges capture orographic rainfall. Grains, cotton, and sugarcane are raised in the Nile Valley of Sudan and Egypt, grains and fruit in the Tigris and Euphrates river valleys. Turkish farmers produce cotton, tobacco, sugar beets, and livestock; and Iraqi and Iranian farmers grow grains, nuts, tea, tobacco, and livestock. Citrus fruits, olives, dates, and wine grapes are grown on the northern coasts of Africa for export to Europe. (Alcohol consumption is not permitted by the Qur'an, but it does occur. Most wine is exported.) The farmers of Yemen and Oman grow coffee and grains. Israelis and Palestinians grow vegetables and fruits.

Ambitious irrigation schemes in Libya, Egypt, Saudi Arabia, Turkey, and Iraq are now expanding commercial, primarily export-oriented, agriculture to neighboring areas that had been too dry to support large-scale cropping. These attempts at expansion have not always been well conceived. Often state-sponsored

irrigation projects have damaged soil fertility through **salinization,** a process that occurs when large amounts of water evaporate, leaving behind dissolved salts and other minerals that inhibit plant growth over time (see the discussion of the disappearing Aral Sea in Chapter 5, pages 266–267). It is particularly likely to occur in arid lands because there is little rain to wash away the salt. Israel has developed more efficient techniques of drip irrigation that curtail salinization, but poorer states have been unable to afford the technology, and other wealthy states have been wary of becoming dependent on a technology developed by Israel, a country they deeply distrust.

The region's numerous political tensions have convinced many governments that they should try to be self-sufficient in food production no matter the expense. To that end, some oil-rich governments have wasted huge amounts of money on shortsighted and poorly engineered development schemes. Saudi Arabia and Libya are pumping massive volumes of precious groundwater to the surface at highly unsustainable rates to grow crops for home use and export. Libya is mining an ancient aquifer for irrigation and other purposes at eight times the rate of the natural replenishment. Saudi Arabia has even used the expensive process of desalinating seawater to provide irrigation for wheat fields. Wheat grown in this way costs 10 times the price of wheat on the world market.

Herding. The tending of grazing animals—particularly camels for transport and sheep for wool—was the economic mainstay of the region for thousands of years, but its economic importance has been in decline since the nineteenth century. Traditionally, camel breeders would take their herds from the sandy deserts of the Arabian Peninsula to cooler Syria or Iraq in the summer. Those in the Sahara would take their animals to summer in the southern fringes of the Atlas Mountains. Since the construction of railroads and highways, camels are no longer needed, but demand for sheep's wool continues. Sheep (and goats), however, cannot travel long distances, so the range of nomadic migrations of sheepherders was always smaller than that of camel herders. The settled herding of sheep is relatively common even in urban fringe areas, and many families engaged in nonagricultural employment maintain a small herd of sheep that the boys of the family tend (see the vignette at the beginning of this chapter).

Nomadic herders, such as the Kurds who live at the juncture of Turkey, Iraq, and Iran, have lost financial and spatial independence as the region modernized and national borders became official dividing lines. In establishing large irrigated agriculture projects, many governments seek control of water sources previously available to the herders, or they may require the nomads to settle permanently in order to provide a labor pool for these projects. Herders are also forced to settle down because their tendency to cross national borders in search of grazing lands and water is now perceived as a national security threat. Finally, nomads are hard to tax or to control in other ways; for example, it is difficult to enforce compulsory school attendance on their children or to be sure adults and children get proper health care. The requirement to settle permanently, however, drastically alters the economy of herding communities and their social structure and can result in impoverishment and social strife.

Attempts at Economic Diversification

Greater economic diversification could bring broader prosperity to the region and limit the risk that a drop in the price of one or two commodities on the world market would bring economic disaster. **Economic diversification** is the expansion of an economy to include a wider array of economic activities in an effort to increase GDP and economic stability; but few countries in the region have achieved that goal. Perhaps the most successful is Israel. Compared to other countries in the region, Israel has a large knowledge-based service economy and a particularly solid manufacturing base. Israel's goods, services, and the products of its modern agricultural sector are exported worldwide. Turkey is the next most diversified, with Egypt, Morocco, and Tunisia also starting to move into new economic activities. All however, including Israel, still depend on substantial government investment, and Israel and Egypt rely on major aid from the United States.

Economic diversification has been hindered by limited resources—such as water, arable land, minerals, and forests. Diversification is also limited by the lack of private investment funds. Beginning in the 1950s, governments, such as those of Turkey, Egypt, Iraq, Israel, Syria, Jordan, Tunisia, and Libya, established state-owned enterprises for the production for local consumption (not export) of such items as machinery and metal items, textiles, toilet paper, cement, processed food, paper and printing, and then protected these enterprises from foreign competition with tariffs and other trade barriers. Local elites favored these state-owned and protected industries because, as managers, they could gain personal profit from them. The participation of government made it difficult for local private investors to compete, and European countries, such as France, Spain, Germany, and Italy, eager to sell their own industrial products to the region, often discouraged local private industries. Finally, potential investors from the region, such as Saudi oil barons, failed, until recently, to finance industrial and other development within their home region, finding more lucrative sites in North America, Australia, or Southeast Asia for their investment funds. Whatever the cause, the industrial sector grew only slowly—in Egypt from 24 percent of the economy in 1968 to 34 percent in 2000, in Jordan from 21 percent in 1968 to 25 percent in 2000. In Turkey, industry actually lost its share of the economy over the same period, but there, remittances from Turkish guest workers in Europe supplemented the economy. Industries overprotected from foreign and domestic competition produced shoddy and expensive products that did not meet international standards and could not compete in the global economy. Furthermore, the extension of governmental control into so many parts of the economy nurtured corruption.

Not only did the desired economic diversification not occur, expensive blunders in state-led development saddled the poorer governments of the region with crippling debt burdens. The need to make debt payments has forced governments to cut back their role in the economy. This trend has been under way since the 1970s, when international lending institutions, such as the International Monetary Fund and the World Bank, imposed structural adjustment programs (SAPs). As discussed elsewhere in this book (see Chapter 3, pages 136–138), SAPs require that governments shift away from state-led development and toward a free-market economy by privatizing state-owned industries; streamlining those that remain government-owned; removing import barriers; and reducing subsidies for food, housing, and agriculture. In return, governments get guarantees of additional loans and are allowed to reschedule the payments on existing debts.

Side Effects of Development Efforts

Although SAPs may force economies to be more streamlined and more productive per capita, they often have had a negative impact on the poor majority, and amount to a disinvestment in human capital, when in fact a healthier, more educated populous is what is most needed. Overall, SAPs in this region have produced significant losses. In Egypt in the 1980s, for example, reductions in food and housing subsidies doubled poverty in rural areas and increased it by half in urban areas. To provide for their impoverished families, hundreds of thousands of men migrated to Jordan and Saudi Arabia to take up short-term construction labor contracts. Millions of other rural poor were pushed into Egypt's cities and into the informal economy. The streets of Cairo are now so clogged with vendors that traffic cannot pass. More than 6 million poor live in crowded conditions, without sanitation and other services. Throughout the region, governments in similar situations are depending more and more on international nongovernmental organizations (NGOs) to provide such services as education and health care. (One example is the NGO in Turkey mentioned in the vignette at the beginning of this chapter.) The Aga Khan, a Muslim philanthropist, aims at strengthening civil society through investments in education and health care and by teaching entrepreneurship and self-sufficiency. The resources of NGOs, however, are too small to serve more than a minority of the needy and are unlikely to provide a long-term solution. Although many ordinary citizens are entrepreneurs, they often operate in the informal economy. The amounts they can invest are small, making it difficult for them to eventually register, expand to create more jobs, and pay taxes to support needed services.

Under these less than ideal circumstances, it has been difficult to attract private foreign investment that would contribute to economic diversification and bring technology transfers and efficient management practices. Nonetheless, beginning in the 1990s, a number of multinational corporations opened operations in Egypt (see page 324); and the countries of Lebanon, Egypt, Turkey, Jordan, Saudi Arabia, Israel, and Kuwait drew multinational investment in technology (see the map of Cisco investments in Figure 1.6, page 12), food and beverage delivery, and auto manufacturing. Foreign direct investment in this region fell after the 9/11 attacks and the subsequent wars in Afghanistan and Iraq, except in services related to military activities. Tourism was the sector most negatively affected, but all investment interest cooled and by 2004 had barely begun to rebound. Nonetheless, the post–9/11 circumstances and the war in Iraq did encourage more interest in local and foreign partnerships for industrial investment. By mid-2004, 13 of the 21 countries in the region were discussing new state-of-the-art petrochemical industries.

The Economic and Political Legacy of Outside Influence

Political and economic cooperation in the region has been thwarted by a complex tangle of hostilities between neighboring countries. Many of these hostilities are the legacy of a long history of outside interference in regional politics. In the early twentieth century, Britain and France carved up non-Turkish parts of the Ottoman Empire into a number of small countries dependent on Europe for defense and trade. Then, during the post–World War II cold war era, the region became the site of further geopolitical maneuvering by the West and the USSR. The Iran–Iraq war of 1980–1988, the Gulf War (1990–1991) (described and discussed in the subregion section, page 330), and the Iraq war of 2003 were all, at least in part, instigated by pressures from outside the region. For example, the Iraq war in the spring of 2003 was waged not by Iraq's neighbors but by the United States, with Great Britain and a few other countries as allies.

The Iraq War (2003–present). In searching for an explanation for the September 11 terrorist attacks of 2001, the U.S. administration of George W. Bush focused on Iraq. President Saddam Hussein, by then regarded as diabolical and ruthless, was thought to harbor a lingering hostility toward the United States for the Gulf War and its aftermath. Actual links between the attacks and Iraq were never made, however; and in fact, given that Hussein was a secular leader with his own antipathy for Islamic fundamentalism, it is unlikely that he and Osama bin Laden could have found common ground. The U.S. president was convinced that Iraq had an arsenal of nuclear and other weapons and posed an immediate threat to the United States. Although connections to the 9/11 attacks were never substantiated, nor the arsenal of weapons found, war was declared on March 20, 2003, with the goals of removing Saddam Hussein from power and turning Iraq into a democracy. After only a short time, on May 1, 2003, President Bush declared the war won, but soon terrorist bombs and insurgent attacks by Hussein sympathizers and/or Iraqis who resented the U.S. occupation were taking the lives of U.S. and allied soldiers at the rate of more than one per day. Iraqi deaths were much higher; the Web site www.iraqbodycount.net estimates that from the beginning of the war to October 27, 2004, there were between 14,000 and 16,000 occupation-related Iraqi deaths. Hussein was captured and incarcerated in December 2003.

The United States had planned to slowly turn over the governing of Iraq to a national assembly selected by caucuses of the mostly former Iraqi exiles, who made up the U.S.-influenced interim Governing Council. The United States preferred delegates be selected by caucuses, because this system would be less likely to result in one group (for example, the Shi'ites) dominating the Iraqi government to the detriment of all others (especially the Kurds, who had largely aided the U.S. military effort). However, powerful political figures in Iraq, most notably a Shi'ite cleric, Ali Sistani, began to call for speedy direct elections of delegates to the national assembly, garnering popular support from not just Shi'ites but some Sunnis as well. Thus the United States and its allies were faced with the conundrum of advocating democracy but not much liking the possibility that Iraqis would vote to install a conservative, religiously based, if not Islamist, republic. In the midst of violent reprisals by insurgents, the United States turned over political control of Iraq to Iraqis on June 30, 2004, in hopes that the violence would stop. Instead, it escalated, and at this writing continues unabated.

"As it turned out, winning the war in Iraq was the easy part. Liberating the country from Saddam's brutal legacy of ethnic hatred is something else."
REPORTER SAM KILEY

It is February 2003. Kadijah, a blind, middle-aged Kurdish woman, is talking to reporter Sam Kiley. She explains she is blind as a result of the chemical warfare waged by Saddam Hussein against the Kurds in 1988. Hussein's goal was to eliminate the mostly agricultural Kurds who lived in the northern part of the country atop large oil reserves and use their houses and land for other purposes. The day she was blinded by the chemical attack on her village (chemicals were dropped from low-flying jets), four members of Kadija's family were killed. In all, between 1988 and 2003, 182,000 Kurds went missing. Saddam's troops actually videotaped the mass execution of many; others, like the children and brothers of Nabat, another middle-aged Kurdish woman, whose village was decimated, were killed in the presence of family members, or have simply never been seen again.

Once Kurdish homes and lands were cleared, Hussein had them reoccupied by various minorities who had been displaced elsewhere in Iraq. When, during April 2003, American troops (just 50 Green Berets) at last arrived in Kirkuk, one of four large Kurdish cities, the Kurdish population was ecstatic. But soon their energies were turned to venting their rage on the Arab, Bedouin, and Turkoman occupants of Kurdish homes. For several anxious days, the 50 Green Berets were left with the task of controlling a city of 700,000 feuding people.

When asked if she could manage to live with the Arab and Bedouin settlers, now that Saddam Hussein is gone, Nabat said "With Arab people? No, never. They might not all be responsible, but my heart would never allow it. It is better to live with our own people." An understandable attitude, but one that is making the building of a peaceful, self-governing Iraq especially difficult. This is why reporter Kiley says the hardest job will be resolving the hatred created by Hussein's regime and keeping the peace.

To learn more about the history of the Kurdish people in Iraq, see the FRONTLINE/World video "The Road to Kirkuk."

Understanding the Continuing Israeli-Palestinian Conflict

Warfare between the Israelis and Palestinians began in 1948 on the very day that the last British soldier left. In the ensuing conflict between the Israelis and forces from neighboring Arab countries, Israel prevailed and Palestinian land shrank yet further (Figure 6.15). By 1949, Israel had expanded and the remnants of Palestinian lands were incorporated into Jordan and Egypt (see Figure 6.15C). In the repeated conflicts that followed—such as the Six Day War (1967) and the Yom Kippur War (1973)—Israel not only defeated alliances of its much larger neighbors, including Egypt, Syria, and Jordan, but also expanded Israeli military control into the territories of these neighbors. The Israeli-occupied lands comprised the West Bank (a part of the former Palestine that had become part of Jordan), the Gaza Strip and the Sinai Peninsula (which had become part of Egypt), and the Golan Heights (which had become part of Syria) (see Figure 6.15D).

As a result of continuing hostilities, hundreds of thousands of Palestinians fled the war zones or were removed to refugee camps in nearby countries. Some Palestinians stayed inside Israel and became Israeli citizens, but they have not been treated as equal to Jewish Israelis by the state (see Table 6.1 on page 312). The loss of land and political oppression led to uprisings among the Palestinians (called **intifada**) and to mounting terrorist incidents which were then responded to by the Israeli military.

When Israel occupied Palestinian lands in 1967, the United Nations Security Council passed resolution 242. It required Israel to return these lands in exchange for peaceful relations between it and neighboring Arab states. This resolution, later dubbed the land-for-peace formula, formed the basis of peace talks between Egypt and Israel at Camp David (1979), which returned the Sinai Peninsula to Egypt. Land-for-peace was also the basis for the Madrid peace talks (1991) and the Oslo Accords (1993) between the Palestinians and Israel.

At the 1993 Oslo Accords, the Palestinian Liberation Organization (PLO), an important wing of the Palestinian Authority, acknowledged the right of Israel to exist in return for gaining control over the Gaza Strip and some portions of the West Bank. An important part of the 1993 peace accord was Israel's commitment to stop settling Jews in the occupied territories including the Gaza Strip and parts of the West Bank of the Jordan River. After seizing these areas in 1967, Israel established hundreds of Jewish settlements, mainly on the West Bank, as part of its effort to absorb new Jewish immigrants and also to create a de facto situation that would secure Israel's continued control of the territory, and possibly to gain access to the water in aquifers under the West Bank (see discussion in Environmental Issues section, pages 316–317). Despite agreements to desist, Israel continued building housing units on the West Bank into 2003 and began constructing a wall around the Palestinian enclave. While Israel was consolidating its hold on disputed territories through settlement, the Palestinian people, many of whom had by then lived in refugee camps for 40 years or more, mounted a prolonged uprising against Israel, known as the first intifada (1987–1993) and the second intifada (2000–present). Both periods were characterized by escalating violence, with Israeli military using force to quell demonstrations and to punish the families of Palestinian activists. The Palestinian sympathizers first used primarily stones and sticks against the Israelis, but as frustrations grew and the Israelis extended the controversial settlements, the Palestinians then began using suicide bombs against civilians in order to gain world attention.

A joint team of peace activists, Israeli Simona Sharoni and Palestinian Mohammed Abu-Nimer, writing in the recently published book, *Understanding the Contemporary Middle East* (2004), states that while most Israelis and Palestinians have concluded that violence is not the answer and that diplomacy would be better, the media continue to misrepresent the conflict. Palestinians are presented as unaccountably prone to violence and their acts are routinely referred to as "terrorist attacks." Meanwhile, the illegal extension of settlements by Israel into Palestinian lands, the building of the wall around the Palestinian territory, and the use of deadly force on the part of the state of Israel are depicted as normal actions taken in the name of "national security."

In the summer of 2000, President Bill Clinton brought the Israeli prime minister, Ehud Barak, and the leader of the Palestinian Authority, Yasser Arafat, to Camp David for intensive, marathon talks to reach a final peace agreement between the antagonists. His efforts failed primarily because the parties could not agree on how to share Jerusalem and on how to resolve the Palestinian refugee question. By the summer of 2001, the second intifada, begun in 2000 to protest new settlements on the West Bank, had degenerated into an outright war between Israel and the Palestinians, making peace seem a distant dream. The "Road Map to Peace in the Middle East" proposed by President George W. Bush in 2002 achieved little, partly because attentions were diverted to the war in Iraq, but also because, in the region, the peace attempt was viewed as merely an effort to improve the circumstances for continued U.S. economic interests in the eastern Mediterranean.

Little recognized in the press or among world leaders is the fact that ordinary citizens, both Israeli and Palestinian, have both separately and collaboratively designed ways to end the conflict, exploring scenarios that would acknowledge the national aspirations and the right to land of both parties. Examples of these bottom-up peace initiatives include joint Israeli–Palestinian peace demonstrations and joint women's groups who have tried to end the Israeli occupation of the West Bank and Gaza. Palestinian–Israeli Physicians for Human Rights have joined to address the medical problems of the overwhelmingly poor Palestinians. Groups from both sides hold youth camps, so that Israeli and Palestinian children can, through personal friendship, break the cycle of hatred. Sharoni and Abu-Nimer write that whereas "official representatives of the two [sides] viewed peace mostly as the absence of war and direct violence [or what might be called] *negative peace*, the grassroots activists within both communities envisioned peace as a transformative process grounded in the presence of justice (positive peace)." The advantage of positive peace strategies is that they proactively focus on justice and equality rather than on mere stability and the absence of violence.

The regional impact of the more than half-century-long Israeli–Palestinian conflict has spawned several major wars and innumerable skirmishes, and is a persistent obstacle to political and economic cooperation in the region. Israel has the most

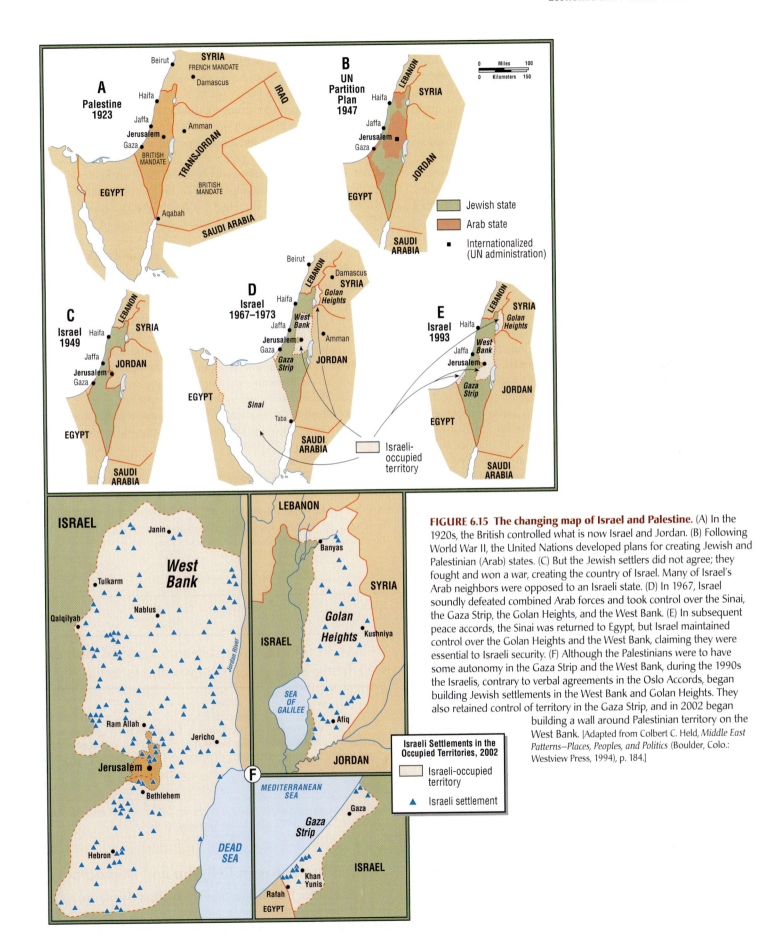

FIGURE 6.15 The changing map of Israel and Palestine. (A) In the 1920s, the British controlled what is now Israel and Jordan. (B) Following World War II, the United Nations developed plans for creating Jewish and Palestinian (Arab) states. (C) But the Jewish settlers did not agree; they fought and won a war, creating the country of Israel. Many of Israel's Arab neighbors were opposed to an Israeli state. (D) In 1967, Israel soundly defeated combined Arab forces and took control over the Sinai, the Gaza Strip, the Golan Heights, and the West Bank. (E) In subsequent peace accords, the Sinai was returned to Egypt, but Israel maintained control over the Golan Heights and the West Bank, claiming they were essential to Israeli security. (F) Although the Palestinians were to have some autonomy in the Gaza Strip and the West Bank, during the 1990s the Israelis, contrary to verbal agreements in the Oslo Accords, began building Jewish settlements in the West Bank and Golan Heights. They also retained control of territory in the Gaza Strip, and in 2002 began building a wall around Palestinian territory on the West Bank. [Adapted from Colbert C. Held, *Middle East Patterns—Places, Peoples, and Politics* (Boulder, Colo.: Westview Press, 1994), p. 184.]

AT THE GLOBAL SCALE   Terrorism as an Economic and Political Strategy

Terrorism is the use, or the threat, of violence, intended to create a climate of fear in a given population. To this we might add that terrorism can be the ultimate effort to bring attention to a cause by those who feel powerless and invisible. State reactions against terrorism can take on some of the qualities of terrorism when police inflict violence on suspected perpetrators before due legal process has been served. As noted below, there are also cases of states that surreptitiously sponsor terror by quietly encouraging and even funding radical elements, but the charge is difficult to prove because citizens of these countries may act without actual government sanction.

In modern times in this region, the terrorist strategy was first successfully used by militant Jewish Zionists (called the Irgun) in 1946, when they bombed the King David Hotel in Jerusalem in an effort to get the British to fulfill their agreement to support the formation of the state of Israel on Palestinian lands; 91 people died. More recently, the word *terrorist* has been attached to Palestinians seeking redress of grievances against the state of Israel for taking their lands in 1948 and thereafter in a series of short wars (started by Arabs, but not without provocation). Groups such as the Islamic resistance movement Hamas, acting on behalf of Palestinians, have launched suicide attacks on Israeli civilians since the late 1980s; they say that they resort to violence only in response to state-sponsored terror: the repeated and overtly brutal Israeli military efforts to quell demonstrations and protect Jewish settlements constructed illegally on the contested West Bank and Gaza Strip territories (see discussion of the Israeli–Palestinian dispute). Hamas, which does not recognize the right of Israel to exist, is thought to get some funding from Iran and Saudi Arabia and some of the Arab Emirates, but primarily it receives donations from Palestinian expatriates and sympathizers around the world.

Terrorism has also been used in North Africa—for example, by Islamists in Algeria protesting the failure of the military government to recognize the results of an election in 1991, and by the Algerian government against those suspected of being Islamists. In Egypt, those protesting the policies and Western alignment of the Sadat and Mubarek governments bombed innocent bystanders, and then were themselves squelched by brutal police tactics. In Saudi Arabia, terrorists launched car bomb attacks in May and November 2003 on upscale residential compounds in Riyadh, where foreign advisers were living.

Terrorism, and its counterpart of violence against civilians by government police, is a major destabilizer of civil society. It gains attention because of its potential to endanger or disrupt the lives of people, even those far from the scene of violence. The September 11, 2001, terrorist attacks in the United States affected economies worldwide by inhibiting travel and commerce and by diverting public funds to security efforts. Aside from its immediate effects of death and destruction and economic recession, terrorism also increases hatred between neighboring groups, causing them to demonize one another—and so the losses proliferate.

Scholars of terrorism point out that terrorist activity is much less likely when people see themselves as having access to timely justice through legal means and through the process of open political participation and free elections.

[*Sources:* [Adapted from "Western team said to be in Libya on anti-weapons mission," *New York Times* (January 1, 2004); "Talk of the nation," National Public Radio (January 21, 2004); BBC news continuous special coverage, *In Depth: Israel and the Palestinians, 2000–2004* at http://news.bbc.co.uk/1/hi/world/middle_east/978626.stm.]

---

modern and diversified economy in the entire region. Since the 1950s (when it took in hundreds of thousands of World War II refugees from Europe), Israel's development has been facilitated by the immigration of relatively well-educated middle-class Jewish settlers from the United States, Russia, and South America; by large aid contributions from the United States and private interests; and by the country's own excellent technical and educational infrastructure. The Palestinian people, on the other hand, are severely impoverished and undereducated (Table 6.1). Were the hostilities to end tomorrow, the majority of Palestinians would remain unequipped to participate in a modern economy. Many Israelis would like to see their country become a major source of investment and technology for the region's poor economies, including Palestine. Although this possibility seemed to be gaining momentum among some Arab elites, in the 1990s it was derailed by mounting violent (terrorist) resistance to Israel's assertive military occupation of southern Lebanon and of territories belonging to the Palestinians, such as the West Bank and Gaza Strip.

TABLE 6.1   Circumstances and state of human well-being among Palestinians and Israelis, 2003

Country	Population, in millions	Infant mortality per 1000 live births	Unemployed (percent)	Percent of population in poverty
Palestine	3.6	26	50	75
Israel	6.7	5	11	18

Sources: Population Reference Bureau Data Sheet, 2003; *United Nations Human Development Report 2003; Arab Human Development Report 2003.*

The Tension between Religion and Democracy

Many people fear that **Islamic fundamentalism (Islamism)** is the greatest threat to the region's political stability and economic development. Islamist movements are grassroots religious revivals that also seek political power. The leaders of these movements seek to take control of national governments currently dominated by secular parties. Islamism has been treated as a major threat since an Islamist revolution led by Ayatollah Ruhollah Khomeini overthrew the secular, though markedly authoritarian and corrupt, government of Iran in 1979, as described more fully on page 329. Although Islamist movements have not displaced any secular governments in the years since 1979, they may be gaining strength in some countries. Several secular governments in the region appear willing to sacrifice even small moves toward democracy in order to suppress Islamist movements. Yet the attempts to suppress these assertive, yet largely peaceful, popular movements may only increase the likelihood of violent revolutionary reaction, as has occurred for some years in Algeria. The United States supports a number of governments (Egypt, Saudi Arabia, Algeria, Morocco, and Turkey) that from time to time have actively suppressed Islamist movements, a move that many in the region see as unwarranted antidemocratic meddling (see discussion of aftermath of a suppressed Islamist election in Algeria in 1991, pages 320–321).

Understanding the popular base of Islamism can yield insights into its strength. Many recruits are young men from lower-class neighborhoods in the region's largest cities. Others are descendants of poor farmers and nomads who were forced off their land and into the city by state-sponsored development programs and imported labor-saving agricultural technologies. Refugee camps in Palestine and elsewhere are also important sources of recruits. Given the crowded, polluted, and often chaotic living conditions in the camps and in such cities as Cairo and Algiers, it is not surprising that many members of the younger generation have come to question the basic philosophy of the governments under which they live. Many Islamist activists, especially the leaders, come from the large pool of recent university graduates who are frustrated by their inability to find employment or to participate in the political process in their own countries and who are disenchanted with foreign interference in the region. Some Islamist leaders are respected religious men without much education who offer seemingly simple solutions to the region's problems. They say that all will be well if people return to fundamentalist versions of Islam and eschew secularism. Perhaps the greatest support for Islamism is gained by appealing to the widespread discontent of millions of increasingly poor people, whose plight has been further worsened by SAPs (see page 308) and economic upheavals brought on by such things as the wars in Iraq. Although there are those who are drawn to the radical alternatives proposed by the Islamists, the movements are still relatively small and based primarily in individual countries; only the most extreme are coordinated internationally (as was apparently the case with the organization Al Qaeda, which carried out the September 2001 air attack on New York City and Washington, D.C.).

Governments in the region have differing relationships with Islamist movements. For example, the Saud family in Saudi Arabia came to power, and remains there, by cooperating with the very conservative Wahhabi school of Islamist thinking. Iran is predominantly Shi'ite and is now governed by what for all practical purposes is an Islamist dictatorship. Other governments, such as those of Egypt and Algeria and, to a lesser extent, Turkey, are using the threat of Islamism as an excuse to keep political power in the hands of a small, wealthy elite that are nonetheless religiously conservative, just not radically so. On the other hand, non-Islamist groups dedicated to moderate reform in such places as Jordan, Bahrain, Oman, Qatar, Tunisia, and Morocco, are attempting to avert political extremism by redistributing political power more broadly.

Reform Efforts from within the Arab Community

In 2002, an independent consortium of Arab governments and Arab scholars, bureaucrats, and activists collaborated to produce a frank and revolutionary report on human development within Arab countries. The conclusion was that three deficits afflicted the larger Arab community (and other Muslim communities, as well): deficits in general human freedoms, in women's rights, and in access to knowledge. Enhancing democratic institutions was identified as the key to erasing these deficits; but it was recognized that domestic democratic reforms were dependent in large part on the development of a better educated electorate. Hence, the *Arab Human Development Report (AHDR) 2003* (pp. 1–26) focused on building a "knowledge society" and urged Arab leaders to introduce their people to the "global knowledge stream." While an educated electorate is essential to democracy, it should be noted that this report carefully skirts the issue of just how present autocratic governments might be reformed into more open, popularly elected entities.

Written in the context of the ongoing U.S.-led war in Iraq with its widening physical and social destruction, the *AHDR 2003* was careful to note both regret over the U.S. occupation of Iraq and recognition that the Hussein regime oppressed the Iraqi people and deprived them of a wide array of rights and freedoms. Observing that, "for most Arabs, the war is seen as an attempt at restructuring the region by outside forces pursuing their own objectives," the report aims at "self-determined change" by inspiring Arab elites to restructure the "region from within, with the ultimate objective of building human development in the Arab world." Self-criticism, the report suggests, is vastly to be preferred to having the region's future mapped from outside. Recognizing that education in Arab countries has long been curtailed by repressive governments, wide disparities in wealth and well-being, conservative interpretations of Islam, and outside influences, the 2003 report says Arab countries must become "sites of knowledge production." They must equip all their people to join the world of science and technology and become world-class innovators.

Although critics say that Arab reformist leaders are doing only a modicum to truly increase democratic participation, there is some evidence to the contrary. The *AHDR 2003* promotes Morocco, Jordan, and Qatar as places that are effectively reforming autocratic institutions and advancing the knowledge society. King Mohammed VI of Morocco is lauded for his efforts to liberalize the social climate for women (see page 302), and in Jordan, King Abdullah II has taken strong measures to improve education, protect women from abuse (such as honor killing), and extend democratic participation through a multiple-party system. In the tiny state of Qatar on the Persian Gulf (the home of Al Jazeera), movement toward the goals expressed in the *AHDR 2003* may be best developed. The emir of Qatar, Sheikh Hamad bin Khalifa Al Thani, has instigated broad-based reform and ambitious education projects.

ENVIRONMENTAL ISSUES

Salam, the Arabic root of the word *Islam*, means peace and harmony. Islam therefore calls on its followers to live in peace and harmony with both human and natural systems. The Islamic scholar and World Bank executive Ismail Serageldin writes that the Qur'an contains numerous references to the role of humans as stewards of the earth. In Muslim societies, these references are typically interpreted to mean that humans have the right to use the earth's resources, but only within the general limitations that Islam places on greed and personal ambition. The Qur'an requires Muslims to avoid spoiling or degrading human and natural environments, to share such resources as water equally with all forms of life, and to conserve natural resources even if they are abundant. Water purity is important in Islamic preparation for worship. A few decades ago, a religious decree in Saudi Arabia approved the human use of wastewater if it had been completely and properly treated and impurities removed from it. This decree was important because it encouraged recycling of water and helped extend the life of existing freshwater aquifers.

In practice, Muslims are like people everywhere: they have not always followed their own religious teachings and cared for the earth. Urban crowding, mechanized agriculture and industry, and the pursuit of material goods have resulted in pollution, species extinctions, and degraded environments. At present rates of use, the region of Southwest Asia and North Africa has a very limited supply of arable land, water, forests, and even minerals. Yet people across this region expect to achieve higher living standards—which, along with population growth, will tax the region's resources mightily. Careful management of the environment will undoubtedly become an important issue over the next several decades. Environmental issues are not yet at the forefront of public debates; it is hard to find much Muslim literature supporting environmentalist policies. Recently, some Islamist critics of modern secular (Western) society have linked materialism to environmental degradation. They label capitalism as a hazard to the environment. In this section, we look at water availability and use and desertification as two environmental issues that are central to the region's future.

Water Availability

Water has always been in short supply in this arid part of the world. Hence, cultural attitudes toward water differ greatly from those of North Americans. Here, people expect to use very little water, and they have devised many strategies to cope with the limited supply. Traditional ways to conserve water include designing buildings to maximize shade and conserve moisture; capturing mountain snowmelt and moving it to dry fields and villages in underground water conduits (*qanats*); bathing in public baths; and practicing countless recycling techniques, such as using bathwater for crop irrigation.

Despite a long tradition of careful water use, several factors have exacerbated the water shortage over the last few decades. Modernization has brought new ways to use water: in household plumbing; sewage treatment; industrial cooling and cleaning; and in mechanized irrigated agriculture. Moreover, the number of people who must share scarce water resources is growing rapidly. If natural supplies of water remain more or less constant, but water use per capita rises, will supplies meet basic needs? Experts have established that for a country to maintain basic human health and to support development, it must have no less than 1000 cubic meters of water available per capita per year. Figure 6.16 shows that, in 1990, many countries in the region already fell well below this standard. By the year 2050, all but three countries will be below the standard.

Given its centrality to human and economic life, water is a strategic resource. Consequently, when demand exceeds supply, countries commonly follow one or a combination of the following strategies: storage of water in reservoirs behind dams, importation of water, water recycling, conservation of the quality and quantity of water, efforts to stimulate rainfall, and desalination of seawater.

The greatest use of water throughout the region is for irrigated agriculture, despite the fact that in most countries agriculture does not contribute significantly to GDP. In a typical country, such as Tunisia (agriculture is 12 percent of GDP and employs 22 percent of the labor force), 88 percent of all the water withdrawn is used for agriculture, 9 percent for domestic purposes, and just 3 percent for industry. (For comparison, the United States uses 42 percent of its water for agriculture, 13 percent for domestic purposes, and 45 percent for industry.) In addition, the water in rivers and streams is being polluted by chemical fertilizers applied to fields. Even strategies to increase the supply of water—damming rivers, desalinization, groundwater pumping—create their own problems. Damming wastes water through evaporation or leakage, deprives downstream environments and people, and floods upstream users. The production and use of energy for desalinization creates air pollution. The pumping of groundwater lowers the water table and causes the land to subside, often resulting in the intrusion of seawater into the aquifer, which then renders the groundwater useless.

Irrigated agriculture is at the heart of the water crisis, and yet it holds the seeds of a solution. Because the overwhelming majority of water is used for irrigation, even small increases in efficiency will yield large increases in the volume of fresh water available for domestic and other uses. Efficiency can be improved through such methods as applying drip, rather than flood, irrigation more widely; reducing reliance on water-intensive crops such as rice, cotton, and citrus fruits; gradually removing government subsidies to large water users; and upgrading water distribution networks to reduce leakage. Another approach, suggested by British geographer Tony Allan, calls for greater reliance on food imports, a controversial idea. Allan argues that importing food for dry areas amounts to importing "virtual water," a reference to the water used in the production of local food and the water content found

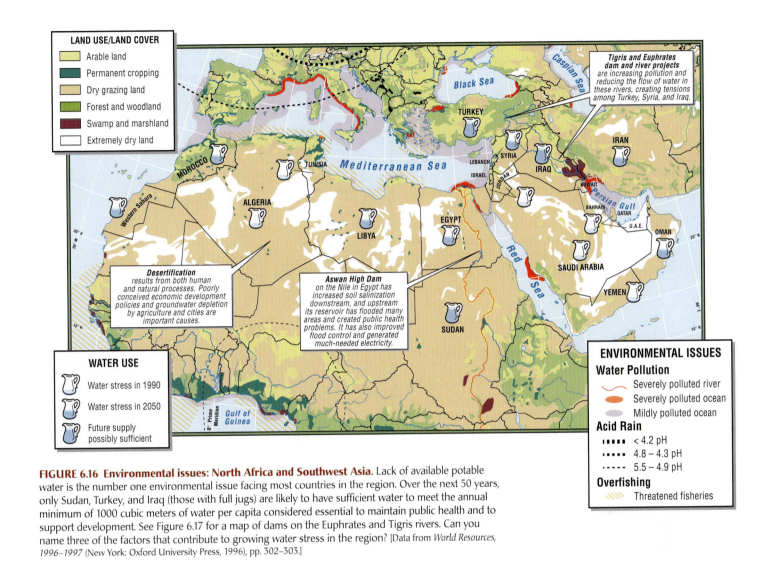

LAND USE/LAND COVER
- Arable land
- Permanent cropping
- Dry grazing land
- Forest and woodland
- Swamp and marshland
- Extremely dry land

Tigris and Euphrates dam and river projects are increasing pollution and reducing the flow of water in these rivers, creating tensions among Turkey, Syria, and Iraq.

Desertification results from both human and natural processes. Poorly conceived economic development policies and groundwater depletion by agriculture and cities are important causes.

Aswan High Dam on the Nile in Egypt has increased soil salinization downstream, and upstream its reservoir has flooded many areas and created public health problems. It has also improved flood control and generated much-needed electricity.

WATER USE
- Water stress in 1990
- Water stress in 2050
- Future supply possibly sufficient

ENVIRONMENTAL ISSUES
Water Pollution
- Severely polluted river
- Severely polluted ocean
- Mildly polluted ocean

Acid Rain
- < 4.2 pH
- 4.8 – 4.3 pH
- 5.5 – 4.9 pH

Overfishing
- Threatened fisheries

FIGURE 6.16 Environmental issues: North Africa and Southwest Asia. Lack of available potable water is the number one environmental issue facing most countries in the region. Over the next 50 years, only Sudan, Turkey, and Iraq (those with full jugs) are likely to have sufficient water to meet the annual minimum of 1000 cubic meters of water per capita considered essential to maintain public health and to support development. See Figure 6.17 for a map of dams on the Euphrates and Tigris rivers. Can you name three of the factors that contribute to growing water stress in the region? [Data from *World Resources, 1996–1997* (New York: Oxford University Press, 1996), pp. 302–303.]

in all foods. Allan suggests that water-poor regions give up water-intensive agriculture and concentrate instead on other economic activities, using the proceeds to import food from water-rich regions. For those who practice this strategy, it could lead to a serious loss of jobs and greater dependency on imported food.

The most ambitious efforts to increase water supplies have involved the construction of dams and reservoirs on major river systems: the Euphrates, the Tigris, the Jordan, and the Nile (see the box "Turkey's Anatolian River Basin Project" on page 316). These projects threaten rights of downstream and upstream water users. Damming a river removes water from huge tracts of land downstream and stops the deposition of fertility-enhancing silt during seasonal flooding. International laws governing water rights along river systems evolved in parts of the world, mainly Europe, where water was plentiful, and these laws are ill-suited to dry regions. For example, under existing laws, it is possible for a country to dam a river that flows through its territory, with little or no consideration for the people living downstream or upstream of the dam. Countries may have little recourse short of war when their lands are flooded or their source of water is cut off by a dam in a neighboring country.

Dams can create a wide range of problems. The artificial reservoir created upstream of a dam not only floods villages, fields, wildlife, and historic sites, but the silt that used to serve as fertilizer downstream will fill up the reservoir behind a dam in just a few years. In a warm climate, the standing water behind a dam becomes a breeding environment for mosquitoes, which carry debilitating diseases such as malaria, and for dangerous parasites such as hookworm and schistosome species (see discussion in Chapter 7, pages 345–346). Also, huge volumes of water are lost from the reservoir through evaporation. Irrigated soil downstream, repeatedly saturated and dried out, will eventually become salty and infertile.

Sharing the water of rivers that flow across national boundaries has never been easy. In fact, the words *river* and *rivalry* come from the same root, and the original sense of "rivalry" was the conflict among users of a watercourse. Naturally, in a disagreement, each country emphasizes the variables that strengthen its case. For example, in debates over who should get Euphrates water, Turkey notes that the rivers start in Turkey and most of the flow in the Euphrates originates there. Meanwhile, Iraq points out that the river travels the longest distance in Iraq; and that the ancient Sumerians of Mesopotamia (meaning "lands between the

AT THE LOCAL SCALE   Turkey's Anatolian River Basin Project

The Euphrates River begins life as trickles of melting snow in the mountains of Turkey, then gathers volume from several tributaries as it flows southeast through the Anatolian Plateau and then through Syria and Iraq. In Iraq, the Euphrates is joined by a second major river, the Tigris, which also originates in Turkey. The two run roughly parallel for hundreds of miles, joining just before they reach the Persian Gulf.

The Anatolian River Basin project (Figure 6.17) was begun as a series of 22 dams to reserve Euphrates waters in Turkey for irrigation and hydropower generation; 19 power plants are planned. The enterprise has since broadened to include reforestation; instruction in farming; the establishment of fisheries; and the provision of such social services as literacy training, health care, and women's centers. But approximately 80,000 people will have to be resettled as waters rise behind the dams, and most will be impoverished Kurds, part of a larger settlement of these once nomadic herders that straddles the borders of Turkey, Iraq, Syria, and Iran.

As important as the Anatolian project may be to Turkey, its effects on Syria and Iraq downstream could be disastrous. If completed as planned, the project will reduce the Euphrates flow to Syria by 30 to 50 percent and to Iraq by around 75 percent. Moreover, the water will be polluted with fertilizers, pesticides, and salts after having served to irrigate crops in Turkey. Syria, whose population will double in just 25 years, depends on the Euphrates for at least half of its water, and Syria also plans to use more of the Euphrates flow for irrigation. Iraq is the last to receive the river water and is the most vulnerable. Iraq and Syria demand a combined total of 700 cubic meters of water per second as their rightful share of the Euphrates flow. Sixty percent of this flow would be eventually released into Iraq. But Turkey currently releases just 500 cubic meters per second.

[*Source:* Stephen Kinzer, "Restoring the Fertile Crescent to its former glory," *New York Times* (May 29, 1997); Christine Drake, "Water resource conflicts in the Middle East," *Journal of Geography* (January/February 1997): 4–12; and Hussein A. Amery, personal communication, August 2001.]

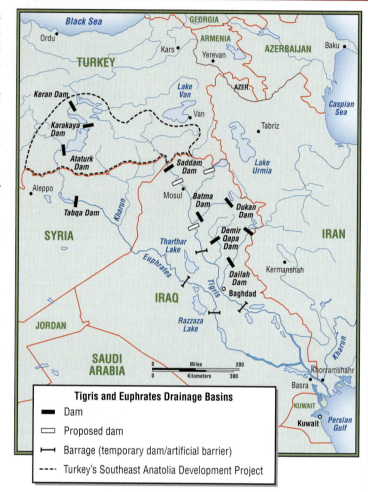

FIGURE 6.17 Dams of the Tigris and Euphrates drainage basins.
[Adapted from Christine Drake, "Water resource conflicts in the Middle East," *Journal of Geography* (January/February 1997): 7.]

two rivers," in both Arabic and Greek) were clearly the earliest major users of the Euphrates water.

A number of factors complicate the chances of reaching water allocation agreements along the Euphrates and the Nile. Most countries in the region are experiencing rapid population growth and a rapid rise in living standards and urbanization. One consequence is the growing consumption of meat, yet raising animals is more water-intensive than growing grains. Urbanites tend to consume more water (indoor plumbing, flush toilets) and use more electricity (often hydroelectricity) than rural residents. Furthermore, the people in these river-basin states have long had territorial and ideological disputes with one another. In the case of the Euphrates, for the time being, the fact that Turkey has the upstream geographic location on the river basin and the strongest

political and military might tilts the balance of power in the basin in Turkey's favor (see the box "Turkey's Anatolian River Basin Project" above).

Water Issues in the Palestine–Israel Dispute. When Palestinians say that according to the 1993 Oslo Accords, Israel must relinquish all of the West Bank and Gaza Strip seized during the 1967 war, what Israelis hear is that they are being told to give up major water rights as well as settlements. The West Bank includes the west side of the Jordan Valley, which has better water resources than the east side. The Palestinians are willing to share some of the water in the aquifers of the West Bank, but they claim the lion's share of the resource because its **recharge area**—the area with the highest precipitation, which recharges the aquifers—is located under the hills of this Palestinian territory.

Israel says it should retain control over water resources in the West Bank and no foreign army will be allowed west of the Jordan River. Furthermore, Israel holds that desalinization projects should be undertaken to provide sufficient water for Israel, Jordan, and the Palestinians, instead of merely redistributing existing resources.

Quietly, Palestinians and Israelis are working to resolve water allotment issues, and here is the quandary: at present, Israel pumps the most water from the Jordan River and from aquifers that underlie the Israeli occupied territories. It uses the water primarily in its technically advanced agricultural sector. But Israeli agriculture employs only 2.6 percent of the labor force, and contributes just 2 percent to the GDP. By contrast, the Palestinian economy remains agriculturally based (33 percent of GDP, 13 percent of labor force) and will be damaged if the Palestinians are not allocated sufficient amounts of water to sustain their production, maintain existing jobs, and create new ones. At present, Palestinians are banned from sinking wells or accessing the water they need.

Desertification. The United Nations defines **desertification** as the ecological changes that convert nondesert lands into deserts. These changes include the loss of soil moisture, the loss of trees or the replacement of plant species by other species that are more adapted to dry conditions, the erosion of soil by wind, and the formation of sand dunes on formerly vegetated land. These alterations can be brought on by long-term climatic change, by relatively short term climate variability, or by human activities that reduce moisture. Desertification is occurring in many places around the globe. Estimates of how much usable land is being lost to desert vary, but the ecologist and author G. Tyler Miller writes that globally, every year, 23,000 square miles (60,000 square kilometers) become new desert. It is not yet known just how much of this loss is due to natural climate change and how much is due to human-induced climate change through the greenhouse effect (the result of burning fossil fuels), human modification of forests and grasslands, and overuse of water resources, but clearly human activity plays a major role.

In the region of North Africa and Southwest Asia, nomadic herders are often blamed for increasing desertification. They are said to overstock rangelands and allow their animals to overgraze. With little grass remaining, the soil is unable to retain moisture. Studies are showing that the real story is more complicated. Geographer and veterinarian Diana Davis found that herders are usually careful to manage their herds so that the grasslands on which they depend are not destroyed. In her research among herders in Morocco, Davis observed that when times are good, herders were more likely to decrease rather than increase their herds, recognizing that it is during good times that rangelands should be allowed to regrow. Davis found that herders' knowledge of range ecology often leads to management practices that accord well with an arid environment. Most likely, a wide array of economic changes has contributed to a general drying of grasslands bordering the Sahara. As groundwater is lowered through urban use and irrigated agriculture, plant roots can no longer reach sources of moisture. Furthermore, the encouragement of settled cattle ranching (rather than nomadic herding) by international development agencies leads to overgrazing of pastureland and to excessive demands on scarce water. When pastoral nomads settle permanently and take up modern life ways their per capita use of water increases.

MEASURES OF HUMAN WELL-BEING

As we have observed elsewhere in this book, the gross domestic product per capita is a less than adequate measure of the actual state of well-being in a region, but it nonetheless gives some indication of the distribution of wealth across the region. In Table 6.2, you can see that the pattern of wealth as measured by GDP per capita (column 2) is uneven. The primary oil-producing nations (Saudi Arabia, Oman, the UAE, Qatar, Kuwait) and Bahrain and Israel have high GDP per capita figures. Nearly all other countries in the region are well below the world average of about U.S. $7376. Even politically influential countries such as Turkey and Egypt have per capita incomes below the world average. Because the GDP per capita figure is only an average, the fact is masked that in this region wealth is unequally distributed (much less so than in the United States or South America, however). Many people live on much less than listed, while a very few are extremely rich.

The United Nations Human Development Index (HDI) (see Table 6.2, column 3) combines three indicators to evaluate well-being in 175 countries and rank them. The indicators are life expectancy at birth, educational attainment, and income adjusted to purchasing power parity (PPP). You may remember that in other chapters we cited the United Arab Emirates' HDI figures in our comparisons of well-being in countries in other regions. Despite its high GDP (highest in this region), the UAE does not rank very high on the HDI scale in comparison to nearly all developed countries in the world. An important reason is its low literacy rates, especially among adult women (just 80 percent have basic literacy, see Table 6.2). Until the past decade, education was available to only a minority of females. Thus, in Kuwait and elsewhere in the Arab states, there is a large population of older illiterate women. Israel is the only country in this region with both a high GDP per capita and a high HDI rank (22). Notice that some of Israel's closest neighbors (Syria, Lebanon, Jordan, and Egypt) rank fairly low on both the GDP index and the HDI. Israel's ranking reflects its ability to provide the basics of a decent life—adequate income, shelter, food, education, and health care—to all its citizens, regardless of ethnicity, sex, or religion. Israeli culture emphasizes education, and Israel has attracted highly educated Jewish immigrants from around the world. Israel, however, has also received help in providing for its citizens through development aid, receiving $172.00 per capita in development assistance from the United States in 1998. Israel's much poorer neighbor, Egypt, which is the next largest recipient of U.S. development assistance after Israel (1998), received $29.00 per capita.

In this chapter, the Gender Empowerment Measure (GEM) and the Gender Development Index (GDI) are combined in Table 6.2 (column 4). For a variety of reasons, the statistics

TABLE 6.2   Human well-being rankings of countries in North Africa and Southwest Asia

Country (1)	GDP per capita, adjusted for PPP[a] in 2001 $U.S. (2)	Human Development Index (HDI) global rankings, 2003[b] (3)	Gender Empowerment Measure (GEM)[c] and Gender Development Index (GDI)[c] global rankings, 2003[d] (4)	Female literacy, (percentage), 2001 (5)	Male literacy, (percentage), 2001 (6)
Selected countries for comparison					
Japan	25,130	9 (high)	44 (GDI = 13)	99	99
United States	34,320	7 (high)	10 (GDI = 5)	99	99
Mexico	8,430	55 (high)	42 (GDI = 52)	89	93
World	7,376	—	—	73	85
The Northeast					
Turkey	5,890	96 (medium)	66 (GDI = 81)	77	94
Iraq[f]	3,197	126 (medium)	— (GDI = 107)	43	64
Iran	6,000	106 (medium)	— (GDI = 86)	70	84
Eastern Mediterranean					
Syria	3,280	110 (medium)	65[f] (GDI = 93)	62	89
Lebanon	4,170	83 (medium)	— (GDI = 70)	81	92
Israel	19,790	22 (high)	23 (GDI = 22)	93	97
Jordan	3,870	90 (medium)	69[f] (GDI = 75)	85	96
Palestinian Territories	NA	98 (medium)	— (GDI = NA)	73	71
The Arabian Peninsula					
Saudi Arabia	13,330	73 (medium)	— (GDI = 68)	68	83
Yemen	790	148 (low)	70 (GDI = 127)	27	69
Oman	12,040	79 (medium)	— (GDI = 71)	64	81
United Arab Emirates	20,530	48 (high)	65 (GDI = 49)	80	75
Qatar	19,844	44 (high)	— (GDI = 41[f])	84	80
Bahrain	16,060	37 (high)	— (GDI = 40)	83	91
Kuwait	18,700	46 (high)	— (GDI = 45)	80	85
Egypt and Sudan					
Egypt	3,520	120 (medium)	68 (GDI = 99)	45	67
Sudan	1,970	138 (medium)	— (GDI = 116)	48	66
The Maghreb					
Western Sahara	—	—	— (GDI)[e]	—	—
Morocco	3,600	126 (medium)	— (GDI = 102)	37	63
Algeria	6,090	107 (medium)	— (GDI = 88)	58	77
Tunisia	6,390	91 (medium)	60[f] (GDI = 76)	62	83
Libya	7,570	61 (medium)	— (GDI = 65[f])	69	91

[a]PPP = purchasing power parity.
[b]Total ranked in world = 175.
[c]Total ranked in world = 70 (no data for many).
[d]Total ranked in world = 144 (no data for many).
[e]Data not available.

[f]In 2000 (*United Nations Human Development Report 2000*). Data not available in 2003.

Source: United Nations Human Development Report 2003 (New York: United Nations Development Programme).

necessary to calculate GEM, which ranks countries on the extent to which women have opportunities to participate in economic and political life, are not available for two-thirds of the countries in this region (only 70 countries are ranked worldwide). The GDI measures only the access of males and females to health care, education, and income, not the extent to which males and females actually participate in policy formation and decision making within a country. As we have already observed, despite the importance of women in the home, only a very small percentage of women across the region participate in policy-making debates or as leaders in the larger society. Many countries have no women in parliament. Turkey has 4.4 percent, Egypt 2.4 percent, and Iran 4 percent. In Jordan, women hold just 3 percent of the seats in its Parliament, but three women have been appointed to the Jordanian cabinet. In Tunisia, women hold 11.5 percent of the parliamentary seats, and other gains are being made: women hold 23 percent of the country's judgeships and head 1500 businesses, managing 13 percent of businesses in greater Tunis. Israel's relatively high GEM rank of 23 is the result of its direct efforts to include females in all aspects of society, including the armed forces. Even in Israel, however, women hold only 15 percent of parliamentary seats.

Quick Review

1. What is shari'a? What role does it play in politics and in daily life?

2. Compare secular Islamic nations with theocratic Islamic nations. In which are women more likely to engage in public activities and occupations?

3. Why did countries in this region organize the oil cartel, OPEC?

4. What are some of the human activities that lead to desertification?

III   SUBREGIONS OF NORTH AFRICA AND SOUTHWEST ASIA

The subregions of North Africa and Southwest Asia present a mosaic of the issues that have been discussed so far in this chapter. Although all countries in the region except Israel share a strong tradition of Islam, they vary in how Islam is interpreted in national life and the extent to which Westernization is accepted. They also vary in prosperity and in the extent to which wealth is evenly distributed in the general population. Conflict is not uncommon in this region, and the closer examination allowed by the subregional perspective may help you understand the factors that have led to the discord and, in some cases, violence.

THE MAGHREB (NORTHWEST AFRICA)

If you are interested in old movies, you have probably already formed intriguing images of North Africa: Berber or Tuareg camel caravans transporting exotic goods across the Sahara from Timbuktu to Tripoli or Tangier; the Barbary Coast pirates; Rommel, the World War II "Desert Fox"; the classic lovers played by Humphrey Bogart and Ingrid Bergman in *Casablanca*. These images—some real, some fantasy, some merely exaggerated—are of the western part of North Africa, what Arabs call the Maghreb ("the place of the sunset"). The reality of the Maghreb, however, is much more complex than popular Western images of it.

The countries of the Maghreb stretch along the North African coast from Morocco to Libya (Figure 6.18). A low-lying coastal zone is backed by the Atlas Mountains, except in Libya. Despite the region's overall aridity, these mountains trigger sufficient rainfall in the coastal zone to support export-oriented agriculture. Except in Morocco, each country's interior reaches into the huge expanse of the Sahara Desert.

The landscapes of the Maghreb reflect the long and changing relationships all the countries have had with Europe. European domination in this area lasted well into the twentieth century. During this time, the people of North Africa became acculturated to Europe, taking on European ways: consumerism, mechanized market agriculture, and manners of dress, language, and popular culture. The cities, beaches, and numerous historic sites of the Maghreb continue to attract millions of European tourists every year, who come to buy North African products—fine leather goods, textiles, handmade rugs, sheepskins, brass and wood furnishings, and paintings—and to enjoy a culture that, despite Europeanization, seems exotic. The agricultural lands of the Maghreb are strategically located close to Europe, where there is a strong demand for Mediterranean food crops: olives, olive oil, citrus fruits, melons, tomatoes, peppers, dates, grains, and fish. Not only is Europe a major market for North African agricultural produce; oil, gas, and petroleum products, which make up over 30 percent of GDP and 95 percent of total exports

Modern high-rise complex in Tripoli, Libya. [P. Robert/CORBIS SYGMA.]

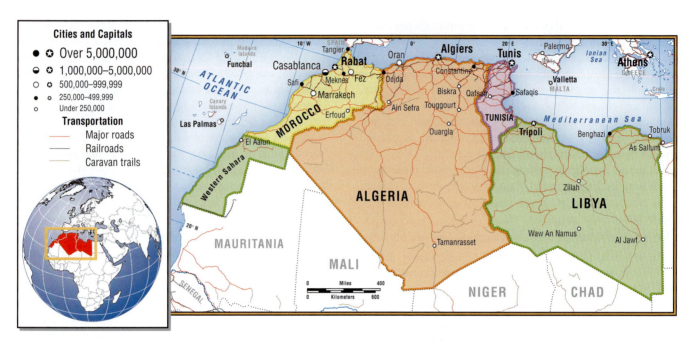

FIGURE 6.18 The Maghreb subregion (Northwest Africa).

in both Algeria and Libya, are also marketed primarily to Europe. Algeria and Libya supply about one quarter of the oil and gas used by the European Union.

Europe is also a source of jobs, both in that local firms are supported by European investment and tourism and in that millions of guest workers have migrated to Europe, many after losing jobs as a result of agricultural modernization. In 1998, *Migration News* estimated that there were half a million North African citizens in Spain and that France had 5 million to 6 million North African migrants, many of them illegal. After the terrorist bombings in Spain in March 2004, sentiment against illegal North African immigrants increased. For the migrants, the trip has become more dangerous; death rates for those who cross the Mediterranean in leaky, flimsy craft are high: in 2003, 77 North Africans died trying to reach the Spanish coast. Their plights once in Europe are reminiscent of illegal Mexican migrants into the United States: low pay, poor housing, and a constant threat of deportation.

In North Africa, settlement and economic activities are concentrated along the narrow coastal zone (see Figure 6.7, page 294). Most people live in the cities that line the Atlantic and Mediterranean shores—Casablanca, Rabat, Tangier, Algiers, Tunis, and Tripoli—and in the towns and villages that link them. The architecture, spatial organization, and lifestyles of these cities mark them as cultural transition zones between African and Arab lands and Europe. The city centers retain an ancient ambience, with narrow walkways fronted by family compounds (see the photo on page 301), but most people live in modern apartment complexes around the peripheries of these old cities. Super-highways, art galleries, shopping malls, and restaurants serving international cuisine serve the citizens. (See the photo of Tripoli, Libya, on page 319.)

Many North Africans are seeking an identity that is less European and retains their own distinctive heritage. In the Maghreb,

the rejection of Europeanization by Islamists has been particularly hostile. As the historian John Ruedy puts it, for those people who were Westernized during the era of European domination and who now make up the educated middle class, independence from Europe meant the right to establish their own secular nations with constitutions influenced by European models. But by the 1990s, the Islamist movement was seeking to challenge independent secular states by linking them in a negative way to European domination. For these religious activists, Islamism is a positive cultural identity. Their interpretation of Islam, including the reassertion of Islamic law, challenges the current, often only quasi-democratic, practices across the Maghreb. Secular leaders and some citizens who are religious but who prefer a secular government are fearful of the outcome.

Violence in Algeria

Algeria, with 28 million people, is arguably the Maghreb's most troubled country. By 1998, civil violence between Islamists and the military-controlled government had resulted in the deaths of at least 100,000 people. The violence disrupted the society and ruined the already strained economy.

After a violent struggle with France for independence, won in 1962, Algeria was ruled by a single-party socialist military dictatorship that repressed all viable opposition. At independence, Algeria's economy retained colonial features, in that Europeans remained in control of the most productive farmland. The new socialist military government expropriated (took over) the large European farms that produced dates, wine, olives, fruit, and vegetables for export, and soon these farms ended up in the control of the Algerian elite, either as private farms or as large cooperatives. Most of the large agricultural population remained on small, infertile plots. By the early 1990s, nearly half the economy, including

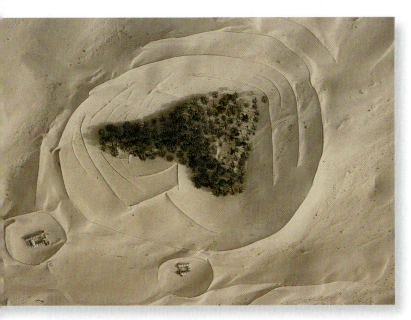

Although modern life in Algeria is centered on the coast, ancient strategies for living with the desert persist in inland areas. At Souf, an oasis in the Algerian Sahara, farmers plant date palms. The roots reach down for the water located near the surface in depressions between the dunes. The farmers also build fences against the prevailing winds, creating wavelike patterns in the sand. Sand collects against the fences, building higher dunes and further sheltering the trees from the desiccating wind. [Georg Gerster/Photo Researchers.]

factories and utilities, was state owned and inefficiently operated. The economy declined to the point that by the late 1990s, unemployment among urban dwellers under the age of 30 was 80 percent or higher. The unequal division of wealth between those Algerians who replaced Europeans in positions requiring skills and education and those who remained poor, with only a few years of education, created social divisions. Those who are well-off and liberal in their attitudes toward economic reform, women's roles, and freedom of speech are often derided by the Islamists as too secular; but a significant number of Islamist leaders are themselves technically and professionally educated people.

During the 1990s, the Algerian military government shifted its policy regarding ownership of industries and undertook major economic structural adjustment. In an attempt to lure foreign investment and create jobs, industries were privatized. During the lag that often occurs between structural adjustment reforms and economic resurgence, average personal income fell by a third. In 1991, under these dismal economic conditions, an Islamist (fundamentalist) party won the first free elections by an overwhelming margin. A peaceful transfer of power could have diffused the pent-up discontent of many Algerians. Instead, the military-influenced socialist government—motivated largely by fears that the newly elected government would pursue an extreme religious path—declared the parliamentary elections to be null and void. Duly elected Islamist officials were jailed, and the country was plunged into a devastating civil war.

Although some Islamist advocates joined religiously based social service agencies and worked to improve life in Algeria's shantytowns, extremist factions of Islamists began brutal terrorist attacks. Particularly targeted were journalists, musicians and artists, and educated working women not observing seclusion or not wearing the veil. Several thousand such women were killed. Also, perhaps as a scare tactic, the populations of whole villages were massacred, with apparently no attention paid to political or religious affiliation. Some studies and a few journalistic reports show that state-sponsored "militias" (troops in civilian clothing) also terrorized civilians to dissuade them from supporting the Islamists.

By mid-1997, public sentiment was turning against extremists in the Islamist movement. In a relatively fair election in June 1997, moderate Islamists, espousing nonviolence, won nearly one-third of the seats in the national assembly. Soon, violence had subsided, and more than 1000 militants surrendered under an amnesty program. In April 2004, with 60 percent of eligible voters participating, President Abdelaziz Bouteflika won a second term with a landslide majority of 83 percent. But shortages of housing and jobs continue to be focal points of resistance among young adults.

THE NILE: SUDAN AND EGYPT

The Nile River begins its trip north to the Mediterranean in the hills of Uganda and Ethiopia in central East Africa. The countries of Sudan and Egypt share the main part of the Nile system (Figure 6.19), and for these countries it is the chief source of water. Although Sudan and Egypt have the Nile River, an arid climate, and Islam (Egypt is 94 percent Muslim, Sudan 70 percent) in common, they differ culturally and physically. Egypt, despite troubling social, economic, and environmental problems, is industrializing and plays an influential role in global affairs, whereas Sudan struggles with civil war and remains relatively untouched by development.

Sudan

Sudan, slightly more than one-quarter the size of the United States, is the largest country in this region and, in fact, in all of Africa (see Figure 6.19). Yet, with 38 million people, it has about half the population of Egypt. The country has three distinct environmental zones that become drier as the altitude decreases. The upland south, called the Sudd (the first zone), consists of vast swamps fed by rivers bringing water from the rainy regions of Central (equatorial) Africa. North of, and somewhat lower than, the Sudd, the much drier steppes and hills of central Sudan (the second zone) are home to animal herders. Yet farther north and lower is the Sahara (the third zone), which stretches to the border with Egypt and then all the way to the Mediterranean. Most Sudanese live in a narrow strip of rural villages along the main stream of the Nile and its two chief tributaries, the White Nile and the Blue Nile (see Figures 6.2 and 6.7). Sudan, then, though primarily a desert and dry steppe country, does have these two important Nile tributaries, and the water they bring from the far upland south irrigates fields of cotton for export. Sudan's only cities are clustered around the famed capital of Khartoum, where the White and Blue Niles join.

There is animosity between southern and northern Sudan that has roots deep in the past and is fed by the politics of oil and religion. Islam is the religion of northern lowland Sudan and did not spread to the upland south until the end of the nineteenth century, when Egyptians, backed by the British, subdued that part of the country, despite fierce resistance. Sudan then became a joint protectorate of Britain and Egypt until independence in 1955. The southern Sudanese people are, for the most part, either Christian or hold various indigenous beliefs, and most are black Africans of Dinka and Nuer ethnicity. In southwestern Sudan (Darfur) there are also Arab-speaking Muslim black Africans. Despite their common religion and language with the north, these southwestern Sudanese, as well as the Christians of the south, fear the Muslim, Arabic-speaking lighter-skinned Sudanese of the north, who for thousands of years raided the upland south for slaves. In 1983, the Muslim-dominated government decided to make Sudan a completely Islamic state and imposed shari'a on the millions of non-Muslim southerners. As a result, Sudan is now home to two civil wars. One is between the Muslim Arab-dominated regime in Khartoum and black, non-Muslim rebels in the south who protest the imposition of shari'a, but also want to stake a claim on Sudan's rich oil reserves. This civil war has claimed 2 million lives since 1983, forced 4 million people to abandon their homes, and orphaned many thousands of children, who were forced to be soldiers for one side or the other. But recently the two sides have come close to a peace pact that would allow for a sharing of power and oil wealth.

This prospect of sharing oil wealth and scarce water has helped to provoke the second civil war in Darfur, where the uniformly Muslim and Arab-speaking population is nonetheless divided between those who identify with the Arab north and those who consider themselves black Africans. Both groups want a share of the oil and both are afflicted by declining per capita supplies of water that upset long-standing cooperative resource-use patterns. The government, leery that concessions will lead to more rebellions, backs an armed

Smoke rises from a house in Tine, on the Chad–Sudan border, after being bombed by a Sudanese government Antonov plane, January 26, 2004. Tens of thousands of Sudanese refugees have poured across the border into Chad since December 2003, fleeing a series of attacks by horse-riding *janjaweed* militiamen and bombardment by government planes. [Antony Njuguna/Reuters/Landov.]

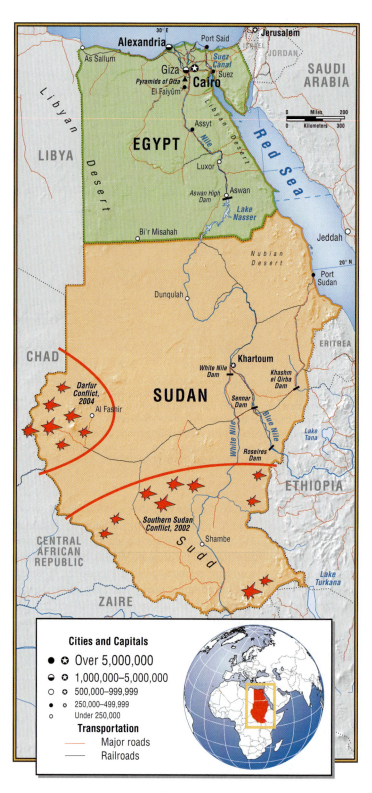

FIGURE 6.19 The Egypt and Sudan subregion. [Sources: USAID@http://www.USAID.gov/locations/sub-saharan_africa/sudan/sudan_bombjuly.pdf; http://www.lib.uTEXAS.edu/maps/AFRICA/darfur_villages_0802_2004.Jpg.]

Arab militia called the *janjaweed*. Samantha Powers, a Harvard University specialist in genocide, reported after a trip to Darfur (PBS, August 2004) that apparently the government gave the *janjaweed* free rein to rape, rob, and kill blacks, and helped by bombing black villages just before the raiders arrived (see the photo at left). The government denies this, but thousands of refugees tell similar stories and Human Rights Watch says documents show beyond a reasonable doubt that the ethnic cleansing (more than 70,000 have died)

PRACTICING GEOGRAPHY   What Does a City Look Like in Egypt?

Dona J. Stewart
Georgia State University

In my research I have tried to understand the forces that shape cities in this region, especially Cairo. Many of us in the United States have an image of North Africa and Southwest Asia (the Middle East) as "frozen in time"; we think of walled cities with minarets and women kept in seclusion. While it is important to understand the historical model of the Middle Eastern city as a foundation for the modern urban system, recent economic, demographic, and political forces have dramatically reshaped cities such as Cairo. Crumbling city walls and gleaming skyscrapers set on artificial islands both are now part of the region's urban mosaic.

Most of the region saw enormous population growth in the 1960s, as infant mortality rates dropped and life expectancy increased. Cities that had been relatively small mushroomed; this was particularly true of capitals, many of which became primate cities. As farmers left the countryside in search of higher income employment in the urban areas, cities grew even more.

I have used a number of methodologies in conducting research in Egypt. In examining the change in Cairo's cultural landscapes over time, and the role of differing political economic forces in reshaping the city, I've used traditional cultural landscape analysis techniques similar to those used by geographers Donald Meinig and Paul L. Knox. In this approach, the landscape can be "read" to identify political economic forces at work. For example, when one sees a McDonald's in Cairo it may indicate the presence of global economic forces where the corporate headquarters in the United States has decided that there is enough potential local demand, and enough people who can afford their product, to open a franchise in Egypt. It can also indicate a change in domestic economic policy, in which the government of Egypt has embraced capitalism (at the prompting of the World Bank) and is now welcoming foreign direct investment. One important research question is how the residents of Cairo perceive the McDonald's. Who is excluded from access by economic constraints? What is the impact, if any, of McDonald's on local fast-food chains? Though McDonald's is a global corporation, the franchise has local investment; how does this impact its "authenticity"?

A colleague with GIS expertise helped me use my local knowledge to assess changes in Cairo's urban structure over time.

Using satellite imagery, historical maps and field checks, we have looked at growth and change in Cairo over the past few decades. Surprisingly, our results indicate that Cairo is showing trends very similar to the North American city. Affluent suburbs, with gated communities and swimming pools, have appeared in the desert around the city. And much like American central cities in the 1960s–80s, population is declining in the old downtown. In the U.S. experience, the poor remain trapped downtown, without access to transportation or the new jobs in the suburbs; it is possible that Cairo may be moving in the same direction, though job creation in the suburbs has not been very significant yet. Comparing research across regions and models is an important way to increase our understanding both of the United States as well as the Middle East.

Of course, my favorite part of conducting research in this region is living in a city like Cairo or Amman and walking the streets or sitting in a coffeehouse watching people go by. One of my favorite memories is of an early morning in the old city of San'a, Yemen. I watched the city wake up, children going off to school, vendors going out to sell their wares, neighbors chatting with each other. As the call to prayer lingered in the background, Toyotas vied for space with pedestrians. Though I have spent many years in the region, I am always conscious of being an outsider of sorts. What for me is a research site and temporary home is the city someone's family has lived in for thousands of years.

[*Sources:* Dona J. Stewart, "Middle East urban studies I: identity and meaning," *Urban Geography*, 22 (2004): 175–181; *The Arab Human Development Report* (New York: United Nations Development Programme); Dona J. Stewart, Z. Y. Yin, S. Bullard, and J. MacLachlan, "Urban decentralization in Cairo: A GIS and imagery analysis approach," *Urban Studies*, 41.]

by the *janjaweed* has Khartoum's blessing. Although a desperately needed international peacekeeping presence was slow to form, refugee aid was sent by Egypt and Saudi Arabia, as well as by numerous other countries, and there was discussion of mobilizing troops from sub-Saharan Africa, under the banner of the African Union (see Chapter 7, page 357) to quell the violence against ordinary people. In October 2004, the African Union sent observers to oversee peace negotiations, but the violence continued.

Egypt

The Nile flows through Egypt in a somewhat meandering track from Egypt's southern border with Sudan north to its massive delta along the Mediterranean. The distance is more than 800 miles (about 1300 kilometers). Egypt is so dry that the Nile Valley and Delta are virtually the only habitable parts of the country (see Figure 6.7, page 294). Ninety-six percent of Egypt is desert. Yet, for thousands of years, agriculture along the banks of the Nile has fed the country and provided high-quality cotton for textiles.

The Nile's flow is no longer unimpeded. At the border with Sudan, the river is captured by a 300-mile-long (483-kilometer-long) artificial reservoir, Lake Nasser. The lake stretches back from the Aswan High Dam, which was completed in 1970 (see Figure 6.19). The Aswan controls flooding along the lower Nile and produces hydroelectric power for Egypt's cities and industries. North of (below) the Aswan, the river environment is greatly modified by the effects of the dam. No longer is the alluvial plain replenished every year by floodwaters carrying a fresh load of sediment from upstream. Irrigation canals keep some fields in nearly continuous production, and fertilizers are used to maintain productivity. The principal crops, now almost all irrigated, are cotton, grains, vegetables, and sugarcane, as well as animal feed. The effects of the dam extend into the Mediterranean: the Nile Delta, no longer replenished with sediment annually, is eroding, and eastern Mediterranean fisheries are affected by the dearth of natural nutrients and the infusion of chemical fertilizers.

Egypt is the most populous of Arab countries (72 million in 2003) and the most politically influential. Its geographic location bridging Africa and Asia, gives it strategic importance, and the country plays an influential role in situations of global importance—for example, debates on world trade, the peace process between Palestine and Israel, and other geopolitical activities, such as the U.S.-led wars in Iraq in 1991 and 2003. Egypt was part of the U.S. coalition in 1991, but opposed the 2003 war in Iraq, saying it was likely to increase terrorism—"Iraq will produce 100 bin Ladens" ran a headline in Cairo. But Egypt is also a major recipient of U.S. aid, which limits its ability to take independent positions. Egypt's limited resources and its high rate of population growth mean that poverty rates are high; 44 percent of the population live on just $2 a day. Although it is the cities that are growing most rapidly (in 2003, Egypt was 43 percent urban), the number of people still living and working on the land has also increased. Despite the increased labor, the Nile fields are no longer sufficient to feed Egypt's people, and food must be imported. Because of increased needs for cash to buy this food, rural men often migrate to neighboring countries where they work to supplement the family income, leaving the women in charge of farming and village life.

PERSONAL VIGNETTE

Mohammed Abdel Wahaab Wad has just spent a week in jail for protesting the loss of the land his family has leased for several generations. In the 1950s, land was redistributed from rich to poor farmers at very low rents so that they would be able to support their families. In the late 1990s, a new law reversed the process—the result of structural adjustment policies (SAPs) imposed by the World Bank and the International Monetary Fund. The purpose of the new law is to increase the land's productivity by cultivating it in large tracts managed with machinery and controlled by the original owners. This reorganization of the agricultural system is expected to boost Egypt's ability to feed itself and grow water-intensive export crops such as cotton and rice. The change in control of the land is expected to affect 900,000 tenant farmers like Mr. Wad, who have continued to use labor-intensive techniques to cultivate crops of vegetables and wheat just sufficient to feed their families.

Mr. Wad, who has 11 children to help him cultivate, has never before even been asked for rent for the 8 acres (3.25 hectares) he occupies on the edge of the Sahara, northwest of the Aswan Dam. Now the new law says that the owner can tell the family to leave.

The landowner is Gamal Azzam, a member of the urban aristocracy, who is adamant that he will recover his land and that the Wad family must go. There are growing worries that in the short run armed conflict will break out and in the long run the result will be another rush to the already crowded cities. Migration to the cities would result in a period of malnutrition for many people while the agricultural system is reorganized around mechanized production, which will undoubtedly produce more food per acre than traditional farmers have been able to produce, but this food will be too expensive for the dispossessed. [*Adapted from Douglas Jehl, "Egypt's farmers resist end of freeze on rents," New York Times (December 27, 1997): A5.*]

Egypt's two main cities, Cairo and Alexandria, are both in the Nile Delta region. Cairo, at the head of the delta, has 14 million people in its environs and is one of the most densely populated cities on earth. Egypt's crowded delta faces a crisis of clean water availability and the threat of disease and pollution. In Chapter 4, we discussed pollution around the Mediterranean. To address its share of this problem, the Egyptian Ministry of the Environment, created in 1997, is seeking international contractors to treat industrial, agricultural, and urban solid and liquid waste that for years has been dumped untreated into the Nile and the Mediterranean. But an example from Cairo illustrates a disturbing pollution linkage between Europe and North Africa. A major industrial complex in Cairo recycles used car batteries sent to Egypt for disposal from all over Europe. The chemicals in car batteries are particularly hard to dispose of safely, and Europe's environmental standards are high. The recycling complex in Egypt, where the Ministry of the Environment has yet to exert its influence, releases lead concentrations 30 times higher than world health standards into the air that blows across Cairo and into the eastern Mediterranean. Exposure to lead affects brain development and, ultimately, intelligence. Some children's playgrounds in Cairo are so polluted they would be considered hazardous waste sites in the United States and Europe.

In the 1990s, Egypt's economy, long plagued with stagnation, inflation, and unemployment, appeared to be turning around. After years of structural adjustment policies, inflation was brought under some control, and multinational companies, such as Microsoft, Owens-Corning, McDonald's, American Express, Löwenbräu of Germany, and three German automakers, opened subsidiaries in Egypt. For the ordinary working people of Egypt, however, this flashy development did not translate quickly into prosperity. By 2003, at the start of the U.S. war in Iraq, Egypt's economy was once again in a recession; unemployment stood at around 20 percent and was expected to increase with the instability caused by war. Protests against the United States were widespread, especially among Egypt's underemployed educated young adults, who, partly because of the general economic and political malaise, but also because of their own personal values and their dismay over Western policies, are attracted to Islamism. Overall, conservative interpretations of Islam seem to be gaining popularity. For example, as already discussed, many young women are choosing to wear the veil as a sign of their piety and their stance against Western influence.

THE ARABIAN PENINSULA

The desert peninsula of Arabia has few natural attributes to encourage human settlement, and today large areas remain virtually uninhabited (Figure 6.20). The land is persistently dry and barren of vegetation over large areas. Streams flow only after sporadic rainstorms that may not come again for years. The peninsula as a whole has one significant resource, oil, which amounts to more than 65 percent of the world's proven reserves. Saudi Arabia has by far the largest portion, with approximately 25 percent of the world's reserves (see Figure 6.14 on page 306).

Traditionally, control of the land was divided among several ancestral groups led by patriarchal leaders called **sheiks.** The sheiks were based in desert oasis towns and in the uplands and mountains bordering the Red Sea. The tribespeople they ruled were either nomadic herders, who traveled over wide areas in search of pasture and water for their flocks, or poor farmers settled where rainfall or groundwater supported crops. Until the twentieth century, the sheiks and those they governed earned additional income by trading with camel caravans that crossed the desert (once the main mode of transport) and by providing lodging and sustenance to those on holy pilgrimages to Makkah.

In the twentieth century, the sheiks of the Saud family, in cooperation with conservative religious leaders of the Wahhabi sect, consolidated the tribal groups to form an absolutist monarchy called Saudi Arabia, which occupies the central part of the peninsula. The Saud family consolidated its power just as reserves of oil and gas were becoming exportable resources for Arabia. The resulting wealth has added greatly to the power and prestige of the Saud family and of their allies, the Wahhabi clerics. Nonetheless, as discussed elsewhere in this chapter (see especially the sections on migration and urbanization and on economic and political issues), the tight control of the ruling family and conservative clerics in Saudi Arabia has inhibited the development of opportunities for young people, many of whom remain undereducated and hence displaced from employment by skilled workers imported from abroad. Discontent among young adults is rising and their discontent is often expressed by embracing fundamentalist Islam. A majority of the suicide hijackers involved in the September 2001 attacks in the United States were Saudi, and Osama bin Laden is a member of a wealthy Saudi family in the construction business. Increasingly, the regional press speculates that sooner or later, political power will shift in this country. Violence like the terrorist attacks in Riyadh in May and November 2003 is likely to increase, in part because there is no civil forum for airing discontent.

Despite the overall conservatism of Arab society, oil money has changed landscapes, populations, material culture, and social relationships across the peninsula. Where once there were mud-brick towns and camel herds, there now are large, modern cities

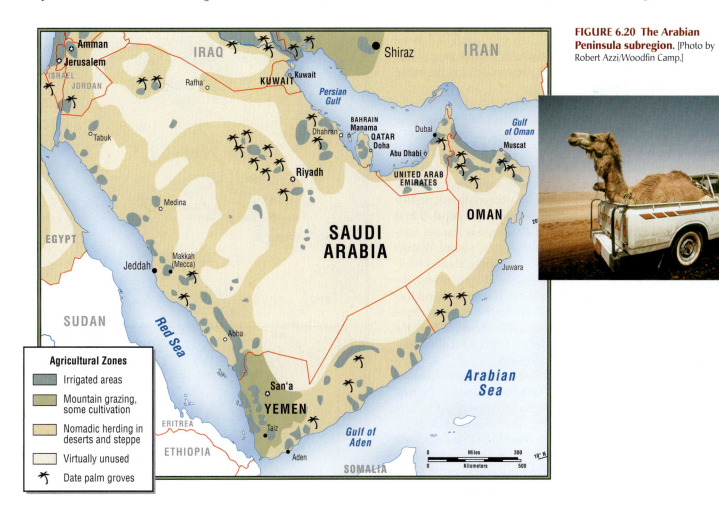

FIGURE 6.20 The Arabian Peninsula subregion. [Photo by Robert Azzi/Woodfin Camp.]

Agricultural Zones
- Irrigated areas
- Mountain grazing, some cultivation
- Nomadic herding in deserts and steppe
- Virtually unused
- Date palm groves

served by airports and taxis. The population of the peninsula burgeoned to 52 million people, half of whom live in Saudi Arabia. Pickup trucks not only have replaced camels but are now used to transport them. Unsustainable irrigated agriculture has made the peninsula nearly self-sufficient in food, in the short term; education is now promoted for young girls as well as boys (though classes are gender-segregated); and among the minority that attend high school and college, women now outnumber men, a fact that will surely influence future politics. Increasing numbers of women work outside their homes, usually in work spaces secluded from men. Despite all this apparent progress, economic and social development for the great majority of peninsula citizens has been slow.

The six smaller nations on the perimeter of the Arabian Peninsula, now ruled by ancestral clans that resisted the Saudi family expansion in the 1930s, have varying profiles. Kuwait and the United Arab Emirates, each with close to 10 percent of the world's oil reserves, are rapidly modernizing and are very affluent; their societies remain conservative. The modest oil and gas reserves of Oman and Qatar are sufficient to generate relatively high standards of living. The emir of Qatar and his wife are generous patrons of education and openly encourage social reform. Bahrain, virtually a city-state, generates income not from its tiny oil reserves but from services related to oil production and transport and from providing entertainment, shopping, and manufactured goods for neighboring wealthy Saudis. Yemen occupies about one-quarter as much space as Saudi Arabia but has a population nearly as large (19 million); it has only small and as yet undeveloped oil reserves. Yemen's standards of education and of living remain by far the lowest on the peninsula and in the entire region (see Table 6.2, page 318).

PERSONAL VIGNETTE   Raufa Hassan is 38 years old and holds a Ph.D. in social communications from the University of Paris. As Yemen's most outspoken feminist, her goal is to help women learn how to vote independently; Yemen's Islamist party, Islah, supports her efforts. Despite having the right to vote, Yemeni women do not participate in political processes. Nearly 70 percent of the population is still rural, and in rural areas only 1 in 10 women is literate. Typically, girls stay home and work rather than go to school. Women perform many essential tasks in the rural economy, such as herding cattle, grinding wheat, and carrying water. After marriage, the average woman bears seven children. Dr. Hassan has found that husbands generally keep their wives' and daughters' voter registration certificates because both men and women believe the women would be likely to lose them. As a result, men often control whether or not, and how, a woman votes.

Sheik Ahmed Abdulrahman Jahaf, whose daughter is running for parliament, worries that what he calls the "backward villagers" will think ill of her. He believes that an educated woman such as his daughter is more attractive as a bride. According to Sheik Jahaf, "The nature of women leads us to the conclusion that a woman's right place is home." In these waters, Dr. Hassan wades carefully, always getting the sheiks' permission before she talks to the village women; but lately the sheiks, too, seem to be changing their views. They see that they must begin to respond to the demands of their

people, both men and women, if they wish to stay in power. They even recognize that if they support the women's right to vote and encourage them to do so, the sheiks' own sons may profit in future elections when female voters are more numerous. [*Adapted from Daniel Pearl, "Yemen steers a path toward democracy with some surprises," Wall Street Journal (March 28, 1997): A1, A11.*]

THE EASTERN MEDITERRANEAN

The part of Southwest Asia and North Africa now occupied by Jordan, Lebanon, Syria, and Israel (Figure 6.21) has been torn by conflict for much of the twentieth century (see discussion on pages 309–312). Yet if tensions were resolved, this area has the potential for economic leadership in the region. One state, Israel, is already technologically advanced and broadly prosperous. Another, Lebanon, was similarly developed and prosperous until disrupted by a civil war between Christians and Muslims in the 1970s. This war ended in 1990, and since then the country has been on an active path of rebuilding its economy and infrastructure. The countries of this subregion are strategically located adjacent to the

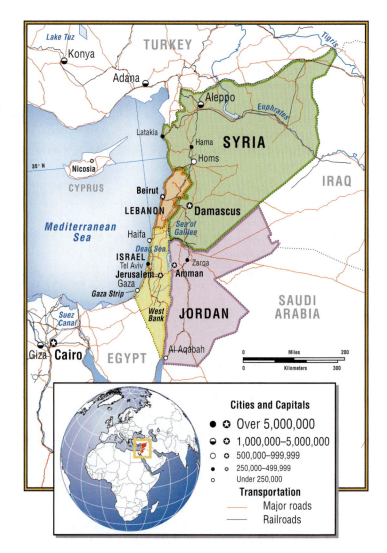

FIGURE 6.21 The Eastern Mediterranean subregion.

potentially lucrative markets of Central Asia and the Gulf states and the rich markets of Europe. Their climate makes them suitable for tourism and semitropical agriculture. Jordan already exports vegetables, citrus fruits, bananas, and olive products; if irrigation water can be found (not an easy matter; see the discussion on desertification and river damming, pages 315–317), Syria could do likewise. Syria lost an important trading partner after the breakup of the Soviet Union; but since then (1991), private investors, especially from the Gulf states, have been helping Syria expand its industrial base to include pharmaceuticals, food processing, and textiles, in addition to gas and oil production.

Jordan, Lebanon, Syria, and the Palestinian people (the Palestinians have no actual political state but live within these three countries and Israel) have all been preoccupied over the last 50 years with political and armed conflict arising out of the establishment of the state of Israel, but encompassing wider issues of access to land and resources and religious rivalry. The Arab–Israeli conflict spilled over Lebanon's borders in the 1970s and exacerbated discord between Christian and Muslim Lebanese. Syria has conflicts with most of its neighbors, notably with Iraq and Turkey over water rights (see page 316) and with Israel over possession of the Golan Heights. The core of the conflict, however, centers on Israel and the Palestinians.

Israel has the most educated, prosperous, and healthy population in the region; a pool of unusually devoted immigrants; and financial resources contributed by the worldwide Jewish community. A range of sophisticated industries are located along the coast, between the large port cities of Tel Aviv and Haifa. Israeli engineering is world renowned, and Israeli innovations in cultivating arid land have spread to the Americas, Africa, and Asia. Some people argue that Israel could be a model of development for neighboring countries because it has managed to develop despite rather meager resources, but this possibility has been hindered by Israel's continual conflict with its neighbors. The years of politically fueled violence have turned Israel into an armed fortress and have discouraged Israelis to the point that about 12 percent of the citizenry have left and returned to Europe or the Americas or Australia.

THE NORTHEAST: TURKEY, IRAN, AND IRAQ

Turkey, Iraq, and Iran (Figure 6.22) are culturally and historically distinct. For example, a different language is spoken in each country: Farsi in Iran, Turkish in Turkey, and Arabic in Iraq. Yet they share some similarities. At various times in the past, each country was the seat of a great empire. Each was deeply affected by the spread of Islam, which is now the religion of more than 97 percent of the region's people. At present, each occupies the attention of Europe and the United States because of location, resources, or potential threats. All three countries share some common concerns, such as how to allocate scarce water and how to treat the large Kurdish population that occupies a zone overlapping all three countries. Moreover, each country has experienced a radical transformation at the hands of idealistic reformist governments, although the paths pursued have varied dramatically.

Turkey

Turkey is more closely affiliated with Europe and the West than any other country in the region, with the possible exceptions of Israel and Lebanon. Turkey has been a strong ally of the West for most of twentieth century, and a strong faction in Turkey seeks to join the European Union. Turkey's affinity to Europe is increased by the fact that many Turks have spent years in Europe as guest workers. Thus far the official reasons that Turkey has not been accepted into the EU are that it has not sufficiently marketized its economy and installed constitutional human rights guarantees, including freedom of religion and protections for minorities, such as the Kurds. Less publicly stated are European worries about absorbing a predominantly Muslim state into (at least nominally) Christian Europe, and recently generated worries about Islamic fundamentalists and the security risk they would pose, were Turkey legally a part of Europe.

Turkey was once the core of the Ottoman Empire, which, as discussed earlier, was dismantled after World War I. At the end of a civil war, in 1923, Turkey undertook a path of radical Europeanization, led by a military officer, Mustafa Kemal Atatürk. Kemal Atatürk is revered as the father of modern Turkey. He and his followers declared Turkey a secular state, modernized the bureaucracy, encouraged women to discard the veil and the custom of seclusion, promoted state-sponsored industrialization, and actively sought to establish connections with Europe. Dramatic transformations took place throughout the twentieth century, but backlashes and simple inertia have been obstacles to change. Although the state has sponsored industrialization and modernization in western

Sheik Rashid Terminal in Dubai's international airport is a futuristic, high-tech facility with clear glass exterior walls—an expensive building to air-condition in the hot, dry climate. On a stopover in June 2000, the authors observed that Filipino young people staffed the dozens of upscale duty-free shops where every imaginable luxury is available—gold jewelry is sold by the ounce, not the piece. Other personnel were from India and Bangladesh. The tall palm trees and other plants that decorate the structure are artificial, thus requiring no water, a particularly scarce commodity in this Arabian Peninsula country. [H. Rogers /TRIP.]

FIGURE 6.22 The Northeast subregion and the Kurds. [Adapted from Edgar O'Ballance, *The Kurdish Struggle 1920–1994* (New York: St. Martin's Press, 1996), p. 235.]

Turkey, much of eastern Turkey remains agricultural and relatively poor. Islam, long deemphasized as a matter of state policy, is still an overriding influence on daily life, and fundamentalist versions of the religion are experiencing a resurgence.

Turkey straddles the Bosporus, a narrow passage from the Black Sea to the Mediterranean that is described (with exaggeration) as the separation between Europe and Asia. During the cold war, Turkey's location made it a strategic member of the North Atlantic Treaty Organization (NATO), an alliance of European countries and the United States whose mission was to contain Russia. In the post-Soviet era, Turkey's location gives it potential advantages if it can establish economic links with Europe, the Caucasus, and Central Asia. Istanbul, already a booming city of more than 10 million, is now the regional headquarters for hundreds of international companies. There are more than 250 U.S.-based companies alone, virtually all of them joint ventures of one kind or another: Coca-Cola, IBM, Eastman Kodak, Kraft Foods, Levi Strauss, and Citibank, to name a few. Turkey offers a large market of potential consumers: 59 percent of Turkey's 71 million people are city dwellers. Many of them once lived in rural areas of Turkey, then traveled to Europe to work in factories and on farms; they have returned eager to pursue a standard of living similar to what they observed in Europe. Also, remittances from Turkish guest workers in Europe add substantially to the country's GDP. The landscape of western Turkey is now dotted with large new houses built with remittances.

Turkey's future depends on its ability to manage relatively abundant water resources. With its mountainous topography, Turkey receives the most rainfall of any country in the region, and the headwaters of the economically and politically significant Tigris and Euphrates rivers are in the mountains of southeastern Turkey. As new irrigation and power generation systems on the Euphrates come on line (see the box "Turkey's Anatolian River Basin Project," page 316), production of cotton, grains, fruits and vegetables, soybeans, and seed oils will increase dramatically. Agriculture employs 40 percent of Turkey's workforce and accounts for 12 percent of its GDP (mostly exports of cotton and luxury food items).

The cheap hydropower will also expand industry, which already accounts for 80 percent of Turkey's exports. Despite a lack of electrical power, the country's diversified economy manages to produce a wide range of goods, including steel, textiles, leather goods, cement, automobiles, and tires, all of which are exported to Europe and Central Asia. Hydropower plus Turkey's already well-trained workforce could spur yet more industrial growth. Turkey's plans to appropriate a lion's share of the flow of the Euphrates River, however, are a source of conflict with its neighbors to the southeast.

Perhaps the greatest obstacle to Turkey's stability is its ongoing conflict with the Kurdish minority. The Kurds are tribal peoples (some are nomadic) who have lived in the mountain borderlands of Iran, Iraq, and Turkey (see Figure 6.22) for at least 3000 years. The division of the Ottoman Empire after World War I by France and Britain dispersed Kurdish lands among Turkey, Iran, Iraq, and Syria, leaving the Kurds without a state of their own. All four countries have had hostile relations with their Kurdish minorities. In Turkey, the conflict has escalated to the point of nearly continuous armed strife and repression of those Kurds who are noncombatant. The war in Iraq has added to the conflict, in part because the loyalties of the transnational Kurdish community are questioned by Turkey, Iraq, Syria, Iran, and the United States.

Iran

Iran occupies a transitional geographic position in this region: Iran's western parts are occupied by people of Arab, Turkish, Kurdish, and Caucasian heritage, its eastern parts by people with roots in South Asia, especially Afghanistan and Pakistan. Because its northern flank borders the Caspian Sea, Iran shares in the debate over the future of this inland sea and its resources, especially water and oil. Iran is also close to Central Asia, and its border regions have common language and ethnic features with Turkmenistan. During the cold war, Iran was a zone of contention between the United States and the Soviet Union.

Like Turkey, Iran has abundant natural resources, a strategic location, and a large population (66 million) that could make it a regional economic power. Iran's large petroleum reserves give it influence in the global debate on energy use and pricing. Yet social and economic turmoil have long held the country back.

Iran's present situation in the region and in global politics results from its geographic location, its recent history, and its status as a theocratic state based largely on Islamism. The theocracy was formed under the leadership of the Islamic fundamentalist Ayatollah Ruhollah Khomeini, a Shi'ite spiritual leader who, in 1979, led a revolution against Shah Reza Pahlavi. Pahlavi's father had seized control of the country in the 1920s and introduced secular and economic reforms patterned after those instituted by Kemal Atatürk in Turkey; he abdicated in favor of his son in 1941. In 1953, the United States helped Shah Reza Pahlavi retain power during a reformist coup attempt. Through U.S. support Pahlavi continued his father's efforts at Europeanization. But his emphasis on military might and on royal grandeur paid for with oil money overshadowed any genuine efforts at agricultural reform and industrialization. The disparity of wealth and well-being in Iran grew ever wider. Finally, with the Khomeini-led Iranian revolution of 1979, political conditions in Iran changed rapidly and radically. Resistance to the new theocratic state was crushed by massive imprisonments and executions. Among those most affected were women. Many upper-class women had lived emancipated lives under the shah's reforms, studying abroad and returning to serve in important government posts. Now all women past puberty had to wear long black chadors, and they could not travel in public alone, drive, or work at most jobs. Yet even some highly educated women supported the return to seclusion, seeing it as a way to counter the unwelcome effects of Western influence.

The turmoil of revolution was characterized by several episodes when Westerners were taken hostage and by general resentment against the West, which in turn led to decades of isolation and conflict. During the 1980s, Iran engaged in a devastating war with Iraq (see the discussion below). As of 2001, Iran began to change both internally and in its relationships with the outside world. Iran sought to rejoin the community of nations by liberalizing internal policies and announcing an end to its active nuclear armament program (the sincerity of this announcement is still debated at the present time). Punitive economic sanctions imposed by the United States and Europe in response to Iran's suspected funding of terrorism in the West were eased for a while; but recently Iran has gained the reputation of being a dangerous, emerging nuclear power deserving of sanctions by the United Nations.

Although many essential products must be imported, the government has begun to work toward economic diversification and industrialization. The country's out-of-date transport system is being updated so that Iran can begin again to participate in trade with its Arab and Central Asian neighbors. Agriculture remains Iran's weakest economic sector: even though it accounts for 25 percent of the GDP and employs one-third of the workforce, Iran must import a large part of its food.

Elections in May 1997 brought the landslide victory of a moderate leader, Mohammad Khatami. His election success in 1997 and again in 2001 is credited to women and young voters who were attracted by his subtle hints that he favored liberalizing reforms. Because of strong conservative resistance, however, the reforms have come slowly and sporadically. In January 2004, the conservative clerics who made up the Council of Guardians and who had final say over who can stand for election, rejected more than 8000 would-be candidates. Hundreds of liberal and moderate candidates were blocked, including many sitting legislators. Protests, increasingly common in Iran especially among university students, were subdued but sufficient to attract the notice of the European Union, to whose disapproval the Grand Council is somewhat sensitive. Measures were taken to partially address the demonstrators' complaints. But in the end, the liberal and moderate candidates were not restored to the ballot and thus were not elected.

Partly because of Iran's location at the borders of several regions undergoing social and economic transitions (Russia and the newly independent states, Southwest Asia, and South Asia), it has the unhappy distinction of being the country with the largest number of international refugees—more than 2.21 million in 2004. More than 600,000 are Iraqis, mostly Shi'ite Arabs from southern Iraq who fled the regime of Saddam Hussein and the Iraq wars; about a quarter of this group (50,000) are Kurds who fled discriminatory treatment in Iraq and Turkey; and at least 1.3 million are Afghans who left during the Soviet invasion of Afghanistan, the subsequent Taliban regime, and most recently the U.S. occupation. For the most part, Iran has at least temporarily absorbed all these people, and only the Iraqis tend to live in refugee camps. Many Afghans are employed at low wages doing the undesirable jobs of curing leather, cleaning wool, and digging ditches. Still, as of 2004, Iran was seeking to repatriate as many refugees as possible, hoping to lower the 15 percent unemployment rate in Iran. In July 2004, 80 Kurds were repatriated to northern Iraq via bus.

Iraq

Iraq, home to one of the earliest farming societies on earth and to the Babylonian Empire of biblical times, was carved out of the dying Ottoman Empire after World War I by Britain and the Arab sheiks. Iraq remained a British mandate until 1932, when it became an independent monarchy. Although about half of the Iraqi people are Shi'ite Muslims, since 1932 the government and most of the wealth have been controlled by non-Kurdish Sunni Muslims. Like many governments in the region, the Iraqi monarchy maintained strong alliances with Britain and the United States and gradually lost touch with its people. A tiny minority monopolized the wealth generated by increasing oil production but invested little in development. The Shi'ites, who live in the south where the oil is located, benefited little from oil earnings.

In 1958, a group of young officers, including Saddam Hussein, overthrew the government and created a secular socialist republic that prospered over the next 20 years despite occasional political disruptions. Hussein founded Iraq's secret police; he assumed leadership of the Ba'ath Party and eventually the presidency of Iraq. Large proven oil reserves (second in size only to those of Saudi Arabia) became the basis of a very profitable state-owned oil industry, and the profits from oil financed a growing industrial base. Agriculture on the ancient farmlands of the Tigris and Euphrates alluvial plain also prospered, even as environmental problems mounted. The proceeds from oil and agriculture produced a decent standard of living for most, who benefited from excellent government-sponsored education and health-care systems, but corruption and political repression increased.

Like several other countries in the region, Iraq has been in a prolonged state of crisis for several decades. Once an oil-rich socialist state, the country has experienced the devastation of multiple wars: the Iran–Iraq war (1980–1988), the Gulf War (1990–1991), and the U.S.-led war in Iraq (2003–present). Ten years of severe economic sanctions imposed by Europe and the United States after the Gulf War crippled the country, and as hardship increased, hundreds of thousands of Iraqis—including an estimated 250,000 children—died due to poor nutrition, poor health-care, and violence instigated by President Saddam Hussein against his own people. By 2000, European countries were lifting sanctions and Iraq was tacitly allowed to sell some oil to buy necessities, but U.S. sanctions remained in place and then, in 2003, war was launched by the United States against Iraq because it was erroneously believed to be a center of Al Qaeda–linked terrorism and dangerous weapons.

Most of Iraq's people (25 million in all) live in the area of productive farmland in the country's eastern half, on the alluvial plain of the Tigris and Euphrates rivers. Sunni Muslims (about 7 million) reside in the northern two-thirds of the plain, around the capital of Baghdad. About 4 million Kurds, who are also Sunnis, live in the northern mountains in the border regions with Turkey, Syria, and Iran. The southern third of the country is occupied by at least 11 million Shi'ite Muslims (see explanation of Shi'ite Islam on page 298) concentrated around Basra at the head of the Persian Gulf. Iraq's main oil fields are also in the south. Friction between Sunnis and Shi'ites has been particularly strong under Saddam Hussein and since his ousting. The Sunni Muslim minority monopolized political power for many decades and it is they who reaped most of the profits of Iraq's oil, despite its location in the Shi'ite south.

The Iran–Iraq War (1980–1988). In 1980, the Iraqi president, Saddam Hussein, invaded Iran, with the intent of acquiring territory along the Shatt al Arab estuary and the fertile lowlands between the Zagros Mountains and the Tigris River. Hussein cal-

culated correctly that other countries would not come to Iran's aid. The historian Bernard Lewis explains that there was an ethnic dimension to this war: Arabs against Persians; a sectarian dimension: Sunni Muslims versus Shi'ite Muslims; and an economic dimension: control over oil extraction and sale. It was also a war between the newly revived traditionalism of Iran and the secularism of Iraq. And it was a geopolitical war: a battle over territory and regional domination. A further geopolitical dimension was that the United States and Arab countries, thinking of Iraq as a buffer zone against the spread of Islamism from Iran, had financed Saddam Hussein's regime and equipped Iraq with sophisticated weapons. But Hussein, expecting a relatively quick victory, had not realized the extent to which Iranian leaders were willing to use the bodies of their religiously inspired young men against the war machine of Iraq. Iran lost hundreds of thousands of people, many of them young boys barely past puberty. Eventually, with heavy financing from wealthy Arab neighbors and with U.S. military and intelligence support, Hussein forced Iran to the peace table. Hussein himself emerged in a marginally better position, but the infrastructures and oil industries of both countries were left in serious disrepair.

The Gulf War (1990–1991). Just two years after the end of the war with Iran, Iraq, using the weapons and knowledge garnered from the West, invaded Kuwait and initiated the **Gulf War.** Kuwait had been unwilling to restrain its oil production in accordance with OPEC guidelines, thereby driving down the price of oil in global markets at a time when Iraq was badly in need of oil revenues to fund its recovery from the war with Iran. It is believed that Iraq's need to boost oil prices was a major factor in the invasion (Iraq also accused Kuwait of stealing Iraqi oil by slant-drilling across the border); but Iraq was also interested in acquiring more territory along the Persian Gulf. Iraq's decision to invade Kuwait ultimately proved disastrous, because it brought retaliation from a coalition of European and Arab countries led by the United States in an operation known as Desert Storm. The result was resounding defeat of Iraq and its withdrawal from Kuwait. More than 100,000 Iraqis died in combat, and 300,000 were wounded. Iraq lost 4000 of its 4230 tanks, whereas the coalition lost just 4 tanks. The United States suffered 148 combat deaths, 121 a result of accidents, and 458 wounded.

Despite its victory, the United States was criticized throughout the region for having rejected several opportunities to solve the conflict peacefully. In the years following the war, Iraq's refusal to give up its lethal chemical weapons, thought by some to be capable of killing on a massive scale, led the UN to impose the most crippling economic sanctions ever leveled against a country. The resulting shortages of food, medicine, and other necessities in Iraq caused massive deaths of civilians, especially children, one in five of whom became malnourished.

The long war with Iran and the subsequent Gulf War impoverished Iraq. According to European private aid agencies that began to enter the country in 2000, the economy remained weak as a result of international sanctions imposed by the United States and Europe after the Gulf War. These measures stopped nearly all sales of the country's oil, and the ensuing economic depression hit the poor, the ill, the elderly, and children the hardest. But even doctors and lawyers, unemployed because of empty state coffers, worked as street vendors. Meanwhile, smugglers formed a new economic elite, many of them being desert people with a knowledge of ancient transport routes into Jordan and Syria. Another group of the newly rich bought up state-owned industries at bargain prices. Money from such sales, political control through extreme repres-

sion of the general populace, imprisonment and execution of enemies enabled the regime of Saddam Hussein to survive until 2003, when the United States invaded again.

For a discussion of the war against Iraq waged by the United States in 2003, see page 309.

Quick Review

1. In Sudan, there are two civil wars in progress. What are the prime issues in each war?

2. What was the reason for returning land to the urban aristocracy in Egypt?

3. How did religion enter into the consolidation of power by the Saud royal family in Saudi Arabia?

4. What are the major sources of fresh water in this region, and how and why do they generate conflict?

REFLECTIONS ON NORTH AFRICA AND SOUTHWEST ASIA

The region of North Africa and Southwest Asia extends across portions of two continents yet demonstrates considerable spatial cohesion. People in this part of the world face common environmental problems defined primarily by water scarcity; settlement patterns reflect the necessity to live close to the few water and moisture sources available. The people also share a common religion: Islam. Although interpreted differently from place to place, Islam nonetheless unites by virtue of its Five Pillars of Islamic Practice. In all parts of the region, there is tension between strict adherence to Islam and more secular ways of life. And gender roles are at the heart of social debate everywhere, because these roles raise such basic questions about how individuals will relate to family and society.

Additionally, most of the region has endured some form of foreign domination over the last several hundred years; and it is presently in transition from outside control to regional or local systems of governance and economic development. Although not united since the Ottoman Empire, the 21 countries described in this chapter share long histories that have intersected repeatedly and often violently. The success of OPEC over the last 30 years has led to other efforts to set up associations that work for com-

mon goals. Although the nations of North Africa and Southwest Asia are far from achieving economic or political union, the possibilities are occasionally discussed within the region.

Just how long the countries of this region will remain in their present relationships is debatable. To some extent, Iran is now reaffirming ancient connections with Central Asia, Afghanistan, and Pakistan. Turkey may become more closely associated with Europe if secular forces dominate; or it may be aligned with Central Asia and Iran if Islamist forces gain power. Though it seems unlikely just now, North Africa might begin to see its future as more closely linked with European countries rimming the Mediterranean, a scenario envisioned by those in southern Europe who see possibilities in organizing countries bordering the Mediterranean as an economic and environmental unit (see Chapter 4).

The trends to look for in the news about this region are a growing environmental movement, political maneuverings over water rights, efforts to develop an educated citizenry capable of participatory democracy, a decline in authoritarian governments juxtaposed with other efforts to establish Islamist governments. Birth rates can be expected to continue in decline as gender roles are slowly but relentlessly redefined. As always, the geographic patterns of these trends will be uneven but related to earlier social and physical patterns.

Chapter Key Terms

Allah, p. 289

cartel, p. 306

Christianity, p. 289

desertification, p. 317

Diaspora, p. 289

economic diversification, p. 308

Fertile Cresent, p. 288

Gulf War (1990–1991), p. 330

hajj, p. 290

intifada, p. 310

Islam, p. 289

Islamic fundamentalism (Islamism), p. 312

Islamists, p. 304

Israeli–Palestinian conflict, p. 309

Judaism, p. 289

monotheistic, p. 289

OPEC (Organization of Petroleum Exporting Countries), p. 306

pogroms, p. 289

polygyny, p. 302

private spaces, p. 300

protectorate, p. 292

public spaces, p. 299

Qur'an (or Koran), p. 289

recharge area, p. 316

salinization, p. 307

seclusion, p. 292

secular states, p. 298

shari'a, p. 298

sheiks, p. 325

Shi'ite (or Shi'a), p. 298

spatial freedom, p. 302

Sunni, p. 298

terrorism, p. 312

theocratic states, p. 298

Zionists, p. 293

CHAPTER 7

SUB-SAHARAN AFRICA

"I never thought the sun could do all this!" exclaimed Melusi Zwane, the principal of Myeka High School, located two hours outside Durban, South Africa.

It was September 2001, and Principal Zwane was referring to the school's new photovoltaic solar panels, which can generate 2.4 kilowatts of power. They were installed in the late 1990s by Solar Electric Light Fund (SELF), a nonprofit group from the United States. SELF, in cooperation with Dell Computer and Infosat Telecommunications, set up the school with computers and a satellite uplink so that the students could have Internet access.

The school lies deep in the Valley of a Thousand Hills, in South Africa, where there is no power grid to supply electricity service (formerly all-white schools in similarly remote areas had electricity decades ago). Students at Myeka High used to read their lessons by candlelight and shared only a few textbooks among themselves. Poverty induced by years of discrimination under the white minority rule of apartheid resulted in less than a third of the students graduating, and those who did had little chance of employment other than agricultural labor. But that changed. By 2001, 70 percent of the students were graduating. The arrival of solar power as well as computers and connections to the Internet gave them access to a world of information and ideas that helped them pass the South African national exams. Being connected to the Internet also inspired the students to test their entrepreneurial skills and to make use of their education, either within the valley or in Durban.

A science teacher at Myeka High, S'busiso Madondo, and two 11th-graders got interested in biogas production, thinking that this form of energy and spin-offs from its production might address even more of the school's needs than solar power. Now waste from the school's toilets, as well as animal dung, are used to make methane, which powers all the school's electrical equipment. The spin-offs are impressive: sanitation problems are solved, and the manure fertilizes school gardens, which grow food for the many students who are AIDS orphans. Teachers and students alike mastered science, math, engineering, and management skills in order to make the system work.

Myeka High School's experience is a vivid illustration of how access to alternative energy sources can positively affect a society, deeply changing its economic and social dynamics. Two billion people, or roughly one-third of the world's population, live in rural areas that are not connected to the energy grid. [*Adapted from David Lipshultz, "Solar power is reaching where wires can't," The New York Times, Business Section, September 9, 2001; Myeka High School Web page (http://www.solarengineering.co.za/myeka_htm1.htm), 2004.*]

After success with solar power that made computers and Internet connection possible for their school, science teacher S'busiso Madondo and two 11th graders got interested in biogas as a source of energy for their school. Here, at left, the biogas digester is being constructed. The students and the people in

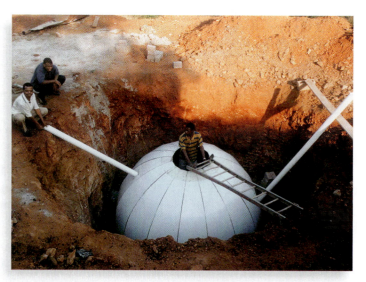

the rural school's community were amused by the idea of cooking and powering their computers with gas generated by cattle manure and human waste, but now they find it works fine. [(*left*) Courtesy of Solar Electric Light Fund. (*right*) Courtesy of Will Cawood and Richman Simelane.]

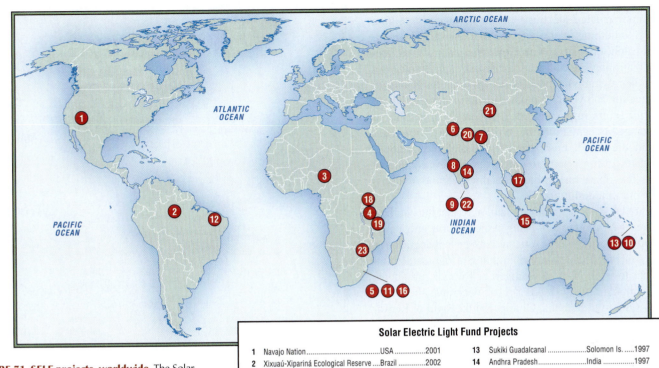

Solar Electric Light Fund Projects

1	Navajo Nation	USA	2001
2	Xixuaú-Xipariná Ecological Reserve	Brazil	2002
3	Jigawa State	Nigeria	Current
4	Gombe Reserve	Tanzania	Current
5	KwaZulu-Natal	South Africa	Current
6	Ladakh	India	Current
7	Phobjikha Valley	Bhutan	Current
8	Gaden Jangtse Tibetan Monastery	India	2002
9	Sri Lanka		Planned
10	Temotu Province	Solomon Is.	Planned
11	Myeka High School	South Africa	2000
12	Itapipoca, District of Marinheiro	Brazil	1998
13	Sukiki Guadalcanal	Solomon Is.	1997
14	Andhra Pradesh	India	1997
15	West Java	Indonesia	1997
16	Maphaphethe, KwaZulu-Natal	South Africa	1996
17	Mekong Delta	Vietnam	1996
18	Uganda		1995
19	Maasai	Tanzania	1995
20	Pulimarang	Nepal	1994
21	Gansu Province	China	1993
22	Sri Lanka		1991
23	Zimbabwe		1991

FIGURE 7.1 SELF projects worldwide. The Solar Electric Light Fund (SELF) helps rural villagers in developing countries improve their lives through clean, renewable energy and modern communications. SELF addresses the digital divide by using the latest in solar and wireless technology to show rural people how to leapfrog into a future based on sustainable energy sources. Those who formerly had to migrate far from home in search of economic and educational opportunities can now stay home and still enjoy life-changing opportunity. The map shows that as of 2004, SELF was active in five world regions. One of SELF's most successful projects, and one that has spawned numerous spinoffs, is that with Myeka High School, in Kwa-Zulu-Natal, located on the eastern shore of South Africa (see opening vignette). [*Source:* http://www.self.org/sou_africasolarschools.asp.]

MAKING GLOBAL CONNECTIONS

The story of Myeka High School illustrates some of sub-Saharan Africa's challenges in the aftermath of colonialism and extreme racial discrimination: poverty, lack of education, and lack of infrastructure needed for economic progress. But it also illustrates the promise of solutions to these problems.

Africa is a diverse continent, with great natural wealth and a long, rich history of peoples and their civilizations. It is the evolutionary birthplace of humankind, and it is one of the world's cultural hearths for agriculture, metalworking, and writing. The world's longest continuous religious and philosophical traditions are also found in Africa.

Sub-Saharan Africa, the region discussed in this chapter, is home to about 700 million people. Several of the fastest-growing economies in the world, and some of the world's richest deposits of oil, gold, platinum, copper, and other strategic minerals, are found here. Yet Africa is not often described in positive terms. The news that comes out of Africa is usually about disease, environmental devastation, war and genocide, political corruption, and poverty.

And, all too often, these reports accurately describe conditions in large areas of the continent.

It has been barely a full generation—less than 50 years—since the first sub-Saharan African country, Ghana, achieved independence from British colonial rule in 1957. During the era of European colonialism (1850s–1950s), wealth flowed out of Africa; and although colonialism officially ended with the granting of political independence in the 1950s, 1960s, and 1970s, wealth is still flowing out of Africa. Investors from the rich countries of the world continue to reap Africa's wealth by extracting minerals and other natural resources under unfair terms of trade—a practice called **neocolonialism.** An example is provided by the fishing industry of Guinea-Bissau, a country that ranks near the bottom on the United Nations Human Development Index (HDI) (166 out of 175). It lies on the west coast of Africa, adjacent to the rich Canary Current. In 1996, 97 percent of the fish caught in the waters of Guinea-Bissau were taken and sold abroad by high-tech foreign fishing fleets from China, Japan, Korea, Russia, the United Kingdom, Italy, Spain, and Portugal. The owners of these fleets paid just U.S. $11 million a year to Guinea-Bissau for licenses to take a catch worth U.S. $130 million annually on the global market. According to a European

Union report of late 2002, this general pattern persists. China pays next to nothing for its share of about half the total commercial catch, but as compensation it has set up some processing plants in Guinea-Bissau that employ local people.

Similar asymmetrical deals have been the rule throughout Africa for 500 years or more. It is not surprising, then, that Africa is now impoverished and often at war with itself. The average per capita income in sub-Saharan Africa is the lowest in the world. From time to time, ethnic conflicts, often arising over access to resources and ill-advised government policies, have turned the generally low standard of living into outright destitution.

But today, there are many hopeful signs that the conflicts can be resolved and the well-being of the majority enhanced. In recent years, a number of Africans have been recognized as among the best leaders and champions of democracy the world has to offer: Mary Balikungeri of Rwanda; presidents Seretse Khama and Q. K. Masire of Botswana, the writer Awes A. Osman of Somalia, Charity Kaluki Ngilu of Kenya, 2004 Nobel Peace Prize winner Wangari Maathai of Kenya, UN Secretary-General Kofi Annan from Ghana; South Africa's former president, Nelson Mandela; former president Julius Nyerere of Tanzania; and the late Nigerian environmentalist and writer Ken Saro-Wiwa, to name but a few. Since independence, African countries and societies have been engaged in determining the appropriate pathways to economic development and progress. Some of them are forging ahead; others are struggling to find a sustainable path.

THEMES TO EXPLORE IN SUB-SAHARAN AFRICA

Several themes are developed in this chapter:

1. **The roots of African poverty.** Sub-Saharan African countries rank among the poorest on earth. The roots of African poverty lie in past exploitative colonial relationships with European countries and in modern global trade patterns that cast Africa as a provider of cheap raw materials and low-cost labor. Colonialism officially ended in the 1950s and 1960s, but modern economic patterns perpetuate the circumstances of colonialism, especially the unsustainable use of resources and an increasing disparity of wealth.

2. **Disadvantageous position in the global economy.** The principal reason Africa interacts with the global economy only as a supplier of resources is because its pervasive poverty means that little is available to invest in upgrading the skills and earning power of its citizens. Large debt payments to governments and banks in the developed world take money from needed social services. Corruption, ethnic clashes, and religious conflict also inhibit economic development.

3. **Population dynamics and disease patterns.** Generally, Africa is not densely populated, but population growth rates are very high, and in some places densities are too high to be supported by

FIGURE 7.2 Regional map of sub-Saharan Africa.

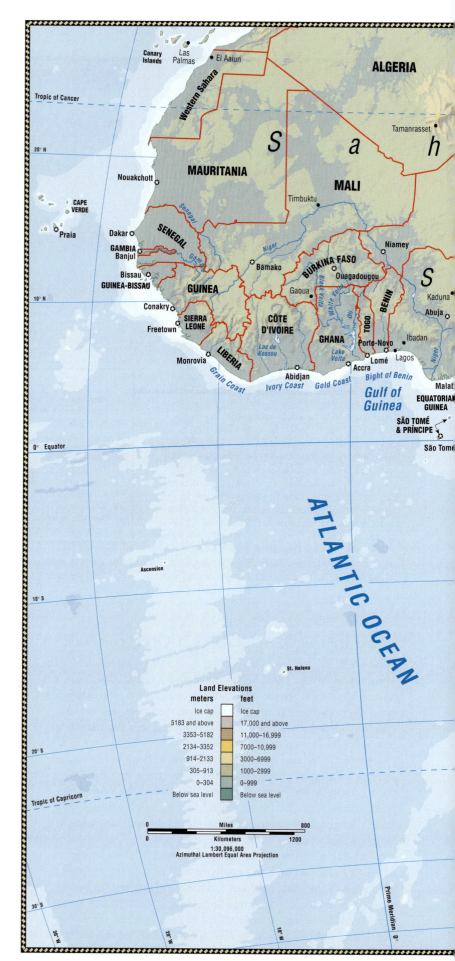

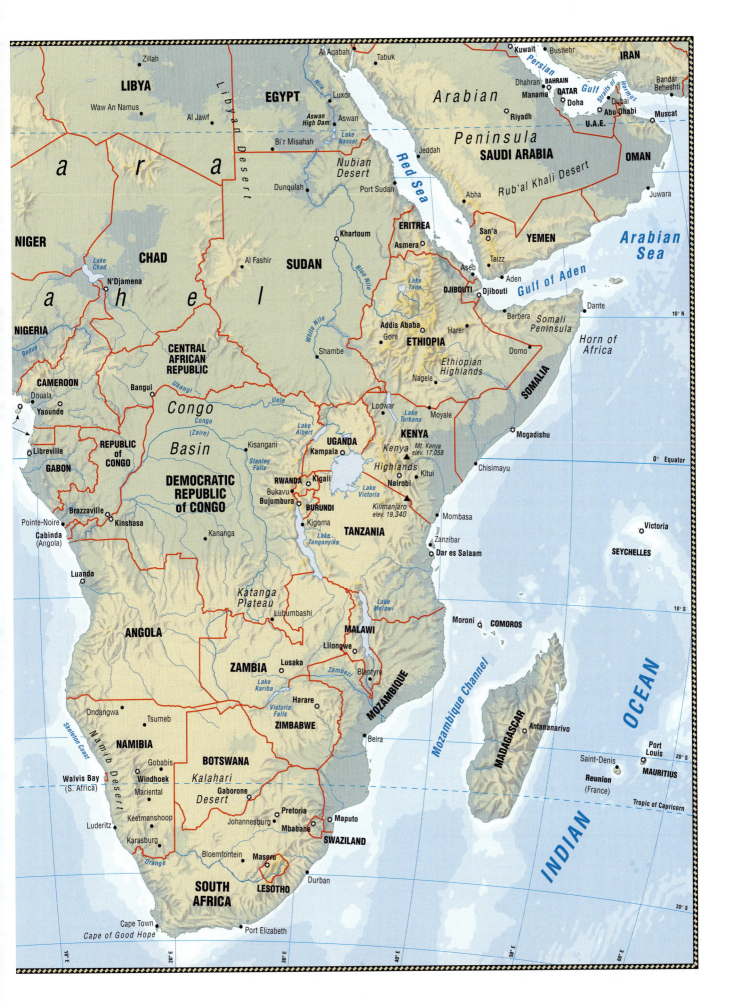

Zillah

LIBYA

Waw An Namus

a   *r*   *a*

NIGER

Lake
Chad

CHAD

a   *h*   *e*   *l*

N'Djamena

NIGERIA

Benue

CAMEROON

Douala
Yaounde

Libreville

GABON

REPUBLIC
of
CONGO

Pointe-Noire
Brazzaville
Kinshasa

Cabinda
(Angola)

Luanda

Ubangi

Congo
(Zaire)

Congo

Basin

DEMOCRATIC
REPUBLIC
of CONGO

Kisangani

Stanley
Falls

Kananga

ANGOLA

Katanga
Plateau

Lubumbashi

Ondangwa

Tsumeb

NAMIBIA

Walvis Bay
(S. Africa)

Skeleton Coast

Namib Desert

Gobabis
Windhoek

Mariental

Luderitz

Keetmanshoop

Karasburg

Orange

Kalahari
Desert

BOTSWANA

Gaborone

ZAMBIA

Lusaka

Lake
Kariba

Zambezi

Victoria
Falls

Harare

ZIMBABWE

Pretoria
Johannesburg
Mbabane

SWAZILAND

SOUTH
AFRICA

Bloemfontein
Maseru

LESOTHO

Cape Town
Cape of Good Hope

Port Elizabeth

Durban

Maputo

Beira

MOZAMBIQUE

Mozambique Channel

MADAGASCAR

Antananarivo

Saint-Denis

Reunion
(France)

Port
Louis

MAURITIUS

INDIAN

OCEAN

Tropic of Capricorn

10° E
20° E
30° E
40° E
50° E
60° E

30° S

20° S

10° S

Equator

0°

10° N

Al Aqabah
Tabuk

EGYPT

Luxor

Aswan
High Dam

Aswan

Nile

Bi'r Misahah

Lake
Nasser

Libyan Desert

Nubian
Desert

Dunqulah

Port Sudan

Khartoum

Al Fashir

SUDAN

White Nile

Blue Nile

Shambe

ERITREA

Asmera

Lake
Tana

Addis Ababa
Gore

ETHIOPIA

Harer

Ethiopian
Highlands

Nagele

Domo

Lodwar

Moyale

SOMALIA

Lake
Turkana

KENYA

Kenya
Highlands

Mt. Kenya
elev. 17,058

Kitui

Nairobi

Kilimanjaro
elev. 19,340

Mombasa

Zanzibar

Dar es Salaam

TANZANIA

Kigoma

Lake
Tanganyika

BURUNDI

Bujumbura

Bukavu

RWANDA

Kigali

UGANDA

Kampala

Lake
Albert

Lake
Victoria

Lake
Malawi

MALAWI

Lilongwe

Blantyre

Moroni

COMOROS

SEYCHELLES

Victoria

Chisimayu

Mogadishu

Horn of
Africa

Somali
Peninsula

Dante

Berbera

Aden

DJIBOUTI

Djibouti

Aseb

Gulf of Aden

Taizz

San'a

YEMEN

Abha

Jeddah

Red Sea

Rub'al Khali Desert

SAUDI ARABIA

Riyadh

Arabian

Peninsula

Arabian
Sea

OMAN

Juwara

Muscat

U.A.E.

Abu Dhabi
Dubai

Straits of
Hormuz

Bandar
Beheshti

IRAN

Busheyr

Kuwait

Persian

Gulf

Dhahran
BAHRAIN
Manama
QATAR
Doha

CENTRAL
AFRICAN
REPUBLIC

Bangui

Uele

local resources. Disease has always been a factor limiting African population growth, but now the HIV-AIDS epidemic is severely reducing growth, affecting social conditions, and slowing improvements in standards of living.

4. Distinctive gender roles. Across Africa's wide range of ethnic groups, it is common for men and women to have distinctly different roles. At least since colonial times, women have performed most of the labor connected with growing and preparing food, most local transport duties, and most marketing of surpluses; men have prepared fields for cultivation, taken responsibility for cash crop production, and often migrated far from their families to work in mines or urban factories.

5. Encouraging developments. Although many hurdles remain, many African countries are joining together to define the region's problems and design solutions for them. In many parts of the continent, the private sector, both informal and formal, is emerging as a vital force in the economy as direct government involvement in agriculture and industry decreases. Political reform remains a difficult process, but democratic elections are a reality in more than half of sub-Saharan African countries.

I THE GEOGRAPHIC SETTING

Terms to Be Aware Of

The language used to describe Africa has been particularly prone to **ethnocentrism,** the often subconscious belief in the superiority of one's own culture or ethnic group. There is now a movement to purge colonial terminology from descriptions of Africa because, too often, these terms are based on European misperceptions of cultures very different from their own. In this book, we use the terms **ethnic group** and **ethnicity,** instead of *tribes* and *tribalism,* to describe groups of people and the cultures they share.

Civilization is another term often misused in relation to Africa. Europeans considered only certain people to be civilized: the people of ancient China, Greece, Mesopotamia, the Indus Valley, the Maya and Inca empires in the Americas, and most of Europe since A.D. 1000. Meanwhile, Africa and many other places were judged to be uncivilized. Today, most scholars recognize the limitations of this European concept of civilization. Africa, for example, has had several ancient societies that fit common definitions of civilization. Moreover, Europeans often grossly misunderstood African cultural practices that they labeled "uncivilized." Furthermore, they seldom evaluated cultural practices in Europe and elsewhere closely enough to determine just how "civilized" they were.

The naming of African countries can often be confusing. There are two neighboring countries called Congo: the Democratic Republic of the Congo and the Republic of the Congo. Because these designations are both lengthy and easily confused, in this text the names will be abbreviated. The Democratic Republic of the Congo (formerly Zaire) will carry the name of its capital in parentheses: Congo (Kinshasa), as will the Republic of the Congo: Congo (Brazzaville). Check the regional map (Figure 7.2) to note the location of these countries and capitals.

PHYSICAL PATTERNS

The African continent is big—in fact, it is the second largest after Asia. At its widest point, it stretches 4000 miles (6400 kilometers) from east to west, and its length from the Mediterranean to its southern tip is nearly 5000 miles (8000 kilometers). But Africa's great size is not matched by its surface complexity. More than one-fourth of the continent is covered by the Sahara Desert, which comprises many thousands of square miles of what, to the unpracticed eye, appears to be a homogeneous landscape. Africa has no major mountain ranges, but it does have several high peaks, including Mount Kilimanjaro (19,324 feet [5890 meters]) just south of the equator and Mount Kenya (17,057 feet [5199 meters]) on the equator. Both have permanent snow and ice at their peaks and altitudinal zonation similar to that on mountains in the tropics of South America (see Figure 3.4 on page 119).

Landforms

Geologists usually place Africa at the center of the ancient supercontinent of Pangaea (see Figure 1.10 on page 23). Over the past 200 million years, several continents broke off from Africa and moved away: North America to the northwest, South America to the west, and India to the northeast. As Africa shed these continent-sized pieces, it readjusted its position only slightly, drifting gently northeast into Southwest Asia. Although Africa remains in approximately the same place on the globe's surface that it has occupied for more than 200 million years, it is still breaking up, especially along its eastern flank. The Arabian Plate has split away from Africa and drifted to the northeast, leaving the Red Sea, which separates Africa and Asia. The Great Rift Valley, another series of developing rifts, curves inland from the Red Sea and extends more than 2000 miles (3200 kilometers) south to Mozambique, near the east coast. At some future time, Africa is expected to break apart along this rift, which is a formation marked by chains of lakes, including two long and narrow ones, Lake Tanganyika and Lake Malawi.

The surface of the continent of Africa can be envisioned as a raised platform, or plateau, bordered by fairly narrow and uniform coastal lowlands and covered by an ancient mantle of rock in various stages of weathering. The platform slopes downward to the north. Thus the southeastern third of the continent is an upland region with several high peaks, whereas the northwestern two-thirds of the continent is a lower-lying landscape, interrupted only here and there by uplands and mountains (see Figure 7.2). Despite their lack of complexity, the landforms of Africa have obstructed transport and hindered connections to the outside world. Routes from the plateau to the coast must negotiate steep escarpments (long cliffs) around the rim of the continent, and the long, uniform coastlines have few natural harbors.

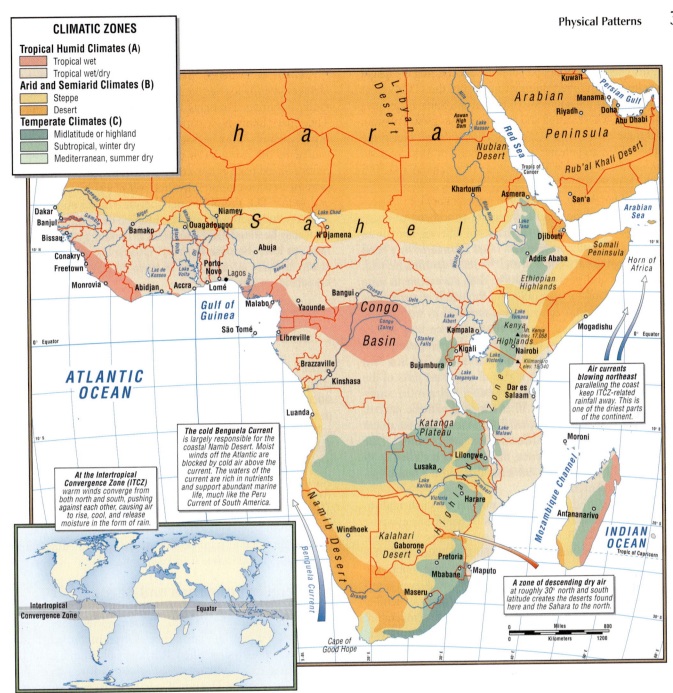

CLIMATIC ZONES

Tropical Humid Climates (A)
- Tropical wet
- Tropical wet/dry

Arid and Semiarid Climates (B)
- Steppe
- Desert

Temperate Climates (C)
- Midlatitude or highland
- Subtropical, winter dry
- Mediterranean, summer dry

At the Intertropical Convergence Zone (ITCZ) warm winds converge from both north and south, pushing against each other, causing air to rise, cool, and release moisture in the form of rain.

The cold Benguela Current is largely responsible for the coastal Namib Desert. Moist winds off the Atlantic are blocked by cold air above the current. The waters of the current are rich in nutrients and support abundant marine life, much like the Peru Current of South America.

Air currents blowing northeast paralleling the coast keep ITCZ-related rainfall away. This is one of the driest parts of the continent.

A zone of descending dry air at roughly 30° north and south latitude creates the deserts found here and the Sahara to the north.

Intertropical Convergence Zone   Equator

FIGURE 7.3 Climates of sub-Saharan Africa. Most of sub-Saharan Africa has a tropical climate. The inset at lower left shows the position and range of the intertropical convergence zone. [Inset adapted from Tom L. McKnight, *Physical Geography* (Upper Saddle River, N.J.: Prentice Hall, 1996), p. 125.]

Climate

Most of sub-Saharan Africa has a tropical climate (Figure 7.3) because 70 percent of the continent lies between the Tropic of Cancer and the Tropic of Capricorn. Average temperatures are generally high, staying above 64°F (18°C) year-round everywhere except at the more temperate southern tip of the continent and in **upland zones** (hills and plateaus). Seasonal climates in Africa differ more by the amount of rainfall than by temperature.

Most rainfall comes to Africa by way of the **intertropical convergence zone (ITCZ)**, a band of atmospheric currents that circles the globe roughly around the equator (see inset, Figure 7.3). At the ITCZ, warm winds converge from both north and south. These winds push against each other, causing the air to rise, cool, and release moisture in the form of rain. The rainfall produced by the ITCZ is most abundant in central and western Africa near the equator. There, in places such as the Congo Basin, the frequent rainfall nurtures the dense vegetation of tropical rain forests.

The ITCZ shifts north and south seasonally, generally following the area of the earth's surface that has the highest average temperature at any given time. The tilt of the earth's axis causes the most direct rays from the sun to fall in a band that sweeps into the Northern Hemisphere during its summer (June–September) and into the Southern Hemisphere during its summer (December–March). Hence, during the height of the Southern Hemisphere summer in January, the ITCZ might bring rain far enough south to water the dry grasslands, or steppes, of Botswana. During the height of the Northern Hemisphere summer in August, the ITCZ brings rain as far north as the southern fringes of the Sahara—an area called the **Sahel**, where steppe and savanna grasses grow. Poleward of both of these extremes, a belt of descending dry air blocks the effects of the ITCZ and creates the deserts found in Africa (and on other continents) at roughly 30°N latitude (the Sahara) and 30°S latitude (the Namib).

Across Central Africa, between the wet equatorial rain forests and the dry scrub of the deserts, there are almost mirror-image bands of ecosystems that reflect the bands of differing rainfall: in both the north and south, the true rain forest is bordered by a band of seasonally dry tropical woodland and the moist **savanna** beyond, where tall grasses and trees intermingle. Both environments have provided suitable land for agriculture for thousands of years. This banded pattern of African ecosystems is modified in many areas. Winds blowing north along the east coast of Africa keep ITCZ-related rainfall away from the **Horn of Africa**, the triangular peninsula that juts out from northeastern Africa below the Red Sea. As a consequence, the Horn of Africa is one of the driest parts of the continent. The Namib Desert of southwestern Africa is the southern counterpart of the Sahara in Africa's banded climatic pattern. Moist air from the Atlantic is blocked from moving over the desert by cold air above the northward-flowing Benguela Current. Like the Peru Current off South America, the Benguela is chilled by its passage past Antarctica. Rich in nutrients, it supports a major fishery up the west coast of Africa.

The climate of Africa presents a number of challenges to human beings. Insects and parasites that breed prolifically in warm, wet climates cause debilitating diseases such as river blindness, schistosomiasis, and malaria. In drier tropical climates, water for drinking, farming, and raising animals is often in short supply; the soils, lacking organic matter, are not particularly fertile, even if irrigated. In humid tropical climates, cultivated soil loses its fertility more easily than in temperate zones. Wherever both temperature and moisture are high, **organic matter** (the remains of any living thing) in the soil decays rapidly, and the nutrients it releases are quickly absorbed by surrounding plants or **leached** (washed) out into groundwater and runoff. In a standing forest, the organic matter shed every day decays and provides a continual source of nutrients for plant growth. But if the forest is removed, the source of nutrients is also gone, and the soil quickly deteriorates. The minerals are leached out, and soil particles wash downslope. The direct rays of the sun bake the soil into a permanently hard surface called **laterite.**

To maintain soil quality, over the ages cultivators in the wet tropics have developed a method of farming called **shifting cultivation.** They clear only small patches, an acre or two at a time. They use the cleared vegetation as "fertilizer," sometimes burning it to release nutrients. And they plant their gardens as a sort of miniature forest, with many species—often 20 or more per garden—of various sizes, shapes, and requirements that cover the soil quickly to prevent it baking hard in the hot sun. Because the soil loses fertility quickly, the small plots produce well for 2 or 3 years and are then allowed to revert to forest for several decades. Typically, commercial agriculture that depends on large cleared fields and long-term production has not succeeded for extended periods in tropical zones because of soil infertility, tropical diseases, and pests.

There is evidence that Africa has experienced repeated changes in climate over the last 20,000 years and that a new, drier cycle is starting. Droughts have been frequent since the 1970s. They may simply be a consequence of the natural cycle, or, as many scientists think, at least partly the result of global warming caused by the rising levels of greenhouse gases produced in the industrialized parts of the world. Scientists expect present climatic patterns in Africa to intensify: hot, wet places may become hotter and wetter; hot, dry places may become hotter and drier. Grasslands on the desert margins may become deserts; low-lying coastal areas may be inundated by seawater as melting glaciers at the poles raise sea level. Because so many Africans live in arid or low-lying zones, as many as one-fourth of them could eventually find their homelands uninhabitable.

HUMAN PATTERNS OVER TIME

Africa's rich past has often been misunderstood and dismissed by people from outside the region—a strategy that helped some outsiders to defend their profit-making ventures. For example, European slave traders and colonizers resisted evidence of the historical achievements of Africa's peoples, justifying their invasion of sub-Saharan Africa by calling Africa the Dark Continent and assuming it was a place where little of significance in human history had occurred. The substantial and elegantly planned city of Benin in West Africa, encountered by European explorers in the 1500s, never became part of Europe's image of Africa, nor did the city of Loango in the Congo Basin. Because of the pervasive influence of Europe, even today, most people are unaware of Africa's internal history or contributions to world civilization.

The Peopling of Africa and Beyond

Africa is the original home of the human species. It was in eastern Africa that the first human species evolved more than 2 million years ago, although they differed anatomically from human beings today. These early humans are known to have ventured far from Africa. Discoveries in Dmanisi, Georgia (between the Caspian and Black seas), in the late 1990s indicate that 1.7 million years ago, early humans, using primitive tools, were living there. Recently analyzed archaeological evidence recovered in Ethiopia shows that anatomically modern humans (*Homo sapiens*) also evolved in Africa, about 200,000 years ago. New research on cranial features of fossil skulls by Andrew Kramer of the University of Tennessee shows that by at least 90,000 years ago, modern humans had reached the eastern Mediterranean and were apparently interbreeding with earlier humans (commonly called Neanderthals). Evidence is mounting, then, that modern humans, like earlier humans, radiated out of Africa, spreading into Asia and eventually into Europe, possibly intermingling with other human populations they encountered, rather than replacing them.

Early Agriculture, Industry, and Trade in Africa

As already observed (in Chapters 1, 3, and 6), humans learned to cultivate plants and domesticate animals at various times after

10,000 years ago and in a number of places: the Fertile Crescent, the Americas, several parts of Asia, and Africa (see Figure 1.15 on page 29). In Africa, people began to cultivate plants as early as 7000 years ago in the southern belt of the Sahara Desert, which was less arid at that time. The highlands of present-day Sudan and Ethiopia in East Africa were another early center of plant and animal domestication and food production. The Bantu peoples seem to have brought plant cultivation to equatorial Africa when they migrated from West Africa to the Congo Basin within the last 2500 years. As early as 1500 years ago, descendants of people from both coastal West Africa and the highlands of East Africa were converging and migrating into Southern Africa, displacing local hunter-gatherers with their farming systems.

Political economist Samir Amin makes the point that before the modern era (A.D. 1500 or so), Africa was neither remote nor backward when compared to Europe or Asia. Africans had complex and varied social and economic systems. Trade routes spanned the continent as well as reaching around the Mediterranean to Rome and east to India and China. For example, the Kush people, who lived in the upper Nile River region 4000 years ago, served as trading middlemen between Africans in the interior of the continent and Egyptians. Gold, elephant tusks, and timber from Africa were exchanged with Mediterranean people for a variety of resources—brass, copper, iron (not yet then available in Africa), olive oil, animals, dyes, food, and manufactured goods such as fine leather products, ceramics, and textiles. In East Africa, in about 300 B.C.E., the kingdom of Aksum in what is present-day Ethiopia took over from the Kush and expanded the trade with Egypt and Assyria to include both Rome and India, and even China.

About 2500 years ago, Africans learned how to smelt iron (probably from the Assyrians in what is now Iraq), and thereafter iron smelting, and eventually steel production in carbon furnaces, spread slowly from northeastern Africa to the highlands of eastern Central Africa and to West Africa (steel was not invented in the West until the 1700s). The slow spread of ironworking is explained in part by the fact that the continent was only lightly populated. Not until about 1500 years ago did the practice diffuse more rapidly. By 1300 years ago, a remarkable iron-producing and trading civilization with advanced agricultural and mining technology had developed in the highlands of southeastern Africa in what is now Zimbabwe. Known now as Great Zimbabwe, this empire traded with merchants from Arabia, India, Southeast Asia, and China, exchanging the products of its mines and foundries for silk, fine porcelain, and exotic jewelry from Asia. Then, for reasons as yet little understood, the Great Zimbabwe empire collapsed. Nineteenth-century British explorers were mystified by their discovery of the ruins of impressive stone structures, tens of thousands of mine shafts, and even fragments of Chinese pottery. The European archaeologists of the time did not believe that indigenous African civilizations could have existed, let alone have had trading links to China, so they mistakenly credited the Great Zimbabwe culture to outsiders, not Africans.

In western Central Africa, several influential centers made up of dozens of linked communities developed in the forest and the savanna. One of several powerful kingdoms was the empire of Ghana (A.D. 700 to 1000), which was located not in the modern country of Ghana, but in present-day Mali north of the headwaters

The archaeological ruins of The Enclosure at Great Zimbabwe are surrounded by the ruins of numerous other structures. [Don L. Boroughs/The Image Works.]

of the Niger River. The Ghana Empire was succeeded by the kingdoms of Mali (1200 to 1400) and Songhay (1400 to 1500), with each succession or takeover enlarging the empire's territorial jurisdiction. The kingdoms' wealthy and prominent converts to Islam periodically sent large entourages on pilgrimages to Mecca, where their opulence was a source of wonder.

In both East and West Africa, Africans traded not only their ivory, iron, animals, and gold, but also fellow Africans. There was a long-standing custom in Africa of enslaving certain classes of people as a result of hostilities between two or more ethnic groups. The treatment of slaves within Africa, governed by local custom, was usually reasonably humane, but by no means egalitarian. At some unknown time in the past, but certainly long before Islam, a slave trade developed with Arab and Asian lands to the east (Figure 7.4). After the spread of Islam (700 A.D.), scholars generally agree that close to 9 million African slaves were exported to parts of Asia (Persia, India, Southeast Asia) and to Islamic areas around the Mediterranean. Protections afforded slaves in Africa were lacking when slaves were traded to non-Africans. Often male slaves were castrated before they were offered for sale to Muslim traders; although there was demand for eunuchs, Muslim law prohibited such treatment on Muslim soil. Despite this practice, numerous sources report that African slaves who worked in the homes and fields of wealthy Mediterranean Muslims were often treated as distant and lesser-status family members and not brutalized.

The Coming of the Europeans

The course of African history shifted dramatically in the mid-1400s, when Portuguese sailing ships began to appear off the West African coast. The names given to stretches of this coast by the Portuguese and other early European maritime powers reflected their interest in acquiring Africa's resources: the Gold Coast, the Ivory Coast, the Pepper Coast, the Slave Coast.

By the 1530s, the Portuguese had organized the slave trade with the Americas. The trading of slaves by the Portuguese, British, Dutch, and French was more widespread and brutal than either the small-scale taking of slaves that had occurred within Africa or

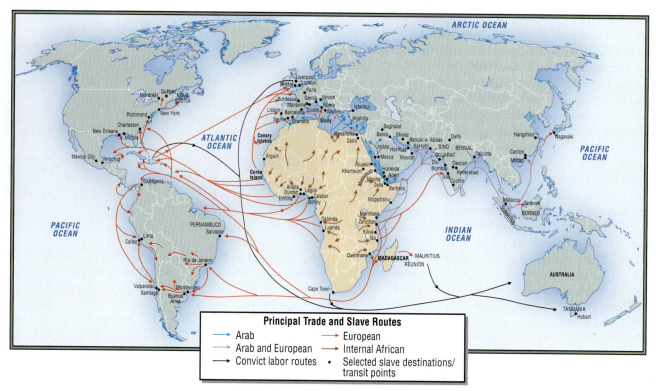

FIGURE 7.4 African diaspora resulting from the slave trade. [Adapted from the work of Joseph E. Harris, in Monica Blackmun Visona et al., eds., *A History of Art in Africa* (New York: Harry N. Abrams, 2001), pp. 502–503.]

the slave trade conducted with Muslims and Asians. Its most distinctive feature was the commercial motivation of the Europeans, who needed cheap labor for their American plantations, which in turn supplied raw materials and money for the industrial revolution in Europe. Slavery as managed by Europeans was more impersonal than slavery as practiced in Africa, and slaves were often treated strictly as a commodity.

To acquire slaves, the Europeans established forts on Africa's west coast and paid nearby African kingdoms with weapons, trade goods, and money to make slave raids into the interior. As in the pre-European slave trade, some of the captives were taken from enemy kingdoms in battle. Many more, however, were kidnapped from their homes and villages in the forests and savannas. Most slaves traded in the international market were male because the raiding kingdoms preferred to keep captured women for their reproductive capacities. From 1600 to 1865, about 12 million captives were packed aboard cramped and filthy ships and sent to the Americas. Up to a fourth of them died at sea. Those who arrived in the Americas went primarily to plantations in South America and the Caribbean; about one-fourth were sent to the southeastern United States (see Figure 7.4).

The European slave trade severely drained the African interior of human resources and set in motion a host of damaging social responses within Africa that are not well understood even today. It made Africans dependent on European trade goods and technologies, especially the guns used by raiding kingdoms. Unfortunately, enslavement of Africans by Africans persists into the present, as recent news stories, primarily from northern Central Africa, reveal.

Among those who were enslaved and shipped far from home, African culture nonetheless survived, and it continues to enrich the language, cuisine, religion, literature, music, agricultural technology, and art of other world regions, especially the Americas.

The Scramble to Colonize Africa

European involvement in Africa gradually increased, culminating in the establishment of formal colonies in the late nineteenth century. The British transatlantic slave trade officially ended in 1807, but other European nations continued trading in slaves until 1865. By that time, Europeans found it more profitable to use African labor *in Africa* to extract raw materials for Europe's growing industries. European interests extended inland to include the exploitation of fertile agricultural zones, areas of mineral wealth, and places with large populations that could serve as sources of labor.

Colonial powers competed avidly for territory and resources. To prevent the eruption of armed conflicts, it became necessary to establish official boundaries between the claimed African territories. The result was the virtually complete seizure and partition of the continent by the time of World War I (Figure 7.5). Africans were not consulted about the partitioning of their continent, and only two African countries retained independence: Liberia on the west coast, because it was populated by former slaves from the United States, and Ethiopia (then called Abyssinia) in East Africa, because its strong monarchy defeated early Italian attempts to colonize it. Otto von Bismarck, the German chancellor who convened the 1884 Berlin Conference at which the competing powers first

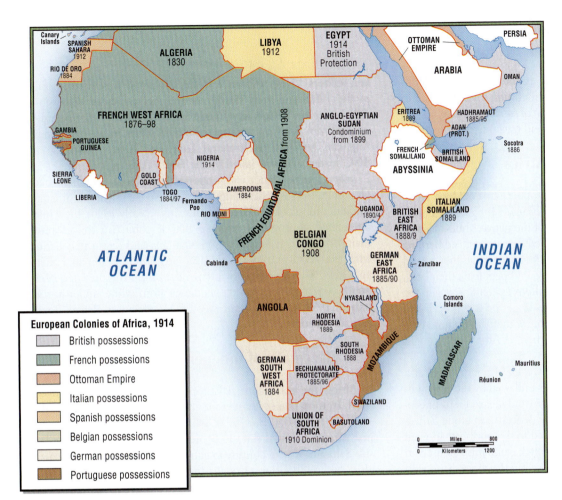

FIGURE 7.5 The European colonies of Africa in 1914. Some countries (those without dates) had been informally occupied by colonial powers for a few centuries. In the latter part of the nineteenth century, as European countries began to compete on a large scale for natural resources, cheap labor, and new markets, most expanded into Africa. The dates on the map indicate the years of officially recognized control by the European colonizing power. [Adapted from Alan Thomas, *Third World Atlas* (Washington, D.C.: Taylor and Francis, 1994), p. 43.]

gerrymandered Africa, declared: "My map of Africa lies in Europe." With some notable exceptions, the boundaries of most African countries today derive from the colonial boundaries set up between 1884 and 1916 by European treaties. These territorial divisions lie at the root of many of Africa's current problems.

There were a few basic geographic patterns in the European domination of Africa during the colonial period that can help us to visualize how the geography of Africa was changed by colonization. The geographer Robert Stock, a specialist in sub-Saharan Africa, identifies three such geographic patterns:

1. European settlers occupied land at relatively high densities in only a few places. These were mainly areas with especially attractive resources or places where Europeans considered the climate comfortable—such as the relatively cool highlands of Kenya in East Africa and the upland plateau of South Africa (see Figures 7.2 and 7.3). In these places, Europeans forced indigenous farmers and herders onto marginal land in reservations or made them labor on European-owned farms and plantations. Areas taken over by Europeans were privileged: taxes were kept low, and roads were constructed to provide ready access to local and foreign markets.

2. Africans continued to farm small plots in agricultural regions and in places considered disagreeable by Europeans. Some of these agricultural zones were quite densely occupied, but Europeans still exercised control. African farmers were often directed to switch from their traditional food crops to export crops. Some complied eagerly, in the hope of becoming wealthier; others changed grudgingly, only because they now needed cash to pay taxes to the colonial regimes and to buy food. Africans suffered malnutrition as more and more land was shifted from food to cash crops or livestock production. Despite the disruption of their economies, these indigenous farming areas received little assistance from Europeans. Transportation was only modestly upgraded, and Africans were often required to pay higher taxes than the Europeans who had taken over their land.

3. Remote areas that were difficult to exploit economically were treated as labor reserves from which young men were often removed and put to work on government projects such as railroad construction. They were separated from their families, subjected to dangerous working conditions, and paid very little. Because they were not given enough food or allowed the time to grow food or hunt their own, many died of malnutrition. In the Belgian Congo, during the reign of King Leopold of Belgium (1865–1909), as many as 10 million people starved. (In 1960, the Belgian Congo became an independent country that was called Zaire until the mid-1990s, when the name was changed to the Democratic Republic of the Congo.)

In all areas colonized by Europe, the main objectives of colonial administrations were to extract as many raw materials as possible, create markets for European manufactured goods, and keep the costs and the commitments of European-administered governments to an absolute minimum. It is useful to examine the case of South Africa to appreciate how the European incursion into Africa led to the expropriation of land, the subjugation of African peoples, and, in this case, the infamous system of **apartheid** (racial segregation).

The Colonization of South Africa

In the 1650s, the Dutch took possession of the Cape of Good Hope at the southern tip of Africa from the Portuguese. By 1673, Dutch settlers were expanding into the lands of the native KhoiKhoi people. The Dutch immigrant farmers, called Boers, practiced herding and farming techniques that used large tracts of land. As they spread into the interior, the Boers pushed indigenous people off their land, despite strong resistance. In 1713, a smallpox epidemic ravaged the remaining KhoiKhoi, thus clearing the stage for yet more European settlement. Lacking sufficient labor for their agricultural ventures, the Boers enslaved laborers elsewhere in Africa and brought them to work on their farms.

The British were interested in the wealth of South Africa and gained control of it in 1795. Slavery was outlawed by the British in 1834, and the Boers felt that this change would leave them in economic straits and unprotected from the Africans. To elude British control, large numbers of Boers migrated to the northeast in what was called the Great Trek (1835). They often came into intense and violent conflict with the Africans in these interior areas. European mining, using African labor, unearthed extremely rich diamond deposits in these areas in the 1860s, and in the 1880s, gold secured the economic future of this land, which became known as the Orange Free State and the Transvaal. Africans were forced to work in the diamond and gold mines under disastrously unsafe conditions and for minimal wages. They lived in unsanitary compounds that travelers of the time compared to large cages.

Britain, eager to claim the wealth of these mines, invaded the Orange Free State and the Transvaal in 1899, waging the bloody Boer War. The war gave them control of the mines briefly, until resistance by Boer nationalists forced the British to grant independence to South Africa in 1910—an independence that applied only to the white Boers and British of the country, a small minority. Black South Africans lacked legal political rights until 1994.

In 1948, apartheid laws were enacted to reinforce the longstanding segregation of Boer society. These laws required everyone except whites to carry passbooks and live in racially segregated rural townships, specific sections of cities, or workers' dormitories attached to mines and industries. Eighty percent of the land was reserved for the use of Europeans, who then made up just 15 percent of the population. Blacks were assigned to ethnicity-based "homelands" that were considered independent enclaves within the borders of, but not part of, South Africa. African people were treated similarly in other areas of the continent, such as Northern and Southern Rhodesia (modern Zambia and Zimbabwe), Tanzania, and Kenya.

The fight to end racial discrimination in South Africa began even before the apartheid laws were formally introduced in 1948. The African National Congress (ANC), the first and most important organization participating in this struggle, was formed in 1912 to work nonviolently for civil rights for black Africans. After the apartheid laws were passed, the ANC began to recruit more outspoken leaders. Overt resistance to the apartheid system intensified in the 1960s and grew steadily despite heavy-handed repression. Among the key leaders who fought for racial justice in South Africa were Steven Biko, an attorney who died while in police custody; Anglican Archbishop Desmond Tutu; and Nelson Mandela, an ANC leader who was jailed by the South African government for 27 years. The resistance was supported by millions of ordinary people: professionals (including a few outspoken whites), migrant workers, and schoolchildren. Finally, in the early 1990s, the white-dominated government realized that it could no longer resist majority rule. It released Nelson Mandela and began negotiating with the ANC and other political organizations for an end to apartheid. Though officially independent from Britain since 1910, South Africa gained its independence from white minority rule only in 1994, when resistance leader Mandela was elected South Africa's president.

The Aftermath of Independence

In Africa, the era of formal European colonialism was relatively short. In most places it lasted for about 80 years, from roughly the 1880s to the 1960s. In 1957, Ghana, in West Africa, became the first African colonial state to achieve its independence. The last sub-Saharan African country to gain independence was Eritrea, in 1993, although Eritrea won its independence not from a European power, but from its neighbor Ethiopia, after waging a 3-year civil war.

Colonial rule has continued to influence the postindependence history of Africa. Many independent African countries have struggled to find the most appropriate economic systems and forms of government. Like their colonial predecessors, most independent African governments became authoritarian, antidemocratic, and dominated by privileged and Europeanized elites. The road to nation building has been a rocky one for these countries, but most of them have been on this road for less than 40 years. Many countries remain economically dependent on their former European colonizers. In the last several years, however, pro-democracy movements have begun to spread across Africa, and by 2001, some 26 countries had democratically elected governments, up from 11 in 1970.

Africa enters the twenty-first century with a complex mixture of enduring legacies from the past and looming challenges for the future. Although Africa has been liberated from colonial domination, it is still strongly influenced by the aftermath of oppressive colonial policies and by neocolonialism exercised by the world's wealthy countries and multinational firms. Because of its reliance on exports and imports, it remains inextricably linked to the global economy, in which it has long operated at a disadvantage. Poverty is expanding rapidly as solutions to Africa's problems are slow in coming. In the recent past, Africa has faced declining economic productivity and rising debt; severe periodic drought and famine; major health problems, including the world's worst HIV-AIDS epidemic; and the challenge of having the fastest-growing population on earth. The brightest spot in Africa's future is that the recent decades of rapid and often wrenching change have made many Africans more willing to consider innovative alternatives to conventional solutions.

POPULATION PATTERNS

A look at the population density map of sub-Saharan Africa (Figure 7.6) will surprise many readers, who may have the erroneous impression that Africa is densely populated. In fact, the population is distributed very unevenly, but generally sparsely, over the continent. Only a few places exhibit the densities that are widespread in Europe, India, and China. Nonetheless, there are serious population problems in Africa. Some countries (for example, Rwanda, Burundi, and Nigeria) have pockets of very high density, but have not developed existing resources to support their people. A number of countries (for example, Chad, Liberia, Mali, Niger, and Madagascar) have annual population growth rates above 3 percent; the populations of these countries could double by 2030. High population growth rates make it difficult for already poor countries to supply adequate educational and health services to their increasing numbers of people. The result is that living standards decline.

Defining Density

The standard of living experienced by a population depends on a region's **carrying capacity**: the maximum number of people it can support sustainably with food, water, and other essential resources. Carrying capacities in Africa vary widely across the continent, and the factors that limit carrying capacity also vary from place to place. In some persistently dry places, such as areas bordering the Sahara Desert, the lack of water limits cultivation and grazing, which, in turn, limits human settlement. In some persistently wet places, such as the Congo Basin, the leached soils cannot sustain long-term cultivation (see page 338). Countries with alternating wet and dry seasons—such as South Africa, Kenya, and other countries in eastern Africa—can support fairly dense populations as long as people use the land and water sustainably. Some otherwise habitable African ecological zones, such as the high plains of the Serengeti southeast of Lake Victoria, are sparsely populated because they are breeding grounds for such diseases as sleeping sickness (trypanosomiasis), borne by tsetse flies, and malaria, borne by mosquitoes (see pages 345–346).

Ultimately, the carrying capacity of any place is affected by cultural, social, economic, and political factors as well as by physical features. For example, in most of Africa, people must depend largely on the local agricultural carrying capacity of the land for a subsistence living. There is insufficient wealth to import such items as food, building materials, and industrial raw materials. Such wealth would allow many more people to be supported. To understand why this is so, compare an American suburban family of four

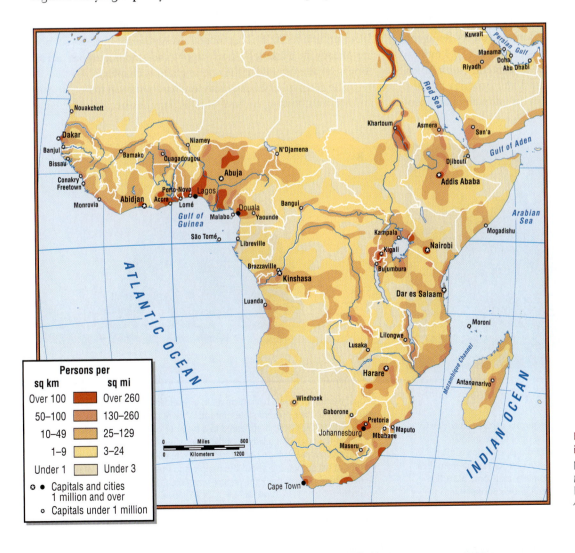

FIGURE 7.6 Population density in sub-Saharan Africa. Africa's population is distributed unevenly, but generally sparsely, over the continent. [Adapted from *Hammond Citation World Atlas* (Maplewood, N.J.: Hammond), 1996.]

living on a half-acre lot, supporting themselves on the adults' annual salaries of $50,000 to $100,000, with a rural African family of seven or eight subsisting almost entirely on 4 acres (1.6 hectares) of land, often with little education and no reliable outside source of cash. The Africans must literally live off their land, whereas the American suburbanites do not subsist on their lawns and flower gardens. The inhabitants of wealthy societies (Europe, North America, Australia, Japan) can tap into resources around the globe (including Africa), often at favorable prices. Their standard of living is only partially linked to the carrying capacity of their country's environment. Thus people in these societies can live in cities that are both dense and affluent. In Africa, on the other hand, because human capabilities are underdeveloped, the vast majority of people, like many in Asia and in Central and South America, have much less access to resources in distant lands because they do not have the skills, technical wherewithal, political power, or cash to acquire them.

Political unrest places additional burdens on a region's carrying capacity. War or oppression often forces people to leave their homes as refugees. Over the last decade of the twentieth century, for example, people in such places as Chad, Sudan, Somalia, Ethiopia, Uganda, Liberia, Sierra Leone, Congo (Kinshasa), Congo (Brazzaville), Rwanda, and Mozambique poured back and forth across borders to escape **genocide**—the deliberate destruction of an ethnic, racial, or political group. Genocide is a tactic usually instigated by the state and carried out by political factions. According to the United Nations High Commission for Refugees, in 2003, Africa hosted about 33 percent of the world's 10.3 million refugees (Asia hosted 41 percent, Europe 20 percent, North America 5 percent, Oceania 0.6 percent, and Latin America and the Caribbean 0.4 percent). The United Nations estimates that if peo-

ple who are displaced within their home countries are counted, Africa has half the world's refugee population, and that about three-fourths of Africa's refugees are women and children. As difficult as life is for these refugees, the burden on the countries that host them is also severe. Even with help from international agencies, the host countries find their own development plans derailed by the arrival of so many distressed people, who must be fed, sheltered, and given health care. Large portions of economic aid to Africa have been diverted to deal with the emergency needs of refugees.

Population Growth

African populations are growing faster than any others on earth. In less than 50 years, the population of sub-Saharan Africa has more than tripled, reaching nearly 657 million in 2000. Between 2000 and 2003, the population increased by 54 million. This rapid growth is the main threat to human well-being in places where the carrying capacity has already been reached or exceeded. The addition of increasing numbers of individuals to feed, educate, house, vaccinate, and employ is outstripping even the best efforts to improve nutrition, education, housing, health care, and employment possibilities. In some places, Africa's already low standards of living are declining further. The UN projects that the number of primary school students (aged 6–11) in sub-Saharan Africa will increase from 76.9 million in 1985 to 144.2 million in 2005 and to 195.3 million in 2025. This increase is significant because schools are already in extremely short supply and underequipped across Africa (see the vignette at the opening of this chapter).

The geographer Ezekiel Kalipeni has found that many Africans are not yet choosing to have smaller families because they view

PRACTICING GEOGRAPHY   **COMPARING LOCAL LANDSCAPES**

Very often refugees are productive, even prosperous, citizens who, due to no fault of their own, lose their homes and the right to live in their accustomed space. Gazing on their stressed and desperate faces often makes the viewer want to find a way to distance him- or herself from their plight. In the early 2000s, civil war gripped Somalia, and some 300,000 refugees fled to neighboring countries. Photo A shows Somali Bantu

people waiting in Kenya to participate in a resettlement program. What can you tell about their present state of mind? Where do you think they may be headed? In 2004, Photo B appeared in a number of newsmagazines. It is an arresting image of a terrible conflict that pitted Muslim Arab-speaking people against each other. Look carefully at the pile of possessions and see if you can tell something about the people and their former state of well-being. Where do you think the picture was taken? How many people can you count? [A, AP/Wide World Photos; B, Oliver Jobard/Sipa.]

children as both an economic advantage and a spiritual link between the past and the future. Childlessness is considered a tragedy. Not only do children ensure a family's genetic and spiritual survival, they still do much of the work on family farms. In regions with high infant mortality, parents have many children in the hope of raising a few to maturity. In short, in most places in Africa, the demographic transition—the sharp decline in births that usually accompanies economic development—is only beginning and is developing slowly (see Figure 1.20 on page 42).

In a handful of countries scattered throughout the region, however, fertility rates are declining. These countries include South Africa, Botswana, Seychelles, and Mauritius (the latter two are tiny island countries off East Africa). In these countries, circumstances have changed sufficiently to make smaller families desirable. First, in all four, the education of women has improved, and gender role restrictions have been relaxed (as measured by the United Nations Gender Empowerment Measure). Hence women are choosing to use contraception because they have life options beyond motherhood. Second, because infant mortality is now relatively low, parents can expect their two or three children to live to adulthood. Nonetheless, the rate of contraception use in sub-Saharan Africa as a whole is less than half of that in all other world regions.

PERSONAL VIGNETTE

Mary, a Kenyan farmer, has just had her third son. The father of Mary's children works in a distant city and visits the family only several weeks a year. He supports himself with his earnings and buys occasional nonessentials for the family. Mary tells an interviewer that three children are enough for happiness, and so today, at age 29, she is having surgery that will prevent conception. She owns only one cow and a small piece of land that can't be further divided, so all she can provide for her children is an education. Mary says that she can afford to educate only three children.

Such attitudes are spreading in Kenya, where food, health care, and jobs are in short supply. Mary plans to augment her farm income by starting a sanitary pit toilet construction business. She has applied for a small loan (U.S. $150) for this purpose. The success of her business could mean that her children will become well educated and that she herself will gain prestige. Studies of the effects of microcredit opportunities have shown that female participants tend to increase their use of birth control markedly. In addition, when women like Mary accomplish their goals, they become role models for other women, who then limit their families so that they can also become self-sufficient owners of small businesses. [Adapted from Jeffrey Goldberg, The New York Times Magazine (March 2, 1997): 39; World Resources, 1996–1997 (New York: Oxford University Press, 1996), p. 5. Information on the role of microcredit in fertility patterns is from Fiona Steele, Sajeda Amin, and Ruchira T. Naved, "The impact of an integrated micro-credit program on women's empowerment and fertility behavior in rural Bangladesh," Policy Research Division Working Paper no. 115 (New York: Population Council), 1998.]

Some factors contribute to reduced family size in any culture. Education helps people understand that under modern conditions, additional children are unlikely to add to a family's wealth as they did in the past. Communication between husband and wife also influences family size. Surveys reveal that when men are included in public health education, they are often amenable to having fewer children and will suggest birth control to their wives. This inclusion of men is especially important in cultures in which men are assumed to make major decisions. Lengthy breast-feeding after childbirth is another long-standing folk birth-control strategy because it retards fertility. In many cultures, including some groups in Africa, the custom of abstaining from sex for several years after a birth is a strategy for limiting family size.

On the other hand, polygynous relationships, which exist in parts of West Africa and elsewhere (see pages 368–369), produce more children than monogamous relationships do. Men who choose polygyny tend to be less well educated and to be especially interested in having many children as a status symbol, because children have traditionally been a source of wealth. Men who choose polygyny also tend to be relatively wealthy (because the husband must provide financial support for his multiple wives), so polygyny itself is sometimes a status symbol.

The population pyramids of Nigeria, Africa's most populous country, and South Africa, the most developed country on the continent (Figure 7.7), demonstrate the contrast between the classic wide-bottomed shape of countries experiencing rapid growth (Nigeria), and countries where growth has slowed markedly (South Africa). South Africa's pyramid has contracted at the bottom in the last several years because the birth rate has dropped from 35 per 1000 total population to 23 per 1000 just since 1990. This drop has undoubtedly resulted partly from the economic and educational improvements and social changes that have come about since the end of apartheid in the early 1990s. But this pattern in the South African pyramid may also be a consequence of the spread of HIV among young adults. HIV-AIDS, the consequences of which are discussed further on pages 347–348, is now the main cause of the slowing of population growth rates in Africa, especially in the Southern Africa subregion. In 2001, 20.1 percent of Southern Africa's adult population between the ages of 15 and 49 was infected with HIV. In other words, one-fifth of those adults who should be in their most productive years, both economically and biologically, are sick and dying.

Population and Public Health

Sub-Saharan Africa has long been troubled by infectious diseases that have been particularly harmful and difficult to control; they include schistosomiasis, sleeping sickness, malaria, river blindness, and cholera. Observers report that infectious diseases (including HIV-AIDS) are by far the largest killers in Africa and are responsible for about 50 percent of all deaths. (Compare the causes of deaths in Africa with those in Europe, as shown in Figure 7.8.) Some of these diseases are linked to particular ecological zones. For example, people living between the 15th parallels north and south of the equator are most likely to be exposed to sleeping sickness (trypanosomiasis). It is within this range that the tsetse fly lives, inhabiting the vegetation near rivers and lakes, in open woodlands, and in grasslands. The disease, spread by the bite of this fly, leaves the victim exhausted and, if left untreated, attacks the central nervous system, resulting in death. Sleeping sickness affects both humans and cattle, and the presence of the tsetse fly has limited animal husbandry

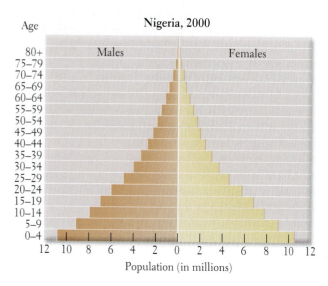

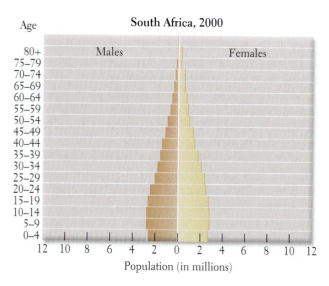

FIGURE 7.7 Population pyramids for Nigeria (134 million) and South Africa (44 million). Nigeria has a population growth rate of 2.8; South Africa's is 0.9. [Adapted from "Population Pyramids for Nigeria" and "Population Pyramids for South Africa" (Washington, D.C.: U.S. Bureau of the Census, International Data Base, at http://www.census.gov/ipc/www/idbpyr.html).]

in some areas. Several hundred thousand people, mostly rural residents, are thought to suffer from sleeping sickness, and most of them are not treated because they cannot afford the expensive drug therapy developed in Germany.

Schistosomiasis (bilharzia) is Africa's most common chronic tropical disease, and malaria its second most common. Both diseases are linked to standing fresh water, and their incidence has increased with the construction of dams and rice paddies. Schistosomiasis develops when a flatworm carried by a particular freshwater snail enters the skin of a person standing in water. The victim first experiences a rash, then a cough and diarrhea, headaches, fever, and general malaise. If the disease is left untreated, the liver and spleen are impaired, and the belly swells and is painful. The condition can lead to bladder cancer. Malaria is spread by the anopheles mosquito, which lays its eggs in standing water. Its victims experience chills and fever alternately. Left untreated, malaria leads to anemia and jaundice and eventually to enlargement of the liver and spleen. In some cases, damage to the brain leads to death. Until recently, little research in Africa was focused on how best to control the most common chronic tropical diseases. Now, several groups based in Kenya are working on control measures for both schistosomiasis and malaria.

The bite of the blackfly introduces a parasite into the human body that causes river blindness (onchocerciasis). This disease afflicts millions of Africans between the 20th parallels north and south, especially those who live along fast-moving rivers, because the blackfly requires fast-moving water to breed. The parasite first causes itching, then disfigurement, and finally eye lesions that lead to blindness. Nineteen countries in the affected zone participate in the Onchocerciasis Control Programme funded by the World Health Organization (WHO), the United Nations Development Program (UNDP), and the World Bank. French scientists and scientists from the member countries have devised ways to control blackfly breeding with insecticides, and a new drug, Ivermectin,

which kills the parasites, has been administered to more than 20 million people since 1997. It appears to act both as a preventative and as a cure.

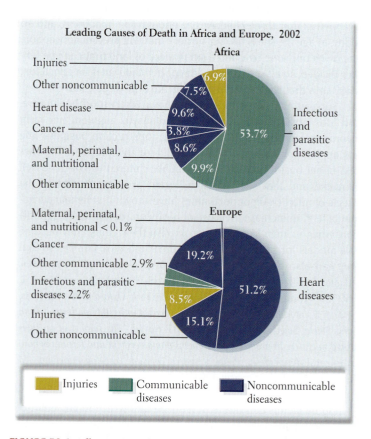

FIGURE 7.8 Leading causes of death in Africa and Europe, 2002. [Adapted from *Global Burden of Disease Estimates, 2002* (Geneva: World Health Organization).]

HIV-AIDS in Africa

The epidemic of human immunodeficiency virus (HIV)-acquired immunodeficiency syndrome (AIDS) is the most severe public health problem in sub-Saharan Africa. Because it debilitates and kills so many people in the prime of life, this disease has devastating social ramifications. According to a joint report by the United Nations and the World Health Organization, in the year 2003, sub-Saharan Africa had an estimated 67 percent of the estimated worldwide total of 40 million HIV-infected people (Figure 7.9). For the world as a whole, only 1.2 percent of the adult population is infected with HIV. In North America, the figure is 0.6 percent, and in East Asia and the Pacific it is just 0.1 percent. In sub-Saharan Africa, however, 8.9 percent of adults are infected, and in certain countries the rate is much higher. The subregion most affected is Southern Africa, where, until the HIV-AIDS epidemic, development prospects were brightest. By 2004, based on improved methods of estimating the figures, 40 percent of adults aged 15 to 49 were infected in Botswana and Swaziland; 25 percent in Zimbabwe; between 20 and 22 percent in Namibia and South Africa; and 16 percent in Malawi and Zambia.

The disease threatens to change sub-Saharan Africa's population growth and life expectancy patterns drastically. According to a Population Reference Bureau report, in Botswana, Africa's second most well-off country after South Africa, HIV-AIDS has already reduced life expectancy from 59 years in 1990 to just 37 years in 2003. Zimbabwe's life expectancy went from 58 years in 1990 to 34 years in 2003 (a reduction of 41 percent). Had the HIV-AIDS epidemic not occurred, Zimbabwe could have achieved a life expectancy of about 70 years by 2005; instead, life expectancy may fall to what it was 200 years ago. Zimbabwe's population growth rate was 3.2 percent in 1993, but by 1997 it had dropped to 2.5 percent, primarily in response to rising standards of living. Then the HIV-AIDS epidemic reached Zimbabwe, and by 2003 the country's population growth rate was only 1.2 percent—no longer because people were choosing to have fewer children, but mostly because of the high rate of death from AIDS. If the epidemic is not stopped, Zimbabwe could lose 20.5 percent of its population by 2010. The demographic effects of HIV-AIDS are comparably dramatic in such countries as Nigeria, Côte d'Ivoire, Burkina Faso, Ethiopia, Uganda, Kenya, Tanzania, Rwanda, Zambia, Namibia, Botswana, and South Africa.

In Africa, HIV-AIDS affects both men and women: 55 percent of HIV-infected adults in sub-Saharan Africa are women, and four-fifths of the world's women infected with HIV are in Africa. Many of them are young mothers infected by their husbands, who may visit sex workers when they travel for work or business. In some cities, virtually all sex workers are infected. In Uganda, one-quarter of married urban women, who are normally considered at low risk, are HIV-positive. The inset in Figure 7.9 shows the dramatic effect of HIV-AIDS on female life expectancy in Southern Africa. Because AIDS develops slowly in its victims, controlling the epidemic is particularly difficult. UN officials suspect that 90 percent of African carriers of HIV do not learn they have been infected until several years after they acquire the virus. During this time, they may pass the disease to other adults and produce several HIV-positive children.

The rapid urbanization of Africa (see the discussion on pages 364–365) has contributed to the swift spread of HIV-AIDS.

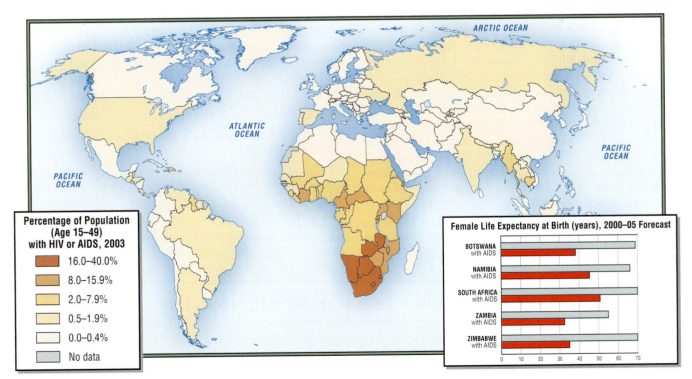

FIGURE 7.9 Global prevalence of HIV-AIDS and its effect on African female life expectancy. [Adapted from *2003 World Population Data Sheet* (Washington, D.C.: Population Reference Bureau); inset adapted from "A survey of sub-Saharan Africa," *The Economist* (January 17, 2004): 11; "AIDS Epidemic Update," UNAIDS and World Health Organization, December 2003, p. 7.]

Transportation between cities and the countryside has improved, and bus and truck drivers are thought to be major carriers of the disease, as are urban migrants who come home for visits. Young men and women encounter each other more easily in urban settings, often without the community pressure that would have precluded intimacy in the countryside. Women may be coerced to have sex, and they have great difficulty insisting that their mates use condoms. According to custom, men often have multiple partners. Where HIV-AIDS education is lacking, some men think that only sex with a mature woman can cause the disease, so very young girls are increasingly sought as sex partners (this is sometimes referred to as "the virgin cure"). For many poor urban women, removed from village support systems, occasional sex work is part of what they do to survive economically.

Education has played a contradictory and changing role in the spread and control of HIV-AIDS in Africa. Early in the epidemic, educated Africans were actually more susceptible to infection, partly because they were likely to live in urban areas and to be free of community-enforced sanctions against multiple sex partners. Now that the epidemic has grown, those who can read and understand explanations of how HIV is spread have an advantage, and rates of infection are now lower among educated people than among people who lack education. In Senegal in West Africa, where the decision to invest heavily in HIV-AIDS prevention was taken in the 1980s (when infection rates across Africa were still low), levels of infection have held at just 1 percent from 1990 through 2002. Through public education and distribution of condoms, Uganda has had great success in lowering the incidence of new HIV infections.

While prevention of HIV-AIDS through education is the first bulwark against the disease, given the number of people that are now infected, treatment is the most effective strategy for reducing AIDS-related illness and death. Unfortunately, the overwhelming majority of Africans with HIV-AIDS cannot afford the costly combination of drugs that keeps victims alive in such places as North America, Europe, and Japan. The antiretroviral "drug cocktail" produced by American and European drug companies costs about U.S. $10,000 a year per patient. In 2000, in an attempt to lower this cost, a global movement, including countries in Latin America, Africa, South Asia, and Southeast Asia, began to challenge the patents of the drug companies. By the end of 2000, the hope for cheaper drug treatments was given a boost when drug firms in Cuba and India began to offer a regimen that cost just $1.00 a day ($365 a year)—a price that is still too high for people in most countries where infection rates are high. Drug companies in the United States and

Europe brought lawsuits against the Cuban and Indian firms producing cheaper copies of the AIDS drugs, arguing that those firms were illegally using drug formulas that had been developed and patented at great cost. Those suits were dropped, however, when the plight of untreated, poverty-stricken AIDS patients in Africa and elsewhere was broadcast worldwide and the drug companies found themselves accused of putting profits ahead of human lives. By 2003, several private foundations were seeking ways to provide affordable drug treatment to the more than 40 million AIDS patients worldwide. In Africa, access to antiretroviral drugs varies by country, but in the region as a whole, as of 2003, barely 0.02 percent of AIDS patients had access to these drugs. The governments of Botswana and South Africa strive to provide antiretroviral drugs to their affected citizens. In 2004, the Central African Republic made antiretroviral drugs available to HIV-AIDS patients at affordable rates with the help of $25 million donated by developed nations, including the United States. These medicines will both improve life expectancy for patients and encourage voluntary HIV testing.

Across the continent, the social consequences of the HIV-AIDS epidemic are enormous. In November 2003, at an HIV-AIDS charity concert in his honor, Nelson Mandela said that he now views HIV-AIDS as a challenge greater than apartheid (see the discussion on page 342). As of 2000, more than 15 million Africans had already died. Young adults, parents, teachers, skilled craftspeople, and trained professionals are lost to a slow, agonizing death in the prime of their lives. Millions of orphans have been created, many without any family left to care for them (80,000 in Botswana alone as of 2003). Elderly grandparents, or just kindly neighbors, must provide for orphaned dependent children (who may also have been infected at birth). Time, money, and energy are being channeled to HIV-AIDS prevention and treatment and away from education and from the control of other diseases such as schistosomiasis, malaria, river blindness, and cholera, which actually threaten a much larger percentage of sub-Saharan Africans. In Kenya, 50 to 70 percent of hospital beds are taken up with AIDS patients.

Quick Review

1. What climatic conditions in Africa incubate diseases such as malaria and schistosomiasis?

2. Name a couple of the places the Aksum reached with their trade routes outside the African continent.

3. What circumstances make heterosexual women in Africa so susceptible to HIV/AIDS?

4. What factors influence the carrying capacity of a region?

II CURRENT GEOGRAPHIC ISSUES

Most countries of sub-Saharan Africa have been independent of colonial rule for less than 50 years. Many are only beginning to build their nations and must come to terms with such issues as arbitrary national boundaries and rapid urbanization. Africans are still developing appropriate systems of government and economic poli-

cies that will lead to sustainable growth and development. Because the effects of the colonial era in Africa run so deep, it is too soon to tell just how some of the African countries will find ways to become politically, economically, and socially viable. Still, some countries are achieving higher standards of living for their citizens, and many

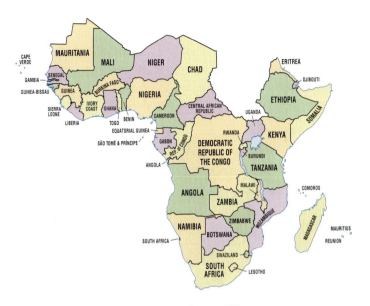

FIGURE 7.10 Political map of sub-Saharan Africa.

competition for land and resources, and few ways to earn income in the globalizing economy or raise capital for industrialization. Too many manufactured products had to be imported. Some countries (Botswana and South Africa, for example) became viable and relatively stable participatory democracies. Both were moving hopefully toward real economic prosperity and were pursuing innovative strategies to strengthen their local and regional economies. Unfortunately, sharply rising oil prices during the early 1970s hit African economies especially hard. Although many were poised to recover in the 1990s, the HIV-AIDS epidemic intervened and appears to be causing major reversals in precisely those countries that otherwise have the brightest futures. Addressing this latest setback requires multiple strategies directed by African leaders, but they are unlikely to succeed without assistance from outside the region.

researchers are optimistic about the future. A broad consensus is developing that Africa's prospects will improve if development policies facilitate efforts by Africans themselves to seek and implement solutions to the problems they face.

ECONOMIC AND POLITICAL ISSUES

Africa cannot continue to produce what it does not consume and consume what it does not produce.

(THE AFRICANS, CO-PRODUCED BY WETA-TV, WASHINGTON, D.C., AND THE BRITISH BROADCASTING CORPORATION)

In the aftermath of colonial rule, Africa was left with underdeveloped human capital, ill-trained leaders, intense and misdirected

Postindependence Economies

We have noted that after years of declining production and shrinking economies, many Africans were poorer at the beginning of the 1990s than they were in the 1960s. Figure 7.11 illustrates this point by comparing GDP per capita in the decades from 1960 to 2000 for selected African and Asian countries. These economies began at roughly similar GDPs per capita in 1960 (dark red bars in Figure 7.11). While GDPs rose modestly, but very steadily, in India (65 percent) and dramatically in South Korea (638 percent) between 1960 and 1990, in Africa, GDPs per capita rose only very modestly in Kenya and Zimbabwe and sank in Ghana and Uganda. Only in Botswana was the pattern of growth over the 40 years spectacular, but Botswana started at a very low base. By 2000, GDPs had risen in nearly all countries shown in Figure 7.11—in several cases quite dramatically (in Africa, see the blue bars for Ghana, South Africa, Botswana, and Zimbabwe).

British development specialist Elizabeth Francis gives several reasons for Africa's difficulties in the postindependence

GDP per Capita, 1960–2000

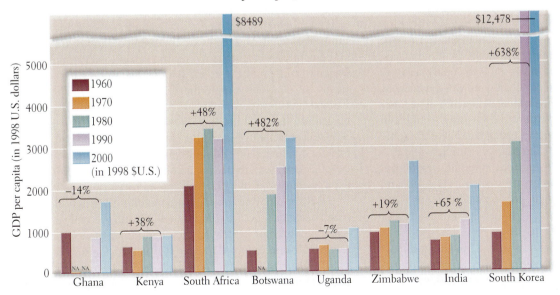

FIGURE 7.11 GDP per capita for selected sub-Saharan African and Asian countries. Notice that some African countries experienced very modest economic growth, or even declines, in 1960–1990. Some grew significantly in 1990–2000, while others did not. [Adapted from Elizabeth Francis, *Making a Living: Changing Livelihoods in Rural Africa* (London and New York: Routledge, 2000), p. 2; data for Botswana, 1960, 1980, 1999: http://www.cepr.net/IMF/Emperor_Table_2.htm.]

(1960–1990) era. First, the long-term impact of colonialism positioned Africa in the global market as a supplier of cheap resources and low-cost labor. That role persists today, leaving African countries in competition with other suppliers of cheap labor and raw materials and with little influence on market prices. Second, political problems followed independence. The political leaders and citizens of the new states had little or no experience in allocating resources, power, and opportunity fairly. Consequently, corruption became widespread, and people were unable to assert their will through democratic channels. Third, decades of civil unrest in many countries, and outright wars in a few, retarded both social and economic progress. Often these wars were exacerbated by outside interference. During the cold war, interventions by the United States, France, and the Soviet Union and its allies intensified the wars in Africa as these world powers attempted to draw new African nations into alliances (see pages 361–362). Fourth, steep oil price increases in the 1970s forced governments into debt to pay for fuel. Finally, in the 1980s and 1990s, the World Bank and the International Monetary Fund promoted ill-advised economic reorganization and structural adjustment programs (SAPs). The belt-tightening SAP strategies, widely prescribed to remedy indebtedness and dysfunctional development (see the discussion of SAPs on pages 353–355), resulted in a fateful **disinvestment** in human capital (removal of financial support for education and health) just when the improvement of human skills and well-being was crucial for further development.

Subsistence Agriculture: A Major Source of Livelihood. Most Africans, 70 percent of whom live in rural areas, produce their own food by farming small plots, raising livestock, or a combination of both. Many also fish, hunt, and gather some of their food. Small-scale cash cropping, adopted during the colonial era to make money to pay taxes, is often part of the system of production. Today people may grow cash crops of peanuts, cacao beans, rice, or coffee to obtain money to pay school fees or purchase a bicycle or other useful items.

Farms are usually 2 to 10 acres (1 to 4 hectares), the size being limited more by lack of available labor than by scarcity of land. Fields given over to food production are often planted with many different species (intercropped) in a planned arrangement that is rotated over the course of the field's use. After a few years of cultivation, a field is allowed to lie fallow (rest) so that the soil can be replenished by the growth of natural vegetation. In some places, either because of situations that restrict ordinary people's access to land or because of localized rapid population growth, the intensity of land use has increased to the point that sufficient fallow times are no longer possible, and soil fertility is decreasing. Farmers in many environments—the tropical forest fringes of Nigeria and the high volcanic plateaus of southern Uganda, for example—now cultivate their fields permanently. Permanent cultivation requires a high degree of specialized knowledge: fields are intercropped, and the needs of particular species are carefully assessed and met. Great care is taken to fertilize the soils continuously and intensively. Sometimes commercial fertilizer is used.

Most African farmers raise some livestock, but they usually limit their holdings to a cow or two, some poultry, and several goats and sheep. Herding is practiced primarily in savannas, on desert

In places where there is no access to arable land, a roof garden can be the solution. In Dakar, Senegal, studies have shown that greening an entire roof can ameliorate the high temperatures created by flat concrete roofs. [Fabio Massimo Aceto/Agency Grazia Neri.]

margins, and in the mixture of grass and shrubs called open bush. Ideally, herders move their animals seasonally to the freshest pasture to minimize damage to the grazable grasses. Increasingly, however, drought or the encroachment of field agriculture has forced pastoralists to reduce or divide their herds. Some herders have even taken up farming, though to do so is considered a step down in status.

Urban farming (see the photo above) is an increasingly significant part of the lives of city dwellers in Africa. Tiny vegetable gardens (with maize, beans, potatoes, yams, greens, and bananas) and fowl pens can be seen even in the hearts of central business districts. A study has shown that in Mombasa, Kenya, 15 percent of households grow food on urban land. These urban gardens provide nutrition to city dwellers; generate employment, especially for women; and put derelict urban land to sustainable, productive use.

Exports of Raw Materials to the Global Economy. After independence, the economies of African countries were still centered around the export of one or two raw materials, and this pattern has continued (Figure 7.12). Southern Africa exports a wide range of minerals and produces 27 percent of the world's gold and 50 percent of its platinum. West Africa exports oil (primarily from Nigeria) and produces about half of the world's cacao beans, nearly all of which supply the European market. (The Africa–Europe link in cacao production and the chocolate trade is discussed in Chapters 1 and 4 and mentioned in this chapter on page 377.) Peanuts and fish are other West African exports. East Africa exports small percentages of the world production of copra (dried coconut, used to make oil), palm oil, and soybeans; Central Africa exports some tropical hardwoods.

Development of commercial agriculture is sorely needed to meet the demands of growing populations and to supply more profitable products for export, yet there are impediments to this goal. Large-scale cultivation of tree crops (such as palm oil, copra, and cacao), rice, and various dessert fruits (such as bananas) is increasingly common as African countries search for crops that will sell

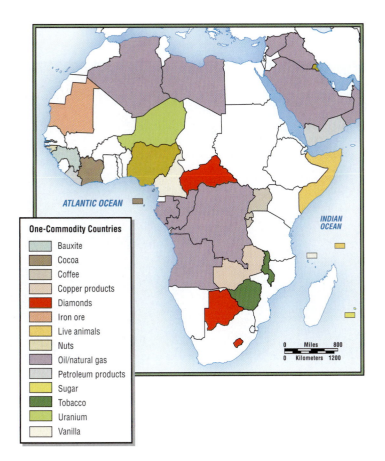

FIGURE 7.12 One-commodity countries. This map depicts those countries that depend on just one commodity for more than 50 percent of their export earnings. [Adapted from George Kurian, ed., *Atlas of the Third World* (New York: Facts on File, 1992), p. 76; data from *United Nations Human Development Indicators 2000* (New York: United Nations Development Programme); source for Congo (Kinshasa): http://www.jubileeplus.org/databank/profiles/congodem.html#3.]

One-Commodity Countries

- Bauxite
- Cocoa
- Coffee
- Copper products
- Diamonds
- Iron ore
- Live animals
- Nuts
- Oil/natural gas
- Petroleum products
- Sugar
- Tobacco
- Uranium
- Vanilla

land to poor farmers. The best and largest of the confiscated farms went to Mugabe henchmen, who then used the land for their own enrichment. The results of this conflict have been disastrous. Food production has dropped so low that a majority of Zimbabweans were on food aid in 2003. The Zimbabwe Horticulture Promotions Board claims that every $10,000 of horticulture export revenue should create four jobs in agriculture and 1.5 jobs in processing. But in 2003, unemployment stood at an astonishing 70 percent, and an estimated 2 million Zimbabweans were working illegally in Botswana and South Africa. On the other hand, while the Zimbabwean economy certainly appeared to be doing better under private, primarily white farm ownership, wages were low, and for the majority there was little chance for upward mobility. Even though elaborate government programs now train former farmworkers as horticulturists and promote crops that will have export appeal in (non-European Union) Eastern Europe and Central Asia, progress has been too slow to make up for the jobs lost when the experienced white farmers were chased off.

Botswana: A Case Study. Since its independence in 1966, Botswana has been hailed as the fastest-rising economic star in sub-Saharan Africa (see Figure 7.11). The GDP per capita was

well in the global marketplace. Getting access to consumers in European and North American markets can be difficult, however, because, despite free trade initiatives, these regions often block African products to protect their own producers. Africa's marginally fertile soils and underdeveloped infrastructure (roads, railroads, irrigation facilities, farm equipment manufacturers, banks) are additional drags on commercial agriculture; still, production has increased thanks to improved crop varieties, more effective environmental management techniques, and greater investment in places such as Botswana and South Africa in the infrastructure needed for agriculture. Commercial farms now produce cashews, cotton, coffee, tea, maize, groundnuts, poultry, and dairy products. But commercial farming requires substantial investment in land, as well as expensive mechanized equipment and chemical fertilizers and pesticides, leaving most small farmers out of the game.

Governments try to help small farmers acquire these costly necessities, but there is much debate over how involved governments should be. In Zimbabwe, the promotion of smallholder horticulture for export is linked with a highly controversial land reform program instituted by the government of Robert Mugabe, wherein wealthy white farmers, many from South Africa, have had their productive farms confiscated and redistributed to poor Zimbabweans, some of whom were fighters in the war of independence in the 1970s. Many Zimbabweans saw these white owners as usurpers of African lands because the holdings were originally taken from African people during the colonial period. Critics of Mugabe point out that while government programs can do much to improve production by smallholders, the goal of Mugabe's land reform scheme was to run off productive white landowners, not to redistribute the

Although not yet one of Mozambique's principal exports, tea is becoming an important cash crop in Gurue, Zambezia Province. For example, Chazeira de Mozambique, the largest tea grower in the region, exports tea to Pakistan and Yemen and plans to expand to other countries. In this photo, workers who have picked tea leaves by hand to preserve their quality and flavor wait in line to have the tea weighed. [Brian Seed.]

U.S. $535 in 1960, and in 1966, Botswana was one of the poorest 25 countries in the world. By 2000, the GDP per capita was U.S. $7184 (adjusted for purchasing power parity, or PPP), second only in sub-Saharan Africa to that of South Africa, and Botswana was a middle-income country by world standards. Between the late 1960s and the early 1990s, Botswana had the highest sustained annual growth rate of GDP in Africa: 6.1 percent per year. How is an arid, landlocked country able to prosper so well and so rapidly? More important, to what extent does the general population share in the country's wealth, and how have recent developments affected the country economically?

In 1966, 40 percent of Botswana's GDP derived from agriculture. Now agriculture, though still important in family economies, accounts for only 6 percent of GDP and 16 percent of employment, while the country gets most of its wealth from mineral resources. Diamond deposits were found shortly after independence, and by 1995 Botswana was the world's third largest diamond producer and ranked first in diamond output value. The country also generates income from deposits of gold, soda ash and potash, copper, nickel, and coal. Diamonds are the mainstay, however. Elsewhere in Africa, the diamond mining industry is extremely exploitative, and the trade is known to fund terrorism and arms for civil wars (so-called blood diamonds), but evidence suggests that in Botswana, the diamond industry has funded infrastructure improvement. Diamonds account for over 63 percent of Botswana's government revenues and 90 percent of its export dollars. When such wealth is statistically apportioned among just 1.5 million people, the per capita GDP is mid-level by world standards.

The actual allocation of wealth in Botswana, however, is highly uneven. A typical diamond worker earns only $85 a month or one-seventh of the average per capita GDP. Meanwhile, income for the richest 10 percent in Botswana is 77 times that of the poorest 10 percent, one of the highest wealth disparities on earth. Perhaps as many as 20 percent of the people have reasonably well paid positions in government or the mining industry; the remaining 80 percent work in agriculture or in primarily low-paid menial, urban jobs, and most suffer serious poverty. So even though there is an appearance of increasing wealth in the statistics, at most a fifth of the population does well, a situation that is reflected in Botswana's rather low global ranking of 125 on the Human Development Index—which, nonetheless, is relatively high for Africa. Still, overall, the trend in Botswana is toward investment in human capital, which should lead ultimately to a more egalitarian society. The country's commitment to educating its children at least through the 10th grade should prepare the population for higher-skilled jobs, if investors can be interested in the country.

Botswana's present preoccupation is with the burgeoning HIV-AIDS crisis, which threatens its prosperity (see the discussion on pages 347–348). While HIV-AIDS is perhaps more widespread in urban areas, the disease has dire consequences in rural areas, where the poor are concentrated, as well. With 40 percent of the total population afflicted, at least 100,000 agricultural workers have been disabled. Studies show that in farm households affected by HIV-AIDS, food consumption drops by 40 percent because the sickened farmers turn to crops that take less work, but are also less nutritious. Production decreases, incomes drop, nutrition worsens,

The Jwaneng diamond mine in Botswana is one of the world's richest. The diamond-laden earth is loaded into trucks and taken to a treatment plant for crushing and sifting. The diamonds are then separated into industrial and gem-quality stones. [Peter Essick/Aurora and Quanta Productions.]

and soon school fees can't be paid. Often Botswana's farming parents die before they can pass on farming and life skills to their children. HIV-AIDS also afflicts agricultural extension agents, with many unable to continue working. To address the issue, Botswana is considering lighter plows and tools that can be used by children and frail adults, seed varieties that require less labor, and garden species that mature at different times of the year so as to spread out the work. And, with the death of so many male farmers, customs restricting equal rights to land for women also are being adjusted. The diamond industry, so important to Botswana's export earnings, employs only about 6000 people, so the effect of HIV-AIDS on this portion of the economy is likely to be less; because of the resources in the diamond sector, affected employees are likely to get access to antiretroviral drugs. President Festus Mogae recognizes that HIV-AIDS could easily derail his country's impressive economic accomplishments. He is open and frank with the Botswana public about the disease, and the Botswana government is setting other priorities aside to provide antiretroviral drugs to as many AIDS sufferers as possible.

South Africa Works! South Africa has managed to create a large and diversified economy, with a total output in 2001 (U.S. $488.2 billion, PPP) that accounted for more than a third of the output of all other sub-Saharan African countries combined. A major reason for South Africa's present state of development is that for centuries it had a well-off minority European population (about 16 percent of the total population) with the skills and external connections to foster economic development, but their energies were invested primarily for their own benefit. The system worked to the great disadvantage of the majority African population, who subsidized the country's prosperity with their low wages.

PRACTICING GEOGRAPHY

What is the destiny of democracy in sub-Saharan Africa?

My field studies span two regions of the continent: the Horn of Africa and Southern Africa. The key issues of concern to me have been government and economic development and class structure as well as leadership, democracy, and global capitalism. In earlier years I was interested in explaining the economic and political crisis that engulfed the continent. I became dissatisfied with the available theories in geography and the social sciences, theories I call "Afro-pessimism." Consequently, I sought out successful African experiences with economic development and public accountability. In my exploration, I found a remarkable story in the Republic of Botswana in Southern Africa, a story filled with wonderful lessons for the rest of the continent. What I learned in Botswana led me to reexamine the role of government in development. The Botswana miracle is Africa's version of the East Asian transformations. The major difference is that the East Asian miracle was guided by ruthless authoritarian government, while a liberal democratic government championed the Africa miracle. In Botswana, I examined the interplay between traditional social structure, colonialism, leadership, and public institutions in the making of this African success.

The Botswana experience has enabled me to revisit two issues in the African political geographic landscape: cultural explanation of political problems and civil wars, and Africa's first postcolonial democratic wave and why it failed. Considering the first question, I compared studies of Somalia and Botswana (a failed state and the most successful state) to reveal that the role of culture and ethnicity in development is a contingent variable

Abdi Samatar,
Department of Geography,
University of Minnesota

rather than a causative one. "Ethnic strife" is the product of, and not the cause of, civil wars. I found out that mismanagement of public resources and their privatization is the key instigator in civil wars. Our work on Africa's first democratic wave examines the internal and external causes of its collapse. Although this research is yet to be concluded, initial findings suggest that two factors jointly undid Africa's economic development: segments of the national elite who were committed to self-aggrandizement and looting of the public purse, and cold warriors who expediently supported those amenable factions. One of the clearest examples is the Somali republic. Here, principal political factions were partially financed by Soviet and American intelligence in 1964 and 1967. The consequence of this alliance was that the first democratic transfer of power in postcolonial Africa was eclipsed by the rise of authoritarian regimes that led to the disintegration of state and civil society in Somalia.

The research methods I use include survey research, case study methodology, archival readings, participant observations, and oral historiography. Countries I worked in include Somalia, Djibouti, Ethiopia, Kenya, Botswana, and South Africa.

In 1910, the Dutch (Boer) and British settlers of South Africa became the rulers of an independent South Africa. During the twentieth century, they established many lucrative industries, such as the refining of minerals, gemstones, and metals from the country's rich mines. These enterprises kept profits in the country, allowing the construction of a large industrial economy, but only a small minority reaped the benefits of this effort.

Although the labor of black workers helped build South Africa's exceptional economy, only a few black South Africans worked in sectors in which wages were relatively high. Very low wages for the majority, a failure to invest in education, housing, and health care, and the apartheid system kept most South Africans poor. Essentially, 100 percent of the population worked to provide a high standard of living for 16 percent, while 84 percent barely subsisted. With the end of apartheid, this pattern has begun to change, but only slowly.

The Current Economic Crisis

By the late 1990s, more than half of sub-Saharan Africa's countries had economic growth rates averaging at least 4 percent a year; yet, this growth has not been enough to erase the effects of the down-

ward spirals of the last two decades. As a result of the low prices paid for their products on the global market and the high prices of imported industrial products, most countries have found it increasingly difficult to repay loans extended for mechanized agriculture, small industries, hydroelectric plants, roads, and so forth (Figure 7.13).

In the early 1980s, the major international banks and financial institutions that had made loans to African governments developed a plan to obtain repayment. Working through the World Bank and the IMF, these institutions threatened to stop all further lending to the debtor countries unless they enacted specific economic reforms, called **structural adjustment programs** (**SAPs**). SAPs try to increase government revenues by reducing state interference in the economy. According to supporters of SAPs, economies guided solely by free-market mechanisms will allocate investment money and resources better, will be more vibrant and entrepreneurial, and hence will produce more tax revenues with which to pay off debts. To raise money, governments are also required to slash their own payrolls by cutting health care, education, training, and other social service programs.

SAPs have accomplished some good. They tightened bookkeeping procedures and thereby curtailed corruption and waste

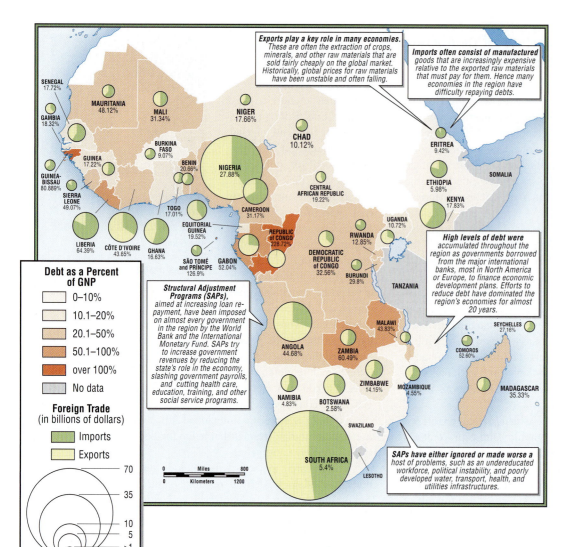

FIGURE 7.13 Economic issues: Sub-Saharan Africa. Increasingly, imports are exceeding exports, even in countries, such as Nigeria and South Africa, with major exportable resources. When imports exceed exports and money is borrowed for large projects, debt rises. [Debt data from *United Nations Human Development Report 2004.*]

Exports play a key role in many economies. These are often the extraction of crops, minerals, and other raw materials that are sold fairly cheaply on the global market. Historically, global prices for raw materials have been unstable and often falling.

Imports often consist of manufactured goods that are increasingly expensive relative to the exported raw materials that must pay for them. Hence many economies in the region have difficulty repaying debts.

High levels of debt were accumulated throughout the region as governments borrowed from the major international banks, most in North America or Europe, to finance economic development plans. Efforts to reduce debt have dominated the region's economies for almost 20 years.

Structural Adjustment Programs (SAPs), aimed at increasing loan repayment, have been imposed on almost every government in the region by the World Bank and the International Monetary Fund. SAPs try to increase government revenues by reducing the state's role in the economy, slashing government payrolls, and cutting health care, education, training, and other social service programs.

SAPs have either ignored or made worse a host of problems, such as an undereducated workforce, political instability, and poorly developed water, transport, health, and utilities infrastructures.

Debt as a Percent of GNP
- 0–10%
- 10.1–20%
- 20.1–50%
- 50.1–100%
- over 100%
- No data

Foreign Trade (in billions of dollars)
- Imports
- Exports

70
35
10
5
>1

in bureaucracies. They reduced the power of the elites to commandeer resources for their own profit. They eliminated some inefficient and corrupt state agencies that were cheating farmers in the process of getting their crops to market. (In Ghana, for example, cacao producers can now negotiate directly with buyers from Europe and no longer have to buy favor with government commodity brokers.) And SAPs stopped the practice of capping food prices to appease urban dwellers; as a result, local food producers received fairer prices for their crops and were encouraged to produce more. Where they were well-managed, SAPs closed corrupt state-owned monopolies in industries and services, and they opened some sectors of the economy to medium- and small-scale business entrepreneurs. They also made tax collection more efficient.

SAPs have also created problems. Researchers on Africa, including Adebayo O. Olukoshi, of the Nordic Africa Institute in Sweden, and Elizabeth Francis, cited previously, agree that SAPs have made it harder for the poor majority to make a decent living and stay healthy. They have also failed to reduce the debt burden. On the contrary, debt has continued to grow, despite the fact that the region as a whole is now spending more on debt payments than

on health care and education combined—more than U.S. $13 billion a year. One reason debt persists is that SAPs have generally eliminated more jobs than they have created. To maximize revenues and achieve greater efficiency, governments have sold off state-owned businesses to the private sector and removed protections for local industries. Both steps usually bring firings and layoffs. Additional jobs have been lost as large state bureaucracies have been significantly reduced.

Moreover, contrary to SAP theory, Africa did not attract foreign direct investment (FDI) once the cuts had been made. As of 2003, foreign investment in the whole of Africa was less than one-fifth that in Asia and slightly less than that in the city of Hong Kong (Figure 7.14). South Africa and Nigeria together accounted for half the direct foreign investment in all of sub-Saharan Africa. Investors have been discouraged by problems that SAPs either ignored or made worse. Loss of public funds for schools perpetuated an underskilled workforce, which in turn led to unemployment, disaffection, and political instability. Deteriorating infrastructure provided poor-quality water, transport, health care, banking, and utility services. The result has been economic decline and rising unemployment, which has left more than two-thirds of the population in

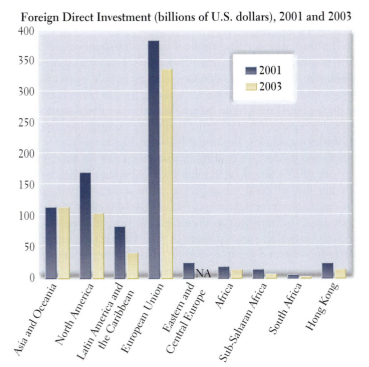

Foreign Direct Investment (billions of U.S. dollars), 2001 and 2003

FIGURE 7.14 Foreign direct investment in several world regions, plus South Africa, 2001–2003. Foreign direct investment (FDI) flows to Africa are notably lower than to other world regions. [Data from *World Investment Report, 2003, FDI Policies for Development*, UNCTAD, http://www.unctad.org/en/docs/wir2003ch1_en.pdf.]

poverty and prone to social unrest. On the other hand, there are indications that conditions for foreign direct investment in Africa may be improving. A 2003 report by the World Bank noted that, small as this investment was, the rate of return on FDI in sub-Saharan Africa was the highest for all world regions. This may have been because the perceived higher risks in the region impelled investors to choose only high-return, quick-turnaround projects. Unfortunately, such "get in, get out" investment is not likely to support stable growth; however, investors who make money in this way may return with larger investments. Another phenomenon may be more helpful: Africans themselves are beginning to invest in Africa. In 2002, for the first time, remittances by African migrant workers amounted to more than private bank loans within the region. Africans are sending money home to start small businesses and to build homes for their families. Both activities have multiplier effects of job creation and reinvestment.

Agriculture and Economic Restructuring. Economic restructuring has affected especially the agriculture sector, which employs 70 percent or more of Africans in one way or another. Modern agricultural development programs provide incentives to encourage investment in cash crops—such as cacao, sugar, bananas, and palm oil—for export. These incentives may include lower taxes on profits, decreased government regulation, and low-cost loans for equipment and land. At the same time that export cropping has been encouraged, however, structural adjustment policies have inadvertently reduced the profitability of food production for Africa's own

consumption. Between 1961 and 1995, per capita food production in Africa decreased by 12 percent, making it the only region on earth where people are eating less well than they used to (see Figure 1.21 on page 44.

For example, the often recommended policy of **currency devaluation** lowers the value of African currencies relative to the currencies issued by other countries. Devaluation promotes the sale of African export crops by making them cheaper on the world market, but it also makes all imports more expensive. Therefore, farmers who grow food for the African market must spend much more on imported seeds, fertilizers, pesticides, and farm equipment; few of these items are as yet manufactured within Africa. Furthermore, with the shift to export production, many farmers no longer have the time or space to grow their own food or food for local markets—so, increasingly, Africans must pay for expensive imported food. Meanwhile, because SAPs have also mandated production increases in many competing countries in other world regions—cacao beans, for example, are now produced in Latin America, Southeast Asia, and several parts of Africa—world market prices for agricultural export crops have declined, eroding farm incomes.

In some countries, land has been taken away from local people outright and made available to foreign-owned export-agriculture ventures. This is one reason that critics equate SAPs with colonial exploitation. In northern Mozambique, for example, where many people used to farm land rented from local elites, Boer plantation owners, who emigrated from South Africa, bought the best of these lands for large, chemical-intensive farms that produce grain, beef, and fruit. Local farmers thereby lost access to lands they had farmed for generations and became dependent on jobs on the new plantations, where they are paid little and endure abusive treatment. The government of Mozambique supports this program because it increases the value of Mozambique's exports and allows it to pay off its debts.

Another issue in agricultural restructuring relates to gender. Agricultural scientists, with their European and North American backgrounds, typically neglected to consider that in Africa, women grow most of the food for family consumption. Hence they did not consult women or include them in development projects, except possibly as field laborers. The displacement of women from the land and food production, as well as the focus on cash crops, contributed to the economic decline in sub-Saharan African countries after 1960 and to loss of sustainable farming practices. Commercial farming techniques often are not sustainable because soils do not remain fertile under continuous use in many moist, humid African environments (see the discussion of African climates and soils on page 338). In Nigeria, Dr. Bede Okigbo, a botanist at the International Institute of Tropical Agriculture (IITA) in Ibadan, had long been aware that traditional farmers (many of them women), growing complex tropical gardens, are highly skilled systems analysts. They can successfully cultivate 50 or more species of plants at one time, sustainably producing more than enough food for their families on a continual basis. "At IITA we were working at a much simpler level than the local farmer," says Okigbo. "We [could manage at most] one or two systems, while they had to deal with something far more complex." By the late 1990s, Dr. Okigbo and other scientists were beginning to persuade international aid agencies to design farm aid projects along lines suggested by experienced local farmers.

Industry and Economic Restructuring. Africa's debt crisis was triggered partly by attempts, prompted by the World Bank, to develop industries for export rather than for domestic markets. For example, in the 1980s, the World Bank financed a shoe factory in Tanzania that was to use that country's large supply of animal hides to manufacture high-fashion shoes for the European market. Machines for making this type of footwear had to be imported at high cost. Due to miscalculations by both Tanzanians and the World Bank, the factory never managed to produce shoes suitable for Europe. Unfortunately, even though there is a large market in Africa for good shoes, the machines could not produce the types of practical shoes Africans would care to wear. The unused factory deteriorated, but Tanzania was still expected to repay the cost to the World Bank.

Whether export-oriented manufacturing industries can succeed is uncertain. The rich industrialized countries place tariffs on African manufactured goods such as footwear and clothing, so Africa lacks access to prosperous markets in Europe and North America. But if Africa sells its products only to other poor regions, the demand for its products may be highly unpredictable because poor people's access to cash fluctuates. Some aspects of SAPs have actually worked against the African export-oriented manufacturing industries they were intended to help. For example, SAPs and belief in the free market have inspired African governments to reduce tariffs that formerly kept out textiles from Asia. The flood of imported cloth has led to factory closings and massive job losses in the textile industries of Tanzania, Zimbabwe, and Uganda. Meanwhile, throughout Africa, currency devaluations have made imported equipment and spare parts for locally focused industries prohibitively expensive.

Economic and industrial recovery began in Africa around 1995, after two decades of stagnation, but the recovery has yet to gain the momentum necessary for Africa to raise its standards of living and improve the well-being of the majority of its people. Investment from both domestic and foreign sources will be required if countries are to broaden and diversify their industrial bases. In its 2001 report on business opportunities in Africa, the Mbendi Group, which provides services for those wishing to invest in Africa, noted that although investment risks are high, the return on successful investments is also higher than in most other world regions, including eastern Europe and Asia—a point reiterated in the 2003 World Bank report cited earlier (page 355). Some observers are now describing African countries such as Ghana, Mozambique, Nigeria, and South Africa as "the last emerging markets." Financial stock markets are opening all over Africa, a sign of increased confidence in the potential for business growth.

The Informal Economy and Economic Restructuring. All in all, the various strategies dictated by SAPs have resulted in hardship for Africa's working people: lower wages, lost jobs, fewer social services, higher school fees, deteriorating working conditions. For many, the escape hatch has been the informal economy. Most people who work in the informal economy perform useful and productive tasks, such as selling craft items, prepared food, or vegetables grown in one of Africa's prolific urban gardens. Others make a living by smuggling scarce or illegal items, such as drugs, weapons, endangered animals, or ivory.

In most African cities, informal trade once supplied perhaps a third to a half of all employment; now it often provides more than two-thirds. This surge of economic activity outside of formal systems is evidence that Africans are working hard to better their own lot during hard times. Informal employment has offered some relief from abject poverty, but it cannot resolve the overarching problems of African economies. Because the informal economy is very difficult to tax, it does not contribute revenue to pay for government services or to repay debt. An informal entrepreneur has trouble building the collateral to expand and create more jobs or to qualify for credit. Some entrepreneurs would gladly register their businesses in order to gain the advantages of being aboveboard, but licensing fees alone are, on average, twice the average annual per capita income, and the bureaucratic snarl can mean that registering a business can take as long as 3 years.

The profits of informal businesses are unreliable, often declining over time as more and more people compete to sell goods and services to people with less and less disposable income. One result is a decline in living standards that is disproportionately severe for women and children. As large numbers of men have lost their jobs in factories or the civil service, they have crowded into the streets and bazaars as vendors, displacing the women and young people who formerly dominated there. Women and children have turned to contract work, manufacturing clothing, shoes, or other items in their homes. Often they work long hours for little pay, using chemicals or techniques that are hazardous to their health. With families disintegrating under the pressure of economic hardship, some women turn to sex work—a growing sector in the informal economy—putting themselves at high risk of contracting HIV-AIDS. More and more children from these disintegrating families must fend for themselves on the streets. Cities such as Nairobi, Kenya, which had very few street children even as recently as 1985, now have thousands.

Alternative Pathways to Economic Development

For many reasons, Africans are seeking alternatives to past development strategies. Regional economic integration along the lines of the European Union, South America's Mercosur, or Southeast Asia's ASEAN is one such strategy. Another is locally designed and locally based grassroots development.

Regional Economic Integration. Many African countries, especially in West Africa, are simply too small to function efficiently in the world economy. Their national markets and their resources cannot nurture a significant industrial base. These countries have remained heavily dependent on Europe; the bulk of their trade is often with former colonial powers. A measure of this dependency is that only 11 percent of the total trade of sub-Saharan Africa was conducted between African countries in 2002. It is difficult to establish regional trade links because the necessary transport and communication networks are not in place. Air travel, and even long-distance telephone calls, from one African country to another often must go by way of Europe, and roads are poor, even between major cities. The forty-plus countries of sub-Saharan Africa each have their own tangle of bureaucratic regulations for business, trade, and work permits, and only a few (Ghana and Tanzania, for example) have attempted to streamline court systems to handle disputes quickly. These impediments make doing business in Africa about 50 percent more expensive than doing business in Asia, so many chose to go elsewhere with their investment money.

Governments in Africa have seen the bad results of surrendering their economic policy authority to distant international lending and finance institutions such as the IMF and World Bank. Therefore, governments are reluctant to surrender what little power they still have to new regional organizations that would implement and streamline economic integration. Nonetheless, there are a number of subregional organizations working toward economic integration. Three of the best known are the **Economic Community of West African States (ECOWAS)**, the **Southern African Development Community (SADC)**, and the **East African Community (EAC)** (Figure 7.15).

ECOWAS, which is dominated by Nigeria, has worked toward removing or reducing tariffs, forming a common currency, forging cooperation among the former British and French colonies of West Africa (about 250 million people), and helping to restore peace in places such as Sierra Leone, Côte d'Ivoire, Liberia, Chad, and Sudan. SADC, which has grown to include 14 countries in Southern and East Africa and includes more than 200 million people, has major projects under way to link transportation and communications infrastructures and to work toward building regional industrial capacity and freer trade. The EAC unites Kenya, Tanzania, and Uganda in an effort to form a common economy, with the possibility of political federation in the future. Participants in this effort, which is supported by the European Union, hope that a vibrant coastal economy in this region will ultimately help the landlocked and still struggling countries of Rwanda, Burundi, Malawi, and Zambia (Congo [Kinshasa] may be added in the future) to find investors for more sustainable development. The **Common Market for Eastern and Southern Africa (COMESA)**, which now includes 20 countries, was established in 1984. Another subregional organization, the **Economic Community of Central African States (CEEAC),** is marginally active because of widespread civil unrest since the early 1990s, but the goal here as well is an economic union.

All of these efforts at subregional integration owe a debt to the Organization of African Unity (OAU), a pan-African movement founded by President Kwame Nkrumah of Ghana in the late 1950s. The OAU, now superseded by the **African Union,** includes all the countries on the African continent, plus nearby islands. It was broadly conceived as an antidote to European dominance; it is now an increasingly active, but still loose, union that promotes economic cooperation and general social welfare. Its activities continue to evolve; in 2003 the African Union Summit in Maputo, Mozambique, voted to form a Pan-African Parliament with five members from each of the countries on the continent. The main goal of the parliament is to develop advisory position papers focused on helping member states, and the continent as a whole, to increase Africa's prosperity and global prominence and to enhance civil society institutions, women's rights, and environmental conservation.

Grassroots Development. Another alternative development strategy for Africa is **grassroots economic development,** aimed at providing sustainable livelihoods in the countryside as an alternative to urban migration as well as coping strategies for poor urban residents. One of the most promising ideas is **self-reliant development,** which consists of small-scale projects primarily in rural areas. These projects use local skills, create local jobs, produce products or services for local consumption, and maintain local control so that participants retain a sense of ownership. One district in Kenya has more than 500 such self-help groups. Most members are women who terrace land, build water tanks, and plant trees. They form farm cooperatives for small-scale production of poultry, honey, fish, dairy products, and crafts. They also build houses, schools, and barns; run bookshops, nurseries, and restaurants; and form credit societies. Nonetheless, there are limitations to such efforts. For one thing, they require numerous facilitators with extraordinary skills in managing people. Moreover, the success of locally funded self-reliant development strategies has tempted governments to not give rural communities their fair share of financial support from tax monies.

Examining the issue of rural transport illustrates how an Africa-centered perspective and a focus on local needs might generate improvements that differ from those that would be advocated by large external development agencies. When non-Africans learn that transport facilities in Africa are in need of development, they usually imagine building and repairing roads, railroads, airports, buses, and the like. But a recent study that analyzed transport on a local level—the time spent, the distances traveled, the loads

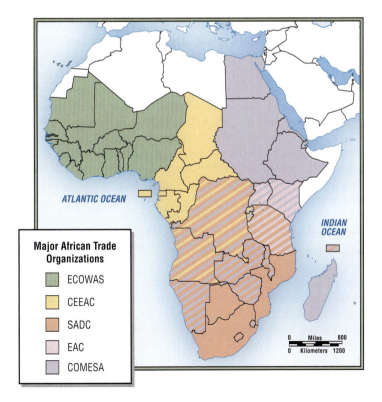

FIGURE 7.15 Principal trade organizations in sub-Saharan Africa.

carried—discovered that women provide a major part of village transport and that most of the goods moved *are carried on their heads*! Women "head up" firewood from the forests, crops from the fields, and water from wells, carrying these loads to their homes, as well as carrying goods bound for rural markets (see the photo below). Eighty-seven percent of this domestic load bearing is on foot; on average, an adult woman moves the equivalent of 44 pounds (20 kilograms) more than 1.25 miles (2 kilometers) a day and spends about 1.5 hours a day doing so. Men tend to carry fewer of these local burdens. In Kasama, Zambia, for example, an adult woman transports the equivalent of 35.7 tons (32.4 metric tons) a year, compared with 7.1 tons (6.4 metric tons) transported by the typical man, no matter what his conveyance. Yet even when development agencies have examined transport at the household and village level, they have focused exclusively on transport by men, suggesting bicycles, wheelbarrows, donkey carts, and pickup trucks to help men out. As forests are depleted and water becomes scarcer, the time spent, loads carried, and distances traveled by women increase. Mothers usually recruit their daughters to help them carry their loads, thus cutting into the girls' time in school.

Analysis of grassroots development shows that tiny changes can make big differences in people's lives—differences that are less disruptive and more sustainable than the changes wrought by large-scale development. Giving women access to low-tech transport such as donkeys and bicycles would free some of their time for education, short training courses, and perhaps small-scale businesses. These improvements may forestall migration to cities.

Technological Development. A generation of modern technological entrepreneurs is emerging in Africa. Well educated and familiar with global communication, they are buoyed by Africa's potential in human and natural resources rather than intimidated by its past problems. These entrepreneurs, who are in their 20s,

American geographer Barbara McDade, a professor at the University of Florida, was inspired by a college course to study Africa. Most recently, she has studied how African entrepreneurs who work in information technology, telecommunications, horticulture exports, retail stores, and electronics form regional networks with one another.

30s, and 40s, were often educated in Europe or North America and first experienced success as young professionals in these wealthy economies. Wishing to avoid becoming part of Africa's brain drain, they developed a desire to find similar opportunities in their own countries. They have established businesses in telecommunications, radio production, graphics and publishing, Internet services, furniture manufacturing, glass manufacturing, bookstore franchises, horticulture, and cut flower exports, among others. In 1994, Africa Online, now the continent's largest Internet Service Provider (ISP) was founded by three young Kenyans who returned from the United States after graduating from MIT, Harvard, and Princeton, respectively. Africa Online has become the leading pan-African Internet service, with operations in Kenya, Uganda, Ghana, Namibia, Zimbabwe, and several other African countries. Although Internet service is still largely confined to the main cities, by the end of 2000, all 54 countries in Africa had achieved some permanent connectivity and had installed some local full-service dial-up ISPs.

Political Issues: Colonial Legacies and African Adaptations

Few places in the world today seem as politically turbulent as Africa. The news media bring us reports of civil wars, genocide in Rwanda, and military despots who execute members of the opposition. We also see the lingering effects of despotic presidents such as Mobutu Sese Seko, who plundered the national treasury of Congo (Kinshasa) for decades while millions of Congolese went without food, shelter, health care, and education (see page 380). Africa's civil wars are partly a legacy of European colonialism and cold war geopolitics and partly the result of Africa's own ethnic and regional

Women returning from market at Kisangani, Congo (Kinshasa), pass by Tshopo Falls, on a tributary of the Congo River. [Tom Friedmann/Photo Researchers.]

FIGURE 7.16 Ethnic groups in sub-Saharan Africa. The map shows the large number of ethnic groups spread across the continent of Africa. Superimposed on the pattern of ethnic groups are the present national boundaries. Very rarely do ethnic group and political boundaries match. [Adapted from James M. Rubenstein, *An Introduction to Human Geography* (Upper Saddle River, N.J.: Prentice Hall, 1999), p. 246.]

Ethnic Group Boundaries

— Ethnic groups

issues, which have been manipulated by unscrupulous elites in efforts to accrue personal power and wealth. Fortunately, there are many reasons for optimism for a more peaceful and prosperous future. Political and economic changes are reducing the power of entrenched elites and forcing them to allow democratic freedoms that, in the future, may bring more responsible government and better economic potential.

Origins of Conflict. In 1990, 45 percent of the population of sub-Saharan Africa was considered at risk of suffering harm resulting from armed conflict of some sort. In many ways, the prevalence of conflict is related to Africa's experience with European colonial rule. Traditional systems of governance were dismantled, and the continent's resources were diverted to other world regions. National borders, which were established by Europeans during the colonial era, split some ethnic groups; others resulted in very different and sometimes hostile groups sharing the same country (Figure 7.16). In too many cases, years of carnage followed independence as some African governments, though not racist in the way of their colonial predecessors, nonetheless practiced ethnic discrimination. Civil wars arose from attempts to crush ethnic or regional separatist movements in places such as Nigeria and Congo (Kinshasa), and from general repression of disadvantaged minorities. The state even encouraged interethnic violence bordering on genocide in Congo (Kinshasa), Uganda, Rwanda, and Burundi. In places such as Congo (Kinshasa), Nigeria, Liberia, and Sierra Leone, conflicts were worsened by military establishments that were only loosely

under the control of civilian governments or that seized control through coups d'état. (The conflict in Nigeria is examined below.)

The European colonial strategy of using local elites to control local populations often created or exacerbated ethnic tensions within a colony and led to minority rule after independence. The British in Uganda, the Germans in Tanzania, and the Belgians in Rwanda and Burundi gave local African elites the responsibility for collecting taxes and drafting laborers for various colonial construction projects. In Rwanda and Burundi, for example, the elite Tutsi were empowered to take advantage of the more numerous Hutu. This practice destabilized long-standing relations between these two groups and was partly responsible for ethnic bloodshed in the 1990s.

The Case of Conflict in Nigeria. Nigeria is the dominant country in West Africa economically because of its large population, its oil reserves, and its relatively diversified economy. It is the most populous country in Africa, and its 123 million people outnumber the total of those who live in the rest of West Africa (see Figure 7.6 on page 343). Like the other countries of West Africa, Nigeria is a culturally complex nation. It presents an interesting case study of how colonial and postcolonial rule have led to increasing tensions, which are too often attributed to conflict among the various ethnic groups that have long occupied the area that is now Nigeria.

As geographer Robert Stock observes, "Nigeria was, and still remains, a creation of British imperialism." When Europeans drew the borders of African countries, they did not place them along natural physical boundaries, or along traditional borders between different ethnic groups. Instead, they deliberately plotted borders to split up some ethnic groups (see Figure 7.16). Conversely, the Europeans often forced disparate groups into one unusually diverse country. That is what the British did with Nigeria, which has 395 indigenous languages in 11 language groups (Figure 7.17).

In present-day Nigeria, there are four main ethnic groups. The Muslim Hausa (21 percent of the population) and Fulani (9 percent) have, until recently, lived mostly in the north, where the moist grassland grades into dry savanna and then semidesert. The Hausa are traders, dryland farmers, and animal herders. Since independence in 1960, they have dominated both the government of Nigeria and its military. The Yoruba (20 percent) live in southwestern Nigeria in the moister woodland and grassland environments. The Igbo (17 percent) are centered in the southeast in and near the tropical rain-forest zone. Hundreds of other ethnic groups, such as the Ogoni, live among and between these larger groups.

Colonial administration created a north-south dichotomy, especially in education. Among the Hausa, the British let local Muslim leaders administer community life. These leaders did not encourage public education. In the south, among the Yoruba and Igbo, Christian missionary schools were common. At indepen-

dence, the south had more than ten times as many primary and secondary school students as the north. This disparity in education led to a disparity in income, with southerners holding most government civil service positions, yet the poorer northern Hausa continued to dominate politically. Bitter disputes have erupted over the distribution of development funds and jobs. Accusations of rigged census data and election fraud usually fall out along ethnic and north-south divisions.

In 1966, after an Igbo-led military coup and a Hausa-led countercoup, the slaughter of 30,000 Igbos ignited the 3-year Biafran War. The Igbo portion of southeastern Nigeria tried to secede (unsuccessfully). More than 200,000 people died as a result of the war and ensuing food shortages.

But an even more important actor in Nigerian politics is oil, a major resource that has brought what political geographer Michael Watts calls "rapacious frontier capitalism," and environmental violence. Oil is Nigeria's primary resource: 90 percent of its foreign trade earnings come from the sale of oil. Nigeria's oil reserves have the potential to make it the richest country in Africa, but this resource has been mismanaged to enrich local military leaders and their associates and foreign oil companies, especially Shell Oil. A Nigerian commission reported that between 1990 and 1994, the military regime in power stole $12 billion in oil money, and a sub-

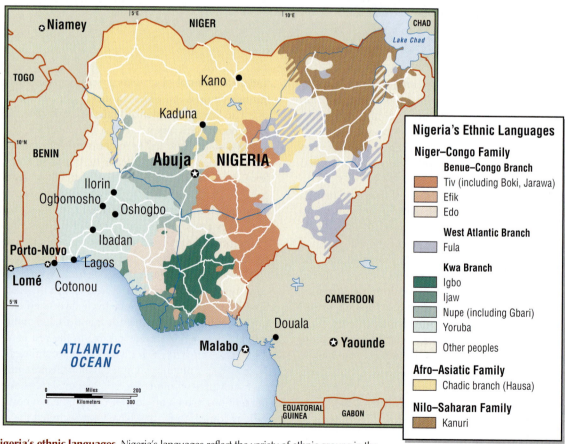

FIGURE 7.17 Nigeria's ethnic languages. Nigeria's languages reflect the variety of ethnic groups in the country. The 11 main languages are indicated in color, but many smaller groups speak distinct languages or Creole dialects not shown on this map. [Adapted from James M. Rubenstein, *An Introduction to Human Geography* (Upper Saddle River, N.J.: Prentice Hall, 1999), p. 167.]

sequent regime is thought to have taken another $12 billion. Officially, Nigeria earns about $10 billion a year from oil, so corrupt officials robbed the Nigerian people of an amount equal to half of the official earnings.

During the 1990s, ethnic and regional differences continued to mold Nigerian affairs, but were further complicated by oil politics. Much of Nigeria's oil is located on lands occupied by the Ogoni people, which lie along the northeastern edges of the Niger River delta. Yet although the state in which the Ogoni live produces about 44 percent of Nigerian oil, virtually none of the profits from this production, and very little of the oil itself, go to the state as a whole, much less to Ogoniland. Although Ogoniland has received few benefits from oil extraction, it has suffered from the resulting pollution. Oil pipelines crisscross Ogoniland, and spills and blowouts are frequent. Oil spills many times the size of the *Exxon Valdez* spill in Alaska have happened repeatedly. There have been 111 spills between 1985 and 1994. Natural gas, an untapped byproduct, is burned off, even though it could be used to generate electricity—something many Ogoni go without.

In the 1990s, a prominent Ogoni writer and businessman, Ken Saro-Wiwa, began to help the Ogoni organize a protest movement. The movement proposed major changes in Nigerian federal organization, alliances with pro-democracy movements in Nigeria, and direct action against Shell and Chevron Oil installations. By its own admission, in 1996, Shell Oil netted $200 million in profits yearly from Nigeria, but in 40 years gave only $2 million to the Ogoni community, whose oil they appropriated. In 1995, the world community was outraged when Ken Saro-Wiwa was summarily executed with nine other Ogoni by the Nigerian government.

 *It is a vicious circle—you cannot get investment while there is violence and killing and disturbance and then if you don't get investment which [would create] employment then the killing, the violence will go on. So it is a very, very bad vicious circle.*

PRESIDENT OLUSEGUN OBASANJO OF NIGERIA,
INTERVIEW WITH THE BBC, 2002

In 2002, Amina Lawal, then pregnant, was sentenced to death by stoning for adultery. She lived in a Muslim village in northern Nigeria, where shari'a had replaced the corrupt secular legal system 3 years before, when the military dictatorship ended. The democratic government in Abuja, in the Christian south, declared she would never be executed. Her case exacerbated tensions between the different ethnic groups in the north and south. This simmering conflict erupted during preparations for the Miss World pageant to be held in Abuja that year. When a Christian fashion writer suggested that Muhammad would surely have chosen one of the beauty contestants for his wife, the northern Muslims were offended. Riots between Muslims and Christians broke out in the northern city of Kaduna, and the pageant, which Christian Nigerian officials had hoped would help to burnish Nigeria's image globally, packed up and left.

Amina Ladan-Baki, a banker and world rights activist, believes that the discord between ethnic and religious groups is manipulated to suit the agendas of politicians. "The politicians mislead people. They use religion. They use diverse cultures to unite or disunite. 'Vote for me because I am of your stock.'"

President Olusegun Obasanjo, quoted above, believes the violence is due to poverty.

In February 2004, a shari'a court of appeals overturned Amina Lawal's conviction. However, other women have been charged with adultery in the 12 Nigerian states that follow shari'a, and they face the possible sentence of death by stoning. Meanwhile, tensions between the north and the south make it unlikely that the north will relax its conservative approach to Islam.

To learn more about Amina Lawal and about the beauty pageant, watch the FRONTLINE World video, "The Road North: What the Miss World Riots Reveal about a Divided Country."]

Can geographic strategies resolve the tensions in Nigeria? While enduring these divisive and violent civil disputes, some Nigerians have continued to pursue solutions to unwieldy ethnic and political spatial patterns. One strategy has been to create more political states (Nigeria now has 30) and thereby reallocate power to smaller local units, something Ken Saro-Wiwa had advocated during his reform protests. Recently, large, wealthy states have been subdivided to ensure more equitable distribution of development money and oil profits and to prevent any one state from seizing control of the central government. Although dividing the country into more states has increased administration costs, it also seems to have eased regional ethnic and religious hostilities. Another strategy to defuse ethnic conflict has been to relocate the capital city from Lagos on the Atlantic coast to Abuja, an interior village north of the confluence of the Niger and Benue rivers, in neutral territory away from the heartlands of the Hausa, Yoruba, and Igbo. Abuja is a **forward capital**: a capital city built to draw migrants and investment to a previously underdeveloped area. The cost of establishing Abuja has increased Nigeria's debt by $4.5 billion, and it turns out that most civil servants are unwilling to move from the vibrant cosmopolitan city of Lagos.

The Cold War in Africa. The cold war between the United States and the former Soviet Union deepened and prolonged many conflicts in sub-Saharan Africa. After independence, some African governments sought to redress the problems of economic underdevelopment and exploitation during the colonial period by turning to socialism, often receiving economic and military aid from the Soviet Union. Still gripped by bitter anticommunism, the United States (with its allies) tried to undermine these regimes by arming and financing rebel groups—as it did in Angola, Congo (Kinshasa), and Namibia—or by pressuring other governments in the region to intervene. In the 1970s and 1980s, the United States encouraged South Africa's apartheid government, which itself was facing increasing political pressure from the Soviet-backed African National Congress of Nelson Mandela, to carry out military interventions against socialist governments in Namibia (considered at the time a sort of colony of South Africa), Angola, and

During the Mozambique civil war, the world's arms makers supplied enough weapons to arm nearly every person in the country with a rifle. More than 6 million AK-47s alone entered Mozambique (a country with only 18 million people). In a recent move to reduce guns among the population, the Anglican Church began buying back the weapons in a Guns for Plowshares program. Guns are exchanged for tools and agricultural implements, sewing machines, bicycles, schoolbooks, and other useful items. The weapons are immediately destroyed. On any given day, one might encounter the unlikely sight of a church employee sawing guns in half in the churchyard. [João Silva/ CORBIS SYGMA.]

African traditions of patronage (wherein the leader is expected to look after the needs of the people) to create huge, inefficient government bureaucracies, with corruption penetrating to the highest levels. For example, in Nigeria, which has sub-Saharan Africa's second largest economy (after South Africa's), political and economic elites appropriated immense wealth from the country's large oil industry. Hence, today, despite Nigeria's oil wealth, most Nigerians are no better off than any other Africans. When poor Nigerians took actions that could have reduced oil profits—for example, by asking for compensation for damages created by oil spills—they suffered brutal government repression. But again, some optimism is in order: as of the mid-2000s, several countries (South Africa, Botswana, Mozambique, and even Nigeria) enjoy responsible leadership that has steered them through difficult times and has still managed to achieve economic growth and social stability. The pan-African movements discussed on page 357 now focus on fostering transparent government, and the movement itself practices and reinforces participatory democracy.

Shifts in African Geopolitics

Signs of progress toward democratic systems are especially visible in the southern part of the region. For years, the white-dominated government of South Africa tried to prolong white rule throughout Southern Africa. Now that white political control has ended in South Africa, the country's leaders are pursuing friendlier and more cooperative relations with neighboring countries, bringing greater stability to Southern Africa as a whole.

Across Africa, legal opposition parties have been able to express dissent more openly and freely. There is hope for a "second independence" as autocratic, corrupt ruling elites are replaced or moderated by political parties that enjoy the electoral support of the majority of the people. Botswana, Kenya, and South Africa are examples of this trend. At least one encouraging statistic supports hopes for more democratic government across Africa. In 1990, only four countries in sub-Saharan Africa were pluralist democracies. During the 1990s, 47 states promised or allowed multiparty elections (Figure 7.18). By the end of the twentieth century, even Nigeria, long under a military dictatorship, had held elections. The trend toward democracy promises greater prosperity for the region, because democracy allows for greater transparency in government proceedings, public criticism of leaders, chances for the opposition to speak, and opportunities to replace leaders through elections. In these ways those who rule are held accountable for the consequences of bad or corrupt policies.

Despite this trend, there remain several states—including Sierra Leone, Liberia, Ethiopia, Rwanda, Malawi, and Congo (Kinshasa)—where violence or open warfare make fair elections impossible. Two countries, Côte d'Ivoire and Zimbabwe, have devolved from democracies with growing economies to autocratic governments sponsoring economic decline and degenerating well-being. Côte d'Ivoire suffered a military coup d'état in 1999. Zimbabwe struggled against a long downward slide—the result of inappropriate economic strategies, increasing social unrest due to spreading poverty, and authoritarian government—culminating in rigged elections in 2002 that left Robert Mugabe in power.

Mozambique. Another major area of cold war tension was the Horn of Africa, where the governments of Somalia and Ethiopia alternately received aid from the United States and the Soviet Union, much of which went to fund fighting between those two African nations.

Elite Rule. For a time after independence, African elites in a number of countries continued colonial traditions of authoritarian and corrupt rule. Often a single political party monopolized power, and leaders repressed any viable political opposition. The elites argued that the open political debates of Western democracies were incompatible with African traditions of communal identity and collective decision-making. Similarly, they tended to warp

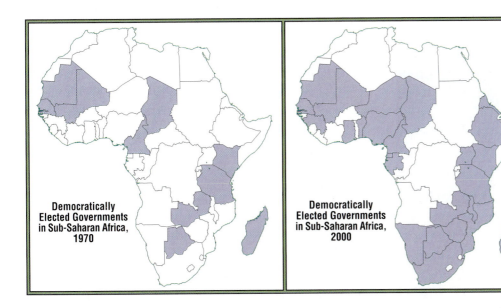

FIGURE 7.18 Democratically elected governments in sub-Saharan Africa. Between 1970 and 2000, the number of democratically elected governments (color) in sub-Saharan Africa rose and fell, but by 2000, the number had increased nearly two and one-half times, from 11 to 26. [From Barbara E. McDade.]

SOCIOCULTURAL ISSUES

A majority of sub-Saharan Africa's people appear to live traditional lives in rural villages. A closer look, however, reveals the influences of the Western colonizers and the modern world on such aspects of traditional culture as religion, gender roles, and community life. The growing numbers of people who are moving to the cities find their lives transformed still further as they exchange the isolated village for crowded shantytowns. Yet even in the cities, traditional African culture is visible in religious patterns, domestic life, and especially in the plentiful urban subsistence gardens.

Settlement Patterns

Human settlements can take many forms, from single isolated agricultural homesteads to modern high-rise cities. These various types of living arrangements reflect how people relate to one another economically, politically, and socially. This section discusses the two main settlement patterns found in sub-Saharan Africa: small rural villages and dense low-rise cities. There is a trend toward more urban settlement and the high-rise model of European, Asian, and American cities.

Rural Settlements. More than 70 percent of sub-Saharan Africa's people still live in villages, making this region, along with South Asia, a very rural area. There are thousands of versions of the African village (see the photos on page 364), with many different types of houses and village arrangements, all depending on the cultural heritage of the inhabitants. But there are some constants in African villages, too. People usually live in extended family compounds, consisting of several houses arranged around a common space in which most activities take place. These compounds may stand alone, dispersed across an agricultural landscape, or they may be grouped with other compounds and surrounded by fields

---

AT THE REGIONAL SCALE   African Women in Politics

Charity Kaluki Ngilu will never forget the day she became a professional politician. She was washing dishes in her kitchen in Kitui, about 75 miles (120 kilometers) east of Nairobi, Kenya, when she saw a group of women approaching her back door. The women had worked with her to build better waterworks and health clinics in the town. Mrs. Ngilu answered the knock on the door, drying her hands on an apron. The women said they wanted her to run for parliament in Kenya's first multiparty elections. She assumed they were joking.

That was in 1992. Ngilu not only beat the governing party's incumbent, but then became a major advocate for women's issues. On July 6, 1997, Ngilu became the first woman to run for president in Kenya. Her very candidacy was a blow to convention because, as in most of Africa, Kenyan women are not likely to vote at all, and the majority of men are not likely to support female candidates. Never-theless, many men supported her because they felt she was capable of making bigger changes than a man could. Although she didn't win (she came in fifth), her campaign focussed attention on old colonial-era laws that crippled those in opposition to the government by allowing them to be jailed. In 1999, Ngilu became a member of the board of Parliamentarians for Global Action–a group of legislators focused on environmental planning efforts at the global scale. In 2000, she was in her second term in parliament. By 2004, she was also Minister of Health, in charge of Kenya's greatest challenge, the response to the HIV-AIDS epidemic.

[*Source:* Adapted from James C. McKinley, Jr., "A woman to run in Kenya? One says, 'Why not?'" *New York Times* (August 3, 1997): 3; Kennedy Graham, ed., *The Planetary Interest: A New Concept for the Global Age* (London: Rutgers University Press, 1999); Deutsche Stiftung für International Entwicklung, http://www.dse.de/ef/ded/ngilu.htm, February 2, 2004.]

These photos show three of the many types of villages found in sub-Saharan Africa. *Top:* One of the very colorful Ndebele villages in Zimbabwe. *Middle:* Cattle are returned to a Masai village compound in East Africa after a day's grazing. *Bottom:* Granaries with removable thatch roofs sit alongside houses in this Dogon village in Mali. [*top:* Walter Bibikow/Index Stock Imagery/PictureQuest; *middle:* Frans Lanting/Minden Pictures; *bottom:* Henning Christof/Das Fotoarchiv.]

designated for each family. Villages, and the compounds that make them up, are economic units. Subsistence agriculture is almost always an important component of village life, with labor carefully apportioned by custom. Increasingly, rural people also work seasonally on commercial farms for pay or have several other ways of earning small amounts of cash.

Urbanization. Although the majority of Africans live in rural areas, cities are not a recent settlement pattern in Africa. Many cities are long established, with histories extending back to the age of the great empires of West Africa and the kingdoms of East and Southern Africa. In the past, the cultural and social institutions of these cities focused on maintaining social and economic continuity with the surrounding rural areas. Although the village remains the predominant settlement type, African cities have grown very quickly indeed in recent decades, and their links with rural areas are now more tenuous. The average annual growth rate of African cities—more than 5 percent—exceeds even the growth rates of cities in Asia (3 percent) and Latin America (2.5 percent). In the 1960s, only 15 percent of Africans south of the Sahara lived in cities; now about 30 percent do, but the percentage of the population that is urban varies considerably among countries (Figure 7.19).

African cities have attracted massive migration for at least two reasons. First, life in rural villages and towns is perceived as offering few jobs and little opportunity for upward mobility. Second, there is a widely held misconception that life in the cities offers quick access to money and prestige. In reality, as discussed on page 356, a majority of Africa's city dwellers work mostly at very low wages in the informal economy. On the other hand, urbanization can add a new dynamism to economic development in Africa by attracting

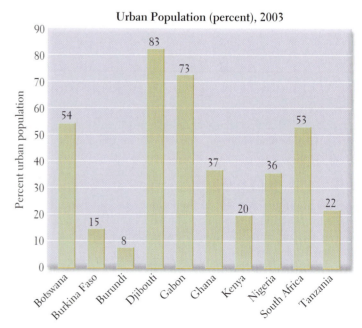

Urban Population (percent), 2003

FIGURE 7.19 **Urbanization in selected countries, 2003.** [Data from *2003 World Population Data Sheet* (Washington, D.C.: Population Reference Bureau).]

African urban landscapes vary a great deal, ranging from contrasting patterns of shantytowns and modern mid-rise office buildings to single-family homes and apartment towers to elegant villas set in landscaped grounds. Marketplaces, both formal and impromptu, are common features and are the sites of important social interactions. *Above:* Closely packed houses in a shantytown near downtown Lagos, Nigeria, leave little room for ventilation in the tropical heat. The dwellings surround a parking lot for the cars of office workers. *Right:* An urban view of Nairobi, Kenya, a major economic center in East Africa. [*Above:* L. Gilbert/CORBIS SYGMA; *right:* B. E. McDade/University of Florida.]

talented young people to educational institutions and then to jobs in industry and services. African cities also are attempting to recruit those who have migrated to Europe and North America by advertising high-tech or managerial jobs and urban lifestyles now available in Africa.

As is often the case in developing countries, most African countries have one very large primate city, usually the capital, which attracts virtually all migration. (The primate city phenomenon is discussed in Chapter 3, page 128.) Kampala, Uganda, for example, is almost ten times the size of Uganda's next largest city. The population of another typical primate city, Kinshasa in the Democratic Republic of the Congo, has soared from 450,000 in 1960 to an estimated 4.6 million in 2000. The population of Lagos, Nigeria, was 48 times larger in 2000 (11 million) than it was in 1950 (230,000). Such rapid and concentrated urbanization usually overwhelms a country's urban infrastructure: housing, sewage treatment, schools, utilities, transport, and government bureaucracy.

Typically, governments have paid little attention to the housing needs of the throngs of poor urban migrants. Most migrants have had to construct their own shanties on illegally occupied land, using materials that would be discarded as trash in North America, Europe, Singapore, or Japan. They live in vast squatter settlements surrounding the older urban centers. Transport in these huge and shapeless settlements is a jumble of government buses and private vehicles. People often have to travel long hours to reach distant jobs, getting most of their sleep while sitting on a crowded bus. They spend a good deal of their travel time moving at a snail's pace through extremely congested traffic. Africa's mid-level officials are aware of the transport problem because they deal with it themselves.

For some members of the small but growing middle class, new homes and jobs may be part of a single project. In South Africa and Botswana, a few industries are providing housing or locating close to housing developments, thus resolving the home-to-job transport problem for at least some of their workers. The housing NGO, Habitat for Humanity, active in many parts of Africa, aims to build not just houses, but self-sustaining communities. It has been active in dozens of African urban areas for several decades.

Religion

Religion is particularly important in African daily life. The region's rich and complex religious traditions derive from three main sources: the indigenous belief systems, Islam, and Christianity. Hinduism, like Islam and Christianity, was introduced from outside Africa, but its influence is found primarily in specific locations along Africa's east coast and on the eastern coastal islands, where there are significant East Indian or mixed African-Indian populations.

Indigenous Belief Systems. Traditional African religions, sometimes classed as **animism,** probably have the most ancient heritage of any religion on earth and are found in every part of Africa. Map

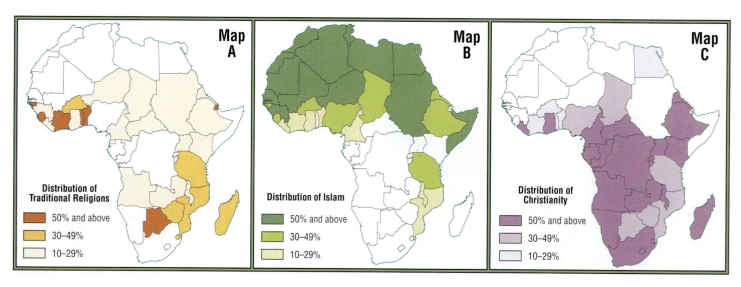

FIGURE 7.20 Religions in Africa. [Adapted from Mathew White, "Religions in Africa," in *Historical Atlas of the Twentieth Century* (October 1998), at http://users.erols.com/mwhite28/afrorelg.htm.]

A in Figure 7.20 shows the countries in which traditional beliefs remain particularly strong. According to these beliefs, all beings, including plants and animals, are part of a whole that is situated in, and part of, the physical environment. African beliefs and rituals seek to bring the vast host of departed ancestors into contact with people now alive—who, in turn, are the connecting links in a timeless spiritual community that stretches into the future. The future is reached only if present family members procreate and perpetuate the family heritage in other ways, such as by storytelling and making offerings to the spirits. The people in a living community are usually led by a powerful man, or occasionally a powerful woman, who combines the roles of politician, patriarch or matriarch, and spiritual leader.

According to traditional African beliefs, the spirits of the deceased are all around—in trees, streams, and art objects, for example. In return for respect (expressed through ritual), these spirits offer protection from life's vicissitudes and from the ill will of others. Rosalind Hackett, a scholar of African religions and art, writes that "African religions are far more pragmatically oriented than Western religions, being concerned with explaining, predicting and controlling misfortune, sickness and accidents." She says they give believers a way "to make sense of and act upon adverse forces in their public and private lives." African religions remain fluid and adaptable to changing circumstances. For example, Osun, the god of water, traditionally credited with healing powers, is now invoked also for those suffering economic woes.

Religious beliefs in Africa are not static, but evolve continually as new influences are encountered. If Africans who practice traditional beliefs convert to Islam or Christianity, they commonly retain parts of their indigenous religious heritage and blend them with aspects of the new faith, creating a fresh entity—a process sometimes referred to as **syncretism** (or **fusion**). The maps in Figure 7.20 show an overlap of beliefs, but they do not convey the blending of two or more belief systems, which is widespread. In the Americas, the African Diaspora has participated in the creation of new belief systems developed from the fusion of Roman Catholicism and African beliefs: Voodoo in Haiti; Obeah in Jamaica; Condomble in Brazil; and, more recently, Santeria in Cuba, Puerto Rico, and North America. In Africa, Christian missionaries of the established churches—Roman Catholic, Episcopal, Baptist, and so forth—often refused to allow Africans to incorporate their folk beliefs. This rejection helped to spawn

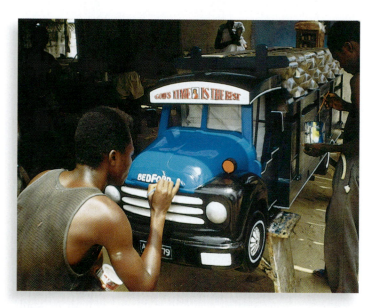

Coffin art in Ghana. Ghana's coffin carvers accomplish an apparently seamless interweaving of religion, art, and modern life. They are experts at interpreting the life of the deceased artistically: a Mercedes-Benz for a chauffeur, a plane for a pilot, a boat for a fishing fleet captain. Both rich and poor prize the coffin carver's art; people may commission a coffin long before death comes, often spending their life savings to buy a niche in the cultural memory of the community. [Carol Beckwith and Angela Fisher/Robert Estall Photo Library, U.K.]

AT THE REGIONAL SCALE The Gospel of Success

We first encountered the "gospel of success" in Chapter 3. In Africa, this adaptable version of evangelical Christianity combines certain interpretations of Christianity and capitalism with Central and West African beliefs in the importance of sacrificial gifts to ancestors and in the power of miracles. The message is simple: according to preachers such as those at the Miracle Center in Kinshasa, Congo, "The Bible says that God will materially aid those who give to Him. . . . We are not only a church, we are an enterprise. In our traditional culture you have to make a sacrifice to powerful forces if you want to get results. It is the same here." The supplicants accept the message that generous gifts to such churches will bring divine intervention to alleviate their miseries, whether physical or spiritual. They donate food, television sets, clothing, and money—one woman gave three months' salary in the hope that God would find her a new husband. The collection plates are large plastic bags. By combining spiritual ministration with a touch of hucksterism, the leaders of this largely urban movement receive the respect and adulation that Africans used to reserve for rural village leaders.

Like all religious belief systems, the gospel of success is best understood within its cultural context. Many of the believers, new to the city, feel isolated and are seeking a supportive community to replace the one they left behind. People view their material contributions to the church as similar to the labor and goods they previ-

Worshipers at the Miracle Center in Kinshasa, Congo. [Robert Grossman/CORBIS SYGMA.]

ously donated to maintain their standing in a village compound. In return for these dues, members receive social acceptance and community assistance in times of need.

---

independent, charismatic Christian sects that better suited the need of Africans for religions that specifically represented their culture and their experience. These independent churches have given rise to "gospel of success" churches, which have grown quickly in recent decades (see the box "The Gospel of Success" above).

Islam and Christianity in Africa. Islam began to extend into Africa south of the Sahara, especially East Africa, soon after Muhammad's death in 632 (see the discussion of Islam and Christianity in Chapter 6). As Figure 7.20 shows, Islam is now the predominant religion throughout North Africa and is important in much of West and East Africa, including the central coastal zone on the Indian Ocean, where Islamic traders were especially active. When the British began to colonize West Africa formally in the late 1800s, they obtained the help of Islamic leaders in governing the countryside and towns, especially in the drier northern areas. For this reason, the British did not encourage Christian missionaries in the interior away from the coast; hence, today, the descendants of these Islamic administrators are still politically powerful in Nigeria and some other parts of West Africa.

Christianity came to northeastern Africa before it spread to Europe. In the fourth century, Christianity was adopted by the people of the kingdom of Aksum, the antecedent of modern-day Ethiopia. Today, the Ethiopian Coptic Church maintains a tradition of Christian practices derived from this long heritage. Christianity began to spread into sub-Saharan Africa in the nineteenth

century, when Methodist missionaries from Europe and America became active along the coast of West Africa. Roman Catholic missions followed, and today Christianity is prevalent along the coast of West Africa and in Central and Southern Africa. Christian missionaries in Central and Southern Africa found a niche as providers of the education and health services that colonial administrators had neglected. In East Africa, Christians are either the majority or an important minority. There, as in West Africa, colonial administrators in the British-held territories discouraged Christian missionaries from going north into Muslim territory (for example, northern Sudan) because Islamic leaders were already aiding the colonization process.

In the 1980s, old-line established churches began to gain adherents in East and West Africa. For example, the Church of England (Anglican Church) has been growing so rapidly in Kenya, Uganda, and Nigeria that by 2000 there were more Anglicans in these countries than in the United Kingdom. African Anglicans appear to be conservative on social issues—they are against female priests and the acceptance of homosexuals—but they are often liberal on economic issues. For example, in 1998, African Anglican bishops persuaded the worldwide Anglican Communion to oppose holding the world's poor countries to crippling debt payments. The Anglican Church in Africa attracts the educated urban middle class, whereas modern evangelical versions of Christianity appeal to the less educated, more recent urban migrants—the most rapidly growing cohort of African populations.

Gender Relationships

Long-standing African traditions dictate a fairly strict division of labor and responsibilities between men and women. The exact allocation of work, power, and family wealth by gender varies from subregion to subregion. In general, however, women are responsible for domestic activities, including rearing the children, tending the sick and elderly, and maintaining the house. Women carry about 90 percent of the water, collect 80 percent of the firewood, and produce and prepare nearly all the food. In those countries where rice is popular (and it is increasingly the staple of choice for urban working women with families to feed), rural women are the primary cultivators of rice. Labor studies show that women also handle about 50 percent of the care of livestock, and that when there are small agricultural surpluses or handcrafted items to trade, it is women who transport and sell them in the market. Throughout Africa, married couples tend to keep separate accounts and manage their earnings as individuals, so when a wife sells her husband's produce at market, she usually gives the proceeds to him.

Men are usually responsible for preparing land for cultivation, a job that often entails clearing large trees and heavy brush with hand tools. Men then stack the refuse, let it dry and burn it, and turn the soil, a job that is harder than it looks. In the fields intended to produce food for family use, women sow, weed, and tend the crops as well as process them. In the fields where cash crops are grown, men perform most of the work. Women may help to weed and harvest these fields, but it is usually understood that any earnings belong to the men. When husbands in search of cash income migrate to work in the mines or in urban jobs, women take over nearly all agricultural work. Studies have shown that the majority of agricultural laborers in Africa are women, and that they contribute about 70 percent of the total time spent on African agriculture. They tend to work with hand tools in the fields and in the home, and during their reproductive years, they often do field labor with a child strapped on their backs.

In rural areas, African men, on average, do not have as many responsibilities as women, nor do they work as hard or for as many hours. The reasons for this situation are complicated. Over the last century, many men have migrated to the cities or the mines to work, at least seasonally. There they may endure long hours of hard labor, difficult commutes, and crude living conditions. But men's labor has been mitigated more often than women's by labor-saving machines. As the geographer Robert Stock observes, rural African women do not have the "double day" that women in industrial societies have, but rather a "double double day," in that so much of their work entails backbreaking physical labor and time-consuming treks for water and fuel, usually without the aid of even a wheelbarrow or bicycle (see the previous discussion of women and transport on pages 357–358).

The retreat of men from virtually all tasks directly related to supporting domestic life seems to have started with European colonialism. At that time, men became increasingly engaged in activities to earn cash, as laborers or as cash-crop cultivators. At first, this change took place to meet the colonial administration requirement that taxes be paid in cash; later, men worked for cash to pay for children's school fees, basic electricity, certain consumer goods, and even food, when families lost access to land due to agricultural

modernization. European Christians colonized Africa during the Victorian era, when the prevailing attitude in Europe was that women were lesser creatures who should remain in the home and let their husbands deal with the outside world. This idea influenced the colonial policy in Africa of recruiting men for cultivating cash crops and doing work for wages, while leaving women to shoulder all the domestic work as well as what had formerly been the shared work of subsistence agriculture.

In the precolonial past, there were social controls that tempered the effects of male domination over women's lives. Most marriages were social alliances between families; therefore, husbands and wives spent most of their time doing their tasks with family members of their own sex, rather than with each other. Traditional gender relationships, however, were modified wherever Muslim or Christian religious and cultural practices were introduced. In most cases, it seems that men gained power and women lost freedom. Muslim women, restricted to domestic spaces, could no longer move about at will, seize opportunities to trade in the markets, or engage in public activities. In Islamic Africa, the strong patriarchy left many women in a sort of servitude to male wishes, with ambiguous legal rights. While Christian women operate in the public sphere more than Muslim women, they too are socialized to restrict their activities to the home. It is important to note that having multiple wives—the practice of **polygyny**—is more common in sub-Saharan Africa than in Muslim northern Africa, as Figure 7.21 shows, although, overall, Africans in polygynous marriages are a small minority. Those countries with the highest rates of polygyny

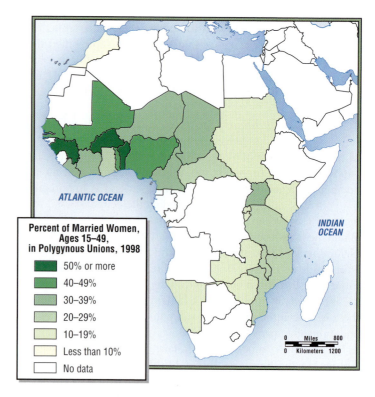

FIGURE 7.21 Polygyny in Africa. Polygyny is less common in East Africa than in West Africa, and it is uncommon in Southern African families. [Adapted from *The World's Women, 2000: Trends and Statistics* (New York: United Nations, Department of Economic and Social Affairs, 2000), pp. 27–28.]

—Senegal, Burkina Faso, Mali, and Cameroon—have relatively small populations.

Female Circumcision

A practice known as **female circumcision** is widespread in at least 27 countries throughout the central portion of the African continent (Figure 7.22). Often mistakenly thought to be a Muslim practice, the custom was reported by Herodotus in 500 B.C. and predates both Christianity and Islam. It is now practiced by some African Muslims, Christians, and followers of ancient traditional African religions. Circumcision is probably intended to ensure that a female is a virgin at marriage (and hence not pregnant by another male), and that she thereafter has a low interest in intercourse, but it also has many complex symbolic meanings. In the procedure, which is a much more radical operation than male circumcision and usually performed without anesthesia, the labia minora and the clitoris are removed and, in the most extreme cases (called *infibulation*), the vulva is stitched nearly shut. Female circumcision eliminates any possibility of sexual stimulation for the woman. Because the practice results in the permanent removal of a healthy organ, many physicians refer to it as *female genital mutilation* or *female genital excision*. For those who have undergone female genital excision, urination and menstruation are difficult, intercourse is painful, and childbirth is particularly devastating because the scarred flesh is inelastic. Recent research shows that the practice also leaves women exceptionally susceptible to HIV infection.

In some culture groups, such as the Kikuyu of Kenya, nearly all females would have been "circumcised" 40 years ago; today only about 40 percent of Kikuyu schoolgirls have been. But in Africa as a whole, as many as 130 million girls and women presently living have undergone the procedure (2 million a year, 6000 per day). The practice has spread beyond Africa, also taking place surreptitiously among African émigrés in Europe and North America. In Boston, at Brigham and Women's Hospital, Dr. Nawal Nour heads a center that handles the special needs of women who have undergone the procedure.

Defenders of the custom cite its symbolic importance as ritual purification; often girls who have not undergone excision are considered unclean and unmarriageable. The *New York Times* reporter Celia W. Dugger interviewed a 12-year-old in Côte d'Ivoire, who said that she wants more than anything to be "cut down there," gesturing to her lap. All her friends have had it done, and afterward they were showered with gifts and money and there was a huge celebration for relatives and friends. The father of this girl said that if he did not have it done to her, he would not be allowed to speak in village meetings, and no man would marry her. Nonetheless, the girl's mother, who had undergone excision, said she hated the custom because it deprives a woman of sexual sensitivity.

Many African and world leaders, both men and women, have concluded that the custom constitutes an extreme human rights abuse, but because it is so deeply ingrained in some value systems, the most successful eradication campaigns have emphasized the threat it poses to a woman's health. Eliminating female genital mutilation will require enforcement of laws against it already in place in many African countries (there have been few arrests thus far) and the passing of new legislation elsewhere. There have been

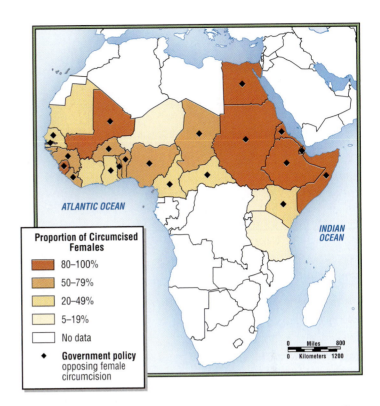

FIGURE 7.22 Prevalence of female circumcision. This practice occurs all across the center of the African continent in spite of government policies against it. [Adapted from Joni Seager, *The State of Women in the World: An International Atlas* (London: Penguin, 1997), p. 53; updated with figures from Amnesty International: http://www.amnesty.org/ailib/intcam/femgen/fgm9.htm; http://www.crlp.org/pub_fac_fgmicpd.html; *The World's Women, 2000: Trends and Statistics* (New York: United Nations, Department of Economic and Social Affairs, 2000), pp. 159–161.]

successful efforts to change people's attitudes toward the practice and to create acceptable alternative practices. A growing number of Kenyan families, for example, are turning to a new rite known as "circumcision through words." This new ritual involves a week-long program of counseling, ending with a community celebration and affirmation of a girl's passage to adulthood. In Ghana, one of the most successful programs encourages community leaders to perform alternative rite-of-passage ceremonies. In Uganda, a health education program seeks to show the public that some cultural practices can change without compromising the society's values. It is reported that in just 2 years (1994 to 1996), female genital mutilation in Uganda declined by more 30 percent, largely as a result of this program; in 2003, according to Amnesty International, only 5 percent of Ugandan girls were cut.

Ethnicity and Language

Ethnicity refers to the shared language, cultural traditions, and political and economic institutions of a group. The term refers primarily to culture, not to race, but because race often carries social significance, ethnicity can include it.

The map of Africa presented in Figure 7.23 shows a rich and complex mosaic of ethnic languages. Yet despite its complexity, this map does not adequately depict Africa's cultural diversity. Most ethnic groups have a core territory in which they have traditionally

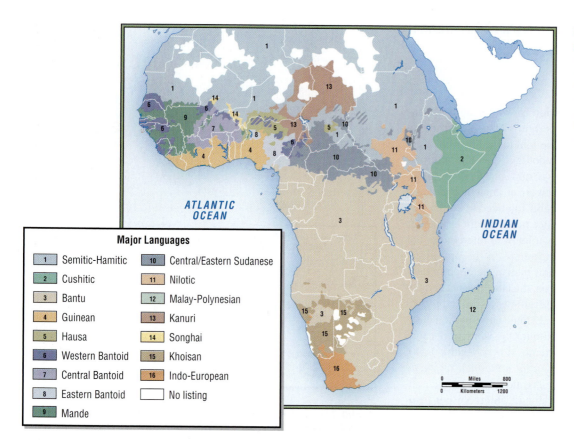

FIGURE 7.23 Major language groups of sub-Saharan Africa. [Adapted from Edward F. Bergman and William H. Renwick, *Introduction to Geography—People, Places, and Environment* (Englewood Cliffs, N.J.: Prentice Hall, 1999), p. 256; Titus Didactica, Frankfurt, Germany, http://titus.uni-frankfurt.de/didact/karten/afr/afrikam.htm.]

Major Languages

1	Semitic-Hamitic	10	Central/Eastern Sudanese
2	Cushitic	11	Nilotic
3	Bantu	12	Malay-Polynesian
4	Guinean	13	Kanuri
5	Hausa	14	Songhai
6	Western Bantoid	15	Khoisan
7	Central Bantoid	16	Indo-European
8	Eastern Bantoid		No listing
9	Mande		

lived, but very rarely do groups occupy discrete and exclusive spaces. Often, several groups share a space, perhaps practicing different but complementary ways of life and using different resources. For example, one ethnic group might be subsistence cultivators, another might herd animals on adjacent grasslands, and a third might be craft specialists working as weavers or smiths. On the other hand, people can share an ethnic identity, yet have little in common culturally or in the spaces they occupy. In Kenya, for example, Kikuyu villagers in the hinterland live very different daily lives from Kikuyu urbanites who reside in Nairobi. People may also be very similar culturally and occupy overlapping spaces, but identify themselves as being from different ethnic groups. Hutu and Tutsi cattle farmers in Rwanda share an occupation, similar languages, and similar ways of life. As a result of Belgian, German, and eventually British colonial policy, however, which exaggerated ethnic differences, Hutus and Tutsis now think of themselves as having very different ethnicities. In the 1990s (and again in 2004), the Hutu and Tutsi people engaged in several episodes of mutual genocide. Now those seeking reconciliation note that cultural training must play a central role in changing the way Rwandans think about each other. Not surprisingly, deciding which language shall be used in school instruction is a highly politicized and crucial debate.

Some countries in Africa have only one or two ethnic groups; others—such as Nigeria, Tanzania, and Cameroon—have many. Nigeria's three largest groups, the Hausa, Igbo, and Yoruba, are of roughly equal size, each with about 25 million people; but there are hundreds of smaller groups in the country. The vast majority of African ethnic groups have peaceful and supportive relationships with one another.

To a large extent, language correlates with ethnicity. More than a thousand languages, falling into more than a hundred language groups, are spoken in Africa. Some are spoken by only a few dozen people; others, such as Hausa, by several million. Language use is always in flux, so some African languages are now dying out. They are being replaced by languages that better suit people's needs or have become politically dominant. Increasingly, a few **lingua francas** (languages of trade) are taking over, such as Swahili and Arabic in East Africa and Hausa in West Africa. The new Pan-African Parliament now organizing under the auspices of the African Union (see page 357) is debating whether to use Swahili or English as its official language. Swahili, also called Kiswahili, is an Arabic-influenced Bantu language. Former colonial languages such as English, French, and Portuguese are also widely used in commerce, education, on the Internet, by pan-African congresses, and in the exchange of scientific knowledge.

ENVIRONMENTAL ISSUES

Africa is home to many unique animal and plant species that warrant protection and preservation. But attempts to ensure the well-being of Africa's human citizens often threaten the habitats that support endangered animals and plants. For example, Africans who cannot afford kerosene or gas to cook food must use wood—but in some densely populated areas, or where access to wood is restricted by large land monopolies, so many people are collecting wood that the forests have disappeared, and even shrubs are constantly pruned for firewood. The wisest African planners design strategies that both provide for basic human needs and address Africa's environmental issues. This section presents a sample of the environmental issues facing Africa today (Figure 7.24).

Desertification

Climatologists think that parts of Africa are now in a natural cycle of increased aridity. The effects are most dramatic in the region called the Sahel (Arabic for "shore" of the desert), a band of arid grassland 200 to 400 miles (320 to 640 kilometers) wide that runs east-west along the southern edge of the Sahara (see Figure 7.3 on page 337). Over the last century, the Sahel has shifted to the south in a process known as **desertification,** by which arid conditions spread to areas that were previously moist. For example, the *World Geographic Atlas* in 1953 showed Lake Chad situated in a tropical rain forest well south of the southern edge of the Sahel. By 1998, the Sahara itself was encroaching on Lake Chad. This shift of aridity to the south—and its opposite, the shift of moister conditions to the north—is part of a long-term natural cycle, but human activity may be speeding up the cycle and broadening the area affected.

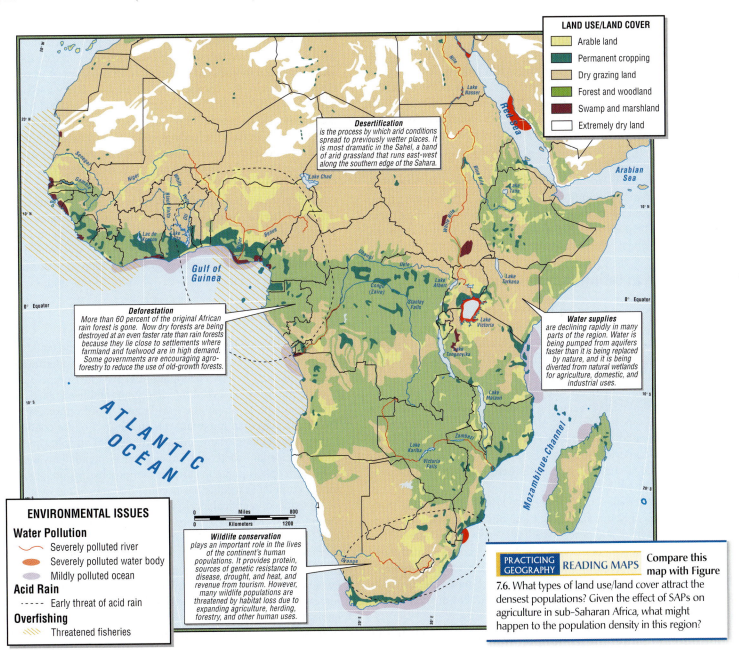

LAND USE/LAND COVER
- Arable land
- Permanent cropping
- Dry grazing land
- Forest and woodland
- Swamp and marshland
- Extremely dry land

Desertification
is the process by which arid conditions spread to previously wetter places. It is most dramatic in the Sahel, a band of arid grassland that runs east-west along the southern edge of the Sahara.

Deforestation
More than 60 percent of the original African rain forest is gone. Now dry forests are being destroyed at an even faster rate than rain forests because they lie close to settlements where farmland and fuelwood are in high demand. Some governments are encouraging agroforestry to reduce the use of old-growth forests.

Water supplies
are declining rapidly in many parts of the region. Water is being pumped from aquifers faster than it is being replaced by nature, and it is being diverted from natural wetlands for agriculture, domestic, and industrial uses.

ENVIRONMENTAL ISSUES
Water Pollution
- ⟋ Severely polluted river
- ⬭ Severely polluted water body
- ⬭ Mildly polluted ocean

Acid Rain
- ----- Early threat of acid rain

Overfishing
- ⬚ Threatened fisheries

Wildlife conservation
plays an important role in the lives of the continent's human populations. It provides protein, sources of genetic resistance to disease, drought, and heat, and revenue from tourism. However, many wildlife populations are threatened by habitat loss due to expanding agriculture, herding, forestry, and other human uses.

PRACTICING GEOGRAPHY   READING MAPS   Compare this map with Figure 7.6. What types of land use/land cover attract the densest populations? Given the effect of SAPs on agriculture in sub-Saharan Africa, what might happen to the population density in this region?

FIGURE 7.24 Environmental issues: Sub-Saharan Africa. Africa's environmental issues are varied and are less connected to urbanization and industrialization than those of other regions, but they are no less worrisome. The drying out of environments (desertification) is probably the result of both natural and human-induced processes, some stemming from bad development advice. Deforestation and losses of plant and wildlife species are also related to global market demands as well as to local circumstances that result in forest clearing for fuelwood. In some places, water resources are being depleted or polluted by growing urban demands and irrigated export agriculture.

The Sahel and other dry grasslands in Africa are what geographers call **fragile environments** because they exist in zones that are barely sufficient for the needs of native plants and animals. Rainfall is already low in these grasslands, at 10–20 inches (25–50 centimeters) per year, and there are only low levels of organic matter in the soil to provide nutrients. Consequently, any further stress, such as fire, plowing, or intensive grazing, may cause dry-adapted native grasses to give way to introduced species, such as tough bunchgrass and thorny shrubs, that are more tolerant of stress. But these alien species do not cover the soil as well as the native plants, so they allow rain to evaporate more quickly, and they are not useful for grazing. Soon, the dry, denuded soil is blown away by the wind, and the remaining sand piles up into dunes as the grassland becomes more like a desert.

Indigenous animal herders are often blamed for desertification. Although overgrazing can trigger this process, natural cycles are often the underlying cause. Moreover, economic development activities can advance desertification. For example, agencies such as the World Bank have sometimes advised animal herders to raise cattle instead of their traditional animals, such as camels and goats, because cattle can be sold as meat in distant urban markets. But raising cattle places greater stress on native grasslands than more traditional herding. Irrigated agriculture can also lead to desertification when minerals from the water build up in the soil over time, an effect called salinization (see description in Chapter 5 on pages 266–267. The overharvesting of trees and shrubs for fuel also leads to drier soil and advances desertlike conditions.

Forest Vegetation as a Resource

Africans have long made extensive use of forest products for construction, furniture, tools, fencing, medicines, religious carvings, fuel, and food. Like forests all over the world, the forests of Africa are disappearing. It is estimated that by 2000, 1.15 million square miles (3 million square kilometers)—more than 60 percent—of the original African rain forests were gone. Much of the loss of African forests is attributed to the demand for farmland and fuel by growing populations, especially in West Africa (see Figure 7.24). In recognition of the rapid loss of forests, governments in some countries (such as Ghana, Uganda, and Tanzania) are encouraging **agroforestry**—the raising of economically useful trees—to take the pressure off old-growth forests.

In October 2004, the Nobel Peace Prize was awarded to Wangari Maathai of Kenya for her contributions to sustainable development, democracy, and peace. Nearly 30 years ago, she founded the Green Belt Movement through which poor women planted more than 30 million trees, thus protecting their environment, promoting their own environmental awareness, and improving their self-esteem. Maathai is known for saying that a healthy environment is essential for democracy to flourish.

So far, logging for export is only a minor part of African economies, but it is responsible for the severe depletion of forests in easily accessible countries such as Côte d'Ivoire in West Africa. The harvested logs are sold as raw wood, even though the financial return would be much greater if the trees were harvested sustainably and processed in Africa into refined wood products. The most extensive remaining tracts of African rain forest are in the Congo River basin, most of which lies in the country of Congo (Kinshasa), where sustainable forest management is not practiced at all. Ironically, when the armed conflict that has engulfed the country and its neighbors since the mid-1990s subsides, commercial logging is expected to expand and the loss of forested land to increase.

Dry forests—those that lose their leaves during the dry season—once covered nearly twice as much territory as rain forests. Now dry forests are being destroyed even faster than rain forests because they lie close to settlements and their products are in great demand by growing populations. Their greatest use is as fuel. Africans still use wood or charcoal—a prepared wood product that burns hotter than wood—to supply nearly all domestic and industrial energy. Making and selling charcoal is an important industry in the informal economy of most African countries. Much of the wood used is harvested free because Africans traditionally have considered forests to be a resource held in common. In urban areas, wood and charcoal remain the cheapest fuels available. Even in Nigeria, which is a major oil producer, most people use fuelwood because they cannot afford petroleum products (kerosene or bottled gas) as fuel. In fact, Nigeria is the leading producer of fuelwood in Africa.

For a decade or more, many Africans have recognized the need to use fuelwood more sustainably. Fast-growing trees such as Leucana are farmed for use in making charcoal. Economic development planners have tried to promote fuel-efficient stoves, but users find that the stoves lack the convenience and nostalgic associations of old-fashioned cooking fires. Some Africans are turning to alternative energy sources—hydropower, bottled gas, solar power, and wind power—to meet their growing energy needs, but the trend is not yet strong.

Wildlife and National Parks

Africa's wildlife has long played a role in the global and continental economy. African animal products such as skins, taxidermy specimens, and ivory have been important exports to Europe, China, and the Americas for several hundred years, and live animals from Africa continue to supply the world's zoos. Furthermore, wildlife has always played an important role in the lives of the continent's human populations. Many Africans still depend on wild game and fish as their main sources of protein, but as populations increase faster than opportunities to earn cash for purchasing food, pressure on the continent's wildlife has become extreme. Some species are being reduced to critical levels, as is the case with mountain gorillas in Congo (Kinshasa), where the ongoing war has nearly eliminated this large primate. Reports documenting the diminishing numbers of gorillas, elephants, lions, zebras, and giraffes and the decline of their rain forest and grassland habitats have raised public concern about Africa's diminishing wildlife heritage. Although this increasing international interest has been helpful, it may not be enough to save many animals. The cheetah, for example, is very near extinction, with only 12,000 individuals left. About 7000 have been killed in the last 10 years, mostly by farmers on the fringes of nature reserves who kill cheetahs to protect their livestock. The use of dogs to protect livestock, an old African strategy, has peacefully resolved the conflict between cheetahs and some farmers. Nevertheless, as more and more areas are turned over to agriculture, the loss of habitat will further reduce the number of cheetahs.

Several strategies are emerging to address this many-sided crisis. To improve domestic food supplies and reduce expensive imports, there is growing interest in game farming. Wild animals are resistant to African diseases, are adjusted to the climate, and can withstand occasional droughts or heat waves better than domesticated animals. Another strategy to conserve African wildlife and use it in economic development relates to national parks and tourism. The governments of Kenya, Tanzania, South Africa, Zimbabwe, Botswana, and Zambia have all recognized **ecotourism** as a promising source of income for Africa, and national parks exist in all these countries. In fact, Africa has one-third of the globe's preserved national parkland, covering about 10 percent of the earth's surface. Africa is unique in that its national parks are home to the only remaining concentrations of migratory plains animals in the world, and the potential to develop natural-area tourism is large in all of the countries listed above. But the idea that African national parks can both conserve wildlife and earn income will need careful nurturing. The parks are often islands within a sea of rural poverty, and because they are underfunded, it is easy for poachers to intrude. Because there has been too little effort to explain how preservation of wildlife could benefit local people or contribute to their income, hostility toward the parks has produced a number of violent attacks on foreign tourists. News of these attacks eclipses, at least temporarily, the potential of the parks to attract more tourists. In addition, Africa's economic crisis has further reduced the revenues available to maintain parks. At the World Parks Conference in Durban, South Africa, in 2003, there was recognition that African parks are unlikely to be self-sustaining, much less profitable, in the near future, but that nonetheless they are worth the investment because of the valuable assets they protect. One African initiative seeks to establish a $250 million fund to be supported by African governments, the business community, local people, and international donors to ensure that protected areas provide clear benefits to African communities.

Water

In a growing number of African countries, the amount of water per capita is declining rapidly. In many parts of arid Africa, water is being pumped from aquifers faster than it is being replaced by nature; it also is being diverted from natural wetlands. Large economic development projects may redirect water for industrial, agricultural, or urban uses, leaving rural inhabitants of arid lands with insufficient water and causing the loss of complex ecosystems. Kenya, for example, has only 830 cubic yards (635 cubic meters) of water per capita per year, compared to the accepted global standard of 1300 cubic yards (1000 cubic meters); its water supply is expected to drop to just 250 cubic yards (190 cubic meters) per capita per year by the year 2050.

Safe water is scarce even in the moist areas of the continent. Water is being polluted by salinization, human wastes, and chemical poisons (see Figure 7.24). In rural areas, most households must carry all their water from springs, wells, pools, or streams; in urban areas, water is often drawn from standpipes. It is a long-standing tradition that women are the procurers of water. As sources become depleted or polluted, women must walk farther and farther to collect safe water. The difficulty of carrying water limits the amounts available for use, and illness is spread when insufficient water is used to wash dishes, diapers, clothing, and other materials. Because plumbing and sewage treatment are usually not available, even in cities, human wastes often find their way into water sources.

Saving what is now being wasted is perhaps the greatest source of "new" water for most places on earth. Agriculture accounts for about 70 percent of global water use. In Africa, many large agricultural projects are irrigated, and most modern irrigation systems lose a great deal of water through leakage and evaporation because water often runs through unlined open ditches and is sent into fields via sluices or sprinklers. In addition, standing pools of water are sources of disease (see page 346). Culturally aware development planners say that the solutions to Africa's water problems lie less in high-tech applications than in rediscovering folk ideas about water conservation that fit with local customs. A decade or two ago, West African women who delivered irrigation water directly to the roots of their plants by using human water brigades were seen as inefficient by foreign development planners. Today, the water these practices save (and the camaraderie they foster) are better appreciated. The installation of "old-fashioned" sanitary roof catchments and cisterns in place of complex piped delivery systems can save construction costs as well as water, as can the use of sanitary hand-flushed pit toilets in place of public sewer systems. Remember that at Myeka High School, toilet refuse, rather than being flushed with water, is turned into valuable assets: methane that produces energy for the school and fertilizer for its gardens (see the chapter-opening vignette).

MEASURES OF HUMAN WELL-BEING

The very low gross domestic product (GDP) per capita figures in Africa (Table 7.1, column 2) indicate a desperately poor continent. However, GDP per capita does not include economic activities that take place outside the formal market system: sharing, bartering, participation in the informal economy, raising food in family plots, and conserving resources. Within the low range of GDP per capita figures, there is still considerable variation. Botswana, Gabon, Mauritius, Cape Verde, Namibia, South Africa, Swaziland, and Seychelles all have per capita GDPs above U.S. $3000 per year, but together they account for only 51 million people, less than 10 percent of Africa's population. Moreover, within each of these countries, there is a wide range of incomes. Many people live on much less than the official per capita GDP.

The Human Development Index (HDI) combines three measures—life expectancy at birth, educational attainment, and adjusted real income—to arrive at a ranking of 175 countries that is sensitive to more factors than just income (see Table 7.1, column 3). Most African countries rank in the lowest third of world nations on the HDI. Because GDP figures are so low, there is little tax money to invest in education, basic health care, and sanitation. The literacy figures (columns 5 and 6) reflect both the generally low level of education and discrimination against women. During the 1980s and 1990s, per capita government spending fell, often in response to cutbacks mandated by SAPs, and the introduction of fees limited the number of people who could afford services such as education and health care. One result has been that more mothers (Text continues on page 376.)

TABLE 7.1 Human well-being rankings of countries in sub-Saharan and other selected countries

Country (1)	GDP per capita, adjusted for PPP[a] in 2001 $U.S. (2)	Human Development Index (HDI) global rankings, 2003[b] (3)	Gender Development Index (GDI) global rankings, 2003[c] (4)	Female literacy (percentage), 2001 (5)	Male literacy (percentage), 2001 (6)
Selected countries for comparison					
Japan	25,130	9 (high)	13	99	99
United States	34,320	7 (high)	5	99	99
Mexico	8,430	55 (medium)	52	89.5	93
West Africa					
Benin	980	159 (low)	131	24.6	54
Burkina Faso	1,120	173 (low)	143	14.9	35
Cape Verde Islands	5,570	103 (medium)	82	67	85
Côte d'Ivoire	1,490	161 (low)	134	38.4	60
Gambia	2,050	151 (low)	123	30.9	45
Ghana	2,250	129 (medium)	104	64.5	81
Guinea	1,960	157 (low)	—	19[d]	36[d]
Guinea-Bissau	970	166 (low)[b]	137	24.7	55
Liberia	1,100[d]	—	—	22[d]	54[d]
Mali	810	172 (low)	142	16.6	37
Mauritania	1,990	154 (low)	125	30.7	51
Niger	890	174 (low)	144	8.9	25
Nigeria	850	152 (low)	124	57.7	73
Senegal	1,500	156 (low)	128	28.7	48
Sierra Leone	470	175 (low)	—	18[d]	45[d]
Togo	1,650	141 (medium)	118	44	73
Central Africa					
Cameroon	1,680	142 (low)	114	65.1	79
Central African Republic	1,300	168 (low)	138	36.6	60
Chad	1,070	165 (low)	135	35.8	53
Congo (Brazzaville)	970	140 (medium)	111	75.9	88
Congo (Kinshasa)	680	167 (low)	136	51.8	74
Equatorial Guinea	1,817[d]	116 (medium)	109[d]	76	95
Gabon	5,990	118 (medium)	—	53[d]	74[d]
São Tomé and Príncipe	1,469[d]	122 (medium)	—	62[d]	85[d]

TABLE 7.1 (*Continued*)

Country (1)	GDP per capita, adjusted for PPP[a] in 2001 $U.S. (2)	Human Development Index (HDI) global rankings, 2003[b] (3)	Gender Development Index (GDI) global rankings, 2003[c] (4)	Female literacy (percentage), 2001 (5)	Male literacy (percentage), 2001 (6)
Horn of Africa					
Djibouti	2,370	153 (low)	—	55.5	68
Eritrea	1,030	155 (low)	126	45.6	68
Ethiopia	810	169 (low)	139	32.4	48
Somalia	600[d]	—	—	14[d]	36[d]
East Africa					
Burundi	690	171 (low)	141	42	57
Kenya	980	146 (low)	115	77.3	85
Rwanda	1,250	158 (low)	129	61.9	75
Tanzania	520	160 (low)	130	67.9	85
Uganda	1,490	147 (low)	117	58	78
East African Islands					
Comoros	1,870	134 (medium)	108	48.8	63
Madagascar	830	149 (low)	121	60.6	74
Mauritius	9,860	62 (medium)	59	81.7	88
Seychelles	10,600[d]	36 (high)	—	60[d]	56[d]
Southern Africa					
Angola	2,040	164 (low)	—	28[d]	56[d]
Botswana	7,820	125 (medium)	101	80.6	75
Lesotho	2,420	137 (medium)	110	93.9	73
Malawi	570	162 (low)	132	47.6	76
Mozambique	1,140	170 (low)	140	30	61
Namibia	7,120	124 (medium)	100	81.9	84
South Africa	11,290	111 (medium)	90	85	87
Swaziland	4,330	133 (medium)	107	79.4	81
Zambia	780	163 (low)	133	72.7	86
Zimbabwe	2,280	145 (low)	113	85.5	92

[a]PPP = purchasing power parity.
[b]The high, medium, and low designations indicate where the country ranks among the 175 countries classified into these three categories by the United Nations.
[c]Only 144 out of a possible 175 ranked in the world; no data for many.
[d]Data is for 1998; data not available for 2003.

Sources: United Nations Human Development Report 2000 (New York: United Nations Development Programme), pp. 157–168; *United Nations Human Development Report 2003* (New York: United Nations Development Programme).

and infants have died during and after childbirth. Spending on education fell from U.S. $11 billion in 1994 to U.S. $7 billion in 1998. When school fees were introduced, enrollment in elementary schools fell from 77.1 percent to 66.7 percent. School fees are increasingly recognized as counterproductive because they render government incapable of enforcing school attendance and leave the decision to educate up to often illiterate parents who may lack the funds or the understanding of just how education can help their boys and girls out of poverty.

For various reasons, such as the low status of women in African countries and official hesitancy to reveal the data, the statistics necessary to calculate the Gender Empowerment Measure (GEM) are available for only five countries, so this measure is not included in Table 7.1. Nonetheless, there has been progress for women. Women occupy 50 percent or more of professional and technical positions in Botswana (52 percent) and Namibia (55), but all other African countries failed to report data in this category. In South Africa, Namibia, Uganda, Tanzania, Eritrea, and Mozambique, women hold more than 20 percent of the seats in parliament (in the United States, women hold just 14 percent of the seats in Congress). As representatives in parliament, women are able to influence future policies that affect gender equality.

The Gender Development Index (GDI) ranks 36 sub-Saharan African countries among a total of 144 countries (data is missing from many countries; see Table 7.1, column 4). This index measures the access of men and women to health care, education, and income, but not the degree to which they participate in public policy formulation and decision making.

Quick Review

1. How does the policy of currency devaluation both help and hurt African farmers?

2. Name a few of the ways British colonial systems of government contributed to the current conflict in African countries.

3. What parts of Africa are predominantly Muslim? Christian?

4. What are a few of the factors that might prevent African national parks from reaching their full potential for protecting game and earning revenue?

III SUBREGIONS OF SUB-SAHARAN AFRICA

We divide sub-Saharan Africa into four subregions that predominantly reflect geographic location: West Africa, Central Africa, East Africa, and Southern Africa. Such factors as the current geopolitical situation and tradition have also guided the placement of a particular country in a particular subregion. Chad, for example, could have been discussed with the arid countries of northern West Africa, along with Niger and Mali. We have placed it in Central Africa, chiefly because most of Chad's people live in the far southern reaches of the country, are closer in culture to Cameroon and the Central African Republic, and are affected primarily by political and economic circumstances in Central Africa. For each subregion, we provide a short discussion of the physical setting (see Figure 7.2 on pages 334–335 for landform features) and then concentrate on specific circumstances that have led to current conditions.

WEST AFRICA

West Africa, which occupies the bulge on the west side of the African continent, includes 15 countries (Figure 7.25). It is physically framed on the north by the Sahara Desert, on the west and south by the Atlantic Ocean, and on the east by Lake Chad and the mountains of Cameroon. The region can be thought of as a series of horizontal physical zones grading from dry in the north to moist in the south. The north-south division also applies to economic activities—herding in the north, farming in the south—and, to some extent, religions and cultures—Muslim in the north, Christian in the south. Most of these cultural and physical features do not have distinct boundaries, but rather zones of transition and exchange.

Stretching across West Africa from west to east are horizontal vegetation zones that reflect the horizontal climate zones shown in Figure 7.3 (see page 337). Remnants of tropical rain forests still exist along the Atlantic coasts of Côte d'Ivoire, Ghana, and Nigeria, although much forest has been lost as a result of intensifying settlement and agriculture. To the north, this moist environment grades into drier woodland mixed with savanna. Farther north still, as the environment becomes drier, the trees thin out and the savanna dominates. The savanna grades into the yet drier Sahel region of arid grasslands. In the northern parts of Mali, Niger, and Mauritania, the arid land becomes actual desert, part of the Sahara.

Along the coast, and inland where rainfall is abundant, crops such as coffee, cacao, yams, palm oil, corn, bananas, sugarcane, and cassava are common. Farther north and inland, the crop complex is adapted to the drier climate: millet, peanuts, cotton, and sesame grow in the savanna environments. Cattle are also tended in the savanna and north into the Sahel; farther south, however, the threat of tsetse flies prevents cattle raising (see pages 345–346). People in the Sahel and Sahara are generally not able to cultivate the land, except along the Niger River during the summer rains. Food, other than that provided by animals and the fields along the riverbanks, must be traded for with coastal zones.

West African environments have dried out as more and more people clear the forests and woodlands for cultivation and to obtain cooking fuel. When the forest cover is gone, it no longer blocks sunlight and wind. As a result, sunlight evaporates moisture, and desiccating desert winds blow south all the way to the coast. The drying out of the land is exacerbated by the recurrence of a natural cycle of drought. The ribbons of differing vegetation running west to east are not as distinct as they once were, and even coastal zones occasionally experience dry, dusty conditions reminiscent of the Sahel.

The geographies of religion and culture in West Africa also reflect the north-south physical patterns to some extent. Generally,

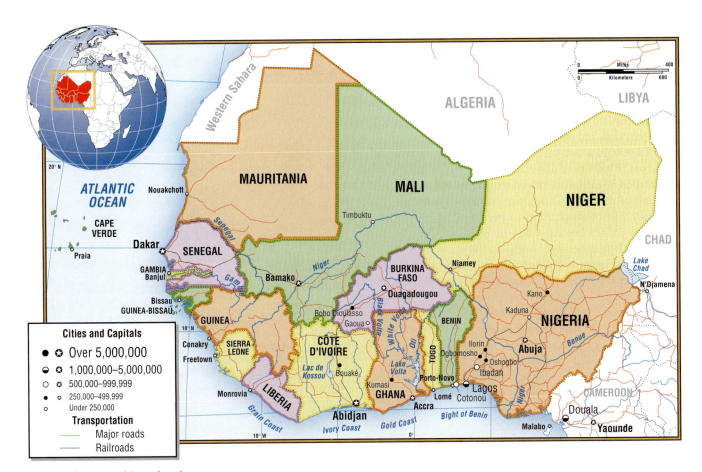

FIGURE 7.25 The West Africa subregion.

the north is Muslim, and people are light-skinned and tend to have angular features. The south is populated by darker-skinned people who practice a mix of Christian and traditional religions. Some large ethnic groups once occupied distinct zones, but as a result of migration, members of specific groups live today in many different areas and in a number of West African countries. There are hundreds of smaller ethnic groups, each with its own language; the map depicting ethnic languages in Nigeria (see Figure 7.17 on page 360) is representative of the cultural complexity of the entire subregion. Nigeria illustrates the potential difficulties faced by a nation attempting to unite a multitude of ethnic groups (see the section "The case of conflict in Nigeria" on pages 359–361).

Conflict in West Africa

The 1990s and early 2000s have seen continuing bloody conflict in the far West African countries of Guinea, Sierra Leone, Liberia, and Côte d'Ivoire. The wars in these four countries are intertwined, but each conflict is fed by local disputes that ultimately can be traced back to poverty and horrendously bad leadership. In the 1980s, Liberia was a country with a reasonably bright future. Then, in 1989, Liberian Charles Taylor, an American-educated, Libyan-trained-and-armed would-be revolutionary, tried to stage a coup. He failed, but then joined with a Sierra Leone rebel who had forced children to form an army, which captured that country's diamond mines. At least 200,000 people were murdered in Sierra

Leone, and the mines became the source of funding for further aggression. Taylor returned to Liberia with his AK-47-armed child soldiers, and in 1997, terrorized Liberians elected him president. He then invaded Guinea under the pretext that Liberian dissidents were ensconced there. With American military aid, Guinea pushed Taylor back, but only after many more deaths.

In 2002, the conflict spilled into Côte d'Ivoire, formerly one of the most stable West African countries, but one nonetheless troubled by regional disparities. Abidjan, the southern coastal capital, has fine roads, skyscrapers, French restaurants, and a modern seaport, but the north is much poorer and full of cacao plantations that employ immigrant agricultural workers at very low wages, and, in some cases, enslaved children from nearby Burkina Faso and Mali. Investigators say the slavery and abusive wages resulted when unscrupulous Côte d'Ivoire cacao planters started looking for a way to cut costs in the midst of a depressed world cacao market. But the civil conflict in Côte d'Ivoire is only tangentially related to the issue of child slavery. The extreme poverty of the north and the large immigrant labor population led to rebellion against the south, with support for the rebels coming from Burkina Faso, the home of many workers in Côte d'Ivoire. Support for the southern faction came from Liberia, already a scene of conflict. Again, child soldiers were armed and told to support themselves by looting, which in relatively wealthy Côte d'Ivoire was a tempting proposition for poor, illiterate, hungry boys and young men. In 2003, UN peacekeepers were belatedly sent to Liberia. By 2004,

AT THE REGIONAL SCALE   Water Management in Mali and Niger

The Niger is the most important river of West Africa. It carries summer floodwaters northeast into the normally arid, clay-lined lowlands of Mali and Niger (see Figure 7.25). There the waters spread out into lakes and streams that nourish wetlands. For a few months of the year (June through September), these wetlands ensure the livelihoods of millions of fishers, farmers, and pastoralists, who share the territory in carefully synchronized patterns of land use that have survived for millennia. These desert wetlands along the Niger produce eight times more plant matter per acre than the average wheat field and provide seasonal pasture for millions of cattle, goats, and sheep—the highest density of herds in all of Africa.

Because of mounting population pressure, international experts are advising the governments of Mali and Niger to dam the river and channel it into irrigated agriculture projects that will help to feed the more than 20 million overwhelmingly poor people of Mali and Niger. Even though 80 percent of the people are farmers, the two countries already must rely on imported food. But the dams will forever change the seasonal rise and fall of the river that has supported an intricate mix of wildlife and human uses for thousands of years. The impact will be felt in Europe as well, because the Niger is a seasonal migration stopover for birds that spend the summer in Europe, north of the Mediterranean.

Every rainy season, the Bani River, a major tributary of the Niger, floods and turns the city of Djenne, Mali, into a series of islands. Djenne, with a population of 20,000, is the oldest known urban settlement in sub-Saharan Africa (about 2200 years old). The annual flood revitalizes the surrounding soil and the river's marine life and permits the population to feed itself. A proposed upstream dam threatens the city's self-sufficiency. [Sarah Leen/Matrix.]

Charles Taylor was in exile in Nigeria. Guinea, Liberia, and Sierra Leone were calm and rebuilding under the watch of UN peacekeeper troops sent from Pakistan, Bangladesh, and Ethiopia, but conflict in Côte d'Ivoire continued. Rebuilding Liberia alone was expected to cost $500 billion. The United States promised $200 million to disarm, demobilize, and rehabilitate the child armies. Meanwhile, human rights groups urged chocolate eaters in Europe and elsewhere to be mindful of the slave labor that keeps chocolate prices low for the consumer.

CENTRAL AFRICA

If you were to look at a map of Africa without any prior knowledge, you might reasonably expect the countries of Central Africa—Cameroon, the Central African Republic, Chad, Congo (Brazzaville), Equatorial Guinea, Gabon, São Tomé and Príncipe, and Congo (Kinshasa)—to be the hub of continental activity, the true center around which life on the continent revolves (Figure 7.26). This is not the case. Paradoxically, Central Africa plays a peripheral role on the continent. Its dense tropical forests, tropical diseases, and difficult terrain make it the least accessible part of the continent, and recent armed conflict over resources and power has left it in a deteriorating state that discourages visitors and potential investors.

Central Africa consists of a core of wet tropical forestlands surrounded on the north, south, and east by bands of drier forest and then savanna (see Figure 7.3 on page 337). The core is the drainage basin of the Congo River, which flows through this area and is fed by a number of tributaries. The forests of Central Africa, although rich in resources, form a barrier that has isolated Central Africa from the rest of the continent for centuries and has impeded the development of transportation, commercial agriculture, and even mining. Moreover, the soils of the forest are not suitable for long-term cash-crop agriculture (see page 338). Most Central Africans are subsistence farmers who live in rural settings along the river courses and the occasional railroad line. Cities are only beginning to grow; too often, the newcomers are people escaping civil violence in the countryside.

Although Central Africa possesses significant mineral and forest wealth, a long string of corrupt governments has pocketed the profits, and smugglers of minerals and other resources have leaked away the potential wealth to foreign interests. Persistent warfare between rival groups has discouraged most legitimate investors, Africans or outsiders.

The region's colonial heritage has contributed to these difficulties. In what is now known as the Democratic Republic of the Congo, for example, the Belgian colonists left a terrible legacy of violence. In the 1880s, Belgium was a small but wealthy and industrialized European country, and its king, Leopold II, had a personal fortune to invest. After failing to establish a colony in Latin America or Asia, King Leopold obtained international approval in 1885 for his own personal colony, the Congo Free State, with the promise that he would end slavery, protect the native people, and guarantee free trade. He did none of these things. The Congo Free State consisted of 1 million square miles (2.5 million square kilometers) of tropical forestland rich in ivory, rubber, timber, and copper, with a population of 10 million people, whose labor King Leopold would soon appropriate to extract Congo's resources for his own profit.

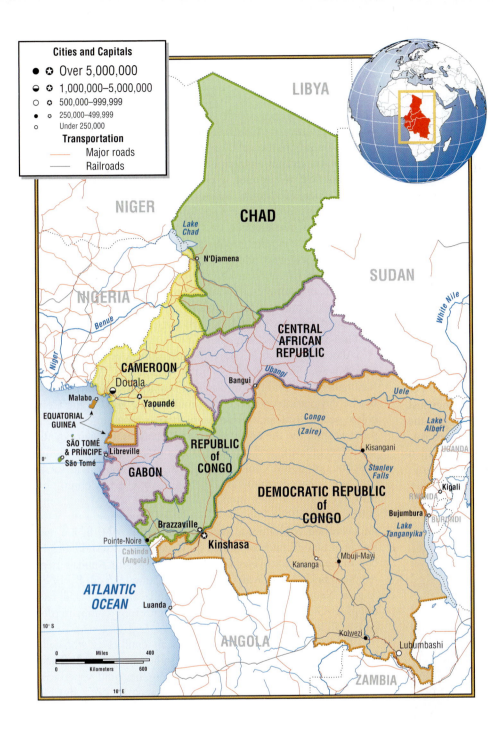

FIGURE 7.26 The Central Africa subregion.

The Belgians used physical violence and terror to control the people of Congo and gain access to their resources. The cruel behavior of Belgian colonial officials was witnessed by the young writer Joseph Conrad when he obtained a job on a steamer headed up the river in 1890. Conrad's novella, *Heart of Darkness* (published in 1902), and a pamphlet by Mark Twain, "King Leopold's Soliloquy" (published in 1905), awakened Europeans and Americans to Leopold's perfidy, and efforts began to end his regime. French colonists in what is now Congo (Brazzaville), Gabon, and the Central African Republic; the Germans in Cameroon; and the Portuguese in neighboring Angola used similar tactics. For example, a chronicler of the 1890s recorded that after a group of Africans rebelled against the French, their heads were used as a decoration around a flower bed in front of a French official's home.

The indigenous people of Central Africa form an overlapping ethnic mix, and they speak some 700 languages. The artificial polit-

ical borders imposed by the Europeans, and their policy of deliberately pitting Africans against one another, exacerbated tensions between ethnic groups. Indigenous governance systems were disrupted when Europeans removed chiefs and installed new leaders who would do their bidding. Long-standing traditional restraints on unwise leadership were lost, thus opening the door for the type of power abuses that developed in the twentieth century.

Most of the countries of Central Africa received their independence in the 1960s, but the Europeans stayed on to continue their economic domination. Tax-free profits were sent home to France, Belgium, or Portugal; rarely were profits reinvested in Central Africa. With no money to develop an infrastructure, no experience with multiparty democracy, and no educated constituency, many Central African countries, such as Congo (Kinshasa), fell into the hands of corrupt leaders. Political upheavals that resulted when some groups were left out of representation in government were accompanied by downward-spiraling economies. As in other parts of sub-Saharan Africa, the average person's purchasing power has decreased over the last few decades. These countries have little tax revenue because of tax concessions to foreigners as well as government corruption. So maintenance of roads, railroads, waterways, telephone networks, electricity grids, and buildings is negligible. Industries, both government and private, cannot survive for lack of electricity and raw materials. People are so poor that there is little market for any type of consumer product.

Congo (Kinshasa): A Case Study of Devolution

PERSONAL VIGNETTE   For women in Kisangani, Congo (Kinshasa), the saying, *"Se débrouiller"* — "Make do with what you have"—is a way of life. Mauwa Funidi, 45, the first college graduate in her once prominent family, now sells small bags of charcoal on the street in Kisangani to earn a few pennies to help support her extended family. She still holds her job as a university librarian, which paid U.S. $300 per month in 1976, but after two decades of inflation, paid her just U.S. $11 per month in 1997. The university library is empty, the windows are broken, and books and other publications are rotting in the humidity. Thirty years ago, education was celebrated in Kisangani. Now there are no funds for the university, and the government has stopped funding kindergarten through twelfth grade. The majority of children do not finish primary school because their families cannot afford the tuition of U.S. $7 a month.

Mauwa Funidi in her defunct library, Kisangani, Congo (Kinshasa). [Stephen Crowley/NYT Pictures.]

Many of the 47 million inhabitants of Congo (Kinshasa) are in circumstances similar to those of Mauwa Funidi's family. The average GDP per capita is U.S. $680 (PPP) a year. Yet the country has an enormous wealth of untouched natural resources: copper, gold, diamonds, and oil. Congo (Kinshasa) is sub-Saharan Africa's fourth largest oil producer, and oil is its major foreign trade commodity. It also has enough undeveloped hydroelectric resources to supply power to every household in the country and to those of some neighboring countries as well.

The problems in Congo (Kinshasa) have resulted from European colonization, the cold war, and the desire of local leaders to stay in power and enrich themselves. When independence came in 1960, political and economic collapse was imminent as Belgian colonial officials withdrew, leaving untrained Congolese in charge. The new government of Patrice Lumumba tried to correct the inequalities of the colonial era by nationalizing foreign-owned companies so that the profits could be kept in the country. Lumumba's strategies, adopted during the height of the cold war, raised the red flag of communism to the West. The Soviet Union supported and armed Lumumba's pro-socialist movement. The Western allies arranged for the killing of Lumumba and supported and armed opposing politicians, such as the extraordinarily corrupt Mobutu Sese Seko.

After five years of violent political struggle, Mobutu Sese Seko seized power in 1965. Once firmly established, Mobutu also nationalized the country's institutions, businesses, and industries—but he did not intend to use the profits for the good of the Congolese people. Instead, he and his supporters began to extract personal wealth from the newly reorganized ventures. Foreign experts, trained technicians, and capital left the country. The Congo's infrastructure, economy, educational system, food supply, and social structure declined rapidly. As conditions worsened, Mobutu and his supporters used their money and their paid army to hold onto power. By the 1990s, the vast majority of Congolese people were impoverished. Mobutu was finally overthrown in 1997 by Laurent Kabila, a new dictator supported by Uganda and Rwanda, neighbors to the east who coveted Congo's mineral wealth. Kabila's inept regime quickly led to widening strife in Central Africa. By 1999, Angola, Namibia, Zimbabwe, Chad, Rwanda, and Uganda all had troops in Congo (Kinshasa) and were aggressively competing for parts of its territory and resources.

To some extent, the Central African conflict is ethnically based. For example, Kabila's Rwandan allies are primarily Tutsi. They have attacked Congolese Hutus, whom they hold responsible for genocidal attacks against Tutsis in Rwanda in 1997. But the prospect of economic and political power is an even stronger incentive for conflict. In January 2001, Laurent Kabila was assassinated, reportedly by previous supporters from neighboring countries. His son, Joseph Kabila, was installed as head of state. The level of violence subsided, but there have been no meaningful changes in social or economic policies. A tiny minority continues to appropriate profits from the sale of the country's mineral wealth.

EAST AFRICA

East Africa (Figure 7.27) is a land of relatively dry coastal zones that contrast with moister interior uplands. A vital and productive farming economy has long existed in the uplands. On the coast, there is a lively trading economy with links across the Arabian Sea and Indian Ocean. This commerce has especially influenced the culture and economy of the southern half of the subregion.

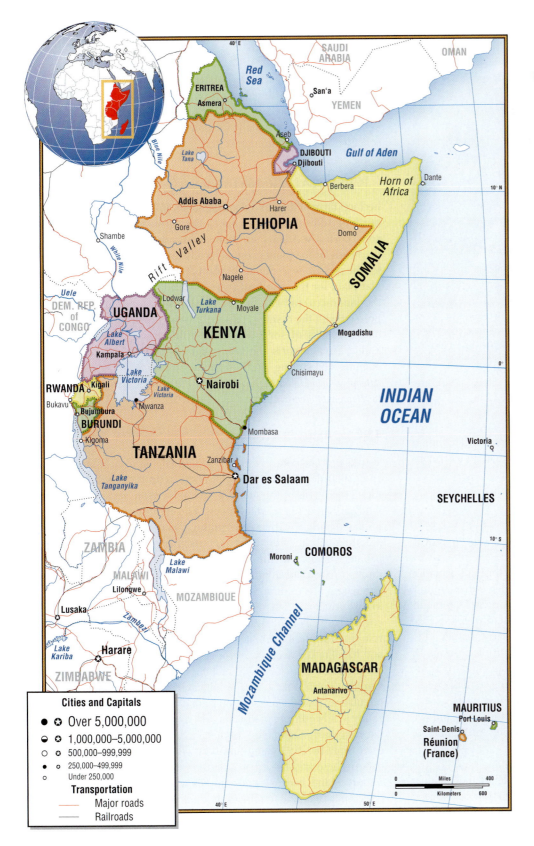

FIGURE 7.27 The East Africa subregion.

In the northeast, the Horn of Africa reaches along the southern shore of the Gulf of Aden, across from the southwestern tip of the Arabian Peninsula. This more northerly section of East Africa includes the countries of Djibouti, Eritrea, Ethiopia, and Somalia. South of the Horn of Africa are the countries of Kenya, Uganda, and Tanzania; the tiny interior highland countries of Rwanda and Burundi in the Great Rift Valley; and the islands of Madagascar, Comoros, Seychelles, and Mauritius in the Indian Ocean across the Mozambique Channel.

Africa's ability to support its people is being tested in East Africa. The entire Horn of Africa has suffered periodic famine since the mid-1980s as a result of naturally occurring drought and

In this Italian-built factory in Asmera, Eritrea's capital, women make shoes for foreign markets. Eritrea has new laws guaranteeing gender equality: women now have the right to divorce, to vote, to acquire and own land, and to work outside the home. [Robert Caputo/Aurora & Quanta Productions.]

closely adapted to particularly fragile environments. Pastures and fields are rotated systematically. Agronomists have recently begun to appreciate the precision of these ancient and specialized mixed-agriculture systems, which are now threatened by development schemes and population pressure.

PERSONAL VIGNETTE Kadija and Hassain Saide Adem have a 10-acre (4-hectare) farm in the Ethiopian Highlands 150 miles (400 kilometers) north of Addis Ababa, the capital. They grow mostly sorghum and some corn. The Adems' sorghum plants are 15 feet (4.5 meters) tall; the heads of the plants contain seeds that are *landraces*, the term for the genetic descendants of plants carefully nurtured in a particular place for thousands of years. Hassain says he prefers to grow the landraces, rather than purchasing seeds, because he is impressed with the results. His seeds are highly productive: he once had 29 stalks of sorghum sprout from a single seed. His wife, Kadija, usually sorts the seeds, determining which are best for planting, for storing, and for cooking. [*Adapted from Pattie Lacroix, "Ethiopia's living laboratory of biodiversity" (Ottawa, Ontario: International Development Research Centre of Canada, 1995).*]

ineffective government that has at times fostered civil conflict. Even in the moister southern portion of East Africa, growing consumption is straining the region's carrying capacity.

The Interior Highlands

The landforms of East Africa's interior are dominated by the Great Rift Valley. This irregular series of developing tectonic rifts curves inland from the Red Sea and extends southwest through Ethiopia and into the highland "Great Lakes" country, which curves around the western edges of the subregion and extends into Mozambique to the south. The rift valley, which contains numerous archaeological sites related to the evolution of humans and large mammals, is today covered primarily with temperate and tropical savanna, although patches of forest remain on the high slopes and in southwestern Tanzania. Most rural people make a living by cattle herding, although herding and cultivation are combined in the wetter central uplands of Ethiopia and Kenya and in the woodlands of southern Tanzania. A mixture of shifting subsistence and plantation cultivation sustains populations in the south.

In the north of East Africa, near the rift valley, is an area of mountains and plateaus called the Ethiopian Highlands, which range in elevation from 4000 feet (1200 meters) to over 10,000 feet (3000 meters). The Eritrean and northern Ethiopian zone of these highlands is occupied by the ethnic Tigre. The highlands are drained by deep river valleys, including the valley where the headwaters of the Blue Nile lie. Rain falls during the summer in the semiarid highlands, and the area includes a mixture of environments that vary with elevation. The Oromo, Amhara, and Gurage live in the southern reaches of the highlands. The various ethnic groups of the highlands are known for their complex and productive systems of **mixed agriculture,** meaning that they raise cattle and use the manure to fertilize a wide variety of crops that are

The Coastal Lowlands

To the east and south of the Ethiopian Highlands is a broad, arid apron that descends to the Gulf of Aden and the Indian Ocean. Along the coasts of the Horn of Africa, there are large stretches of sand and salt deserts and few rivers. The country of Somalia, which is shaped like a 7 with its point jutting into the Indian Ocean, occupies the majority of the low, arid coastal territory in the Horn. The people who live in this area can eke out only a meager living through oasis cultivation and dry-grass herding. Somalia's Muslim population of 8 million, virtually all ethnically Somalis, are aligned in six principal clans. Warlords and political adventurers have engaged in a violent rivalry for power over the last decade. The resulting turmoil has created refugees, who cannot be productive, and has blocked access to some lands that are crucial to the food production system. Thus Somalian cultivators must overuse the land that is available. For this political reason, not because of rapid population growth, pressure on arable land has increased significantly in all countries of the Horn of Africa. Nonetheless, at present rates of growth, populations in the Horn are expected to double and even triple by 2050. Such numbers will be difficult to accommodate in these arid lands because of the scarcity of water and other resources and because the complex agricultural systems require skilled management to remain productive.

East Africa's shores from southern Somalia southward have attracted traders from around the Indian Ocean—from China, Indonesia, Malaysia, India, Persia (Iran), Arabia, and elsewhere in Africa. The result has been a grand blending of human genetics, human cultures, plants, animals, and languages. In the sixth century, Arab traders brought Islam to East Africa, where it remains an important influence in the north and along the coast. Traders, speaking the lingua franca of Swahili, established networks that linked the coast with the interior. These networks penetrated deeply into the farming and herding areas of the savanna and into the semifeudalistic cultures of the highland lake country. Today, trade across the Indian Ocean continues: Arab countries are the most important trading partners, but Asia, China, India, and Japan are increasingly important to East African trade, while trade with Indonesia and Malaysia is increasing. East Africa exports mostly agricultural products (live animals, hides, bananas, coffee, cashew nuts, tobacco, and cotton) and imports machinery, transportation equipment, industrial raw materials, and consumer goods.

The Islands

Madagascar, which lies off the coast of East Africa, is the fourth largest island in the world; only the islands of Greenland, New Guinea, and Borneo are larger. Its unique plant and animal life—the result of *diffusion* (transmission by natural or cultural processes) from mainland Africa and Asia and subsequent evolution in isolation from the mainland—is highly prized by biologists and greatly threatened by environmental degradation. The other East African island nations—Comoros, Seychelles, and Mauritius—are much smaller and less physically complex than Madagascar. All the islands have a cosmopolitan ethnic makeup, the result of thousands of years of trade across the Indian Ocean. The people and languages of Madagascar, for example, have ancient origins in Southeast Asia. Further cultural mixing in the islands took place during European (primarily French and British) colonization. During the colonial era, these islands supported large European-owned plantations, worked by laborers brought in from Asia and the African mainland. More recently, islanders have added tourism to the economic mix. Most visitors come from the African mainland, from Asian countries surrounding the Indian Ocean, and from Europe.

Kenya: A Modern Case Study in Contested Space

Kenya is the wealthiest and most industrialized country in East Africa, producing tea, coffee, cacao, beer, cement, petroleum products, horticultural products, and processed food. It is a regional hub of trade, yet it remains one of the poorest countries in the world (see Table 7.1 on pages 374–375).

In ancient times, traders in this part of Africa exchanged commodities—slaves, ivory, rare animals—that can no longer be legally traded. By the early nineteenth century, the depletion of elephant herds (killed for their ivory tusks) and changing attitudes toward slavery had pushed the region into a period of stagnation that made it easier for the British to gain control. In the reorganization of East African economies under colonialism, the British relocated native highland farmers in Kenya, such as the Kikuyu, to make way for European coffee plantations; and they displaced herders, such as the Masai in Kenya and Tanzania, from the savanna. Some of the lands of the Masai became game reserves designed to protect the savanna wildlife for the use of European hunters and tourists.

Today, the former colonial game reserves of Kenya and Tanzania are among the finest national parks in Africa, offering protection to wild elephants, giraffes, zebras, and lions, as well as many less spectacular but no less significant species. Tourism, mostly in the form of visits to wildlife parks, accounts for 20 percent of foreign exchange earnings. The future of the parks is precarious, however, because of competing demands for the land by herders and cultivators and the tendency to view wild animals as a useful consumable resource. In Kenya, a living elephant is worth close to $15,000 a year in income from tourists, and tourism brings in about $200 million a year to the country. Nonetheless, some still see hunting elephants for their ivory as a profitable activity, even though the profits to a hunter may be no more than $1000 per dead animal. The Masai complain that elephants destroy their grazing lands and gardens, and hence they see no reason to protect elephants, which, when protected, they complain, quickly overpopulate an area and stress its environment. Modern park policy in Kenya now advocates public education that clearly points out the long-term value to local communities of saving endangered species.

After Kenyan independence from Britain in 1963, government policies resettled black African farmers on highland plantations formerly owned by Europeans, giving the country a strong base in export agriculture through the production of coffee, tea, sugarcane, corn, and sisal. Under postcolonial policies, nomadic peoples have fared less well than farmers. Government strategies for managing the national wildlife parks have inadvertently resulted in the spread of tsetse fly infestations to the local pastoralists' herds. Disputes over scarce water supplies have broken out between cultivators who wish to irrigate their crops and herders who need regular sources of water for their animals. So far, the government has sided with the farmers, even to the point of policing the nomadic herders from armed helicopters.

Agriculture, especially export agriculture, is important to Kenya's economy, but the country's dependence on this sector has also put people's livelihoods and the famous national parks at risk. One problem is related to population growth. In the mid-twentieth century, populations throughout East Africa began to grow extremely rapidly. In 1946, Kenya had just 5 million people; by 2003, it had 31 million trying to eke out an existence from the same land and resource base. In the past, the primarily arid land could adequately support relatively small populations when managed according to traditional herding and subsistence cultivation customs. Today, the land's carrying capacity is severely stressed. Competition for cultivable land has pushed thousands of small farmers out of the more fertile uplands into the drier, more fragile savannas that had been both communal rangelands and habitats for wild animals. The small plots they have plowed in the savanna are only marginally productive and have displaced nomads, elephants, impala, and giraffes alike. While Kenyans try desperately to increase agricultural production, volatile world market prices for some export crops can send the national economy plunging; world coffee prices fell sharply in 2001, for example. These problems are threatening both Kenyan and Tanzanian game reserves and parks. Population pressure has forced farmers to plow and overgraze wildland buffer zones surrounding the parks. Hungry people have begun to poach park animals for food, and there is now a large underground trade in bush meat.

SOUTHERN AFRICA

At the beginning of the new millennium, both optimism and worry are in the voices heard across Southern Africa (Figure 7.28). In the case of South Africa, the cause for optimism is that it is now free of the scourge of apartheid, racial conflict has abated, and the peace dividend is substantial economic growth. South Africa is the largest and wealthiest country, and perhaps the best-run country, in the whole of sub-Saharan Africa. Other countries in Southern Africa—Namibia, Botswana, Lesotho, Swaziland, and Angola—are enjoying new stability and economic growth as well. Only Zimbabwe, once thought of as having great promise, is quickly becoming a worrisome failed state because of unwise development advice and bad governance (Zimbabwe is discussed on page 351 and at several other locations in this chapter). But the main cause for worry in the region is the HIV-AIDS epidemic, which by 2003 threatened from 20 to 30 to 40 percent or more of the population in every country in Southern Africa (see the discussion on pages 347–348).

Southern Africa is a plateau ringed on three sides by mountains and a narrow lowland coastal strip (see Figure 7.2 on pages 334–335). At its center is the Kalahari Desert, but for the most part, Southern Africa is a land of savannas and open woodland. Population density is low; the highest densities are found in the southeastern part of South Africa (see Figure 7.6 on page 343). Africa's mineral wealth is concentrated in Southern Africa, where there are rich deposits of diamonds, gold, chrome, copper, uranium, and coal.

Angola and Mozambique are both former Portuguese colonies. Their wars of independence evolved into civil wars, with Marxist governments on one side and rebels supported by South Africa and the United States on the other. Mozambique ended its civil war in 1994. By 1996, the once devastated economy was experiencing one of the highest rates of growth on the continent. In Angola, where fighting continued into 2002, stability and economic growth are still elusive. Liberation movements fighting to free Angola from Portuguese rule split into factions in the early 1970s. A Marxist faction took over the government and held the coastal regions, where most Angolans live and where the country's oil and fishing

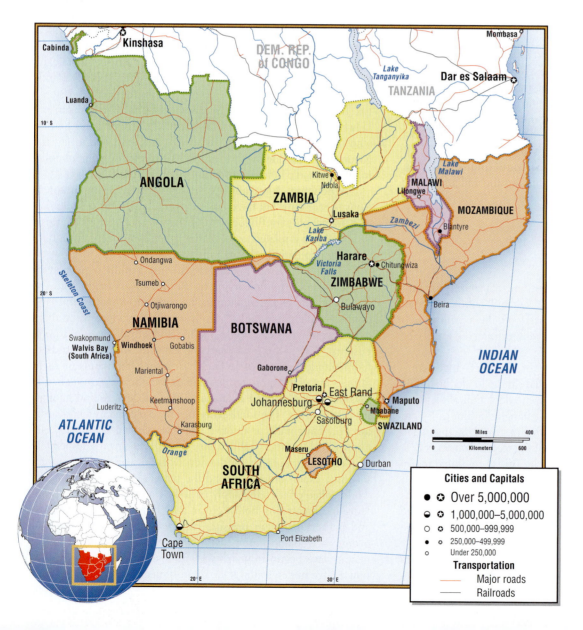

FIGURE 7.28 The Southern Africa subregion.

CULTURAL INSIGHT The Thumb Piano

In Zimbabwe, the Shona people have considered the mbira (thumb piano) a core element of their culture for several hundred years. The mbira has both a sacred and a secular role in society. Practitioners use it in *bira* ceremonies to contact the spiritual realm. Missionaries viewed the music as pagan and tried to ban it. In the 1970s, the mbira was one of the principal instruments used in the subtle songs and traditional parables of protest against white rule. These songs infuriated members of the white government because they could not understand them. Now the mbira is once again used extensively, both for entertainment and to lift the spirit.

Musicians play the mbira by depressing and releasing the ends of the finely tuned metal keys with their thumbnails. The keys vibrate against the carefully crafted wooden resonator and produce warm, metallic tones that add mysterious, melodious sounds to any composition. There are many different styles of mbiras, and the number of keys can range from 8 to 52. In contrast to many modern instruments, which are box-shaped, traditional mbiras are made of resonant calabash bowls.

A mbira, or thumb piano. [Jose Azel/Contact/Woodfin Camp & Associates.]

resources are located. Another faction, backed by the United States until 1993, known as Union for the Total Independence of Angola (UNITA), drew its strength from among the Ovimbundu-speaking people of the south-central plateau, where gold, diamonds, uranium, and manganese are located. UNITA held the upper hand for years, controlling 70 percent of the country. For a while, UNITA garnered support from the notoriously corrupt Mobutu regime in Congo (Kinshasa) and from the sale of diamonds and other resources. In early 1998, UNITA and the Angolan government agreed to a peace treaty, but episodic violence continued into 2002 in the countryside.

The three interior countries of Malawi, Zimbabwe, and Zambia are still overwhelmingly rural. Agriculture and related food-processing industries form the basis of the economies, and their products are a major portion of exports. Over the last two decades, however, agricultural productivity has fallen. The case of Malawi is illustrative. Malawi had enough food for its people and surpluses for export until the 1980s, when land productivity began to decline as a result of overuse. After independence from Britain in 1964, the government chose to focus on large-estate agriculture to produce cash crops (tobacco, tea, sugar) for export. Fragile tropical soils that ordinarily need years of rest between plantings were expected to produce continually. After a few years, the soil was depleted, and production on plantations declined. Smallholders, who had lost land to commercial producers, found themselves with insufficient land to allow for necessary fallow periods. Their productivity, too, began to fall. The historian F. Jeffress Ramsay writes that, by the 1990s, 86 percent of Malawi's rural households had less than 5 acres (2 hectares) of land on which to grow food for themselves. Yet these smallholders produced nearly 70 percent of the food for rural and urban consumption in Malawi. In addition, the country had to feed more than 600,000 refugees from Mozambique's civil war. Meanwhile, Malawi's government, for reasons that are not clear, held food reserves under lock, while the people went hungry.

Although Zambia and Zimbabwe are struggling to reconstitute their cultural heritage after generations of British colonialism (see the box "The Thumb Piano" above), their economies remain focused on long-established extractive industries. In Zimbabwe, chromium, gold, coal, nickel, and silver mining accounts for 40 percent of the country's exports; a platinum mine opened in 1994 was expected to make Zimbabwe the world's second largest producer, after South Africa. But the economy has failed because of misplaced advice from international lending agencies and the inept and now illegitimate government of Robert Mugabe (see the discussion on page 351). Zambia has rich copper resources and has been a major world producer, but the industry lost money in the 1990s. The loss resulted from a combination of high debt, falling world prices in the 1980s, inefficient management, and the low productivity of miners hampered by poor working conditions. Botswana's situation was discussed at some length earlier in this chapter (see pages 351–352).

Majority Rule Comes to South Africa

South Africa is a country of beautiful and striking vistas. Just inland of its long coastline lie uplands and mountains that afford dramatic views of the coast and the ocean. The interior is a high rolling plateau reminiscent of the Great Plains in North America. South Africa is entirely within the midlatitudes and has consistently cool temperatures that range between 60°F and 70°F (16°C and 20°C). The extreme southern coastal area around Cape Town has a Mediterranean climate: cool and wet in the winter, dry in the summer.

About twice the size of Texas, South Africa has a population of 44 million people: 75 percent are of African descent, 14 percent of

European descent, and 2 percent of Indian descent; the rest have mixed backgrounds. After more than 300 years of colonization by Europeans and rule by the white minority, free elections were held in 1994, and black Africans gained control of the government. The end of white minority rule came after nearly 50 years of strict segregation by color enforced brutally by the apartheid system. When the former political activist against apartheid, Nelson Mandela, was elected president in 1994, he insisted that in order to heal the racial divide, the country face the reality of what had happened under apartheid, through Truth and Reconciliation Commissions. Those who had been brutal on both sides were asked to step forward and admit what happened in detail, then they were formally forgiven for their misdeeds.

The country is now considered to be one of the world's top ten emerging markets. The country has a balanced economy that accounts for 45 percent of the entire continent's production. It has built competitive industries in communications, energy, transport, and finance (it has a world-class stock market) and supplies goods and services to neighboring countries. For example, 75 percent of Namibia's and Botswana's imports come from South Africa. In return, products from all over Africa are sold in South Africa's markets, and many people throughout Southern Africa are dependent on jobs in South Africa.

The Angolan peace treaty, economic successes in Namibia and Botswana, and racial reconciliation and political and economic progress in South Africa, as well as in the tiny countries of Swaziland and Lesotho, have prompted forecasters to speculate that Southern Africa, more than any other subregion, holds the potential to lead sub-Saharan Africa into a more prosperous age. Certainly, poverty is still widespread, resources are overstressed, and the HIV-AIDS epidemic, which is particularly severe in this region (see pages 347–348), is bringing terrible losses. But with creative, indeed courageous, leadership and outside help, South Africa and a few of its neighbors may be able to move directly from agrarian economies into the information age of the twenty-first century, bypassing the industrial phase experienced by Europe, North America, and Japan.

Quick Review

1. What did Charles Taylor use to finance his takeover of Liberia?

2. How is the political situation in Somalia restricting food production?

3. In Kenya, under what conditions is an elephant worth more alive than dead?

4. Describe the features of the South African economy that account for its regional influence.

REFLECTIONS ON SUB-SAHARAN AFRICA

Late one evening in a restaurant in central Europe, after a lengthy conversation that touched on some of the earth's perplexing problems, my colleague leaned across the elegant white linen tablecloth and asked, "But don't you think that Africa is, after all, better off for having been colonized by Europeans?"

How does one reply to such a question? Africa, the ancient home of the human race, is of all regions on earth the one in greatest need of attention. Africa is poor and, by some measures, getting poorer. Yet the reasons for Africa's poverty are not immediately apparent to the casual observer. Africa is blessed with many kinds of resources: agricultural, mineral, and forest. Nonetheless, it does not get much income from these resources; instead, the income goes elsewhere. And although most of Africa is not densely occupied, rapid population growth is thwarting efforts to improve standards of living. Africa's people work hard, but their productivity, at least as measured by the standards of the developed world, is low.

An understanding of the reasons for Africa's present poverty and its social and political instability begins to emerge only through an exploration of its history over the last several centuries of European colonialism and through a careful analysis of how that history is still affecting the organization of African economies and societies. Colonialism methodically removed Africans from control of their own societies and turned Africa's peoples and resources to the service of distant countries. Even today, with colonialism officially dead for more than three decades, outsiders in the global marketplace continue to view African resources as available for the taking at less than a fair price. In view of all this, it is hard to believe that anyone would think Africa is better off for having been colonized.

Africans are only beginning to devise new economic development strategies and political institutions to replace the ones imposed by outsiders over the last 500 years. Africa remains a continent of countries created out of foreign perceptions of how Africans should be organized politically, but now African leaders are articulating wider visions. Pan-African movements for economic integration and regional cooperation are emerging.

It is tempting to suggest that the rest of the world should leave Africa to the Africans for the next era. But that view may be too simplistic. Although Africa had plenty of help getting into its current predicament, and although outsiders have prospered from its wealth, it has received very little foreign aid. Europe, North America, and Japan together give about U.S. $16.2 billion a year in aid to sub-Saharan Africa, of which the United States gives only about 4 percent, or U.S. $665 million, annually. Egypt alone receives four times more U.S. foreign aid ($2.4 billion). U.S. citizens give less in aid to sub-Saharan Africa than they reap in profits from private investment in Africa. For example, in 2000, the Coca-Cola Company *alone* took profits of roughly $148 million from its African operations.

Many free-market economists, such as Jeffrey Sachs, director of the Center for International Development at Harvard University, think wealthy nations should provide financial support for Africa's comeback so that Africans can themselves become a market for imported products. Experts suggest several strategies. First, African debt to foreign governments, international lending agencies, and some private lenders should be cancelled so that tax money can once again support schools, health care, and social services. As of January 2001, American Baptists, Presbyterians, and Methodists and the international Anglican and Roman Catholic leadership were publicly on record as supporting this idea. Soon thereafter, the leaders of the G8 countries (the world's most developed economies) agreed in principle to forgive debt owed their governments by the poorest countries in Africa. What remained to be decided was how to ensure that the debt repayment money would go instead to badly needed health, education, and social programs. A second suggestion is that the developed world lower tariffs against African manufactured products to foster development of African industries. This proposal is opposed by people who worry that jobs in the developed world will be lost if cheaper African consumer goods can enter the developed countries free of tariffs. A third suggestion is that future aid to Africa be designed to take advantage of indigenous skills and knowledge and that development planning and aid money address local needs as defined and managed by local experts. This fundamentally conservative perspective acknowledges the need to help, but puts its faith in the ultimate ability of the African people to look out for themselves.

Chapter Key Terms

African Union   357
agroforestry   372
animism   365
apartheid   342
carrying capacity   343
Common Market for Eastern and Southern Africa (COMESA)   357
currency devaluation   355
desertification   371
disinvestment   350
dry forests   372
East African Community (EAC)   357
Economic Community of Central African States (CEEAC)   357

Economic Community of West African States (ECOWAS)   357
ecotourism   373
ethnic group   336
ethnicity   336
ethnocentrism   336
female circumcision   369
forward capital   361
fragile environment   372
genocide   344
grassroots economic development   357
Horn of Africa   338
intertropical convergence zone (ITZC)   337
laterite   338
leaching   338

lingua franca   370
mixed agriculture   382
neocolonialism   333
organic matter   338
polygyny   368
Sahel   338
savanna   338
self-reliant development   357
shifting cultivation   338
Southern African Development Community (SADC)   357
structural adjustment programs (SAPs)   353
syncretism (or fusion)   366
upland zones   337

CHAPTER 8

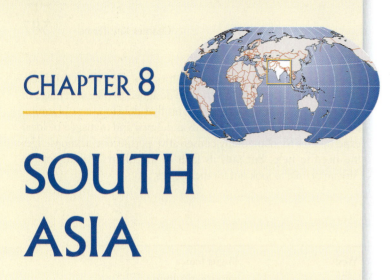

SOUTH ASIA

PERSONAL VIGNETTE Ramesh Seshadri leans back from his computer at the Indian Institute of Technology, a prestigious institution of higher learning in Chennai (Madras), South India. He stares out his window, which faces the Bay of Bengal. The beach is strewn with litter and dotted with the makeshift tents of poor people who call the beach home. This reminder of his country's increasing population and decreasing environmental quality makes Ramesh think about a job opportunity in the United States as a network technician that a friend has just written him about. Many of Ramesh's classmates dream of a job in the United States, where salaries and living standards far surpass those in India. Alternatively, Ramesh could take a job in India's own booming information technology sector, which would allow him to remain close to his family and to Hindu religious traditions.

Then there is an opportunity offered by Lal Krishnan, a wealthy businessman of Indian descent who is now minister of environment for a Caribbean island nation. On a recent trip to Chennai, Krishnan visited a student group that Ramesh leads. The group organizes demonstrations urging India's government to base its rule on Hindu principles and not, as they feel it has in the past, on the secular ideals of Western-style democracy. Krishnan took note of a campaign that Ramesh organized to ban the thin plastic shopping bags that are commonly discarded in the streets. The cows that roam the streets of Chennai and most urban areas in India often eat the bags and develop lethal digestive problems. Devout Hindus like Ramesh, who hold cows to be sacred, are concerned about this problem. After the campaign, Krishnan offered Ramesh a position as an organizer and fundraiser for the Caribbean branch of the World Council of Hindus. This global organization works with Indians living abroad to raise awareness about issues of concern to Hindus all over the world, including newly emerging religion-based concerns about environmental deterioration. The organization also seeks financial support for Hindu political parties in India. Even though the salary is much lower than what he could earn in a computer job, Ramesh is excited about this opportunity because it would enable him to work against what he sees as the degradation of Hindu culture and the Indian environment by foreign ways and ideas. Nevertheless, a guaranteed job in the United States and a well-paid technical career are very tempting. [*Adapted from the field notes of Alex Pulsipher, India, 1999.*]

MAKING GLOBAL CONNECTIONS

Ramesh's concerns and opportunities typify the global forces influencing South Asia today as well as the more local problems South Asians face. This region is the second poorest in the world, after sub-Saharan Africa, and the most densely populated. These pressures have motivated many South Asians to emigrate to other nations. Figure 8.1 shows the distribution of South Asians throughout the world (also see the box "The Indian Diaspora" on page 395). Should Ramesh choose to go abroad, he will be joining a migration that began more than 2000 years ago as traders from the Indian subcontinent ventured to Arabia, Africa, and Southeast Asia. The migration continued under British colonization. In some places, people of South Asian origin have worked their way into positions of power and prosperity and can aid contemporary South Asians who wish to migrate.

South Asia is a relatively easy region to define. It is bordered by the Indian Ocean to the south and by extraordinarily high mountains to the north. Although it is complex culturally and politically, it has a number of unifying features: the village is one; the common experience of British colonialism is another.

Physically, South Asia is far smaller than Africa: it would fit into the African continent five times. Yet, with nearly 1.44 billion people, it has twice Africa's population. The countries that make

Children of ethnic Tamil immigrants from South India prepare for the annual Hindu temple celebration at Sri Kamadschi Ambal Temple in Deu, Germany. [Yavuz Arsian/Peter Arnold.]

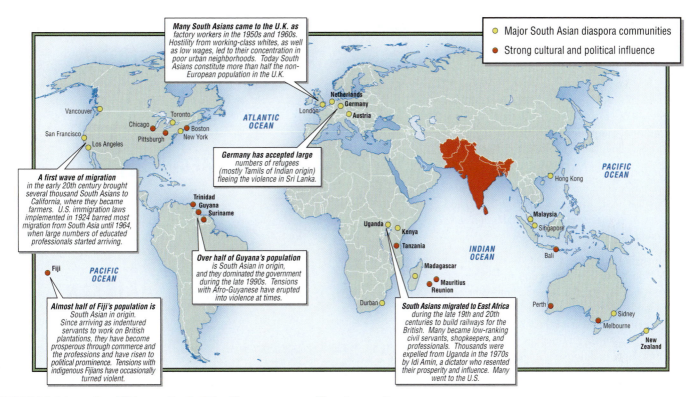

Many South Asians came to the U.K. as factory workers in the 1950s and 1960s. Hostility from working-class whites, as well as low wages, led to their concentration in poor urban neighborhoods. Today South Asians constitute more than half the non-European population in the U.K.

● Major South Asian diaspora communities
● Strong cultural and political influence

Germany has accepted large numbers of refugees (mostly Tamils of Indian origin) fleeing the violence in Sri Lanka.

A first wave of migration in the early 20th century brought several thousand South Asians to California, where they became farmers. U.S. immigration laws implemented in 1924 barred most migration from South Asia until 1964, when large numbers of educated professionals started arriving.

Over half of Guyana's population is South Asian in origin, and they dominated the government during the late 1990s. Tensions with Afro-Guyanese have erupted into violence at times.

Almost half of Fiji's population is South Asian in origin. Since arriving as indentured servants to work on British plantations, they have become prosperous through commerce and the professions and have risen to political prominence. Tensions with indigenous Fijians have occasionally turned violent.

South Asians migrated to East Africa during the late 19th and 20th centuries to build railways for the British. Many became low-ranking civil servants, shopkeepers, and professionals. Thousands were expelled from Uganda in the 1970s by Idi Amin, a dictator who resented their prosperity and influence. Many went to the U.S.

FIGURE 8.1 Interregional linkages: South Asian diaspora communities. As a result of several waves of South Asian emigration, communities of ethnic Indians may be found throughout the world (see the box "The Indian Diaspora" on page 395).

up the region are Afghanistan and Pakistan in the northwest; the Himalayan states of Nepal and Bhutan; Bangladesh in the northeast; India; the island country of Sri Lanka; and several groups of islands in the Arabian Sea and the Bay of Bengal (Figure 8.2). Because its clear physical boundaries set it apart from the rest of the Asian continent, the term **subcontinent** is often used to refer to the entire Indian peninsula, including southeastern Pakistan and Bangladesh.

I   THE GEOGRAPHIC SETTING

Terms to Be Aware Of

South Asians are reclaiming the original names of many of their important places. Under British colonial rule, which lasted until 1947, many of the ancient place names were changed or modified. In the last decade, Indians, especially, have begun to use the old names. The city of Bombay, for example, is now officially *Mumbai*, the city of Madras is *Chennai*, and the Ganges River is the *Ganga River*. In the spirit of returning to indigenous names, the country neighboring far eastern India is referred to as *Burma* rather than Myanmar.

PHYSICAL PATTERNS

The physical geography of South Asia was created by ancient and continuing shifts in the earth's crust. Many of the landforms and

even the climatic features of this region are the result of forces that positioned the Indian subcontinent along the southern edge of the Eurasian continent, where the warm Indian Ocean surrounds it and the Himalayas, a massive mountain range, shield it from cold air flows from the north.

Landforms

The Indian subcontinent and its surrounding territory comprise some of the most spectacular landforms on earth and illustrate dramatically what can happen when two tectonic plates collide. About 180 million years ago, the Indian-Australian Plate, which carries India, broke free from the eastern edge of the African continent and drifted to the northeast. It began to collide with the Eurasian Plate about 60 million years ago, and India became a giant peninsula jutting into the Indian Ocean. As the relentless pushing from the south continued, both the leading (northern) (*Text continues on page 392.*)

THEMES TO EXPLORE IN SOUTH ASIA

Here are some themes to follow as you read this chapter.

1. The ancient and layered pattern of cultural influences. This region has experienced multiple waves of cultural and religious influences since pre-historic times. The most recent is globalization.

2. The importance of village life. South Asia has many large cities, but 70 percent of the region's people live in hundreds of thousands of villages, and rural modes of spatial organization and interaction persist even in the cities.

3. The lingering influence of British colonization. Although British colonization ended more than 50 years ago, it has left its mark on the landscape and culture. South Asians are struggling to define national identities that can accommodate some British and other outside influences while retaining strong South Asian traditions and moral values.

4. Extremes between rich and poor. Disparities of wealth and startling contrasts between traditional and technically advanced ways of life are commonplace across the region and pose dangers to the continuation of democracy in these plural societies.

5. The challenge of continuing population growth. South Asia now has a population comparable to that of China. High population growth puts stress on resources, and the region faces a number of challenges in providing its people with the basics of life, such as clean drinking water.

6. Environmental concerns and conflicts. As in other densely populated regions experiencing economic growth and rising consumerism, environmental issues lie at the forefront of people's minds. In addition to the increasing problems of land degradation, shortage of water, and pollution, a number of conflicts have emerged over how resources are to be used and the impact of large-scale development on local people.

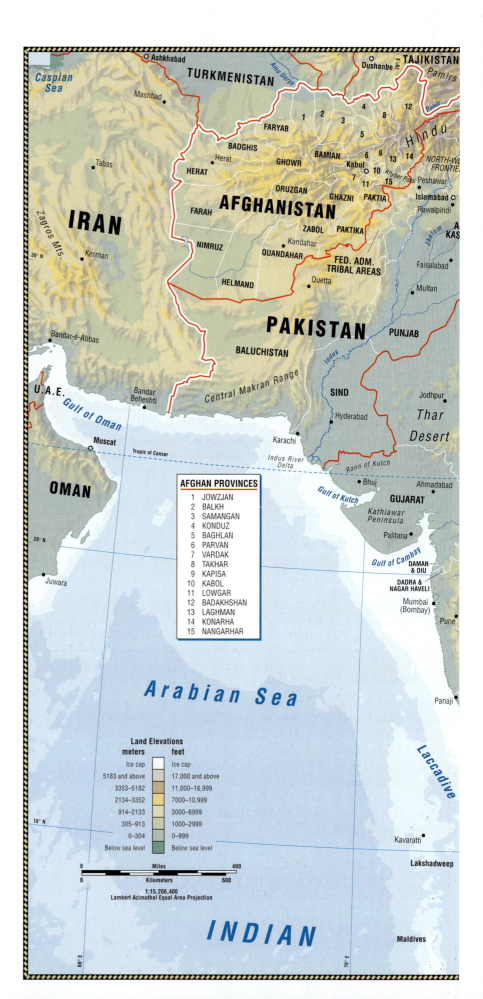

FIGURE 8.2 Regional map of South Asia.

edge of India and the southern edge of Eurasia crumpled and buckled. The massive Himalayas, rising more than 28,000 feet (8500 meters), were formed from this collision, but the effects extended far beyond it. To the west and east of the Indian landmass, the pressure pushed curved crinkles into the "fabric" of Eurasia. In the west, these crinkles became the Hindu Kush, the Pamir, and the other mountains of Pakistan and Afghanistan; in the east, they became the mountains of far eastern India and adjacent Burma, China, and Thailand. As a result of the continuous compression, the Plateau of Tibet rose up behind the Himalayas to more than 15,000 feet (4500 meters) in some places (see Figure 8.2). Elsewhere in Asia, as far north as Siberia, the land arched, bent, and cracked.

South and southwest of the Himalayas are the Indus and Ganga river basins (also called the Indo-Gangetic Plain). Still farther south, near the tip of the Indian subcontinent, is the Deccan Plateau, an area of modest uplands 1000 to 2000 feet (300 to 600 meters) high, interspersed with river valleys. This upland region is bounded on the east and west by two moderately high mountain ranges, the Eastern and Western Ghats. They descend to a long but narrow coastline interrupted by extensive river deltas and alluvial plains. The river valleys and coastal zones are densely occupied; the uplands only slightly less so. Because of the high degree of tectonic activity and deep crustal fractures, South Asia is prone to devastating earthquakes, such as the magnitude 7.7 quake that shook the state of Gujarat in western India in January

2001. In August 2001, seismologists announced that measured pressures in the earth's crust indicate that a very large earthquake is overdue in the Himalayan region.

Climate

The end of the dry [winter] season [April and May] is cruel in South Asia. It marks the beginning of a brief lull that is soon overtaken by the annual monsoon rains. In the lowlands of eastern India and Bangladesh, temperatures in the shade are routinely above a hundred degrees; the heat causes dirt roads to become so parched that they are soon covered in several inches of loose dirt and sand. Tornadoes wreak havoc, killing hundreds and flattening entire villages. Even the wind provides little relief, as it whips up sandstorms, making it impossible to see farther than six feet in any direction. Inhaling the sand and dust leads to widespread respiratory problems that cause many to spend long stretches of the summer ill.

This is also a time of hunger, as with each passing day thousands of rural families consume the last of their household stock of grain from the previous harvest and join the millions of others who must buy their food. Each new entrant into the market nudges the price of grain up a little more, pushing millions from two meals a day to one, from 90 percent of the minimal caloric intake needed to sustain life, to 70 percent.

(Alex Counts, *Give Us Credit*, 1996, p. 69.)

FIGURE 8.3 Winter and summer monsoons in South Asia. (A) In the winter, cold, dry air blows from the Eurasian continent south across India toward the ITCZ, which is much farther south. (B) In the summer, the ITCZ moves north across India, picking up huge amounts of moisture from the ocean, which it then deposits over India. (C) Landforms also affect rainfall. In summer, moisture-laden air from the ocean rises and cools when it reaches the coastal mountains, releasing copious rain. The rising and cooling process is repeated when the wet air reaches the Himalayas.

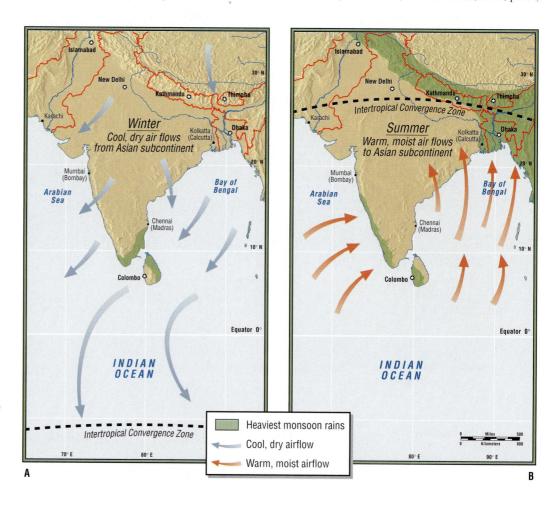

From mid June to the end of October [summer] is the time of the river. Not only are the rivers full to bursting, but the rains pour down so relentlessly and the clouds are so close to village roofs that all the earth smells damp and mildewed, and green and yellow moss creeps up every wall and tree. . . . Cattle and goats become aquatic, chickens are placed in baskets on roofs, and boats are loaded with valuables and tied to houses. Cooking fires are impossible . . . so the staples [are] precooked rice, a dry lentil called dal, and jackfruit, a large smelly melon that ripens on trees during this season. . . . Because most villages are built on artificial mounds raised above the fields, as the floods rise villages become tiny islands, . . . self-sustaining outpost[s] cut off from civilization . . . for most of three months of the year.

(James Novak, *Bangladesh: Reflections on the Water*, 1993, pp. 24–25.)

These two passages highlight the contrasts between South Asia's winter and summer **monsoons** (dominant wind patterns) (Figure 8.3). In winter, cool, dry air flows from land (the continent of Asia) to the ocean; in summer, warm, moisture-laden air flows from the Indian Ocean over the Indian subcontinent, bringing with it heavy rains. The abundance of rainfall is amplified by another process taking place near the equator, where air masses moving south from the Northern Hemisphere converge with those moving north from the Southern Hemisphere. Warm, moisture-laden air masses from the north and south come together and rise, forming clouds that produce copious precipitation as they rise and cool. This belt of warm, rising air that circles the earth around the equator is called the intertropical convergence zone (ITCZ) (see Figure 7.3, page 337). In June, July, and August, when it is summer in the Northern Hemisphere, the ITCZ shifts north of the equator; in November, December, and January, when it is summer south of the equator, the ITCZ shifts south of the equator. Climatologists now think that the powerful "inhalation" of air over the Eurasian landmass in summer actually bends the ITCZ belt even farther north. After the ITCZ rushes over the Indian Ocean, picking up enormous

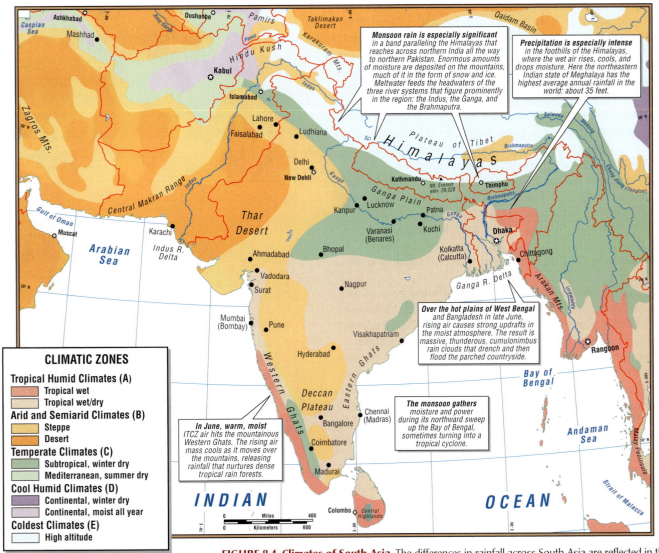

FIGURE 8.4 Climates of South Asia. The differences in rainfall across South Asia are reflected in the varying climatic zones of the region.

amounts of moisture, the pull of the Eurasian "inhalation" draws it farther north toward Asia, where its moisture is deposited on the Indian subcontinent.

The monsoons and the ITCZ are major influences on South Asia's climate (see Figure 8.3 on page 392). In early June, the ITCZ- generated warm, moist air first reaches the mountainous Western Ghats and Sri Lanka's central highlands. The rising air mass cools as it moves over the mountains, releasing rainfall that nurtures dense tropical rain forests. After this dumping of huge amounts of rain, the moisture in the air mass is reduced, and somewhat less rain (but still significant amounts) falls to the east of the Western Ghats. Once on the other side of India, the monsoon gathers additional moisture and power in its northward sweep up the Bay of Bengal, sometimes turning into a tropical cyclone. As the monsoon system reaches the hot plains of West Bengal and Bangladesh in late June, warm rising air causes strong updrafts in the moist atmosphere. These updrafts create massive, thunderous cumulonimbus rain clouds that drench and then flood the parched countryside. Precipitation is especially intense in the foothills of the Himalayas. The northeastern Indian state of Meghalaya has the highest average annual rainfall in the world: about 35 feet (10.6 meters). Rainfall diminishes to the west of Nepal, but seasonal rain is still significant in a band parallel to the Himalayas that reaches across northern India all the way to northern Pakistan by July. These differences in rainfall are reflected in the varying climatic zones (Figure 8.4) and agricultural zones (see Figure 8.15 on page 413) of South Asia. Rice is generally grown in the wettest areas, and cereals—especially wheat—are the dominant crops in the drier areas. Pasture and nonagricultural land dominate in the driest areas and in the areas where altitude inhibits plant growth.

By November, the cooling Eurasian landmass sends cooler, drier air over South Asia. This heavier air from the north pushes the warm, wet air back south to the Indian Ocean. Although very little rain falls during this winter monsoon, parts of South India and Sri Lanka receive winter rains as the ITCZ drops moisture picked up on its now southward pass over the Bay of Bengal.

Monsoon rains deposit large amounts of moisture over the Himalayas, much of it in the form of snow and ice. Meltwater feeds the headwaters of the three river systems that figure prominently in the region: the Indus, the Ganga, and the Brahmaputra. All three rivers begin within 100 miles (160 kilometers) of one another in the Himalayan highlands near the Tibet-Nepal-India borders (see Figure 8.2 on pages 390–391). The Ganga flows generally south through the mountains and then east across the Ganga Plain to the Bay of Bengal. The Brahmaputra flows first east, then south, and then west to join the Ganga in forming the giant delta of Bengal. The Indus, in the far west, takes a southwesterly course across arid Pakistan and empties into the Arabian Sea. These rivers, and many of the tributaries that feed them, are actively wearing down the surface of the Himalayas; they carry an enormous load of sediment, especially during the rainy season when their volume increases. Their velocity slows when they reach the lowlands, and much of the sediment settles out as silt. It is then repeatedly picked up and redeposited by successive

floods. As illustrated in the diagram of the Brahmaputra River (Figure 8.5), the seasonally replenished silt nourishes much of the agricultural production in the densely occupied plains of Bangladesh. The same is true on the Ganga and Indus plains.

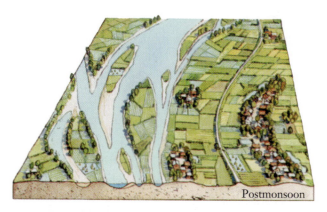

FIGURE 8.5 The Brahmaputra River in Bangladesh at various seasonal stages. In the premonsoon stage (*top*), the river flows in multiple channels across the flat plain. During peak flood stage (*middle*), the great volume of water overflows the banks and spreads across fields, towns, and roads. It carves new channels, leaving some places cut off from the mainland. In the postmonsoon stage (*bottom*), the river returns to its banks, but some of the new channels persist, changing the lay of the land. As the river recedes, it leaves behind silt and algae that nourish the soil. New ponds and lakes form and fill with fish. The people have learned to adapt their farms to a changing landscape. Throughout much of the country, farmers are able to produce rice and vegetables nearly year-round. [Adapted from *National Geographic* (June 1993): 125.]

AT THE GLOBAL SCALE The Indian Diaspora

From the tropical island of Bali in Indonesia to Queens in New York City, one can hear South Asian temple bells ringing and catch the pungent fragrance of spices being ground for curry and kabob dishes. Both the Balinese and the South Asian Americans, though separated by thousands of miles, belong to what is called the **Indian diaspora,** the set of all people of South Asian origin living (and often born) abroad. Some are the descendants of groups who have been living outside India for a thousand years or more. The term *Indian diaspora* is used because for many centuries, India was the popular name for the whole region of South Asia.

The earliest migrations of South Asians were to eastern Africa and the islands of Southeast Asia. Archaeological evidence shows that people from the Indus Valley in what is today Pakistan traded with people in eastern Africa and Mesopotamia as early as 4000 years ago. The Balinese trace their cultural roots to trade networks between southern India and Southeast Asia established over 2000 years ago. Their primarily Hindu culture and religious practices are a remnant of that time.

The next major wave of South Asian emigration occurred from about 1850 to 1920 under British colonial rule. Hundreds of thousands of people were recruited, and sometimes even kidnapped, from various parts of the Indian subcontinent to labor as indentured servants on British plantations in such places as Trinidad and Guyana in the Caribbean, Fiji in the Pacific Ocean, and Malaysia in Southeast Asia. Some were recruited to build railroads in East Africa (especially in Kenya and Uganda) or to serve with the British army in other parts of the world. Many of these migrants remained in their new homes permanently.

After World War I, another wave of migrants from the west coast of India (mainly merchants and civil servants) arrived along the coast of East Africa, where they became successful merchants and exporters. They were pushed out of countries such as Uganda during African nationalistic movements in the 1970s. Their descendants live in the United Kingdom, the United States, Canada, and Australia. Members of these communities are called "twice-migrants." A substantial number of twice-migrants in the United States and elsewhere have opened shops and restaurants or purchased small mom-and-pop motels. The films *Mississippi Masala* (1992) and *Miss India Georgia* (1998) illustrate South Asians' struggles to adapt to life in the United States; *East Is East* (1999) and *Bend It Like Beckham* (2000) portray South Asian immigrant life in England.

Over the past 30 years, another great wave of migrants has left South Asia. Since the oil price rises of the early 1970s, tens of thousands of Muslims from India, Pakistan, Bangladesh, and Sri Lanka have found work in the oil-rich nations of the Arabian Peninsula and Southeast Asia. Others have found jobs in England's factory towns (Oldham, Leeds, Manchester) or as taxi drivers and construction workers in Canadian and U.S. cities. Probably the most socially visible members of South Asian diaspora communities are those professionals in medicine and information technology who have settled in North America, Australia and New Zealand, and Europe. In the United States, citizens and residents of South Asian origin now form the wealthiest ethnic group, as defined by the U.S. census. Recent migrants, who maintain particularly close contact with their homelands, remit millions of dollars back to their families and make regular return visits.

Although migration allows people to escape from the relative poverty and environmental degradation of South Asia, it also contributes to the **brain drain**—the flight of the best and brightest South Asians to wealthier regions. Trends that may counter the brain drain are increasing foreign investment in India and greater access to foreign markets by Indian companies, both of which create desirable jobs in India, but also intensify the foreign influence that alarms many Indians. Ironically, some of the strongest reactions against foreign influences in South Asia are funded by members of the South Asian diaspora in the United States and the Caribbean. Surrounded on a daily basis by what they see as degrading Western influences, some diasporic South Asians try to strengthen the cultural and religious traditions they cherish back home by financially supporting religion-based political parties and activism of the type that Ramesh Seshadri practices (see the vignette that opens this chapter).

[*Source:* Adapted from material contributed by geographer Carolyn Prorok.]

HUMAN PATTERNS OVER TIME

A variety of groups have migrated into South Asia over the millennia, many of them as invaders who conquered peoples already there. The merging and blending of different cultural, religious, social, and political elements has given South Asia a unique and rich heritage. One result of that heritage is a level of cultural diversity that few other world regions can equal. Another is that South Asia is one of the most politically dynamic and contentious places in the world.

The Indus Valley Civilization

There are indications of early humans in South Asia as long as 200,000 years ago, but the first evidence of modern humans is dated at about 38,000 years before the present. The first substantial settled agricultural communities, known as the **Indus Valley civilization** (or Harappa culture), appeared about 4500 years ago along the Indus River in modern-day Pakistan and along the Saraswati River in modern-day India. The architecture and urban design of this very early civilization were quite advanced for the time: there were multistory homes with piped water and sewage disposal; planned towns with wide, tree-lined boulevards laid out in a grid; and a high degree of consistency in building materials over a region that covered more than 1000 square miles (2500 square kilometers). Fine beadwork and jewelry, evidence of a trade network that extended to Mesopotamia and eastern Africa, continue to fascinate students of this region, as does a writing system that has yet to be deciphered. Moreover, much of the Indus Valley civilization's agricultural system survives to this day, including techniques for storing monsoon rainfall to be used for irrigation in dry times; methods for cultivating wheat, barley, oilseeds, cotton, and other crops adapted to arid conditions; and the use of wooden plows drawn by oxen.

The reasons for the decline of the Indus Valley civilization after about 800 years (3700 years before the present) are debated. Some scholars believe that complex ecological changes brought about a gradual demise; others argue that foreign invaders brought a swift collapse. Although the civilization disappeared, many aspects of its culture blended with subsequent foreign influences. Thus the Indus Valley civilization is the foundation of modern South Asian religious beliefs, social organization, linguistic diversity, and cultural traditions.

A Series of Invasions

The richness of the Ganga and Indus river valleys and the agricultural societies that lived there attracted people from the drier surrounding areas of Central and Southwest Asia. The first recorded invaders of South Asia, who called themselves **Arya** (Aryan is the contemporary term), moved into the Indus Valley and Punjab from Central and Southwest Asia probably as early as 3500 years ago. Many scholars believe that the Aryans, in conjunction with the indigenous Harappa culture and other indigenous cultures, instituted the remarkably influential caste system and

The great fortress at Agra, built in 1565–1571, has walls 72 feet (22 meters) high enclosing an area about 1.5 miles (2.4 kilometers) in circumference. [J. H. C. Wilson/Robert Harding Picture Library.]

brought some of the early elements of classical Hinduism, the major religion of India (see pages 406–408). Thus contemporary Hinduism and the caste system represent more than 3000 years of the blending of ritual concepts and practices from many communities. The caste system divides members of society into a social status hierarchy (see the discussion of caste on page 409), but throughout history the strength and makeup of this system have varied. Periods in which caste was rigidly enforced have alternated with periods in which caste hierarchies were relatively loose.

A distinctive feature of the cultural tradition of South Asia has been its ability to accommodate immense diversity as new influences have blended with those already present. In addition to the Aryans, other invaders who arrived from Central Asia included the Persians and Alexander the Great, the Greek warrior-explorer who penetrated as far as Punjab in 326 B.C. but was stopped by a mutiny of his own soldiers. After Alexander, waves of Turkic and Mongolian peoples continued to enter the region from the northwest, while Arab traders peacefully settled along the Malabar coast of western India and in Sri Lanka. Starting about 1000 years ago, these groups introduced the religion of Islam into the region, initiating a period of Hindu-Muslim alliance building, cultural accommodation, and reciprocity.

The invasion in 1526 by the **Mughals,** a group of TurkicPersian people from Central Asia, intensified the spread of Islam by giving it imperial stature and artistic grandeur. The Mughals reached the height of their power and influence in the seventeenth century (see Figure 6.5 on page 290) and controlled the north central plains, but farther south their influence was much weaker. Here other Muslim and Hindu kingdoms and chiefdoms were to be found. When the last great Mughal ruler (Aurangzeb) died, in 1707, several groups both within and outside areas of Mughal control began to challenge Mughal authority. Over the next century, a number of regional states and kingdoms rose in power and competed with one another for territory and political control (Figure 8.6). At the same time, several European trading companies competed with one another to gain a foothold and expand their interests in the region. Of these, the British were the most successful. Through intrigue, manipulation of political rivalries, strategic alliances, and military conquest, they defeated their European rivals, subdued most of the regional powers, and supplanted the last of the Mughals by 1857.

One legacy of the Mughals is the more than 420 million Muslims now living in South Asia. The Mughals also left a unique heritage of architecture, art, literature, and linguistics that includes the Taj Mahal, the fortress at Agra, miniature painting, and the tradition of lyric poetry. The Mughals also helped to produce the **Hindustani** language, which became the lingua franca (language of trade) of the northern Indian subcontinent. Hindustani is still used by more than 400 million people (see pages 405–406). In the northern reaches of South Asia, the Islamic ideals of the Mughals influenced, and in some cases supplanted, the largely Hindu cultural complex, making a lasting mark on aesthetics, social interactions, gender roles, and religious architecture. In all these ways, the Mughals contributed important elements to modern definitions of what it means to be Indian, Pakistani, or Bangladeshi.

FIGURE 8.6 Precolonial historical map of South Asia. After the death of Mughal ruler Aurangzeb, in 1707, the ability of the Mughals to assert strong central rule throughout South Asia declined. A number of regional states emerged that competed with one another for territory and power. Among the strongest was the Maratha Confederacy, but it also was composed of a number of smaller states dominated by the Marathas. Weakness at the center and constant rivalry paved the way for British conquest by the end of the eighteenth century. [Adapted from William R. Shepherd, *The Historical Atlas* (New York: Henry Holt, 1923–1926), p. 137; and Gordon Johnson, *Cultural Atlas of India* (New York: Facts on File, 1996), p. 111.]

The Legacies of Colonial Rule

Great Britain was the most recent influential invader of South Asia. Britain was already a strong trading presence in the 1750s, when the British East India Company established its control first over a number of coastal trading centers and then over interior regions. The British controlled most of South Asia from the 1830s through 1947, profoundly influencing the region politically, socially, and economically. Even areas not ruled by the British felt the influence of colonialism. Afghanistan was able to blunt British attempts at military conquest, but had to contend with

British meddling in its role of "buffer state" against expanding Russian interests. Even though Nepal remained nominally independent during the colonial period, large numbers of one of its ethnic groups—the Gurkhas—served in the British imperial army. And Bhutan became a protectorate of the British Indian government.

Economic Influence. British colonial policy in South Asia, as elsewhere, was to use the region's resources primarily for the benefit of Britain. This policy often resulted in detrimental effects on South Asia. A typical example was the textile industry in

Bengal (modern-day Bangladesh and the Indian state of West Bengal), where Britain made one of its first inroads into South Asian economies. Bengali weavers, long known for their high-quality muslin cotton cloth, initially benefited from the greater access British traders gave them to overseas markets in Asia, in the Americas, and on the European continent, as well as in Britain itself. The high quality of Bengali cloth and the great demand for it reflected the advanced manufacturing economy of South Asia, which in 1750 produced 12 to 14 times more cotton cloth than Britain alone and more than all of Europe combined. However, as Britain's own highly mechanized textile industry developed during the second half of the eighteenth century, cheaper British cloth replaced Bengali muslin, first in world markets and eventually throughout South Asia. The British East India Company further supplanted indigenous textile manufacturing by instituting a rule against looming. Those who continued to run their own looms rather than work in British-owned plants were severely punished. Thus, as one British colonial official put it, while the mills of Yorkshire prospered, "the bones of Bengali weavers bleached the plains of India."

The British-induced reversal of India's fortunes in the eighteenth century profoundly affected its people's lives in the nineteenth century. Many people who were pushed out of their traditional livelihood in textile manufacturing were compelled to find work as landless laborers in an economy that already had an abundance of agricultural labor. Others migrated to emerging urban centers. Increasingly, bandits (called *dacoits* in India) roamed the countryside, while small landowners lost their land to large landowners as a consequence of highly corrupt and ill-conceived land revenue (tax) systems instituted by the British East India Company in some parts of India. In the 1830s, a drought exacerbated the privations of colonial rule, and more than 10 million people starved to death in the heart of the Ganga Plain. It was during these trying times that South Asian laborers were pressed into joining the stream of indentured laborers emigrating to other British colonies in the Americas, Africa, Asia, and the Pacific (see the box "The Indian Diaspora" on page 395).

What little economic development did take place in South Asia was allowed only if it benefited Britain. The use of tariffs is an example. Whereas Europe and the United States, and later Japan, used protective tariffs to provide a shield against cheaper imported manufactured goods, the British prohibited protective tariffs in South Asia. Hence, imported British products ruined many established South Asian industries and channeled their workers into agricultural pursuits. For the remainder of the colonial period, Britain encouraged the production of tropical agricultural raw materials, such as cotton, jute (a fiber used in making gunnysacks and rope), tea, sugar, and indigo (a blue dye). These products were intended to supply Britain's own growing industries and to fit in with British consumption patterns.

The economic historian Dietmar Rothermund argues that this British tendency to inhibit and reverse Indian industrial development also encouraged population growth, now such a prominent issue in all of South Asia. Poor farmers produced as many children as possible, both for farm labor and as insurance

against destitution in their old age. Had they moved from being farmers to being industrial workers and business owners, they might have favored smaller families, like their industrial counterparts elsewhere, and they might simply have saved their earnings to provide for their old age.

Nonetheless, the British Empire did bring some benefits. Trade with the rest of the empire brought prosperity to a few areas, especially the large British-built cities on the coast, such as Bombay (Mumbai), Calcutta (Kolkatta), and Madras (Chennai). It built a railroad system that boosted trade within South Asia and greatly eased the burden of personal transport. In addition, English became a common language for South Asians of widely differing backgrounds, assisting both trade and cross-cultural understanding. Moreover, most of today's South Asian governments retain institutions put in place by the British to administer their vast empire. South Asian governments have also inherited many of the shortcomings of their colonial forebears, such as highly bureaucratic procedures, a resistance to change, and a tendency to remain distant and aloof from the people they govern. Nonetheless, these governments have proved functional. In particular, democratic government, though it was not instituted on a large scale until the final years of the empire, has given people an outlet for voicing their concerns and has enabled many peaceful transitions of elected governments since 1947.

Independence and Partition. Perhaps the most enduring and damaging outcome of colonial rule was the partition of British India into the independent countries of India and Pakistan in 1947 (Figure 8.7). The idea of two nations was first suggested by some Muslim political leaders who were concerned about the fate of a

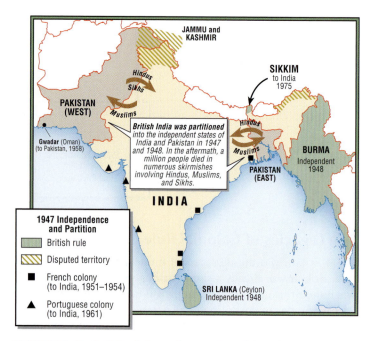

FIGURE 8.7 South Asian independence and Partition. The European colonies in South Asia were divided up in 1947 with the partition of India and Pakistan. [Adapted from *National Geographic* (May 1997): 18.]

minority Muslim population in a united India with a Hindu majority. Although more a political strategy to ensure the interests of Muslims than a real political demand, the idea of a separate state for Muslims took hold and became part of the independence agreement that the British and the Indian National Congress (India's principal nationalist party) eventually sanctioned. It was decided that northwestern and northeastern India, where the population was predominantly Muslim, would become a single country, Pakistan. This new country thus consisted of two parts, known as West and East Pakistan, separated by all of north central India. Although both India and Pakistan maintained secular constitutions, with no official religious affiliation, the general understanding was that Pakistan would have a Muslim majority and India a Hindu majority. Fearing that they would be persecuted if they did not move, some 4 million Hindus and Sikhs migrated from their ancestral homes in Pakistan to India; similarly, another 4 million Muslims left their homes in India for Pakistan. In the process, families and communities were divided, looting and rape were widespread, and more than a million people were killed in innumerable skirmishes between religious groups.

Many historians argue that Partition could have been avoided had it not been for deliberate British efforts throughout the colonial era to heighten tensions between South Asian Muslims and Hindus, thus creating a role for themselves as indispensable and benevolent mediators. There is much evidence for these so-called divide-and-rule tactics. British administrators at the local level commonly favored the interests of minority communities in order to weaken the power of majority communities, which could have threatened British authority. At the national level, the British worked out special political agreements with Muslim leaders as a way of weakening Muslim support for the Indian National Congress (Congress party), which led the independence movement against the British and claimed to represent the interests of both Muslims and Hindus. Thus Partition spawned repeated wars and skirmishes, and relations between India and Pakistan remain strained in the best of times. Sporadic armed conflict between the two countries over Kashmir, in the northwest of India, continues today.

Since Independence. In the more than 50 years since the departure of the British, South Asians have experienced both progress and setbacks. Democracy has expanded steadily, albeit somewhat slowly. India has maintained its status as the world's most populous democracy and is gradually dismantling age-old traditions that hold back poor, low-caste Hindus and other disadvantaged groups. Agricultural advances have brought relative prosperity to some rural areas. Industrialization has advanced as well: it now constitutes a larger share of the GDP than agriculture, and a small but increasing number of educated Indians are reaping the economic benefits of the global growth in information technology. Nonetheless, economic development in all South Asian countries continues to face a number of challenges. As of 2003, these countries have a collective annual gross domestic product per capita of about U.S. $2110, adjusted for PPP—only sub-Saharan Africa has a lower average regional per capita GDP ($1710)—and poverty persists in both rural and urban areas throughout the region.

Although threats from outside the region have diminished, the danger of violence originating from within has steadily increased. After 1947, East and West Pakistan, despite their common religion, struggled to overcome their stark cultural differences, economic disparity, and geographic separation. East Pakistan was the more populous and was the chief earner of the country's foreign exchange through the export of raw materials. Yet East Pakistan saw much of its earnings spent by officials in West Pakistan. West Pakistan, on the other hand, dominated government and military positions. In 1970, the first general election in 12 years gave a leader from East Pakistan a landslide victory. The leaders of West Pakistan refused to turn over the government. After protracted negotiations, West Pakistan attacked East Pakistan, which resulted in a cry for independence. Thus, as the result of a bloody civil war and intervention by India in 1971, Pakistan was divided, and East Pakistan became the independent country of Bangladesh.

In the years since 1971, civil wars have plagued Sri Lanka, Afghanistan, and parts of India, and the possibility of nuclear confrontation between Pakistan and India has loomed ominously whenever the dispute over Kashmir has heated up (see pages 419–420). Notably, however, India has managed to hold together despite many pressures favoring its disintegration, and the region's disputing parties have been known to lay aside conflict in emergencies. Such was the case after the earthquake in Gujarat in 2001, when Pakistan sent much-appreciated aid to India. Future threats to the region's stability may result from the stresses that a huge and still growing population places on the region's already precarious natural environment and from the uneven distribution of resources and economic prosperity.

POPULATION PATTERNS

Although Africa has the fastest-growing population on earth, South Asia is one of the most densely populated regions in the world (Figure 8.8, compare with Figure 7.6 on page 343). In cities, commuters cling to the outside of packed buses as they travel lopsidedly through the streets of Peshawar, New Delhi, and Mumbai, and crowds of people pack urban sidewalks. Though less than 30 percent of the population is urban, South Asia has several of the world's largest cities: Mumbai with 16 million, Kolkatta with 13 million, Delhi with nearly 13 million, Dhaka with 13 million. People come to South Asian cities for all the reasons that cities attract people everywhere. Some hope for business opportunities or better jobs with higher pay; others come for an education or training. Some seek anonymity and individualism—perhaps even a rise in caste status (see page 409)—that are not possible in close-knit rural communities. Many are pushed into urban migration by agricultural modernization that reduces rural incomes and employment opportunities and by population growth that makes access to land and resources ever more difficult. Some are refugees who have left drought-stricken or flooded countryside.

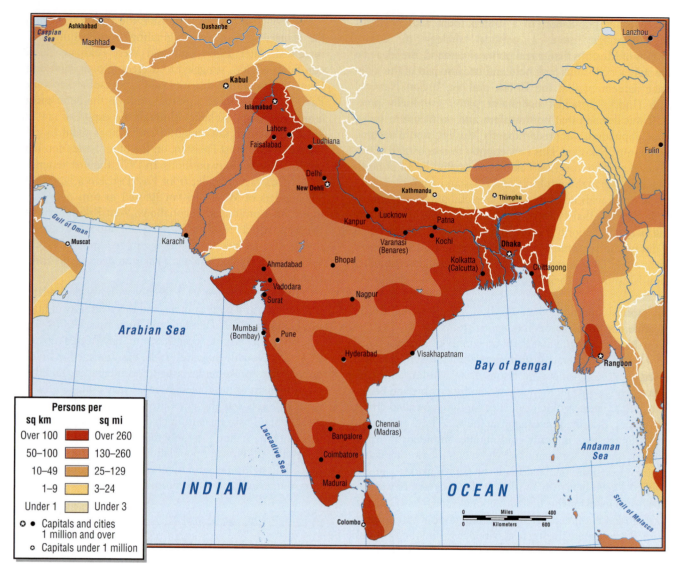

FIGURE 8.8 Population density in South Asia. [Adapted from *Hammond Citation World Atlas* (Maplewood, N.J.: Hammond, 1996.]

PERSONAL VIGNETTE   One consequence of South Asia's rapid urban growth is that employment and availability of affordable housing have not kept pace, and many people simply live on the streets. In a National Public Radio interview in August 1997, an Indian journalist recalled once impulsively asking a bicycle rickshaw driver about himself as he peddled her through Delhi. He replied that his belongings—a second set of clothes, a bowl, and a sleeping mat—were under the seat where she was sitting. He had come to Delhi from the countryside 14 years before, and he had never found a home. He knew virtually no one, he had few friends and no family, and no one had ever inquired about him before. He worked virtually around the clock and slept here and there for 2 hours at a time. [*Gagan Gill, "Weekend Edition," National Public Radio (August 16, 1997).*]

Because South Asia's population is so young—more than a third of the region's people are under age 15—rapid population growth will continue to strain efforts to improve life for South Asians. As a

A bicycle rickshaw and driver in Delhi. Bicycle rickshaws are just one form of public transportation one finds in South Asia. [Lindsay Hebberd/Woodfin Camp & Associates.]

Decline in Total Fertility Rates

Births per woman

	1960	1993	2003
Bangladesh	6.7	4.9	3.2
India	5.9	3.9	3.1
Nepal	5.8	5.6	4.7
Pakistan	6.9	6.7	4.6
Sri Lanka	5.3	2.3	2.0

Percent decline, 1960–2003

Bangladesh	India	Nepal	Pakistan	Sri Lanka
52%	47%	19%	33%	62%

FIGURE 8.9 Decline in total fertility rates.
Some South Asian countries have experienced large declines in fertility rates in the last 30 years.
[Adapted from *A Demographic Portrait of South and Southeast Asia* (Washington, D.C.: Population Reference Bureau, 1994); and *World Population Data Sheet*; Population Reference Bureau, 2003.]

region, South Asia already has more people (1.44 billion) than China (1.26 billion). By 2020, India alone is expected to overtake China, where the rate of natural increase (0.6 percent) is less than half that of India (1.7). Although the Indian economy is not particularly efficient, it has grown at a steady annual rate of 5–6 percent for several decades. The size of the middle class has expanded to over 250 million people—nearly as large as the total U.S. population. Each year, though, India adds nearly 18 million people to its ranks. To accommodate these new Indians adequately, every single year the country would need to build 127,000 new village schools, hire 373,000 new schoolteachers (50 students per teacher), build 2.5 million new homes (7 people per home), create 4 million jobs, and produce 180 million new bushels of grain and vegetables. Food production in India has kept pace with population growth, but its ability to continue doing so is not secure. The technological

advances in agriculture responsible for increased yields may not be sustainable because of their high costs and negative effects on the environment and soil (see pages 412–414.)

South Asia has been trying to reduce births since 1952. India spends over a billion dollars a year on population control programs. Unlike many other countries, which rely on financial aid from developed countries to fund such programs, India pays for nearly all of this effort on its own. Fertility rates have indeed declined significantly in India and Bangladesh, and especially in Sri Lanka, but much less in Pakistan and Nepal (Figure 8.9). Why does population continue to boom despite such efforts? The answers are similar to those we have given in earlier chapters.

In Pakistan, for example, a look at the population pyramid (Figure 8.10) shows that a significant portion of the population is in the early reproductive years, so even a one-child-per-couple policy

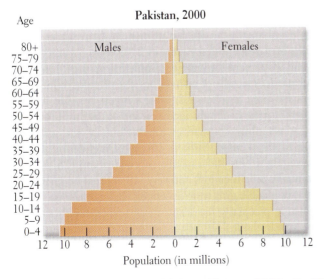

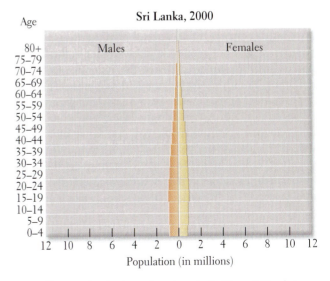

FIGURE 8.10 Population pyramids for Pakistan and Sri Lanka, 2000.
Both Pakistan and Sri Lanka have experienced decreasing fertility rates since 1960 (see Figure 8.9), but the great size and youth of Pakistan's population ensure that it will continue to grow for years. Sri Lanka has a much smaller

population, and its growth rates have slowed markedly. Hence Sri Lanka's pyramid no longer has a pyramidal shape. [Adapted from "Population Pyramids for Pakistan" and "Population Pyramids for Sri Lanka" (Washington, D.C.: U.S. Bureau of the Census, International Data Base, May 2003).]

FIGURE 8.11 Contraceptive use by level of education. The positive relationship between education and the use of contraception is virtually universal. In South Asia, women with at least some education use contraception at higher rates than do women with no education. In countries such as Pakistan, where the overall use of contraception is low, education drastically changes a woman's attitude toward birth control. In countries where education and contraceptive use are already fairly high, some education still increases the rates of use significantly (from 54 to 64 percent in Sri Lanka), but the effects of even more education—through secondary school, for example—are negligible. [Adapted from *A Demographic Portrait of South and Southeast Asia* (Washington, D.C.: Population Reference Bureau, 1994), p. 18.]

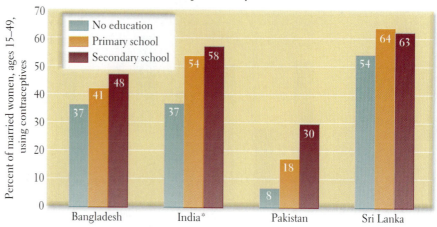

Contraceptive Use by Level of Education

*Ages 15–44

would result in population growth for years to come. Of equal importance, poor, rural, uneducated people see children as their only source of wealth. Babies not only bring joy, but also quickly become productive family members: 3-year-olds are already able to perform some useful tasks, and 10- or 12-year-olds pull their weight, quite literally, in the fields or even perform such highly skilled tasks as carpet weaving (see the box "Should Children Work?" on the next page). Furthermore, grown children are the only retirement plan that most South Asians will ever have. Because access to health care is limited, infant mortality rates in the region average 67 per 1000 live births (only sub-Saharan Africa's are higher); hence couples often choose to have more than two or three children to ensure that at least some will reach maturity.

The pyramid for Sri Lanka, on the other hand, shows a significant decline in fertility. Sri Lanka has a much lower infant mortality rate (13 per 1000 live births) than Pakistan (91 per 1000). Sri Lanka's low rate is related to high rates of education for women, discussed below, and development policies aimed at more egalitarian wealth distribution, which have fostered the growth of the middle class. As families achieve middle-class status, lifestyles change, child survival rates increase, and the role of children shifts from economic asset (worker in the fields) to economic liability (student). Hence the incentive to have a lot of children diminishes with economic development (Figure 8.11).

Another reason families favor a large number of offspring in regions like South Asia is that they view sons as more beneficial than daughters. Sons, it is believed, contribute more to a family's wealth than daughters, and many couples want at least two sons to provide for them in their old age. Daughters are thought to have little economic value, in part because the domestic work they do rarely earns cash. In patriarchal societies, women are rarely educated, and most women have few ways to achieve fulfillment except as prolific mothers. A popular toast to a new bride is "May you be the mother of a hundred sons."

Even the middle class becomes caught up in the desire for sons. Many couples who wish sons patronize high-tech labs that specialize in identifying the sex of an unborn fetus. Their intention is to abort the fetus if it is female. Several Indian states have banned this use of technology, but enforcement is difficult. The practice of selective abortion and the practices of neglecting girl children and of female infanticide have resulted in an odd circumstance throughout India: among adults, men outnumber women. (As a result of their longer natural life span, adult women outnumber adult men in most places.) The 2001 Indian census showed 933 females for every 1000 males. In India, women outnumber men only in the state of Kerala—where there are 1058 females for every 1000 males. (In the United States, there are about 1038 females per 1000 males.)

Indian social scientists explain the exception of Kerala by noting that its entire population is more educated. The literacy rate in Kerala is 90 percent, versus 58 percent in India as a whole. In Kerala, the elected communist state government has for many years provided education and broad-based health care for all. Education is credited with the fact that 63 percent of women in Kerala use contraception; in India as a whole, the rate is only 48 percent. Women who limit the number of children they bear have healthier children and are less likely to die as a result of childbirth. The country of Sri Lanka, like the Indian state of Kerala, has also succeeded in bringing down population growth rates while enhancing health for women by focusing on basic literacy. In Sri Lanka, literacy rates in 2003 were 89 percent for women and 95 percent for men (see Table 8.1 on page 426).

Education, especially of women but also of men, reduces the incentives for large families. Educated women are able to contribute significantly to family incomes and to the well-being of elderly parents, thus countering the preference for sons. Yet according to the *United Nations Human Development Report 2000*, during the 1990s, India spent an average of just U.S. $17 per capita annually on education, and very little of that on girls (only 44 percent of eligible girls were in school). By comparison, South Korea—where prosperity has leapt upward while family size has shrunk—spends about $320 per capita annually, and the state of Tennessee (ranked 48th in the United States in education spending per capita) spent about $3000 per capita annually in the late 1990s. Paradoxically, although education has the power to reduce fertility rates quickly, the increasing costs of rapid population growth make adequate education harder and harder to provide.

AT THE REGIONAL SCALE   Should Children Work?

Beginning in the mid-1990s, a number of reports questioned the exploitation of child labor in Asia, especially in the hand-woven carpet industry, which has been the subject of reports of outright enslavement of kidnapped children ("Weekend Edition," National Public Radio, March 17, 2001). Carpet weaving is an ancient artistic and economic enterprise in Central and East Asia as well as in South Asia. Traditionally, it has been a family-run enterprise, with women and children the weavers and men the merchants. International trade has increased the demand for fine handwoven carpets, but it is South Asian middlemen and foreign traders from Europe and America, not the weavers, who profit, and some unscrupulous carpet merchants have apparently resorted to forced labor to produce carpets.

Children have always participated in the home-based carpet industry. For thousands of years, young children have learned weaving skills from their parents and have become proud members of the family's production unit. Yet, in today's society, children must balance their role as family workers with the need to attend school, where they learn the skills that will enable them to survive in a modern economy.

Should children be allowed to work? It is clear that kidnapping and enslavement must be stopped, but is there room for different cultural views of what childhood should be like? When consumers buy goods made by children, are they supporting family values or greedy factory owners and middlemen? The United Nations and South Asian governments are now addressing these questions. They have instituted an active program to curb child labor abuses, while remaining open to the positive experience that learning a skill and being part of a family production unit can be for a child. In India, for example, there is now a national system to certify that exported carpets are made in shops where the children go to school, have an adequate midday meal, and receive basic health care. Such carpets will bear the label "Kaleen."

If abuses can be eliminated, the custom of child labor is likely to continue because of its significant psychological and economic benefits to children and their families. Until the modern era, such work was considered an important part of a child's training for adulthood, even in the United States and Europe. Some experts argue that it is the lack of meaningful roles for children that leads to juvenile crime.

Two young boys tie knots at a carpet loom in India. [Cary Wolinsky/Stock, Boston.]

Quick Review

1. Which two seasonal wind patterns contribute to South Asia's summer rains?

2. In what ways did Islam spread to South Asia?

3. Although both India and Pakistan have secular constitutions, they are each assumed to have a majority population of a particular faith. Which religion is associated with India, and which with Pakistan?

4. Which European country presided over independence and Partition?

5. Generally speaking, what is the relationship between education and the use of contraception for women in the region?

II   CURRENT GEOGRAPHIC ISSUES

FIGURE 8.12 Political map of South Asia.

For the Westerner first learning about South Asia, the size of the region and the immense diversity of its people and landscapes may seem overwhelming. The writer Gordon Johnson notes that once "confronted by the massive physical scale of the subcontinent, the size of the population, and the long reach of its traditions," the outsider is wise to try to "catch a glimpse of the whole by pursuing the specific and studying the particular." That is the approach we take here. In the section on sociocultural issues, we start with the feature that most characterizes the region and makes it distinctive: village life. We then examine language, religion, caste, and gender relationships—for the most part, within the context of South Asian village life. These cultural characteristics touch all lives, yet vary greatly in practice across the region. The sections on economic and on political issues touch on those matters that are receiving coverage in the local South Asian and international press: economic change and income disparities, relative degrees of democratization, agricultural change, technologically advanced industrialization, and innovative economic development strategies.

SOCIOCULTURAL ISSUES

Many travelers to South Asia say that life there is best understood if it is observed in the intimate setting of the village, where relationships among individuals, and among and within groups, are easier to discern than they would be in the city. We begin our coverage of sociocultural issues by visiting a few villages. We then explore how religion, caste, and gender influence life in the village and how some people escape from traditional religious, caste, and gender restrictions by leaving the village.

Village Life

Joypur. The writer Richard Critchfield, who has studied village life in more than a dozen countries, writes that the village of Joypur (Bangladesh) in the Ganga Delta is set in "an unexpectedly beautiful land, with a soft languor and gentle rhythm of its own," where in the early evening, mist rises above the rice paddies and hangs there, "like steam over a vat." In the heat of the day, the village is sleepy: naked children play in the dust, women meet to talk softly in the seclusion of courtyards, and chickens peck for seeds. Here and there, under the trees, people ply their various trades: one may, for example, encounter the village tailor sewing school uniforms on his hand-cranked sewing machine. But it is at dusk that the village comes to life. The men return from the fields, and after a meal in their home courtyards, they come "to settle in

Several generations of a northern Indian family, from the village of Dharamkot in Himachal Pradesh province, enjoy one another's company. [David Morgan.]

groups before one of the open pavilions in the village center and talk—rich, warm Bengali talk, argumentative and humorous, fervent and excited in gossip, protest and indignation" as the men discuss their crops, an upcoming marriage, or national politics.

Ahraura. The anthropologist Faith D'Aluisio and her colleague Peter Menzel give us another peek into village life as night falls. This village is in the state of Uttar Pradesh in north central India. In the enclosed women's quarters of the village, lamps flicker, and the chirping of cicadas rises from the surrounding fields. In one walled compound, Mishri is finishing up her day by the dying cooking fire as her year-old son tunnels his way into her sari to nurse himself to sleep. Mishri, who is 27, lives in a tiny world bounded by the walls of the courtyard she shares with her husband and five children and several of her husband's kin. Almost all her activity takes place in seclusion. Like most villages of northern India, her village observes the practice of purdah, in which women keep themselves apart from male gazes. Mishri feels fortunate that her husband does not need her help in the fields, making possible her seclusion, a mark of status. Within the compound, she works from sunup to sundown, only taking an hour off in the early afternoon to chat with two women friends who cover their faces and move quickly from their own courtyards to hers for the short visit. Mishri is devoted to her husband, who was chosen for her by her family when she was 10; out of respect, she never says his name aloud.

The vast majority—about 70 percent—of South Asians live in hundreds of thousands of villages like Joypur and Ahraura. The village way of life unites the entire region. Even many of those now living in South Asia's giant cities were born in a village or visit one regularly. In fact, were one to observe city life in places as widely separated and culturally different as Chennai, Mumbai, Kathmandu, and Peshawar, one would discover that they are hardly cities at all. Rather, each comprises thousands of tightly compacted, reconstituted villages, where daily life is intimate and familiar, not anonymous as in Western cities.

Language and Ethnicity

Within the life of one village, there is often room for considerable cultural variety. Differences based on caste, economic class, religion, and even language are usually accommodated peacefully by longstanding customs that guide cross-cultural interaction. One equalizing factor is that everyone in South Asia is, in one way or another, a minority. As the Indian writer and diplomat Shashi Tharoor writes:

A *Hindi-speaking male from the Gangetic Plain state of Uttar Pradesh might cherish the illusion that he represents the "majority community," . . . but he does not. As a Hindu he belongs to the faith adhered to by some 82 percent of the population, but a majority of the country does not speak Hindi; a majority does not hail from Uttar Pradesh; and if he were visiting, say, Kerala, he would discover that a majority is not even male. . . . Our archetypal Hindu has only to step off a train and mingle with the polyglot polychrome crowds thronging any of India's five major metropolises to realize how much of a minority he really is. Even his Hinduism is no guarantee of majorityhood, because his caste automatically places him in a minority as well: if he is a Brahmin, 90 percent of his fellow Indians are not; if he is [of the] Yadav [caste], 85 percent of Indians are not, and so on.*

(Shashi Tharoor, *From Midnight to Millennium*, 1997, p. 112.)

There are many distinct ethnic groups in South Asia, each with its own language, dialect, and subdialect. In India alone, 18 languages are officially recognized, but there are actually hundreds of separate languages and literally thousands of dialects. Figure 8.13 shows the complexity of the distribution of languages in

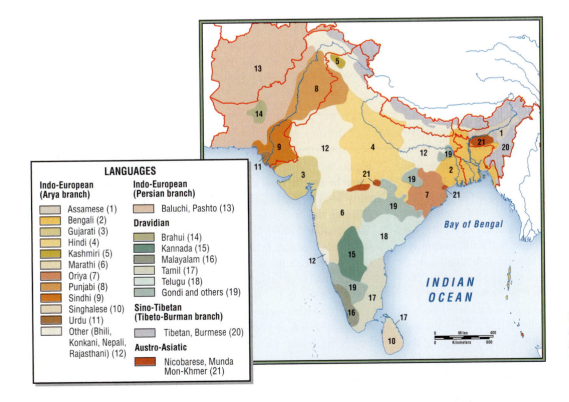

LANGUAGES

Indo-European (Arya branch)
- Assamese (1)
- Bengali (2)
- Gujarati (3)
- Hindi (4)
- Kashmiri (5)
- Marathi (6)
- Oriya (7)
- Punjabi (8)
- Sindhi (9)
- Singhalese (10)
- Urdu (11)
- Other (Bhili, Konkani, Nepali, Rajasthani) (12)

Indo-European (Persian branch)
- Baluchi, Pashto (13)

Dravidian
- Brahui (14)
- Kannada (15)
- Malayalam (16)
- Tamil (17)
- Telugu (18)
- Gondi and others (19)

Sino-Tibetan (Tibeto-Burman branch)
- Tibetan, Burmese (20)

Austro-Asiatic
- Nicobarese, Munda Mon-Khmer (21)

FIGURE 8.13 Languages of South Asia. [Adapted from Alisdair Rogers, ed., *Peoples and Cultures* (New York: Oxford University Press, 1992), p. 204.]

South Asia. That complexity results from the history of multiple invasions from outside, the relentless movement and rearrangement of people, and the long periods of isolation experienced by particular groups. In Figure 8.13, the patches of lavender (20) and the patches of dark red (21) indicate some of the most ancient culture groups in the region. The dark red patches represent remnants of Austro-Asiatic languages that were once more widely distributed and were left as isolated pockets despite the sweeping cultural changes brought by invaders. These aboriginal languages are distantly akin to others found farther east in Southeast Asia. The languages represented by numbers 1–12 are linked to various groups of Aryan people who entered South Asia from Central Asia during prehistory and brought with them distinctive languages and other cultural features.

The Dravidian language-culture group (identified by numbers 14–19 in Figure 8.13) is another ancient group that predates the Aryan invasions by a thousand years or more. Today, Dravidian languages are found mostly in southern India, but an ancient version of Dravidian was widely used across the subcontinent more than 4000 years ago. A small remnant of the Dravidian past can still be found in the Indus Valley in south central Pakistan (number 14 in Figure 8.13), where a language derived from Dravidian (Brahui) is still spoken. Dravidians appear to have had a matrilineal social organization, meaning that descent and inheritance were figured through the female line and important family decisions were made by the mother and mother's brother. This female-centered family model was modified, but not completely erased, by Aryan patriarchal models coming from the north. Remnants of these matrilineal ideas are still to be found among various groups in southern and northeastern India.

By the time of British colonialism, Hindustani was the lingua franca of all of northern India and what is today Pakistan. Hindustani is an amalgam of Persian (an Indo-European language the Mughals used as their court language) and Sanskrit-based northern Indian languages. The Muslims wrote Hindustani in a form of Arabic script and called it Urdu, whereas the Hindus and other groups wrote it in a script derived from Sanskrit and called it Hindi. Today, Urdu is the national language of Pakistan and Hindi the national language of India, but people who speak the common forms of these languages can understand one another in the same way as speakers of American and British English do. Hindi, because of its origins in Hindustani and the popularity of romantic Hindi-language films, is understood by most Pakistanis and by about 50 percent of India's population. Only Chinese and English are spoken by more people worldwide. But Hindi is not the first language of the majority: as Tharoor noted, no Indian language can make that claim.

English is a common second language throughout the region. For years, it was the language of the colonial bureaucracy, and it remains a language used at work by professional people. From 10 to 15 percent of South Asians speak, read, and write English.

Religion

The main religious traditions of South Asia are Hinduism, Buddhism, Sikhism, Jainism, Islam, and Christianity. Buddhism and Jainism originated as reformist movements in Hinduism. The geographic distribution of the adherents of these various faiths is uneven, as Figure 8.14 shows.

Hinduism. Hinduism is a major world religion practiced by approximately 900 million people, 800 million of whom live in India. It is a complex belief system, full of seeming contradictions that often make it difficult for outsiders to understand. These contradictions result largely from the fact that Hinduism includes a broad range of beliefs and practices that have their roots in highly localized folk traditions (known as the Little Tradition) as well as a classical system (known as the Great Tradition). These traditions are based on the amalgam of Harappan and Aryan ritual beliefs recorded in ancient Sanskrit texts. For example, most Hindus worship a number of gods and goddesses, but many of these are found only in one region, in one village, or even in one family. Scholars explain this diversity as a product of Hindu notions of divinity, which are quite flexible. Hence, although local areas have retained their own deities, over time some local gods have become incorporated into the classical Hindu pantheon.

CULTURAL INSIGHT   Food Customs in South Asia

Customs governing food and its preparation demonstrate how religion and caste influence everyday life in South Asia. The best known of India's food rules is that Hindus do not eat beef. This prohibition may have grown from the fact that, historically, cattle have been more valuable as living resources than as food because they were the primary source of labor, fuel, and fertilizer and are thus a symbol of life. In Afghanistan, Pakistan, and Bangladesh, it is pork that is proscribed, a prohibition that predates Islam. But there are countless other restrictions that vary from place to place.

The long tradition of highly seasoned, meatless cuisine in South Asia is particularly strong in southern India. Many notions of purity and pollution surround food preparation throughout the region. According to tradition, Hindus should not accept food prepared by anyone from a lower caste, and only the right hand is to carry food to the mouth because the left hand is reserved for cleaning the body after defecating and hence is ritually impure. Menstruating women are also not allowed to prepare food. These taboos are fading in urban areas where caste and gender rules are now less important.

In rural areas, where the vast majority of South Asians still reside, the serving of food is also culturally prescribed. Women and girls eat after men and boys, a practice that often leaves them undernourished in poorer families and partly accounts for the higher death rate of female children than of male children.

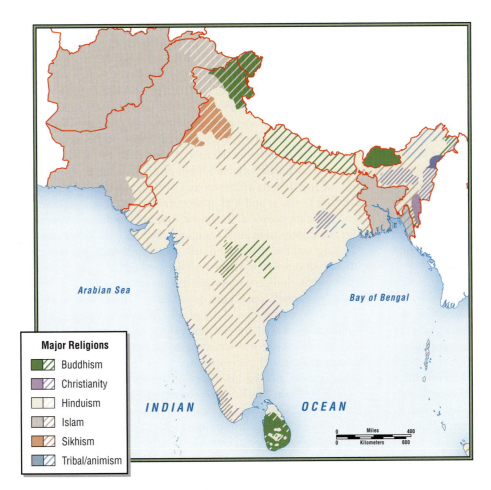

FIGURE 8.14 Religions in South Asia. The map attempts to convey the overlapping patterns of religion in the region, showing brownish gray where Islam is dominant, light tan where Hinduism is dominant, and so forth. The color of the hatching indicates which two or more religious traditions are practiced in a particular part of the region. [Adapted from Gordon Johnson, *Cultural Atlas of India* (New York: Facts on File, 1996), p. 56.]

Major Religions
- Buddhism
- Christianity
- Hinduism
- Islam
- Sikhism
- Tribal/animism

A major tenet of classical Hindu philosophy, as described in the 4000-year-old Hindu scriptures called the Vedas, is that all gods are merely illusory manifestations of the ultimate divinity, which is formless and infinite. Many devout Hindus worship no gods at all. These Hindus may engage in meditation, yoga, and other spiritual practices designed to liberate them from illusions and bring them closer to the ultimate reality, described as infinite consciousness. The average person, however, is seen as needing to personify divinity in the form of a god.

Nevertheless, there are some things that almost all Hindus have in common, such as the belief in reincarnation—the idea that any living thing that desires the illusory pleasures (and pains) of life will be reborn after it dies. Hindus usually observe caste identities, which are deeply interwoven with classical Hindu ritual; dietary rules commonly distinguish caste groups. A reverence for cows, which are seen as only slightly less spiritually advanced than humans, binds all groups together.

Geographic Patterns in Religious Beliefs. There is an overlapping distribution pattern of religions in South Asia. Hindus, as we have observed, are the most numerous, and they are found mostly in India. The Ganga Plain is considered the **hearth** (place of origin) of Hinduism, and every 12 years, during India's largest religious festival, millions of Hindus converge on the city of Allahabad to bathe at the confluence of the Ganga and Yamuna rivers as an act of devotion.

Other religions are important in various parts of the region. **Buddhism** began as a reform and reinterpretation of Hinduism. Its origins are in northern India, and it flourished there early in its history before spreading eastward to East and Southeast Asia. Only 1 percent of South Asia's population, or about 10 million people, are Buddhists; they are a majority in Bhutan and Sri Lanka.

Muslims are followers of **Islam**, which is discussed in Chapter 6 (pages 289–290 and 298–300). The 420 million Muslims in the region form the majority in Afghanistan, Pakistan, Bangladesh, and the Maldives. Muslims are also a large and important minority in India (numbering about 120 million there); they live mostly in the northwestern and central Ganga Plain but are also scattered throughout the country.

The 18 million **Sikhs** in the region combine the beliefs of Islam and Hinduism. **Sikhism** was founded in the fifteenth century by Guru Nanak as a challenge to contemporary socioreligious systems, including Hinduism and Islam. The new religious philosophy was inspired by both Hindu and Islamic ideals. Sikhs espouse belief in one God, high ethical standards, and meditation. Philosophically, Sikhism rejects the idea of caste, but accepts the Hindu idea of reincarnation. (In everyday life, however, caste continues to play a role in people's identity.) Sikhs live mainly in Punjab, in northwestern India, but are also found elsewhere. Their influence in India is greater than their numbers because many Sikhs hold positions in the military and police.

More Sikhs live in diaspora communities than live in India itself (see the box "The Indian Diaspora" on page 395), and some have financed a Sikh separatist movement (see the section on the conflict in Punjab, pages 419–420).

The faith tradition of **Jainism** is more than 2000 years old. **Jains** (about 10 million people, or 1 percent of the region's population) are found mainly in western India and in large urban centers. They have more influence than their numbers would suggest, especially within the Hindu community. They are known for their nonviolence and strict vegetarianism.

Parsis, though few in number, are a highly visible minority in India's western cities, where they have distinguished themselves in business, politics, and the arts. Zubin Mehta, a famous symphony conductor, is a Parsi. Parsis are descended from Persian migrants from Iran who did not give up their traditional religion of Zoroastrianism when Iran became Muslim.

The first **Christians** in the region are thought to have arrived in the far southern Indian state of Kerala with St. Thomas, the Apostle of Christ, in the first century A.D. (Christianity is discussed in Chapter 6 on page 289.) Today, Christians are an important minority along the west coast of India. In some places in northeastern India, more than half the descendants of the ancient aboriginal inhabitants are Christian, the result of British colonial mission efforts in the late nineteenth and twentieth centuries. At that time, ethnic minorities adopted Christianity, partly as a hedge against the encroachment of Hindu power. Christians are also a small minority elsewhere in India, as well as in Sri Lanka, Pakistan, and Bangladesh.

Small communities of **Jews** are found along the Malabar Coast and in such major cities as Mumbai, Kolkatta, and Ahmadabad. (Judaism is discussed in Chapter 6, page 289.) Some are thought to be the descendants of ancient migrants who arrived perhaps 2000 years ago or even earlier. Others came via Europe in the modern period.

Animism is practiced throughout South Asia, especially in central and northeastern India, where there are indigenous people whose occupation of the area is so ancient that they are considered aboriginal inhabitants. (Animism is discussed in Chapter 7, pages 365–367). In South Asia, animist beliefs often incorporate aspects of Hinduism, Islam, Buddhism, or Christianity.

The Hindu-Muslim Relationship. The different religious traditions of South Asia have influenced one another, and this is especially true of the two largest faiths. Where Hindus have lived in close association with Muslims—often within the same villages—they have absorbed Muslim customs, such as honoring Muslim saints. For their part, Muslims have adopted some of the Hindu ideas of caste. Although Muslims make up only 12 percent of the Indian population, the political and social relationships between Hindus and Muslims are enormously complex. The great independence leaders Mohandas Gandhi and Jawaharlal Nehru both emphasized the common cause that once united Muslim and Hindu Indians: throwing off British colonial rule. Since independence, members of the Muslim upper class have been prominent in Indian national government and the military. Muslim generals served India willingly, even in its wars with Pakistan after Partition. In the upper echelons of society, Hindus

and Muslims often socialize amicably, live in the same neighborhoods, share recreational facilities, and join the same clubs. In urban areas, middle-class Hindus and Muslims go to school together and occasionally marry one another.

But there is a darker side to the Hindu-Muslim relationship. Especially in Indian villages, some upper- and middle-caste Hindus regard Muslims as members of low-status castes. Religious rules about food are often the source of discord because dietary habits (such as differing degrees of vegetarianism) are a primary means of distinguishing caste. Hindus regard the cow as sacred and use its products with reverence: milk for food and dried excrement for fertilizer and fuel. Although Hindus do not kill cows for food or hides, people at the lowest rungs of Hindu society process and consume individual animals that die by other means. Muslims, on the other hand, run slaughterhouses and tanneries (though discreetly), eat beef, and use cowhide to make shoes and other items. Indigent cows, picked up off the streets of India's cities, are sometimes exported to nearby Muslim countries. So, due to their work with and use of cows, Muslims appear similar to members of a low-status caste to Hindus. Also fueling this perception is the occasional conversion of entire low-caste Hindu villages to Islam. Those who convert seek to escape the hardships of being members of a low caste.

The Hindu-Muslim relationship is no less complex in Bangladesh. After the separation of Bangladesh from Pakistan, most upper-class Muslims moved to Pakistan, leaving Bangladesh with a preponderance of poor Muslim farmers. Meanwhile, many of the Hindu landowners remained, and some lower-caste Hindus converted to Islam. In Bangladeshi villages, the Muslims are usually a majority, but the Hindus are often somewhat wealthier and disdainful of the Muslims. Although the two groups may coexist amicably for many years, they view themselves differently, and **communal conflict**—that resulting from religious differences—can erupt over seemingly trivial events.

PERSONAL VIGNETTE The sociologist Beth Roy, a specialist in conflict resolution who studies communal conflict in South Asia, recounts an incident in the village of Panipur, Bangladesh, in her book *Some Trouble with Cows.* The incident started when a Muslim farmer carelessly allowed one of his cows to graze in the lentil field of a Hindu. The Hindu complained, and when the Muslim reacted complacently, the Hindu seized the offending cow. By nightfall, Hindus had allied themselves with the lentil farmer and Muslims with the owner of the cow. More Muslims and Hindus converged from the surrounding area, and soon there were thousands of potential combatants lined up facing each other. Fights broke out. The police were called. In the end, a few people died in the ensuing riot, and relationships in the village were deeply affected by the incident. In the words of Roy, the dispute "delineat[ed] distinctions of caste, class, and [religious] culture so complex they intertwine[d] like columbines climbing on an ancient wall." [*Adapted from Beth Roy,* Some Trouble with Cows—Making Sense of Social Conflict *(Berkeley: University of California Press, 1994), pp. 18–19.*]

Caste

Caste, the ancient system of dividing society into hereditary hierarchical categories, seems alien to many people outside South Asia, and yet social division and inequality are common in all societies, including Europe and America. In fact, all human groups have deeply ingrained concepts of relative social status. The old often have more authority than the young; or conversely, as in U.S. pop culture, the young sometimes have greater influence than the old. In many situations in Europe and America, men still have more power and prestige than women. Nearly everywhere on earth, social difference is indicated by clothes, hairstyle, body decorations, manner of speaking, material possessions, residential location, and religion. In America, something so simple as the use of "ain't" or "youse" or a certain kind of headgear (baseball cap, motorcycle helmet, straw hat) conveys information about one's origins, role, or point of view. In some quarters, race still carries overtones closely akin to those of caste in South Asia.

Caste is a custom associated primarily with Hindu India, but all religious communities in the region (including Christians, Buddhists, and Muslims) have incorporated elements of the system into their cultures. One is born into a given subcaste, or community (called a *jati*), and that happenstance largely defines one's experience for a lifetime—where one will live, where and what one can eat and drink, with whom one will associate, one's marriage partner, and sometimes one's livelihood (see the box "Food Customs in South Asia" on page 406). The classical caste system has four main divisions or tiers, called *varna*, within which are many hundreds of *jatis*, which vary from place to place; there are also important groups that fall outside the caste system.

The four *varna* include **Brahmins**, the priestly caste, who are the most privileged in ritual status and thus must conform to behaviors that are considered most ritually pure (for example, strict vegetarianism, abstention from alcohol, and abstention from certain types of work). Then, in descending rank, are **Kshatriyas**, who are warriors and rulers; **Vaishyas**, who are landowning farmers and merchants; and **Sudras**, who are low-status laborers and artisans. A fifth group, the **Harijans** (also called Dalits—"the oppressed"—or untouchables), are actually considered to be so lowly as to have no caste. Harijans perform tasks that caste Hindus consider the most despicable and ritually polluting: killing animals, tanning hides, sweeping, and cleaning. A sixth group, also outside the caste system, is the *adivasis,* who are thought to be descendants of the region's ancient original inhabitants (for example, see numbers 20 and 21 in Figure 8.13, page 405).

Within each *varna* are numerous *jatis* and sub-*jatis*. Although *jatis* are associated with specific occupations, in the modern economies of today this aspect of caste is more symbolic than real. Members of a particular *jati* do, however, follow the same social and cultural customs, dress in a similar manner, speak the same dialect, and tend to live in particular neighborhoods or villages. This spatial separation between caste groups arises from the higher-caste communities' fears of ritual pollution through physical contact or sharing water or food with lower castes. When one stays in the familiar space of one's own *jati*, one is enclosed in a comfortable circle of families and friends that becomes a mutual aid society in times of trouble. This cohesion within *jatis* and the attachment to place help to explain the persistence of a system that seems to put such a burden of shame and poverty on the lower ranks.

Although it seems rigid, the caste system is quite dynamic. The particular hierarchy of *jatis* in a given locale is often disputed by the various groups themselves and changes over time. *Jatis* are constantly jockeying with one another for position and status. Caste identity has also asserted itself in different ways at different times throughout history. Many scholars argue that British policies entrenched and politicized caste (and religious) identity in ways that are still influential today. Most recently, caste has become almost "ethnicized" and has taken on a renewed importance in both national and local politics. It is important to note that caste and class are not the same thing. Class refers to economic status, and there are class differences within caste groups because of differences in wealth. Because historically, upper-caste groups (Brahmins and Kshatriyas) owned or controlled most of the land and lower-caste groups (Sudras) were the laborers, caste and class tended to coincide with one another, though there were exceptions. Today, as a result of expanding educational and economic opportunities, caste and class status are less connected. Some Vaishyas and Sudras have become large landowners and extraordinarily wealthy businesspeople, while some Brahmin families live in poverty or struggle to achieve a middle-class standard of living. By and large, however, Harijans remain very poor.

In the twentieth century, Mohandas Gandhi, one of India's leaders during the fight for independence, began an official effort to eliminate untouchability (discrimination against "untouchables"). As a result, India's constitution bans caste discrimination. In the late 1940s, India began an affirmative action program that reserves a portion of government jobs, places in higher education, and parliamentary seats for Harijans (referred to as "Scheduled Castes") and *adivasis* ("Scheduled Tribes"). Together, Scheduled Castes and Scheduled Tribes now constitute approximately 23 percent of the Indian population and are guaranteed 22.5 percent of government jobs. In 1990, this program of affirmative action was extended to include other socially and educationally "Backward Castes" (low-caste groups), reserving an additional 27 percent of government jobs for these groups.

Among educated people in urban areas, the campaign to eradicate caste-ism (discrimination on the basis of caste) has been remarkably successful. Some Dalits throughout the country are now powerful officials. Members of high and low castes now ride the city buses side by side, eat together in restaurants, use the same restrooms, drink from the same water fountains, and attend the same schools and universities. More remarkably, for some urban Indians—educated professionals who meet in the workplace—caste is disappearing as the crucial factor in finding a marriage partner. Nonetheless, it would be incorrect to conclude that caste is now irrelevant in India. Less than 5 percent of registered marriages cross *jati* lines. Nearly everyone notices social clues that reveal an individual's caste; and in rural areas, where the majority of Indians still reside, the divisions of caste remain prevalent.

Geographic Patterns in the Status of Women

Within South Asia there are a number of highly successful women in such occupations as business, the media, academia, medicine, politics, law, and law enforcement—some in very high positions of power. But the overall status of women in the region is notably lower than the status of men. Despite this general condition, the particular experiences of women vary considerably across region, caste, religion, class, and age. Women's status and welfare are lowest in the belt that stretches from the northwest in Afghanistan across western India and the Indo-Gangetic Plain into Bangladesh. Literacy rates for women here are among the lowest in the region, and women's access to education and property rights are more restricted than elsewhere. Women fare better in eastern, central, and southern India and in Sri Lanka, where different marriage and inheritance practices have given women greater access to education and resources.

Urban women enjoy greater individual freedom than rural women, but middle- and upper-caste women are more restricted in their movement than are lower-caste women. Lower-caste women have greater mobility, but they must contend with sexual harassment and exploitation from upper-caste men. Young women today have significantly expanded educational and employment opportunities compared with those of a generation ago, and they are entering the skilled workforce in large numbers.

The socioeconomic status of Muslim women is notably lower than that of their Hindu and Christian counterparts. A recent national survey in India reported that Muslims on the whole have an average standard of living well below that of most Hindus. This disparity translates into educational levels for Muslim women that are significantly below the national average. Muslim women's workforce participation rates also tend to be the lowest in the country. Similar patterns prevail in Muslim-dominated countries such as Pakistan and Afghanistan, but the arrival of export-processing industries in Bangladesh (similar to maquiladoras in Mexico; see Chapter 3, pages 160–163) may be expanding opportunities for low-income women there.

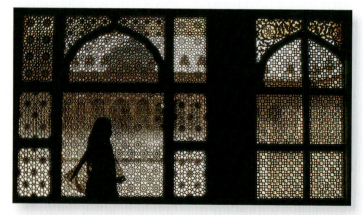

Lattice screens known as *jalee* are an architectural feature in South Asian areas where women are in purdah. Like the louvers and latticed bay windows of Saudi Arabia (see the photo on page 302 in Chapter 6), *jalee* allow ventilation and let in light, but shield women from the view of strangers. [Lindsay Hebberd/Woodfin Camp & Associates.]

Women in Afghanistan arguably have had the most difficult lives since an archconservative Islamist movement, the **Taliban**, gained control of the government there from the mid-1990s to November of 2001. The Taliban (meaning "God's students") supported strict (some argue distorted) interpretations of Islamic law (see the discussion of the Taliban on pages 420–422). Both men and women had to follow strict dress codes (men had to wear beards and wear robes), and females had to live in seclusion: girls and women were not allowed to work outside the home or attend school, and they had to wear a completely concealing heavy veil, called a *burqa*, whenever they were out of the house. (*Burqa* is the name used for this kind of garment throughout South and Southeast Asia; *chador* is the Persian name, and *abiya* is the Arabic name.) The Taliban even decreed that women whisper and not make noise as they walked, because the sound of their footsteps was distracting and potentially erotic to men. Although the Taliban are no longer in power, cultural and religious conservatism continues to adversely affect Afghani women, although concerted efforts to change their lives are under way.

PERSONAL VIGNETTE From behind her microphone at Radio Sahar (Dawn), Nurbegum Sa'idi speaks to a female audience on a wide range of topics. Radio Sahar, located in Herat, is the latest in a network of independent women's community radio stations to spring up in Afghanistan since early 2003. Radio Sahar has a broadcast radius of 50 to 70 km and provides 2 hours of daily programming consisting of educational items that address cultural, social, and humanitarian matters as well as music and entertainment. Radio Sahar is supported by a community radio advisory board composed of a variety of people from the local community, and the women members—none of whom are radio professionals—run the radio station themselves.

According to local activists, initiatives such as Radio Sahar are vital to improving the condition of women in this region. Given the high percentage of illiterate women in Afghanistan with little or no access to education, radio provides one of the most powerful ways to reach and educate women, allowing them to connect with one another in this conservative, male-dominated society. As Sa'idi attests, "It's great when you feel you can bring about change. The feedback we have been getting from listeners tells us that Sahar is providing new hope for the women in Herat." [*Adapted from United Nations Office for the Coordination of Humanitarian Affairs Integrated Regional Information Network, "Afghanistan: New Radio Station to Tackle Women's Problems," November 20, 2003.*]

Throughout South Asia, most women are partners in marriages arranged for them, often without their wishes being consulted. Usually a bride (who may be as young as 12) goes to live in her husband's family compound, where she becomes a source of labor for her mother-in-law. Most brides work at domestic tasks for many years until they have produced enough children to have their own crew of small helpers, at which point they gain some prestige and a measure of autonomy. Mothers in South Asia actually enjoy more respect than in the West. Women's power and mobility increases when they become mothers-in-law themselves. But in some communities, the death of a husband is a disgrace to a woman and can completely deprive her of all support and even of her home, children, and reputation. Widows are often ritually scorned and blamed for their husbands' deaths. Widows of higher caste rarely remarry, and in some areas, they become bound to their in-laws as household labor or are asked to leave the family home.

Purdah is the practice of concealing women from the eyes of nonfamily men, especially during their reproductive years. It is a custom observed in various ways across the region. The practice is strongest in the northwest and across the Indo-Gangetic Plain, where it takes the form of seclusion of women and of veiling or head covering in both Muslim and Hindu communities (see the story of Mishri in the village of Ahraura, on page 405). Purdah is weaker in central and southern India, but even here, separation between unrelated men and women is maintained in public spaces. The custom is generally not observed by the *adivasis* (native peoples) of the region or by low-caste Hindus, but this is changing. In recent decades, as low-status households increase their economic standing, they emulate upper-caste practices such as purdah in an effort to upgrade their overall social status (a process called *Sanskritization*). To these people, the ability to seclude women signals surplus wealth and increased ritual purity. Unfortunately for women, the effect of this trend is to limit their economic independence and autonomy in the long run.

Bride Burning and Female Infanticide. For some years now, a rare practice in India called bride burning or dowry killing, in which a husband and his relatives stage an "accidental" kitchen fire that kills his wife, has been increasing. The wife's death enables the widower to marry again and collect the dowry that, by custom, comes from a wife's family. The government of India released figures in late 1987 that affirmed 1786 such deaths in that year alone. Although not widespread—India has hundreds of millions of married women—this practice speaks to the problem of domestic violence that is pervasive throughout the region. In some cases, the threat of bride burning is used to extort further dowry, in the form of cash or durable goods, from a wife's family. More commonly, a young bride may be made to feel undeserving of her status in the groom's family if the family feels cheated in the marriage negotiations. A related practice is that of female infanticide, in which girl babies, deemed unaffordable because of the dowry investment they will require, are killed.

Changing customs regarding dowry appear to be a cause of the growing incidence of both bride burning and female infanticide in India. Until the last several decades, it was the custom among the lower castes to pay a **bride price,** not a dowry: a groom paid the family of the bride a relatively small sum that symbolized the loss of their daughter's work to her family's economy and the gain of her labor to his. **Dowry,** on the other hand, is wealth provided by the bride's family at the time of marriage; it originated as an exchange of wealth between landowning, high-caste families. With her ability to work reduced by purdah, an upper-caste female was considered a liability. The dowry that went with a bride to her new husband gave her dignity as a wife and leverage in her husband's household, because the dowry had

to be returned if the marriage dissolved. As the practice of purdah spread to poorer, lower-caste families wanting to upgrade their status, the custom of dowry spread with it. Moreover, the payment of a substantial dowry could further increase the status of a bride's family by attracting a groom of higher status. A good dowry settlement for the first child in a family (male or female) set the tone for future marriage settlements of siblings. Thus caste and class became conflated as dowry was redefined in practice as a non-refundable endowment paid by the bride's family to a higher-status groom and his family.

Oddly, increasing education for males and increasing family affluence reinforced the custom of dowry. Young men, even those of low caste, came to feel that their diplomas increased their worth as husbands and put them in the category of those who deserved a bride with a substantial dowry. This upgrade in status through education gave them the power to demand larger and larger dowries. Soon the practice spread through the lower castes, and now the poorest of families are crippled by the dowries they must pay to get their daughters married, which in turn influences the matches they can make for their sons.

Despite the fact that the Indian government bans the practice of dowry, the social duty to provide a dowry for a daughter is taken extremely seriously, because the stigma of having an unmarried daughter is huge. Some ambitious families are giving their daughters a graduate school education, with its promise of earning power, in lieu of a dowry. But for an increasing minority of families, the birth of more than one daughter threatens the family with hopeless impoverishment, whereas the birth of sons promises future daughters-in-law who will bring dowry wealth with marriage. A village proverb captures this inequitable relationship: "When you raise a daughter you are watering another man's plant." Some families view the birth of a daughter as such a calamity that they are led to the desperate act of poisoning second and third daughters soon after birth.

Education and the Status of Women. Several agencies (such as Oxfam, India Literacy Project, and South Asian Women's Network) are seeking to improve the status of women and their earning opportunities and to diminish the influence of purdah, which keeps women illiterate and confined to their houses. These organizations believe that freeing women from purdah and other ancient strictures will increase the educational attainments of all of South Asia's children and improve the health and nutrition of families. Research has shown that even with as little as 2 years of education, women choose to have fewer children (see Figure 8.11 on page 402) and begin to make decisions that enhance the well-being of their children. United Nations researcher Martha Nussbaum found that women who can read often seek a way to earn some income and generally invest their earnings in food, medicine, and schooling for their children. Over time, freeing women from purdah will also encourage lower fertility and greater economic growth because women will channel less of their energies into reproduction and more into such activities as business, innovative agriculture, recycling of resources, teaching, and other service occupations.

The Internet has become a forum for discussion within South Asian communities about issues related to the low status of

In this Mumbai neighborhood Internet café, people of all ages and both sexes are welcome and receive instruction and assistance as they need it. Virtually every city of more than 10,000 people now has such cafés. Internet technology itself may empower women: many of the workers in India's Internet cafés are young women, who meet many international travelers in the course of a day and learn to use the technology themselves because they must help customers. [Dinodia.]

women: low levels of literacy, health care, purdah, the preference for sons, bride burning, and female infanticide. For example, descriptions of daughters performing skilled mechanized agricultural work and conducting Hindu death ceremonies—both tasks typically done only by sons—have been posted on the India Family Net Talk Web site at http://indiafamily.net/ (home page). Also posted on this site are testimonies about the value of educated daughters and personal accounts of young women who have experienced son preference and the low status of being a daughter. Web sites like this one, and access to global Internet connections in general, lend strength to social change efforts.

Gender Equality at the Village Level and Beyond. Although women in India face a number of constraints and disadvantages, a strong activist movement in India has led to fairly enthusiastic enforcement of constitutional protections, at least after the fact. In the 1980s, Prime Minister Rajiv Gandhi introduced *panchayati raj* (village government) to encourage gender equality in village life. Thirty percent of the seats on these local councils must be reserved for women during a given election cycle. Furthermore, since 1993, there has been training for women *panchayati raj* members to improve their effectiveness on the councils. Support is growing for legislation that would reserve one-third of the seats in the lower house of the Indian parliament and in state assemblies for women for a 15-year trial period. After that time, it is hoped, women will have achieved the political experience to win elections without the aid of quotas.

A minority of upper- and middle-class urban professional women in India, Pakistan, Sri Lanka, and Bangladesh may have more in common with their counterparts in Europe and America than they do with village women in their own countries. In such cities as Karachi, Delhi, Mumbai, Bangalore, Chennai, Dhaka, and Colombo, there are growing numbers of highly successful businesswomen, female directors of companies, highly qualified female technicians, high-ranking female academics, and women who serve prominently in government. India, Bangladesh, Sri Lanka, and Pakistan have all had women heads of state.

ECONOMIC ISSUES

South Asia is a region of startling economic contrasts, where a country (India) that is home to hundreds of millions of desperately poor people can also foster a growing computer software industry and a space program. These extremes reflect the propensity of South Asian economies to favor the interests of a privileged minority over those of the poor majority. Although the British colonial system deepened the extent of South Asia's poverty and widened the gap between rich and poor, the current wealth disparities in the region result mostly from economic policies favored by post-independence leaders. Despite India's celebrated democratic traditions, for example, the poor have often been left out of the political process and bypassed or hurt by economic reforms.

Agriculture remains the basis of South Asian economies, but rapid industrialization and self-sufficiency have been the dream since independence. Recent strategies adopted to encourage economic development include an emphasis on information technology in India and innovative finance strategies pioneered in Bangladesh; and the service sector has expanded more rapidly than either agriculture or industry in almost all South Asian countries over the past decade.

Agriculture and the Green Revolution

Well over 60 percent of the region's population is still rural and engaged in agricultural labor. In tiny Nepal, 90 percent of the population is so occupied. Nonetheless, the contribution of agriculture to most national economies (GDP) hovers below 25 percent. Although production per unit of land has increased dramatically over the past 50 years, agriculture is the least efficient sector of the regional economy, meaning that it gets the lowest return on investment of land, labor, and cash.

Figure 8.15 shows the distribution of agricultural zones in South Asia. Except in the far northwest and in highland areas, double-cropping is common throughout the region: crops adapted to dry conditions are planted in the winter, and those adapted to wet conditions are planted in the summer when monsoon rains are plentiful. Farmers in some areas of Bangladesh can harvest three crops a year. Rice is the main crop wherever rainfall is plentiful, especially in the flat Ganga Plain and around the eastern and southern rim of India, and occupies about one-third of the total land planted in grain. Wheat, grown in the west central and northwestern parts of the region, is the second most important crop, and its cultivation is spreading with the use of irrigation systems. Other grains grown include millet and grain sorghum. Cotton remains an important cash crop and is often grown with the aid of irrigation. Animal grazing is predominant in the dry areas of Pakistan, Afghanistan, and northwestern India.

Until the 1960s, agriculture in South Asia was based largely on traditional small-scale systems that managed to feed families in good years, but often left them hungry in years of drought or flooding. These systems did not produce sufficient surpluses for the region's growing cities, which even now rely on imported food. Large-scale mechanized agriculture, where it existed, was aimed at producing export crops such as cotton, flax, jute, tea, and rice. Much of South Asia's agricultural land is still cultivated by hand, and overall agricultural development has been neglected in favor of industrial development, especially in India. Nonetheless, by the 1970s, important gains in agricultural production had begun.

Beginning in the late 1960s, a so-called **green revolution,** promoted primarily by international agricultural research agencies, boosted grain harvests dramatically through the use of seeds selected for high yield and for resistance to disease and wind damage. Other components of the revolution were fertilizers, mechanized equipment, irrigation, pesticides, herbicides, double-cropping, and, to a lesser extent, an increase in the amount of land under cultivation. As a result, Pakistan is now self-sufficient in wheat, rice, and sugarcane and produces surpluses for export.

India has become one of the world's leading producers of grain and now exports rice to the Americas, Asia, and Africa. Where the new techniques were used, yield per unit of farmland improved by more than 30 percent between 1947 and 1979. International loans financed the building of dams to store monsoon water for irrigation and to create hydroelectric power. These projects, in turn, boosted industrial growth and created jobs. India was able to repay its green revolution–related loans, which boosted its reputation for creditworthiness.

The green revolution, however, has not made India reliably self-sufficient in food. Food production on a per capita basis has increased since 1960; however, some scholars and policy makers are concerned about India's long-term ability to maintain this trend. In addition, the benefits of the green revolution have been very uneven. Some states, such as Punjab and Haryana, which have extensive irrigation networks, have gained tremendously while other states have lagged behind. Many poor farmers who were unable to afford special seeds, fertilizers, pesticides, and new equipment were forced off rented or borrowed land and into low-wage farm labor in the early decades of the green revolution (1960s and 1970s). Although a number of such farmers were able to buy into the green revolution in the 1980s, and rural incomes did expand, those farmers who lost access to land have migrated to the cities, where they have difficulty finding jobs of any sort.

Although the new technologies have resulted in dramatically higher production rates, the increased food supplies have not

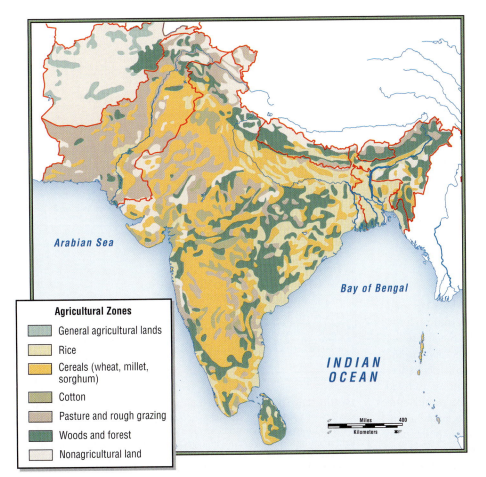

Agricultural Zones

- General agricultural lands
- Rice
- Cereals (wheat, millet, sorghum)
- Cotton
- Pasture and rough grazing
- Woods and forest
- Nonagricultural land

FIGURE 8.15 Agricultural zones in South Asia. Agricultural zones in this region form a complex pattern influenced by landforms, climate, cultural customs, and recent development theories. Note especially the distribution of rice zones as opposed to cereal zones. [Adapted from Gordon Johnson, *Cultural Atlas of India* (New York: Facts on File, 1996), p. 34; "Afghan Economy (map)," SESRTCIC (Statistical, Economic and Social Research and Training Centre for Islamic Countries) InfoBase at http://www.sesrtcic.org/members/afg/afgmapec.shtml.]

PRACTICING GEOGRAPHY READING MAPS Compare this map to Figures 8.4 and 8.8. Is there any correlation between the distribution of agriculture and population densities? How does climate affect agriculture?

eliminated hunger and malnutrition. Between 1970 and 2001, there was an 18 percent increase in the amount of food produced per capita in South Asia. During the same time, the proportion of undernourished people dropped from 33 percent of the population to 22 percent; however, the total number of hungry people increased 23 percent. The hungry still could not afford food, while the increased supplies of food were sold to those who could pay for it or were exported. Some South Asian rice, for example, finds its way to specialty stores in the United States. Another factor in food distribution is that rural areas are bypassed during food scarcities. South Asian governments place a high priority on ensuring a sufficient supply of food to the cities because urban unrest poses a greater threat to the interests of government and to the middle and upper classes than does rural unrest.

Meanwhile, increasing soil salinity and other kinds of environmental damage created by chemical fertilizers, pesticides, and high levels of irrigation are reducing yields in many areas. One such area is the Pakistani Punjab, that country's most productive, but highly irrigated, agricultural zone.

Green revolution technologies also inadvertently reduced the utility of many crops for the rural poor, especially women. For example, the new varieties of rice and wheat yield more grain, but less of the other components of the plant previously used by women, such as the wheat straw used to thatch roofs, make brooms and mats, and feed livestock. Moreover, women's already low status in agricultural communities often erodes further as their contribution to household production is supplanted by new technology, such as small tractors and mechanized grain threshers, which are usually controlled by male members of the family.

A potential remedy for some of the failings of the green revolution style of agriculture is **agroecology**: the use of traditional methods to fertilize crops and natural predators to control pests. Unlike green revolution techniques, the methods of agroecology are not disadvantageous to poor farmers because the necessary resources are readily available in most rural areas. To participate, the farmers do not need access to cash, but only to knowledge, which can be taught orally to small groups and over the radio. Studies in South India that compared agroecology techniques with those of the green revolution found their productivity and profitability to be equal, but agroecology techniques reduced soil erosion and loss of soil fertility.

Industry over Agriculture: A Vision of Self-Sufficiency

After independence from Britain in 1947, South Asia's new leaders favored industrial development over agriculture. Influenced by socialist ideas, especially the model of the Soviet Union, they concluded that agriculture was incapable of supplying the growth and technological innovation that poor countries needed. Government involvement in industrialization was considered necessary to ensure the levels of job creation that would cure poverty. Another motive, especially in India, was to create a self-sufficient economy that did not need to import manufactured goods from the industrialized world. The new South Asian leaders engineered government takeovers of the industries they

believed to be the linchpins of a strong economy: steel, coal, transport, communications, and a wide range of manufacturing and processing industries.

For the most part, South Asian industrial policies failed to meet their goals. There was enough economic growth in the early years of independence to make India the eighth most industrialized country in the world, in terms of the relative output of the industrial sector compared to that of the agricultural and service sectors. But the emphasis on self-sufficiency in industry was ill suited to countries that had been primarily agricultural for years. In India, for example, governments invested huge amounts of money in a relatively small industrial sector that even today employs only 17 percent of the population (agriculture employs 60 percent). Only a small portion of the population directly benefited from this investment, so industrialization failed to increase South Asia's overall prosperity significantly.

Another problem was that the measures governments took to boost employment often contributed to inefficiency and ignored market incentives. One policy encouraged industries to employ as many people as possible, even if they were not needed. So, for example, it still takes 250,000 Indian workers to produce the same amount of steel as 8000 Japanese workers; consequently, Indian steel costs much more to produce than Japanese steel. In addition,

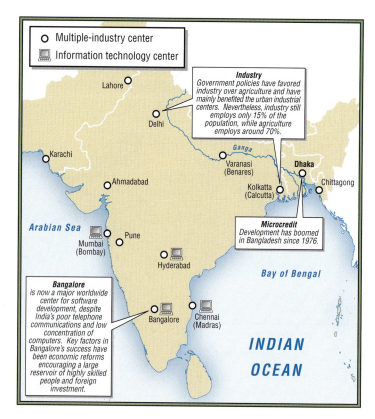

FIGURE 8.16 Economic issues: Industrial and information technology centers in South Asia. The multiple-industry symbols mark locations where three or more major industrial activities (heavy industry, extractive industry, light manufacturing, or service industry) take place. [Adapted from Gordon Johnson, *Cultural Atlas of India* (New York: Facts on File, 1996), pp. 184–185, 190–191, 198–199, 204–205, 210–211, 217, 219–220, 222.]

decisions about which products particular manufacturing industries should produce were made by ill-informed government bureaucrats rather than being driven by consumer demand, as in the former USSR. Until the 1980s, items that would improve daily life for the poor majority were produced in only small quantities. Such items included cheap cooking pots, buckets, cheap yet sturdy bicycles, and simple tools. At the same time, there is a relative abundance of such items as vacuum cleaners, watches, TVs, kitchen appliances, and cars—but only a very few can afford to purchase them (although their numbers are increasing). Figure 8.16 shows the distribution of manufacturing and service industries in South Asia as of the mid-1990s.

Economic Reform: Achieving Global Competitiveness

During the 1990s, much of South Asia began to undergo a wave of economic reforms (structural adjustment programs, or SAPs) intended to make national economies more efficient and productive by cutting back on government expenditures and opening up economies to foreign investment and trade (see the explanation of SAPs in Chapter 3, pages 136–138). In contrast to other regions in which structural adjustment has been mandated by the International Monetary Fund (IMF) and World Bank, India's economic reforms were initiated by the Indian government itself in response to a financial crisis that emerged in the 1980s. Begun in 1991, they were aimed at privatizing industries, removing government regulation, and opening up the economy to foreign goods and foreign investment. Investment, trade, and finance were the principal targets for reform; specific measures included devaluation of the Indian rupee, reduction of tariffs and restrictions on foreign investment, and a relaxing of industrial licensing requirements. Although privatization of public-sector industries and banks has not proceeded very far, liberalization of the economy has been

significant—evidenced in part in investment by international companies such as Coca-Cola, Microsoft, Cisco, Sony, Union Carbide, and Lever Brothers, and by the wide range of foreign goods that are now available.

In this more competitive environment, productivity has increased in the export and industrial sectors of the economy. India, in particular, has achieved spectacular successes in high technology. But overall, India's rate of economic growth remained slow as of 2002–2003 (4.3 percent), in large part as a result of actual declines in agricultural production due to drought. In 2003, India's economy showed greater growth (7.6 percent), but a large deficit, poor-quality infrastructure, slow growth in employment, and increasing disparities between regions of the country remain major concerns. Most of the new growth that has taken place has occurred in the Indian Ocean states of Maharashtra, Gujarat, and Karnataka, where the cities of Mumbai, Ahmadabad, and Bangalore, respectively, are growth centers (see Figure 8.16). The rest of the country lags behind, although it should gain ground if a strong global economy persists. Pakistan and Bangladesh, with growth rates of 5.4 and 5.3 percent, respectively, in 2003, also have been slow to show overall economic improvement.

Like SAPs in Middle and South America and Africa, however, the new policies are producing wider disparities in income. The urban and industrial elites have received most of the gains, while little regard has been paid to the effects of reform on the poor, many of whom have lost jobs and access to social services. The devaluation of the Indian rupee has also led to inflation, and the poor find it increasingly difficult to make ends meet.

South Asia's service economies have been expanding as the agricultural and industrial sectors have gone through ups and downs. As a whole, between 20 and 40 percent of South Asian workers are employed in the service sector, but this sector's contribution to the GDP in India, Pakistan, Sri Lanka, and

---

AT THE GLOBAL SCALE   The Export of High-Skilled Jobs to India

One of the effects of globalization has been a migration of jobs from industrialized countries in the West to countries in Asia, Africa, and Latin America in search of lower production costs. Manufacturing jobs were the first to go in the 1980s. Since the early 1990s, an increasing number of call center, back-office, and even high-skilled jobs have found their way to countries such as India that have large, college-educated, low-cost workforces. India's pharmaceutical and high-tech industries have attracted a number of American jobs in recent years, and now Wall Street firms such as J. P. Morgan, Goldman Sachs, and Lehman Brothers are beginning to seek out India's highly skilled workers.

According to the *New York Times*, both J. P. Morgan and Morgan Stanley planned to hire dozens of researchers in Mumbai in 2003–2004, and Goldman Sachs planned to hire 250 employees to work in its new Indian unit. In fact, total outsourcing of business-processing jobs by American companies is expected to

reach $136 billion over the next decade. Of the 3.3 million jobs that will be created by American companies, 1 million will move abroad. India, along with China, Russia, and the Philippines, is expected to gain the most.

A downturn in capital markets and stiff global competition makes cost cutting imperative for Wall Street firms. India's relatively low real estate costs and salaries seem to provide a solution. For example, whereas a junior analyst from an Ivy League school costs $150,000 in the United States, an Indian graduate from a top business school costs only $35,000 a year. Yet, for that Indian employee, this amount translates into a salary of over 1 million rupees, which buys a much higher standard of living than the U.S. employee would enjoy.

[*Source:* Saritha Rai, "As it tries to cut costs, Wall Street looks to India," *New York Times* (October 8, 2003).]

Bangladesh is over 50 percent. Hence the service sector is seen as having the best chance of competing successfully in the global economy. Within the service economy, facilities that engage in trade, transport, storage, and communication (including information technology) show the most growth; finance, insurance, real estate, business services, and tourism have also grown quickly. All these activities are connected in some way with international commerce and benefit from India's success at developing information technology. Tourism has considerable potential as South Asia's own middle class expands and outsiders are attracted to the region. Tourism sites abound in both coastal and interior mountain locations, especially in India.

Differing Views of Globalization. The growth of the service sector and information technology in South Asia seems to foretell increasing connections to the global marketplace. The development of such connections is both desired and feared across the region. This ambivalence is particularly strong in India, the country with the strongest links to the global economy. Globalization is desired because it is expected to increase the number of high-paying jobs, raise standards of living, and fuel local production of consumer products and services (see the box "The Export of High-Skilled Jobs to India" on page 415). Cities throughout the region have already experienced these benefits.

Some small rural places see information technology as the answer to their geographic isolation. For example, the small community of Dhab (40,000 people), located on an island in the Ganga River downstream of Varanasi (Benares), has no electricity and few jobs. The shifting course of the Ganga between the rainy and dry seasons has isolated the island community and inhibited development. Dhab's citizens are lobbying for dikes that would control the river's course and, more important, would support electricity and Internet cables and thus bring Dhab from the era of candles and oil lamps to the information age in just a few months.

Those who worry about globalization include environmental activists and conservative community leaders, who occasionally find common cause in opposing projects like the one proposed for Dhab. Environmentalists warn against unduly modifying the river's natural seasonal oscillations with dikes, and community leaders fear that access to information on the Internet will cause a breakdown of traditional social and economic relationships and bring a flood of Westernization. Television is another agent of globalization that is having a dramatic cultural impact, as the following vignette illustrates.

The tiny kingdom of Bhutan, couched in the Himalayas between China and India, had for centuries been more or less secluded from the rest of the world. That all changed in June 1999, when a royal decree legalized television, making Bhutan the last country to "plug in." Within a short time, several entrepreneurs, such as Rinzy Dorji, whom the Bhutanese call "The Cable Guy," were in business. For $5 a month, the price of a bag of chilies, Rinzy provides Bhutanese households with 45 cable TV channels—everything from the BBC to *Baywatch*.

Although Rinzy's business is booming, not everyone welcomes the new technology. As Kinley Dorji, editor of Bhutan's only newspaper, describes, "Soon after television started, we started getting letters to the editor of the newspaper from children, children who seemed very hurt. The letters actually specifically asked about this World Wrestling Federation program, 'Why are these big men standing there and hitting each other? What is the purpose of it?' They didn't understand; they were very hurt. Now, a few months later my son jumps on me one morning and says, 'I am Triple H, and you can be Rock.' And, suddenly we are fighting. Suddenly these are new heroes for our children."

Foreign Minister Lyonpo Jigma Thinley thoughtfully observes, "People have suddenly realized that there are so many things they desire, which they were not even aware of before. The truth is that most of these television channels are commercially driven. And some of the Bhutanese people are driven towards consumerism. And, that is inevitable. It's unfortunate, but inevitable." On the other hand, he has also heard people saying, "My God! We didn't know that we are living in a peaceful country. There seems to be violence and crime everywhere in the world." He concludes, "So, in a way, the positive thing is that people realize how good a life they are living in this country."

To learn more about the impact of satellite and cable TV in Bhutan, see the FRONTLINE/World Video "Bhutan; The Last Place."

Economic Development and Poverty Rates. Poverty continues to decline in South Asia, but poverty rates have decreased much more slowly since 1991 than they did from about 1968 to 1990. In India, the poor are defined as that portion of the population that cannot afford a diet with the minimal caloric intake needed to sustain life. Population growth over the last 50 years has outstripped economic growth. Despite all the policy efforts that have reduced poverty from nearly 45 percent of the population in 1952 to less than 25 percent by 2000, there are now nearly half again as many people who qualify as poor—simply because there are now so many more people overall. In 2000, there were 256 million people who could not afford to eat the minimum diet, whereas 50 years ago there were only 180 million.

Innovative Help for the Poor. In recent years, South Asians have developed some promising strategies for helping poor people. One of these is **microcredit,** a program that makes very small loans available to poor would-be business owners in both rural and urban areas. Throughout South Asia, and indeed, in much of the world, poor people have a difficult time obtaining loans. They do not have much collateral, so lending to them seems risky; nor do they need large sums, so lending to them is unprofitable. Hence the poor must rely on small-scale moneylenders who often charge extremely high interest rates of 30 percent or more *per month.*

In the late 1970s, Mohammed Yunnis, an economics professor in Bangladesh, responded to this problem by starting the Grameen Bank, or "People's Bank," which makes small loans, mostly to people in rural villages who wish to start businesses. The loans often pay for the start-up costs of small enterprises such as chicken raising, small-scale egg production, or construction of pit toilets. The problem of collateral is resolved by having potential borrowers organize themselves into small groups that are collectively responsible for paying back the loans. If one member fails to repay a loan, then the group will be denied loans in the future. This system, reinforced with weekly meetings, creates incentives for mutual support among group members as well as peer pressure to repay the loans. The weekly meetings are often the only time that women in purdah leave the confines of their homes and may be their first contacts with women (and sometimes men) who are not their kin. The repayment rate on the loans is extremely high, averaging around 98 percent—much higher than most banks achieve. Hence the Grameen Bank can afford to charge interest rates of around 13 to 14 percent per year, much lower than those of village moneylenders.

So far, the Grameen Bank has been an enormous success in Bangladesh, where it has loaned over U.S. $1.8 billion to more than 2 million borrowers. Similar microcredit projects have been established in India and Pakistan and throughout Africa, Middle and South America, North America, and Europe.

PERSONAL VIGNETTE   In a small hamlet in Bangladesh not too far from the Indian border is the house of Mosamad Shonabhan, a 32-year-old married woman whose life has been changed by her 11-year participation in the Grameen Bank. Everyone agreed that she had been the smartest of her brothers and sisters, but because her father earned only 50 cents a day as a farm laborer, she could not go to school, and was instead married at the age of 14 to a young barber. For a year she lived in her father-in-law's house, but financial problems soon forced her to move back into her father's house. There she faced increasingly dire circumstances as his health deteriorated. After a few years, a local political leader suggested that she join the Grameen Bank's lending program. She was afraid to go, as she had never handled money and had heard a local rumor that the bank's real purpose was to convert people to Christianity. Nevertheless, she went, and eventually took out a loan for $40.00 that would allow her to set up a small rice-husking operation in her father's backyard. Eleven years and eleven loans later, she earns about $1.50 every day—three times what her father had made—and is a pillar of the local community. Her main source of income is a small shop inside her father's old house, which she bought from her siblings after his death. She also leases an acre of land, which produces enough rice to feed her family and the numerous guests and friends who now come by to see her. She plans to open another shop that her husband will run. She is being taught to read by her 15-year-old daughter, whom she plans to send to university. [*Adapted from the field notes of Alex Pulsipher, 2000.*]

POLITICAL ISSUES

Since independence in 1947, South Asia has peacefully resolved many conflicts, smoothed numerous potentially bloody transfers of power, and nurtured vibrant public debate over the issues of the day. India is often cited as a bastion of democracy that serves as an example of political enlightenment to the rest of the developing world. In recent years, however, there have been increasing signs that corruption, demagogic leadership, and violence are eroding democracy in the region. Shifting patterns of authority have increased tensions between upper- and lower-caste groups. The rise of religious nationalism also threatens peaceful relations both within and between the region's nations.

Caste and Democracy

Since the beginning of broad-based democracy in India in the 1930s, caste has been a defining yet contradictory factor in both local and national politics. At the local level, most political parties design their vote-getting strategies to appeal to subcaste (*jati*) loyalties. They often secure the votes of entire *jatis* with such

In this village in Bangladesh, women borrowers gather weekly at the local Grameen Bank, usually in someone's house, to pay their loan installments. [Alex Pulsipher.]

political favors as bringing in new roads, schools, or development projects to communities dominated by those groups. These arrangements fly in the face of the official ideologies of the major political parties, which deny any caste loyalties, and of Indian government policies, which actively work to undermine discrimination on the basis of caste. Currently, the role of caste in politics seems to be increasing, as several new political parties that explicitly support the interests of low castes have emerged and formed new group alliances to increase their political influence. This assertion of political rights by low castes has been met with a backlash from upper-caste groups, resulting in a number of violent clashes in recent years. Hence, caste has been woven into the political system in ways that create and maintain tension and conflict. We will see shortly that the same is true of religious identity.

Religious Nationalism

Increasingly, people frustrated by government inefficiency, recent scandals surrounding corruption, and the failure of governments to deliver on their promises of broad-based economic development and prosperity are joining religious nationalist movements. Although people in South Asia live under legally secular governments, religious nationalism has long been a reality, shaping relations between people and their governments. **Religious nationalism** is the belief that a particular religion is strongly connected to a particular territory, perhaps even to the exclusion of other religions, and that those who share a belief system should have control over their own political unit—be it a neighborhood, part of a country, or a separate country. Although both India and Pakistan were formally created as secular states, India is increasingly thought of as a Hindu state and Pakistan and Bangladesh as Muslim states. Many people in the dominant religious group strongly associate their religion with their national identity.

Hindu nationalism in India is predominantly supported by men from middle- and upper-caste groups who fear the erosion of their castes' political influence and who resent, in particular, the extension of the quota system for government jobs and seats in universities to lower-caste groups (see page 409). This sentiment has been fueled by the threat middle- and upper-caste groups see in a politically mobilized lower-caste alliance. Previously, the dominant (upper and middle) castes were able to direct the votes of the lower castes. Now, however, lower-caste groups are no longer willing to follow the dictates of the dominant castes.

Political parties based on religious nationalism have gained popularity throughout South Asia, often fueling conflicts between religious majorities and minorities within a particular state. Although their members think of these parties as forces that will purge their country of corruption and violence, they are, in fact, usually only slightly less corrupt, and certainly no less violent, than other parties (see the box "Babar's Mosque" below). The Ganga River basin has emerged as the center of Hindu nationalist activism because of the sacred status of the Ganga River, the plethora of Hindu sacred places in the basin, and the long history of Muslim rule there. Periodically, there is violence against Muslims, who make up about 15 percent of the population in this part of India.

Regional Political Conflicts

The most intense armed conflicts in South Asia today are **regional conflicts,** in which nations dispute territorial boundaries or in which a minority actively resists the authority of a national or state government (Figure 8.17). The resisting group sees government officials as depriving it of a voice in its own governance. Two regional conflicts in the neighboring Indian states of Punjab and

---

AT THE REGIONAL SCALE Babar's Mosque: The Geography of Religious Nationalism

Proponents of religious nationalism often try to gain mass support through political campaigns that interweave South Asian history, mythology, and landscape. In late 1992 and early 1993, a series of riots occurred throughout South Asia that were the culmination of a long campaign waged by India's leading proponents of Hindu nationalism, the Vishwa Hindu Parishad (VHP—World Hindu Council). The riots were triggered by the destruction of a Muslim mosque in the town of Ayodhya, in Uttar Pradesh on the Ganga Plain (see Figure 8.17). The Mughal emperor Babar had built the mosque during his early invasions of India in the sixteenth century. It supposedly stood on the ruins of a Hindu temple believed by local residents to mark the birthplace of the Hindu god Ram. After years of campaigning for the mosque's destruction, the VHP finally succeeded in late 1992, when a highly organized Hindu mob of 300,000 razed the structure. The destruction led to riots that ripped through most major northern Indian cities. Mobs, commanded by urban Hindu nationalist political parties and

social organizations (many of them funded by Indian Hindus living abroad), burned and looted selected Muslim businesses and homes, often with the complicity of the police. Nearly 5000 people died.

The VHP's selection of the Ayodhya mosque, an important Muslim heritage site, as a locus of protest was calculated to elicit violent reactions from Muslims. The locations of the riots also reflected the urban base of Hindu nationalist parties, and their highly organized criminal actions reflected the violent gangster-style tactics that were common on the South Asian political scene in the 1990s. The violence soon spread across international borders: in retaliation, Muslim mobs in Bangladesh and Pakistan harassed Hindu communities and destroyed their temples.
[*Sources*: Alex Pulsipher, field notes, India, 1993; and Stuart Corbridge and John Harriss, *Reinventing India* (Malden, Mass.: Blackwell-Polity Press, 2000).]

FIGURE 8.17 Political issues: Conflicts in South Asia. Well-known conflicts in the region include the dispute between India and Pakistan over Kashmir (see inset); the civil war and the ongoing "war on terrorism" in Afghanistan; the internationally funded Hindu nationalist movement, for which Ayodhya is a symbol; the Assam separatist rebellion in far northeastern India; the Sikh secession movement in Punjab; and the conflict between Tamils and Singhalese in Sri Lanka. [Inset adapted from *National Geographic* (May 1997): 19.]

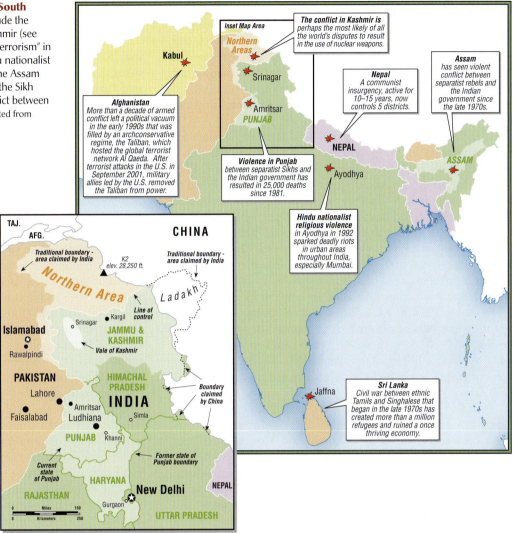

Kashmir can serve as examples. Both states are in far northwestern India along the Pakistan border, and both conflicts have religious and political components. Although the Punjab crisis has faded in recent years, Kashmir continues to be a source of unrest that has the potential to destabilize the region.

Conflict in Punjab. Punjab is the ancestral home of the Sikh community. When India and Pakistan were partitioned in 1947, Punjab was divided between the two countries. Caught between majority populations of Muslims and Hindus during the violence that followed Partition, large numbers of Sikhs chose to live in the Indian part of Punjab or elsewhere in India because they thought the secular Indian constitution would better allow them to preserve their unique identity. The Sikh community has since pressed India for greater recognition of its distinct religious and ethnic identity and for greater regional political autonomy. Despite a long history of peaceful coexistence and intermarriage between Hindus and Sikhs in Punjab, conflicts between rural (Sikh) and urban (Hindu) interests over access to land and water, over development policies and funds, and over the control of religious sites have created communal tensions.

After independence, Sikh political leaders successfully agitated for the creation of two states from Punjab: a Sikh majority state called Punjab and a Hindu majority state called Haryana (see the inset in Figure 8.17). The border between the two states had to be drawn on the basis of language (Punjabi and Hindi) because division based on religious difference would have been unconstitutional. Since then, however, Sikhs have felt alienated from the rest of India. In 1973, the Akali Dal, the moderate Sikh political party, issued demands for religious concessions, equitable water rights, and greater political autonomy, which were ignored by the central government of India. In an attempt to keep the Akali Dal fractured and prevent them from gaining control of the Punjab State Assembly, the Congress party cultivated relations with Sikh extremists.

This strategy backfired in the early 1980s as the extremists gained strength and barricaded themselves in the holiest Sikh shrine, the Golden Temple in Amritsar, which they then used as a base of operations to agitate for an independent Sikh nation to be called "Khalistan." In 1984, government forces (headed by a Sikh general) attacked the shrine, damaged the temple, and killed the militants as well as numerous innocent pilgrims caught in the crossfire. This incident deeply alienated and further radicalized the Sikh community. Shortly thereafter, Prime Minister Indira Gandhi, who had called for the attack, was assassinated by two of her Sikh bodyguards. Riots and organized mob violence spread throughout India over the next few days, resulting in the deaths of more than 2700 Sikhs, who are a wealthy and influential minority in many cities outside of Punjab.

In 1985, an agreement acceding to many Sikh demands for water rights and control of religious sites was signed between a moderate Sikh leader, Harchand Singh Longowal, and then Prime Minister Rajiv Gandhi, and the Akali Dal was elected to govern Punjab. In 1987, however, the central government (ruled by the Congress party) replaced the Akali Dal officials with their own appointed leaders, and the terms of the 1985 agreement were never implemented. This further alienated Sikhs and led to more political violence, which continued into the early 1990s; 25,000 people have died from political violence in Punjab. The Congress party controlled the state government for much of the 1990s, but in 1997 the Akali Dal won the elections and formed a coalition government with the Hindu nationalist party, the BJP. In the 2002 elections, however, voters returned the Congress party to power. Although the violence has subsided, the original demands expressed by Sikhs have yet to be met, so tensions could reemerge.

Conflict in Kashmir. Of all the armed disputes in South Asia, many experts fear that the situation in Kashmir most threatens global peace. In Kashmir, Hindu and Muslim differences date back to before Partition in 1947. At that time, Kashmir was one of hundreds of independent states that had surrendered much of their autonomy to the British, although their leaders retained nominal control. All these states were eventually absorbed into either Pakistan or India. Kashmir has long been a Muslim-dominated area, but its ruler in 1947 was a Hindu king (maharaja). Pakistan's leaders believed that Kashmir should be turned over to them based on the rules for Partition established by the British. But the kings and princes of the autonomous states also wanted a say in where their territories would be allotted. Kashmir's Hindu king wanted Kashmir to remain independent; its most prominent popular leader—a Muslim—favored constitutional rule and accession to India. In fact, many Muslims rejected the idea of a separate nation for Muslims. They saw themselves as Indians and favored a united, secular, and democratic India. When Pakistan-sponsored raiders invaded western Kashmir in 1947, the maharaja quickly agreed to join India. A brief war between Pakistan and India resulted in a cease-fire line (line of control) that became a tenuous boundary (see Figure 8.17 inset).

A popular vote for or against joining India that was supposed to decide Kashmir's fate was never held. India argued that because Kashmiris had already voted in Indian elections, they had, in effect, voted to be part of India. Pakistan attempted another invasion of Kashmir in 1965, but was defeated. The two countries are technically still waiting for a UN decision on where the final border will be, but Pakistan effectively controls the thinly populated mountain areas north and west of the densely populated Vale of Kashmir. India holds nearly all the rest, where it maintains an ominous presence with more than 500,000 troops. China claims the Ladakh part of Kashmir (see Figure 8.17 inset).

As many as 20,000 people have been killed in the conflict over Kashmir. Civil war has erupted repeatedly over the years because many Kashmiris support independence from both India and Pakistan. As in Punjab, much of the conflict has centered around the right of Kashmiris to elect their own state leaders and their demand for greater political autonomy within India's federal political system; the national government in New Delhi has often appointed its own favorites in an attempt to maintain strong central control. Anti-Indian Kashmiri guerrilla groups, equipped with weapons and training from Pakistan, have carried out many bombings and assassinations. Blunt counterattacks launched by the Indian government have killed large numbers of civilians. The deaths have alienated more Kashmiris, including the local police force, which is now seen as sympathetic to the militants. Sporadic fighting between India and Pakistan continues along the boundary line, where at one location the two countries intermittently clash in the world's highest battle zone, at an altitude of 20,000 feet (6000 meters). Another development in the Kashmir dispute is the increasing involvement throughout the 1990s of Islamic militant *mujahedeen*, soldiers who took part in Afghanistan's war of resistance against the Soviet Union in the 1980s. The mujahedeen are a powerful destabilizing force because, although they receive weapons and training from Pakistan (as do the Kashmiri militants), it is difficult to tell whether their actions are always controlled directly by Pakistan or perhaps by global terrorist networks.

A further complication in the Kashmir conflict is that both India and Pakistan have nuclear weapons. Since the testing of nuclear weapons by both countries in 1998, India and Pakistan have come close to war on two occasions. In the 1999 Kargil crisis, over 1000 Pakistani troops infiltrated Indian-held Kashmir at Kargil, and in 2002, India deployed hundreds of thousands of troops along its border with Pakistan following an attack on the Indian parliament in December 2001. Although international security analysts believe that the conflict in Kashmir is the one most likely to result in the use of nuclear weapons, because of the nationalistic fervor of the protagonists, some argue that the existence of nuclear weapons on both sides of the border has actually deterred an all-out war in much the same way the United States and the Soviet Union were deterred from engaging in armed conflict during the cold war period.

War and Reconstruction in Afghanistan

In the 1970s, political debate in Afghanistan became polarized between urban elites, who favored industrialization and democratic reforms, and rural conservative religious leaders, whose positions as landholders and ethnic leaders were threatened by the proposed reforms. Some urban elites allied themselves with the Soviets, who, fearing that a civil war in Afghanistan would destabilize the Central Asian states of the Soviet Union, invaded Afghanistan in 1979. This was still the cold war era, and the United States, Pakistan, and Iran gave enormous support to the anti-Soviet movement. The anti-Soviet forces (mujahedeen) were formed by rural conservative men—a collection of ethnic leaders (often labeled "warlords") and their followers who were strongly influenced by militant Islamist thought. The mujahedeen were tenacious fighters, and in 1989, after heavy losses, the Soviets gave up and left the country. Anarchy prevailed for a time as the Afghan factions fought each other, but the rural conservatives eventually defeated the reformist urban elites.

In the early 1990s, a radical religious-political-military movement, called the Taliban, emerged from among the mujahedeen. For the most part, the Taliban are young men from remote villages, many in southern and eastern Afghanistan. They are led by students from the *talibs* (Islamist schools of philosophy and law), but most followers have had little chance even to learn to read. The Taliban saw their role as controlling corruption and bringing stability and peace by strictly enforcing the **shari'a,** the Islamic social and penal code (see the explanation of the shari'a in Chapter 6, page 298). They viewed themselves as guardians of the social order and saw Western influence as so fundamentally

corrupting that it had to be rooted out. In the 1990s, the Taliban took particular aim at urban professional women, whom they saw as symbolic of Westernization. All women were forced into domestic seclusion; whenever they appeared briefly in public, they had to wear the completely concealing *burqa* (see the discussion of veiling in Islam in Chapter 6, page 303). Other efforts by the Taliban to purge their society of non-Muslim influences included restricting education for everyone, especially girls and young women; destroying 1500-year-old Buddhist sculptures in the Bamiyan Valley; and banning the production of opium, to which many Afghan men are addicted.

By 2001, the Taliban controlled 95 percent of the country, including the capital, Kabul, and were moving north, where an alliance of non-Pashto ethnic groups (the Northern Alliance) was mounting a counteroffensive. Meanwhile, an active resistance movement of Afghan exiles, many of them educated Pashto, labored against the Taliban in Pakistan, Europe, and America. The events of September 11, 2001, greatly assisted the resistance

---

PRACTICING GEOGRAPHY   ## Have You Seen the State?

Over the past few years I have become interested in how people see the state. This might seem like an odd topic, but think for a moment about how varied our sightings of anything can be. Sometimes we see a person or object directly. At other times we might see them on TV or in a photograph. We form other sightings from radio or the Web, as well as from conversations we hold with people. We also draw on rumors and memory. And most of the time we are forming our accounts through prior images, perceptions, and biases. Tom Stoppard had one of the characters in his play *Jumpers* ask out loud what the sun looked like after Copernicus. Did it still look round and yellow? Yes, of course. But it had come to look like a stationary object around which the earth moved in orbit, and no longer looked like an object in motion around the earth.

But what about the state? Ask yourself if you have seen the state today. Perhaps you have seen the state if you have been sitting in a government building, or on a college campus that is funded mainly by the taxpayers. Maybe you have seen or met a police officer, immigration officer, or social worker. Or maybe you have seen the state through a form you have to fill out. What I have become interested in is what we might call the microgeographies of the state, particularly in rural eastern India, where I have been working off and on for 25 years. Poorer men and women there see the state in all sorts of ways. They might find employment in a government food-for-work scheme, they might be harassed by a forest guard, and they might press their rights to a minimum wage in a court of law. Perhaps most of all, though, people tend to see the state in bureaucratic offices that bring together the policing, revenue-collection and developmental functions of government. The state can have an everyday and

localized quality to it, and it is often approached with a degree of concern, if not fear, and with the help of intermediaries.

It seems to me that if we look carefully at how people see and meet the state we come to learn a good deal about ideas of hierarchy, power, and citizenship in a given society. And the research tip here is simply to observe closely. Take something that is a commonplace and make

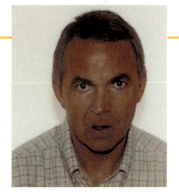

Stuart Corbridge
*Department of Geography and Environment,
London School of Economics*

it strange. Take, for example, standing in a queue. If I go to a barbershop in Miami or Delhi on a Saturday morning I might see eight other men in line before me. If I try to sit in the barber's chair one turn too early, I might politely be told by someone, "I'm sorry, it's my turn next." But if I step in two or three ahead of my turn someone will say, "Who do you think you are?" The queue, in other words, signals a resource that is scarce (ratio of chairs to customers) and requires an ethic of equality and respect for others. So what should we make of queue-jumping in India? Why do groups of men routinely push in front of women in office queues? And where else do we see this happening? (Check out Grand Central Station in New York City or London's King's Cross at rush hour, and you'll see that the problem is not confined to poorer or non-Western countries.) Is there a geography to waiting in line? Observations like these help illuminate broad patterns of interactions, including politics at the personal, regional, and national levels.

movement in overthrowing the Taliban. In an effort to root out Osama bin Laden and his Al Qaeda network, who were believed responsible for the terrorist attacks, the United States and its allies set out to topple the Taliban regime through a series of air strikes supported by Northern Alliance ground forces. By late 2001, the Taliban were overpowered. The United Nations stepped in to help establish an interim coalition government under the leadership of Hamid Karzai. In June 2002 a Loya Jirgha (national assembly) was convened to designate a new government, and Hamid Karzai was appointed head of state. In October 2003, the new government unveiled a draft constitution that was approved by another Loya Jirgha in early January 2004. Ratification of the new constitution marks an important step in the reconstruction of Afghanistan.

Postwar reconstruction and the establishment of democratic government in Afghanistan face a number of challenges, including the country's ethnic and linguistic diversity (see Figure 8.20 on page 429), regional disparity, its wealth, resources, and infrastructures, and a landscape devastated by war. Furthermore, the ability of the new government to ensure security and meet the needs of people outside Kabul has been thwarted by the continuing influence of the warlords. A report released by ACTIONAID (an antipoverty NGO based in the United Kingdom) in November 2003 indicated that the majority of Afghanistan's rural poor favor democratic government based on Islam, but feel that their concerns have not been adequately addressed. There is a critical need for basic services in rural areas that is not being met. Rural people are also concerned that corruption, crime, and factional fighting may hijack the democratic process.

The Future of Democracy

Although there are many political hot spots in South Asia as a whole, countries such as India and Sri Lanka have elected democratic governments regularly since their independence. Signs that democracy is expanding, thus making better government a possibility, include the recent peaceful creation of three new states in India and the fact that a more competitive multiparty system has been taking shape in that country. There, voters are increasingly intolerant of corruption and violence. Although expanding democracy has resulted in greater influence for Hindu nationalists in the short run (they were voted out of office in the spring 2004 elections), it has also led to fairer and more peaceful elections and a clearer focus on providing opportunities for women and other disadvantaged groups.

Although Pakistan has had elections, it functions as a military dictatorship. In Bangladesh, after years of military dictatorship, democratic elections have occurred with some regularity. In Nepal, a more freely elected legislature and multiparty democracy, intended to reduce the king's power and give ordinary citizens a greater political voice, were introduced in 1990. These political reforms, however, resulted in neither improved economic conditions for the Nepalese people nor real political change. So, since 1996, revolutionaries inspired by Maoist ideals have waged a "people's war" against the Nepalese state in an effort to destroy the monarchy. Peace talks between the rebels and the

government have twice broke down between 2001 and 2003, and the ensuing violence since August 2003 has terrorized the Nepalese. Eight thousand people have been killed, and another 100,000 to 200,000 have been displaced, rendering moot their new political rights.

In 2002, the tiny Himalayan kingdom of Bhutan granted its people the right to elect local representatives in government, and the next elections are scheduled for November 2005.

Finally, in Afghanistan, the first general elections following the removal of the Taliban were held in October 2004. Hamid Karzai became Afghanistan's first popularly elected president, winning 55.4 percent of the vote. The elections mark a significant step toward the construction of a democratic society.

ENVIRONMENTAL ISSUES

In 1973, in the Chamoli district of Uttar Pradesh, India, a sporting-goods manufacturer planned to cut down a grove of ash trees so that his factory, in the distant city of Allahabad, could use the wood to make tennis racquets. The trees were sacred to nearby villagers, however, and when their protests were ignored, a group of local women took a dramatic action that became a symbol of the struggle to protect South Asia's environmental quality. When the loggers came, they found the women hugging the trees and refusing to let go until the threat to their grove ended. Soon the manufacturer located another grove. The women's action grew into the **Chipko** (or social forestries) **movement,** which has spread to other forest areas, slowing deforestation and increasing ecological awareness.

Deforestation

Deforestation is not new to the Indian subcontinent. The spread of agriculture at the time of the Indus Valley civilization, as well as the expansion of newly developing Hindu kingdoms after the Aryan migrations, came at the expense of forested lands more than 3000 years ago. Ecological historians Madhav Gadgil and Ramachandra Guha have shown that the western regions of the subcontinent (from India to Afghanistan) became drier and drier as the forests vanished. They suggested that the early success of Buddhism and Jainism in these areas resulted, in part, from the focus of these two religions on vegetarianism, which possibly served as an adaptive response to a decline in game supplies as habitat was lost to deforestation. Under British colonialism, deforestation intensified once again. Starting in the mid-nineteenth century, perhaps a million trees a year were felled for use in building the railroad alone. Laxman Satya, another ecological historian, has shown that the massive deforestation of the nineteenth century contributed to the increasing aridity and heat of twentieth-century Deccan Plateau environments.

In the twenty-first century, the subcontinent's forests are still shrinking due to commercial logging and village populations expanding into forestlands for living and cultivation space. Tourism is also a factor. Forests are cleared to make trails for tourist trekkers in mountainous forest zones, such as those in Nepal, Kashmir, and the Nilgiri Hills of southern India; and trekkers consume wood for cooking and heat. Among the results of forest clearing are massive landslides that often close railroad lines and roads during the rainy season. Landslides increase erosion, and the eroded mountain environments are less able to absorb rainfall during future summer monsoons. Another result of forest clearing in mountainous areas is increased flooding and silt deposition in lowland places such as Bangladesh.

Unlike China and many African nations facing similar problems, South Asia has a healthy and vibrant climate of environmental activism that brings the consequences of deforestation to the attention of the public. Activism focused on saving forests is a reaction to a pattern found throughout South Asia, in which the resources of rural areas are channeled to urban industries without consideration of the needs of local rural people. The proponents of the social forestry movement argue that management of forest resources should be turned over to local communities. They say that people living at the edges of forests possess complex local knowledge of these ecosystems, gained over generations—knowl-

edge about which plants are useful for building materials, for food, for medicinal use, and for fuel. These people have the incentive to manage carefully because they want their progeny to benefit from forests for generations to come. In contrast, the government forestry departments typically shut local people out of their traditional forestlands, forcing them to depend on smaller and more marginal common-access lands that may now contain only a very few of the useful species.

The fact is, though, that burgeoning local populations themselves contribute to deforestation as they try to obtain ever more firewood for themselves and fodder for their animals. It is difficult to convince impoverished people to conserve resources. Their need for income predisposes them to collaborate with poachers of rare forest products. Moreover, the powerful industrial and government interests that monopolize forest reserves are not likely to yield control to local people easily. Despite these problems, several natural reserves have been established in India, and local residents are gaining increasing influence in decisions about control of these areas (see the box "A Visit to the Nilgiri Hills" below).

AT THE LOCAL SCALE A Visit to the Nilgiri Hills

As part of their research trip through southern India in the summer of 2000, Lydia and Alex Pulsipher visited several natural reserves in the Nilgiri Hills. These hills, part of the Western Ghats south of Mysore in Karnataka, harbor some of the last scraps of forest in southern India. One of the most successful of the Nilgiri reserves is Longwood Shola, a tiny remnant of ancient tropical evergreen forest. Here, two local naturalists, members of ethnic minorities native to the Nilgiri Hills, manage the 287-acre (116-hectare) park, which harbors 13 mammal species, 52 bird species, and 118 plant species, many of them found only in the Nilgiris. One bird species, the Grey Jungle Fowl, is ancestor to the domestic chicken. Tiny as it is, the forest of Longwood Shola protects three perennial streams that supply fresh water to 16 downstream villages. Among the projects the naturalists have initiated is a reforestation effort for which they themselves are growing the seedlings from local native species.

Phillip Mulley, a naturalist, Christian minister, and leader of the Badaga ethnic group, helps visitors to understand the issues that face the people native to the Nilgiri Hills. He noted that in addition to being a region of wildlife reserves, this part of India is the homeland of indigenous peoples (Badaga, Toda, Kota, Kurumba) who must now compete for space with a burgeoning tourist industry (1.2 million visitors a year in 1991) and huge tea plantations that have been established where forests stood until recently. Some of the newest tea plantations were cut out of forestlands by the state government to provide employment for Tamil refugees from the Tamil-Singhalese conflict in Sri Lanka (see page 440). So while one branch of the state government (the forestry department) and citizen naturalists try to preserve forest-

lands, another branch (social welfare), faced with a huge refugee population, is cutting them down.

[*Source:* Lydia and Alex Pulsipher, field notes, Nilgiris, June 2000.]

Increasingly, geographers are using digital photography and video in their research. This rare photo of a Bengal tiger encountered unexpectedly in the Mudumalai Wildlife Sanctuary in the Nilgiri Hills in southwestern India is from a mini–digital video filmed by Alex Pulsipher from the back of an elephant in June 2000. The Bengal tiger is an endangered species. [Alex Pulsipher.]

Water Issues

One of the most controversial environmental issues in South Asia today is the use of water. South Asia has more than 20 percent of the world's population, but only 4 percent of its fresh water. It is not surprising, then, that conflict exists between India and Bangladesh over access to the waters of the Ganga River (Figure 8.18).

Conflict over Ganga River Water. In recent years, during the dry season, India has diverted 60 percent of the Ganga's flow to Kolkatta to flush out channels where silt is accumulating and hampering river traffic. India's policies, however, deprive Bangladesh of normal water flow. Less water is available for irrigation, causing crop yields to fall, and salt water from the Bay of Bengal penetrates inland, ruining fields. The diversion has also caused major alterations in Bangladesh's coastline, damaging its small-scale fishing industry. Thus, to serve the needs of Kolkatta's 13 million people, the livelihoods of 40 million rural Bangladeshis have been put at risk, triggering protests in Bangladesh. In the late 1990s, India signed a treaty promising to reduce the scale of the diversions, but as of 2001, they had not been reduced. As with other major environmental problems, a solution has been hard to achieve because the population adversely affected is not only poor and rural, but is located in a different region—in this case, in a different country—from the politicians and bureaucrats who are in a position to respond to protests.

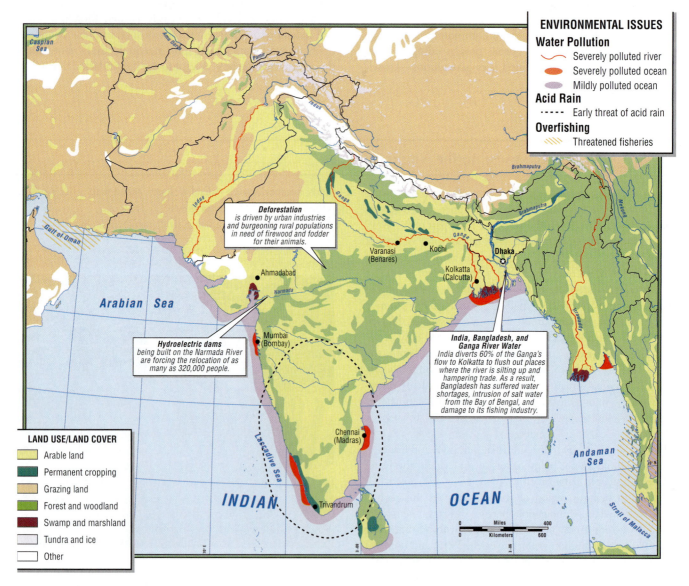

FIGURE 8.18 Environmental issues: South Asia. Environmental issues in South Asia are closely related to the ways in which land and water resources are used by an ever more densely settled population. Forests are cleared for industrial logging and for agriculture, increasing the likelihood of erosion. Modern agricultural methods and industrialization increase pollution in the water, soil, and air. This pollution, in turn, affects coastal zones, including fisheries. Government efforts to clear rivers for navigation, harness hydropower, and impound water for irrigation all affect downstream users, who are often poor and politically powerless. Yet, on a per capita basis, resource use is far less than in industrialized countries such as the United States, where each individual consumes 17 times the resources a South Asian does.

Similar water use conflicts occur between states within India and between the wealthier and poorer sectors of the population. Just 17 five-star hotels in Delhi use about 210,000 gallons (800,000 liters) of water daily, enough to serve the needs of 1.3 million slum dwellers. At the state level, Harayana diverts water from the Ganga River, depriving farmers downstream in Uttar Pradesh of the means to irrigate their crops.

Conflict over Dams. Hydroelectric dams are planned for many of India's rivers to supply the electricity needed to support a modern infrastructure and India's new information technology industries. An example is the Sardar Sarovar hydroelectric dam, one of 30 planned for the Narmada River in Madhya Pradesh (see Figure 8.18). The environmental problems posed by these dams have drawn national and international attention. As many as 320,000 people will have to be relocated to make room for new reservoirs. The law requires government agents to provide land of equal or better value to displaced people, but popular resistance hardened in the late 1980s when the land given to farmers to be displaced by the Sardar Sarovar Dam turned out to be barely arable. Facing starvation on these new lands, the farmers and their families returned to their old villages and fields in 1989 and eventually marched, 80,000 strong, on the capital in New Delhi; there they demanded that the project be halted. This protest and others culminated in a 1993 decision by India's Supreme Court to turn down World Bank loans of U.S. $450 million for the Narmada project. Nonetheless, after a brief hiatus, construction resumed, although legal challenges intermittently interrupted work on the dam as late as 2001. In resuming construction, the Indian government was bowing to pressure from the state government of Gujarat. Although the disputed dam is in Madhya Pradesh, most of the technology industries that would use the electric power and the farmers who stand to benefit from the irrigation waters provided by the project live in Gujarat (see Figure 8.24 on page 437). The "Save the Narmada" movement has persuaded the international funders of such large dam development projects to demand environmental impact studies more frequently. No such study was provided for the Narmada hydroproject.

Water Purity of the Ganga River. Water purity is an issue in historic religious pilgrimage towns such as Varanasi (Benares), where each year millions of Hindus come to die, be cremated, and have their ashes scattered over the Ganga River. The Ganga is believed to be a goddess that literally pours life and purifies the world. As the number of such final pilgrimages has increased, wood for cremation fires has become scarce, and incompletely cremated bodies are being dumped into the river, where they pollute water used for drinking, cooking, and bathing. In addition, dead cows are often consigned to the river. Recently, the government installed an electric crematorium with its own generator on the banks of the river. The cost is only 70 rupees per funeral, compared to the 2000 rupees for a traditional funeral pyre. The line at the electric facility is much longer than the one at the pyres; thus there is hope for reduced levels of unburned human remains in the river.

Of greater concern, however, is the amount of industrial waste and sewage dumped into the river. Most sewage enters the river in raw form because Varanasi's sewage system (built by the British early in the twentieth century) long ago exceeded its capacity. The

Women who must wash clothes in the watercourse flowing through Mumbai's Dharavi section, Asia's largest slum, have a difficult time because garbage and raw sewage pollute the water daily. [Steve McCurry/National Geographic Image Collection.]

danger posed by sewage is measured in terms of the river's **biochemical oxygen demand,** a figure that indicates the degree of oxygen consumption caused by decomposing matter such as fecal material or animal remains. The Ganga's biochemical oxygen demand is 375,000 times the level of safety. Pumps have been installed to move the sewage up to a new and expensive sewage processing plant built on the Varana, a small stream that enters the Ganga on the city's southern boundary. There the sewage is processed and released back into the Ganga. Unfortunately, the plant is so overwhelmed by the volume of water during the rainy season that it can process only a small fraction of the city's sewage. During the dry season, the municipal electrical supply is haphazard at best. Thus it seems that traditional Western technologies for sewage treatment are unsuited to the extreme conditions of India.

Veer Bhadra Mishra, a Brahmin priest and professor of hydraulic engineering at Banaras Hindu University in Varanasi, is on a mission to clean up the Ganga using unconventional methods.

He is working with engineers from the United States to build a series of processing ponds that will use India's heat and monsoon rains to clean the river at half the cost of more technologically sophisticated methods. In addition, he preaches a contemporary religious message to the thousands who visit his temple on a bank of the sacred Ganga. The belief that the Ganga purifies all it touches leads many Hindus to think that it is impossible to damage this magnificent river. Mishra reminds them that because the Ganga is their symbolic mother, it would be a travesty to smear her with sewage and industrial waste.

Industrial Pollution

Elsewhere in the region, the air as well as the water may be endangered by industrial activity. Emissions from vehicles and coal-burning industries are so bad that breathing Delhi's air is equivalent to smoking 20 cigarettes a day. The acid rain caused by industries up and down the Yamuna and Ganga rivers is destroying good farmland and such great monuments as the Taj Mahal — which, scientists say, has "marble cancer": its soft, intricately carved surfaces are becoming increasingly pockmarked. M. C. Mehta, a Delhi-based lawyer, became an environmental activist partly in response to the condition of the Taj Mahal. For more

than 20 years, he has successfully promoted environmental legislation that has removed hundreds of the worst polluting factories from the river valleys. His efforts are also a response to a horrible event that took place in central India in 1984, when an explosion at a pesticide plant in Bhopal produced a gas cloud that killed at least 3000 people and severely damaged the lungs of 50,000 more. The explosion was largely the result of negligence on the part of the U.S.–based Union Carbide Corporation, which owned the plant, and the local Indian employees who ran it. In response to the tragedy, the Indian government launched an ambitious campaign to clean up poorly regulated factories.

MEASURES OF HUMAN WELL-BEING

We have emphasized throughout this book that gross domestic product (GDP) per capita (Table 8.1, column 2) is at best a crude indicator of well-being and is best used in conjunction with other measures. GDP per capita in South Asia is nearly as low as it is in Africa (and in some cases lower), and a large proportion of people in South Asia are extremely poor. Most people in this region,

TABLE 8.1   Human well-being rankings of countries in South Asia and other selected countries

Country (1)	GDP per capita, adjusted for PPP[a] in 2001 $U.S. (2)	Human Development Index (HDI) global rankings, 2003[b] (3)	Gender Empowerment Measure (GEM) global rankings, 2003[c] (4)	Female literacy (percentage), 2001 (5)	Male literacy (percentage), 2001 (6)	Life expectancy, 2001 (7)
Selected countries for comparison						
Japan	25,130	9 (high)	44	99	99	81
United States	34,320	7 (high)	10	99	99	77
Mexico	8,430	55 (medium)	42	89.5	93.5	73
South Asia						
Afghanistan[d]	700	—	—	21	51	47
Bangladesh	1,610	139 (medium)	69	30.8	49.9	61
Bhutan	1,536[e]	136 (medium)	—	28[f]	56[f]	63
India	2,840	127 (medium)	—	46.4[g]	69	63
Maldives	2,082	86 (medium)	—	97	97	67
Nepal	1,310	143 (low)	—	25.2	60.5	59
Pakistan	1,890	144 (low)	58	28.8	58.2	60
Sri Lanka	3,180	99 (medium)	67	89.3	94.5	72

[a]PPP = purchasing power parity.
[b]The high, medium, and low designations indicate where the country ranks among the 175 countries classified into these three categories by the United Nations.
[c]Total ranked in world = 70; no data for many.
[d]Statistics for Afghanistan from *CIA World Fact Book*, 2003. Data for female literacy based on 1999 estimate; GDP based on 2002 estimate.

[e]In 1998. Data not available for 2001.
[f]In 1995. Data not available for 2001.
[g]The 2001 *Census of India* provisionally reports a female literacy rate of 54 percent.

Source: United Nations Human Development Report 2003.

however, are frugal and resourceful. Because they recycle nearly everything, their villages tend to be extremely clean. They are entrepreneurs in the informal economy, they grow their own food whenever possible, and they strictly limit cash expenditures through reciprocal exchange agreements with one another. Through such efforts, the people of South Asia manage to give themselves a somewhat higher standard of living than the GDP figures would indicate.

South Asia does not exhibit the wide variations of GDP among countries seen in some world regions (East Asia, Southeast Asia, and Oceania, for example). In fact, the one country with even a marginally higher GDP, the Maldives, has only a tiny population, whose income is increased by tourism. Sri Lanka's slightly higher GDP per capita results from a physical environment that is favorable to agriculture (tea is a chief product) and from the presence of exportable minerals. More important, a history of investing in the education and health of its citizens has helped Sri Lanka to equalize wealth distribution among its 19 million people. The country would undoubtedly be even more prosperous if conflict between the Tamil and Singhalese ethnic groups had not hindered development for several decades (see page 440).

The United Nations Human Development Index (HDI) (see Table 8.1, column 3) combines life expectancy at birth, educational attainment, and adjusted real income to arrive at a ranking of 175 countries. As of 2003, India (with 1 billion people) and

Bangladesh (with 147 million) had advanced to the lowest ranks of the medium range (those countries ranking between 56 and 141). This is a notable accomplishment because of the huge populations involved (more than one-sixth of the earth's population). Again, the rankings are based on averages, so there are still staggering numbers of people who remain very poor. With the exceptions of the Maldives and Sri Lanka, both literacy (columns 5 and 6) and life expectancy (column 7) are very low in this region, especially for women. And, as in Africa, governments have not provided even the most basic services. According to the United Nations, only 60 percent of the eligible children in India are in secondary school; in Pakistan, the figure is only one-fourth (1995); and in Afghanistan, just one-eighth (1995). In India, Bangladesh, and the Himalayan states of Nepal and Bhutan, clean water is available to only about 80 percent of the population. In the entire region, less than half of the people have access to sanitary toilets, including outhouses and pit toilets.

The United Nations Gender Empowerment Measure (GEM) (see Table 8.1, column 4) ranks countries by the extent to which women have opportunities to participate in economic and political life. For all but Bangladesh (ranked 69 out of 70), Pakistan (ranked 58), and Sri Lanka (ranked 67), these figures were not available in 2003 (worldwide, only 70 of 175 countries reported these data). It is reasonable to assume that the missing figures for the unranked countries would also be low.

Quick Review

1. Which two written languages were derived from Hindustani?

2. What is the difference between a dowry and a bride price?

3. What were a few of the techniques and technologies that brought about the green revolution?

4. Describe the role of Kashmir in India–Pakistan relations.

5. Why has the Taj Mahal been said to have "marble cancer"?

III   SUBREGIONS OF SOUTH ASIA

In this section, we subdivide the region of South Asia into subregions for closer examination. The subregions are grouped roughly according to their physical and cultural similarities. In several cases, parts of India are grouped with adjacent countries. This is true in the Himalayan region, in northeastern South Asia, and in the southernmost South Asian region, where parts of India and the country of Sri Lanka are treated as a subregion.

AFGHANISTAN AND PAKISTAN

Afghanistan and Pakistan (Figure 8.19) share location, landforms, and history, and they have both been involved in recent political disputes over global terrorism. Many cultural influences have passed through these mountainous countries into the rest of South Asia: Aryan migrations, Alexander the Great and his soldiers, the continual infusions of Turkish and Persian peoples, and the Turkish-Mongol influences that culminated in the Mughal invasion of the subcontinent at the beginning of the sixteenth century. Today, both countries are primarily Muslim and rural; 80 percent of Afghanistan's people and 67 percent of Pakistan's live in villages and hamlets. Both countries must cope with arid environments, scarce resources, and the need to find ways to provide rapidly growing populations with higher standards of living. Both countries are home to conservative Islamist movements. Afghanistan was ruled until late 2001 by a fundamentalist religious group, the Taliban; Pakistan is ruled by a military dictatorship that needs the support of fundamentalists to stay in power. Afghanistan's more extreme poverty is made worse by the armed strife from which it has suffered for more than 20 years.

FIGURE 8.19 The Afghanistan and Pakistan subregion. In addition to location and landforms, Afghanistan and Pakistan share a number of cultural influences and contemporary challenges to development.

The landscapes of Afghanistan and Pakistan are best imagined in the context of the ongoing tectonic collision between India and Eurasia. At both ends of the Himalayas, the collision uplifted curved crinkles in the Eurasian landmass. The lofty Hindu Kush, Pamir, and Karakoram mountains of Afghanistan and Pakistan are the western manifestation of this buckling (see Figure 8.2 on pages 390–391). This system of high mountains and intervening valleys swoops away from the Himalayas and bends down to the southeast toward the Arabian Sea. Landlocked Afghanistan, bounded by Pakistan on the east and south, Iran on the west, and Central Asia on the north, is entirely within this mountain system. Pakistan has two contrasting landscapes: the north, west, and southwest are in the mountain and upland zone just described; the central and southeastern sections are arid lowlands watered by the Indus River and its tributaries.

Afghanistan

The Hindu Kush Mountains in the north of Afghanistan fan out into lower mountains and hills and then into plains to the north, west, and south. In these gentler but arid landscapes, characterized by steep, sparsely vegetated meadows and pasturelands, most of the country's people struggle to earn a subsistence living from grazing animals and some cultivation. The main food crops are wheat, fruit, and nuts. Historically, opium poppies, native to this region, have been an important cash crop.

As of mid-2003, there were more than 28 million Afghans, 43 percent of whom were 14 years of age or younger. Life expectancy is only 46 years, yet the population is growing by 2.4 percent per year. Literacy in Afghanistan is among the lowest on earth. Only 20 percent of women and 50 percent of men over age 15 can read (see the vignette on page 410). Women bear six children on average. Generally, low life expectancy, low literacy, and a high birth rate are the markers of an extremely poor country.

The less mountainous regions in the north, west, and south are associated with Afghanistan's three main ethnic divisions (Figure 8.20). In the northwest are the ethnic groups who share culture and language traditions with the Turkmen and Uzbeks of Central Asia. In the northeast are the Tajiks and in the west the Hazara, both of whom are closely aligned with Iranian culture and languages. In the south, the Pashto-speaking Pathans (Pashtuns) are culturally akin to groups farther south across the Pakistan border. The various ethnic groups have remained separate and competitive, but there is significant variation within ethnic (often called tribal) groups, especially regarding views toward religion, education, and gender roles.

The ethnic diversity of Afghanistan and its neighbors has thwarted many previous efforts to unite the country under one government. Contrary to media reports, Afghanistan's ethnic groups have not been at continuous war with one another, but the events of the last several decades (see discussion on pages 420–422) have disturbed long-standing relationships and feelings of trust. More important, the sheer devastation of two decades of war has left the country crippled. Although the traditional rural subsistence economy of Afghanistan proved remarkably resilient and self-sufficient in the face of ongoing civil strife, the sufficiency of that system has been compromised by an ongoing drought that threatens over a million people with starvation.

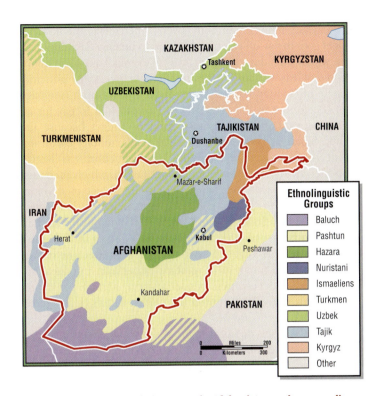

FIGURE 8.20 Ethnolinguistic groups in Afghanistan and surrounding countries. This map of the Central Asian vicinity of Afghanistan illustrates how, over the millennia, particular groups have been divided and moved to different places. In the twentieth century, the Soviet Union moved people to facilitate agricultural and industrial projects. The Taliban caused further dislocations in the 1990s. Those now seeking to form a democratic government in Afghanistan must deal delicately with the leaders of these different ethnic groups, often labeled warlords. [Phillippe Rekacewicz and Cecile Marin, "A tangle of nations," *Le Monde diplomatique,* January 2000, www.monde.diplomatique.fr.]

Pakistan

Although not much larger in area than Afghanistan, Pakistan has five times as many people (149 million). Pakistanis also live primarily in villages. Some are sprinkled throughout the arid mountain districts associated with herding and subsistence agriculture, but it is the lowlands that have attracted most settlement. Here, the ebb and flow of the Indus River and its tributaries during the wet and dry seasons form the rhythm of agricultural life. The river brings fertilizing silt during floods and provides water to irrigate millions of cultivated acres during the dry season.

Pakistan enjoyed an overall annual economic growth rate of 6 percent in the 1980s and early 1990s. This growth was fueled substantially by agriculture in the Indus Valley basin, especially in the Pakistani Punjab area, where cash crops such as cotton, wheat, rice, and sugarcane are grown on large tracts of irrigated land. Pakistan has had irrigation systems for many thousands of years, but recent irrigation projects, encouraged by international funding, have overstressed the system. By the mid-1990s, waterlogged and salinized soil had drastically reduced production, and some fields were entirely barren.

Other problems are related to financial issues. In 1996, the International Monetary Fund withheld U.S. $70 million of a loan because of governmental mismanagement and apparent fudging on budgetary figures. Currently, over 70 percent of the national budget goes to pay off external debt and to finance Pakistan's military, which includes a standing army of over half a million troops and a nuclear weapons program, thought necessary primarily because of Pakistan's disputes with India over Kashmir (see page 420). Beginning in the early 1990s, Pakistan had to pay for an increasing proportion of its own military budget because the United States, which had funded Pakistan's military expansion for decades, reduced its support. This reduction was largely motivated by the collapse of the Soviet Union, which increased U.S. friendliness with India, and Pakistan's insistence on developing its own nuclear bomb. Following September 11, 2001, however, the United States sought to repair relations with Pakistan, offering significant military and social aid in return for Pakistan's support against the Taliban in Afghanistan. Partly because of Pakistan's huge investment in its military, rather than in education, health care, and economic modernization, annual per capita GDP is only U.S. $1890—lower than either India's (U.S. $2840) or China's (U.S. $4020).

Most Pakistanis do not reap the benefits of the country's overall economic growth. Two-thirds of the population live on less than U.S. $2 a day. The productive farmland of Punjab is owned by a small elite. Although textile and yarn-making industries are growing around the cities of Lahore and Karachi, providing jobs for rural workers, the wealth generated is not passed on to the workers in the form of higher wages, and the reinvestment of profits in Pakistan is low. In addition, international drug dealing is a major activity, and corruption among officials is widespread. Arif Nizami, editor of the Lahore newspaper *The Nation*, is quoted by *New York Times* correspondent John Burns as saying: "Pakistan is a country where millions cannot get two square meals a day, yet the prime minister has a fleet of planes, flies to his place in the country in a personal helicopter, and lives in a palace [that would] shame the White House."

HIMALAYAN COUNTRY

The Himalayas form the northern border of South Asia (Figure 8.21). The regional map in Figure 8.2 (page 390) shows the mountainous zone running from west to east through the northern borderlands of India and continuing through Nepal and Bhutan. Notice that the country of India actually extends east in a narrow corridor running between Nepal and Bhutan to the north and Bangladesh to the south, then balloons out in a far eastern lobe containing many small states. Physically, this mostly mountainous region grades from relatively wet in the east to dry in the west because the main monsoons move up the Bay of Bengal and strike the east first and hardest (see Figure 8.3 on page 392).

Life in this region is dominated by the spectacular mountain landscape. This strip of Himalayan territory can be viewed as having three zones: (1) the high Himalayas, (2) the foothills and lower mountains to the south, and (3) a narrow strip of southern lowlands running along the base of the mountains. This band of lowlands is actually the northern fringe of the Indo-Gangetic Plain in the west and the narrow Assam Valley of the Brahmaputra River in the east. Although some people manage to live in the high Himalayas, most of the population lives in the foothills, where they cultivate subsistence strips and terraces in the valleys or herd sheep and cattle on the hills. This area is so rural in character that the Bhutan capital city of Thimphu, with just 27,000 inhabitants, is the largest town in that country; the capital city of Nepal, Kathmandu, has just 52,000 people.

FIGURE 8.21 The Himalayan country subregion. The Himalayas stretch across several state and national boundaries and link diverse cultural groups in a system of reciprocal trade.

AT THE LOCAL SCALE Salt for Grain and Beans

The Dolpo-pa people (pictured here) are yak herders and caravan traders who live in the high, arid part of Nepal. In that difficult environment, they can produce only enough barley and corn to feed themselves for half the year. Through trade, the Dolpo-pa parlay a half-year's supply of grain into enough food to feed them for a whole year. At the end of the summer harvest, they load a portion of their grain onto yaks and head for Tibet. There, they trade grain to Tibetan nomads in return for salt, a commodity in short supply in Nepal. They leave some of the salt in their village for their own use; the rest they carry on to Rong-pa villages in the central Nepal foothills. There they trade this salt for enough grain to last them through the winter. Bargaining is fierce, and it can take days before a price is agreed upon; prices average 3 to 5 measures of corn for 1 measure of salt. The trading season ends in November, so the Dolpo-pa usually winter over in a convenient field, paying for the privilege.

The Rong-pa people are sheep and goat herders, and salt is a necessary nutrient both for them and for their animals. In recent years, because of salt shortages in Tibet, the Dolpo-pa have not brought enough salt to meet all the Rong-pa's needs. The Rong-pa, therefore, load goats and sheep with bags of red beans and set out for Bhotechaur, where they meet Indian traders at a large bazaar and trade their beans for iodized Indian salt; a good price is 1 measure of beans for 3 measures of salt. Often they will also sell a sheep or two to buy cloth, or perhaps a copper pan, to take home. [*Source:* Adapted from Eric Valli, "Himalayan caravans," *National Geographic* (December 1993): 5–35.]

The Dolpo-pa people trading with Tibetan nomads. [Eric Valli.]

Culturally, the region is Muslim in the west and Hindu and Buddhist in the middle; indigenous beliefs are important throughout, but are especially strong in the far eastern portion. Throughout the subregion, but especially in valleys in the high mountains and foothills, indigenous cultures continue to live in traditional ways, isolated from daily contact with the broader culture (see the FRONTLINE/World video vignette on page 416). There are many of these groups in such places as the Indian state of Arunachal Pradesh, at the eastern end of the region; there, a population of less than 1 million speaks more than 50 languages. Despite language differences, the Himalayan people have learned to survive in their difficult mountain habitat by relying on one another. An example of this reciprocity between cultures comes from central Nepal, where two indigenous groups engage in a complicated cycle of trade that links them with both Tibet and India (see the box "Salt for Grain and Beans" above).

Most people in the Himalayan subregion are very poor. Statistics for the various Indian states in this region hover near those of Nepal, which has an annual per capita GDP of just U.S. $1310. In Nepal, the average life expectancy is 59, literacy is about 43 percent, and some 65 percent of the children under 3 years of age are malnourished. At present, only 18 percent of the land is cultivated, and the country's high altitude and convoluted topography make agricultural expansion unlikely. The government's strategy, therefore, is to improve crop productivity and lower the population growth rate.

The Indian state of Arunachal Pradesh is one of the region's more prosperous areas. Moist air flowing north from the Bay of Bengal brings plentiful rain as it lifts over the mountains. This is one of the most pristine regions in India. Forest cover is abundant, and a dazzling array of flora and fauna occupies habitats in descending elevations: glacial terrain, alpine meadows, subtropical mountain forests, and fertile floodplains. Conditions at different altitudes are so good for cultivation that lemons, oranges, cherries, peaches, and a variety of plants native to South America—pineapple, papaya, guava, beans, maize, and potatoes—are now grown commercially for shipment to upscale specialty stores.

Over the past 20 years, the Himalayan subregion has experienced numerous changes brought by the increasing numbers of tourists trekking and climbing the mountains and seeking spiritual enlightenment at the numerous holy sites. Tourism in Nepal, in particular, has been a mixed blessing. As in Southeast Asia (see the box "Tourism Development in the Greater Mekong Subregion" in Chapter 10, pages 518–519), it has created economic opportunities for some people, but its impact on the landscape has been devastating in other ways.

NORTHWEST INDIA

Northwest India stretches almost a thousand miles from the Punjab-Rajasthan border with Pakistan to somewhat east of the famous Hindu holy city of Varanasi on the Ganga (Figure 8.22). It is dry country, yet contains some of the wealthiest and most fertile areas in India.

In the western part of this subregion, there is so little rainfall that houses can safely be made of mud with flat roofs. Widely spaced cedars and oaks are the only trees, and the landscape has a dusty khaki color. Yet filling the landscape between the trees are fields of barley and wheat, potatoes, and sugarcane, plowed by villagers using oxen and humpbacked cattle. Along the northern reaches of this region (Punjab and Uttar Pradesh), the rivers descending from the Himalayas compensate for the deficient rainfall. The most important is the Ganga; its many tributaries flow east and water the whole of Uttar Pradesh, bringing not only moisture but fresh silt from the mountains.

The western half of the region contains one of India's poorest states—Rajasthan—as well as its wealthiest—the agriculturally productive Indian part of Punjab. Rajasthan, with only a few fertile valleys, is dominated by the Thar (Great Indian) Desert in the west, which covers more than a third of the state. Historically, small kingdoms were established wherever water could be contained (Rajasthan means "land of kings"). Seminomadic herders of goats and camels still cross the desert with their animals. Perhaps the best known are the Rabari (see the photo on the next page). Originally a caste of camel herders and dung gatherers, today the Rabari, about 250,000 strong, continue their annual migrations during the dry season in search of green pastures. As they travel, they sell or trade animal dung for grazing rights, thus keeping farmers' fields fertile and their fires burning. By the 1990s, however, there were more and more small farmers occupying former pasturelands. Although these farmers want the dung, they cannot afford to lose one bit of greenery to the passing herds, so increasing population pressure means that the benefits of reciprocity between herders and farmers are being lost.

In arid Rajasthan, less than 1 percent of the land is arable, yet agriculture, poor as it is, produces 50 percent of the state's domestic product. It is not surprising that the annual per capita GDP is just U.S. $500. The crops include rice, barley, wheat, oilseeds, peas and beans, cotton, and tobacco. A thriving tourist industry, focused on the palaces and fortresses constructed by warrior princes of the past who fought off invaders from Central Asia, accounts for much of the other half of the GDP.

Punjab, one of India's most prolific agricultural states, has an annual per capita GDP of U.S. $1105, double that of Rajasthan. The differences in wealth result partly from differences in physical geography and the introduction of green revolution technology. Rajasthan is a desert; Punjab receives water and fresh silt carried down by rivers from the mountains. Although Rajasthan is about three times the size of Punjab, only 1 percent of its land is cultivated. Nearly 85 percent of Punjab's land is cultivated, and it is particularly productive. In 1994 and 1995, Punjab alone provided 62 percent of India's food reserves. The typical crops are maize, potatoes, sugarcane, peas and beans, onions, and mustard.

Although agriculture is the most important economic activity throughout the region and employs about 75 percent of the people, industry is also important in the union territory (city-state) of

FIGURE 8.22 The Northwest India subregion. This subregion encompasses the Indian states of Rajasthan, Punjab, Haryana, and Uttar Pradesh, and the union territory (city-state) of Delhi. It stretches almost a thousand miles, and, although dry, contains some of the most fertile agricultural land in India.

In the Rabari village of Bhopavand, near the border of Rajasthan and Gujarat, camels are taken out to graze. Many Rabari still lead a partially nomadic life, traveling northward through Rajasthan and across the Thar Desert in the dry season. [Dilip Mehta/Contact Press Images.]

Delhi and on the Ganga Plain. Here, as elsewhere in the region, typical industries are sugar refining and the manufacture of cotton cloth and yarn, cement, and glass. People also make craft items and hand-knotted wool carpets. Urban areas have more modern forms of industry. In the city of Chandigarh, for example, there are 15 medium-sized and large-scale industrial facilities that produce electronic and biomedical equipment, household appliances, tractor parts, and cement tiles and pipes.

The city of New Delhi, India's capital, has about twice the population of Washington, D.C., and is located approximately in the center of Northwest India. The city center was built by the British in 1931 just south of the old city of Delhi—an important Mughal city—amidst the remains of seven ancient cities. It has all the monumental hallmarks of an imperial capital—and all the problems one might expect of a big city in a country as poor as India. The Delhi metropolitan area (including the old and new cities and outlying suburbs) has more than 13 million people and attracts a continuing stream of migrants. Most of the new arrivals have left nearby states to escape conflict or poverty, or both. The city is also home to refugees who fled Tibet to escape Chinese oppression and to others who fled the Afghan-Soviet war of the 1980s. Annual per capita GDP in Delhi (U.S. $1335) is below average for India, but actual incomes for most people are lower still; Delhi's tiny minority of the extremely wealthy pulls up the average. The low literacy rate for an urban area—just 82 percent—is partly a result of the continual arrival of migrants from poor rural areas, but Delhi also has far fewer schools than it needs for its population. In fact, the city has difficulty providing even the most basic services: there are insufficient water, power, and sewer facilities, and 75 percent of the city structures violate local building standards. Many people have no buildings to inhabit at all; like many of the poor worldwide, they live in shanties constructed from refuse.

Pollution is a particular problem: Delhi has been designated the world's fourth most polluted metropolitan area. The annual death rate from pollution-related illness has reached 7500. Most of the pollution comes from the more than 3 million unregulated motor vehicles. Taxis, trucks, buses, motorized rickshaws, and scooters, most without pollution control devices, all compete for space, for cargo, and for passengers.

Despite this grim picture, life in Delhi is vibrant and upbeat, in part because the middle class is growing and many of the poor sense that there is a chance for upward mobility. The global market economy, evident across the nation, is especially visible in cities like New Delhi, where clubs, department stores, boutiques, cinemas, and video parlors share space with traditional businesses, and where brand names such as McDonald's, Louis Vuitton, Nike, Baskin-Robbins, and Compaq are common.

NORTHEASTERN SOUTH ASIA

Northeastern South Asia, in strong contrast to Northwest India, has a wet tropical climate. This subregion bridges national and state boundaries, encompassing the states of Bihar, Jharkhand, and West Bengal in India; the country of Bangladesh; and the far eastern provinces of India—all clustered at the north end of the Bay of Bengal (Figure 8.23). Its dominant shared features are two rivers, the Ganga and the Brahmaputra, and the giant drainage basin and delta region created by these rivers. The wet climate and fertile land have nourished a population that is now among the densest on earth, often struggling to support itself in the overcrowded conditions.

The Ganga-Brahmaputra Delta is the largest delta on earth. Every year, the two rivers deposit enormous quantities of silt, building up the delta so that it extends farther and farther out into the bay. The rivulets running into the bay repeatedly change course, until the bay is periodically flushed out by a huge tropical storm (a cyclone in 1991 left 130,000 delta dwellers dead). The people of the delta have learned never to regard their land as permanent (see Figure 8.5 on page 394). Their dwellings, means of transport, and livelihoods are adapted to drastic seasonal changes in water level and shifting deposits of silt. Villages on river terraces are clusters of houses on mounds, surrounded by lush green forests and fields. The houses are intricately woven from bamboo and palm material, and have sloping thatched roofs able to withstand heavy rainfall. In the lowlands, the houses are usually raised on posts above the high-water line. Transport is by small boats; people fish during the wet season and farm when the land emerges from floods. Even beyond the delta, most people have to cope with flooding at some time of the year, either when the summer monsoon rains come or, in the foothills of the Himalayas, when the spring snowmelt in the mountains swells rivers beyond their banks.

The Case of West Bengal

With a population of 80 million packed into an area slightly larger than Maine, West Bengal is India's most densely occupied state: 2342 people per square mile (904 people per square

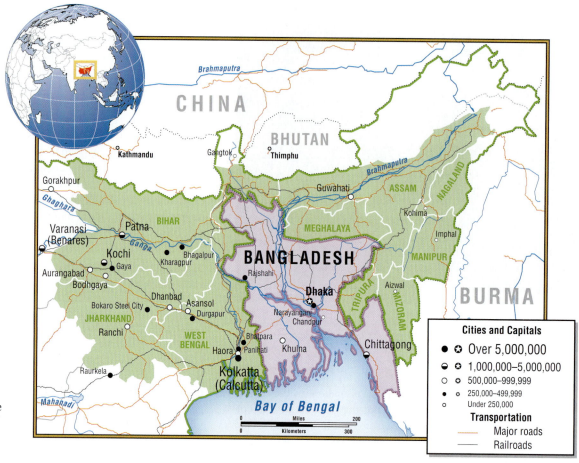

FIGURE 8.23 The Northeastern South Asia subregion. This subregion bridges national and state boundaries, encompassing the states of Bihar, Jharkhand, and West Bengal in India; the country of Bangladesh; and the far eastern provinces of India—all clustered at the north end of the Bay of Bengal.

kilometer). Its population has been swelled by refugees from eastern Bengal (now part of Bangladesh) since Partition in 1947, and even more immigrants have come in since the Chinese takeover of Tibet and the Pakistani civil war that gave Bangladesh its independence. Today, better employment opportunities in West Bengal and the possibility of farming in the eastern Indian states draw a continual flow of Bangladeshi migrants (most of them illegal). The result is a mix of cultures that are frequently at odds: there are often Hindu and Muslim religious demonstrations and disputes with indigenous people over land occupied by new migrants.

Nearly 75 percent of the people in this crowded state earn a living in agriculture, yet agriculture is not a major contributor to West Bengal's GDP, accounting for only 35 percent of domestic production. In addition to growing food for their own consumption, many people work as rice and jute cultivators or tea pickers—all labor-intensive but low-paying jobs. Twenty-five percent of India's tea comes from West Bengal; the plant is grown in the far north of the state around Darjeeling, a name well known to tea drinkers. Each leaf must be selected and picked by hand, and it is often women and children who do this work. One result of a dense population dependent on agriculture is that continuous intensive cultivation often overstresses the land severely and soil fertility declines over time. In addition, surrounding woodlands are depleted by impoverished farm laborers who must gather firewood because they cannot afford kerosene.

Kolkatta (Calcutta), one of the most famous cities on earth, is in the delta region of West Bengal. This giant, vibrant city of 13 million people is known in the West for its legendary beggars and for Mother Teresa's ministrations to the poor at Nirmal Hriday (Home for Dying Destitutes). Bengalis, however, are proud of their two Nobel laureates (Rabindranath Tagore and Amartya Sen) and Academy Award–winning filmmaker Satyajit Ray. Kolkatta was built on a swampy riverbank in 1690 and served as the first capital of British India. But its sumptuous colonial built environment, substantially upgraded in the nineteenth century, has become lost in squatters' settlements that surround the city and have invaded its parks and boulevards. It is often characterized as a city in such a state of decline that nothing can be done to improve conditions. The reasons for Kolkatta's decline are economic and social, brought on by massive immigration since 1947. In addition, outmoded regulations limit incentives to start businesses and to improve private property. For example, apartment owners have little reason to make improvements when, under rent controls presently in place, a tenant can pay as little as U.S. $1.40 a month for a four-room apartment and pass the lease down to the next generation. On the other hand, for those only precariously in the middle class, rent hikes would be disastrous. Such conundrums produce an inertia in governance that the city's educated young people find stifling, and many are leaving Kolkatta—a situation that makes it difficult to revive the city.

This street scene in Kolkatta shows a small marketplace and several modes of transportation. Count them. Relics of European influence are visible in the architecture. Note the mosque in the background. [Steve McCurry/National Geographic Image Collection.]

Somraj Kundu, a 20-year-old who has been accepted for graduate studies at Oxford University in England, identifies four reasons he is leaving Kolkatta: the breakdown of traditional extended families; the much too rapid increase in population as immigrants pour in; corruption at many levels (one can bribe an official to get a passport or a professor to avoid receiving a low grade in a course); and, perhaps most important, the lack of opportunity for economic advancement for him and others like him.

Eastern India

Eastern India has two physically distinct regions: the river valley of the Brahmaputra as it descends from the Himalayas, and the mountainous uplands stretching south of the river between Burma on the east and Bangladesh on the west. Although migrants have recently arrived from across South Asia, traditionally this region has been occupied by ancient indigenous groups that are related to the hill people of Burma, Tibet, and China.

The Indian state of Assam encompasses the river valley of the Brahmaputra, eastern India's most populated and most productive area. Hindu Assamese make up two-thirds of the population of 29 million, and indigenous Tibeto-Burmese ethnic groups make up another 16 percent; the rest are recent migrants. To reduce the proportion and influence of the Assamese people, who have continually objected to being under central Indian control, the Indian government has made large tracts of land available to outsiders: for example, Bengali Muslim refugees fleeing the civil war in East Pakistan (now Bangladesh) in 1971, Nepali dairy herders, and Sikh merchants. There were violent disputes in the late 1970s and early 1980s between Assamese and the new settlers and between Assam and India. These disputes continue; in the fall of 2004, new outbreaks of violence resulted in more than 50 deaths.

More than half the people in Assam work in agriculture by growing and producing food; another 10 percent are employed on tea plantations or in forestry (forests cover about 25 percent of the land area). Two-thirds of the cultivated land is in rice, but tea is the main cash crop—Assam produces half of India's tea. By the 1990s, Assam's oil and natural gas accounted for more than half that produced in all of India. Given India's shortage of energy, this alone could explain India's efforts to dominate the Assamese politically.

Colorful names such as "Land of Jewels" and "Abode of the Clouds" convey the exotic beauty of the emerald valleys, blue lakes, dense forests, carpets of flowers, and undulating azure hills in the mountainous sections of eastern India that surround Assam. In these uplands, occupied largely by indigenous ethnic groups, people produce primarily for their own consumption, with 80 percent or more of the inhabitants making a living from the land. Many practice a particularly complex version of traditional tropical horticulture specifically adapted to the region; others cultivate rice on permanently terraced fields.

Although literacy rates are not high in most of the region, the state of Mizoram ranks second in India, at 88 percent, because of the influence of Christian missionary schools.

Bangladesh

The American James Novak, who has lived and worked in Bangladesh, writes:

Bangladesh is not so much a land upon water as water upon a land. One-third of Bangladesh's physical space of fifty-five thousand square miles is comprised of water in the dry season, while in the rainy season up to 70 percent is submerged. Water is the central reality of Bangladesh, just as its shortage is the central reality of Saudi Arabia. At least 10 percent of the people live in boats, up to 40 percent depend on the sea and rivers for a livelihood, and 100 percent depend on rain and floods for food. Water is the main source of protein [fish], the major provider of crop fertilizer and transport, and unquestionably the greatest source of wealth. Bangladesh's main crops—rice, jute, and tea—cannot exist without huge amounts of water.

(James J. Novak, *Bangladesh: Reflections on the Water*, 1993, pp. 22–23.)

Today, Bangladesh is one of South Asia's poorest countries. It is also one of the region's most populous democracies and the world's most densely populated agricultural nation. More than 146 million people live in an area slightly smaller than Alabama. Population density is 2639 people per square mile (1019 per square kilometer)—compared to 78 people per square mile (30 per square kilometer) in the United States—yet all but 21 percent of the people live in rural areas, trying to manage as farmers on severely overcrowded land. Some 50 million people live below the poverty line, meaning that their daily caloric intake is below 2122, the minimum standard for adults.

Although desperately poor, Bangladesh is better off today than it was just a few years ago. The percentage of rural people living in poverty dropped 9 percent in the years from 1989 to 1994, according to the U.S. Agency for International Development. Higher literacy and the availability of contraceptives (54 percent of women use some form of birth control), both funded

not continuous; areas between patches are densely occupied by people in dispersed rural villages. The isolation of small populations of wild plants or animals in separate forest patches weakens their gene pools, and extinction becomes inevitable, even if it is somewhat delayed by protection. In recognition of this problem, there are now plans to reconstitute forest corridors between parks. Estimates of future human population growth, however, do not bode well for the future of wildlife anywhere in this subregion.

Central India is notable for its several industrial areas (see Figure 8.16 on page 414). The state of Gujarat in the west, on the Arabian Sea, has light industries in all of its major cities. Gujarat's service and industrial sectors account for 75 percent of its domestic production; the rate is even higher, at 83 percent, in the neighboring state of Maharashtra. Even the central plateau and eastern areas of Central India, although more rural in character, have pockets of industrial activity. An example is the city of Indore, a commercial and industrial hub whose residents think of it as a mini-Mumbai. Cotton textiles are the city's main product, although that industry is currently in decline. Nearby, the town of Pithampur, known as India's Detroit, houses several automobile plants, a steel plant, a container plant, and small-appliance factories. Industry and urbanization are clearly connected, and in fact, the western part of the region is exceptionally urbanized for India. In Gujarat, about 37 percent of the people live in urban areas; in

Called *nodi bhanga lok* ("people of the broken river"), the people who occupy the constantly shifting silt of the Ganga Delta region are looked down upon by more permanent settlers on slightly higher ground. Because they must often flee rising floodwaters, they are known to be less secure financially and are thought to lack the qualities of thrift and good citizenship that come from living in one place for a lifetime. The delta floods come from the Brahmaputra and from storms that sweep up the Bay of Bengal. [James Blair/National Geographic Image Collection.]

primarily by foreign aid, have brought a significant reduction in fertility rates, from 7 children per woman in 1974 to 3.6 in 2003. Infant mortality has also dropped, from 128 per 1000 in 1986 to 66 per 1000 in 2003. Economically, there are signs that the textile industry, once the source of considerable wealth in Bangladesh, is reviving. Bangladesh now ships more than U.S. $2 billion worth of garments to the United States and Europe annually. Although Bangladesh remains in distress, there is no reason to conclude that further progress is not possible. In fact, Bangladeshi innovation in microcredit is contributing to progress among the poor in many countries through the model provided by its Grameen Bank (discussed on pages 416–417).

CENTRAL INDIA

The Central India subregion stretches across the widest part of India, from Gujarat in the west to Orissa in the east (Figure 8.24). It contains India's last untouched natural areas as well as much of its industrialized area. The Narmada River, site of many hydroelectric dams (see page 425), flows across the region and empties into the Gulf of Cambay.

Some of India's most significant environmental battles are being fought in the central highlands of this subregion. Central India has most of India's remaining forest cover and is home to a concentration of national parks and sanctuaries (most notably the tiger reserves). Yet already the forest cover is merely patchy,

One day during a recent visit to India, Lydia and Alex Pulsipher visited Koli, a fishing village located on the bay in the heart of Mumbai. At sunset, the fishing crews often retire to their boats to celebrate the day's events. The Koli were the fisherfolk living on the seven islands in 1600 when the city was founded. The village is now a crowded low-rise labyrinth between a busy urban thoroughfare and the bay. Many people in the village work at bureaucratic jobs in the city. Although the village had the look of a shantytown and was one of the more densely occupied places we visited, its 400 years on the site and the relative affluence of the inhabitants gave the village an air of permanence. More than a few living quarters boasted TVs and white marble floors, and we saw an occasional computer. [M. M. Navalkar/Dinodia.]

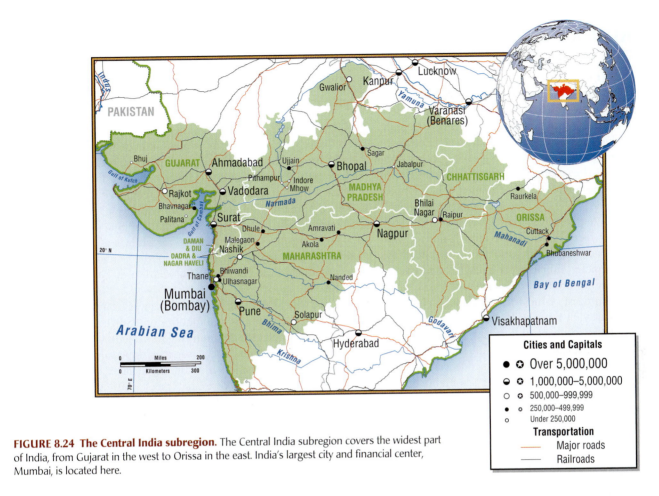

FIGURE 8.24 The Central India subregion. The Central India subregion covers the widest part of India, from Gujarat in the west to Orissa in the east. India's largest city and financial center, Mumbai, is located here.

Maharashtra, the location of the megacity Mumbai, the figure is about 43 percent.

Even though this region might be considered newly industrializing, it too can lose industries if it fails to compete effectively in the mobile global economy. For example, the city of Ahmadabad in Gujarat, with 4.5 million people, used to be known as the Manchester of India because of its large textile mills. Although the Arvind denim mills are still in operation, most of the old mills are closed today, having lost out to newer mills and cheaper labor elsewhere in Asia. But another aspect of globalization may help Gujarat keep its economic edge. Many Gujaratis live abroad in a worldwide diaspora of merchants and businesspeople who maintain ties with their homeland. The diaspora could prove to be a major source of foreign investment for Gujarat.

Bombay is the name by which most Westerners know Maharashtra's capital, but since 1995 its official name has been Mumbai, after the Hindu goddess Mumba. In the sixteenth century, when a local sultan gave the bay, with its seven small islands, to the Portuguese, it became known as Bom Bahia in Portuguese, or "Beautiful Bay." Eventually the British joined the islands with bridges and landfill and built the largest deepwater harbor on India's west coast. Mumbai, with over 16 million people, is now India's most prosperous city. It hosts India's largest stock exchange and the nation's central bank. It pays about a third of the taxes collected in the entire country and brings in nearly 40 percent of

India's trade revenue; its annual per capita GDP is three times that of Delhi. Yet this wealth is not readily visible. By some estimates, lack of housing has reduced half the population to living on the streets. Real estate developers have a booming trade in high-rise condominiums built for the city's rapidly growing middle class on land reclaimed from the sea.

Mumbai is also home to Asia's largest slum. The community of Dharavi houses more than 600,000 people on less than 1 square mile, and most of those people have no plumbing. People work at three or four jobs, and the community is known for its inventive entrepreneurs. One young man, for example, collects and sells aluminum cans that once held ghee, India's form of butterfat. He says he makes about 15,000 rupees a month (U.S. $480), nearly twice the salary of the average college professor in India and much more than he made as a truck driver. Hence, despite widespread poverty and ethnic and religious tensions, Mumbai has more than a few success stories.

Mumbai is known popularly in India as "Bollywood" because it produces popular Hindi movies portraying love, betrayal, and family conflicts. The stories, played out on lavish sets and accompanied by popular music and dance, serve to distract the huge audiences from the physical difficulties of daily life, at least temporarily. Mumbai produces many more films than Hollywood, on much smaller budgets. The stars make six or more films a year and are so popular that movie posters are everywhere, in public and private spaces alike.

SOUTHERN SOUTH ASIA

Southern South Asia (Figure 8.25) resembles the rest of South Asia in that the majority of people work in agriculture—ranging from about 70 percent in the west to somewhat more than 50 percent in the east. But the region is set apart by its relatively high proportion of well-educated people, its advanced technology sectors, its strong tradition of elected communist state governments (primarily in Kerala), its focus on environmental rehabilitation and preservation, and the higher status of women. The cultural mix here also sets it apart. It is the center of ancient Dravidian cultures and languages that predate Aryan influences.

This part of India receives consistent rainfall and is well suited for growing tobacco and rice; also grown are peanuts, chilies, limes, cotton, spices (cinnamon and cloves), and castor-oil plants (the source of an intestinal medicine and a skin lubricant). The southwestern coast of India (the Malabar Coast) is a narrow coastal plain backed by the Western Ghats. The sea-facing slopes of these mountains, just back of the coast, are some of the wettest in India; they support teak, rosewood, and sandalwood, all highly valued furniture woods. Small parts of the Deccan Plateau, a series of uplands to the east, are also forested (see Figure 8.2 on page 391). Here, dry deciduous forests yield teak, eucalyptus,

cashew, and bamboo (see the box "A Visit to the Nilgiri Hills" on page 423). Several large rivers and numerous tributaries flow toward the east across this plateau and form rich deltas along the lengthy and fertile coastal plain facing east to the Bay of Bengal.

The Case of Kerala

Kerala is a well-watered, primarily coastal state in far southwestern India. It is often cited as an aberration on the Indian subcontinent because its people enjoy a higher standard of well-being than the rest of the country. Literacy rates are the highest in the whole of South Asia, at close to 90 percent (see the discussion of Kerala in the section on population patterns, page 402). Women enjoy relatively high status and outnumber men in the population—there are no "missing females" as elsewhere in India. Moreover, women actively participate in the economy and are far less secluded than elsewhere. Standards of health are also better in Kerala.

Just why Kerala stands out in these ways is not entirely understood, but the state has a unique history. Since the 1950s, it has had a series of elected communist governments that have strongly supported broad-based social services, especially education for both males and females. In addition, it has a long history of a traditional family structure that gives considerable power to women; for example, a husband often resides with his wife's

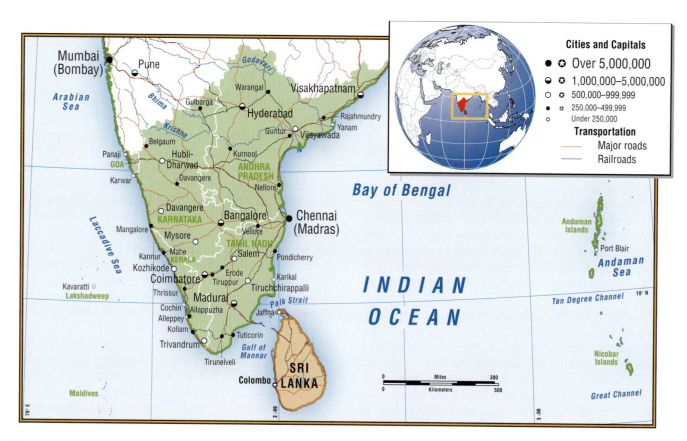

FIGURE 8.25 The Southern South Asia subregion. The Indian states of Andhra Pradesh, Karnataka, Kerala, and Tamil Nadu, together with Sri Lanka, constitute this subregion. As the center of ancient Dravidian cultures and languages that predate Aryan influences, this subregion stands apart from the rest of South Asia.

AT THE LOCAL SCALE   Kerala's Fishing Industry

Geographer Holly Hapke has studied how recent structural changes in the fishing industry of Kerala have altered family economics among Christian fisherfolk in and around Trivandrum, in the far south. Before independence in 1947, fishing in Kerala was carried out almost entirely on a small scale. Thousands of fishers worked from boats that held crews of 1 to 40 men. The catches were small and consisted of multiple species of fish. Once the boats had returned, the wives of the fishermen sorted and cleaned the fish, then sold them on the beach to wholesalers or in town at the market. Unlike most women elsewhere in South Asia, Kerala fishwives regularly dealt with the public as businesswomen; at the same time, they performed their multiple household and child-rearing tasks.

In the 1950s, the Indian government hoped to improve the incomes of Kerala's fishers by increasing the catch for the local market. It introduced large mechanized vessels to villages north of Trivandrum. Very soon, the project shifted to harvesting prawns for the world market, but mechanization also spread to other fisheries along the coast. By the mid-1970s, the mechanized fleet was encroaching on traditional fishers, damaging their gear, and competing for their catch. The intensification of fishing has led to declining fish stocks and pronounced decreases in income for Trivandrum's fishing villages, which continue to use traditional technology. Men no longer harvest sufficient fish, so the women no longer have fish to clean and sell. Although the men occasionally find wage labor in the mechanized fishing industry, fishermen are no longer bringing in sufficient income. So families have become increasingly dependent on the women's earnings. The women, therefore, have started buying fish from male wholesalers and then reselling them in the market or to regular customers, all the while keeping up with their household duties. To compete as fish vendors against the larger operators, the women must themselves deal in larger volumes; this requires investment cash, which they must borrow at the high rates charged by moneylenders. The exorbitant interest depletes the women's earnings, leaving them operating on the margins, with their family economies at risk.

Hapke concludes that state planners ignored women's roles in processing and marketing the fish catch; when the system was reorganized, women were left out entirely. They must now bear more of the responsibility for their family incomes, but they have little access to institutional and societal supports. Despite their considerable entrepreneurial spirit, their situation is likely to worsen as the traditional sector disappears.

[*Source*: Adapted from Holly M. Hapke, *Fish Mongers, Markets, and Mechanization: Gender and the Economic Transformation of an Indian Fishery* (Syracuse, N.Y.: Syracuse University, 1996).]

Women sell fish in a Trivandrum neighborhood market. [Holly Hapke.]

---

family (a pattern also found in Southeast Asia), instead of the reverse. The concept of female seclusion, while present, is less stringently practiced; women are often on the street, and it is not unusual to see a lone woman going about her shopping duties. Agriculture employs less than half the population; instead, remittances from migrants to the Middle East figure prominently in the state's economy, and fishing is an important economic sector as well (see the box "Kerala's Fishing Industry" above).

Another feature of life in Kerala that may have added to the openness of society and greater freedom for women is its significant contact with the outside world. Beginning thousands of years ago, traders from around the Indian Ocean, and especially Southeast Asia, made calls along the Malabar Coast, possibly bringing new ideas about gender roles. Then, in the first century A.D., the Apostle Thomas is said to have come to spread Christianity. Eventually, Muslim traders brought their faith by sea as well. The state's 31.8 million people have a history of cultural and religious pluralism that seems to have led to somewhat more tolerance than exists in neighboring states. Still, some groups, especially fisherfolk, experience a pervasive sort of discrimination. Fisherfolk are one of the very lowest-caste communities, and there is a broad unwritten assumption that they will always be an underclass.

Sri Lanka

Sri Lanka, known as Ceylon until 1972, is a large island country off India's southeast coast known for its beauty. From the coastal plains, covered with rice paddies and with nut and spice trees, the land rises to hills where tea and coconut plantations are common. At the center is a mountain massif that reaches nearly 8200 feet (2500 meters) at its highest elevation. The southwest monsoons bring abundant moisture to the mountains, giving rise to forests

Jewelry stores line a portion of Sea Street in Colombo, the capital of Sri Lanka. Gemstones have long been one of Sri Lanka's resources, and polishing and setting them is a major industry. Other businesses include textiles and clothing. Tourism would be a substantial industry but for the conflict between Singhalese and Tamils. Britain, Germany, and the United States are Sri Lanka's primary export trading partners. [Steve McCurry/National Geographic Image Collection.]

and several unnavigable rivers that produce hydroelectric power. Close to 30 percent of the land is cultivated, and just over 25 percent remains forested. Despite its natural beauty, Sri Lanka has suffered particularly violent civil unrest that grew out of ethnic rivalry, postcolonial politics, and high unemployment.

The original hunter-gatherers and rice cultivators of Sri Lanka, today known as Veddas, now number less than 5000. They were joined several thousand years ago by settlers from northern India who built numerous city-kingdoms. Today known as Singhalese, these descendants of northern Indians make up about 74 percent of the population of 18.7 million. The Singhalese brought Buddhism to Sri Lanka, and today 70 percent of Sri Lankans (most of them Singhalese) are Buddhist.

About a thousand years ago, Dravidian people from southern India, known as Tamils, began migrating to Sri Lanka; by the thirteenth century they had established a Hindu kingdom in the northern part of the island. Later, in the nineteenth century, the British imported large numbers of poor Tamils from Tamil Nadu in southeastern India to work on British-managed tea, coffee, and rubber plantations. Known as "Indian" Tamils, the plantation-based population shares linguistic and religious traditions with the "Sri Lankan" Tamils of the north and east, but each community

considers itself distinct because of its divergent historical, political, economic, and social experiences. Together, the two Tamil populations make up about 18 percent of the total population of Sri Lanka. The Sri Lankan Tamils have done well, dominating the commercial sectors of the economy, whereas the Indian Tamils have remained isolated and largely poverty-stricken laborers on interior plantations.

Sri Lanka once had a thriving economy, led by a vibrant agricultural sector, and a government that made significant investments in health care and education. It was thought to be poised to become one of Asia's most developed economies. But conditions in rural areas took a drastic turn for the worse in the 1960s. Declining prices for the chief agricultural exports of tea, rubber, and coconuts prompted the government to shift investment away from rural development and toward urban manufacturing and textile industries, dominated by Singhalese. This shift exacerbated a political conflict that already existed because the legal status of the plantation laborers imported by the British had not been guaranteed at the time that the British withdrew as a colonial government. The Singhalese alienated all of Sri Lanka's Tamils by calling into question the status of the Indian Tamils as citizens of Sri Lanka and threatening to bar them from participation in elections. The Singhalese also claimed that only Singhala, not Tamil or English, was an official language. Moreover, Buddhist Singhalese were privileged in obtaining university admissions, job opportunities, and other social services. In news stories, protests against this favoritism by Tamil plantation workers have been confused with brutal acts of terrorism by guerrilla forces, known as the Tamil Tigers, operating mostly in Sri Lanka's far northeast. In 1987, Indian troops intervened at the request of the Sri Lankan government, but that action failed to end the violence.

Meanwhile, resentment grew in southern Sri Lanka in response to the government's lack of attention to growing poverty and corruption. In 1989, Sri Lanka appeared to be near collapse as the result of an insurrection by southern Singhalese guerrillas and the withdrawal of the Indian forces in the north. Armed conflict flared repeatedly throughout the 1990s, and the civil war has produced more than a million refugees and severely impeded Sri Lanka's economic development. Urban industrial development has lagged despite efforts to create a free trade zone to attract industries. Repeated terrorist bombings have frightened off international investors. As the conflict continues in the first decade of the new millennium, security remains tight on major highways and at the airport.

Quick Review

1. The collision between which two tectonic plates has created the mountains of Afghanistan and Pakistan?

2. What are a few of the reasons New Delhi attracts migrants despite its pollution and lack of services?

3. What nickname was given to Mumbai in recognition of its film industry?

4. Why do the people of Kerala enjoy a generally higher standard of living than others in India?

5. Kolkatta (Calcutta) is suffering from brain drain. Why are some young people leaving the city?

REFLECTIONS ON SOUTH ASIA

Conflicting images of life in South Asia abound. A common expression in South Asia is that any statement one may make about a region as complex as South Asia can be matched by an opposite statement that is equally true. For example, South Asia has some of the largest and most crowded cities on earth, yet more than 70 percent of the population lives in rural villages. Poverty is endemic, and the gaps between rich and poor are widening, but South Asia is also the home of one of the most potentially empowering and far-reaching development strategies ever conceived—microcredit. Although telephones are missing from most homes, information technology is flourishing. And while religious conflict between Muslims and Hindus threatens to precipitate nuclear war between Pakistan and India, the region is a mecca for those seeking spiritual enlightenment. It is also the home of the highly effective strategy of nonviolent social resistance, begun by Mohandas Gandhi and adopted by advocates for human rights all over the world, from the American South to South Africa to Beijing's Tiananmen Square. Another contrast is that India is the world's largest democracy, yet it is also a place where religious intolerance has led to thousands of deaths in the last few years. In South Asian democracies, a person's sex significantly determines whether that person has access to sufficient food, education, income, opportunity, and, indeed, to life itself. Yet in no other region have more women been elected head of state, and South Asian women are among the most articulate supporters of women's rights in the global forum.

Perhaps one of the most provocative characteristics of South Asia is that it has spawned eloquent and prolific writers who frequently publish their work in English. Many Indian and Bengladeshi writers make the best-seller lists in Europe and North America (see the bibliography on this textbook's Web site: www.whfreeman.com/pulsipher). Most have something to add to the global conversation about poverty, human rights, and development. This literary tradition is exemplified by Bengali Nobel laureate Rabindranath Tagore, who wrote prophetically before his death in 1941 about the need for South Asia to assess carefully its Western-led development paths:

We have for over a century been dragged by the prosperous West behind its chariot, choked by the dust, deafened by the noise, humbled by our own helplessness and overwhelmed by the speed. We agreed to acknowledge that this chariot-drive was progress, and the progress was civilization. If we ever ventured to ask, "progress towards what, and progress for whom," it was considered to be peculiarly and ridiculously Oriental to entertain such [reservations] about the absoluteness of progress. Of late, a voice has come to us to take count not only of the scientific perfection of the chariot but of the depth of the ditches lying in its path.

(Rabindranath Tagore, "Crisis of Civilization," in *Collected Works of Rabindranath Tagore*, vol. 18.)

It took more than three decades for Tagore's critique of development policy to gain wider acceptance. The "progress for whom" question is only now being asked in the highest halls of policy formulation at the World Bank and the United Nations. Strategies for aiming development at the poorest, rather than at those who are already reasonably well off, are finally being invented; as we have seen, South Asians are some of the most innovative creators of these strategies. Readers have only to check the newspapers in their towns and cities to discover small groups of cooperative borrowers meeting regularly to support and encourage one another in entrepreneurial ventures, emulating their million-plus counterparts in the Grameen Bank of Bangladesh, where microcredit got its start. At the close of a chapter about a region that is arguably one of the poorest on earth, it is illuminating to note that this region is also a leader in inventive ideas for development as well as in provocative thought about the present trajectory of human society.

Chapter Key Terms

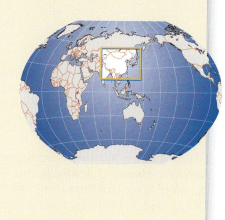

CHAPTER 9

EAST ASIA

something so I can come back and change things." These sentiments echo those of migrants everywhere.

On the fifth day, the bus reached Changan. When Xiaohui arrived, in 1996, the city's population had grown from 30,000 to 180,000 in less than a decade. The Mattel Barbie doll factory, in the middle of town, is in a modern white building known locally as "Paradise of Girls" because nearly all 3500 employees are female. Within a day, Xiaohui had completed her training, signed a 3-year contract, and learned how to put brown or blonde hair into Barbie doll heads with varying skin colors. The pay, 200 yuan a month (about U.S. $24), was considerably less than the 350 yuan the recruiters had promised. Xiaohui learned by asking around that no one was earning more than 400 yuan.

When a reporter inquired about the women's pay, a Mattel official replied, "We're confident that we provide these workers with a better standard of living than they would have without Mattel." [Adapted from Kathy Chen, "Boom-town bound," Wall Street Journal (October 29, 1996): 1, A6.]

PERSONAL VIGNETTE Eighteen-year-old Hong Xiaohui (Hong is her family name) and two friends eagerly signed up when the labor officials came to her remote farming village in Sichuan Province. They had come to hire workers for Mattel's Barbie doll plant in Changan, a factory town between Guangzhou (Canton) and Hong Kong in the southern coastal province of Guangdong.

When the bus left a few days later, it held 100 young women. Most of them had never left their home villages before, but now they were on a 1200-mile (1900-kilometer) trip that would take 5 days, most of it along bumpy, winding roads. Exhaust fumes and a lack of ventilation would bring constant nausea. Although nearly all the women were short of cash, most had a supply of food provided by their anxious mothers: sugarcane, bags of oranges, salted duck eggs, peanuts, and bottles of orange soda. While chatting with her friends, Xiaohui said, "I've always itched to get away from home. Life there is empty. I want to be independent. I want to learn

MAKING GLOBAL CONNECTIONS

Hong Xiaohui's first trip to China's booming southeastern coast illustrates some of the major transitions under way in the vast region of East Asia. She and other rural villagers from China's interior agricultural provinces are among the millions who are making East Asia's coastal cities some of the biggest and most rapidly growing cities on earth. Not only will these young women work for wages that are among the lowest in the global economy, but many will make sacrifices to send remittances home to poor relatives. Although the remittances are helpful, migration deprives poor rural interior provinces of the energy, initiative, and leadership of their young adults. Thus the migration of rural people from the interior to the coast contributes to the growing disparity in wealth between China's poor interior and its booming coastal provinces (Figure 9.1A).

Hong Xiaohui's story also hints at how gender roles are changing throughout East Asia. Although many of the old restrictions on women persist, increasingly it is young women who are the workers in the factories of multinational firms such as Mattel. The pattern of young women leaving home to find work in distant factories and offices is now global, as illustrated by Figure 9.1B. Most sacrifice their own comfort and well-being to send most of their wages home to needy relatives. Some inadvertently get caught up in human trafficking and become enslaved in the sex trade. Only a few have the financial freedom to truly seek their own fortunes. (Similar trends are discussed in the chapters on Middle and South America, Europe, Russia and the newly independent states, South Asia, and Southeast Asia.)

East Asia, home to nearly one-fourth of humanity, is a vast territory that stretches from the Taklimakan Desert in the far west to Japan's rainy Pacific coastline and from the frigid mountains of Mongolia in the north to the steaming subtropical forests of China's southeastern coastal provinces. The East Asia world re-

Many women have migrated from rural areas to work in the global factories of China's special economic zones. In this photo, women workers are inspecting Reebok sneakers at the Kong Tai shoe manufacturing plant in Shenzhen, China. [Forrest Anderson/Time Life Pictures/Getty Images.]

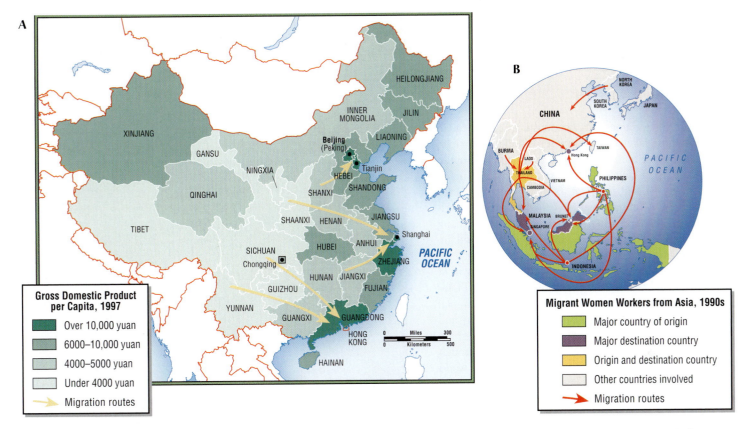

FIGURE 9.1 Economic changes in China. (A) Regional disparity in wealth in China. U.S. $1.00 = 8.28 yuan (May 2004). Note the significant differences in economic prosperity between the eastern coastal provinces and most of China's interior. The arrows show patterns of migration. (B) The pattern of young women leaving home to find work in distant factories and offices is now global; this trend is especially strong in East and Southeast Asia. [A, adapted from *The Economist* (January 16, 1999): 40; B, adapted from Joni Seager, *The State of Women in the World Atlas* (New York: Penguin, 2003), p. 73.]

gion consists of the countries of China, Mongolia, North Korea, South Korea, Japan, and Taiwan. East Asia also includes Hong Kong and its smaller neighbor, Macao; both are now parts of China, but were previously European colonies. Several of the world's most rapidly growing economies of the recent past are in East Asia. China, especially, will provide a huge market for products from other world regions. Japan has achieved economic success and a high standard of living as a supplier of consumer products to the United States and much of the developed world. East Asia is also the most populous world region and is among the most precarious environmentally. Burgeoning rates of consumption in this region will add to local environmental stress and to such problems as the worldwide accumulation of greenhouse gases. For all these reasons, East Asia will figure prominently in global relationships for years to come.

THEMES TO EXPLORE IN EAST ASIA

Current trends in the economy, migration, population, and environment of East Asia are reflected in the following themes, which appear repeatedly throughout this chapter:

1. **Rapid economic growth.** Since implementing economic reforms, China has become one of the fastest-growing economies in the world. Previously, state-aided market economies, pioneered by Japan, also experienced rapid growth, but this growth has slowed in recent years.

2. **Influence of ideas from traditional Chinese thought.** Ancient Chinese forms of government and philosophy have influenced the entire region. One can see these influences in the development of economies, in the dominance of bureaucracies, and in gender roles. Yet countries such as Japan and China have incorporated these influences in very different ways.

3. **Population concentration on coasts and in lowlands.** In all parts of East Asia, the population is unevenly distributed, with the greatest concentrations in coastal areas and other lowlands.

4. **Environmental stress.** Large, densely settled populations pursuing modern, consumerist lifestyles are straining the environment, causing floods, dangerous levels of pollution, and overcrowding.

5. **Urban-rural disparities.** Throughout the region, there are disparities in wealth and well-being between rural and urban areas. This pattern is especially dramatic in China. The disparities are growing as cities gain special economic concessions and attract millions of rural migrants.

I THE GEOGRAPHIC SETTING

Terms to Be Aware Of

East Asian place names can be unusually complicated, especially those taken from Chinese. Whenever possible, we will give place names in English transliterations of the appropriate Asian language. We also avoid redundancies. For example, *he* and *jiang* are both Chinese words for *river*. Hence the Yellow River (so called because it is yellowed by the heavy sediment load it carries) is the Huang He, and the Long River is the Chang Jiang (also called the Yangtze in its lower course). It is incorrect to add the term *river* to either name, because it is already there. Pinyin (a spelling system based on Chinese sounds) versions of Chinese place names are now commonplace. For example, the city once called Peking in English is now Beijing, and Canton is Guangzhou. The region known as Manchuria is here referred to simply as China's Far Northeast. Although China refers to Tibet as Xizang, people around the world who support the idea of Tibetan self-government avoid using that name. This text uses Tibet for the region (with Xizang in parentheses) and Tibetans for the people who live there.

PHYSICAL PATTERNS

A quick look at the regional map of East Asia (Figure 9.2) reveals that its topography is perhaps the most rugged in the world. Powerful tectonic forces have produced a wide range of complex landforms. East Asia's varied climates result from a dynamic interaction between huge warm and cool air masses and the land and sea. Rapidly expanding human populations have affected the variety of ecosystems that have evolved over the millennia and that still contain many important and unique habitats.

Landforms

There are few flat surfaces to be found in the rugged landscapes of East Asia. Many of those that do exist are too dry or too cold to be very useful to humans. The large numbers of people who occupy the region have had to be particularly inventive in creating spaces for agriculture. They have cleared and terraced entire mountain ranges using only simple hand tools, irrigated drylands with water from melted snow, drained wetlands using elaborate levees and dams, and applied their complex knowledge of horticulture and animal husbandry to help plants and animals flourish in difficult conditions.

A simple way to visualize the varied landforms of East Asia is to think of them as analogous to the shapes that would be formed in a

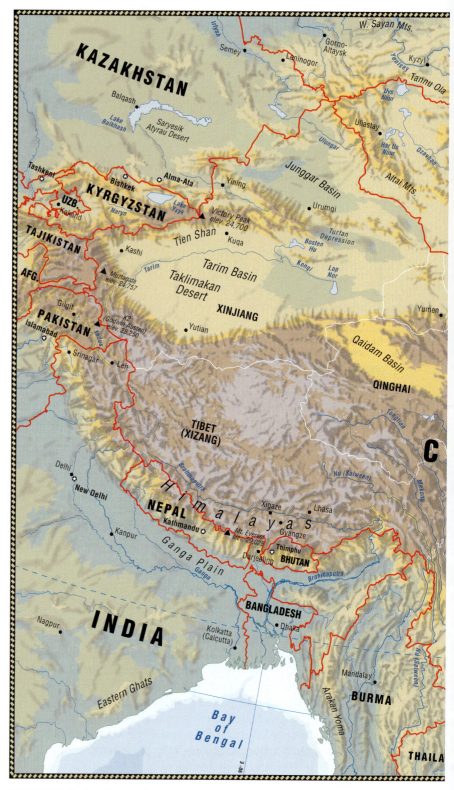

FIGURE 9.2 Regional map of East Asia.

huge carpet if a grand piano (representing the Indian subcontinent) were shoved deeply into one side of it. The mountain ranges, plateaus, depressions, fissures, and bulges across Eurasia have resulted from the slow-motion collision of the Indian-Australian Plate (which carries the Indian subcontinent) with the southern edge of Eurasia (the Eurasian Plate) that began roughly 50 million years ago and continues today. The Himalayas are the most dramatic result of

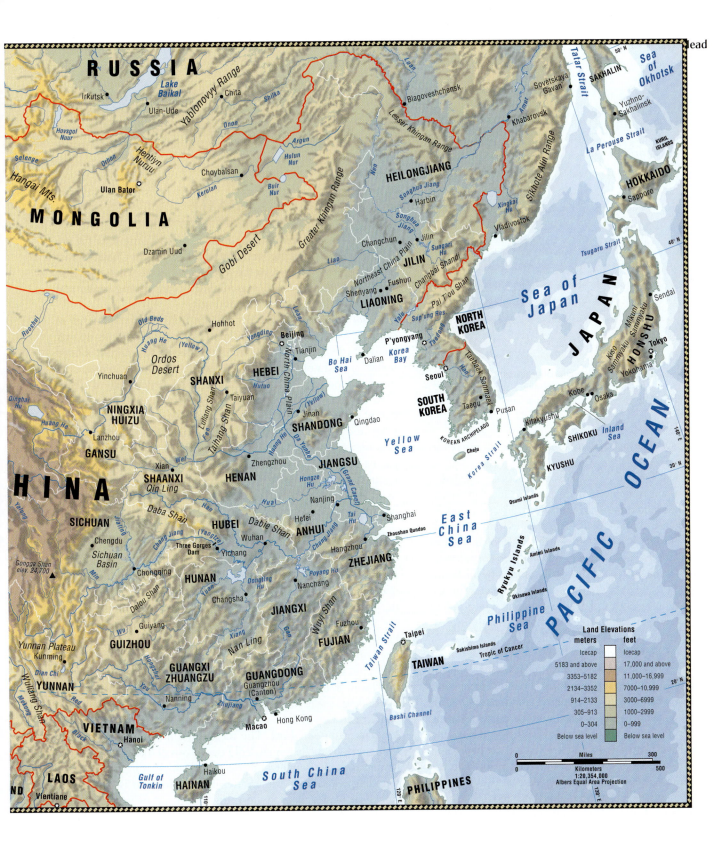

this collision. The area on the north side of the Himalayas absorbed some of the pressure by bulging upward and spreading outward to the northeast, forming the Plateau of Tibet (also known as the Xizang-Qinghai Plateau or Northern Plateau). It is depicted in rust and lavender in Figure 9.2. To the north of the plateau, in Xinjiang, the land responded to the tectonic forces by sinking to form basins, including the Qaidam Basin and the sea-level Tarim Basin in the

Taklimakan Desert. Farther north, it buckled again to form the mountains on either side of the Junggar Basin and those of northern Mongolia and southern Siberia. To the east and west of the Himalayan impact zone, wrinkled mountain and valley formations curve away to the southwest and southeast from the Plateau of Tibet.

The landforms of East Asia form four steps. The top step is the Plateau of Tibet. Many of the rivers of China and the Southeast

In the midst of the Taklimakan Desert, international petroleum company trucks are driving to a remote oil field that has potential reserves far greater than those in the United States. [Reza/National Geographic Image Collection.]

Asian mainland have their headwaters along the eastern rim of this plateau.

The second step down is a broad expanse of basins, plateaus, and low mountain ranges (depicted in deep yellow and light greens in Figure 9.2). These landforms include the deep, dry basins and deserts of Xinjiang, to the north of the Plateau of Tibet, and the broad, rolling highland grasslands and deserts of the Mongolian Plateau northeast of Xinjiang. East of Xinjiang, this step also includes the upper portions of China's two great river basins, through which flow the Huang He and Chang Jiang. To the south is the rugged Yunnan Plateau. The topography of this plateau is dominated by a system of deeply folded mountains and valleys that bends south through the Southeast Asian peninsula. The middle portions of the Nu (Salween), Mekong, and Red rivers are found here.

The third step, directly east of this upland zone, consists mainly of broad coastal plains and the deltas of China's great rivers (shown in green in Figure 9.2), with intervening low mountains and hills (shown as yellow patches on the green areas). Starting from the south is a series of three large lowland river basins: the Zhujiang (Pearl River) basin, the massive Chang Jiang basin, and the Huang He lowland basin on the North China Plain. The Far Northeast of China, the Korean Peninsula, and the westernmost parts of southern Siberia are also part of this third step.

The fourth step consists of the continental shelf, covered by the waters of the Yellow Sea, the East China Sea, and the South China Sea. Numerous islands—including Hong Kong, Hainan, and Taiwan—are anchored on this continental shelf; all are part of the Asian landmass. The islands of Japan have a different geological origin: they are volcanic, rather than being part of the continental shelf. These islands lie along a portion of the Pacific Ring of Fire (see Figure 1.11 on page 24). They rise out of the waters of the northwestern Pacific in the highly unstable zone where the Pacific, Philippine, and Eurasian plates grind against one another. The volcanic Mount Fuji is perhaps Japan's most recognizable symbol.

The entire Japanese island chain is particularly vulnerable to disastrous eruptions, earthquakes, and seismic sea waves (**tsunamis**). Mount Fuji last erupted in 1707. However, in 2001, there were deep internal rumblings, suggesting that this period of dormancy may be ending.

CLIMATE

East Asia has two contrasting climatic zones (Figure 9.3): the dry continental west and the monsoon east. Recall from Chapter 1 that *monsoon* refers to the seasonal reversal of surface winds that flow from the continent to the surrounding oceans during winter and from the oceans inland during summer.

The Dry Interior. The dry western zone lies in the interior of the East Asian continental landmass. Because land heats up and cools off more rapidly than water, locations in the middle of large bodies of land in the midlatitudes tend to experience intense cold in winter and intense heat in summer. The western part of China is an extreme example of a midlatitude continental climate. This part of China includes the Mongolian Plateau, the basins of Xinjiang, and the Plateau of Tibet—all of which are very dry and cold in winter and very hot during the day in summer. Summer nights can be cold because there is too little vegetation or cloud cover to retain the warmth of the sun after nightfall. Heat is lost so rapidly that the difference between summer daytime and nighttime temperatures may be as much as 100°F (55°C).

Grasslands and deserts of several varieties cover most of the land in this dry region. Only scattered forests grow on the few relatively well watered mountain slopes and in protected valleys supplied with water by snowmelt. In all of East Asia, humans and their effects are least conspicuous in the large, uninhabited portions of the deserts of Tibet (Xizang), the Tarim Basin in Xinjiang, and the Mongolian Plateau. The grasslands of Mongolia and the basins of Far Northwest China traditionally supported only scattered groups of nomadic herders. However, they are increasingly being put to more intensive uses, such as irrigated agriculture in the Tarim and Junggar basins. In Mongolia, many nomads now live on large-scale, stationary livestock cooperatives, also supported with irrigation.

The Monsoon East. The monsoon climates of the east are influenced by the extremely cold conditions of the huge Eurasian landmass in the winter and the warm temperatures of the surrounding seas and oceans in the summer. In what is commonly called the winter monsoon, dry, frigid arctic air sweeps south and east through East Asia, producing long, bitter winters on the Mongolian Plateau, on the North China Plain, and in Far Northeast China. The winter monsoon also causes occasional freezes in southern China. Central and southern China have shorter, less severe winters because they are protected from the advancing arctic air by the east-west ranges of the Qin Ling Mountains and because they lie close to the warm waters of the South China Sea.

In the summer, continental warming pulls in wet tropical air from the Pacific Ocean and its adjacent seas (see Figure 8.1 on page 000). The warm, wet air from the ocean deposits moisture on the land as seasonal rains. As the summer monsoon moves northwest, it

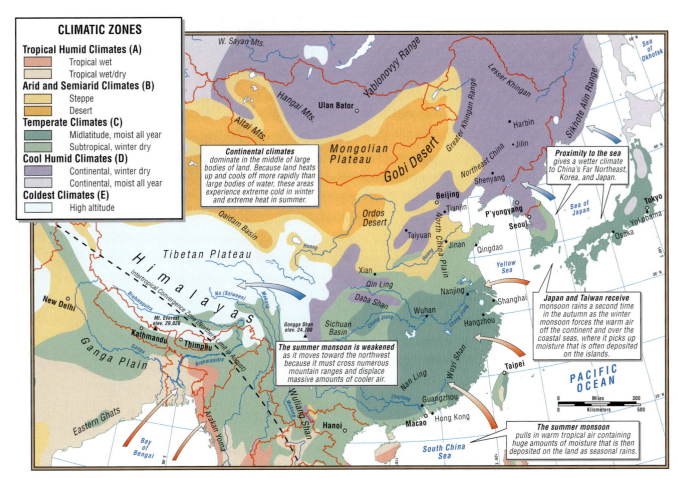

CLIMATIC ZONES

Tropical Humid Climates (A)
- Tropical wet
- Tropical wet/dry

Arid and Semiarid Climates (B)
- Steppe
- Desert

Temperate Climates (C)
- Midlatitude, moist all year
- Subtropical, winter dry

Cool Humid Climates (D)
- Continental, winter dry
- Continental, moist all year

Coldest Climates (E)
- High altitude

Continental climates dominate in the middle of large bodies of land. Because land heats up and cools off more rapidly than large bodies of water, these areas experience extreme cold in winter and extreme heat in summer.

Proximity to the sea gives a wetter climate to China's Far Northeast, Korea, and Japan.

Japan and Taiwan receive monsoon rains a second time in the autumn as the winter monsoon forces the warm air off the continent and over the coastal seas, where it picks up moisture that is often deposited on the islands.

The summer monsoon is weakened as it moves toward the northwest because it must cross numerous mountain ranges and displace massive amounts of cooler air.

The summer monsoon pulls in warm tropical air containing huge amounts of moisture that is then deposited on the land as seasonal rains.

FIGURE 9.3 Climates of East Asia. Two contrasting climatic zones characterize much of East Asia: the dry continental west and the moist (monsoon) east.

must cross numerous mountain ranges and displace cooler air. Consequently, its effect is weakened toward the northwest. Thus the Zhujiang basin in the far southeast is drenched with rain and enjoys warm weather for most of the year, whereas the Chang Jiang basin, which lies in central China to the north of the Nan Ling Mountains, receives only about 5 months of summer monsoon weather (see Figure 9.2). The North China Plain, north of the Qin Ling and Dabie Shan ranges, receives only about 3 months of monsoon rain. Very little monsoon rain reaches the southern Mongolian Plateau or Inner Mongolia (in China). The Far Northwest also gets very little rain.

China's Far Northeast is wet in summer; Korea and Japan have wet climates all year because of their proximity to the sea. But all of these areas still have hot summers and cold winters because of their northerly location and their exposure to the continental effects of the Eurasian landmass. Japan and the more southerly Taiwan actually receive monsoon rains twice: once in spring, when the main monsoon moves toward the land, and again in autumn, as the winter monsoon forces warm air off the continent. This retreating warm air picks up moisture over the coastal seas, which is then deposited on the islands. Japan, however, has a much longer and more severe winter than subtropical Taiwan. Much of Japan's autumn precipitation falls as snow.

The eastern areas of the Asian landmass, watered by seasonal rains, once supported rich forests with diverse plant and animal life, much of it unique to East Asia. The character of the forests changed from north to south: coniferous forests covered the Far Northeast and graded to deciduous broad-leafed forests in Korea, Japan, and the North China Plain. Broad-leafed evergreen forests and even tropical rain forests grew in the south. Over the past two or three millennia, however, agriculture has transformed the landscape. Many lowland ecosystems were wiped out as farmers expanded into any well-watered or irrigable land, be it flat or hilly. Hills and low mountainsides were rarely left covered with undisturbed forest; they were logged continuously, planted with orchards, or completely cleared and terraced for agriculture. Today, the few undisturbed natural areas in the humid zone are remote, deeply convoluted, and increasingly threatened by development.

HUMAN PATTERNS OVER TIME

East Asia is home to some of the most ancient civilizations on earth. Archaeological evidence indicates that settled agricultural societies have flourished in China for over 7000 years. A little more than 2000 years ago, the basic institutions of government that still exist across the region today were established in eastern China. Until the twentieth century, China was the main source of wealth, technology, and culture for the people of East Asia.

Until recently, the hearth of East Asian culture was thought to be the North China Plain, a rich agricultural area that was unified under successive empires and has come to be regarded as **China proper.** Archaeological discoveries in other parts of China—most notably the southwestern province of Sichuan—now indicate that Chinese civilization evolved from several hearths. Bordering China proper to the north and west were the lands of inner Asia, inhabited by Mongolian peoples considered alien and uncivilized by the Chinese. The huge, dry Mongolian territories were unsuitable for crop farming, but ideal for nomadic animal herding. The nomads, for their part, pitied the Chinese farmers who were tied to a particular place for a lifetime. On East Asia's eastern fringe, the Korean Peninsula and the islands of Japan and Taiwan were profoundly influenced by the culture of China; nonetheless, they were isolated enough that each developed a distinctive culture and usually maintained political independence. These characteristics proved crucial during the twentieth century, when these areas (except for North Korea) leapt ahead of China economically and militarily, partly by integrating European influences that China disdained.

The Beginnings of Chinese Civilization

Although humans and their ancestors have lived in East Asia for hundreds of thousands of years, the earliest complex civilizations appeared in various parts of China about 4000 years ago. Written records exist only from the civilization that was located in north central China. There, a small, militarized, feudal aristocracy controlled vast estates on which the mass of the population worked and lived as impoverished farmers and laborers. The landowners usually owed allegiance to one of the petty kingdoms that dotted northern China. These kingdoms were relatively self-sufficient and well-defended with private armies, and they often proved insubordinate to central authority.

Between 400 and 221 B.C., a new order emerged that eradicated feudalism and laid the foundations for the great Chinese empires that dominated East Asia for more than 2000 years. Following a long period of war between the petty kingdoms of northern China, a single dominant kingdom emerged. What gave this kingdom, known as the Qin dynasty, its advantage was the use of a trained and salaried bureaucracy to extend the monarch's authority into the countryside. The old feudal alliance system was scrapped, and the estates of the aristocracy were divided into small units and sold to the previously semi-enslaved farmers. The kingdom's agricultural output increased because the people worked harder to farm land they now owned. In addition, the salaried bureaucrats who replaced their former masters were more responsible about building and maintaining levees, reservoirs, and other public works that reduced the threat of floods and other natural disasters. Although the Qin dynasty was short-lived, subsequent empires maintained its bureaucratic ruling methods, which have proved essential to the governing of China as a whole right up to the present.

Confucianism

Closely related to China's bureaucratic ruling tradition is the philosophy of **Confucianism.** Confucius, who lived about 2500 years

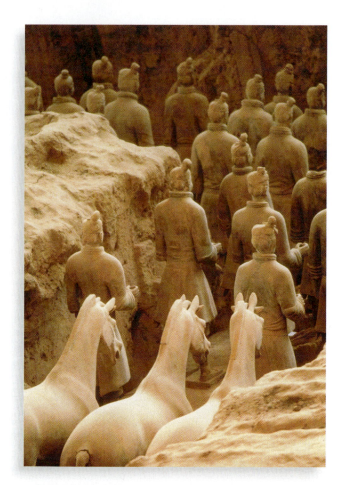

After unifying China in 211 B.C., Qin Shi Huang became its emperor. One of his many public works, which included China's Great Wall, was the creation of a life-size terra-cotta army of thousands of soldiers in a vast underground chamber that became his mausoleum. Since 1974, archaeologists have been excavating, restoring, and preserving this site, near modern-day Xian, which attracts more than 2 million visitors a year. [Wally McNamee/Woodfin Camp & Associates.]

ago (prior to the Qin dynasty), was an idealist interested in reforming government and eliminating violence from society. He espoused the view that relationships between people were the basis of a civilized society and that all human relationships involved a set of defined roles and mutual obligations. If each person understood his or her proper role and acted accordingly, a stable, uniform, and enduring society would result. Confucian values include respect for parents (filial piety) and government officials, courtesy, loyalty (to family and government), knowledge (moral wisdom), and integrity. Saving face and giving face (allowing others to maintain their personal dignity) is an important cultural norm in East Asia that stems from Confucian values.

The foundation of Confucian philosophy was the patriarchal extended family, in which the oldest male held the seat of authority as well as the responsibility to provide for the well-being of everyone in the family. All other family members were aligned under him according to age and sex and owed their allegiance and obedience to the patriarch who attended to their material needs. Extending these conventions, Confucianism held that commoners must obey imperial officials, and that everyone must obey the emperor, who, in turn, was obliged to ensure the welfare of society. As the supreme human being, the emperor was seen as the source of all order and civilization—in a sense, the grand patriarch of all China.

Over the centuries, Confucian philosophy penetrated all aspects of Chinese society, altering its social, economic, and political geography and circumscribing the lives of the masses at the base of the hierarchical pyramid, who were marshaled as a labor force. An early literary interpretation of the ideal woman demonstrates how women were almost always placed under the authority of others and confined to the domestic spaces of the home, where they performed many economic tasks. A student of Confucius wrote: "A woman's duties are to cook the five grains, heat the wine, look after her parents-in-law, make clothes, and that is all! When she is young, she must submit to her parents. After her marriage, she must submit to her husband. When she is widowed, she must submit to her son." The system also rested on the labor of large numbers of low-status men and children, who worked in agriculture and as servants.

Confucian ideology had a lasting effect on China's human geography because of the way it was utilized by elite groups to maintain power and position. Elites emphasized the Confucian ideal of obedience and loyalty to maintain the status quo when it suited their interests, even at the expense of others in society. Elites, for example, claimed the produce raised by impoverished farmers and the wealth produced by artisans and craft workers. To block the emergence of competing powerful segments of society, elites attempted to limit merchants' wealth wherever possible. In parable and folklore, merchants were characterized as a necessary evil, greedy and disruptive of the social order. At certain points over the past 2000 years, the expropriation of farmers' and artisans'

wealth through high taxes and the curtailment of merchants' prosperity left these groups little incentive or investment capital for agricultural improvements, industrialization, or entrepreneurship. At other times, however, the anti-merchant aspect of Confucianism was less influential, and trade and entrepreneurship flourished. For instance, Confucianism was never as strong in the southern coastal area as it was in the North China Plain. Here, a vibrant maritime society in which trade was important thrived.

Although the Confucian bureaucracy allowed empires to expand (Figure 9.4), its rigidities often led to periods of imperial decline. The heavy tax burden on farmers led to periodic revolts that weakened imperial control. In addition, invasions by nomadic peoples from what is today Mongolia and western China dealt the final blow to several successive Chinese empires—despite the defensive walls, such as the Great Wall of China, built along China's northern border. So weakened were the Chinese in the 1200s that the Mongolian military leader Genghis Khan and his descendants were able to conquer all of China and then push west as far as Hungary and Poland (see Chapter 5, page 242). It was during this dynasty (Yuan) that the first direct contacts between China and Europe took place.

China's Preeminence

In the tenth century, China proper was the world's most developed region: it had the world's wealthiest economy, its largest cities, and the highest living standards. Improved strains of rice allowed dense farming populations to expand throughout southern China and sup-

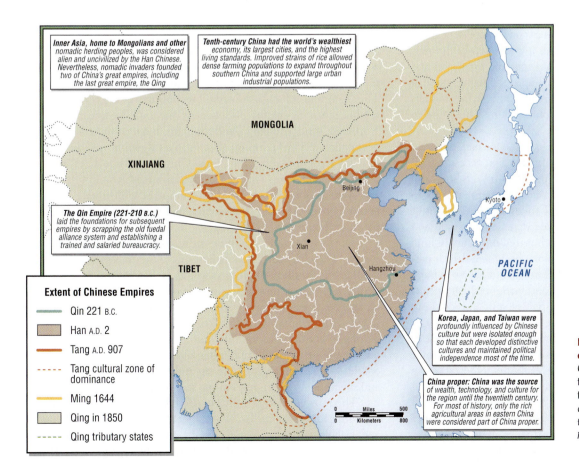

Inner Asia, home to Mongolians and other nomadic herding peoples, was considered alien and uncivilized by the Han Chinese. Nevertheless, nomadic invaders founded two of China's great empires, including the last great empire, the Qing

Tenth-century China had the world's wealthiest economy, its largest cities, and the highest living standards. Improved strains of rice allowed dense farming populations to expand throughout southern China and supported large urban industrial populations.

The Qin Empire (221–210 B.C.) laid the foundations for subsequent empires by scrapping the old fuedal alliance system and establishing a trained and salaried bureaucracy.

MONGOLIA

XINJIANG

Beijing

Kyoto

Xian

TIBET

Hangzhou

PACIFIC OCEAN

Korea, Japan, and Taiwan were profoundly influenced by Chinese culture but were isolated enough so that each developed distinctive cultures and maintained political independence most of the time.

China proper: China was the source of wealth, technology, and culture for the region until the twentieth century. For most of history, only the rich agricultural areas in eastern China were considered part of China proper.

Extent of Chinese Empires
- Qin 221 B.C.
- Han A.D. 2
- Tang A.D. 907
- Tang cultural zone of dominance
- Ming 1644
- Qing in 1850
- Qing tributary states

Miles 500
Kilometers 800

FIGURE 9.4 The extent of Chinese empires, 221 B.C.–A.D. 1850. The Chinese state expanded and contracted throughout its history. The colors on the map indicate the extent of its dominion at various times. [Adapted from *Hammond Times Concise Atlas of World History*, 1994.]

ported large urban industrial populations. Metallurgy flourished. As early as 1078, people in northern China were producing twice as much iron as people in England did 700 years later. Nor was innovation lacking: Chinese inventions included papermaking, printing, paper currency, gunpowder, and improved shipbuilding techniques. All these advances contributed to the growth of remarkable urban areas. When the Venetian trader Marco Polo wrote about his travels in China between 1271 and 1295 (during the Yuan, or Mongol, dynasty), Europeans were stunned to learn that the imperial city of Hangzhou (near modern Shanghai) was 100 miles in circumference; had a population of 900,000; was dotted with parks and served by a sewer system; had a large canal system; and supported several principal markets, each frequented daily by 40,000 to 50,000 people. Trade flourished, and a strong mercantile economy had developed. Wares included rhinoceros horns from Bengal, ivory from India and Africa, coral, agate, pearls, crystal, rare woods, and spices.

Although the Chinese empire experienced unprecedented economic expansion under the Ming dynasty (1368–1644), its rulers rigidly pursued a series of policies that favored elite interests at the expense of ingenuity and change and ultimately left China ill-prepared to respond to the challenge posed by Europe after 1600. Despite the high level of technological know-how that China attained early on, European technology and military might eventually surpassed that of China in the eighteenth century.

European Imperialism in East Asia

By the mid-1500s, Spanish and Portuguese traders had found their way to East Asian ports. They were interested in acquiring the silks, spices, and ceramics of East Asia, and they brought a number of new food crops from the Americas, such as corn, peppers, peanuts, and potatoes, to offer in exchange. The introduction of the new American food crops initiated a spurt of economic expansion and population growth. By the mid-1800s, China's population stood at over 400 million, and many Chinese were migrating to the frontiers of the empire: to Yunnan in central southwestern China and to Manchuria in the northeast.

By the nineteenth century, European influence had increased markedly as European merchants attempted to gain access to Chinese markets. Between 1839 and 1860, the Opium Wars broke out between Britain and China over China's unsuccessful attempt to crack down on the illegal importation of opium from India. British merchants exchanged opium for Chinese wares, such as silks and fine ceramic dinnerware, that were much prized in Britain. China lost these wars and paid dearly. Hong Kong became a British possession; British merchants were granted unfettered access to trade in key coastal cities such as Shanghai; and foreigners (Europeans and Americans) were granted extensive rights to establish communities and conduct business on Chinese soil.

The final blow to China's preeminence in East Asia came in 1895, when a rapidly modernizing Japan won a spectacular naval victory over China in the Sino-Japanese War. After its defeat by the Japanese, the Qing dynasty made only halfhearted attempts at modernization, and it collapsed in 1912 after a coup d'état. Until China's communists took control in 1949, much of the country was governed by provincial warlords in rural areas and by a mixture of Chinese and Western administrative agencies in the major cities. During this era, radical ideologies gained popularity in new Western-style universities as intellectuals searched for a new basis of political authority to replace Confucianism. Of particular interest were various forms of socialism and communism.

China's Turbulent Twentieth Century

Two rival reformist groups arose in China in the early twentieth century. One was the Nationalist party, known as the Kuomintang (KMT), led by Chiang Kai-shek. The KMT united the country in 1924 and at first reorganized it politically and economically according to the socialist views of Lenin. After 1927, however, the KMT increasingly became the party of the urban upper and middle classes. The rival group, the Chinese Communist party (CCP), appealed most to the far more numerous rural laborers. Japan took advantage of these internal struggles to seize control of China's Far Northeast (Manchuria) in 1931. Then, in 1937, while the Western powers were distracted by European fascism, Japan invaded China, conquering most of its major cities and coastal provinces. For a while, the KMT and CCP tried to unite against this common enemy, but once Japan surrendered to the Allied forces at the end of World War II (in 1945), the two Chinese reformist parties no longer cooperated. Although the KMT was backed by the United States (which was then increasingly preoccupied with opposing communist movements), it was quickly pushed out of the country by the now very popular CCP, led by Mao Zedong. The KMT and many of its urban supporters fled to Taiwan. There they formed a government-in-exile opposed to the mainland communists. Even today Taiwan is governed separately from the rest of China.

On the Chinese mainland, Mao Zedong's revolutionary mobilization of the rural majority resulted in the historic proclamation of the People's Republic of China on October 1, 1949, in Beijing's Tiananmen Square. Before long, Mao's government became the most powerful and expansionist government China had ever had. The Communist party assumed total control over the economy. It established a dominant presence in the outlying areas of the Northeast, Inner Mongolia, and Xinjiang, and it launched a brutal occupation of Tibet (Xizang). The People's Republic of China was in many ways similar to a traditional Chinese dynasty. The Confucian bureaucracy was replaced by the Chinese Communist party, and Mao Zedong became a sort of emperor with unquestioned authority over the party, the military, and virtually all other activities of government.

The communist revolution substantially changed Chinese society. The chief early beneficiaries were the masses of Chinese farmers and landless laborers. It would be hard to exaggerate the desperate plight of the Chinese people on the eve of the revolution in 1949. Huge numbers lived in abject poverty, there were frequent famines, massive floods regularly devastated villages and farmland, and infant mortality rates were high. For the millions outside the small middle and upper classes, life was short and miserable. The vast majority of women and girls held low social status and spent their lives in unrelenting servitude.

The revolution drastically changed this picture. All aspects of economic and social life became subject to central planning. Land and wealth were reallocated, often—although certainly not always—resulting in improved standards of living for those who needed it most. Heroic efforts were made to improve agricultural production and to ameliorate the age-old twin afflictions of floods and droughts. Huge numbers of Chinese, regardless of age, class, or sex, were mobilized to construct massive public works projects almost entirely by hand. Chinese laborers terraced and reterraced mountainsides with hand tools. Bucket brigades unclogged silted waterways. People built roads across difficult terrain with virtually no machinery. Industries were founded in many regions with an eye to providing jobs to a broad segment of society. The famed "barefoot doctors"—people with rudimentary medical training—dispensed basic medical care, midwife services, and nutritional advice to the remotest locations. Schools were built in the smallest of villages. Opportunities for women opened up, and some of the worst abuses against them—such as the crippling binding of women's feet to make them small and childlike—stopped. Until recent decades, famine had occurred somewhere in the country every few years. Since the mid-1970s, however, China has achieved a remarkable ability to feed its people. The vast majority of Chinese who are old enough to have witnessed these changes say that, materially, life is now better by far than before the revolution.

Nonetheless, the progress made under Communist party rule has come at enormous human and environmental costs. During the **Great Leap Forward** (a government-sponsored program of massive economic reform initiated in the 1950s), millions died needlessly, persecuted for dissenting from party policies or killed by famine or other avoidable disasters that resulted from the rush to fulfill poorly planned development objectives. Similarly, many of China's considerable natural resources have been sacrificed to its vision of progress.

In the aftermath of famine and other disasters, some Communist party leaders tried to correct the inefficiencies of the centrally planned economy, only to be purged themselves as Mao Zedong attempted to stay in power. One early ploy to keep political control was the initiation in 1966 of the **Cultural Revolution,** a series of highly politicized and destructive mass campaigns to force the entire population to support the continuing revolution. Educated people and intellectuals were a main target because they were thought to instigate dangerously critical evaluations of Communist party central planning. Many Chinese scientists and scholars were summarily sent to labor in mines and industries or to jail, where they often died. The public was encouraged to purge suspected dissidents. People who had engaged in any type of petty capitalism were severely punished, as were those who adhered to any type of organized religion. The Cultural Revolution so disrupted Chinese society that by Mao's death in 1976, the communists had been thoroughly discredited.

Two years later, a new leadership formed around Deng Xiaoping. In the early 1980s, he initiated a series of reforms to give China's economy some characteristics of open-market economies while at the same time maintaining Communist party political control. The pace and direction of reform, however, did not keep up with popular expectations. In April and May of 1989, more than 100,000 students and workers agitating for greater freedom of expression and government accountability marched on Tiananmen Square in Beijing. On June 3 and 4, the government suppressed the demonstrations by firing into the crowd, killing hundreds and injuring thousands. Arrests, hasty trials, and executions

followed. By 2000, many observers were concerned that, although remarkable levels of economic growth had been achieved, the strains of rapid movement to a market economy might be creating further political instability in China that could lead the government to abridge civil and human rights once again.

The Transformation of Japan

The Japanese islands were probably first inhabited about 20,000 years ago. Modern Japanese populations are descended from migrants from the Asian mainland, the Korean Peninsula, and the Pacific islands. By A.D. 300, Japanese society was divided into military clans that had established their rule over most of the islands that are now part of Japan. These islands included the central and largest island of Honshu, the island of Hokkaido to the north, and several smaller islands to the south. Settlement was concentrated in central Honshu and the southern islands. Leaders of the Yamato clan, who claimed divine approval, ruled until about 700. During the Yamato period, ideas and material culture imported from China and the Korean Peninsula transformed the everyday lives of Japan's people and their use of the land. These imports included Buddhism, Confucian bureaucratic organization, architecture, Chinese writing, the arts, and agricultural technology.

Between 800 and 1300, Japan turned inward, and influences from the Asian mainland waned. The Japanese successfully fended off invasion by the Mongols, whose empire in the 1300s stretched from coastal China to Europe. During the ensuing period of isolation, a feudal system with a rigid class structure evolved, and Japan was divided into mini-kingdoms. This was the situation when Portuguese sailors landed in Japan in 1543, followed by Christian missionaries. Active trade with Portugal brought new ideas and technology that strengthened the wealthier feudal lords, who then seized control from the emperor and unified Japan under a military bureaucracy. Between 1600 and 1868, a hereditary line of elite military rulers (known as **shoguns**) imposed a strict four-tier social class system, expelled foreigners, and imposed isolation once again, especially rejecting all European influence. The shoguns ruled from a castle in Edo, the city that was later to be renamed Tokyo, while the emperor, stripped of real power and authority, remained in the imperial capital, Kyoto.

In 1853, a small fleet of naval vessels commanded by U.S. Commodore Matthew C. Perry arrived unexpectedly near the mouth of Tokyo Bay. Japan was quickly forced to recognize that more than two centuries of self-imposed isolation from the rest of the world would have to end. The foreigners, who were stronger in military might and much more advanced in technology, were in a position to dictate terms of trade and foreign relations. The threat was compounded by domestic unrest over various political and economic issues that had been building for decades. Consequently, by 1868, the family line of shoguns who had ruled Japan since 1600 lost power to a group of reformers, the Meiji Oligarchs, who restored symbolic authority to the emperor and moved his seat from the ancient capital Kyoto to Tokyo. In the emperor's name, they set Japan on a crash program of modernization and industrial development. They sent Japanese students abroad and recruited experts from around the world, especially from Western nations, to teach everything from foreign languages to modern technology.

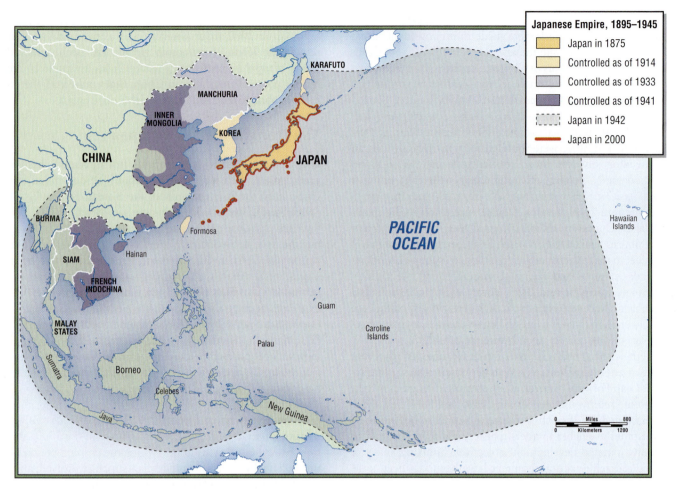

FIGURE 9.5 Japan's expansions, 1875–1942. Japan colonized Korea, Taiwan (Formosa), Manchuria, China, and parts of Southeast Asia to further its program of economic modernization and to fend off European imperialism in the early twentieth century. [Adapted from *Hammond Times Concise Atlas of World History*, 1994.]

Meanwhile, Japanese settlement expanded from central Honshu and the southern islands to northern Honshu and Hokkaido, where new farmlands and cities were developed and strategic resources, such as coal, were exploited for heavy industry. To expand its resource base further and to gain a labor force for its mines and factories, Japan colonized Korea, Taiwan (then known as Formosa), and Manchuria, and then pushed farther into China and into Southeast Asia (Figure 9.5).

Japan's imperial ambitions ended with its defeat in World War II (1941–1945), and the country was occupied by U.S. military forces until 1952. The U.S. government imposed many social and economic reforms and required Japan to create a democratic constitution. Japan's postwar constitution allows only a very limited military, and for more than 50 years the country has relied primarily on U.S. forces based in Japan, South Korea, and the western Pacific to protect it from attack. Some observers believe that China could one day pose a military threat to Japan and motivate the country to reassess both its strategic alliance with the United States and the post–World War II constitutional provisions that prohibit an active military. The alliance with the United States is strained by Japanese rightists, who, for reasons of national pride, want to close American bases and rearm Japan. An additional stress on the alliance has been caused by a number of serious crimes, including

rape and murder, committed by U.S. military personnel based in Okinawa against Japanese citizens.

Japan rebuilt rapidly after World War II, aided in the early years by the United States. It eventually became a giant in global business, exporting automobiles, electronic goods, and many other products to the United States, Western Europe, and other parts of the world. By 2001, Japan remained the world's most technologically advanced economy, after the United States; but China had taken over as the world's second largest economy.

Conflict and Transfers of Power in East Asia

The history of the entire region is to some extent grounded in what transpired in China and Japan. All East Asian countries, for example, have been affected by ancient Confucian philosophy, the relics of which can be found in present-day folk beliefs and social policy. More recently, the 1949 communist revolution in China has had ramifications for the rest of the region, as has the economic rise of Japan. North Korea and Mongolia have joined China in the communist experiment. Taiwan and South Korea have chosen to model themselves after Japan's state-aided market economy.

The Korean Peninsula. Korea was a unified country until a cold war clash led to its division into North and South Korea in the

Thinking Geographically: ON THE WEB

years after World War II. From 1910 to 1945, Japan occupied Korea and treated it as a colony. Korea became a source of cheap minerals and agricultural raw materials for Japan's own growing manufacturing centers. Resentment toward the Japanese grew in Korea, and nationalist sentiment increased. Western ideas, including communism, flooded into Korea, competing with the more traditional Confucian and Buddhist beliefs.

As World War II was drawing to a close in August 1945, the Soviet Union declared war against Japan and invaded Manchuria and the northern part of Korea. Rather than allowing the entire country to come under Soviet control, the United States proposed dividing Korea at the 38th parallel. The United States took control of the southern half of the peninsula and the Soviet Union the northern half. Over the next few years, the United States held elections in the south, recognized South Korea's independence, and withdrew its troops. Meanwhile, in the north, Korean and Soviet communists established a military government. In June 1950, North Korean forces attacked South Korea. The United States soon came to the south's defense, leading a United Nations force that fought a 3-year war against North Korea, the Soviet Union, and China. The Korean War, which ended in a truce in 1953, caused huge losses of life on both sides and devastated the peninsula's infrastructure. Since then, North Korea has closed itself off from the rest of the world and remains impoverished. South Korea has evolved a prosperous state-aided market economy, with U.S. assistance in its early years and the model of Japan as a guide afterward.

Taiwan. In the late nineteenth century, Taiwan (once known as Formosa) was a poor agricultural island on the periphery of the Chinese Qing dynasty. In 1895, as the Qing dynasty weakened, Japan annexed the island as a colony and exploited its resources for more than half a century. Then, in 1949, the Chinese Nationalists (the Kuomintang) lost to the Chinese communists and had to flee the mainland. They chose to settle in Taiwan, despite opposition from the indigenous people and earlier migrants from Fujian, and renamed the island the Republic of China (ROC). Taiwan became home to the exiled government of Chiang Kai-shek and more than a million migrants from the mainland. Over the next 50 years, it became a modern, crowded, and highly industrialized society and played a leading role, through investment and technology diffusion, in the rapid transformation of East Asian and Southeast Asian economies.

Mongolia. Mongolia is traditionally a land of nomadic herders. It long posed a threat to China because its horsemen periodically moved south into the territory of the Chinese empire. China colonized the region in 1691 and controlled it until the early twentieth century. A communist revolution followed soon after independence from China in the 1920s, and Mongolia continued to follow a communist system under Soviet guidance until the breakup of the Soviet Union in 1989, which launched it on a difficult road to a free-market economy.

POPULATION PATTERNS

East Asia is the most populous world region. China, with 1.3 billion people, has more than one-fifth of the world's population. Yet, throughout East Asia, people are having markedly fewer children than in the past. Chinese couples residing in urban areas are limited by law to one child (rural families may have two), and the Chinese are now reproducing at a rate considerably lower than the world average: 13 births per 1000 people yearly, as opposed to the world average of 22 per 1000. Nevertheless, China's population will continue to grow for years to come, simply because so many Chinese are just entering their reproductive years.

If the one-child family pattern continues, there is some hope of curtailing population growth significantly. With relatively fewer births, the average age of the population will rise fairly quickly. As the population ages, China's growth will eventually slow and possibly even stop—but not until perhaps the year 2050, and only after millions more people are added. However, a halt in growth depends on continued and improved compliance with control measures. Couples who are more affluent often choose to have two or even three children and pay the fines imposed for having more than one child. If this trend continues, China may double its population in less than the presently estimated 117 years. On the other hand, if China follows Japan's lead and very small families—some with no children at all—become common, then China will face another problem: a very large elderly population that must be supported by a relatively small group of people of working age.

Elsewhere in East Asia, population growth rates hover around China's rate of 0.6 percent. Japan has the lowest rate in the region, at 0.1 percent (one-sixth the rate in the United States). In fact, Japan's population is growing so slowly that it will not double for 700 years. Thus Japan's population, like China's, is rapidly aging (see the box "An Aging Population and the Immigration Debate" on page 467). By 2025, Japan will have a ratio of one pensioner for every two workers—comparable to that of several European countries with aging populations. Only in Mongolia (with 2.5 million people) and North Korea (with 22.7 million) are women still averaging more than two children each. But even in those two countries, increasing opportunities for women outside the home mean that family size is shrinking, family life and gender roles are changing, and the numbers of dependent elderly are growing.

As you can see in the population map of East Asia (Figure 9.6), people are not evenly distributed on the land. In fact, many parts of Japan, the Koreas, Taiwan, and China are very lightly settled—either because the terrain is very rugged or (in the case of western China and Mongolia) because the climate is extremely dry or cold. Ninety percent of China's people are clustered on only one-sixth of the total land area. They extract a very high level of agricultural production from this land, though at considerable environmental cost. People are concentrated especially densely in the eastern third of China: in the North China Plain, the middle and lower Chang Jiang basin, the delta of the Zhujiang in Southern China, and the Sichuan Basin. South Korea is also densely settled, as are northern and western Taiwan. In Japan, settlement is concentrated in a band stretching from the cities of Tokyo and Yokohama on Honshu Island through the coast of the Inland Sea to the islands of Shikoku and Kyushu. This urbanized region is one of the most extensive and heavily populated metropolitan zones in the world, accommodating well over half of Japan's total population. The rest of Japan is mountainous and more lightly settled. Because the overall population of East Asia is so large, and because population issues are especially

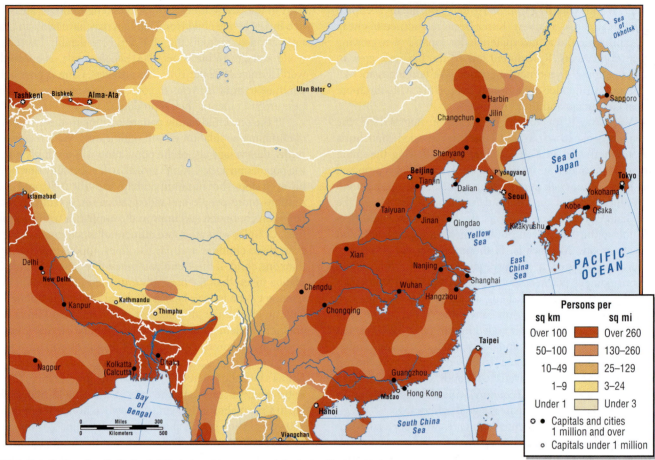

FIGURE 9.6 Population density in East Asia. [Adapted from *Hammond Citation World Atlas,* 1996.]

important to this part of the world, a number of demographic topics are discussed in the sociocultural issues section below.

Population and Health Issues: HIV-AIDS and SARS

According to the Joint United Nations Program on HIV-AIDS, East Asia is experiencing a concentrated and potentially explosive HIV-AIDS epidemic. By the end of 2003, between 700,000 and 1.3 million adults and children in the region were infected with HIV. Although rates of infection are relatively low by global standards—less than 1 percent of the population are infected—this figure masks a large number of infections. During 2003, an estimated 150,000 to 270,000 adults and children in the region were newly infected with HIV, and at least 32,000 people died as a result of the disease. The majority of HIV infections are in China, where concentrated epidemics have been under way for many years in certain regions (Yunnan, Xinjiang, Guangxi, Sichuan, Henan, and Guangdong) and are poised to take off in several others. The number of reported HIV-AIDS cases has increased significantly in recent years, particularly among injecting drug users and sex workers. Japan is also witnessing a steady increase in the rate of new HIV infections. The number of new HIV-AIDS cases reported

annually has doubled since the 1990s to more than 900 in 2002. This rise has been accompanied by an increase in other sexually transmitted infections over the same period and may be related to evidence of more widespread sexual activity among Japanese youth. The incidence of HIV-AIDS in South Korea is somewhat lower than in Japan, and in Mongolia, only about 100 adult cases have been reported.

The appearance of another virus in the region captured the world's attention in November 2002, when the first cases of a previously unknown, but highly contagious, disease called Severe Acute Respiratory Syndrome (SARS) appeared in southeastern China. The epicenter of the epidemic was Guangdong Province. From there the disease quickly spread across the Zhujiang Delta to Hong Kong and then to the rest of the world. Soon cases were reported in Singapore, Vietnam, Hong Kong, and Toronto, Canada, and a smaller number in the United States and Europe. In total, SARS infected 8400 people in 29 countries in less than 6 months. The rapidity with which the disease spread and the potentially fatal consequences of infection—nearly 800 people died—caused global alarm. In March 2003, the World Health Organization issued a global health alert and travel advisory that brought international travel to and from East Asia to a virtual standstill and led governments to screen travelers and quarantine infected individuals.

Renata Simone, reporter: "That SARS spread so quickly around the world is a warning about the dark side of globalization, a fact of life in the new world of international trade and fast, easy travel."

The airport in Hong Kong is one of the busiest in the world, but it was eerily quiet the day Dr. David Ho, world-renowned HIV-AIDS researcher, arrived in Hong Kong in June 2003. Global alarm over the outbreak of SARS cut flights from 1000 a day to a mere 40. Dr. Ho joined an international group of researchers who teamed up at the University of Hong Kong to tackle SARS. In the field of HIV-AIDS research, Ho developed a protease inhibitor treatment that keeps HIV-AIDS patients healthy for years. He thinks a similar mechanism may help with SARS. When Ho, who lives in New York, got the call asking for help, he couldn't say no. "It's sort of like you see a fire in the next town, and you're very anxious, but folks over there say, 'please come and help us,' and you definitely cannot say 'no.' And so here we are."

The SARS virus is able to attach to a healthy human cell, penetrate it, and then replicate, creating copies of itself that will attack other cells, causing more and more damage to a victim's lungs. Dr. Ho's idea is to stop this powerful virus by using peptides (pieces of protein) to block the initial attachment to the host cell. His team tests this idea in the lab using combinations of 12 specially formulated peptides in cell cultures containing the SARS virus. The experiment takes 48 hours to incubate, a painful wait for the scientists. Finally, something interesting shows up. Five of the 12 peptides have protected the cells from the SARS virus.

The next day, Dr. Ho announces the results at a press conference. Praising the diligence of his team, he states, "It's been a heroic effort, and I say some of the heroes are here in front of you." Through the extraordinary effort and brilliance of scientists like Dr. Ho, SARS was successfully contained. On May 18, 2004, the World Health Organization declared that the virus had been contained, but warned that living virus particles were still contained in research labs around the world, and that these specimens posed a health risk. Indeed, the last known cases of SARS were linked to a research lab in Hong Kong.

Learn more about SARS and its potential global impact in the FRONTLINE/World video "Chasing the Virus."

Quick Review

1. In which season(s) does Japan receive monsoon precipitation?

2. What innovation allowed the Qin dynasty to unite China under one government?

3. Which two political parties struggled for control of China after the fall of the Qing dynasty in the early twentieth century?

4. What are some of the possible downsides of the slowing population growth rate in China and Japan?

II CURRENT GEOGRAPHIC ISSUES

East Asia, like other world regions, faces the difficult task of providing a decent standard of living for its growing population—and doing so equitably and without causing environmental damage. Although East Asia's countries adopted new economic systems after World War II, most of them are making progress in creating a better life for their citizens. In fact, Japan has one of the highest standards of living in the world. The great challenge in this part of the world will be to maintain improved standards of living for large numbers of people in a manner that does not place too much stress on the environment. As economic prosperity expands, people's consumption of goods and resources increases, and environmental degradation is usually the result. How to balance economic prosperity with environmental integrity is an issue every region faces. Yet East Asia's particularly large population makes finding solutions to this problem especially imperative.

POLITICAL AND ECONOMIC ISSUES

After World War II, the countries of East Asia adopted new economic systems. The communist regimes of China, Mongolia, and North Korea relied on central planning to set production quotas and to allocate goods among their citizens. In contrast, Japan, Taiwan, and South Korea established **state-aided market economies**, with the assistance and support of the United States and Europe. This economic system is similar to the free-market economies of the West in that market forces, such as supply and demand and

FIGURE 9.7 Political map of East Asia.

competition for customers, determine many economic decisions. However, the government maintains a much more interventionist role in the economy than it does in countries such as Britain or the United States. More recently, the differences among East Asian countries have diminished as China and Mongolia have adopted reforms that allow market forces, rather than government planning quotas, to set production quantities and prices. This transition has caused hardship for many people because rapid growth under free-market reform has been uneven, bringing income inequality and the loss of social services. The older market economies of East Asia are in a recession that has lasted since the mid-1990s.

The State-Aided Market Economy Countries

Throughout the nineteenth century, the economies of Japan, Korea, and Taiwan were minuscule compared to China's. Then, during the twentieth century, all three grew tremendously. Despite economic setbacks in the 1990s, Japan's economy remains the second largest in the world, as measured by the total value of goods and services produced, not adjusted for purchasing power parity. It is more than three and a half times larger than China's economy, even though Japan's population is only one-tenth as large as China's. But if the value is adjusted for purchasing power parity (PPP), China's economy ranks second. It is about twice as large as Japan's, and growing so rapidly that it may overtake the U.S. economy in size in a decade or so. South Korea, with one-twenty-sixth the population of China, has the world's thirteenth largest economy. Tiny Taiwan's economy is one-sixth as large as mainland China's adjusted for PPP. The credit for the economic success of Taiwan and South Korea belongs mostly to Japan, which—as far back as the late nineteenth century—began to develop the model of a state-aided market economy that all three countries have used to achieve prosperity.

Japan's Economic Rise. Japan is renowned for its successful export industries, which produce automobiles, cameras, computers, video and audio equipment, and many other products. Yet, less than a century and a half ago, the country was one of the most isolated in the world, its doors closed tightly to all trade and international travel. At a time when the industrial revolution and urbanization were transforming Europe and North America, Japan was primarily a nation of poor farmers subsisting in a feudalistic society. Its industrial technology was primitive and far behind that of other leading countries. Thus Japan's rapid rise to world prominence in trade and technology since the middle of the nineteenth century is one of the most remarkable tales in modern history—all the more so because of its limited base of natural resources and its crippling defeat at the end of World War II.

It would be hard to overstate the extent of devastation in Japan at the end of the war. Except for Kyoto, which was spared because of its historical and architectural significance, all major cities had been mercilessly bombed and destroyed by the United States—most notably Hiroshima and Nagasaki with nuclear weapons and Tokyo with incendiary bombs. Factories, roads, bridges, rail lines, power stations, and other infrastructure were in ruins. The people were hungry because food production and distribution were disrupted, and they were homeless and jobless because of the destruc-

tion wreaked by the bombing. At the same time, several million Japanese returned from Japan's former colonies as refugees. According to a description in the Japanese press, the condition of the country immediately after the war was "one hundred million people in a state of trauma."

Thus it was literally from ashes and despair that Japan rebuilt itself in the second half of the twentieth century into one of the world's wealthiest and most influential nations. The key ingredients of its success were the incredibly hard work and sacrifice of the Japanese citizenry and a relatively benign period of occupation (1945–1952) during which the United States imposed democratic and economic reforms on the country. Procurements for the Korean War (1950–1953) were a particular stimulus for Japanese industry. Also contributing to rapid recovery were the opening of markets abroad for Japanese products, guaranteed lifetime employment for most urban workers, and cozy relations between Japanese business and political institutions. A strong state bureaucracy helped wealthy private capitalists to start industrial enterprises that became the foundation of early economic development in Japan—a strategy that was later replicated in Taiwan and South Korea. The government provided advice, financial assistance, and protection from internal and external competition.

Japanese innovation in manufacturing was another factor in Japan's success. Japanese manufacturers organized firms in highly efficient spatial arrangements, and their practices have been imitated globally. Firms that supply components such as engine parts for automobiles are often clustered around a single internationally renowned company where final assembly takes place. This proximity allows suppliers to deliver their products "just in time," literally minutes before they are needed. Defects in production runs can be spotted more quickly, so fewer defective parts are produced, and less space is taken up by warehouses. This **kanban system** has influenced economic geography globally. Rather than outsourcing parts to cheap labor pools in developing countries, many firms all over the world pursue an alternative strategy of locating the assembly of both components and final products in a single area, either abroad or at home.

For much of the post–World War II period, Japan's economy grew roughly 10 percent a year, with heavy industry and manufacturing the leading sectors. By 1964, when it hosted the Summer Olympics in Tokyo, Japan was eager to engage in trade. The Olympics gave Japan a chance to show off its event-planning abilities and its new international hotels, light rail system, high-speed bullet train linking Tokyo with other major cities, and ultramodern Olympic sports facilities. Japan's economic growth continued steadily through the next three decades, with the result that Japanese brand names such as Sony, Panasonic, Nikon, and Toyota have become household words around the world.

More recently, Japan's economy has centered on the service sector. Since 1992, employment has shifted significantly from manufacturing to services, such that the service sector now employs 70 percent of the workforce and constitutes 68 percent of Japan's gross domestic product. Banking and financial services are especially important industries that have contributed to Tokyo's status as a global city.

Japan's "economic miracle" has had, and continues to have, an immense worldwide effect. Resources from all parts of the

In Kyushu, Japan, a highly mechanized Toyota plant turns out cars for the domestic market. Here precision robot arms weld car bodies. [Mike Yamashita/Woodfin Camp & Associates.]

world are shipped to Japan: minerals and ore from Africa; coal and steel from Australia; timber from the Philippines, Indonesia, Malaysia (see Figure 10.17 on page 531), Canada, and the United States; and oil from Southwest Asia. Japan's purchases of resources are mainstays of local economies in the producing countries and make Japan something of a global employer. Japanese know-how and technology turn the imported raw materials into inexpensive yet high-quality manufactured goods that are in demand throughout the world. Profits from these industries have often been invested abroad in factories, hotels, and resorts, thus securing multiple markets for the country.

Recession. In the early 1990s, however, the Japanese economy showed signs of strain, and by the end of the decade all three of East Asia's state-aided market economies had experienced significant economic downturns. Japan's economy was the weakest, shrinking by several percentage points a year. Its economic problems are linked to the economic crisis in Southeast Asia that began in the mid-1990s and became apparent by 1997 (discussed in Chapter 10). Japan's trade with Southeast Asia shrank. In 1999, 45 percent of its exports were to that region, but by 2003, Japan was shipping just 30 percent of its exports to South Korea, Taiwan, and Southeast Asia combined. In the late 1990s, Japan was the single largest investor in Southeast Asian ventures, including agriculture, forestry, manufacturing, mining, and tourism. When companies in Southeast Asia began to close because of bad debts, Japanese investors lost money. Similarly, when millions of Southeast Asians lost their jobs, they could no longer afford to buy Japanese products. As a result, many Japanese companies began to experience significant business losses.

Corruption also contributed to the Japanese economic crisis. The long-standing close relationship between government and industry had nurtured favoritism. Politicians often encouraged economic development projects that were not needed because the projects brought money and jobs to voters in their home districts. For example, many rural areas in Japan now have elaborate transportation systems far beyond the needs of their small rural popula-

tions. Another problem was that Japan overexpanded its productive capacity. Between 1988 and 1992, Japan increased its productive capacity by an amount equal to the entire economy of France—just when the demand for its products began to contract at home and abroad. Bureaucrats who had nurtured enterprises with advice and financial support were often unwilling to let them go bankrupt, even when they were being poorly managed or no longer had the potential to make profits. The Japanese government forced banks to make large loans to troubled firms. Eventually, both the firms and the banks went bankrupt. South Korea and Taiwan have experienced financial crises similar to Japan's, though Taiwan has suffered less, largely because it has been more willing to let its weak enterprises go bankrupt.

Some experts on Japan link the Japanese financial crisis to overprotected Japanese workers who have little incentive to perform at their full potential. Good workers are rewarded with job security rather than higher pay, a policy that has encouraged conformity at the expense of innovation and creativity. Another consequence of the relatively low wages is that middle- and lower-class Japanese consumers cannot buy many of the high-quality items they themselves produce, whereas their counterparts in other industrialized economies form a large internal market for locally made wares. Japanese high-end goods are sold instead to consumers in North America and Europe. Many experts argue that these characteristics of the Japanese economy make it inherently unstable because they render the country dependent on a few large external (foreign) markets. Should the economies of North America and Europe falter, so would the economies of Japan and all the countries following its economic policies. It is therefore argued that workers in Taiwan, South Korea, and Japan should be paid more, with raises contingent on performance, so that there will be incentives for these workers to generate more of their own technological innovations. In addition, more prosperous workers would create an internal market for domestic products, thus strengthening the economy.

Economic Challenges. Experts on Japan now debate whether the economic and political model that served the country so well for most of the second half of the twentieth century can continue to be successful. As a result of the economic recession, purchasing power has declined, and urban populations—though still affluent by the standards of most other countries—are growing uneasy. The Japanese people, long accustomed to tolerating extended workdays, overcrowding, pollution, high prices, and only modest buying power in the interest of rapid economic growth, are agitating for changes that will improve their living standards. The greater participation of women in the workforce and the need for employees to be more creative and innovative are leading to changes in the workplace (see the discussion on pages 492–493). Moreover, increased life expectancy is producing an aging population. The ever larger proportion of the elderly will soon require more social spending, which will further burden the working-age population (see the box "An Aging Population and the Immigration Debate" on page 467).

Internationally, China's rising economic power promises opportunity for Japanese investors, but China could also pose a competitive threat in the marketplace. Japan is also experiencing increasing economic competition from neighboring South Korea and Taiwan and the newly industrializing countries of Southeast

Asia—all countries with strong export economies. Meanwhile, the United States has been pushing Japan to remove its barriers against foreign imports. Since 1950, Japan's access to North American markets has been a major factor in its economic success, but it has consistently blocked incoming trade goods.

The Communist Command Economy

The communist economic systems of China, Mongolia, and North Korea emerged in the mid-twentieth century. After World War II, all three countries abolished private property, and the state took control of virtually every aspect of the economy: agricultural and industrial production; construction; service industries such as transportation and distribution, utilities, social services, and education; and sales of food and consumer goods. Many leaders saw this course as the only way to salvage societies that had been badly damaged by a grossly self-serving elite, by Japanese and European colonial domination, and by civil war. The idea was to centralize the management of all facets of the economy and thereby achieve maximum efficiency. Economies directed by central government planning agencies in this way are called **command economies** (see the discussion of the Soviet command economy in Chapter 5, page 248). These sweeping changes ultimately proved less successful than was hoped.

The Commune System. When the Communist party first came to power in China in 1949, its top priority was to make great improvements in both agricultural and industrial production. Its first strategy was land reform, which to the communists meant taking large, unproductive tracts out of the hands of landlords and putting them into the hands of the millions of landless farmers. By the early 1950s, much of China's agricultural land was divided into tiny plots. But it soon became clear that this system was inefficient. The farmers needed to produce enough to feed people who were being drawn out of agriculture to work in the many expanding industries. Communist leaders decided to band small landholders together into cooperatives so that they could share their labor and pool their resources to acquire tractors and other equipment to increase agricultural production. In time, these cooperatives were expanded into full-scale communes with an average of 1600 households each. The communes, at least in theory, took over all aspects of life. They were the basis of political organization (all within the Communist party), and they provided health care and education. In addition, the communes were responsible for building rural industries to supply themselves with such items as fertilizers, gunnysacks, pottery, and small machinery. They also had to fulfill the ambitious expectations of the leaders in Beijing for better flood control, expanded irrigation systems, and the generation of surplus funds for investment elsewhere.

The commune system had several difficulties. Farmers were required to spend so much time building roads, levees, and drainage ditches or working in the new rural industries that they had much less time to farm. Local Communist party administrators often compounded the problem by reporting harvests that were larger than they actually were, to impress their superiors in Beijing. The leaders in Beijing responded by requiring larger grain shipments to the cities, which created food shortages in the countryside.

These measures pushed the system past its limits in the late 1950s. Thirty million people starved to death during a massive famine between 1958 and 1961.

Focus on Heavy Industry. Industrial production beyond that of small farm-linked industries was focused on heavy industry rather than on consumer goods. The leadership believed that China needed first to improve its infrastructure—to build roads, railways, and dams—all of which required heavy equipment. So the government emphasized mining for coal and other minerals, producing iron and steel, and building heavy machinery. The state obtained funds for industrial development from the already inefficient agricultural sector in several ways. Farmers were required to sell their crops to the state at artificially low prices, and the state kept any profits gained in reselling the food. In addition, the state diverted funds that had been earmarked for agricultural improvements to industrialization. Other funds for industry came from profits in mining and forestry and from savings on workers' wages, which were kept very low.

By design, most people in communist China were not allowed to consume more than the bare necessities. On the other hand, the "iron rice bowl" policy guaranteed nearly everyone a job for life and the means to obtain the essentials. As was the case in Japan, the emphasis on job security, rather than on rewarding creativity and hard work, led to overwhelming conformity and a lack of innovation. Hence productivity remained low. Central planners in Mongolia and North Korea also favored giving workers job security rather than higher wages.

Regional Self-Sufficiency in China. One objective of China's leaders was to reduce regional variations in prosperity. For centuries, agricultural and herding areas in the interior, especially those in more remote areas to the west, had been much poorer than the coastal areas to the east, where industries and trade were concentrated. After the revolution, the communist leadership instituted a policy of **regional self-sufficiency.** It encouraged each region to develop independently, building agricultural and industrial sectors of equal strength, in the hope of creating jobs and evening out the national distribution of income. Government funds were used to set up industries in nearly every province, regardless of practicality. For example, a steel industry was established in Inner Mongolia at great expense, when other provinces already had steel industries that could have been improved with a far greater payoff. Similarly, agriculture was extended into marginal physical environments far from markets, often in areas where continuous irrigation was necessary or where growing seasons were very short. Huge mechanized grain farms were developed on cleared forestlands in far northern Heilongjiang, when the same effort elsewhere would have yielded better results and allowed the forests to be saved for other purposes. This policy of regional self-sufficiency also encouraged individual communities to develop backyard smelters, kilns, tool and die factories, or their own tiny tractor factories. Certainly these efforts were a tribute to community ingenuity, but ultimately they wasted time and resources while producing relatively little.

Cold war tensions also encouraged the dispersion of industry across China and regional self-sufficiency. Communist leaders believed that a dispersed pattern of industrial development and the ability of each region to feed itself would foil an enemy's effort to destroy China's productive capacity quickly.

Globalization and Market Reforms in China

Dramatic changes have occurred in China in recent decades, and their effects have reverberated throughout the world. In the late 1970s, the failure of China's command economy to make dramatic improvements in living standards motivated China's leaders to enact market reforms. Initially, these reforms were made domestically at the local level, and they changed China's economy in three ways. First, economic decision making was decentralized. Second, farmers and small businesses were permitted to exchange their produce and goods in markets. The markets encouraged competition and improved the efficiency with which food was produced, and goods were manufactured and distributed to consumers. Finally, regional specialization, rather than regional self-sufficiency, was encouraged.

Eventually, the new markets expanded to allow foreign investment in Chinese enterprises and the sale of foreign products in China. This shift to a market-based economy has transformed East Asia and the surrounding areas as China has become a participant in the global economy. In the 1990s, it emerged as a significant producer of manufactured goods, and it represents a market of more than 1 billion potential consumers. Nearly every major company in the world is eager to sell its goods to customers in China.

The Reforms in Overview. The key change in China's economic policies was the decentralization of decision making, which enabled the market system to function, encouraged regional specialization, and subsequently led to the opening up of Chinese markets. To decentralize decision making, the government introduced **responsibility systems,** which gave managers of many state-owned enterprises the right and responsibility to improve the efficiency of their operations. Managers set production quotas and prices for goods and services according to supply and demand. To some extent, anyone could create a new enterprise. Those who earned money could either spend it or reinvest it. The reforms increased regional specialization as managers and entrepreneurs (sometimes with assistance from the state) took advantage of the different resources and opportunities offered by different areas of the country. For example, most new manufacturing has been located in coastal areas, such as the city of Shenzhen (see the box "Pearl River Megalopolis," page 488), that are more accessible to ocean transport and the global economy. In contrast, interior areas have refocused on agriculture and raw materials industries.

Regional Trends in Agriculture. The market reforms and integration into the global economy have brought new opportunities to some agricultural sectors. Throughout China, farmers have been encouraged to organize family-sized or larger operations that meet the local food demand. Because agricultural potential varies greatly across China, regional specialization in agriculture has increased. Those farmers who live close to cities are producing food for urban markets. Instead of selling the grain they grow, increasing numbers of farmers are feeding it to animals to satisfy a rising demand for meat among the affluent. In places where the

growing season is long enough, such as the far southern provinces of Yunnan and Guangdong, farmers are now growing fresh produce for distant markets in northern cities as well as overseas. China has become a major exporter of several high-value, labor-intensive agricultural products to other countries in Asia, particularly Japan and South Korea. Exports of fresh and preserved fruits, vegetables, fish, animal products, and manufactured foods are likely to increase as China expands its comparative advantage in such products. At the same time, China has recently increased its imports of bulk grains to fill domestic demand, and this trend is also likely to continue in the near future.

While changing patterns of food production, consumption, and trade may be expected in any country undergoing the economic transformations that China is experiencing, they nevertheless raise new challenges for China's **food security:** its ability to supply sufficient basic food to all its people consistently. Several factors are putting pressure on the food production system and restricting China's ability to increase grain production in particular. Most notably, these factors include the reduction of arable land and shortages of water (see the discussion on pages 471–473). Only a portion of China's vast territory can support agriculture because much of the land is too cold, too dry, or too steep to cultivate (Figure 9.8). China's huge population has already stretched the productive capacity of many fertile zones well beyond sustainability. And demands are increasing—in some parts of China, there are six times more people living on productive farmland today than there were just 50 years ago. As urban populations grow more affluent, their taste for meat and other animal products will place an added burden on China's agricultural land. The raising of animals for meat requires more land and resources than the plant-based national diet of the past. On the other hand, like Japan, China now has the economic capacity to buy the food and grain it needs from other countries. (For a comparison, see the discussion of Japan's food security on pages 489–491.) While some researchers and policy makers are concerned about China's increasing dependence on foreign suppliers, others view the

Nearly half of China's crop production takes place along the fertile banks of the Chang Jiang (Yangtze River). Among the crops grown are rice, wheat, barley, corn (maize), beans, cotton, and hemp. Note the terrace-style fields throughout the landscape. [Michael S. Yamashita/CORBIS.]

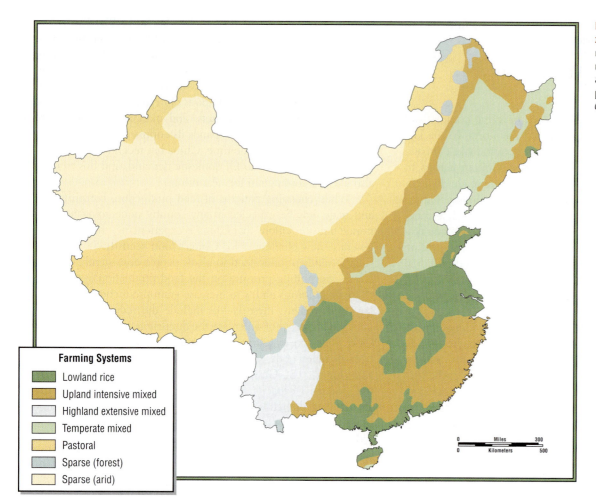

FIGURE 9.8 China's agricultural zones. As part of the economic reforms instituted in China, greater regional specialization in agricultural products is taking place. [Adapted from *Hammond Citation World Atlas,* 1996.]

Farming Systems
- Lowland rice
- Upland intensive mixed
- Highland extensive mixed
- Temperate mixed
- Pastoral
- Sparse (forest)
- Sparse (arid)

Miles 300
Kilometers 500

availability of food from multiple sources as a positive development in China's ability to maintain its food security.

A Market Focus for Rural Enterprises. One of the most remarkable recent economic developments in China is the growth of rural enterprises that provide a wide variety of goods and services—from operating coal mines to making plastic flowers to assembling electronic equipment. These enterprises have become the mainstay of many rural economies, especially in the eastern and southern coastal provinces of Jiangsu, Fujian, and Guangdong. Such enterprises may still be township or village collectives, although some are privately owned. All of them leave major decisions to managers (rather than to the collectives) and price their products according to market demand. Rural enterprises have bloomed so quickly and successfully that they now constitute a quarter of the Chinese economy, and they may actually produce over half of the country's industrial output and 40 percent of its exports. Statistics are not easy to come by, but it is known that these enterprises now employ more than 125 million people—more than the Chinese government itself—and account for more than 30 percent of farm household income.

Nonetheless, environmental pollution and corruption have accompanied the growth of these rural enterprises. In the mountains west of Beijing, for example, where many such industries are located, pollution exceeds that in the city. Clouds of exhaust from trucks that link the rural industries with their markets contaminate the air. Paper mills pollute waterways, farmers' fields, and aquifers. Some managers (perhaps even a majority) steal funds from their own enterprises, try to evade taxes, and pay officials to look the other way when they breach environmental regulations. The government, although moving to enact legislation that could lessen these problems, is wary of driving rural enterprises out of business because they have provided important new sources of income and economic growth.

The Persistence of Regional Disparities. Since the reforms, the Chinese economy has become more productive overall, and many citizens have achieved greater prosperity. The per capita gross domestic product has doubled several times, from U.S. $300 in the mid-1980s to U.S. $4020 (PPP) by 2001. But the reform of state-owned enterprises has proceeded slowly, largely because new private enterprises have not emerged fast enough to employ those who would lose their jobs if state-owned factories reduced their workforces. In addition, long-term regional disparities in wealth are once again increasing (see Figure 9.1A on page 443). Rural workers' wages are falling far behind the wages of workers in urban areas, thus increasing the disparity in standards of living between urban and rural regions.

Furthermore, remote provinces, such as Tibet (Xizang) and Yunnan, and those with little mineral or agricultural wealth, such as Inner Mongolia and Qinghai, have lost special support from the central government, yet have had difficulty generating rural enterprises. Increasingly, so many young people and skilled laborers are leaving the interior provinces (often without official permission) that the only people left in some rural communities are children, some women, and the very old. Meanwhile, the eastern and southern coastal provinces—with their larger cities, more developed infrastructure, skilled workforces, and ready access to ocean transport—are enriching themselves through manufacturing industries and foreign trade.

International Trade and Special Economic Zones. Central to China's new market reforms are the **special economic zones (SEZs)** and related **economic and technology development**

zones (**ETDZs**) that now stretch into the interior of China (Figure 9.9). In the early 1980s, China's government, wary of the disruption that could result from a rapid opening of the economy to international trade, selected five coastal cities to function as free trade zones: Zhuhai, Shantou, and Shenzhen in Guangdong Province (near Hong Kong); Xiamen in Fujian; and Haikou on Hainan Island. Industries in these cities were allowed to practice management methods that were not permitted in the rest of the country until recently. For example, foreign investors and their companies were given income tax breaks, reductions in import duties, and the freedom to take their profits out of China. In the years since they were established, these SEZs have succeeded in attracting international investors and industry. Industry, in turn,

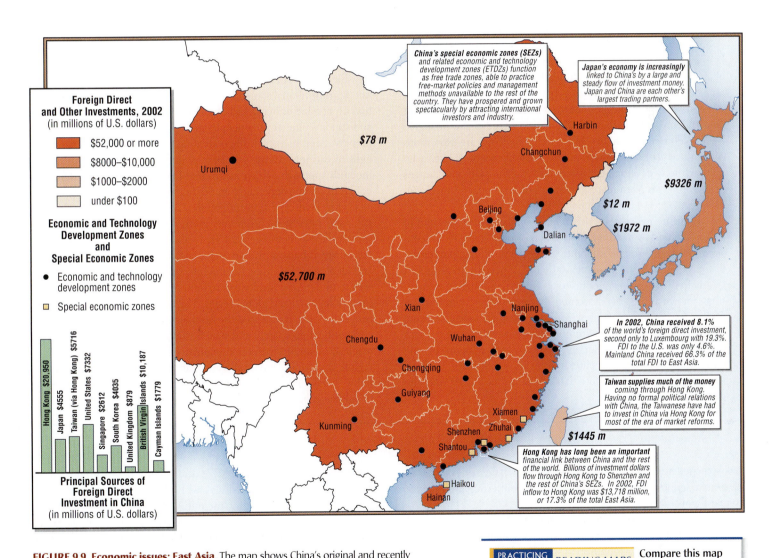

FIGURE 9.9 Economic issues: East Asia. The map shows China's original and recently designated special economic zones (SEZs) and economic and technology development zones (ETDZs). The colors on the map reflect levels of foreign investment (direct and otherwise) in each of the countries of the region. Also shown are the principal countries from which the investments in China came. Although surely high, there is some evidence that China's figures are greatly inflated due to "round-tripping" (in which capital leaves the country for another, usually a tax haven, and then returns as "foreign" investment). [Adapted from http://www.harper.cc.il.us/ ~mhealy/geogres/maps/eagif/chsez.gif; http://www.sezo.gov.cn/; http://www.UNCTAD.org, and http://www.fdimagazine.com/ news/fullstory.php/aid/215/China%92s_FDI_merry-go-round.html.]

PRACTICING GEOGRAPHY   READING MAPS   Compare this map with Figures 9.6 and 9.8. What are the likely reasons for dense population along the coast? What relationship do you see between population density and patterns of foreign direct investment? What effects do you think the growth of China's cities in both size and population may have on the practice of agriculture?

PRACTICING GEOGRAPHY   How does the changing Chinese city reflect the changing Chinese economy?

Rapid industrialization in East Asia in the final quarter of the twentieth century has had a range of impacts on people and places across the region, as forces of globalization have combined with local conditions to produce complex new cultural and economic landscapes. In China, enormous amounts of capital investment have poured into the coastal cities and provinces with strong ties to the world economy. My research has been based in the cities of the South China coast, which is also China's historic maritime trading region.

China's cities are places of dramatic contemporary transformation and spatial restructuring. Entire districts of some cities have been moved and rebuilt, while some cities are entirely new, like Shenzhen. One of my long-standing research interests is urban landscape transformation, which expresses cultural and economic change in the Asian city. Under rapid industrialization, large-scale real estate development projects have characterized many cities in East Asia. Such megaprojects, like the new city center under construction in Shenzhen, or the development of the Shanghai Pudong district, the site of China's tallest building, are expressions of both real economic power and the ideas of economic elites who guide China's development under reform. The plans and designs of such projects often represent aspects of traditional Chinese culture, reinterpreted in contemporary terms.

The emergence of an urban labor market is also apparent in Chinese cities, as male migrant workers typically engage in construction work. Women migrants also make up a large proportion of the population in many coastal cities, especially in Southern China in the special economic zones. Indeed, Shenzhen is sometimes known as the "city of women" because of the enormous size of the female migrant labor force working in manufacturing enterprises. So the built environment of the new Chinese city represents the unevenness emerging in society and economy: on one hand, the new economic elite, and on the other hand, the new low-wage workers on whose labor China's economy and the global economy depends.

Carolyn Cartier
University of Southern California

I also work on understanding rapid regional growth in China in geographic terms. For example, how should we understand growth spatially? It is interesting that one way the Chinese state continues to be substantially involved in the economy is by managing its administrative geography. For example, the central government can redesignate rural counties as cities, or make counties come under the jurisdiction of cities, which makes their land subject to urban leasing and real estate development, thus propelling rapid growth.

Fieldwork in East Asia has always defined my research, but not by conventional standards. Rather, I see fieldwork as a contextual experience of living and observing events and places in cities and regions, interviewing people and assessing political economic processes, and then critically evaluating such empirical realities in light of existing history, scholarship, and theory. I take a mixed methods approach, which means that I collect and analyze different kinds of data depending on the research questions at stake. In analyzing landscape transformation, I interview architects, planners, and interested community groups as well as people on the street to assess their divergent views about how places should be transformed. I assess primary materials as representations of ideas and ideologies, including documents, newspapers, and plans, and seek to understand local views in light of political economic pressures at national and global scales. This is an approach to human geography that transcends the subfields of cultural, economic, and political geography and which is also interdisciplinary in its reach. It maintains the core disciplinary concerns of place, space, region, and landscape, and ways that spatial processes are centrally at stake in explaining landscape transformation as well as the global economy.

---

has spawned related activities in these free trade zones, such as insurance, general commerce, information gathering and processing, and consulting services.

Since 1990, many more cities along the coast and inland have become ETDZs. Shanghai, for example, long a window on the outside world for China, was chosen to spearhead the expansion of free trade zones along the Chang Jiang all the way to Chongqing. By the late 1990s, the Chinese authorities had designated 32 provincial locations in the interior of China as ETDZs, a distribution that stretches all the way to Urumqi in Xinjiang. These ETDZs provide footholds for international investors and multinational companies eager to establish operations in a country with such a huge pool of cheap labor and such a large number of potential con-

sumers. Today the original SEZs and related development zones (there are now at least 48 SEZs and ETDZs altogether) are major **growth poles** in the country, meaning that their success is drawing yet more investment and migration. Coastal cities such as Shenzhen, Qingdao, Wenzhou, and Changan are attracting skilled people from the rest of the country. In just 20 years, many coastal cities have grown from mid-sized towns or even villages into some of the largest urban areas in the country (see the box "Pearl River Megalopolis" on page 488).

Life in the Growing Cities. In the vignette that opens this chapter, we met Hong Xiaohui, who left her home in Sichuan Province to work in Mattel's factory in Changan. She is typical of the millions of young migrants who are leaving sheltered rural village life to work in the highly competitive industries of the SEZs. Many arrive intending not only to succeed as individuals, but to send money home to their impoverished families and possibly even return one day to improve their home communities. We return now to Xiaohui's story because her triumphs and travails over the first year after she left home demonstrate what life is like for migrants from the hinterlands to China's booming coastal cities.

PERSONAL VIGNETTE   When Xiaohui first arrived in Changan, she learned that, in addition to her pay being lower than what she had been promised, the shifts at Mattel were usually 12 hours long, with no overtime pay and just one day off a month. More important, when Xiaohui rapidly mastered her task of putting hair on Barbie dolls, she began to entertain dreams of becoming a supervisor—only to discover that such a position might require sexual favors to the plant manager. With an income equaling just U.S. $24 a month, Xiaohui and her fellow workers could afford only a crowded dormitory room and canteen food. There were 11 women in one small concrete room with a tiny bath.

In less than 2 months, Xiaohui broke her contract with Mattel and returned home to her village in Sichuan, where—using ideas and the assertiveness she had gained from her time in the south—she opened a kebab stand. In just one month, she made ten times her investment of U.S. $12. Soon she also opened a Hot Pot soup shop in her family home. But Xiaohui yearned to return to the southern coast to seek employment on her own. So, leaving the kebab shop to her sister and brother-in-law, she eventually returned to Changan.

In one year, the city had grown by 20 percent and now had 1400 foreign companies trying to hire workers. Xiaohui asked one of her old friends to help her find work. In a matter of days, she found a job in a Japanese-owned factory—and also one for her sister, who rushed to Changan, leaving her husband and baby in Sichuan. The sisters quickly made friends with several young women from various parts of China, and together they enjoyed an active urban social life, hampered only by a tight budget and their generally conservative village ways. Just a year after Xiaohui's first trip to Changan, she was making enough to live relatively comfortably, though still in a crowded dormitory. She was sending money home to her parents and once again saving to start a business. This time she dreamed of opening a karaoke bar back in her village. [*Adapted from Kathy Chen, "Life Lessons", Wall Street Journal (July 9, 1997).*]

Hong Xiaohui's story illustrates why China's central government in Beijing thinks of southeastern China as "the golden coast" and hopes it will become a model for the rest of the country. Yet the golden coast is not without problems. Policing the activities of thousands of foreign firms is already an impossible task for the Beijing government. Finding ways to tax businesses without discouraging small capitalists like Hong Xiaohui is difficult. Just housing the flood of migrants and providing them with sanitation and transportation services is expensive and poses environmental problems. Developing the infrastructure of the coastal zone fast enough to accommodate all these new residents will divert resources from rural places where life is still unbearably hard. As a result, yet more people will move to the cities from the interior, and the regional disparity of wealth will increase.

One of the most visible results of China's economic reforms is the growing number of impoverished and unemployed people now crowding into its cities. During the Maoist era, the Chinese government established the **hukou system,** a permanent residence registration system in which people were classified by their place of residence (rural-urban) and their occupation, or entitlement status (agricultural-nonagricultural). Although this system was not originally intended to restrict the movement of people around the country, by the 1970s it became a means by which the state could control the flow of people from rural to urban areas. (Migration from urban to rural areas was not restricted.) This was possible because most employment opportunities were in state-owned enterprises that required an appropriate hukou registration. Without an "urban-nonagricultural" classification, people could not get jobs in cities, and so they were not able to move from rural to urban areas unless they could get their status changed, which was not easy to do. The economic reforms of the 1980s and 1990s, however, have created non-state employment opportunities in urban areas, thereby increasing people's mobility. An estimated 80 million to 100 million people have left economically depressed rural areas for cities without official permission. Because they lack official urban resident documentation, the members of this **floating population** do not have access to housing, subsidized food, or even medical care in the cities. Although some find work and contribute to the support of rural relatives, many others are unemployed and homeless. Some large industrial towns already have 20 percent unemployment, a figure that is likely to rise. This floating population may become a source of social instability as social and economic disparity between peasant migrants and full-status urbanites intensifies. Mongolia is facing similar dilemmas as it, too, implements market reforms (see page 498).

China and the World Trade Organization China's dramatic economic changes over the last 20 years, its strong growth rate, and its present ability to attract investment while maintaining a low ratio of external debt to GNP led to China's admission to the **World Trade Organization** (**WTO**) in 2001. (Taiwan was admitted the same year.) The WTO, the world's preeminent trade body, seeks to remove barriers to global trade. A poor and relatively isolated country in 1980, China is increasingly prosperous and linked to the rest of the world. As of 2004, it had one of the fastest-growing economies in the world (outstripping Japan's economy in size), partly because it received more foreign direct investment than any other developing country (see Figure 9.9 on page 461). Many governments and large multinational corporations see China's entry into the WTO as essential to the goal of global economic integration.

The entrance of China into the global economy has had far-reaching effects. For example, 80 percent of toys sold in the United States are now manufactured in China. The sheer size of its population and demand for raw materials gives China tremendous clout in global financial and commodities markets. When Chinese premier Wen Jiabao announced in April 2004 that the Chinese government would take strong action to slow the excessive growth of its

AT THE REGIONAL SCALE   The Information Highway in China: Bringing Openness to Government?

Before the economic reforms of the past two decades, the central government in China controlled the news media. By the late 1990s, however, the expanding use of electronic communication devices was loosening central control over information. Over 23 million people are now connected to the Internet, and many more have access to telephones (Table 9.1). Many more people now have personal computers, which provide access to the Internet. Just a few years ago, only a very few people had telephones, but now anyone can buy temporary cellular phone access without showing identification. As a result, many more people have anonymous access to an international network of information. Greater public access to information—which many people believe is essential in a market economy—has had a number of important social effects.

One such effect is that it is difficult for the government to promulgate inaccurate explanations for disasters or problems caused by inefficiency and corruption. In March 2001, for example, the government first attributed an explosion in a rural primary school, which killed at least 38 schoolchildren, to the act of a suicide bomber. But soon reporters in Beijing randomly dialed numbers in the village area code to reach some of the children's parents directly. The parents explained that their dead children had been forced to work in a fireworks factory during and after school hours. Proceeds from the factory went to the school budget and to school officials. The Chinese prime minister apologized publicly on March 15. He acknowledged that there had been an "industrial accident" that was

causing him to reflect on his own work, perhaps meaning that he felt implicated. Analysts inside and outside China see this apology as a watershed event that demonstrates the power of technology to make information more accessible. [*Source:* Craig S. Smith, "China arrives at a moment of truth," *The New York Times* (April 1, 2001): 5.]

TABLE 9.1 Changes in access to telecommunications in China, 1990 to 1998–2001

Telecommunications method	1990	1998–2001
Rural telephone lines	1.4 million	34 million
Internet hosts	0	1264
Internet users	0	23.4 million
Mobile phone subscribers	0	110 million
Regular phone lines	7.5 million	137 million
Personal computers	0	11.3 million

Sources: United Nations Human Development Report 2000 and *2003* (New York: United Nations Development Programme), Table 12; R. Benewick and S. Donald, "Access to information flows," in *The State of China Atlas* (New York: Penguin, 1999), pp. 198–200; Craig S. Smith, "China arrives at a moment of truth," *New York Times* (April 1, 2001): 5.

economy, the value of shares of U.S. mineral companies immediately plummeted on fears that this would decrease demand for items such as copper, aluminum, and iron ore. The following day, U.S. shipping company shares also dropped. China is a huge importer of oil—it accounts for 30 percent of the growth in global demand for oil—and its growing demand for gasoline has already translated into higher prices at the pump. This influence is likely to increase when the number of automobiles on its roads reaches a projected 100 million in 2014.

The inclusion of China in the WTO became highly controversial as awareness spread that much of China's growth is based on environmentally destructive activities and on abuses of workers. Moreover, China has brutally suppressed separatist movements in Tibet (Xizang) and Xinjiang and has committed human rights abuses against citizen groups agitating for democracy and greater political freedom. In 2001, most governments of the world and some business leaders were willing to ignore these problems to avoid angering the Chinese government and jeopardizing trade. They argued that once China was in the WTO, increased trade would result in more openness and prosperity, which would ultimately improve environmental and labor standards. In fact, there may already be signs of a new openness (see the box "The Information Highway in China" above). Some consumers, environmentalists, and workers outside China counter that increased trade may only result in more widespread abuses. They suggest

that China's entry into the WTO should have been conditional upon improvements in human rights, worker safety, and environmental protection. Workers in the United States and Europe, some of whom are losing their jobs as a result of competition from China, occasionally led demonstrations against China's admission into the WTO.

SOCIOCULTURAL ISSUES

East Asia's economic progress has led to social and cultural change throughout the region. Successful attempts to control population growth, made possible by prosperity, are creating new problems—such as an aging society, an imbalance in the numbers of males and females, and controversies over immigration. Modernized economies are changing work patterns and family structures. In this section, we look at how countries in the region are responding to these changes.

Population Policies and the East Asian Family

Population patterns in East Asia began to change after World War II. By the 1970s, government policy, aided by rapid urbanization, had reduced fertility in Japan and Taiwan to just over 2 children per woman aged 15 to 45. South Korea's fertility rate was still high, at 4.3 children per woman. By 2000, however, women in Japan, Taiwan, and South Korea were bearing fewer than 2 children on average. In China, Mongolia, and North Korea, fertility rates remained very high into the 1970s: in China and North Korea, nearly 5 children per woman, and in Mongolia, 7 children per woman. In China, leaders realized that the rapidly rising population was sapping the country's ability to make economic progress. Strict population control measures were deemed necessary, and by 1980, China had adopted a policy of one child per family.

The One-Child Policy in China. Chinese demographers and the Chinese people at large soon realized that the one-child policy would dramatically affect family life and the social fabric. For example, within two generations, the kinship categories of sibling, cousin, aunt and uncle, and sister- and brother-in-law disappeared from all families that complied with the policy—a major loss in a society that has long placed great value on the extended family. But it is the prospect of the one child being a daughter, with no possibility of having a son in the future, that has caused the most despair. Despite strong conditioning by the communist government to consider the sexes equal, the preference for sons remains strong. Sons are preferred partly because of the belief that only a son can inherit property, pass on the family heritage, and provide sufficient income for aging parents (even though daughters or daughters-in-law are expected to provide daily personal care of the aged). For years, the makers of Chinese social policy have sought to eliminate the preference for males by empowering women economically and socially. They believe that female children will be just as desirable as males when it is clear that well-trained, powerful daughters can bring honor to the family name and earn sufficient income for their families.

The offspring in one-child families, whether male or female, have few kinfolk with whom to share tasks when they reach adulthood. They have increased responsibilities for both child care and elder care. And in China, the elderly are especially dependent on the younger generation. There was never a nationwide pension system. Pensions are even less common now, since the closing of many collectives during the economic reforms of the 1980s and 1990s. Hence, as the single-child family spreads through the generations, elderly parents will remain dependent on their one child.

The one-child policy has indeed reduced population growth rates: between 1950 and 1990, the Chinese birth rate dropped dramatically, and the one-child family is now the most common family form. Nevertheless, the policy has never been very popular—for the reasons just discussed and because the policy is enforced unevenly. Rural families are exempt from the policy, and from time to time, the government has offered incentives to limit family size to young couples in some parts of the country, but not in others. In some urban areas, couples who have only one child receive a monthly subsidy and a housing allowance. In other areas, complying couples receive special chances for promotion. In some cases, those who have additional children may lose these benefits, receive demotions, and have to pay a fine; or a pregnant mother may be forced to abort her second child. In other cases, the infractions may be ignored. Some Chinese—those newly affluent in the market economy and some private-enterprise farmers—may have additional children and simply pay the substantial fines. Some ethnic minorities are officially exempted from family-size restrictions, presumably to answer political charges that the majority Han group is too dominant. Still, there is evidence that family-size limits have been imposed on ethnic Tibetan women, some of whom have been forced to have abortions.

There is no doubt that population control has been successful in reducing births, but some of the drop in fertility is simply the result of modernization of life in China and would have happened without the one-child policy. In almost all situations, people who learn to read, move to the city, and take up a lifestyle based on cash income will choose to have fewer children because children are expensive in the new urban environment. They do not help to produce income as they did on the farm; in fact, they require a major investment of time and money to prepare them for urban adulthood.

The population pyramid on the left in Figure 9.10 reflects two sharp past declines in birth rates. The first took place during the Great Leap Forward (1959–1961), when policy errors resulted in famine and malnutrition led to lower fertility. After that era, China's birth rate grew rapidly. The second decline began 30 years ago and continued as the result of strict population control efforts in the early 1980s. In the mid-1980s, birth rates increased when market reforms were instituted. Increasing prosperity prompted some Chinese to have two or more children despite official sanctions. By the mid-1990s, another significant decline in birth rates was under way, perhaps inspired more by the realities of raising children in urban situations than by the one-child rule. The pyramid on the right illustrates the long-term effect of current trends on China's population 50 years from now. A smaller proportion of young people will be supporting a relatively larger elderly population, as is currently the situation in wealthy countries in the West (see Chapters 2 and 4).

Gender Values and Population Control. The left-hand pyramid in Figure 9.10 shows another interesting phenomenon. If you look carefully, it appears that for 50 years or more, the number of

This army officer enjoys some moments with his only child, a son. [Alain le Garsmeur/Panos Pictures.]

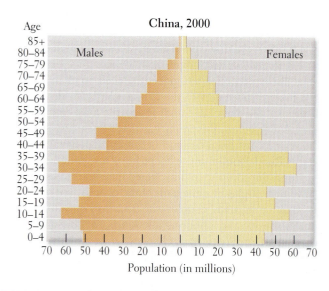

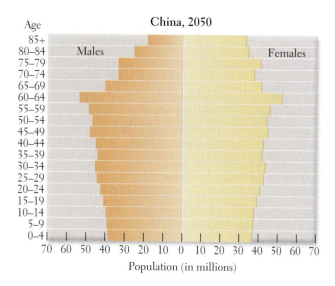

FIGURE 9.10 Population pyramids for China, 2000 and 2050. [Adapted from U.S. Bureau of the Census, International Data Base, at http://www.census.gov/cgi-bin/ipc/idbpyry.pl.]

male infants born has significantly exceeded the number of female infants. The pyramid shows that in 2000, among those 0–4 years of age, there were roughly 44 million girls and over 50 million boys, indicating a deficit of about 6 million girls, or a ratio of 114 boys to every 100 girls. (Without intervention, the normal sex ratio at birth is 105 boys to 100 girls.) What happened to the missing girls? There are several possible answers. Given the preference for male children, the births of these girls may simply have gone unreported as families hoped to conceal their daughter and try again for a son. There are many anecdotes of girls being raised secretly or even disguised as boys. Adoption records indicate that girls are given up for adoption much more often than boys. Or the girls may have died in early infancy, either through benign neglect or actual infanticide. Finally, some parents have access to medical tests that can identify the sex of a fetus, and there is evidence that in China, as elsewhere around the world, some of these parents choose to abort a female fetus. A side effect of the preference for sons is that there is now a growing shortage of women of marriageable age throughout East Asia. In China alone, between 1991 and 1996, 88,000 young women and girls were freed by the police from kidnappers who, for a fee, provide mates for men who have had difficulty finding one.

Elsewhere in East Asia, the cultural preference for sons persists, with the deficit of girls showing up on the 2000 population pyramids for Japan, the Koreas, Mongolia, and Taiwan. Nonetheless, evidence shows that attitudes are changing. Increasingly, girls are being educated as well as or even better than boys: in Japan, South Korea, and Mongolia, the percentage of women receiving higher education equals or exceeds that of men. In China, only half as many women as men receive higher education. Data is not available for North Korea and Taiwan.

Family and Work in Industrialized East Asia

In Japan, Taiwan, and South Korea, the institution of marriage must meet the needs of urban life because close to 80 percent of the population now lives in cities. In China, roughly 60 percent of the people still live in small towns and rural areas, but urbanization is increasing rapidly. Life in a small city apartment is different from life among a host of relatives in a farming village. Nonetheless, in urban situations throughout East Asia, the wife still performs most domestic duties and dedicates herself primarily to looking after her husband and children. Often she also has responsibility for the care of elderly parents or parents-in-law. Surveys show that East Asian women who do work outside the home do not yet desire a job with full responsibilities. Although there is a small but growing group of trained young women who seek careers and are as ambitious as men, even many university-educated married women say they wish to earn only supplementary income for the household.

One reason that the primary duties of family life have been assigned to women is that jobs in East Asian industrial economies are particularly demanding. Workdays are long, and commuting often adds more than 3 hours to the time away from home. Even more important is the "culture of work," which in Japan, Taiwan, and South Korea (and increasingly in China) is based on male camaraderie and demands an especially high level of loyalty to the firm. In Japan, for example, a firm will have a corporate song, a corporate exercise program, and a strict dress code. After work, leisure time often must be spent with business colleagues. Management is a vertical hierarchy that is accessible only to those who are perceived as diligent and socially interactive after hours. Rivalry between firms is often so strong that it elicits a siege mentality among workers. It is considered disloyal to refuse overtime, and a man who takes off early to spend time with his wife or to help her with care of the children or elderly is not a "real man." Women, when hired, are there to support men as secretaries and assistants, not to participate as full members of corporate teams.

By 2000, the East Asian urban work ethic and family structure were being publicly challenged, not least by men themselves. In Tokyo, for example, a group calling itself Men Concerned about Child Care meets regularly to discuss ways to participate more fully in home and community life. Another group has formed to lobby for 4-hour workdays for both men and women, so that both can spend

AT THE LOCAL SCALE   An Aging Population and the Immigration Debate in Japan

With increasing longevity and falling fertility, Japan faces a new social and economic environment. Japan's population has the largest proportion of elderly people among the industrialized nations, while the nation's birth rate is the lowest. Japanese women now have an average of 1.3 children, well below the level needed to maintain the current population of 127.5 million. If this trend continues, Japan's population will plummet to just over 100 million by 2050, shrinking the country's labor pool by more than a third and dragging down the country's national wealth.

The solution seems obvious: throw open the country to immigrants from Asia, Africa, and Latin America. A recent United Nations report estimates that Japan will have to import over 640,000 immigrants per year just to maintain its present workforce and avoid a 6.7 annual drop in gross domestic product. Many business executives support importing foreign workers in the near future.

But standing in the way of such a policy decision is a political culture that seeks to keep Japan ethnically pure and avoid the social problems that many associate with immigration in Western countries. "It's been obvious for a while now that Japan desperately needs an influx of workers from outside," says Tony Lazlo, director of an organization that researches multicultural issues in Japan. "People in business have been forced to recognize this, but it's not really widely known in Japanese society." According to Lazlo, resistance to higher levels of immigration is likely to take two forms. One is simply resistance to having a lot of foreigners around—prejudices against foreigners are deep-rooted. The other will come from people who believe that Japan is not really ready for immigration and needs more time.

In the meantime, foreigners dribble into Japan in a multitude of legal (and many illegal) guises to fill the most dangerous and low-paying jobs. The integration of these workers—who come from places such as Peru, Bolivia, Pakistan, Sri Lanka, the Philippines, and Africa—has been less than perfect and points to the challenges Japan faces ahead.

[Adapted from Asia Society, "Japan's aging population: A challenge for its economy and society" (October 7, 2003), http://www.asiasource.org; Howard W. French, "Insular Japan needs, but resists, immigration," *New York Times* (July 24, 2003; David McNeill, "Time running out for shrinking Japan," *Japan Times* (September 2, 2003).]

time with the family. This group has filed lawsuits against employers who routinely require employees to accept job transfers to distant locations where they cannot take their families. These movements may eventually result in some important changes in gender and work roles throughout the region. It is particularly significant that urban men are the ones challenging the extremes of the lockstep East Asian work ethic. They are the group that has been most regimented and most deprived of personal time and family life, and they are also the group that has the power to change the system. If work rules are liberalized and workdays shortened, these policies will add to the cost of East Asian products and lessen their competitiveness in the marketplace. But if employers do not make work situations more attractive to middle managers, efficiency and innovation may suffer, which will also reduce competitiveness.

Indigenous Minorities

Cultural diversity exists throughout East Asia, even though most countries have one dominant ethnic group. In China, for example, 93 percent of the population call themselves "people of the Han." The name harks back about 2000 years to the Han dynasty, but it gained currency only in the early twentieth century, when nationalist leaders were trying to create a mass Chinese identity. The term *Han* does not denote an ethnic group, but rather connotes a general and unspecified pride in Chinese culture and a sense of superiority over ethnic minorities and outsiders. The main language of the Han is Mandarin, although there are many dialects. (All versions of Chinese use the same writing system.)

Because China is such a populous country (1.3 billion people in 2003), the non-Han minorities, though only 9 percent of the population, still number about 117 million people. There are more than 55 different minority groups scattered across the vast expanse of China; as Figure 9.11 shows, most live outside the Han heartland of eastern China. They are located in the hinterlands: near the

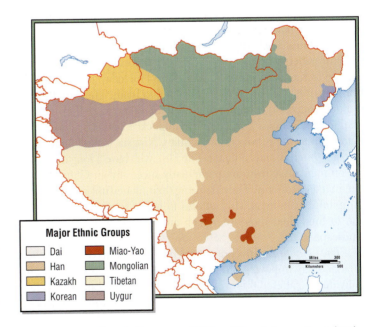

FIGURE 9.11 Major ethnic groups of China. Although the most populated areas in the east and southeast are dominated by the Han, China is home to several distinct ethnic groups. This map does not show recent migrations by Han into far western China (Xinjiang) and Tibet (Xizang); nor does it show the Hui, who consist of people from many ethnic groups who have converted to Islam and who are found in disparate locations along the old Silk Road and in coastal southeast China. [Adapted from Chiao-min Hsieh and Jean Kan Hsieh, *China: A Provincial Atlas* (New York: Macmillan, 1995), p. 12.]

borders with Mongolia and Russia in the north and with Central Asia and Pakistan in the west; in Xinjiang and Tibet (Xizang); and in the southern mountains that lead to Southeast Asia. Some of these areas have been designated *autonomous regions*, where these groups theoretically can manage their own affairs. In practice, however, the Han-dominated Communist party has not allowed self-government. The central government in Beijing controls the fate of the minorities, especially those who live in zones that are believed to pose security risks or that have resources of economic value.

Turkic-Speaking Peoples. One such zone is Far Northwest China, where Turkic-speaking peoples, such as the Uygurs and Kazakhs, live. Many of these peoples remain nomadic. The Beijing government has sent troops and hundreds of thousands of Han settlers to this area of China. The Han settlers manage mineral extraction and run military bases, facilities for nuclear testing, and power generators. Communist party officials are frank in their assertion that an important role of these Han settlers is to dilute the power of minorities within their own lands and to assist these minorities in ridding themselves of "unacceptable" cultural practices and distinctive national identities. In the case of the Turkic-speaking peoples, cultural assimilation to Han ways would include giving up Islam. Although assimilation may be the long-term outcome, for now there has been a rebirth of Islamic culture, inspired in part by the resurgence of Central Asia in the aftermath of the Soviet decline. Turkic people have revived trade across the borders. The old Silk Road market of Kashgar (now Kashi) is back in business. (The **Silk Road** was the ancient trading route between eastern Asia and regions to the west, including Europe; see Figure 9.19, page 485.) Islamic prayers are once again heard five times a day, Muslim women are again wearing Islamic dress, and Islamic architectural traditions are being revived.

The Hui. The original Hui are the descendants of ancient Muslim traders who plied the Silk Road across Central Asia from Europe to Kashgar (Kashi) to Xian. Most Hui today are descended from other ethnic groups who converted to Islam. They still live as distinct groups, some numbering a million or more, on the western margins of the Loess Plateau, in small pockets of the North China Plain, and in the provinces of Sichuan, Ningxia, Yunnan, Hunan, Gansu, Qinghai, and Xinjiang. The Hui, who have a long tradition of commercial activity, are particularly successful in China's new free-market economy. They are using their money not only for consumption of luxury goods, but also to revive religious instruction and to fund their mosques, which are now more obvious in the landscape.

The Tibetans. In contrast to the prosperous Hui are the Tibetans, an impoverished ethnic minority of nearly 5 million individuals scattered thinly over a huge and high mountainous region in western China. Their homeland is widely known as Tibet, but since Chinese troops invaded in 1950, the territory has been divided and assigned to what the Chinese government designates as Xizang and Qinghai provinces. The Chinese government also suppressed the Tibetan Buddhist religion, long the mainstay of most Tibetans' daily lives, by destroying thousands of temples and monasteries and massacring many thousands of monks and nuns. The spiritual leader of Tibet, the Dalai Lama, was forced into exile

At the market in Kashgar (now called Kashi), an ancient Silk Road city, you can still buy a camel, a fine Oriental rug, the original bagel, or the skin of an endangered animal. You can also get a haircut or get your teeth fixed, as this person is doing. [Reza/National Geographic Image Collection.]

in India in the 1950s, along with thousands of his followers. The Potala, his palace in Tibet's ancient capital Lhasa, is now a museum run by the Chinese government. To dismantle traditional Tibetan society further, hundreds of thousands of Han Chinese were resettled in Tibet, where they control the economy and major cities, exploit mineral and forest resources, and force native Tibetans to adopt Chinese ways. We discuss Tibet in greater detail on pages 484–485.

Indigenous Population Groups of Southern China. In China's far southeast, in Yunnan Province, there are more than 20 groups of ancient native peoples, including the Bai, Yi, Li, Miao, Yao, Dai, and Zhuang. These peoples speak many different languages and live in remote areas tucked away in the deeply folded mountains. Many of these groups have ethnolinguistic connections to the Tibeto-Burmese people, others to the Thai and Cambodian people. Interestingly, women are treated more equally here than in Han culture areas. A crucial difference may be that among several groups, most notably the Dai, the husband moves in with the wife's family at marriage and provides her family with labor or income. A husband inherits from his wife's family rather than from his birth

PRACTICING GEOGRAPHY   COMPARING LOCAL LANDSCAPES

Study the landscapes in these two photos. Photo A is from Beijing, China. Photo B was taken in Melaka, Malaysia. Can you determine what type of building appears in each landscape? What do the architectural features of the buildings tell you about what they are? What clue does the man's attire in Photo A provide? Finally, what cultural influences link the two buildings? Why? [A, Peter Sanders Photography; B, Kevin Bubriski.]

family. In contrast, in most of China, a woman traditionally leaves her birth family and moves to her husband's family compound, where she works under the direction of her mother-in-law. Anthropologists think it is not a coincidence that the minorities in Yunnan and neighboring provinces, unlike the majority of Chinese, value female children just as highly as males. Similar patterns of family structure and gender values are noted among indigenous peoples in Southeast Asia, as discussed in Chapter 10.

Aboriginal Peoples in Taiwan. In Taiwan and the adjacent islands, the Han account for 95 percent of the population, but this region is also home to 60 minority groups. Nine of these are aboriginal groups that have lived in Taiwan for thousands of years. They have some cultural characteristics—languages, certain types of weaving, the making of bark cloth, iron smithing methods, and agricultural and hunting customs—that indicate a strong connection to ancient cultures in far Southeast Asia and the Pacific. Mountain dwellers have resisted assimilation better than plains dwellers; both are now protected and may live in mountain reserves if they choose. But increasingly, these Taiwanese aboriginal peoples, who numbered 365,000 in 1994, are attending schools and participating in mainstream urbanized and modernized Taiwan life. Hence native cultures—embodied in languages, beliefs, skills, and modes of family support—are dying out. Like those of native groups in other societies worldwide, the average incomes of the Taiwanese aboriginals lag far behind those of the majority. In addition, they experience the acute social problems typical of the underclass everywhere (alcoholism, low self-esteem, unemployment, under-education, and poverty).

The Ainu in Japan. Japan has a particularly strong sense of cultural solidarity and a tendency to suppress difference. Nonetheless, there are several indigenous cultural minorities in Japan. Not surprisingly in a country that prizes sameness, they have suffered considerable discrimination, although the topic is sensitive and rarely discussed. A small and distinctive minority group is the **Ainu.** Now numbering only about 16,000, the Ainu are a racially and culturally distinct group who are thought to have migrated many thousands of years ago from the northern Asian steppes. The Ainu, who lived by hunting, fishing, and some cultivation, once occupied the northern parts of Honshu and Hokkaido islands. They are being displaced by forestry and other development activities. Today there are very few full-blooded Ainu because, despite prejudice, they have been steadily assimilated into the mainstream population. Some now make a living by demonstrating traditional crafts to tourists visiting "Ainu villages." Among some Ainu there is now a revival of "Ainu spirit," with new attention given to relearning traditional ways and teaching children the Ainu language.

ENVIRONMENTAL ISSUES

The countries of East Asia share a number of environmental concerns. Some result from high population density and rapid economic development, others from long-term patterns of resource management or ineffective planning in more recent times. China has the widest array of environmental problems, some of which also affect Mongolia and North Korea. Japan, Taiwan, and South Korea are affected by issues that result primarily from their high degree of industrialization, their efforts to overcome limited resource bases, and their high rates of urbanization.

China's remarkable record of improved well-being is now slipping, as a result of the very activities that produced its dramatic economic turnabout. It is now apparent that China has the most severe environmental problems on the planet—so severe that they could

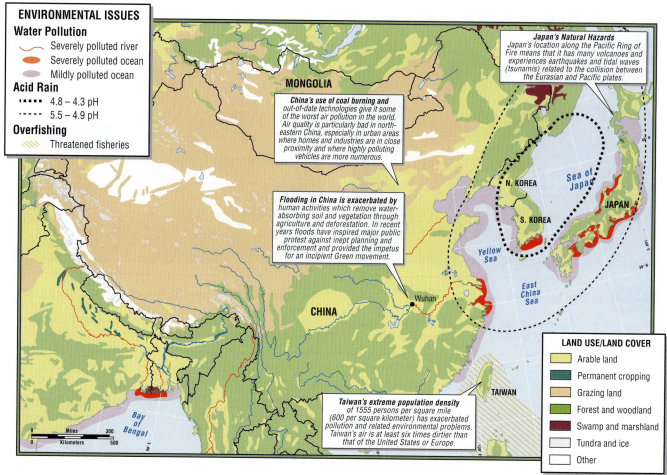

FIGURE 9.12 Environmental issues: East Asia. Environmental issues in East Asia are extensive and include air and water pollution, water scarcity and flooding, and deforestation (25 percent of China's forests have been clear-cut in just three decades). Air pollution in this region often results in acid rain (the pH of acid rain is explained in Figure 2.22 on page 92), part of which originates in China but is blown to the east by prevailing winds.

prevent future economic progress (Figure 9.12). Accomplishments such as industrialization, improved transport, increased agricultural production, better housing, widespread home heating, and urbanization are producing air and water pollution, water scarcity, and other negative environmental effects. Pollution alone causes degenerative health conditions such as asthma, chronic bronchitis, intestinal diseases, and various cancers. Adding to the problem is population growth that is still significant, despite vigorous efforts at birth control. Of the several environmental problems that exist in China, including resource depletion and erosion, we focus here on just two: air pollution and water availability and quality.

Air Pollution in China

Since the late 1970s, industrial output in China has been growing at an annual rate of 18 percent. China's rapidly developing industries require large amounts of energy. China's consumption of fossil fuel energy has risen sharply over the last 20 years: between 1977 and 1999, coal consumption doubled, and coal now fulfills 75 percent of China's energy needs. The use of coal is likely to increase despite more use of oil, gas, and other sources of energy, such as hydroelectricity generated by the new Three Gorges Dam (see the

box "The Three Gorges Dam "on page 472). China has considerable reserves of coal.

Coal burning is the primary cause of China's poor air quality and is blamed for a critical toll in respiratory ailments. The combustion of coal releases high levels of two pollutants: suspended particulates and sulfur dioxide (SO_2). In Chinese cities, these emissions can be ten times higher than World Health Organization guidelines. Figure 9.13 shows emission levels of particulates and sulfur dioxide for a number of Chinese cities and compares them with those for other large cities worldwide.

The geographic pattern of air pollution in China correlates with population density and urbanization. In the lightly populated plateaus and mountains of the north, northwest, and west, air pollution occurs primarily in urban areas where industries are concentrated and where there are large numbers of homes to be heated. In the Sichuan Basin and in most of eastern China, air pollution is prevalent in both cities and rural areas. Air pollution is the worst in China's Northeast because homes and industries are in close proximity and both depend heavily on coal for fuel. Less heating is necessary in the warmer climate of southeastern China, but the air is still laden with sulfur and suspended particles because there are many industries in this region and they burn a

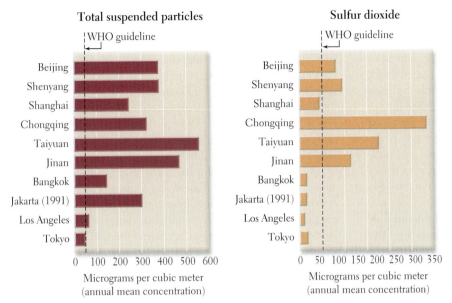

Ambient Concentrations of Air Pollutants, 1995

Total suspended particles

Sulfur dioxide

Micrograms per cubic meter
(annual mean concentration)

Micrograms per cubic meter
(annual mean concentration)

FIGURE 9.13 Air pollution concentrations in selected Chinese and other large cities, 1995. The vertical dashed lines indicate World Health Organization safe guidelines. [Adapted from *World Resources 1989–99* (New York: Oxford University Press, 1998), p. 117.]

softer, highly polluting coal. Because precipitation rates are higher in the south, air pollution also becomes water pollution. The rain absorbs sulfur in the air and becomes acid rain, which is displaced to the east by prevailing winds (see Figure 9.12).

Throughout eastern China, large numbers of people are exposed to high levels of gaseous emissions from vehicles. The use of personal cars in China has hardly begun: Beijing has just one-tenth the vehicles that Los Angeles has. Nevertheless, pollution attributable to vehicles in Beijing already equals that of Los Angeles. The reason Beijing has such high auto emissions is that most of China's vehicles are fueled with leaded gasoline, and lead emissions can reach 14 to 25 times the suggested maximum per vehicle. Chinese cars also emit 6 to 12 times more carbon monoxide than foreign vehicles.

Water in China: Too Much, Too Little, Polluted

Flooding. Like many places in the world, China suffers from having too much water at certain times and locations and too little at others. During the summer monsoon, huge amounts of rain are deposited on the land, often causing catastrophic floods along the rivers and their tributaries. Flooding is a common and particularly severe problem along the Chang Jiang and Huang He (see pages 477 and 481). Engineers have constructed elaborate systems of dikes, dams, reservoirs, and artificial lakes to help control flooding and to provide water for irrigation and power for industry and urban populations. Since 1949, for example, the government has built 120,000 miles (193,000 kilometers) of dikes.

Despite these efforts, the flood control system failed repeatedly in the late 1990s. Along the Chang Jiang, heavy rains in June and July of 1996 caused some of the worst flooding in two centuries. Several thousand people drowned, and millions were still homeless months later. Crops that were ready for harvest rotted in the fields. Much of the farmland and many essential fish-farming

facilities remained unusable for months. This catastrophe was repeated with even more severity in 1998.

Farther north, along the Huang He, the rains are less heavy, but levels in water channels are still reaching the highest marks ever recorded because silt is building up in the riverbed. The situation is so precarious that the silt must be removed by hand to prevent levees from bursting—a disaster that could kill more than a million people.

Flooding is a natural phenomenon, but in China, as in many other places, it has been exacerbated by human activities. Many environmental scientists attribute the sharp rise in worldwide deaths and damage due to floods in the last three decades to deforestation (25 percent of China's forests were clear-cut during that time), land cultivation, urbanization, mining, and overgrazing—all human activities that remove water-absorbing vegetation and soil. Another suggested factor is a change in rainfall patterns caused by global warming, the result of greenhouse gases emitted by the burning of fossil fuels.

Drought. At the other end of the water availability scale, droughts occur somewhere in China every year and often cause more suffering and damage than any other natural hazard. Droughts are triggered by periods of abnormally low rainfall and abnormally high temperatures, with attendant higher evaporation rates. They are made worse by many of the same human factors that contribute to flooding: overgrazing, irrigation, deforestation, overpopulation, and urbanization. When people begin to live or farm in dry environments, as many millions have done in China in the twentieth century, they clear natural vegetation and tap underground aquifers for water. Removing vegetation increases evaporation rates, and the more intense demand for water depletes the aquifers. Ecological forecasters believe that northern China, parts of the western United States, and other naturally dry areas of the world that are subject to increased human occupation and use of water for irrigation are mining groundwater from aquifers at grossly

AT THE LOCAL SCALE   The Three Gorges Dam

The Chang Jiang carries about two-thirds of all goods shipped on China's inland waterways. Before the first locks and dams were built at Gezhouba above Yichang in the 1980s, barges, junks, and tugs traveling upstream to Sichuan could not handle the grade and the current at certain points, so they had to be winched upriver with human power. Crews of as many as 200 men were harnessed to long towlines of braided bamboo extending from the bow of the boat to the shore. "For the whole month's work we were paid three silver dollars in addition to our food," reported one tow crewman when interviewed in the 1980s, "but we had to furnish our own towing harness and sandals."

A workforce of more than 100,000 laborers, most working only with their hands, constructed the Gezhouba lock and dam. The lock improved navigation, and the 21 dam turbines generate three times more electricity than the entire Chinese electrical system supplied in 1949. Despite these improvements, the amount of electricity supplied is now woefully insufficient.

All this will change after mid-2009, when the Three Gorges Dam is scheduled for completion (see the map at beginning of the chapter and Figure 9.17 on page 480). The Three Gorges Dam is the largest engineering project in history: it will be 600 feet (183 meters) high and 1.4 miles (2.3 kilometers) wide. It is designed to improve navigation on the Chang Jiang, generate electricity equivalent to at least one-tenth of the national total, and control flooding. The Three Gorges Dam will transform Sichuan and other provinces in the heart of China by opening the way for industrialization and global trade. The city of Chongqing has already benefited considerably from economic activities associated with the dam and has emerged as the largest city in China and its leading metropolis of the interior.

The Three Gorges Dam will also have enormous costs. Not only is its construction expensive, but there are also incalculable costs associated with relocating the 1.3 million people who lived where the dam will form a reservoir some 370 miles (600 kilometers) long. Thirteen major cities, 140 large towns, hundreds of small villages, as many as 1600 factories, and at least 62,000 acres (25,000 hectares) of farmland will be submerged. The new lake will also destroy important archaeological sites, as well as some of China's most spectacular natural scenery.

Other concerns include the possibility of disaster brought on by a major earthquake, the problem of siltation of the reservoir, and

disruptions of the ecology of the river itself. One example of likely ecological loss is the giant sturgeon, a fish that can weigh as much as three-quarters of a ton. Huge numbers of sturgeon used to swim more than 1000 miles (1600 kilometers) up the Chang Jiang past the Three Gorges location to spawn, but now the reproductive process of the sturgeon is being irretrievably interrupted. Hundreds of other species of plants and animals will also be affected, but there is little money allotted to fund research or salvage efforts.

Consequently, the dam has many critics, both in China and abroad. In 1992, many Chinese and foreign scientists and engineers tried to dissuade the Chinese government from beginning the project. Moreover, international funding sources such as the World Bank withdrew support because of the risks. However, construction of *Da Ba* (the Big Dam) is a top government priority and is proceeding quickly, even ahead of schedule.

Known as the Pearl of the Three Gorges (and considered to be one of the world's eight unique structures), Shibaozhai Temple is a 12-story wooden pagoda built about 1545 during the Ming dynasty (1368–1644). The temple stands atop a huge rectangular rock, hugging the side of a rock face some 720 feet tall. Shibaozhai Temple will just barely survive the construction of the Three Gorges Dam; the water in the dam reservoir will lap at the temple's base, and the town of Shibaozhai, 300 feet below, will be completely submerged. [Wolfgang Kaehler.]

unsustainable rates. These areas are likely to confront severe water shortages in the next several decades.

Water shortages threaten China's food security and economic growth. In the North China Plain, where half of China's wheat and one-third of its corn are produced, the water table is rapidly falling: in 2000 alone, it dropped 10 feet (3 meters). Agricultural production fell short of need in 2000, so coming shortages of irrigation water raise questions of food security. Because China's population is projected to grow by another 126 million over the next 10 years, and because of urbanization and rising water consump-

tion, the country's urban water demand is expected to increase from 65 billion cubic yards (50 billion cubic meters) in 2001 to 105 billion cubic yards (80 billion cubic meters) by 2010. Industrial water needs are expected to expand by 62 percent. The United Nations reports that in China's cities, water shortages are curtailing industrial production by more than U.S. $10 billion a year. Thus water shortages are retarding development.

With its aquifers being depleted, China has limited options: find new sources of water or find ways to conserve. Conservation is the less expensive option by far and is much likelier to reap results.

The question remains: can conservation save enough water to make a difference?

Water Pollution. Another problem related to shortage of water is water pollution. The United Nations reports that pollution of China's water, as well as the overall scarcity of water, causes health problems costing U.S. $4 billion a year. Fully one-third of the population does not have safe drinking water. Raw sewage and chemical pollution have contaminated lakes, reservoirs, and 29,000 miles (47,000 kilometers) of China's waterways. Much of the pollution comes from synthetic nitrogen fertilizers, but some of the worst water polluters are the village enterprises that have been so successful in creating jobs and alleviating rural poverty since 1978. Furthermore, only 24 percent of the Chinese population has access to adequate sanitation. In Sichuan Province, only 37 percent of the 9 million people who live in the city of Chengdu are connected to any sort of sewer system. In rural areas, 90 percent of the people have only latrines or more primitive facilities. In the east, coastal cities have so depleted underground water reserves that new construction is collapsing on subsiding land and salt water is intruding into water supplies. Some Chinese scientists estimate that one-half of the groundwater supplying Chinese cities is contaminated.

Efforts to Improve Environmental Health. The desire for improved environmental health is fueling efforts at both the national and local levels to alleviate environmental problems. The 1998 floods along the Chang Jiang, coming on the heels of the 1996 floods, prompted a major public protest against inept planning and inadequate enforcement of flood-control regulations, both acknowledged by the government. One result was the beginning of a Green movement. High literacy rates and a freer press are creating an informed public that is beginning to press for environmental cleanup. Although only 30 percent of urban water is now recycled, many of China's cities are taking steps to improve recycling procedures. Ultimately, this processed water will be safe to use elsewhere. A system of permits, incentives, and penalties aimed at removing pollutants from industrial wastewater discharges is being imposed on Chinese industry and farming at all levels.

But the proposed water management solutions have their own negative environmental effects. One example is a plan to supply water to cities, which includes conservation incentives, but also includes river diversions. Some of China's rivers are being diverted (at least partially) to provide water for growing metropolitan areas. In the Loess Plateau, for example, water from the Huang He is being diverted to several growing cities. River diversion has enabled stricter monitoring of groundwater usage: those who use more than their planned allotment are fined, those who save water are rewarded. But river diversion can lower the quality of water downstream and deprive downstream users, such as farmers and small villages, of sufficient water. In addition, river diversion destroys natural biological habitats.

Damming rivers to generate electricity and control flooding also solves some problems while creating others. China has built 22,000 to 24,000 large dams—half of all the large dams in the world—most of them since 1949. The most spectacular and potentially damaging such project is the Three Gorges Dam on the Chang Jiang (see the box "The Three Gorges Dam"). The problems it is creating include most of the typical environmental concerns that arise with dam projects.

Environmental Problems Elsewhere in East Asia

Many environmental problems found in China also exist elsewhere in East Asia. Although Japan, the Koreas, and Taiwan do not suffer water deficits because they are located in the wet northwestern Pacific, North Korea did suffer flooding in the late 1990s as an immediate consequence of deforestation, and suffered soil depletion as a long-term consequence. That country has also recently suffered a series of devastating crop failures related to environmental mismanagement, and it is thought that many thousands of people have starved. Mongolia, like northern and northwestern China, has always had to cope with arid conditions; much of the country is desert and grassland. Mongolia's forests are made up of species especially adapted to cold, dry environments; in the late 1990s, hundreds of thousands of acres of these forests were lost to fires.

All these countries have experienced the ills of water and air pollution connected with modern agriculture, industrialization, and urbanized living, but to varying degrees. Mongolia and North Korea have lower overall levels of pollution only because they are not yet heavily industrialized. Air pollution in the largest cities of Japan (Tokyo and Osaka), Taiwan (Taipei), and South Korea (Seoul) and in adjacent industrial zones is severe enough to pose public health risks. Antipollution legislation is being passed, and enforcement is increasing, but in these cities, as elsewhere around

A family masked against air pollution goes for a drive in a park in Kaohsiung, southwestern Taiwan. [Jodi Cobb/National Geographic Image Collection.]

the globe, high population densities and rising expectations for better standards of living make it difficult to improve environmental quality. Taiwan's case is an example.

Taiwan's extreme population density of 1555 people per square mile (600 per square kilometer) and its high rate of industrialization have exacerbated pollution and related environmental problems. In the far north, around the metropolitan area of Taipei, seven cities house a total of 3.26 million people, and densities can rise to 5000 people or more per square mile (1930 per square kilometer). As standards of living have increased, so has per capita consumption of water, sewage facilities, energy, and material possessions of all sorts.

Automobile ownership has been rising rapidly in Taiwan, to the point at which there are now four motor vehicles (cars or motorcycles) for every five residents. This adds up to more than 16.5 million exhaust-producing vehicles on this small island. In addition, there are nearly eight registered factories for every square mile (three per square kilometer), all emitting waste gases. As a result, Taiwan has some of the dirtiest air on earth—publicly

acknowledged by the government as six times dirtier than that of the United States or Europe (see the photo on page 473).

In the category of natural hazards, Japan is a special case. Its location along the northwestern edge of the Pacific Ring of Fire means that it has many volcanoes, and it also experiences earthquakes and tsunamis. These natural hazards are a constant threat in Japan, and the heavily populated zone from Tokyo southwest through the Inland Sea is particularly endangered. Earthquakes are also a critical natural hazard in Taiwan.

MEASURES OF HUMAN WELL-BEING

More than in most other regions on earth, there is extreme variation from country to country in East Asia on all indices of human well-being. Table 9.2 combines several indicators that are widely used by international agencies to compare how different countries are doing in providing basic health care and welfare for their citizens.

TABLE 9.2 Human well-being rankings of countries in East Asian and other selected countries

Country (1)	GDP per capita, adjusted for PPP[a] in 2001 $U.S. (2)	Human Development Index (HDI) global rankings, 2003[b] (3)	Gender Development Index (GDI) global rankings, 2003[c] (4)	Female literacy (percentage), 2001 (5)	Male literacy (percentage), 2001 (6)	Life expectancy (years), 2001 (7)
Selected countries for comparison						
United States	34,320	7 (high)	5 (GEM = 10)	99	99	77
Mexico	8,430	55 (medium)	52 (GEM = 42)	90	94	75
East Asia						
China	4,020	104 (medium)	83	79	93	71
Hong Kong	24,850	26 (high)	26	90	97	81
Japan	25,130	9 (high)	13 (GEM = 44)	99	99	81
North Korea[d]	1,000	—	—	99	99	63
South Korea	15,090	30 (high)	30 (GEM = 63)	97	99	76
Macao[d]	19,400	—	—	92	97	82
Mongolia	1,740	117 (medium)	95	98	99	65
Taiwan[e]	16,100	—	—	94	94	76

[a]PPP = purchasing power parity.
[b]The high, medium, and low designations indicate where the country ranks among the 175 countries classified into three categories by the United Nations.
[c]Only 144 out of a possible 175 ranked in the world.
[d]Data constructed from several sources.
[e]2000 estimate.

Sources: Human Development Report 2003 (New York: United Nations Development Programme);
2003 World Population Data Sheet (Washington, D.C.: Population Reference Bureau).

Human well-being is a complicated assessment that includes social, cultural, and political values as much as economic measures. The most general indicator of well-being is gross domestic product (GDP) per capita adjusted for purchasing power parity (PPP) (see Table 9.2, column 2). This value ranges from U.S. $1000 per capita per year in North Korea to U.S. $25,130 per capita per year in Japan. This is the widest spread between countries (25 to 1) in any world region. Yet this apparently enormous regional disparity of wealth is somewhat misleading. Socialist governments in North Korea, Mongolia, and China have attempted for half a century to keep abject poverty at bay by providing basic necessities for their citizens. In fact, mass abject poverty such as that found in India or Bangladesh, or during famine in Africa, has not existed in China or Mongolia for several decades. Mongolia and China now fall into the medium category in the Human Development Index ranking. Figures for North Korea are not available. Although that country is known to be extraordinarily poor, the life expectancy value indicates that at least until recently, some basic needs were being met (North Korea experienced a devastating famine from 1995 to 2001). Nonetheless, North Korea's infant mortality rate is five times higher than South Korea's and significantly higher than those for China and Mongolia. Complete figures for Taiwan are not available because of its unresolved political status vis-à-vis China. Nonetheless, Taiwan has a relatively high GDP per capita and provides many social services for its citizens. Were Taiwan ranked on the HDI, it would doubtless be in the high category.

United Nations Gender Empowerment Measure (GEM) figures are missing for all East Asian countries except South Korea, Japan, and Mongolia, and Gender Development Index (GDI) figures are not available for all countries. As a result, it is difficult to assess progress toward gender equity; the patterns appear to be irregular at best. In China, nearly 100 percent of young girls and boys attend primary school, but only 65 percent of eligible girls attend secondary school, compared to 75 percent of boys. Only 0.003 percent of women go to college, compared to 0.006 percent of men. Women workers earn about two-thirds of what men workers earn. In Mongolia, women are 21 percent more likely than men to receive a high school education, and almost twice as many women as men go beyond high school. Yet, despite being more qualified, Mongolian women, like Chinese women, earn about two-thirds of what men earn. In the industrialized societies of Japan and South Korea, nearly all women attend high school, and more than two-thirds attend college or receive training beyond the high school level. Yet Japanese and South Korean women earn less than half of what men earn.

China has made the most spectacular improvements in human well-being. Since the 1970s, China has reduced infant mortality by 50 percent, and life expectancy has risen from 63 to 71 years. These advances are the result of efforts to improve basic health care delivery, including massive immunization and family planning programs, programs to control infectious and parasitic diseases, and improved nutrition. Also contributing to well-being are better housing, improved water quality, and literacy training. These advances were mirrored for a time in Mongolia and North Korea, but both countries have lost ground in the 1990s. In Japan, Taiwan, and South Korea, basic health indicators such as life expectancy and infant mortality are comparable to, if not higher than, those for Europe and the United States.

Quick Review

1. How has the hukou system been used to control rural-to-urban migration? How is it being circumvented?

2. What issues provoked resistance to the idea of China joining the WTO?

3. How has the one-child family affected the social landscape of China?

4. What are the sources of pollutants that contribute to China's poor air quality?

III SUBREGIONS OF EAST ASIA

This section discusses Japan, Taiwan, North and South Korea, and Mongolia as four distinct subregions of East Asia. The remaining country, China, is so large in area and population, and contains so many physically and culturally distinctive parts, that it is divided into four subregions (Figure 9.14). The focus here is mainly on the modern period in each of these countries (from the twentieth century on); brief historical summaries were given earlier in the chapter (see pages 447–453). Although China dominates East Asia in size and population and wields great influence, all the countries have their own unique physical environments and social circumstances. Taiwan, South Korea, and Japan remain far more developed than China, and their experiences with rapid economic growth have influenced China, which is catching up to them, as well as many other developing countries. Mongolia and North Korea are both markedly less developed and less prosperous than the other countries in the region.

CHINA'S NORTHEAST

China's Northeast consists of the Loess Plateau, the North China Plain, and the Far Northeast (Figure 9.15, page 477). The Loess Plateau and the North China Plain are the ancient heartland of China proper. By the eighth century, the city of Chang'an (not to be confused with the town of Changan in Southern China)—near modern Xian in the southeastern Loess Plateau—was an imperial capital and cultural center for peoples of both the Loess Plateau and the North China Plain. At that time, with 2 million inhabitants, Chang'an may have been the largest city on earth. After A.D. 900, the center of Chinese civilization shifted from the plateau down and east to the North China Plain, but the Loess Plateau remained a crucial part of China. Chang'an and Xian served as the eastern terminus of the lucrative Silk Road trade that connected

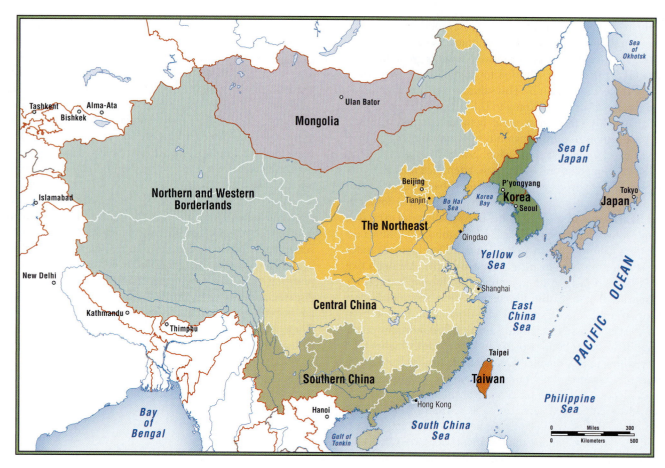

FIGURE 9.14 Subregions of East Asia. This map shows the four subregions of China as well as the other four subregions of East Asia: Taiwan, Japan, North and South Korea, and Mongolia.

China with Central Asia and Europe by the time of the Roman Empire, and perhaps earlier. In addition, the Loess Plateau acted as a **buffer zone**—a neutral area that serves to prevent conflict—between the more populous plain in the east and the arid areas to the west and north, occupied by often hostile nomadic herders.

The Loess Plateau and the North China Plain are linked physically as well as culturally. Both are covered by fine yellowish **loess,** or windblown soil. For millennia, dust storms have picked up loess from the surface of the Gobi and other deserts to the north and west, then carried it east. The loess has drifted into what used to be deep mountain valleys in Shanxi and Shaanxi provinces, creating an undulating plateau. The Huang He, in an earth-moving process that might be even more massive, has transported enormous amounts of loess sediment from the Loess Plateau to the coastal plain. Over the millennia, this river has created the North China Plain by depositing its heavy load of loess sediment in what was once a much larger Bo Hai Sea (see Figures 9.15 and 9.16).

The Loess Plateau

The Loess Plateau is sheltered from the moisture-bearing monsoon rains by higher surrounding mountains. An unusually diverse mixture of peoples from all over Central Asia and East Asia has found it

a fertile, though challenging, place to farm and herd. Among the many culture groups that share this populous and productive plateau are the Hui and the Han of China and, particularly in the western reaches, Mongols, Tibetans, and Kazakhs. Many of the inhabitants farm cotton and millet in irrigated valley bottoms or raise sheep on the drier grassy uplands. China's largest coal reserves, also found here, supply energy for the nation's industries.

Soil erosion is a particularly severe problem in the Loess Plateau. Thousands of years of human occupation have stripped the land of ancient forests, leaving only grasses to anchor the loess. Because the loess is so thick—hundreds of feet deep in many areas—deep gullies form after torrential rains. Such gullies now cover much of the landscape. Although hillsides have been terraced for agriculture, they are prone to landslides, so many of them are no longer farmed. For decades, the government has maintained a reforestation campaign to stabilize slopes, and the new forests are spreading. But cropland is too badly needed to restore all the land to forests.

The North China Plain

The North China Plain is the largest and most populous expanse of flat, arable land in China. Since the sixth century, it has been home

FIGURE 9.15 China's Northeast subregion. The Loess Plateau and the North China Plain comprised China proper in early Chinese history. The Far Northeast was formerly known as Manchuria.

to most of the imperial dynasties. From this region, the Han Chinese have dominated the empire and the country.

Today the North China Plain is one of the most densely populated spaces in the country: it has an average of more than 1400 people per square mile (540 per square kilometer). More than 370 million people—more than live in the whole of North America—live in a space about the size of France. The majority are farmers who work the relatively small fields that cover the plain. Most produce wheat, which grows well in the relatively dry climate. The food his-

torian E. N. Anderson says of the North China Plain that there is not one square inch of natural cover left. The forests that once blanketed the plain were cut down millennia ago, and now the only trees that survive are those around temples and those planted as windbreaks.

The Huang He (Figure 9.16) is the most important physical feature of the North China Plain. The river serves as a major transport artery; it also provides some irrigation water, but is particularly famous for its disastrous floods. After the river descends from the

In some places in Shanxi Province, the loess is so thick and firm that people build energy-efficient, cavelike houses in it. [Wolfgang Kaehler.]

Loess Plateau, its speed slows dramatically on the flat surface of the plain, and its load of silt begins to settle out. The silt steadily raises the level of the riverbed year after year. To contain the river, people dredge the channel and bank the silt into levees along both sides of the streambed. Over time, deposited silt has built up the riverbed until, in some places, it lies higher than the plain. During the spring surge, the river occasionally breaks out of the levees and rushes across the surrounding plain in a destructive flood. The spring surge has helped the Huang He cut many new channels over time, as can be seen in Figure 9.16. In one way, the spring surge is a gift that endows the plain with a new layer of fertile soil as much as a yard (a meter) thick. But over 3000 years of recorded history, some 1500 floods have wiped out crops, brought famine to millions of people, and destroyed whole cities. For these reasons, the river is referred to as both the Mother of China and China's Sorrow.

China's Far Northeast (Manchuria)

The northeastern corner of China, once known as Manchuria, has long been considered a peripheral region, partly because of its location and partly because of its harsh climate. The winters are long and bitterly cold, the summers short and hot. The frost-free season is often less than 120 days, so the types of crops that can be grown are limited. Nonetheless, China's Far Northeast has become increasingly important for its agriculture and natural resources. The region is endowed with fertile plains, rain that falls primarily in the summer, thickly forested uplands, and mineral resources of oil, coal, gas, gold, copper, lead, and zinc. Heilongjiang, the most northerly province, is now covered with huge state farms that grow wheat, corn, soybeans, sunflowers (for oil), and beets. With the discovery of oil at Daqing, the province has become the country's

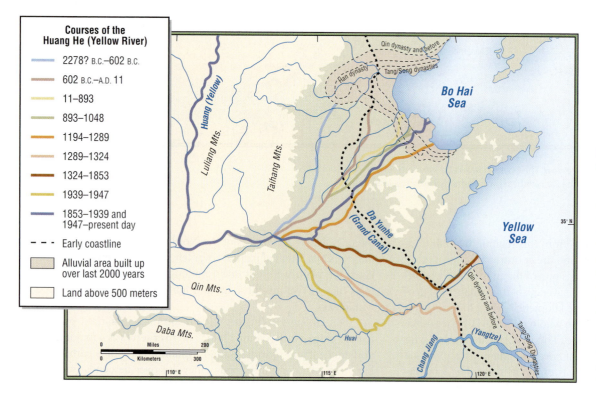

Courses of the Huang He (Yellow River)

— 2278? B.C.–602 B.C.
— 602 B.C.–A.D. 11
— 11–893
— 893–1048
— 1194–1289
— 1289–1324
— 1324–1853
— 1939–1947
— 1853–1939 and 1947–present day
- - - Early coastline
▢ Alluvial area built up over last 2000 years
▢ Land above 500 meters

FIGURE 9.16 The Huang He's changing course. The lower course of the Huang He has changed its direction of flow many times over the last several thousand years. In 2000 B.C., it flowed north and entered the Bo Hai Sea, south of Beijing. Then it repeatedly shifted to the east, then southeast, like the hand of a clock, until it joined the Huai, and finally cut south all the way to the Chang Jiang delta at Shanghai. Now it once again flows into the Bo Hai Sea. [Adapted from Caroline Blunden and Mark Elvin, *Cultural Atlas of China,* rev. ed. (New York: Checkmark Books, 1998), p. 16.]

The cityscape of Beijing is adapting to the market economy. Advertisements swathe old buildings from the communist era, cranes mark the locations of future skyscrapers of glass and steel, and bicycles compete with pedestrians, who must pick their way around hazardous potholes and building materials. Meanwhile, dust and exhaust mix to create a constant polluted haze. [Stuart Franklin/Magnum Photos.]

leading oil and gas producer. The rich mineral base is the foundation for a growing industrial sector. The huge iron and steel complex in Anshan, in Liaoning Province, produces more than 20 percent of the national output.

Development in Far Northeast China has been helped by its extensive rail and road network and is increasingly linked to river traffic on the Heilong Jiang (Amur River), which runs along more than 1000 miles (1600 kilometers) of the border between Russia and China and is the conduit for increasing trade among Russia, China, Japan, and the United States. In the nineteenth century, the Russians built a part of the Trans-Siberian Railroad through the fishing village of Harbin in Heilongjiang. Harbin, located on the Songhua Jiang, is now an industrial city of more than 4 million. Ports on the Heilong Jiang already handle more than 20 percent of the trade between Russia and Japan. U.S. trade passing through Heilong Jiang ports doubled several times over between 1995 and 2000. Although China has ample resources (fish, petroleum, and wheat) to sell along the Heilong Jiang, trade alliances along the river have been fragile so far, and disputes with Russia over fishing rights have been frequent. Nonetheless, through the efforts of the World Wildlife Federation, Russia and China have agreed to species preservation along the Heilong Jiang, where huge and diverse wetlands provide habitat for endangered storks, leopards, and the Siberian tiger.

Beijing and Tianjin

Beijing, China's capital city, lies between the North China Plain and the Far Northeast. It is the administrative headquarters of the People's Republic of China, has several of the nation's most prestigious universities, and, with the nearby port city of Tianjin, is becoming an industrial and transport center. Some of the grandest Chinese architectural masterpieces have been preserved in Beijing, notably the Forbidden City, the former imperial headquarters. Nonetheless, Beijing has lost some of its character under the communists, who have torn down many older monuments. Supposedly, the demolition was intended to improve the flow of traffic, but it also served to blot out some of China's politically out-of-fashion history. Nowadays, in the context of China's acceptance of capitalism, much of Beijing is being reconstructed, this time as a center for international commerce, with new neighborhoods of high-rise office towers and hotels replacing older neighborhoods of tile-roofed single homes and communist-era apartment blocks.

CENTRAL CHINA:

Central China consists of the upper, middle, and lower portions of the Chang Jiang (Yangtze River) basin (Figure 9.17). Like many of Eurasia's rivers, the Chang Jiang starts in the Plateau of Tibet. It then skirts the southeastern rim of the Red Basin in Sichuan Province (this is the river's upper basin, also known as the Sichuan Basin). From there, the Chang Jiang flows rapidly through the Witch Mountains (Wu Shan) to China's densely occupied central plain (the river's middle basin). Finally, it winds through the coastal plain (its lower basin) to the Pacific Ocean, exiting in a huge delta on which sits the famous trading city of Shanghai.

Sichuan Province

Sichuan is China's most populous province, with 107.2 million people, and one of its richest in resources. It has fertile soil, a hospitable climate, inventive cultivators, and sufficient natural resources to support diversified industry. Of all the provinces of China, Sichuan perhaps best embodies the revolutionary ideal of complementary agricultural and industrial sectors.

The heart of the province is the Sichuan (or Red) Basin, so named because of the underlying red sandstone. It is a fertile

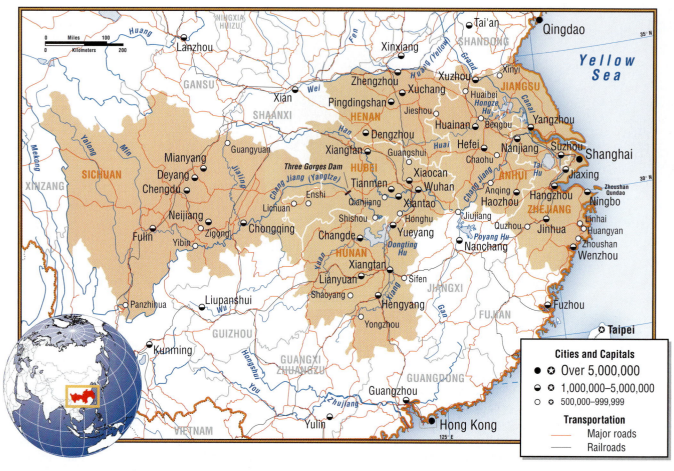

FIGURE 9.17 The Central China subregion: The Chang Jiang Basin. Central China consists of the upper, middle, and lower portions of the Chang Jiang (Yangtze River) basin.

region of hills and plains crossed by many rivers that drain south toward the Chang Jiang. The shape and angle of the Sichuan Basin result in a milder and wetter climate than is usual at such a midlatitude, continental location. Because the basin is tilted toward the south, and therefore receives comparatively direct sunlight during much of the year, the overall temperatures are higher than would be expected. Furthermore, the basin is surrounded by mountains that are highest in the north and west (the latter rising to 24,000 feet [7300 meters]), which form a barrier against the arctic blasts of winter. The mountains also trap and hold the moist, warm air moving up from the southeast all year round. For these reasons, the Sichuan climate is generally mild and humid, and the basin is so often cloaked in fog or low cloud cover that it is said: "A Sichuan dog will bark at the sun."

For the last several thousand years, the basin has been home to a dense population of relatively affluent farmers. They have cleared the native forests, manicured the landscape, and channeled the rivers into an intricate system of irrigation streams. Only a few patches of old-growth forest are left in the uplands and mountains. The geographer Chiao-Min Hsieh vividly describes how the wet "rice fields in [the lowlands] are shaped like squares on a chessboard. Everywhere one can hear water gurgling like music as it brings life and growth to the farms."

The use of irrigation in such a wet climate may seem surprising, but it is an effort to maximize production. By diverting the river water into many small sluiceways, the farmers can control flooding and make water available during the winter dry season, when temperatures are still high enough for some crops to be cultivated. Moreover, this is a region where **wet rice cultivation** is practiced. This type of rice production, which results in particularly prolific yields, requires that the roots of the plant be submerged in water for part of the growing season. It would be risky to depend solely on rainfall to keep the rice paddies filled.

Most of the people of Sichuan live in rural areas in the basin, where the population density is more than 800 people per square mile (300 per square kilometer). Their crops include rice, wheat, and corn. Most farmers raise animals such as silkworms, pigs, and fowl as a sideline; in the new market economy, it is often these sidelines that earn them the most cash income (see the box "Silk and Sericulture" on the next page). In the mountain pastures to the west of the basin, farmers raise cattle, yaks, sheep, and horses, which produce meat, hides, wool, and draft animal power. Many of the mountain herders are Tibetan (Sichuan has the largest group of Tibetans outside Tibet—close to 1 million).

The two main cities of the Sichuan Basin are Chengdu, the capital, and Chongqing. Chengdu, with a current population ap-

AT THE LOCAL SCALE Silk and Sericulture

Silk has long been an important crop in central and southern China. Silk cloth has figured prominently in China's interactions with the outside world—indeed, it gave its name to the Silk Road trade route between China, Central Asia, and Europe. Silk production spread to Japan in the seventh century, and it is now a major industry in many parts of South and Southeast Asia. Although the overall economic importance of silk has been reduced, both because China has turned to other industries and because fabrics made of artificial fibers and cotton have largely replaced silk fabrics, silk is still an important commodity in Chinese commerce.

The silk fiber is made by caterpillars (of the *Bombyx* moth family) that produce a fine, strong, continuous fiber from which they spin a cocoon. The caterpillars are raised from eggs and are fed on the leaves of the mulberry tree. The caterpillars that hatch from one ounce of eggs will consume the leaves of 25 to 30 trees! To make silk, the completed cocoons, with the caterpillars inside, are plunged into hot water to kill the insects. The cocoons are then carefully dried and unspun so that the fiber can be used to weave cloth.

The process of raising the caterpillars and harvesting the cocoons is called *sericulture*. People who tend the cocoons must take great care to control temperature, humidity, and disease. Losses of 70 percent of the product are not uncommon, although technical assistance has improved production. Sericulture is primarily a sideline cottage industry run by women—one of the ways that Chinese women, and now women throughout East Asia, have gained access to their own cash.

---

proaching 9 million, was an ancient trading center. It remains a main hub for national rail and highway service. Chengdu is also home to light industry, especially food processing and the manufacturing of textiles and precision instruments. Some of its finer-quality products are sold in the global market, as a check of labels in gourmet food stores or discount clothing stores will attest. Chongqing has recently achieved Provincial City status (equal to a province), like Beijing, Shanghai, and Tianjin. With a population of 30 million, it is the largest city in China, and indeed the world (depending on the index of measurement). Chongqing has a high concentration of heavy industry. Its thousands of iron and steel manufacturers and machine-building industries take advantage of Sichuan's rich deposits of iron, coal, copper, and lead. And, as a result of the construction of the Three Gorges Dam (see the box on page 472), it has become a hub for shipping and other new economic activities.

Sichuan's historic ability to support its dense population has made it a productive region, but with an increasing population and growing expectations for yet more affluence, degradation of the environment may cause intolerable pollution levels. This side effect is a result of the basin's tendency to retain warm, moist air, which allows industrial and vehicle emissions to build up. For example, sulfur dioxide emissions in Chongqing in 1995, measured in micrograms per cubic meter, were five times greater than the guidelines recommended by the World Health Organization (see Figure 9.13 on page 471).

The Central Plain (the Middle and Lower Basins and Shanghai)

After the Chang Jiang emerges from its concrete shackles in the Three Gorges Dam, it traverses an ancient, undulating lake bed that is interrupted in many places by low hills. This lake bed forms the middle basin of the Chang Jiang. It and the river's lower basin, leading to the coast, are rich agricultural regions dotted with industrial cities.

The lake bed of this middle basin and the bed of the lower basin (a coastal plain) have been filled with **alluvium** (river-borne sediments) carried down from the Sichuan Basin. Other rivers entering the Chang Jiang Basin from the north and south also bring in loads of silt, along with significant volumes of water, which are added to the main river channel. The Chang Jiang system carries a huge amount of sediment—as much as 186 million cubic yards (142 million cubic meters) per year—past the large industrial city of Wuhan. This sediment, which under natural conditions was deposited on the floor of the basins during annual floods, formerly enriched agricultural production. But because the floods often destroyed people, animals, and crops, the Chang Jiang, like the Huang He, now flows between levees. Still, it occasionally breaches the levees and floods both rural and urban areas, as it did twice in the late 1990s.

The climate of the middle and lower basins, though not as pleasant as Sichuan's, is milder than that of the North China Plain. The Qin Ling Mountains and other, lower hills that extend eastward across the northern limits of the basin block some of the cold northern winter winds. They also trap warm, wet southern breezes, so the basins retain significant moisture during most of the year. The growing seasons are long: 9 months in the north and 10 in the south. The natural forest cover has long since been removed to make agricultural land for the ever-increasing rural population. Summer crops are rice, cotton, corn, and soybeans; winter crops include barley, wheat, rapeseed, sesame seed, and broad beans. As the river approaches the Pacific, it deposits the last of its sediment load in a giant delta. The old and rapidly reviving trading city of Shanghai sits at the delta's outer limits (see the box "Shanghai's Urban Environment" on page 482).

There are more than 400 million people in the hinterland drained by the lower Chang Jiang and its tributaries. In the past, many of them were farmers, growing wheat and rice, but every

AT THE LOCAL SCALE   Shanghai's Urban Environment

Shanghai has a long history as a trendsetter. The opening of its gates to Western trade in the early nineteenth century spawned a period of phenomenal economic growth and cultural development that led it to be called "the Paris of the East." That process seems to be repeating itself today. As a result of the opening of China to the global economy, Shanghai is undergoing a rigorous rejuvenation. In less than a decade, Shanghai has seen the construction of over one thousand skyscrapers, and several other architectural projects are remaking Shanghai's urban landscape faster than planners can think. In addition to the skyscrapers, a subway line, highway overpasses, bridges, and tunnels have been constructed, and several of the city's neighborhoods have received a facelift.

One of the most impressive new developments is Pudong, the new financial center of Shanghai, which sits across the Huangpu River from the famous Bund—the elegant row of big brownstone buildings that served as the financial capital of China until half a century ago. Previously a nondescript stretch of dirt paths and sprawling neighborhoods of gray houses with tiled roofs, Pudong is being rapidly transforming as 140 high-rises soar to the sky. The grandest of these is Jin Mao Tower, which ascends over 420 meters (1380 feet) high, making it the third largest building in the world. Atop the Jin Mao Tower, Hyatt has opened its first grand hotel in China, and some 45 banks have settled into the building's Lujiazui Finance and Trade Zone. The Jin Mao Tower is the first of three mega-towers that will mark the corners of the Shanghai World Financial Center.

In addition to the new buildings shaping Pudong's skyline, a U.S. $1.6 billion airport project is to be operational in October 2004, as will be the world's longest underwater pedestrian tunnel, which will link Pudong to central Shanghai. Shops, restaurants, boutiques, and open-air cafes will line Century Avenue, "China's Champs Elysées." Shanghai is also increasing the area of green space in the city to improve the local environment for its residents. All of this activity is part of the local government's plan to turn Shanghai into an international metropolis. Government officials hope these efforts will attract further investment and economic development, which Shanghai will need to accommodate the millions of new migrants it expects to receive over the next decade. [Adapted from Ron Gluckman, "21st century city," *Time Asiaweek* (July 1999; http://www.gluckman.com/Pudong.htm); Asko Ahokas, "Lost in modern Shanghai," *Design Forum* (December 19, 2002).]

The Jin Mao Tower looms high in the skyline of Pudong, Shanghai's new financial and commercial center. [Liu Liqun/CORBIS.]

year more people migrate to the many industrialized cities of the region. Shanghai overshadows them all in importance. For many centuries, Shanghai's location facing the East China Sea, with the whole of central China at its back, positioned the city well to participate in whatever international trade was allowed. When the British forced trade on China in the 1800s, Shanghai became their base of operations. Before the communist revolution, Shanghai was among the most cosmopolitan cities on earth, home to well-educated literati, wealthy traders, connoisseurs of fine antiquities—and a goodly portion of underworld figures as well. After the revolution, the communists designated the city as the nexus of capitalist corruption, and it fell into disgrace.

Shanghai was well situated to take advantage of the changes that began in the 1980s. In the early twenty-first century, the Chinese government has given developers incentives (such as reduced taxes, help in preparing sites, and permission to take profits out of the country) to build factories, workers' apartment blocks, international banks, luxury apartment houses, shopping malls, and elegant postmodern-style villas. Today, some of the Overseas Chinese (see the box on page 487 and Chapter 10, page 525)—particularly those who fled the repression of the Communist party to places such as Taiwan, Singapore, and Kuala Lumpur—are returning to invest in everything from television stations, production companies, and discotheques to factories, shopping centers, and entertainment parks. Europeans and both North and South Americans have been attracted to Shanghai too, hoping to find joint ventures with Chinese partners so they can tap into the huge pool of consumers that is emerging in China.

CHINA'S FAR NORTH AND WEST

The northern and western interior provinces have long been considered the periphery of China (Figure 9.18). Here the land is especially dry and often cold, and today's cultures retain influences from a long history of nomadic pastoralism. Settlements are widely dispersed, and crop agriculture usually requires irrigation. Much of this

FIGURE 9.18 China's Northwest subregion.

large interior zone is designated as autonomous regions, rather than as provinces, because of the high percentage of ethnic minority populations. As mentioned in the earlier discussion of minorities in China, however, the designation of an autonomous region has not meant that the region's people have ultimate power over their own affairs, nor can they even count on being the majority population in the future. The central government in Beijing retains control over economic and political policy and over the region's industrial assets. Furthermore, millions of Han from eastern China are being encouraged to immigrate to the Far North and West, where they are offered the best jobs. In comparison to eastern (coastal) China, the native citizens of this region are very poor—for example, farmers here earn one-third the income of farmers in the eastern regions.

Xinjiang

The Xinjiang Uygur Autonomous Region, in Far Northwest China, is the largest of all the provinces and autonomous regions. It accounts for one-sixth of China's territory. Once considered

extraordinarily remote from eastern China, it is now, paradoxically, becoming a region of lively trade and enterprise.

Xinjiang has only 17.5 million inhabitants, and their roots lie mainly in Central Asia. The most numerous, at 8 million, are the Turkic-speaking Uygurs. There are also Mongols, Persian-speaking Tajiks, Kazakhs, Kirghiz, Manchu-speaking Xibe, and Hui. New migrants of Han Chinese origin number about 6 million. The peoples native to the region once made their living as nomadic herders and animal traders, moving with their herds and living in **yurts** (or **gers**)—round, heavy felt tents stretched over collapsible willow frames. These cozy houses can be folded and carried on horseback or in horse-drawn carts. In the last few decades, many nomadic people have taken jobs in the emerging oil industry and now live in apartments provided for workers. The Han migrants live primarily in the cities. They work as bureaucrats; in the oil, gas, or nuclear power industries; in state-owned agricultural colonies; and on highway and railroad projects.

Xinjiang consists of two dry basins: the Tarim Basin, occupied by the Taklimakan Desert, and the smaller Junggar Basin to the

In Yining, northwest Xinjiang, a Kazakh girl plays "kiss the maiden." In the game, a suitor gallops after the girl and tries to steal a kiss. If he succeeds, she chases him and tries to beat him with a whip. The contest tests the riding skills of both future spouses. [Jay Dickman.]

northeast. Both are virtually surrounded by 13,000-foot-high (4000-meter-high) mountains topped with snow and glaciers. In this distant corner of China, far from the world's oceans, rainfall is exceedingly sparse. Snow and glacial meltwater from the high mountain peaks are important sources of moisture. Much of the meltwater makes its way to underground rivers, where it is protected from the high rates of evaporation on the surface. Long ago, people built conduits called *qanats* deep below the surface to carry groundwater dozens of miles to areas where it was needed. *Qanats* have made some of the hottest and driest places on earth productive. The Turfan Depression, situated between the Junggar and the Tarim basins, is a case in point. Temperatures often reach 104°F (40°C), and evaporation rates are extremely high. But because the *qanats* bring irrigation water, this region produces some of China's best luxury foods: melons, grapes, apples, and pears. The produce is destined for urban populations in eastern China.

Xinjiang once occupied a central place in the global economic system. Traders carried Chinese and Central Asian products (such as silk, rugs, spices and herbs, and ceramics) over the Silk Road (Figure 9.19) to Europe, where they were exchanged for gold and silver. Today, the province is recovering some of its ancient economic vitality, but just who will profit is not yet clear.

Xinjiang is now one of a growing number of places on earth where the local and the global, the very traditional and the very modern, confront each other daily. Herdspeople still living in yurts and *gers* dwell under high-tension electric wires that supply new oil rigs. Tajik women weave traditional rugs that are sold to merchants who fly from as far as New Jersey in the United States to the ancient trading city of Kashgar (now Kashi) for the Sunday market, which can draw more than 100,000 shoppers (see the photo on page 468). With the breakup of the Soviet Union, citizens of the

new republics of Central Asia are eager to return to their trading heritage, and China is welcoming them and attracting outside investors from Europe and the Americas by establishing SEZs in cities such as Kashgar and Urumqi.

The Uygur and other ethnic leaders of Xinjiang are wary of Beijing's intentions, especially regarding Xinjiang's oil and gas resources. Uygur and other minority leaders have posted Web pages on the Internet in which they join Uygurs in Kazakhstan and Kyrgyzstan in warning that Beijing's true intent is not to improve life for the local people of Xinjiang, but rather to exploit their oil and gas resources and to appropriate settlement land for eastern China's excess population. A young Uygur man in Urumqi, Xinjiang's capital, spoke of his discontent: "I am a strong man, and well-educated. But [Han] Chinese firms won't give me a job. Yet go down to the railroad station and you can see all the [Han] Chinese who've just arrived. They'll get jobs. It's a policy to swamp us."

The Plateau of Tibet (Xizang and Qinghai): The Tibetan Culture Region

Situated in the Far West of China, the Plateau of Tibet is the traditional home of the Tibetan people. Officially, it includes the Xizang Autonomous Region and Qinghai Province. Tibet (Xizang) and Qinghai average 13,000 feet (4000 meters) and 10,000 feet (3000 meters) above sea level, respectively. They are surrounded by mountains that soar thousands of feet higher. They have rather cold, dry climates (late June can feel like March does on the American Great Plains) because of their high elevation and because the Himalayas to the south block warm, wet air from moving in. Across the region, but especially along the northern foothills of the Himalayas, snowmelt and rainfall are sufficient to support a short growing season for barley and such vegetables as peas and broad beans. Snowmelt also forms the headwaters of some major rivers: the Indus, Ganga (Ganges), and Brahmaputra begin in the western Himalayas; the Nu (Salween), Irrawaddy, Mekong, Chang Jiang, and Huang He all begin along the eastern reaches of the Plateau of Tibet.

Traditionally, the economies of Tibet (Xizang) and Qinghai have been based on the raising of grazing animals. The yak is the main draft animal, and it also provides meat, milk, butter, cheese, hides, fur, and hair, as well as dung and butterfat for fuel and light. Other animals of economic importance are sheep, horses, donkeys, cattle, and dogs. Animal husbandry on the sparse grasses of the plateau has required a mobile way of life so that the animals can be taken to the best available grasses at different times of the year. Yet, for several decades, the Chinese government has pressured Tibetan herdspeople to give up their portable houses (yurts and *gers*) and settle in permanent locations so that their wealth can be taxed, their children schooled, their sick cared for, and dissidents curtailed. Still, throughout the region, many native (non-Han) peoples continue to live mobile yet solitary lifestyles as they have for centuries, occasionally adopting some of the accoutrements of modern life and adapting to its restrictions.

The history of Tibet's political status vis-à-vis China is long and complex, characterized by both cordial relations and conflict. During China's imperial era (prior to the twentieth century), China

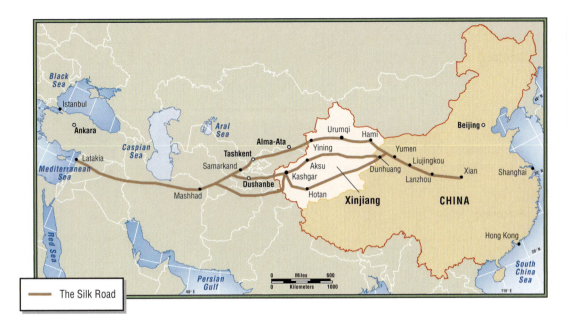

FIGURE 9.19 The ancient Silk Road. Merchants who plied the Silk Road rarely traversed the entire distance. Rather, they moved back and forth along only part of the road, trading with other merchants to the east or west. [Adapted from *National Geographic* (March 1996): 14–15.]

and Tibet entered a "patron-priest" relationship, in which Tibet was for the most part politically autonomous. It maintained its own government and sent representatives to the Chinese imperial court, but faced the constant threat of invasion and Chinese meddling in its affairs. In the early 1900s, as China was collapsing under European pressure and embroiled in civil war, Tibet declared itself separate and free from China and conducted its affairs as an independent country. It was able to maintain this position until 1949–1950, when the Red Army "liberated" Tibet, promising a "one-country two-systems structure." A Tibetan uprising in 1959 led China to abolish the Tibetan government and violently reorder its society. Tibet's spiritual-political leader, the Dalai Lama, was forced into exile, and dissidents faced military crackdown.

By the 1990s, the Beijing government's strategy was to overwhelm Tibetans with secular social and economic modernization rather than military force. China is attempting to bring its economic boom to Tibet by spending hundreds of millions of dollars on housing and on roads and other infrastructure, by trying to attract trade and tourism to this remote region, and by capitalizing on European and American interest in Tibetan culture. China sees its actions in Tibet as part of its overall strategy to integrate the entire country economically and socially. Schools are being built and job markets opened up to young Tibetans. Increasingly, Tibetans are accepting the continued Chinese presence and are channeling their Tibetan cultural pride into efforts to preserve the Tibetan language and religion.

Women have always had a relatively high position in traditional Tibetan society. Among the nomadic herders, they were free to have more than one husband, just as men were free to have more than one wife. In addition, as with some other minority groups in China, traditional marriage customs call for the husband to join the wife's family. This custom alone allowed women to attain a higher status than they had among the Han. Buddhism introduced patriarchal attitudes from outside Tibet, yet encouraged female independence. At any given time, up to one-third of the male population was living a short-term monastic life, so Tibetan Buddhist women have been particularly self-sufficient,

often spending days herding on horseback as well as performing most other daily duties that support community life. Chinese culture has typically regarded the women of the western minorities as barbarian, precisely because their roles were not clearly defined: they rode horses and worked alongside the men in herding and agricultural activities.

SOUTHERN CHINA

Southern China has two distinct sections. The first is made up of the mountainous and mostly rural provinces of Yunnan and Guizhou to the southwest. The second includes the coastal provinces of Guangxi, Guangdong, and Fujian to the southeast, where the booming cities of China's evolving economic revolution are located.

The Yunnan-Guizhou Plateau

The provinces of Yunnan and Guizhou share a plateau noted for its natural beauty and mild climate and for being the home of numerous indigenous groups that are culturally distinct from the Han Chinese (see Figure 9.11 on page 467). The plateau is a rough land of deeply folded mountains that trend north-south and carry the headwaters of the Nu (Salween) and Mekong rivers (Figure 9.20). The heavily forested valleys may be as deep as 5000 feet (1500 meters), yet only 1300 feet (400 meters) across. Although people can call to one another across the valleys, it may take more than a day's difficult travel to reach the other side. In some places, rope and bamboo bridges have been slung across the chasms. The landforms here are unstable, and earthquakes cause heavy damage to the carefully constructed rice paddy terraces, which must be absolutely level to receive gravity-fed irrigation water.

The valuable natural resources of the Yunnan-Guizhou Plateau are primarily biotic. Yunnan Province is one of the most heavily forested regions in the country. Although Yunnan is required to

FIGURE 9.20 The Southern China subregion.

supply wood for China's industrialization, it still has many untouched zones. Yunnan is called "The National Botanical Garden" because it is home to many of China's native plant species and one-third of its bird species (400 or more). Both flora and fauna are now threatened by recent human disturbance of the environment. A British team of biologists traveling in Yunnan in the mid-1980s reported that by then "birds were absent even in the reserves." Until the 1970s, the tropical forests of far southern Yunnan (close to Burma, Thailand, and Laos) harbored elephants, bears, porcupines, gibbons, and boa constrictors.

Kunming, capital of Yunnan Province, lies at the heart of a booming heroin trade that flows from major producers in the mountains of Burma and Laos to China's northern and eastern coastal cities, where the heroin is shipped to the global market. Kunming police say that heroin can be bought for 100 yuan (U.S. $12.50) per gram in Yunnan, a price that has left many local youths addicted. Kunming is now also a center of drug treatment facilities, where herbal medicine is the chief therapy.

The Southeastern Coast

The southeastern coastal zone of China has long been a window to the outside world. Its ports began launching ships that journeyed as far as the Persian Gulf, and possibly the north and east coasts of Africa, during the Tang dynasty (618–907). Arab traders began to visit this part of China around the same time. By the fifteenth century, some of the first Europeans in the region described a string of flourishing trading towns all along the coast. People

from this region have long ventured out to seek their fortunes throughout Southeast Asia and beyond. The overwhelming majority of Overseas Chinese (see the box on the next page) have their roots along China's southeastern coast. Even today, fisherfolk sail

The lower-lying, eastern parts of the Yunnan-Guizhou Plateau are composed of karst (limestone) deposits that are eroded by water to form fantastic landscapes of jagged peaks jutting out from surrounding flat valleys. This phenomenon is found in its most extreme form in Guilin, Guizhou Province, where there is a karst "forest" of 90-foot-high (27-meter-high) jagged limestone columns interspersed with lakes. [Keren Su/China Span.]

AT THE GLOBAL SCALE The Overseas Chinese

One of the ways in which China has had an impact on the rest of the world is through the migration of its people to nearly all corners of the earth. The first recorded emigration by the Chinese took place over 2200 years ago, and after the establishment of the Han dynasty (202 B.C.), China's contact with foreign countries was continuous. Contact spread eastward to Korea and Japan; westward into central Asia and the Middle East via the Silk Road; and south and southwest over a maritime route that led from northern Vietnam through Thailand and Burma to India.

Aside from diplomatic emissaries and religious pilgrims, trade was probably the most significant early impetus for Chinese migration. Just as merchants from western and southern Asia came to China, Chinese traders traveled by sea around the Indian Ocean to western Asia and eastern Africa. Eventually, in all of Southeast Asia and parts of South Asia, Chinese merchants, artisans, sailors, and workers played such a significant economic role that the region came to be known in Chinese as *Nanyang*, the South Seas. Most of the travelers to Nanyang hailed from the southern coastal provinces of Fujian, Guangdong, and Zhejiang. Taking their families with them, some settled in Nanyang permanently and came to be known as the "overseas Chinese."

The arrival of the Europeans in East Asia initiated further migration of Chinese overseas. Artisans and craftspeople (such as carpenters, shipbuilders, weavers, and masons) traveled to Nanyang in increasing numbers, as did unskilled workers who worked for Europeans in construction, transportation, mining, and plantation agriculture. Large-scale international migration of Chinese laborers began after the defeat of China in the First Opium War (see page 450) in 1842. Nanyang became the destination of a large number of unskilled workers, but the Chinese also went to other parts of the world that were new destinations for them. Over the next century, economic hardship in China and a growing international demand for labor spawned the migration of as many as 10 million Chinese to countries in all the inhabited continents. As a result, Chinese communities are present in places as widely scattered as Singapore; London; Central Europe; São Paulo, Brazil; the Caribbean; San Francisco; and Toronto, Canada.

[Adapted from Wei Djao, *Being Chinese: Voices from the Diaspora* (Tucson: University of Arizona Press, 2003).]

their legendary junks far out to sea to meet officially forbidden trading partners from Taiwan, Vietnam, and elsewhere. In the 1980s, the central government of China decided to take advantage of this long tradition of outside contact by designating several of the old coastal fishing towns as SEZs, anticipating that they would quickly grow into major trading cities. This designation gives them special rights to conduct business with the outside world and allows them to create conditions that will attract foreign investors (see Figure 9.9 on page 461).

The principal river of southern China, the Zhujiang (Pearl River), is joined by several tributaries to form one large delta below the city of Guangzhou (Canton), called the Pearl River Estuary. The lowlands along the rivers and delta have a subtropical climate and a perpetual growing season. The rich delta sediment and the interior hinterland are used to cultivate sugarcane, tea, fruit, vegetables, herbs, mulberry trees for sericulture (the raising of silkworms), and timber—products that are sold in the various SEZs and exported to global markets.

Guangzhou has long been the most important trading center in Southern China, but since the nineteenth century, Hong Kong (Xianggang)—which until 1997 was a British crown colony—has posed stiff competition. Now there are several other trading and manufacturing centers along the southern coast that have SEZ status: Xiamen in Fujian Province; Shantou in far eastern Guangdong Province; Shenzhen, adjacent to Hong Kong; Zhuhai, close to the old Portuguese trading colony of Macao, across the Pearl River estuary from Hong Kong; and Haikou, on Hainan Island (see Figure 9.9).

Rainy and warm Yunnan Province is increasingly important nationally for its market gardening and raising of small animals. Here a family transports its wares in the produce baskets typical of the subregion. [Eastcott/Momatiuk/ Woodfin Camp & Associates.]

AT THE REGIONAL SCALE Pearl River Megalopolis

The megalopolis along the eastern seaboard of the United States, described in Chapter 2 (see page 80), is but one of the world's great concentrations of urban settlement. Another has formed along the lower Pearl River (Zhujiang) around Guangzhou, Hong Kong, Macao, and the SEZ cities of Shenzhen and Zhuhai, and it is now one of the world's largest and fastest-growing metropolitan areas. The total population is probably greater than 25 million. This region is often called the South China Megalopolis or the Pearl River Megalopolis.

The SEZ city of Shenzhen has experienced particularly dramatic growth. Located just north of Hong Kong, at the neck of the Hong Kong Peninsula, Shenzhen has received spillover investment from Hong Kong and has emerged as a leading center of labor-intensive manufacturing plants that make an array of products for the global export market. As recently as the late 1980s, it was no more than a local farmers' market center with a population in the range of tens of thousands. Now there are 7 million or more residents, as mi-

grants have poured in from a wide area of China to find work and seek their fortunes. Rice and vegetable fields have been paved over and now sprout some of Asia's tallest high-rise buildings. Densities run as high as 23,500 people per square mile (9000 per square kilometer). Shenzhen has grown so quickly that the official population is 3 million, while migrant workers make up another 4 million.

Many problems accompany such rapid growth. Although Shenzhen and other SEZ cities represent vast improvements in living conditions for many migrants, many others find poor working conditions, exploitative pay, and poor, overcrowded, and overpriced housing. As in other cities around the world, there are social problems such as crime, prostitution, and alcoholism, as well as a general decline of traditional values and culture. Some people see what they consider an unhealthy emphasis on money and material possessions. Environmental problems include traffic congestion, air and water pollution, and poor building construction, which has led to some disastrous fires and building collapses.

Hong Kong

Hong Kong, one of the most densely populated cities on earth, has packed its 6.4 million people into just 23 square miles (60 square kilometers) of the territory's total of 380 square miles (985 square kilometers). This very small place has the world's eighth largest trading economy and the world's largest container port. It is also the world's largest producer of timepieces. Hong Kong's populace has achieved an annual gross domestic product per capita of more than U.S. $20,000, with an annual economic growth rate of 10 percent at the end of 2000.

In July 1997, Hong Kong's niche as a British trading enclave ended when Britain's 99-year lease ran out and Hong Kong became a special administrative region (SAR) of China. There had been considerable worry that China would absorb Hong Kong and no longer allow it the economic and political freedoms it had enjoyed. However, Hong Kong has long played an important role as China's unofficial link to the outside world of global trade. Some 60 percent of foreign investment in China was funneled through Hong Kong before 1997. Over the years since 1997, Hong Kong's situation has shifted a bit toward less autonomy; however, dramatic changes in its fortunes are unlikely. Given the very obvious success of many cities along China's Pacific coast, from Macao to Shanghai, Hong Kong will probably continue its role as a world financial hub in the development of this very rapidly growing region. Its new state-of-the-art airport, called Hong Kong International or Chek Lap Kok, which opened in 1999 on a nearby artificial island, is evidence of expectations that the city will continue to play a major role in global trade.

Macao

Macao (also spelled Macau) is the oldest permanent European settlement in East Asia. Portuguese traders first arrived in about 1516,

and by 1557 they had established a colonial trading center. Macao remained a Portuguese colony until the Chinese regained control of it at the end of 1999. The city is built on a series of three islands that lie adjacent to the southern China coast across the Pearl River Estuary from Hong Kong. Macao grew rapidly after 1949 with the influx of refugees from the mainland communist revolution; today, most of its 1.7 million people are Chinese from nearby southern provinces. Many earn a living from the manufacture of clothing, textiles, toys, and plastic products and from the tourism industry. Macao is known as a center of casino and dog-track gambling, and every weekend thousands of visitors from the mainland and Hong Kong cross the water to Macao by high-speed ferry. Although it is not quite as prosperous as Hong Kong (see Table 9.2 on page 474), the standard of living in Macao vastly exceeds that of the mainland.

JAPAN

The Japanese archipelago is a chain of four main islands and hundreds of smaller ones (Figure 9.21). Most are volcanic and mountainous. Prone to severe earthquakes, tsunamis, and disastrous tropical storms, and so mountainous that only 18 percent of its land can be cultivated, Japan might seem an unlikely place to find 126 million people living in affluent comfort. Yet its citizens have learned to cope with these limitations and have made the most of crowded conditions in cities and countryside alike.

Counting its tiny southernmost islands, Japan stretches over a range of latitudes roughly comparable to that between Canada's province of Nova Scotia and the Florida Keys (46°N–24°N). Although half a world away, it has a climate similar to the east coast of North America. The climate is cool on the far northern island of Hokkaido; temperate on the islands of Honshu, Kyushu, and Shikoku; and tropical in the southern Ryukyu Islands. Despite the

FIGURE 9.21 The Japan subregion. The Japanese archipelago is a chain of four main islands and hundreds of small ones. Most of these islands are volcanic in origin and so mountainous that only 18 percent of Japan's land can be cultivated.

moderating effect of the surrounding oceans, the great seasonal climatic shifts of nearby continental East Asia give Japan a more extreme seasonal variation in temperature than would be the case if it lay farther out to sea.

The large central island of Honshu, the most densely populated of Japan's islands, also has the most mountains and forests. Because it has gained access to forest products in Southeast Asia and other parts of the world, Japan has been able to keep much of its own forestland in reserves. Today the interior mountains of Honshu (and Hokkaido) remain largely forested, although many of the slopes are planted with a monoculture of *sugi*, Japanese cedar, to be harvested commercially. The few flat lowlands and coastal areas of northern and southern Honshu were once intensively cultivated and supported a dense rural population. Industrial cities now fill many of these same lowlands, especially in the south. Tokyo-Yokohama is one of the largest urban agglomerations in the world, with 26.2 million people. It lies on the east-central coast of Honshu on the Kanto Plain, Japan's largest flatland. Kobe-Osaka, Japan's second largest metropolitan area with 10.6 million people, is one of several major urban areas located on the flatlands that ring the Inland Sea.

Japan's other three main islands are Hokkaido, the most northerly and least populated island, referred to as Japan's northern frontier, and Kyushu and Shikoku, two smaller southern islands. The northern slopes of Kyushu and Shikoku, facing the Inland Sea, have narrow strips of land put to agricultural, industrial, and urban uses. The southern slopes are more rural and agricultural. Off Kyushu's southern tip, the very small, mountainous Ryukyu Islands stretch out in a 650-mile (1000-kilometer) chain, reaching almost to Taiwan. Although many areas are densely settled and are important for tourism and specialty agriculture, such as the growing of pineapples, the Ryukyu Islands are considered a poor rural backwater of the four big islands. Okinawa, the site of the strategic U.S. military base that has recently been the source of U.S.-Japanese tensions, lies at the southern end of this island chain.

Food Production and the Challenge of Limited Resources

With 126 million people and only 18 percent of its land suitable for agriculture, Japan has nearly the largest ratio of people to farmland in the world. More than 7000 people depend on each square mile (more than 2700 on each square kilometer) of cultivated land. Thus Japanese agriculture depends heavily on high-yield varieties of rice and other crops as well as on irrigation, fertilization, and

PRACTICING GEOGRAPHY   How do you get to know Tokyo?

Roman Cybriwsky
Temple University

As a young faculty member I was an urban geographer interested in neighborhood change and downtown renovation in American cities, working mostly in Philadelphia. However, in 1984, Temple University asked me to be one of the first professors at its new extension campus in Tokyo. I had never been to Japan or anywhere else in Asia before and accepted the offer as an exciting challenge. I moved to Tokyo with my family and have been alternating academic year assignments between Philadelphia and my new city ever since. Tokyo became my new principal study area. I also came to know other cities in Japan, China, and several countries of Southeast Asia quite well.

Geographers often fall in love with places and want to know them intimately. In my early Philadelphia years I studied that city's various nooks and crannies, hidden neighborhoods, local histories, and newest developments. But Tokyo was a wholly new experience, with a vastly different culture, different ways of living, working, and playing, and with vastly different urban form and transportation. The geographer in me made me want to know it. To do so, I walked and walked and walked, as I had previously done in Philadelphia, with maps, guide books, and new friends who were experts. As best I could at middle age, I struggled to learn Japanese. I shot roll after roll of film—images of what, to me, were unfamiliar urban scenes that I needed to make sense of. I started to publish and to present what I had learned at academic conferences, where I faced the criticisms and comments of academic peers. I am now proud to claim that I know the city quite well. I also know that there is no end to learning a city, particularly one so complex. Tokyo is one of the most interesting and exciting places in the world, with aspects that reflect the many deep traditions of Japan and others that are truly global and leading-edge.

I have used my access to Asia to also learn more about cities and cultures outside Japan. A few years ago good fortune introduced me to a small island of Indonesia named Batam. There, I have gotten to know something more about the global economy and its impact on environment and society in a developing country.

I first saw Batam when it was mostly covered by tropical jungle and most of its several thousand residents either farmed or fished for a living. But in 20 years the island has grown to some 500,000 people drawn from all over Indonesia to take jobs in the American, Japanese, and European electronics factories that have sprouted there, in construction, and in hotels and golf courses. There is also a huge underground economy, including a bustling red-light nightlife to serve men from nearby Singapore. I find Batam's changes fascinating and have followed them over nearly two decades with repeated fieldwork. I study Batam like I studied Philadelphia and Tokyo: I walk a lot, even in the equatorial sun, I read, I take photos, and I ask questions of the people I meet. Almost all are poor, young migrants from impoverished regions of Indonesia who have come to this new place with high hopes for improving their lives and providing for their families. Once again, a new language is necessary, although I struggle with Indonesian even more than I did with Japanese. But I've gotten to know and care about still another place, albeit one with many problems, and my life is that much richer.

Who knows where life will take me next? My geography dreams are of Shanghai, now perhaps the most dynamic city in the world, Buenos Aires, called the Paris of Latin America and famous for its tango clubs, and Kiev, the beautiful, gold-domed capital of my ethnic homeland. The photo shows me in Peru. I'm a happy person because, as a professional geographer, my job is to go places and then share my experiences with students in classes and in publications.

Look me up if you are ever in Tokyo.

mechanization. Because flat land is scarce, Japanese farmers have had to make efficient use of hill slopes by planting tea bushes and orchards and by terracing for rice paddies.

Although the agricultural way of life plays a large part in Japanese national identity, today only 4 percent of the population farms for a living. Japanese farms are increasingly mechanized, and they are highly productive in terms of output per acre. But the produce is some of the most expensive in the world. A single cantaloupe can cost as much as U.S. $10, a head of lettuce U.S. $5, a pound of beef more than U.S. $50. These high prices are partly a reflection of the very superior quality of the produce, but they are also the result of government efforts to support Japanese farmers, who are awarded subsidies to persuade them to stay on the land.

The seas surrounding Japan, where warm and cold ocean currents mix, are an especially important source of food. There are some 4000 coastal fishing villages and tens of thousands of small craft that work these waters and bring home great quantities of tuna, halibut, mackerel, salmon, and other fish. Japan also sends out larger fishing ships, complete with canneries and other processing equipment, to harvest oceans around the world. Tokyo's early morning fish market is the largest fish market in the world. In addition, a large aquaculture industry produces oysters, seaweed, and other foods in shallow bays and freshwater fish in artificial ponds.

Nevertheless, Japan is one of the largest consumers of foreign agricultural products and has one of the lowest levels of food self-sufficiency (the percentage of food consumed daily that is supplied by domestic production) in the world. In 2001, Japan's food imports totaled U.S. $43 billion. The Japanese government has expressed concern over increasing food imports and their implications for food security. Underlying its concern is the realization that Japan's food supply can be substantially affected by unpredictable circumstances, such as poor harvests, both at home and abroad. Hoping to limit Japan's dependence on food imports, the government has set a goal of raising its food self-sufficiency from the current level of 41 percent to 45 percent by the year 2010.

Living in Japan

Perhaps more than any other country on earth, Japan is renowned for its ability to confound foreigners. In an attempt to penetrate some of Japan's cultural complexity, we will look at several aspects of life in Japan: its unique cultural patterns, its urban life, the home, and its elderly population.

Distinctive Culture. Japan is a land of fascinating cultural combinations. The cultural base is a substantial foundation of long-standing Japanese customs and traditions, but atop that is a plentiful mix of borrowings from around the world, creating a unique blend of cultural contrasts and apparent contradictions. We see the mix in many ways: in the kinds of clothes people wear, in the foods they eat, in the music they listen to, in the sports they enjoy, and even in how they practice religion. Thus, among other contrasts, Japan is a land of kimonos *and* blue jeans, sushi *and* hamburgers, the koto (a traditional stringed instrument) *and* rap music, sumo wrestling *and* baseball. When they marry, Japanese couples often change their clothing after the ceremony from traditional wedding kimonos to a Western-style tuxedo and a white bridal dress for the reception, and they will have posed during the day for formal photographs in both outfits.

There are many attempts to explain Japan's unique cultural patterns. Some experts trace the contrasts to the forced opening of Japan by U.S. Commodore Matthew Perry in 1853. In the years that followed, many Japanese, particularly in Tokyo, took on exaggerated international trappings to accommodate foreign visitors and protect what was truly Japanese. There is often a distinction in Japanese life between what is displayed on the outside, *omote*, and what is private on the inside, *ura*. The former often protects the latter, as in the contrasts between *omote-ji*, the outer layer of cloth in a kimono, and *ura-ji*, the layer closest to the skin, and between *Omote-Nippon*, the urban-industrial eastern (Pacific Ocean) side of Japan that trades with the world, and *Ura-Nippon*, the more traditional and secluded western (Sea of Japan) side of the country. Still other observers of Japan emphasize that the country's penchant for copying foreign fashions, trends, and technologies, often slavishly, is a response to cultural forces—such as a stodgy educational system (see the box "Japanese College Life" on page 492)—that stifle creativity and reward conformity. There are no easy ways to explain Japan's culture; the only guarantee is that a visit to Japan will be both fascinating and mystifying.

Urban Japan. Nearly all of Japan's major cities are located along the coastal perimeter. The cities are where Japan's rapid industrialization has taken place, and their coastal location facilitates the import of raw materials and the export of finished products. Ideas from the outside world can penetrate easily, aiding technological advancement and cultural synthesis. Tokyo, for example, has the world's largest stock exchange, numerous centers for research and development, and some of the world's most beautiful modern architecture. It is also a major international cultural center: museum exhibitions, concert tours, and leading intellectuals and other dignitaries from around the world all visit the city regularly.

Nonetheless, Japanese cities also suffer from overcrowding and pollution. Overcrowding is especially severe. In Tokyo, it is not uncommon for a middle-class family of four, plus a grandparent or two, to live in a one-bedroom apartment. Japanese cities cannot expand to relieve crowding because they are limited by surrounding mountains and the ocean, by building codes minimizing potential damage from earthquakes, and by regulations protecting Japan's scarce agricultural lands. But perhaps the greatest causes of overcrowding are corruption and a lack of competition in the construction industry, which keep housing in short supply. Although there is growing pressure for change, most Japanese stoically endure minuscule, expensive apartments and long commutes to work. In the Tokyo metropolitan area, it is common to travel 2 or even 3 hours one way in trains that are so overcrowded that stations employ "shovers," who physically push as many passengers as possible into a single car.

The greatest source of discomfort for urban dwellers is pollution. Several cities have endured major episodes of poisoning from mercury and polychlorinated biphenyls (PCBs), and all large cities have chronic air pollution from automobiles and factories, as well as noise pollution. In the 1970s, local grassroots campaigns led to the first serious government efforts to limit pollution and to establish a moderately effective environmental agency. During Japan's boom years in the 1980s and early 1990s, many cities, including badly congested Tokyo, were able to expand public park space and open other recreational amenities such as artificial beaches, sports fields, and new aquariums.

The Home. Japanese homes are known for their simple, elegant aesthetics and their highly functional use of space. Traditional homes are made of wood and are usually roofed with heavy

The suburbs of Japanese cities are densely packed with family homes. During the day, housewives go about their many duties: shopping, laundry, child care, and care of the elderly. [Michal Heron/Woodfin Camp & Associates.]

CULTURAL INSIGHT   Japanese College Life

Being a college student in Japan is a very different experience from being one in the United States. In Japan, once a student is admitted to a university, he or she is almost certain to graduate on schedule after 4 years, regardless of grades or class attendance. Students simply don't flunk out, as they sometimes do in North America. What they don't learn in college, they learn on the job after graduation. Employers, who normally expect young workers to stay with them until retirement, prefer in-house training and often shuffle new employees from position to position to expose them to all aspects of their operations. What they look for during job interviews is personality: the ability to get along with others, be part of a team, and have a sense of responsibility. Students hone these traits in college by being active in various groups: clubs for sports such as tennis or golf or interest clubs ranging from photography to foreign affairs to environmental issues. Thus the first weeks at a Japanese university are spent joining various groups and building new friendships that will carry students through their 4 years.

Employers believe that the score a student receives on the university entrance exam is evidence of that student's potential to learn. Therefore, the biggest academic pressure on students is not in college, but near the end of high school, when they must take a test that will determine the kind of university they can enter, as well as what employers will think of them after graduation. The most prestigious universities in Japan are Tokyo University, Keio University, and Waseda University, all in Tokyo. Studying at one of these universities provides entry into important business networks and to employment in big companies. Many Japanese students who don't like their country's approach to higher education elect to study abroad in the United States, Europe, Australia, or elsewhere. Alternatively, they can enroll in a foreign university operating in Japan. The largest of these is Temple University, an American institution based in Philadelphia with a campus in central Tokyo.

tiles. Even in larger rural homes, interior space is much smaller than most Westerners are accustomed to. The three or four rooms of a typical middle-class home are used for many purposes during the course of the day: they are transformed as needed by sliding doors of paper and wood or by folding decorative room dividers. Furniture is simple, with much of family life centered on a low table and floor cushions. Sleeping is done on a firm mattress, called a futon, which is often rolled out on whatever floor space is available. However, such patterns are changing in many newer houses and apartments in favor of Western-style arrangements of rooms and furniture. Most Japanese kitchens are generously equipped with a full range of electrical appliances, including the all-important automatic rice cooker, and almost every Japanese family has a color TV, video or DVD player, and sound system. Many Japanese also have automobiles, personal computers, cell phones, and other consumer items.

An Aging Society. Because of Japan's healthy, low-fat, high-protein diet, good medical care, and perhaps even lucky genes, among other factors, Japanese life expectancy is the longest in the world. For men, life expectancy is now 78 years; for women, 85 years. This longevity, combined with a low birth rate, has given Japan's population a large proportion of elderly people. According to estimates for 2003, 19 percent of the Japanese population is 65 years old or older. This figure compares to 16 percent for the United Kingdom, 13 percent for the United States, and 7 percent for China. Thus Japan has a disproportionately large social security obligation that is becoming larger and harder to fund over time. The country also faces a challenge in providing a full range of services and facilities for the aged: affordable medical care, recreational and cultural opportunities, and easier physical access to mass transportation and public buildings.

It has long been the practice in Japan that most older people who cannot fully care for themselves are cared for by their adult children. This pattern is being eroded, however, as younger gener-

ations find that they lack space in their dwellings for their parents or grandparents, or prefer privacy to traditional family obligations. Thus a relatively new industry of nursing homes and other care facilities is growing in Japan.

Working in Japan

Japan's distinctive culture is reflected in the working lives of its people, as well as in their attitudes toward their own work and toward outsiders who perform many of the menial jobs.

Traditional Work Patterns. The Japanese are renowned for their hard work and high-quality output, and most are devoted to their jobs. The lives of many people are shaped by the companies that employ them. The employer provides lifetime employment, regular paid vacations, a pension upon retirement, and, often, subsidized housing. Some companies even have a graveyard for employees. Job-hopping is considered reprehensible, and loyalty is reinforced by personnel management that instills camaraderie and conformity. Individualism is regarded as selfish, and innovation is discouraged. Often, employees who have good ideas will not present them for several years or will give credit to someone else. The Japanese language does not even have a word for "entrepreneur"; only recently has the English term become a buzzword in Japanese.

Traditional work patterns are changing in the face of the current economic downturn, as many companies have had to reduce their workforces and cut back employee benefits. Moreover, some younger college graduates have turned away from the old pattern of lifetime employment in favor of temporary employment. These workers, called "freeters," move from job to job, earning what they need, then taking a break to pursue other interests. They settle for a lower standard of living and give up job-related benefits, but they are unencumbered by contracts and do not feel compelled to acquiesce to their superiors or work overtime. This freelance approach to the world of work allows them time to start their own businesses. Whether by choice or necessity, accepting a temporary or part-time job is risky. The old loyalty system makes it difficult to find a new job. Firms worried about industrial security are often unwilling to hire someone who has worked for a competing company.

Japan's Dual Economy. Japan's kanban system (see page 456) has given the country a sort of dual economy. In this system, firms that supply components are clustered around a single final assembly plant. The high-status final assembly firms offer better wages and benefits and more secure employment than do the supplier firms. Supplier firms tend to hire more women, who are almost always paid at lower rates. Often, even though these women actually work more than full time, they are formally classified as part-time workers and hence receive fewer or no benefits, and can be laid off and rehired relatively easily.

Foreign Workers. Foreign workers from poor countries are also on the wrong side of the dual economy. They come from China, the Philippines, Bangladesh, Iran, Brazil, and other places as guest workers to take some of the hard, dirty, dangerous, and low-paying jobs that today's Japanese workers avoid. Many of the migrants from Brazil and other Latin American countries are descendants of Japanese immigrants who went abroad in search of work some generations ago, when Japan was a poor country. Foreign men often work in construction, in factories, and as dishwashers in restaurants; women are hotel maids, cleaners, and other low-status workers. There are also many foreign women employed as hostesses in bars and as sex workers. The recruitment of women for sex work has become a controversial issue because unscrupulous "agents," who are actually gangsters, force many of these women into sex-industry jobs even though they had initially promised them other kinds of work (see the box "The Trade in Women" in Chapter 5, page 263).

Because Japan's population is aging, there is a shortage of Japanese workers of prime employment age. Foreign workers have become necessary to keep Japan's economy productive. Consequently, the country is debating whether to begin allowing permanent immigration from poor countries (as opposed to admitting temporary guest workers). Permanent immigrants would eventually change Japanese society, making it less homogeneous and less distinctive (see the box "An Aging Population and the Immigration Debate in Japan" on page 467).

TAIWAN

Taiwan is located a little over 100 miles (160 kilometers) off China's southeastern coast (Figure 9.22). With an area of 14,000 square miles (36,000 square kilometers) and almost 23 million people, Taiwan is a crowded place. The island, which is shaped somewhat like a leaf, has a mountainous spine running from the northeastern corner to the southern tip. A rather steep escarpment faces east, and a long, gentler slope faces west. Most of the population lives on the western side, especially in lower elevations along the coastal plain. Increasingly, as farming declines, people are concentrated in a few urban centers. The greatest concentration is in the far north, in the area surrounding Taipei, the capital.

Taiwan today is a geopolitical hot spot. Its status is ambiguous: the United Nations does not recognize it as a country, because of China's opposition, yet Taiwan operates as an independent country

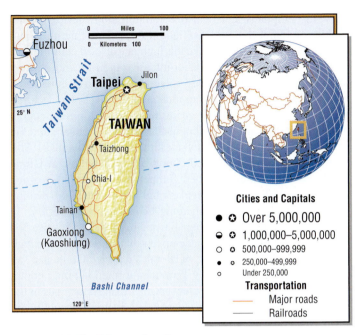

FIGURE 9.22 The Taiwan subregion.

in nearly every way, and its economic significance in East Asia cannot be ignored. Taiwan's existence is a source of political tension between it and China. China views Taiwan as a renegade province and wishes to regain control of it, much as it has done with Hong Kong and Macao (see page 488), and it has maintained the right to use force to do so. The sentiments of the Taiwanese are mixed. Pro-independence president Chen Shui-bian was only very narrowly re-elected in March 2004, and a referendum on boosting defenses against China held at the same time failed to pass. The United States tacitly supports Taiwan and sells it arms, but its strong interest in trade with mainland China and its desire to avoid military entanglements keep it from taking an overt stand on behalf of Taiwan. The degree to which the United States would support Taiwan should armed conflict erupt appears to vary with U.S. administrations.

Taiwan, despite its small size and small population, is one of the most prosperous countries of the **Asia-Pacific region,** a huge trading area that includes all of Asia and the countries around the Pacific Rim. It is fourteenth globally in the size of its foreign trade, and its GDP per capita in 2000 was U.S. $16,100. After the communist revolution in China, Taiwan took advantage of its ardent anticommunist stance, its burgeoning refugee population, and its geographic location close to the mainland to draw aid and investment from Europe and America. The island's economy quickly changed from overwhelmingly rural and agricultural to mostly urban and industrial. Many industries made products for Taiwan's impressive export markets. Then, slowly, the economic emphasis changed again, from labor-intensive industries to high-tech and service industries requiring education and skill. By 1999, only 8 percent of the people were still working in agriculture.

By the mid-1990s, the domestic economy was expanding rapidly, and local buyers were absorbing a major proportion of Taiwan's own production: home appliances, electronics (including the latest in computers and related communications devices), automobiles, motorcycles, and synthetic textiles. The Taiwanese performed a neat pirouette to make this high rate of local consumption possible. As Taiwan lost its labor-intensive industries to cheaper labor markets throughout Asia—including mainland China—Taiwanese entrepreneurs built computer chip and electronics factories in the very places that were giving them such stiff competition: Thailand, Indonesia, the Philippines, Malaysia, Vietnam, and the new SEZs in Southern China. It is estimated that in the mid-1990s, Taiwanese businesses invested about U.S. $25 billion in China alone. Hence Taiwan remained competitive as a rapidly growing economy with strong export markets and also profited from dealing with the less advanced, but emerging, economies in the region.

Life for Taiwanese women has changed radically since the 1950s. At that time, women made up less than one-third of the high school students, and their low social status reflected Confucian values. After World War II, many Taiwanese women left their homes in towns and villages and took low-paying urban factory jobs in the textile and garment industries that fueled Taiwan's economy. By the 1990s, Taiwanese women had equal access to education, at least through high school. In addition, 40 percent worked outside the home, and most were choosing to have only one or two children. Despite these gains, women are still greatly underrepresented in supervisory and management positions throughout the government bureaucracy and in private firms, and very few are politicians.

Today, Taiwan's likely future role in the wider world is a hotly debated question. As mainland China emerges as a world power with a huge population and enormous market potential, tiny Taiwan can no longer promote the idea that it speaks for (or is) China. On the other hand, Taiwan is perfectly located to participate in the development of mainland markets and in related efforts to integrate the Asia-Pacific economy. Young Taiwanese voters have recently elected officials who support an independent status for Taiwan rather than the long-held hope of eventual reunification with a noncommunist China.

KOREA, NORTH AND SOUTH

Some of the most enduring international tensions in East Asia have focused on the Korean Peninsula. As discussed earlier in this chapter, after centuries of unity under one government, the peninsula was divided in 1953. Communist, inward-looking North Korea does not participate in the global economy, while more cosmopolitan, newly democratic South Korea has pursued a model of state-aided capitalist development.

Physically, Korea juts out from the Asian continent like a down-turned thumb (Figure 9.23). Two rivers, the Yalu (Amnok in Korean) and the Tumen, separate the peninsula from the Chinese mainland and a small area of Russian Siberia. Low-lying moun-

Cities and Capitals
- ● ✪ Over 5,000,000
- ◖ ✪ 1,000,000–5,000,000
- ○ ✪ 500,000–999,999
- ● ○ 250,000–499,999
- ○ Under 250,000

Transportation
- —— Major roads
- —— Railroads

FIGURE 9.23 The Korea subregion. The Korean Peninsula was once a unified country. Since 1953, it has been divided into North and South Korea.

tains cover much of North Korea and stretch along the east into South Korea, covering nearly 70 percent of the peninsula. There is little level land for settlement in this mountainous zone. The rugged terrain disrupts ground communications from valley to valley. Along the western side of Korea, alluvial plains slope toward the Yellow Sea, and most people live on these western slopes and plains. Although the peninsula is surrounded by water on three sides, its climate is essentially continental because it lies so close to the huge Asian landmass. The same cyclic monsoons that "inhale" and "exhale" over the Asian continent (see pages 446–447) bring hot, wet summers and cold, dry winters.

The Korean Peninsula was a unified country as early as A.D. 668. Scholars think that present-day Koreans are descended primarily from people who migrated from the Altai Mountains in western Mongolia. The Korean language appears to be most closely related to languages from this region. Other groups—Chinese, Manchurians, Japanese, and Mongols—have invaded the peninsula, sometimes as settlers, other times as conquerors. The Chinese Confucian values and Buddhism brought by some of these groups have influenced Korea's educational, political, and legal systems. Korea is noted for early advances in mathematics and medicine and for a system of printing by movable type. This form of printing, developed by 1234, predated the Gutenberg press in Europe by 200 years.

Contrasting Political Systems

Although an armistice brought the Korean War to an end in 1953, North Korea and South Korea each assumed a strong nationalist stance that has resulted in 50 years of hostile competition between the two. Both governments adopted the Korean concept of *juche*, which means self-reliance or "the right to govern yourself in your own way." In North Korea, *juche* was interpreted as unquestioned loyalty to the Great Leader, Kim Il Sung, and later to his son, Kim Jong Il, who succeeded him in 1997. In North Korea, there is no pretense of democracy, and the government remains a military dictatorship. North Korea limits its involvement with its continental neighbors, China and Russia, and declines to trade with them. Its restrictive internal and external economic policies make it one of the poorest nations in the world. Despite its apparent desire for isolation, North Korea has occasionally threatened South Korea and its allies, especially Japan.

In 2002, North Korea provoked global concern when it announced its intent to resume production of plutonium, forced UN nuclear monitors to leave the country, and then, in early 2003, withdrew from the Treaty on the Non-Proliferation of Nuclear Weapons. As of 2004, most intelligence assessments indicated that North Korea had assembled nuclear weapons, but the extent of North Korea's nuclear weapons capacity is unknown; U.S. officials suggest that the number of weapons is perhaps one to six. It is also unclear why North Korea is claiming to have greater nuclear weapons capability than what it may actually possess. Perhaps its leader, Kim Jong Il, is using the nuclear issue as a hard-line ploy to negotiate a nonaggression pact and improved aid from the United States. Perhaps North Korea is preparing for an attack it believes the United States plans to launch. Regardless, it is generally agreed that the presence

On this April morning in Pyongsung City, about 20 kilometers (12.5 miles) from North Korea's capital, P'yongyang, one person is doing tae kwon do exercises while others sweep the sidewalk along the stream. [Hyungwon Kang.]

of nuclear arms in North Korea is cause for concern. North Korea's political system is believed to be in terminal decline, and who will monitor the weapons if the government disintegrates?

In South Korea, *juche* has come to mean vigorous individualism, coupled with pride in and loyalty to one's own people and nation. Social criticism and even violent protest by labor unions and other social movements are allowed, with the understanding that one's ultimate loyalty is still to South Korea. Politically, South Korea allied itself with the United States, Japan, and Europe shortly after World War II and sought economic revival through foreign aid and capitalist development. But despite its economic success, South Korea was governed by a series of military dictatorships until 1997, when democratic elections were held. Kim Dae-jung, an outspoken critic of the military who endured years of harassment and imprisonment for his beliefs, was elected president.

Gender Roles

The cultural history of Korea bears a certain resemblance to that of China, primarily because both countries have been shaped by Confucian thought. Society is organized hierarchically. Elder males have the greatest authority, and women are expected to be subservient to men in all public situations. Parents prefer sons, partly because, like the Chinese, they believe that sons will provide better income for them in their old age and that sons' wives will provide the necessary personal care. Within the household, women wield financial power, even to the point of controlling men's spending. Yet this responsibility often means that women must make insufficient funds stretch so that the earning power of the husband will not be questioned and the family lose face. Women are responsible for most of the household work, even though most women in both North and South Korea now work outside the home. Although data are not available for North Korea, women who work in the formal economy of South Korea earn less than half what their male counterparts earn. Few managers and administrators are women, and women hold just 6 percent of the seats in parliament. Men are expected to maintain the family's public social and economic connections. After work, they meet along the streets and in shops to drink, eat, tell stories, play games, and make personal connections that can facilitate everyday life for the entire family.

Contrasting Economies

Economically, the two countries differ dramatically. North Korea has better physical resources for industrial development, including forests and deposits of coal and iron ore. Its many rivers, which descend from the mountains, have considerable potential to generate hydroelectric power. South Korea has better resources for agriculture because its flatter terrain and slightly warmer climate make it possible to double-crop rice and millet. Nevertheless, in recent decades, South Korea has surpassed North Korea not only in agricultural production, as expected, but also in industrial development. Its manufacturing industries now export goods throughout the world.

A major reason for South Korea's economic success has been the formation of huge corporate conglomerates known as *chaebol*. These conglomerates include such companies as Samsung, Hyundai, and Daewoo. South Korea's governments have assisted the *chaebol* by making credit easily available to them and by assisting them to purchase foreign patents so that Koreans could focus on product quality and marketing, rather than on inventing. Today there are numerous facilities in South Korea that produce consumer electronic equipment (televisions, stereos, computers, videocassette recorders, and microwave ovens) as well as cars, ships, vehicles, clothing, shoes, and iron and steel. Nevertheless, the *chaebol* have come under increasing criticism in recent years because their close connections to the government have led to corruption. Unwise loans and investments have disrupted the South Korean economy. Laws restricting *chaebol* finance deals were enacted in 2000. One of the largest *chaebol*, Daewoo, was dismantled because it was deeply in debt and had too many layers of management.

Despite these problems, the South Korean economic system has worked well enough that today South Koreans, who were extremely poor at the end of the Korean War, can now afford the products they formerly only exported. For example, even in rural areas where incomes are comparatively low, 90 percent of households have a refrigerator, an electric rice cooker, and a propane gas range, and many have telephones and color television sets.

North Koreans have benefited from some communist economic policies, particularly access to basic health care and education. In general, however, as the economy has deteriorated, these services have also declined. North Korea's industries are inefficient and its workers poorly motivated; furthermore, since the demise of the Soviet Union, the country has suffered a loss of markets, raw materials, fuel, and technical training. Poor harvests and recurrent cycles of floods and droughts brought extensive famines beginning in 1995 and continuing through 2001. Although the government has been unwilling to release much information, it is possible that up to 2 million people died of hunger over this 6-year period. Life expectancy has declined, and the infant mortality rate increased between 2001 and 2003. The government has responded ineffec-

tively to this dire situation. In general, North Korean farms are organized as centrally controlled collectives, and these collectives have not been able to meet the demand for food. Only very small plots are allowed for the private production of vegetables and fruit and for raising small animals. In August 2001, a United Nations World Food Program visitor to North Korea reported "no significant improvement in the country's ability to feed itself" in the past few years, and warned that present UN programs, which feed 7.6 million North Koreans, would have to continue for years to come.

The future of the Koreas remains uncertain. Closer cooperation and even reunification are increasingly discussed, not only because so many families have been painfully split by the long dispute, but also because closer cooperation would bring obvious economic benefits to both sides. For example, South Korea needs more energy to grow its industrialized economy so that it can better compete with Japan, Taiwan, and eventually China. North Korea also needs energy to begin its development, and it has numerous sites at which hydroelectric dams could be built, especially if South Korea were to provide the investment capital. But reunification would require much sacrifice from South Koreans, just as it has from West Germans as they attempt to merge their economy with that of much poorer, formerly communist East Germany. Because of the stark difference in wealth between the two Koreas, reunification would mean that money that might have been used to develop South Korea further would have to be used to pay for basic improvements in North Korea. Poor refugees might flood into South Korea, lowering standards of living there even more.

MONGOLIA

PERSONAL VIGNETTE In the summer of 1991, as their truck descended onto a broad plain within the high Altai Mountains of Mongolia, the visitors saw two horsemen herding hundreds of sheep. One of the horsemen, a suntanned man in his sixties, rode over, dismounted, and greeted the two Americans (and their interpreter) with a smile and a query about their state of health. He wore a traditional knee-length coat, fastened at the side and the shoulder with buttons and belted with a silk sash, and Western-style black leather boots. Anthropologists Cynthia Beall and Melvyn Goldstein were to discover that his easy, confident demeanor was typical of Mongolians. He seemed amused to have found these strangers popping up among the sheep on this normally lonely sweep of grazing land.

They had spoken for only a few moments when the Mongolian herder said cheerfully, "You know, I heard on the radio that your Foreign Minister Baker has visited our capital and that our two countries are now friends. That is good. Please come to visit my camp later. It's not far from [where you are headed]. We can talk more then. I have many questions to ask you about America, and I have a lot to say about [how things are changing in Mongolia]." Goldstein and Beall admit that this encounter shattered their assumptions that Mongolians would be wary of outsiders, hesitant to express themselves, and unaware of the outside world. [*Adapted from Melvyn C. Goldstein and Cynthia M. Beall, The Changing World of Mongolia's Nomads (Berkeley: University of California Press, 1994).*]

With just 2.5 million people, Mongolia occupies a territory so large (Figure 9.24) that the average population density is nearly the lowest on earth, at 4 people per square mile (1.5 per square kilometer). For thousands of years, the economy has been based on the nomadic herding of sheep, goats, camels, horses, and yaks. Today, nearly half the people are still engaged in rural activities in some way related to this traditional lifestyle. The other half now live in cities, primarily the capital of Ulan Bator. Urban workers are employed in a wide range of services and in industries related to the processing of mined minerals and such animal products as hides, fur, and wool. In the 1990s, Mongolia's attempted transition from communist central planning to a market economy was interrupted by an economic recession that forced some people working in urban jobs to return to the countryside.

The Mongolian Plateau lies in the heart of Central Asia, directly north of China and south of Siberia. It is high, cold, and dry, with an extreme continental climate. The physical geography of Mongolia can be broken down into four major zones. In the far south and extending to the border with China is the Gobi Desert —actually a very dry grassland that grades into true desert in especially dry years or where it is overgrazed. To the east and northeast of the Gobi is a huge, rolling, somewhat moister grassland. The remaining two zones are Mongolia's two primary mountain ranges: the forested Hangai (Khangai) in north-central Mongolia and the grass- and shrub-covered Altai, which sweep around west and south and into the Gobi Desert.

History

Scientists are not sure where the ancestors of the Mongolians came from, but it is reasonable to assume that they descended from groups of people who have occupied the massive mountains and plains of Central Asia for more than 40,000 years. **Nomadic pastoralism**—a way of life centered on the tending of grazing animals—is one of the most complex agricultural traditions and dates at least as far back as the cultivation of plants (8000 to 10,000 years). Nomadic herders must understand the complex biological requirements of the animals they breed, and they must also know the ecological and seasonal vegetation changes in the landscapes they traverse with their herds. Mongolian pastoralists have perfected management practices and equipment, such as the portable *ger*, that make it possible to move whole communities and hundreds of animals frequently over the course of the year.

The present country of Mongolia is the northern part of what was once a much larger culture area in eastern Central Asia known by the same name. The height of Mongolian influence was between 1235 and 1366. During this period, the Mongols created a land-based empire that stretched from the eastern coast of China to central Europe, including parts of modern Russia and Iran. While in control of China, the Mongols improved the status of farmers, merchants, scientists, and engineers; fostered international trade; and broke the control of traditional elites by abolishing their automatic access to privilege. They refined the manufacture of textiles, jewelry, and blue and white porcelain, and trade in these wares along the Silk Road flourished. Although the Mongols brought many important changes to China, they alienated many people with their authoritarian rule and discrimination against ethnic Chinese. The Mongols were deposed in 1366. They retreated to the north and, over the next 300 years, lost control of territory in Asia and Europe. Eventually, the southern part of Mongolia became part of China.

By 1900, after long years under Chinese colonization, Mongolians were among some of the poorest people in Asia, despite their herding skills. Profits from herding went to the feudal elite, and trade was monopolized by Chinese merchants who sent their

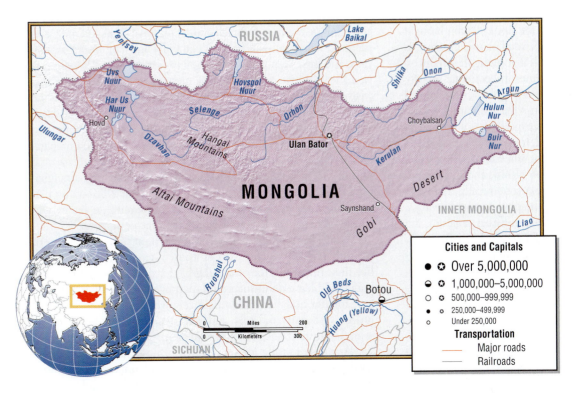

FIGURE 9.24 The Mongolia subregion.

profits home rather than investing in Mongolian development. In 1921, after considerable turmoil, Mongolia declared independence from China, and 3 years later a communist revolution took place.

The Communist Era in Mongolia

From 1924 to 1989, Mongolia was a communist country that sought guidance from the Soviet Union. Russian advisors and technicians wielded considerable influence in setting up a system of central planning that organized the economy and allocated resources. In time, the nomadic pastoral economy was reorganized according to socialist policies, but without drastic disturbance of traditional lifeways. The old rural households made up of extended families became economic collectives. Some Mongolians continued to herd animals, others engaged in sedentary livestock production, and others farmed crops. During this time, Mongolia was partially industrialized and urbanized. By 1985, nomadic herding and forestry accounted for less than 18 percent of the Mongolian national income and somewhat less than 30 percent of the labor force, whereas employment in mining and industry, communications, transport, and construction accounted for about 48 percent of the national income and more than a third of the labor force.

The Soviets subsidized an elaborate system of social services. By 1989, there was nearly universal education through middle school, and adult literacy reached 93 percent. Health care became available to all, and life expectancy rose dramatically. The previously high infant mortality rates had dropped, and family size had decreased. That nomadic herder who impressed the American anthropologists Melvyn Goldstein and Cynthia Beall in the early 1990s was undoubtedly a beneficiary of these educational and social reforms.

Recent Economic Issues

At the end of the 1980s, the collapse of the Soviet Union brought drastic social and economic downturns. Several natural disasters, as well as lower world prices for the minerals that Mongolia produces, led to increased national debt and declines in economic growth, human well-being, and social order. Few Mongolians were trained to replace the Soviet managers who had run the economy. Because of layoffs and factory closings, everyone experienced a decline in living standards. Between 1990 and 1992, total public expenditures dropped by 58 percent, and education was cut by 69 percent. Kindergartens and rural boarding schools closed, and many older students left school to help their families economically. Throughout the 1990s, young adults remained unemployed and rootless, some turning to substance abuse for solace. Street children, the outcasts of disintegrating families, proliferated in Ulan Bator. By 2003, about one-third of the population lived in poverty. Nonetheless, there were signs of a better future. Mongolia's new membership in the World Trade Organization was bringing increased international attention and development assistance, and private foreign investors were beginning to take an interest in the country.

Mongolia apparently has vast reserves of copper, gold, molybdium, and coal. The Oyu Tolgoi deposit, in the southern Gobi Desert, may hold the world's fifth largest deposits of gold and copper. Several global mining companies are currently conducting

The Batsuur family lives in a *ger*, the traditional tentlike Mongolian house that can easily be dismantled, packed up, and moved to another place. Located on the outskirts of Ulan Bator, the house has electricity (for the hot plate and television), a coal-burning stove, and a variety of furniture and decorative and functional items for the six people who live in it. [Leong Ka Tai and Peter Menzel/Material World.]

exploratory drilling across Mongolia. One reason for this intense interest is that China needs the minerals for it huge and growing industrial sector.

Gender Roles

Traditionally, Mongolian women have enjoyed a status approximately equal to that of men, but outside forces, such as Lamaist Buddhism in the sixteenth century and Chinese rule in the seventeenth and eighteenth centuries, eroded their status. Nevertheless, in today's nomadic pastoral economy, men and women share many tasks. Women usually breed animals—a highly specialized task—and tend the mothers and their young, preserve meat and milk, and produce many essentials from animal hides and hair, including felt, yarn, and furs. Men also perform many of these tasks. Women provide the materials for house (*ger*) construction and also dye and weave beautiful furnishings for the interiors. Both boys and girls are taught to ride horses at an early age, and both help with the herding. As they mature, women become more responsible than men for the routines of daily life in the home: child care, food preparation and serving, home maintenance, and care of the elderly and infirm. Men engage primarily in pasturing the herds and arranging the marketing of the animals and animal products. Men also perform other tasks—such as the occasional planting of barley and wheat—that are carried out beyond the home compound.

Under communism, much of this system remained intact or was enhanced. The daily work of women herders was valued, and, like their husbands, they were eligible for old-age pensions. A woman could leave an unsuccessful marriage because there was support for her and her children. A few women became teachers, judges, and party officials, and soon women were better represented in institutions of higher education.

As a result of the jolting changes of the past decade, women have lost some of the status they enjoyed under communism. Nonetheless, women are regaining legal protections and access to education, jobs, health care, housing, and credit. Women entrepreneurs are emerging. For example, an out-of-work truck mechanic received a small loan and put herself and her brothers to work repairing trucks. A woman who used to make sausages in a state-run meat plant now makes them in her kitchen, employing 14 of her former fellow workers. The sausages are peddled on the streets. As in so many other countries that are moving from communism to a market economy, the informal economy of street peddlers and small-scale manufacturing and service people has been an essential component of economic transition.

Quick Review

1. Which part of China has the largest expanse of cultivable land?

2. Why was Shanghai unpopular with the Communists?

3. How have the Han commonly thought of traditional Tibetan women?

4. Describe the Japanese relationship between workers and employers.

5. What is Taiwan's status vis-à-vis China?

REFLECTIONS ON EAST ASIA

The opening of China to the global economy and to outside influences presents huge economic opportunities for other East Asian countries, but it also means stiff competition for their economies. Most of China's East Asian competitors have already begun to invest in the rapidly growing SEZs in Chinese coastal areas. But the expectation of benefits from China's debut into capitalism could be premature. It is not at all clear yet how the rapid social changes occurring in China will be worked out politically, either within the country or internationally. Will economic liberalization lead to more political freedom for ordinary people and more public input into such issues as environmental policy? Will China's emerging policies toward Hong Kong and Taiwan bring continued rapid economic growth, social change, and outside contact? Or will they result in a reversion to greater control of people and economies and a break with the international community?

Leadership within the region is in question. Residual animosities toward Japan linger because of the excesses it committed during its imperial period before and during World War II. Whether Japan can earn the confidence of fellow Asians and play a leadership role in East and Southeast Asia may depend on how it solves its own internal financial crises and on whether it opens its own markets to its neighbors. China's rapidly expanding economy will surely enhance its position as a regional leader. If it seriously seeks a leadership role, China would be wise to woo the Overseas Chinese, many of whom were forced to leave China during the revolution but have since prospered in Taiwan, Singapore, and Malaysia. Politically, there is no clear leader in the effort to bring more participatory democracy to East Asia. Although variations on parliamentary democracy appear to be gaining ground in East Asia, in every country in the region, actual power overwhelmingly tends to remain with a male elite.

A combination of factors related to population control will profoundly affect the future. These factors include shrinking family size, the increase in the proportion of retired and elderly people, and the fact that women are no longer available to stay home to care for children and elders. To cope with these changes, governments may have to spend more on social services and be content with slower economic growth. The population issue most emphasized in East Asia today is the need to curtail population growth. Yet the issue that may most influence the future is finding a way to care for large numbers of aging people.

There is also the question of how East Asia will deal with the world's cultural diversity. Barring a major reversal of policy by China, both personal and electronic contact between all of East Asia and the rest of the world will increase significantly. Not only will outsiders from all backgrounds have more influence, but women and minority groups within each country are bound to become more politically assertive. A possible outcome is that East Asian society will become more welcoming to other cultures and to the idea of women acting in expanded roles.

Finally, East Asia, along with the rest of the industrialized world, will be considering how to balance the desire for a clean, safe environment with the desire for consumer lifestyles. In this regard, the tremendous talents of innovation and synthesis found throughout East Asia may be called upon even more in the future than they have been in the past.

Chapter Key Terms

Ainu 469
alluvium 481
Asia-Pacific region 493
buffer zone 476
chaebol 496
China proper 448
command economy 458
Confucianism 448
Cultural Revolution 451
economic and technology development zones (ETDZs) 461

floating population 463
food security 459
ger 483
Great Leap Forward 451
growth poles 462
hukou system 463
kanban system 456
loess 476
nomadic pastoralism 497
qanats 484

regional self-sufficiency 458
responsibility systems 459
shogun 451
Silk Road 468
special economic zones (SEZs) 461
state-aided market economy 455
tsunami 446
wet rice cultivation 480
World Trade Organization (WTO) 463
yurt 483

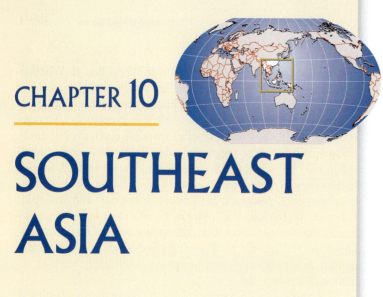

CHAPTER 10

SOUTHEAST ASIA

| PERSONAL VIGNETTE   It is breakfast time, and the food stand in the village in northeastern Thailand is crowded. The middle-aged woman cooking the food is friendly, her portions large, and the price right. For the equivalent of five cents, she serves up a huge mango leaf filled with rice, fish paste, and fried beetles.

One of the half-dozen people sitting on a bench eating his breakfast is a sinewy, bare-chested laborer in his late 30s. It is a
hot, lazy day, so the two reporters from the *New York Times* take the opportunity to chat idly with the laborer about the food and their families. Soon he mentions his daughter, who is 15. His voice softens as he speaks of her. She is beautiful and smart, and he has invested his hopes in her.

"Is she in school?" the reporters ask.

"Oh, no," the father replies, mildly amused. "She is working in a factory in Bangkok . . . making clothing for export to America." He explains that she makes U.S. $2.00 a day for a nine-hour shift, six days a week. It is dangerous work; twice, the needles of her sewing machine went right through her hands. But the managers bandaged her hands and she soon went back to work. He is pleased that she has kept her job despite widespread factory closings, and he worries about what she would do if her factory shut down.

The reporters noted that the man was not indifferent to his daughter's injuries; he appeared to care for her deeply. Her persistence in the face of injury made him proud. From his point of view, the long hours were an advantage because they meant more pay for her, and her income promised a better future for their entire family. The reporters' subsequent conversations with factory managers revealed that although some enforce particularly long work days, others find workers' enthusiasm for long hours exasperating because it is hard to provide supervisors and security guards around the clock. [*Adapted from the writings of Nicholas D. Kristof and Sheryl Wu Dunn, "Two cheers for sweatshops,"* New York Times Magazine *(September 24, 2000): 70–71.*]

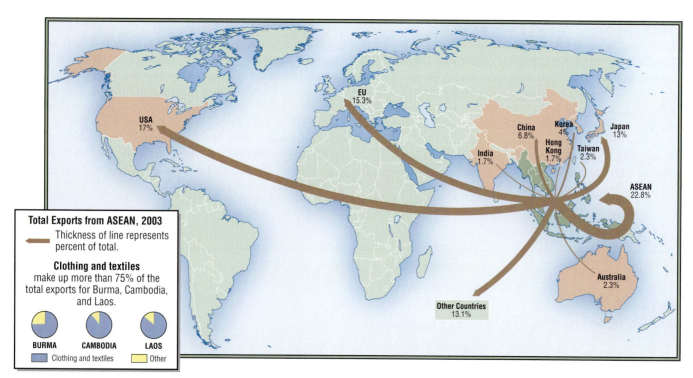

FIGURE 10.1 Total exports from ASEAN, 2003. The countries of Southeast Asia export a wide variety of goods around the world (see the discussion of the ASEAN trade organization on page 521). Nearly one-third of the exports go to the European Union and the United States; 31.8 percent goes to ASEAN's nearest neighbors; 22.8 percent is trade within ASEAN; and 13.1 goes to the rest of the world. Clothing and other textiles make up 75 to 85 percent of the total exports for Cambodia, Laos, and Burma. For the richer Southeast Asian countries like Thailand and Malaysia, these consumer goods are only a small percentage (less than 10 percent) of the total exports. Can you think of any reasons why this is so? [Adapted from *ASEAN Statistical Yearbook 2003* (Jakarta: ASEAN Secretariat, 2003), at http://www.aseansec.org/13100.htm; ASEAN Secretariat, at http://www.aseansec.org/64.htm; *International Trade Statistics 2003*, World Trade Organization, at http://www.wto.org/english/res_e/statis_e/statis_e.htm; and Australian Government Department of Foreign Affairs and Trade, at http://www.dfat.gov.au/geo/.]

Rapid industrialization in Southeast Asia has created contrasting scenes in urban landscapes. *Left:* Money from profitable industries and eager outside investors has been invested in sleek buildings, such as the upscale shopping mall at the base of the Petronas Towers in Kuala Lumpur, Malaysia. *Center:* Widespread layoffs force many former urban factory workers to fall back on other skills to earn a living. Here, a couple in Vientiane, Laos, makes simple iron tools in a street-side blacksmith shop. *Right:* After years of working under difficult conditions, thousands of factory workers began to demonstrate at trade meetings, like this one in Bangkok, Thailand, in Spring 2000. [Mac Goodwin (left), Alex Pulsipher (center and right).]

MAKING GLOBAL CONNECTIONS

The country of Thailand, where the laborer and his daughter live, and several other Southeast Asian countries—Burma, Laos, Cambodia, Vietnam, and Malaysia—occupy a substantial peninsula that extends from the southeast corner of Asia. The rest of the region consists of thousands of islands, most of which belong to one of two island countries: the Philippines and Indonesia. Southeast Asia is easily connected to neighboring regions by sea and has long provided its spices and other agricultural products to the rest of the world (see Figure 10.2). In the past half-century, the region has become a key link in the global economy and supplies many of the consumer and electronic goods sold at inexpensive prices in the malls and galleries of the world.

Beginning in the 1960s, Southeast Asia became a center of rapid industrialization when governments and local investors began to build factories that employed thousands of willing workers. The workers' wages were low and the raw materials they used inexpensive, so the prices of the finished products made them highly competitive in the world market. By the 1970s, businesses and governments were courting investors from Japan, Europe, Australia, and North America, and by the 1990s, industrial facilities had greatly expanded. Millions of rural workers were attracted to urban areas throughout the region. Although the wages of a few dollars a day were very low by Western standards, they were enough to make a substantial difference to people's standard of living. Parents could afford simple medicines, better shelter, and better nutrition for their families. With increasing government revenues and private profits, a building boom ensued. Magnificent banks, hotels, and office buildings, such as Petronas Towers in Kuala Lumpur, began to rise in capital cities across the region.

By the early 1990s, Southeast Asia was regarded as a model of rapid development that other regions might emulate. But in the late 1990s, the region began to appear less adequate as a model of success. First, Western consumers became aware of the unhealthy and exploitative conditions under which the products they were buying from Southeast Asia were made. The factories became known as sweatshops. Soon protesters appeared on U.S. college campuses and at meetings of the World Bank and World Trade Organization. These agencies were seen as encouraging the growth of such factories (see the photos on this page). Second, the environmental costs of rapid industrialization were revealed: deforestation to make room for rubber trees and oil-producing palms, soil erosion caused by loss of forest cover, and heavy air pollution from the burning of fossil fuels. Finally, in 1997, a widespread economic recession hit. Thousands of factories closed, at least temporarily, and families that depended on factory wages had to turn to the informal economy to pay for necessities.

By 2001, demand for Southeast Asian products was reviving and factories were increasing production. But by that time it was clear that world events would continue to have an impact on the job security of Southeast Asia's factory workers.

THEMES TO EXPLORE IN SOUTHEAST ASIA

This chapter discusses the characteristics of Southeast Asia and the successes and problems the region has faced in recent years. Our discussion will include the following themes.

1. **Distinct culture groups.** Many distinct culture groups in Southeast Asia have lived side by side for centuries, absorbing a spectrum of cultural influences from the outside world, yet retaining their uniqueness.

502

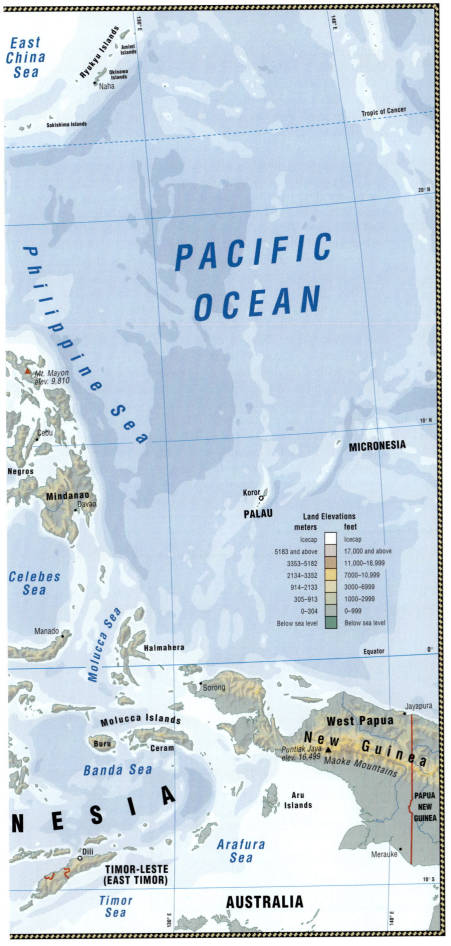

FIGURE 10.2 Regional map of Southeast Asia.

2. **Disparities in well-being.** Cultural and physical diversity have contributed to stark disparities in wealth and well-being, often within a given country.

3. **Management of the tropical environment.** Social and economic factors have influenced human strategies for managing physical resources, but so has the tropical environment. The rich soils of river valleys and deltas, and the much poorer soils of tropical rain forests in the uplands and islands, are being farmed intensively. Farms and plantations support large numbers of people, but there have been environmental consequences, including soil erosion, flooding, and loss of biodiversity.

4. **Availability of labor and resources for consumer product manufacturing.** The countries of the region range from agricultural societies that are just modernizing (Laos and Cambodia) to fiercely competitive industrialized societies that are active in the global marketplace (Indonesia, Malaysia, Thailand, and Singapore). Throughout the region, workers receive low wages and work under unhealthy conditions while producing products for consumers around the globe who are usually much wealthier and healthier. Resources are extracted at unsustainable rates.

5. **Challenges to national unity.** Authoritarian governments throughout the region—some fairly elected, some not—are encountering increasing difficulty in maintaining control over countries that have conflicting interest groups. Separatist movements are challenging several governments, which are accused of violating human rights in responding to the threats.

6. **Persistence and change in family structure.** Although traditional family structures are patriarchal, they have some unique features that give women more power and freedom than they have in many other world regions. Furthermore, as a result of recent economic changes and urbanization, the extended family is giving way to the nuclear family. These changes help to account for the generally higher levels of education among women in the region and for reductions in fertility.

I THE GEOGRAPHIC SETTING

Terms to Be Aware Of

Many of Southeast Asia's inhabitants choose to dispense with place names that connote their colonial past. As a result, many names that are familiar to Westerners are now not the names preferred by the people of the region. In this chapter, we generally use the place name officially preferred by each locality. The older, more familiar Western name is given in parentheses the first time that a place name appears. In the case of Burma, however, we made an exception to this practice: Burma, the old name for Myanmar, is now being revived popularly; hence, *Burma* is used in this text.

PHYSICAL PATTERNS

The physical patterns of Southeast Asia have a continuity that is not immediately apparent when looking at a map of the region. On a map, one sees a unified mainland region that is part of the Eurasian continent and a vast and complex series of islands arranged in chains and groups. The apparent contrast between mainland and islands obscures the fact that these landforms are related in origin. Moreover, despite covering a large territory, most of the region has a tropical or subtropical climate.

Landforms

Southeast Asia is a region of peninsulas and islands (see the regional map, Figure 10.2). Although the region stretches over an area nearly as extensive as Europe, most of that space is ocean. In fact, the area of all the land in Southeast Asia amounts to less than half that of the contiguous United States. There is a large mainland peninsula, sometimes called Indochina, that extends to the south of China. It is occupied by Burma, Thailand, Laos, Cambodia, and Vietnam. This peninsula itself sprouts a long, thin penin-

sular appendage that is shared by outlying parts of Burma and Thailand, a part of Malaysia, and the city-state of Singapore, which is actually built on a series of islands at the southern tip. This long, thin peninsula is usually considered part of the archipelago that lies to the south and east of the mainland. The Southeast Asia **archipelago** is a series of large and small islands, fanning out over an area larger than the continental United States. These islands are grouped into the countries of Malaysia, Indonesia, and the Philippines. Indonesia alone has some 17,000 islands and the Philippines 7000. The independent country of Brunei shares the large island of Borneo with Malaysia and Indonesia. Timor-Leste (East Timor), which gained independence in 2002, shares the island of Timor with Indonesia. The entire region, except for the northernmost part of Burma, lies in the tropics, with the islands located on and near the equator.

The irregular shapes and landforms of the Southeast Asian mainland and archipelago are the result of the same tectonic forces that were unleashed when India split off from the African Plate and crashed into Eurasia (discussed first in Chapter 8, pages 389–392; see also Figure 8.2). The collision, which is still under way, has pushed up the Himalaya Mountains and has forced up folds in the land to the east and west of the Himalayas. The eastern folds are a series of high mountainous ridges and intervening gorges that bend out of the Tibetan Plateau and turn south, then fan out to become the peninsula of Indochina. The gorges widen into valleys that stretch toward the sea, each valley containing a river or two that begins in the mountains of China to the north. The major rivers are the Irrawaddy and the Salween in Burma;

FIGURE 10.3 Sundaland, Sahulland, and Wallacea 18,000 years ago. The now-submerged shelf of the Eurasian continent that extends under the region's peninsula's and islands was above sea level during recurring ice ages.

[Adapted from Victor T. King, *The Peoples of Borneo* (Oxford: Blackwell, 1993), p. 63.]

the Chao Phraya in Thailand; the Mekong, which flows through Laos, Cambodia, and Vietnam and forms the border of Thailand and Laos for more than 620 miles (1000 kilometers); and the Black and Red rivers of northern Vietnam. (See the regional map in Figure 10.2.) On the mainland, the mountain ranges descend from the high peaks of the north, which reach heights of almost 20,000 feet (6100 meters), to lower ranges of 2000 to 3000 feet (600 to 900 meters) toward the south, and even more lower hills.

The curve formed by the islands of Sumatra, Java, Bali, Timor and the other islands of the Lesser Sunda group, and New Guinea conforms approximately to the shape of the Eurasian Plate's leading edge (see Chapter 1, page 23). As the Indian-Australian Plate plunges beneath the Eurasian Plate along this curve, hundreds of volcanoes have been created, especially on the islands of Sumatra and Java. Volcanoes are also being created in the Philippines, where the Philippine Plate is pushing against the eastern edge of the Eurasian Plate. The volcanoes of the Philippines are part of the Pacific Ring of Fire (see Chapter 1, page 24). Volcanic eruptions, and the mudflows and landslides that occur in their aftermath, endanger and complicate the lives of many Southeast Asians. In the long run, though, the volcanic material creates new land and provides minerals that can enrich the soil for farmers.

The now-submerged shelf of the Eurasian continent that extends under the Southeast Asian peninsulas and islands was above sea level during recurring ice ages, when much of the world's water was frozen as glaciers (Figure 10.3). The exposed shelf, known as Sundaland, allowed ancient people and Asian land animals (elephants, tigers, rhinoceros, proboscis monkeys, and orangutans) to travel south into what became the islands of Southeast Asia as the ice age ended and the sea level rose. Sundaland remained exposed until about 16,000 years ago. A second exposed continental shelf, known as Sahulland, was attached to Australia and New Guinea; its western boundary stopped near the islands of Timor in the south and the Moluccas in the north. Sundaland and Sahulland never met because they were, and are, separated by a deep ocean trench. This trench extends north from Bali through the Makassar Strait and along the western side of the Philippines. The trench (and the nearby islands) is known as **Wallacea**. Wallacea is a **biogeographical transition zone** between Asian and Australian flora and fauna. Within this narrow zone there is some mixing of the two groups; but beyond it they remain distinct. Humans using watercraft were probably the only major land creatures to move across to New Guinea and Australia; dogs apparently made the trip with them. But Balinese tigers, for example, were not able to spread east, and Australian kangaroos could not move west.

Climate

The tropical climate of this region is distinguished by continuous warm temperatures in the lowlands—consistently above 65°F (18°C)—and heavy rain (Figure 10.4). The rainfall is the result of two major processes: the monsoons (seasonally shifting winds) and the intertropical convergence zone (ITCZ), a band of rising warm air that circles the earth roughly around the equator (see discussion in Chapter 8, pages 393–394 and Figure 8.3). The wet summer season extends from May to October, when the warming of the Eurasian landmass sucks in moist air from the surrounding seas.

Dayak farmer Abdur Rani plants cassava in a section of Kalimantan rain forest ravaged earlier by loggers. The haze—like the one that covered most of Indonesia, Malaysia, and parts of other Southeast Asian countries in 1997—is smoke from forest fires. [Michael Yamashita/Woodfin Camp & Associates.]

Between November and April, there is a long dry season on the mainland, when the seasonal cooling of Eurasia causes dry air from the inner continent to blow out toward the sea. On the many islands of Southeast Asia, however, the winter can also be wet because the air that blows out from the continent picks up moisture as it passes south and east over the seas. The air releases its moisture as rain after ascending high enough to cool down. With rains coming from both the monsoon and the ITCZ, the island part of Southeast Asia is one of the wettest regions of the world. Its copious year-round rainfall supports dense tropical vegetation over much of the landscape.

Periodically, the normal patterns of rainfall are interrupted, especially in the islands, and severe droughts result. These periodic droughts, which occur irregularly every 2 to 7 years, are part of the El Niño phenomenon (see Figure 11.5, page 557). In an El Niño event, the usual patterns of air and water circulation in the Pacific are reversed. Ocean temperatures are cooler than usual in the western Pacific near Southeast Asia. Instead of warm, wet air rising and condensing as rainfall, cool, dry air sits at the ocean surface. In 1997, an El Niño event brought drought and cool temperatures to the countries of Malaysia and Indonesia. Crops failed; springs and streams dried up; and the heavy, cool air prevented air contaminants from venting into the upper atmosphere—especially the pollution caused by automobile and industry exhaust and forest fires. The result was several weeks of the worst widespread air pollution ever experienced in the region.

The soils in Southeast Asia are typical of the tropics. Although not particularly fertile, they will support dense and prolific vegetation when left undisturbed for long periods. The high temperatures and damp conditions promote the rapid decay of **detritus**—dead plant material and insects—and the quick release of useful minerals.

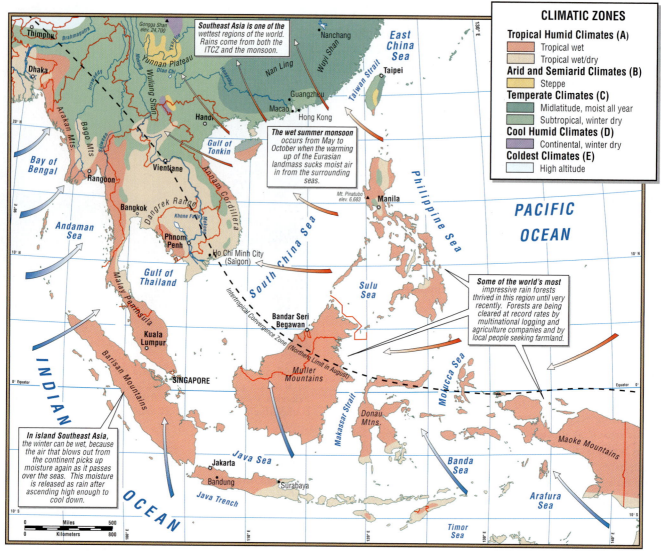

FIGURE 10.4 Climates of Southeast Asia. Southeast Asia experiences few periods of prolonged dryness. The islands are washed by rains that result both from monsoon patterns originating on the Eurasian continent and from rising warm, wet air along the intertropical convergence zone (ITCZ). The arrows on the map show airflows during the northern summer.

These minerals are taken up directly by the living forest rather than enriching the soil. Some of the world's most impressive rain forests thrived in this region until very recently. Today, the forests are being cleared at record rates, especially by multinational logging companies; but also by local people seeking farmland (see the section on deforestation later in this chapter, pages 531–532). Often the refuse is burned, creating smoke that contributes to severe air pollution under certain conditions. The fires that played a role in the severe pollution episode of 1997 had been set in Indonesia to clear land for commercial purposes, most notably for oil palm plantations. Winds carried the copious smoke to Malaysia, Singapore, and the Philippines (see the photo on page 505).

HUMAN PATTERNS OVER TIME

Southeast Asia's position as a group of islands and peninsulas surrounded by seas has made it easily accessible to ocean trade and to outside cultural influences. First settled by migrants from the Eurasian continent, it later was influenced by Arab, Indian, and Chinese traders. Later still, it was colonized by Europe (1500s to early 1900s) and Japan (World War II). In the late twentieth century, the region emerged from a long period of colonial domination ready to profit from selling manufactured goods to its former colonizers. Its economic success has not yet been sufficient to catapult the majority of its people out of poverty, but signs of the region's development include declining birth rates and longer life expectancy (the demographic transition), rapid industrialization, growing urbanization, and rising rates of literacy.

The Peopling of Southeast Asia

The modern indigenous populations of Southeast Asia arose from two migrations widely separated in time. In the first, a group of hunters and gatherers called **Australo-Melanesians** moved from the present northern Indian and Burman parts of southern Eurasia into the exposed landmass of Sundaland about 40,000 to 60,000 years ago. They were the ancestors of the indigenous peoples of New Guinea, Australia, and Indonesia's easternmost islands. Very small numbers of their descendants still live in small pockets in upland areas elsewhere, notably in the Philippines, Borneo, Java, Sumatra, the Malay Peninsula, and the Andaman Islands.

In the second migration, people from southern China began moving into Southeast Asia about 10,000 years ago, at the end of the ice age that closed the Pleistocene epoch. Their migration gained momentum about 5000 years ago, when a culture of skilled farmers and seafarers from South China, named **Austronesians,** migrated first to Taiwan, then to the Philippines, and then into island Southeast Asia and the Malay Peninsula. Other groups of Austronesians appear to have traveled southward along the coast of China and Vietnam. Some of these sea travelers eventually moved westward to southern India and as far as Madagascar off the east coast of Africa. They also moved eastward to the far reaches of the Pacific islands (see Chapter 11). Today there are no clear biological or geographic boundaries between the descendants of the early Australo-Melanesians and the more recent Austronesians and others who drifted south, southeast, and southwest from China.

The migrations of many different peoples to this highly accessible region, together with its extremely fragmented physical geography, have led some geographers to compare Southeast Asia to other regions that also have a highly complex cultural geography: the Balkans of southeastern Europe and the islands of the Caribbean.

Other Cultural Influences

Southeast Asia has been and continues to be shaped by a steady stream of cultural influences, both internal and external. These influences have been so numerous partly because the sea that touches or surrounds all the countries except Laos brought ships and travelers. The shifting winds of the monsoon also played a role. In spring and summer, winds blowing from the west brought seaborne traders, religious teachers, and occasionally even invading armies from the coasts of India and Arabia. These newcomers brought religions—Hinduism, Buddhism, and Islam—and trade goods such as cotton textiles, and food plants such as mangoes and tamarinds deep into the Indonesian archipelago. The traders, religious teachers, and warriors penetrated as far as Cambodia, Laos, Vietnam, and China on the mainland as well as reaching the islands farther south. The merchant ships sailed home on the monsoon winds of autumn and winter that blow from the northeast. They carried Southeast Asia's people—as well as spices, bananas, sugarcane, root crops, silks, domesticated pigs and chickens, and other goods—to the wider world.

Islam came to the region mainly through South Asia, after India fell to Muslim conquests in the thirteenth century. Today Islam is a major religion in Southeast Asia, and Indonesia has the world's largest Muslim population. China was also influential, especially in northern and central Vietnam, which was part of the Chinese Empire for about a thousand years until the year 939. After the tenth century, the Vietnamese expelled the Chinese, becoming a major bulwark that kept China's empire from expanding into the rest of the region. The blending of these and other cultural influences is obvious today in such places as Singapore and parts of Malaysia, where Buddhist, Taoist, Confucian, and Hindu temples exist side by side with Islamic mosques. Other areas are more homogenous, such as Burma, Thailand, and Laos, which are largely Buddhist, and Indonesia, which is mostly Muslim.

As important as external influences have been, indigenous cultural characteristics also have shaped the character of the region. Several powerful urban empires emerged in Southeast Asia, including Angkor (800–1400), Pagan (800–1100), and Srivijaya (600–1000). The cities of these kingdoms were impressive for their size and architecture. At its zenith in the 1100s, the city of Angkor was among the largest in the world, and its ruins are still an important heritage site in Cambodia (see Figure 10.11 on page 519).

As we shall see, ideas about family structure and gender roles that appear to date from the earliest inhabitants of the region can still be observed. These ideas may account for the relatively high status of women in the region (compared to parts of South Asia and China, for example)—a status that declined during the period of Western colonialism.

Colonialism

Over the last five centuries, Europe, Japan, and the United States have also influenced Southeast Asia. All three established colonies or quasi colonies in the region (Figure 10.5). The Portuguese were drawn to Southeast Asia's fabled spice trade. By sailing around Africa, they reached India in 1498 and by 1511 they established the first permanent European settlement at the port of Malacca on the southwest coast of the Malay Peninsula. Although better ships and weapons gave the Portuguese an advantage, their anti-Islamic and pro-Catholic policies provoked strong resistance in Southeast Asia. Only in their Timor-Leste (East Timor) colony did the Portuguese establish Catholicism, and today the people of Timor-Leste are predominantly Catholic.

By 1540, the Spanish had established trade links across the Pacific between the Philippines and their colonies in the Americas, especially Mexico. Like the Portuguese, they practiced a style of colonial domination grounded in Catholicism. The Spanish met less resistance, however, because their attitude toward non-Christians was somewhat more tolerant. The Spanish ruled the Philippines for 350 years, and as a result the Philippines are the most deeply Westernized and certainly the most Catholic part of Southeast Asia.

The Dutch were the most economically successful of the European colonial powers in Southeast Asia. From the sixteenth to the nineteenth centuries, under the auspices of the Dutch East India Company, they extended their control of trade over most of what is today called Indonesia. Their privately chartered trading company was headquartered in Batavia (now Jakarta) on the island of Java. Dutch colonists were far less interested in territory than in profits, and they avoided direct administrative rule by placing local leaders in charge. In exchange for granting this local control, the Dutch obtained exclusive trading rights, and they were renowned for their often bloody extermination of Southeast Asian trade competitors. Eventually, however, the Dutch became interested in export cash crops, primarily coffee, sugar, and indigo. From 1830

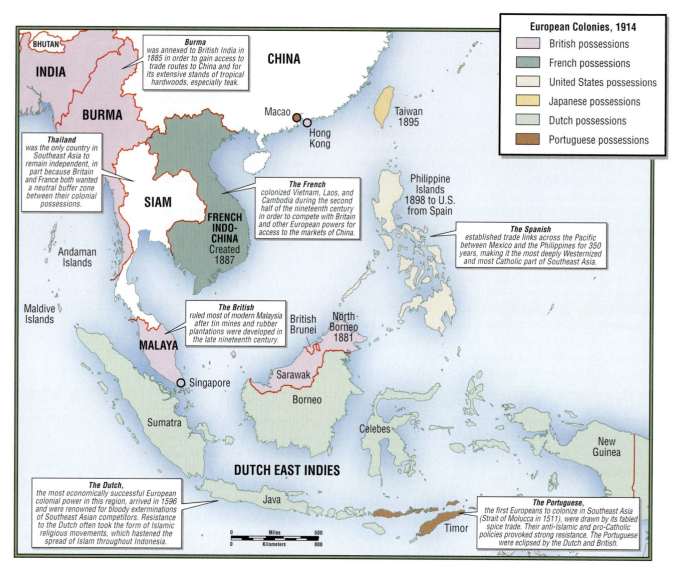

European Colonies, 1914
- British possessions
- French possessions
- United States possessions
- Japanese possessions
- Dutch possessions
- Portuguese possessions

BHUTAN

INDIA

CHINA

BURMA

Burma was annexed to British India in 1885 in order to gain access to trade routes to China and for its extensive stands of tropical hardwoods, especially teak.

Macao

Hong Kong

Taiwan 1895

Thailand was the only country in Southeast Asia to remain independent, in part because Britain and France both wanted a neutral buffer zone between their colonial possessions.

SIAM

FRENCH INDO-CHINA Created 1887

The French colonized Vietnam, Laos, and Cambodia during the second half of the nineteenth century in order to compete with Britain and other European powers for access to the markets of China.

Philippine Islands 1898 to U.S. from Spain

The Spanish established trade links across the Pacific between Mexico and the Philippines for 350 years, making it the most deeply Westernized and most Catholic part of Southeast Asia.

Andaman Islands

Maldive Islands

The British ruled most of modern Malaysia after tin mines and rubber plantations were developed in the late nineteenth century.

MALAYA

British Brunei

North Borneo 1881

Sarawak

Borneo

Singapore

Sumatra

Celebes

New Guinea

DUTCH EAST INDIES

The Dutch, the most economically successful European colonial power in this region, arrived in 1596 and were renowned for bloody exterminations of Southeast Asian competitors. Resistance to the Dutch often took the form of Islamic religious movements, which hastened the spread of Islam throughout Indonesia.

Java

The Portuguese, the first Europeans to colonize in Southeast Asia (Strait of Molucca in 1511), were drawn by its fabled spice trade. Their anti-Islamic and pro-Catholic policies provoked strong resistance. The Portuguese were eclipsed by the Dutch and British.

Timor

Miles 500
Kilometers 800

FIGURE 10.5 European colonies in Southeast Asia, 1914. Of the present-day countries in Southeast Asia, only Thailand (Siam) was never colonized. [Adapted from *Hammond Times Concise Atlas of World History*, 1994, p. 101.]

to 1870, the Dutch diverted farmers from producing their own food to working without pay in Dutch enterprises, especially plantations. Disruption of local food-producing systems caused periodic severe famines. Resistance to the Dutch often took the form of Islamic religious movements, and the rise of these movements hastened the spread of Islam throughout Indonesia. As in South Africa, the Dutch settlers made little effort to spread Christianity.

Like their Dutch rivals, the British were commercially motivated. In the early nineteenth century, the British East India Company, operating with government and military powers sanctioned by the British Crown, controlled a few key ports on the Malay Peninsula. The British held these ports both for their trade value and to protect the Strait of Malacca, through which passed trade between China and Britain's empire in India. In the nineteenth century, Britain extended its rule over the rest of modern Malaysia to benefit from the tin mines and plantations in the interior. Once Malaysia came under their rule, the British argued that Malays, and in fact all "Orientals," were incapable of governing themselves

effectively and needed a "benevolent" European power to rule over them. Similar justifications were offered for the violent annexation of Burma to British India in 1885. By controlling Burma, Britain controlled access to trade routes to China and to extensive stands of tropical hardwoods.

In the early seventeenth century, French Catholic missionaries began actively seeking converts in the eastern mainland area of Southeast Asia—the modern states of Vietnam, Cambodia, and Laos. The French eventually colonized the region in the late nineteenth century, spurred by rivalry with Britain and other European powers for greater access to the markets of nearby China.

In all of Southeast Asia, the only country to remain independent was Thailand (formerly Siam). Like Japan, it protected its sovereignty by undergoing a massive drive toward European-style modernization. In addition, Thailand was adept at using diplomacy to prevent European colonization. Another factor in Thailand's ability to maintain its sovereignty was that Britain and France both wanted Thailand to serve as a neutral **buffer zone,** a territory separating adversaries. The existence of a neutral zone between their colonial possessions on mainland Southeast Asia lessened the possibility of direct confrontation between the two colonial powers.

Struggles for Independence

Agitation against colonial rule began in the late nineteenth century, when Filipinos fought first against Spain and then, against the Americans, after the United States took control of the Philippines in 1898. However, independence for the Philippines and the rest of Southeast Asia had to wait until the end of World War II. By then, Europe's ability to administer its colonies was weakened, partly because its attention was diverted by the devastation of the war. Also, during the war, Japan had exploded the myth of European superiority by conquering most of European-held Southeast Asia. At first, the Japanese were welcomed as liberators from European domination. But the Japanese soon imposed harsh administrative policies, and Southeast Asians turned their attention to ridding themselves of all outside control. By the mid-1950s, the colonial powers had granted self-government to most of the region. Only Singapore had to wait until the mid-1960s to become independent.

The most bitter battle for independence took place in the French-controlled territories of Vietnam, Laos, and Cambodia. Although all three became nominally independent in 1949, France retained political and economic power. Various nationalist leaders led resistance movements against continued French domination. Although these leaders did not begin as communists, and, in fact, shared ancient antipathies toward China for its previous efforts to dominate the region, the resistance leaders accepted military assistance from communist China and the Soviet Union when the French refused to release control. Thus, the cold war was brought to mainland Southeast Asia.

In 1954, the French lost their war against Vietnamese guerrillas when they were defeated by Ho Chi Minh at Dien Bien Phu. The United States, increasingly worried about the spread of international communism, stepped in and eventually replaced the French in Vietnam. This began what was known as the Vietnam War in the United States and, in Vietnam, the American War. The resistance leaders controlled the northern half of Vietnam, and attempted to wrest control of the southern half from the United States and a U.S.-supported South Vietnamese government. The pace of the war accelerated in the mid-1960s. Many people in the United States were turned against the war by nightly broadcasts showing gruesome pictures of burning children and village massacres, by the loss of their own sons, and by revelations of the ineptness of U.S. officials. After several years of brutal conflict, public opinion in the United States forced the country to withdraw from the conflict in 1973. The war finally ended in 1975. During the Vietnam War, more than 4.5 million people died in Indochina, including more than 58,000 Americans. Another 4.5 million on both sides were wounded, and bombs, napalm, and defoliants ruined much of the Vietnamese environment; land mines continue to be a hazard to this day.

The U.S. withdrawal from Vietnam in 1973 ranks as one of its most profound defeats. After the war, the United States imposed economic sanctions against Vietnam that lasted until 1993 and crippled its recovery. In Cambodia, where the war had spilled over the border, a particularly violent revolutionary faction called the Khmer Rouge seized control of the government in the mid-1970s. Inspired by a vision of rural communist society, they attempted to destroy virtually all traces of European influence. They targeted Western-educated urbanites in particular, forcing them into labor camps where many died of starvation or were summarily executed. Over 2 million Cambodians, or one-quarter of the population, were killed during Khmer Rouge rule. In 1978, Vietnam invaded Cambodia in an effort to drive out the Khmer Rouge and ruled the country through a Cambodian puppet government until 1989. The country then fell into a period of civil war. Despite a massive United Nations effort to establish multiparty democracy in Cambodia throughout the 1990s, the country remains plagued by political tension between rival factions and government corruption, and none of the Khmer Rouge leaders have ever been tried or held accountable for the massacre that took place under its rule.

Former Khmer Rouge village chief, Choch: "It's not true. If I had done those things, how could I live here now?" In Cambodia, victims of the Khmer Rouge reign of terror now often live as close neighbors to those who tortured and killed their loved ones. The perpetrators have never had to stand trial or face punishment because little forensic work has been done on the murders and massacres. Those responsible can blandly deny having been involved.

One night in 1977 soldiers came to the home of Samrith Phum and took her husband away. She thought he was just going to a meeting, but he never came home. Samrith was then only 20 years old and had three young children, one a newborn. With her infant in arms she went to talk to the Khmer Rouge village chief, a man named Choch and asked him, "Brother, do you know where my husband is?" But the village chief told her not to worry about other people's business. Says Samrith, "I didn't ask him any more after that. I was hopeless. I knew my husband was dead." A short while later, Choch appeared at Samrith's door and said he would take her to see her husband. Instead, he drove to the nearby prison and locked her up with her baby. She was released a year later only after the Vietnamese drove the Khmer Rouge from power. Today, Samrith still lives just down the street from Choch—an all-too-common horror in Cambodia. For his part, Choch denies any involvement in the killings or even ever being at the prison saying, "It's not true. If I had done those things, how could I live here now?"

To hear more about the aftermath of the Khmer Rouge, watch "Cambodia: Pol Pot's Shadow" and read Amanda Pike's "Reporter's Diary: In Search of Justice" at http://www.pbs.org/frontlineworld/ stories/cambodia/diary03.html.

Although the independence era has brought violence to some areas, it has also brought relative peace and economic development to much of Southeast Asia. Since the 1960s, the economies of Thailand, Malaysia, Singapore, Indonesia, and the Philippines have grown considerably, aided largely by industries that export manufactured products to the rich countries of the world. Some critics regard this situation as a form of neocolonialism because the countries of the region remain dependent on markets in distant wealthy nations and often compete with one another to sell their products to their former colonizers. This stiff competition often results in wages that are too low to provide a decent standard of living and in weak pollution controls and worker safety regulations.

POPULATION PATTERNS

Today more than half a billion people occupy the peninsulas and islands of Southeast Asia. Twice as many people as live in the United States are packed into a land area half its size. The population map in Figure 10.6 reveals, however, that several parts of the region are only lightly populated. Relatively few people live in the upland reaches of Burma, Thailand, and northern Laos, where the land is particularly rugged and difficult to traverse. Much of

Cambodia is also lightly settled. In some islands of Southeast Asia, wetlands, dense forests, mountains, and geographic remoteness have inhibited settlement. Small groups of indigenous people have lived here for thousands of years, supported by shifting subsistence cultivation and by hunting, gathering, and small-plot permanent agriculture. Over the last several decades, however, in previously lightly settled places such as Kalimantan in Borneo, Sulawesi (the Celebes), the Moluccas, and West Papua (until recently called Irian Jaya), thousands of new settlers have been lured by promises of farmland and jobs in commercial agriculture and in extraction of forest products and minerals.

Most of the people of Southeast Asia (about 60 percent) live in patches of particularly dense rural settlement along coastlines, on the floodplains of major rivers, and in the river deltas of the mainland. In the islands, settlement is concentrated in the northern Philippines and on Java. These places are attractive because people can pursue several ways of making a living: the rich and

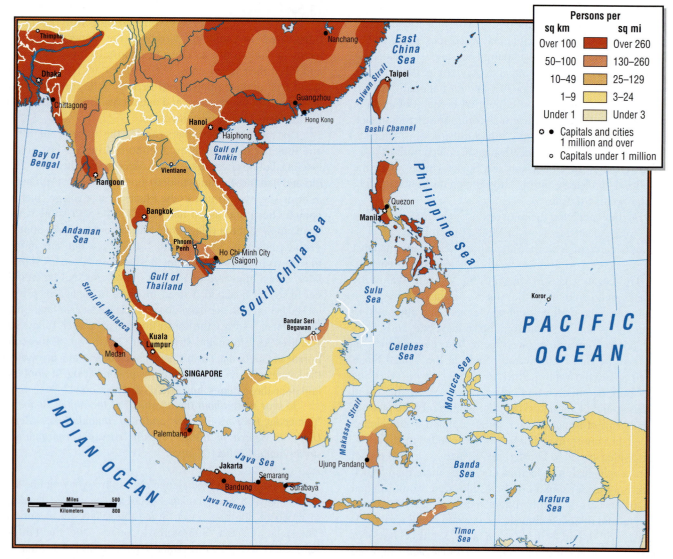

FIGURE 10.6 Population density in Southeast Asia. Population densities vary considerably throughout the region. Although twice as many people as live in the United States are packed into a land area half its size, several parts of the region are only lightly populated. [Adapted from *Hammond Citation World Atlas*, 1996.]

TABLE 10.1 Southeast Asia's largest metropolitan regions

Rank (2000)	Metropolitan region	Population, in thousands (1980)	Population, in thousands (2000)	Average annual growth rate, 1980–2000, percent[a]
1	Jakarta, Indonesia	6503	13,000	3.46
2	Manila, Philippines	5926	11,200	3.18
3	Bangkok, Thailand	4960	7,700	2.20
4	Ho Chi Minh City (Saigon), Vietnam	2701	4,600	2.66
5	Rangoon, Burma	2200	4,500	3.58
6	Singapore	2410	4,000	2.53
7	Hanoi, Vietnam	820	3,100	6.65
8	Bandung, Indonesia	1463	2,900	3.42
9	Kuala Lumpur, Malaysia	938	2,200	4.26
10	Medan, Indonesia	1379	2,100	2.10
11	Palembang, Indonesia	787	1,550	3.39
12	Ujung Pandang, Indonesia	709	1,250	2.84
13	Phnom Penh, Cambodia	530	1,200	4.09

[a]Growth rates calculated using natural logarithm method: $\dfrac{\ln\,(2000\ \text{population}/1980\ \text{population})}{20}$.

well-watered soils allow intensive agriculture on small plots, and there are also opportunities for small business, petty trade, wage labor, and enjoyable social interaction. About 37 percent of the region's people live in cities, and the cities of Southeast Asia are among the most crowded on earth. The most populated metropolitan areas are Jakarta, Manila, and Bangkok (Table 10.1). These and other cities of the region are growing very rapidly (see page 527, for a discussion of the reasons for this growth). Rural migrants stream into the increasingly dense slum and squatter areas that exist in all of the region's cities except Singapore. Indeed, in some cities, as much as one-third of the population resides in these temporary, usually self-built, settlements.

Population Dynamics

Overall, Southeast Asians have smaller families today than they did in the past, but the patterns vary geographically. In some places, fertility rates have declined sharply since the 1960s (Figure 10.7).

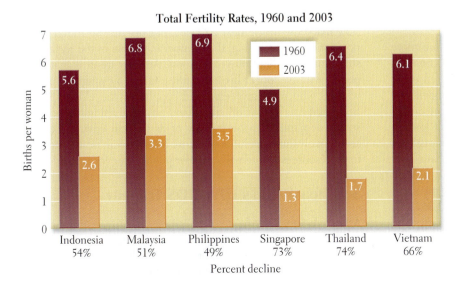

Total Fertility Rates, 1960 and 2003

Births per woman

	1960	2003
Indonesia 54%	5.6	2.6
Malaysia 51%	6.8	3.3
Philippines 49%	6.9	3.5
Singapore 73%	4.9	1.3
Thailand 74%	6.4	1.7
Vietnam 66%	6.1	2.1

Percent decline

FIGURE 10.7 Total fertility rates, 1960 and 2003.
Total fertility rates declined from 1960 to 2003 for several Southeast Asian countries. The percentage decline for each country is indicated below the name of the country.
[Adapted from *A Demographic Portrait of South and Southeast Asia* (Washington, D.C.: Population Reference Bureau, 1994), p. 9; and *World Population Data Sheet,* 2004 (Washington, D.C.: Population Reference Bureau).]

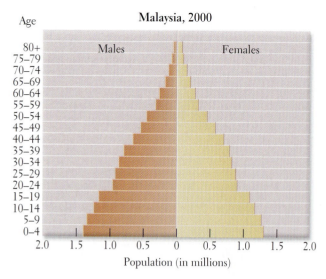

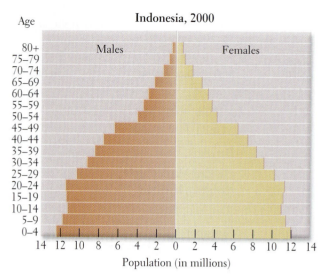

FIGURE 10.8 Population pyramids for Malaysia and Indonesia. Note that the scale for the population axis is roughly eight times greater for Indonesia than for Malaysia. In 2000, Indonesia had 206.1 million people while Malaysia had just 22.7 million people.

Singapore, an unusually wealthy city-state, had a fertility rate in 2003 below replacement level: 1.4 children per adult woman. The Singapore government is so concerned about the low fertility rate that it offers young couples various incentives for marrying and procreating and priority for the best public housing. Singapore's population continues to grow despite low birth rates because Singapore attracts a steady stream of highly skilled immigrants and less skilled illegal immigrants. Thailand, with one of the region's most successful family planning programs, has the next lowest fertility rate in the region: 1.7 children per adult woman. Compared to Singapore, Thailand is very poor, with just one-fourth the income per capita. Nevertheless, rapid economic change has made couples feel that a large number of children are a burden. Although wages are low and working conditions often exploitative, Thai citizens have many employment opportunities and are allowed to move freely throughout the country to find work. High literacy rates for both men and women and Buddhist attitudes that accept the use of contraception have also been credited for the decline in Thailand's fertility rate.

Two extremely poor countries in the region show the more usual correlation between poverty and high fertility. In Cambodia and Laos, women average between four and five children each, and infant mortality rates are 95 per 1000 births for Cambodia and 102 per 1000 for Laos. Vietnam, where per capita incomes are as low as in Cambodia and Laos, nonetheless has a relatively low fertility rate (2.3) and infant mortality rate (26). Vietnam's lower rates are explained by the fact that this socialist state provides basic education and health care to all people, regardless of income. In Vietnam, literacy rates are more than 90 percent for both men and women, whereas in Cambodia only 58 percent and in Laos only 54 percent of women can read. Also, Vietnam's rapidly growing economy is attracting foreign investment; hence employment for women is expanding, replacing child rearing as the central role in their lives.

The use of modern birth control by women in Southeast Asia (51 percent) is significantly more prevalent than it is in South Asia (42 percent). Generally, couples tend to use birth control when access to health services, education, and jobs—especially for women—improves. As educational levels increase so does the use of contraceptives.

Southeast Asia's population is young and hence is likely to keep growing even as family sizes shrink. Those 15 and younger make up 31 percent of the population and have not yet begun to reproduce. Figure 10.8 shows the population pyramids for Malaysia, the most developed country in the region (except for the city-state of Singapore), and for Indonesia, the country with the largest population (220.5 million). Malaysia's pyramid remains wide at the bottom, indicating that the population is young and thus will continue to grow even though family size is shrinking. Indonesia's birth rate slowed sharply 25 years ago but then increased again about 5 years ago, giving the lower slopes of its pyramid a blocky, irregular shape. Analysts anticipate that population growth will slow markedly in the next 25 years, but because so much of the population is now young, the number of people in the region will still double in about 40 years or less. Another trend is that more people will live longer lives, possibly a mixed blessing for younger workers who must give care and support to the elderly. Given Southeast Asia's relatively small land area, the rising level of consumer demand, and the fragility of its tropical environments, it seems likely that population pressure on land, soil, fresh water, and forest resources will remain a major public issue for many years.

Southeast Asia's HIV/AIDS Tragedy

As in sub-Saharan Africa (see Chapter 7), human immunodeficiency virus (HIV) and acquired immunodeficiency syndrome (AIDS) constitute a significant public health issue in Southeast Asia. Infection rates are growing across the region: at present, Cam-

CULTURAL INSIGHT Caring for the Ill

In Thailand, Buddhist monks often provide intermittent care for the gravely ill in temples, but the high incidence of AIDS in recent years has strained this custom. In the crematorium in Lopburi, Thailand, dozens of fist-sized white cotton bags sit unclaimed at the foot of the gilded statue of Buddha because the victims' families fear getting sick themselves. In 1998, the Thai government announced a plan to have the Ministry of Agriculture and Cooperatives and Buddhist agencies cooperate to provide care for AIDS patients, because the traditional temple care system is overwhelmed. An AIDS community rehabilitation center was built at Lopburi, 75 miles north of Bangkok; it shelters approximately 10,000 patients and also serves as a public education center for HIV/AIDS. Thousands of students visit the patients, participate in their care, and learn that HIV is transmitted by drug use and sexual activity. They also see art exhibits made of the bones of victims, an acceptable way of paying homage to the dead in Thai culture. Thailand maintains several Web sites on HIV/AIDS research and care. See, for example, http://www.pattayamail.com/500/features.shtml and http://www.aids2004. org/.

bodia, Thailand, and Burma have the highest rates. In Thailand, AIDS is the leading cause of death, overtaking accidents, heart disease, and cancer, and more than a million people are thought to be infected. Men between the ages of 20 and 40 are the most common victims, but the number of women victims is rising. Many observers think that the disease may soon increase rapidly in places throughout the region that are only lightly affected so far—rural areas and secondary cities just being brought into closer economic and social contact with the region's big cities. The growth of the sex industry and especially of sex tourism is also a factor in the spread of AIDS (see the discussion on pages 518–519).

There are several reasons for the gloomy forecast of increasing HIV/AIDS and related infections. Southeast Asian societies have a number of religious traditions, especially Islam and Catholicism, which strictly limit both expressions and discussions of sexuality. At the same time, there are ancient, popularly accepted traditions that allow a fair amount of latitude in sexual practices that facilitate the spread of the disease. Conservative religious leaders restrict public sex education and AIDS prevention programs because these are viewed as unseemly topics, but popular customs support sexual experimentation, at least among some segments of the population. In Thailand, for example, a visit to a brothel has long been a rite of passage for young Thai men, and brothels are found in most neighborhoods. Studies in 1991 showed that more than 60 percent of female sex workers in urban locations were infected with HIV and that 80 percent of young male soldiers visited brothels in that year. Also, just as in South Asia and Africa, truck drivers in Southeast Asia commonly spread HIV because they travel far from home and may have sexual partners in several different locales. The UN AIDS Epidemic Update for 2000 noted that casual sex (between those who do not know each other well) is the most significant way in which HIV is spread. A newly infected carrier is particularly contagious, yet is unaware of his or her condition and is relatively unconcerned about spreading the disease to a stranger. In Thailand, a rigorous government education campaign to use condoms is reducing infection rates. Neighboring Malaysia, in contrast, has a rapidly growing HIV/AIDS problem that is made worse by religious prohibitions on public discussion of sexuality, condom use, and drug abuse.

Quick Review

1. What climatic circumstances made it possible for animals and plants to diffuse into the islands of Southeast Asia?

2. What cyclical conditions can cause air pollution to build up over the island countries in Southeast Asia?

3. What are some reasons why the United States stepped in when the French were defeated in Vietnam?

4. How does increasing rural to urban migration affect the prevalence of AIDS in Southeast Asia?

II CURRENT GEOGRAPHIC ISSUES

Not surprisingly, some themes identified elsewhere around the world are also found in Southeast Asia. Like the Americas, Africa, and South Asia, Southeast Asia has a history of colonial rule; has experienced uneven development resulting in wide disparities of wealth; is expanding its links to the global economy through manufacturing and migration; and is home to a wide array of culture groups and religions, with social conflict an increasing reality. Environmental concerns—especially those related to urbanization and to unsustainable extraction of forest and soil resources—are also a theme. Nonetheless, there are some surprises. Southeast Asia has an economic, political, cultural, and social repertoire with a distinctly different flavor from all other regions; and, of course, the space it occupies is unique.

Because Southeast Asia has developed rapidly since World War II, the region is often cited as a proving ground for development strategies that may be usable elsewhere. But recent economic

FIGURE 10.9   Political map of Southeast Asia.

recession and political instability have cast some doubt on the stability of Southeast Asian systems and on their usefulness as a model for other regions.

ECONOMIC AND POLITICAL ISSUES

The economic and political situation in the region has changed dramatically in recent years. From the mid-1980s to the late 1990s, Southeast Asian countries had some of the highest economic growth rates in the world. Several countries earned a reputation as economic "tigers" by aggressively embracing capitalism and achiev-

ing a remarkable level of economic growth and reduction in poverty. Many countries that once struggled to feed themselves exported rice as well as a range of increasingly valuable manufactured goods. But beginning in 1997, economic growth stagnated and political instability increased. The foreign investors who had supported much of the region's growth withdrew. In the late 1990s, poverty was again an increasingly painful reality for many people, and thwarted hopes for prosperity led to rising political violence. By 2003, though, economic recovery helped reduce the number of people living in poverty, and, according to the World Bank, the number of people living on less than $2 per day in several countries fell to all-time lowest levels.

Although popular images of Southeast Asia link the region to industrialization, in fact the economies of all the countries in this region are grounded in agriculture and services as well as industry. It is not uncommon for individuals to earn a living by plying skills in all three sectors at once.

Agriculture

Compared to services and industry, agriculture's role in the region's economy has been declining. It accounts for less than one-sixth of total regional output. However, about 60 percent of Southeast Asians still live in rural settlements. These people continue to depend on agriculture for part of their support (Figure 10.10), even though many of them also fish, make and sell crafts, perform services, or work part-time for wages. Agriculture also yields important products for export (especially rice, rubber, sugar, coconut products, and palm oil) and provides important components of urban diets (rice, fruits and vegetables, fish, meat, and dairy products).

Several forms of agriculture are practiced in Southeast Asia. So-called **slash-and-burn** (**shifting** or **swidden**) **cultivation** is an ancient form of multicrop gardening using small plots, often on rain-forest lands and in uplands that do not support more intensive types of agriculture. This system requires intimate knowledge of

PRACTICING GEOGRAPHY   COMPARING LOCAL LANDSCAPES

When you examine these two photos carefully (see photo interpretation guidelines in Chapter 1, page 6), what observations can you make about the nature of agriculture and women's economic roles in the situations depicted in photos A and B? Speculate on where the photos were taken. [A, Paul A. Souders/CORBIS; B, Keren Su/CORBIS.]

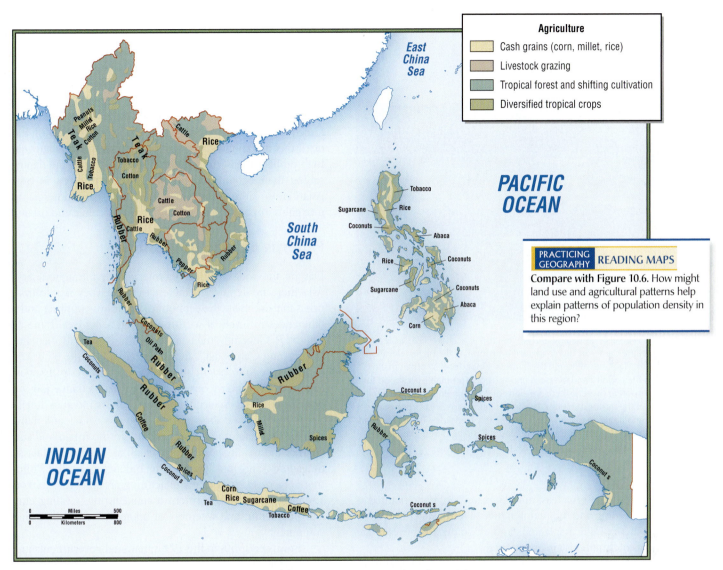

Agriculture

☐ Cash grains (corn, millet, rice)
☐ Livestock grazing
☐ Tropical forest and shifting cultivation
☐ Diversified tropical crops

PRACTICING GEOGRAPHY   READING MAPS
Compare with Figure 10.6. How might land use and agricultural patterns help explain patterns of population density in this region?

FIGURE 10.10 Agricultural patterns in Southeast Asia. Tropical forests and crops, rice production, and shifting cultivation dominate the agricultural patterns of Southeast Asia. [Adapted from *Hammond Citation World Atlas*, 1996, pp. 74, 83, 84.]

plant and soil characteristics. The farmer intercrops multiple species with careful attention to local soil and climate and attends to the needs of individual species and individual plants with hand tools (see Chapter 3, pages 121–122, and Chapter 7, page 350 for similar cultivation systems in the Americas and Africa). This type of cultivation is practiced today by subsistence farmers in the hills and uplands of mainland Southeast Asia and in coastal Sumatra, Borneo, Sulawesi, the southern Philippines, and West Papua—that is, in areas where population densities are relatively low (see Figure 10.6 on page 510). Shifting agriculture is possible only in low-density areas because farmers move their fields every three years or so, as soil fertility is depleted. Where populations are sparse, plots can be abandoned for decades before they are used again, allowing the forest to regrow and the soil to be replenished.

Another venerable and highly productive form of agriculture in the region is **wet rice** (or **paddy**) **cultivation.** Wet rice cultivation is permanent, as opposed to shifting. It entails the hand-planting (usually by women) of rice seedlings in flooded fields that are first cultivated (usually by men) with hand-guided plows pulled by water buffalo. First, laborers terrace and bank the soil by hand to create multiple adjacent ponds on a hillside or plain. Then, they apply animal manure to the soil to maintain fertility and deliver water to the paddies by directing natural runoff or by pumping water from streams. Wet rice cultivation has transformed landscapes throughout Southeast Asia (see the photo on page 524). It is practiced in river valleys and deltas, where rivers bring a yearly supply of silt, and on rich volcanic soils on such islands as Java and Sumatra in Indonesia and Luzon in the Philippines. Although wet rice farming has extensive environmental effects—in the form of forest clearing, landscape manipulation, and waterlogging of soils—this type of agriculture is sustainable over long periods and yields are fairly high. It is often cited as one of the better examples of nature being harnessed effectively in the service of humans.

Many farmers are abandoning paddy cultivation for work in the cities. The rice fields are now worked mostly on the weekends. Increasingly, they are prepared quickly with a type of tractor ("iron

Large-scale vegetable gardening and single-crop agriculture, like this tea plantation in the Cameron Highlands, are transforming the higher elevations of peninsular Malaysia. Although the terraces supply food to urban markets both within the country and overseas and provide a profitable livelihood for farm and corporate owners, there are signs of environmental damage amid the beauty of the pastoral settings. Erosion in fields and construction sites (upper left) and along roadsides is causing deadly landslides. Because the land is less able to hold moisture, there is increased flooding during the rainy season and drought during the dry season. [Alex Pulsipher.]

buffalo"), and rice is broadcast into the paddy, not hand-planted as seedlings. Weeding and the maintenance of water systems and terraces are increasingly neglected, and in some areas yields are declining as old methods are abandoned.

During the recent decades of prosperity, some of Southeast Asia's farming families left the countryside after selling their small farms to corporate plantations. The result has transformed rural ecology. Small farms that were once operated by families have been combined into large commercial farms owned by local or multinational corporations and are producing such crops as rubber, African oil palm trees, bananas, pineapples, and rice. This style of agriculture entails combining huge tracts of land into one system; clear-cutting patches of forest that once occupied the fringes of farms; deep plowing the soil; usually planting one species of crop plant over many square miles; bolstering soil fertility with chemicals; using mechanized equipment; and, in the case of rice, using large quantities of water. Commercial farming reduces the need for labor, and its goal is large-scale production for cities and for export. Quick profit, not long-term sustainability, is the objective. Many commercial farmers have achieved dramatic boosts in harvests (especially of rice) by using high-yield crop varieties, the result of green revolution research that has been applied in many parts of the world. Although *all* types of farming, except some forms of small-scale shifting cultivation, have significant negative effects, the impact of large-scale commercial farming appears to be wider and longer lasting. Commercial farming can lead to displacement of traditional farmers and forest dwellers, loss of wildlife

habitat and plant and animal species, soil erosion and loss of soil fertility, flooding, and the chemical pollution and depletion of land and groundwater resources.

Poor farmers usually cannot afford to become green revolution farmers because the new technologies are too expensive for them. Often they cannot compete with green revolution farms and are forced to migrate to the cities to look for work. In contrast, the farmers and corporations who can afford the new technologies can earn high profits, at least until soil resources and water are depleted, by selling their agricultural products to the millions in Southeast Asian cities. Here as elsewhere in the world, commercial agriculture gains importance as economies industrialize and urbanize.

Patterns of Industrialization

Despite the widespread economic growth throughout Southeast Asia, there are significant differences in the types of industrial enterprises that dominate the economies in various countries and related differences in levels of wealth and well-being (see Table 10.3 on page 535). Generally, the poorer countries—such as Vietnam, Laos, Burma, and Cambodia—and the rural parts of the other countries depend primarily on agriculture, fishing, and forestry. Even in some rural areas, however, construction and maintenance jobs on such projects as new hydroelectric dams are raising incomes. In cities and towns, garment and shoe making, food processing, and many types of light manufacturing are expanding. In the urban and suburban areas of the wealthier countries—such as Thailand, Malaysia, and Singapore—and in a few areas in Indonesia and the Philippines, more elaborate types of manufacturing are important: automobile assembly, textile and garment manufacturing, refining chemicals and petroleum, and assembly of computers and other electronic equipment.

From the 1960s to the 1990s, national governments in Southeast Asia restricted and controlled foreign investment in the region. Using their own tax revenues, along with a modest amount of regulated foreign investment, governments tried to nurture economic development that would steadily create new jobs and be shielded from the ups and downs of global economic cycles. Export agriculture was favored, as were import substitution industries. These are local industries that produce consumer goods for local use; the use of tariffs excludes competing goods made outside the region. For the most part, these policies resulted in strong and sustained economic growth. Standards of living increased markedly, especially in Malaysia, Indonesia, and Thailand. Unlike similar efforts in Middle and South America that have been less successful (see Chapter 3, pages 134–135), these countries were aided by a higher proportion of local (versus foreign) capital; strong support of government and economic elites, whose interests were tied to successful industrial development; and a focus on cheap mass consumer goods rather than expensive luxury consumer goods, all of which created a broad-based domestic market that could fuel further economic expansion.

EPZs. In the 1970s, the region adopted an additional strategy for encouraging economic development, this one was aimed at manufacturing products for export, primarily to developed countries. Foreign multinational corporations were attracted through the establishment of **export processing zones (EPZs),** or free trade zones. These are specially designated areas, similar to the maquiladoras of Mexico, where foreign companies can set up manufacturing plants using inexpensive labor and, sometimes, inexpensive local raw materials to produce items only for export. Subcontractors, often locally owned, also may set up factories to supply needed parts. Taxes are eliminated or greatly reduced. The first export processing zone was established in Malaysia in 1971 and attracted electronics industries. Since then, EPZs have been established in Indonesia, Vietnam, and the Philippines for the production of garments, textiles, and shoes and are now a major strategy to expand economic development.

EPZs generally employ an almost entirely female labor force. Between 80 and 90 percent of the workers in these new factories are women, not only in Southeast Asia but in other regions as well. So widespread is this development around the world that scholars and activists view the **feminization of labor** as a distinct characteristic of globalization over the past three decades (see Chapter 3, pages 160–162, and Chapter 9, pages 442 and 443). The geographer Jonathan Rigg has analyzed several studies of this phenomenon in Southeast Asia and concluded that bosses prefer to hire young, single women because they are perceived as the least troublesome employees. Statistics do show that, at least for now, women will work for lower wages than men, will not complain about poor and unsafe working conditions, will accept being restricted to certain jobs on the basis of sex, and are not as likely as men to agitate for promotions. But the situation varies from country to country. For example, female Filipino workers are more assertive than Malaysians. Supervisors are somewhat more likely to be female in U.S.- and European-owned firms than in Japanese- or Korean-owned ones, though both male and female bosses in this region generally tend to be authoritarian, and factory workers usually have little say in workplace policies.

Growth Triangles. Singapore is part of a new trend that is taking shape in the region: the emergence of larger transnational economic regions called **growth triangles** or economic zones. The Singapore-Johor-Riau (or Southern) Growth Triangle (see Figure 10.13 on page 523) was created in about 1989 when low-cost manufacturing began to be relocated from Singapore to Johor, just across the border in southern Malaysia, and eventually to the nearby Indonesian Riau archipelago. The move was prompted by Singapore's relatively high labor costs, land shortages, and expensive fresh water. Multinational corporations locating in the Southern Growth Triangle strategically pick the advantages of Singapore's efficient business and port infrastructure and highly skilled labor force (workers earned about U.S. $600 per month in 1990) for some tasks, and for other tasks choose the lower costs of the cheaper land and water and semiskilled labor in nearby Johor and Riau. (Labor costs are U.S. $220 per month in Johor and U.S. $90 per month in Riau.) A single multinational firm can locate core high-technology telecommunications, managerial, and shipment activities in Singapore, less technical activities (for example, textiles and electronics manufacturing) in Johor, and assembly operations in Riau.

Not surprisingly the working conditions and general well-being of workers in Singapore are better than in Johor or Riau; and, in general the benefits of Southeast Asia's "economic mira-

This view from atop the Raffles Hotel captures the active and prosperous spirit of Singapore. In the foreground, you can see the old colonial town center and, across the river, the skyscraper banking district. The pleasure boat harbor is to the left, port facilities at upper left. Singapore is a sleek and modern planned environment, but here and there historic buildings have been adapted to modern uses, usually connected with tourism. Singapore is the world's largest port, and from the hotel's penthouse restaurant one can see dozens of container ships waiting far out in the open ocean for permission to come into the docks for loading or unloading. [D. Saunders/TRIP.]

cle" have been unequally apportioned. In the region's new factories and other enterprises, it is not unusual for assembly-line employees to work 10 to 12 hours a day, 7 days a week for less than the legal minimum wage and with no benefits. Labor unions that would address working conditions and wage grievances are rare, and international consumer pressure to improve working conditions has been only partially effective. For example, in the late 1990s, the U.S. shoe manufacturing firm Nike became the focus of worldwide outrage over the treatment of its workers in Vietnam, Indonesia, and elsewhere across Asia; the employees were frequently exposed to hazardous chemicals, as well as to physical abuse and psychological cruelty on the job. In 1997, Nike laborers in Vietnam were paid $10 per week for 65 hours of work. The production cost for a pair of Nike shoes was U.S. $5.60, while the retail price in America was $90 to $145. Nike's labor tactics became widely known to American university students when student activists revealed that many campus stores were selling equipment and garments with the Nike logo that were made under these substandard conditions in Southeast Asian factories. Student activist groups succeeded in finding out where Nike's factories were located in Asia, and investigations of factory working conditions revealed the abuses. Although the students' exposure of abuses in 1997 resulted in some improvements in working conditions, succeeding reports by U.S. university students and labor and human rights groups continued. In April 2000, Nike workers were still earning less than U.S. 20 cents an hour, far less than a living wage, and were often required to work 72-hour weeks. Managers still

AT THE REGIONAL SCALE   Tourism Development in the Greater Mekong Subregion— Economic Boom or Cultural Bust?

By 5:30 A.M., the Buddhist monks, clad in saffron robes, were already walking down Sisavangvong Road, which leads toward the old town of Luang Prabang, the ancient capital of a kingdom that once covered present-day Laos, southern China, and northeastern Thailand. In the gloomy light of dawn, they looked like a strip of moving orange silk. White-haired old men of the town, kneeling and bowing to show respect, distributed glutinous rice to the monks as they went through their morning ritual of collecting alms from the community. The harmonious moment was broken when a young girl invited tourists to buy some rice wrapped in a banana leaf to offer to the monks—commercializing a Buddhist religious ritual.

Luang Prabang is known as the seat of Lao culture, rich with monasteries and monuments. It is one of several stops along a newly developed heritage route for tourists in the Greater Mekong Subregion that includes Cambodia's Angkor Wat and Vietnam's Hue historic sites (Figure 10.11). Since being declared a World Heritage Site by the United Nations Educational, Scientific and Cultural Organization (UNESCO) in 1995, Luang Prabang has become one of the tourist industry's new "hot" spots in Southeast Asia. Airline routes, boat tours from Thailand, and planned road links bring more than 500 tourists a day to the town, making it a key contributor to the tourism dollars that Laos earns.

The opening up of such destinations in mainland Southeast Asia is part of a large international initiative to develop the Greater Mekong Subregion (GMS) and enhance economic relations among the six Mekong countries (Burma, Cambodia, China, Laos, Vietnam, and Thailand). The scheme is financed by the Asian Development Bank with support from tourism promoters such as the World Tourism Organization and the Pacific Asia Travel Association.

The promotion of tourism in Southeast Asia is viewed as a major opportunity for economic development and the best way to get income to the people; but critics asking the "development for whom?" question argue that the people themselves will have little choice in deciding the pace or direction of its development. It is unclear how a small geographic area that has been isolated for decades can deal with a large and sudden influx of thousands of tourists who will strain the infrastructure and services, damage cultural assets, and perhaps make local people feel unwelcome in their own home spaces. In Luang Prabang it is already apparent that jarring changes are underway. Every second building along Phothisarat Street in the center of Luang Prabang's old town is a restaurant, shop, bakery, or guest house, interspersed with a spattering of Internet cafés.

Pollution, too, has become a problem. Plastic bottles and garbage litter the area, and shreds of plastic cling to the tall weeds lining the bank of the Mekong River that leads out of town. Within the town itself, tensions are rising between those who make tourism bucks and those, like farmers on the other side of the river, who do not. And, whereas once young people in the town may have aspired to be doctors or teachers, they now want to be tour guides.

[Adapted from Teena Amrit Gill, "Locals lose out as tourism booms," *Asia Times* Online, March 12, 2002, at http://www.atimes.com/se-asia/DC12Ae02.html; Bui Nguyen Cam Ly, "As hordes of tourists come, heritage goes," *Inter Press Service News*, 2003, at http://www.ipsnews.net/mekong/stories/heritage.html; and Gulfer Cezayirli, "Fast-growing Asian tourism should enlist help of the urban poor," *Asian Development Bank*, 2003, at http://www.adb.org/Media/Articles/2003/2009_Regional_Asian_Tourism_Should_Enlist_Help_of_the_Urban_Poor/default.asp.]

physically abused workers and called them degrading names, and workers still worked in environmentally hazardous conditions. When questioned, Nike placed the blame on subcontractors in Taiwan and Korea who set up and administer Nike factories throughout Southeast Asia. Subcontracting is a growing strategy for reducing costs and transferring responsibility. As of 2004, the same abuses by other sportswear corporations were being documented by the student-founded NGO, Educating for Justice, which was also lobbying the United States Olympic Committee (USOC) to require companies that obtain Olympics licenses to observe international labor standards.

Tourism Development. The fastest-growing industry in Southeast Asia, and indeed the world, is tourism. Many countries view tourism as an opportunity to broaden the base of their economies and to spread industrial development more evenly throughout their territories. Since the 1970s, tourism has been viewed by many Southeast Asian countries as playing a key role in economic development, and organizations such as the Asian Development Bank

and the Association of Southeast Asian Nations (ASEAN) have promoted tourism in the region in a big way. Between 1991 and 2001 the number of international visitors to the region doubled to over 42 million, which accounted for over 40 percent of the world's total. Earnings from tourism contribute tens of billions of U.S. dollars to Southeast Asian economies. While the surge of tourism in Southeast Asia has created a number of opportunities for several local communities, it has also raised a number of concerns. Chief among these is the threat that uncontrolled mass tourism poses to the long-term survival of cultural heritage sites and local cultures. (See the box "Tourist Development in the Greater Mekong Subregion" above.)

Tourism and the Sex Industry. Related to the growth of the tourist industry in general in Southeast Asia is the expansion of **sex tourism.** Sex tourism is in some ways an extension of the sexual entertainment industry that served foreign military troops stationed in Asia after World War II and during and after the conflicts in Korea and Vietnam. Now civilian men arrive from around the

FIGURE 10.11 **World Heritage Sites in the Greater Mekong Subregion.** [Adapted from UN World Heritage Convention, at http://unesco.org/; and the Mekong River Commission, at http://www.mrcmekong.org/shopping/productsbycategory.asp?incatalogID=58str Catalog_name-maps.]

globe to live out their fantasies during a few weeks of vacation in Thailand, Cambodia, Indonesia, Vietnam, or the Philippines. In 2000, 9 million tourists visited Thailand alone, up from 250,000 in 1965, and some observers estimate that as many as 70 percent are looking for sex. In recent years, Thai government officials have encouraged sex tourism to create jobs, and they praise the sector's role in helping the country weather the economic crisis of 1997. Corrupt public officials also support sex work because it provides them with a source of illegal, untaxed income from bribes.

One result of the "success" of sex tourism and high local demand for sex workers (see the discussion of HIV/AIDS on pages 512–513) is that organized crime has entered the field, and girls and women are being coerced into sex work. Some research indicates that girls may have been sold by their families to pay off family debt. Others have been kidnapped, often at a very young age. Demographers estimate that 20,000 to 30,000 Burmese girls taken against their will (some as young as 12) are working in Thai brothels; their wages are too low to make buying their own freedom pos-

sible. In the course of their work, they must service more than 10 clients a day and are routinely exposed to physical abuse and sexually transmitted diseases, especially HIV. The industry is found throughout the region, although perhaps not to the extent it is in Thailand.

PERSONAL VIGNETTE Twenty-five-year-old Watsanah K. (not her real name) awakens at 11 every morning, attends afternoon classes in English and secretarial skills, and then goes to work at 4 P.M. in a bar in Patpong, Bangkok's red light district. There she will meet men from Europe, North America, Japan, Taiwan, Australia, Saudi Arabia, and elsewhere who will pay to have sex with her. She leaves work at about 2 A.M., studies for a while, and then goes to sleep.

Watsanah was born in northern Thailand to an ethnic minority group whose members, like many others in the area, are poor subsistence farmers who have recently become involved in the global drug trade by growing opium poppies. Watsanah married at

15 and had two children shortly thereafter. Several years later, her husband developed an opium addiction. She divorced him and left for Bangkok with her children. There she found work at a factory that produced seat belts for a nearby automobile plant. In 1997, Watsanah lost her job when a financial crisis ripped through Thailand and the rest of Southeast Asia. She became a sex worker to feed her children.

Although the pay, between $400 and $800 a month, is much better than the $100 a month she earned in the factory, the work is dangerous and demeaning. Sex work, though widely practiced and generally accepted in Thailand, is illegal and the women are looked down on, so Watsanah must live in constant fear of going to jail and losing her children. Moreover, she cannot always make her clients use condoms, which puts her at high risk of contracting AIDS or other sexually transmitted diseases (STDs). "I don't want my children to grow up and learn that their mother is a prostitute," says Watsanah, "that's why I am studying. Maybe by the time they are old enough to know, I will have a respectable job." [Adapted from the field notes of Alex Pulsipher and Debbi Hempel, 2000; "Sex industry assuming massive proportions in Southeast Asia," International Labor Organization News (August 19, 1998); coverage of the HIV/AIDS conference in Thailand, July 11–16, 2004, by the Kaiser Family Foundation, at http://www.kaisernetwork. org/aids2004/kffsyndication.asp?show=guide.html.]

Economic Crisis and Recovery: The Perils of Globalization

Beginning in 1997, a financial crisis swept the region, forcing millions of people back into poverty and creating tensions that disrupted Southeast Asia's political order. A major cause of the crisis was the rapid shift of most Southeast Asian economies away from government regulation and toward the free market. Another factor was widespread corruption.

By the late 1980s, most governments were opening up their national economies. One change was a greater willingness to allow the marketing of foreign products in Southeast Asia. Foreign governments in Japan, North America, Europe, and elsewhere negotiated with Southeast Asian governments to gain local markets for their countries' products. Multinational corporations, such as automobile manufacturers, saw opportunities to save on labor, land, transportation, and resource costs by locating their facilities and selling their products within the region. Thus, factories manufacturing products for export in the EPZs were joined by foreign-owned factories manufacturing products they hoped to sell in Southeast Asia.

In opening national economies to the free market, governments focused especially on the crucial financial sector. Emboldened by their years of success in creating jobs and raising incomes and living standards, Southeast Asian firms were eager to expand; to do so, they needed investment capital. Southeast Asian banks were given greater freedom to determine the kinds of investments they would make and were allowed to accept more money from foreign investors. Soon, these banks were flooded with money from investors in the rich countries of the world, who hoped to profit as the banks invested in the region's growing economies. As a result, the stock prices of many public companies were inflated beyond their value. Also, to make quick profits, the bankers made

risky loans to real estate developers, often for high-rise office building construction. As a result, by the mid-1990s, many cities had a glut of office space—in Bangkok alone, there was U.S. $20 billion worth of unsold office space.

However, lifting the controls on investments might not have been such a problem had the region not been plagued by a kind of corruption called **crony capitalism.** In most Southeast Asian countries, high-level politicians, bankers, and wealthy business owners often have close personal relationships or are part of the same extended family, and these personal connections can encourage corruption. For example, in Indonesia, the most lucrative government contracts and business opportunities were for decades reserved for the children of former President Suharto, who ruled Indonesia from 1967 to 1997. His children became some of the wealthiest people in Southeast Asia. With the opening of Southeast Asian economies to foreign investors, this kind of corruption expanded considerably as foreign investment money was easily diverted to bribery or unnecessary projects that brought prestige to political leaders.

The cumulative effect of crony capitalism and the pursuit of quick returns on risky investments was that many ventures failed to produce profits at all. In response, foreign investors withdrew their money on a massive scale. Before the crisis, in 1996, there was a net inflow of U.S. $96 billion to Southeast Asia's leading economies as well as to South Korea, which was also affected by the crisis. In 1997, there was a net outflow of U.S. $12 billion. The panicked withdrawal of investment money was the immediate cause of the economic crisis.

The International Monetary Fund (IMF) made a major effort to keep the region from sliding deeper into recession. In 1998, the IMF implemented a bailout package worth some U.S. $65 billion, much of it used to rescue banks. The IMF considers the banks essential to the region's economic stability, but critics argue that the bailout gives these banks little incentive to improve their disastrous record. Also the IMF's bailout packages were designed more to protect foreign investors than local economies and often came at the expense of local residents and laborers (see below). The packages consisted of structural adjustment programs, which, like those discussed in Chapters 3 (pages 136–138) and 7 (pages 353–355), required countries to reduce tariffs and abandon other policies intended to protect domestic industries. The goal was to lessen corruption and increase efficiency in order to win back the confidence of foreign investors. Critics point out, however, that dependence on foreign investment leaves the region open to a recurrence of a panicked withdrawal of funds. And, if foreign investors take over the most profitable sectors of the economy, local enterprise will suffer, and profits made by foreign investors are likely to be invested outside the region.

Nevertheless, by 2000, the economic recovery of Southeast Asia was apparently under way. Between 2001 and 2003, economies registered growth rates between 3 and 7 percent. This marked an improvement over the negative growth rates of 1997–1999 but still fell short of pre-crisis levels. Local currencies had begun to reappreciate, and several nations experienced an expansion in trade. Southeast Asian countries were successfully expanding older strategies, such as the establishment of export processing zones to attract additional multinational corporations to Southeast Asia.

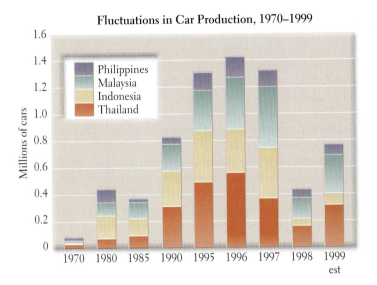

Fluctuations in Car Production, 1970–1999

FIGURE 10.12 Car production in Southeast Asia, 1970–1999. The sudden decline in car manufacturing in 1998 is an illustration of the depth of the sudden economic decline in the region. [From "Car making in Asia: politics of scale," *The Economist* (June 22, 2000): 68.]

Consequences of the Economic Crisis

The Southeast Asian economic crisis had a number of disturbing consequences. Low-income people suffered the most. Many lost their jobs as a result of the rapid withdrawal of foreign investment money, which shut down many foreign-backed businesses. (Figure 10.12 shows the effect of the crisis on car producers in four countries in the late 1990s.) Moreover, the crisis sparked a decline in the value of most Southeast Asian currencies between 1997 and 2001. As a result, prices rose for the imported food and consumer goods on which many urban people depend and for the fertilizers and pesticides that commercial agriculture requires. As prices rose, consumption fell and firms no longer operated profitably. Workers faced layoffs and firings. The 1998 IMF bailout is to be repaid with tax dollars, so low-income people are likely to suffer cuts in health care, education, and food subsidies for years to come. In Indonesia, for example, rates of poverty during the crisis climbed from roughly 11 to 60 percent. Frequent outbreaks of rioting, looting, and general violence troubled urban and rural areas.

With economic recovery under way since 2002, however, inflation rates dropped, incomes increased, and poverty rates dropped to near, if not below, pre-crisis levels. In Thailand the number of people living in poverty dropped to an all-time low of less than 10 percent. While this is encouraging, economic recovery has not been even across the region. The greatest recoveries have been achieved in Thailand and Malaysia, while in Indonesia and the Philippines economic recovery has been mixed and less impressive. And the external debt of all the affected countries remains high.

The Association of Southeast Asian Nations

Southeast Asian countries trade more with the rich countries of the world than they do with one another (see Figure 10.1 on page 500).

They export food, timber products, minerals, processed commodities, and manufactured finished goods and components to Japan, Australia, New Zealand, Europe, and the United States. They import manufactured products that they do not make themselves, including high-fashion consumer products, industrial materials, machinery, and parts. Trade among countries within the region has been inhibited by the fact that they all export similar goods and traditionally have imposed tariffs against one another. It is in this context that a region-wide free trade zone was created by the **Association of Southeast Asian Nations (ASEAN)**, an increasingly important bulwark of both economic growth and political cooperation in the region. Started in 1967 as an anticommunist, anti-China association, ASEAN now includes all the states in the region except Timor-Leste, and focuses on nonconfrontational accords that strengthen regional cooperation. An example is the Southeast Asian Nuclear Weapons-Free Zone treaty signed in December 1995 by all ten Southeast Asian nations.

ASEAN is assembling a free trade association patterned after the North American Free Trade Agreement and the European Union. This trade alliance, known as the **ASEAN Free Trade Association (AFTA)**, was launched in 1992. According to current plans, most tariffs will be reduced to 0 to 5 percent by 2010 for the older ASEAN members and 2015 for Cambodia, Laos, Burma, and Vietnam. The main objective of AFTA is to help ASEAN members compete better in the world marketplace. Eliminating tariffs and other trade barriers between countries will lower production costs and make ASEAN's manufacturing industries more efficient. This will allow ASEAN products to be priced lower and hence to gain a competitive edge in the global market. At the same time, the region's consumers will be more likely to buy the more efficiently produced regional products, thus expanding trade among ASEAN countries.

Intraregional trade in ASEAN is growing, albeit slowly. In 1990, 18 percent of the region's total trade was within the ten ASEAN countries; by 2003, it had increased to nearly 23 percent. But participation in intraregional trade remains skewed; Singapore alone accounts for 41 percent of the total. The original ASEAN member states (Thailand, Indonesia, Singapore, the Philippines, and Malaysia) account for over 95 percent of the total intraregional trade. Burma, Laos, Cambodia, and Vietnam hardly participate at all.

Pressures for Democracy and Self-Rule

Although there are strong demands for a greater public voice in the political process in Southeast Asia, significant barriers to democratic participation exist. Undemocratic socialist regimes control Laos and Vietnam, for example, and a military dictatorship runs Burma. There are few democratic institutions even in slightly wealthier countries, such as Indonesia, or in the rich city-state of Singapore. Some Southeast Asian leaders, such as Singapore's former prime minister Lee Kuan Yew, have argued that Asian values are not compatible with Western ideas of democracy. Yew and other leaders have argued that Asian values are grounded on the Confucian view that individuals should be submissive to authority. Hence, Asian countries should avoid the highly contentious public discourse of electoral politics. (For more on this view of Asian

AT THE LOCAL SCALE *Pancasila:* Indonesia's National Ideology

Indonesia is the largest country in Southeast Asia and the most fragmented—it comprises 3000 inhabited islands stretching over 3000 miles (8000 kilometers) of ocean. It is also the most culturally diverse, with dozens of ethnic groups and multiple religions. Until the end of World War II, Indonesia was not a nation but rather a loose assemblage of distinct island cultures, which Dutch colonists managed to hold together as the Netherlands East Indies. When Indonesia became an independent country in 1945, its first president, Sukarno, hoped to forge a new nation out of disparate parts. To that end, he articulated a famous and controversial national philosophy known as *Pancasila,* based on tolerance, particularly in matters of religion.

The five precepts of *Pancasila* are belief in God and the observance of conformity, corporatism, consensus, and harmony. It is easy to see that all four of these words could be interpreted to discourage dissent or even loyal opposition; and they seem to require a perpetual stance of boosterism. For some people, the strength of *Pancasila* is that it ensures there will never be either an Islamic or a communist state. Others note the chilling effect that the philosophy has had on participatory democracy (the two political parties are state controlled) and on criticism of the president and the army. And, in fact, there had been no orderly democratic change of government until 2004. Sukarno was in power from 1945 until he was deposed by Suharto in 1965. Suharto declared himself president and was in power until forced to step down in 1998. A series of appointed successors followed him, including Megawati Sukarnoputri, Sukarno's daughter, in 2001.

Indonesia held its first presidential election in 2004. Susilo Bambang Yudhoyono won with just over 60 percent of the 140-plus million votes cast.

An example of how *Pancasila* becomes part of the daily lives of people and counters egalitarian ideals is provided by the state-sponsored women's organization Dharma Wanita (Women's Duty). The organization is modeled on an organization of American military wives, and membership is obligatory for all female government employees and the wives of male government employees. Its purpose is to teach women to serve as good wives and to support their husbands' careers, regardless of the women's actual marital or career status. There is a chapter in every government office, and a woman holds office in Dharma Wanita according to her husband's position in an agency or firm. The ideals of Dharma Wanita have spread with immigrant Indonesian women to other parts of the region and to North America.

[*Sources:* Adapted from Saskia Wieringa, "The politicization of gender relations in Indonesia," *The Colonial Widow* (The Hague: The Digital City Project, 1995), http://www.indonesiachicago.org/dharmawanita/events.htm; and *The Economist* (October 7, 2004).]

Thinking Geographically: ON THE WEB

values, see the box "*Pancasila:* Indonesia's National Ideology" above.) But this position conveniently overlooks the Confucian expectation that governments rule with justice and the needs of all society in mind (see Chapter 9) and that citizens act against corruption and misuse of power to restore the ideal social order. In fact, many people living under undemocratic, authoritarian regimes in the region have rebelled. People in Timor-Leste, Aceh, the Molucca Islands, Sulawesi, and elsewhere in Indonesia risked their lives to resist the authoritarian and exploitative policies of President Suharto.

Is Indonesia Breaking Up? In recent years, rising violence and instability in Indonesia have many people wondering if this multi-island country of 220 million might be headed for disintegration. Separatist movements have sprouted in four distinct areas (Figure 10.13). One movement has been successful, resulting in the formation of the independent country of Timor-Leste (2002), but only after a long, bloody conflict following the invasion of Timor-Leste by Indonesia in 1978. The large far eastern province of West Papua has a growing separatist movement that formed in response to discrimination against the local Melanesian population by people coming from densely populated Java in hopes of finding cheap plentiful land. Just to the west in the Moluccas, in 2002, clashes between Christians and Muslims killed thousands and encouraged the revival of a Christian-led separatist movement. In the far western province of Aceh, separatists continue to battle Indonesian security forces, as they have for decades, largely because they resent that most of the wealth yielded by their province's oil industry goes to the central government in Jakarta on the island of Java.

Jakarta tries sporadically to appease these movements by giving people in separatist areas greater control over their own affairs (see the discussion of devolution in Chapter 4, page 220), but it is also wary of strengthening separatism by allowing too much autonomy. An alternate strategy is to bypass the provincial governments, which are the most prone to separatism, and instead give more power to local and municipal authorities. Some say this strategy will only increase corruption at the local level while doing little to improve the behavior of the Indonesian armed forces, whose often brutal tactics have alienated locals throughout Indonesia, not just in separatist areas.

Progress Toward Democracy. Pressure for democracy has been rising in several Southeast Asian countries, in response to both economic development and economic crisis. In Thailand, years of economic growth have increased literacy levels and awareness of the democratic privileges that exist in many other countries. In the late 1990s, Thais successfully agitated for constitutional provisions that reduce corruption and provide for regular elections. In Malaysia, increased education and prosperity have similarly pushed the government to act in a somewhat more democratic and less authoritarian manner. In Indonesia, after three decades of semidictatorial rule by President Suharto, massive demonstrations in the wake of the region's economic crisis led to Suharto's resignation and a series of democratic parliamentary and presidential elections. People in Burma, on the other hand, have been protesting the rule of a military regime for more than a decade. The regime refused to step aside when the people elected Aung San Suu Kyi to lead a civilian reformist government in 1990.

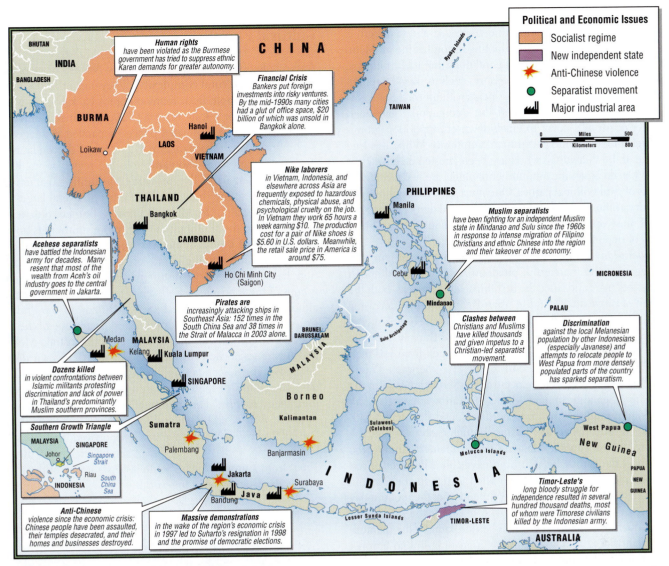

FIGURE 10.13 Political and economic issues: Southeast Asia. Economics, politics, and culture interact to create particularly complex circumstances in Southeast Asia. While rapid industrialization has improved standards of living, incomes vary widely. Some countries seek development with socialist policies; others pursue capitalist policies and democracy, with varying degrees of success. Indonesia has experienced the most violence and political unrest in the recent past, as the boxes in the map indicate. Thailand, Malaysia, and Singapore are, for now, the most stable politically and the most prosperous.

She has been under and out of house arrest for more than a decade; and the military remains in control.

International Terrorism. During the late 1990s, a series of bombs exploded in the Philippines and across Indonesia, and small terrorist cells were discovered in Malaysia, far southern Thailand, and Singapore. Until the bombing of the Sari Hotel in Bali (Indonesia) in October 2002, which killed nearly 200 foreign tourists, terrorist activity in the region was local in nature. It was carried out by local militant groups pursuing domestic political agendas and grievances—most often revenge for government campaigns against Muslim separatists (see Figure 10.13). The Bali bombing, the bombing of the Marriott hotel in Jakarta in August 2003, and the attack in April 2004 on a police station in Muslim southern Thailand, however, have drawn attention to apparent connections between local groups and international terrorist networks. Of particular concern is the Indonesian-based organization Jemaah Islamiah

(JI), which is believed to have links to Al Qaeda (though the nature of these is debated) as well as its own extensive network of terrorist cells. *The Economist* (Global Agenda, August 7, 2003) reported that Jemaah Islamiah has roots in radical Muslim movements of the 1940s and 1950s that sought to create an Islamic superstate in Southeast Asia. These groups survived various crackdowns by the Indonesian government and then re-emerged after the fall of President Suharto in 1998; but elections of secular political parties in Malaysia, Indonesia, and Thailand in 2004 showed that the citizenry of Southeast Asia, who are accustomed to a tolerant version of Islam, do not support the militants.

Despite heightened efforts of state governments in Southeast Asia to combat the rise of terrorism in the region, the task they face is daunting and complicated. Distinguishing between local and international organizations is difficult, and the inefficiency of underfunded security forces makes it relatively easy for terrorists to slip back and forth across borders undetected. Another concern is the strategy that governments are adopting to combat the problem. Rather than trying to address local conditions that fuel militancy, states are resorting to repressive measures such as illegal deportations and indefinite detention of dissidents without trial. State governments in the region have used the international campaign against terrorism as a cover to crack down on opposition groups that merely challenge the government's authority. While it is true that some of the individuals who have been arrested do have clear ties to Al Qaeda, a significant number of others are just prominent members of rival political parties. One must ask what might be the long-term ramifications for political stability and democracy in the region, if those opposed to a particular government were routinely detained or deported for being dissenters.

Modern-Day Pirates. Loosely related to the rise of terrorism in the region is the increase in piracy in Asian waters. Although piracy has been around since the first ships sailed the seas, both the escalation of pirate attacks in the Asian Pacific and the increasing violence associated with these attacks have become a global concern. According to the International Maritime Bureau, the total number of pirate attacks in the world has jumped from less than 50 in 1994 to over 450 in 2003. In recent years, between 35 and 50 percent of these attacks have occurred in Southeast Asia (in the South China Sea and Strait of Malacca). Narrow channels, shallow reefs, thousands of tiny getaway islands, and slow traffic with some 900 commercial vessels passing through each day make the waters around Singapore, Malaysia, and Indonesia among the world's most dangerous for maritime navigation. Political unrest and economic recession have been cited as reasons behind the wave of pirate attacks in the region. While most of the modern-day pirates are essentially petty thieves who make for the safe in the captain's quarters, organized crime is behind some of the marauding in Asian waters, and an apparently increasing link to terrorist organizations has become a matter of concern.

One of the factors that make prevention of piracy so difficult is that more and more ships sail under so-called "flags of convenience." That is, to avoid taxation in their own country, owners register ships in other countries, such as Liberia, sail under their flags, and hire crews from around the world (see the PBS *FRONTLINE/World* program "The Lawless Sea," highlighted in Chapter 4, page 209). Unlike the colonial era when the British navy patrolled the seas, nowadays national governments have little vested interest in protecting ships from attacks by pirates. In addition, many pirates operate in collusion with powerful businessmen and corrupt government officials who profit from the enterprise.

SOCIOCULTURAL ISSUES

The cultural complexity of Southeast Asia supplies people in the region with an array of strategies for responding to rapid and pervasive change. But culturally complex societies can often be in precarious balance if national mores fail to help citizens relate well across ethnic and religious lines. We will also see that attitudes about family size and gender roles are changing in response to new economic circumstances.

Cultural Pluralism

Anthropologists like to say that Southeast Asia is a place of **cultural pluralism** (or **cultural complexity**). They mean that it is inhabited by groups of people from many different backgrounds who have lived together for a long time yet have remained distinct, partly because they lived in isolated pockets separated by rugged topography and spans of ocean. Small groups of descendants of the Australo-Melanesians who came into the region some 40,000 to 60,000 years ago still live primarily by hunting, gathering, and swidden agriculture in the interior forests of several larger islands, primarily Sumatra, the Malay Peninsula, Borneo, Sulawesi, New Guinea, and the Philippines. Far more numerous today are descendants of the Austronesians who migrated from the South China mainland into the islands about 5000 years ago. Past isolation has contributed to distinctive cultural differences among groups of these people as well. They tend to live in coastal zones, although some, such as the Igorots of Luzon in the Philippines, live in the mountainous interiors. Language is one of the clearest markers of ethnicity among groups of both Australo-Melanesian and Austronesian descent; each group speaks its own distinctive language. The descendants of these two major divisions of the earliest inhabitants of the region are usually referred to together as *indigenous peoples*. They are distributed in small patches from northern Burma and Thailand to southern Indonesia. Over the last 2000 years, newer immigrants have arrived, first from Arabia, Central Asia, India, China, Japan, and Korea, and then (after 1500) from Europe. Today, Southeast Asia is one of the most ethnically diverse regions in the world.

Look very closely at the bag the young woman is carrying and the drink the young man is holding as they pause amid manicured rice fields in Indonesia. Even in remote areas it is difficult to escape the symbols of globalization in Southeast Asia. [Ian Lloyd/Black Star.]

Southeast Asia entered an era of increasing interaction with the outside world when Europeans began colonizing the region after 1500. Most recently, Southeast Asia's cultural mix has been broadened by ideas from around the globe: free-market capitalism, communism, nationalism, consumerism, and environmentalism. These ideas and the material culture associated with them—from advertising to motorbikes to spandex to green politics—have also modified the life and landscapes of Southeast Asia.

Given the introduction of such new ideas, one might expect that by now Southeast Asia would be a melting pot. To some extent, notably in the cities, it is showing some signs of moving in that direction. In some of the region's countries, however, dozens of different languages are still spoken. (Indeed, it is estimated that 1000 of the world's 6000 or so languages are spoken here.) People practice many different religions, and neighboring families may trace their roots to remarkably different racial and ethnic origins. Today, the barriers separating groups are falling as modernization and industrialization draw large numbers of people into the cities and resettlement projects move several ethnic groups together onto newly opened lands. Migration and cross-cultural marriage are creating ethnically mixed groups who are unsure of their heritage and may never have seen their ancestral landscapes.

The Overseas Chinese. One group, prominent beyond its numbers, is the so-called Overseas (or Ethnic) Chinese. Small groups of traders from southern and coastal China have been active in Southeast Asia for thousands of years; some got their start as fishermen and women who met their customers on the high seas. Over the centuries, there has been a constant trickle of these emigrants out of China. The ancestors of most of today's Overseas Chinese, however, began to arrive in large numbers during the nineteenth century, when the European colonizers needed labor for their plantations and mines. Many of those who fled China's communist revolution (1949) sought permanent homes in Southeast Asia's trading centers. Today, over 26 million Ethnic Chinese live in Southeast Asia and work as artisans, merchants, middlemen, bankers, technology engineers, transport owners, developers, and laborers.

Despite the fact that most have modest incomes, the Chinese in Southeast Asia have the reputation of being rich and influential in government and commerce, yet strongly loyal to their own group, often importing kin as employees rather than hiring local people. Some in the region think the Overseas Chinese should be forced to change their habits, assimilate, and—especially—hire non-Chinese workers. Occasionally, the proponents of these ideas have turned to violence.

Many low- and middle-income workers who have been hurt by the recent financial crisis (see the discussion on pages 520–521) blame their sorrows on the Overseas Chinese. The Chinese are industrious and more prosperous than the average; with their region-wide connections and access to start-up money, they are able to take advantage of the new growth sectors of modernizing economies. Sometimes new Chinese-owned enterprises have put out of business older, more traditional establishments that many local people depend on. For example, in the town of Klaten in central Java, an open-air bazaar where local Indonesians once sold an array of goods has been shut down to make way for a government-sponsored, air-conditioned shopping mall. Only the Chinese can afford the expensive store rents. Since the economic crisis, resentment against the Chinese has brought a wave of violence. Chinese people have been assaulted, their temples desecrated, their homes and businesses destroyed. Indeed, since the 1970s, conflicts involving the Chinese have occurred in Vietnam, Malaysia, and many places in Indonesia (Sumatra, Java, Kalimantan, and Sulawesi).

Despite recent hostilities, Chinese commercial activity throughout the region has reinforced the perception that the Chinese are diligent, clever, and civic-minded businesspeople, often working especially long hours. Some Overseas Chinese, increasingly aware that the global economy presents dangers as well as opportunities to their group, are attempting to alleviate their situation through public education. There are Overseas Chinese study centers in Thailand, Vietnam, Malaysia, Singapore, Indonesia, and the Philippines. They maintain a higher profile as civic participants within their local communities or establish more formal international connections and services, such as investment banking companies, that will give them financial flexibility and spatial mobility should life in one particular place become too inhospitable.

Religious Pluralism

The major religious traditions of Southeast Asia include Hinduism, Buddhism, Confucianism and Taoism, Islam, Christianity, and animism (Figure 10.14; for further discussion of these religions see Chapter 6, pages 298–300, Chapter 7, pages 365–367, Chapter 8, pages 406–408, and Chapter 9, pages 448–449). Of these, only animism is native to the region. Animism is the traditional belief system of the indigenous peoples. For animists, such natural features as trees, rivers, crop plants, and the rains all carry spiritual meaning and are the focus of festivals and rituals to give thanks for bounty and to mark the passing of the seasons.

The other religions were brought by immigrants, traders, colonists, and missionaries, and their patterns of distribution reveal an island-mainland division (see Figure 10.14). Buddhism is dominant on the mainland, especially in Burma, Thailand, and Cambodia. In Vietnam, people practice a mix of Buddhist, Confucian, and Taoist beliefs that originated in China. Islam is dominant on the southern Malay Peninsula and in the islands of Malaysia and Indonesia, and it is growing in the southern Philippines. Christianity is the predominant religion only in Timor-Leste and in the Philippines. Catholicism was imported to the Philippines by the Spanish colonists who arrived there to open transpacific trade with Mexico in the 1540s. Hinduism first arrived with Indian traders thousands of years ago and was once much more widespread; now it is found only in small patches, chiefly on the islands of Bali and Lombok, east of Java. Recent Indian immigrants who came as laborers in the twentieth century during the later part of the European colonial period have reintroduced Hinduism to Burma, Malaysia, and Singapore.

All of Southeast Asia's religions have undergone change as a result of exposure to one another and to the traditional animist beliefs. Filipino Catholics and other Christians in Burma and Thailand (the Karen and Wa hill people, for instance) believe in spirits, and they practice rituals that have their roots in animism. Often, missionaries for Christianity or Islam deliberately pointed

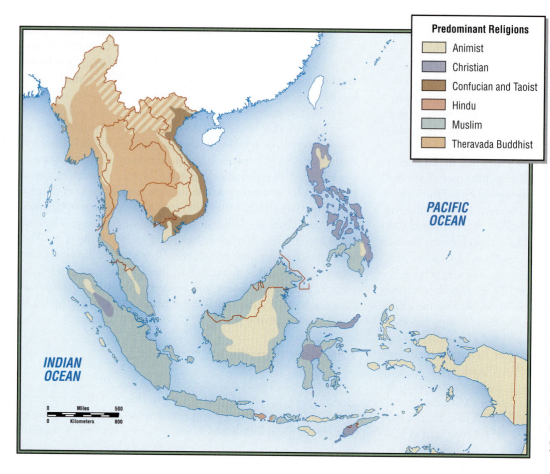

PACIFIC OCEAN

INDIAN OCEAN

Predominant Religions
- Animist
- Christian
- Confucian and Taoist
- Hindu
- Muslim
- Theravada Buddhist

0   Miles   500
0   Kilometers   800

FIGURE 10.14 Religions of Southeast Asia. Southeast Asia is religiously very diverse—five of the world's six major religions are practiced in Southeast Asia. Buddhists are the largest group on the mainland and Indonesia is the largest Muslim country in the world. Malaysia is also Muslim. Christians are a minority across the region. Animism is the oldest belief system and is found in both island and mainland places. [Adapted from Alisdair Rogers, ed., *Peoples and Cultures* (New York: Oxford University Press, 1992), p. 222.]

out ways in which the new faith was similar to indigenous beliefs. Hindus and Christians in Indonesia, surrounded as they are by Muslims, have absorbed ideas from Islam, such as the seclusion of women. On the other hand, Muslims have absorbed ideas and customs from indigenous belief systems, especially ideas about kinship and marriage, as illustrated in the following vignette.

PERSONAL VIGNETTE Marta is a *dukun*, a woman who prepares a bride for a traditional Javanese wedding. Nearly all Javanese are Muslims, but Islam does not have elaborate marriage ceremonies, so colorful animist rituals have survived. Usually, the bride and groom have never met, their families having worked out the match. Hence, the first meeting at the wedding ceremony is surrounded by a great deal of mystery. Marta's job is to reinforce this mystery and to prepare the couple for a life together. She bathes and perfumes the bride, puts on her makeup, and dresses her, all the while making offerings to the spirits of the bride's ancestors and counseling her about how to behave as a wife and how to avoid being dominated by her husband.

The groom, meanwhile, is undergoing ceremonies that prepare him for marriage. Each will sign the marriage certificate before they meet. In the traditional Javanese view, it is best that a young couple not be in love—that way, they will not fall out of love. Rather, from the start they will hold each other at arm's length, not investing all their emotional capital in a relationship that is bound to change over the course of the decades as they mature and their family grows up. Marta's role is to create a magically reinforced

bond between these two strangers. After marriage, as before, a man's closest friends will be his male age-mates and kin, and a woman's will be her female friends and kin.

Divorce in Malaysia and Indonesia is fairly common among Muslims, who often go through one or two marriages early in life before they settle into a stable relationship. Although the prevalence of divorce is lamented by society, it is not considered outrageous. Apparently, ancient indigenous customs allowed for mating flexibility early in life, and this attitude is still tacitly accepted. [*Adapted from Walter Williams,* Javanese Lives *(Piscataway, N.J.: Rutgers University Press, 1991), pp. 128–134.*]

Family, Work, and Gender

Family organization in Southeast Asia is variable, as one might expect in a region of so many cultures. There are some patterns that are remarkably different from the patriarchal patterns found elsewhere around the world. Nonetheless, there is a patriarchal overlay that derives from many sources: Islam, Hinduism, and Christianity, to name a few. In this section, these differences in family life in Southeast Asia are described in greater detail.

Family Patterns. Throughout the region, it is common for a newly married couple to reside with the wife's parents, or at least to be most closely associated with them. This ancient family structure survives today even where Islam is the dominant religion. Along with this custom there is a range of behavioral rules that empower the woman in a marriage, despite some basic patriarchal attitudes. For example, a family is headed by the oldest surviving male, usually the wife's father. Only when he dies does he pass on his wealth and power to his son-in-law, the husband of his oldest daughter, not to his own son. (A son goes to live with his wife's parents and inherits from them.) Hence, a husband may live for many years as a subordinate in his father-in-law's home. Instead of the wife being the outsider under the hegemony of her mother-in-law, as in Southwest Asia, China, India, and elsewhere, it is the husband who must kowtow. There is an inevitable tension between the wife's father and the son-in-law, which is resolved by the custom that the two men must practice ritual avoidance. The wife manages communication between them by passing messages and even money back and forth. Consequently, she has access to a wealth of information crucial to the family and has the opportunity to influence each of the two men. This role gives her power that wives in Muslim families in India or Southwest Asia could not even imagine.

Urban families tend to be nuclear in the early years of a marriage. In urban families influenced by modernization, the young couple may choose to live apart from the extended family. This arrangement takes the pressure off the husband in daily life. Because this nuclear family unit is often entirely dependent on itself for support, wives usually work for wages outside the home. The drawback of this compact family structure, as many young families are now finding out in Europe and United States, is that there is no pool of extended kin (grandparents, siblings, aunts, and uncles) available to help working parents with child care. There are not even any kin or neighbors to be hired for child care and housework because most women work, and grandmothers often live back in the village. By the same token, no one is left to help elderly parents maintain the old rural family home. So, increasingly, tiny urban apartments must accommodate a young family and one or more aged grandparents in need of care. On the brighter side, working grandmothers in their vigorous middle years can often help their adult children financially.

"Women are good with money" is a common saying in Indonesia. The social researchers Hannah Papanek and Laurel Schwede have written about working women and money in Indonesia. Women in that country must take into account the high divorce rate: already, in the 1970s, 38 percent of urban women over the age of 35 and 43 percent of all rural women had been married more than once. Thus, Indonesian women often start planning at the beginning of their marriage for the eventuality of having to support themselves. They tend to keep their spending low. In 70 percent of the families studied by Papanek and Schwede, women were the family money managers. Women are customarily the managers of family budgets elsewhere around the region as well.

Work Patterns. As the economies of Southeast Asia are transformed, so are the work lives of its people. The changing pattern of work in Southeast Asia is more complex than a simple shift from agricultural work to factory work or from rural to urban occupations. Agriculture is changing from small-plot family farming to large-scale commercial operations. The larger farms need fewer, but more highly skilled, workers. Rural communities, which used to be inward-looking and self-sufficient, now produce crops primarily for export: rice, palm oil, or bananas, for example. People cycle in and out of rural areas, spending a few years in urban factory work, then returning for an interlude of village life. Statistics indicate that many people remain registered as rural residents even when they are actually living in the city. For example, in 1992, 72 percent of the people of Indonesia identified themselves as rural, yet only 56 percent of the labor force worked in agriculture in that year. Households counted as rural in the census may include several individuals who live and work in cities and return only on weekends, if then.

Men continue to work in agriculture, construction, and services as well as in manufacturing. They have a higher rate of employment than women throughout the region, but women are catching up in the number of paid jobs they hold (Table 10.2), and they have always played an important role in traditional agriculture. Women's wages, however, average only about half those of men (Table 10.2). In Brunei, Burma, Malaysia, and the Philippines, significantly more women than men are completing training beyond secondary school. Hence, if training qualifications were the sole consideration, women would appear to have an advantage for future employment in Southeast Asian industries, such as information technology, where education is an advantage. It will be interesting to see if women eventually do become the majority in this industry and if their wages are commensurate with those of men with similar qualifications.

Migration

Migration has long been a common feature of life in Southeast Asia. Well before the era of European colonization, outsiders arriving from Arabia, India, and China contributed major components to the region's cultural mix. Today, many people leave the region in search of work or move to new areas within the region as part of vast resettlement schemes. Recently, rural-to-urban migration within the region has been particularly significant, as is the case elsewhere in the world.

Rural-to-Urban Migration. Southeast Asia as a whole is just 37 percent urban, but the balance between rural and urban employment is changing quickly in response to global market forces. Malaysia is already 57 percent urban, the Philippines 47 percent, and Singapore 100 percent. Often, the focus of migration is the capital of a country, which may quickly become a primate city— one that is two or three times the size of the second largest city and overwhelmingly dominates the economic and political life of the country. Indeed, Bangkok is nearly 30 times larger than Thailand's next largest metropolitan area (Chang Mai), and Manila is 10 times larger than Cebu, the Philippines' second city.

Individuals and families leave rural areas to escape poverty in the countryside where multinational firms now cultivate oil palms, rubber trees, or fruit for urban and export markets. Commercial agriculture often displaces subsistence cultivators who must then obtain cash to buy what they previously provided for themselves. These are the **push factors** in migration. On the other hand, people are attracted to the city by the higher living standards they expect to find there, perhaps working for multinational producers of shirts or sneakers such as Gap, Inc., or Nike, or for a semiconductor manufacturer such as Applied Materials in Singapore. These are the **pull factors.** Migrants often leave behind young and aged relatives whom they must then support with earnings sent home (**remittances**). Rarely, however, can primate cities such as Jakarta, Kuala Lumpur, Bangkok, or Manila provide sufficient

TABLE 10.2 Gender comparisons for Southeast Asia: GDP, education level, and labor force participation

Country (1)	Estimated earned income (adjusted for PPP in U.S. $) Female	Male (2)	Female secondary school enrollment as percentage of male enrollment (3)	Female tertiary school enrollment as percentage of male enrollment (4)	Labor force participation (%), 2000 Female	Male (5)
Brunei	11,716	26,122	105[a]	196	52	81
Burma	1,011[a]	1,389	95	175	68	90
Cambodia	1,621	2,113	59	38	85	86
Indonesia	1,987	3,893	99	77	58	85
Laos	1,278	1,962	81	59	78	90
Malaysia	5,557	11,845	111	108	50	81
Philippines	2,838	4,829	118	110	51	83
Singapore	14,992	30,262	98[a]	81[a]	55	84
Thailand	4,875	7,975	97[a]	111[a]	78	90
Vietnam	1,696	2,447	97[a]	74	78	84
United States	26,389	42,540	102	132	70	82

[a] In 2000. Data not available in 2003.

Source: United Nations Human Development Report 2003, Tables 22 and 24; Human Development Report 2000, Tables 2 and 28 (New York: United Nations Development Programme); Population Reference Bureau, 2002, Women of Our World, pp. 19–20; http://www.prb.org/pdf/WomenOfWorld2002.pdf.

jobs, housing, and services for the new arrivals. Of all the primate cities in Southeast Asia, only Singapore provides well for nearly all of its citizens, but even there uncounted illegal immigrants live on the margins.

PERSONAL VIGNETTE Mak (age 33) and Lin (age 27), husband and wife, left their two sons in the care of her parents in a village north of Bangkok, Thailand. They could not afford the school fees for even one of their sons on their rural wages; yet they hoped to educate both boys, even though educating just the eldest is the custom. So Mak and Lin traveled by bus for 10 hours to reach Bangkok, where—after several anxious days—they both found grueling work unloading bags of flour from ships in the harbor.

Each day when they finished their work, they walked miles to their quarters in one part of a tiny houseboat anchored, with thousands of others, on the Chao Phraya River. That river is Bangkok's low-income housing site, its source of water, its primary transport artery, and its sewer. Mak and Lin had to step gingerly across dozens of boats to get to theirs, intruding repeatedly on the privacy of their fellow river dwellers. They bathed in the dangerously polluted river, washed their sweaty, flour-covered work clothes by hand, and cooked their dinner of rice over a Coleman stove. Their only entertainment was provided by the private lives of their too numerous and too near-at-hand neighbors, who were often drunk on cheap liquor. For two years, Mak and Lin sent remittances to their family and managed to save enough to buy a bicycle, to pay the school fees, and to tide them over for several months at home. Although they were glad to be out of Bangkok, both nevertheless agreed that they would eventually go back if they could not find work near their village.

Resettlement. To alleviate rural unemployment, to forestall migration to the crowded cities, and for many other reasons, governments in Southeast Asia have developed **resettlement schemes** to move large numbers of rural people from one part of a country to another. (**Transmigration** is the local term for resettlement in Indonesia.) Beginning in 1904, Dutch colonists in Indonesia began to move people from the islands of Java, Bali, and Madura to land only lightly occupied by indigenous people in the outer islands of Sumatra, Kalimantan in Borneo, Sulawesi, and West Papua in New Guinea. These immigrants supplied labor on plantations growing such export crops as rubber, coconut, and palm oil. Much larger resettlement efforts on these same islands began after 1950. More than 5 million people have been relocated, making the Indonesian effort one of the largest land resettlement schemes ever.

The reasons for resettlement in Indonesia have changed over time, from promoting food production, especially national self-sufficiency in rice, to more general goals of regional development, national integration, and population redistribution. Resettlement also dovetails with government policies to assimilate Indonesia's outlying indigenous culture groups into mainstream society. Such policies are carried out by relocating people, especially Javanese,

Designated SP6, this new settlement carved out of the West Papua rain forest is part of the government's resettlement program. People who elect to move here receive a one-way air ticket, a house, 5 acres (2 hectares) of land, and a year's supply of rice. Estimates are that 5 million to 6.5 million people have participated in transmigration in Indonesia since 1950. [George Steinmetz.]

two. Severe environmental damage, ethnic discord, and breaches of human rights have been common.

In 2000, the indigenous people began to fight back in the Molucca and Sulawesi islands and in Kalimantan. Violent clashes occurred on several occasions. For example, the Dayak people of Kalimantan, many of whom are now Christians but continue to live in traditional ways, launched an attack on settlers from Madura (near Java), who happened to be Muslim. The Dayaks complain that their lands were taken without their consent and for far too little payment. Dayaks have killed hundreds of Madurese settlers. In the southern Philippines, Muslim settlers have been overwhelmed by Christians resettled to Mindinao by the central government in a political move to control Muslim dissidents.

Migration out of the Region. **Extraregional migration,** or migration to countries outside the region, is also important in Southeast Asia because of its effects on families, income, and employment. Increasingly in Southeast Asia, remittances from such migration are a major contributor of **foreign exchange**—foreign currency such as U.S. dollars that countries need to purchase imports. Sometimes, as in the case of the Philippines, they are the number one source. According to the Migration Policy Institute, Filipinos working abroad sent home over U.S. $ 6 billion in 2002 while workers from Indonesia sent home U.S. $ 1 billion. In addition to contributing to foreign exchange earnings, these remittances provide an important lifeline to many families in this region.

Extraregional migration may be either short-term or permanent. Recently, women have constituted a major proportion of these migrants (60 to 80 percent). In the Philippines, women emigrants have outnumbered men since 1992, and many are skilled nurses and technicians who work in European and North American cities and in Southwest Asia. Southeast Asian Muslim women (chiefly from Indonesia) have become part of the "maid trade" of 1 million or more who work in the homes of wealthy citizens in the Arab states of Saudi Arabia, Kuwait, and United Arab Emirates (Figure 10.15). Whereas about 10,000 Indonesian women migrated to Saudi Arabia each year in the 1980s, by 1998 this number

from the densely occupied cultural heartland to areas occupied by indigenous people (see the photo above). It is also Indonesian policy to disperse people from the crowded core in western Java to avoid political unrest. Many thousands of people not part of the formal resettlement schemes have joined the stream of people to the outer islands. The results have been unfortunate. In some cases, thousands of informal migrants moving onto the lands of indigenous peoples have built shantytowns and destroyed native habitats for plants and animals. Many thousands of acres of forest have been cleared for agricultural resettlement in areas where the tropical soils are too fragile to sustain cultivation for more than a year or

Thinking Geographically: ON THE WEB

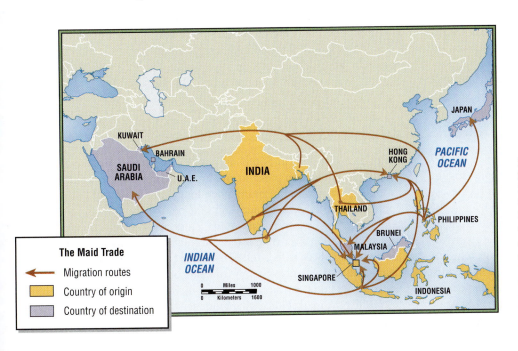

The Maid Trade

→ Migration routes

▢ Country of origin

▢ Country of destination

FIGURE 10.15 Interregional linkages: the "maid trade." In the 1990s, between 1 million and 1.5 million Southeast Asian women were working elsewhere in Asia (Japan, Malaysia, Brunei, Singapore, Kuwait, Bahrain, the United Arab Emirates, and Saudi Arabia) as foreign domestic workers. Educated women from Southeast Asia also migrate and send remittances home. Wealthy countries within Southeast Asia, such as Malaysia and Singapore, are, in turn, destinations for women from India, Sri Lanka, Indonesia, and the Philippines. Thailand both sends and receives workers. [Adapted from Joni Seager, *The Penguin Atlas of Women in the World*, 2003), p. 73.]

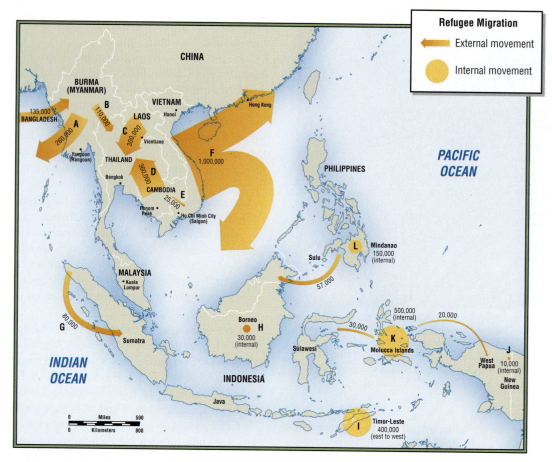

FIGURE 10.16 Flow of refugees in Southeast Asia. Refugees fled violence in Burma, Laos, Cambodia, and Vietnam in the decades following the end of the Vietnam War. (A) 1992–1995: Rohingya Muslims fled from Burma to Bangladesh, but about half have since returned. (B) 1995: Burman refugees in Thailand. (C) 1973–1975: Laotians fled communism. (D) 1979–1993: Cambodian refugees in Thailand. (E) 1995: Cambodian refugees in Vietnam. (F) 1975–1989: Boat people from Vietnam. (G) 2000–2001: Aceh refugees in Sumatra. (H) 2000: Resettled Madurans displaced by indigenous Dayak. (I) 1999–2000: Timor/Timor-Leste, postcolonial conflict. (J) 2000: West Papua. (K) 2000: Molucca. (L) 1978–2004: Philippines internal conflict resulted in some fleeing to Malaysia, others were displaced within the Philippines. [Adapted from Jonathan Rigg, *Southeast Asia* (London: Routledge, 1997), p. 130; and http://www.refugees.org/.]

rose to over 380,000 per year. This exponential increase has been fueled in part by the economic crisis (see pages 520–521) when many women lost their jobs and had to seek alternative employment opportunities overseas.

Skilled male workers from Southeast Asia are especially well known in the world merchant marine (see the vignette on page 30, Chapter 1). They typically serve in the lower echelons of shipboard occupations, working as seamen, cooks, or engine mechanics. In recent years, some have advanced to officer status.

Refugees from Conflict. In this region, as in others, **forced migration** still plays a role in the movement of people (Figure 10.16), and war is perhaps its most prevalent cause. During the last half of the twentieth century, many millions of mainland Southeast Asians fled into neighboring states to escape protracted conflict. Thailand, Laos, and Cambodia, in particular, received many refugees during and after the Vietnam War. More recently, refugees have fled discord on many of the islands, especially those of Indonesia. On the surface, it may appear that the discord is between ethnic or religious groups because the groups in conflict happen to be of different ethnicities or religions. In fact, disputes often arise over such issues as resettlement (Molucca, Sulawesi, and Kalimantan), political autonomy (Timor-Leste), and unsustainable extraction of resources from peripheral provinces by the central government (West Papua and Aceh, both provinces of Indonesia).

ENVIRONMENTAL ISSUES

Virtually everywhere in Southeast Asia, environments are deeply affected by population pressure and by such commercial activities as mining, logging, and export agriculture. Most Southeast Asian countries are rich in resources: oil and natural gas, timber and other forest products, minerals (gold, gemstones, tin, and copper), and fertile soils in river valleys and volcanic areas. These resources are being rapidly depleted—often extracted by foreign-owned companies and sold abroad, with only a fraction of the profits being reinvested in the host countries. In this region, as elsewhere, it is useful to consider the market forces that influence the rate of resource use and to ask whether or not the rate is sustainable and who benefits: Are the citizens of a region the ones benefiting from the development of their resources? Are those resources perhaps being sold off at unsustainably low prices?

People in Europe and North America may inadvertently be part of the problem. We, may, for example, have beautiful tropical hardwood paneling in our dens, purchased at an attractively low price at the local branch of a multinational home improvement warehouse, or we may use computer paper by the case. We are on the demand side of the Southeast Asian resource equation. Though we live far from Southeast Asia, we often support our lifestyles with its resources. In the United States, for example, we consume 690 pounds (313 kilograms) of paper per capita (the highest rate in the world). Japan's rate of paper use is 500 pounds (225 kilograms); and most of Europe consumes 220 to 440 pounds (100 to 200 kilograms) per person. Most Southeast Asian countries consume less than 67 pounds (30 kilograms) per capita. Singapore is an exception, consuming 450 pounds (218 kilograms) of paper per person. Much of the wood fiber in paper products comes from Brazil, the Philippines, Indonesia, or other tropical pulpwood producers. The map in Figure 10.17 shows the destinations for Indonesia's timber products. Although we have paid for our paneling and paper, we must ask if the price was fair, especially in light of the environmental damage that results from logging.

Deforestation

Environmentalists estimate that 13 to 19 square miles (34 to 50 square kilometers) of Southeast Asia's rain forests are destroyed each day, a rate 50 percent faster than in the Amazon. About 20 percent of the forests are destroyed by legal logging, road and pipeline construction, and utilities installation. Another 25 percent are destroyed for resettlement schemes. Fifty-five percent disappear to gain land for commercial agriculture and shifting cultivation. Companies that have legal rights to log the land over a period of 25 years sometimes choose to "cut and run" in just a few months and thus avoid fulfilling already woefully ineffective conservation agreements. Moreover, illegal logging often outpaces legal logging, and its impact on the environment is even worse because it is done secretly and in a matter of hours, using heavy equipment, and with no attention to preserving natural habitat.

Indirectly, much of the deforestation is linked to population growth and poverty. Traditional, nonintensive tropical swidden or shifting agriculture techniques on forested lands were especially effective and sustainable in Southeast Asia when the population was only about 10 million, as it probably was in about 1800. Until the early twentieth century, population density remained low enough to allow the use of small cultivation plots and long fallow periods, during which the forest could fill in the clearings. But recently, popula-

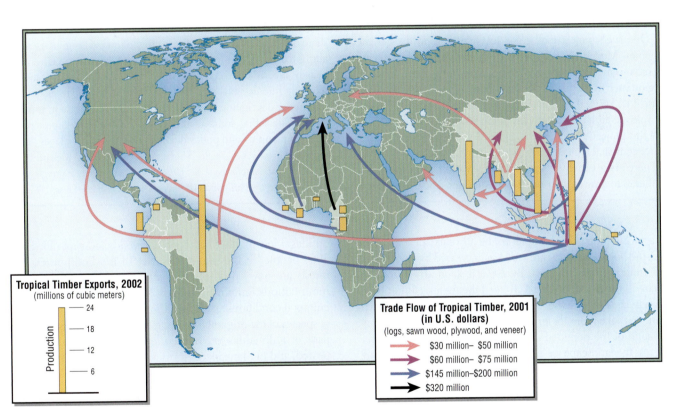

Tropical Timber Exports, 2002
(millions of cubic meters)

Production

— 24
— 18
— 12
— 6

Trade Flow of Tropical Timber, 2001
(in U.S. dollars)

(logs, sawn wood, plywood, and veneer)

→ $30 million– $50 million
→ $60 million– $75 million
→ $145 million–$200 million
→ $320 million

FIGURE 10.17 Interregional linkages: tropical timber production and exports, 2001–2002. East Asia and the European Union are the main importers of tropical timber products (logs, sawn wood, plywood, and veneer) while the main exports come from Southeast Asia. China and Japan lead the world in imports of these woods, and China's share is growing while Japan's is decreasing. On the map, the columns represent the millions of cubic meters of tropical timber produced in a given area. The arrows show the direction and value of the timber trade flows. [Adapted from *Commodity Atlas*, United Nations Conference on Trade and Development (New York: United Nations, 2004), pp. 44–45. http://www.unctad.org/Templates /webflyer.asp?docid=5221&intItemID=1397&lang=1.]

PRACTICING GEOGRAPHY

Can Medium-Sized Cities Play an Important Role in Developing Countries?

Richard Ulack
Department of Geography,
University of Kentucky

Ever since I was a boy, I have been interested in exotic, faraway places, especially in Asia. Reading constantly about exotic destinations, a fascination with maps, and collecting stamps were among the things that whetted my appetite for visiting and studying such places. As an undergraduate student I majored in geography (and history) and when I entered graduate school (at Penn State) I knew I had finally gotten to the point where I could pursue these earlier interests by conducting research on and (hopefully!) in "faraway" places. Several faculty members in my graduate program had had research experiences in Southeast Asia and, more specifically, in the Philippines. Thus, it was partly through their influence that Southeast Asia became the region in which I specialized.

In college I began to develop an interest in Third World development, and thus early in my career I specialized in the subfield of geography called "population geography." Within that broad area I was most interested in questions revolving around urbanization, which included of course internal migration. "Why do people move?", "Where do they move to?", and "What are the consequences of the moves to individual migrants, as well as to the places they came from and their destinations?" were but a few of the questions that I addressed in my research projects.

I spent nearly three years at various times in my career conducting research (and teaching) in the Philippines on such questions. One such project took place in Cebu City in the central Philippines. Most studies at the time which dealt with Southeast Asian urbanization focused on the very largest, or "primate," cities of the region such as Manila, Jakarta, and Bangkok. I chose Cebu City because it is classified as a "medium-sized" city and I was interested in the role medium-sized cities play in the development process. (As part of this project, two other colleagues were doing similar studies on other comparable-sized cities elsewhere in Southeast Asia.)

There are a variety of ways that information and data could be gathered to answer the research questions posed; among them are secondary analysis of existing studies, as well as utilizing available national and local data sources (e.g., censuses, surveys). The technique I used to gather the "data" involved the questionnaire method wherein over 1000 households were randomly selected in the Cebu metro area and several outlying small villages and rural areas in Cebu Province. The purpose of including these smaller areas was to help in assessing the migration, economic, and other ties that the city had with these rural areas. Detailed questions were asked in each household that included information about the composition of the household, employment and economic characteristics, place of birth of each household member, migration history, reasons for moving, and so forth. Upon returning from the field, the data were then subjected to various statistical analyses which were interpreted and reported on in a number of journal articles and a monograph.

Very rapid urbanization is of course a major characteristic (some would say "problem") in developing countries and a large share of the population growth in cities, often over one-half, can be attributed to rural-to-urban migration. And, again, the cities most heavily affected are usually the very largest. One of our purposes in selecting middle-sized cities for study was to be able to assess their role in the development process and to suggest that there is potential for alleviating some of the growth in the very largest places through planning and policy directed toward smaller places.

[*Source:* Michael A. Costello, Thomas R. Leinbach, and Richard Ulack, *Mobility and Employment in Urban Southeast Asia: Examples from Indonesia and the Philippines* (Boulder, Colo.: Westview Press, 1987).]

---

tion has grown rapidly. For example, in the 19 years between 1973 and 1992, the population of Southeast Asia increased by 48 percent, from 317 million to 469 million. By 2003, it stood at 544 million (more than 50 times that of 1800). To feed the increased numbers, cultivation is expanding to the steepest of slopes. Trees are cut down for sale and to make room for farmland. Erosion is drastically changing the shape of the denuded land. Without a forest cover, water moves across the landscape too rapidly to be absorbed, carrying away precious topsoil, which is very thin in tropical environments. Watercourses cut new gullies and ravines in the uplands and deposit tons of silt in the lowlands and surrounding oceans.

The Philippines serve as an example of widespread deforestation. Logging is the primary culprit, though clearing for commercial agriculture is next. In 1900, 75 percent of the Philippines was covered with **old-growth forests,** which are forests dominated by large trees in the mature stages of their life cycle. By 1990, only 2.3 percent of the country was so forested. In 1991, floods blamed on illegal logging that had been allowed by corrupt officials killed 3000 people on the island of Leyte. In the Philippines, original forests are now confined to small ribbonlike patches on higher mountainous terrain. **Secondary forest** will regrow after the first cutting if conditions are right, although it will not have the diversity of species of the original forest. On Sumatra, and in Sarawak and in parts of Kalimantan on Borneo, even secondary forest has disappeared, and in many places former forestlands are so degraded that they are useless for agriculture. The map in Figure 10.18 shows a range of environmental concerns in the region.

Mining

Mechanical strip-mining is probably the extractive activity that most disrupts the land. This technique is increasingly used in Southeast Asia to extract such minerals as copper, silver, and gold. The land is cleared of forest, and heavy equipment then peels away

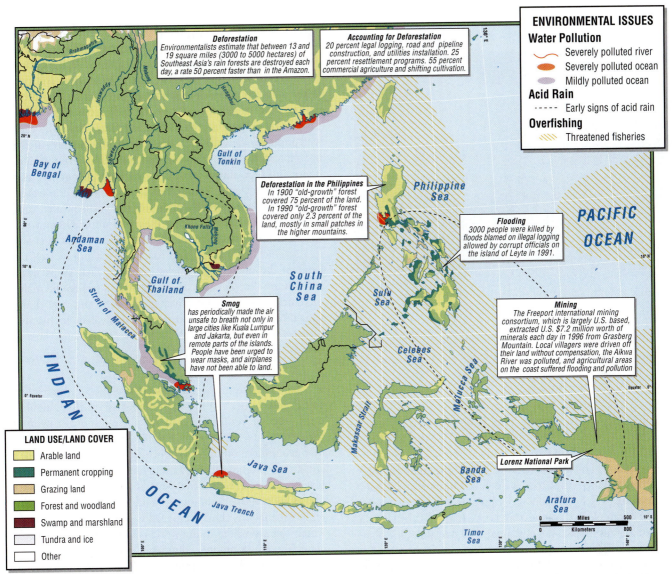

ENVIRONMENTAL ISSUES

Water Pollution
- Severely polluted river
- Severely polluted ocean
- Mildly polluted ocean

Acid Rain
- Early signs of acid rain

Overfishing
- Threatened fisheries

Deforestation
Environmentalists estimate that between 13 and 19 square miles (3000 to 5000 hectares) of Southeast Asia's rain forests are destroyed each day, a rate 50 percent faster than in the Amazon.

Accounting for Deforestation
20 percent legal logging, road and pipeline construction, and utilities installation. 25 percent resettlement programs. 55 percent commercial agriculture and shifting cultivation.

Deforestation in the Philippines
In 1900 "old-growth" forest covered 75 percent of the land. In 1990 "old-growth" forest covered only 2.3 percent of the land, mostly in small patches in the higher mountains.

Flooding
3000 people were killed by floods blamed on illegal logging allowed by corrupt officials on the island of Leyte in 1991.

Smog
has periodically made the air unsafe to breath not only in large cities like Kuala Lumpur and Jakarta, but even in remote parts of the islands. People have been urged to wear masks, and airplanes have not been able to land.

Mining
The Freeport international mining consortium, which is largely U.S. based, extracted U.S. $7.2 million worth of minerals each day in 1996 from Grasberg Mountain. Local villagers were driven off their land without compensation, the Aikwa River was polluted, and agricultural areas on the coast suffered flooding and pollution.

Lorenz National Park

LAND USE/LAND COVER
- Arable land
- Permanent cropping
- Grazing land
- Forest and woodland
- Swamp and marshland
- Tundra and ice
- Other

FIGURE 10.18 Environmental issues: Southeast Asia. Most Southeast Asians live frugally and use far fewer resources per capita than people in Europe or North America. Nonetheless, rapid industrialization and dependence on income from extractive activities such as export agriculture, forestry, and mining have led to environmental crises over the last two decades. The observant traveler in Southeast Asia, seeing evidence of environmental stress, may wonder to what extent consumers in the developed world are implicated.

the layers of soil and rocks until the desired mineral is exposed. Until 2004, companies in the United States were required to rehabilitate the landscapes they bulldozed and to dispose of mining and processing waste without polluting watercourses. Strip-mined land can never be truly restored, however, and despite regulations, many mining operations in the United States are still heavy polluters. One reason that international mining companies are interested in Southeast Asia is that environmental regulations do not exist or are not enforced.

The most publicized mine in the region is the 13,000-foot-high (4000-meter-high) "Mine in the Clouds," an open-pit multi-mineral mine on Grasberg Mountain in West Papua (see the photo on page 534). The mountain contains the world's largest known gold reserve. Every day in the late 1990s, the Freeport-McMoRan international consortium (a group with strong ties to firms in the United States) extracted U.S. $7.2 million worth of copper, gold, and silver from this mine. Freeport, Indonesia's largest foreign taxpayer, now has contracts to mine over 9 million acres (3.6 million hectares) in the mountain range. Mining will continue here for at least another century, according to Freeport president George Mealey. Local villagers, however, complain that they have been driven off their land without compensation. And the Komoro, who live on the coast, have demonstrated that tailings from the mines have polluted the Aikwa River, causing flooding and ruining their stands of sago palm. Mining by Freeport has also threatened the nearby Lorenz National Park, in direct violation of Indonesian law.

In Indonesia (Kalimantan and West Papua) and the Philippines (Mindanao), national armed forces have been called out to quell protests against loss of native habitats to mining operations. Such clashes reflect the conflict between the interests of local, often remote, politically weak forest dwellers and traditional

The Freeport-McMoRan Grasberg open-pit mine. [George Steinmetz.]

cultivators on the one hand and national governments focused on rapid economic development on the other.

Air Pollution

Usually, we think of places surrounded by the sea as pristine and their air as invigorating. But twice in the 1990s, the islands of Southeast Asia have been covered with a putrid and persistent cloud of pollution. The most recent instance occurred in the fall of 1997. Although large cities such as Kuala Lumpur and Jakarta experienced the worst pollution, air pollutants surged beyond safe levels even in remote parts of the islands (see Figure 10.18). People were urged to wear masks, and airplanes could not land. One plane crash is thought to have been caused by the smog.

The immediate cause of the poisonous smog was smoke from fires set on forestland that had been recently logged and was being prepared for the planting of palm oil trees and luceana, a fast-growing tree used for paper pulp production. Unfavorable winds, drought, and a delayed monsoon combined to spread the fires. A high-pressure air mass (possibly linked to El Niño) kept the smoky haze and the region's normal industrial air pollution from diffusing, as it usually does, into the upper atmosphere. The ASEAN environmental ministers meeting to deal with the crisis had to move their meeting to avoid the unhealthy air.

Proliferating motor vehicles and a lack of emissions control also make the air dangerous to breathe. Few cars, trucks, or buses have pollution-control devices, and the respiratory effects of breathing exhaust are felt by everyone who walks along streets or rides in vehicles, even in small towns and rural areas. A brown haze of pollution obscures the skylines of all cities most of the time and many pedestrians choose to wear masks. The following vignette demonstrates one way in which Jakarta's effort to control automobile emissions has intersected with the city's informal economy. Like Washington, D.C., New York, and some other U.S. cities, Jakarta tries to relieve pollution and traffic congestion by requiring car-pooling.

PERSONAL VIGNETTE Traffic regulations in Jakarta require that on some streets, all cars must carry at least three occupants. At the entrances to such streets wait the "jockeys," young boys who will provide a driver with the necessary number of passengers. The driver pays the boys 500 rupiah for the ride and then drops them on the other side of the restricted area, where they may make the return trip with another needy driver. From time to time the boys are rounded up, banned, fined, imprisoned, and sometimes even beaten by the police. But they always return. In fact, the traffic generates a great deal of employment. Young men and women at the traffic lights, wearing masks against the pollution, sell bottled water, candy, and cigarettes. Most are financing their education. Other young men earn a living by stopping the traffic on main roads to enable vehicles to enter from small side roads. By placing himself bodily in front of the oncoming vehicles, a young man stops traffic long enough to allow the cars from side streets to join the mainstream of vehicles. The grateful driver gives the "traffic cop" 100 rupiah, sometimes as much as 500 rupiah (equivalent to twenty cents). [*Adapted from Jeremy Seabrook,* In the Cities of the South—Scenes from a Developing World *(London: Verso, 1996), p. 296.*]

Some people suggest that the repeated occurrence of preventable disasters, such as the air pollution crisis of 1997, highlights the weaknesses of Southeast Asian social institutions. Environmental regulations are not strongly enforced, and those who pay bribes may avoid them entirely. Meanwhile, the political systems do not encourage public debate. Throughout the region, great value is placed on gentle, consensual persuasion as embodied in the principles of *Pancasila* in Indonesia (see the box on page 522) rather than on legal sanctions. This is especially true for environmental issues, which—as elsewhere in the world—consistently take a backseat to economic development. Ultimately, success in addressing Southeast Asia's environmental problems will depend both on consumers outside the region giving serious thought to the environmental consequences of their purchases and on the Southeast Asian general public becoming more aware of how they can change unsustainable development policies.

MEASURES OF HUMAN WELL-BEING

Human well-being is a complicated measure that is determined by sociocultural and political factors as much as by economic ones. As stated in earlier chapters, gross domestic product (GDP) per capita (see Table 10.3, column 2), the most common index used to compare countries, is not an accurate indicator of well-being, so additional measures are included here. Southeast Asia's countries, unlike those of sub-Saharan Africa and South Asia, have greatly varying GDP per capita figures. The GDP figures for some countries (Singapore and Brunei) are very high, for others (Malaysia, Thailand, and the Philippines) are only moderate, and for still others (Burma, Cambodia, Laos, and Vietnam) are very low. Singapore and Brunei enjoy widespread prosperity, but elsewhere in the

TABLE 10.3 Human well-being rankings of countries in Southeast Asia

Country (1)	GDP per capita, adjusted for PPP^a in 2001 $U.S. (2)	Human Development Index (HDI) global rankings, 2003^b (3)	Gender Empowerment Measure (GEM) and Gender Development Index (GDI) global rankings, 2003^c (4)	Female literacy (percentage), 2001 (5)	Male literacy (percentage), 2001^d (6)
For comparison					
Japan	25,130	9 (high)	44 (GDI = 13)	99	99
United States	34,230	7 (high)	10 (GDI = 5)	99	99
Mexico	8,430	55 (medium)	42 (GDI = 52)	89	93
Southeast Asia					
Brunei	18,600^e	31 (high)	N/A^f (GDI = 31)	88	95
Burma	1,199^g	131 (medum)	N/A (GDI= 102)^h	81	89
Cambodia	1,860	130 (medium)	64 (GDI = 105)	58	81
Indonesia	2,940	112 (medium)	N/A (GDI = 91)	83	91
Laos	1,620	135 (medium)	N/A (GDI = 109)	54	77
Malaysia	8,750	58 (medium)	45 (GDI = 53)	84	91
Philippines	3,840	85 (medium)	35 (GDI = 66)	95	95
Singapore	22,680	28 (high)	41 (GDI = 28)	89	96
Thailand	6,400	74 (medium)	55 (GDI = 61)	94	97
Vietnam	2,070	109 (medium)	N/A (GDI= 89)	91	95

^{a}PPP = purchasing power parity.

^{b}The high, medium, and low designations indicate where the country ranks among the 175 countries classified into three categories (high, medium, low) by the United Nations.

^{c}Total ranked in the world = 104 (no data for many).

^{d}Figures computed by authors based on female literacy as percent of male literacy.

^{e}2002 estimate, *CIA Factbook*, 2003.

^{f}N/A = data not available.

^{g}In 1998, data not available in 2001.

^{h}In 2001, data not available in 2003.

Sources: United Nations Human Development Report 2003, Tables 1, 22, 24 (New York: United Nations Human Development Programme); and *2003 World Population Data Sheet* (Washington, D.C.: Population Reference Bureau).

region GDP figures mask wide variations in well-being. To survive, those with the lowest incomes depend on subsistence cultivation, reciprocal exchange of labor and services, and the informal economy. But often these very poor people live in close proximity to a vastly better-off minority who live in fine houses, travel widely, and consume at high levels. Protracted war in Vietnam, Laos, and Cambodia has hindered development in those countries. Their economies, especially Vietnam's, are expected to improve as low wages attract foreign investment. In Burma, a military dictatorship and widespread corruption have been major factors in the continuing poverty of this country, which is among the richest in the region in natural resources.

As with GDP, the countries of Southeast Asia vary in their rankings on the United Nations Human Development Index (HDI) scale (see Table 10.3, column 3). Singapore's HDI rank is very high indeed (28), whereas Cambodia (130), Laos (135), and Burma (131) all rank in the bottom third, globally. In those countries with the lowest HDI rankings (Laos and Cambodia), literacy figures (columns 5 and 6) are exceptionally low, especially for women. Governments in these countries have not provided even the most basic education or health-care services. United Nations Gender Empowerment Measure (GEM) figures (see Table 10.3, column 4) are missing for all Southeast Asian countries except Malaysia (45), the Philippines (35), and Singapore (41). Because only 70 of the possible 175 countries are ranked on GEM, these scores place all three in the bottom half of the rankings. (But notice that even very rich countries can have relatively low GEM rankings—as does Japan, with a ranking of 44.) Women in Singapore have many opportunities for employment, though at lower wage scales than men (see Table 10.2, page 528). In Malaysia and the Philippines, a few well-connected women have access to some high-ranking jobs as administrators and as professional and technical workers. Many Filipino women obtained experience working as bureaucrats and in private companies that served the U.S. military until several large American military bases were closed. Now, many well-educated and experienced Filipino women, especially those in health-related fields, are migrating to work in the United States or other wealthy nations. Elsewhere in the region, most women lag well behind men in education level, wages, and access to jobs. Still, compared to South Asia and Africa, women are gaining some ground toward more equal treatment; Table 10.2 shows that young women are now staying in school, sometimes longer than young men.

Quick Review

1. Why is a growth triangle such as Singapore-Johor-Riau attractive to corporations?

2. Describe one way that opening national markets to the free market helped to bolster growing Southeast Asian economies. In what ways did this strategy backfire?

3. Why is Southeast Asia said to be a place of cultural and religious pluralism?

4. Name a few of the "push" and "pull" factors that encourage rural-to-urban migration.

5. Explain why gold mining in Southeast Asia is a threat to environments and people.

III SUBREGIONS OF SOUTHEAST ASIA

The subregions of Southeast Asia all have tropical environments that can support dense rain forests as well as rice paddy and other cultivation practices wherever the soil is rich enough and water is sufficient. All the subregions are ethnically diverse, retain vestiges of European colonialism, and have dominant religions—Buddhism on the mainland, Islam and Catholicism in the islands. In all subregions, especially in cities and towns, Overseas Chinese minorities are especially active in commerce. One obvious difference among the subregions is the pace of modernization. Another interesting difference is the response of various countries to the ethnic diversity within their borders. In most countries, cultural differences are tolerated; in the tiny Muslim state of Brunei, people are encouraged to conform to one standard. In one or two countries, the various ethnic groups are more or less equal in wealth and power; in most, one or two ethnic groups dominate. Countries also differ in their approaches to achieving national unity, a major concern throughout Southeast Asia.

MAINLAND SOUTHEAST ASIA: Thailand and Burma

Burma and Thailand occupy the major portion of the Southeast Asian mainland and share the long, slender peninsula that reaches south to Malaysia and Singapore (Figure 10.19). Although the two countries are adjacent and share similar physical environments, Burma is poor, depends on agriculture, and has a repressive military government, whereas Thailand is rapidly industrializing and has a more open, less repressive society. Both countries trade in the global economy, but in different ways: Burma supplies raw materials; Thailand provides low-wage labor and hosts multinational manufacturing firms, and a thriving tourism industry.

The landforms of Burma and northeastern Thailand consist of a series of ridges and gorges that bend out of the Tibetan Plateau and descend to the southeast, spreading out across the Indochina Peninsula. Rivers such as the Irrawaddy and the Salween, which originate far to the north on the Tibetan Plateau, flow south through narrow valleys to the huge delta at the southern tip of Burma or to the large plain in central Thailand. Most agriculture takes place in the Burmese interior lowlands around Mandalay and in Thailand's large central plain.

Ancient migrants from southern China, Tibet, and eastern India settled in the mountainous northern reaches of these two countries. The rugged topography has protected these indigenous peoples—the Shans, Karens, Mons, Chins, and Kachins are among the largest such groups—from outside influences. Hence, they follow traditional ways of life and practice animism. The valleys and lowlands to the south in Burma, and the large central plain in Thailand, are places of urban development, Buddhism, and modernization. Burma is named for the Burmans, who constitute about 70 percent of the population and live primarily in the lowlands. Thailand is named for the Thais, a diverse group of indigenous people who also originated in southern China.

From 70 to 80 percent of the world's remaining teakwood still grows in the interior uplands of Burma. A single tree can be worth U.S. $200,000. Other resources include oil, tin, antimony, zinc, copper, tungsten, limestone and marble, precious stones, and natural gas. Although Burma is rich in natural resources, in part because of corruption, it ranks as one of the region's poorest countries, with an estimated per capita annual income of U.S. $1700 and an HDI of 131. About 75 percent of the people still live in rural villages, and the economy is based primarily on rural activities: the cultivation of paddy rice, corn, oilseed, sugarcane, and legumes and the logging of teak and other tropical hardwoods. Burma's opium production (opium is the source of heroin) rises and falls from year to year, but Burma and Afghanistan vie for first- and second-largest producer. Burma is estimated to supply over 60 percent of the heroin market in the United States and it is also a major source of amphetamine-type stimulants. The U.S. General Accounting Office reported in the late 1990s that the global supply of opium was so high that the cost of heroin in New York City was lowered to less than one-quarter of the cost in the 1980s—a price drop that fueled increased usage. Opium poppy cultivation

FIGURE 10.19 Mainland Southeast Asia: Thailand and Burma subregion. The major portion of mainland Southeast Asia is occupied by Thailand and Burma.

is the major source of income for a number of indigenous ethnic groups, especially the Wa people. This group also manufactures synthetic drugs, especially methamphetamine. The Wa, who number about 700,000, are said to protect their territory in northern Burma with surface-to-air missiles. Those who are persuaded to stop growing poppies experience a 90 percent drop in income.

The military government of Burma has manipulated the drug traffic and has even harvested poppies, especially in the lightly settled uplands and mountains near the Thai border. When indigenous inhabitants have protested, they have been silenced by assassination, plundering, resettlement, and the abduction of their young women into the sex industry (see the discussion on pages 518–520). There are more than 100,000 refugees from Burmese military violence in camps in Thailand and Bangladesh. Since 1997, the international community has applied economic sanctions against the military government, including a U.S. embargo on investment in Burma.

Is Thailand an Economic Tiger?

During the boom years of the mid-1990s, Thailand was known as one of the Asian "tiger" economies, meaning that it was rapidly approaching widespread modernization and prosperity. Thailand takes pride in having avoided colonization by European nations and in having transformed itself from a traditional agricultural society into a modern nation. But this self-image of Thailand is not entirely accurate. Although Thailand's rapid industrialization and phenomenal urban growth are impressive, more than 50 percent of working people are still farmers who, according to official statistics, produce less than 10 percent of the nation's wealth. (Remember, though, that much of what farmers, especially women, produce is not counted in the statistics.) Just 31 percent of the population, or 20 million people, live in cities, and most of them endure crowded, polluted, and often impoverished conditions.

According to the United Nations, between 1975 and 1998, Thailand had the world's fastest growing economy, averaging an annual growth in GDP per capita of 4.9 percent (U.S. annual growth then averaged 1.9 percent). At the same time, the country kept unemployment relatively low for a developing country—at 6 percent—and inflation in check at 5 percent or lower. The growth rate climbed as a result of the rapid industrialization that took place in Thailand because the government provided attractive conditions for large multinational corporations from Europe, Japan, Taiwan, and the United States. These firms located in Thailand to take advantage of its literate, yet low-wage, workforce and its lenient environmental laws and other laws affecting manufacturing and trade.

The Bangkok metropolitan area, which lies at the head of the Gulf of Thailand, contributes 50 percent of the country's wealth and has 14 percent of its population. Still, cities elsewhere in Thailand are also growing and attracting investment: Chiang Mai in the north, Khon Kaen in the east, and Surat Thani on the southern peninsula. And despite the significant downturn in the economy and rise in inflation during the regional economic recession (1997–2002), the country still has a growing middle class with money to spend and invest.

The life circumstances of Buaphet Khuenkaew in the vignette following illustrate a standard of living common among Thai people who live on the urban fringe. Many of them retain some agricultural ways of life and at the same time take up opportunities for employment in the city. Also illustrated are several points made earlier about typical family organization, residence patterns, and the relative autonomy of women in Southeast Asia, as compared to other places in Asia.

PERSONAL VIGNETTE Buaphet Khuenkaew, 35, lives in Ban Muang Wa, a village near the northern city of Chiang Mai. She married at 18 and has two children: a son, 10, and a daughter, 17. A Buddhist with a sixth-

Buaphet shares the morning meal with her family before setting off to work. [Joanna Pinneo/Material World.]

grade education, she is both a homemaker and a seamstress. Six days a week she drives the family motor scooter 30 minutes to her job in Chiang Mai, where she sews buttonholes in men's shirts for 2800 baht (U.S. $118) a month. The children perform weekday household chores when they return from school.

Buaphet's husband, Boontham, is a farmer who is about five years older than she. The couple knew each other before they were married. He would visit her at her parents' home, and eventually they fell in love. One day he and his parents came to the house with a bride price of 10,000 baht (U.S. $420) in gold, and he asked her to marry him. She accepted. They used the gold to build their house on land owned by her mother, across the street from where she was born. They have electricity and a small television set. Drinking water comes from a well, is filtered through stones, and is stored in ceramic jars. Many household activities, including bathing, washing clothes, and washing dinner dishes, take place in small shelters outside, but meals are eaten inside. Although her husband feels that men are rightly regarded as superior in Thai society, Buaphet reports that she and her husband have an egalitarian marriage in which all decisions are made jointly. As is common for married couples in the region, they do not spend much of their leisure time together. She regularly spends time with her female friends and relatives, he with his male friends and family. Buaphet says she is happy with her life but also regularly complains about not having appliances and more up-to-date furnishings as her friends have. [Adapted from Faith D'Alusio and Peter Menzel, Women in the Material World (San Francisco: Sierra Club Books, 1996), pp. 228–239.]

Industrialization has had unanticipated detrimental side effects in Thailand. Thousands of rural people have been drawn to the cities seeking jobs, status, and an improved quality of life. Many of them, like the Bangkok migrants Mak and Lin described earlier (see page 528), arrive only to find a difficult existence and meager earnings. Catherine Shepherd and Julian Gearing, reporters for *Asiaweek*, a regional weekly newsmagazine, tell the story of Luan Srisongpong, a 33-year-old farmer from the southeast province of

AT THE LOCAL SCALE The Yadana Project: A Natural Gas Joint Venture

Bangkok, Thailand, needs power. Neighboring Burma has natural gas reserves. The Yadana natural gas field, located some 43 miles (70 kilometers) off Burma's coast in the Andaman Sea, contains reserves of some 5 trillion cubic feet (142 billion cubic meters). These reserves are being tapped in a joint venture led by TOTAL of France (holding a 31 percent interest), UNOCAL of California (28 percent), the Burmese government oil company (15 percent), and the Petroleum Authority of Thailand (26 percent). The liquefied petroleum gas (LPG) will be piped to the port at Daminseik, Burma; from there it will pass through 155 miles (250 kilometers) of buried pipeline to a power plant outside Bangkok. From 2000 to 2030, the gas from the Yadana field is expected to generate up to 2800 megawatts of electricity for Bangkok per day. (A megawatt is the amount of electricity required to light 10,000 100-watt lightbulbs.) By comparison, San Francisco, Berkeley, and Oakland, California, together require about 1300 per day megawatts for residential use.

Burma will gain a natural gas facility near Rangoon that will provide 125 million cubic feet (3.5 million cubic meters) a day for domestic consumption. It will also gain an electricity-generating plant that uses the gas and an accompanying fertilizer plant (fertilizer is a by-product of LPG production). Burma will also receive U.S. $200 million annually from sales of natural gas to Thailand. In addition, Thailand will pay another U.S. $200 million a year to TOTAL and UNOCAL. The project has additional benefits for Burma. It will create 2000 high-wage jobs; provide much-improved medical care and renovated hospitals; pay for new schools; rebuild roads; improve village water systems; provide electricity; and fund several village-scale development projects, primarily the breeding of pigs, chickens, shrimp, and cattle for profit. Burmese villagers displaced by the pipeline are to be compensated.

As good as all this sounds for Burma, a major problem remains. Burma is ruled by a military government that has committed numerous human rights violations. In acquiring land for the pipeline, the Burmese government is thought to have moved whole villages forcibly and then used forced labor to prepare the land for the pipeline. Meanwhile, the environment is being altered by clearing and bulldozing for the pipeline construction. At shareholder meetings, critics in the United States protested UNOCAL's position. They filed court cases suggesting that the company indirectly participated in human rights abuses that the Burmese government may have committed against its own citizens in connection with the Yadana pipeline project. UNOCAL's position is that it is simply an oil company and is not involved in, nor should it be held accountable for, political problems. The company argues that its expenditures to bring jobs and improvements in quality of life to local villagers should outweigh other, negative effects.

[*Source:* Adapted from "Country Report: Myanmar," *The Economist* (November 11, 1999): 16; http://www.earthrights.org/burma.shtml.]

Ubon Ratchathani. He, along with 20,000 others, is a member of the Assembly of the Poor, a protest group that has convened three times since 1995 on the streets of Bangkok to protest living and working conditions. Many are protesting large development programs, such as the Yadana pipeline (see the box "The Yadana Project" above), that have forced them off the land and into the city. Some protesters are farmers who suffer from food price controls meant to quiet hungry urban workers. Others are objecting to factory layoffs, which take place when industries produce more than they can sell or move to parts of Asia that have even cheaper labor. The Assembly of the Poor is effective: half the problems on its list have been resolved to its satisfaction. The fact that public protest is permitted in Thailand is noteworthy, because protest is forbidden in Burma and many other parts of the region.

The geographer Jim Glassman challenges the idea that Thailand is an Asian tiger. Its infrastructure is not nearly as developed as other middle-income countries. It has only one international airport, and there are only about ten telephone lines per 100 people. (But, as in rapidly modernizing parts of the old Soviet Union, the cellular phone industry is leaping into the breach; there are now more than 3 million cellular phones in Thailand.) Whereas Korea and Taiwan have devoted significant funds to alleviating inequalities in education, basic health care, and housing, Thailand has only recently begun to address some of these needs. Even though Thailand has literacy rates above 94 percent, true education levels are minimal. Only 33 percent of the children attend secondary school, and a tiny minority go on to higher education. As a result, Thailand's workers are less able to compete in the global wage market, and there is a deficit of highly trained scientists and engineers, compared to such countries as the Philippines and Malaysia. Glassman adds that Thailand has received much less foreign aid than, for example, Korea and Taiwan, which were given significant U.S. aid during the cold war era to fund education, infrastructure improvements, and some land reform.

MAINLAND SOUTHEAST ASIA: Vietnam, Laos, Cambodia

On the map in Figure 10.20, the countries of Vietnam, Laos, and Cambodia appear to be ideally suited for peaceful cooperation, sharing as they do the Mekong River and its delta on the southeastern peninsula of Eurasia. Nonetheless, the three have suffered a disruptive half-century of war that pitted them against one another. Until the end of World War II, all three were colonies, known collectively as French Indochina. All three have fought a long struggle to transform themselves from colonies controlled by the French to independent nations. In the early 2000s, they remain essentially communist states; yet free-market capitalism has been allowed in measured doses. Visitors and investors began arriving in the 1990s,

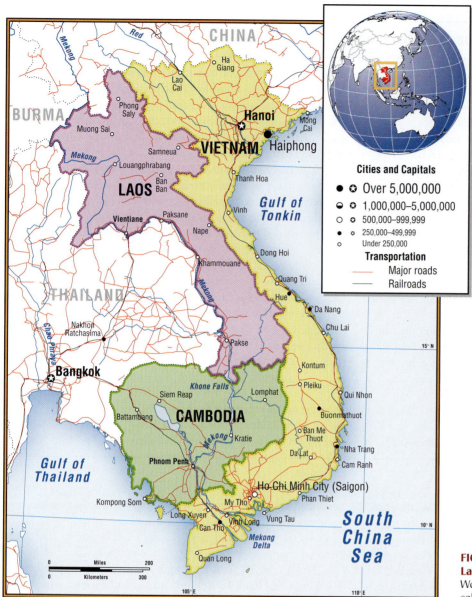

FIGURE 10.20 Mainland Southeast Asia: Vietnam, Laos, and Cambodia subregion. Until the end of World War II, Vietnam, Laos, and Cambodia were collectively known as French Indochina.

and by 2000 once-somber city streets were abuzz with people and enterprise.

Vietnam is a long, slender country with nearly 1000 miles of coastline on the South China Sea. Both the northern and southern extremities of the country are marked by deltas that are important rice-growing regions: the Red River delta in the north and the Mekong River delta in the south. A long, curved spine of mountains runs through Laos and into Vietnam (see Figure 10.2, page 502). In the south, these mountains are flanked on the west by the broad alluvial plain of the Mekong that is occupied by Cambodia and Thailand and on the east by the long coastline and fertile river deltas of Vietnam. The rugged mountainous territory in the north is the least densely occupied area in mainland Southeast Asia. Laos, almost entirely mountainous, has only 61 people per square mile (38 per square kilometer). Most of the subregion's population is in the Mekong River delta and coastal zones of Vietnam. There, people take advantage of the wet tropical climate and flat terrain

to cultivate rice. Throughout the subregion, at least 75 percent of the people support themselves as subsistence farmers; in Cambodia, only 16 percent of the people live in cities.

The Mekong River is the region's major transport artery. It originates in China, high on the Tibetan Plateau, and flows south along the border of Burma and Laos and then Thailand and Laos. It then crosses central Cambodia and far southern Vietnam. The southern part of the subregion is a large wetland formed by the Mekong Delta. Here, people live in houses on stilts to accommodate the seasonal floods, as in the Ganga Delta in India. Until recently, the Mekong River had but one bridge for automobile traffic and one dam, located in Yunnan, China, and completed in 1995. China now plans eight Mekong dams and several bridges on its side of the border.

The main trading partners of Laos and Cambodia are Thailand and Vietnam (sources of imports) and the United States. and Thailand (destinations for timber, vegetables, and inexpensive

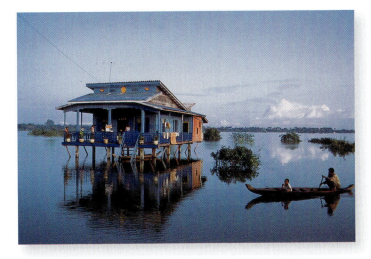

People who live along the Mekong River in Cambodia and southern Vietnam usually build their houses on stilts. During the rainy season, the river can rise 10 feet (3 meters) or more, broadening into a lazy flow 2 to 3 miles (3 to 5 kilometers) wide. [Michael Yamashita/Woodfin Camp & Associates.]

manufactured goods). Thai investors wield considerable power because they have broad investment interests in such resources as timber, gypsum, tin, gold, precious gems, and hydropower and in the production of rice, vegetables, coffee, sugarcane, and cotton for export. Some investors are hoping to promote the Mekong Basin as a secondary destination for European tourists who visit Thailand. (See the box "Tourist Development in the Greater Mekong Subregion" on pages 518–519.) Despite the regional economic downturn in the late 1990s, all three countries, and especially Vietnam, appear to be entering a period of rapid economic growth.

Vietnam

Vietnam is by far the most populous of the three eastern countries of the Southeast Asian mainland and the most developed. It has a population of 81 million people, 25 percent of whom live in cities. In contrast, Laos has 5.6 million people and Cambodia 12.6 million.

After the United States withdrew its military forces in 1973 (see page 509), the communist government, assisted by the Soviet Union, began to invest aggressively in health care, basic nutrition, and basic education. In the 1990s, only about 214 people out of every 100,000 attended college, but most completed elementary school and literacy was above 90 percent for both women and men. (Cambodia and Laos have much lower rates of well-being; see Table 10.3.) Because 63 percent of the Vietnamese people are cultivators living in rural areas, they have extensive knowledge of such practical matters as useful plants and animals, home building and maintenance, fishing, medicinal herbs, and related subjects. The Vietnamese have developed a national flair for excellent cuisine. When times are good, the dinner table is filled with artfully prepared fish, vegetables, herbs, rice, and fruits.

Vietnam has mineral resources (phosphates, coal, manganese, offshore oil) and at one time had a lush forest cover, much of which was destroyed by defoliants in the later years of the Vietnam War. Despite its resources, Vietnam's economy languished for nearly two

decades after the war, partly because the United States imposed an economic blockade and partly because of the inefficiencies of communism. In the mid-1980s, the Hanoi leadership began to introduce elements of a market economy in a program called **doi moi**, meaning economic and bureaucratic restructuring (very similar to perestroika in Russia or even the SAPs imposed on so many countries by international lending agencies in recent decades). With the lifting of the U.S. embargo in 1994, firms from all over the world began to open branches in Vietnam. By the late 1990s, about 90 percent of the country's industrial labor force worked in the

Vietnam's capital city of Hanoi is home to more than 3 million people. The city's tree-lined streets and architecture retain a French colonial flavor. The 1911 opera house, in a central square at the end of the boulevard, serves today as the municipal theater. [David Allen Harvey/National Geographic Image Collection.]

private sector producing such exportable products as textiles and clothing, cement, fertilizers, and processed food.

Wages remain extremely low, partly because of an oversupply of workers. Thirty percent of Vietnam's population is under the age of 15, which means that more than a million new workers enter the labor force every year. They are joined by several hundred thousand displaced agricultural workers. The Vietnamese government hopes that local entrepreneurs and multinational firms investing in Vietnam will provide jobs for the expanding workforce. Multinationals that are exploring new or expanded investment in Vietnam include Boeing, Citigroup, Coca-Cola, General Electric, General Motors, Cisco Systems, J. Walter Thompson, Nike, and Proctor & Gamble.

Although economic growth in Vietnam has ranged between 7 and 8 percent per year throughout the 1990s and early 2000s, the factors that brought growth caused considerable dislocation elsewhere in society. One focus of the government's market economy program has been the privatization of land and other publicly held assets. The farmers who own newly privatized land have a greater security of tenure and hence pay more attention to conservation and efficient production. But the process often leaves out the poorest people, who do not get land. In Vietnam, there was an attempt to accommodate the landless by generating employment in non-farm rural enterprises, but the downturn in Asian economies in the late 1990s put a halt to this effort. Those left without land or agricultural work ended up in the informal service economy, and their well-being has declined dramatically. Also, new enterprises that promise to bring in direct foreign investment have sometimes taken precedence over local interests. For example, in the late 1990s, small-scale Vietnamese farmers felt the squeeze when their taxes were raised and some of their lands were confiscated to be used in such foreign-funded enterprises as tourist hotels, golf courses, and oil refineries.

ISLAND AND PENINSULAR SOUTHEAST ASIA: Malaysia, Singapore, Brunei

Malaysia and neighboring Singapore and Brunei are the most economically successful countries in Southeast Asia. Malaysia was created in 1963 by combining the previously independent Federation of Malaya (the lower portion of the Malay Peninsula) with Singapore (at the peninsula's southernmost tip) and the territories of Sarawak and Sabah (on the northern coast of the large island of Borneo to the east) (Figure 10.21). All had been British colonies since the nineteenth century. The city-state of Singapore became independent from Malaysia in 1965. The tiny and wealthy sultanate of Brunei, also on the northern coast of the island of Borneo, refused to join Malaysia and remained a British colony until independence in 1984. Brunei does not release statistics, but the estimated per capita income is U.S. $18,600, and virtually all citizens have a high standard of living supported by wealth from oil and natural gas. A look at Table 10.3 (page 535), on human well-being, will confirm that Singapore is one of the richest countries on earth, with Brunei not far behind. And Malaysia, though much less wealthy than the other two, still ranks relatively high (at 58) in human well-being.

Malaysia

Malaysia is home to 25 million ethnically diverse people. Most live in West Malaysia, a section at the end of the Malay Peninsula that has 40 percent of the country's land area and 86 percent of its people. Nearly 60 percent of Malaysia's people are Malay, and nearly all Malays are Muslims. Ethnic Chinese make up 24 percent of the population, and they are mostly Buddhist. Eight percent are Tamil-

FIGURE 10.21 The Island and Peninsular Southeast Asia subregion. Malaysia occupies part of peninsular Southeast Asia as well as a portion of the island of Borneo. Singapore became independent from Malaysia in 1965.

Members of some of Malaysia's culture groups can be seen in the capital city, Kuala Lumpur, which has slightly more than 1 million inhabitants. [Stuart Franklin/Magnum Photos.]

and English-speaking Indian Hindus, and 2 percent are forest-dwelling indigenous peoples who live primarily on Borneo. Until the 1970s, conflicts among these groups divided the country socially and economically. Malaysians have worked hard to improve relationships among these groups, and they have had noteworthy success.

Throughout most of its history, West Malaysia was inhabited by Malays and a small percentage of Tamil-speaking Indians. The Indians were traders who plied the Strait of Malacca, the ancient route from India to the South China Sea. Even before the eastward spread of Islam in the thirteenth century, Arab traders of the ninth and tenth centuries began using this route, stopping at small fishing villages along the way to replenish their ships. By 1400, virtually all ethnic Malay inhabitants had converted to Islam.

During the colonial era, the British brought in Chinese Buddhists to work as laborers on West Malaysian plantations. The Chinese eventually became merchants and financiers; the Indians achieved success in the professions and in small businesses. The far more numerous Malays remained poor village farmers and plantation laborers. The British maintained control over their Southeast Asian colonies by promoting ethnic, religious, and economic rivalries. As economic disparities widened, antagonisms between the groups increased. After independence, animosities exploded with the onset of widespread rioting by the poor in 1969. Political rights were suspended, and it took two years for the situation to calm down. The violence so shocked and frightened Malaysians that they agreed to address some of the fundamental social and economic causes.

After a decade of discussions among all groups, Malaysia launched a long-term affirmative action program, called "Bumiputra," in the early 1980s. Its core was a new economic policy designed to help the Malays and indigenous peoples gain economic advancement. The policy required Chinese business owners to have Malay partners. It set quotas that increased Malay access to schools and universities and to public jobs. As the plan evolved in the 1980s, the goal became to bring Malaysia to the status of a fully developed nation by the year 2020, a plan known as Vision 2020. The program has succeeded in narrowing some social inequalities; and *Asiaweek*, a regional newsmagazine, reported that in late 1997, the real standard of living in Malaysia—adjusted for per capita purchasing power—was more than double what it had been at the beginning of the 1990s.

A sense of nationhood and national pride in diversity has developed, with many Malaysians now making a point of noting their cross-cultural origins. In 1999 at the end of the affirmative action reform period (1980s and 1990s), a tour guide in Malacca proudly itemized his personal heritage as Tamil, Portuguese, Dutch, British, Malay, Chinese, and Sarawakan.

Nonetheless, disparities between rich and poor remain wider in Malaysia than in any other Southeast Asian country. The capital, Kuala Lumpur, boasts beautifully designed skyscrapers like those in Singapore, including the two tallest buildings in the world in 2000—Petronas Towers I and II. Yet at least one-fifth of the city's total population, which amounts to several hundred thousand people, are squatters. In most of Malaysia, squatters reside on land, but in crowded urban areas that are close to water, many of them live in raft houses on rivers or bays. Rafts are an ingenious way of overcoming the acute housing shortage. In 1993, the geographer Asmah Ahmad studied urban raft dwellers in Malaysia and found that the typical domicile is a wood structure of 36 by 24 feet (11 by 7.3 meters), with a zinc roof. A rear platform provides space for domestic chores, and a wooden plank gives access to the riverbank. People often cultivate fast-growing food crops such as sweet potatoes on the riverbank, so that if they must move, not too much will be lost.

The economy of Malaysia has traditionally relied on export products such as palm oil, rubber, tin, and iron ore. In contrast, much of the growth of the 1990s has come from Japanese and U.S. investment in manufacturing, especially of electronics, and from offshore oil in the South China Sea. Timber exports are also important, and the cleared land is developed into additional palm oil and rubber tree plantations. Now Malaysia is hoping to benefit from a new, multibillion-dollar, high-tech manufacturing corridor 10 miles wide by 30 miles long (16 kilometers wide by 48 kilometers long) just south of Kuala Lumpur. Called the Multimedia Supercorridor (or MSC), it includes two new cities: the new national capital, called Putrajaya, and another city named Cyberjaya. The MSC is expected to help the nation spring fully prepared into a prosperous twenty-first century.

Like many Asian economic tigers, Malaysia is subject to volatile economic patterns. Currency values and stock market prices may fluctuate widely. Lavish building projects may be halted for years, as they were in Kuala Lumpur from 1998 through 2001. Overall, however, it does seem that Malaysia is likely to achieve increasing prosperity and continue to foster cultural diversity while discouraging extremist factions, religious or ethnic.

Singapore

Singapore occupies a hot, flat, humid island just off the southern tip of the Malay Peninsula. Its 4.2 million totally urban inhabitants live at a density of over 17,528 people per square mile (10,955) per square kilometer), yet it is one of the wealthiest countries on earth. Singapore's wealth is derived from the manufacture of highly technical pharmaceutical, biomedical, and electronic products; financial services; oil refining and petrochemical manufacturing; and oceanic transshipment services. Unemployment is low and poverty unusual, except among temporary and often illegal immigrants, primarily from Indonesia. Singaporeans seem to share the notion that financial prosperity is a worthy goal. The free-market economy reigns, but there is strict government control over virtually all aspects of society. Nonetheless, daily life for most people is secure, relaxed, and affluent.

Ethnically, Singapore is overwhelmingly Chinese (76.7 percent). Malays account for 14 percent of the population, Indians for 7.9 percent, and mixed or "other" for 1.4 percent. Singapore's people are religiously diverse: 42 percent are Buddhist or Taoist, 18 percent Christian, 16 percent Muslim, and 5 percent Hindu; the rest include Sikhs, Jews, and Zoroastrians. The government encourages unity among this diversity. Singapore officially subscribes to a national ethic not unlike *Pancasila* in Indonesia. The nation commands ultimate allegiance. Loyalty is to be given next to the community and then to the family, which is recognized as the basic unit of society. Individual rights are respected but not championed. The emphasis is on shared values, community consensus rather than conflict, and racial and religious harmony.

Singapore has a meticulously planned elegant cityscape with safe, clean streets and little congestion. Permits are required for virtually any activity: having a car radio, owning a copier, working as a journalist, having a satellite dish, being a sex worker, or performing as a street artisan or entertainer. Eighty percent of the people live in government-built housing estates, and workers are required to contribute up to 25 percent of their wages to a government-run pension fund. Home care of the elderly, however, is the responsibility of the family and is enforced by law. Each child, on average, receives ten years of education and can continue further if grades and exam scores are high enough. Literacy averages close to 95 percent. Law and order is strict. A few years ago, a U.S. teenager was sentenced to a caning for spray painting graffiti. Drug users are severely punished, and drug dealers are sentenced to life imprisonment or death. But virtually all citizens of Singapore seem to have accepted this control and strictness in return for a safe city and the next highest per capita income in Asia after Japan.

INDONESIA

Indonesia is a recent amalgamation of island groups, inhabited by people who before the colonial era never thought of themselves as a national unit (Figure 10.22). Rivalry between island peoples remains strong, and many resent the dominance of the Javanese in government and business. This resentment and other burgeoning issues of identity and allegiance are serious threats to continued unity (see the section "Is Indonesia Breaking Up?" on page 522).

The archipelago of Indonesia consists of the large islands of Sumatra and Java; the Lesser Sunda Islands east of Java; Kalimantan, which shares Borneo with Malaysia and Brunei; West Papua on the western half of New Guinea; Sulawesi; and the Moluccas.

FIGURE 10.22 The Indonesia subregion. Many of the Indonesian island names have changed in the past three decades. We are using the Indonesian names in this book, but here include in parentheses the name better known in the West—for example, Sulawesi (Celebes).

Two members of the Dani, an indigenous group in West Papua, try to adjust to urban life in the highland trading town of Wamena. The Dani have little in common with recent Muslim migrants from Java, who were promised acreage carved out of the Dani's forested lands. [George Steinmetz.]

In all, there are some 17,000 islands, but many of them are small uninhabited bits of coral reef. The term *Indonesia* was coined by James Logan, an Englishman residing in Singapore, in 1850. He combined two Greek words, *indos* (Indian) and *nesoi* (islands), to create the name. Indonesians themselves now refer to the archipelago as *tanah air kita*, meaning "Our Land and Water." This name conveys a sense of the archipelago environment but falsely indicates a universal feeling of togetherness and environmental concern.

Java and Sumatra are the two most economically productive islands and are home to 80 percent of Indonesia's population. Java alone has 60 percent of the nation's 221 million people, but only 7 percent of the country's available land. Over thousands of years, the ash from Java's 17 volcanoes has made the soil rich and productive, capable of supporting large numbers of people. Another 21 percent of the people live on the adjacent volcanic island of Sumatra. Recently, nearly 3 million people were resettled from Java to Sumatra, most under government sponsorship (see the discussion on pages 528–529). The resettlement of Javanese to other islands in Indonesia is an increasingly contentious issue because indigenous people resent being inundated with Javanese culture and concepts of economic development. They see the resettlement as a Javanese effort to gain access to and profit from indigenously held natural resources, such as oil, precious metals, and forestlands.

Timber harvesting in Kalimantan and West Papua is a representative example of timber operations throughout Southeast Asia and illustrates the manner in which development initiatives clash with the rights of indigenous peoples. Kalimantan, the southern part of the island of Borneo, has about 10.5 million people and about one-third of Indonesia's forests. In the mid-1980s, East Kalimantan province supplied 30 percent of the country's timber. West Papua, with a population of about 2 million, is the western section of the island of New Guinea. Some 85 percent of West Papua is covered with rain forests and thick mangrove swamps. These two places are wet, warm, and mountainous. Such tropical environments, like those in the Americas and Africa are at their most productive and sustainable when supporting their natural vegetative cover. They do not easily sustain large human populations, nor can they support export-oriented agriculture, forestry, or mining without being damaged.

Logging companies have obtained leases to huge tracts of rain forest for timber operations on Kalimantan and West Papua. These companies are usually owned by foreigners, sometimes in partnership with Javanese companies controlled by political leaders. In clear breech of official government policy, the favored method of operation is clear-cutting, which means that loggers cut every tree in an area instead of taking only the valuable species and preserving the forest in some form. Clear-cutting is fast and gives quick returns. It lets companies avoid the surveillance of human rights advocates and environmentalists that would come with legal forestry efforts. The indigenous peoples who live in the forest are quickly driven off the land and are often resettled near state-owned factories affiliated with the timber interests or on newly developed state rubber plantations. In either case, they become a source of cheap, compliant labor (see the photo of Dani people in a West Papua town above left). Sometimes settlers from overcrowded rural and urban areas are also brought in to subdue the indigenous peoples more quickly. These transmigrants are expected to cultivate the newly cleared land, work for which they may have no prior experience.

Shantytowns, like this one on the outskirts of Jayapura, the capital of West Papua, rise on the outskirts of most Indonesian cities as thousands of people flock to urban areas looking for jobs. Notice the mosque: native Papuans are rarely Muslim. Jayapura is a city of more than 310,000 people. On any given day, an interisland ferry may arrive with another 2000 people—peddlers, contract laborers, students, sex workers, and soldiers—looking for work. [George Steinmetz.]

In this way, the now deforested land becomes contested space between displaced indigenous peoples and incoming settlers (see the discussion of violence against transmigrants by Dayak forest people in Kalimantan, page 529). The government's argument that resettlement provides jobs is true, strictly speaking. But the habitat is degraded beyond repair, and people are left in impoverished circumstances, in tiny shacks on muddy tracts of land, with little sense of community. The central government invokes the principles of *Pancasila*, arguing that resettlement unifies the people as Indonesians by spreading modernization and eliminating ways of life that differ from the government's vision of the norm.

THE PHILIPPINES

The Philippines, lying at the northeastern reach of the Malay archipelago, encompasses more than 7000 islands spread over about 500,000 square miles (1.3 million square kilometers) of ocean (Figure 10.23). The two largest islands are Luzon in the north and Mindanao in the south. Together, they make up about two-thirds of the total land area, which is about the size of Arizona. The Philippine Islands, part of the Pacific Ring of Fire, are volcanic. The violent eruption of Mount Pinatubo in June 1991 devastated 154 square miles (400 square kilometers) and blanketed most of Southeast Asia with ash. Volcanologists had predicted the eruption and precautions were taken, so although more than a million people were threatened, only 250 died. Over time, volcanic eruptions have given the Philippines relatively fertile soil and rich mineral deposits (gold, copper, iron, chromate, and several other elements).

The Philippines had 7 million people when it became a U.S. protectorate in 1898. Just over 100 years later, it has almost 82 million and is home to one of the largest and most densely settled urban agglomerations on earth. Close to 50 percent of Filipinos live in cities, a high proportion compared to Southeast Asia as a whole. The metropolitan area of Manila, the country's capital, has a population of at least 11 million. Urban densities exceed

FIGURE 10.23 The Philippines subregion. The Philippines encompasses over 7000 islands spread over an area of 500,000 square miles.

50,000 people per square mile (31,250 per square kilometer). Many people in the cities are unemployed squatters living in shelters built out of scraps. Some of the newest urban dwellers are people displaced by the eruption of Mount Pinatubo in 1991. Others were displaced by rapid deforestation, dam projects, and the mechanization of commercial agriculture. The masses of urban poor in the Philippines (estimates of those classed as poor by local standards range between 28 and 40 percent of the population) represent a particular political and civil threat, because the people's faith in the government has not yet been restored after the excesses of the dictator Ferdinand Marcos, who maintained a flamboyant and brutal regime from 1965 until he was deposed in 1986.

The radical increase in population density over the last century might well have caused social problems by itself, but Philippine society is also complex culturally. Within the Malay majority—almost 96 percent of the population—there are at least 60 distinct ethnic groups. The Chinese make up about 1.5 percent of the population, and the remaining 3 percent include Europeans, Americans, other Asians, and non-Malay indigenous peoples. Most people in the Philippines are Roman Catholic (83 percent), a result of nearly 400 years as a Spanish colony (1520–1898). In contrast, the southern and central portions of the island of Mindanao and the Sulu Archipelago are predominantly Muslim. Over the last decades the central government has been resettling many thousands of Catholics in Mindanao, apparently to dilute the Muslim population. Possibly in retaliation, in the 1990s, Muslim extremists, perhaps linked to international terrorist movements, began operating in the southern Philippines and bombs killed a number of people. In 2002, the United States sent Special Forces soldiers to help train Philippine military personnel in combating the suspected terrorists.

In the Philippines, as elsewhere in Southeast Asia, wealth is apportioned by ethnicity and religion. The vast majority of the wealthy are descendants of Spanish and Spanish-Filipino plantation landowners or of Chinese business and financial people. It is estimated that 30 percent of the top 500 corporations in the islands are controlled today by ethnic Chinese-Filipino people (those of Chinese extraction but resident in the Philippines for generations). This group is less than 1 percent of the population. Meanwhile, the poor are overwhelmingly Malay people, and, in the southern islands, they are often Muslim as well.

By 1972, family conglomerates controlled 78 percent of all corporate wealth in the Philippines. Unlike Malaysia, the Philippine government under Marcos did not embark on a new economic policy to give the more disadvantaged ethnic groups access to jobs and education, nor were there consistent efforts to improve roads, ports, railways, and other infrastructure. Instead, after violent protests erupted in the early 1970s, President Marcos declared martial law in 1972 and managed to hang on to power for another 14 years until he was overthrown. During that time and since, there has been little progress in infrastructure development, job creation, or wealth redistribution. The Philippines still spends very little on education for its populace, just U.S. $25 per capita in the 1990s.

Violent social unrest has been very much a part of life in the Philippines. Filipinos are keenly aware that a better social and political system is possible. The long association of the Philippines with the United States (beginning in 1898 as a protectorate, and after World War II as the site of large U.S. military installations) gave the Philippine people experience with U.S. institutions and culture. But some of the worst aspects of U.S. culture were also imported (drug use, sex work, environmental abuse, wasteful consumerism), especially around the six U.S. military bases. Nonetheless, the Subic Bay Naval Base alone employed 32,000 local people and indirectly created 200,000 jobs. Many of the workers, especially women office workers, were exposed to information, education, and opportunities to travel and migrate that other Southeast Asians have not had. Furthermore, some of the urban protest strategies employed in the United States have been brought to the Philippines by liberation theology Catholics (see the discussion of liberation theology in Chapter 3, page 148) and by community developers.

Geographers are often interested in the strategies used to gain and maintain control of contested space. Denis Murphy, coordinator of Urban Poor Associates, is an American who helps urban squatters resist government eviction plans in Manila. He notes that the urban poor in cities such as Manila form ad hoc organizations to take care of one another and to challenge those who would displace them—be it government, developers, the police, or the military. In an area of Manila called Bulaklak, meaning "flower," squatters foiled urban developers by building their own houses and naming all the streets after flowers. Then they embarked on a self-financed project to develop a drainage system and to mark individual lots, even though ultimate landownership remains questionable. Each household contributes 5000 pesos (U.S. $200) for the project, apportioned over a 15-year period. As of 1996, construction was under way on the drainage system. Nonetheless, Bulaklak risks violent demolition, a common fate of squatter settlements in Manila and elsewhere. Murphy believes that although the squatters' movement is not motivated by socialist ideology, they represent a threat because they demonstrate their ability to construct autonomous, egalitarian alternatives to the status quo.

In their efforts to settle social unrest and attract investment, Marcos's successors have been hampered by the triple economic disasters of the Mount Pinatubo eruption, the closing of the American military bases, and a drastic cut in U.S. foreign aid in 1993. One of the few bright spots in the Philippine economy is the success of microcredit programs that encourage the founding of small businesses by the poor (see the discussion of the Grameen Bank in Chapter 8, pages 416–417). The successes of these small entrepreneurs illustrate the spirit that Denis Murphy talks about and that too often remains untapped in Philippine society.

PERSONAL VIGNETTE When Jesusa Ocampo made her first batch of *macapuno* candy for sale 14 years ago, she had no idea that the venture would one day grow into a full-fledged business. Neighbors snapped up all 20 packets of her coconut-based sweet. It wasn't long before Ocampo was making over 100 packets of candy daily. There was just one problem: she was too poor to pay for the expansion needed to meet rising demand. Then a friend told her about a microcredit program financed by the Philippine government. Beginning with a loan of just U.S. $145 obtained through a local cooperative, Ocampo gradually built up her tiny operation. She flourished, and so did her credit rating. In the spring of 1997, she obtained a loan of U.S. $3000, her nineteenth loan. She and her family are now building their own home on land they have purchased outside Manila.

The case of the Philippines under Marcos shows how poor leadership and corrupt government can hold a country back. Despite the rampant sale of its once-rich timber resources, the economy has been slow to grow. The agricultural sector—rice, coconuts, corn, sugarcane, bananas, pineapples, mangos, fish, and animals—has dropped in productivity over the past 20 years, sinking from 25 percent of the total economy in the 1980s to just 15 percent in 2003, yet agriculture employs at least 45 percent of the labor force. Meanwhile, in 2001, industry accounted for 31 percent of the economy but employed just 15 percent of the labor force. Perhaps the most meaningful figure, however, is that although officially only 10 percent of the population is unemployed or underemployed, real unemployment figures may be as high as 30 percent.

Paradoxically, because the Philippines remained underdeveloped longer than its neighbors, the country was less affected than other countries across the region by the financial crisis and turmoil of the late 1990s. Foreign investors continued to build facilities, construction in Manila grew, and an industrial park rose on the former U.S. naval base at Subic Bay. Although the country is less developed than most of its neighbors, between 2001 and 2003 the Philippines' economy grew a steady 4 percent, which was comparable to Malaysia (4.5 percent). Because of population growth, however, GDP per capita only increased 2 percent in the Philippines. Poverty decreased and inflation remained low, but the Filipino peso depreciated 4 percent, while other currencies in the region appreciated in value, and export growth remained slow. Importantly, it is remittances from workers abroad that helped to keep domestic consumption strong during this period.

Democratic processes are found nearly everywhere in the region, but democracy flourishes in some countries more than others.

Quick Review

1. Why did the U.S. place an embargo on investment in Burma?

2. Describe one way Malaysia is addressing the effects of economic disparity.

3. In what ways is Singapore's understanding of individual rights different from Western concepts of those rights?

4. Name some events that have hampered progress towards improved well-being in the Philippines.

REFLECTIONS ON SOUTHEAST ASIA

For years, Southeast Asia has been touted as a model of fast, efficient development, following the peerless example of Japan. Many countries in Southeast Asia—especially Malaysia, Thailand, Singapore, and Indonesia—have grown more quickly and consistently than countries in any other developing region. Table 10.4 shows that the distribution of wealth in Southeast Asia has been far more equitable than in Latin America.

Southeast Asia is often held up as a model to regions such as Latin America and sub-Saharan Africa, in terms of rate of growth, distribution of wealth, and rate of personal savings. It is important to note, though, that what is often called the miracle of Southeast Asian development is almost always calculated in economic terms. Those who do not think only in those terms (geographers, political scientists, anthropologists) have advocated a more thorough analysis of the "miracle." They would also look at how rapid economic progress affects the environment, cross-cultural relations, human rights and political participation, investment in social programs and education, families, and the status of women.

For a time, it appeared that human and environmental issues were not likely to receive much attention in Southeast Asia. Then, in the late 1990s, the situation began to change. A major economic downturn in 1997, in conjunction with an alarming environmental episode—a persistent poisonous haze throughout island Southeast Asia—turned the region's attention to factors that had escaped notice for some time. The threatening economic recession precipitated appalling revelations of bad loans based on widespread corruption in high places. Under scrutiny, the poisonous smog was shown to have resulted from rampant environmental exploitation, including the clear-cutting of timber and the subsequent burning of the resulting debris for plantation cultivation, as well as from industrial and vehicle emissions. Soon it seemed as if the region's competitive edge had been founded on wildly unsustainable prac-

TABLE 10.4 Income spread ratio of selected countries in Middle and South America and Southeast Asia

Country	Income spread ratio[a]
Brazil	29.7
Peru	11.7
Chile	19.3
Mexico	17.0
Costa Rica	11.5
Malaysia	12.4
Singapore	9.7
Thailand	8.3
Philippines	9.7
Vietnam	5.6
Indonesia	5.2

[a] The higher the number, the greater the spread between the income of the wealthiest 20 percent and the income of the poorest 20 percent.

Source: *United Nations Human Development Report 2003*, Table 13 (New York: United Nations Human Development Programme).

tices. None of the information was really new; it was just that the supposed economic prowess of the region had always overridden indications that all was not well.

Starting in the late 1990s and into the mid-2000s, ethnic and religious hostilities broke out in numerous places in Indonesia and also in the Philippines and northern Burma and in Thailand. Most often, these problems were linked to resource allocation, with indigenous people complaining either about the loss of their ancestral lands to new settlers moved in by the government or about the appropriation of their resources (especially minerals and forestlands) by multinational corporations with the collusion of government officials. But in at least some cases (Mindinao and Southern Thailand), Islamic extremists (Jemaah Islamiah) gained local support by taking up the causes of people who previously had not been particularly militant.

The question now is, will the economic, environmental and now extremist crises in Southeast Asia lead to a different definition of development, to a recognition that the means of development—democratic participation and widespread access to opportunity and information—are more important than such ends as quick profits and high mass consumption? Certainly the people of the region have amply demonstrated that they have the fortitude to work hard, but it seems possible that the national philosophies of conformity and harmony may thwart deep reforms and too easily gloss over real issues between ethnic and religious groups. Elections in the mid-2000s indicate that there is little pubic support for religious extremism and, if the social, economic, and environmental issues are addressed, the region may very well pull off another miracle. So far, no other region in the world has achieved development that is both socially and physically sustainable.

Chapter Key Terms

archipelago   504

ASEAN Free Trade Association
(AFTA)   521

Association of Southeast Asian Nations
(ASEAN)   521

Australo-Melanesians   506

Austronesians   507

biogeographical transition zone   505

buffer zone   508

crony capitalism   520

cultural pluralism   524

detritus   505

doi moi   541

export processing zones
(EPZs)   517

extraregional migration   529

feminization of labor   517

forced migration   530

foreign exchange   529

growth triangles   517

old-growth forests   532

Pancasila   522

pull factors   527

push factors   527

remittances   527

resettlement schemes   528

secondary forests   532

sex tourism   518

slash-and-burn (shifting or swidden)
cultivation   514

transmigration   528

Wallacea   505

wet rice (or paddy) cultivation   515

CHAPTER 11

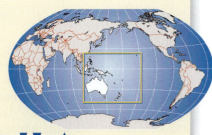

OCEANIA:
AUSTRALIA, NEW ZEALAND, AND THE PACIFIC

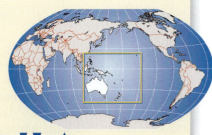

PERSONAL VIGNETTE

It is not the koalas or the Olympics which make Australia such a wonderful country. What makes Australia the best place in the world is the diversity of the people who live here. . . . There are indigenous Australians, Japanese migrants, transgender[ed people], middle aged hippies, and many others, who all belong to subcultures.

MAYU KANAMORI

Mayu Kanamori was born and raised in Tokyo and educated in English and Japanese. She moved to Australia in 1981 and has since become a much-needed cultural mediator between Japan and Australia. Today she lives in Sydney and works as a freelance photographer and writer for Australian and Japanese news-

The Buchannon family lives on Eveleigh Street, Redfern, commonly known as "the Block," in Sydney, Australia. In her exhibit in Tokyo, photographer Mayu Kanamori had this caption for her photo: "The Block is where many of the aboriginal people in inner Sydney live. It is considered a slum, and the media have called it a 'police no go' area. There are many problems of poverty, alcoholism, and drugs on the Block, and many white people will not dare to go near. Despite media reports, however, there are aboriginal families who live on the Block with hopes, dreams, and aspirations like the rest of Australia. This photo is when granddad said, 'Moggi, look at the camera.'" [Mayu Kanamori.]

papers and other publications. Much of her work focuses on helping Australians and Japanese better understand and appreciate each other.

The quote cited above and the photograph of the Buchannon family were presented at an exhibition of Kanamori's work in Tokyo titled "Australia Without Make-Up." The exhibition was directed, in part, at the many Japanese people who visit Australia but see little more than the friendly facade of blue skies and the Sydney Opera House. Kanamori similarly tries to reveal an unfamiliar side of Japan to Australians through exhibitions such as "Unseen Faces of Japan," which features images—poor Japanese laborers, Japanese Christians, and a gay pride parade in Tokyo— that break down Australian stereotypes about Japan. Another of Kanamori's exhibitions illustrates the environmental and cultural consequences of uranium mining for Australia's indigenous people. (Japan buys Australian uranium for nuclear-generated electricity.) [*Adapted from http://www.mayu.com.au/exhibitions/index.html.*]

MAKING GLOBAL CONNECTIONS

Cultural mediators like Mayu Kanamori are increasingly important in Oceania because dramatic shifts are reorienting the region as a whole toward Asia and away from old colonial powers and allies in Europe and North America. Oceania, which comprises Australia, New Zealand, Papua New Guinea, and the myriad Pacific islands, has been dominated politically and economically by people of mainly European descent for more than 400 years. Beginning in the early 1800s, thousands of Europeans made the long ocean voyage to Australia and New Zealand (far fewer to the Pacific islands)— most of them motivated by the simple desire to make a new life for themselves. Almost everywhere they went, they dominated, often violently, the indigenous people of the region, some of whom had been living there for perhaps as long as 60,000 years.

Indigenous people throughout Oceania, as well as minority immigrants such as the Japanese, are challenging European domination today by taking steps to revive traditional cultures and by agitating for greater recognition of their rights. The people of the Pacific have developed an attitude or point of view they call the Pacific Way. This concept has different meanings in different parts of the region, but it includes the sense that traditional methods of reaching consensus provide a way for Pacific people to work together to address the environmental, economic, and other issues of their society.

Asia poses a much greater challenge to European influence in the region than do the movements for indigenous or migrant rights. Though three-fourths of Oceania's people are of European descent, and migration rates show no signs of changing that fact in the near future, Asian influences are becoming more prominent primarily because the economies of the region are becoming more dependent on Asian markets and Asian investment. For example, trade with the European Union has shrunk markedly as a proportion of total trade. Australia now exports more to Asia than it does to Europe and the United States combined (see Figure 11.11 on page 573). Tourism is an important component of the economies of most countries in Oceania, and the Japanese are the largest group of tourists. Asia may never dominate Oceania politically (as

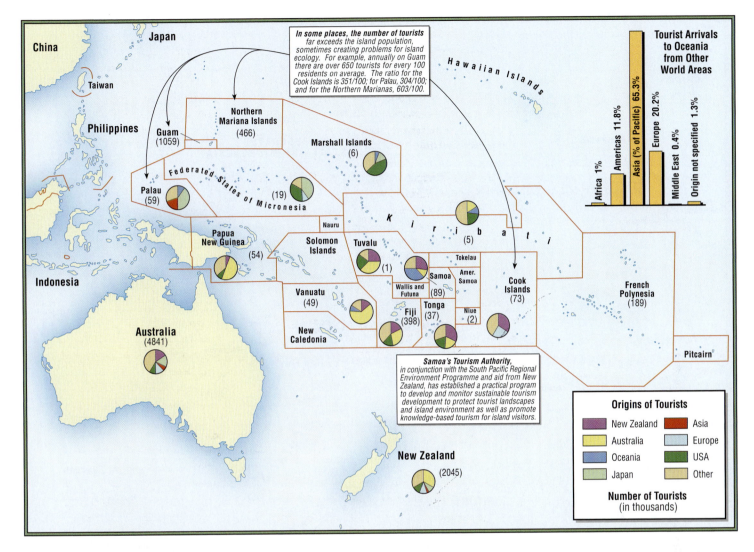

In some places, the number of tourists far exceeds the island population, sometimes creating problems for island ecology. For example, annually on Guam there are over 650 tourists for every 100 residents on average. The ratio for the Cook Islands is 351/100; for Palau, 304/100; and for the Northern Marianas, 603/100.

Samoa's Tourism Authority, in conjunction with the South Pacific Regional Environment Programme and aid from New Zealand, has established a practical program to develop and monitor sustainable tourism development to protect tourist landscapes and island environment as well as promote knowledge-based tourism for island visitors.

Tourist Arrivals to Oceania from Other World Areas

Africa 1% · Americas 11.8% · Asia (% of Pacific) 65.3% · Europe 20.2% · Middle East 0.4% · Origin not specified 1.3%

Origins of Tourists
- New Zealand
- Australia
- Oceania
- Japan
- Asia
- Europe
- USA
- Other

Number of Tourists (in thousands)

FIGURE 11.1 Global issues map. Tourism plays a major role in the economies of all countries in Oceania. The numbers of tourists (in thousands) are listed with place names, and in some cases, small-island inhabitants are outnumbered many times over by their visitors (see inset textbox). The origins of the tourists reflect changing trade patterns in the region. Most, by far, come from Asia (65.3 percent), and most of these visitors are from Japan (light green wedges). Can you explain why so many Japanese visit Oceania? Who do you think might be the tourists of the future? [Adapted from World Tourism Organization, at http://www.world-tourism.org/facts/tmt.html; *Financial Times World Desk Reference* (New York: Dorling Kindersley), 2004 (various pages); Cook Islands, at http://166.122.164.43/archive/2004/March/tcp-ck.htm; and Guam, at http://www.travelweeklyeast.com/articles/standard.asp?isArticle=1924&pCat=6&rmenu=articles.]

Japan attempted to do during World War II) or demographically (although Asians are now a majority on some Pacific islands, such as Hawaii). However, Oceania's 32 million people can no longer ignore the almost 4 billion people to the north in Asia, as they have tried to do in the past.

A look at the globe reveals that the Pacific Ocean covers more than a fourth of the earth's surface. Oceania is located in the central and southwestern Pacific. This part of the Pacific contains the continent of Australia and more than 20,000 islands, many of which are uninhabited atolls barely rising above the surface of the sea. Setting Australia aside for a moment, the single island of New Guinea and the two islands of New Zealand account for 90 percent of the land area in the Pacific. The other 10 percent is shared among thousands of islands lying mostly in the central and southern Pacific. Oceania's relatively small population (32 million) has prompted us to omit the usual treatment of its subregions—Australia, New Zealand, and the Pacific islands—at the end of the

chapter. Instead, we have written about these subregions in the section on current geographic issues.

THEMES TO EXPLORE IN OCEANIA

Several themes characterize Oceania:

1. The growth of Pacific regional consciousness. People of the Pacific are forging a regional identity by reviving diverse traditional cultures while encouraging cooperation that brings people of diverse backgrounds into close contact.

2. The increasing role of Australia as the regional "superpower." Reliance on outside (U.S. or European) military forces has diminished in Oceania and in Southeast Asia as Australia has taken on greater responsibility for regional peacekeeping operations—recently, for example, in the Solomon Islands and in Timor-Leste.

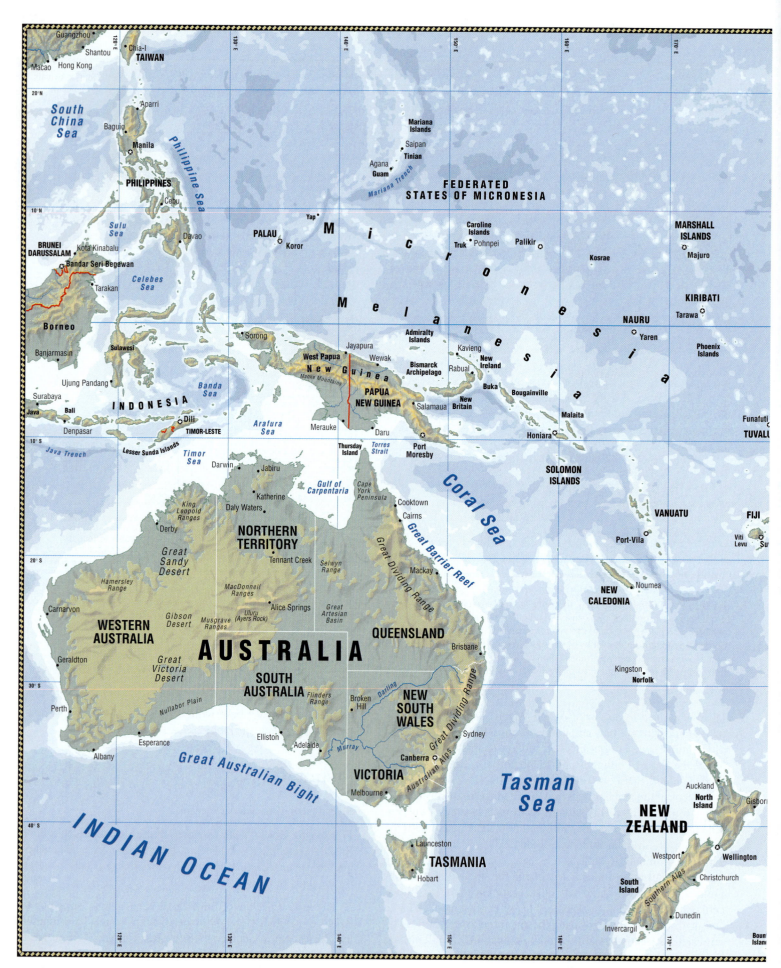

South
China
Sea

TAIWAN

Guangzhou
Shantou
Chia-I
Macao Hong Kong

20°N

Aparri

Baguio
Manila

PHILIPPINES

Cebu

BRUNEI
DARUSSALAM
Bandar Seri Begawan

Kota Kinabalu

Davao

Sulu
Sea

Tarakan

Borneo

Banjarmasin

Celebes
Sea

Sulawesi

Ujung Pandang

Surabaya

Bali

Java

Denpasar

INDONESIA

Banda
Sea

Sorong

Dili
TIMOR-LESTE

Lesser Sunda Islands

Java Trench

Timor
Sea

Arafura
Sea

Darwin

Jabiru

10°S

10°N

Yap

PALAU

Koror

M
i
c
r
o
n
e
s
i
a

Truk

Caroline
Islands

Pohnpei

Palikir

Kosrae

FEDERATED
STATES OF MICRONESIA

Mariana
Islands

Saipan
Tinian

Agana
Guam

Mariana Trench

MARSHALL
ISLANDS

Majuro

KIRIBATI

NAURU
Yaren

Tarawa

Phoenix
Islands

M
e
l
a
n
e
s
i
a

Admiralty
Islands

West Papua
New Guinea
Madke Mountains

Jayapura

Wewak

PAPUA
NEW GUINEA

Merauke

Daru

Thursday
Island

Torres
Strait

Port
Moresby

Kavieng

New
Ireland

Rabaul

Buka

Bismarck
Archipelago

Salamaua

New
Britain

Bougainville

Malaita

Honiara

SOLOMON
ISLANDS

Funafuti
TUVALU

VANUATU

FIJI

Viti
Levu

Su

Port-Vila

NEW
CALEDONIA

Noumea

Coral Sea

Gulf of
Carpentaria

Cape
York
Peninsula

Cooktown

Cairns

Great Barrier Reef

King
Leopold
Ranges

Derby

Katherine

Daly Waters

NORTHERN
TERRITORY

Tennant Creek

Selwyn
Range

Great
Sandy
Desert

Hamersley
Range

MacDonnell
Ranges

Alice Springs

Uluru
(Ayers Rock)

Great
Artesian
Basin

Great Dividing Range

Mackay

Carnarvon

WESTERN
AUSTRALIA

Gibson
Desert

Musgrave
Ranges

AUSTRALIA

QUEENSLAND

Brisbane

Geraldton

Great
Victoria
Desert

SOUTH
AUSTRALIA

Flinders
Range

Broken
Hill

Darling

NEW
SOUTH
WALES

Kingston
Norfolk

Perth

Nullarbor Plain

Elliston

Adelaide

Murray

Canberra

Sydney

Australian Alps

Great Dividing Range

Esperance

VICTORIA

Albany

Great Australian Bight

Melbourne

Tasman
Sea

Auckland
North
Island

Gisbon

NEW
ZEALAND

INDIAN OCEAN

Launceston

TASMANIA

Hobart

Westport

South
Island

Southern Alps

Wellington

Christchurch

Dunedin

Invercargil

Boun
Island

120°E
130°E
140°E
150°E
160°E
170°E

552

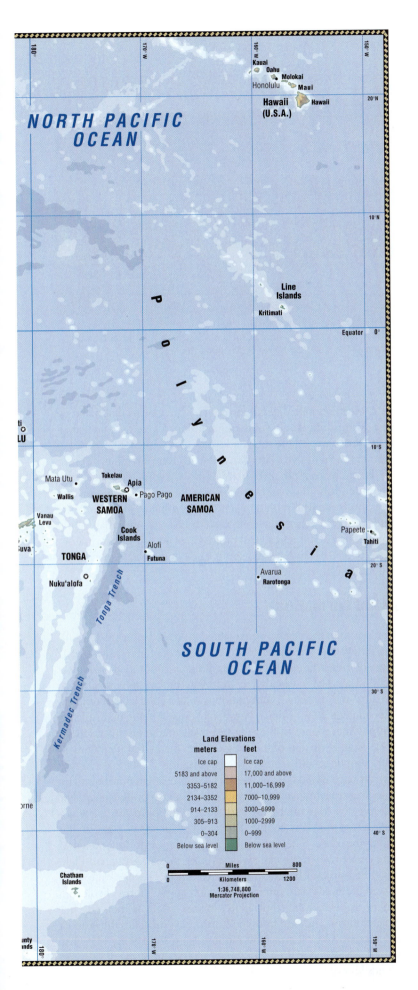

FIGURE 11.2 Regional map of Oceania.

3. **Reorientation toward Asia.** Trade and social interaction with Asia is increasing as the emphasis on European roots declines.

4. **Growing recognition of environmental issues.** Pacific Islanders are cooperating to address nuclear and other types of pollution. Nevertheless, some of the most dangerous environmental risks to the region—future rising sea levels and the depletion of the ozone layer—are the result of global trends beyond the control of the people of the Pacific.

5. **Increasing attention to the rights of indigenous peoples.** Australia and New Zealand are beginning to recognize indigenous claims to lands and resources as people of both indigenous and European descent learn to take pride in their countries' heritage.

6. **Familiarity with mobility.** Travel among the islands for education, sports, cultural events, and trade is facilitating unity and cooperation among residents of formerly isolated places.

I THE GEOGRAPHIC SETTING

Terms to Be Aware Of

Island place names in the Pacific often can be confusing. Groups of islands were given names by European seafarers, mapmakers, and colonizers. Individual islands within those groups had their own names. In the twentieth century, some islands became independent and formed countries. For example, the Caroline Islands, located north of New Guinea, still go by that name on maps and charts, but they have divided into two countries: Palau (a small group of islands and atolls at the western end of the Caroline Islands) and the Federated States of Micronesia, which extends over 2000 miles west–east from Yap to Kosrae. Other island groups (the Hawaiian Islands and French Polynesia, for example) became states of larger countries. In addition, there are the three larger island groupings—Micronesia, Melanesia, and Polynesia—that are not political units but are based on ethnic and cultural links. This is the reason that you may see different place names in similar locations on some maps in this chapter.

PHYSICAL PATTERNS

The huge expanse of the Pacific Ocean is of supreme importance in this region as both an avenue and a barrier. For living things, the Pacific serves as a link between widely separated lands. Plants and animals have found their way from island to island by floating on the water; and humans have long used the ocean as a conduit for communication: visiting, trading, and raiding. Sea life is a source of food, and today it is also a source of income, as catches are sold in

Uluru (Ayers Rock, shown here) and the nearby Olgas of central Australia are smooth remnants of ancient mountains that have resisted erosion. These sites are held sacred by central Australian Aborigines. They are also among Australia's most popular tourist destinations. [Susan Metros.]

the global market. The movement of the water and its varying temperatures influence climates in all the region's landmasses. But the wide expanses of water have also acted as a barrier, profoundly limiting the natural diffusion of plant and animal species and keeping Pacific Islanders relatively isolated from one another. The vast ocean has imposed solitude and has fostered self-sufficiency and subsistence economies well into the modern era.

Continent Formation

The largest landmasses of the Pacific are the frozen continent of Antarctica (not considered here) and the ancient continent of Australia at the southwestern perimeter of the region. The continent of Australia is partially composed of some of the oldest rock on earth and has been a more or less stable landmass for more than 200 million years, with very little volcanic activity and only an occasional mild earthquake. Australia was once a part of the great landmass, called **Gondwana,** that formed the southern part of the ancient supercontinent Pangaea (see Figure 1.10 on page 23). What became present-day Australia broke free from Gondwana and drifted until it eventually collided with the part of the Eurasian Plate on which Southeast Asia sits. The impact created the mountainous island of New Guinea to the north of Australia; downwarping created the lowland that is now filled with ocean between New Guinea and Australia.

Australia is shaped roughly like a dinner plate with a lumpy, irregular rim and two portions missing: one in the north (the Gulf of Carpentaria) and one in the south (the Great Australian Bight). The center of the plate is the great lowland Australian desert. The lumpy rim is composed of uplands: the Eastern Highlands (labeled "Great Dividing Range" in Figure 11.2) are the highest and most complex of these. Over millennia, the forces of erosion—both wind and water—have worn Australia's landforms into low, rounded formations, some quite spectacular.

Off the northeastern coast of the continent lies the **Great Barrier Reef,** the longest coral reef in the world. A *coral reef* is an intricate structure composed of the calcium-rich skeletons of tiny living creatures called coral polyps. The tiny polyps that make up the reef are abundant beyond enumeration. These extremely fragile creatures are unable to live in water that is too cold, too full of sediment, or polluted. The Great Barrier Reef stretches in an irregular arc for more than 1000 miles (1600 kilometers). The giant reef interrupts the westward-flowing ocean currents in the mid–South Pacific circulation pattern, shunting warm water to the south, where it warms the southeastern coast of Australia. In and around the reef, there is a great profusion of aquatic environments, some protected and warm, some cool and wildly active with pounding surf.

Island Formation

The islands of the Pacific were created (and are being created still) by a variety of processes related to the movement of tectonic plates. Some of these islands are remnants of the great southern landmass Gondwana. These islands are found in the western reaches of Oceania and are large, mountainous, and geologically complex. They include New Guinea, New Caledonia, and the main islands of Fiji. Other islands in the region are volcanic in origin and form. These islands are part of the Ring of Fire, the belt of frequent volcanic eruptions and earthquakes circling the Pacific (see Figure 1.11 on page 24.) Many of these islands are situated in boundary zones where tectonic plates are either colliding or pulling apart. For example, the Mariana Islands east of the Philippines are volcanoes that were formed when the Pacific Plate plunged beneath the Philippine Plate. The two much larger islands of New Zealand were created when the eastern edge of the Indian-Australian Plate was thrust upward by its convergence with the Pacific Plate. The Hawaiian Islands were produced through another form of volcanic activity associated with **hot spots**—places where particularly hot magma moving upward from the Earth's core breaches the crust in tall plumes. Over the past 80 million years, the Pacific Plate has moved across one of these hot spots, creating the string of volcanic islands known as Hawaii (Figure 11.3).

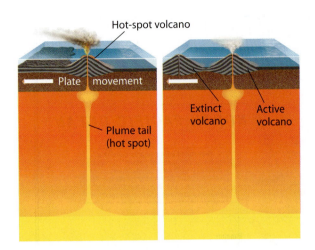

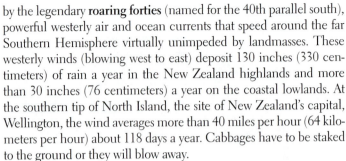

FIGURE 11.3 Formation of "hot-spot" islands. Some islands in the Pacific, such as the Hawaiian chain, are formed by magma bursts from a hot spot onto the surface of a tectonic plate. Wide, thin layers of lava spread from the point of eruption, slowly building undersea mountains that emerge as islands. As a tectonic plate moves over a hot spot, a chain of volcanic islands is formed. [Adapted from Frank Press, Raymond Siever, John Grotzinger, and Thomas H. Jordan, *Understanding Earth*, 4th ed. (New York: W. H. Freeman, 2004), p. 133.]

The islands that are volcanic in origin exist in three forms: volcanic high islands, low coral atolls, and raised or uplifted coral platforms known as *makatea*. **High islands** are usually volcanoes that rise above the sea into mountainous, rocky formations that contain a rich variety of environments. New Zealand, the Hawaiian Islands, Tahiti, the Samoa Islands, and Easter Island are among the many examples of high islands. An **atoll** is a low-lying island formed of coral reefs that have built up on the circular or oval rim of a submerged volcano. These reefs are arranged around a central lagoon that was once the volcano's crater. As a consequence of their low elevation, these islands tend to have only a small range of environments and very limited supplies of fresh water. Subsequent uplift of an atoll may result in the formation of a *makatea*. These raised coral platforms may have cliffs over 65 feet (20 meters) high.

Climate

Although the Pacific Ocean stretches nearly from pole to pole, Oceania is situated within a primarily tropical and subtropical region of that ocean. Oceania is widest through its tropical zones, and most of its islands lie within, or close to, the tropics. The tepid water temperatures of the central Pacific bring year-round mild climates to nearly all the inhabited parts of the region (Figure 11.4). The seasonal variability in temperature is greatest in the southernmost reaches of Australia and New Zealand.

Moisture and Rainfall. With the exception of the arid interior of Australia, much of Oceania is warm and humid nearly all the time. New Zealand and the high islands of the Pacific receive copious rainfall and once supported dense forest vegetation. Travelers approaching New Zealand, either by air or by sea, sometimes notice a distinctive long, white cloud that stretches above the two islands. A thousand years ago, Polynesian settlers called the Maori also noticed this phenomenon, and so they named the place *Aotearoa*, "land of the long white cloud." The distinctive mass of moisture is brought in

by the legendary **roaring forties** (named for the 40th parallel south), powerful westerly air and ocean currents that speed around the far Southern Hemisphere virtually unimpeded by landmasses. These westerly winds (blowing west to east) deposit 130 inches (330 centimeters) of rain a year in the New Zealand highlands and more than 30 inches (76 centimeters) a year on the coastal lowlands. At the southern tip of North Island, the site of New Zealand's capital, Wellington, the wind averages more than 40 miles per hour (64 kilometers per hour) about 118 days a year. Cabbages have to be staked to the ground or they will blow away.

Australia's relatively low landforms and dry climate are remarkably different from New Zealand's mountainous topography and moist temperate climate. Two-thirds of the continent of Australia is overwhelmingly dry. The Eastern Highlands block the movement of moist easterly winds (blowing east to west) and deflect them to the south, so rain does not reach the interior. As a result, the large, dry portion of Australia receives less than 20 inches (50 centimeters) of rain a year (see Figure 11.4), and humans have found rather limited uses for this territory. But the eastern windward slopes of the highlands receive abundant moisture. This moist eastern rim of Australia was favored as a habitat by the Aborigines as well as by the Europeans who displaced them after 1800. During the southern summer, the fringes of the monsoon that passes over Southeast Asia and Eurasia (see Chapter 10, page 506) bring moisture across the Australian north coast. There rainfall varies in a given year from 20 to 80 inches (50 to 200 centimeters).

Overall, Australia is so arid that there is only one major river system—in the temperate southeast, where most Australians live. There, the Darling and Murray rivers drain one-seventh of the continent and flow into the ocean at Adelaide. A measure of the overall dryness of Australia is that the entire average *annual* flow of the Murray-Darling river system is equal to just one day's average flow of the Amazon in Brazil.

An atoll in the Tuamotu Archipelago of French Polynesia. As is the case here, usually the land area of an atoll is not continuous, but instead forms a sort of necklace of flat islets, known as *motu*, around a central lagoon. Often the necklace surrounds one or more islands that are the remnants of the old volcanic core. Some of these lagoons are home to *Pinctada margarifera*, the black-lipped pearl oyster, which produces the famous and expensive black pearls. [David Doubilet.]

Thinking Geographically: ON THE WEB

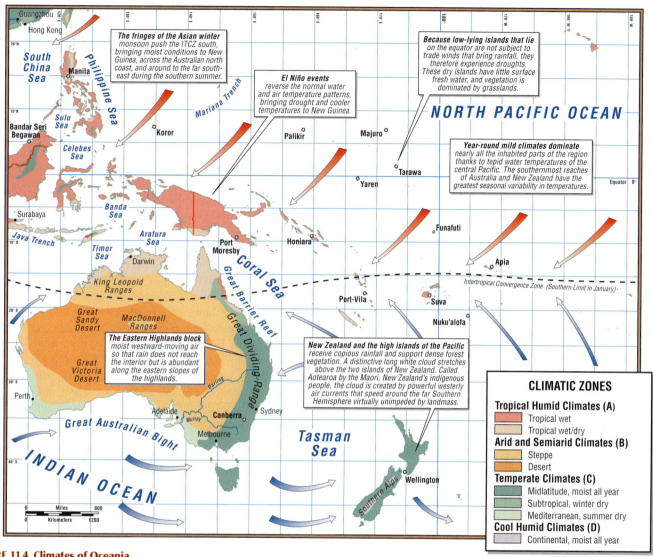

FIGURE 11.4 Climates of Oceania.

Many of the mountainous high islands in the Pacific exhibit orographic rainfall patterns, with a wet windward side and a dry leeward side (see Figure 1.13 on page 25). Low-lying islands across the region vary considerably in the amount of rainfall they receive. Some of these islands lie directly in the path of trade winds that deliver between 60 and 120 inches of rain a year on average. These islands support a remarkable variety of plants and animals. Other islands, particularly those on the equator, receive considerably less rainfall and often experience extreme droughts. As a result, these dry islands are dominated by grasslands and are unable to support much animal life.

El Niño. In an irregular pattern every 2 to 7 years, a number of changes take place in the circulation of air and water in the Pacific (Figure 11.5). These cyclical changes (also known as oscillations), which are not yet well understood, have been given the popular name **El Niño** by Peruvian fishers. Normally (see Figure 11.5A), water in the equatorial western Pacific (the New Guinea–Australia side) is warmer than water in the eastern Pacific (the Peru side). Warm air rises in the west, and rain clouds form. In the east,

cool air descends, and there is little rainfall. As an El Niño event develops (see Figure 11.5B), water temperatures (and the temperatures of the air above the water) are cooler than usual in the west and warmer than usual in the east. Instead of warm, wet air rising over the mountains of New Guinea and condensing as rainfall, cool, dry air sits at the ocean surface. The result is less cloud cover and less rainfall in the west, but more in the east (see Figure 11.5C).

The El Niño event of 1997–1998 illustrates the effects of this phenomenon. By December 1997, New Guinea had received very little rainfall for almost a year. Crops failed, springs and streams dried up, and fires broke out in tinder-dry forests. The cloudless sky allowed heat to radiate up and away from elevations above 7200 feet (2200 meters), so temperatures at high elevations dipped below freezing at night for stretches of a week or more. Tropical plants died, and people unaccustomed to chill weather sickened. Meanwhile, along the coasts of North, Central, and South America, the warmer than usual weather brought unusually strong storms, high ocean surges, and damaging wind and rainfall. Recently, an opposite mode, in which normal conditions become

A. Normal Equatorial Conditions

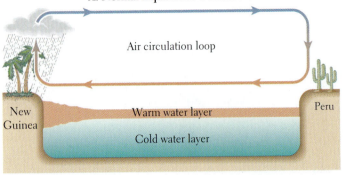

B. Developing El Niño Conditions

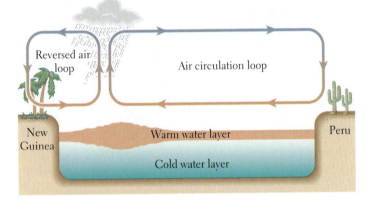

C. Fully Developed El Niño

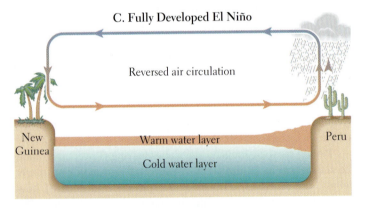

FIGURE 11.5 The El Niño phenomenon. [Adapted from Environmental Dynamics Research, Inc., 1998; Ivan Cheung, George Washington University, Geography 137, Lecture 16, October 29, 2001: http://www.gwu.edu/~/geog137/download/lecture16.ppt.]

unusually strong, has been identified and named *La Niña*, though scientists have barely begun to study it.

Flora and Fauna

The fact that Oceania comprises an isolated continent and numerous islands has had a special effect on its animal life (*fauna*) and plant life (*flora*). Many of its species are **endemic**: they exist in a particular place and nowhere else on earth. This is especially true of Australia, but many Pacific islands also have endemic species.

Plant and Animal Life in Australia. The uniqueness of Australia's plant and animal life is the result of the continent's long physical isolation, its large size, its relatively homogeneous landforms, and its strikingly arid climate. Since Australia broke away

from the southern supercontinent of Gondwana more than 65 million years ago, its plant and animal species have evolved in isolation (see Figure 1.10 on page 23 and the discussion of Wallacea on page 505 in Chapter 10). One spectacular result of this long isolation has been the development of more than 144 living species of marsupial animals. **Marsupials** are mammals that give birth to their young at a very immature stage and then nurture them in a pouch equipped with nipples. The best known marsupials are the kangaroos; other species include wallabies, wombats, phalangers, the tiger cat, possums, the koala, numbats, and bandicoots. These various marsupials fill ecological niches that in other regions of the earth are occupied by rats, badgers, moles, cats, wolves, ungulates (grazers), and bears. The **monotremes**, egg-laying mammals that include the duck-billed platypus and the spiny anteater, are exclusive to Australia and New Guinea. Birds are also unusually varied, and parrot species are especially diverse. Some of the 750 species of birds known on the continent migrate in and out, but more than 325 species are endemic.

Most of Australia's endemic plant species are adapted to dry conditions. Many of the plants have deep taproots to draw moisture from groundwater and small, hard, shiny (sclerophyll) leaves to reflect heat and to hold moisture. Much of the continent is grassland and scrubland, with bits of open woodland; there are only a few true forests, found in pockets along the Eastern Highlands, the southwestern tip, and in Tasmania (Figure 11.6). Two plant genera account for nearly all the forest and woodland plants: *Eucalyptus* (450 species, often called "gum trees") and *Acacia* (900 species, often called "wattles").

Plant and Animal Life in New Zealand and the Pacific Islands. The prehuman biogeography of the Pacific islands has long interested those who study evolution and the diffusion of

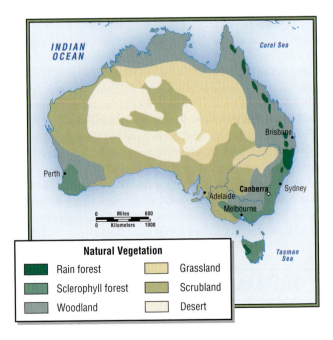

FIGURE 11.6 Australia's natural vegetation. Much of Australia is grassland and scrubland; a few forests can be found in the Eastern Highlands, in the far southwest, and in Tasmania. [Adapted from Tom L. McKnight, *Oceania* (Englewood Cliffs, N.J.: Prentice Hall, 1995), p. 28.]

plants and animals, including the great scientist Charles Darwin, who formulated many of his ideas about evolution after visiting the Galápagos Islands of the eastern Pacific. Islands have to snare their plant and animal populations from the sea and air around them. Birds and large storms drop seeds and spores. Dead logs and other floating debris from storms carry small animals, seeds, and plant shoots from larger islands and continents to distant atolls and volcanic islands. Some seeds—such as large, buoyant coconuts—can float to a random destination on their own. Then, in a complicated process, these organisms "colonize" their new home. For example, coconut trees accomplish the long process of spreading inland from the beach by dropping the ripe fruit of each succeeding generation onto ever slightly higher ground. Over time, isolated plant and animal colonizers may evolve into new species that are endemic to one island. High, wet islands generally contain more and more varied species because their more complex environmental mosaics provide niches for a wider range of wayfarers and thus a greater range of circumstances for evolutionary change.

The flora and fauna of islands are also modified by human inhabitants once they arrive. In prehistoric times, Asian explorers in oceangoing sailing canoes brought plants such as taro, bananas, and breadfruit and animals such as pigs, chickens, and dogs. Today, human activities from tourism to military exercises to city building continue to change the plants and animals of Oceania.

Generally, the distribution of land animals and plants is richest in the western Pacific, near the larger landmasses, which are the sources of species that diffuse to the islands. Biological diversity thins out to the east, where the islands are smaller and farther apart and hence less likely to be encountered by plant or animal wayfarers drifting on the wind or water. Although the natural rain-forest flora of New Zealand, New Guinea, and the high islands of the Pacific is abundant, the fauna of New Zealand and the Pacific islands is often described as simply "absent" because there are no indigenous land mammals, almost no indigenous reptiles, and only a few indigenous species of frogs. The reason for the absence of large animals is that New Zealand and the islands were never connected to the landmass of Eurasia as were the islands of Southeast Asia. There was never a land bridge that animals (and eventually humans) could cross. On the other hand, indigenous birds (especially waterfowl) in New Zealand are numerous and varied. New Zealand is the home of the kiwi and (in the past) the huge moa. Some species of the latter bird grew 12 feet (3.7 meters) high. The moa was a major source of food until the Maori people hunted it to extinction before Europeans arrived. Today, New Zealand may well be the country most characterized by introduced species of mammals, fish, and fowl (sheep are one example). Nearly all were brought in by European settlers.

HUMAN PATTERNS OVER TIME

PERSONAL VIGNETTE   *With courage, you can travel anywhere in the world and never be lost. Because I have faith in the words of my ancestors, I'm a navigator.*

MAU PIAILUG

In the 1990s, the *Hokule'a* sailed from Hawaii to Easter Island and back, a 14,000-mile round trip, guided by traditional navigational techniques, such as the reading of wave patterns. [Cary Wolinsky.]

In 1976, Mau Piailug made history by navigating a traditional Pacific island voyaging canoe through the 2400 miles (3860 kilometers) of deep seas that separate Hawaii and Tahiti (149°E, 17°S). He did so without a compass, charts, or other modern instruments, using methods passed down through his family. He relied mainly on observations of the stars, the sun, and the moon to find his way. When clouds covered the sky, he used the patterns of ocean waves and swells, as well as the presence of sea birds, to tell him of distant islands over the horizon.

Piailug reached Tahiti 33 days after leaving Hawaii and made the return trip in 22 days. His voyage settled a major scholarly debate. For centuries, scholars and explorers had debated how people had managed to settle the many remote islands of the Pacific without navigational instruments, thousands of years before the arrival of Europeans. Many scholars did not believe that navigation without instruments was possible and argued that would-be settlers might have simply drifted about on their canoes at the mercy of the winds, most of them starving to death on the seas, with a few happening on new islands by chance. It was hard to refute this argument because local navigational methods had died out almost everywhere. However, in isolated Micronesia, where Piailug is from, indigenous navigational traditions had survived.

Piailug learned his methods from his grandfather in secret because the Germans who first colonized Micronesia, and later the Japanese, banned long-distance navigation. The colonizers had wanted to prevent their Micronesian forced laborers from sailing away. After World War II, however, Piailug was free to sail, and he began making small voyages of several hundred miles. Eventually he attracted the attention of some Hawaiians who were building a traditional voyaging canoe, *Hokule'a* (see the photo above), with the aim of proving that traditional Pacific Islander sailing and navigational technologies were adequate for long-distance Pacific

travel. Since the successful voyage of 1976, Piailug has trained several students in traditional navigational techniques. Throughout the Pacific, traditional sailing and navigational techniques have become a symbol of rebirth and a source of pride. [*Adapted from Richard Nile and Christian Clerk,* Cultural Atlas of Australia, New Zealand, and the South Pacific *(New York: Facts on File, 1996), pp. 63–65.*]

The story of humans in Oceania began in ancient times with a series of sea voyages across huge distances. Then, in historical times, the region experienced a process of European colonization that resembled colonialism in other world regions in its land use, settlement patterns, exploitation of resources, and treatment of indigenous people. Although a minority of the Pacific islands remain under the jurisdiction of a Western power, the majority of the independent nations of Oceania are interacting less with their former colonizers and more with Asia, their nearest neighbor.

The Peopling of Oceania

The longest-surviving inhabitants of Oceania are Australia's **Aborigines.** As mentioned in Chapter 10, the ancestors of the Aborigines were among the Australoids, who migrated from Southeast Asia possibly as early as 60,000 years ago. Amazingly, some memory of this ancient journey may be preserved in Aboriginal oral traditions, which recall mountains and other geographic features that are now submerged under water. At about the same time that the Aborigines were settling Australia, related groups of Australoids were settling nearby areas. These were the **Melanesians,** so named for their relatively dark skin tones, a result of high levels of the protective pigment melanin. The Melanesians spread throughout New Guinea and other nearby islands, giving this region its name, Melanesia. They lived in isolated pockets. One indication of the great age and isolation of the Melanesian settlement of New Guinea is the existence of hundreds of distinct yet related languages on that island. Like the Aborigines, the Melanesians survived mostly by hunting, gathering, and fishing, although some groups—especially those inhabiting the New Guinea highlands—also practiced agriculture. There, agricultural features excavated in the terrain of a swampy basin have been dated at 9000 years ago, making this region one of the earliest places on earth where plant domestication has been firmly documented.

Much later, about 5000 to 6000 years ago, groups of linguistically related **Austronesians** migrated out of Southeast Asia and continued the settlement of the Pacific. By about a thousand years ago, some Austronesians had reached most of the remaining far-flung islands of the Pacific. They sometimes mixed with the Melanesian peoples they encountered (Figure 11.7). These Austronesians were renowned for their ability not only to survive at sea, but also to navigate over vast, featureless distances by using the stars and reading the waves. They were fishers, hunter-gatherers, and cultivators who developed complex cultures and maintained trading relationships among their widely spaced islands.

In the millennia that have passed since first settlement, humans have continued to circulate throughout Oceania. Some apparently set out because their own space was too full of people and conflict, food reserves were declining, or they wanted a life of

greater freedom. It is also likely that Pacific peoples were enticed to new locales by the same lures that later attracted some of the more romantic explorers from Europe and elsewhere: sparkling beaches, magnificent blue skies, beautiful people, scented breezes, and lovely landscapes.

The vast area settled by these people encompasses three distinct subregions. **Micronesia** refers to the small islands lying east of the Philippines and north of the equator. **Melanesia** includes New Guinea and the islands south of the equator and west of Tonga (the Solomon Islands, New Caledonia, Fiji, and Vanuatu). **Polynesia** refers to the numerous islands situated inside a large irregular triangle formed by New Zealand, Hawaii, and Easter Island (a tiny speck of land in the far eastern Pacific). Polynesia is the most recently settled part of the Pacific; some Polynesian influence remains in Melanesia in such places as Fiji and the Solomon Islands (see Figure 11.7).

Arrival of the Europeans

The earliest recorded contact between Pacific peoples and Europeans took place in 1521, when the first Europeans to cross the Pacific, led by the Portuguese explorer Ferdinand Magellan, landed on the island of Guam in Micronesia. That encounter ended badly. The islanders, intrigued by European vessels, tried to take a small skiff. For this crime, Magellan had his men kill the offenders and burn the village to the ground. By the 1560s, the Spanish had set up a lucrative trade route between Manila in the Philippines and Acapulco in Mexico. Explorers from other European states followed, first taking an interest mainly in the region's valuable spices. The British and French explored extensively in the eighteenth century, including three separate scientific expeditions by the region's greatest European voyager, James Cook (see discussion on the "Age of Exploration" in Chapter 4).

The Pacific was not formally divided among the colonial powers until the nineteenth century, and by that time, the United States, Germany, and even Japan had joined France and Britain in taking control of various island groups. European colonization of Oceania proceeded according to the models developed in Latin America, Africa, South Asia, and Southeast Asia, with the major emphasis being on extractive agriculture and mining. Native people were often displaced from their lands or exposed to exotic diseases to which they had developed no immunity, and their populations declined. This invasive early contact between Europeans and Pacific Islanders has been termed "the fatal impact."

Many enduring notions about the Pacific arose from the European explorations of the eighteenth and nineteenth centuries. During this time, Europeans debated whether or not civilization actually improves the quality of life for human beings. Some Europeans argued that civilization corrupts and debases people. This point of view grew partly out of news filtering home to Europe about the negative effects of colonization and partly out of the experience with industrialization in Europe, where crowded, dirty, impersonal, often crime-ridden cities were proliferating. Romanticists glorified what they termed "primitive" people living in distant places supposedly untouched by corrupting influences; they coined the term **noble savage** to describe such people. These ideas influenced explorers of

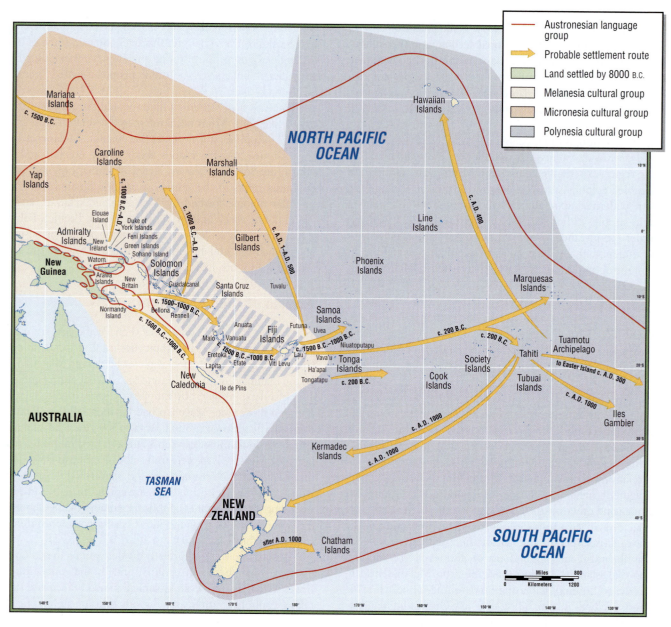

FIGURE 11.7 Primary culture groups in the island Pacific. Modern *Homo sapiens*, scholars agree, reached coastal New Guinea around the same time they reached Australia, about 50,000 years ago. By about 25,000 years ago, people were spread across a large part of New Guinea and had even begun moving across the ocean to nearby Pacific islands. Archaeologists have found a site on Buka in the northern Solomon Islands dated to about 26,000 years ago. Movement into the more distant Pacific islands apparently began with the arrival of the Austronesians, who went on to inhabit the farthest reaches of Oceania. This map shows the division of Pacific islands into Melanesia, Micronesia, and Polynesia. [Adapted from Richard Nile and Christian Clerk, *Cultural Atlas of Australia, New Zealand, and the South Pacific* (New York: Facts on File, 1996), pp. 58–59.]

the Pacific, who often spoke of the people they encountered as being in a pristine state of nature that was more conducive to virtuous moral conduct. Europeans were caught off guard when, from time to time, the very same islanders rebelled, armed themselves, and attacked those who were taking their lands and resources. Usually the surprised Europeans quickly revised their opinions and relabeled the "noble savages" as brutish and debased.

The realities of life in Oceania were much more balanced than the positive and negative extremes that Europeans perceived. In Australia and on New Guinea and the other larger islands of Melanesia, a relatively plentiful resource base made it possible for people to live in small, simple societies, less subject to the stratifi-

cation and class tensions of so much of the world. On the smaller islands of Micronesia and Polynesia, in contrast, land and resources were scarcer. On these islands, many societies were **hierarchical,** with layers of ruling elites at the top and undifferentiated commoners at the bottom. Moreover, many of the peoples of Oceania coexisted in a state of moderate antagonism. Although warfare occurred, hostilities were often settled ritualistically and by means of annual tribute-paying ceremonies, rather than by resorting to mortal violence. Individual rulers rarely amassed large territories or controlled them for long.

Perhaps the most enduring myth Europeans created was their characterization of the women of the Pacific islands as gentle, simple, compliant love objects. (Tourist brochures still promote this notion.) Although there is ample evidence to suggest that Pacific Islanders did have more sexual partners in a lifetime than Europeans did, the reports of unrestrained sexuality related by European sailors were no doubt influenced by the exaggerated fantasies one might expect from all-male crews living at sea for months at a time. The notes of Captain James Cook are typical: "No women I ever met were less reserved. Indeed, it appeared to me, that they visited us with no other view, than to make a surrender of their persons."

Over the years, such notions about Pacific island women have been encouraged by the paintings and prints of Paul Gauguin, the writings of the novelist Herman Melville (*Typee*), and the studies of the anthropologist Margaret Mead (*Coming of Age in Samoa*), as well as by movies and musicals such as *Mutiny on the Bounty* and *South Pacific*. In reality, gender roles in the Pacific varied considerably from those in Europe, but not in the ways European explorers imagined. Women often exercised a good bit of power in family and clan, and their power increased with motherhood and advancing age. In Polynesia, a woman could achieve the rank of ruling chief in her own right, not just as the consort of a male chief. In everyday Polynesian life, men were the primary cultivators of food as well as the usual cooks. Women were primarily craftspeople but also contributed to subsistence by gathering fruits and nuts and by fishing for shellfish in reefs and lagoons near the shore. And in some places—Micronesia, for example—lineage was established through women, not men.

The Colonization of Australia and New Zealand

Though all of Oceania has experienced European or American rule at some point, the most Westernized parts of the region are Australia and New Zealand. The colonization of these two countries by the British has resulted in many parallels with North America. In fact, the American Revolution was the major impetus for "settling" Australia because, once the North American colonies were independent, the British needed somewhere else to send their convicts. In early nineteenth-century Britain, a relatively minor theft—for example, of food or a piglet—might be punished with a term of 7 years' hard labor in Australia. After their sentences were served, most former convicts chose to stay in the colony. A steady flow of English and Irish convicts arrived in Australia until 1868, and they are given credit for Australia's rustic self-image and egalitarian spirit. They were joined by a much larger group of voluntary immigrants from the British Isles who were attracted by the availability of inexpensive farmland. Waves of these immigrants arrived until World War II. New Zealand was settled somewhat later, in the mid-1800s. And, although its population also mostly derives from British immigrants, New Zealand was never a penal colony.

Another similarity among Australia, New Zealand, and North America was the treatment of indigenous peoples by European settlers. In both Australia and New Zealand, native peoples were killed outright, annihilated by infectious diseases to which they had little immunity, or shifted to the margins of society. The few who lived on territory deemed undesirable by Europeans were able to maintain their traditional way of life, but the vast majority who survived lived and worked in grinding poverty, either in urban

Arearea ("Amusement") by Paul Gauguin, 1892. The painting shows Tahitian women rendered in a European Romantic pastoral style that emphasizes their gentle, compliant demeanor. [Musée d'Orsay, Paris. Photograph by Erich Lessing, Art Resource, New York.]

slums or on cattle and sheep ranches. Today, native peoples still suffer from pervasive discrimination and maladies such as alcoholism and malnutrition that are common in an **underclass,** the lowest social stratum, composed of the disadvantaged. Even so, some progress is being made toward improving their lives, as discussed on pages 565–566. New Zealand has made great strides in appreciating the culture and rights of its native Maori population. Australia is beginning to do the same with regard to the Aborigines, although in Australia there is vocal resistance to this project among some people of European background.

Oceania's Growing Ties with Asia

During the twentieth century, Oceania's relationship with the rest of the world changed at least three times: from a predominantly European focus to identification with the United States and Canada to its currently emerging linkage with Asia. Up until roughly World War II, the colonial system gave the region a European orientation. In most places, the economy depended largely on the export of raw materials to Europe. Thus, even when a colony gained independence from Britain, as Australia did in 1901 and New Zealand did in 1907, people remained strongly tied to their "mother countries"— even today the Queen of England remains the head of state in both countries. During World War II, however, the European powers could provide only token resistance to Japan's invasion of much of the Pacific and its bombing of northern Australia.

After the war, the United States became the dominant power in the Pacific and throughout much of Asia. Although U.S. dominance did not sever the region's economic linkages with Europe, U.S. investment became increasingly important to the economies of Asia and Oceania. Australia and New Zealand joined the United States in a cold war military alliance known as ANZUS. Both fought alongside the United States in Korea and Vietnam, suffering considerable casualties and experiencing significant anti-war movements at home. U.S. cultural influence was strong, too, as its products, technologies, movies, and pop music penetrated much of Oceania.

By the 1970s, another shift was taking place as Oceania became steadily drawn into the growing economies of Asia. One could consider this development long overdue, given Asia's proximity to Oceania, its potential as a market for Oceania's raw materials, and some cultural connections. Many Pacific islands have significant Chinese, Japanese, Filipino, and Indian minorities, and even the small Asian minorities of Australia and New Zealand are increasing. Since the 1960s, Australia's thriving mineral export sector has become increasingly geared toward supplying Japan's burgeoning manufacturing industries. Similarly, since the 1970s, New Zealand's wool and dairy exports have gone mostly to Asian markets. As we shall see, these transformations are accompanied by considerable cultural and economic strain. Nonetheless, despite

occasional backlashes against "Asianization," Australia, New Zealand, and the rest of Oceania are becoming more open to Asian influences.

POPULATION PATTERNS

Oceania occupies a huge portion of the planet, yet it has few people compared to other world regions. Whereas Southeast Asia has more than half a billion people and East Asia well over a billion, just over 32 million people are spread throughout Oceania—from New Guinea, Australia, and New Zealand across the Pacific to Hawaii and Easter Island. The Pacific islands, including Hawaii, have somewhat over 8 million people; Australia has 20 million, and New Zealand 4 million.

As Figure 11.8 indicates, the people of this region are unevenly distributed. Most Australians live in a string of cities along

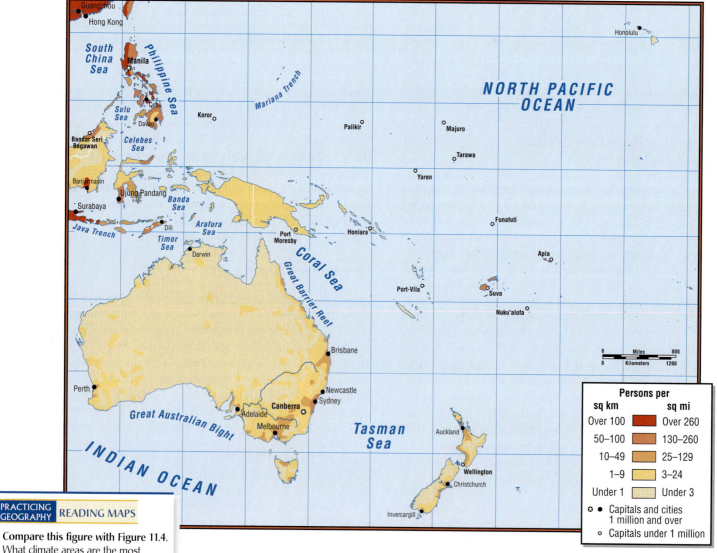

Persons per		
sq km		sq mi
Over 100	🟥	Over 260
50–100	🟧	130–260
10–49	🟨	25–129
1–9	🟨	3–24
Under 1	⬜	Under 3
⬕ ●	Capitals and cities 1 million and over	
⬕	Capitals under 1 million	

PRACTICING GEOGRAPHY   READING MAPS

Compare this figure with Figure 11.4. What climate areas are the most heavily populated? In what ways does climate influence human settlement patterns?

FIGURE 11.8 Population density in Oceania. Oceania is sparsely populated compared to other world regions. Its 32 million inhabitants are unevenly distributed and tend to be concentrated on a few islands and along the fertile coastlines of Australia and New Zealand. [Adapted from *Hammond Citation World Atlas* (Maplewood, N.J.: Hammond, 1996).]

AT THE REGIONAL SCALE Urbanization in Island Oceania

All over the world, increasing numbers of people are migrating from the countryside to towns and cities. This trend is no less visible in Oceania's islands, where a steadily growing proportion of the population lives in urban areas. The image of the Pacific islands as a place of palm-fringed lagoons populated with rural people celebrating ancient cultural traditions contrasts starkly with reality. Throughout the Pacific, urban centers have transformed national landscapes, and in some small states, such as Palau and the Marshall Islands, they have become the new landscape. Although cities are places of opportunity, they are also sites of both cultural change and conflict. The growth of urban areas has created several social, economic, and environmental problems in the Pacific islands. Some of these are unique to the region; many are similar to the problems faced in other parts of the developing world.

The great majority of Pacific island towns, and all the capital cities, are located in ecologically fragile coastal settings. Many of these towns were established during the colonial era and were situated in places suitable for only limited numbers of people. Consequently, little land is available for development, and access to

housing is limited. Squatter settlements have been a visible feature of the region's urban areas for several decades. The discharge of untreated sewage and other wastes into coastal waters and lagoons has damaged marine environments, has reduced the productivity of subsistence fisheries, and has periodically led to the outbreak of diseases such as cholera. Air pollution is a new phenomenon, as is noise pollution. Urban unemployment is on the rise, as is crime, and low economic growth has reduced the revenue available to governments to manage urban development.

Cultural changes, too, have resulted from urbanization. Although many urban residents were born outside the towns and maintain close connections to their rural homes, some urban islanders have sought to disavow rural life, ethnic identity, and cultural commitments. Increasingly, people are marrying in town and across language divisions, creating new patterns of social alliances and networks. This, along with the adoption of urban lifestyles, is creating new social tensions and changing the very nature of social life in the region.

[Adapted from John Connell and John Leah, *Urbanisation in the Island Pacific* (London and New York: Routledge, 2002).]

---

the country's well-watered and relatively fertile eastern and southeastern coasts—Brisbane, Newcastle, Sydney, Canberra, Melbourne, Adelaide—and on the southwestern tip of the continent in and around the city of Perth. Australia and New Zealand have among the highest percentages of city dwellers outside Europe: 85 percent for Australia and 77 percent for New Zealand in 2003. The vast majority of people in these two countries live in modern, affluent cities and work in a range of occupations typical of highly industrialized societies. New Zealand has one very large city, Auckland, and several medium-sized cities on both North Island and South Island. Nonetheless, overall densities in both countries are low: Australia averages just 7 people per square mile (3 per square kilometer) and New Zealand 38 per square mile (15 per square kilometer).

The smaller islands of the Pacific are generally much less urbanized (but see the box "Urbanization in Island Oceania" above), and the population density of these islands is often linked to specific features and to their history of settlement and development. Some are sparsely settled or uninhabited; others—including some of the smallest Pacific islands, such as the Marshall Islands and Tuvalu—have 800 to 1000 people per square mile (307 to 386 per square kilometer). The highest density is found on the tiny island of Nauru, with 1360 people per square mile (525 per square kilometer). In the last half of the twentieth century, the people of Nauru lived in relative prosperity on the proceeds from the mining of phosphate (used in the manufacture of fertilizer), derived from eons of bird droppings. But the phosphate reserves are now depleted, and the government, to avoid financial disaster, has been exploring new, high-tech ways of making money, such as offshore banking.

The urban poor of Oceania, whether they live in the developed economies of Australia and New Zealand—where they tend

to have Aboriginal or Maori origins—or in cities across the Pacific–such as Port Moresby in Papua New Guinea, Suva in Fiji, Papeete in Tahiti, and Honolulu in Hawaii—may have higher cash incomes than rural Pacific Islanders, but much lower standards of living. In the cities, there is little they can do to supplement their incomes with self-produced food or other necessities (see the discussion of subsistence affluence on page 570).

There are two different population profile patterns in Oceania. The Pacific islands tend to have high fertility rates, between 2.5 and 5.7 children per woman. Their populations are young, with 27 to 44 percent under the age of 15 and just 3 to 5 percent elderly. In contrast, the population profiles of Australia and New Zealand closely resemble those of North America, Europe, and Japan. There, fertility rates average less than 2 children per woman, and people tend to live into their mid- to late seventies. The overall trend throughout the region, however, is toward smaller families and aging populations—although this trend is just beginning in the Pacific islands and is much more obvious in Australia and New Zealand.

Quick Review

1. If you discovered an island shaped like a ring on top of high cliffs, what would you call it, and how would you explain its formation?

2. Why is New Zealand known for its numerous and distinctive birds but has no indigenous land mammals?

3. Explain the adoption and subsequent rejection of the term "noble savage" by nineteenth-century Europeans to describe the people of the Pacific.

4. Explain why land is limited in the cities of Oceania's Pacific islands.

11 CURRENT GEOGRAPHIC ISSUES

Many current geographic issues in Oceania are related to the transition now under way from European to Asian and inter-Pacific cultural influences. This change in cultural orientation, in turn, is related to the larger transition to a global economy now under way everywhere. The old relationships were built on historical factors, such as the settlement of Australia and New Zealand by Europeans. The new relationships, in contrast, are influenced by practical financial and geographic considerations, such as physical proximity to Asia and recently opened markets—especially in China, but also in South and Southeast Asia. Hence Australia now trades more with Asia and less with Europe, and New Zealand, while continuing to trade mostly with Australia, finds increasing opportunities in Asia and the Pacific.

Emerging technologies such as e-mail and the Internet and rapid air travel have a special effect on the far-flung locations of Oceania. They encourage cultural sensitivity and the modification of old prejudices because people have greater access to information about unfamiliar ways of life, can correspond instantly with distant colleagues and friends, and have the chance to experience distant places personally. One result of this greater interaction is that New Zealand and the Pacific islands are finding philosophical grounds for a closer mutual identity, including the fostering of greater public awareness of environmental issues and the renewal and acceptance of their common Polynesian as well as European cultural heritage.

SOCIOCULTURAL ISSUES

The cultural sea change away from Europe and toward Asia and the Pacific has been accompanied by new respect for indigenous peo-

FIGURE 11.9 Political map of Oceania. The reader is advised that showing and naming all of the political units of Oceania on this map is an impossibility. See Figures 11.1 (page 551) and 11.2 (pages 552–553) and the text discussion on the complexities on page 553.

ples: the Aborigines of Australia; the Maori of New Zealand; and the Melanesians, Micronesians, and Polynesians of the Pacific islands. In addition, the growing sense of common economic ground with Asia has heightened awareness of the attractions of Asian culture.

Ethnic Roots Reexamined

Weakening of the European Connection. Until very recently, most people of European descent in Australia and New Zealand thought of themselves as Europeans in exile. Many considered their lives incomplete until they had made a pilgrimage to the British Isles or the European continent. In her book *An Australian Girl in London* (1902), Louise Mack wrote: "[We] Australians [are] packed away there at the other end of the world, shut off from all that is great in art and music, but born with a passionate craving to see, and hear and come close to these [European] great things and their home[land]s."

These longings for Europe encouraged Australians and New Zealanders to think of themselves as transients in their own countries, separate from the region in which they resided; such feelings were accompanied by racist attitudes toward the Aborigines, Maori, and Asians. The historian Stephen H. Roberts managed to write a history of settlement in Australia without even mentioning the Aborigines. In a later book, *The Squatting Age in Australia*, published in 1935, he noted of the Aborigines: "It was quite useless to treat them fairly, since they were completely amoral and usually incapable of sincere and prolonged gratitude." The prevailing idea was that both Australia and New Zealand should preserve European culture in this nether region of the Southern Hemisphere, warding off not only indigenous peoples but also Pacific Islanders and Asians in general. Hence, in the 1920s, migrants from Asia, Africa, and the Pacific were legally barred. Trading patterns further reinforced connections to Europe: until World War II, the United Kingdom was the primary trading partner of both New Zealand and Australia.

When migration from the British Isles slowed in the mid-1940s, after World War II, both Australia and New Zealand began to lure immigrants from southern and eastern Europe, many of whom had been displaced by the war. Hundreds of thousands came from Greece, parts of the former Yugoslavia, and Italy. The arrival of these non-English-speaking people began a shift toward a more multicultural society. The election of more liberal governments led to a loosening of restrictions against Asian migrants in the late 1950s and early 1960s. In the 1960s, the whites-only immigration policies were set aside, and migration from China, India, Vietnam, South Africa, and elsewhere in the Pacific was allowed. The greatest increase in Asian migrants took place in the early 1970s after the United States withdrew from Vietnam, when many Vietnamese refugees sought a safe haven.

As a result of the new immigration policy, the ethnic makeup of the populations of Australia and New Zealand is changing. Australia is now among the most ethnically diverse countries on earth

Percentage of Foreign-born Australians from Various Countries

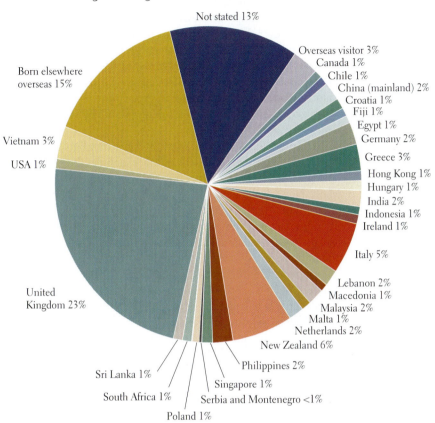

FIGURE 11.10 Australia's cultural diversity. About 74 percent (13.2 million) of Australia's people were born in the country; the rest were born in other places, making Australia one of the world's most culturally diverse nations. The pie chart, based on the 1996 census, shows the birthplaces of the 26 percent of Australia's residents born elsewhere. [Data from 1997 Commonwealth of Australia.]

Thinking Geographically: ON THE WEB

(Figure 11.10). In both Australia and New Zealand, immigrants from Asia, especially from Vietnam and China, are a significant portion of new arrivals. Nonetheless, Asians remain a small percentage of the total population in both countries (less than 6 percent in Australia). People of European descent are declining as a proportion of the population, but are expected to remain the most numerous segment throughout the twenty-first century.

Strengthening of the Connections with Indigenous Peoples. The population makeup of Australia and New Zealand is also changing in another respect. In Australia, for the first time in recent memory, the number of people who claim indigenous origins is increasing. Before Europeans arrived at the end of the eighteenth century, the sole inhabitants of Australia were about 750,000 Aborigines. In the intervening two centuries, the overall population has grown to about 20 million, but the Aboriginal population, just 2.2 percent of that total, is only half what it was in 1800. Therefore, when those claiming Aboriginal origins rose by 33 percent between 1991 and 1996, it was a surprise. In New Zealand, those claiming a Maori background rose by 20 percent. These increases are thought to be linked in two ways to changing attitudes toward indigenous peoples in the region. First, more people now understand that colonial attitudes are largely responsible for the low social standing and impoverished state of indigenous peoples. Hence, those who have some Aborigine or Maori ancestors are more willing to claim them. Another factor is that relationships between European and indigenous peoples are more common and more open, and so the number of people with mixed heritage is increasing.

In 1988, during a bicentennial celebration of the founding of white Australia, a contingent of some 15,000 Aborigines protested that they had little reason to celebrate. During the same 200 years, they had been excluded from their ancestral lands, from basic civil rights, and even from the national consciousness. Into the 1960s, Aborigines had only limited rights of citizenship, and it was even illegal for them to have a drink of alcohol. Until 1993, Aborigines were thought to have no prior claim to any land in Australia. During settlement in the late eighteenth century, the British had deemed all Australian lands to be available for British use because the Aborigines were thought to be too primitive to have concepts of land tenure, since their nomadic cultures had "no fixed abodes, fields or flocks, nor any internal hierarchical differentiation." After the Australian High Court declared this position void in 1993, Aboriginal groups began to win some land claims, mostly for land in the arid interior previously claimed only by the Australian government.

In New Zealand, relations between the majority European-derived population and the indigenous Maori have proceeded somewhat more amicably. Today, New Zealand proudly portrays itself as a harmonious multicultural society. Yet there is deep ambivalence about the foundations of this supposed harmony. The geographer Eric Pawson writes that when the Maori signed the Waitangi Treaty with the British in 1840 and accepted European "guns, goods and money," they thought they were granting rights of land usage only in return for a trading relationship. The Maori did not regard land as a tradable commodity, but rather as an asset of the people as a whole, used by families and larger kin groups to fulfill their needs. Pawson writes: "To the Maori the land

There were many different groups of Aborigines in various parts of Australia when the Europeans arrived, but all seem to have based their moral laws and daily customs on the idea that the spiritual and physical worlds are intricately related. The dead are everywhere present in spirit, and they guide the living in how to relate to the physical environment. *Dreamtime* refers to the time of creation when the human spiritual connections to rocks, rivers, deserts, plants, and animals were made clear. This representation (in Kakadu National Park, Northern Territory) is of Djawok, a creator who left his image of a cuckoo on this rock. Aboriginal people who have remained close to their heritage still read the landscape as a complex sign system conveying spiritual meaning. Territory is recognized and made familiar by following the tracks of ancient "song lines." Particular tribal groups are associated with specific animals or landscape features, from which they gather solace and inspiration. When people die, they are said to "go into the country." [Belinda Wright/National Geographic Image Collection.]

in the early 1800s to 42,000 in 1900, and they came to occupy the lowest and most impoverished rung of New Zealand society. In the 1990s, however, the Maori began to reclaim their culture, and they established a tribunal that forcefully advances Maori interests through the courts. New Zealand formalized its efforts to right past wrongs and bring greater equality and social participation to the Maori and other minority groups in the country in 1996, when the government agreed to settle several long-standing Maori claims to land and fishery rights.

During the twentieth century, Maori numbers rebounded, reaching 532,000 by 2002. This number includes those who previously hid their Maori origins but are now proud to claim them. Many New Zealanders now embrace the Maori component of their cultural heritage; overall, New Zealand may lead the world in addressing past mistreatment of indigenous peoples. Nonetheless, the Maori still have higher unemployment, lower educational attainments, and poorer health than the population as a whole.

Balancing Indigenous Rights. The Fiji island group in the southern Pacific exemplifies an issue that is typical in the Pacific islands: how to balance the rights of indigenous people with those of "outsiders" who may have lived on Pacific islands for several generations. Fiji, one of the first colonial domains in the Pacific to achieve independence, has a population that is about evenly divided between indigenous Fijians and the descendants of indentured sugar plantation workers brought in by the British from India more than a century ago. Fijians of Indian origins (Indo-Fijians) have flourished as the owners of tourism facilities and other businesses and as growers of sugarcane, one of the country's leading exports. They hold significant economic and political power, especially in western urban centers and in areas of tourism and sugar cultivation. On the other hand, the indigenous Fijians, who are governed by traditional chiefs, tend to live on rural islands in the east of the island group and to be less prosperous. Complicating matters is a codified system established by the British whereby land rights are held by indigenous clans and land cannot be alienated or sold. Indo-Fijians who grow sugarcane or run businesses do so on leased land. In 1987 and again in May 2000, rivalry between these two main groups of Fijians led indigenous Fijians to attempt coups d'état against legally elected governments dominated by Indo-Fijians. The conflict has retarded the tourism economy and set back economic and social development. Many Indo-Fijians have left the islands, fearing that indigenous rights will weaken egalitarian policies. Political leaders around the region, from Australia to Hawaii, have reminded indigenous Fijians that indigenous rights cannot be held superior to fundamental human rights.

Forging Unity in Oceania

In embracing its Maori roots, New Zealand has in effect also begun to embrace its connection to the wider Pacific, especially Polynesia. Although the majority of New Zealand immigrants continue to come from the United Kingdom, the second largest influx has come from the Pacific islands. Auckland now has the largest Polynesian population of any city in the world.

A sense of unity with Oceania as a whole is developing throughout the region as people begin to appreciate the region's

was sacred ... [and] the features of land and water bodies were woven through with spiritual meaning and the Maori creation myth." The British, on the other hand, assumed that the treaty had transferred Maori lands to them, giving them *exclusive* rights to settle the land with British migrants and to extract wealth through farming, mining, and forestry. Therein lay the grounds for a conflict that has resulted in significant losses for the Maori.

In the 50 years after 1840, the Maori lost control of more than 84 percent of their former lands; by 1950, they had lost all but 6.6 percent. European settlers and the government owned and occupied the rest. Maori numbers shrank from a probable 120,000

cultural complexity and cooperate in activities ranging from schooling to sports to environmental activism. One way in which this unity is manifested is through interisland travel. Today, people usually fly rather than traveling by sea. They often travel in small planes from the outlying islands to hubs such as Fiji, where jumbo jets can be boarded for Auckland, Melbourne, or Honolulu. Cook Islanders call these little planes, which carry 5 to 30 passengers, "the canoes of the modern age." New Zealanders migrate to Australia to teach or train. A businessman from the Kiribati group in Micronesia flies to Fiji to take a short course at the University of the South Pacific. A Cook Islands teacher takes graduate training in Hawaii.

Who funds all this travel, especially in the Pacific islands? Many travelers do so themselves. A person may earn money by selling food or handmade crafts at public events. A popular way for a group to collect plane fares is to sell raffle tickets for baskets of food or a bicycle. (There may be plenty of customers in one's own extended family.) Governments help, too, with scholarships to the various universities in Australia, New Zealand, Hawaii, and Fiji. Because there is often a waiting list for these grants, a college education may come later in life, when one is 30 or 40.

The arts have become a powerful medium for forging a regional identity through events such as the Festival of Pacific Arts. First held in 1972 in Fiji, this weeklong gathering takes place once every 4 years in a different location. The festival features indigenous music, dance, and other traditional forms of artistic expression, such as wood carving and tattooing. Recently, these events have also included the skills of oceangoing navigation and canoe voyaging. Several other trends are encouraging a growing sense of unity throughout the region: the use of pidgin languages, the movement called the Pacific Way, and the popularity of sports.

Languages in Oceania. The Pacific islands, and most notably Melanesia, have a rich variety of languages. In some cases, islands in a single chain have several different languages. A case in point is Vanuatu, a chain of 80 mostly high volcanic islands to the east of northern Australia (see Figure 11.2 on pages 552–553). At least 108 languages are spoken by a population of just 180,000—an average of one language for every 1600 people! New Guinea is the largest, most populous, and most ethnically diverse island in the Pacific. No less than 800 languages are spoken on New Guinea by a populace of 5.5 million. Languages are both an important part of a community's cultural identity and a hindrance to cross-cultural understanding.

In Melanesia, and elsewhere in the Pacific, the need for communication with the wider world is served by several **pidgin** languages that are sufficiently similar to be mutually intelligible. Pidgins are made up of words borrowed from the several languages of people involved in trading relationships. Over time, they can grow into fairly complete languages, capable of fine nuances in expression. When a particular pidgin is in such common use that mothers talk to their children in it, then it can literally be called a "mother tongue." In Papua New Guinea, pidgin English is the official language.

The imposition of European languages in Oceania extinguished a number of native tongues. For example, there were 500 or more Aboriginal languages in Australia at the time of first

The Australian Aboriginal Embassy is a collection of colorful wooden structures erected on the lawns in front of Old Parliament House in Canberra as a protest against the seizure of Aboriginal lands by European settlers at the end of the eighteenth century and the continuing refusal of the government to acknowledge that injustice. The protest began in 1972 and was continuing in 2001, when the authors visited. Nearby is a permanent encampment of Aboriginal people who staff the facility and welcome the many visitors. [Mac Goodwin.]

contact; most are now gone. Attempts to revive and preserve indigenous languages have crystallized an effort at regional solidarity known as the Pacific Way.

The Pacific Way. The **Pacific Way** is a term used since 1970 to convey the idea that Pacific Islanders and their governments have a regional identity growing out of their own particular social experience. It includes the empowering idea that Pacific Islanders have the ability to control their own development and solve their own problems. The concept of the Pacific Way first emerged when Pacific peoples were grappling with the problems presented by school curricula imposed by the colonial powers, which suppressed the use of native languages and often depicted the region as peripheral and backward. They developed the Pacific Way as a philosophy to guide them in writing their own school texts. It embodied the idea that Pacific island children should first learn about their own cultures and places before they studied Britain, France, or the United States.

Increasingly, the Pacific Way has grown into an integrated approach to economic development and environmental issues, with an emphasis on consensus as a traditional approach to problem solving. In some ways, the concept resembles *Pancasila* in Southeast Asia (see Chapter 10), and it has some of the same potential for abuse. But there have been constructive steps to implement the Pacific Way. The South Pacific Regional Environmental Program (SPREP), in existence since 1980, emphasizes regional cooperation and grassroots environmental education. The geographer Carolyn Koroa, formerly of the University of the South Pacific in Fiji, reports that the organization is extraordinarily egalitarian and that SPREP programs routinely incorporate traditional environmental knowledge into efforts to promote sustainable livelihoods, often based on traditional crafts.

Sports as a Unifying Force. Sports and games are a major feature of daily life throughout Oceania, and the region has both shared them with and borrowed them from cultures around the world. Long-distance sailing, now a world-class sport, was an early skill in this region. Surfing evolved in Hawaii from ancient navigational customs that matched human wits against the power of the ocean. Sporting traditions introduced from outside have been embraced and modified to suit local needs. The Scottish game of golf is a favorite of Japanese and South Korean tourists in Hawaii, Tahiti, and Guam. On hundreds of Pacific islands and in Australia and New Zealand, the rugby field, the volleyball court, the soccer field, and the cricket pitch (see the photo below) are important centers of community activity. Baseball is a favorite in the parts of Micronesia that were U.S. trust territories. Women compete in the popular sport of netball, which is played a little like basketball but

In the Trobriand Islands near New Guinea, the British game of cricket has been reformulated to include local traditions. Village teams of as many as 60 men dress in traditional garments and decorate their bodies in ways that are reminiscent of British cricket uniforms, such as the painting-on of white shin pads, yet also include elements of magical decoration. Chants and dances are part of Trobriand cricket matches. The matches can go on for weeks, and they attract crowds of young onlookers, as seen in this picture. [Robert Harding Picture Library, London.]

without a backboard. Australia or New Zealand often fields the world champion netball team.

Sports competitions, including native dance, are the single most common and resilient link among the countries of Oceania. Such competitions encourage regional identity, and they provide opportunities for ordinary citizens to travel extensively around the region and to sports venues in other parts of the world. The South Pacific Games—featuring soccer, boxing, tennis, golf, and netball, among other sport—are held every 4 years. The Micronesians hold periodic games that incorporate many traditional tests of skill, such as spearfishing, climbing coconut trees, and racing outrigger canoes.

In New Zealand, the Maori and Samoan cultures are central to that country's world dominance in rugby. At the beginning of a game, the players perform the Maori war dance, the Haka, to arouse feelings of aggression. Many of the New Zealand players are Samoan in origin; Samoans are Polynesians, noted for their size, strength, and acuity in games of skill. In Oceania, as in societies the world over—in North America, the Caribbean, Africa, Europe, and South Asia—sports have smoothed the way for acceptance of diverse groups, especially at the Olympic level. Nonetheless, here as elsewhere, there is a danger that a particular group can be considered good at sports, yet still find full participation in society blocked.

Women's Roles in Oceania

There is great variation in gender roles in Oceania, and they are changing significantly throughout the region. In Australia and New Zealand, women's access to jobs and policy-making positions has improved over the last few decades. Among the results is that women bureaucrats in Australia have been credited with materially improving the living conditions of the poorest segments of Australian society. Increasingly, young women are choosing careers and postponing marriage until their 30s. In New Zealand, 58 percent of women aged 24 to 29 are not married, nor are 38 percent of those aged 30 to 34. Many of these women have mates, but choose not to marry them. Nonetheless, Australian and New Zealand societies continue to reinforce the housewife role for women in a variety of ways. For example, the expectation is that women, not men, will interrupt their careers to stay home to care for young or very old family members.

In the Pacific, recent research has shown that, historically (and prehistorically), gender roles and relationships varied greatly from island to island and were in a continual state of flux. Moreover, gender roles changed over the course of life, and this is still very much the case. Today, many young women fulfill traditional roles as wives and mothers and practice a wide range of domestic crafts, such as weaving and basketry. Then, in middle age, they may return to school and take up careers. Some Pacific women, with the aid of government scholarships, pursue higher education or job training that takes them far from the villages where they raised their children, yet their incomes are often shared with children and grandchildren. Accumulating age and experience may boost Pacific women into community positions of considerable power.

The New Zealand geographer Camilla Cockerton reports that throughout their lives, Pacific women contribute significantly

On the Cape York Peninsula, Queensland, a ringer, or station hand, takes a break before heading home at the end of the workday. [Sam Abell/National Geographic Image Collection.]

to family assets through the formal and informal economies. Most traders in marketplaces are women, and the items they sell are usually made and transported by women. Yet, like women everywhere, they have trouble obtaining credit for their businesses. For example, of the 2039 loans approved by the Agricultural Bank of Papua New Guinea in January 1991, only 4 percent (or 91) went to women. Cockerton also reports that Pacific women who migrate play a huge role in Pacific economies, sending larger remittances to their home countries than men do, despite their lower incomes.

Being a Man: Persistence and Change

Because of its cultural diversity, Oceania encompasses many acceptable roles for men, but images of supermasculinity are popular throughout the region. In the Pacific, men traditionally were cultivators, deepwater fishermen, and masters of seafaring. In Polynesia, they also were responsible for many aspects of food preparation, including cooking. Men fill many positions in the modern world, but idealized male images continue to be associated with vigorous activities such as rugby, cricket, and canoe building and navigating.

In Australia and New Zealand, the cult of the supermasculine, white, working-class settler has long had prominence in the national

mythologies. Only recently have the stories of ordinary men, female immigrants, Aborigine laborers, and the many early Chinese settlers come to public attention. In New Zealand, the classic male settler was a farmer and herdsman; in Australia, he was more often an itinerant laborer—stockman, sheep shearer, cane cutter, or digger (miner). This many-skilled drifter, who possessed a laconic, laid-back sense of humor, went from station (large farm) to station or from mine to mine, working hard but sporadically, gambling, and then working again, until he had enough money or experience to make it in the city. There, he often felt ill at ease and chafed to return to the wilds. Now immortalized in songs, novels, films such as the *Crocodile Dundee* series, and American TV auto advertisements, these men are portrayed as a rough and nomadic tribe. Although they may have had families somewhere, they traveled as loners. Not surprisingly, male camaraderie, a demonstrated loyalty to their "mates" (male friends), and frequent brawls dominated their social life. No small part of this characterization derived from the fact that many of Australia's first immigrants were convicts who seized the opportunity for a new identity and measure of freedom "down under."

Today, as part of larger efforts to recognize the diversity of Australian society, new ways of life for men are emerging and are breaking down the national image of the tough male loner. Nonetheless, the old model persists and remains prominent in the public images of Australian businessmen, politicians, and movie stars.

ECONOMIC AND POLITICAL ISSUES

Although Oceania's reorientation away from Europe and toward Asia and the Pacific has many cultural ramifications in the region, the process is driven largely by global economic forces such as trade, tourism, and migration.

Export Economies

For decades, natural resources and agricultural products have supported the national economies of Australia and New Zealand. Australia, for example, supplies about 50 percent of the world's wool exports; it is the world's largest supplier of coking coal used in steel manufacturing; and it is first in bauxite production, third in gold, and fourth in nickel. New Zealand specializes in dairy products, meat, fish, wool, and timber products. But neither country has been a major supplier to the world market of more profitable manufactured goods.

There are several reasons for this pattern of exports. First, during the early and mid-twentieth century, Australia and New Zealand used tariffs to protect their manufacturing industries from outside competition so that jobs could be created for new immigrants. At the same time, domestic markets in both countries were (and still are) too small to support multiple competitors. Thus, with limited outside sources of manufactured goods and few domestic ones, the type of competition among manufacturers that would result in quality improvement and price reductions rarely occurred. As a result, imported products were more popular than those produced locally, despite their higher costs. Second, most of the major trading partners of Australia and New Zealand purchase only raw materials to supply their own manufacturing industries. Japan, for example, now buys 18 percent of Australia's exports (Korea and China together buy 15 percent), and almost all of those purchases are raw materials. A third factor that has discouraged the development of manufacturing is the restrictive import policies of Japan and other Asian countries, which mean that Australia cannot sell many of its own manufactured goods in those markets.

New Zealand has attempted to create more profitable and reliable international markets for its agricultural products by lowering transportation costs and by producing luxury products. Milk is sold as whole milk powder, whey, and casein because these forms are cheap to ship. However, New Zealand also produces finished dairy products such as butter and cheese; and exotic gourmet foods such as venison and the bright green, lemon-banana-tasting kiwifruit have found markets in Europe, Asia, and North and South America. New Zealand is now the world's leading exporter of fine dairy products, and more than 90 percent of its production is sold overseas.

The Pacific islands depend on raw natural resources for their exports, and there is a major difference between their economies and those of Australia and New Zealand: on many of the islands, subsistence lifestyles are still common. On the island of Fiji, for example, part-time subsistence agriculture engages more than 60 percent of the population. But a desire for modern amenities has made it imperative to find methods of generating cash. In Fiji, sugar and other cash crops—copra, cassava, ginger, and rice, along with timber—make up about 17 percent of the economy, down from 22 percent in 1999–2000. Gold mining, however, is still a significant sector of the Fijian economy, and tourism has assumed additional importance. Today, tourism is the primary source of income in many areas of the Pacific, and fisheries and seafood products are second.

Security and Vulnerability in the Pacific

In the tiny islands of the Pacific, economic growth is hindered by small resource bases, small populations, and remote locations. Although many islands have economies based on tourism, mineral resources, and sugar, coconut, and other cash crops, few have any significant manufacturing industries. Hence they must import manufactured goods from Asia. What few Pacific manufacturers there are have to mount aggressive "buy Pacific" appeals to maintain markets for their wares, which almost always carry higher prices than imports.

Many Pacific islands are cushioned to some extent from the stresses of reorientation to Asia and to the global economy by customs that contribute to self-sufficiency. Official income figures (see Table 11.1 on page 580) do not reflect the fact that many households still rely on fishing and subsistence cultivation for much of their food supply, nor do they include income from thriving informal economies and from remittances sent regularly by family members overseas. These resources mean that Pacific Islanders can have a safe and healthy life with relatively little formal income. Home-grown food can be more than adequate and nutritious; islanders who can be self-sufficient while saving extra cash for travel and occasional purchases of manufactured goods are sometimes said to have achieved **subsistence affluence.** If there is poverty, it is related to geographic isolation, which often means lack of access to information and opportunity. But computers and the revolution in global communication networks (satellite media, cell phones, and the Internet), although not yet widely available in the Pacific islands, have the potential to alleviate this isolation for some island groups.

Some places—such as Papua New Guinea, the Solomon Islands, and Tuvalu—find that they are not cushioned by local customs. Once they too had vibrant local cultures, but in comparison to other parts of Oceania, they are now poor and underedu-

These two photos illustrate traditional economic activities in two regions of the world. Can you determine where these activities are taking place? What does it appear each person is doing? What inferences can you draw about the nature of these traditional economic activities and the particular ways in which they interact with local environments? [A, Owen Franken/CORBIS; B, Holly Hapke.]

AT THE LOCAL SCALE Native Hawaiian Rights versus Residential Tourism

Although tourism is important to Pacific economies, it can threaten the local people's way of life. An example is provided by a recent effort to build a retirement home for U.S. mainlanders near Honolulu, Hawaii. In the early 1990s, the national officials of a major U.S. Protestant denomination voted to build a retirement home for their church members on the island of Oahu. The idea was to acquire land and build a multilevel care facility to which church members from the mainland could retire, living independently until they needed nursing home care. This type of retirement relocation to sunny climes is often called "residential tourism." The church officials proceeded to look for affordable land close to Honolulu, yet with landscapes of rural tropical beauty. They found a suitable tract in Pauoa Valley, one of the last valleys near Honolulu where rural people of Native Hawaiian origins still live in extended family compounds and grow their traditional gardens.

Were the Native Hawaiians the only inhabitants of Pauoa Valley, their solidarity and Hawaiian laws would protect the land against sale; but new immigrants from North America and Asia now share the valley, and some were eager to sell. Yet, if the retirement home were built, life would change irreversibly, even for those residents who chose not to sell. The ecology and ambience of the valley would be transformed by the introduction of a large complex with 141 apartments and a 106-bed nursing home, large concrete parking lots, and manicured grounds. Once the church officials understood the issues raised by the Native Hawaiians,

they decided to build the retirement home in the mainland United States rather than in Pauoa Valley.

Archaeologists Tom Dye and Mac Goodwin assist Pauoa Valley residents to establish the archaeological features of the valley. The documentation of such sites established long-standing occupation of the valley by Native Hawaiian ancestors and helped residents fend off the acquisition of their land by a mainland-based church. [Mac Goodwin.]

cated (see Table 11.1 on page 580), in large part because of the ways in which colonialism affected their societies. They have high birth rates (a fertility rate of 4.1 children per woman) and low literacy rates and life expectancies. Their economic potential is limited to tourism or extractive activities such as mining, export agriculture, fishing, or forestry. Conditions on many of the smaller Pacific islands typify what has been termed a **MIRAB economy**—one based on migration, remittance, aid, and bureaucracy. Many families depend on the remittances sent by family members living and working overseas, and foreign aid from former or present colonial powers supports bureaucracies that supply employment for the educated and semiskilled.

Tourism in Oceania: The Hawaiian Case

Tourism is a growing part of the economy throughout Oceania, but perhaps nowhere in the region are the issues raised by this industry clearer than in Hawaii. Travel and tourism is the largest industry in Hawaii, and in 2003 it produced 22.4 percent of the gross state product (GSP). By comparison, travel and tourism account for 10.4 percent of GDP worldwide. In 2003, this segment of Hawaii's economy provided 171,200 jobs, employed almost 22 per-

cent of Hawaii's workforce, and accounted for U.S. $10.7 billion in output. Tourism also accounted for 24.1 percent of Hawaii's tax receipts.

In the late 1990s, the Hawaiian economy began to falter, partly because of a decline in tourism. The downturn in the Japanese economy was the chief factor contributing to this decline. Although the total number of visitors to Hawaii increased in the early 1990s, the point of origin of these visitors had been shifting from North America to Asia, particularly to Japan, the Philippines, Indonesia, and South Korea. By 1995, 40.3 percent of all visitors to Hawaii came from Asia, mostly from Japan. Thus the dramatic slump in Asian economies in the late 1990s had a major effect on the economy of Hawaii. The decline in tourism hurt not only the tourist industry (travel companies, airlines, hotels, restaurants, tour companies, and so forth), but also the construction industry, which had been thriving on the building of office towers, condominiums, hotels, and resort and retirement facilities. By December 2000, arrivals of tourists had returned to record levels because of a surge in Japanese visitors, and arrivals from the U.S mainland were up 5.6 percent. But by the summer of 2001, arrivals were down again, because of a continuing recession in Japan and a slowdown in the United States. The terrorist attacks of September 11, 2001, sharply affected Hawaii's economy. During the fall of 2001, hotel occupancy ran at 50 percent of expected rates, and the Hawaiian economy contracted by an estimated 1.5 percent for the year. Although the industry had recovered by June 2002, these fluctuations illustrate the vulnerability of tourism to economic downturns and political events.

Many other islands in Oceania also depend on tourism for a significant part of their income, so global economic ups and downs, especially in Asia, cause declines and surges in their economies, too. Guam, for example, depends on middle- and lower-class Japanese and South Koreans—laborers, clerks, students, midlevel managers, and assembly-line workers—for its tourism clientele. More than 1 million tourists visit Guam annually, of which 90 percent are from Japan and a large proportion of the remainder are from South Korea. During the recession in Asia in the late 1990s, several hundred thousand Japanese and South Koreans cut the yearly Guam vacation from their budgets—a loss to Guam of several hundred million dollars per year.

New Asian Orientations

The increasing influence of Asia in Oceania is seen most clearly in the realm of economics. Throughout the Pacific, Asians dominate the tourist trade, both as tourists and as investors, making a vital contribution to Pacific economies and affecting land use significantly. For example, an important segment of the Honolulu tourist infrastructure—hotels, golf courses, specialty shopping centers, import shops, nightclubs—is geared to visitors from Japan, and many such facilities are actually owned by Japanese investors (although Japan's economic downturn in the 1990s forced some to sell their holdings). In addition, increasing numbers of Asians are taking up residence in the region. In Hawaii, for example, the strongest force favoring reorientation toward Asia in the future may be the simple fact that Asians, many whose families have lived there for generations, now make up 58 percent of the population.

The rise in Asian influence is also seen in the increasing exports of raw materials to Asia (Figure 11.11). In the Pacific, coconut, forest, and fish products sell mostly to Asian markets; increasingly, Asian companies control these industries. Fishing fleets from Asia are able to purchase licenses, and they regularly ply the offshore waters of Pacific island nations. The products of Australian and New Zealand farms—wool, dairy products, meat, and hides—once found a ready market in Britain, but now supply Asian consumers and factories. Australian coal and iron supply the rapidly growing Asian manufacturing industries.

Asia buys 60 percent of Australia's exports and 35 percent of New Zealand's. Increasingly, the islands of the Pacific sell to Japan and Southeast Asia and buy much of what they need from Japan, South Korea, Taiwan, Singapore, Malaysia, Thailand, and Indonesia. Although Asia's rising influence in Oceania is associated with growing prosperity, the region is increasingly vulnerable to economic downturns in Asia.

The Stresses of Reorientation to Asia. The shift toward greater trade with Asia has posed challenges for Oceania. Throughout the region, local industries that used to enjoy protected trade with Europe have lost that advantage because new European Union regulations stemming from membership in the World Trade Organization prohibit such preferential trade agreements. Now these industries face competition from much larger firms in Asia.

Australia and New Zealand are experiencing considerable stress as a result of reorientation to Asia. Their export sectors—mainly wool, cattle, dairy products, coal, and metals—have maintained a competitive edge largely through mechanization; thus they employ only a small proportion of the population. Most people work in other sectors of the economy, such as services or manufacturing. In the past, the manufacturing and service sectors grew, despite relatively small domestic demand, partly because they benefited from government subsidies, tariffs, and other barriers to foreign investment. Today, as trade barriers fall worldwide and as Asian manufacturing industries continue to become more competitive, Australia and New Zealand are under increasing pressure to reform their economies by reducing the numbers of employees and streamlining operations, while also cutting social services and eliminating subsidies and tariffs.

Reform measures have been especially traumatic for the Australian and New Zealand labor movements, which historically have been among the world's strongest. Australian coal miners' unions successfully agitated for the world's first 35-hour workweek. Other labor unions won a minimum wage, pensions, and aid to families with children long before such programs were enacted in many other industrialized countries. For decades, these arrangements were highly successful: both Australia and New Zealand enjoyed living standards comparable to those in North America, but with a more egalitarian distribution of income. Since the 1970s, however, competition from Asia has meant that increasing numbers of workers have lost jobs and many hard-won benefits. In Australia, unemployment has risen from 2 percent in the 1960s to between 8.5 and 9.5 percent in the 1990s; it hovered around 6 percent by 2002. Meanwhile, previously high rates of social spending have been cut across the board to address mounting government deficits. The loss of social support, especially for those who have lost jobs, has contributed to rising income disparity in recent years. Nonetheless, Australia had about half the poverty rate of the United States throughout the 1990s.

Social tensions related to the reorientation toward Asia are especially pronounced in New Zealand. Because of its small size and limited resource base, New Zealand has always found it harder to compete in world markets than has Australia. For decades, it managed well by selling its major exports of wool, meat, and dairy products to the United Kingdom on preferential trading terms. When the United Kingdom joined the European Common Market in 1973, however, the other members insisted that it curtail its protected agricultural imports from New Zealand, which France felt competed unfairly with its own agricultural products. The loss of the U.K. market sent New Zealand's economy into decline. By the mid-1980s, the government was forced to cut agricultural subsidies, funding for the welfare state, and the government payroll. At the same time, New Zealand's whites-only immigration policies were dropped and the doors were opened to Asian immigrants, especially to wealthy professionals, who, it was hoped, would reinvigorate the economy with their savings and invest-

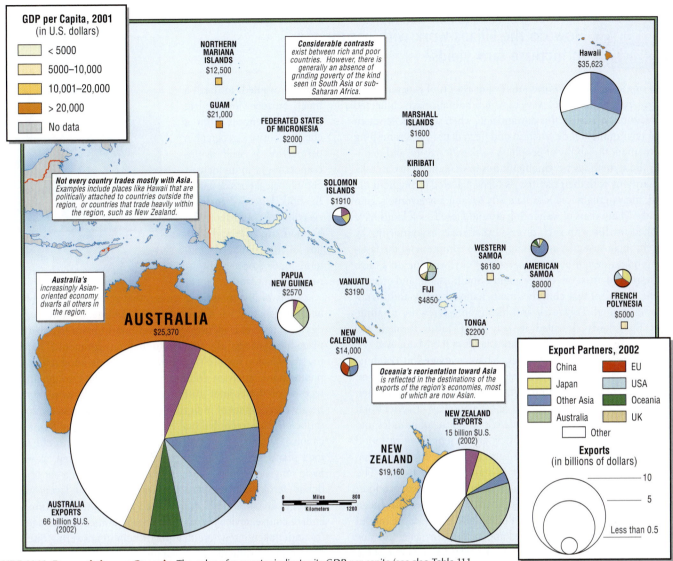

GDP per Capita, 2001
(in U.S. dollars)

- < 5000
- 5000–10,000
- 10,001–20,000
- > 20,000
- No data

Considerable contrasts exist between rich and poor countries. However, there is generally an absence of grinding poverty of the kind seen in South Asia or sub-Saharan Africa.

NORTHERN MARIANA ISLANDS $12,500

GUAM $21,000

FEDERATED STATES OF MICRONESIA $2000

MARSHALL ISLANDS $1600

KIRIBATI $800

Hawaii $35,623

Not every country trades mostly with Asia. Examples include places like Hawaii that are politically attached to countries outside the region, or countries that trade heavily within the region, such as New Zealand.

SOLOMON ISLANDS $1910

Australia's increasingly Asian-oriented economy dwarfs all others in the region.

PAPUA NEW GUINEA $2570

VANUATU $3190

FIJI $4850

WESTERN SAMOA $6180

AMERICAN SAMOA $8000

FRENCH POLYNESIA $5000

AUSTRALIA $25,370

NEW CALEDONIA $14,000

TONGA $2200

Oceania's reorientation toward Asia is reflected in the destinations of the exports of the region's economies, most of which are now Asian.

AUSTRALIA EXPORTS 66 billion $U.S. (2002)

NEW ZEALAND $19,160

NEW ZEALAND EXPORTS 15 billion $U.S. (2002)

Export Partners, 2002

- China
- Japan
- Other Asia
- Australia
- EU
- USA
- Oceania
- UK
- Other

Exports (in billions of dollars)
- 10
- 5
- Less than 0.5

Miles 800
Kilometers 1200

FIGURE 11.11 Economic issues: Oceania. The color of a country indicates its GDP per capita (see also Table 11.1, page 580), and the colors of the pie chart indicate its export partners. The white sections of the pie charts ("Other") can include trade with Canada, Mexico, the Caribbean, the European Union, sub-Saharan Africa, and other locales, some of them new trade partners.

ments. The new immigration policy has resulted in considerable social tension, as some longtime residents of New Zealand resent the presence of wealthy newcomers at a time when they themselves are jobless or strapped for cash. According to the United Nations, income disparity in New Zealand during the 1990s was roughly twice that in Australia.

The Future: A Mixed Asian and European Orientation? Despite the powerful forces pushing Oceania toward Asia, important factors still favor a strong Western influence. The economic recession that swept through Asia in the late 1990s emphasized the need for Oceania to maintain broad contacts with economies outside Asia, especially in Europe and the United States, that seem to be more stable. Another factor is the lingering fear of Chinese aggression, justified to some extent by expansionist moves that China has made in the South China Sea toward Taiwan and the potentially oil-rich Spratly Islands. Despite a recent move by China to expand diplomatic and cultural relations with Australia, and despite increasing trade links between the two countries, it is

likely that both Australia and New Zealand will retain a cultural affinity with the West for generations to come.

In the Pacific, several islands, especially those in Micronesia, may be strongly drawn into the Asian sphere. But lingering vestiges of colonialism will prevent others from straying very far from a Western orientation. Both the United States and France maintain possessions in Polynesia and Micronesia, and any desire for independence in these possessions has not been sufficient to override the financial benefits of aid, subsidies, and investment money from the former colonial powers. Hence, as Oceania is being steadily drawn closer to Asia, strong ties to the United States and Europe remain.

ENVIRONMENTAL ISSUES

A case could be made that there is more public awareness of environmental issues in Oceania than in any other world region. Despite its relatively small population, Oceania faces many environmental

PRACTICING GEOGRAPHY

How do the Fijians work with nature to increase taro yields?

For me, one of the more interesting aspects of geography has always been the variety of ways in which humans have enhanced their natural setting. In this day and age, when we learn so much about environmental destruction and degradation, it is refreshing to remember the ability of people to work *with* nature, and to discover those methods by which cultures have come to live in relative harmony with their habitat. As I have traveled throughout the world, the vast realm of Oceania easily became my favorite region to study. Many years of wandering around these far-off lands have enriched my life with such diverse experiences as journeying by train through Australia, and taking time to comprehend the fascinating urban design of its young capital city; working on a farm on the rugged coast of New Zealand's South Island, and "tramping the tracks" of that country's many national parks; and island-hopping through Polynesia, and hiring on as crew for a sailing yacht that slowly threaded its way from Tahiti through the Tuamotu Archipelago to the distant islands of the Marquesas.

As a professional geographer, I have managed to combine these regional and topical interests with a very practical focus on agriculture, beginning with my Ph.D. dissertation research, when I became fascinated with the art and technology of agricultural terracing on Pacific islands. The careful sculpting of hillsides into a series of level garden platforms is an ancient practice found throughout many regions of the earth, from the Mediterranean across the mountainous areas of Asia and into the Pacific. The technique perhaps reached its fullest elaboration in two locations: in the Andes of what is now southern Peru under the Incan empire, and in the islands of the Philippines, where entire mountains have been intricately carved and carefully irrigated for rice production. Terracing has been viewed as a visible outcome of the process of **agricultural intensification**, whereby additional inputs of labor to a given area of land typically lead to an increase in crop yields. This process can occur for a number of reasons, but usually implies population pressure on resources. Terracing can increase the natural carrying capacity of the land in a way that is sustainable, conserving both soil and water required for food production. Scholars have studied these agrosystems in many places, but there was a gap in our knowledge regarding terraces in the Fiji Islands, where they were once widely used for growing the dietary staple taro, a starchy tuber, so that is where I decided to conduct my research.

Unlike other studies of the subject, which largely emphasize the archaeological aspects of single instances of abandoned sys-tems, I wanted to take a much broader view, and to investigate terraces from a comprehensive geographic perspective as well as incorporating a more focused chronological consideration. In short, I needed to learn everything I could about the location, structure, and function of these artificial landforms once so prevalent on many Fijian islands.

Robert Khulken
*Department of Geography,
Central Washington University*

One of the reasons I chose to become a geographer is because geography is a synthesizing discipline, bringing together disparate lines of evidence to discover how humans in a particular place have successfully adapted to their environment. Therefore, my study had to involve the following methods: conducting aerial photographic analysis; drafting maps; searching through archival records of the former British colonial officials who regulated nearly every aspect of indigenous life; seeking permission to transcribe officially recorded oral histories regarding land tenure allocations; living in remote villages while learning the rudiments of food plants, cropping techniques, and garden design, as well as basic language skills; excavating trenches through a set of abandoned terraces to diagram the constructed soil layers and to obtain charcoal samples that allowed me to date their formation; and, of course, reviewing any literature documenting terrace studies undertaken elsewhere that might help in answering questions I had raised during the course of my own work.

Finally, I was most fortunate to have the unique opportunity to document a fully functional system of irrigated taro terraces still operating on the remote southern island of Kadavu. I first asked permission of local villagers to make a video documentary of their gardens, a project in which they enthusiastically agreed to participate. This videotape, which was narrated in the standard dialect, is now being used by the Ministry of Agriculture to educate Fijians from other areas of the country in the methods of constructing and maintaining irrigated terraces for growing taro. This is one small example of how basic geographic research can contribute to the preservation of traditional cultural-ecological knowledge and may benefit a small island nation struggling to achieve greater self-sufficiency and security in domestic food production.

---

problems (Figure 11.12), and they are often discussed in public forums such as the press, exhibits, and the broadcast media. The human introduction of many nonnative species has ravaged the region's unique ecosystems, as has the expansion of the human population itself. Even remote areas have been subjected to intense pollution from mining and the testing of nuclear weapons, especially during the cold war era. Finally, global environmental crises related to climate change and ozone depletion are particularly threatening to this region, and urbanization is creating new pressures, especially in the Pacific islands (see the box "Urbanization in Island Oceania" on page 563).

Despite the increasingly severe environmental problems facing the Pacific islands, there are a few bright spots on the horizon. In many areas, a revived pride in cultural heritage has encouraged

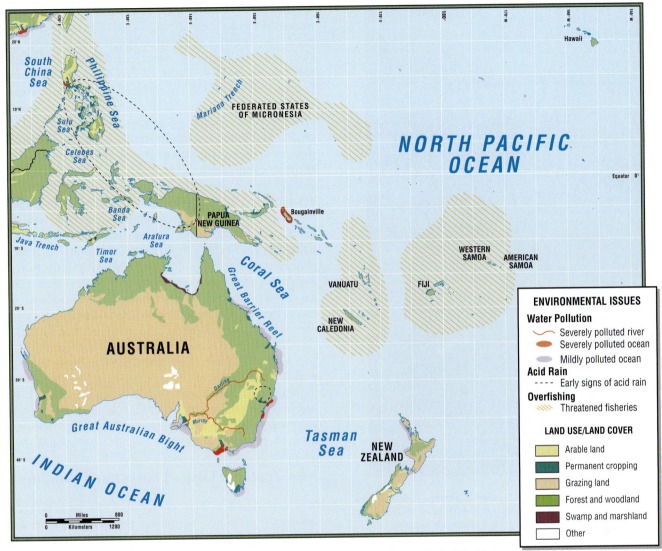

FIGURE 11.12 Environmental issues: Oceania. In Oceania, environmental issues are often the center of public debate. Deforestation and losses of species diversity are common issues. Overfishing is widely recognized as a growing problem, often because of abuse by fishing fleets that come from outside the region. Air pollution is locally a problem in Australia and New Zealand, but overall is less problematic than in neighboring Southeast Asia.

Pacific Islanders and other indigenous peoples to reestablish more sustainable lifestyles and to become active in Green movements, especially as cultural allegiances shift away from distant Europe. Because environmental conditions vary around the region, we will address Oceania's environmental issues by subregion.

Australia: Human Settlement in an Arid Land

Unique endemic organisms occupy many ecological niches in Australia's unusually arid environments (see page 555). Since settlement by Europeans, nonnative species have displaced or even driven to extinction many of Australia's native plants and animals. At least 41 bird and mammal species and more than 100 plant species have become extinct. In all likelihood, many more disappeared before they were biologically classified.

European rabbits are among the most destructive of Australia's introduced species, and their story illustrates the complex problems these species can cause. Rabbits were brought to Australia by early British settlers who enjoyed hunting and eating them. With no natural predators in Australia, the rabbits multiplied quickly, browsing on many local plants and destroying the food supply for indigenous species. The holes they dug to live in created considerable soil erosion. Attempts to control the rabbit population by introducing European foxes and cats only increased the threat to native species. The vigorous European predators drove many native predator species to extinction without having much effect on the rabbit population. In recent years, intentionally introduced diseases have destroyed as much as 90 percent of the target rabbit population. Yet, without rabbits to eat, the cats and foxes now prey more heavily on indigenous species. Hence the eradication of feral cats and foxes has been stepped up as well.

The dingo fence. The world's longest fence snakes 3200 miles (5800 kilometers) between Yalata in South Australia and Jandowae, near the coast of Queensland. Made of tough wire attached to wooden posts, it attempts to separate the dingo, Australia's indigenous dog, from sheep herds. A single dingo can kill as many as 50 sheep in one night of marauding. The fence requires constant patrolling and mending, but it has reduced the dingoes' killing of sheep dramatically. Kangaroos, the dingoes' natural prey, have learned to live on the sheep side of the fence, where their population has boomed beyond sustainable levels. Kangaroos compete with sheep for the same scarce grass and water. [Medford Taylor/National Geographic Image Collection.]

Moreover, some rabbits have proved resistant to disease, and the rabbit population may rebound.

Agriculture has also had a huge impact on Australian ecosystems. Because the climate is arid and the soils are infertile, the dominant land use in Australia is the grazing of introduced species—primarily sheep, but also cattle. More than 15 percent of the land is given over to grazing; increasingly, ranchers are using irrigation, herbicides, and pesticides to extend the seasonal availability of pastureland. Crop agriculture is limited to just 6 percent of the land, most of it in the southeast. Crops, too, now rely on irrigation and chemicals. One result is that natural underground water reserves are being depleted. Another is that many rivers and lakes have become polluted by agricultural chemicals and the algae that feed on them.

In a place where the overwhelming majority of the people live in cities, urban activities are the most likely to create environmental problems. Australia's major cities form a series of nodes along the eastern and southeastern coasts from Brisbane to Adelaide. The nodes themselves are of fairly low density, and they are connected by dispersed towns and coastal developments. As a result, commuters typically travel long distances daily and depend primarily on private cars for transport (although urban streetcars and light rail systems and interurban commuter lines are more common than they are in the United States). The dependence on cars has a variety of environmental impacts. Streets and parking lots not only consume space but also seal the soil surface. Paving thus reduces the amount of rainfall that the ground can absorb, decreasing the retention of scarce rainfall and increasing flooding and erosion. Because of the great distances traveled, more automobile fuel is used in Australia than elsewhere in this region, and because of inefficiencies, per capita energy consumption is higher in Australia than in most western European countries. Consequently, per capita greenhouse gas emissions are higher than in any other industrialized countries except the United States, Luxembourg, and Singapore, and photochemical smog and chemical haze are common in the major cities.

Water scarcity is another problem associated with large urban populations. Domestic per capita water consumption has risen markedly over the last 30 years, due mostly to lifestyle changes: more flushing toilets, more washing machines, more showers, more irrigation of ornamental plants and crops. Australians have had to build giant water catchment and storage facilities, which have displaced natural habitats and diverted the natural flow of streams.

New Zealand: Loss of Forest and Wildlife

New Zealand shares many of Australia's environmental problems: loss of endemic plants and animals, pollution, and water shortages. There are some differences, however, which result from New Zealand's different set of physical and human features. There were no human beings in New Zealand until about 1000 years ago, when Polynesian Maori people settled there. When they arrived, dense midlatitude rain forest covered 85 percent of the land. The Maori were cultivators who brought in yams and taro, as well as other nonindigenous plants and birds. By the time of European contact, forest clearing and overhunting by the Maori had already degraded the environment and driven several species of moa birds to extinction. Archaeological evidence also indicates that intense competition for increasingly scarce resources led to warfare.

European settlement in New Zealand drastically intensified land use well beyond Maori levels (Figure 11.13). The settlers made concerted efforts to re-create the British open landscape and farming system in their new land. Today, just 23 percent of the country remains forested. The Europeans used most of the land they cleared for export agriculture: pastures and cropland. Grazing has become so widespread that today there are 15 times as many sheep as people, and 3 times as many cattle. The Europeans also brought many new crops and inadvertently introduced damaging pests. Soils exposed by the clearing of forests proved thin and acidic, and ranchers buttressed them with chemicals to maintain pastures. As in Australia, chemicals have polluted waterways. New

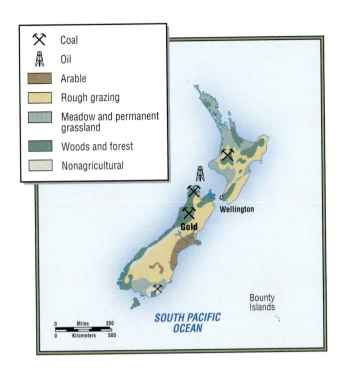

FIGURE 11.13 **Land use and natural resources of New Zealand.** As a result of European settlement and the clearing of land for farming, only 23 percent of New Zealand remains forested. [Adapted from Richard Nile and Christian Clerk, *Cultural Atlas of Australia, New Zealand, and the South Pacific* (New York: Facts on File, 1996), p. 194.]

Zealand's environment has actually become hostile to many native species. Ranches, farms, roads, cities, and towns claim more than 90 percent of the lowland area, resulting in unusually high extinction rates for endemic plants and animals. Hydroelectric schemes have disrupted the natural flow of rivers, particularly on South Island.

The Pacific Islands: At Risk from Global Trends

The Pacific islands, though widely dispersed, share a number of environmental concerns and opportunities. Their small sizes, often tiny populations, and insignificant influence on the global economy give them little say about international forces that nonetheless affect their environments. Mining, nuclear pollution, ozone depletion, global warming, economic globalization, and increasing tourism all have environmental effects on the Pacific islands.

Resource Extraction and Pollution. Pollution from mining has disrupted traditional lifeways in Papua New Guinea. Two mines in particular have had devastating effects on the country's environment. Both were operated by Australian mining companies that took advantage of poorly enforced or nonexistent environmental codes to dump large amounts of mine waste into New Guinea's rivers. As a consequence, entire river systems have been devastated and indigenous subsistence cultivators displaced.

In the case of the Ok Tedi mine on mainland Papua New Guinea, 30,000 local villagers sued the Australian parent mining company, BHP, for U.S. $4 billion. Two villagers, Rex Dagi and

Alex Maun, traveled to Europe and America to explain their cause and meet with international environmental groups. They and their supporters convinced American and German partners in the Ok Tedi mine to divest their shares. The international pressure also resulted in objections to BHP's plans to mine diamonds in Canada's Northwest Territories. The parties reached an out-of-court settlement that sets several important precedents: (1) the plaintiffs won U.S. $125 million in trust funds for damage mitigation; (2) the plaintiffs won a 10 percent interest in the mine; and (3) the settlement provides that any further disputes will be heard in Australian rather than New Guinea courts, making it less likely that the mining company can manipulate the justice system.

The second case involved a copper mine on the island of Bougainville (administered as part of Papua New Guinea, but lying in the Solomon Islands; see Figure 11.2, page 552). Soon after an Australian-owned company began mining in Bougainville in the 1960s, on land deforested and stripped for the purpose, the erosion of mine waste into rivers poisoned downstream cultivation plots, forestlands, and fisheries. People who lost the resources on which they had subsisted were forced into the cash economy, many of them fleeing to new mining market towns where their subsistence skills were of little use (see the photo of a similar situation below). By the 1990s, protests against the environmental and social damage done by the copper mine had developed into a war for the independence of Bougainville Island from Papua New Guinea, in which 1 in 8 Bougainvilleans (some 20,000 people) lost their lives. Australia and New Zealand aided the Papua New Guinea government by supplying helicopters and munitions to use against the rebels, until public pressure in Australia put a stop

Supermarkets, such as this one in Tari in the Papua New Guinea highlands, are rearranging the ways in which people have traditionally obtained food and other goods. Among the customers are forest dwellers displaced by mining operations. The stores are doing away with long-established personal trading networks and are rapidly bringing the impersonal global economy into one of the last frontiers on the planet. Pacific people are happy to have access to the goods, though many worry about the changes that result. [Sassoon/Robert Harding Picture Library, London.]

to this support. Then the Papua New Guinea government hired Executive Systems, a mercenary group (soldiers for hire) of white South Africans, to suppress the secessionist movement. This inflammatory action convinced the Australian government to support the closing of the mine and to supply peacekeeping forces. By January 2001, peace negotiations were under way, with Bougainville seeking to become an autonomous unit within Papua New Guinea with its own judiciary and police force.

Another major environmental issue for the Pacific islands is radioactive pollution from nuclear weapons tests and reactor waste. During World War II, U.S. nuclear bomb experiments destroyed Bikini atoll in the Marshall Islands and caused cancer among the islanders. Similarly, Mururoa atoll in French Polynesia has been the site of 180 nuclear weapons tests and the recipient of numerous shipments of nuclear waste from France. The result has been widespread cancer, infertility, birth defects, and miscarriages among the native populations of nearby islands. One response to this pollution was the establishment of the South Pacific Nuclear Free Zone by the Treaty of Rarotonga in 1985. Most independent countries in Oceania have signed this agreement, which bans weapons testing and waste dumping on their lands. As a result of political pressure from France and the United States, French Polynesia and U.S. territories such as the Marshall Islands have not signed the agreement. Waste dumping and weapons testing continue. Japan, North Korea, South Korea, France, and the United States have all explored the possibility of depositing their radioactive waste in the Marshall Islands. Since the 1990s, Pacific leaders have coordinated their efforts against nuclear testing and waste dumping through the United Nations Working Group on Indigenous Populations (UNWGIP).

The United Nations Convention on the Law of the Sea. The 1994 UN Convention on the Law of the Sea (called the Law of the Sea Treaty, for short) establishes rules governing all uses of the world's oceans and seas. (The treaty has been signed by 157 countries; the United States is not one of them.) There is no overarching enforcement agency, however. The treaty, based on the sound idea that all problems of the oceans are interrelated and need to be addressed as a whole, is an example of how the globalization of Pacific island economies has thwarted environmental protection for the islands. The treaty allows islands to claim rights to ocean resources 200 miles (320 kilometers) out from the shore, and island countries can now make money by licensing privately owned foreign fishing fleets from Japan, South Korea, Russia, and the United States to fish within these offshore limits. However, protecting the fisheries from overfishing by these rich and powerful licensees has turned out to be an enforcement nightmare for tiny island governments with few resources. Similarly, it has proved difficult to monitor and control the exploitation of seafloor mineral deposits by foreign companies. Even mining operations conducted legally outside the 200-mile limit have the potential to pollute fisheries and other ocean resources.

Multiple Effects of Tourism. Even tourism, which until recently was considered a "clean" industry, has now been shown to create environmental problems. Foreign-owned tourism enterprises have often accelerated the loss of wetlands and worsened beach erosion. Tourism increases the use of scarce water resources, the production of sewage, and the consumption of environmentally polluting products such as gasoline, kerosene, fertilizers, plastics, and paper. The widely promoted concepts of ecotourism and cultural tourism, discussed in other chapters, are now common elements of development in the Pacific, but they have quickly become commercialized by mass tourism. All tourist ventures, no matter how ecologically aware, have some impact on the local environment, and the invasive presence of visitors invariably makes demands on local cultures.

Global Warming and Ozone Depletion. Global warming, which may raise sea levels by melting land ice and thermally expanding ocean water (see Chapter 1, page 46), is of obvious concern to islands that already barely rise above the waves (Figure 11.14). If sea levels rise the 4 inches (10 centimeters) per decade predicted by the International Panel on Climate Change, many of the lowest-lying atolls will simply disappear under water. Other islands, some with already very crowded coastal zones, will be severely reduced in area and will become more vulnerable to storm surges and cyclones. (The highest point on the densely occupied island of Nauru, for example, is 200 feet, or 61 meters.) The peoples of the Pacific have little direct control over the forces that appear to be producing these precarious circumstances because they themselves produce only a very small percentage of the earth's greenhouse gases. The Association of Small Island States (AOSIS), whose membership is largely from Oceania, is an important organization that was involved in the international climate change negotiations that resulted in the Kyoto Protocol (see Chapter 1, page 47).

People in the whole of Oceania are particularly at risk from depletion of the ozone layer. Ozone is a type of oxygen molecule that is normally heavily concentrated in a layer of the upper atmosphere. This ozone layer filters out biologically harmful ultraviolet radiation emitted by the sun. Since the mid-1980s, the ozone layer has been thinning, and a hole in the layer has appeared periodically over Antarctica. The cause is almost certainly a buildup of a class of manufactured chemicals called chlorofluorocarbons (CFCs), which float up to the ozone layer and destroy it. Increasing amounts of ultraviolet radiation reaching the surface of the earth through the damaged ozone layer are likely to increase the prevalence of skin cancer. Australia already has the highest incidence of skin cancer in the world, which is explained by the fact that its largely white, transplanted population has little protective skin pigment (melanin) yet lives under intense sunlight. Higher levels of ultraviolet radiation are also likely to increase the incidence of eye disease and weaken the immune systems of a variety of organisms, including humans. Increased ultraviolet radiation could also decrease agricultural and marine productivity, and it is one of Australia's most significant natural hazards.

MEASURES OF HUMAN WELL-BEING

The overall status of human well-being in Oceania is probably higher than the statistics indicate, especially for the Pacific islands, because subsistence agriculture and reciprocal exchange remain important in the everyday economy. Statistics for Australia, New Zealand, and several of the larger islands of the Pacific are easy to obtain, but data for many of the smaller island groups is missing, or

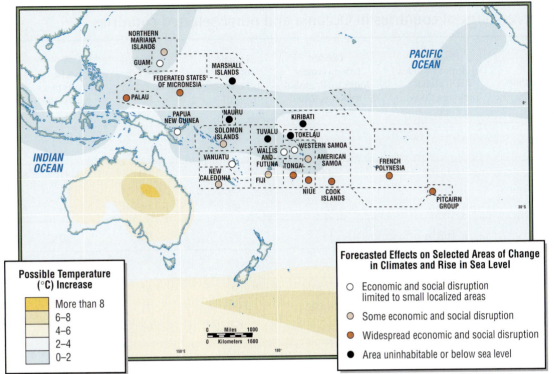

FIGURE 11.14 Global warming as it may affect Oceania. Scientists studying global warming have forecast its effects on land temperatures and sea levels. Should these forecasts materialize, how will living conditions in various parts of Oceania change? [Adapted from Richard Nile and Christian Clerk, *Cultural Atlas of Australia, New Zealand, and the South Pacific* (New York: Facts on File, 1996), p. 223.]

only partial data is available. Keep in mind that the sizes and populations of many of the island states are very small.

As discussed throughout the text, gross domestic product per capita (Table 11.1, column 2) is at best a crude indicator of well-being. Oceania, like most other regions, has a wide variation in GDP per capita. Australia and New Zealand provide their citizens with a high standard of living that is further enhanced by publicly financed social programs such as health care and subsidized housing. The high standard of living in these societies is demonstrated by their long life expectancies and high literacy rates. Further evidence of their high living standards is the fact that 67 percent of the dwellings in Australia are owned by their occupants. Generally, the Pacific islands have relatively low GDP figures. Still, even the lowest figures of U.S. $800–$1000 probably support an adequate standard of living for most people. The GDP statistics do not reflect income from the informal economy, remittances from family members overseas, or the fact that many households practice subsistence fishing and agriculture.

The United Nations Human Development Index (HDI) (see Table 11.1, column 3) is not particularly useful for Oceania. Notably, Australia and New Zealand rank high—fourth and twentieth, respectively. Hawaii is not ranked because it is part of the United States. If it were ranked, Hawaii would have a higher score than the United States as a whole (which is ranked seventh) because it ranks near the top of the U.S. states in life expectancy, educational attainment, and income. Fiji's rank dropped from 66 in 2000 to 81 in 2003, and all other places ranked by the United Nations are in the lower range of the medium category on the HDI index. These rankings are partly the result of the difficulty of providing services such as medical care and advanced training to very small, remote populations. Changing diets (more sugar and fat) and lifestyles (more sedentary) have provoked an increase in health problems such as diabetes and heart disease, resulting in significantly lower life expectancies. Specialized, high-tech health-care facilities are generally not to be found outside New Zealand, Australia, and Hawaii.

The United Nations Gender Empowerment Measure (GEM) rankings (see Table 11.1, column 4) are also missing for most of the Pacific islands, but, as we have observed, women participate actively in island life and enjoy increasing prestige as they age. New Zealand's high rank (12) results partly from the high percentage of women in parliament (29 percent of the seats) and from the fact that more than one-third of administrators and managers, and more than half the professional and technical workers, are women. In 1997, New Zealand acquired its first woman prime minister when the transportation minister, Jenny Shipley, became head of the ruling coalition. In 1999, Helen Clark was elected prime minister after 25 years of leadership in parliament as a member of the Labour (liberal) party.

Literacy rates are high throughout Oceania, with the exception of Papua New Guinea and the Solomon Islands in Melanesia. This statistic is particularly meaningful because it indicates that most of the citizenry is equipped to participate in democratic processes and in public debate on regional issues.

Quick Review

1. Does the subsistence affluence lifestyle support the idea of the Pacific Way? How?

2. In the late 1990s, how did the economic slump in Japan come to affect Hawaii and other Pacific islands?

3. Describe an example of the environmental havoc created by nonnative species in Australia.

4. Discuss how decisions and actions originating far from Oceania nevertheless influence environmental conditions on Pacific islands.

TABLE 11.1 Human well-being rankings of countries in Oceania and other selected countries

Country (1)	GDP per capita, adjusted for PPP[a] in 2001 $U.S. (2)	Human Development Index (HDI) global rankings, 2003[b] (3)	Gender Empowerment Index (GDI) global rankings, 2003 (4)	Female literacy (percentage), 2001 (5)	Male literacy (percentage), 2001 (6)	Life expectancy (years), 2001 (7)
Selected countries for comparison						
Japan	25,130	9 (high)	44	99	99	81
United States	34,320	7 (high)	10	99	99	77
United Arab Emirates	20,530	48 (high)	65	80	75	74
Australia	25,370	4 (high)	11	99.0	99.0	80
Melanesia (selected units)						
Fiji	4,850	81 (medium)	—(61 in 2000)	91	95	67
Papua New Guinea	2,570	132 (medium)	—(91 in 1998)	55	71	57
Solomon Islands	1,910	123 (medium)	—	62[c]	62[c]	71
Vanuatu	3,190	128 (medium)	—	—	—	67
New Caledonia (French Overseas Territory)	25,200	—	—	90	90	73
Micronesia (selected units)						
Guam (U.S. dependency)	21,000	—	—	99	99	78
Federated States of Micronesia	2,000	—	—	88	91	68
Kiribati	800	—	—	—	—	62
Marshall Islands	1,600	—	—	94	94	68
Northern Marianas	12,500	—	—	96	97	76
Nauru	5,000	—	—	—	—	61
Palau	9,000	—	—	90	93	69
Polynesia (selected units)						
French Polynesia (French Overseas Territory)	5,000	Shares French rank of 17 (high)	— Shared French rank of 31 in 1998	98	98	72
Hawaii (U.S.)	35,623	Shares U.S. rank of 7 (high)	Shares U.S. rank of 10	99	99	77
Tonga	2,220	—	—	99	99	68
Tuvalu	1,100	—	—	—	—	66
Western Samoa	6,180	70 (medium)	—	99	99	69
American Samoa	8,000	—	—	97	98	76
New Zealand	19,160	20 (high)	12	99.0	99.0	78

[a]PPP = purchasing power parity.
[b]The high and medium designations indicate where the country ranks among the 175 countries classified into three categories (high, medium, low) by the United Nations.
[c]In 1998. Data not available for 2003

Sources: United Nations Human Development Report 2003 (New York: United Nations Development Programme); 2003 World Population Data Sheet (Washington, D.C.: Population Reference Bureau); CIA World Factbook 2003 (Washington, D.C.: Central Intelligence Agency).

REFLECTIONS ON OCEANIA

One way to assess Oceania and its prospects is to think of it in relation to its nearest neighbors. In other chapters, we have noted the economic vigor of East and Southeast Asian economies, the energy and potential they seem to have for future expansion, and the possibility that they will either subvert or pull along other economies around the globe. We have also considered the many challenges these regions face, not the least of which are cronyism, corruption, inattention to environmental issues, and lack of commitment to enhancing democratic processes. Will Oceania's economic relationships with Asia leave it wealthier and more secure—or more precarious economically, environmentally, and socially?

Some people have suggested that Oceania might implement the Pacific Way effectively and, perhaps, teach it to others. These observers are encouraged by Oceania's somewhat superior environmental record, its leadership in the movement to ban nuclear weapons, and the strength of Pacific indigenous cultures. Although the concept of the Pacific Way is only now emerging, it seems to hold considerable promise in the post–cold war era, when many people are in a mood to demilitarize, to find less wasteful lifestyles, and to discover something grander than national chauvinism. The Pacific Way is based on ancient techniques for conflict resolution through cooperation and consensus. It is, in fact, more of an ideal and a legend than a reality, because traditional Pacific cultures were not particularly democratic. Many of the region's multi-island nations are not accustomed to collective decision making for the whole, and the disruptive forces of political factionalism and ethnic conflict have yet to be overcome. But it appears that most parts of Oceania are at a stage where cooperation and consensus may seem desirable. Most Pacific islands are redefining their relationships with former colonial governments and are choosing independence of some sort; at the same time, they recognize that they can hardly go it alone as tiny states.

Meanwhile, Australia and New Zealand are newly aware of their "Pacificness." As New Zealand political scientist Steven Levine has noted: "If in time a genuinely consensual 'Pacific Way' does emerge, it will no doubt be based on the extension of the values of empathy and social harmony found in some island cultures to the ethnically distinct yet native-born 'newcomers' in their midst." Why not, then, become a region with a splendid mission—a place where value is placed on sustainable development, where affluence is defined in nonmaterial ways that focus on the quality of essentials rather than the quantity of nonessentials, where the primary goal is to enhance human and social capital, and where collaboration and common consent are honored?

Chapter Key Terms

Aborigines 559
agricultural intensification 574
atoll 555
Austronesians 559
El Niño 556
endemic 557
Gondwana 554
Great Barrier Reef 554
hierarchical societies 560

high islands 555
hot spots 554
makatea 555
marsupials 557
Melanesia 559
Melanesians 559
Micronesia 559
MIRAB economy 571
monotremes 557

noble savage 559
Pacific Way 567
pidgin 567
Polynesia 559
roaring forties 555
subsistence affluence 570
underclass 561

ANSWERS TO THE QUICK REVIEW

These answers are brief responses to the review questions throughout the book. They are not intended to be definitive or comprehensive; please refer to the pages indicated for a more complete discussion of the topic.

CHAPTER 1

Quick Review I (p. 30)

1. While lines of latitude run parallel to the equator, lines of longitude run from pole to pole. Also, lines of latitude diminish in length as they approach the poles; lines of longitude are all the same length (p. 6).

2. Local, subregional, regional, and global (p. 10)

3. Digital divide (p. 13)

4. Religion, language, material culture, technology, ideas about gender and race (pp. 15–22)

5. Internal processes that drive the movement of plate tectonics are linked to the formation of mountain ranges, volcanoes, and earthquakes, while rivers, floodplains, and deltas are formed by the external processes associated with climate (pp. 22–24).

Quick Review II (p. 50)

1. National governments can impose tariffs, import quotas, and capital controls (p. 34).

2. A population pyramid shows the number of people who live in a particular country or region at a particular time according to age and gender. It can indicate a typical life expectancy, gender imbalances in different age groups, and the proportion of younger and older residents (p. 40).

3. Overgrazing, deforestation, mismanagement of farmland, as well as irrigation (salinization) (p. 44)

4. When forests are cut down, carbon is released from the biomass (whether it is burned or decomposes), and the carbon eventually combines with oxygen to become carbon dioxide. It is the buildup of carbon dioxide in the atmosphere that is responsible for gradually increasing average temperatures (p. 46).

5. The United States is not a nation-state, according to Rousseau's definition, because its citizens are culturally, linguistically, and politically diverse (p. 49).

CHAPTER 2

Quick Review I (p. 67)

1. Glaciers scoured out depressions in the central lowlands (p. 56).

2. The Gulf of Mexico (p. 58)

3. See page 56 for a description.

4. The U.S. population shifted to the South and West in the twentieth century (pp. 64–65).

5. As a result of the 9/11 attacks, the economic recession that had already begun deepened as investors lost confidence and U.S. cities lost business; 79,000 people in New York City lost their jobs; and airlines suffered from lower levels of air travel. See pages 66 and 67 for a discussion of the economic effects of 9/11 and for ways in which the U.S. national identity was redefined.

Quick Review II (p. 95)

1. See pages 68 and 69 for a discussion of U.S. and Canadian asymmetries, similarities, and interdependencies.

2. The goal of NAFTA is to increase trade but not necessarily balance it (p. 76).

3. The United States spends more than Canada per capita on health care (p. 77), but Canada insures all and has lower death rates and longer life expectancies.

4. About 10 million immigrants, mostly from Europe, came to the United States between 1820 and 1880. Immigration rates increased after that as immigrants came from a broader range of European countries and Russia. From 1881 to 1920, around 23 million people immigrated. Rates then tapered off from 1921 to 1960 (only 8 million immigrants in that period), although immigrants started arriving from Canada and Mexico. Since 1960, more than 24 million immigrants have arrived. Some still come from Europe, though the vast majority are now from Asia, Mexico, and Middle and South America (pp. 81–84).

5. Often, rural land is less expensive to develop, builders have fewer prior development projects to contend with, and the land is less polluted (p. 80).

6. Compared with extended families, the nuclear family is more mobile (p. 89).

Quick Review III (p. 110)

1. New England has had abundant fish and timber (p. 96).

2. Population is decreasing in small towns in the Great Plains; soil is eroding 16 times faster than it can be replaced; and irrigation water is being mined from aquifers at unsustainable rates (pp. 102–103).

3. See pages 84–85, 100–101, 104, 110 for a discussion of immigration.

4. Logging (especially clear-cutting) and its effects on the animal and plant life of the forest is a topic of much discussion, as are the effects of hydroelectric dams (pp. 107–108).

5. The influence of French colonization is still evident in Québec and the U.S. South, while Spain left its mark on Florida, California, and the Southwest. German, Slavic, and Scandinavian people were prominent in the settlement of the Great Plains. England may have had the most pervasive influence, as English has become the most common language spoken across the continent, but the English colonies were founded along the Atlantic coast from New England to Georgia (pp. 95–104).

CHAPTER 3

Quick Review 1 (p. 131)

1. During an El Niño year, the Peru Current changes direction, so instead of bringing a cold ocean current and drought to the west coast of South America, it brings warm water and rain. This means poor catches for fishers because warm water is less nutrient-rich than cold water (p. 120).

2. Tropical rain forests grow in the *tierra caliente*, the hot lowlands, as do domesticated crops like bananas, pineapples, and cacao. The slightly higher altitude *tierra templada* (between 3000 and 6500 feet) has a springlike climate year-round, and is suitable for crops like corn, beans, wheat, and coffee. *Tierra fria* (between 6500 and 12,000 feet) is cooler yet, and suitable for midlatitude crops like root vegetables and cool-weather vegetables like cabbage and broccoli. In the *tierra helada*, the coldest zone in the highest altitudes (above 12,000 feet), vegetation gives way to glaciers and snow-topped mountains (pp. 118–119).

3. Until recently, most people in the region lived in agricultural areas, where children were a source of wealth because they provided labor. Also, the high infant mortality rate led parents to have many children, the Catholic Church discouraged family planning, and ideas of gender in the region encouraged both men and women to have large families (p. 127).

4. See pages 128–131 for a discussion of migration.

Quick Review II (p. 156)

1. Countries that borrowed large amounts of money from international banks in order to finance their industrialization were unable to repay the money when prices for their raw material exports dropped in the global market. To get out of this "debt crisis," they adopted structural adjustment policies recommended by the IMF. They sold government-owned industries, removed tariffs on imported goods, encouraged free-market activity, and cut back on government spending and essential services to free up money for debt payments (pp. 135–137).

2. Citizens of countries that funnel money into debt payments instead of education or health care may be poorly equipped to take jobs that require skills. For poor families, it may be necessary for children to work instead of attend school, making it more difficult for them to find decent-paying jobs when they are older (pp. 137–138).

3. Assuming that the digital divide can be overcome, the Internet may be able to make education and entrepreneurial opportunities more available to all parts of society, two goals that can strengthen democracy (pp. 144–145).

4. Acculturation is the borrowing of different and diverse cultural traits; assimilation entails the loss of one's own cultural practices while learning those of others (p. 146).

5. Deforestation in the Amazon contributes significantly to global warming by reducing carbon dioxide absorption and the amount of free oxygen released by forest plants (p. 151).

6. Standard of living depends on factors that are not reflected in the GDP, such as the availability of health care and education and the equality of wealth distribution. Participation in the informal economy and social support customs can also improve quality of life, additional factors that are not reflected in the GDP (pp. 154–156).

Quick Review III (p. 175)

1. While Haiti is still undeveloped and rural, Barbados has developed diverse industries. The literacy rate is much higher in Barbados, and there are employment opportunities in Barbados for skilled workers. Cuba is an independent country with a socialist (communist) regime, whereas Puerto Rico is a capitalist commonwealth within the United States. Since the demise of the Soviet Union in 1991, Cuba's economy has been in decline while Puerto Rico's economy, with U.S. support, has been more stable. By the mid-2000s, both islands showed signs of stagnation (pp. 158–160).

2. Poorly regulated factories in maquiladora zones contribute to air and groundwater pollution. The factories draw large numbers of low-paid laborers who must live in housing with inadequate water and sewage facilities (pp. 161–162).

3. The United States wanted the contras to destabilize the Sandinistas because they were thought to be communists (p. 166).

4. A communist guerrilla movement in Peru (p. 170)

5. The culture of Middle and South America has been influenced by native cultures and immigrants from Europe, Africa, East Asia, South Asia, and Southeast Asia (p. 146).

CHAPTER 4

Quick Review I (p. 193)

1. Dikes, landfill, and wind-powered water pumps (p. 183; see also p. 216)

2. The extensive coastline of Europe permits much of the region to feel the temperature-moderating effect of the surrounding oceans and the warming effect of the North Atlantic Drift (p. 188).

3. The techniques that were used by the Romans to colonize Europe were later used by Europeans in North and South America (p. 185).

4. Its colonies, principally in the Caribbean (p. 188)

5. Germany (p. 189)

6. While U.S. cities tend to decline at the center and to spread out as more and more land is developed for suburban housing and commercial zones, European cities tend to expand in concentric circles around the old central city (p. 192).

Quick Review II (p. 213)

1. Increased efficiency (no exchange fees), stronger voice in the creation of EU economic policy, and consistent value of currency relative to other EU members (pp. 195–196)

2. This prevents producers of GMOs from competing in the European market (p. 196).

3. In Europe, immigrants are expected to embrace the culture of Europe and abandon their traditional ways of doing things. This is assimilation. Immigrants to the United States, on the other hand, find that acculturation (learning the basics of American culture, like language) is enough to settle in. They may even be rewarded for retaining some aspects of their cultural heritage, like music and food (p. 204).

4. The conservative welfare system (p. 206)

5. From Europe (p. 209)

Quick Review III (p. 233)

1. Although tourism provides substantial revenues for popular destination countries like France, it also leads to overdevelopment and pollution (pp. 200, 217, 228, 230).

2. Scotland and Wales have created their own governing bodies, thus taking for themselves powers that used to belong to a tightly unified United Kingdom (p. 220).

3. The countries of Eastern and Central Europe are adapting to capitalism at different rates. Generally, the countries closest to Western Europe such as Poland and the Czech Republic have developed democratic institutions and market economies. Countries farther to the east, like Ukraine, Romania, and Bulgaria, are struggling through privatization and have not been able to maintain the standard of living that people were used to under communist governments (pp. 227–230).

4. This nickname reflects the shady politicians and government indiscretions that have come to characterize Italian domestic politics (p. 223).

5. Norway has focused on advancing gender equality. Men are allowed paternity leave, and children's games that promote male violence are discouraged. Employment ads are required to be gender neutral, and at least 40 percent of the public board and committee member seats must be filled by each of the sexes (p. 225).

6. Many people from former British colonies now live in Britain (p. 220).

CHAPTER 5

Quick Review I (p. 247)

1. Permafrost is permanently frozen soil that prevents precipitation from sinking into the ground; in the summer, rain forms swamps and marshes on the soil surface (p. 239).

2. The Silk Road connected China and the Mediterranean (p. 242).

3. After the breakup of the Soviet Union, the widespread loss of jobs, health care, and social networks (often associated with the workplace), and rising rates of alcoholism took their toll on life expectancy (pp. 244–245).

4. A command economy is one in which a central authority owns all the real estate, resources, and means of production, and determines which products will be manufactured and how much they will cost. In the USSR, Stalin used this system to foster development; once developed, in theory, the country would then be able to become truly communist, using no government, no money, and producing goods that would be shared by all for the welfare of all (p. 243).

Quick Review II (p. 269)

1. The government cannot collect tax revenue from the informal economy (p. 250).

2. The great wealth amassed by the oligarchs in the transition from communism to capitalism affords them extraordinary political power (p. 250).

3. Chechnya is an internal republic of Russia that has been fighting for independence since the USSR was dissolved (pp. 258–259).

4. The Soviets, believing that nature was valuable only insofar as it could be mined for resources, overexploited resources, polluted environments, and made few efforts at conservation (pp. 264–265).

Quick Review III (p. 280)

1. European Russia was where early Slavic people established the Russian Empire. It has the best climate, agricultural land, most of Russia's industry, and most of the people (p. 269).

2. In the past, the resources of the Russian Far East were too far from developed areas to be useful, but now port facilities on the coast have made more resources accessible to Japan, Korea, and China. New rail links and oil pipelines may also help to develop the area (p. 273).

3. Warmer climate and rich soils help boost agricultural capability in Ukraine and Moldova (p. 275).

4. The mountain slopes and plains of Caucasia are warm and wet compared with the rest of the land in this region, making it uniquely suited for warm-weather crops like citrus fruits and bananas. More than 50 ethnic groups live in a patchwork of cultural enclaves, a pattern that can lead to disputes over the region's land and natural resources, such as its vast oil reserves. Disputes are often manipulated by outside entities like Russia, Turkey, and Iran that seek access to resources (pp. 276–277).

CHAPTER 6

Quick Review I (p. 297)

1. Europe (p. 284)

2. A belt of dry air circles the earth between 20° and 30° North and South latitudes (p. 286).

3. The Fertile Crescent (p. 288)

4. Their expulsion from the Eastern Mediterranean by the Romans, beginning in A.D. 73 (p. 289).

5. Many refugees are displaced by earthquakes or droughts; many others have fled from conflict in Iraq and Afghanistan. Two million Palestinians are living in camps that were set up when Israel was created (p. 297).

Quick Review II (p. 319)

1. Shari'a is Islamic religious law, intended to guide daily life according to the principles of the Qur'an. The role it plays in politics and daily life varies country by country according to whether the state is theocratic or secular. Muslim individuals in the region may be conservative or liberal, and may interpret and observe shari'a loosely or strictly (pp. 298–299).

2. Islam is the official religion in the theocratic states and the leaders must be Muslim and ensure a conservative legal system based on shari'a. In secular states, there is no state religion and no direct religious influence in state affairs. Women are more likely to occupy public spaces and work in the public sphere in secular Islamic countries (pp. 298 and 301).

3. To ensure better returns from the exploitation of their main resource by controlling the volume of oil production and hence oil prices (pp. 305–306).

4. Humans contribute to desertification through settled cattle-ranching (overgrazing), overuse of irrigation (soil salinization), pumping of groundwater, and diversion of streams for urban populations, and by the burning of fossil fuels (global warming) (p. 317).

Quick Review III (p. 331)

1. The politically dominant Muslim northern Sudanese have imposed shari'a law on the non-Muslim southern Sudanese. This conflict is complicated by rivalry over oil resources. In Darfur (southeastern Sudan), the conflict is between groups of Arab-speaking Muslims over oil, land, and water. The government in Khartoum failed to quell, and instead instigated, vigilante violence against ordinary citizens in Darfur (pp. 322–323).

2. This was intended to encourage investment in modern industrialized agriculture (p. 324).

3. The Saud royal family allied themselves with the conservative Wahabi clerics in order to consolidate their power and avoid wealth-sharing (p. 325).

4. Fresh water provided by rivers in this region, such as the Nile and the Euphrates, is often a source of conflict as countries place dams that restrict a river's flow into other countries. The aquifers under the West Bank are a source of conflict between the Israelis and the Palestinians, each of whom claim the lion's share of the resource (pp. 316–317, 323–324, 327–328).

CHAPTER 7

Quick Review I (p. 348)

1. Warm and wet conditions enhance the conditions for the spread of malaria and schistosomiasis (p. 346).

2. Rome, India, and China (p. 339)

3. Increased rural to urban migration has made women more susceptible to HIV: men may visit HIV positive sex workers and transmit the virus to their wives, and urban women separated from village social networks are more likely to have sex more often and with more partners. Poverty may force women to turn to sex work in order to make ends meet; and women are not empowered to demand safe sex (pp. 347–348).

4. The number of people a region can support sustainably depends on the availability of water, the fertility of the land, and the quantity of natural resources. Low local carrying capacities can be supplemented by importing resources from other regions (pp. 343–344).

Quick Review II (p. 376)

1. Currency devaluation allows African agricultural products to sell more cheaply on the world market, but it makes farming equipment and pesticides more expensive for Africans to buy, since such aids must still be imported (p. 355).

2. The British imposed national boundaries that grouped many different cultural groups within one country; sometimes they used one cultural group to govern others in an area; and they pitted cultural groups against each other. The animosities created by these practices linger today (p. 359).

3. The northern half of Africa has a majority of Muslims, and the central and southern portions have a majority of Christians. However, there are Muslims, Christians, and animists throughout the continent (pp. 366–367).

4. Africa's national parks are underfunded, and they don't have the monetary resources they need to guard the animals and other resources in the parks from poachers (pp. 372–373).

Quick Review III (p. 386)

1. Profits from the diamond mines of Sierra Leone (p. 377)

2. The large numbers of refugees are blocking access to land that has been used in the past, refugees cannot raise food themselves, and the land that is left over is being overused (p. 382).

3. Elephants are worth much more alive if they can earn money as tourist attractions, but they are still being poached for their ivory (p. 383).

4. South Africa has a well-rounded slate of industries, and supplies goods and services to surrounding countries. Many people in the area are dependent on South African jobs (p. 386).

CHAPTER 8

Quick Review I (p. 403)

1. The summer monsoon and the intertropical convergence zone, or ITCZ (p. 393)

2. Islam was first brought to the region by Turkic and Mongolian invaders and Arab traders. Then the Mughals, who invaded in 1526, established an Islamic empire (p. 396).

3. India has a majority of Hindus, and Pakistan has a majority of Muslims (p. 399).

4. Britain (p. 399)

5. Higher education correlates with increased use of contraception (p. 402).

Quick Review II (p. 427)

1. Urdu and Hindi (p. 406)

2. A dowry is wealth provided by the bride's family upon her marriage to provide for her care; a bride price is wealth given to a bride's family by the groom's family to compensate them for the loss of her labor (p. 411).

3. Fertilizers, mechanized equipment, irrigation, pesticides, herbicides, and double-cropping all contribute to the green revolution (p. 413).

4. India and Pakistan both claim the territory of Kashmir, which is currently split unequally between the two. Kashmiris themselves are divided over which country they would rather be part of. Many would prefer to be independent (p. 420).

5. Acid rain, caused by industrial smog, is eating away at the marble surface of the monument (p. 426)

Quick Review III (p. 440)

1. Indian and Eurasian (p. 429)

2. Many people immigrate to New Delhi to escape conflicts in nearby states and widespread poverty. The middle class in New Delhi is growing; the poor have hope of upward mobility (p. 433).

3. Bollywood (p. 437)

4. The government has historically supported social services; women have power in the family structure and seclusion is practiced less often; and both males and females have high literacy rates (pp. 438–439).

5. Bureaucratic inertia and corruption, population pressure, the breakdown of families, and the lack of opportunities have all fueled the emigration of educated youth (pp. 434–435).

CHAPTER 9

Quick Review I (p. 455)

1. In the spring and fall (p. 447)

2. A trained and salaried bureaucracy was used to extend the ruler's influence into the countryside (p. 448).

3. The Chinese Communist party and the Kuomintang party were the contenders. The Kuomintang lost, and its supporters were eventually forced to flee to Taiwan (p. 450).

4. Tax revenue may drop, and it will be more difficult to support the elderly (p. 453).

Quick Review II (p. 475)

1. Without the appropriate hukou residence registration, it was impossible to get a job in government-owned urban enterprises. With the rise of private enterprise, migration has become more feasible, because private employers do not require a particular hukou designation (p. 463).

2. China's economic growth was linked to environmental and labor abuses (p. 464)

3. Kinship categories, such as siblings, aunts, uncles, and cousins, are disappearing, and the burden of caring for elderly parents falls entirely on the single child (p. 465).

4. Coal burned for both industry and home-heating is the primary pollutant, but auto emissions, particularly from leaded fuel, are increasingly a problem (pp. 470–471).

Quick Review III (p. 499)

1. The North China Plain (p. 476)

2. The communists thought of Shanghai historically as a nexus of capitalist corruption (p. 482).

3. The Han considered Tibetan women barbaric because their gender roles were more loosely defined and less limiting (p. 485).

4. In Japan, workers are encouraged to be loyal to their employers, which provide such benefits as lifetime employment, regular paid vacations, a pension upon retirement, and subsidized housing (p. 492).

5. Taiwan's status is ambiguous. The Chinese consider Taiwan a province; the Taiwanese consider themselves independent (p. 493).

CHAPTER 10

Quick Review I (p. 513)

1. During the last ice age, sea levels dropped, making it possible for animals and plants to travel across land that is now covered by the oceans (p. 505).

2. Periodic El Niño events interrupt normal air circulation patterns (p. 505).

3. The United States was worried about the spread of international communism in Vietnam and elsewhere in Southeast Asia (p. 509).

4. Migrants to cities are at increased risk because the disease is more prevalent there, hence individuals are more likely to encounter HIV positive partners. Returning urban migrants may unknowingly carry the disease back to their rural communities (p. 513).

Quick Review II (p. 536)

1. They can take advantage of Singapore's highly skilled workforce while having the option of cheaper resources and labor for manufacturing in less developed cities nearby (p. 517).

2. Once national economies were opened to the free market, Southeast Asian companies were free to solicit outside investors to help finance their expansion; when those investors withdrew their funds, however, national economies collapsed (p. 520).

3. Southeast Asia is inhabited by groups of people from many different backgrounds who have lived together for a long time yet have remained distinct (p. 524).

4. Push factors include spreading poverty in rural areas and the necessity of participating in the cash economy. Pull factors include the prospect of higher living standards and economic opportunities in the city (p. 527).

5. Gold mining in Southeast Asia is not regulated, so after companies strip the forest and disrupt the earth to remove minerals, they are not required to restore the landscape, control erosion, or to protect waterways from pollution. Local villagers are driven from their lands, national parks are threatened, and agricultural viability is lost (pp. 532–533).

Quick Review III (p. 548)

1. The military government of Burma has committed atrocities against the citizens, reaped the benefits of opium production, and is a major protector of methamphetamine production (pp. 536–537).

2. An affirmative action program called Bumiputra has reduced some social inequalities and increased national respect for diversity (p. 543).

3. Whereas Western culture places the most emphasis on the rights of the individual, in Singapore the rights of an individual are respected, but the emphasis is on community consensus, shared values, and religious and racial harmony (p. 544).

4. The eruption of Mt. Pinatubo, the closing of U.S. military bases, and drastic cuts in U.S. aid in the 1990s (p. 547)

CHAPTER 11

Quick Review I (p. 563)

1. Atolls form when coral reefs build up on the rim of a submerged volcano, creating a ring-shaped island that may be uplifted above the surface of the ocean. A *makatea* is an uplifted atoll that extends well above the ocean surface (p. 555).

2. New Zealand has always been surrounded by ocean. Birds flew or drifted to its shores, but land mammals never had the opportunity to travel there, other than with the help of humans (p. 558).

3. As the Pacific islands were discovered, some Europeans and explorers romanticized the islanders, whom they thought were uncorrupted by civilization. After those same islanders attacked Europeans for taking their land and resources, they were relabeled uncivilized brutes (pp. 559–560).

4. Many cities were established where colonizing countries found suitable land for relatively small numbers of people along fragile coastal environments. These locations do not support the growing numbers of urban dwellers (p. 563).

Quick Review II (p. 579)

1. Subsistence affluence is consistent with the Pacific Way, which includes the idea that Pacific Islanders have the ability to control their own development and solve their own problems (pp. 567 and 570).

2. Tourism in Hawaii and other islands dropped substantially (p. 571).

3. Rabbits imported from Europe depleted the food sources of native species and contributed to soil erosion with their digging (p. 575).

4. The Pacific islands have little political power and have been subjected to damaging treatment by corporations (which have stripped natural resources in mining operations) and countries (France and the United States have used Pacific islands as nuclear test sites and nuclear waste dumps). They are also particularly sensitive to global climate change, since global warming will lead to higher sea levels (pp. 577–578).

SUGGESTED ANSWERS TO READING LOCAL LANDSCAPES

Chapter 2 (page 80)

Photo A was taken on the peninsula of Southwest Seattle, just across the bay from Seattle proper. The fact that some of the buildings look Asian may be related to the fact that there are many Asian immigrants in Seattle. There are a host of fanciful items and spaces in the picture: wind socks, wind chimes, steps and platforms, including an observation deck at the top of the hill from which one has a panoramic view of downtown Seattle. Photo B was taken in Malacca, Malaysia, which lies across the Strait of Malacca from Sumatra, Indonesia. This settlement lies along the Malacca River and was taken from a tour boat that plies the river every hour or so. People use the river for transport and a footpath follows the course of the river, but they also have cars. Like the folks in Seattle, those who created these spaces are given to making "yard art" constructions. Can you spot some?

Chapter 3 (page 123)

In Photo A, an Ecuadorian woman is barbecuing a whole guinea pig, or *cuy*, in a small-town restaurant in South America. Cooking over hot coals dates back to the Incas, and the technique was used throughout the Americas in pre-Columbian times. Barbecuing is now a worldwide custom. In Photo B, a North American man is barbecuing a whole chicken that is resting on an open can of beer. The boiling beer keeps the bird moist while it cooks over the hot coals.

Chapter 4 (page 210)

The potential for nuclear contamination figures in both pictures. In Photo A, Denisa Kukucova is gathering flowers for her wedding near Slovakia's oldest nuclear power plant. The plant was built by a Soviet designer and has been upgraded with Western technology. Still, nuclear power is now out of favor in Europe, and many worry that living near such cooling towers might be dangerous. Photo B shows Clement Strauss, a Russian of German descent, with his family at their home in Kazakhstan. An affluent engineer, he nevertheless suffers extensively from nuclear contamination and worries about the future of his children. He is one of an estimated 1.2 million people living in an area that was contaminated by repeated nuclear test explosions beginning in 1949. When the Soviet Union dissolved, Kazakhstan, a relatively poor country, was left with many deteriorating nuclear facilities.

Chapter 5 (page 259)

Native economies in the Arctic regions often center on just one or two species of animals. The Nenets reindeer herder in Photo A lives in a settlement at the mouth of the Taz River in northern Siberia. Because of the remoteness of Siberia, and, until recently, a dearth of resources that others might covet, Siberians have continued to live a traditional lifestyle. Reindeer supply food, clothing, shelter, and transportation. In Siberia, oil exploration by European Russians now threatens native Siberian lands. Photo B shows a group of Inuit near Barrow, Alaska, as they slice blubber from a bowhead whale. From whales they derive oil, meat, and bone. Unlike most Native Americans, who live on reservations, the Inuit still occupy their native lands in the United States and Canada. There they hunt mammals, including whales. They also fish and maintain other traditional ways of life. Oil exploration by corporations in the "Lower 48" and Canada threatens to force some of these people off their lands. The store-bought clothes suggest jobs in the cash economy now supplement traditional lifeways.

Chapter 6 (page 291)

Nearly half the ingredients for a Mexican Mole sauce, shown in Photo A, came to Mexico via the Spanish, who in turn had borrowed from the Arabs of North Africa. The chocolate, tomatoes, pumpkin seeds, and peppers were contributed by the Americas; but the garlic, onions, sesame, almonds, cinnamon, cloves and raisins came from the Mediterranean and Asia and entered Europe courtesy of Arab traders who plied the Indian Ocean and the Sea of China. Photo B shows the ingredients for an Indonesian vegetable curry, including four essentials from the Americas: pumpkin, peppers, tomatoes, and potatoes. Indonesians first learned of these ingredients from Arab traders, who brought news of new American food plants they learned about in the Mediterranean shortly after 1500.

national government in Khartoum encouraged "ethnic cleansing" by mounted warriors (*janjaweed*) to drive dark-skinned, Muslim, Arab-speaking peoples from the southwestern Darfur region. This family did not start out impoverished: they have many possessions, including some fine handwoven rugs that can be seen on top of the pile.

Chapter 8 (page 427)

Photo A depicts a relatively new suburban neighborhood in Bangalore, India, where a number of return migrants from the United States live. Photo B shows a home in Tustin Ranch, California. Migration and return migration have long been means by which material culture, in this case, housing styles, diffuses from one place to another. Return migrants to India have built houses similar to those they lived in in the United States.

Chapter 7 (page 344)

Photo A shows Somali Bantu standing in line to be processed for resettlement in the United States; this 2002 resettlement program was the largest ever attempted from Africa. Some of the refugees were settled first in Atlanta, Georgia, and then moved to cheaper Lewiston, Maine, or Tucson, Arizona. Others went to Minneapolis, Minnesota. Twenty-two U.S. cities are now home to Somali Bantu refugees. Photo B shows a family of at least six in flight from Darfur. They are huddled next to a shelter constructed of their possessions. In 2004, hundreds of thousands of people in Sudan (discussed in Chapter 6) became refugees when the

Chapter 9 (page 469)

The buildings in both photos are mosques. The head covering of the man in Photo A and the minaret, or tower, in the Photo B provide clues to identifying these as Islamic structures. In Photo A, a sexton is watering potted shrubs in one of the courtyards of the Niujie ("Oxen Street") Mosque in Beijing, China. The mosque dates from 997, and its minaret is in the Chinese pagoda style, indicating the manner in which Islam melds with local customs and forms. The minarets of the Kampung Keling Mosque in Malacca, Malaysia, shown in Photo B, blend elements of Malay and Chinese architecture with Islamic influences.

Chapter 10 (page 514)

Photo A shows women peasants transplanting rice seedlings in a paddy field at Banaue, in the northern region of Luzon Island, in the Philippines. The women shown in Photo B are harvesting grapes in Turpan, Xinjiang Ugyur Autonomous Region, China. Both types of agriculture are labor intensive, and these activities are not easily adapted to mechanization. Women in many regions of the world are often heavily involved in, if not entirely responsible for, many harvesting activities and intermediary agricultural tasks such as transplanting and weeding. The surrounding hills visible in Photo A suggest that the paddy field is likely part of a terrace system. Photo B appears to be in a ravine where grapevines are protected from the drying winds of Xinjiang and where vine roots can access stored moisture in the ground.

Chapter 11 (page 570)

The fisherman shown in Photo A is using the once-traditional method of throwing stones to guide fish into nets off the island of Moorea in French Polynesia. Photo B shows fishermen using the traditional *kattumaram* (thought to be an invention of Pacific Islanders) to fish off the shores of Kerala State on India's southwest coast. Artisanal fishermen all over India face increasing competition from commercial fleets of mechanized trawl boats that encroach upon their fishing grounds and damage their craft and gear. In French Polynesia, subsistence fishermen also face pressures from tourism and the environmental destruction it generates.

PRONUNCIATION GUIDE

Aborigine	ab-uh-RIHJ-uh-nee	Argentina	ahr-juhn-TEE-nuh	Belo Horizonte	BEH-loo aw-rih-ZAWN-teh ["oo" as in "book"]
Abuja	ah-BOO-jah	Arias, Oscar	AH-ree-ahs		
Addis Ababa	AH-dihss AH-buh-bah	Armenia	ahr-MEE-nee-uh		
Adelaide	AD-uh-layd	Arunachal Pradesh	ah-ROON-ah-chahl prah-DAYSH	Benares	beh-NAHR-ehss
Aden, Gulf of	AH-duhn			Benelux	BEHN-uh-luhks
Adirondack Mountains	ad-uh-RAHN-dak	Arya	AIR-ee-uh	Bengal	behn-GAHL
		Assam	ah-SAHM	Benin	beh-NEEN
adivasis	ah-di-VAH-sees	Aswan High Dam	ahss-WAHN	Berber	BURR-burr
Adriatic Sea	ay-dree-AT-ihk	Atacama Desert	ah-tuh-KAH-muh	Bering Strait	BAIR-ing
Afrikaner	af-rih-KAH-nurr	Athens	ATH-ihnz	Bhopal	boh-PAHL
Agra	AH-gruh	Auckland	AWK-luhnd	Bhopavand	boh-pah-VAHND
Ahmadabad	AH-muh-duh-bahd	Australia	aw-STRAYL-yuh	Bhotechaur	boh-teh-CHAWR
Aikwa River	IKE-wah	Austria	AW-stree-uh	Bhutan	boo-TAHN
Ainu	IE-noo	Austronesian	aw-stroh-NEE-zhuhn	Biafra	bee-AH-fruh
Al Salaam	ahl sah-LAHM	Ayatollah	eye-yah-TOH-lah	Bihar	bee-HAHR
Albania	ahl-BAY-nee-ah	Ayers Rock	AIRS	Bo Hai Sea	BOH-HYE
Alexandria	al-ihk-SAN-dree-uh	Ayodhya	ah-YOHD-yah	Boer	BOHR
Algeria	ahl-JEER-ee-uh	Azerbaijan	ah-zair-bye-JAHN	Bogotá	boh-goh-TAH
Algiers	ahl-JEERZ	Aztec	AZ-tehk	Bolivia	boh-LIHV-ee-ah
Algonquin	al-GAHN-kwuhn	Baghdad	bahg-DAHD/BAG-dad	Bolshevik	BOHL-shuh-vihk
Alhambra	ahl-HAHM-brah	Bahamas	buh-HAH-muhz	Bom Bahia	bohm bah-EE-ah
Allah	AH-luh	Bahrain	bah-RAYN	Bombay	bawm-BAY
Allahabad	ah-luh-huh-BAHD	Bai	BYE	Bordeaux	bohr-DOH
Altai Mountains	AHL-tye	Baikal, Lake	bie-KAHL	Borneo	BOHR-nee-oh
altiplano	ahl-tee-PLAH-noh	baksheesh	bak-SHEESH	Bosnia	BAWZ-nee-uh
Amazon River	AM-uh-zahn	Bali	BAH-lee	Bosporus	BAWSS-puh-ruhss
Amhara	ahm-HAHR-uh	Balkans	BAHL-kuhnz	Botswana	bawt-SWAH-nah
Amnok River	AHM-NAWK	Baltic Sea	BAWL-tihk	Bougainville	boo-gahn-VEEL
Amritsar	ahm-RIHTZ-surr	Ban Muang Wa	BAHN MWAHNG WAH	Brahmaputra River	brah-mah-POO-truh
Amsterdam	AM-sturr-dam			Brahmin	BRAH-mihn
Amu Darya	ah-moo DAHR-yah	Bandare Abbas	BAHN-dahr AH-buhss	Brasília	bruh-ZIHL-yuh
Amur River	ah-MOOR	Bandung	BAHN-doong ["oo" as in foot]	Brisbane	BRIHZ-buhn
Anasazi	ahn-uh-SAH-zee			Brundtland	BROONT-lahnd
Anatolia	an-uh-TOH-lee-uh	Bangalore	bang-guh-LOHR	Brunei	broo-NYE
Andalucía	ahn-dah-loo-THEE-ah	Bangkok	bang-KAWK	Brussels	BRUH-suhlz
Andaman Sea	AN-duh-muhn	Bangladesh	bahng-gluh-DEHSH	Bucharest	BOO-kuh-rehst
Andes	AN-deez	Bantu	BAN-too	Budapest	BOO-duh-pehsht
Angkor	AHNG-oor	Barbados	bahr-BAY-dohss	Buenos Aires	BWAY-nohss IE-rehss
Angola	ang-GOH-luh	Barbary Coast	BAHR-buh-ree	Buka	BOO-kah
Antigua	an-TEE-guh	Barbuda	bahr-BOO-duh	Bukhara	boo-KAH-rah
Antwerp	AHN-twurrp	Basque	BASK	Bulgaria	buhl-GAIR-ee-uh
Aotearoa	ah-oh-tay-ah-ROH-ah	Basra	BAHZ-rah	Burkina Faso	burr-KEEN-uh FAH-soh
apartheid	uh-PAHRT-hyt	Batavia	buh-TAY-vee-uh		
Appalachia	ap-uh-LACH-uh	Bedouin	BEHD-oo-ihn	Burma	BURR-muh
Appalachian Mountains	ap-uh-LACH-uhn	Beijing	BAY-JYIHNG	Burqa	BURR-kuh
		Belarus	byeh-lah-ROOS	Burundi	boo-ROON-dee [both "oo" as in "book"]
Arabia	uh-RAY-bee-uh	Belém	buh-LAYM		
Arabian Sea	uh-RAY-bee-uhn	Belfast	BEHL-fast	Byzantine Empire	BIHZ-uhn-teen
Arafat, Yasser	AHR-uh-faht, YAH-seer	Belgium	BEHL-juhm	Cahokia	ku-HOH-kee-uh
		Belgrade	BEHL-grahd	Cairo	KYE-roh
Aral Sea	AHR-uhl	Belize	buh-LEEZ	Cambodia	kam-BOH-dee-uh

Cameroon	kahm-uh-ROON	Côte d'Ivoire	KOHT deev-WAHR	Fujian	FOO-jee-ehn
Canaan	KAY-nuhn	Croatia	kroh-AY-shuh	Fulani	foo-LAH-nee
Canberra	KAN-burr-uh	Curitiba	koor-ee-TEE-bah	Gabon	gah-BOHN
Canton	KAHN-tawn	Cuzco	KOOZ-koh	Galicia	gah-LEESH-ee-ah
Cape Verde Islands	VAIR-day	czar	ZAHR	Gambia	GAHM-bee-uh
Caracas	kah-RAH-kahs	Czech Republic	CHEHK	Gandhi, Indira	GAHN-dee, ihn-DEER-ah
Caribbean	kuh-RIH-bee-uhn	Czechoslovakia	chehk-oh-sloh-VAH-kee-ah	Gandhi, Mohandas	GAHN-dee, moh-HAHN-dahss
Carpentaria, Gulf of	kahr-puhn-TAIR-ee-uh	Dai	DIE	Ganga River	GAHNG-gah
Casablanca	kah-suh-BLAHNG-kuh	Dalai Lama	DAH-lye LAH-mah	Ganges River	GAN-jeez
Caspian Sea	KASS-pee-uhn	Dalit	DAH-liht	Gauguin, Paul	goh-GA(N)
Castro, Fidel	KAH-stroh, fee-DEHL	Damascus	duh-MASS-kuhs	Gaza Strip	GAH-zah
Catalonia	kaht-uh-LOH-nee-uh	Danube River	DAN-yoob	Genghis Khan	JEHNG-gihss kahn/GEHNG-gihss kahn
Caucasia	kaw-KAY-zhuh	Daqing	DAH-CHIHNG	Genoa	JEHN-oh-uh
Caucasus Mountains	KAW-kuh-suhss	Darjeeling	dahr-JEE-lihng	Georgia	JOHR-jah
Cebu	say-BOO	Dayak	DYE-ak	ger	GURR
Celebes	seh-LAY-behss	Deccan Plateau	DEHK-uhn	Gezhouba	GUH-JOH-BAH
Cerro San Cristóbal	SEHR-roh, sahn, krees-TOH-vahl	Delhi	DEHL-ee	Ghana	GAH-nah
Ceylon	say-LAWN	Deng Xiaoping	DUHNG SHYAU-PIHNG	Ghats Mountains	GAHTSS
chador	chah-DOHR	Dhaka	DAHK-uh	ghee	GEE
chaebol	CHYE-BOHL	Dharamkot	duh-RUHM-koht	Gibraltar	jih-BRAHL-turr
Chamoli	chuhm-OH-lee	Dharavi	duh-RAH-vee	Gobi Desert	GOH-bee
Chamorro, Violetta	chah-MOHR-oh, vee-oh-LEHT-ah	Dharma Wanita	DAHR-muh wah-NEE-tuh	Goiás	goy-AHSS
Chandigarh	CHUHN-dih-gurr	Dien Bien Phu	dyehn-byehn-FOO	Golan Heights	GOH-lahn
Chang Jiang	CHAHNG JYAHNG	Dinka	DIHNG-kuh	Gorbachev, Mikhail	gohr-buh-CHAWF, mee-KILE
Changan	CHAHNG-AHN	Djibouti	jih-BOO-tee		
Chao Phraya River	chau prah-YAH	Dnieper River	DNYEH-purr	Granada	grah-NAH-dah
Chechnya	chyehch-NYAH	doi moi	DOY MOY	Grasberg Mountain	GRASS-burg
Chelyabinsk	chyehl-yah-BEENSK	Dolpo-pa	DOHL-poh-pah	Greater Antilles	an-TIL-eez
Chengdu	CHUHNG-DOO	Donetsk	duh-NYEHTSK	Grenada	gruh-NAY-duh
Chennai	chehn-IE	Dravidian	drah-VIHD-ee-uhn	Grenadines	GREHN-uh-deenz
Chernobyl	chyair-NOH-bihl	Dubuque	duh-BYOOK	Gro Harlem	GROH HAHR-lehm ["oo" as in "book"]
Cherokee	CHAIR-uh-kee	Duvallier, François	doo-VAHL-yay, frahnts-WAH		
Chiang Kai-shek	JYAHNG KYE-SHEHK	East Timor	TEE-mohr	Guadeloupe	gwah-duh-LOOP
Chiang Mai	CHYAHNG MYE	Ecuador	EHK-wah-dohr	Guangdong	GWAHNG-DOONG
Chiapas	chee-AH-pahss	Edo	AYD-oh	Guangxi	GWAHNG-SHEE
Chile	CHEE-leh	El Niño	ehl NEEN-yoh	Guangzhou	GWAHNG-JOH
Chipko Movement	CHIHP-koh	El Paso	el-PAH-soh	Guatemala	gwah-teh-MAH-lah
Choctaw	CHAWK-taw	El Salvador	ehl sahl-vah-DOHR	Guayaquil	gwah-yah-KEEL
Chongqing	CHOHNG-CHIHNG	Eritrea	air-ih-TREE-uh	Guiana Highlands	gee-AHN-ah
Ciudad Juárez	see-oo-DAHD HWAHR-ehss	Essen	EHSS-uhn	Guinea	GIH-nee
		Estonia	ehss-TOH-nee-uh	Guinea-Bissau	GIH-nee bih-SAU
Colombia	koh-LOHM-bee-ah	Ethiopia	ee-thee-OH-pee-uh	Guizhou	GWAY-JOH
Colombo	koh-LOHM-boh	Euphrates River	yoo-FRAY-teez	Gujarat	GOO-juh-raht
Comoros	KOHM-uh-rohz	Eurasia	yoo-RAY-shuh	Guyana	gye-AHN-uh
Confucianism	kuhn-FYOO-shuhn-ihz-uhm	Faroe Islands	FAHR-oh	hacienda	hah-see-EN-dah
		Farsi	FAHR-see	Hague, The	HAYG
Congo	KAWNG-goh	Fiji	FEE-jee	Haikou	HYE-KOH
Constantinople	kawn-stan-tih-NOH-puhl	Filipino	fihl-ih-PEE-noh	Hainan	HYE-NAHN
		Finland	FIHN-luhnd	Haiti	HAY-tee
Copenhagen	KOH-puhn-hay-guhn	Florence	FLOHR-uhntss	hajj	HAHJ
Coptic Christians	KAWP-tihk	folkhem	FOHLK-hehm	Haka	HAH-kah
Córdoba	KOHR-doh-vah	Fortaleza	fohr-tah-LAY-zah	Han	HAHN
Costa Rica	KOHSS-tah REE-kah	French Guiana	gee-AHN-ah	Hangai Mountains	HAHNG-GIE
				Hangzhou	HAHNG-JOH

Term	Pronunciation
Hanoi	hah-NOY
Harbin	HAHR-BIHN
Harijan	HAH-ree-jahn
Haryana	hahr-YAH-nah
Hausa	HAU-zuh
Havana	ah-VAHN-ah/huh-VAN-uh
he	HUH
Heilongjiang	HAY-LUNG-JYAHNG
Helsinki	hehl-SIHNG-kee
Herzegovina	hairt-suh-goh-VEE-nuh
hijab	hee-JAHB
Himachal Pradesh	hih-MAH-chahl prah-DAYSH
Himalaya	hih-MAHL-yuh/hih-muh-LAY-uh
Hindi	HIHN-dee
Hindu Kush	HIHN-doo KOOSH [second "oo" as in "book"]
Hindu	HIHN-doo
Hindustan	hihn-doo-STAHN
Hiroshima	hih-ROH-shih-muh
Hispaniola	hihs-puhn-YOH-luh
Ho Chi Minh City	HOH CHEE-MIHN
Hohokam	HOH-HOH-kahm
Hokkaido	haw-KYE-doh
Honduras	ohn-DOO-rahss/hawn-DOOR-uhss
Honshu	HAWN-shoo
Huang He	HWAHNG HUH
Hui	HWAY
Hukou	HOO-koh
Hungary	HUHNG-guh-ree
Hussein, Saddam	hoo-SAYN, sah-DAHM
Hutu	HOO-too
Iberia	eye-BEER-ee-uh
Igbo	IHG-boh
Inca	ING-kuh
Indonesia	ihn-doh-NEE-zhuh
Indore	ihn-DOHR
Indus River	IHN-duhss
intifada	ihn-tih-FAH-duh
Iquitos	ih-KEE-tohss
Iran	ih-RAHN
Iraq	ih-RAHK
Irian Jaya	EER-ee-ahn JYE-yah
Irrawaddy River	eer-uh-WAWD-ee
Islam	ihz-LAHM
Israel	IHZ-ree-uhl
Istanbul	ihss-STAHN-bool / ihss-tahn-BOOL ["oo" as in "book"]
Jainism	JYE-nihz-uhm
Jakarta	juh-KAHR-tuh
jalee	jah-LEE
Java	JAH-vuh
Jayapura	jye-yuh-POOR-uh
Jeddah	JEHD-uh
Jerusalem	juh-ROO-suh-lehm
ji	JEE
jiang	JYAHNG
Jiangsu	JYAHNG-SOO
Jilin	JEE-LIHN
Johor	juh-HOHR
Jönköping	YEHN-keh-pihng
Jordan	JOHR-duhn
Joypur	joy-POOR
Junggar Basin	DZOONG-GAHR ["oo" as in "book"]i
Jwaneng	JWAY-nehng
Kadavu	KAHN-da-voo
Kalahari Desert	kah-lah-HAH-ree
Kalimantan	KAH-LEE-MAHN-TAHN
Kaliningrad	kah-LYIH-nihn-graht
Kalmykia	kuhl-MIHK
Kampala	kahm-PAH-luh
Kanto Plain	KAHN-TOH
Kaohsiung	GAU-SHYOONG ["oo" as in "book"]
Karachay	kahr-ah-CHYE
Karachi	kuh-RAH-chee
Karakoram Mountains	kahr-uh-KOHR-uhm
Karen	kuh-REHN
Karnataka	kahr-nah-TAH-kah
Karzai, Hamid	Kahr-zi, HAM-id
Kasama	kah-SAH-muh
Kashgar	kahsh-GAHR
Kashmir	kash-MEER
Kazakhstan	kah-zahk-STAHN
Kazan'	kuh-ZAHN
Kerala	KAIR-uh-luh
Khartoum	kahr-TOOM
Khatami, Mohammad	kah-TAH-mee
Khmer Rouge	KMAIR ROOZH
Khomeini, Ruholla	koh-MAY-nee, roo-HAWL-uh
Khon Kaen	KAWN GAN
Kiev	KEE-ehf
Kikuyu	kih-KOO-yoo
Kinshasa	kihn-SHAH-suh
Kiribati	keer-ih-BAH-tee
Kisangani	kee-sahn-GAH-nee
Kitui	kee-TOO-ee
Klaten	KLAH-tehn
Kobe	KOH-bay
Kohtla-Järve	KOHT-luh YAIR-vuh
Kolkatta	kohl-KAH-tah
Kootenai	KOOT-uh-nee
Koran	koo-RAHN ["oo" as in "book"]
Kosovo	KOHSS-uh-voh
Krajina	kreye-EE-nuh
Krtova	krah-TOH-vuh
Kshatriya	kih-SHAH-tree-uh
Kuala Lumpur	KWAH-luh loom-POOR [first "oo" as in "book"]
Kunming	KOON-MIHNG ["oo" as in "book"]
Kuomintang	GOAH-mihn-TAHNG
Kurd	KURRD
Kuwait	koo-WAYT
Kyrgyzstan	keer-gihz-STAHN
Kyushu	KYOO-shoo
Lagos	LAH-gohss
Lahore	luh-HOHR
Lake Nasser	NAH-surr
Laos	LAH-ohss
Laredo	luh-RAY-doh
Latino	lah-TEE-noh
Latvia	LAT-vee-uh
Lebanon	LEH-buh-nuhn
Lee Kuan Yue	LEE KWAHN YWEH
Lenin, Vladimir	LYEH-nyihn, vlah-DEE-meer
Lesotho	leh-SOO-too
Leyte	LAY-tay
Liberia	lye-BEER-ee-uh
Libya	LIH-bee-uh
Lille	LEEL
Lima	LEE-muh
Lithuania	lihth-oo-AY-nee-uh
Ljubljana	lyoo-BLYAH-nuh
llanos	YAH-nohs
Loango	loh-AHNG-goh
Loess Plateau	LUHS
Lombok	lawm-BAWK
Lopburi	LAWP-BOO-REE
Louisville	LOO-uh-vuhl/LOO-ee-vihl
Lumumba, Patrice	loo-MOOM-buh, pah-TREESS [both "oo" as in "book"]
Luxembourg	LOOK-suhm-boork [first "oo" as in "book"]
Luzon	loo-ZAWN
Macao	muh-KAU
Macedonia	mass-ih-DOH-nee-uh
machismo	mah-CHEEZ-moh
Madagascar	mad-uh-GASS-kurr
Madras	muh-DRAHSS
Madrid	mah-DRIHD
Madura	mah-DOOR-uh
Mafia	MAH-fee-ah
Maghreb	MAH-gruhb
Magnitogorsk	mahg-nyee-tuh-GOHRSK

Maharashtra	mah-hah-RAHSH-trah	Milan	mih-LAHN	Nicaragua	nee-kah-RAH-gwah
Makassar Strait	muh-KASS-urr	Milosevič, Slobodan	Mee-LOHse-vitch, SLOH-boh-dahn	Nicholas II	NEE-koh-lahss
Makkah	MEH-kuh			Niger	NYE-jurr
Malabar	MAL-uh-bahr	Mindanao	mihn-dah-NAU	Nilotic	nye-LAW-tihk
Malawi	muh-LAH-wee	minifundios	mih-nee-FOON-dee-ohss ["oo" as in "book"]	Nirmal Hriday	NEER-mahl HREE-day
Malay	muh-LAY				
Malaysia	mah-LAY-zhuh	Mizoram	mee-ZOH-rahm	nodi bhanga lok	NOH-dee BAHNG-gah LAWK
Maldives	mawl-DEEVZ	Mobutu Sese Seko	moh-BOO-too SAY-see SAY-koh		
Mali	MAH-lee			Nogales	no-GAH-layss
Managua	mah-NAH-gwah	Mogadishu	moh-guh-DEE-shoo	Noriega, Manuel	nohr-ee-AY-gah, mahn-WEHL
Manaus	mah-NAUS	Molucca islands	maw-LOOK-uh ["oo" as in "book"]		
Manchu	MAN-CHOO			Norilsk	nuh-REELSK
Manchuria	man-CHOOR-ee-uh	Molucca Sea and Islands	maw-LOOK-uh ["oo" as in "book"]	Novgorod	NAWV-guh-ruht
Mandarin	MAN-duh-rihn			Novosibirsk	noh-voh-sih-BEERSK
Manila	mah-NIHL-uh	Mombasa	mawm-BAH-suh	Nswazi	uhn-SWAH-zee
Mao Zedong	MAU DZUH-DOONG	Mon	MOHN	Nubian	NOO-bee-uhn
		Mongol	MAWNG-gohl	Nuevo Laredo	NWAY-voh lah-RAY-doh
Maori	MAU-ree	Mongolia	mawng-GOH-lee-uh		
Maputo	muh-POO-toh	Montego Bay	mawn-TEE-goh	Oahu	oh-AH-hoo
maquiladora	mah-kee-lah-DOH-rah	Montenegro	mawn-tih-NEH-groh	Ob River	AWB
Marcos, Ferdinand	MAHR-kohss, FAIR-dee-nahnd	Montserrat	mohnt-srat	Obeah	OH-bee-uh
		Morocco	muh-RAW-koh	oblast	OH-blahsst
Marianas	mahr-ee-AH-nahss	Moscow	MAWSS-kau	Oceania	oh-shee-AN-ee-uh
marianismo	mah-ree-ah-NEEZ-moh	Mount Fuji	FOO-jee	Ogallala	oh-gu-LAHL-uh
		Mozambique	moh-zuhm-BEEK	Ogoni	oh-GOH-nee
Marie Galante	muh-REE guh-LAHNT	Mt. Vesuvius	veh-SOO-vee-uhss	ohana	oh-HAH-nah
		mugam	MOO-gahm	Okhotsk	oh-KAWTSK
Marquesas	mahr-KAY-zahss	Mughal	MOO-guhl	okrug	OH-kroog ["oo" as in "book"]
Marseille	mahr-SAY	Muhammad	moo-HAHM-ihd ["oo" as in "book"]		
Martinique	mahr-tih-NEEK			Olgas	OHL-guhss
Marx, Karl	MAHRKS, KAHRL	mujahedeen	moo-jah-hu-DEEN	Oman	oh-MAHN
Masai	mah-SYE	mulatto	moo-LAH-toh	Ordos Desert	OHR-dohss
Mauna Loa	MAU-nah LOH-ah	Mumbai	moom-BYE	Orinoco River	ohr-ee-NOH-koh
Mauritania	mawr-ih-TAY-nee-uh	Muslim	MOOZ-lihm ["oo" as in "book"]	Orissa	oh-RIH-sah
Mauritius	maw-RIHSH-uhss			Osaka	oh-SAH-kah
mbira	uhm-BEER-uh	Myanmar	myahn-MAHR	Ottawa	AW-tuh-waw
Mecca	MEH-kuh	Nagasaki	nah-guh-SAH-kee	Ottoman	AW-tuh-muhn
Medan	may-DAHN	Nagorno Karabakh	nah-GOHR-noh kah-rah-BAHK	Ottoman Empire	AW-tuh-muhn
Medina	meh-DEE-nuh			Ovimbundu	oh-vihm-BOON-doo [first "oo" as in "book"]
Mediterranean Sea	meh-dih-tuh-RAY-nee-uhn	Nairobi	nye-ROH-bee		
		Nakhodka	nah-KAWD-kuh		
Megalopolis	meh-guh-LAW-puh-lihss	Namibia	nuh-MIHB-ee-uh	Pahlavi, Reza	PAH-luh-vee, RAY-zah
		Nan Ling Mountains	NAHN LIHNG	Pakistan	pah-kih-STAHN
Meghalaya	mehg-AH-lah-yah	Nanjing	NAHN-JYIHNG	Palau	pah-LAU
Meiji	MAY-jee	Naples	NAY-puhlz	Palembang	pay-lehm-BAHNG
Mekong River	MAY-KAWNG	narghile	NAHR-guh-leh	Palestine	PAL-uh-stien
Melanesia	mehl-uh-NEE-zhuh	Narmada River	nahr-MAH-duh	Pamir Mountains	pah-MEER
Mesa Central	MAY-zah sehn-TRAHL	Nauru	NAU-roo	pampas	PAHM-pahss
		Nazca Plate	NAHZ-kuh	Panama	PA-nuh-mah
Mesabi	muh-SAH-bee	Nazi	NAHT-see	Pancasila	pahng-kuh-SEE-luh
Mesopotamia	meh-suh-puh-TAY-mee-uh	Neapolis	nay-AH-poh-leess	panchayati raj	pahn-chah-YAH-teh RAHJ
		Nehru, Jawaharlal	NAY-roo, jah-wah-HAHR-lahl		
mestizo	mehs-TEE-zoh			Pangaea	pan-JEE-uh
Mexicali	meh-hee-KAH-lee	Nepal	neh-PAHL	Panipur	pahn-ee-POOR
Mexico	MEH-hee-koh/ MEHK-sih-koh	Netherlands Antilles	an-TIL-eez	Papeete	pah-pee-AY-tay
		New Delhi	DEHL-ee	Papua New Guinea	PAH-poo-ah
Miao	MYAU	New Guinea	GIH-nee	Paraguay	PAHR-uh-gwaiy
Micronesia	mie-kroh-NEE-zhuh				

Paraná River	pah-rah-NAH	Riau	REE-au	Shevardnadze, Edvard	shyeh-vahrd-NAHD-zyeh, EHD-vahrd
pardo	PAHR-doh	Riga	REE-guh	Shi'a	SHEE-uh
Pashto	PAHSH-toh	Rio de Janeiro	REE-oh dih zhuh-NAY-roh	Shi'ite	SHEE-eyt
Patagonia	pa-tuh-GOHN-yuh			Shikoku	SHE-kaw-koo
Pathan	puh-TAHN	Riviera	rih-vee-AIR-uh	shish kebab	SHEESH kuh-bahb
Pauoa Valley	PAU-oh-ah	Riyadh	ree-YAHD	Shona	SHOH-nuh
perestroika	pyeh-ryih-STROY-kah	Romania	roh-MAY-nee-uh	Siberia	sye-BEER-ee-uh
Persia	PURR-zhuh	Rondônia	rawn-DOH-nyuh	Sichuan	SSI-CHWAHN
Peru	peh-ROO	Rong-pa	RAWNG-pah	Sicily	SIHSS-uh-lee
Peshawar	peh-SHAH-wurr	Rotterdam	RAWT-urr-dam	Sierra Leone	see-AIR-uh lee-OHN
Petén	peh-TEHN	Rub'al Khali	roob ahl KAH-lee	Sierra Madre	see-AIR-ah MAHD-ray
Philippines	FIHL-uh-peenz	Rwanda	roo-AHN-duh	Sikh	SEEK
Phnom Penh	NAHM pehn	Ryukyu Islands	RYOO-kyoo	Sikkim	SIHK-ihm
Piazza del Plebiscito	PYAHT-suh dehl pleh-bih-SHEE-toh	Sabah	SAH-bah	Silesia	sih-LEE-zhuh
		Sahara	suh-HAHR-uh	Sinai Peninsula	SYE-nye
pilaf	PEE-lahf	Sahel	suh-HAYL	Singapore	SIHNG-uh-pohr
Pinatubo, Mount	pee-nah-TOO-boh	Saigon	sye-GAWN	Singhalese	sihn-ghuh-LEEZ
Pithampur	pee-tahm-POOR	Salinas, Carlos	sah-LEE-nahss, KAHR-lohss	Sioux Falls	SOO
Pizarro, Francisco	pee-ZAHR-oh, frahn-SEESS-koh			Slovakia	sloh-VAH-kee-uh
		Salish	SAY-lihsh	Slovenia	sloh-VEE-nee-uh
pogrom	puh-GRAWM	Salvador	sahl-vah-DOHR	Sofia	soh-FEE-uh
Pointe-Noire	PWANT NWAHR	Salween River	SAHL-ween	Somalia	soh-MAHL-ee-uh
polygyny	puh-LIHJ-uh-nee	Samarkand	sah-mahr-KAHNT	Somoza, Anastacio	sah-MOH-zah, ah-nah-STAH-see-oh
Polynesia	pawl-ih-NEE-zhuh	Samoa	sah-MOH-ah		
Port Moresby	MOHRZ-bee	San Diego	san dee-AY-goh	Songhua	song-HUAH
Pôrto Alegre	POHR-too ah-LEHG-reh	San José	sahn hoh-SAY	sottogoverno	soh-toh-goh-VAIR-noh
		San Juan	sahn HWAHN	Soviet Union	SOH-vyeht
Portugal	POHR-chuh-guhl	Sandinista	sahn-dee-NEESS-tah	Sri Lanka	shree LAHNG-kah
Potosí	poh-toh-SEE	Sanskrit	SAN-skriht	St. Croix	KROY
Pudong	POO-DOONG	Santeria	sahn-tuh-REE-uh	St. Kitts and Nevis	KIHTSS; NEH-vihss
Puerto Rico	PWAIR-toh REE-koh/POHR-toh REE-koh	São Paulo	sau PAU-loh	St. Lucia	LOO-shuh
		Sarajevo	sah-rah-YAY-voh	St. Martin	MAHR-tuhn
Punjab	poon-JAHB	Sarawak	suh-RAH-wahk	Stalin, Joseph	STAH-lihn, YOH-sehf
purdah	PURR-duh	Saro-Wiwa, Ken	SAHR-oh WEE-wah	Strait of Malacca	mah-LAHK-uh
Qaidam Basin	CHYE-DAHM	Saud	sah-OOD	Subic Bay	SOO-bihk
qanat	GAH-naht	Saudi Arabia	sah-OO-dee uh-RAY-bee-uh	Sudan	soo-DAN
Qatar	KAH-tahr			Sudd	SOOD
Qin Ling Mountains	CHIHN LIHNG	Sault Ste. Marie	SOO SAYNT muh-REE	Sudra	SOO-druh, SHOO-druh
Qin	CHIHN	Scandinavia	skan-dih-NAY-vee-uh		
Qingdao	CHIHNG-DAU	Scotland	SKAWT-luhnd	Suez Canal	SOO-ehz
Qinghai	CHIHNG-HYE	Seattle	see-AT-uhl	Suharto	SOO-HAHR-toh
Québec	kay-BEHK	Semey	seh-MAY	Sulawesi	soo-lah-WAY-see
Québecois	kay-behk-WAH	Senegal	SEHN-ih-gahl	Sulu Archipelago	SOO-loo
Quito	KEE-toh	Seoul	SOHL	Sumatra	soo-MAH-truh
Qur'an	koo-RAHN ["oo" as in "book"]	Serbia	SURR-bee-uh	Sunda Islands	SOON-duh ["oo" as in "book"]
		Seychelles	say-SHEHLZ		
Rabari	rah-BAHR-ee	Shaanxi	SHAH-AHN-ZHEE	Sundaland	SOON-duh-luhnd ["oo" as in "book"]
Rabat	rah-BAHT	shah	SHAH		
rai	RYE	Shanghai	SHAHNG-HYE	Sunni	SOO-nee
Rajasthan	rah-jah-STAHN	Shantou	SHAHN-TOH	Surat Thani	suh-RAHT TAH-nee
Ramadan	rahm-uh-DAHN	Shanxi	SHAHN-SHEE	Suriname	SOO-rih-NAHM
Rangoon	rahn-GOON	Shari'a	shah-REE-ah	Suva	SOO-vah
Recife	reh-SEE-feh	Shatt al Arab	SHAHT ahl AHR-uhb	Svalbard	SVAHL-bahr
reketiry	ryeh-kyeh-TEE-ree	sheik	SHAYK	Swahili	swah-HEE-lee
Renaissance	REHN-uh-sahnss	Shenzhen	SHEHN-JUHN	Swaziland	SWAH-zee-land
Reynosa	ray-NOH-sah			Syr Darya	SEER DAHR-yah
Rhine River	RINE				

Syria	SEER-ee-uh	Transvaal	tranz-VAHL	Venice	VEH-nihss
Tahiti	tah-HEE-tee	Trinidad and Tobago	TRIH-nih-dad, toh-BAY-goh	Veracruz	va-rah-CROOSS
taiga	TYE-guh			Vienna	vee-EHN-uh
Taipei	TYE-PAY	Tripoli	TRIH-puh-lee	vilas rural	VEE-lahss roo-RAHL
Taiwan	TYE-WAHN	Trivandrum	trih-VAHN-druhm	Vistula River	VIHSS-choo-luh ["oo" as in "book"]
Tajik	tah-JEEK	Trobriand Islands	TROH-bree-uhnd		
Tajikistan	tah-jee-kyih-STAHN	tsetse	TSEH-tsee	Vitória	vee-TOHR-ee-ah
Taklimakan Desert	TAHK-luh-muh-KAHN	Tshopo Falls	TCHOH-poh	Vladivostok	vluh-dyuh-vuh-STAWK
		Tuamotu Archipelago	too-uh-MOH-too		
Taliban	TAHL-ee-bahn			Volga River	VOHL-guh
Tamil Nadu	TAH-mihl NAH-doo	Tuareg	TWAH-rehg	Volgograd	vuhl-guh-GRAHT
Tamil	TAH-mihl	Tumen River	TOO-MUHN	Wa	WAH
Tangier	tahn-JEER	Tunis	TOO-nihss	wadi	WAH-dee
Tanzania	tan-zuh-NEE-uh	Tunisia	too-NEE-zhuh	Waitangi Treaty	wie-TAHNG-gee
Tarim Basin	TAH-REEM	Turfan Depression	TOOR-FAHN	Wallacea	woh-LAH-see-yuh
Tashkent	tahsh-KEHNT	Turin	TOOR-ihn	Warsaw	WOHR-saw
Tassili 'n Agger	tah-see-LEE nah-JAIR	Turkmenistan	turrk-mehn-ih-STAHN	Wenzhou	WUHN-JOH
Tatar	TAH-turr	Turkoman	TURR-kuh-mahn	West Papua	PAP-uh-wah
Tehran	teh-RAHN	Tutsi	TOOT-see	Wu Shan	WOO SHAHN
Tehuantepec	teh-WAHN-teh-pehk	Tutu, Desmond	TOO-too	Wuhan	WOO-HAHN
Tel Aviv	tehl ah-VEEV	Tuva	TOO-vah	Xi Jiang	JOO JYAHNG
Tenochtitlán	teh-nawch-teet-LAHN	Tuvalu	too-VAH-loo	Xiamen	SHYAH-MEN
Thailand	TYE-land	Ubon Ratchathani	OO-buhn RAH-chayt-nee	Xian	SHEE-AHN
Thar Desert	TAHR			Xianggang	SHYAHNG-GAHNG
Thimphu	THIHM-boo	Udaipur	oo-dye-POOR	Xibe	SHEE-BUH
Tiananmen Square	TYEHN-AHN-MEHN	Ujung Pandang	OO-joong pahn-DAHNG [second "oo" as in foot]	Xinjiang Uygur	SHIHN-JYAHNG WEE-gurr
Tianjin	TYEHN-JIHN				
Tibet	tih-BEHT			Xizang	SHEE-DZAHNG
tierra caliente	tee-AIR-ah cah-lee-EHN-tay	Ukraine	yoo-KRAYN	Yalu River	YAH-LOO
		Ulan Bator	OO-lahn BAH-tawr	Yamato	YAH-mah-toh
Tierra del Fuego	tee-AIR-ah dehl FWAY-goh	Uluru	oo-LOO-roo	Yamuna River	yah-MOO-nah
		Umbanda	oom-BAHN-dah	Yangtze River	YAHNG-TSUH
tierra fria	tee-AIR-ah FREE-uh	ummah	OO-muh	Yemen	YEH-muhn
tierra helada	tee-AIR-ah ay-LAH-dah	United Arab Emirates	EHM-uh-ruhts	Yi	YEE
				Yichang	YEE-CHAHNG
tierra templada	tee-AIR-ah temp-LAH-dah	Ural Mountains	YOOR-uhl	Yokohama	yoh-kaw-HAH-mah
		Uruguay	OO-roo-gwaiy	Yoruba	YOH-roo-bah ["oo" as in "book"]
Tigris River	TYE-grihss	Urumqi	OO-ROOM-CHEE [both "oo" as in "book"]		
Tijuana	tee-HWAH-nah			Yugoslavia	yoo-goh-SLAH-vee-uh
Timbuktu	tihm-book-TOO [first "oo" as in "book"]	Uttar Pradesh	OO-turr prah-DAYSH	Yunnan	YOO-NAHN
		Uygur	WEE-gurr	yurt	YURRT
Togo	TOH-goh	Uzbekistan	ooz-behk-ih-STAHN	Zabbaleen	zah-bah-LEEN
Tokarp	TOH-kahrp	Vaishya	VYE-shyuh	Zagros Mountains	ZAH-grohss
Tokyo	TOH-kee-oh	Valdez	val-DEEZ	Zaire	zah-EER
tolkuchka	tohl-KOOCH-kah	Vancouver	van-KOO-vurr	zakat	zah-KAHT
Tonga	TAWNG-gah	Vanuatu	vah-noo-AH-too	Zambia	ZAM-bee-uh
Tordesillas	tohr-day-SEE-yahss	Varanasi	vuh-RAH-nuh-see	Zhuang	JWAHNG
Toronto	tuh-RAWN-toh	Vedda	VEHD-uh	Zhuhai	JOO-HYE
Toulouse	too-LOOZ	Venezuela	veh-neh-ZWAY-lah		
Transcaucasia	tranz-kaw-KAY-zhuh				

GLOSSARY

Aborigines (p. 559) the longest surviving inhabitants of Oceania, whose ancestors, the Australoids, migrated from Southeast Asia possibly as early as 50,000 years ago, over the Sundaland landmass that was exposed during the ice ages

acid rain (p. 91) acidic precipitation that has formed through the interaction of rainwater or moisture in the air with sulfur dioxide and nitrogen oxides emitted during the burning of fossil fuels

acculturation (pp. 146, 204) adaptation of a minority culture to the host culture enough to function effectively and be self-supporting; cultural borrowing

adivasis (p. 409) a social group outside the caste system thought to be descendants of the ancient original inhabitants of South Asia

African Union (p. 357) a political organization consisting of all the countries on the African continent and some nearby islands promoting economic cooperation and social welfare

age distribution or **age structure** (p. 40) the proportion of the total population in each age group

Age of Exploration (p. 186) a period of accelerated global commerce and cultural exchange facilitated by improvements in European navigation, shipbuilding, and commerce in the fifteenth and sixteenth centuries

agribusiness (p. 71) the business of farming conducted by large scale operations that produce, package, and distribute agricultural products

agricultural intensification (p. 574) additional inputs of labor to a given area of land that typically lead to an increase in crop yields

agroecology (p. 409) the practice of traditional, nonchemical methods of crop fertilization and the use of natural predators to control pests

agroforestry (p. 372) the raising of economically useful trees

Ainu (p. 469) an indigenous cultural minority group in Japan characterized by their light skin, heavy beards, and thick, wavy hair, who are thought to have migrated thousands of years ago from the northern Asian steppes

air pressure (p. 24) the force exerted by a column of air on a square foot of surface

Allah (p. 289) the Arabic word for "God," used by Arabic-speaking Christians and Muslims in their prayers

alluvium (p. 481) river-borne sediment

altiplano (p. 169) an area of high plains in the central Andes of South America

animism (pp. 365, 409) a belief system in which natural features carry spiritual meaning

apartheid (p. 342) a system of laws mandating racial segregation in South Africa, in effect from 1948 until 1994

aquifer (p. 94) a natural underground reservoir

archipelago (p. 504) a group, often a chain, of islands

Arya (p. 396) an ancient people of Southwest Asia; the first recorded invaders of South Asia

ASEAN Free Trade Association (AFTA) (p. 521) a free-trade association of Southeast Asian countries launched in 1992 by ASEAN and patterned after the North American Free Trade Agreement and the European Union

Asia-Pacific region (p. 493) a huge trading area that includes all of Asia and the countries around the Pacific Rim

assimilation (pp. 146, 204) the loss of old ways of life and the adoption of the lifestyle of another culture

Association of Southeast Asian Nations (ASEAN) (p. 521) an organization of Southeast Asian governments established to further economic growth and political cooperation

atoll (p. 555) a low-lying island, formed of coral reefs that have built up on the circular or oval rims of submerged volcanoes

Australo-Melanesians (p. 506) a group of hunters and gatherers who moved from the present northern Indian and Burman parts of southern Eurasia into the exposed landmass of Sundaland about 40,000 to 60,000 years ago

Austronesians (pp. 507, 559) a Mongoloid group of skilled farmers and seafarers from southern China who migrated south to various parts of Southeast Asia between 10,000 and 5000 years ago

autonomous regions (p. 257) territories within Russia and China designated as the homelands of indigenous peoples

autonomy (p. 15) the capacity of a people or a country to control their own affairs

average population density (p. 38) the average number of people per unit area (for example, per square mile or square kilometer)

Aztecs (p. 121) native people of central Mexico noted for advanced civilization before the Spanish conquest

baby boomer (p. 90) a member of the largest age group in North America, the generation born in the years after World War II, from 1947 to 1964, in which a marked jump in the birth rate occurred

barriadas (p. 128) see **favelas**

barrios (p. 128) see **favelas**

bimodal (p. 73) having or requiring two means of operation. An economic sector that is bimodal may require both unskilled and highly skilled employees.

biochemical oxygen demand (p. 425) the amount of dissolved oxygen consumed by decomposing biological waste or remains

biogeographical transition zone (p. 505) an area in which there is some mixing of two or more distinct types of flora and fauna

birth rate (p. 39) the number of births in a given population per unit time, especially per year

Bolsheviks (p. 242) a faction of communists who came to power during the Russian Revolution

boreal forest (p. 104) northern coniferous forests

Brahmins (p. 409) members of the Hindu priestly caste, which is the most privileged caste in ritual status

brain drain (pp. 131, 395) the migration of educated and ambitious young adults to cities or foreign countries, depriving the sending communities of talented youth in whom they have invested years of nurturing and education

bride price (p. 411) a price paid by a groom to the family of the bride; the opposite of dowry

brownfields (p. 79) old industrial sites whose degraded conditions pose obstacles to redevelopment

Buddhism (p. 407) a religion of Asia that originated in northern India in the sixth century B.C. and emphasizes modern living and peaceful self-reflection leading to enlightenment

buffer zone (pp. 476, 508) a neutral area that serves to prevent conflict

Canadian Shield (p. 104) the vast glaciated territory in Canada lying north of the Great Plains; characterized by thin soils, innumerable lakes, and large rivers

capital (p. 33) wealth in the form of money or property used to produce more wealth

capital controls (p. 34) restrictions on business investment by foreigners requiring that investment capital stay in the host country for a minimum length of time or that companies be partially controlled by local investors

capitalists (p. 243) a wealthy minority that owns the majority of factories, farms, businesses, and other means of production

carrying capacity (p. 343) the maximum number of people that a given territory can support sustainably with food, water, and other essential resources

cartel (p. 306) a group that is able to control production and set prices for its products

cash economy (p. 42) an economic system in which the necessities of life are purchased with monetary currency

caste (p. 409) an ancient Hindu system for dividing society into hereditary hierarchical classes

chaebol (p. 496) huge corporate conglomerates in South Korea that receive government support and protection

chain migration (p. 81) a pattern in which immigrants to a new country encourage their family and friends to join them, thus creating a community of culturally similar immigrants in a particular place

chernozem (p. 241) black, fertile soils of the region stretching south from Moscow toward the Black and Caspian seas

China proper (p. 448) the North China Plain, a rich agricultural area that was unified under successive empires and that is considered one of the hearths of East Asian culture

Chipko movement (p. 422) a grassroots Indian environmental movement that attempts to slow down deforestation and increase ecological awareness

Christianity (p. 289) a monotheistic religion based on belief in the resurrection and on the teachings of Jesus of Nazareth, a Jew, who described God's relationship to humans as primarily one of love and support, as exemplified by the Ten Commandments

Christians (p. 408) followers of Christianity

clear-cutting (p. 61) the cutting down of all trees on a given plot of land, regardless of age, health, or species

climate (p. 24) the long-term balance of temperature and precipitation that characteristically prevails in a particular region

cold war (p. 189) the contest that pitted the United States and Western Europe, espousing free-market capitalism and democracy, against the former USSR and its allies, promoting a centrally planned economy and a socialist state

cold war era (p. 48) the period from 1946 to the early 1990s, when the United States and its allies in Western Europe faced off against the Union of Soviet Socialist Republics and its allies in Eastern Europe

colonias (p. 128) see **favelas**

colonizing (or **mother**) **country** (p. 123) the country from which the people of a colony or former colony derive their origin

colony (p. 32) a (usually) distant land acquired by a more powerful country for economic gain

command economy (pp. 243, 458) an economy in which government bureaucrats plan, locate, and manage all production and distribution

Common Market for Eastern and Southern Africa (COMESA) (p. 357) a subregional organization that promotes economic cooperation among 20 African states

communal conflict (p. 408) a euphemism for religion-based violence in South Asia

communism (p. 242) an ideology, based largely on the writings of the German revolutionary Karl Marx, that calls on workers to unite to overthrow capitalists and establish an egalitarian society where workers share what they produce

Communist party (p. 243) the political party that ruled the former Russian Empire from 1917 to 1991

Confucianism (p. 448) a Chinese philosophy that teaches that the best organizational model for the state and society is a hierarchy based on the patriarchal family

contested space (p. 140) any area that several groups claim or want to use in different ways, such as the Amazon or Palestine

contiguous regions (p. 11) regions that lie next to each other

conurbation (p. 129) an interconnected group of urban areas and suburban developments; a metropolitan area; see megaopolis

Corridor Five (p. 200) a planned and partially constructed major transportation roadway across southern Europe, stretching from Barcelona, Spain, to Kiev, Ukraine

Creole (p. 124) a person of Spanish descent born in the Americas

Creole cultures (p. 116) distinctly American cultures that emerged from blending immigrant cultures such as those of Spain, Africa, Holland, Germany, Britain, China and Japan.

crony capitalism (p. 520) a type of corruption in which politicians, bankers, and entrepreneurs, sometimes members of the same family, have close personal as well as business relationships

cultural diversity (p. 15) differences in ideas, values, technologies, and institutions among culture groups

cultural homogeneity (p. 15) uniformity of ideas, values, technologies, and institutions among culture groups

cultural identity (p. 15) a sense of personal affinity with a particular culture group

cultural marker (p. 15) a characteristic that helps to define a certain culture group

cultural pluralism (p. 524) the cultural identity characteristic of a region where groups of people from many different backgrounds have lived together for a long time, yet have remained distinct

Cultural Revolution (p. 451) a series of highly politicized and destructive mass campaigns launched in 1966 to force the entire population to support the continuing revolution in China

culture (p. 14) everything we use to live on earth that is not directly part of our biological inheritance

culture group (p. 14) a group of people who share a set of beliefs, a way of life, a technology, and usually a place

currency devaluation (p. 355) the lowering of a currency's value relative to the U.S. dollar, the Japanese yen, the European euro, or other currency of global trade

czar (p. 242) title of the ruler of the Russian Empire; derived from the word "caesar," the title of the Roman emperors

death rate (p. 39) the ratio of total deaths to total population in a specified community

delta (p. 23) the triangular-shaped plain of sediment that forms where a river meets the sea

democratic institutions (p. 188) institutions—such as constitutions, elected parliaments, and impartial courts—through which the common people gained a formal role in the political life of the nation

demographic transition (p. 42) the change from high birth and death rates to low birth and death rates that usually accompanies a cluster of other changes, such as change from a subsistence to a cash economy, increasing education rates, and urbanization

demography (p. 37) the study of population patterns and changes

dense nodes (p. 79) small regions in an urban area that have particularly dense populations

deposition (p. 23) the settling out of sand and soil particles carried by wind or moving water when the speed of their flow slows

desertification (pp. 317, 371) a set of ecological changes that converts nondesert lands into deserts

detritus (p. 505) dead organic material (such as plants and insects)

development (p. 35) usually used to describe economic changes like greater productivity of agriculture and industry that lead to better standards of living.

"Development for whom?" (p. 35) a question that asks whether a particular development strategy will truly improve average standards of living for most local people, or if it will raise production levels and increase profits for only a few, those who are already well off

devolution (p. 220) the weakening of a formerly tightly unified state

Diaspora (p. 289) the dispersion of Jews around the globe after they were expelled from the eastern Mediterranean by the Roman Empire beginning in A.D. 73

diffusion (p. 28) the process by which agriculture and the domestication of animals spread around the world from a few places

digital divide (pp. 13, 24, 145) the discrepancy in access to information technology between small, rural, and poor areas and large, wealthy cities; major government research laboratories; and universities

disinvestment (p. 350) removal of financial support

doi moi (p. 541) economic and bureaucratic restructuring in Vietnam

domestication (p. 28) the process of developing plants and animals through selective breeding to live with and be of use to humans

double day (p. 205) the longer workday of women with jobs outside the home who also work as family caretakers, housekeepers, and cooks at home

dowry (p. 441) a price paid by the family of a bride to the groom; the opposite of bride price

dry forests (p. 372) forests that lose their leaves during the dry season

dumping (p. 201) the cheap sale on the world market of overproduced commodities, lowering global prices and hurting producers of these same commodities elsewhere in the world

early extractive phase (p. 133) a phase in Central and South American history, beginning with the Spanish conquest and lasting until the early twentieth century, characterized by colonial policies such as mercantilism that resulted in unequal trade

earthquake (p. 22) a catastrophic shaking of the landscape, often caused by the shifting and friction of tectonic plates

East African Community (EAC) (p. 357) an organization formed by Kenya, Tanzania, and Uganda to promote economic links among the countries of East Africa

Economic Community of Central African States (CEEAC) (p. 357) a subregional organization that promotes economic cooperation among 11 Central African states

Economic Community of West African States (ECOWAS) (p. 357) an organization of West African states working toward forming an economic union

economic core (p. 60) the dominant economic region within a larger region; in nineteenth-century North America the core included southern Ontario and the north-central part of the United States (chiefly Illinois, Indiana, Ohio, New York, New Jersey, and Pennsylvania)

economic diversification (p. 308) the expansion of an economy to include a wider array of economic activities

economic integration (p. 190) the free movement of people, goods, money, and ideas among countries

economic restructuring (p. 114) reorganization of an economy to encourage economic growth in markets free of government controls

economic and technology development zones (ETDZs) (p. 461) zones in China with fewer restrictions on foreign business, established to encourage foreign investment and economic growth

economies of scale (p. 194) reductions in the unit costs of production that occur when goods or services are produced in large amounts, resulting in a rise in profits per unit

economy (p. 31) the forum in which people make their living, including the spatial, social, and political aspects of how resources are recognized, extracted, exchanged, transformed, and reallocated

ecotourism (pp. 53, 373) nature-oriented vacations often taken in endangered and remote landscapes, usually by travelers from industrialized nations

El Niño (pp. 120, 556) periodic climate-altering changes, especially in the circulation of the Pacific Ocean, now understood to operate on a global scale

endemic (p. 557) belonging or restricted to a particular place

erosion (p. 23) the process by which fragmented rock and soil are moved over a distance, primarily by wind and water

ethnic cleansing (p. 190) the systematic removal of an ethnic group or people from a region or country by deportation or genocide

ethnic group (pp. 14, 336) a group of people who share a set of beliefs, a way of life, a technology, and usually a geographic location

ethnicity (pp. 86, 336) the quality of belonging to a particular culture group

ethnocentrism (p. 336) the belief in the superiority of one's own cultural perspective

euro (p. 195) the official currency of 12 of the 15 European Union countries, as of January 1, 1999

European colonialism (p. 11) the practice of taking the human and natural resources of distant places in order to produce wealth for Europe

European Economic Community (EEC) (p. 194) an economic community created in 1958, when Belgium, Luxembourg, the Netherlands, France, Italy, and West Germany agreed to eliminate certain tariffs and promote mutual trade and cooperation in the interest of achieving stronger economic ties within Europe

European Union (EU) (p. 178) a supranational institution including most of West, South, and North Europe, established to bring economic integration to member countries

evangelical Protestantism (p. 148) a Christian movement that focuses on personal salvation and empowerment of the individual through miraculous healing and transformation; some practitioners preach the "gospel of success" to the poor: that a life dedicated to Christ will result in prosperity of the body and soul

exchange or service center (p. 32) barter or trade for money, goods, or services

exclave (p. 226) a portion of a country separated from the main part and constituting an enclave in respect of the surrounding territory

export processing zones (EPZs) (or free trade zones) (pp. 137, 517) specially created legal spaces or industrial parks within a country where, to attract foreign-owned factories, duties and taxes are not charged

extended family (p. 146) a family consisting of related individuals beyond the nuclear family of parents and children

external debts (p. 114) the debts a country owes to other countries or international financial institutions

external processes (geophysical) (p. 23) landform-shaping processes that originate at the surface of the earth, such as weathering, mass wasting, and erosion

extraction (p. 32) the acquisition of a material resource through mining, logging, agriculture, or other means

extractive resource (p. 31) a resource such as mineral ores, timber, or plants that must be mined from the earth's surface or grown from its soil

extraregional migration (p. 529) short-term or permanent migration to countries outside a region

favelas (p. 128) Brazilian urban slums and shantytowns built by the poor; called colonias, barrios, or barriadas in other countries

female circumcision (p. 369) the removal of the labia and the clitoris and sometimes the stitching nearly shut of the vulva

feminization of labor (p. 517) the increasing representation of women in both the formal and informal labor force

Fertile Crescent (p. 288) an arc of lush, fertile land formed by the uplands of the Tigris and Euphrates river systems and the Zagros Mountains, where nomadic peoples began the earliest known agricultural communities

feudalism (p. 185) a politico-social system once prevalent in Europe and Asia and elsewhere in which a class of professional fighting men, or knights, defended the monarch and the peasants or serfs, who cultivated the lands of their protectors.

floating population (p. 463) jobless or underemployed people who have left economically depressed rural areas for the cities and move from place to place looking for work

floodplain (p. 23) the flat land around a river where sediment is deposited during flooding

food security (p. 459) the ability of a state to supply a sufficient amount of basic food to the entire population consistently

forced migration (p. 530) the movement of people against their wishes

foreign direct investment (FDI) (p. 133) the amount of money invested in a country's business by citizens, corporations, or governments of other countries

foreign exchange (p. 529) foreign currency such as U.S. dollars that countries need to purchase imports.

formal economy (p. 32) all aspects of the economy that take place in official channels

formal institutions (p. 14) associations such as official religious organizations; local, state, and national governments; nongovernmental organizations; and specific businesses and corporations

forward capital (pp. 175, 361) a city built to draw migrants and investment for economic development

fragile environment (p. 372) an area that contains barely enough water, soil nutrients, or other resources essential to meet the needs of plants and animals; human pressure in such environments may result in long-term or irreversible damage to plant and animal life

free trade (p. 34) the movement of goods and capital without government restrictions

free trade zones (p. 137) see **export processing zones**

frontal precipitation (p. 28) rainfall caused by the interaction of large air masses of different temperatures and densities

gender structure (p. 40) the proportion of males and females in each age group of a population

genetic engineering (p. 71) the genetic manipulation of crops to produce strains resistant to pests or disease

genetically modified organisms (GMOs) (p. 196) plants or animals whose genetic code has been modified to make them more attractive as commodities

Geneva Conventions (p. 66) treaties concerning the conduct of war, some of which protect the rights of prisoners of war

genocide (p. 344) the deliberate destruction of an ethnic, racial, or political group

gentrification (p. 79) the renovation of old urban districts

geopolitics (p. 48) the use of strategies by countries to ensure that their best interests are served

ger (p. 483) see **yurt**

glasnost ("openness") (p. 244) an opening up of public discussion of social and economic problems, which occurred in the Soviet Union under Mikhail Gorbachev in the late 1980s

global economy (p. 30) the worldwide system in which goods, services, and labor are exchanged

global scale (p. 10) the level of geography that encompasses the entire world as a single unified area

global warming (p. 46) the predicted warming of the earth's climate as atmospheric levels of greenhouse gases increase

globalization (p. 13) the growth of interregional and worldwide linkages and the changes they are bringing about

Gondwana (p. 554) the great landmass that formed the southern part of the ancient supercontinent Pangaea

government subsidies (p. 75) amounts paid by the government to cover part of the production costs of some products and to help domestic producers to sell their goods for less than foreign competitors

grassroots economic development (p. 357) economic development projects designed to help individuals and their families achieve sustainable livelihoods

Great Barrier Reef (p. 554) the longest coral reef in the world, located off the northeastern coast of Australia

Great Basin (p. 105) the dry, mountainous region in North America between the Rocky Mountains and Pacific Coastal zone

Great Leap Forward (p. 451) an economic reform program under Mao Zedong intended to raise China to the industrial level of Britain and the United States in a short time

Green parties (p. 208) environmentally conscious political parties

green revolution (p. 413) increases in food production brought about through the use of new seeds, fertilizers, mechanized equipment, irrigation, pesticides, and herbicides

gross domestic product (GDP) (p. 32) the market value of all goods and services produced by workers and capital within a particular nation's borders and within a given year

gross domestic product (GDP) per capita (p. 35) the market value of all goods and services produced by workers and capital within a particular nation's borders and within a given year; dividing the value by the number of people in the country results in the per capita value

Group of Eight (G8) (p. 254) an organization of highly industrialized countries: France, the United States, Britain, Germany, Japan, Italy, Canada, and Russia

growth poles (p. 462) zones of development whose success draws more investment and migration to a region

growth rate (p. 39) see **rate of natural increase**

growth triangles (p. 517) large transnational economic regions in which firms attempt to find the best skills and resources for the best prices

guest workers (p. 201) immigrants from outside Europe, often from former colonies, who came to Europe (often temporarily) to fill labor shortages; Europeans expected them to return home when no longer needed

Gulf War (1990–1991) (p. 330) a conflict in which a military coalition of the United States and European and Arab countries quickly defeated Iraq after it invaded oil-rich Kuwait in 1990

hacienda (p. 134) a large agricultural estate in Middle or South America, more common in the past; usually not specialized by crop and not focused on market production

hajj (p. 290) the pilgrimage to the city of Makkah (Mecca) that all Muslims are encouraged to undertake at least once in a lifetime

Harijans (or Dalits or untouchables) (p. 409) members of a social group in Hindu India considered so lowly as to have no caste

hazardous waste (p. 91) nuclear, chemical, or industrial wastes that can have damaging environmental consequences

hearth (p. 407) place of origin

hierarchical societies (p. 560) social structures in which groups are ranked according to power, prestige, and wealth, with layers of ruling elites at the top and undifferentiated commoners at the bottom

high islands (p. 555) volcanoes that rise above the sea into mountainous, rocky formations that contain a rich variety of environments

Hindustani (p. 396) during Mughal rule, the lingua franca (language of trade) of all of northern India and what is today Pakistan; the precursor of modern Urdu and Hindi

Hispanic (p. 56) a loose ethnic term that refers to all Spanish-speaking people from Latin America and Spain; equivalent to **Latino**

homogenization (p. 13) the process of becoming more nearly uniform

Horn of Africa (p. 338) the part of Africa that juts out from East Africa and wraps around the southern tip of the Arabian Peninsula

hot spot (p. 554) an individual site of upwelling material (magma) that originates deep in the mantle of the earth and arrives at the surface in a tall plume; hot spots tend to remain fixed relative to migrating plates

hub-and-spoke network (p. 72) the organization of air service in North America around hubs, or strategically located airports used as collection and transfer points for passengers and cargo traveling from one place to another

hukou system (p. 463) China's system by which its citizens' permanent residence is registered

human geography (p. 3) the study of various aspects of human life that create the distinctive landscapes and regions of the world

human well-being (p. 35) the ability of people to obtain for themselves a healthy life in a place and community of their choosing

import quota (p. 35) a limit on the amount of a given item that may be imported into a country over a period of time

import substitution industrialization (ISI) (p. 134) a form of industrialization involving the use of public funds to set up factories to produce goods that previously had been imported

Incas (p. 121) Native American people who ruled the largest pre-Columbian state in the Americas, with a domain stretching from southern Colombia to northern Chile and Argentina

income disparity (p. 133) a dramatic gap in wealth and resources between the rich elite and the poor majority of a country or region

Indian diaspora (p. 395) the set of all people of South Asian origin living (and often born) abroad

Indus Valley civilization (p. 395) the first substantial settled agricultural communities in South Asia, which appeared about 4500 years ago along the Indus River in modern-day Pakistan and along the Saraswati River in modern-day India

industrial production (p. 32) manufacture of goods for sale

industrial revolution (p. 33) a series of innovations and ideas that allowed manufacturing to be mechanized

informal economy (pp. 32, 139) all aspects of the economy that take place outside official channels

informal institutions (p. 14) ordinary or casual associations, such as the family or a community

information technology (IT) (p. 73) the part of the service sector that relies on the use of computers and the Internet to process and transport information, including banks, software companies, medical technology companies, and publishing houses

institutions (p. 14) all of the associations, formal and informal, that help people get along together

internal processes (geophysical) (p. 23) processes, like plate tectonics, that originate deep beneath the surface of the earth

internal republics (p. 257) more than 30 ethnic enclaves in Russia whose political status and relationship to the Russian state are continuously being renegotiated republic by republic

International Monetary Fund (IMF) (p. 50) a financial institution funded by the developed nations to help developing countries reorganize, formalize, and develop their economies

Internet (p. 73) a computer network that allows for the electronic transfer of all kinds of information in seconds

interregional linkages (p. 11) economic, political or social connections between contiguous or widely separated regions

Interstate Highway System (p. 72) the federally subsidized network of highways in the United States

intertropical convergence zone (ITZC) (p. 337) a band of atmospheric currents circling the globe roughly at the equator; warm winds from both north and south converge at the ITCZ, pushing air upward and causing copious rainfall

intifada (p. 310) a prolonged Palestinian uprising against Israel

iron curtain (p. 189) a fortified border zone that separated Western Europe from Eastern Europe during the cold war

Islam (pp. 289, 407) a monotheistic religion that emerged in the seventh century A.D. when, according to tradition, the archangel Gabriel revealed the tenets of the religion to the Prophet Muhammad

Islamic fundamentalism (Islamisno) (p. 312) a grassroots movement to replace secular governments and civil laws with governments and laws guided by Islamic principles

Islamists (p. 304) fundamentalist Muslims who favor a religion-based state that incorporates conservative interpretations of the Qur'an and strictly enforced Islamic principles into the legal system

Israeli–Palestinian conflict (p. 309) a long-running conflict over the control of territory between Israel and Palestinians who live on lands now occupied by Israel

Jainism (p. 408) a faith tradition that is more than 2000 years old; Jains are found mainly in western India and large urban centers and are known for their strict vegetarianism

Jains (p. 408) followers of Jainism

jati (p. 409) in Hindu India, the subcaste into which a person is born, which largely defines the individual's experience for a lifetime

Jews (p. 408) followers of Judaism; people of Jewish heritage and culture

Judaism (p. 289) a monotheistic religion characterized by the belief in one God, Yaweh, a strong ethical code summarized in the Ten Commandments, and an enduring ethnic identity

kanban system (p. 456) an efficient Japanese manufacturing system in which suppliers are located in close proximity to the main factory where final assembly takes place

knowledge economy (p. 73) the markets based on the management of information, such as those of finance, media, and research and development

Kshatriyas (p. 409) members of the Hindu warrior and ruler caste

ladino (p. 161) a local term for mestizo used in Central America

laterite (p. 338) a permanently hard surface left when minerals in tropical soils are leached away

Latino (p. 56) a loose ethnic term that refers to all Spanish-speaking people from Middle and South America and Spain; equivalent to Hispanic

leaching (p. 338) the washing out into groundwater of soil minerals and nutrients released into soil by decaying organic matter

liberation theology (p. 148) a movement within the Roman Catholic church that uses the teachings of Christ to encourage the poor to organize to change their own lives and the rich to promote social and economic reform

lingua franca (pp. 17, 370) a language used to communicate by people who do not speak one another's native languages; often a language of trade.

livestock ranch (p. 134) a farm that specializes in raising cattle and sheep

living wages (p. 34) minimum wages high enough to support a healthy life

local scale (p. 10) the level of geography that describes the space occupied by an individual such as a city, village or rural area

loess (pp. 56, 476) windblown dust that forms deep soils in North America, central Europe, and China

long-lot system (p. 97) a system of long, narrow plots of land stretching back from the edge of the St. Lawrence River, which gave French Canadian settlers access to resources extending inland from the river

low-till (p. 71) a low-impact method of plowing that reduces the loss of topsoil

machismo (p. 147) a set of values that defines manliness in Middle and South America

makatea (p. 555) raised coral platforms formed when atolls are uplifted

maquiladoras (pp. 110, 137) foreign-owned factories, often located in Mexican towns just over the border from U.S. towns, that hire people at low wages to assemble manufactured goods that are then sent elsewhere for sale; now also used for similar firms in other parts of Mexico and Middle America

marianismo (p. 146) a set of values based on the life of the Virgin Mary that defines the proper social roles for women in Middle and South America

marketization (p. 249) the process of developing a free market economy

marsupials (p. 557) mammals that give birth to their young at a very immature stage and then nurture them in a pouch equipped with nipples

material culture (p. 14) all the things, living or not, that humans use

medieval period (p. 185) the period in Europe A.D. 450–1300 during which civil society declined and commerce ceased as the Roman Empire collapsed. By 1250, town life, trade, and commerce revived and diffusion from outside and European innovation encouraged the flourishing of the arts, philosophy, and architecture

megalopolis (p. 80) an area formed when several cities grow to the extent that their edges meet and coalesce

Melanesia (p. 559) New Guinea and the islands south of the equator and west of Tonga (the Solomon Islands, New Caledonia, Fiji, Vanuatu)

Melanesians (p. 559) a group of Australoids named for their relatively dark skin tones, a result of high levels of the protective pigment melanin; they settled throughout New Guinea and other nearby islands

mercantilism (p. 186) the policy by which the rulers of Spain and Portugal, and later of England and Holland, sought to increase the power and wealth of their realms by managing all aspects of production, transport, and trade in their colonies

Mercosur (p. 139) a free trade zone created in 1991 that links the economies of Brazil, Argentina, Uruguay, and Paraguay to create a common market

mestizo (p. 125) a person of mixed European and Native American descent

metropolitan areas (p. 79) cities with population of 50,000 or more and their surrounding suburbs

microcredit (p. 416) a program that makes very small loans available to poor entrepreneurs

Micronesia (p. 559) the small islands lying east of the Philippines and north of the equator

Middle America (p. 116) in this book, a region including Mexico, Central America, and the islands of the Caribbean

migration (p. 128) the movement of people from one place to another

MIRAB economy (p. 571) an economy based on migration, remittance, aid, and bureaucracy

mixed agriculture (p. 382) the raising of a variety of crops and animals on a single farm, often to take advantage of several environmental niches

Mongols (p. 242) a loose confederation of nomadic pastoral people centered in eastern Central Asia, who by the thirteenth century established by conquest an empire stretching from Europe to the Pacific

monotheistic (p. 289) pertaining to the belief that there is only one god

monotremes (p. 557) egg-laying mammals such as the duck-billed platypus and the spiny anteater

monsoon (pp. 25, 393) opposing winter and summer pattern of atmospheric and moisture movement between continents and oceans, in which warm, wet air coming in from the ocean brings copious rainfall during the summer months; in winter, cool, dry air moves from the continental interior toward the ocean

mother country (p. 123) see **colonizing** (or **mother**) **country**

Mughals (p. 396) dynasty of Central Asian origin that ruled India from the sixteenth to the nineteenth century

multiculturalism (p. 15) the state of relating to, reflecting, or being adapted to diverse cultures

multinational corporation (or transnational) (p. 33) a business organization that operates extraction, production, and/or distribution facilities in multiple countries

Muslims (p. 407) followers of Islam

nation (p. 49) a group of people who share a language, culture, political philosophy, and usually a territory

nation-state (p. 49) a political unit, or country, formed by people who share a language, a culture, and a political philosophy

nationalism (p. 188) devotion to the interests or culture of a particular country or nation (cultural group); the idea that a group of people living in a specific territory and sharing cultural traits should be united in a single country to which they are loyal and obedient

neocolonialism (pp. 134, 333) modern efforts by dominant countries to control economic and political affairs in other countries to further their own aims

neoliberalism (p. 137) the idea that social justice—meeting basic nutritional, shelter, and educational needs for everyone—can best be achieved through free-market economic development

New Urbanism (p. 79) the growing popularity of urban areas

noble savage (p. 559) a term coined by European Romanticists to describe what they termed the "primitive" peoples of the Pacific, who lived in distant places supposedly untouched by corrupting influences

nodes (p. 79) see **dense nodes**

nomadic pastoralism (p. 497) a way of life and economy centered on the tending of grazing animals who are moved seasonally to gain access to the best grasses

nongovernmental organization (NGO) (p. 50) an association outside the formal institutions of government, in which individuals from widely differing backgrounds and locations share views and activism on political, economic, social, or environmental issues

nonmaterial resources (p. 31) skills and knowledge of economic value

nonpoint sources of pollution (p. 266) diffuse sources of environmental contamination, such as untreated automobile exhaust, raw sewage, or agricultural chemicals that drain from fields into the urban water supplies

North American Free Trade Agreement (NAFTA) (p. 139) a free trade agreement created in 1994 that added Mexico to the 1989 agreement between the United States and Canada

nuclear family (p. 89) a family consisting of a father and mother and their children

oblasts (p. 257) the provinces of Russia

Ogallala aquifer (p. 94) the largest North American natural aquifer, which underlies the Great Plains

old-growth forests (p. 532) forests that have never been logged and therefore contain diverse ecosystems

oligarchs (p. 250) people who acquired great wealth during the privatization of Russia's resources, who use that wealth to exercise power

organic matter (p. 338) the remains of any living thing

Organization for Economic Cooperation and Development (OECD) (p. 47) the highly industrialized countries of North America, Europe, East Asia and Oceania (Australia, Austria, Belgium, Canada, Czech Republic, Denmark, Finland, France, Germany, Greece, Hungary, Iceland, Ireland, Italy, Japan, Korea, Luxembourg, Mexico, Netherlands, New Zealand, Norway, Poland, Portugal, Spain, Sweden, Switzerland, Turkey, United Kingdom, United States)

Organization of Petroleum Exporting Countries (OPEC) (pp. 67, 306) a cartel of oil-producing countries—including Algeria, Indonesia, Iran, Iraq, Kuwait, Libya, Nigeria, Oman, Qatar, Saudi Arabia, the United Arab Emirates, and Venezuela—that was established to regulate the production, and hence the price, of oil and natural gas

orographic rainfall (p. 25) rainfall produced when a moving moist air mass encounters a mountain range, rises, cools, and releases condensed moisture that falls as rain

outsourcing (p. 67) a strategy whereby employers lower their costs by relying on short-term contracts with workers provided by a third-party company

Pacific Rim (Basin) (p. 65) all of the countries that border the Pacific Ocean on the west and east

Pacific Way (p. 567) the idea that Pacific Islanders have a regional identity and a way of handling conflicts peacefully, which grows out of their own particular social experience

Pancasila (p. 522) the Indonesian national philosophy based on tolerance, particularly in matters of religion; its precepts include belief in God and the observance of conformity, corporatism, consensus, and harmony

Pangaea hypothesis (p. 22) the proposal that about 200 million years ago all continents were joined in a single vast continent, called by geologists Pangaea

Parsis (p. 408) a highly visible religious minority in India's western cities; Parsis are descendants of Persian migrants who did not give up their traditional religion of Zoroastrianism when Iran became Muslim

perestroika (p. 244) a restructuring of the Soviet economic system in the late 1980s in an attempt to revitalize the economy

permafrost (p. 239) permanently frozen soil a few inches or feet beneath the surface

permeable national borders (p. 110) borders subject to easy flow of people and goods

physical geography (p. 3) the study of the earth's physical processes to learn how they work, how they in turn affect humans, and how they are affected by humans

pidgin (p. 567) a language used for trading made up of words borrowed from the several languages of people involved in trading relationships

plantation (p. 134) a large estate or farm on which a large group of resident laborers grow (and partially process) a single cash crop

plate tectonics (p. 22) a theory proposing that the earth's surface is composed of large plates that float on top of an underlying layer of molten rock; the movement and interaction of the plates create many of the large features of the earth's surface, particularly mountains

pluralistic state (p. 49) a country in which power is shared among groups, each defined by common language, ethnicity, culture, or other characteristics

pogroms (p. 289) episodes of persecution, ethnic cleansing, and sometimes massacre, especially conducted against European Jews

political ecologist (p. 43) a geographer who studies the interactions among development, human well-being, and the environment

polygyny (pp. 302, 368) the taking by a man of more than one wife at a time

Polynesia (p. 559) the numerous islands situated inside an irregular triangle formed by New Zealand, Hawaii, and Easter Island

population pyramid (p. 40) a graph that depicts age and gender structures of a country

populist movements (p. 148) popularly based efforts seeking relief for the poor

price supports (p. 200) legal minimum prices set as a means of maintaining high prices despite an overabundance of supply

primate city (p. 128) a city that is vastly larger than all others in a country and in which economic and political activity is centered

private spaces (p. 300) spaces in the home relegated to women

privatization (p. 249) the sale of industries formerly owned and operated by the government to private companies or individuals

producer services (p. 199) a section of the service economy tailored to the needs of governments and businesses for advice, information, testing, licensing, and strategic planning

protectorate (p. 292) a relationship of partial control assumed by a European nation over a dependent country, while maintaining the trappings of local government

Protestant Reformation (p. 186) a European reform (or "protest") movement that challenged Catholic practices in the sixteenth century, resulting in the establishment of Protestant churches

public spaces (p. 299) social spaces such as the town square, shops, and markets, in contrast to the domestic (home and private) sphere

pull factors (p. 527) positive features of a place that attract people to move there

purchasing power parity (PPP) (p. 35) the amount of goods or services that U.S. $1 will purchase in a given country

purdah (p. 411) the practice of concealing women, especially during their reproductive years, from the eyes of nonfamily men

push factors (p. 527) negative features of the place where people are living that impel them to move elsewhere

pyroclastic flow (p. 117) a type of volcanic eruption characterized by blasts of superheated rocks, ash, and gas that move with great speed and force

qanats (p. 484) ancient underground conduits that carry groundwater for irrigation

Québecois (p. 56) French Canadian ethnic group or members of that group; also, all citizens of Québec, regardless of ethnicity

Qur'an (or Koran) (p. 289) the holy book of Islam, believed by Muslims to contain the words Allah revealed to Muhammad through the archangel Gabriel

racism (p. 21) the negative assessment of people, often those who look different, primarily on the basis of skin color and other physical features

rain shadow (p. 25) the dry side of a mountain range, facing away from the prevailing winds

rate of natural increase (growth rate) (p. 39) the rate of population growth measured as the excess of births over deaths per 1000 individuals per year, without regard for the effects of migration

recharge area (p. 316) the area with the highest precipitation, which recharges the aquifers

region (p. 9) a unit of the earth's surface that contains distinct patterns of physical features or of human development

regional conflict (p. 418) a conflict created by the resistance of a regional ethnic or religious minority to the authority of a national or state government

regional geography (p. 8) the analysis of the geographic characteristics of particular places

regional self-sufficiency (p. 458) an economic policy that encouraged each region to develop independently in the hope of evening out the national distribution of production and income in communist China

regional trade bloc (p. 34) an association of neighboring countries that have agreed to lower trade barriers for one another

religious nationalism (p. 418) the belief that a certain religion is strongly connected to a particular territory and that adherents should have political power in that territory

remittances (pp. 128, 527) earnings sent home by immigrant workers

Renaissance (p. 186) a broad European cultural movement in the fourteenth through sixteenth centuries that drew inspiration from the Greek, Roman, and Islamic civilizations, marking the transition from medieval to modern times

resettlement schemes (p. 528) government plans to move large numbers of people from one part of a country to another to relieve urban congestion, disperse political dissidents, or accomplish other social purposes

resource (p. 31) anything that is recognized as useful, such as mineral ores, forest products, skills, or brainpower

resource base (p. 191) the selection of raw materials and human skills available in a region for domestic use and industrial development

responsibility systems (p. 459) economic reforms that gave the managers of Chinese state-owned enterprises the right and the responsibility to make their operations work efficiently

Ring of Fire (p. 23) the tectonic plate junctures around the edges of the Pacific Ocean, characterized by volcanoes and earthquakes

roaring forties (p. 555) powerful westerly air and ocean currents that speed around the far Southern Hemisphere virtually unimpeded by landmasses

Russian Federation (p. 257) Russia and its political subunits, which include 30 internal republics and more than 10 so-called autonomous regions

Rust Belt (p. 99) the economic core region of North America that is characterized by obsolete industries and abandoned factories

Sahel (p. 338) a band of arid grassland that runs east–west along the southern edge of the Sahara Desert

salinization (p. 307) damage to soil caused by the evaporation of water, which leaves behind salts and other minerals

savanna (p. 338) grassland in tropical or subtropical regions

scale (p. 2) the proportion that relates the dimensions of a map to the dimensions of the area it represents

Schengen Accord (p. 201) an agreement signed in the 1990s by the European Union and many of its neighbors that called for free movement across common borders

seclusion (p. 292) the requirement that a woman stay out of public view, a regional cultural practice that predates Islam

secondary forests (p. 532) forests that grow back after the first cutting, often with decreased diversity of species

secular states (p. 298) countries that have no state religion and in which religion has no direct influence on affairs of state or civil law

secularism (p. 17) a way of life informed by values that do not derive from any one religious tradition

self-reliant development (p. 357) small-scale development schemes in rural areas that focus on developing local skills, creating local jobs, producing products or services for local consumption, and maintaining local control so that participants retain a sense of ownership

serfs (p. 242) persons legally bound to live on and farm land owned by the lord

service sector (p. 72) economic activity that amounts to doing services for others

sex tourism (p. 518) the sexual entertainment industry that services men who arrive in Southeast Asia (and other places) from around the globe to live out their fantasies during a few weeks of vacation

shari'a (pp. 298, 421) literally, "the correct path"; Islamic religious law that guides daily life according to the principles of the Qur'an

sheiks (p. 325) patriarchal leaders of tribal groups on the Arabian Peninsula

shifting cultivation (pp. 120, 338) a productive system of agriculture in which small plots are cleared in forestlands, the dried brush is burned to release nutrients, and the clearings are planted with multiple species; each plot is used for only two or three years and then abandoned for many years of regrowth

Shi'ite (or **Shi'a**) (p. 298) the smaller of two major groups of Muslims with different interpretations of shari'a; Shi'ites are found primarily in Iran and southern Iraq

shogun (p. 451) a member of the military elite of feudal Japan

Sikhism (p. 407) a religion of South Asia that combines beliefs of Islam and Hinduism

Sikhs (p. 407) adherents of Sikhism

Silk Road (p. 468) the ancient trading route between China and the Mediterranean

silt (p. 117) fine soil particles

Slavs (p. 241) a group of farmers who originated between the Dnieper and Vistula rivers, in modern-day Poland, Ukraine, and Belarus

slash-and-burn (**shifting** or **swidden**) **cultivation** (p. 514) an ancient form of multicrop gardening using small plots, often on lands that do not support more intensive types of agriculture

smog (p. 91) a combination of industrial air pollution and car exhaust (*smoke* + *fog*)

social welfare (p. 206) in Europe, elaborate tax-supported systems that serve all citizens in one way or another

socialism (p. 243) a social system in which the production and distribution of goods are owned collectively, and political power is exercised by the whole community

South America (p. 116) the continent south of Central America

Southern African Development Community (SADC) (p. 357) an organization of 14 countries in Southern and East Africa working together for regional development and freer trade

sovereignty (p. 49) the capability of a country to govern its own affairs

Soviet Union (p. 234) see **Union of Soviet Socialist Republics**

spatial analysis (p. 2) the study of how people, objects, or ideas are, or are not, related to one another across an area

spatial freedom (p. 302) the ability to move about without restrictions

special economic zones (SEZs) (p. 461) free trade zones within China

state-aided market economy (p. 455) an economic system based on market principles, such as private enterprise, profit incentives, and supply and demand, but with strong government guidance; in contrast to the free market (limited government) economic system of the United States and Europe

steppes (p. 239) semiarid, grass-covered plains

structural adjustment policies (SAPs) (p. 34) requirements for economic reorganization toward less government involvement in industry, agriculture, and social services, sometimes made by the World Bank and the International Monetary Fund as conditions for giving loans to borrowing countries

structural adjustment programs (SAPs) (pp. 137, 353) policy changes designed to release money for loan repayment to international banks

subcontinent (p. 389) term used to refer to the entire Indian peninsula, including Pakistan and Bangladesh

subduction (p. 116) the sliding of one lithospheric (tectonic) plate under another

subregions (p. 10) smaller divisions of the world regions delineated to facilitate study of patterns particular to the area

subsidence (p. 129) sinking land

subsidies (p. 200) monetary assistance granted by a government to an individual or group in support of an activity, such as farming or housing construction, that is viewed as being in the public interest

subsistence affluence (p. 570) the ability of Pacific Islanders who rely primarily on a home-grown diet, informal local economies, and remittances from overseas to maintain a safe and healthy lifestyle on relatively little formal income

subsistence lifestyle (p. 42) a way of life in which each family produces its own food, clothing, and shelter

Sudras (p. 409) members of the Hindu caste of low-status laborers and artisans

Sunni (p. 298) the larger of two major groups of Muslims with different interpretations of shari'a

sustainable agriculture (p. 44) farming that meets human needs without poisoning the environment or using up water and soil resources

sustainable development (p. 43) efforts to improve standards of living in ways that will not jeopardize those of future generations

syncretism (or **fusion**) (p. 366) the blending of elements of a new faith with elements of an indigenous religious heritage

taiga (pp. 104, 239) subarctic forests

Taliban (p. 410) an archconservative Islamist movement that gained control of the government of Afghanistan in the mid-1990s

tariff (p. 34) a tax imposed by a country on imported goods, usually intended to protect industries within that country

technology (p. 18) an integrated system of knowledge, skills, tools, and methods upon which a culture group's way of life is based

temperature-altitude zones (p. 118) regions of the same latitude that vary in climate according to altitude

terrorism (p. 312) the use, or the threat, of violence, intended to create a climate of fear in a given population

theocratic states (p. 298) countries that require all government leaders to subscribe to a state religion and all citizens to follow certain rules decreed by that religion

thermal inversion (p. 91) a warm mass of stagnant air that is temporarily trapped beneath heavy cooler air

thermal pollution (p. 266) the return of unnaturally hot water to the environment, as by industrial processes or hydro and nuclear power plants

tierra caliente (p. 118) low-lying "hot lands" in Middle and South America

tierra fria (p. 119) "cool lands" at relatively high elevations in Middle and South America

tierra helada (p. 119) very high elevation "frozen lands" in Middle and South America

tierra templada (p. 119) "temperate lands" with year-round springlike climates at moderate elevations in Middle and South America

trade winds (p. 119) winds that blow west across the Atlantic

transmigration (p. 528) the local term used for resettlement in Indonesia

tsunami (p. 446) a large sea wave caused by an earthquake

tundra (pp. 104, 239) a treeless area between the ice cap and the tree line of arctic regions, which has a permanently frozen subsoil

underclass (p. 561) the lowest social stratum, composed of the disadvantaged

Union of Soviet Socialist Republics (USSR) (p. 234) the nation formed from the Russian empire in 1922 and dissolved in 1991

United Nations (UN) (p. 50) an assembly of 185 member states that sponsors programs and agencies that focus on scientific research, humanitarian aid, planning for development, fostering general health, and peacekeeping assistance

upland zones (p. 337) areas of higher altitude

urban sprawl (p. 80) the encroachment of suburbs on agricultural land

Vaishyas (p. 409) members of the Hindu landowning farmer and merchant caste

varna (p. 409) the four hierarchically ordered divisions of society in Hindu India underlying the caste system: Brahmins (priests), Kshatriyas (warriors/kings), Vaishyas (merchants/landowners), and Sudras (laborers/artisans)

volcano (p. 22) an area between plates or a weak point in the middle of a plate where gases and molten rock, called magma, can come to the earth's surfaces through fissures and holes in the plate

Wallacea (p. 505) a trench extending from Bali through the Makassar Strait and along the western side of the Philippines; within this zone there is some mixing of Asian and Australian flora and fauna

weather (p. 24) the short-term (day-to-day) expression of climate

weathering (p. 33) the process by which rocks are physically or chemically broken up

welfare state (p. 188) a social system in which the state accepts responsibility for the well-being of its citizens

wet rice (or **paddy**) **cultivation** (pp. 480, 515) a prolific type of rice production that requires the submersion of the plant roots in water for part of the growing season

World Bank (p. 34) a global lending institution that makes loans to countries that need money to pay for development projects

world cities (p. 187) cities of worldwide economic or cultural influence

world region (p. 10) a part of the globe delineated to facilitate study of patterns particular to the area

World Trade Organization (WTO) (pp. 36, 468) a global institution whose stated mission is the lowering of trade barriers and the establishment of ground rules for international trade

yurt (or *ger*) (p. 483) a round, heavy felt tent stretched over collapsible willow frames used by nomadic herders in northwestern China and Mongolia

Zionists (p. 293) European Jews who worked to create a Jewish homeland (Zion) on lands once occupied by their ancestors in Palestine

INDEX